The Second Hutton Symposium on the Origin of Granites and Related Rocks

Proceedings of a Symposium held at the Australian Academy of Science, Canberra
23–28 September 1991

Edited by

P. E. Brown
Department of Geology
University of St. Andrews
St. Andrews, KY16 9ST, Scotland

and

B. W. Chappell
Department of Geology
Australian National University
GPO Box 4, Canberra City, ACT 2601, Australia

Copublished in volume format by arrangement with, and with the permission of, the Royal Society of Edinburgh

THE GEOLOGICAL SOCIETY OF AMERICA · 1888 ·

SPECIAL PAPER

272

1992

The papers in this volume were originally published together as *Transactions of the Royal Society of Edinburgh: Earth Sciences,* Volume 83, Parts 1 and 2 (ISSN 0263-5933). Copies of that issue are available from the Royal Society of Edinburgh, 22 George Street, Edinburgh EH2 2PQ, Scotland.

Copublished as a limited edition in volume format in the United States by The Geological Society of America, Inc., 3300 Penrose Place, P.O. Box 9140, Boulder, Colorado 80301

Printed in Northern Ireland and bound in the U.S.A.

Library of Congress Cataloging-in-Publication Data

Hutton Symposium on the Origin of Granites and Related Rocks (2nd : 1991 : Australian Academy of Science)
Proceedings of a symposium held at the Australian Academy of Science, Canberra, 23–28, September 1991 : Second Hutton Symposium on the Origin of Granites and Related Rocks / edited by P.E. Brown and B.W. Chappell.
p. cm. — (Special paper ; 272)
"Originally published together as Transactions of the Royal Society of Edinburgh: Earth sciences, Volume 83, parts 1 and 2"--T.p. verso.
Includes bibliographical references and index.
ISBN 0-8137-2272-1
1. Granite—Congresses. I. Brown, P. E. (Peter E.) II. Chappell, B. W. III. Title. IV. Series: Special papers (Geological Society of America) ; 272.
QE462.G7H88 1991
552'.3—dc20 92-18911
CIP

Cover Photo: Taken by P. E. Brown at Qernertog in the Proterozoic belt of S. Greenland, this photograph shows the flat-lying contact between rapakivi granite forming the upper part of the cliff and migmatized metasediments beneath. Cliff height approximately 500 m (1,650 ft).

10 9 8 7 6 5 4 3 2 1

Second Hutton Symposium
The Origin of Granites and Related Rocks

Proceedings of a Symposium held at the
Australian Academy of Science, Canberra
23–28 September 1991

Edited by P. E. Brown and B. W. Chappell

CONTENTS

BROWN, P. E.
Preface v

CHAPPELL, B. W. and WHITE, A. J. R.
I- and S-type granites in the Lachlan Fold Belt 1

BACON, C. R.
Partially melted granodiorite and related rocks ejected from Crater Lake caldera, Oregon 27

MILLER, C. F., HANCHAR, J. M., WOODEN, J. L., BENNETT, V. C., HARRISON, T. M., WARK, D. A. and FOSTER, D. A.
Source region of a granite batholith: evidence from lower crustal xenoliths and inherited accessory minerals 49

MAHOOD, G. A. and CORNEJO, P. C.
Evidence for ascent of differentiated liquids in a silicic magma chamber found in a granitic pluton 63

NORMAN, M. D., LEEMAN, W. P. and MERTZMAN, S. A.
Granites and rhyolites from the northwestern U.S.A.: temporal variation in magmatic processes and relations to tectonic setting 71

SANDIFORD, M., FODEN, J., ZHOU, S. and TURNER, S.
Granite genesis and the mechanics of convergent orogenic belts with application to the southern Adelaide Fold Belt 83

ELLIS, D. J. and OBATA, M.
Migmatite and melt segregation at Cooma, New South Wales 95

ZEN, E-an
Using granite to image the thermal state of the source terrane 107

CHAMPION, D. C. and CHAPPELL, B. W.
Petrogenesis of felsic I-type granites: an example from northern Queensland 115

BARBERO, L. and VILLASECA, C.
The Layos Granite, Hercynian Complex of Toledo (Spain): an example of parautochthonous restite-rich granite in a granulitic area 127

ZECK, H. P.
Restite-melt and masic–felsic magma mixing and mingling in an S-type dacite, Cerro del Hoyazo, southeastern Spain 139

BARBARIN, B. and DIDIER, J.
Genesis and evolution of mafic microgranular enclaves through various types of interaction between coexisting felsic and mafic magmas 145

KILPATRICK, J. A. and ELLIS, D. J.
C-type magmas: igneous charnockites and their extrusive equivalents 155

HAAPALA, I. and RÄMÖ, O. T.
Tectonic setting and origin of the Proterozoic rapakivi granites of southeastern Fennoscandia 165

BROWN, P. E., DEMPSTER, T. J., HARRISON, T. N. and HUTTON, D. H. W.
The rapakivi granites of S Greenland—crustal melting in response to extensional tectonics and magmatic underplating 173

ALLEN, C. M.
A nested diapir model for the reversely zoned Turtle Pluton, southeastern California 179

STEPHENS, W. E.
Spatial, compositional and rheological constraints on the origin of zoning in the Criffell Pluton, Scotland 191

WYBORN, L. A. I., WYBORN, D., WARREN, R. G. and DRUMMOND, B. J.
Proterozoic granite types in Australia: implications for lower crust composition, structure and evolution 201

HILL, R. I., CHAPPELL, B. W. and CAMPBELL, I. H.
Late Archaean granites of the southeastern Yilgarn Block, Western Australia: age, geochemistry, and origin 211

MEIGHAN, I. G., FALLICK, A. E. and McCORMICK, A. G.
Anorogenic granite magma genesis: new isotopic data for the southern sector of the British Tertiary Igneous Province 227

WILLIAMSON, B. J., DOWNES, H. and THIRLWALL, M. F.
The relationship between crustal magmatic underplating and granite genesis: an example from the Velay granite complex, Massif Central, France 235

RECIO, C., FALLICK, A. E. and UGIDOS, J. M.
A stable isotopic ($\delta^{18}O$, δD) study of the late-Hercynian granites and their host-rocks in the Central Iberian Massif (Spain) 247

SHIMURA, T., KOMATSU, M. and IIYAMA, J. T.
Genesis of the lower crustal garnet–orthopyroxene tonalites (S-type) of the Hidaka Metamorphic Belt, northern Japan 259

SEWELL, R. J., DARBYSHIRE, D. P. F., LANGFORD, R. L. and STRANGE, P. J.
Geochemistry and Rb–Sr geochronology of Mesozoic granites from Hong Kong 269

WEAVER, S. D., ADAMS, C. J., PANKHURST, R. J. and GIBSON, I. L.
Granites of Edward VII Peninsula, Marie Byrd Land: anorogenic magmatism related to Antarctic–New Zealand rifting 281

RAPELA, C. W., PANKHURST, R. J. and HARRISON, S. M.
Triassic "Gondwana" granites of the Gastre district, North Patagonian Massif 291

BLEVIN, P. L. and CHAPPELL, B. W.
The role of magma sources, oxidation states and fractionation in determining the granite metallogeny of eastern Australia 305

CANDELA, P. A.
Controls on ore metal ratios in granite-related ore systems: an experimental and computational approach 317

EWART, A., SCHON, R. W. and CHAPPELL, B. W.
The Cretaceous volcanic-plutonic province of the Central Queensland (Australia) coast—a rift related 'calc-alkaline' province 327

ESPERANÇA, S. and MOYER, T. C.
Rhyolites as probes of the crust: examples from Tertiary volcanism in Arizona, southwestern U.S.A. 347

LAW, R. D., MORGAN, S. S., CASEY, M., SYLVESTER, A. G. and NYMAN, M.
The Papoose Flat Pluton of eastern California: a reassessment of its emplacement history in the light of new microstructural and crystallographic fabric observations 361

HUTTON, D. H. W.
Granite sheeted complexes: evidence for the dyking ascent mechanism 377

HUTTON, D. H. W. and INGRAM, G. M.
The Great Tonalite Sill of southeastern Alaska and British Columbia: emplacement into an active contractional high angle reverse shear zone 383

BURNHAM, C. W.
Calculated melt and restite compositions of some Australian granites 387

NEKVASIL, H.
Feldspar crystallisation in felsic magmas: a review 399

HOLTZ, F., JOHANNES, W. and PICHAVANT, M.
Peraluminous granites: the effect of alumina on melt composition and coexisting minerals 409

JOHANNES, W. and HOLTZ, F.
Melting of plagioclase in granite and related systems: composition of coexisting phases and kinetic observations 417

BLENCOE, J. G.
A two-parameter Margules method for modelling the thermodynamic mixing properties of albite-water melts 423

GREEN, T. H.
Experimental phase equilibrium studies of garnet-bearing I-type volcanics and high-level intrusives from Northland, New Zealand 429

DePAOLO, D. J., PERRY, F. V. and BALDRIDGE, W. S.
Crustal versus mantle sources of granitic magmas: a two-parameter model based on Nd isotopic studies 439

WILLIAMS, I. S.
Some observations on the use of zircon U–Pb geochronology in the study of granitic rocks 447

PATERSON, B. A., STEPHENS, W. E., ROGERS, G., WILLIAMS, I. S., HINTON, R. W. and HERD, D. A.
The nature of zircon inheritance in two granite plutons 459

PIDGEON, R. T. and COMPSTON, W.
A SHRIMP ion microprobe study of inherited and magmatic zircons from four Scottish Caledonian granites 473

ABSTRACTS 485

List of Authors' names and addresses 504

INDEX – Authors of papers 507

Preface

P. E. BROWN

The Second Hutton Symposium on The Origin of Granites and Related Rocks was held in Canberra on the 23rd to 28th of September 1991, the time and place having been decided during the first symposium in Edinburgh in 1987. In Canberra, Bruce Chappell chaired an organising committee of C. M. Allen, P. C. Blevin (Secretary), I. H. Campbell, A. Ewart, R. H. Flood, R. Hine, D. A. Steele, A. J. R. White and D. Wyborn. The superb meeting facilities of the Australian Academy of Science provided an ideal venue on the campus of the Australian National University and were enjoyed by an international representation of some 200 geologists. The smooth organisation of the meeting in such distinguished surroundings contributed greatly to its success.

Immediately prior to the symposium, Bruce Chappell, Allan White and Ian Williams led a week long field excursion to study the granites of the Lachlan Fold Belt. This excursion took a transverse section through the fold belt, starting in Melbourne and finishing in Canberra and was made particularly memorable by a blizzard in the Snowy Mountains, which made the crossing of Dead Horse Gap an anxious time for the leaders. Those participating received a definitive and beautifully produced guide book written by the leaders and supplemented en route by voluminous isotopic data, much of which served to confirm the important role of zircon studies to petrogenetic interpretations. There is a vast amount of integrated field, petrographic, chemical and isotopic information on the Lachlan Fold Belt and those people attending the excursion were much indebted to the leaders for their skilful presentation of its complexities. Debate on the outcrops was lively and frequently stimulated by the invaluable observations of W. S. Pitcher. Edge was added by the not infrequent polarisation of viewpoints into Aussie versus Poms, though reference to the Lachlan boulder belt was a purely British (Pom) postgraduate allusion to an unaccustomed style of outcrop. A mid-symposium excursion to examine enclaves in the Cowra granodiorite was led by D. Wyborn, A. J. R. White and B. Chappell and proved to be a delightful social occasion, as well as scientifically stimulating.

The quality of oral and poster contributions to the symposium was high, though a general desire for more discussion time became evident as the meeting progressed, this despite the high level of discipline exercised by speakers in not exceeding their allocated time. The contributions covered, as would be expected, the wide field of granite geology, but if I- and S-types occupied a large part of the debate then, given the locale of the meeting, that was to be expected. In particular, both in the auditorium and in the field, discussion of the I-type concept aroused strong opinions, with attention frequently focused on the heat source problem. The role of abduction, however, received comparitively brief consideration. Experimental and theoretical studies were well represented and an important contribution from Wayne Burnham on the justification of the White and Chappell I- and S-type model is published here in full. No one attending the meeting could be in any doubt of the advances made by isotope geochemists towards understanding the origins of granite. In particular the role of the SHRIMP instrument at the Australian National University received much attention, while on the basis of Nd isotopes, D. J. DePaolo gave some penetrating insights into the crust vs mantle debate.

The papers included in this volume, plus the abstracts, give an almost complete record of everything presented. The majority of manuscripts were available at the time of the symposium and referees were approached from among the delegates present, with a view to immediate action. Though the procedure was somewhat unconventional the refereeing was rigorous and ensured the rapid production of this volume. Both referees and authors are acknowledged for entering into the spirit of a rapid turn around while the quality of presentation and production owes much to the efforts of Bruce Chappell, Charlotte Allen and Beth Ingpen.

Following Edinburgh in 1987, the scientific and social success of the Canberra symposium ensured the continuity of the Hutton cycle of meetings, the next of which will be in the U.S.A. in 1995 to be hosted by the Department of Geology at Maryland University.

For the 1991 meeting the Canberra committee gratefully acknowledges the following sponsors: Newmont Australia Ltd., Normandy Poseidon Ltd., Petro Services Ltd., The Shell Company of Australia, Ansett Australia, Australian Tourism Commission, BHP-Utah Minerals International, Bureau of Mineral Resources, Plutonic Resources Ltd.

Transactions of the Royal Society of Edinburgh: Earth Sciences, **83,** 1–26, 1992

I- and S-type granites in the Lachlan Fold Belt

B. W. Chappell and A. J. R. White

ABSTRACT: Granites and related volcanic rocks of the Lachlan Fold Belt can be grouped into suites using chemical and petrographic data. The distinctive characteristics of suites reflect source-rock features. The first-order subdivision within the suites is between those derived from igneous and from sedimentary source rocks, the I- and S-types. Differences between the two types of source rocks and their derived granites are due to the sedimentary source material having been previously weathered at the Earth's surface. Chemically, the S-type granites are lower in Na, Ca, Sr and Fe^{3+}/Fe^{2+}, and higher in Cr and Ni. As a consequence, the S-types are always peraluminous and contain Al-rich minerals. A little over 50% of the I-type granites are metaluminous and these more mafic rocks contain hornblende. In the absence of associated mafic rocks, the more felsic and slightly peraluminous I-type granites may be difficult to distinguish from felsic S-type granites. This overlap in composition is to be expected and results from the restricted chemical composition of the lowest temperature felsic melts. The compositions of more mafic I- and S-type granites diverge, as a result of the incorporation of more mafic components from the source, either as restite or a component of higher temperature melt. There is no overlap in composition between the most mafic I- and S-type granites, whose compositions are closest to those of their respective source rocks. Likewise, the enclaves present in the more mafic granites have compositions reflecting those of their host rocks, and probably in most cases, the source rocks.

S-type granites have higher $\delta^{18}O$ values and more evolved Sr and Nd isotopic compositions, although the radiogenic isotope compositions overlap with I-types. Although the isotopic compositions lie close to a mixing curve, it is thought that the amount of mixing in the source rocks was restricted, and occurred prior to partial melting. I-type granites are thought to have been derived from deep crust formed by underplating and thus are infracrustal, in contrast to the supracrustal S-type source rocks.

Crystallisation of feldspars from felsic granite melts leads to distinctive changes in the trace element compositions of more evolved I- and S-type granites. Most notably, P increases in abundance with fractionation of crystals from the more strongly peraluminous S-type felsic melts, while it decreases in abundance in the analogous, but weakly peraluminous, I-type melts.

KEY WORDS: aluminium saturation, fractionated, granite, infracrustal, I-type, Lachlan Fold Belt, metaluminous, source rocks, S-type, supracrustal.

Granites and related volcanic rocks occur extensively in that part of southeastern Australia known as the Lachlan Fold Belt (LFB). The Palaeozoic rocks of this belt cover an area of close to 300,000 km^2 and granites comprise a little over 20% of that area (61,000 km^2). Related volcanic rocks occur over a further 5% of the LFB, and are more abundant in the eastern half of the belt. The general geological setting of these rocks has been discussed by White and Chappell (1983) and Chappell *et al.* (1988) and a coloured map of the granites and volcanic rocks in the belt has been published recently (Chappell *et al.* 1991). Where it is exposed over its full width, the belt is some 750 km wide. Radiometric age data on the granites are sparse, but point to most of the granites of this large area being emplaced within a restricted period of time from 420 to 390 Ma with some plutons as young as 360 Ma in the central part of the belt north of Melbourne. In the most easterly part of the belt 2800 km^2 of granite of Carboniferous age (~320 Ma) are present (references to ages in Chappell *et al.* 1988).

Before the opening of the Southern Ocean, the dominantly Caledonian-age granites of this belt were continuous with those of the northern part of the Trans-Antarctic Mountains. They probably also continue north from the exposed LFB beneath younger cover rocks, to link with granites of that age in North Queensland (see fig. 1 of Chappell & Stephens 1988). Therefore, prior to the separation of Antarctica, these granites probably extended over a distance of at least 3600 km, with a width of up to 750 km. The observations that have been made on the granites of the LFB in southeastern Australia and the inferences that have been drawn from them should be seen in the context of their occurrence within this very extensive 'granite belt'.

Our studies of the granites of the LFB, which led to the recognition of the I- and S-types of granite, commenced with detailed mapping of part of the Berridale Batholith in 1963 (Lambert & White 1965), which was extended to cover all of that batholith and parts of the neighbouring Kosciusko Batholith (White *et al.* 1976; Hine *et al.* 1978) during the following decade. It is fortunate that this work commenced in that particular area since many of the features critical to our understanding of the genesis of these granites are particularly well shown in that part of the LFB. Specifically, the more mafic granites of the Berridale and Kosciusko batholiths show a distinct bimodality in many properties and it became apparent that these differences were a reflection of differences in source rock composition, with one group of granites derived from older igneous rocks, the other from older sedimentary rocks. These I-type and S-type groups were formally proposed by Chappell and White (1974).

The purpose of this contribution is to review the concept of I- and S-type granites in the light of much more extensive studies throughout the LFB in the last 20 years. We will conclude that this separation into two types is viable throughout the belt, and will show that it can be extended to volcanic rocks and to crystal fractionated granites, both of which were absent from the region where our studies of the LFB granites commenced. A summary of the criteria that can be used to distinguish the two types in the LFB is given in Section 16.

1. Two contrasting granite types in the Lachlan Fold Belt—historical review

The widespread occurrence of cordierite-bearing granites in the LFB, that were unequivocally derived from sedimentary source rocks, is fundamental to the recognition in that area of what we have called S-type granites. The occurrence of cordierite granites was recognised by Baker (1940) in the Strathbogie Batholith. Tattam (1925) had suggested that cordierite in the Bulla granite near Melbourne was due to contamination and Hills (1938), in an interesting contribution, had argued that the andalusite and sillimanite occurring in some granites of the LFB is of magmatic 'pyrogenetic' origin. The occurrence of andalusite, sillimanite and cordierite in the Cooma Granodiorite has long been known (Joplin 1942). Since the LFB is also home to many varieties of hornblende-bearing granites, in retrospect it is clear that the information pointing to a diversity of source rock compositions for the granites of the LFB has long been available.

The occurrence of two quite distinct types of granites in the Kosciusko region of SE Australia was apparently first recognised by Dallwitz (in Ball *et al.* 1948) who studied the petrographic features of granites along the proposed water tunnel through the Snowy Mountains from Waste Point to Khancoban. Dallwitz showed that the majority of granites in the area could be placed in a 'foliated' group in which, among other features, the biotite was pleochroic from red-brown to buff. The second group included small 'massive' granite bodies, which in all cases intruded the foliated granites, and in which the biotite was pleochroic from very dark chocolate-brown to brownish yellow. Dallwitz suggested that the two types may have originated in different ways, with the foliated granites resulting from the granitisation of sediments and the massive granites forming as mobilised fractions of that earlier type.

Browne (1929) had earlier distinguished the gneissic granite of the Cooma Complex, with its assemblage of andalusite, cordierite and sillimanite, from the granites of the Kosciusko region to the west where there had been superimposed a foliation of varying intensity. On a regional scale, Browne (in David 1950) recognised three types of granite based on their structural features and the characteristics of their associated metamorphic rocks. He proposed a classification that related each type to a specific orogenic event, that is the 'Ordovician' gneissic granites, the 'Silurian' foliated granites and the 'post-Silurian' massive granites. Joplin (1962) referred to the three types as gneissic, foliated and massive, and although she agreed that the gneissic granites are the oldest and the massive granites the youngest, she considered that each type could not be correlated with a specific orogeny. In his review of plutonism in the LFB, Vallance (1969) again referred to these three types, naming them the Cooma type, occurring in areas of medium- to high-grade regional metamorphism, the Murrumbidgee type, which are often foliated and occur in areas of low-grade metamorphism, and the Bathurst type, which are mainly massive and which have imposed medium- to high-grade metamorphic features on their contact rocks. Following Browne (1931), Vallance (1969) stated that members of the first two groups have been closely associated with deformation, whereas the granites of the third group are typically 'post-kinematic'.

Joplin (1962) demonstrated that there are chemical differences between these three granite types; the gneissic or Cooma type has higher K_2O/Na_2O ratios and lower CaO abundances than the other two types, while the massive or Bathurst type has the highest CaO and Na_2O contents and lowest K_2O/Na_2O ratios. Joplin (1962) related these chemical features to the progressive development of granites during a single magmatic cycle.

Kolbe and Taylor (1966) published a geochemical study of granites from the Snowy Mountains and adjacent areas. They found that the gneissic granites have a distinctive and unusual chemical composition in which a high concentration of K, Fe, Mg and other ferromagnesian elements is coupled with very low Ca, Sr and relatively low Na. They concluded that the chemical data are consistent with the view that the gneissic granites could have formed entirely from the melting of the psammopelite sediments in the area. This conclusion was consistent with the field data (Joplin 1942) and the Sr isotopic data of Pidgeon and Compston (1965) who found an initial $^{87}Sr/^{86}Sr$ ratio of 0·718 for the Cooma granite, a value very similar to that of the surrounding high-grade metasediments. The second group recognised by Kolbe and Taylor (1966), the granodiorites and adamellites, included both foliated and massive granites, which they could not distinguish chemically, except for the two locations of Island Bend and Khancoban where the rocks contained lower Fe, Mg and other ferromagnesian elements, and higher Na, Sr and Ba. This second group includes both the I- and S-type granites recognised in later work. It is probable that the distinction between those two types was not made at that earlier time because they are not correlated simply with the massive and foliated groups. Kolbe and Taylor (1966) also noted that the overlap in chemical composition between foliated and massive types is parallelled in their overlapping K–Ar ages (Evernden & Richards 1962). Kolbe and Taylor (1966) concluded that the composition of these rocks is consistent with an origin by mobilisation and melting at depth, without significant differentiation, of varying proportions of geosynclinal sediments consisting on average of equal parts of normal shale, greywacke, and the exposed clay-rich psammopelites. These authors also recognised a third group of rocks in the region, the 'leucogranites', which include both aplites and crystal fractionated I-type granites of a type that will be considered below (Section 14).

Commencing in 1966, Joyce carried out a PhD study of the Murrumbidgee Batholith (Joyce 1970, 1973). The granites of that batholith include the large Clear Range Granodiorite that was grouped with the foliated granites and which is now classified as S-type. Joyce's conclusion that this unit was derived from the partial melting of sedimentary rocks was an important contribution to the development of such ideas in the LFB and this is discussed below (Section 3).

We began a field study of the granites of the Kosciusko

Batholith near Jindabyne in 1968 and that work showed that the massive granites form small plutons that were intruded into larger and older plutons of foliated granite. The petrographic distinctions made by Dallwitz were confirmed and, in addition, the massive granites were found to be generally hornblende-bearing, whereas the foliated granites contained cordierite, most of which is extensively altered to muscovite. In a BSc thesis project, Hine (1971) obtained chemical data on these granites. He showed that in comparison with the foliated granites, the massive granites are higher in Al_2O_3, CaO, Na_2O and Sr, and lower in K_2O, Pb, Rb and Zr, at any given SiO_2 content. These studies were continued to the north in another BSc thesis by Williams (1973). That study confirmed the earlier observations of Hine. By that time, the I- and S-type granite groups had been recognised, and using data provided by W. Compston, Williams (1973) showed that the time interval between the older S-type and younger I-type granites was no more than 5 Ma. Initial Sr isotopic compositions were also significantly different between the two groups, with the S-type granites having an initial $^{87}Sr/^{86}Sr$ ratio of about 0·711, while the one examined I-type granite has an initial ratio of 0·707. In a slightly later study, Williams *et al.* (1975) showed that all of the 20 lithological units of the Berridale Batholith were emplaced over an interval of about 10 Ma, and that there is no clear sequence in time from S- to I-type.

The studies of Hine and of Williams were brought together in the publication of Hine *et al.* (1978) which documented the strongly contrasting chemical and mineralogical characteristics of the I- and S-type granites at the eastern side of the Kosciusko Batholith. Most of the I-type rocks of that area were grouped together in the Jindabyne Suite and that paper is also an important reference for the suite concept as it has been applied to studies of the granites of the LFB.

The presence of foliation in many of the granites of the Kosciusko Batholith, and in some other parts of the LFB, dominated petrological thinking about those rocks for many years. The detailed studies in the Jindabyne region resolved these problems, and an important facet of the recognition of the I- and S-types was the realisation that this compositional subdivision does not coincide with the structural classification of the massive and foliated groups. Thus in the Kosciusko Batholith, the S-type granites are often foliated but are in fact more often massive, while the I-type granites are generally massive but may show a weak foliation. Elsewhere, for example only a short distance to the east in the Berridale Batholith, both types are mostly massive, although rocks in the vicinity of major faults may show petrographic evidence of dislocation. The historical interpretation of these structural features is that the massive granites were intruded after the older granites obtained their foliation, and to an extent, that is true. However, in general, it is thought that the S-type granites are more frequently foliated because their high quartz contents (e.g. KB32 in Table 3) means that they have a framework of that mineral, which is susceptible to deformation, whereas the tonalitic I-type rocks have a stronger framework of plagioclase crystals (e.g. KB22 in Table 2) which is resistant to deformation. This was first pointed out by White *et al.* (1977).

As we have noted, foliated rocks are much less common in the Berridale Batholith, despite the presence of large areas of S-type granites. This is true to such an extent that Vallance (1969) included that batholith among his examples of massive or Bathurst-type granites. As a result of the massive character of these granites, the twofold subdivision that became recognised in mapping that area was among the more mafic rocks, between those that contained hornblende and those that were either biotite-rich without hornblende, or contained cordierite, frequently altered to secondary muscovite. That work, extending into the Kosciusko Batholith, led to the formalisation in 1972 of the I- and S-type subdivision. That concept was presented at a meeting of the IGCP Circum-Pacific Plutonism Project in Santiago, Chile, in September 1973 and was published as an extended abstract the following year (Chappell & White 1974). It was published in a fuller account in the proceedings of the 1982 Nanjing meeting on 'The Origin of Granites and their Metallogenic Relations' (Chappell & White 1984), and there was a limited discussion of the topic in the memoir on Circum-Pacific plutonism (White & Chappell 1983).

The geological features of most of the Berridale Batholith and the eastern part of the Kosciusko Batholith have been published in the detailed NSW Geological Survey reports of White *et al.* (1977) and of White and Chappell (1989) and the associated 1:100 000 geological maps (White *et al.* 1976, 1989). Isotopic studies of granites of the area have been reported by O'Neil and Chappell (1977), Compston and Chappell (1979) and McCulloch and Chappell (1982). Work by D. Wyborn extended the I- and S-type subdivision to volcanic rocks, and three S-type volcanic suites were described by Wyborn *et al.* (1981). The source rocks of some specific I- and S-type granite suites were discussed by Chappell (1984) and examples of both types of granite were included as examples in discussing the restite model (Chappell *et al.* 1987). Three S-type granite suites were discussed in some detail by White and Chappell (1988).

2. Distribution of the I- and S-type granites in the Lachlan Fold Belt

We have noted earlier that granites make up a little over 20% of the total exposed area of Palaeozoic rocks of the LFB, with the abundance rising to about 35% in the eastern part of the belt. S-type granites make up a little more than half the total area of exposed granite, but unlike the I-type granites, they are not found throughout the fold belt. Chappell *et al.* (1988) recognised that the granites of the LFB have provincial characters, which they took to be a reflection of regional differences in the compositions of their source rocks, and hence of the deep crust. On that basis they subdivided the LFB into nine basement terranes, in each of which the granites have a distinctive character. S-type granites occur in six of these terranes, those of the central parts of the LFB, that is Kosciusko, Wagga, Melbourne, Stawell, Bassian, and Taswegia (Chappell *et al.* 1988).

The granites of the LFB have been subdivided on the basis of their associated rocks into regional aureole, contact aureole, and subvolcanic groups (White *et al.* 1974) and there are S-type granites representative of those three groups, and I-types granites of the last two groups. The Cooma Granodiorite is the classic example of a regional aureole granite in the LFB, being at the centre of a small low pressure regional metamorphic complex (Joplin 1942; Pidgeon & Compston 1965; Chappell & White 1976; Munksgaard 1988).

Most of the granites, both I- and S-type, occur as plutons within Ordovician turbidites of low regional metamorphic grade, surrounded by contact aureoles up to 1 km wide. These contact aureole granites are the dominant group in the LFB, and we stress that most S-type granites are of this type and are not regional aureole types. These plutons are often associated in large batholiths, where subvolcanic granites associated with related volcanic rocks may also be present. Related volcanic rocks are particularly abundant in the eastern part of the LFB where they make up about 15% of the exposed Palaeozoic rocks. The largest I-type complex is the Bega Batholith, in the eastern part of the belt, with a total area of 8600 km^2 and the poorly exposed Wagga Batholith close to the central part of the belt has an area of 17,500 km^2, about three-quarters of which is S-type.

3. The petrogenic basis of the I- and S-type granite subdivision

We will see below that the most mafic granites of the LFB can be assigned to either the I- or S-type group in a completely unambiguous way using the chemical and mineralogical properties of the rocks, which are the basis of this classification system. We have called these two groups I- and S-type, to imply that these two groups were derived from different types of source rocks. Hence the strongly peraluminous biotite-rich and normally cordierite-bearing mafic S-type granites, and the corresponding volcanic rocks, are presumed to have been derived from source rocks that were themselves peraluminous, as a result of their components having been through at least one cycle of weathering at the Earth's surface. On the other hand, the hornblende-bearing granites are considered to have been derived from metaluminous source, that is one that is igneous and ultimately mantle-derived, and which has never been through a weathering cycle. At this point we examine our basis for extrapolating from the composition of the granites themselves to the composition of their source rocks.

The 'granite controversy' concluded in the mid-1950s when it became widely recognised, mainly as a result of experimental studies (Tuttle & Bowen 1958), that granites are dominantly magmatic in origin. By that time it was generally thought that granites formed by partial melting of metasedimentary material in the crust, and that all stages in the evolution of granites could be observed, from those formed at an earlier time deep in the crust, to later-formed high level intrusive plutons. These concepts were embodied in the phrase 'granites and granites' by Read (1948) and more formally in his *granite series* (Read 1949), which sought to relate all granites in space and time. In our terminology, the granite series corresponds to the range of S-type granites, but it does not account for the I-type granites. The publication by Hurley *et al.* (1965) of Sr isotopic data for the Sierra Nevada Batholith (I-type in our terminology) showed that the Sierran granites could not have been derived from old source rocks of sedimentary composition, and that paper marked the beginning of the now widely-held view that many, if not most, granites are of mantle derivation.

In the early 1960s, one of us was involved in a study of the late-Permian hornblende-bearing granites of the Moonbi district of the New England Fold Belt of eastern Australia (Chappell 1966, 1978). As part of that study, it was proposed that the mafic enclaves in those rocks were remnants from the partial melting of mafic material. Such a possibility had been suggested by Bateman *et al.* (1963) for the mafic hornblende-bearing enclaves in some granites of the Sierra Nevada batholith. Furthermore, Presnall and Bateman (1973) had considered the most likely source for most mafic enclaves in the Sierra Nevada batholith to be 'refractory' material that remained after partial fusion. The new concept in the study of Chappell (1966, 1978) was that variation in the amount of material contributed to the magma from the enclaves, was responsible for the systematic linear chemical variation within certain granite suites of that area. The suggestion that mafic enclaves could be residual material from the source of the granite magma, and the proposal by Chappell (1966) that the variation in composition of some granite suites resulted from differing proportions of restite, formed the basis of the *restite model* presented by White and Chappell (1977) and later amplified by Chappell *et al.* (1987). While this model is not widely accepted, it did lead to the important conclusion, for which there is stronger support (see Section 5), that granites form 'images of their source rocks' (Chappell 1979). Hence, in the Moonbi district, for example, the hornblende-bearing granites are presumed to have been derived by the partial melting of source rocks that initially had that same character. At the northern edge of those granites of the Moonbi district there are peraluminous 'two-mica' granites of the Banalasta Suite, now known to contain pseudomorphs after cordierite (Flood & Shaw 1975). In a statement that foreshadowed the full development of the I- and S-type subdivision in the LFB, Chappell (1966) stated that those peraluminous granites were probably produced by the partial melting of 'sialic crust'.

Mafic peraluminous granites that contain abundant cordierite (unaltered or altered), up to 30% biotite, and never hornblende, are developed extensively in that part of the LFB that we have assigned to the Kosciusko and Wagga basement terranes (Chappell *et al.* 1988), and comprise the more mafic parts of the Bullenbalong and Koetong supersuites. The first of these groups also has extensive volcanic equivalents (Wyborn *et al.* 1981) that, when sufficiently unaltered, contain cordierite + orthopyroxene + biotite ± almandine. For the whole of the Bullenbalong Supersuite (White & Chappell 1988) and the more mafic part of the Koetong Supersuite (Chappell *et al.* in prep.) we ascribe the compositional variation to varying degrees of separation of melt and restite. The Clear Range Granodiorite of the Murrumbidgee Batholith (Joyce 1970, 1973) includes the most mafic (Fe-rich) S-type granite yet analysed from the LFB (sample ER62 of Table 3). In his study, Joyce (1973) concluded that the composition of this mafic granodiorite to tonalite unit resulted from extensive reaction of parental magma that formed by the partial melting of psammopelitic rocks, with relict solid material at, or near, the source of the magma. In proposing such an origin, Joyce was applying the model of Chappell (1966) that had been developed for hornblende-bearing granites to the analogous situation in which the granites were strongly peraluminous and contained abundant metasedimentary enclaves. In the earlier study, the granites were presumed to have been derived from older igneous rocks; in the later study the source materials were older sedimentary rocks.

The application by Chappell (1966, 1978) and Joyce (1970, 1973) of an early form of what was later developed as the *restite model* by White and Chappell (1977) was fundamentally important in recognising the significance of the source rock differences for I- and S-type granites. The occurrence of hornblende-rich enclaves (Chappell 1966) or biotite-rich enclaves (Joyce 1970) was important in the original recognition of restite in many granites and in distinguishing the I- and S-type groups. However, we do not now assign to enclaves a major role in the restite model (Chappell *et al.* 1987; Chappell & White 1991). Most of the restite in granites is carried up from the source as single crystals or small aggregates and not as large fragments or enclaves. Whatever its physical form, it is such entrained restite that carries with it a chemical signature of the source rocks of many granites in the Lachlan Fold Belt.

4. A source for the Lachlan Fold Belt granites within the crust

It is of fundamental importance to recognise that the source rocks of the LFB granites were located in the crust. It is apparently necessary to give the reasons why we consider this to be so, to judge, for example, from the comment of Barbarin (1990, p. 228), from which the present authors might be seen as standing alone in advocating a crustal origin for most granites. The arguments for a crustal derivation for the LFB granites were stated by Chappell (1984, p. 695) and they bear repeating here.

"It is asserted that the Lachlan granites are of crustal derivation for the following reasons. First, as we shall see, approximately half of them are derived by partial melting of sedimentary rocks, with little or no evidence for any mantle component. Second, many of the granites, including mafic ones, involved melting at 'minimum' temperatures in equilibrium with quartz and two feldspars, so that the depth limit of feldspar stability was not exceeded. Third, most of the rocks are potassic, in contrast to the calcic granites of the Cordilleran and other young fold belts that might contain a component of mantle or subducted oceanic crust origin. Fourth, all oxygen isotopic compositions measured are significantly higher than mantle values and this must be regarded as an intrinsic or primary feature. Fifth, most Sr and Nd isotopic compositions are more evolved than mantle values."

The arguments for a crustal source for the LFB granites will be developed further in Section 12. For the S-type granites of the LFB, the amount of any mantle component is limited by the strongly peraluminous nature of the more mafic S-type granites, and because of the lack of metaluminous enclaves (Section 12.1). The general chemical features of these granites can be accounted for without invoking the presence of any mantle component (Sections 6.1 and 12.1) and the differences between S-type suites can likewise be accounted for completely by earlier weathering processes (Section 12.2). Since a mantle component is not required, and because there is no evidence for such material, we take the view that the S-type granites of the LFB were of crustal origin. Hence these rocks correspond to Read's granite series (Section 3), with the addition of volcanic rocks.

I-type granites are produced either by partial melting or by fractional crystallisation of mantle-derived source material. There are two important chemical constraints to these processes. First, that source material must be sufficiently mafic to itself be a possible mantle product produced, in many cases, in large quantities; this effectively restricts the source materials generally to basalt through andesite compositions. Second, to produce large volumes of granite magmas, the source material from the mantle must be 'fertile' in terms of the granite-forming elements, that is Si, Al, Na and K; this means that rocks of andesitic rather than basaltic composition are the more likely source materials. In the LFB we favour the production of I-type granites by partial melting, rather than by fractional crystallisation, of mantle-derived material, for two main reasons. First, the amount of gabbro associated with the granites is very limited, less than 0·1% of the total area of the plutonic rocks. Second, the granites are often isotopically distinct from mantle materials, and the evidence supports a process of aging of these primitive materials in the crust prior to partial melting, as for the Jindabyne Suite (Section 12.3.2), rather than mixing with isotopically evolved older crust. In our view, the I-type granites of the LFB are of *indirect* mantle origin and the production of felsic granites from much more mafic materials by a sequence of partial melting events, as proposed by Chappell and Stephens (1988, section 7), still seems to us to be valid. Such an evolution may be more generally applicable in other areas that contain I-type granites like those of the LFB, which are dominantly granodioritic and contain on average close to 70% SiO_2 and 3·5% K_2O (Table 1). A more direct mantle origin may be appropriate for dominantly tonalitic I-type granites such as those of the Cordillera (e.g. Silver & Chappell 1988), and would be required for the quartz dioritic (M-type) rocks of island arcs (e.g. Whalen 1985).

5. Granite suites

Granites of the Lachlan belt can be grouped into suites (Griffin *et al.* 1978; Hine *et al.* 1978) where each suite has a distinctive chemical character consistent with its derivation from source rocks of fixed composition. This does not mean that suites themselves have a fixed or uniform composition. It does mean that throughout their range of composition, suites have distinctive properties, e.g. high Na, or low Cr, or a certain range of Mg/Fe, that are reflecting similar features in their source rocks. In this way, granites form an 'image' of their source rocks (Chappell 1979). Granite magmas are produced by partial melting, so that a melt phase and a residual or restite phase are produced. Magmas result when the partial melt product that is a mix of melt and restite becomes mobile, which occurs at a melt content of about 30%, the critical melt fraction of van der Molen & Paterson (1979). Granite magmas must originally contain restite at their source and many never rid themselves of that component. According to the restite model (White & Chappell 1977; Chappell *et al.* 1987), most of the variation within those suites of rocks in which some restite is retained as the magma leaves its source region can be ascribed to varying degrees of separation of restite from melt before solidification of the granites. Fractional crystallisation, as a mechanism of producing variation within suites, will be of relatively minor importance while restite is still present, within the restite fractionation regime of White and Chappell (1983). Crystal fractionation may become important after all restite has been removed. In the case of high temperature magmas, such as those of the Boggy Plain Supersuite of the LFB (Wyborn *et al.* 1987), efficient separation of melt and restite must have occurred at or near the source as there is no evidence for restite even in the most mafic members of that suite. In such a case there will be a less direct but still detectable correlation between the composition of the granite and its source rock, as a result of the partitioning of elements between melt and segregated restite (Chappell & Stephens 1988). In such a case, high concentrations of an element in the source will result in high concentrations in both the melt and the restite, and conversely for low concentrations. Hence, in the Boggy Plain Supersuite, all mafic and moderately mafic rocks are high in K, Sr and Ba.

Suites must also have distinctive characters. Also, largely as a result of their distinctive chemical properties, they generally have characteristic mineralogical, modal, and petrographic features.

Differences between suites therefore are interpreted to result from differences in source-rock compositions. In the LFB, the first-order subdivision into suites is between those granites formed from igneous and sedimentary sources, the I- and S-types. Both these types have a distinctive chemical, modal, mineralogical and isotopic characters that reflect the differences in the features of these two types of sources. We will now examine the distinctive characteristics of the I- and S-type granites in more detail.

6. Chemical compositions of I- and S-type granites

6.1. Average chemical compositions for the Lachlan Fold Belt

Average chemical compositions of 1074 I-type and 704 S-type granites from the LFB are listed in Table 1. Rocks containing less than 57% SiO_2 have been omitted from this average; such an arbitrary limit applies only to the I-type rocks as the S-type granites have a natural lower limit, with the lowest SiO_2 measured being 63·42% and only four values being less than 65% SiO_2. The number of analyses used to calculate the averages is a little more than twice that used for previous published average values (White & Chappell 1983; Chappell & White 1984). The additional data include some analyses that were previously not assigned as I- or S-type, but are mainly analyses obtained after November 1982, and which come largely from rocks of the central, western and southern parts of the belt. Both averages are a little more felsic than those previously reported (by 1·52% SiO_2 for the I-type and 1·83% SiO_2 for the S-type average), and the average compositions are substantially higher in Rb which has increased from 132 ppm to 164 ppm for the I-type granites and from 180 to 245 ppm for the S-type granites. These changes mainly follow from the occurrence of crystal fractionated granites in the more westerly parts of the belt and also partly from the inclusion of many previously unassigned felsic granites that are now classified as either I- or S-type (Section 14). For the I-type average, a lesser factor is that analyses down to 53% rather than 57% SiO_2 were included in the earlier calculations.

Table 1 Average analyses of I-type and S-type granites from the Lachlan Fold Belt

Number of samples	I-type 1074	S-type 704
SiO_2	69·50	70·91
TiO_2	0·41	0·44
Al_2O_3	14·21	14·00
Fe_2O_3	1·01	0·52
FeO	2·22	2·59
MnO	0·07	0·06
MgO	1·38	1·24
CaO	3·07	1·88
Na_2O	3·16	2·51
K_2O	3·48	4·09
P_2O_5	0·11	0·15
total FeO	3·12	3·06
at. $Fe^{3+}/(Fe^{2+}+Fe^{3+})$	0·290	0·153
Trace elements (ppm)		
Ba	519	440
Rb	164	245
Sr	235	112
Pb	19	27
Th	20	19
U	5	5
Zr	150	157
Nb	11	13
Y	31	32
La	31	27
Ce	66	61
Sc	13	11
V	57	49
Cr	20	30
Co	10	10
Ni	8	11
Cu	9	9
Zn	48	59
Ga	16	18
Sn	6	10

For I-type granites in the LFB, there is a complete gradation in composition from granites to gabbros, although mafic rocks are very much less abundant than the more felsic ones. As in other fold belts, mafic rocks in the LFB occur in two ways, both as mafic phases of generally more felsic plutons, and as separate mafic bodies. For the S-type granites, on the other hand, the low-SiO_2 limit of 63·4% is well within the granite range of composition. The mean SiO_2 values are $70{\cdot}91 \pm 2{\cdot}82\%$ (1σ) for the S-type and $69{\cdot}50 \pm 4{\cdot}62\%$ (1σ) for the I-type granites. Again, for total iron as FeO, the mean values are very close with a greater range for the I-types ($3{\cdot}12 \pm 1{\cdot}55\%$) and the S-types ($3{\cdot}06 \pm 1{\cdot}23\%$). The more restricted compositional range of S-type granites is an important difference between the two types. For the elements listed in Table 1, the standard deviations for I-type granites are greater for all elements except Rb, Zn, Ga and Sn. S-type granites equivalent in composition to the most mafic I-type granites, and transitional to gabbros, are not found in the LFB. There is no S-type analogue to the high temperature I-type Boggy Plain Supersuite. Mafic S-type melts were not present and the compositions of all the more mafic S-type granites that do occur in the LFB appear to be determined by the quantity of restite that was entrained in a relatively low-temperature melt phase; this imposes a lower SiO_2 limit for S-type granites that corresponds to the mean SiO_2 content of the sedimentary source rocks, which is closer in composition to the melt than are the I-type sources to the I-type melts.

The first significant feature of the averages in Table 1 is that for many elements the values are very similar. For those elements that are significantly more abundant in the average S-type granite, some differences are partly due to the greater abundance of crystal fractionated granites among the S-type granites of the LFB; such rocks are high in P, Rb, Ga and Sn. Likewise, the lower S-type values for Ba and Sr are partly due to the effects of feldspar fractionation (Section 14). Data from the granites of the eastern part of the LFB, where crystal fractionation is not common, such as the averages of White and Chappell (1983), show that of those five elements most affected by crystal fractionation, only for Rb and Sr do there appear to be intrinsic differences between the I- and S-types. Hence, for P, Ga, Sn and Ba most or all of the differences result from more extensive crystal fractionation among the S-type granites.

Relative to the I-type granites, the S-types contain significantly less CaO, Na_2O and Sr. These three elements are distinct from all others listed in Table 1 that they occur in feldspars but not in common clay minerals. Thus, when feldspars are weathered to clay minerals at the Earth's surface, these three elements are lost in solution and go to form limestone (Ca and Sr) or accumulate in sea water and may form evaporites (Na). Thus granite source rocks that include a clay mineral component will produce granites low in these elements (S-types) on partial melting. Much of the more detailed variation between S-type suites is caused by variations in the abundance of these three elements, reflecting varying amounts of a clay mineral component in the source rocks.

Relative to the I-types, the S-type granites contain more K, Rb and Pb and also contain significantly more Cr and Ni if the effects of greater crystal fractionation among the S-type granites are discounted. For Cr and Ni this contrast results from the I-type granites being largely derived from source rocks low in these two elements, that is from previously fractionated rocks. The S-type granites, on the other hand, come from sources containing more of these elements, which are typically relatively abundant in shales. That K, Rb and Pb are high in the S-type granites is not difficult to explain, since these three elements are incorporated in clay minerals but it is perhaps surprising that the abundances of these elements are higher in S-type than in I-type granites. This must imply either that the crust that was weathered to produce the S-type source rocks contained more of these three elements than did the source rocks of the I-type granites, or that they were concentrated in the clay minerals produced by weathering; this may be a provincial feature for the LFB.

While both granite types have very similar average total Fe contents, the S-type granites have a distinctly lower Fe^{3+}/Fe^{2+} ratio. The average values of the atomic ratio $Fe^{3+}/(Fe^{3+}+Fe^{2+})$ are 0·153 (S-type) and 0·290 (I-type), but there is overlap between the two types. The generally more reduced character of the S-type granites is thought to be due to the presence of graphite in the sedimentary source rocks, as proposed for late Palaeozoic granites in the adjacent New England Fold Belt by Flood and Shaw (1975), but it could be due to the intrinsic nature of the structures of metaluminous and peraluminous melts (e.g. Mysen & Virgo 1989).

6.2. Aluminium saturation and undersaturation

As a result of their lower Ca and Na (but not higher Al, see Table 1) contents, S-type granites generally have a higher molecular $Al_2O_3/(Na_2O+K_2O+CaO)$, the *aluminium saturation index* (ASI) of Zen (1986). For S-type granites, this ratio is always greater than 1·0, that is the rocks are always saturated in Al, or peraluminous. Almost one-half of the I-type granites of the LFB are likewise peraluminous, but weakly so, with a ratio generally less than 1·1. Metaluminous and rare peralkaline granites (ASI $<$ 1) are exclusively I-type. We note that Avila-Salinas (1990) 'defined' ASI = 1·1, the boundary between I- and S-type granites used by Chappell and White (1974), as the boundary between metaluminous and peraluminous granites. We follow the original definition of 'peraluminous' by Shand (1927) as a rock "characterised by the presence of alumina in molecular excess of the sum of potash, soda, and lime". As we emphasise here, many I-type granites are peraluminous in that original sense.

For a rock to be fertile in terms of producing granite melt in any quantity at temperatures that can be achieved in the crust, it must contain quartz, feldspars and water, the last component being either free or available from the breakdown of hydrous minerals. The source rocks of granites must all have these features in common but can differ in other ways. For example, those sources that are igneous rocks will generally contain amphibole and therefore will be undersaturated in Al (ASI $<$ 1). Sedimentary rocks that contain significant amounts of clay components will be saturated in Al (ASI $>$ 1) and at the source of granites will contain Al-rich minerals such as biotite, muscovite and cordierite. Compositionally very similar felsic granites can be produced from these two quite different types of source rocks since the initial stages of partial melting, at least, will consume dominantly quartz and feldspars. This will produce a melt of near haplogranite composition that is close to the thermal minimum composition of Tuttle and Bowen (1958). Very similar granites could also be produced from diverse sources if melting continued to higher temperatures and produced more mafic melts that underwent fractional crystallisation. Under both circumstances, the granites derived from a diversity of source rocks could be both chemically and mineralogically similar, and under these circumstances, the only clue to the nature of a granite protolith might well be isotopic compositions. This is a limiting situation, but one that is not uncommon, and one that presents difficulties in classifying granites according to their source rocks on the basis of their chemical or mineral compositions.

We have raised this consideration partly to make the point that diverse source rocks can produce very similar granite compositions. We also want to point out that the processes of evolution of granite magmas will trend towards removing chemical differences within those magmas that reflect specific features of the compositions of the source rocks. The separation of restite, either at the source or subsequently, and fractional crystallisation will move the total composition of any magma towards that of low-temperature melts relative to the source rock composition. In that way, granite compositions will converge towards that of a haplogranite as they become more felsic. Conversely, it is the more mafic rocks whose compositions will be closer to that of the source materials because they contain additional source material, either entrained as restite, or melted from the source at higher temperatures.

The ASI values for the LFB granites are shown in Figure 1. Most values of ASI for the LFB granites lie between 0·85 and 1·35 and there is a continuous range of values with a peak at an ASI slightly greater than 1·0 (Fig. 1A); there is no separation into two groups in that plot but there is an asymmetry to the peak with a longer tail on the high-ASI side, due to a weak peak being superimposed on that side. This tail becomes more pronounced if the ASI values for those rocks that we have classified as S-type are scaled up by a factor of 1074/704, that is the number of I-type divided by the number of S-type analyses, so that representatives of both types cover the same area on the histogram; this is plotted in Figure 1B, where the second frequency peak at ASI ~1·15 becomes more evident. Figures 1C and 1D show the ASI values for rocks that we have classified as I-type and as S-type, based on other criteria. The S-type data in Figure 1D are also scaled so that the total area of the S-type histogram is equal to that for the I-type granites in Figure 1C.

Several observations can be made from the information presented in Figure 1.

(1) For the granites of the LFB, there is a broad range in aluminium saturation, with most values between ASI ~ 0·85 and 1·35, with a maximum near ASI = 1·0.

(2) There is no separation in ASI values between the complete data sets for I- and S-type granites, but those values have an asymmetric distribution towards the higher side, caused by a broad maximum in the S-type values between about ASI = 1·1 and 1·2. Chappell and White (1974) placed a boundary value of ASI = 1·1 between the two types, a value that does apply in the Kosciusko

Batholith (fig. 4 of Hine *et al.* 1978), but the additional data now show a substantial overlap. A significant implication of that earlier value of 1·1 was that *many I-type granites are weakly peraluminous,* a point that we emphasise in this paper, and which has not always been appreciated by other workers.

(3) ASI values for the I-type granites ($SiO_2 > 57\%$) extend down to a value of 0·695 but samples below 0·85 make up only 2·3% of the total number of I-type rocks with >57% SiO_2 and these are cumulate rocks. Some rocks thought to be I-type have ASI > 1·1, ranging up to 1·20, and these make up 2·8% of the total. The mean ASI for the I-type granites is 0·985 and the median value is 0·990 with 45% of the values being greater than ASI = 1·0, so that almost half of the I-type granites of the LFB are peraluminous. For a smoothed frequency curve in Figure 1C, the modal value is 1·01.

(4) ASI values for the S-type granites form a broader, less regular, and more asymmetrical distribution on the ASI frequency histogram (Fig. 1D). There is a lower limit at ASI = 1·01, a broad maximum, and the values extend up to 1·99. The mean value is ASI = 1·179.

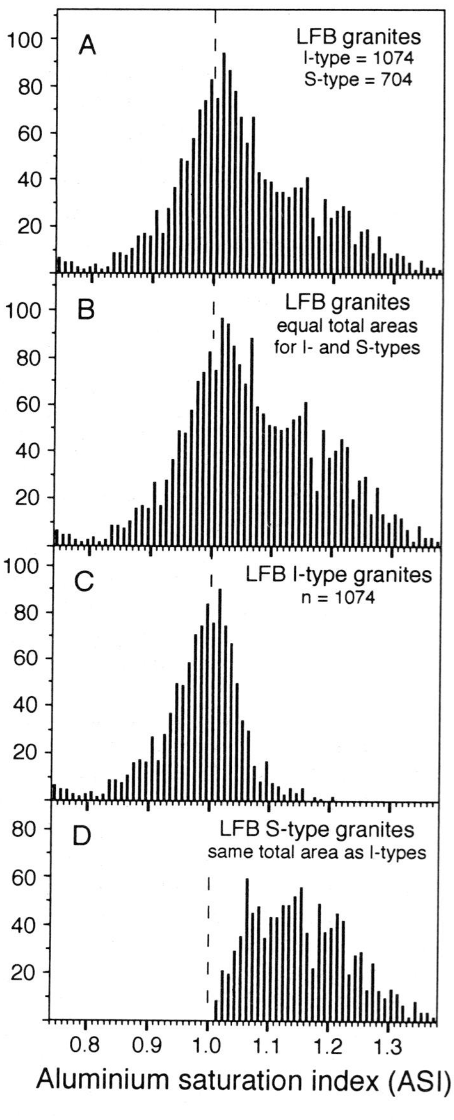

Figure 1 Histograms of ASI values for LFB granites. (A) All LFB granites between ASI = 0·74 and 1·37. Six I-type granites ($SiO_2 >$ 57%) have values below 0·74 down to a minimum of 0·69; all of these samples are cumulate rocks of the Boggy Plain Supersuite. There are 42 S-type values between 1·37 and 1·99 which are not shown. Samples with ASI values outside the range from 0·74 to 1·37 are included in the total number of 1074 I-type and 704 S-type granites analysed from the LFB. (B) All LFB granites with the frequency of the S-type granites scaled up by a factor of 1074/704 so that the total area of each type is the same. (C) I-type granites of the LFB. (D) S-type granites of the LFB scaled to the same area as the I-type granites.

There are three reasons why the overlap between the I- and S-type granites in Figure 1 might be exaggerated relative to that of the primary magmatic compositions.

(1) In assessing the data represented in Figure 1, it must be appreciated that the ASI values are sensitive to the effects of hydrothermal alteration which can lead to the destruction or modification of feldspars and the mobilisation of Na, K and Ca, with consequent changes in the ASI. All of the 1778 samples of granite represented in Figure 1 are unaltered; samples that were obviously altered or weathered, based on field and petrographic criteria, were discarded. There must, however, be some effects of alteration within the ASI data and more so for the felsic granites which are more likely to have formed from melts having high activities of H_2O; these are the rocks closest to the main interval of overlap from ASI = 1·0 to 1·1 in both groups. Later discussion (Section 14.4) will show that small amounts of such post-magmatic alteration are the principal cause of the overlap between the two groups.

(2) The measurement of consistent and accurate ASI values is not a trivial procedure. The analytical error in the data plotted in Figure 1 is close to ±0·02 with 95% confidence, which is not significant in the present context.

(3) There are undoubtedly some cases in Figure 1 in which the granites have been assigned to the wrong group. In particular, this is so for those felsic granites that have not undergone crystal fractionation and which also cannot be placed in suites with more mafic rocks. In such felsic rocks, the chemical compositions are always very similar and the distinctive minerals of the more mafic rocks, hornblende or cordierite, are not present.

Figure 1 shows that the compositions of primary 'minimum-temperature' melts are always mildly peraluminous. The significant difference between the two groups is that as the rocks become more mafic, I-types *always* become progressively more metaluminous and S-types *always* progressively more peraluminous. This is because more mafic rocks will always be closer to the composition of their source rocks than 'minimum-temperature' melts, having either entrained material from that source as restite, or incorporated it in solution in the melt phase, or both. Clearly then, to confirm whether or not the granites of the LFB were derived from source materials of two distinct compositions, one igneous, the other sedimentary, it will be most instructive to examine the compositions of the more mafic granites. The more mafic rocks form the better image of their source and we will therefore look at the ASI frequency distribution of the most mafic of the granites of the LFB.

6.3. Bimodal chemical compositions of mafic granites of the Lachlan Fold Belt

We have calculated the ASI values of the most mafic (Fe-rich) 10% of the analysed granites, that is of 107 I-type granites and 70 S-type granites. The results are shown in Figure 2 and several points arise from that diagram. The

ASI values are bimodal, with a maximum value for I-type granites of ASI = 1·010, and a minimum value for the S-types of 1·056. None of the more mafic granites under consideration here have values between those two limits, despite the fact that the average ASI value for all of 177 mafic granites under consideration is 1·021, which is within that gap. For these most mafic compositions there is a clear separation between metaluminous and peraluminous groups that can reasonably be taken to correspond to source rocks of corresponding character, based on earlier arguments.

For the I-type compositions represented in Figure 2, all ASI values below 0·8 are cumulates from higher temperature liquid magmas, mainly from the Boggy Plain Supersuite of Wyborn *et al.* (1987). At higher ASI levels such cumulate rocks are still represented and they are joined by mafic restite-rich granites.

Two groups can be recognised within the 70 mafic S-type granites. The major group of 62 rocks have ASI values between 1·056 and 1·360 and these are shown on Figure 2. These are the 'batholithic' S-type granites of White and Chappell (1983), considered to have been derived from unexposed pre-Ordovician sedimentary source rocks. The other eight rocks are much more strongly peraluminous and range in ASI from 1·55 to 1·99, are not shown on Figure 2, and are representatives of the Cooma Supersuite. These two S-type groups are discussed below (Section 6.4.2).

Because melting within the crust initially involves dominantly quartz and feldspar components, all 'minimum-temperature' melts will have ASI values close to 1·0, and higher temperature melts will also be closer to 1·0 than their source rocks. Thus I-type melts will be less metaluminous, and S-type melts will be less peraluminous, than their source materials. Therefore granites, except those that are cumulates comprising either strongly metaluminous or strongly peraluminous mineral assemblages, will have ASI values closer to 1·0 than their source materials. Hence, even among the relatively mafic compositions represented in Figure 2, there would have been some movement of both the metaluminous and peraluminous modes towards ASI = 1. The lack of peak overlap among the mafic I- and S-type granites in Figure 2 clearly shows that there were two distinct types of granite source rock compositions in the LFB. There was clearly a significant range in ASI within the S-type sources, and probably within the I-type group as well. It is possible that these ranges in composition were partly produced by the mixing of igneous and sedimentary source materials, with the limitation that in no case did such mixing produce source rock compositions close to ASI = 1,

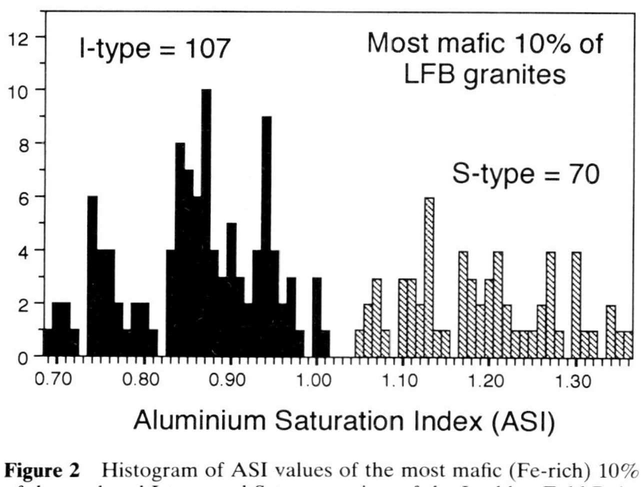

Figure 2 Histogram of ASI values of the most mafic (Fe-rich) 10% of the analysed I-type and S-type granites of the Lachlan Fold Belt.

and certainly no continuum of compositions, at least for the suites represented by samples in Figure 2. If limited mixing did occur then these data show that the sources retained a dominantly 'igneous' or 'sedimentary' character. There are other restrictions on such mixed source rocks in the LFB and these will be discussed elsewhere (Section 12).

I- and S-type 'minimum-temperature' melts near ASI = 1 were in equilibrium with hornblende and with a peraluminous mineral such as cordierite, respectively. Both are peraluminous because hornblende can coexist with peraluminous melt (e.g. Holloway & Burnham 1972). However, because of the solubility of aluminosilicate minerals in haplogranite melts, S-type melts should have higher ASI values. This prediction is supported by observations on felsic granite compositions derived by fractional crystallisation of feldspars (Section 14.4).

6.4. Variations in chemical composition within the I- and S-type granite groups

Both the I- and S-type granites show a significant ranges in composition, which reflect both variations in source rock compositions for each type, and the degree of fractionation, by whatever mechanism, of the granite composition away from that of its source. To illustrate the variations in composition within the two groups, 11 analyses of each type are listed in Tables 2 and 3, with CIPW norms, and modal analyses where available. Such a restricted number of analyses cannot approximate all of the possible compositions, but they do serve to illustrate most of the important and distinctive features of the two types.

6.4.1. Chemical composition of the I-type group. In the following discussion of I-type granite compositions, the sample numbers are those of the 11 analyses in Table 2. These rocks include a range of more mafic compositions with one volcanic rock (BV5), two felsic granites that are considered to have not undergone crystal fractionation (MG20 and BB21) and two granites whose compositions show the effects of feldspar fractionation (VB230 and TB12). Six compositions are diopside-normative and five are corundum-normative, illustrating the point that I-type granites can range from strongly metaluminous to weakly peraluminous (Section 6.2).

The most mafic and most felsic rocks that have been analysed from the Moruya Suite (Griffin *et al.* 1978) are samples MG42 and MG20. The first of these is thought to have a composition close to that of the source rocks of that suite (Chappell 1984; Chappell & Stephens 1988), and the second represents a felsic low temperature melt derived from that source composition. Other rocks of the Moruya Suite represent mixture of the MG20 melt composition and varying amounts of entrained restite, up to the maximum amount present in MG42 (White & Chappell 1977; Burnham 1992). Both of these rocks are high in Na_2O, and in that and other respects, MG42 resembles some Cordilleran I-type tonalites (Silver & Chappell 1988). In the LFB, such high Na is characteristic of rocks at the eastern margin, at the eastern side of the Bega Batholith, including the small Moruya Batholith. The Bega Batholith shows some striking lateral compositional asymmetries (Section 11), which among the major elements is most marked for Na and sample AB6 from the Glenbog Suite near the western edge of the Bega Batholith shows that character, with a much lower Na_2O content than does MG42, despite the much higher SiO_2. Sample BV5 is from a volcanic rock, the Kadoona Dacite, which is closely related to the plutonic rocks of the Bega Batholith and is similar to AB6 in composition.

Table 2 Chemical and modal analyses of some I-type granites and a volcanic rock from the Lachlan Fold Belt

	MG42	MG20	AB6	BV5	KB139	KB22	BP16	CV87	BB21	VB230	TB12
SiO_2	60·13	74·91	67·39	66·17	55·50	62·29	62·26	63·73	76·56	76·73	75·71
TiO_2	0·99	0·15	0·53	0·59	0·70	0·54	0·54	0·88	0·19	0·08	0·07
Al_2O_3	16·42	13·37	14·48	14·52	18·63	16·46	15·42	15·14	12·60	12·44	12·31
Fe_2O_3	2·04	0·38	1·58	1·80	2·03	1·85	2·24	0·87	0·43	0·37	0·40
FeO	3·73	1·16	3·02	3·40	4·98	3·41	3·57	3·77	0·43	0·47	0·84
MnO	0·11	0·07	0·09	0·09	0·12	0·10	0·10	0·07	0·02	0·04	0·02
MgO	3·27	0·24	1·72	2·19	4·81	2·85	3·35	2·40	0·21	0·22	0·02
CaO	5·91	1·10	4·32	4·72	8·10	6·05	5·94	3·74	0·97	0·45	0·64
Na_2O	3·94	4·21	2·63	2·46	2·22	2·60	3·02	2·65	3·52	3·46	3·42
K_2O	1·41	3·50	2·99	2·74	0·95	2·01	2·14	4·08	4·16	4·78	4·83
P_2O_5	0·28	0·06	0·11	0·11	0·10	0·11	0·17	0·24	0·03	0·02	0·01
H_2O+	1·03	0·58	0·95	0·68	1·66	1·20	1·00	1·02	0·47	0·52	0·45
H_2O-	0·18	0·18	0·13	0·16	0·18	0·20	0·12	0·20	0·15	0·17	0·12
CO_2	0·28	0·02	0·08	0·31	0·06	0·10	0·20	0·32	0·11	0·05	0·56
rest	0·18	0·13	0·15	0·16	0·14	0·16	0·23	0·31	0·14	0·10	0·19
TOTAL	99·90	100·06	100·17	100·10	100·18	99·93	100·30	99·42	99·99	99·90	99·59
Trace elements (ppm)											
Ba	325	580	415	440	215	355	555	1150	545	56	6
Rb	36·5	90	135	124	38·5	89	72	215	214	367	765
Sr	487	123	148	168	256	254	578	301	99	24·0	6·5
Pb	6	15	15	26	6	9	14	22	17	42	60
Th	5·8	9·0	16·8	16·2	4·4	9·2	11·2	26·0	19·8	32·5	86
U	1·8	3·4	3·0	3·4	0·8	1·4	2·2	5·8	3·6	15·2	26·0
Zr	168	102	145	162	68	112	115	297	115	79	157
Nb	14·0	12·0	9·0	9·0	4·5	6·5	5·5	20·5	16·5	27·0	32·5
Y	23	25	32	29	18	21	19·	27	26	72	180
La	20	19	26	26	13	23	28	39	33	15	42
Ce	50	44	62	61	30	47	63	85	66	38	111
Sc	19	4	20	21	32	23	24	15	5		4
V	113	6	101	106	169	131	142	115	7	3	<1
Cr	33	3	9	17	37	25	70	71	<1	<1	<1
Co	20	3	13	16	26	17	21	14	1	1	1
Ni	26	<1	4	5	15	7	18	35	<1	<1	<1
Cu	23	2	3	9	8	13	38	39	<1	<1	<1
Zn	68	37	55	60	76	62	74	72	17	14	55
Ga	17·4	15·0	15·8	16·6	17·4	17·4	15·8	18·6	12·0	14·0	30·0
Sn	5	4	4		4	5	5	5	3	11	40
CIPW norms											
Q	14·35	33·71	27·93	26·99	12·05	21·17	18·17	19·56	37·79	36·85	35·44
C		0·80						0·14	0·61	0·80	0·32
or	8·33	20·68	17·67	16·19	5·61	11·88	12·65	24·11	24·58	28·25	28·54
ab	33·34	35·62	22·25	20·82	18·79	22·00	25·55	22·42	29·79	29·28	28·94
an	22·95	5·07	18·87	20·48	38·06	27·31	22·20	16·99	4·62	2·10	3·11
di	3·66		1·51	1·81	1·19	1·62	4·98				
hy	10·11	2·30	7·07	8·52	17·94	10·34	9·93	10·86	0·68	1·05	1·18
mt	2·96	0·55	2·29	2·61	2·94	2·68	3·25	1·26	0·62	0·54	0·58
il	1·88	0·28	1·01	1·12	1·33	1·03	1·03	1·67	0·36	0·15	0·13
ap	0·65	0·14	0·26	0·26	0·23	0·26	0·40	0·56	0·07	0·05	0·02
ASI	0·908	1·064	0·956	0·947	0·972	0·958	0·872	1·009	1·051	1·069	1·026
Modal analyses											
Quartz	7·4	34·6	30·1		16·6	25·2	18·0	29·4	34·3	31·5	32·1
Kfspr	0.4	26·2	13·8		0·6	3·0	7·0	17·5	35·0	36·9	36·1
Plag	59·8	34·7	36·1		52·4	49·9	48·0	36·5	28·3	27·3	25·9
Musc									0·2	0·2	0·2
Biot	16·3	4·5	13·4		8·6	14·7	12·0	13·5	1·9	4·1	5·0
Hbld	14·2		6·6		19·7	6·7	12·0	2·2			
Cpx					<0·1		2·2				
Opaq	1·2				2·1	0·4	0·8	0·2			
Sphene	0·3										
Apat	0·4					0·1		0·3			
Topaz											0·3
Fluorite											0·4

Notes:

MG42	Tuross Head	quartz diorite	36°00·87′S	150°09·70′E
MG20	Bodalla	monzogranite	36°08·01′S	150°05·38′E
AB6	Glenbog	granodiorite	36°36·49′S	149·21·21′E
BV5	Kadoona	dacite	35°46·75′S	149·38·78′E
KB139	Grosses Plain	quartz diorite	36°32·98′S	148°50·19′E
KB22	Jindabyne	tonalite	36°25·50′S	148°30·64′E
BP16	Boggy Plain	quartz monzodiorite	35°50·69′S	148°34·88′E
CV87	Lysterfield	granodiorite	37°57·59′S	145°15·37′E
BB21	Maffra	monzogranite	36°32·09′S	148°58·56′E
VB230	Thologolong	monzogranite	35°58·02′S	147°27·28′E
TB12	Coles Bay	monzogranite	42°08·93′S	148°16·47′E

The above four samples are from a part of the LFB, the Bega Basement Terrane, in which S-type granites of batholithic dimensions, derived from deeper levels of sedimentary crust, are absent. The other four relatively mafic compositions in Table 2, KB139, KB22, BP32 and CV87 are from parts of the belt in which I- and S-type granites occur together. The first two of these are from the Jindabyne Suite, comprising the small and generally massive plutons of the eastern side of the Kosciusko Batholith that intrude the more extensive bodies of S-type granite (Section 1 and Hine *et al.* 1978). The quartz diorite KB139 is the most mafic rock that has been found in that suite, containing the maximum possible amount of restite and hence like MG42 it is thought to be close to the source composition for the suite. The tonalite KB22 is a more typical member of the Jindabyne Suite. Unlike the Moruya Suite, this does not extend to very felsic compositions, with a maximum SiO_2 of 67·15%, in a rock whose composition is thought to be close to the primary melt composition (Chappell 1984; Burnham 1992). In this suite the compositional variation resulted dominantly from variation in the amount of a restite component in a higher temperature and more mafic melt. In contrast to such an origin for KB22, sample BP16 in Table 2 is a cumulate composition from a high-temperature and completely molten magma that separated from its restite at a very early stage, from the Boggy Plain pluton (Wyborn *et al.* 1987). KB22 and BP16 are texturally very dissimilar, but are alike compositionally, with the cumulate rock BP16, however, being lower in Al and richer in Mg, Cr, Ni, Cu, P, Ba, Rb, Sr and Pb. Sample CV87, from the Lysterfield Granodiorite on the eastern outskirts of the city of Melbourne, shows the characteristics of I-type granites of the Melbourne Basement Terrane, which are ilmenite-bearing, have very low Fe^{3+}/Fe^{2+} ratios and very high Ba contents. CV87 also has much higher Ti, K, P, Rb, Pb, Th, U, Zr and Ga contents than any of the other more mafic granites in Table 2.

The remaining three compositions in Table 2 are of very felsic granites. Sample BB21, like MG20 from the Moruya Suite, is considered to be a primary 'minimum-temperature' melt composition, but unlike that other rock it is not associated in a suite with more mafic compositions. A comparison of the composition of BB21 with the crystal fractionated granite compositions VB230 in TB12 in Table 2, is very instructive, as also is a comparison with the S-type analogues in Table 3.

The last three samples in Table 2 show a sequence of I-type compositions from that of a felsic melt that has not undergone fractional cyrstallisation (BB21), to one from which a moderate amount of quartz and feldspar fractionation has occurred (VB230), to an extremely fractionated composition (TB12). There are marked increases in the contents of Rb, Pb, Th, U, Nb, Y, REE, Ga and Sn, while the transition metals, Ba and Sr decrease in abundance. These changes can be accounted for by dominant fractional crystallisation, with perhaps a small amount of superimposed hydrothermal alteration. These compositional features will be discussed in more detail and contrasted with those of S-type granites in Section 14.

6.4.2. Chemical composition of the S-type group. The 11 S-type granite compositions listed in Table 3 are all corundum-normative and are all more strongly so than those I-type granites of Table 2 that are weakly peraluminous. These S-type analyses include two from the Cooma Supersuite (VB221 and CC1), a mafic granite from the Clear Range Suite (ER62), two granites (KB32 and KB102) and a volcanic rock (SV51) from the Bullenbalong Supersuite, two relatively felsic granites of the Dalgety Suite (BB11 and BB2), a granite from the Melbourne Basement Terrane (CV115) and two granites that have undergone significant feldspar fractionation (VB140 and TB83).

The first five analyses in Table 3 show very clearly the distinctive features of more mafic S-type granites in the LFB. They are low in Na and Sr and, except for ER62, also low in Ca, and are consequently strongly peraluminous. K and Rb are generally high for such Fe-rich compositions, and the Fe^{3+}/Fe^{2+} ratios are much lower than those of the mafic I-type granites in Table 2. Samples VB221 and CC1 from the Cooma Supersuite show the S-type characteristics in an extreme form and are typical of the very strongly peraluminous rocks (ASI > 1·5) similar to the feldspar-poor Ordovician sediments widely exposed in the Lachlan belt. The most distinctive feature of the rocks of that supersuite is the low CaO and Na_2O contents, with these components averaging 1·17% and 1·45%, respectively, in 14 analyses.

Sample ER62 from the Clear Range Suite of the Murrumbidgee Batholith (Reid 1980) is the most Fe-rich S-type granite that has been analysed from the LFB. This rock has a low Na_2O content (1·73%) which for the most mafic batholithic S-type granites is typically less than 2% but not as low as in the Cooma Supersuite, and ER62 is also very low in K. Unlike most other S-type granites it is high in Ca, a specific feature of the Clear Range Suite. Another very mafic S-type granite is KB32 from the Jillamatong Granodiorite (Hine *et al.* 1978) and this restite-rich composition (Burnham 1992) from the Bullenbalong Suite (White & Chappell 1988) is more typical of the LFB than is ER62. The composition of the volcanic rock SV15 (Wyborn *et al.* 1981) is quite close to that of KB32. Sample KB102 is a felsic representative of the Bullenbalong Suite and is very close to a primary 'minimum-temperature' melt in composition, with little if any entrained restite, and also, with its Rb content of 187 ppm, showing no evidence of feldspar fractionation.

Samples BB11 and BB2 are from the Dalgety Suite of the Berridale Batholith. That suite is restricted to more felsic compositions than is the Bullenbalong Suite and displays less pronounced S-type characteristics. For example, in rocks of similar Fe contents, Dalgety contains more Ca and Na and is distinctly less peraluminous than the Bullenbalong Suite, a feature reflected by the presence of only small amounts of cordierite in the least felsic samples. The possible causes for this less peraluminous character will be discussed later (Section 12). BB2 is thought to be very close to a primary restite-free S-type 'minimum-temperature' melt that has not undergone crystal fractionation, while BB11 contains a significant amount of restite (Burnham 1992).

Sample CV115 from the Strathbogie Batholith (White & Chappell 1988) is the rock that was studied experimentally by Clemens and Wall (1981). This is the only rock from the Melbourne Basement Terrane listed in Table 3 and it illustrates many of the distinctive features of the S-type granites of that area. Like the associated I-type granites referred to above (CV87), these rocks are high in Ba and have very low Fe^{3+}/Fe^{2+} ratios. CV115 is relatively high in SiO_2, and it is probably close to a primary S-type melt in composition. Its petrographic features suggest that it might have contained a small amount of restite.

Sample VB140 is representative of the Koetong Supersuite (Chappell *et al.* in prep.). It is included here both as an example of a S-type granite that has undergone moderate crystal fractionation and because it has a composition very close to that of the average composition of the biotite granites of the Cornubian Batholith (Chappell &

Table 3 Chemical and modal compositions of some S-type granites and a volcanic rock from the Lachlan Fold Belt

	VB221	CC1	ER62	KB32	SV51	KB102	BB11	BB2	CV115	VB140	TB83
SiO_2	67·87	72·00	65·04	67·68	68·05	73·81	70·89	73·48	73·65	72·48	73·37
TiO_2	0·73	0·54	0·73	0·64	0·61	0·30	0·35	0·22	0·36	0·25	0·01
Al_2O_3	14·80	13·72	14·82	14·70	14·41	13·22	13·92	13·41	13·08	14·50	15·09
Fe_2O_3	0·38	0·59	0·91	0·68	1·15	0·62	0·72	0·44	0·16	0·29	0·03
FeO	4·30	3·48	4·65	4·03	3·31	1·65	2·26	1·47	2·27	1·33	0·65
MnO	0·06	0·06	0·06	0·07	0·06	0·04	0·06	0·04	0·05	0·03	0·03
MgO	2·01	1·76	3·27	2·22	2·05	0·60	1·00	0·71	0·61	0·47	0·02
CaO	1·05	0·95	3·98	2·26	2·68	1·68	2·31	1·70	1·20	0·66	0·29
Na_2O	1·38	1·49	1·73	1·92	2·03	2·97	2·56	2·63	2·58	2·80	4·62
K_2O	4·37	3·73	2·91	3·60	3·58	4·05	3·96	4·63	4·60	5·20	4·09
P_2O_5	0·19	0·13	0·12	0·15	0·15	0·09	0·09	0·08	0·17	0·33	0·29
H_2O+	1·84	1·28	1·18	1·51	1·21	0·46	1·39	0·74	0·90	1·03	0·33
H_2O-	0·26	0·14	0·11	0·22	0·15	0·16	0·15	0·14	0·16	0·16	0·11
CO_2	0·28	0·13	0·11	0·12	0·24	0·10	0·29	0·11	0·09	0·16	0·53
rest	0·21	0·21	0·20	0·19	0·19	0·15	0·18	0·13	0·19	0·13	0·18
Total	99·73	100·21	99·82	99·99	99·87	99·90	100·13	99·93	100·07	99·82	99·64
Trace elements (ppm)											
Ba	610	765	475	475	530	495	625	350	755	220	2
Rb	186	153	142	183	170	187	178	236	244	419	1400
Sr	126	127	160	139	138	89	198	97	79	65	8·0
Pb	35	35	20	27	25	29	33	37	25	36	4
Th	26·5	21·5	19·4	19·2	19·6	18·6	18·0	19·2	15·6	14·2	11·8
U	2·6	3·8	1·2	4·0	3·8	3·6	4·2	6·4	5·8	10·8	3·4
Zr	245	201	168	187	189	139	136	100	183	99	22
Nb	18·0	9·0	11·0	13·0	10·5	7·5	10·5	10·0	11·5	20·0	39·5
Y	24	39	15	27	32	40	40	33	35	14	13
La	37	31	43	32	32	24	31	23	24	20	2
Ce	82	68	88	69	74	58	70	49	58	45	7
Sc	13	12	20	15	15	7	12	7	7	4	2
V	71	64	123	87	93	29	42	25	25	15	<1
Cr	54	56	118	61	57	8	13	10	13	10	<1
Co	17	15	21	17	16	7	8	6	8	4	<1
Ni	24	24	32	20	22	5	6	4	7	4	<1
Cu	16	11	23	17	14	5	6	3	8	11	<1
Zn	95	71	83	82	73	39	47	28	67	77	63
Ga	20·4	15·0	18·2	18·2	18·2	14·8	15·2	13·8	17·8	22·0	40·2
Sn	3	2		8		6	6	7	12	25	48
CIPW norms											
Q	35·45	42·43	27·73	32·31	32·24	35·84	33·25	35·05	36·43	34·27	30·46
C	6·35	5·82	1·88	3·89	2·68	1·11	1·44	1·17	2·08	3·86	3·23
or	25·83	22·04	17·20	21·27	21·16	23·93	23·40	27·36	27·18	30·73	24·17
ab	11·68	12·61	14·64	16·25	17·18	25·13	21·66	22·25	21·83	23·69	39·09
an	3·97	3·86	18·96	10·23	12·32	7·75	10·87	7·91	4·84	1·12	0·46
di											
hy	11·49	9·51	14·84	11·44	9·34	3·59	5·58	3·82	5·05	3·02	1·26
mt	0·55	0·86	1·32	0·99	1·67	0·90	1·04	0·64	0·23	0·42	0·04
il	1·39	1·03	1·39	1·22	1·16	0·57	0·66	0·42	0·68	0·47	0·02
ap	0·44	0·30	0·28	0·35	0·35	0·21	0·21	0·19	0·40	0·77	0·67
ASI	1·751	1·736	1·145	1·360	1·229	1·092	1·115	1·096	1·189	1·362	1·272
Modal analyses											
Quartz	26·5	48·3	31·4	34·0		37·9	37·5	34·4	21·8	41·8	
Kfspr	19·3	8·8	0·2	8·6		16·2	24·5	29·5	45·9	24·6	
Plag	10·8	18·9	37·0	29·3		33·8	27·1	27·1	20·1	21·3	
Musc	2·7	4·0		2·6				0·2		4·0	
Biot	16·8	17·4	31·4	21·5		12·1	10·7	8·7	10·2	7·9	
Andal		2·4									
Cord	19·9	tr		4·0				tr	1·5		
Garn	2·1								tr		
Opaq	0·3	0·1		tr			0·1	0·1			
Apat	0·5	0·1		0·2			0·1		0·2	0·4	

Notes:

VB221	Bethanga	granite	36°06·67′S	147°03·20′E
CC1	Cooma	granodiorite	36°14·26′S	149°06·99′E
ER62	Murrumbidgee	tonalite	36°02·11′S	149°06·07′E
KB32	Jillamatong	tonalite	36°31·35′S	148°35·76′E
SV51	Hawkins	dacite	34°40·15′S	148°46·22′E
KB102	Happy Jacks	granodiorite	36°03·52′S	148°29·60′E
BB11	Dalgety	monzogranite	36°29·46′S	148°50·19′E
BB2	Numbla Vale	monzogranite	36°39·54′S	148·49·22′E
CV115	Strathbogie	granite	36°49·57′S	145°51·71′E
VB140	Granya	monzogranite	36°07·51′S	147°19·20′E
TB83	Lottah	monzogranite	41°13·32′S	148°00·13′E

Hine in prep.). Relative to less evolved rocks of the Koetong Supersuite, and more generally in comparison with the compositions KB102, BB2 and CV115, sample VB140 has a distinctive trace element composition, with more P, Rb, Nb, Ga and Sn, and less of the transition metals, Ba, Sr, Zr, Y, REE and Th. It is strongly peraluminous (3·86% normative corundum), a feature resulting from the fractionation of quartz, feldspars and a small amount of biotite (an assemblage with ASI slightly greater than 1) from a less peraluminous composition similar to rock BB2. An example of an extremely fractionated S-type granite is sample TB83 from the topaz-bearing Lottah Granite of the Bassian Batholith of northeastern Tasmania. For that rock, most of the changes noted above for VB140 have gone to more extreme values. The origin of the extreme compositional characteristics of this unit is controversial, with a magmatic origin proposed by MacKenzie *et al.* (1988, 1990) and a metasomatic-hydrothermal origin being favoured by Higgins (1990). In our view, the composition of TB83 is *dominantly* magmatic, as MacKenzie *et al.* (1988, 1990) argued, and in most respects it is a continuation of the crystal fractionation processes that operated to produce VB140. However, there are features that must have resulted from superimposed hydrothermal alteration, in particular the very low Ba content, and the low amounts of Ca, Sr and Pb, all of which result from the destruction of feldspars by hydrothermal solutions, so that the amounts present in the bulk rock are less than those that can be attained by fractional crystallisation alone. The chemical features of TB83 that can be used to characterise it as a highly fractionated S-type granite are the combination of features such as high Rb, Nb and Sn, and low Ba and Sr, with the very high P_2O_5 content (0·29%) for such a felsic rock. In Section 14 these features will be discussed further and contrasted with those of I-type granites.

7. Some petrographic and mineralogical features of the I- and S-type granites of the Lachlan Fold Belt

The various mineralogical differences between the two granite types correspond with the chemical differences discussed in Section 6. Thus S-type granites are always peraluminous and contain an Al-rich biotite often with other Al-rich minerals such as muscovite, cordierite, andalusite, sillimanite or garnet. The most felsic I-type granites are also peraluminous, as all lowest temperature granite melts have that property (Section 6.2). More mafic I-type granites are metaluminous and consequently contain hornblende. The generally more reduced Fe in the S-type granites corresponds to the common occurrence of ilmenite in those rocks rather than the magnetite of the I-types. Magnetite-bearing S-types are rare in the LFB and virtually confined to the Dalgety Supersuite (e.g. BB11 in Table 3). Ilmenite-bearing I-type granites are not so uncommon (e.g. CV87 in Table 2).

There are many mineralogical and modal differences between I- and S-type granites that correlate with their differences in chemical composition. The most important of these arise from the more peraluminous nature of the S-types. In the mildly peraluminous S-type rocks, for example the Dalgety Granodiorite of the Berridale Batholith (White *et al.* 1977), an Al-rich biotite is present (Tetley 1978), cordierite is rare, and pseudomorphs after orthopyroxene are present (e.g. BB11). In more strongly peraluminous rocks (ASI > 1·1), another Al-rich mineral apart from biotite normally occurs. Such minerals include cordierite, sillimanite and, less commonly, almandine garnet or andalusite. All of these minerals may be of either restite or precipitate origin. Muscovite is generally present, but with the exception of a few granites (e.g. VB140) it is secondary and formed by subsolidus alteration. Cordierite is commonly partly or completely altered to muscovite + green biotite. Peraluminous I-type granites may contain Al-rich biotite + muscovite ± Mn-garnet ± andalusite. Metaluminous I-types contain hornblende and less commonly clinopyroxene, both of which are diagnostic. Mn-rich garnets can occur in either type, when the rocks are felsic with high Mn/Fe ratios. The occurrence of the different Al-bearing minerals can be demonstrated using an ACF diagram (Section 10; fig. 5 in Hine *et al.* 1978). Those rocks (e.g. VB140) containing primary muscovite crystallised at higher H_2O contents resulting from the fractionation of anhydrous phases, and probably at a deeper level.

As a consequence of their lower Ca and Na contents, S-type granites are generally more quartz-rich. Since hornblende is unstable in peraluminous rocks, mafic S-type granites may be very biotite-rich, up to 30% of the rock. Such biotite-rich rocks may be low in K-feldspar, even when moderate amounts of K are present, and are tonalities (e.g. ER62).

We have seen above that S-type granites are as a group more reduced than I-types. S-type granites normally contain ilmenite and no magnetite, and the oxygen fugacity was therefore below the quartz-fayalite-magnetite (QFM) buffer, and probably near the magnetite-wustite (MW) buffer. Some S-type granites contain magnetite (Whalen & Chappell 1988) and must have been close to the QFM buffer. Many-I-type granites contain the assemblage magnetite + sphene and therefore formed above the ilmenite-hedenbergite (or amphibole)-magnetite-sphene-quartz (SMQHIL) buffer (Wones 1989) and below the hematite-magnetite (HM) buffer. Some I-types contain ilmenite instead of magnetite + sphene (Whalen & Chappell 1988), and therefore crystallised below the SMQHIL buffer (e.g. CV87). In summary, I-type granites generally have higher fO_2 than S-types. These features are thought to be inherited from the source. Consequently, I-types often contain magnetite + sphene and less commonly ilmenite without magnetite. S-types generally contain ilmenite and rarely magnetite and they never contain primary sphene, although secondary sphene, formed by alteration of biotite, is common in both types and is non-diagnostic. Thus, the I- and S-type subdivisions are often the same as the magnetite and ilmenite series of Ishihara (1977), but the correspondence is not exact (Takahashi *et al.* 1980). A further indication of the difference in oxidation between the two types is the presence of sulfur as S^{2-} in S-type granites, generally in pyrrhotite (Whalen & Chappell 1988), and as S^{6+} in I-type granites in apatite (our unpublished data).

The difference in oxidation of the two types gives rise to some useful discriminants. The K-feldspar in S-type granites is always white in colour, never pink, if the rock is fresh and not weathered or hydrothermally altered. In I-type granites, on the other hand, the K-feldspar crystals are frequently pale pink in colour, but sometimes white. This is a useful field criterion which reflects the different oxidation states of the rocks. Biotites in the two types often have distinctive chemical compositions and Nockolds (1947) observed that the composition of biotite in granites is a function of its paragenesis. Chappell (1966) showed that biotite in granites always contains close to 2·5 atoms of Al atoms in tetrahedral coordination per 24 (O, OH, F), and that biotite coexisting with hornblende contains negligible amounts of Al in octahedral coordination, whereas biotite occurring with muscovite contains about 0·6 atoms of Al in octahedral coordination per formula unit. Fe^{3+} and Ti are generally less abundant in the biotites of S-type granites (Tetley 1978). Oxidised I-type biotites are typically pleochroic from chocolate to straw-coloured, whereas S-type biotites usually have γ = foxy red-brown with very prominent pleochroic haloes. Reduced I-type granites such as CV87 also contain red-brown biotite.

8. Associated volcanic rocks

In some parts of the LFB, suites of volcanic rocks can be matched in chemical composition and certain mineral characteristics with suites of granites of the same region (e.g. Wyborn *et al.* 1981). The following assemblages are those commonly found in corresponding plutonic (left) and volcanic rocks (right):

Qz + Plag + Kf + Bio + Cd	Qz + Plag + Opx + Bio + Cd + Kf in matrix
Qz + Plag + Kf + Bio	Qz + Plag + Opx + Bio + Kf in maxtrix
Qz + Plag + Kf + Bio + Hbe	Qz + Plag + Opx + Cpx ± Bio ± Hbe + Kf in matrix

The first two pairs of assemblages are peraluminous and the third pair are metaluminous. Those in the first line are S-type, and those in the third are I-type, while those of the second pair may be either I- or S-type. These will be discussed further in Section 10.

The assemblages in the volcanic rocks are anhydrous or more anhydrous than those in corresponding plutonic rocks. In the terminology of Chappell *et al.* (1987) the large crystals (or phenocrysts) in the volcanic rocks are often primary restite, with the hydrous assemblages of the granites resulting from the reaction of these minerals with the OH-bearing melt phase as the magma cools to granite (modified or secondary restite).

Those granites that are closely related to volcanic rocks commonly have miarolitic cavities, graphic intergrowths of feldspar and quartz, and porphyritic fabrics typical of near-surface intrusions regardless of whether they are I- or S-type; some are pressure-quenched. These, and other data (experimental and theoretical), indicate that the amount of H_2O in the melt component of both the I- and S-type magmas was much the same (3–4 wt%).

9. Enclaves in I- and S-type granites of the Lachlan Fold Belt

Accidental xenoliths are rare in granites of the LFB and are mostly confined to the margins of the plutons. The more abundant enclaves display the following features.

(1) The minerals present in the enclaves generally match those of the enclosing granites. Thus enclaves in the I-type granites are the common hornblende-bearing microgranular variety. Those in the S-type granites are more varied and comprise a variety of rocks, such as biotite-rich surmicaceous enclaves, fragments of high-grade gneiss, orthopyroxene-rich rocks, and fragments of milky quartz, which cannot be matched to any exposed wall rocks (Chappell & White 1991).

(2) The chemical composition of the enclaves reflects that of the enclosing granites. Hence those in the I-type granites are metaluminous, whereas those in the S-type granites, while more diverse, contain an important group with strongly peraluminous compositions. Among the I-type enclaves, the abundances of both major and trace elements reflect those in the host granites within a particular suite, although they do not necessarily follow the precise variation defined by the host granites on Harker diagrams (Chen *et al.* 1991).

(3) Except for one unit, the Murrumbucka Tonalite (Section 12), enclaves are either distinctly I- or S-type, and except for Murrumbucka, mixtures of hornblende-bearing and varied metasedimentary enclaves are not seen.

(4) There is a general correlation between the number of enclaves and the composition of the enclosing rock, so that more mafic granites, both I- and S-type, contain more enclaves. In detail, this correlation varies from suite to suite. Chen *et al.* (1990) made a quantitative estimate of the abundance of mafic enclaves in the granites of the Glenbog Supersuite and showed that their abundance in the most felsic unit (Nimmitabel) is less than 0·1%, whereas in the Blue Gum Tonalite, the most mafic unit, it is up to about 6%. On the other hand, mafic rocks of the Boggy Plain Supersuite, which comprise a large fraction of the most mafic I-type granites of the LFB, are poor in enclaves. For example, for one unit of that supersuite, the zoned Boggy Plain pluton, Wyborn (1983) states that macroscopic enclaves are absent from the diorite, are exceedingly rare in the granodiorite, and very rare in the monzogranite.

It can be seen that there are similarities in the occurrence of enclaves in both I- and S-type granites, but the distinct differences in chemical and mineral composition reflect those features of the host granites. When present, enclaves appear to be a reliable guide to the nature of the host rock.

The lack of enclaves in mafic rocks of the Boggy Plain Supersuite, along with other petrographic and compositional evidence, is taken to indicate that the source magmas for that supersuite were liquids that contained no restite (Wyborn *et al.* 1987; Chappell *et al.* 1987). Since none of the mafic S-type granites of the LFB formed from completely liquid magmas, a lack of mafic enclaves in mafic granites can be taken to indicate I-type character. As noted above, the S-type granites of the LFB contain a distinctive but diverse group of enclaves that can be regarded as restite. We favour an analogous origin for most of the mafic enclaves in the I-type granites, but this is more difficult to establish. Hence, probably in most cases the enclaves not only reflect the composition of the host granite, but also the source rocks of those granites. Irrespective of the origin of these enclaves, they do possess distinctly different characteristics in I- and S-type granites of the LFB and they can be used empirically to distinguish between the more mafic granites of the two groups.

10. Relationship between chemical and mineral compositions

The relationship between the chemical and mineral compositions of I- and S-type granites and volcanic rocks, discussed in Sections 6, 7 and 8, are shown on a modified ACF diagram in Figure 3A. Positions of the major mineral phases occurring in these rocks are shown in that figure, and also the boundary between metaluminous and peraluminous compositions, that is the line joining anorthite and hypersthene, for which ASI = 1. The distinctive minerals of I-type plutonic and volcanic rocks, hornblende and augite, are below that line, while the minerals characteristic of S-type rocks, andalusite, sillimanite, cordierite and almandine garnet lie above it. Hypersthene, which occurs commonly in volcanic rocks and rarely, at least as an unaltered mineral, in plutonic rocks of both types, is on the ASI = 1 boundary. Biotite, the mafic mineral shared by both granite types, lies within the peraluminous field, consistent with the observation that the compositions of felsic I-type granites that contain no hornblende are slightly peraluminous (Section 6). The variable content of Al in biotite, shown in Figure 3A, corresponds to the lower and upper limits of Al content of the biotite described in Section 7.

Three typical and distinct plutonic rock and corresponding volcanic rock mineral assemblages were given in Section 8. These correspond to (a) the area above the anorthite to Al-rich biotite join, (b) the area within the anorthite-biotite field, and (c) the area below the anorthite to Al-poor biotite join. The assemblages of minerals for plutonic rocks present in Figure 3A for each of these, are (a) Plag + Bio ± Cd ± Mu, (b) Plag + Bio ± altered Opx, (c) Plag + Bio + Hbe. For volcanic rocks the assemblages are (a) Plag + Cd + Opx ± Bio ± Gn, (b) Plag + Opx ± Bio, (c) Plag + Cpx + Opx ± Hbe ± Bio. For granites, primary cordierite is generally partly or completely altered to a fine-grained mixture of muscovite and green biotite, and pseudomorphs consisting of biotite and quartz in some granites are probably altered hypersthene.

The mineral assemblages of granite enclaves listed in Section 8 are also represented by Figure 3A. The mafic enclaves of I-type granites contain assemblage (c) above. The more compositionally diverse enclaves in S-type granites plot over a wide area of the triangle, but compositions below the anorthite to hypersthene join are rare. Almandine-rich garnet is more common in S-type enclaves than it is in the host granites, and in both cases it is generally surrounded by a rim of cordierite or altered cordierite. Likewise, pyroxenes are more common as mineral cores in enclaves, than in the associated granites.

ACF diagrams are useful for illustrating the way in which the mineral and chemical compositions both of granites and volcanic rocks can be correlated. Also, with good data, they show that there is a precise correlation between chemical and mineral compositions. Hine *et al.* (1978, fig. 5) shows the clear separation between I- and S-type granites of the Kosciusko Batholith on an ACF diagram, and it also successfully discriminates between hornblende-bearing and hornblende-free I-type granites. We have plotted selected analyses from Tables 2 and 3 in Figure 4. That diagram shows the very mafic I-type granites MG42 and KB139, and the somewhat less mafic sample KB22, lying within the hornblende-bearing field, while the strongly peraluminous S-type samples VB221, CC1 and KB32 are in the cordierite field. The mafic S-type granite that does not contain cordierite, ER62, plots within the anorthite-biotite field. The two felsic 'minimum-temperature' compositions, those of the I-type BB21 and S-type BB2, lie just above the anorthite to Al-rich biotite join, consistent with the presence in those rocks of muscovite and cordierite, respectively.

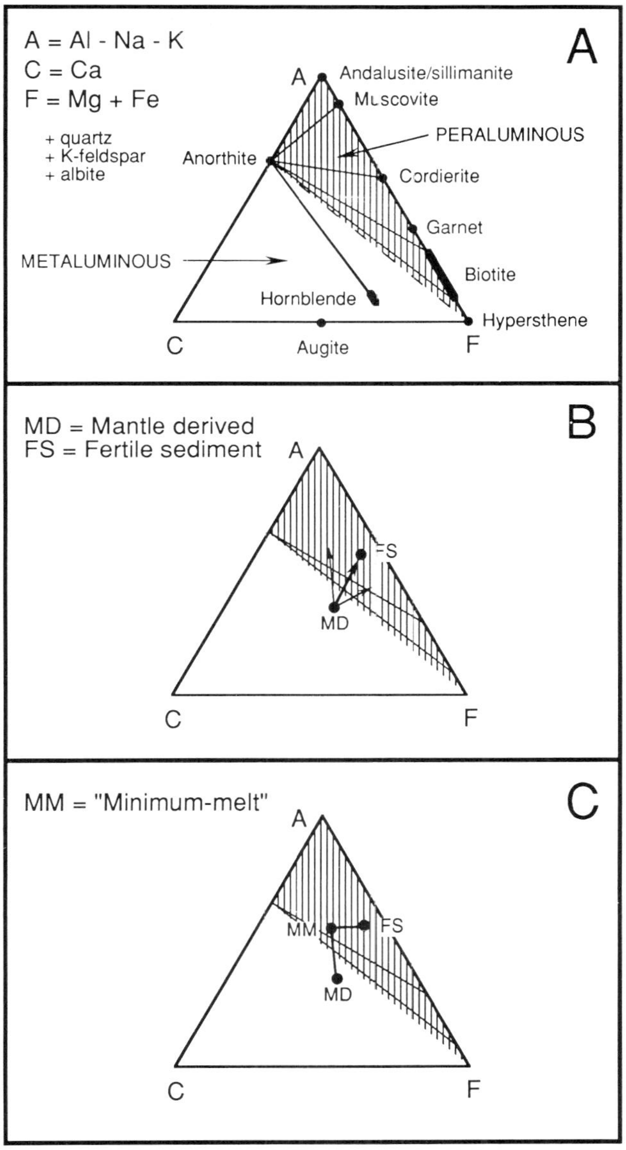

Figure 3 ACF diagrams showing: (A) the relationship between chemistry and mineralogy of I- and S-type granites and volcanic rocks; (B) the evolution of granite source rocks; and (C) the relationship between source rock compositions and derived granite compositions. The hatched area of the diagram corresponds to peraluminous rock compositions. See Sections 10 and 13 of text for a detailed discussion of this figure.

11. Isotopic features of the I- and S-type granites of the Lachlan Fold Belt

Differences in oxygen isotopic composition between the I- and S-type granites of the Berridale Batholith were found by

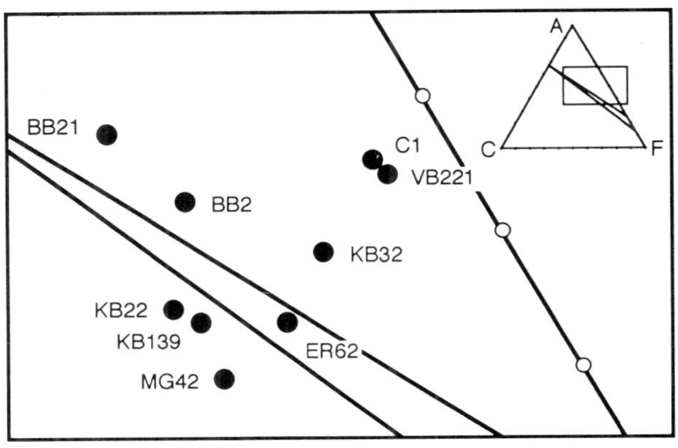

Figure 4 Chemical compositions of some LFB granites from Tables 2 and 3 plotted on an enlarged portion of an ACF diagram.

O'Neil and Chappell (1977), with the two types having $\delta^{18}O$ values less than and greater than 10‰ respectively, relative to SMOW. This division is supported by additional unpublished data on granites of the LFB. Oxygen isotopic compositions could be the best single discriminant between the two granite types in that area, but data are unfortunately limited in areal distribution. As is often the case, this parameter might not appear to be such a good discriminator between the two granite types if comprehensive data were available.

Compston and Chappell (1979) showed that the I-type granites of the Kosciusko, Berridale and Moruya batholiths have a wide range of initial $^{87}Sr/^{86}Sr$ ratios, from 0·704 to 0·712. McCulloch and Chappell (1982) obtained ε_{Nd} compositions for I-type granites from Kosciusko and Berridale between +0·4 and −8·9, with an inverse correlation between the initial Sr and Nd isotopic compositions. S-type granites from the same region extended that correlation, with $^{87}Sr/^{86}Sr$ between 0·7094 and 0·7180 and ε_{Nd} between −5·8 and −9·2. The most evolved values were from the Cooma Granodiorite, for which the data closely matched those from the surrounding high grade metasedimentary rocks (see also Pidgeon & Compston 1965), whereas the isotopic compositions of the batholithic S-type granites (Section 6.3) were always less evolved than the country rocks, with a maximum value of initial $^{87}Sr/^{86}Sr$ of 0·7153 and a minimum ε_{Nd} of −8·8.

Roddick and Compston (1976, 1977) measured Sr isotopic compositions on several bodies of S-type granite in the Murrumbidgee Batholith, north of the Cooma Granodiorite. We have recalculated their total rock $^{87}Sr/^{86}Sr$ compositions to initial values at 415 Ma. For the Shannons Flat Monzogranite, the range of initial $^{87}Sr/^{86}Sr$ is between 0·7083 and 0·7096 for 26 samples, with a mean value of 0·7087; these are less radiogenic values than those of S-type granites referred to above. For the Clear Range Suite, the range in initial $^{87}Sr/^{86}Sr$ values is from 0·7107 to 0·7152, with a mean value for 29 samples of 0·7126; these results exclude one duplicated outlier of 0·7076. As Roddick and Compston (1977) pointed out, some of the total rock isochrons established for rock units of the Murrumbidgee Batholith show significant internal scatter and give apparent ages about 65 Ma older than the maximum possible emplacement age. These observations show that the Sr was not isotopically uniform in the source rocks, and was not well mixed during partial melting and emplacement. Such isotopic heterogeneity is consistent with an origin of these granites from a large volume of metasedimentary source material.

The Bega Batholith is the largest I-type granite complex in the LFB (8620 km^2) and it comprises longitudinal strips that are relatively very uniform in composition, that we refer to as supersuites. Transverse to these there are some strong compositional asymmetries. Towards the west, away from the present continental margin, the average ε_{Nd} initial isotopic compositions determined by M. T. McCulloch for the various supersuites changes from +3·5 to −6·6, and for $^{87}Sr/^{86}Sr$ from 0·7040 to 0·7094 (McCulloch *et al.* 1982; Chappell *et al.* 1990). At the same time, the $\delta^{18}O$ values also increase from near 8‰ to 9–9·5‰. These changes correlate with marked decreases in Na and Sr and lesser ones for Al and P. Ca and Sc increase significantly towards the west, while Rb and V increase slightly. None of the other elements that have been measured show any systematic changes. All of these changes take place in rocks that we regard as I-type and their significance is discussed briefly in Section 12.

Gray (1990) examined Sr isotopic compositions in a traverse across the LFB, using both previously published information and a considerable body of new Sr isotopic data for granites of the central and western parts of the belt. For 30 I-type plutons, his new initial $^{87}Sr/^{86}Sr$ values are in the range from 0·7042 to 0·7121, essentially the same as that established by Compston and Chappell (1979). For nine batholithic S-type granites examined by Gray (1990), the initial ratios were in the interval from 0·7082–0·7166, and two granites of the Cooma Supersuite from locations almost 190 km W of Cooma yielded values of 0·7188 and 0·7197.

The above data for the initial composition of the Sr and Nd isotopic systems in the LFB can be summarised as follows. For I-type granites the range in initial $^{87}Sr/^{86}Sr$ is from 0·704 to 0·712 and for ε_{Nd} from +3·5 to −8·9. For the batholithic S-type granites the corresponding values are 0·708–0·717 and −5·8 to −8·8. For the granites of the Cooma Supersuite the values are 0·718–0·720, with one ε_{Nd} measurement of −9·2. There is a large overlap between the I- and S-type initial isotopic compositions, with a continuity in $^{87}Sr/^{86}Sr$ from 0·704 to 0·720, and a strong negative correlation between initial Sr and Nd isotopic values. Gray (1984) proposed that these transitional isotopic features resulted from the mixing of two components in differing proportions. This was a possibility examined earlier by McCulloch and Chappell (1982) who did not favour such a process in the case of the granites that they examined because of difficulties in accounting for the chemical composition of the granites using such a model. These difficulties will be examined further in Section 12.

12. Infracrustal and supracrustal source rocks

We have seen (Section 11) that there is a significant overlap in the initial Sr and Nd isotopic compositions of the I- and S-type granites of the LFB, and a strong negative correlation between initial Sr and Nd isotopic values. Gray (1984) proposed that these transitional isotopic features resulted from the mixing of two components of different ages, in differing proportions, to produce the source rocks of the granites. He argued that all Sr and Nd isotopic data published at that time for the granites of the LFB lie on a mixing line between basalt and a granite magma derived from metasediments. Gray (1990) presented additional isotopic data and again argued for the mixing of a granite component produced by melting the Ordovician basement with basalt, to produce the observed isotopic compositions. This is one possible interpretation of the isotope data, but it is not a unique solution, and we will show that it not consistent with the chemical data available for the LFB granites.

12.1. Evidence from the chemical compositions of the S-type granites

White and Chappell (1988) pointed out that in mafic S-type granite suites, the rocks become more peraluminous as they become more mafic and such a trend is clearly shown by figure 9 of that paper which plots ASI against total Fe content for granites of the Bullenbalong Suite. Such variation is, of course, not consistent with the more mafic rocks containing a greater amount of basalt component, and as White and Chappell (1988) noted, it is precisely the opposite to what the mixing model of Gray (1984) would predict. Gray (1990) has argued that the data plotted by White and Chappell (1988) in fact define two trends, one with a negative correlation between ASI and total FeO at high Fe, produced by mixing, and the other a positive correlation at lower Fe contents, produced by fractional crystallisation. However, if fractional crystallisation was important in producing the range of rocks of the Bullenbalong Suite, then the more mafic rocks would have to be regarded as cumulates, otherwise liquidus temperatures would be extremely high. If they are cumulate rocks, then they would contain a greater proportion of any recrystallised basalt component. However, there are no hornblende-bearing enclaves in these granites and the chemical and mineralogical compositions of the granites are not consistent with the presence of a basalt component. For example, sample KB32, a mafic member of the Bullenbalong Suite (Table 3), contains 4·64% total FeO but only 2·26% CaO and 1·92% Na_2O, and 3·89% normative corundum. These data are consistent with the presence of a large amount of a sedimentary component, but not of basalt, in this rock. These features can also be correlated with the presence of abundant metasedimentary enclaves, including lumps of quartz. Field, petrographic and chemical data on these mafic S-type granites indicate that the dominant cause of variation between high Fe and ASI granites, and low Fe and ASI granites, is the presence of variable amounts of restite, which to account for the compositional variations, must be Fe-rich and strongly peraluminous.

12.2. Chemical and isotopic differences between S-type suites

A critical test of the mixing hypothesis for the generation of the range of S-type granites will be to examine in detail the isotopic variation within individual suites as a function of chemical composition, which has not yet been done. However, data do exist that enable such variation to be examined between S-type suites, which show that there is a broad correlation of chemical and isotopic composition between suites and this will now be examined.

Consider the chemical and isotopic compositions of the Cooma, Bullenbalong and Dalgety suites, represented by samples CC1, KB32 and BB11 in Table 3. In chemical composition, these suites represent a sequence of increasing Na, Ca and Sr contents for rocks containing comparable amounts of total Fe. The data in Table 3 show that while BB11 contains only slightly more Ca (and much more Sr) than KB32, it is a substantially more felsic rock. Because of the increasing Na and Ca contents, this is also a sequence of decreasing normative corundum, with values of 5·82%, 3·89% and 1·44% in these three samples. The data of McCulloch and Chappell (1982) show that this is also a sequence of progressively less evolved initial $^{87}Sr/^{86}Sr$ compositions, with Cooma ~0·718, Bullenbalong ~0·715 and Dalgety ~0·710. Cooma, with the lowest Ca, Na and Sr contents, is the most isotopically evolved, with initial $^{87}Sr/^{86}Sr$ values matching those of the surrounding Ordovician metasediments (Pidgeon and Compston 1965). In the absence of any other information, there would be two ways in which the production of these differences in both chemical and isotopic composition between these three suites could be interpreted, and in both cases the Cooma compositions represent an end-member. These data alone are consistent with the model of Gray (1984, 1990) in which the differences between Cooma and the other two suites would be accounted for by the addition of different amounts of more isotopically primitive material containing greater abundances of Ca and Sr, that is a basaltic component. The alternative explanation is that the differences are the result of the partial melting of aged sources of different sedimentary maturities and feldspar contents reflected in differences in Ca and Sr contents and hence also of Rb/Sr, and therefore of $^{87}Sr/^{86}Sr$ at the time of granite formation. The second model is preferred because it is consistent with the observed chemical variations *within* suites (Section 12.1), and because there is no evidence for any basaltic component as enclaves in the less isotopically evolved suites. This model implies that in those parts of the LFB in which the large S-type batholiths occur, there is a metasedimentary basement to the exposed Ordovician sediments, that is less mature and less isotopically evolved than those rocks. Such a pre-Ordovician basement has also been proposed by Wyborn and Chappell (1983) as the source for the detritus in the Ordovician flysch deposits.

12.3. Chemical and isotopic compositions of I-type granites

12.3.1. Bega Batholith. The overall chemical and isotopic characteristics of the Bega Batholith have been summarised in Section 11. Because of its regional compositional patterns and asymmetries, this large I-type granite complex is critical to understanding I-type source rocks in the LFB. Most of the compositional changes that are observed, that is the more evolved Sr and Nd, higher $\delta^{18}O$, and lower Na and Sr, but not the higher Ca, are consistent with the incorporation of a progressively increasing amount of a sedimentary component in the granites towards the western parts of the batholith. The higher Ca contents in the granites on the western side restrict the amount of a sedimentary component similar to the Ordovician country rocks (average CaO content ~0·1%) that can be present. Williams *et al.* (1988) report that inherited zircons are common in the granites of the Bega Batholith and that the relative and absolute abundances of the oldest inherited components (between 1150 and 3350 Ma) increase westwards, but in virtually all cases it is the youngest inherited component (<650 Ma) that predominates.

The more isotopically evolved compositions of the I-type suites on the western side of the Bega Batholith cannot be simply the result of mixing varying proportions of materials derived from older igneous and sedimentary components, because the isotopic variation within those suites is not correlated with changes in chemical composition. For the Glenbog Suite, comprising twelve separate plutons that extend over a distance of 265 km close to the western edge of the batholith, with a range of average Fe contents, Chappell and McCulloch (1990) report that initial $^{87}Sr/^{86}Sr$ ranges from 0·7083 to 0·7108 and total FeO from 1·7% to 4·7%, with no correlation between chemical and isotopic compositions of individual samples. These data imply that the amount of sedimentary material in the Glenbog Suite magmas was either too small to be detected isotopically, or was rather uniformly mixed into the extensive source rocks prior to partial melting.

12.3.2. Jindabyne Suite. The Jindabyne Suite comprises nine plutons of tonalite with minor amounts of granodiorite and quartz diorite, on the eastern side of the Kosciusko Batholith (Hine *et al.* 1978). The plutons were emplaced into Ordovician sedimentary rocks and older S-types granites. In Table 2, sample KB139 is the most mafic rock that has been analysed from this suite, and KB22 is a typical tonalite. The significance of the Jindabyne Suite is that despite the strong I-type chemical and mineralogical features, the rocks have rather evolved isotopic compositions. Compston and Chappell (1979) reported initial $^{87}Sr/^{86}Sr$ values in the range 0·7066 to 0·7070. For the samples KB139 and KB22, McCulloch and Chappell (1982) measured initial $^{87}Sr/^{86}Sr$ values of 0.7065 and 0.7077 and ε_{Nd} of −3·7 and −5·0. McCulloch and Chappell (1982) showed that if these isotopic compositions resulted from mixing two components with the compositions of depleted mantle, and continental crust as represented by the Cooma Granodiorite, these would be represented in about equal amounts in the source. They argued that the chemical composition of the Jindabyne Suite did not allow the presence of such a large fraction of sediment in the source rocks, as the abundance of K, Rb, Cr and Ni in the granites is too low, and the amount of Ca is too high. McCulloch and Chappell (1982) concluded that the Jindabyne Suite was produced from material of predominantly mantle derivation that had aged in the crust prior to generation of the granites. This supported the earlier conclusion of Compston and Chappell (1979) that the relatively evolved isotopic character of the Jindabyne Suite resulted from the aging of primitive source materials for ~700 Ma prior to partial melting at about 410 Ma BP.

The chemical data on the Jindabyne Suite would allow the presence of a small component of sedimentary material in the source rocks. Such a component is required by the zircon U-Pb ages for sample KB22 measured by I. S. Williams (pers. comm.), who in a sample of approximately forty grains, purposely biased towards older cores, found ten cores with ages in the range from 600 Ma to 1200 Ma, and three that were older. The older cores are unequivocally of sedimentary origin while the younger inherited crystals could either be from sediments, or from older igneous rocks of the type postulated by Compston and Chappell (1979) for the source materials of the suite. Quantification of the amount of sedimentary material based on the zircon inheritance ages is fraught with problems stemming from the question of zircon survival. However, their low abundance probably restricts the amount to less than 10% of the total source material, since in mafic S-type granites known to be of sedimentary origin, for example KB32 in Table 3, every zircon crystal contains an inherited core (I. S. Williams, pers. comm.). Hence, the combined chemical, isotopic and age data for the Jindabyne Suite lead to a similar conclusion to that reached for the western suites of the Bega Batholith, namely that there is a component of sediment in the source rocks of these isotopically evolved suites, but that the amount is not sufficient alone to account for the evolved isotopic compositions.

12.4. The bimodal nature of chemical composition of the mafic I- and S-type granites

It was shown in Section 6 that the compositions of the most mafic granites of the LFB are bimodal. These rocks are closer in composition to the source rocks than the more felsic granites, implying that the compositions of the source rocks of the granites were also bimodal. When these differences were expressed as the ASI parameter, it was seen that there was a significant range in ASI within the S-type sources, and probably also within those of the I-type granites. In Section 6 it was noted that these ranges could have been partly produced by the mixing of igneous and sedimentary source materials, with the limitation that in no case did such mixing produce source rock compositions close to ASI = 1, so that if mixing did occur the sources retained a dominantly 'igneous' or 'sedimentary' character. We have argued in Section 12.2 that the differences in ASI for the source rocks of the S-type granites resulted from different degrees of weathering of the source materials. For the I-type source rocks, the inherited zircon ages show that there was a component of sedimentary material in some of the source rocks, but the chemical data and their relationships with the isotopic information imply that such material was much less abundant than the other components in the source rocks. In the case of isotopically evolved I-type suites such as Glenbog and Jindabyne, this restriction on the amount of sediment implies that the other components in the source had aged prior to partial melting, as proposed by Compston and Chappell (1979).

12.5. A special case—the Murrumbucka Suite

The Murrumbucka Tonalite (115 km^2), a unit of the Murrumbidgee Batholith, has properties that elsewhere are distinctively I- or S-type (Reid *et al.* 1983), and this rock was perhaps derived from a mixed igneous and sedimentary source. It contains both hornblende-bearing and metasedimentary enclaves (Section 9). It has a metaluminous composition and contains hornblende, which are I-type characters, but has low Na and Sr and high Cr and Ni, contains abundant biotite (>15%), and has evolved isotopic compositions ($^{87}Sr/^{86}Sr \sim 0{\cdot}71$, $\varepsilon_{Nd} < -8{\cdot}6$), that are typical of S-type granites (Reid *et al.* 1983). The Arable Granodiorite (60 km^2) in the Berridale Batholith, some 20 km SW of the Murrumbucka Tonalite, shares some of the chemical features of the latter unit and the two are grouped in the Murrumbucka Suite. These two granites are the only candidates for significantly mixed source rocks in the LFB, but they make up only 0·3% of the total area of granites in the belt.

12.6. A further contrast between I- and S-type granites—homogeneity versus heterogeneity

For those I-type granites in which the variation is due to separation of melt and restite, element variation diagrams show high degrees of correlation. For S-type granites, the correlations are much poorer, and this is ascribed to their derivation from more heterogeneous source rocks. The implication is that the I-type source rocks were relatively homogeneous through volumes large enough to produce plutons up to 1000 km^2 in area in the Lachlan Fold Belt and also over distances of up to a few hundred kilometres, in cases where there are well-defined suites that consist of several plutons.

12.7. Infracrustal and supracrustal source rocks

The homogeneity of I-type source rocks was one of several arguments put forward by Chappell and Stephens (1988, section 5.1) for I-type granites having been derived from deeper levels than S-types, and from material produced by earlier underplating of the crust, rather than from supracrustally deposited igneous rocks. We therefore suggest that the I- and S-type granites were derived from source rocks of fundamentally different origin, one formed by deposition on the crust, the other by accretion beneath the crust. Following the conclusion of White and Chappell (1984), we therefore have:

S-type = Sedimentary or Supracrustal

I-type = Igneous or Infrastructural

It is because the source rocks of I- and S-type granites are so fundamentally different in their origins, that granites showing transitional characteristics are so rare in the LFB.

The amount of igneous supracrustal material incorporated in the S-type source rocks of the LFB was very small. A component of sedimentary material was present in the source rocks of some I-type granites, but the igneous component was always dominant. Chappell and Stephens (1988) suggested that the sedimentary component in I-type sources probably represents material that had previously been subducted, which would be consistent with the uniformity of I-type source rocks over large areas.

13. A sequence of granite sources

The final stage in the development of the granites of the LFB involved the partial melting of crust (Section 4). At an earlier stage of evolution, all of that crust was produced by partial melting of the mantle, with the S-type granites being derived after that more primitive crust was modified by weathering. The various stages involved in the evolution of I- and S-type granite sources can be illustrated using an ACF diagram, and this is done in Figure 3B. Such a diagram can also be used to illustrate the relationship between the compositions of granites and their source rocks, as is shown in Figure 3C.

13.1. Production of granite source rocks

Irrespective of whether I-type granites are produced more directly by fractional crystallisation of mantle-derived liquids, or less directly by partial melting of mantle-derived solids, the source materials will be broadly basaltic to andesitic in composition and their position on an ACF diagram can be represented by the composition MD in Figure 3B. The production of I-type granites from composition MD will be discussed in Section 13.2.

The ability of a sedimentary rock to produce a body of granite magma, or its fertility, is primarily dependent on the presence of sufficient Na_2O, K_2O and SiO_2 to provide a significant fraction of melt (White & Chappell 1988). H_2O is also necessary and this may be provided by biotite. These sandy shales or feldspathic greywackes will be peraluminus and are represented by the position FS in Figure 3B. All sedimentary rocks are themselves ultimately produced from mantle-derived solid rocks corresponding to composition MD, and in terms of the parameters represented in Figure 3B there are two main processes that achieve this result, both involving the conversion of feldspars to clay minerals (Section 6.1). The loss of Ca and Na produces ‘vectors’ away from the C component for Ca, and towards the A component for Na. The effect of these changes is to move the bulk composition of the weathered system from MD to FS in Figure 3B.

A sequence of source compositions between MD and FS could be produced in two ways. First, the position along the MD to FS join could be a function of the degree of weathering, so that more strongly peraluminous compositions result from more intense weathering, leading to points in Figure 3C closer to and then extending beyond FS. We favour this mechanism for the production of the source rocks of various S-type suites, within the peraluminous field of Figure 3B (Section 12.2). Second, a similar sequence of compositions would result from mixing mantle-derived and sedimentary components. Our observations in the LFB show that the source rocks of the granites are bimodal (Section 6.3) and we consider that the amount of mixing of these two end-members is limited (Section 4).

13.2. Production of granite magmas by partial melting within the crust

Granite magmas are produced when the amount of melt exceeds the critical melt fraction of van der Molen & Paterson (1979). If that melt is all produced at ‘minimum’ temperatures then its composition will be slightly peraluminous, both for I- and S-type source rocks, and its composition will correspond to the position MM in Figure 3C. The magma, or ‘minimum melt’ plus restite in this case, will initially have the same composition as that of the source material, and with fractionation of those two components, compositions between MM and either MD (I-type) or FS (S-type) will be produced. This is the situation with, for example, the I-type Moruya Suite (Griffin *et al.* 1978), and apparently with all S-type suites in the LFB.

If melting continues to higher temperatures then the composition of the melt will change and move towards that of the source rock. Again, a series of compositions will be generated, but felsic compositions corresponding to MM will only result when all restite has been removed and further crystal fractionation has occurred. Such high temperature and generally more mafic I-type granites are not uncommon in the LFB, for example with the Jindabyne Suite (Hine *et al.* 1978 and Section 12.3.2 and Table 2). If restite separates from a high temperature melt at a very early stage, then a completely liquid magma that formed in the crust and with a composition close to MD can produce a series of compositions between MD and MM, as is the case for the Boggy Plain Supersuite (Wyborn *et al.* 1987).

The fractional crystallisation of a mantle-derived melt would produce a series of compositions along the line from MD to MM, similar to the products of melts such as Boggy Plain derived from the crust.

Figure 3C shows why all S-type granites are peraluminous. It is also consistent with the observation that felsic I-type granites are peraluminous and the more mafic varieties metaluminous. Also, since the S-type granites of the LFB were derived from source rocks of different maturity, that is within the peraluminous part of the line from MD to FS in Figure 3B, the compositions of S-type granites are dispersed through an area from those that are strongly peraluminous and cordierite-rich, to others that plot within the anorthite-biotite field, and are cordierite free; this is illustrated by the compositions VB221 and ER62 in Figure 4.

14. Strongly fractionated granites

The original subdivision into I- and S-type granites by Chappell and White (1974) was made using rocks from the Berridale–Kosciusko region (Section 1). We have seen in Section 6.2 that in other parts of the LFB there is overlap in composition between the two types, particularly among the more felsic rocks, but not among the most mafic rocks. Overlaps are not seen in the original area (e.g. Hine *et al.* 1978) and thus the criteria proposed by Chappell and White (1974) to separate granites in the Berridale–Kosciusko region are not always precisely applicable elsewhere in the LFB, or in places outside the LFB. While this overlap arises because of the presence of more felsic rocks for which it might be difficult to make the I- and S-type discrimination, the concept that the rocks have come from two distinct types of source remains valid.

The restricted range in composition of ‘minimum-temperature’ felsic granites can make it difficult to assign an individual rocks to either the I- or S-type group (Section 6.2). If such a composition has been further modified by fractional crystallisation, then the resulting bulk rock composition will be moved even further away from the more mafic compositions for which the original criteria of Chappell and White (1974) were proposed. We now examine the effects of such processes during crystallisation and show that new criteria can be erected that discriminate between the two granite types in crystal fractionated felsic compositions.

14.1. The geochemical effects of fractionation of feldspar

Some felsic granites within the LFB have undergone crystal fractionation of quartz and feldspars at near 'minimum-temperature' conditions. Such a process moves the major element composition of the melt into the haplogranite field where I- and S-type granites overlap, but at the same time produces distinctive changes in trace element compositions that can be used to discriminate between the two types.

The effect of prolonged feldspar fractionation is to concentrate in the melt those elements with bulk partition coefficients less than one, such as Rb, and to deplete those with coefficients greater than one, for example Ba and Sr. Hence the abundance of those three elements can be used as an indicator of magmatic crystal fractionation. We recognise that similar changes can result from hydrothermal alteration, which removes feldspars and some of their trace elements in another way, that is by post-magmatic alteration. However in the general case we consider that fractional crystallisation has been the dominant process if the Na:K ratio of the rock is close to those of a 'minimum-temperature' melt and if the abundances of Ba, Sr and Rb, and also Ca, are within the limits that can be reached when reasonable assumptions are made about the magnitude of partition coefficients and the degree of fractional crystallisation (Chappell *et al.* in prep.). For example, for the two I-type (VB230 and TB12) and two S-type (VB140 and TB83) compositions in Tables 2 and 3, that show the effects of crystal fractionation, those of VB230 and VB140 can be reached by moderate amounts of feldspar fractionation and any hydrothermal alteration that might have occurred cannot be detected in those compositions. The most evolved I-type sample TB12, has a magmatic Na:K ratio (compare with VB230), but the very low Ba and Sr levels do indicate some alteration. For the S-type sample TB83, the Ba content of 2 ppm requires alteration, and this is supported by the extremely high Rb and very low Sr, Pb and CaO contents, and also the very high Na:K ratio (see also Section 6.4.2).

To assist in examination of the effects of feldspar fractionation on granite compositions, averages of 'unfractionated' and 'fractionated' felsic granites, both I- and S-type, are listed in Table 4. We use the term 'unfractionated felsic granite' to refer to felsic granites formed from magmas that had lost all or most of their restite, but from which there had been no separation of precipitating crystals. The term 'fractionated felsic granite' refers to compositions resulting from the separation of feldspars from a crystallising felsic granite melt. We take the boundary between unfractionated and fractionated felsic granites of the LFB at 250 ppm Rb. That element is used rather than Ba or Sr or some combination of the three, because in the LFB it has the least variable concentration in granite source rocks and primary melts. This value of 250 ppm chosen as the limiting concentration is arbitrary, because the actual value will vary from one suite to another and also because the processes of restite and crystal fractionation are not completely mutually exclusive, with a transitional region where both may operate. Nevertheless, it is useful to make this distinction and the value of 250 ppm Rb that has been chosen represents a concentration which is probably near the upper limit of the transition from restite to crystal fractionation in most cases. For example, in the unfractionated felsic granites of the Bullenbalong Suite (White & Chappell 1988) the highest Rb content is 233 ppm. Samples BB2 and CV115 in Table 3, with Rb contents of 236 and 244 ppm, are thought to be close to primary S-type 'minimum-temperature' melts that have not undergone crystal fractionation. Also, the two most felsic but unfractionated felsic I-type granites in Table 2, CV87 and BB21, contain 215 and 214 ppm Rb.

14.2. General compositional effects of fractionation

As representatives of clearly fractionated felsic granite compositions, the averages of all 64 I-type and 119 S-type granites from the LFB containing more than 300 ppm Rb, are separately listed in Table 4. The unfractionated felsic granite compositions also listed in that table are averages of all granites containing between 200 and 250 ppm Rb, that is 131 I-type and 160 S-type samples. A general view of the differences in composition between unfractionated and fractionated felsic granites, both I- and S-type, and between both types after crystal fractionation, can be obtained by inspection of the data in Table 4, and also by further examination of the last three analyses in Tables 2 and 3.

Except for P, Pb, Th, Y, La and Ce, the elements listed in Table 4 change in concentration in the same direction with crystal fractionation in both I- and S-type granites. For the major elements, the changes are dominated by the removal of biotite or orthopyroxene, and plagioclase. For the trace elements, Rb, U, Nb, Ga and Sn increase, and Ba, Sr and Zr and the transition metals decrease in abundance with crystal fractionation. These changes are also seen for the last three I-type granites in Table 2 and, apart from U, for the last three S-type granites in Table 3. We note that while there is a distinct tendency for U to increase in amount with fractional crystallisation, its abundance in highly felsic granites is always erratic, apparently as a result of post-magmatic mobilisation even in rocks that show no other sign of late stage alteration.

14.3. The distinction between fractionated felsic I- and S-type granites

Five of the elements that are listed in Table 4 are significantly different in their pattern of abundance in moving from unfractionated to fractionated I- and S-type granites. P_2O_5 moves in opposite directions with fractional crystallisation, decreasing in the I-type and increasing in the S-type granites. Th and Y increase substantially in amount with crystal fractionation in the I-type group, while remaining fairly constant in the S-types. La and Ce decrease in the S-type while remaining steady in the I-type granites. All of these changes are also shown strongly by the last three analyses in both Tables 2 and 3. In addition to these changes, there are differences for Na_2O and Pb. The former increases more strongly with crystal fractionation in the S-type averages. Pb increases in abundance in the I-type averages while decreasing slightly in the S-types, but these changes in concentration are small and may not be significant.

Table 4 Average compositions of some felsic granites from the Lachlan Fold Belt

	Unfractionated I-type	Fractionated I-type	Unfractionated S-type	Fractionated S-type	All A-type
ppm Rb	200–250	>300	200–250	>300	
No. of samples	131	64	160	119	43
SiO_2	72·90	76·17	71·58	74·40	73·47
TiO_2	0·30	0·10	0·42	0·16	0·30
Al_2O_3	13·48	12·51	13·83	13·50	12·88
Fe_2O_3	0·54	0·32	0·45	0·28	0·90
FeO	1·47	0·71	2·38	1·14	1·63
MnO	0·05	0·04	0·05	0·04	0·06
MgO	0·66	0·12	1·02	0·27	0·30
CaO	1·63	0·61	1·74	0·67	1·06
Na_2O	3·27	3·37	2·57	3·06	3·50
K_2O	4·42	4·92	4·33	4·84	4·62
P_2O_5	0·09	0·02	0·14	0·18	0·07
FeO total	1·96	1·00	2·79	1·39	2·44
Trace elements (ppm)					
Ba	488	99	512	150	547
Rb	219	424	221	475	188
Sr	147	31	114	43	96
Pb	29	35	28	25	27
Th	25	47	19	17	24
U	6	16	4	11	5
Zr	151	116	168	92	322
Nb	14	21	12	19	26
Y	38	75	34	28	71
La	35	35	28	16	55
Ce	74	79	63	37	130
Sc	8	6	10	5	11
V	25	3	41	7	9
Cr	10	<1	23	7	2
Co	6	3	9	4	3
Ni	5	<1	10	2	2
Cu	6	2	7	3	5
Zn	38	29	53	46	95
Ga	16	19	17	21	22
Sn	7	13	8	23	8
Normatic C	0·57	0·58	2·09	2·44	0·36

Explanation of Table 4: The term "unfractionated felsic granite" refers to felsic granites formed from magmas that had lost all or most of their restite, but from which there had been no separation of precipitating crystals. The term "fractionated felsic granite" refers to compositions resulting from the separation of feldspars from a crystallising felsic granite melt, as indicated by the Rb content.

The most distinctive difference between the compositions that result from the crystal fractionation of felsic I- and S-type melts is that P decreases in abundance in the I-type and increases in the S-type compositions. To illustrate this, in Figure 5 we have plotted P_2O_5 against Rb, separately for both types, for rocks containing more than 250 ppm Rb. The least fractionated S-type granites contain a range of P_2O_5, between 0·03 and a little more than 0·20% P_2O_5, with the main body of samples being above 0·08%. With increasing Rb, the P_2O_5 content of these rocks immediately starts to rise and continues to do so as Rb increases to more than 1000 ppm. The P_2O_5 data for the more fractionated rocks show a broad scatter, imaging the variable abundance in the least fractionated rocks. For the I-type granites, the pattern is dramatically different. Below 320 ppm Rb, most P_2O_5 values (44 analyses) are from 0·01 to 0·06% with only 11 values of 0·08% and above. Above 320 ppm Rb, such higher values are no longer present and the P_2O_5 levels fall progressively so that the six samples containing more than 600 ppm Rb contain a maximum of 0·01% P_2O_5.

Figure 5 illustrates this dramatic difference between the behaviour of P_2O_5 in I- and S-type felsic magmas, that is between mildly peraluminous (~0·5% normative C) and strongly peraluminous (>2% normative C) melts, during crystal fractionation. This difference results from the high solubility of P in strongly peraluminous melts (Montel *et al.* 1988). It is clear from the data presented here that the solubility of P is very low in weakly peraluminous (~0·5% normative C) felsic I-type melts. In contrast, we have observed a maximum P_2O_5 concentration of 0·42% (Fig. 5), in two felsic granites containing 4·04 and 4·87% normative C. Similarly elevated P_2O_5 concentrations (0·5%) have been found in the strongly peraluminous (~5% normative C) Macusani glasses (Pichavant *et al.* 1988).

A quarter (49 of 197) of the S-type points in Figure 5 are from the Koetong Supersuite. This is the largest such unit in the LFB, with an area of a little over 10,000 km^2 (16% of the total granites), extending over a distince of 410 km in the Wagga Batholith in the centre of the LFB. Within that supersuite there is a complete progression from mafic granites containing metasedimentary enclaves, through to 'two mica' or topaz-bearing granites with high Rb and P abundances, with all rocks being peraluminous (Chappell *et al.* in prep.). All rocks of this supersuite are clearly S-type, and are probably derived from source rocks with a composition close to the mafic granites. This observation links the granites in which there is a positive correlation between Rb and P contents, very clearly to sedimentary source rocks. An analogous sequence of I-type rocks through from metaluminous mafic compositions to slightly peraluminous felsic granites high in Rb and low in P is provided by the Boggy Plain Supersuite (Wyborn *et al.* 1987) which provides 9 of the 106 I-type points in Figure 5.

While the average I-type compositions in Table 4 confirm the earlier observation in Section 6.2 that felsic I-type granites are slightly peraluminous, they also show that they remain that way with crystal fractionation. The normative C contents of the two felsic granite averages in Table 4 of 0·57% and 0·58% imply that fractionation involved the separation of phases with about the same average bulk normative C content. The major mineral phases were mainly quartz and feldspars (ASI = 1) and lesser biotite (ASI > 1) which together would have had the slightly peraluminous composition required by the I-type averages. The separation of that group of minerals from a more peraluminous liquid would increase the normative C content of the derived liquids, which is precisely what is observed in the two average S-type compositions of Table 4.

The felsic granites of North Queensland discussed by Champion and Chappell (1992) are an excellent example of a regional development of strongly fractionated I-type granites. Likewise, the Cornubian granites are a classic example of fractionated S-type granites and we note that the 26 Cornubian granites analysed by Chappell and Hine (in prep.) contain between 346 and 675 ppm Rb and 0·18 and 0·31% P_2O_5 and are thus firmly in the crystal fractionated S-type granite field of Figure 5. Their average Cornubian composition CNG, which also incorporates previously published data on those rocks, contains 483 ppm Rb, 0·25% P_2O_5 and 3·23% normative C.

14.4. The implications for unfractionated felsic granite compositions

The observation that for granites containing more than 500 ppm Rb, the P contents are distinctly bimodal, has important implications for the primary compositions of less fractionated felsic granites. The gap between P concentrations at high Rb contents strengthens the view that the variation observed in Figure 5 must have been produced at high temperatures under magmatic conditions. If those granite compositions were principally the result of post-magmatic alteration, then the gap that is present between the I- and S-types, that is between the P contents of mildly and strongly peraluminous compositions, should not be present, since such alteration would involve the mobilisation of Rb, P and other elements. The fact that the gap is consistent with data on the solubility of P in silicate liquids established experimentally *at high temperatures,* confirms a dominantly magmatic origin for these compositions. The compositions are completely magmatic (Section 6.2), and the data represented in Figure 5 in fact support the operation of widespread but in most cases minor hydrothermal alteration in the felsic granites of the LFB, which will now be discussed.

The observation that the fractionated felsic I- and S-type granites of the LFB have distinctive P contents that correlate with the degree of Al oversaturation, implies that the unfractionated felsic liquids from which they were originally derived also differed in Al oversaturation. Hence the overlap that is observed in ASI values of the felsic granites (Section 6.2) cannot have been a feature of the primary melt compositions, because if that had been the case the more strongly fractionated liquids would not have separated into two distinct groups. Development of the observed ASI overlap must have occurred after the magmatic stage which established the fractionation directions, and must therefore have occurred as the result of alteration of feldspars and the partial mobilisation of Na, K and Ca, as pointed out in Section 6.2.

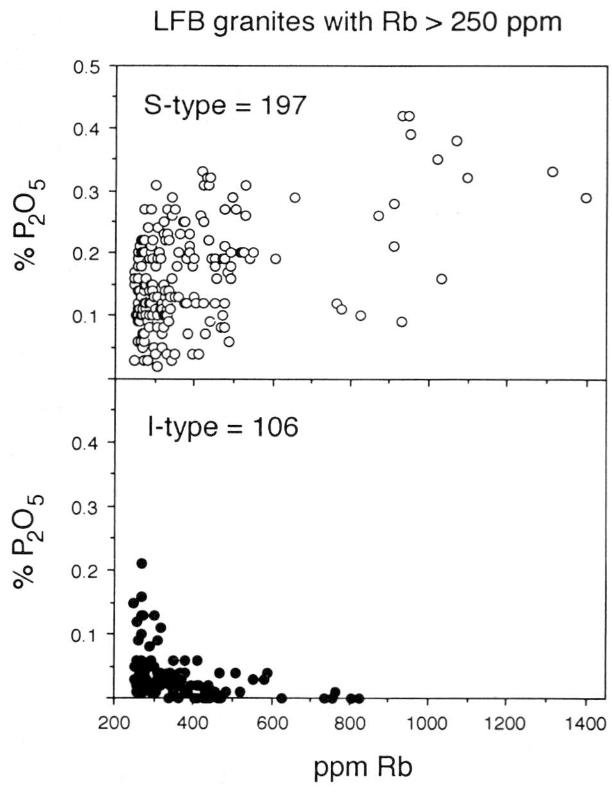

Figure 5 Variation diagrams of P_2O_5 against Rb for all LFB granites containing more than 250 ppm Rb, 197 S-type and 106 I-type.

14.5. A comparison of felsic I- and S-type granites with A-type granites

The average composition of the 43 A-type granites from the LFB that we have analysed is listed in Table 4. Collins *et al.* (1982) compared the compositions of the Gabo and Mumbulla A-type suites, comprising 17 of these analyses, with I-type granites of the adjacent Bega Batholith and showed that it was the higher contents of a group of highly charged cations in the A-type group that is most distinctive, including Y, the REE and Ga, and also Zr and Nb in the less felsic Gabo Suite. With additional data, and a fuller understanding of the granites of the LFB, we can briefly explore these differences further. A detailed account of the A-type granites of the LFB will be presented by King *et al.* (in prep.).

The A-type average in Table 4 is rather close in major element composition to the I- and S-type averages. Most major elements in the A-type average are within the interval for a given element for the two I-type compositions, except that the A-type average contains more total FeO, MnO and Na_2O. Compared with the S-type compositions, in addition to those differences, the A-type average contains less Al_2O_3 and P_2O_5.

The distinctive feature of the A-type granites become clear when the trace element contents listed in Table 4 are considered. We note first, however, that for some of these elements (Ba, Pb, Th, U and Cu) the A-type average is very close to both the unfractionated felsic I- and S-type averages, and for Rb and Sr it would be very close if a slightly less Rb-rich interval and not one between 200 and 250 ppm, had been chosen for calculating the unfractionated felsic granite averages. For other elements (V, Cr, Co and Ni) the A-type average is similar to that for the fractionated I-types. However, the A-type average values are much higher than all I- and S-type compositions in Table 4 for Zr, Nb, Y, La, Ce, Sc, Zn and Ga. For Nb and Ga, the fractionated I- and S-type averages move towards the A-type values, relative to the unfractionated values, as also does Y for the I-types, but the concomitant changes lead to distinct differences with the A-type average, that is a composition with higher Rb, U and Sn, and lower Ba and Sr.

The important feature of the A-type granites in this context is that most individual samples have not experienced crystal fractionation, as evidenced by their low Rb and high Ba and Sr contents, yet these rocks have distinctively high abundances for several trace elements that the other two granite types only reach for a few of those elements in highly fractionated compositions. For these trace elements (Zr, Nb, Y, REE, Sc, Zn and Ga) there is a discontinuity between A- and I-type compositions, at least in the LFB, and for that reason the A-type granites are regarded as a separate granite type.

It is possible that some of the crystal fractionated granites of the LFB that we regard as being fractionated I-types, might in fact be fractionated A-types. However that could only be so in some cases, because generally the average Nb and Ga contents for the fractionated felsic I-type granites (Table 4) are less than those of the average A-type composition which is unfractionated, and Nb and Ga always increase in amount with crystal fractionation in granites of the LFB. Also, in some cases, those fractionated felsic granites can be traced back through more mafic compositions, for example in the Boggy Plain Supersuite (Wyborn *et al.* 1987), and thus the character of their parental magmas can be firmly established. Further, while either I- or A-type sources might be appropriate for production of some of the most fractionated granites, the former are almost two orders of magnitude more abundant in the LFB ($\sim$30,000 km^2 *vs* $\sim$365 km^2), and are therefore the more probable source.

15. Metallogeny of the two granite types

Chappell and White (1974) recognised that mineralisation is different between the two granite types, with Sn mineralisation appearing to be confined to highly silicic S-type granites while tungsten and porphyry-type Cu and Mo deposits are associated with I-types. The metallogenic aspects of the two types are considered in detail by Blevin and Chappell (1992) and that paper should be consulted for full details. They conclude that the mineral deposits related to granites in eastern Australia are mainly determined by the oxidation state and the degree of fractional crystallisation of the associated granites. Since the oxidation state correlates fairly closely with the nature of the source rocks (Sections 6.1 and 7), there is a good but not perfect correlation between the type of mineralisation and the I- or S-type character of the associated granite.

Major Sn mineralisation is associated with both I- and S-type granites that are reduced and have undergone fractional crystallisation, but the major hard-rock deposits are associated with S-type granites while the mineralisation associated with I-type granites is more dispersed. Cu and Au are associated with oxidised I-type granites of intermediate composition, while Mo is associated with similar granites that are more fractionated and oxidised. W mineralisation is not related to any particular granite type.

16. Summary

Granites comprise a little more than 20% of the area of Palaeozoic rocks of the LFB and related volcanic rocks are also an important component of the belt. A minor group of granites with distinctive trace element compositions (A-types) make up 0·6% by area of the granites (Section 14.5). The Murrumbucka Suite (Section 12) has properties that may have resulted through a derivation from a mixed igneous and sedimentary source (? IS-type); this suite comprises 0·3% of the LFB granites. The remaining granites of the LFB (>99%) have properties that are a result of derivation from either dominantly or completely igneous or sedimentary sources. These are the I-type or infracrustal granites, and the S-type or supracrustal granites (Section 12).

The compositional features that disciminate between the two granite types are seen most clearly in the more mafic rocks (Section 6.3). Many of these features converge as the compositions of the rocks become felsic, and it may be difficult to discriminate between felsic rocks of the two types when they are not associated with more mafic rocks in the same suite, although the concept of a separation into I- and S-types remains valid (Section 6.2). When felsic magmas undergo fractional crystallisation, the compositions of some elements again diverge, and the two types can again be easily distinguished.

The more important features that discriminate between the I- and S-type granites are the following.

(1) Relative to I-type granites, the S-types contain significantly less CaO, Na_2O and Sr, three elements that are lost when feldspars are altered to clay minerals by weathering or hydrothermal alteration, and the S-types usually have lower Fe^{3+}/Fe^{2+} ratios (Section 6.1).

(2) As a consequence of the lower Ca and Na contents, the S-type granites are always peraluminous, and the observed range is from ASI = 1·01 to 1·99. I-type granites are about equally divided between metaluminous and peraluminous compositions, lying mostly between ASI = 0·69 and 1·10, with others thought to be I-type up to ASI = 1·20 (Section 6.2). Mafic I-type granites are relatively metaluminous and the mafic S-types more peraluminous than felsic S-types so that when only the most mafic granites are considered, there is a clear separation of ASI between the two types (Section 6.3).

(3) Mineralogical differences between the I- and S-type granites correspond with the chemical differences. Hence, S-type granites contain Al-rich biotite often with other Al-rich minerals. The most felsic I-type granites, which are also peraluminous, may contain small amounts of muscovite in addition to biotite, and the more mafic varieties contain hornblende. S-type granites in almost all cases contain ilmenite rather than magnetite; I-type granites mostly contain magnetite, but ilmenite-bearing varieties are not uncommon (Section 7).

(4) When fresh, the chemical compositions of volcanic rocks can be matched with granite suites. The volcanic rocks contain pyroxene-bearing assemblages equivalent to the hydrous assemblages of the granites (Section 8.1).

(5) Most of the more mafic rocks contain enclaves, whose abundance can be correlated generally with the colour index of the enclosing granite. The enclaves share the distinctive chemical and mineralogical features of the associated granite, with those in the I-type granites being metaluminous and hornblende-bearing, while the S-type granites contain more diverse assemblages of metasedi-

mentary material. Those mafic granites that do not contain enclaves are I-type (Section 9).

(6) With the restricted amount of data available, oxygen isotopic compositions are a good discriminant, with I-type and S-type granites having values less than and greater than 10‰ respectively. I-type granites range in $^{87}Sr/^{86}Sr$ from 0·74 to 0·712 and in ε_{Nd} from +3·5 to −8·9. For the S-type granites, the corresponding values are 0·708 to 0·720 and −5·8 to −9·2 (Section 7). There is a strong negative correlation between initial Sr and Nd isotopic values of the two granite types, which is not the result of a simple mixing of isotopically primitive and evolved end members (Section 12).

(7) The fractionation of feldspars from felsic granite melts results in significant changes in the abundance of some trace elements. Of particular interest is the behaviour of P, which increases in abundance with fractionation of the more strongly peraluminous S-type melts, and decreases to very low levels in fractionated I-type melts, so that fractionated felsic granites can be separated into I- and S-types using a Rb–P_2O_5 variation diagram (Section 14.3).

17. Acknowledgements

Our work on the granites of the Lachlan Fold Belt has benefited from discussions with numerous persons over a long period, particularly in the field. We are grateful to Charlotte Allen, Phil Blevin, David Steele and Ian Williams for constructive comments on this paper. The final diagrams were prepared by Ian Williams. This research has been supported by the Australian Research Council and its antecedents, and by a group of mineral exploration companies operating through the Australian Mineral Industries Research Association Ltd. Long-term support from both The Australian National University and La Trobe University has been essential to this research.

18. References

Avila-Salinas, W. A. 1990. Tin-bearing granites from the Cordillera Real, Bolivia; a petrological and geochemical review. GEOL SOC AMERICA SPEC PAP **241,** 145–59.

Baker, G. 1940. Cordierite granite from Terip Terip, Victoria. AM MINERAL **25,** 543–548.

Ball, C. W., Dallwitz, W. B. & Noakes, L. C. 1948. Geological reconnaissance of the proposed hydro-electric works in Kosciusko area between Waste Point and Khancoban. REC BMR GEOL GEOPHYS 1948/7.

Barbarin, B. 1990. Granitoids: main petrogenetic classifications in relation to origin and tectonic setting. GEOL J **25,** 227–38.

Bateman, P. C., Clark, L. D., Huber, N. K., Moore, J. G. & Rinehart, C. D. 1963. The Sierra Nevada Batholith: a synthesis of recent work across the central part. US GEOL SURV PROF PAP 414D.

Blevin, P. L. & Chappell, B. W. 1992. Some controls on granite metallogeny: an eastern Australian perspective. TRANS R SOC EDINBURGH EARTH SCI **83,** 305–16.

Browne, W. R. 1929. An outline of the history of igneous activity in New South Wales till the close of the Palaeozoic Era. PROC LINN SOC NSW **54,** 9–34.

Browne, W. R. 1931. Notes on bathyliths and some of their implications. J R SOC NSW **65,** 114–16.

Burnham, C. W. 1992. Calculated melt and restite compositions of some Australian granites. TRANS R SOC EDINBURGH EARTH SCI **83,** 387–97.

Champion, D. C. & Chappell, B. W. 1992. Petrogenesis of felsic I-type granites: an example from northern Queensland. TRANS R SOC EDINBURGH EARTH SCI **83,** 115–26.

Chappell, B. W. 1966. Petrogenesis of the granites at Moonbi, New South Wales. Unpublished PhD thesis, The Australian National University.

Chappell, B. W. 1978. Granitoids from the Moonbi district, New England Batholith, eastern Australia. J GEOL SOC AUST **25,** 267–83.

Chappell, B. W. 1979. Granites as images of their source rocks. GEOL SOC AMERICA PROG ABST **11,** 400.

Chappell, B. W. 1984. Source rocks of I- and S-type granites in the Lachlan Fold Belt, southeastern Australia. PHILOS TRANS R SOC LONDON **A310,** 693–707.

Chappell, B. W. & Hine, R. (in prep.) Geochemistry and petrogenesis of the Cornubian Batholith.

Chappell, B. W. & McCulloch, M. T. 1990. Possible mixed source rocks in the Bega Batholith: constraints provided by combined chemical and isotopic studies. ICOG-7, ABST GEOL SOC AUST **27,** 17.

Chappell, B. W. & Stephens, W. E. 1988. Origin of infracrustal (I-type) granite magmas. TRANS R SOC EDINBURGH EARTH SCI **79,** 71–86.

Chappell, B. W. & White, A. J. R. 1974. Two contrasting granite types. PACIFIC GEOL **8,** 173–174.

Chappell, B. W. & White, A. J. R. 1976. *Plutonic rocks of the Lachlan Mobile Zone.* International Geological Congress 25, Field Guide Excursion 13C.

Chappell, B. W. & White, A. J. R. 1984. I- and S-type granites in the Lachlan Fold Belt, southeastern Australia. *In* Xu Keqin & Tu Guangchi (eds) *Geology of granites and their metallogenic relations,* 87–101. Beijing: Science Press.

Chappell, B. W. & White, A. J. R. 1991. Restite enclaves and the restite model. *In* Didier, J. & Barbarin, B. (eds) *Enclaves and Granite Petrology,* 375–81. Amsterdam: Elsevier.

Chappell, B. W., White, A. J. R. & Wyborn, D. 1987. The importance of residual source material (restite) in granite petrogenesis. J PETROL **28,** 1111–38.

Chappell, B. W., White, A. J. R. & Hine, R. 1988. Granite provinces and basement terranes in the Lachlan Fold Belt, southeastern Australia. AUST J EARTH SCI **35,** 505–21.

Chappell, B. W., Williams, I. S., White, A. J. R. & McCulloch, M. T., 1990. Granites of the Lachlan Fold Belt. ICOG 7 Field Guide Excursion A-2. REC BMR GEOL GEOPHYS 1990/48.

Chappell, B. W., English, P. M., King, P. L., White, A. J. R. & Wyborn, D. 1991. *Granites and related rocks of the Lachlan Fold Belt (1:1 250 000 scale map).* Canberra: BMR Geol. Geophys.

Chappell, B. W., Wyborn, L. A. I., White, A. J. R., Burnham, C. W. & Wyborn, D. (in prep.) S-type granites of the Wagga Basement Terrane: an example of compositional variation resulting from sequential restite fractionation, fractional crystallization and hydrothermal alteration.

Chen, Y. D., Price, R. C., White, A. J. R. & Chappell, B. W. 1990. Mafic inclusions from the Blue Gum and Glenbog granite suites, southern Australia. J GEOPHYS RES **95,** 17757–85.

Chen, Y. D., Chappell, B. W. & White, A. J. R. 1991. Mafic enclaves of some I-type granites of the Lachlan Fold Belt, southeastern Australia. *In* Didier, J. & Barbarin, B. (eds), *Enclaves and granite petrology,* 113–24. Amsterdam: Elsevier.

Clemens, J. D. & Wall, V. J. 1981. Crystallization and origin of some peraluminous (S-type) granite magmas. CAN MINERAL **19,** 111–32.

Collins, W. J., Beams, S. D., White, A. J. R. & Chappell, B. W. 1982. Nature and origin of A-type granites with particular reference to south-eastern Australia. CONTRIB MINER PETROL **80,** 189–200.

Compston, W. & Chappell, B. W. 1979. Sr-isotope evolution of granitoid source rocks. *In* McElhinny, M. W. (ed.) *The Earth: its origin, structure and evolution,* 377–426. London: Academic Press.

David, T. W. E. 1950. *The Geology of the Commonwealth of Australia.* London: Edward Arnold.

Evernden, J. F. & Richards, J. R. 1962. Potassium-argon dates in eastern Australia. J GEOL SOC AUST **9,** 1–49.

Flood, R. H. & Shaw, S. E. 1975. A cordierite-bearing granite suite from the New England Batholith, Australia. CONTRIB MINER PETROL **52,** 157–64.

Gray, C. M. 1984. An isotopic mixing model for the origin of granitic rocks in southeastern Australia. EARTH PLANET SCI LETT **70,** 47–60.

Gray, C. M. 1990. A strontium isotope traverse across the granitic rocks of southeastern Australia: petrogenetic and tectonic implications. AUST J EARTH SCI **37,** 331–49.

Griffin, T. J., White, A. J. R. & Chappell, B. W. 1978. The Moruya Batholith and geochemical contrasts between the Moruya and Jindabyne Suites. J GEOL SOC AUST **25,** 235–47.

Higgins, N. C. 1990. Comment on "Origin of alkali-feldspar

granites: an example from the Poimena Granite, northeastern Tasmania, Australia". GEOCHIM COSMOCHIM ACTA **54,** 2305–11.

Hills, E. S. 1938. Andalusite and sillimanite in uncontaminated igneous rocks. GEOL MAG **75,** 296–303.

Hine, R. 1971. Granite studies of the Kosciusko Batholith. Unpublished BSc thesis, The Australian National University.

Hine, R., Williams, I. S., Chappell, B. W. & White, A. J. R. 1978. Contrasts between I- and S-type granitoids of the Kosciusko Batholith. J GEOL SOC AUST **25,** 219–34.

Holloway, J. & Burnham, C. W. 1972. Melting relations of basalt with equilibrium water pressure less than total pressure. J PETROL **13,** 1–29.

Hurley, P. M., Bateman, P. C., Fairbairn, H. W. & Pinson, W. H. 1965. Investigations of initial $^{87}Sr/^{86}Sr$ ratios in the Sierra Nevada plutonic province. BULL GEOL SOC AM **76,** 165–74.

Ishihara, S. 1977. The magnetite-series and ilmenite-series granitic rocks. MINING GEOL **27,** 293–305.

Joplin, G. A. 1942. Petrological studies in the Ordovician of N.S.W. I. The Cooma Complex. PROC LINN SOC NSW **67,** 156–96.

Joplin, G. A. 1962. An apparent magmatic cycle in the Tasman Geosyncline. J GEOL SOC AUST **9,** 51–69.

Joyce, A. S. 1970. Geochemistry of the Murrumbidgee Batholith. Unpublished PhD thesis, The Australian National University.

Joyce, A. S. 1973. Petrogenesis of the Murrumbidgee Batholith, A.C.T. J GEOL SOC AUST **20,** 179–97.

King, P. L., White, A. J. R. & Chappell, B. W. (in prep.) Characterization and origin of the A-type granites of the Lachlan Fold Belt.

Kolbe, P. & Taylor, S. R. 1966. Geochemical investigations of the granitic rocks of the Snowy Mountains area, New South Wales. J GEOL SOC AUST **13,** 1–25.

Lambert, I. B. & White, A. J. R. 1965. The Berridale Wrench Fault: a major structure in the Snowy Mountains of New South Wales. J GEOL SOC AUST **12,** 25–33.

McCulloch, M. T. & Chappell, B. W. 1982. Nd isotopic characteristics of S- and I-type granites. EARTH PLANET SCI LETT **58,** 51–64.

McCulloch, M. T., Chappell, B. W. & Hensel, H. D. 1982. Nd and Sr isotope relations in granitic rocks of the Tasman Fold Belt, eastern Australia. ABST ICOG **5,** 246–7.

MacKenzie, D. E., Black, L. P. & Sun, S. S. 1988. Origin of alkali-feldspar granites: an example from the Poimena Granite, northeastern Tasmania, Australia. GEOCHIM COSMOCHIM ACTA, **52,** 2507–24.

MacKenzie, D. E., Sun, S. S. & Black, L. P. 1990. Reply to N. C. Higgins' comment on "Origin of alkali-feldspar granites: an example from the Poimena Granite, northeastern Tasmania, Australia". GEOCHIM COSMOCHIM ACTA **54,** 2313–22.

Montel, J. M., Mouchel, R. & Pichavant, M. 1988. High apatite solubility in peraluminous melts. TERRA COGNITA **8,** 71.

Munksgaard, N. C. 1988. Source of the Cooma Granodiorite, New South Wales—a possible role of fluid-rock interactions. AUST J EARTH SCI **35,** 363–77.

Mysen, B. O. & Virgo, D. 1989. Redox equilibria, structure, and properties of Fe-bearing aluminosilicate melts: relationships among temperature, composition, and oxygen fugacity in the system Na_2O–Al_2O_3–SiO_2–Fe–O. AM MINER **74,** 58–76.

Nockolds, S. R. 1947. The relation between chemical composition and paragenesis in the biotite micas of igneous rocks. AM J SCI **245,** 401–20.

O'Neil, J. R. & Chappell, B. W. 1977. Oxygen and hydrogen isotope relations in the Berridale Batholith. J GEOL SOC LONDON **133,** 559–71.

Pichavant, M., Kontak, D. J., Briqueu, L., Herrera, J. V. & Clark, A. H. 1988. The Miocene-Pliocene Macusani Volcanics, SE Peru. II. Geochemistry and origin of a felsic peraluminous magma. CONTRIB MINER PETROL **100,** 325–338.

Pidgeon, R. T. & Compston, W. 1965. The age and origin of the Cooma granite and its associated metamorphic zones, New South Wales. J PETROL **6,** 193–222.

Presnell & Bateman, P. C. 1973. Fusion relations in the system $NaAlSi_3O_8$–$CaAl_2Si_2O_8$–$KAlSi_3O_8$–SiO_2–H_2O and generation of magmas in the Sierra Nevada batholith. BULL GEOL SOC AM **84,** 3181–202.

Read, H. H. 1948. Granites and granites. *In* Gilluly, J. (ed.) *Origin of Granite,* GEOL SOC AMERICA MEM **28,** 1–19.

Read, H. H. 1949. A contemplation of time in plutonism. Q J GEOL SOC LONDON **105,** 101–56.

Reid, E. J. 1980. The geology and geochemistry of the Murrumbucka region. Unpublished MSc thesis, The Australian National University.

Reid, E. J., Stewart-Richardson, D. & Chappell, B. W. 1983. Petrogenesis of the Murrumbucka Tonalite: a granitoid with both I- and S-type character. ABST GEOL SOC AUST **9,** 179–81.

Roddick, J. C. & Compston, W. 1976. Radiometric evidence for the age of emplacement and cooling of the Murrumbidgee Batholith. J GEOL SOC AUST **23,** 223–33.

Roddick, J. C. & Compston, W. 1977. Strontium isotopic equilibration: a solution to a paradox. EARTH PLANET SCI LETT **34,** 238–46.

Shand, S. J. 1927. *Eruptive rocks.* London: Murby.

Silver, L. T. & Chappell, B. W. 1988. The Peninsular Ranges batholith: an insight into the evolution of the Cordilleran batholiths of southwestern North America. TRANS R SOC EDINBURGH EARTH SCI **79,** 105–21.

Takahashi, M., Aramaki, S. & Ishihara, S. 1980. Magnetite-series/Ilmenite-series vs I-type/S-type granitoids. MINING GEOL SPEC ISSUE **8,** 13–28.

Tattam, C. M. 1925. Contact metamorphism in the Bulla area. PROC R SOC VICTORIA **37,** 230–47.

Tetley, N. W. 1978. Geochronology by the $^{40}Ar/^{39}Ar$ technique using HIFAR reactor. Unpublished PhD thesis, The Australian National University.

Tuttle, O. F. & Bowen, N. L. 1958. Origin of granite in the light of experimental studies in the system $NaAlSi_3O_8$-$KAlSi_3O_8$-SiO_2–H_2O. GEOL SOC AM MEM **74**.

Vallance, T. G. 1969. Southern and Central Highlands fold belt: plutonic and metamorphic rocks. J GEOL SOC AUST **16,** 180–200.

van der Molen, I. & Paterson, M. S. 1979. Experimental deformation of partially melted granite. CONTRIB MINER PETROL **70,** 299–318.

Whalen, J. B. 1985. Geochemistry of an island arc plutonic suite: the Uasilau Yau Yau Intrusive Complex, New Britain, P.N.G. J PETROL **26,** 603–32.

Whalen, J. B. & Chappell, B. W. 1988. Opaque mineralogy and mafic mineral chemistry of I- and S-type granites of the Lachlan Fold Belt, southeast Australia. AM MINERAL **73,** 281–96.

White, A. J. R. & Chappell, B. W. 1977. Ultrametamorphism and granitoid genesis. TECTONOPHYSICS **43,** 7–22.

White, A. J. R. & Chappell, B. W. 1983. Granitoid types and their distribution in the Lachlan Fold Belt, southeastern Australia. *In* Roddick, J. A. (ed.) *Circum-Pacific plutonic terranes,* GEOL SOC AM MEM **159,** 21–34.

White, A. J. R. & Chappell, B. W. 1988. Some supracrustal (S-type) granites of the Lachlan Fold Belt. TRANS R SOC EDINBURGH EARTH SCI **79,** 169–81.

White, A. J. R. & Chappell, B. W. 1989. *Geology of the Numbla 1:100 000 Sheet (8624).* Sydney: New South Wales Geological Survey.

White, A. J. R., Chappell, B. W. & Cleary, J. R. 1974. Geologic setting and emplacement of some Australian Palaeozoic batholiths and implications for intrusive mechanisms. PACIFIC GEOL **8,** 159–71.

White, A. J. R., Williams, I. S. & Chappell, B. W. *Berridale 1:100 000 Geological Sheet.* Sydney: New South Wales Geological Survey.

White, A. J. R., Williams, I. S. & Chappell, B. W. 1977. *Geology of the Berridale 1:100 000 Sheet (8625).* Sydney: New South Wales Geological Survey.

White, A. J. R., Chappell, B. W., Williams, I. S. & Glen, R. A. 1989. *Numbla 1:100 000 Geological Sheet.* Sydney: New South Wales Geological Survey.

Williams, I. S. 1973. Granites north of Jindabyne, New South Wales. Unpublished BSc thesis, The Australian National University.

Williams, I. S., Compston, W., Chappell, B. W. & Shiriahase, T. 1975. Rubidium-strontium age determinations on micas from a geologically controlled, composite batholith. J GEOL SOC AUST **22,** 497–505.

Williams, I. S., Chen, Y., Chappell, B. W. & Compston, W. 1988. Dating the sources of Bega Batholith granites by ion microprobe. ABST GEOL SOC AUST **21,** 424.

Wones, D. R. 1989. Significance of the assemblage titanite + magnetite + quartz in granitic rocks. AM MINERAL **74,** 744–49.

Wyborn, D. 1983. Fractionation processes in the Boggy Plain zoned pluton. Unpublished PhD thesis, The Australian National University.

Wyborn, L. A. I. & Chappell, B. W. 1983. Chemistry of the Ordovician and Silurian greywackes of the Snowy Mountains, southeastern Australia: an example of chemical evolution of sediments with time. CHEM GEOL **39,** 81–92.

Wyborn, D., Chappell, B. W. & Johnston, R. M. 1981. Three S-type volcanic suites from the Lachlan Fold Belt, southeast Australia. J GEOPHYS RES **86,** 10 335–348.

Wyborn, D., Turner, B. S. & Chappell, B. W. 1987. The Boggy Plain Supersuite: a distinctive belt of I-type igneous rocks of potential economic significance in the Lachlan Fold Belt. AUST J EARTH SCI **34,** 21–43.

Zen, E-an. 1986. Aluminum enrichment in silicate melts by fractional crystallization: some mineralogic and petrographic constraints. J PETROL **27,** 1095–117.

B. W. CHAPPELL and A. J. R. WHITE, Department of Geology, The Australian National University, Canberra ACT 2601, Australia.

Transactions of the Royal Society of Edinburgh: Earth Sciences, **83**, 27–47, 1992

Partially melted granodiorite and related rocks ejected from Crater Lake caldera, Oregon

Charles R. Bacon

ABSTRACT: Blocks of medium-grained granodiorite to 4 m, and minor diabase, quartz diorite, granite, aplite and granophyre, are common in ejecta of the ~6,900 yr BP caldera-forming eruption of Mount Mazama. The blocks show degrees of melting from 0–50 vol%. Because very few have adhering juvenile magma, it is thought that the blocks are fragments of the Holocene magma chamber's walls. Primary crystallisation of granodiorite produced phenocrystic pl + hyp + aug + mt + il + ap + zc, followed by qz + hb + bt + alkali feldspar (af). Presence of fluid inclusions in all samples implies complete crystallisation before melting. Subsolidus exchange with meteoric hydrothermal fluids before melting is evident in $\delta^{18}O$ values of −3·4–+4·9‰ for quartz and plagioclase in partially melted granodiorites (fresh lavas from the region have $\delta^{18}O$ values of +5·8–+7·0‰); $\delta^{18}O$ values of unmelted granodiorites from preclimatic eruptive units suggest hydrothermal exchange began between ~70 and 24 ka. Before eruption, the granitic rocks equilibrated at temperatures, estimated from Fe–Ti oxide compositions, of up to ~1000°C for *c*. 10^2–10^4 years at a minimum pressure of 100–180 MPa. Heating caused progressive breakdown or dissolution of hb, af, bt, and qz, so that samples with the highest melt fractions have residual pl + qz and new or re-equilibrated af + hyp + aug + mt + il in high-silica rhyolitic glass (75–77% SiO_2). Mineral compositions vary systematically with increasing temperature. Hornblende is absent in rocks with Fe–Ti oxide temperatures >870°C, and bt above 970°C. Oxygen isotope fractionation between qz, pl, and glass in partially fused granodiorite also is consistent with equilibration at $T \geq 900$°C ($\Delta^{18}O_{qz\text{-}pl} = +0{\cdot}7 \pm 0{\cdot}5$‰). Element partitioning between glass and crystals reflects the large fraction of refractory pl, re-equilibration of af and isolation or incomplete dissolution of accessory phases. Ba and REE contents of analysed glass separates can be successfully modelled by observed degrees of partial melting of granodiorite, but Rb, Sr and Sc concentrations cannot. Several samples have veins of microlite-free glass 1–5 mm thick that are compositionally and physically continuous with intergranular melt and which apparently formed after the climactic eruption began. Whole-rock H_2O content, microprobe glass analysis sums near 100% and evidence for high temperature suggest liquids in the hotter samples were nearly anhydrous. The occurrence of similar granodiorite blocks at all azimuths around the 8 × 10 km caldera implies derivation from one pluton. Compositional similarity between granodiorite and pre-Mazama rhyodacites suggests that the pluton may have crystallised as recently as 0·4 Ma; compositional data preclude crystallisation from the Holocene chamber. The history of crystallisation, hydrothermal alteration, and remelting of the granitic rocks may be characteristic of shallow igneous systems in which the balance between hydrothermal cooling and magmatic input changes repeatedly over intervals of 10^4–10^6 years.

KEY WORDS: igneous petrology, granite, partial melting, mineralogy, glass, geothermometry, magma

Partial fusion of igneous rocks and the separation and migration of melts is widely believed to be the major process in reworking and differentiation of the crust. Intermediate and silicic magmas typically are formed by a combination of fractionation of mantle-derived magma and assimilation of crustal rocks or their partial melts. Theoretical aspects of partial melting, and melt segregation and transport, have focused on the mantle (e.g. McKenzie 1984), with relatively few treatments of the crustal environment (Wickham 1987). Laboratory experiments have addressed both physical (Arzi 1978; Van der Molen & Paterson 1979; Jurewicz & Watson 1985) and phase equilibrium (e.g. Wyllie 1977) aspects of partial melting of crustal rocks. Phase equilibrium studies, however, are hampered by the small size of experimental charges, the difficulty of attaining equilibrium without large amounts of H_2O present (Johannes 1985), and run times that are short in comparison to melting events in nature. Studies of natural examples of glassy partially fused granitic wall rocks of volcanic necks (Al-Rawi & Carmichael 1967; Dodge & Calk 1978; Kaczor *et al.* 1988) and of xenoliths in lava flows (Maury & Bizouard 1974; Grove *et al.* 1988) and pyroclastic deposits (Le Maitre 1974) have attracted interest, yet these features reflect transient heating with little chance for equilibration of phases in the affected rocks.

Natural examples of partially fused rocks that represent equilibrium, or at least thermal steady-state conditions, can provide valuable information on crystal and liquid compositions. Such a case exists at Crater Lake where variably melted granodiorite, diabase, quartz diorite, granite, aplite and granophyre blocks were ejected in the ~6,900 yr BP caldera-forming eruption of Mount Mazama (Bacon 1983). Diller and Patton (1902, p. 40) first recorded the occurrence of the blocks, and Patton elegantly described their petrography (pp. 133–9), although the origin of the "granophyric secretions" as he called the blocks, was not known at that time. Williams (1942) noted partially melted granitic fragments in the postcaldera lavas of Wizard Island, but it was Taylor (1968) who first pointed out that partially fused granitic blocks were widespread in deposits of the caldera-forming (climactic) eruption. Lidstrom (1971) and Ritchey (1979) also reported on the granitic blocks in their studies of the climactic ejecta.

This paper describes the petrography of the granitic blocks and documents their bulk, mineral and glass compositions. These blocks, particularly the abundant granodiorites, show all stages of fusion from unaffected rock to 50 vol% melt. Mineralogical thermometry and orderly disappearance of hornblende and biotite indicate that melt fraction increases with increasing pre-eruptive temperature under approximately closed-system conditions. The granitic and related rocks apparently were torn from the walls of a shallow, zoned magma body (Bacon & Druitt 1988), where they had existed in quasi-equilibrium for 10^2–10^4 years, brought to the surface virtually instantaneously and rapidly cooled so that no microlites are present in their remarkably fresh glass. They are samples from an epizonal pluton, compositionally very like the Quaternary eruptive rocks above it, that had suffered ^{18}O depletion through exchange with meteoric hydrothermal fluids before being heated by the late Pleistocene/Holocene Mazama magmatic system (Bacon *et al.* 1989). The granitic blocks at Crater Lake appear to be an unparalleled example of a natural partial melting experiment.

1. Geological background

Pyroclastic eruptions bring samples of the contents of magma chambers rapidly to the surface where they cool quickly and preserve a record of magmatic conditions. Fragments from conduit walls, and even from the environs of magma chambers, are transported along with juvenile magma. The largest pyroclastic eruptions result in collapse of the roofs of chambers, forming calderas. Crater Lake, in the southern Oregon Cascade Range, partially fills a relatively small but spectacular Holocene caldera (8 × 10 km rim, ~5 km diameter subsidence) formed in the andesite/dacite edifice known as Mount Mazama in the High Cascades volcanic arc.

Mount Mazama is composed of a cluster of calc-alkaline shield and stratovolcanoes, each of which was active for a comparatively short time during the period ~420–~40 ka (Bacon 1983; Bacon & Lanphere 1990). Monogenetic cones and shield volcanoes erupted basalt, basaltic andesite and andesite in the vicinity of Mount Mazama before and during its construction (Bacon 1990). Beneath Mount Mazama to the S and E is an extensive field of ~0·4 Ma rhyodacitic lava flows underlain by Pleistocene dacite and basaltic andesite. Altered pyroclastic rocks that occur as lithic fragments in the climactic ejecta and below basaltic andesite flows in a drillhole may be coeval with Tertiary rocks of the Western Cascades that crop out ~20 km W of Crater Lake.

The climactic magma chamber of Mount Mazama was in existence by ~30 ka, when the first rhyodacitic lava was erupted from it (Bacon & Druitt 1988). At least five rhyodacitic lava flows and a group of domes were erupted between then and the caldera-forming event, the last flow still hot when the caldera collapsed. The climactic eruption occurred at 6,845 ± 50 yr BP (^{14}C; Bacon 1983), when ~50 km^3 of magma vented in at most a few days, producing (1) a rhyodacitic pumice-fall deposit, followed by (2) welded rhyodacitic ignimbrite, and (3) a zoned deposit of rhyodacitic ignimbrite overlain by andesitic to "basaltic" (actually, gabbroic crystal mush) ignimbrite in valleys and a lithic "lag breccia" on the upper flanks of Mount Mazama (Druitt & Bacon 1986).

The petrology and geochemistry of the juvenile components of the climactic ejecta have been the subject of numerous investigations, most recently those of Bacon and Druitt (1988) and Druitt and Bacon (1988, 1989). The rhyodacitic pumice is uniform in composition and mineralogy, averaging 70·4 wt% SiO_2 (anhydrous). Fe–Ti oxide phenocryst compositions indicate pre-eruptive temperatures of 880–886°C. Analysis of melt inclusions in plagioclase phenocrysts indicates a pre-eruptive H_2O content of ~4 wt% in the melt and a minimum depth to the chamber of 5 km (Bacon *et al.* 1992). The eruption tapped many levels of the chamber concurrently; however, andesite and more mafic clasts ("scoria") are not dominant except in the latest deposits. Changes in proportions of juvenile clasts with stratigraphic height reflect sequential tapping of a layered chamber in which rhyodacitic magma overlay andesitic magma and cumulate mush that increased in temperature and crystal content with depth.

The granitic and related rocks are those medium- to fine-grained block types that have plutonic or hypabyssal textures. They occur in all deposits of the climactic eruption but are most readily collected and best preserved at the top of the lithic breccia within a few kilometres of the caldera rim. The blocks do not show evidence of physical contact or interaction with juvenile (climactic) magma with the following rare exceptions: (1) a granodiorite clast which has adhering mafic scoria with a razor-sharp interface, suggesting syneruptive contact; and (2) a partially melted diabase fragment in a mafic scoria clast in which crystals from the diabase were being dispersed. Xenoliths of some of the same materials occur in postcaldera andesitic lavas of Wizard Island.

The medium to fine grain-size and plagioclase prophyritic texture of the granitic rocks suggest that they are fragments of a subvolcanic pluton. Rock textures are analogous to those of the Tatoosh Pluton in Mount Rainier National Park, Washington (Fiske *et al.* 1963). The pluton beneath Crater Lake may be Pleistocene, or possibly late Tertiary, in age; Cordilleran Mesozoic batholithic rocks are markedly coarser grained. Two samples of granodiorite that resemble blocks from the climactic deposits were obtained from late Pleistocene rocks: an unmelted xenolith (sample 1677) from the 24 ka rhyodacite of Redcloud Cliff and an incipiently fused clast (sample 268) from near the base of the ~70 ka Plinian dacite of Pumice Castle. Therefore, if all of these rocks are from the same pluton, it must have solidified before 70 ka (see also Section 2.2).

Granitic and related clasts in the lithic breccia of the climactic eruption are rhombohedral to tabular blocks whose faces tend to intersect at ~90° angles. Blocks may be dense to pumiceous. Breadcrusted surfaces are common. A 30 cm chlorite + epidote-bearing unmelted quartz diorite block (sample 1092) has radial, "prismatic" jointing reflecting *in situ* cooling following deposition. The largest clasts are granodiorite (commonly to 1 m; a vesicular 4 m block (samples 799 & 800) with 50 vol% melt is the largest known); many contain ellipsoidal quartz diorite microgranular enclaves as large as 20 cm across (sample 800). Recovered aplite and granophyre clasts are ≤25 cm, diabase <20 cm, and most are poorly vesiculated to dense. Rare granodiorite and granite samples have glass veins, or have crystal-free glass on one surface, that are continuous with intergranular glass. Crystals derived from the granitic rocks, or, in one aplite, from climactic magma, occur in tensional

Table 1 Chemical analyses of representative rocks and glasses

Sample no. Rock type Vol% glass	1016 diab n.d.	776 diab n.d.	800 incl n.d.	1092 aqd 0	452 grd 7	466 grd n.d.	799 grd 50	1023 grd 50	1038 grd 30	463 grd 0	1037 grd 12	432 apl 10	799G gls (100)	1023G gls (100)	1038G gls (100)	1326G gls (100)	432G gls (100)	1037G gls (100)	1239G gls (100)	452G gls (100)	1239 gphr 36
SiO_2	51·0	52·5	61·2	61·2	63·6	65·6	67·3	67·7	67·8	68·1	68·8	73·4	75·5	75·8	76·2	76·3	76·4	76·6	76·6	77·6	79·3
Al_2O_3	21·8	18·7	17·3	17·3	16·7	15·9	16·2	15·9	15·8	15·8	15·4	13·7	12·2	11·9	11·9	12·1	12·1	12·0	12·2	11·8	11·5
Fe_2O_3	2·78	2·49	1·76	1·77	1·54	1·34	1·08	1·09	1·11	1·12	1·07	0·71	0·71	0·67	0·64	0·54	0·46	0·52	0·47	0·30	0·24
FeO	5·83	5·22	3·70	3·71	3·23	2·80	2·27	2·30	2·33	2·34	2·26	1·50	1·50	1·42	1·35	1·14	0·97	1·10	1·00	0·62	0·51
MgO	4·28	6·20	3·03	3·30	2·59	1·91	1·42	1·52	1·48	1·40	1·33	0·66	0·23	0·24	0·21	0·15	0·18	0·16	0·13	0·10	<0·10
CaO	8·68	9·14	5·12	5·76	5·21	4·00	3·51	3·40	3·46	3·28	3·87	1·86	0·67	0·66	0·54	0·46	0·53	0·48	0·17	0·31	0·31
Na_2O	3·95	3·92	5·25	4·24	4·52	4·59	5·15	5·13	5·21	4·91	5·00	3·90	3·92	4·00	3·96	3·86	3·37	3·86	3·90	3·15	4·03
K_2O	0·11	0·37	1·64	1·67	1·68	2·98	2·26	2·21	2·02	2·30	1·45	3·76	4·65	4·68	4·71	5·09	5·62	4·79	5·28	5·87	4·03
TiO_2	1·16	0·95	0·73	0·80	0·63	0·64	0·53	0·57	0·57	0·57	0·54	0·41	0·59	0·49	0·47	0·42	0·36	0·42	0·24	0·21	0·12
P_2O_5	0·23	0·34	0·19	0·18	0·13	0·15	0·13	0·13	0·14	0·13	0·12	0·06	0·07	0·03	0·04	0·03	0·02	0·03	0·02	0·02	<0·05
MnO	0·11	0·14	0·13	0·09	0·09	0·07	0·06	0·05	0·06	0·05	0·07	0·03	0·03	0·02	0·04	n.d.	n.d.	0·03	0·03	n.d.	<0·02
LOI	0·00	0·30	0·94	1·50	0·61	0·35	0·90	0·65	0·57	0·64	0·39	0·47	n.d.	n.d.	n.d.	n.d.	n.d.	n.d.	n.d.	n.d.	0·78
Total	100·95	99·15	98·60	99·60	100·11	100·40	99·02	100·46	100·32	100·08	100·33	99·53	99·25	99·30	99·39	98·87	98·76	99·18	99·23	96·69	100·49
Fe^{2+}/Fe	n.d.	n.d.	0·65	0·50	0·45	0·50	0·65	0·56	0·63	0·34	0·57	n.d.	n.d.	n.d.	n.d.	n.d.	n.d.	n.d.	n.d.	n.d.	0·43
H_2O^+	n.d.	n.d.	0·65	1·60	0·24	0·38	0·62	0·48	0·51	0·25	0·32	n.d.	n.d.	n.d.	n.d.	n.d.	n.d.	n.d.	n.d.	n.d.	0·33
Cl	0·089	n.d.	0·076	0·007	0·046	0·058	0·079	0·140	0·075	0·034	0·027	n.d.	0·180	0·330	0·230	n.d.	n.d.	n.d.	0·18	n.d.	0·084
F	0·01	n.d.	0·02	0·03	0·04	0·03	0·03	0·04	0·04	0·04	0·02	n.d.	0·09	0·10	0·11	n.d.	n.d.	n.d.	0·06	n.d.	0·03
Ba	157	311	542	525	548	626	674	671	669	690	552	856	547	567	426	n.d.	n.d.	n.d.	178	n.d.	873
Co	28·6	31·4	14·6	18·8	15·3	11·9	7·8	7·8	14·0	7·2	7·7	6·0	4·0	4·2	4·1	n.d.	n.d.	n.d.	4·0	n.d.	0·4
Cr	29·6	197	45·9	46·3	30·4	20·2	12·7	12·9	13·4	14·0	13·6	4·7	<8	1·8	<2	n.d.	n.d.	n.d.	<2	n.d.	<2
Cs	<0·6	0·2	2·7	0·5	2·7	3·5	3·0	2·5	2·4	3·3	1·7	3·1	7·9	7·7	9·8	n.d.	n.d.	n.d.	9·0	n.d.	3·3
Hf	2·5	1·0	3·7	3·7	4·2	5·0	5·0	5·1	5·1	5·4	4·8	7·5	10·6	5·9	11·2	n.d.	n.d.	n.d.	6·5	n.d.	5·4
Nb	4·1	2·3	4·2	4·4	4·7	6·8	5·2	5·0	6·3	6·5	6·5	n.d.	12	10	12	n.d.	n.d.	n.d.	16	n.d.	4·2
Rb	7	16	33	32	48	50	47	41	37	42	25	94	113	117	151	n.d.	n.d.	n.d.	192	n.d.	67
Sc	25·7	20·7	14·8	14·5	12·7	10·7	8·95	9·32	8·99	9·26	8·58	5·39	9·03	7·48	8·46	n.d.	n.d.	n.d.	6·80	n.d.	2·25
Sr	948	1405	486	547	538	437	379	390	389	389	446	224	52	54	52	n.d.	n.d.	n.d.	14	n.d.	26
Ta	0·17	0·11	0·30	0·35	0·32	0·44	0·46	0·43	0·39	0·50	0·43	0·58	0·95	0·70	1·12	n.d.	n.d.	n.d.	1·63	n.d.	0·51
Th	1·1	0·5	2·6	3·3	3·6	4·7	4·7	4·4	4·4	4·9	4·5	10·0	11·3	8·2	15·6	n.d.	n.d.	n.d.	16·2	n.d.	6·2
U	0·4	<1·0	1·4	1·3	1·6	2·1	2·2	1·8	1·7	1·9	1·7	3·9	5·0	3·5	6·7	n.d.	n.d.	n.d.	5·4	n.d.	2·4
Y	15	15	21	17	17	21	25	25	21	19	19	24	38	30	43	n.d.	n.d.	n.d.	56	n.d.	21
Zn	66	93	85	70	63	52	57	38	36	41	47	20	50	40	36	n.d.	n.d.	n.d.	39	n.d.	15
Zr	113	82	144	147	165	187	197	195	190	193	178	260	367	202	375	n.d.	n.d.	n.d.	191	n.d.	179
La	11	11	15	15	14	16	19	17	18	19	17	22	28	21	33	n.d.	n.d.	n.d.	41	n.d.	17
Ce	22	23	33	29	26	33	38	35	35	37	34	44	63	46	72	n.d.	n.d.	n.d.	90	n.d.	35
Nd	16	20	19	16	13	18	22	19	18	20	19	23	29	22	34	n.d.	n.d.	n.d.	44	n.d.	13
Sm	3·2	3·2	4·9	3·7	3·4	4·0	4·9	4·2	4·2	4·2	4·0	4·9	7·4	5·6	8·2	n.d.	n.d.	n.d.	9·9	n.d.	3·3
Eu	1·11	1·10	1·20	0·98	0·86	0·85	0·90	0·90	0·91	0·95	0·90	0·74	0·49	0·50	0·44	n.d.	n.d.	n.d.	0·14	n.d.	0·38
Gd	2·5	2·1	3·9	n.d.	3·4	4·7	4·4	4·1	4·4	4·2	4·9	5·0	n.d.	n.d.	n.d.	n.d.	n.d.	n.d.	n.d.	n.d.	3·5
Tb	0·42	0·35	0·57	0·50	0·51	0·64	0·63	0·51	0·58	0·59	0·57	0·65	1·03	0·83	1·20	n.d.	n.d.	n.d.	1·57	n.d.	0·61
Yb	1·3	0·9	2·2	1·8	1·7	2·0	2·1	2·0	2·0	2·1	1·9	2·3	4·1	3·1	4·6	n.d.	n.d.	n.d.	5·6	n.d.	2·0
Lu	0·24	0·14	0·30	0·36	0·26	0·30	0·32	0·33	0·31	0·32	0·32	0·36	0·60	0·45	0·70	n.d.	n.d.	n.d.	0·76	n.d.	0·32

Notes: Major elements (oxide wt%) by X-ray fluorescence in rocks (K. C. Stewart, analyst) and electron microprobe in glasses (C. R. Bacon, analyst), recalculated to sum to 100% with $Fe^{2+}/Fe = 0·7$; analysed Fe^{2+}/Fe reported where available. LOI is loss on ignition at 900°C. Total is original total with all Fe as Fe_2O_3. H_2O^+ determined coulometrically with a Karl Fischer titration. F and Cl (wt%) by specific ion electrode, Nb (ppm) by spectrophotometry and ICP (P. J. Aruscavage and S. T. Pribble, analysts). Rb, Sr, Y, and Zr (ppm) by energy-dispersive X-ray fluoresence (P. E. Bruggman, analyst). Other elements (ppm) by INAA (L. J. Schwarz and J. N. Grossman, analysts). n.d. = not determined, diab = diabase, incl = inclusion in granodiorite, aqd = altered quartz diorite, grd = granodiorite, apl = aplite, gls = glass, gphr = granophyre. Vol% glass determined by point counting; (100) indicates glass separate.

fractures in some unmelted to incipiently fused samples. The above features and the field occurrence of the blocks indicate that they were rapidly quarried and brought to the surface. The granitic rocks may be fragments of large xenoliths in, or may have formed the deep walls of, the climactic magma chamber. The nearly complete lack of evidence for interaction with juvenile magma and the extreme rarity of xenoliths in any of the lavas of Mount Mazama, including preclimactic rhyodacites, suggest that the blocks came from the walls of the chamber.

2. Rock compositions

2.1. Compositional variation

The granitic and related rocks are diabase, quartz diorite, granodiorite and granite, and granitic aplite and granophyre (Table 1; names based on modes, CIPW norms and textures); granodiorite is most common. These groups are evident in Figure 1, where representative trace elements are plotted against SiO_2.

The intermediate composition samples are chemically similar to Quaternary calc-alkaline eruptive rocks (see Section 2.3). Incompatible element concentrations increase with SiO_2 from quartz diorite to aplite in the same manner as shown by Rb, suggesting these rocks are related by crystal/liquid separation processes with only minor involvement of accessory minerals (see Section 6.3). The increase in Ba from granodiorite to aplite is subdued because of alkali feldspar fractionation. Compatible elements decrease systematically, as illustrated by Sr and Sc. The granophyre sample is unique and, if related to the intermediate rocks, has experienced accessory mineral, biotite and oxide fractionation that has depleted it in Rb, and LRE and HFS elements in comparison to the aplites.

The granitic rocks are more oxidised than fresh glassy eruptive rocks of Mount Mazama, which have Fe^{2+}/Fe ratios $\approx 0{\cdot}7$ (Bacon & Druitt 1988). Fourteen samples from quartz diorite to granophyre have Fe^{2+}/Fe ratios of 0·34–0·66; high-melt-fraction samples tend to have ratios $\geq 0{\cdot}6$ (Table 1). Presumably, Fe^{2+}/Fe ratios were lowered by interaction with hydrothermal fluids that also were responsible for lowering $\delta^{18}O$ values (see Section 2.2). Major element data in Table 1 were recalculated assuming $Fe^{2+}/Fe = 0{\cdot}7$ for ease of comparison with analyses of eruptive rocks in Bacon & Druitt (1988).

The diabases have a range of mantle-incompatible elemental abundances that indicates derivation from a variety of parent magmas. Cumulate textures, and the aluminous nature of these rocks, suggest crystal accumulation. Presence of intergranular partial melt implies pre-eruptive association of diabases and the granitic rocks, but the geochemical data are insufficient to establish a genetic link between these rock types. The diabases may have been dykes or sills that cut the plutonic rocks, or may be unrelated to the source pluton.

2.2. Strontium and oxygen isotopes

Variation in $^{87}Sr/^{86}Sr$ among eight analysed granitic samples is not significantly greater than the standard deviation of the weighted mean value of $0{\cdot}70372 \pm 2$ (Bacon *et al.* 1989). The range in $^{87}Sr/^{86}Sr$ ($0{\cdot}70365 \pm 8$–$0{\cdot}70375 \pm 4$) overlaps values for dacites and rhyodacites of Mount Mazama and for pre-Mazama rhyodacites.

The granitic rocks have remarkably low $\delta^{18}O$ values for igneous rocks (Bacon *et al.* 1989). Fresh lavas of Mount Mazama and the surrounding region have $\delta^{18}O = +5{\cdot}8$–$+7{\cdot}0$‰. Quartz, plagioclase and glass in the partially melted rocks have $\delta^{18}O = -3{\cdot}4$–$+4{\cdot}9$‰. Oxygen isotope fractionation between quartz, plagioclase and glass in the partially melted samples is consistent with equilibrium at $\geqslant 900$°C. The plutonic rocks apparently exchanged oxygen with meteoric hydrothermal fluids, which variably lowered their $\delta^{18}O$ values from an estimated initial whole-rock value of $\sim +7$‰. Their oxygen isotopes re-equilibrated during partial melting so that the range in Δ_{qz-pl} is now $-0{\cdot}2$–$+1{\cdot}2$‰. Samples with lower Fe–Ti oxide equilibration temperatures (see Section 5.1) and smaller melt fractions tend to have larger values of Δ_{qz-pl}, although this correlation is crude because the analytical uncertainty in Δ_{qz-pl} is $\pm 0{\cdot}2$–$0{\cdot}3$‰.

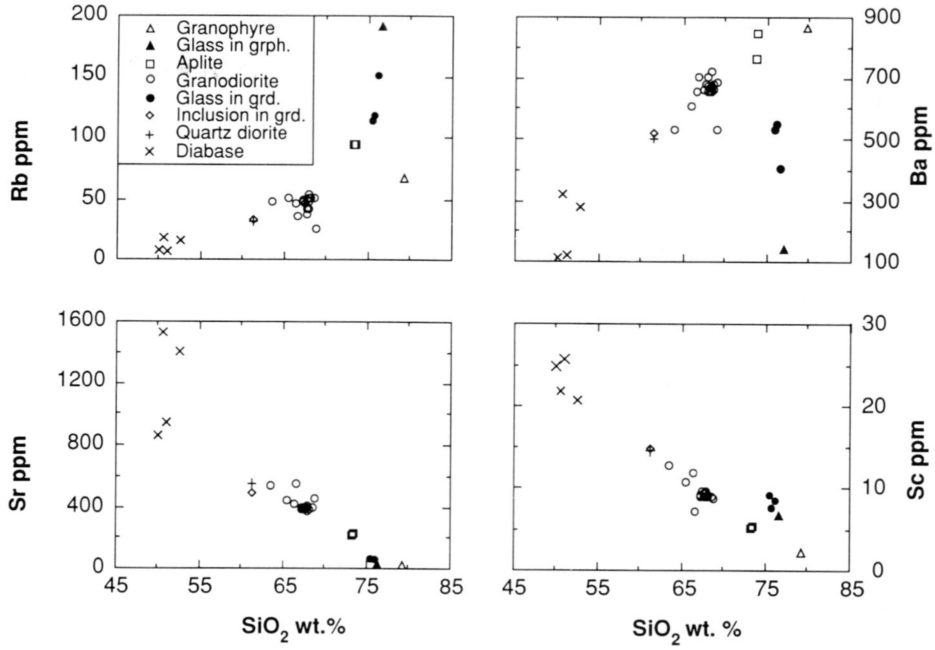

Figure 1 Representative variation diagrams for whole rock and glass separate compositions.

Table 2 Oxygen isotope analyses of mineral separates from granodiorites

Sample no.	Unit	Age	Mineral	$\delta^{18}O_{SMOW}$	Av. $\delta^{18}O_{SMOW}$
461	Lag breccia, climactic eruption	7 ka	Quartz Plagioclase	7·80, 7·71 6·25, 5·98	7·8 6·1
1627	Lag breccia, climactic eruption	7 ka	Quartz Plagioclase	5·11, 5·05 3·25, 3·40	5·1 3·3
1677	Redcloud Cliff rhyodacite flow	24 ka	Quartz Plagioclase	7·72, 7·74 4·33, 4·19	7·7 4·3
268	Tephra of Pumice Castle	~70 ka	Quartz Plagioclase	7·89, 8·16 6·65, 6·64	8·0 6·5

Notes: Analyses by L. H. Adami using techniques described in Bacon *et al.* (1989); all samples unmelted except 268, which is incipiently fused.

New oxygen isotope analyses of quartz and plagioclase from three unmelted and one incipiently fused granodiorite samples (Table 2) may constrain the time of hydrothermal exchange. One of the two samples from the climactic ejecta has low $\delta^{18}O$ values for both quartz and plagioclase. The granodiorite xenolith from the 24 ka Redcloud Cliff flow contains low-$\delta^{18}O$ plagioclase. In contrast, the incipiently fused granodiorite clast from the ~70 ka tephra of Pumice Castle, whose vent is virtually coincident with that of the Redcloud Cliff flow and therefore should have sampled the same volume of granodiorite, has "normal" $\delta^{18}O$ for both quartz and plagioclase; the grain-boundary fusion of this sample may be syneruptive. These limited data (granodiorite samples 268 and 1677 are the only ones that have been found in preclimactic units) are consistent with hydrothermal exchange having begun sometime between ~70 ka and 24 ka, presumably driven by the growing climactic magma chamber.

2.3. Eruptive equivalents of granodiorites?

The granodiorites and quartz diorites are compositionally rather similar to dacites and silicic andesites of Mount Mazama. Comparison of complete analyses reveals no exact matches (most lavas of Mount Mazama contain too much K, Ca, Sr, Ti, and Zr at comparable Si) and establishes no direct relation between the granitic rocks and products of the climactic magma chamber. The ~50 ka dacite of The Watchman and the ~70 ka dacite of Pumice Castle bear a resemblance to the granodiorites but the dacites' incompatible element contents and ratios are inappropriate (Fig. 2).

Another possibility is that the granodiorites are the accumulative complement to more differentiated liquids that erupted to form pre-Mazama rhyodacites. Mineral analyses are permissive of such a relation because compositional ranges of phenocrysts in pre-Mazama rhyodacites (S. Nakada, *unpublished data* 1991) and early-crystallised phases in granodiorites (those which do not appear to have re-equilibrated) generally overlap, although there are some differences in minor element contents of Fe–Ti oxides. The range of incompatible element ratios of pre-Mazama rhyodacites is enclosed by that of the granodiorites and quartz diorites (Fig. 2). The possible relation was tested with Pearce element ratio plots (Russell & Nicholls 1988) and H/M vs H diagrams (Allègre & Minster 1978; highly incompatible/moderately incompatible vs highly incompatible, e.g. Th/Ta vs Th). The majority of such tests were permissive. Only Th–Ta–Nb covariation appears to be inconsistent with such a relationship: in comparison with the rhyodacites (and dacites) the granodiorites have low Ta contents, relative to Th and Nb. Because most of the granodiorites are less differentiated than the rhyodacites, and in fact may be accumulative, the HFSE data appear to be inconsistent with the two groups of rocks being related by crystal/liquid separation processes. Apart from this objection, however, there is no compelling evidence against consanguinity.

2.4. Aplites and granophyre have no eruptive equivalents

The same plots used for comparison of granodiorites with eruptive rocks indicate that the aplites and the granophyre could have been derived from the same magma body as the quartz diorites and granodiorites. However, the differentiated rock types have no eruptive equivalents.

2.5. Relation of diabases to Quaternary lavas

The partially fused diabases show a range of incompatible element contents that requires several parent magmas. In comparison to basalts and basaltic andesites erupted from monogenetic volcanoes near Mount Mazama (Bacon 1990), the diabases have low Mg, K, Rb and Ba, and high Al and Sr, as though they contain accumulative plagioclase. Texturally, they are fine-grained plagioclase mesocumulates. Compositional data are permissive of a relation to the calc-alkaline basaltic andesites, possibly by loss of intergranular liquid through filter-pressing of magma rich in plagioclase microphenocrysts.

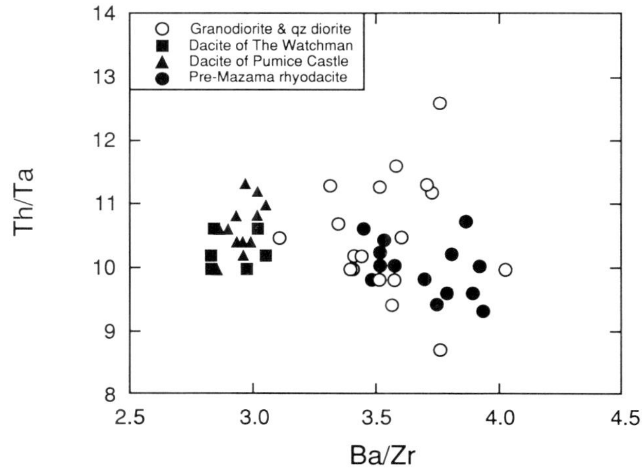

Figure 2 Incompatible element weight ratios for granodiorite and quartz diorite, and for silicic eruptive rocks of similar composition.

3. Petrography

3.1. Quartz diorite and granodiorite

The granodiorites show all stages of partial fusion. Representative samples containing few vesicles, so that melt fractions are easily compared, are shown at uniform magnification in Figure 3. The progressive destruction of the matrix between crystals with increasing melt fraction is evident. In Figure 3(E), at 37 vol% glass, crystals appear interlocking but in Figure 3(F), at 48 vol% glass, contiguity may have been destroyed and randomly oriented crystals appear to swim in glass as in porphyritic lava.

3.1.1. Unmelted samples. These rocks contain euhedral, oscillatory zoned plagioclase (≤6 mm), with interstitial quartz (≤1·5 mm), alkali feldspar, and sodic plagioclase (Fig. 3(A)). Granophyric or micrographic quartz + alkali feldspar + sodic plagioclase intergrowths are common (Fig. 4(A)). Some plagioclase crystals show patchy zoning. Plagioclase typically has tiny oriented acicular crystallites of another phase, except near margins of crystals, that make the feldspar appear dark grey in hand specimen. Sub-μm to ~2 μm fluid inclusions are ubiquitous in feldspars and quartz, and tend to be most abundant in alkali feldspar. Augite and more abundant hypersthene occur as small

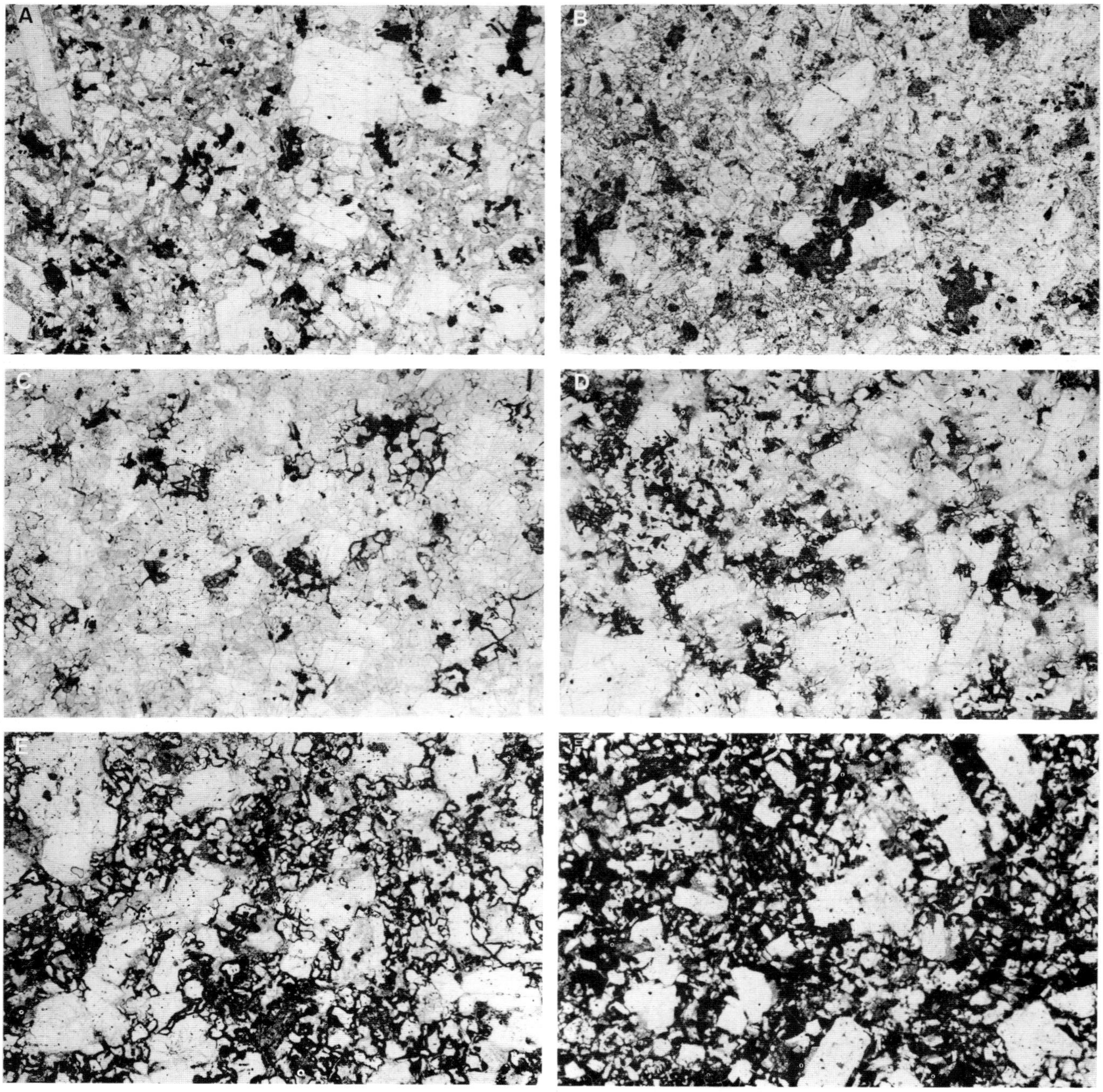

Figure 3 Plane light photomicrographs of granodiorite blocks with increasing degrees of melting: sample numbers, Fe–Ti oxide temperatures (where available), and vol% melt in parentheses; all samples have ≤1% vesicles; width of field = 18 mm for all photomicrographs.

(A) Unmelted mafic granodiorite (463).
(B) Partially melted granodiorite containing hb + bt + vesicular glass (452; 842°C; 7 vol%); colourless vesicular glass is present along grain boundaries.
(C) Partially melted granodiorite containing bt + glass (1423; 5 vol%); brown glass is localised in pools.
(D) Partially melted granodiorite containing bt + glass (1326; 954°C; 17 vol%).
(E) Partially melted granodiorite lacking hydrous silicates (1630; 961°C; 37 vol%).
(F) Partially melted granodiorite lacking hydrous silicates (1631; 1003°C; 48 vol%).

inclusions in plagioclase and as larger discrete crystals. Pyroxenes also replace hornblende, presumably because of reheating. Hornblende is green to brown and forms prismatic crystals that may be replaced by actinolite and biotite in some samples. Brown biotite is present in all rocks. Both biotite and hornblende may occur as inclusions in other phases, notably plagioclase; however, the general impression is that biotite and hornblende were relatively late-crystallising phases. Fe–Ti oxides form discrete crystals and are present as inclusions in silicates. Apatite occurs as large prisms and as acicular inclusions. Zircon has been identified in several samples and sphene is present in one sample. The magmas were saturated with plagioclase, pyroxenes, Fe–Ti oxides and apatite, as were dacites and rhyodacites of Mount Mazama, and biotite, hornblende, alkali feldspar and quartz crystallised on further cooling in the subvolcanic environment.

The rocks do not appear to have been strained before the climactic eruption (see Section 10). Visible hydrothermal alteration typically is limited to formation of fluid inclusions along fracture planes; some fluid inclusions, particularly in alkali feldspar, may be juvenile residual fluid. Quartz diorite sample 1092 has been extensively altered, with epidote,

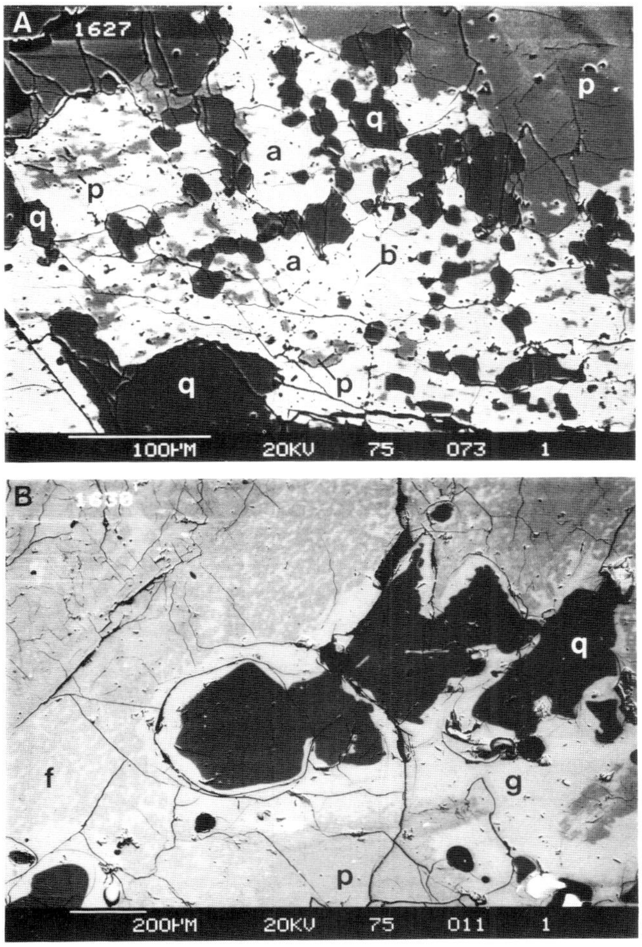

Figure 4 Backscattered electron SEM images of granodiorite textures:

(A) Granophyric groundmass qz + pl + af intergrowth in unmelted granodiorite (1627); letters indicate phases: q = quartz, p = plagioclase (which is strongly albitic in intergrowth), a = alkali feldspar, b = biotite.

(B) High-melt-fraction granodiorite (1630; 961°C; 37 vol% glass); letters indicate phases: q = quartz, p = plagioclase, f = re-equilibrated alkali feldspar ($An_{13}Ab_{70}Or_{17}$; spongy), g = high-silica rhyolite glass; white spots are pyroxene and Fe–Ti oxides.

chlorite, and minor actinolite forming veins and replacing many original minerals.

3.1.2. Partially fused rocks. Low-melt-fraction granodiorites contain quartz, two pyroxenes, two feldspars, two oxides, biotite and hornblende. Some incipiently melted samples contain both hornblende and biotite, others only biotite. As melting increases, first hornblende, then biotite breaks down to pyroxenes + feldspar + oxides; biotite abundance decreases and its composition changes systematically with increasing temperature. Neither hydrous silicate is found in direct contact with glass. The original alkali feldspar apparently persists in some rocks lacking hornblende, but is entirely consumed or transformed in the higher temperature samples. Re-equilibrated alkali feldspar, commonly granophyrically intergrown with quartz, forms dendritic overgrowths on plagioclase and cellular crystals in glass (Fig. 4(B)). The highest-melt-fraction granodiorites contain quartz, one feldspar, two pyroxenes and two oxides. The interiors of plagioclase crystals have not equilibrated with melt (cf. Johannes 1989).

Hypersthene is the most abundant pyroxene in the partially melted rocks. It forms large anhedral crystals that were interstitial to plagioclase before melting, and smaller euhedral inclusions within plagioclase. Primary augite is less abundant. Both pyroxenes replace hydrous silicates and occur as cellular elongated packets of small prisms (mostly hypersthene) in glass.

Titanomagnetite occurs as large crystals that were present in the original rocks. Primary magnetite and ilmenite also form inclusions in silicates. Re-equilibrated magnetite and ilmenite are present in glass and as breakdown products of hydrous silicates. Within oxide grains apatite inclusions are common, and zircon less so.

Colourless to brown glass in many rocks is vesicular but some samples contain dense or poorly vesiculated brown glass. Incipiently melted rocks typically have vesicular colourless glass along grain boundaries. Glass veins (see Section 10 and Fig. 15) may be either dense or strongly vesiculated. Two samples have dense brown glass veins cored by fine-grained, augite + hypersthene-bearing quartz veins. The distribution of intergranular melt appears to be rather uniform within most samples, including those with glass veins. A few low-melt-fraction samples contain melt pools ~0·5–1·0 mm across (Fig. 3(C)). It is uncertain whether the pools resulted from melt migration (Jurewicz & Watson 1985) or from fusion of local granophyric patches (Fig. 4(A)).

Fluid inclusions are present in all samples. Their abundance decreases with increasing temperature.

3.2. Aplite

A few erupted plutonic rocks are fine-grained biotite granite here termed aplite (e.g. 432). These have plagioclase phenocrysts set in an aplitic or micrographic groundmass of plagioclase, alkali feldspar, quartz, biotite, magnetite and ilmenite. Fluid inclusions are present. On partial melting these rocks show the same sequence of mineral reactions as the granodiorites.

3.3. Granophyre

Sample 1239 is partially-melted medium-grained granophyre. It consists of patches of graphically intergrown quartz and alkali feldspar up to 3 mm across which grade outward into 1–2 mm crystals of the same phases. Vesicular brown glass occurs between feldspar + quartz clusters, forming local pools a few millimetres across. Minor plagioclase crystals are comparatively small. Biotite, titanomagnetite, ilmenite, zircon, apatite and monazite are rare.

3.4. Diabase

Fine-grained plagioclase mesocumulates consist of oriented plagioclase laths, fine granular oxides, orthopyroxene and clinopyroxene, with or without poikilitic pyroxenes and(or) hornblende. Plagioclase phenocrysts may be present. Diabase samples are included here because they are associated with the granitic rocks and because their dusty appearance in thin section and local intergranular glass suggest incipient fusion rather than incomplete crystallisation.

4. Mineral compositions

Mineral phases and glass were analysed with an ARL-SEMQ electron microprobe equipped with 9 wavelength-dispersive X-ray spectrometres. Natural and synthetic mineral standards were used. Compositions were calculated with a modified α-factor matrix correction scheme. Glass and Fe–Ti oxides were analysed in 19 and 30 samples, respectively, and all major phases were analysed in a subset of nine representative samples.

4.1. Fe–Ti oxides

Titanomagnetite and ferrian ilmenite occur as equant grains in the crystalline or glassy groundmass and, commonly, are included in plagioclase phenocrysts and mafic silicates of virtually all samples. Several grains of each phase were analysed in most rocks. Up to six point analyses per crystal were averaged, as intragrain compositional variation was within estimated counting error for most elements. Inhomogeneous (oxy-exsolved) magnetite crystals were not analysed. Representative analyses of four magnetite-ilmenite pairs are given in Table 3.

The oxides in unmelted samples have compositions typical of oxide phenocrysts in silicic volcanic rocks (Carmichael 1967). The partially melted samples, however, contain oxides re-equilibrated under high-temperature oxidising conditions, as is evident in the high claculated Fe^{3+} content of the ferrian ilmenite. In the partially melted rocks, oxide minerals in contact with glass have re-equilibrated, whereas those included in silicates appear to record original cooling of the rock (La Tourette *et al.* 1991). In partially melted aplite 432, a large titanomagnetite grain and its coarse ilmenite lamellae have compositions identical to those of discrete titanomagnetite and ilmenite in glass, indicating re-equilibration of an originally more Ti-rich spinel phase. Partitioning of minor elements between touching magnetite and ilmenite grains generally is consistent with equilibrium (e.g. Bacon & Hirschmann 1988), particularly in the high-melt-fraction samples (La Tourrette *et al.* 1991; C. R. Bacon unpublished data 1991; Table 3). Oxide compositions in low-melt-fraction samples are variable, although touching pairs appear to have re-equilibrated. Results of thermometry and oxygen barometry calculations are described below (See Section 5.1).

4.2. Pyroxenes

Pyroxenes (Table 4) were analysed in six samples. Primary augite and more abundant orthopyroxene occur as phenocrysts and as inclusions in plagioclase in the granodiorites. These phases have compositions similar to pyroxene phenocrysts in dacitic to rhyodacitic volcanic

Table 3 Electron microprobe analyses of representative Fe–Ti oxide pairs

Sample no.	463 unmelted	452 hb + bt + gls	1326 bt + gls	1038 gls
Vol% glass	0	7	17	30
Magnetite				
Al_2O_3	0·66	1·10	1·19	1·36
FeO*	85·5	85·2	80·6	79·2
MgO	0·89	1·65	1·96	2·08
CaO	0·01	0·00	0·03	0·03
TiO_2	4·66	5·33	11·4	11·7
MnO	0·91	0·43	0·47	0·45
V_2O_3	0·30	0·61	0·46	0·63
Recalculated				
Fe_2O_3	58·6	57·7	46·7	45·3
FeO	32·8	33·3	38·6	38·5
Total	98·8	100·1	100·8	100·1
X_{usp}	0·131	0·148	0·317	0·331
Ilmenite				
Al_2O_3	0·04	0·19	0·19	0·26
FeO*	49·3	64·6	57·4	58·0
MgO	2·36	1·59	2·22	2·10
CaO	0·01	0·03	0·02	0·03
TiO_2	43·6	29·1	37·7	35·9
MnO	1·55	0·24	0·34	0·26
V_2O_3	0·10	0·35	0·30	0·43
Recalculated				
Fe_2O_3	17·7	46·2	30·9	33·1
FeO	33·4	23·1	29·6	28·2
Total	98·8	100·8	101·3	100·3
X_{ilm}	0·820	0·542	0·695	0·670
T°C	751	(842)	953	980
log f_{O_2}	−13·08	(−10·71)	−9·85	−9·43

Notes: hb = hornblende, bt = biotite, gls = glass; X_{usp} and X_{ilm} are mole fractions of ulvospinel and ilmenite, respectively, calculated according to Stormer (1983). Temperature and oxygen fugacity calculated from Andersen & Lindsley (1988); values in parentheses are provisional because oxide compositions are substantially outside valid range of solution model; values for 1326 and 1038 outside but very near limit of valid range. 463 and 452 oxides are homogeneous, touching grains in groundmass. 1326 and 1038 are averages of two and three compositionally similar, homogeneous, touching oxide pairs, respectively, in glass.

rocks, including those from Mount Mazama (Druitt & Bacon 1989). When in contact with glass, some primary crystals are zoned to relatively magnesian rims. New or re-equilibrated fine-grained pyroxenes in the high-melt-fraction samples have rim compositions that reflect temperatures higher than phenocrysts in juvenile Mazama silicic magmas. There is a trend with increasing temperature toward higher Ca in orthopyroxene and lower Ca in augite in the partially melted rocks (Fig. 5).

4.3. Biotite

Biotite was analysed in six samples (Table 5). Biotite in unmelted sample 463 has the lowest Mg/Mg + Fe (0·66). Biotite in all but one of the partially melted samples has Mg/Mg + Fe = 0·76–0·85, reflecting the oxidising conditions indicated by Fe–Ti oxide compositions (see Section 5.1). In the two samples where biotite and hornblende coexist, biotite has slightly higher Mg/Mg + Fe than hornblende. All of the analysed biotites have high Ti contents (typically >4 wt% TiO_2), the highest being 7·33 wt% TiO_2 in the sample with the highest oxide temperature (968°C) and least biotite of the biotite-bearing rocks.

Table 4 Representative electron microprobe analyses of pyroxenes

Sample no.	463 unmelted	452 hb + bt + gls	1326 bt + gls	1038 gls
Vol% glass	0	7	17	30
Orthopyroxene				
SiO_2	52·8	54·7	52·9	53·4
Al_2O_3	0·57	0·34	0·33	0·22
FeO*	22·2	17·5	21·9	21·1
MgO	21·7	25·8	22·3	22·0
CaO	1·10	0·78	1·13	1·47
Na_2O	0·02	0·01	0·05	0·06
TiO_2	0·17	0·13	0·15	0·15
MnO	0·95	0·80	0·85	0·83
Cr_2O_3	0·00	0·00	0·00	0·00
Total	99·51	100·06	99·61	99·23
Fs	35·6	27·1	34·7	33·9
En	62·1	71·3	63·0	63·0
Wo	2·3	1·5	2·3	3·0
Clinopyroxene				
SiO_2	52·7	53·3	52·4	52·7
Al_2O_3	0·82	0·92	0·94	0·77
FeO*	9·65	7·66	10·06	10·57
MgO	14·3	15·1	14·2	14·2
CaO	21·3	21·3	20·1	19·9
Na_2O	0·40	0·55	0·58	0·57
TiO_2	0·19	0·24	0·41	0·28
MnO	0·42	0·41	0·42	0·43
Cr_2O_3	0·11	0·01	0·02	0·00
Total	99·89	99·49	99·13	99·42
Fs	15·5	12·4	16·5	17·2
En	40·8	43·5	41·4	41·2
Wo	43·7	44·1	42·1	41·5

Notes: 463 opx is inclusion in plagioclase, cpx is relict grain in hornblende; 452 opx and cpx are rims of adjacent relict grains in biotite; 1326 opx and cpx are small grains near each other in glass; 1038 opx and cpx are small adjacent grains in glass; hb = hornblende, bt = biotite, gls = glass.

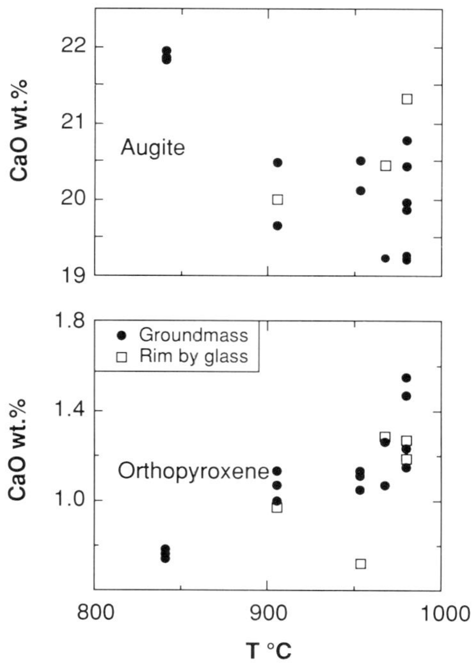

Figure 5 Pyroxene CaO contents versus Fe–Ti oxide temperature for partially melted granodiorites.

Halogen contents of biotite vary substantially from low F and relatively high Cl in unmelted and low-melt-fraction samples, to high F and diminished Cl in the high-melt-fraction rocks. Halogen variation as a function of Mg/Mg + Fe of biotite is plotted in Figure 6. The more magnesian biotites have F contents of over 1 anion per formula unit (up to 4·9 wt% F).

4.4. Amphibole

Hornblende is present in two of the samples chosen for detailed study (Table 5). The hornblende in these rocks is actinolitic (0·54–0·84 Al per formula unit). Hornblende in the low-melt-fraction sample has higher Mg/Mg + Fe (0·73–0·75) than that in the unmelted one (0·64–0·69). Fluorine contents of hornblendes are comparable in the two samples, but Cl is roughly twice as abundant in hornblende in the partially fused rock.

4.5. Feldspars

Feldspars were analysed in ten rocks ranging in composition from quartz diorite to granophyre and in texture from unmelted to the highest melt fraction (Fig. 7, Table 6). Although rim compositions are reported, absolute rim compositions are uncertain because a finely-focused beam could not be used on these sodic feldspars.

Plagioclase in quartz diorite and granodiorite retains its oscillatory zoning, with cores typically $\leq An_{50}$; more calcic cores may be present. Hydrothermally altered quartz diorite (1092) has core, rim and groundmass plagioclase of An_{44}–An_{55} (n = 7). Unmelted granodiorite (463) provides a baseline for monitoring changes in feldspar composition during partial melting. Intermediate, and a few calcic, plagioclase cores are overgrown by oscillatory zones ending in rims as sodic as Ab_{84}. Alkali feldspar in the same rock ranges from Or_{56}–Or_{48} and contains $\leqslant$1% An.

The onset of melting (7 vol% glass) is seen in granodiorite sample 452, which contains biotite + hornblende + intergranular glass (Fe–Ti oxide $T = 842°C$). Rims of feldspars in contact with glass have a range of compositions: An_{22}–An_{25} and Or_{51}–Or_{55} rims probably are relics of original crystallisation; An_{11}–An_{14}, Ab_{69}–Ab_{72}, Or_{20}–Or_{15} and $An_6Ab_{53}Or_{41}$ rims may approximate coexisting equilibrium compositions under melting conditions.

Sample 1326 is representative of medium-melt-fraction biotite-bearing granodiorites (Fe–Ti oxide $T = 954°C$, 17 vol% glass). Plagioclase has rims as sodic as $An_{17}Ab_{73}Or_{10}$ and groundmass alkali feldspar is as calcic as $An_7Ab_{67}Or_{26}$. Higher temperature, biotite-bearing granodiorite sample 1327 (Fe–Ti oxide $T = 968°C$, 40 vol% glass) contains a single equilibrium feldspar of average composition $An_{14}Ab_{71}Or_{15}$. Three high-melt-fraction granodiorites that lack biotite have equilibrium feldspars whose average composition becomes less potassic with increasing Fe–Ti oxide temperature: $An_{13}Ab_{70}Or_{17}$ (1630, 961°C, 37 vol% glass); $An_{12}Ab_{73}Or_{15}$ (1038, 981°C, 30%); $An_{17}Ab_{70}Or_{12}$ (1631, 1003°C, 48%). Equilibrium alkali feldspar in sample 1326 and feldspars in the four samples with Fe–Ti oxide $T > 960°C$ appear to be re-equilibrated alkali feldspar of the original alkali feldspar + plagioclase + quartz intergrowth (compare Fig. 4(A) and (b)).

Low-melt-fraction aplite sample 432 (Fe–Ti oxide $T = 906°C$, 10 vol% glass) contains plagioclase and alkali feldspar that average (rim and groundmass) $An_{23}Ab_{68}Or_9$ and $An_3Ab_{44}Or_{53}$, respectively. Granophyre sample 1239 (36 vol% glass) contains a single feldspar whose rim and groundmass compositions, in contact with glass, average

Table 5 Representative electron microprobe analyses of biotites and amphiboles

Sample no. Vol% glass	463 unmelted 0	463 unmelted 0	452 hb + bt + gls 7	452 hb + bt + gls 7	1326 bt + gls 17	1327 bt + gls 40
	hb	bt	hb	bt	bt	bt
SiO_2	50·5	38·4	52·1	39·9	39·4	38·7
Al_2O_3	4·34	13·4	3·44	12·3	13·0	13·4
FeO*	14·8	14·6	10·5	10·4	8·32	9·22
MgO	15·0	16·1	17·6	18·5	19·5	17·4
CaO	10·7	0·01	11·4	0·01	0·03	0·03
Na_2O	1·31	0·50	1·07	0·83	0·94	0·78
K_2O	0·34	8·68	0·33	8·79	8·66	8·05
TiO_2	0·89	4·16	0·84	4·81	5·22	7·33
MnO	0·43	0·25	0·24	0·10	0·06	0·06
Cr_2O_3	0·00	0·01	0·01	0·00	0·02	0·02
Cl	0·12	0·24	0·12	0·22	0·17	0·13
F	0·60	0·65	0·49	1·75	4·27	2·14
Less O	0·28	0·33	0·23	0·79	1·84	0·93
Total	98·75	96·67	97·88	96·82	97·75	96·33

Notes: 463 amphibole and biotite are cores of adjacent small grains; 452 amphibole and biotite are cores of relatively large grains; 1326 biotite is core of small grain with reaction rim; 1327 biotite is core of small grain with reaction rim; hb = hornblende, bt = biotite, gls = glass.

$An_3Ab_{61}Or_{36}$. Because the glass in these evolved rocks is poor in Ca, it seems possible that their An-poor alkali feldspars have re-equilibrated during melting.

Ba and Sr concentrations were determined in each feldspar analysis ($Sr_{L\alpha}$ intensities were corrected for $Si_{K\beta}$ interference). Sr concentrations in plagioclase cores are <1,100 ppm in granodiorites, 800–1,900 ppm in quartz diorites and mafic granodiorites, and are not simply correlated with An; Ba contents are ⩽900 ppm in plagioclase cores and vary inversely with An. The Sr contents are similar to those of phenocrysts in late Pleistocene and Holocene magmas of Mount Mazama (Druitt & Bacon 1989), and their variation suggests some, but relatively less, variety in parent magmas. Plagioclase rims have decreased Sr and increased Ba contents relative to cores; Ba reaches ~1800 ppm at ~$An_{10}Ab_{83}Or_7$ in the unmelted granodiorite. Alkali feldspar in unmelted sample 463 contains ⩽100 ppm Sr and 200 to 3,000 ppm Ba, the high Ba contents of plagioclase rims and groundmass alkali feldspar reflecting crystallisation of the last liquid. Similarly, alkali feldspar in granophyre 1239 and aplite 432 has ⩽200 ppm Sr and 1,300–2,500 ppm Ba. Alkali feldspars in the other partially melted samples have significantly higher Sr contents, typically several hundreds of ppm, and contain a substantial An component.

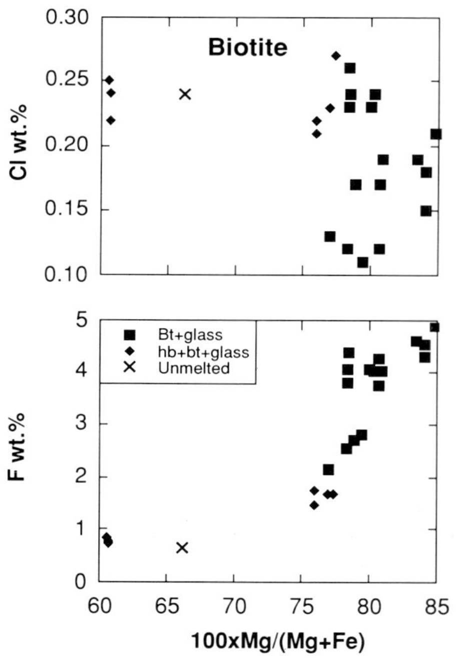

Figure 6 Biotite halogen contents versus 100 × Mg/Mg + Fe; symbols indicate presence of glass and halogen-bearing silicates; unmelted sample contains biotite and actinolitic hornblende; hornblende + biotite + glass points with low Mg/Mg + Fe are for sample with the lower Fe–Ti oxide temperature of two hb + bt + glass samples.

The Ba and Sr contents of rim and groundmass feldspars in the common partially melted granodiorites might be expected to vary systematically with temperature and melt fraction. However, Ba and Sr data for plagioclase do not show notable changes with temperature or melt fraction (Sr data also scatter as a result of poor precision), either because plagioclase has not re-equilibrated with melt or because these elements are compatible in feldspars. Ba in alkali feldspar, however, tends to decrease from ⩾5,000 to ~2,000 ppm with increasing temperature and melt fraction, along with decreasing Or and increasing An. Systematic change in alkali feldspar composition may result from re-equilibration during the melting reaction that consumes much of the original groundmass quartz + alkali feldspar + Na-plagioclase intergrowths.

5. Physical conditions

Previous studies of geothermometry of silicic volcanic rocks have shown that the Fe–Ti oxide thermometer/oxygen barometer yields precise results that allow useful comparisons to be made between quenched samples. Moreover, the oxides typically readjust their compositions to new conditions more rapidly than do silicates. Because virtually all of the samples in this study contain the requisite oxide phases and all samples were rapidly quenched, this thermometer should be the most reliable for the ejected blocks.

Because no precise mineralogical barometers are applicable to the phase assemblages and likely pressure range, pressure is estimated by other means.

5.1. Temperature and oxygen fugacity

Equilibration temperatures and oxygen fugacities have been determined for 27 samples (Fig. 8). Temperature and f_{O_2} are reported only for magnetite and ilmenite pairs that are in mutual contact or are averages of homogeneous populations of each phase. In many cases, plotted values are averages of those for two to four like pairs. Analyses were recalculated according to Stomer (1983); T and f_{O_2} were calculated according to Andersen and Lindsley (1988). Only pairs that passed the Mg/Mn intercrystalline partitioning equilibrium test of Bacon and Hirschmann (1988) were used. Temperatures are given to the nearest degree in order to preserve intersample analytical precision, which may be approximately ±5°C. Accuracy probably is the order of several tens of degrees.

Five unmelted samples were studied, three of which contained biotite and hornblende and two only biotite. Fe–Ti oxide equilibration temperatures of inclusions in silicates and free crystal pairs are 689–834°C, all lower than those of phenocrysts in precaldera rhyodacite lavas and climactic pumice of Mount Mazama (≥847°C; Druitt & Bacon 1989). Oxygen fugacities for the unmelted samples suggest a more oxidised trend than the eruptive rocks (Fig. 8) but there is a range of values for any given temperature. It is unclear whether free oxides in the unmelted samples record original crystallisation conditions or the effects of reheating.

Two low-melt-fraction samples containing hornblende + biotite were studied. Sample 450 contains inclusions in plagioclase and biotite that give 733°C and 801°C, respectively, and oxide pairs in contact with glass that yield 755°C and 785°C. The oxides in glass, and one set of inclusions, plot at notably lower f_{O_2} than data for the other low-melt-fraction and unmelted samples. Presumably, this

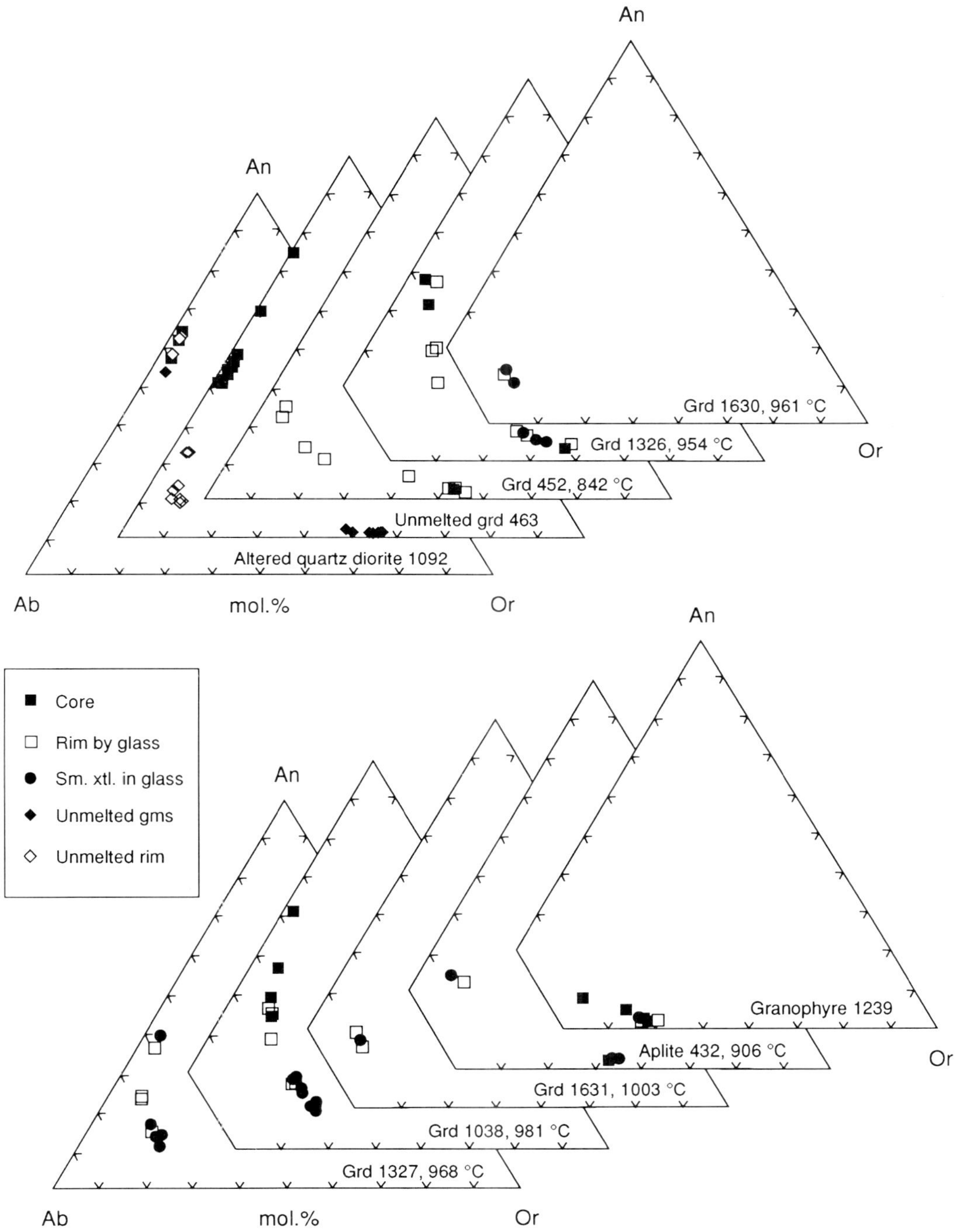

Figure 7 Feldspar compositions.

Table 6 Representative electron microprobe analyses of feldspars

Sample no.	463 unmelted	463 unmelted	463 unmelted	452 hb + bt + gls	452 hb + bt + gls	1038 gls
Vol% glass	0	0	0	7	7	30
	pl core	pl rim	af rim	pl rim	af rim	af rim
SiO_2	55·6	66·6	65·6	63·6	66·0	66·9
Al_2O_3	28·0	20·5	19·1	23·2	19·4	20·0
FeO*	0·26	0·17	0·00	0·18	0·19	0·47
MgO	0·02	0·00	0·02	0·01	0·00	0·02
CaO	10·06	1·89	0·15	4·49	0·59	2·41
Na_2O	5·78	9·35	4·90	8·33	5·10	8·27
K_2O	0·21	1·52	9·65	0·99	8·90	2·59
SrO	0·13	0·04	0·00	0·06	0·08	0·06
BaO	0·03	0·04	0·31	0·07	0·61	0·21
Total	100·09	100·11	99·73	100·93	100·87	100·93
An	48·4	9·2	0·7	21·6	2·9	11·8
Ab	50·4	82·1	43·2	72·7	45·2	73·1
Or	1·2	8·8	56·0	5·7	51·9	15·1

Notes: 463 pl is 3 mm phenocryst, af is groundmass; 452 pl and af are adjacent small grains; 1038 af is skeletal crystal in glass; hb = hornblende, bt = biotite, gls = glass, pl = plagioclase, af = alkali feldspar.

represents the onset of melting, with equilibrium between touching grains but lack of homogenisation of phase compositions throughout the sample. Sample 452, containing hornblende + biotite + 7 vol% glass (Fig. 3(A)), has Fe–Ti oxide inclusions that give 864°C and crystals in glass that give 842°C on a high-f_{O_2} trend with data for rocks with larger melt fractions, but both pairs are outside the calibration range of the thermometer. The data for 452 plot at comparable temperatures to, but much higher f_{O_2} than, those for phenocrysts in preclimactic hornblende-bearing rhyodacites of Mount Mazama (Fig. 8).

Nine granodiorites containing biotite and glass (12–40 vol% glass) yield temperatures of 886–968°C over a small range in f_{O_2} (Fig. 3(C)–(D)). Biotite in these samples becomes rarer with increasing temperature. Because these granodiorites are compositionally like those that contain hornblende, it appears that the maximum stability of (low Al) hornblende under the pressure and f_{H_2O} conditions of melting was ~870°C.

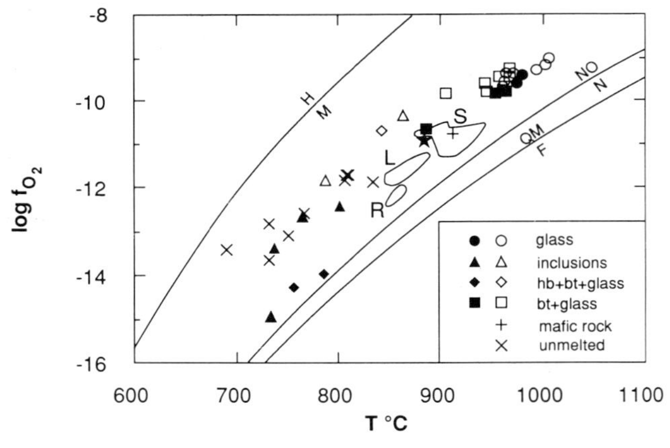

Figure 8 Oxygen fugacity versus temperature calculated from compositions of coexisting Fe–Ti oxides; filled symbols, + and × indicate oxide compositions within the calibration range of the Andersen and Lindsley (1988) model; open symbols are for data at the limit of, or slightly beyond, the calibration range; star and fields are for juvenile magmas of the climactic chamber (Druitt & Bacon 1989): star = climactic rhyodacite (nine climactic pumices + several related preclimactic lava and pumice samples); S = scoria of climactic eruption; L = Holocene preclimactic lava and pumice of Llao Rock; R = late Pleistocene rhyodacite and pumice of Redcloud Cliff, Steel Bay, and Grouse Hill; hematite-magnetite and fayalite-magnetite-quartz buffer curves are from Chou (1978), Ni–NiO from Huebner and Sato (1970).

Ten extensively melted granodiorites (28–50 vol% glass, one 18%) that lack biotite (Fig. 3(E)–(F)) give temperatures of 961–1,006°C at oxygen fugacities on trend with those of biotite-bearing samples (Fig. 8). Overlap in temperature with biotite-bearing granodiorites is only a few degrees, and overlap is not present if only data strictly within the calibration range of the thermometer are considered. The rocks thus record the maximum thermal stability of biotite in this system, which is between 964°C and 976°C; this biotite has Mg/Mg + Fe as high as 0·85 and contains ~5 wt% TiO_2 and 1 afu F (Table 5).

The ranges of temperature and f_{O_2} indicated by the Fe–Ti oxides in the granodiorites are both greater than for the climactic magma chamber of Mount Mazama (Fig. 8). Recognising that some climactic mafic scoria samples lack ilmenite, and that these have other indications of equilibration at temperatures >940°C, it is reasonable that the high temperatures recorded by the partially melted rocks reflect conditions in the deep walls of the chamber. The T–f_{O_2} trend defined by the partially melted rocks is more oxidising than that indicated for the climactic magmas, which themselves are quite oxidised, except for the high-melt-fraction samples where the T–f_{O_2} trend is in line with that of the climactic rhyodacite and most-voluminous scoriae. Thus, the T–f_{O_2} relations converge for the erupted magmas and the partially melted samples that were most likely to have been in pre-eruptive communication with them.

Aplite sample 432, which contains biotite and 10 vol% glass, has inclusions that yield the comparatively low temperature of 787°C at f_{O_2} similar to the higher-temperature holocrystalline samples. Re-equilibrated oxides in this rock give 906°C at the high f_{O_2} of the high-temperature partially-melted granodiorites. This sample may record both its original crystallisation and re-equilibration at a low degree of melting.

One partially melted diabase xenolith was found embedded in an olivine + augite-rich scoria block in the ignimbrite veneer deposit atop the lithic breccia (Druitt & Bacon 1986). In this example, thin veins from the scoria penetrate the xenolith. It is not clear whether this texture reflects pre- or syn-eruption contact. Fe–Ti oxides in the diabase xenolith give 913°C at the relatively low f_{O_2} of late scoriae of the climactic eruption (Fig. 8).

Temperatures were calculated for a few pairs of feldspar compositions that were judged to represent equilibrium (Elkins & Grove 1990). Rims for two pairs in low-melt-fraction granodiorite 452 give 779°C and 816°C (the latter being suspect), whereas the most likely Fe–Ti oxide temperature for this sample is 842°C. Two pairs from aplite 432, which has an Fe–Ti oxide temperature of 906°C, are 846°C and 878°C. As the oxide temperatures for both of these samples are based on compositions beyond the calibration limit, agreement between the two thermometers is reasonable. A feldspar pair in biotite-bearing granophyre 1239, for which no oxide temperature is available, yields 883°C.

5.2. Pressure

Minimum pressure on the partially melted rocks is constrained by data for rhyodacite pumice of the climactic eruption. Dissolved H_2O contents of melt inclusions in plagioclase phenocrysts in the rhyodacite indicate saturation pressures of 100–180 MPa (Bacon *et al.* 1992) for the climactic chamber, which corresponds to a minimum depth of 5–9 km (see also Section 6.1). Unmelted granitic rocks may have come from approximately this depth, and the partially melted rocks from the deeper parts of the erupted volume which is unlikely to have been more than a few hundred metres deeper than the top of the magma chamber. Breakdown of hornblende at lower temperature than biotite is consistent with low pressure (<200 MPa) and H_2O-undersaturation (Hollister *et al.* 1987; Johnson & Rutherford 1989a).

Recently, the Al content of hornblende has been used to infer pressure of crystallisation where hornblende equilibrated with melt + fluid + biotite + quartz + sanidine + plagioclase + sphene + magnetite or ilmenite. Initial crystallisation of the groundmass of the granodiorites, and the early stages of melting (e.g. 452), satisfy the phase assemblage requirement except for the lack of sphene, a condition likely to have a negligable effect on calculated pressure for low-TiO_2 hornblende (Johnson & Rutherford 1989b). Calculated pressures for hornblende rims in these samples are ~0 MPa. This barometer is not strongly temperature-dependent at pressures >200 MPa because the H_2O-saturated solidus for common granodiorite is nearly isothermal; at pressures <200 MPa, however, temperature dependence is likely to be significant (Hollister *et al.* 1987) and the barometer as it is presently formulated (there are no calibration points below 200 MPa) can be expected to give inaccurate results. In any case, the ubiquitously low Al_2O_3 contents of the analysed hornblendes (3·2–4·9 wt%) are consistent with pressures <200 MPa, as long as these crystals represent near-solidus conditions.

6. Glass compositions

Microlite-free glass was analysed in 19 samples by electron microprobe. A defocused beam (typically ~20 μm) and a specimen current of 10 na minimised Na migration, which does not appear to have been significant based on normative compositions and analysis of obsidian standards. Glass was separated from three high-melt-fraction granodiorites and the granophyre, and analysed for halogens and trace elements (Table 1). The glasses have not been affected significantly by secondary hydration (except possibly along fractures; see Section 7).

The vol% of glass was determined in 20 partially melted samples with analysed glass and(or) Fe–Ti oxide temperatures by counting 2,000–3,500 points on the upper surfaces of thin sections. Glass contents are reported vesicle-free and are subject to errors in point counting caused by vesiculation and complex glass/crystal interfaces. Because glass content may be underestimated owing to migration of melt to form veins, serial traverses were counted at 1 mm intervals parallel to veins; no systematic variation in glass content was detected, but results are imprecise because plagioclase crystals are significantly large relative to thin section dimensions. Weight fraction of melt was obtained from vol% glass by assuming a density of 2·3 g/cm^3 for glass and an average density of 2·7 g/cm^3 for crystals.

6.1. Major elements

The glasses have high-silica rhyolitic compositions (75·5–77·6 wt% SiO_2, anhydrous; Table 1). Their major element abundances are not unlike those of the high-temperature rhyolites of the Yellowstone Plateau volcanic field (Hildreth *et al.* 1984), although the latter have lower TiO_2 contents and tend to have higher K/Na.

Normative glass compositions have Q + Ab + Or = 93–97 wt%, An = 0–2·7 wt%. Plotted in the Q–Ab–Or system, the glasses form an array at roughly constant Q extending to compositions more Or-rich than the 100 MPa H_2O-saturated minimum melt, but well below the 50 MPa cotectic (Tuttle & Bowen 1985), so that they lie approximately on the 200 MPa H_2O-saturated cotectic shown by Nekvasil (1988). The projection of cotectic curves from the An-bearing system into the haplogranite system has a neglible effect for the normative An contents of the natural melts; however, reduced a_{H_2O} causes cotectics to shift away from Q such that analysed glass compositions would be consistent with a total pressure of ~100 MPa (Nekvasil 1988).

Changes in glass composition with Fe–Ti oxide temperature and melt fraction are complicated by variation in bulk composition of the rocks. Nevertheless, it can be seen in Figure 9 that SiO_2 and K/Na tend to decrease with increasing temperature, while P_2O_5 and TiO_2 increase. Figure 10 shows Fe–Ti oxide temperature, TiO_2, P_2O_5, and K/Na versus melt fraction. At melt fractions >0·1, there is a large increase in melt fraction with little increase in temperature, which implies a small melting interval as long as quartz remains and plagioclase cores are effectively isolated from melt. Compatible oxide (TiO_2, P_2O_5) contents of granodiorites rise more gently with melt fraction than with temperature. The decrease in K/Na of granodiorites with increasing melt fraction reflects consumption of original alkali feldspar, increasing plagioclase entry into the melt, and formation of K-rich ternary feldspar at higher temperatures.

6.2. Trace elements

Glass separates from the three high-melt-fraction granodiorites have less evolved compositions than many high-silica rhyolites. The more incompatible trace elements, Rb, Cs, Th, U, Nb and Ta, have notably high abundances. Plots of concentrations of these elements continue trends established by bulk rock data (Fig. 11), with the exception of Nb in sample 1038 in which the glass has a notably lower Nb/Ta ratio than the bulk rock.

Because of their compatibility with feldspars, Ba and Sr are present at moderate levels, Sr content being notably lower because of retention in refractory plagioclase. However, Ba and Sr concentrations are far greater in the partial melts than in the most differentiated high-silica rhyolites, which owe their low Ba and Sr to fractional crystallisation (e.g. Halliday *et al.* 1991). Glasses from granodiorite 1023, which gave the lowest Fe–Ti oxide temperature of the three granodiorites, and from granophyre 1239 have low Zr and Hf abundances relative to the other two glasses from granidiorites owing to survival, or isolation of more zircon. The transition metals Co and Cr are present at a few ppm, as in many high-silica rhyolites, but Sc (Fig. 1), like Ti, is relatively more abundant at 7–9 ppm.

The REE in glasses have parallel chondrite-normalised patterns (Fig. 12(A)) with substantial negative Eu anomalies. As a consequence, LREE/HREE ratios are nearly constant despite a two-fold variation in LREE. Bulk-rock analyses, however, show some variation in REE patterns (Fig. 12(B)) and ratios, as the granophyre is affected by monazite fractionation. The aplites, presumed to be differentiates of the granodiorite body, have LREE contents higher than, and HREE similar to, those of granodiorites (Fig. 12(C)); they have notably higher LREE/HREE than glasses. Glass/bulk-rock REE relations are illustrated in Fig. 12(D), where it can be seen that enrichment of HREE is slightly greater than of LREE on melting of granodiorites. The granophyre glass/bulk rock pair shows that La and Ce are less enriched than Nd–Gd because of the presence of monazite.

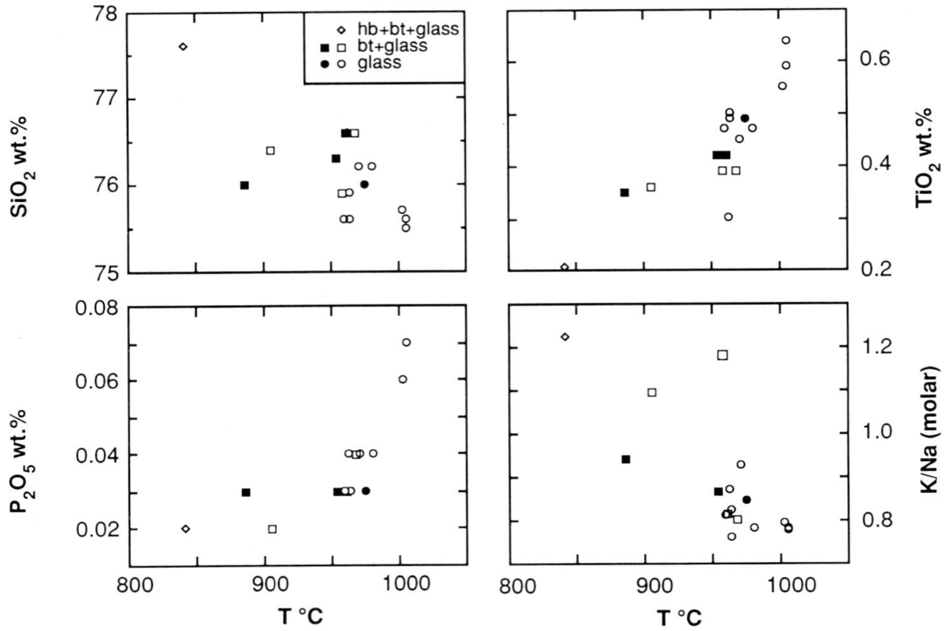

Figure 9 Glass composition versus Fe–Ti oxide temperature; symbols indicate presence of glass and halogen-bearing silicates; filled symbols indicate oxide compositions within the calibration range of the Andersen and Lindsley (1988) model; open symbols are for data at the limit of, or slightly beyond, the calibration range.

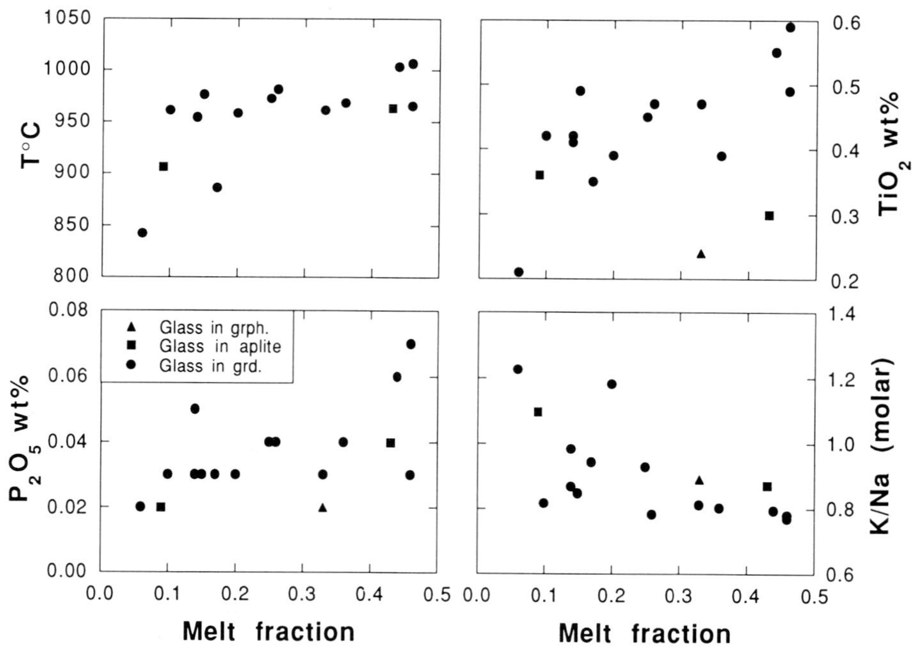

Figure 10 Glass composition and Fe–Ti oxide temperature versus melt fraction (wt fraction of glass); symbols indicate rock types.

6.3. Trace element modelling

Standard models were used to calculate concentrations of Rb, Sr, Ba, REE and Sc in melts formed by crystallisation and partial melting of granodiorite for comparison with analyses of aplites and glass separates, respectively. Unless otherwise specified below, the starting composition was taken as the average of the three granodiorites from which glass was separated. Phase proportions in this average granodiorite were estimated from its normative composition.

Trace element contents of aplitic liquid were modelled by Rayleigh fractionation of granodioritic magma using partition coefficients for phenocrysts in rhyodacitic pumice from Crater Lake (Bacon & Druitt 1988). The Ba, REE (Fig. 13(A)) and Sc contents of liquid resulting from 30% crystallisation of plagioclase + pyroxenes + Fe–Ti oxides + apatite from granodiorite melt (in proportions estimated from the granodiorite norm) match concentrations in aplite sample 432. The major element composition of the calculated liquid is similar to that of the aplite, except that the alpite has 0·8 wt% more K_2O and 0·8 wt% less Na_2O. Rb and Sr contents of the model liquid are ~40% lower than those of 432. Given uncertainties in initial Rb and Sr concentrations and in their behaviour, the trace element model is consistent with aplites being fractionates of the granodiorite pluton.

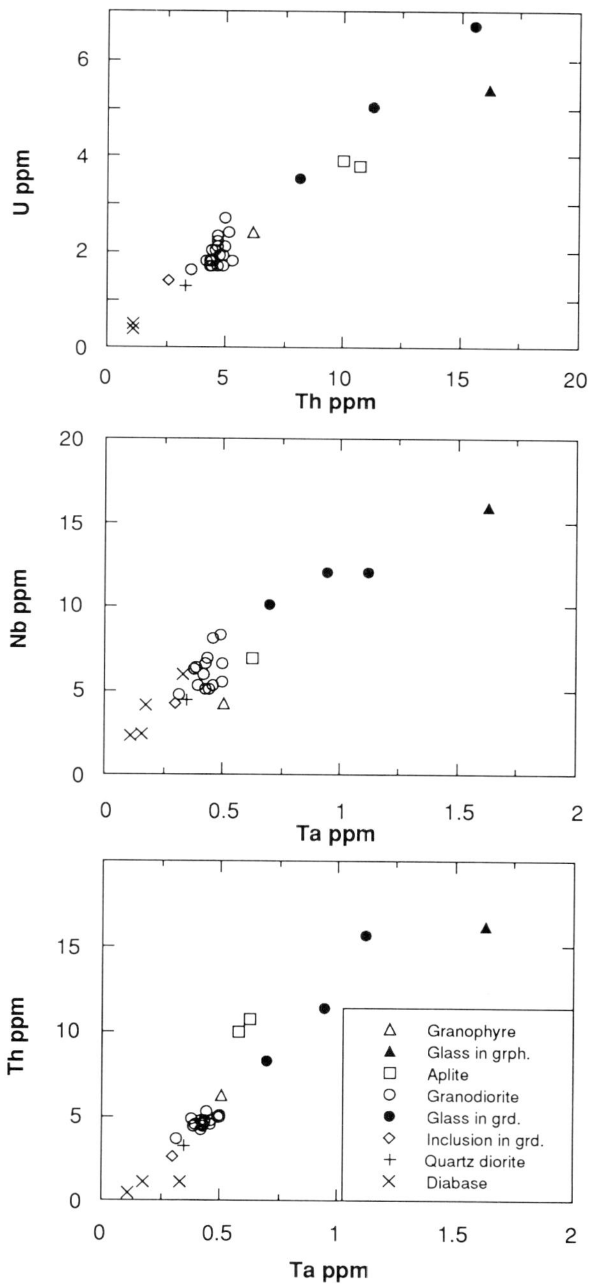

Figure 11 U, Th, Nb, and Ta contents of whole rocks and glasses.

Partial melts of granodiorite were modelled by non-modal equilibrium melting (Shaw 1970, eq. 15) and as accumulated liquids from non-modal incremental melting (Shaw 1970, eq. 14). The average normative composition of the three glasses separated from granodiorites was used to determine proportions of phases entering the melt, except that plagioclase, alkali feldspar and quartz were assumed to melt in equal amounts. Partition coefficeints were chosen from data for phenocrysts in high-silica rhyolitic liquids given by Nash and Crecraft (1985: plagioclase, magnetite, ilmenite), Leeman and Phelps (1981: sanidine), Sisson (1991: pyroxenes), Blundy and Wood (1991: Ba and Sr in plagioclase), and Watson and Green (1981: apatite).

Melting calculations were performed using whole-rock analyses of the three granodiorites as starting compositions. Element concentrations in calculated partial melts were compared with analyses of glass separates at the measured melt fractions. The closest overall approximations to the natural melts are those calculated by equilibrium melting without apatite. These calculated partial melts have REE patterns similar to analysed glasses (Fig. 13(B) & (C)). Modelled Ba contents match analysed values adequately, given uncertainty in partition coefficients. Concentrations of Rb are too low and Sr too high in calculated melts in comparison to the glasses. Adjusting partition coefficients does not eliminate this problem, but appropriate changes in initial concentrations of Rb and Sr, as though early melting domains had fractionated compositions with relatively high Rb and low Sr contents, allow close fits of the models to the glass compositions. Another possibility for Rb is that it is expelled as bioite decreases in abundance with increasing melting. The major failure of the models is that Sc is significantly depleted in modelled liquids relative to granodiorite, whereas Sc contents of rocks and glasses are subequal. A possible explanation for this discrepancy is that pyroxene and oxide partition coefficients used in modelling Sc behaviour are too high.

In summary, conventional models adequately predict the REE and Ba concentrations in partial melts of granodiorite, but are less successful for elements that (1) are compatible and zonally distributed in refractory crystals (Sr in plagioclase), (2) are highly compatible and have uncertain partition coefficients in residual phases (Sc in pyroxenes), or (3) may be enriched in early-melting domains (Rb in late-crystallised matrix and reacting biotite). The "carrier mineral model", in which trace elements enter the liquid as their host minerals melt because melting is more rapid than intracrystalline diffusion (e.g. Bea 1991), is probably more applicable to the partially fused rocks at Crater Lake than the models described above, but its use is precluded by lack of quantitative information on trace element concentrations in mineral phases.

7. Volatiles

Water contents (H_2O^+) of 11 granodiorite blocks, a quartz diorite enclave, and the granophyre are 0·24–0·65 wt% (Table 1). There is no simple correlation between H_2O^+ and temperature, although high-melt-fraction samples tend to have the higher H_2O contents. The hydrothermally altered quartz diorite (1092) has an H_2O^+ value of 1·60 wt%, consistent with its epidote + chlorite + actinolite (+ plagioclase + quartz + oxides) mineralogy. The same samples have CO_2 contents of <0·02–0·15 wt%, again without correlation with temperature. Microprobe analyses of dense glass typically sum to >99 wt%, as though H_2O content were <1 wt%. A single FTIR determination of H_2O in brown glass of high-melt-fraction granodiorite 1038 gave 0·41 ± 0·02 wt% (La Tourrette *et al.* 1991), whereas the whole-rock H_2O^+ content is 0·51 wt%. This observation suggests that the relatively high H_2O^+ of high-melt-fraction bulk rock samples is due to minor secondary hydration of the abundant glass along microcracks. The early breakdown of hornblende and the persistence of quartz to the highest recorded temperatures are consistent with low H_2O contents of melts at upper crustal pressures (Maaløe & Wyllie 1975; Naney 1983; Hollister *et al.* 1987; Johnson & Rutherford 1989a). Finally, the high temperatures recorded in medium- to high-melt-fraction samples indicate melts were H_2O-undersaturated (Johannes 1984). There is no evidence, therefore, for addition of H_2O to granodiorite in the melting zone.

Halogen concentrations in 12 granodiorites and a quartz diorite enclave range from 0·01–0·05 wt% F and from 0·03–0·14 wt% Cl (Table 1). Mass balance calculations indicate that biotite, amphibole and glass can account for virtually all of the F in all bulk samples, that glass contains most of the Cl in high-melt-fraction rocks, and that much of the Cl in unmelted and low-melt-fraction blocks must reside in fluid inclusions. Variation of F content with temperature is not evident, but Cl commonly is more abundant in higher temperature samples so that the Cl/F weight ratio is ~1 in unmelted and low-melt-fraction granodiorites but is typically ~2–4 in high-melt-fraction samples; it is unclear whether Cl was added to high-temperature granodiorites or merely was retained more effectively. Hydrothermally altered quartz diorite 1092 has a normal F content (0·03 wt%) but has Cl/F = 0·2, whereas an otherwise comparable quartz diorite enclave (800) contains an order of magnitude more Cl. Chlorine was removed by hydrothermal solutions capable of depositing epidote but was not as effectively mobilised by fluids responsible for lowering $\delta^{18}O$ values of the granitic rocks (see Section 2.2).

8. Trace element partitioning between melt and crystals

La Tourrette *et al.* (1991) measured U and minor element partition coefficients for Fe–Ti oxides and zircon in medium- to high-melt-fraction samples. In the present study, data are available for Sr and Ba in feldspars, and for effective bulk partition coefficients calculated from analyses of four rocks and glass separates.

8.1. Glass/whole-rock relations

Enrichment and depletion of selected elements during melting is illustrated in Figure 14, where ratios of concentrations in glass/bulk rock are plotted. Of the three granodiorites, sample 1038 has the lowest melt fraction (0·26), and thus its glass shows the greatest relative enrichment in incompatible elements. Of the compatible

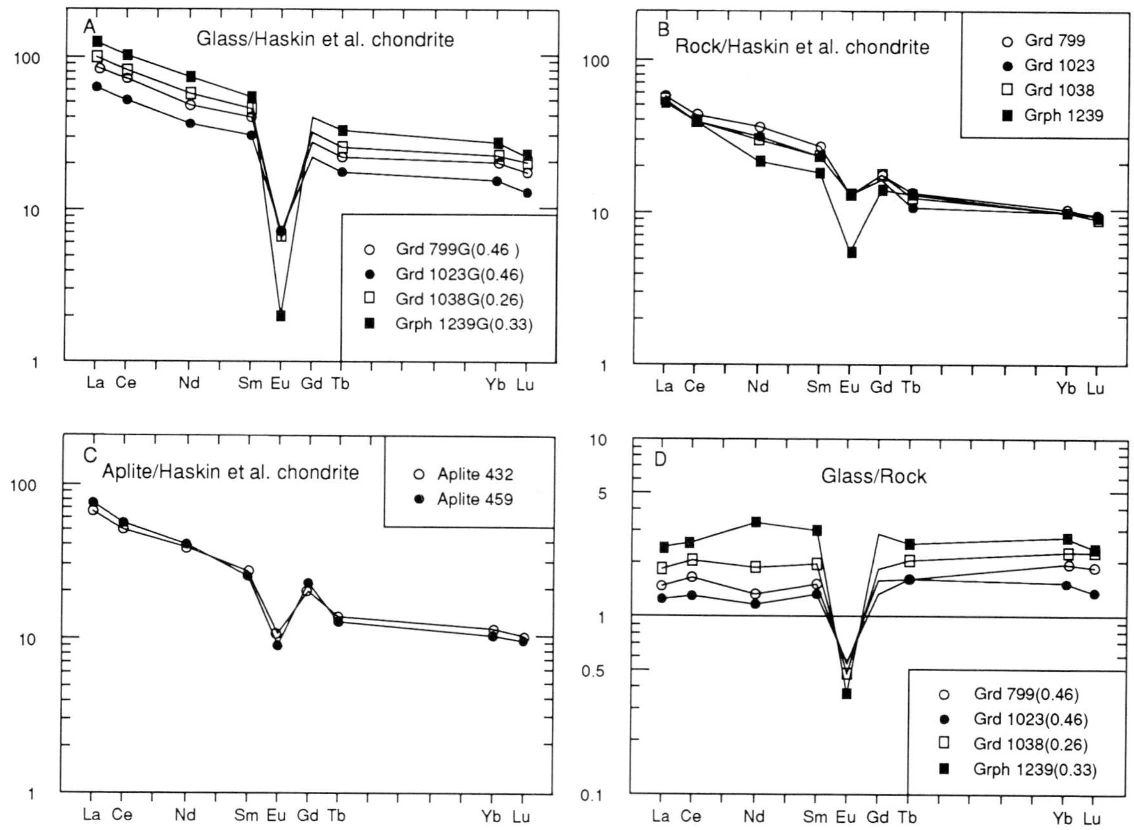

Figure 12 Rare-earth element patterns; legends give rock types (Grd = granodiorite, Grph = granophyre), sample numbers, and melt wt fraction (in parentheses):
(A) Glass separated from granodiorites and granophyre.
(B) Granodiorite and granophyre.
(C) Aplite.
(D) Glass separate data normalised to their respective whole-rock analyses.

elements, Ba, and more strongly Sr, are concentrated in feldspars, P in apatite, Ti in Fe–Ti oxides and Sc in pyroxenes. The behaviour of Zr and Hf is controlled by whether or not zircon was effectively isolated from the melt,

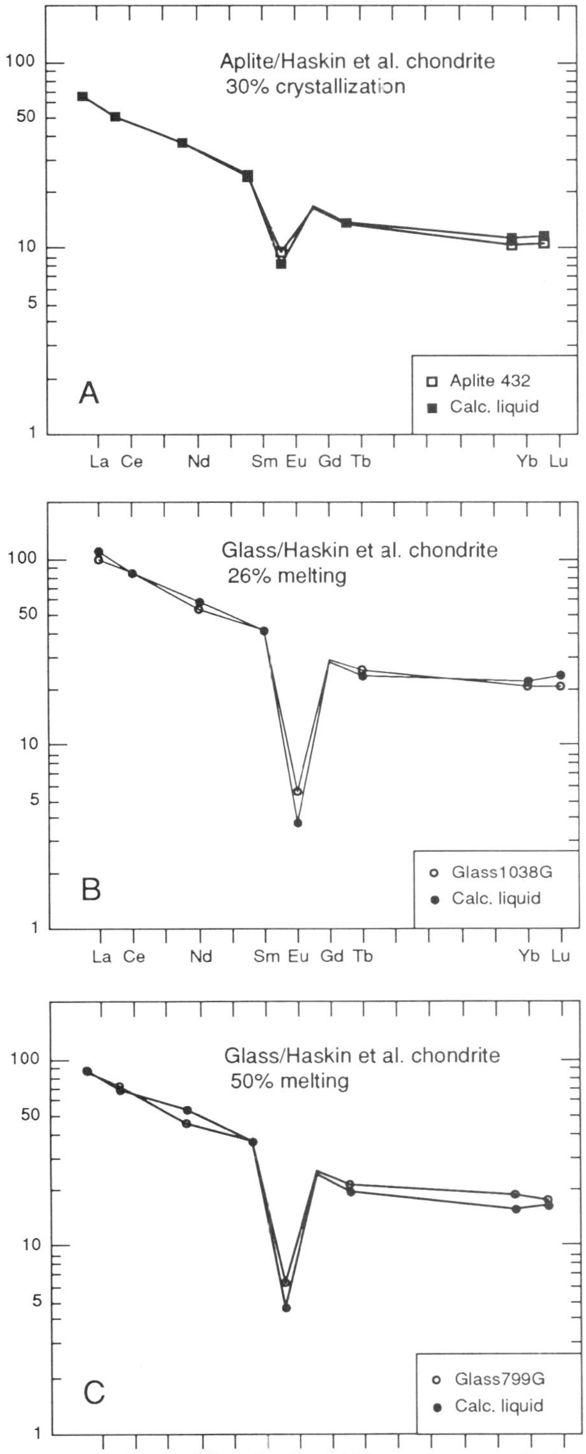

Figure 13 Rare-earth element models:
(A) Calculated REE concentrations in liquid produced by 30% fractional crystallisation of granodiorite compared with composition of aplite sample 432.
(B) Calculated REE concentrations in liquid produced by 26% melting of granodiorite sample 1038 compared with composition of glass separated from the same rock (wt fraction glass = 0·26).
(C) Calculated REE concentrations in liquid produced by 46% melting of granodiorite sample 799 compared with composition of glass separated from the same rock (wt fraction glass = 0·46).

and by zircon dissolution kinetics. La Tourrette *et al.* (1991) noted that, although both phases were present, the melts were undersaturated with zircon (Watson & Harrison 1983) and apatite (Harrison & Watson 1984). This also applies to data for the four glass separates in Table 1 and to all glasses analysed for P by microprobe. The granophyre glass/bulk rock pattern differs in that Nb, Ta, Sc, Co and HREE behave as strongly incompatible elements, and Ba is more compatible than Sr because of the presence of residual alkali feldspar rather than plagioclase.

Determination of melt fractions allows calculation of bulk partition coefficients for the four samples. These partition coefficients are not equilibrium values but represent the *effective* partitioning owing to isolation of inclusions within crystals and slow intracrystalline diffusion, particularly in plagioclase. The following estimates are based on averages for the three granodiorites (melt fractions = 0·26, 0·46 and 0·46; all are biotite free): $D \ll 0{\cdot}1$—Cs, Rb, F, Cl, U, Th; $D \approx 0{\cdot}1$—K, Nb, Ta; $D \approx 0{\cdot}2$–$0{\cdot}3$—HREE; $D = 0{\cdot}3$–$0{\cdot}6$—Zr, Hf, Y, LREE; $D \approx 1{\cdot}0$—$1{\cdot}2$-Sc, Ti, Zn; $D \approx 5$–12 (increasing order)—Ba, Eu, Mn, Co, P, Sr, Cr. In the granophyre (melt fraction = 0·33; rare biotite present), partition coefficients are: $D \ll 0{\cdot}1$—Nb, Ta, Sc, Co; $D \approx 0{\cdot}1$—Cs, Rb, Th, REE, Y, Zn; $D \approx 0{\cdot}2$—$0{\cdot}3$–Ti; $D \approx 0{\cdot}8$–$0{\cdot}9$—Zr, Hf; $D \approx 2{\cdot}3$–7 (increasing order)—Sr, Eu, Ba.

8.2. Sr and Ba in feldspars

Measured Sr and Ba concentrations in glass and alkali feldspar rims in a granodiorite (1038) and the granophyre (1239) allow calculation of partition coefficients. The equilibrium feldspar in the granodiorite, average composition $An_{12}Ab_{73}Or_{15}$, has $D_{Sr} = 7 \pm 3$ and $D_{Ba} = 4 \pm 1$ (1σ of D's calculated from microprobe feldspar analyses and XRF analysis of the glass separate, ignoring uncertainties in analyses). Calculated values for $An_{12}Ab_{88}$ plagioclase at 981°C (the Fe–Ti oxide temperature of sample 1038) from equations of Blundy and Wood (1991) are $D_{Sr} = 9{\cdot}6$ and $D_{Ba} = 1{\cdot}7$. The apparent equilibrium alkali feldspar in the granophyre, $An_3Ab_{61}Or_{36}$, has $D_{Ba} = 13 \pm 1$; feldspar Sr content is too low for accurate analysis.

9. Duration of melting

Compositional variation of glass and crystals, or the lack of it, can be used to constrain the duration of melting using simple isothermal diffusion models. La Tourrette *et al.* (1991) found evidence for Zr concentration gradients in the

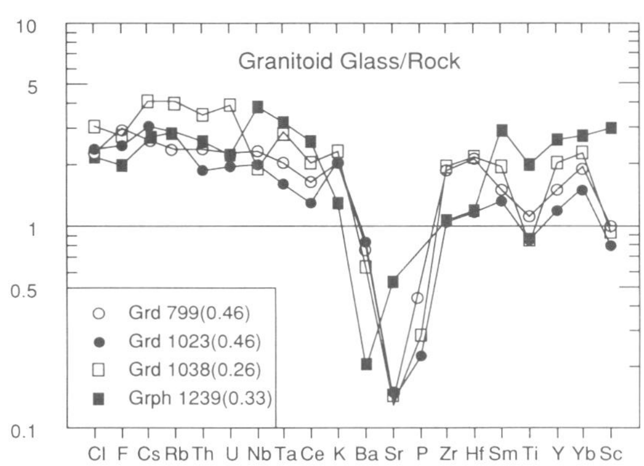

Figure 14 Element abundance diagram of glass separate data normalised to their respective whole-rock analyses; sample numbers and melt wt fractions (in parentheses).

vicinity of zircons in glass in two out of three high-melt-fraction samples investigated. The characteristic Zr diffusion length (l) is between 200 and 500 μm, which implies a time scale of 800 to 8,000 years ($t \approx l^2/D$) when combined with the diffusion coefficients given by Harrison and Watson (1983). This result is consistent with the eruptive history of Mount Mazama, where preclimactic rhyodacite lavas, believed to have been derived from the early climactic chamber, predate the caldera-forming eruption by >20,000 years (Bacon & Druitt 1988).

The *maximum* duration of melting might be constrained by the observation that even in the highest-temperature samples oscillatory zoning is preserved in plagioclase on a scale of 1–10 μm. Calculation of homogenisation times is not definitive, however, because diffusion coefficients for CaAl–NaSi diffusion in plagioclase at 1,000°C are as much as seven orders of magnitude smaller in anhydrous (Grove *et al.* 1984, extrapolated 100°C) than in hydrothermal experiments (Johannes 1985; Liu & Yund 1992). Under anhydrous conditions at 1,000°C 1 μm oscillatory zones would homogenise in 20,000 years versus a day in the presence of H_2O at ≥500 MPa. Because pre-eruptive a_{H_2O} in the partial melts was ≪1, homogenisation times of plagioclase probably would have been of the order of 10^2–10^4 years.

A constraint on the *minimum* duration of heating comes from the observation that titanomagnetite crystals are homogeneous even though their Ti content must have changed in response to heating. Following Lu and Anderson's (1991) study of the Bishop Tuff, the time needed for Ti to diffuse into titanomagnetite was calculated using the spherical model of Crank (1975, eq. 6.19) and diffusivity obtained from eq. 21 of Aragon *et al.* (1984). Isothermal homogenisation of the largest magnetite crystals (~0·2 mm diameter) in the lowest-temperature rocks (sample 450) would occur in a few hundred years at the T and f_{O_2} conditions obtained from Fe–Ti oxides.

The above arguments suggest that the duration of melting was at least 10^2 years and could have been as much as 10^4 years.

10. Veins and fractures

A remarkable feature of the partially melted granitic rocks is the uncommon occurrence of aphyric glass veins. The veins are up to at least 5 mm thick (dense glass) and cut entire blocks or coat single block surfaces that separated along melt veins. Glass veins are dense to highly vesicular. Vein glass is physically continuous and compositionally identical with intergranular glass. Veins occur in granodiorite and granite samples with Fe–Ti oxide temperatures >940°C. In two blocks, glass veins occur on either side of pre-existing 1–2 mm-thick augite + hypersthene-bearing quartz veins (Fig. 15(A)), a result of melt migration into fractures opening along pre-existing planes of weakness. Point counting along serial traverses parallel to veins has not revealed any systematic differences in melt fraction with distance from the veins (see Section 6). Apart from the paired glass veins associated with quartz veins, no blocks have been observed to contain more than one glass vein or have more than one glass-coated surface.

Exceptional conditions must have prevailed in order for viscous intergranular melt to flow through the crystal mush to form veins in a reasonable amount of time. The average velocity of a fluid particle flowing in a porous medium at low Reynolds number is given by Wickham (1987) as

$$\bar{v} = -\frac{k_\phi \cdot \nabla P}{\mu \cdot \phi}$$

where k_ϕ = permeability, μ = fluid viscosity, ∇P = pressure gradient, and ϕ = porosity (=melt fraction = 0·2–0·5). Permeability is calculated as

$$k_\phi = \frac{a^2 \cdot \phi^3}{1000 \cdot (1-\phi)^2},$$

where a = mean radius of solid particles (optimistically taken to be 0·5 mm). Viscosity was calculated from glass composition according to Shaw (1972) and is ~10^6 Pa · s (10^7 poise) for 0·5 wt% H_2O in the melt and T = 940–1,000°C. Calculated flow velocities are of the order of 1 mm · yr^{-1} for $\nabla P = -10$ MPa · m^{-1} (−1 bar · cm^{-1}; Wickham 1987). Seemingly undeformed crystal mush argues against shear-induced melt migration. Alternatively, veins formed during the climactic eruption. Flow velocities that might be appropriate for syneruptive vein formation require very high pressure gradients. Such transient conditions might be achieved during rapid unloading of the chamber walls as the eruption proceeded.

Glass-free, typically orthogonal surfaces of blocks indicate efficient fracturing of unmelted to partially melted rock by

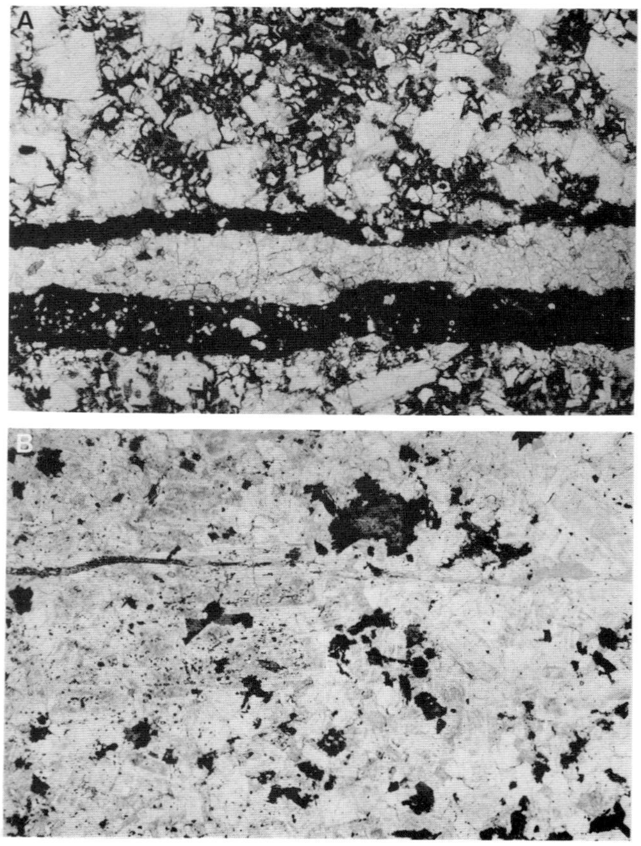

Figure 15 Plane light photomicrographs of granodiorites containing veins and fractures; width of field = 18 mm in both photomicrographs.
(A) Glass veins on either side of pre-existing augite + hypersthene-bearing quartz vein (1038; 981°C); vein glass is compositionally identical and physically continuous with intergranular glass; crystals in vein glass, mostly quartz, were derived from granodiorite; apart from vein, rock contains 30 vol% glass.
(B) Unmelted granodiorite with open tensional fracture containing mineral fragments derived from granodiorite; fracture is one of several parallel fractures visible in same thin section.

the eruptive process. Rarely, unmelted blocks have sets of parallel, open extensional fractures that terminate within single thin sections (Fig. 15(B)). The fractures are $\leq$0·2 mm wide and are incompletely filled with mineral fragments derived from the granodiorite. The open fractures, and glass veins in partially melted rocks, suggest the pluton was "quarried" as a result of decompression accompanying eruption of the climactic magma chamber.

11. Conclusions and implications for melting in the crust

The natural partial melting experiment displayed by the granitic blocks from Crater Lake complements laboratory studies of melting in crustal rocks with data on large samples composed of millimetre-sized crystals that were held at temperatures of up to 1,000°C for of the order of 10^2–10^4 years before rapid quenching. The existence of the partially fused rocks in association with a shallow magma chamber shows that repeated intrusion of intermediate to basaltic magmas can result in high degrees of melting of H_2O-poor crustal rocks. Melting apparently took place in a fluid-absent zone, at least several metres thick, subjected to a conductive thermal gradient between an active magma chamber and brittle rocks with fracture permeability where aqueous fluids circulated.

Bulk compositions of the melted rocks, including radiogenic isotopes, are similar to those of erupted silicic magmas, yet oxygen isotopes indicate widespread subsolidus exchange with meteoric hydrothermal fluids before re-equilibration at high temperature. These data lead to a model in which isotherms rise and fall in the crust with the waxing and waning of mantle imputs to the system so that felsic plutons crystallise, exchange with hydrothermal fluids, and then remelt to contribute to reborn, fractionating magmatic systems. The crust, even upper crust in the present case, is reworked and chemically fractionated by this process.

An important implication of the mineral and glass compositional data is that the behaviour of some trace elements (e.g. Sr) cannot be accurately described by equilibrium partial melting on a timescale of apparently 10^2–10^4 years. Plagioclase retains its original compositional zonation (although it is conceivable that radiogenic isotopes may have homogenised, as oxygen re-equilibrated, because of the greater rate of tracer diffusion than of chemical diffusion) and contains relict fluid inclusions even in the highest-temperature samples. Mineral inclusions in large crystals were not in communication with melt. Although melts were undersaturated in apatite and zircon, these phases failed to dissolve in the time available. Glass compositions suggest that zircon was retained in some residues but apparently not in others formed by equal degrees of melting under similar physical conditions. Despite proximity to an active, H_2O-rich magma body, the partially fused rocks behaved as virtually closed systems, even for volatile components.

Hydrous silicates broke down or reacted with melt systematically with increasing temperature. In the low-pressure environment of the Mazama system, actinolitic hornblende broke down before biotite but was still present at low melt fractions and $T < 870$°C. Biotite reacted with melt so that Mg/Mg + Fe and F increased with increasing temperature, finally disappearing at $T > 968$°C; biotites stable at the highest temperatures are fluor-phlogopites.

The maximum degree of melting known from the ejected blocks is 50 vol%, which may be the maximum melt percentage for maintenance of yield strength under static conditions (Watson 1987; Miller *et al.* 1988). Rocks with this high melt fraction (0·46 on a weight basis) probably were preserved because of the high viscosity of low-H_2O intergranular melt (~10^6 Pa · s) and because they were close to the limit of contiguity of the crystal mush (e.g. Fig. 3(F)). Melting beyond the 50 vol% level would have resulted in viscous flow and convective assimilation of the melt/crystal suspension into the magma chamber. Note that 50 vol% is the upper limit of the range of critical melt fraction (CMF) suggested by Wickham (1987), whereupon bulk viscosity drops rapidly to approximately that of the liquid.

Melt apparently did not segregate from the partially fused granitic rocks under static conditions before onset of the climactic eruption, but then radical depressurisation forced intergranular liquid of moderately- to highly-melted rocks into opening tensile fractures. The latter process was dramatic but probably ineffective at separating and transferring partial melt far from its point of origin.

How can lessons learned from study of Crater Lake granitic blocks be applied to melting in the mid- to lower crust? Conclusions that should affect modelling of crustal melting are: (1) lack of equilibrium between plagioclase and melt in granitic rocks, whose grain size is notably smaller than those of batholithic rocks; (2) isolation and slow dissolution of accessory phases; (3) apparent closed system behaviour evidenced by lack of widespread melt segregation and little or no introduction of fluids during melting, either from external hydrothermal sources or from juvenile magma; (4) similar radiogenic isotopic composition to juvenile magma, but potentially different stable isotope ratios, making an assimilated component difficult to identify through rock composition, mineralogy, or radiogenic isotopes. Critical parameters that might be different in deeper melting zones are the duration of and volume affected by the melting process, the availability of hydrous fluid and the strain regime. Longer time at supersolidus temperatures and increased dissolved H_2O would allow a closer approach to equilibrium melting. Higher concentrations of dissolved H_2O in the melt, which would lower viscosity, and active deformation would aid greatly in segregating melt from residue (Wickham 1987). Rapid melting of granitic rocks in a convecting layer above (and below) a basaltic sill, as envisaged by Huppert and Sparks (1988), requires melt viscosities far lower, and thus higher H_2O contents, than conditions recorded in the partially fused granodioritic walls of the Mazama magma chamber.

12. Acknowledgements

I thank L. C. Calk for ensuring that the Menlo Park electron microprobe has functioned well, and the many USGS chemists (named in footnote to Table 1) who produced the rock analyses used in this paper. L. H. Adami determined $\delta^{18}O$ of mineral separates, S. W. McKnight made the diagrams, C. Kelley helped with manuscript preparation, and R. Oscarson obtained the SEM images. Reviews by T. W. Sisson, G. K. Czamanske, and A. Ewart resulted in substantial improvements in the manuscript.

13. References

Al-Rawi, Y. & Carmichael, I. S. E. 1967. A note on the natural fusion of granite. AM MINERAL **52,** 1086–14.

Allègre, C. J. & Minster, J. F. 1978. Quantitative models of trace element behavior in magmatic processes. EARTH PLANET SCI LETT **38,** 1–25.

Andersen, D. J. & Lindsley, D. H. 1988. Internally consistent solution models for Fe–Mg–Mn–Ti oxides: Fe–Ti oxides. AM MINERAL **73,** 714–26.

Aragon, R., McCallister, R. H. & Harrison, H. R. 1984. Cation diffusion in titanomagnetites. CONTRIB MINERAL PETROL **85,** 174–85.

Arzi, A. A. 1978. Critical phenomena in the rheology of partially melted rocks. TECTONOPHYSICS **44,** 173–84.

Bacon, C. R. 1983. Eruptive history of Mount Mazama and Crater Lake caldera, Cascade Range, U.S.A. J VOLCANOL GEOTHERM RES **18,** 57–115.

Bacon, C. R. 1990. Calc-alkaline, shoshonitic, and primitive tholeiitic lavas from monogenetic volcanoes near Crater Lake, Oregon. J PETROL **31,** 135–66.

Bacon, C. R. & Druitt, T. H. 1988. Compositional evolution of the zoned calcalkaline magma chamber of Mount Mazama, Crater Lake, Oregon. CONTRIB MINERAL PETROL **98,** 224–56.

Bacon, C. R. & Hirschmann, M. M. 1988. Mg/Mn partitioning as a test for equilibrium between coexisting Fe–Ti oxides. AM MINERAL **73,** 57–61.

Bacon, C. R. & Lanphere, M. A. 1990. The geologic setting of Crater Lake, Oregon. *In* Drake, E. T., Larson, G. L., Dymond, J. & Collier, R. (eds.) *Crater Lake—An Ecosystem Study*. PAC DIV AM ASSOC ADV SCI, SAN FRANCISCO, 19–27.

Bacon, C. R., Adami, L. H. & Lanphere, M. A. 1989. Direct evidence for the origin of low-^{18}O silicic magmas: quenched samples of a magma chamber's partially-fused granitoid walls, Crater Lake, Oregon. EARTH PLANET SCI LETT **96,** 199–208.

Bacon, C. R., Newman, S. & Stolper, S. 1992. Water, CO_2, Cl, and F in melt inclusions in phenocrysts from three Holocene explosive eruptions, Crater Lake, Oregon. AM MINERAL (submitted).

Bea, F. 1991. Geochemical modelling of low melt-fraction anatexis in a peraluminous system: The Peña Negra Complex (central Spain). GEOCHIM COSMOCHIM ACTA **55,** 1859–74.

Blundy, J. D. & Wood, B. J. 1991. Crystal-chemical controls on the partitioning of Sr and Ba between plagioclase feldspar, silicate melts, and hydrothermal solutions. GEOCHIM COSMOCHIM ACTA **55,** 193–209.

Carmichael, I. S. E. 1967. The iron-titanium oxides of salic volcanic rocks and their associated ferromagnesian silicates. CONTRIB MINERAL PETROL **14,** 36–64.

Chou, I. M. 1978. Calibration of oxygen buffers at elevated pressure and temperature using the hydrogen fugacity sensor. AM MINERAL **63,** 650–703.

Crank, J. 1975. *The Mathematics of Diffusion*. Oxford: Oxford University Press.

Diller, J. S. & Patton, H. B. 1902. The geology and petrography of Crater Lake National Park. US GEOL SURV PROF PAP **3**.

Dodge, F. C. W. & Calk, L. C. 1978. Fusion of granodiorite by basalt, central Sierra Nevada. J RES US GEOL SURV **6,** 459–65.

Druitt, T. H. & Bacon, C. R. 1986. Lithic Breccia and ignimbrite erupted during the collapse of Crater Lake caldera, Oregon. J VOLCANOL GEOTHERM RES **29,** 1–32.

Druit, T. H. & Bacon, C. R. 1988. Compositional zonation and cumulus processes in the Mount Mazama magma chamber, Crater Lake, Oregon. TRANS R SOC EDINBURGH EARTH SCI **79,** 289–97.

Druitt, T. H. & Bacon, C. R. 1989. Petrology of the zoned calcalkaline magma chamber of Mount Mazama, Crater Lake, Oregon. CONTRIB MINERAL PETROL **101,** 245–59.

Elkins, L. T. & Grove, T. L. 1990. Ternary feldspar experiments and thermodynamic models. AM MINERAL **75,** 544–59.

Fiske, R. S., Hopson, C. A. & Waters, A. C. 1963. The geology of Mount Rainier National Park, Washington. US GEOL SURV PROF PAP **444**.

Grove, T. L., Baker, M. B. & Kinzler, R. J. 1984. Coupled CaAl–NaSi diffusion in plagioclase feldspar: experiments and applications to cooling rate speedometry. GEOCHIM COSMOCHIM ACTA **48,** 2113–21.

Grove, T. L., Kinzler, R. J., Baker, M. B., Donnelley-Nolan, J. M. & Lesher, C. E. 1988. Assimilation of granite by basaltic magma at burnt lava flow, Medicine Lake volcano, northern California: decoupling of heat and mass transfer. CONTRIB MINERAL PETROL **99,** 320–43.

Halliday, A. N., Davidson, J. P., Hildreth, W. & Holden, P. 1991. Modelling the petrogenesis of high Rb/Sr silicic magmas. CHEM GEOL **92,** 107–14.

Harrison, T. M. & Watson E. B. 1983. Kinetics of zircon dissolution and zirconium diffusion in grantic melts of variable water content. CONTRIB MINERAL PETROL **84,** 66–72.

Harrison, T. M. & Watson, E. B. 1984. The behavior of apatite during crustal anatexis: equilibrium and kinetic considerations. GEOCHIM COSMOCHIM ACTA **48,** 1467–77.

Hildreth, W., Christiansen, R. L. & O'Neil, J. R. 1984. Catastrophic isotopic modification of rhyolitic magma at times of caldera subsidence, Yellowstone Plateau volcanic flield. J GEOPHYS RES **89,** 8339–69.

Hollister, L. S., Grissom, G. C., Peters, E. K., Stowell, H. H. & Sisson, V. B. 1987. Confirmation of the empirical correlation of Al in hornblende with pressure of solidification of calc-alkaline plutons. AM MINERAL **72,** 231–329.

Huebner, J. S. & Sato, M. 1970. The oxygen fugacity-temperature relationships of manganese oxide and nickel oxide buffers. AM MINERAL **55,** 934–52.

Huppert, H. E. & Sparks, R. S. 1988. The fluid dynamics of crustal melting by injection of basaltic sills. TRANS R SOC EDINBURGH EARTH SCI **79,** 237–43.

Johannes, W. 1984. Beginning of melting in the granite system Qz—Or—Ab—An—H_2O. CONTRIB MINERAL PETROL **86,** 264–73.

Johannes, W. 1985. The significance of experimental studies for the formation of migmatites. *In* Ashworth, J. R. (ed.) *Migmatites*, 36–85. Glasgow: Blackie.

Johannes, W. 1989. Melting of plagioclase-quartz assemblages at 2 kbar water pressure. CONTRIB MINERAL PETROL **103,** 270–6.

Johnson, M. C. & Rutherford, M. J. 1989a. Experimentally determined conditions in the Fish Canyon Tuff, Colorado, magma chamber. J PETROL **30,** 711–37.

Johnson, M. C. & Rutherford, M. J. 1989b. Experimental calibration of the aluminum-in-hornblende geobarometer with application to Long Valley caldera (California) volcanic rocks. GEOLOGY **17,** 837–41.

Jurewicz, S. R. & Watson, E. B. 1985. The distribution of partial melt in a granitic system· the application of liquid phase sintering theory. GEOCHIM COSMOCHIM ACTA **49,** 1109–21.

Kaczor, S. M., Hanson, G. N. & Peterman, Z. E. 1988. Disequilibrium melting of granite at the contact with a basic plug: a geochemical and petrographic study. J GEOL **96,** 61–78.

La Tourrette, T. Z., Burnett, D. S. & Bacon, C. R. 1991. Uranium and minor-element partitioning in Fe-Ti oxides and zircon from partially melted granodiorite, Crater Lake, Oregon. GEOCHIM COSMOCHIM ACTA **55,** 457–69.

Leeman, W. P. & Phelps, D. W. 1981. Partitioning of rare earths and other trace elements between sanidine and coexisting volcanic glass. J GEOPHYS RES **86,** 10193–9.

Le Maitre, R. W. 1974. Partially fused granite blocks from Mt Elephant, Victoria, Australia. J PETROL **15,** 403–12.

Lidstrom, J. W., Jr. 1971. *A new model for the formation of Crater Lake caldera, Oregon*. Ph.D. Dissertation, Oregon State University.

Liu, M. & Yund, R. A. 1992. NaSi–CaAl interdiffusion in plagioclase. AM MINERAL (in press).

Lu, F. & Anderson, A. T. 1991. Mixing origins of volatile and thermal gradients in the Bishop magma. EOS **72,** 312.

Maaløe, S. & Wyllie, P. J. 1975. Water content of a granite magma deduced from the sequence of crystallization determined experimentally with water-undersaturated conditions. CONTRIB MINERAL PETROL **52,** 175–91.

Maury, R. C. & Bizouard, H. 1974. Melting of acid xenoliths into a basanite: an approach to the possible mechanisms of crustal contamination. CONTRIB MINERAL PETROL **48,** 275–86.

McKenzie, D. 1984. The generation and compaction of partially molten rock. J PETROL **25,** 713–65.

Miller, C. F., WATSON, E. B. & HARRISON, T. M. 1988. Perspectives on the source, segregation and transport of granitoid magmas. TRANS R SOC EDINBURGH EARTH SCI **79,** 135–56.

Naney, M. T. 1983. Phase equilibria of rock-forming ferromagnesian silicates in granitic systems. AM J SCI **283,** 993–1033.

Nash, W. P. & Crecraft, H. R. 1985. Partition coefficients for trace elements in silicic magmas. GEOCHIM COSMOCHIM ACTA **49,** 2309–33.

Nekvasil, H. 1988. Calculated effect of anorthite component on the crystallization paths of H_2O-undersaturated haplogranitic melts. AM MINERAL **73,** 966–81.

Ritchey, J. L. 1979. *Origin of divergent magmas at Crater Lake, Oregon.* Ph.D. Dissertation, University of Oregon.

Russell, J. K. & Nicholls, J. 1988. Analysis of petrologic hypotheses with Pearce element ratios. CONTRIB MINERAL PETROL **99,** 25–35.

Shaw, D. M. 1970. Trace element fractionation during anatexis. GEOCHIM COSMOCHIM ACTA **34,** 237–43.

Shaw, H. R. 1972. Viscosities of magmatic silicate liquids: an empirical method of prediction. AM J SCI **272,** 870–93.

Sisson, T. W. 1991. Pyroxene-high silica rhyolite trace element partition coefficients measured by ion microprobe. GEOCHIM COSMOCHIM ACTA **55,** 1575–85.

Stormer, J. C., Jr. 1983. The effects of recalculation on estimates of temperature and oxygen fugacity from analyses of multi-component iron–titanium oxides. AM MINERAL **68,** 586–94.

Taylor, E. M. 1968. Accidental plutonic ejecta at Crater Lake, Oregon. GEOL SOC AM SPEC PAP **115,** 221 (abstr.).

Tuttle, O. F. & Bowen, N. L. 1958. Origin of granite in the light of experimental studies in the system $NaAlSi_3O_8$—$KAlSi_3O_8$—SiO_2—H_2O. GEOL SOC AM MEM **74**.

Van der Molen, I. & Paterson, M. S. 1979. Experimental deformation of partially-melted granite. CONTRIB MINERAL PETROL **70,** 299–318.

Watson, E. B. 1987. Contiguity and the rheology of partially molten granitoids. EOS **68,** 1143–4.

Watson, E. B. & Green, T. H. 1981. Apatite/liquid partition coefficients for the rare earth elements and strontium. EARTH PLANET SCI LETT **56,** 405–21.

Watson, E. B. & Harrison, T. M. 1983. Zircon saturation revisited: temperature and composition effects in a variety of crustal magma types. EARTH PLANET SCI LETT **64,** 295–304.

Wickham, S. M. 1987. The segregation and emplacement of granitic magmas. J GEOL SOC LONDON **144,** 281–97.

Williams, H. 1942. The geology of Crater Lake National Park, Oregon. CARNEGIE INST WASHINGTON PUBL **540**.

Wyllie, P. J. 1977. Crustal anatexis: an experimental review. TECTONOPHYSICS **43,** 41–71.

CHARLES R. BACON, U.S. Geological Survey MS 910, 345 Middlefield Road, Menlo Park, California 94025-3591, U.S.A.

Transactions of the Royal Society of Edinburgh: Earth Sciences, **83**, 49–62, 1992

Source region of a granite batholith: evidence from lower crustal xenoliths and inherited accessory minerals

Calvin F. Miller, John M. Hanchar, Joseph L. Wooden, Victoria C. Bennett, T. Mark Harrison, David A. Wark and David A. Foster

ABSTRACT: Like many granites, the Late Cretaceous intrusives of the eastern Mojave Desert, California, have heretofore provided useful but poorly focused images of their source regions. New studies of lower crustal xenoliths and inherited accessory minerals are sharpening these images.

Xenoliths in Tertiary dykes in this region are the residues of an extensive partial melting event. Great diversity in their composition reflects initial heterogeneity (both igneous and sedimentary protoliths) and varying amounts of melt extraction (from <10% to >70%). Mineral assemblages and thermobarometry suggest that the melting event occurred at $T \geq 750°C$ at a depth of about 40 km. Present-day Sr, Nd, and Pb isotopic ratios indicate a Mojave Proterozoic heritage, but unrealistic model ages demonstrate the late Phanerozoic adjustment of parent/daughter ratios. A link between these xenoliths and the Late Cretaceous granites, though not fully documented, is probable; in any case, they provide invaluable clues concerning a crustal melting event, recording information about nature of source material (heterogeneous, supracrustal-rich), conditions of melting (moderately deep, moderately high T, accompanied by partial dehydration), and melt extraction (highly variable, locally extensive).

The Old Woman-Piute granites contain a large fraction of inherited zircon and monazite. A SHRIMP ion probe investigation shows that these zircons record a Proterozoic history similar to that which affected the Mojave region. Zonation patterns in zircons, and to a lesser extent monazites and xenotimes, document multiple phases of igneous, metamorphic, and sedimentary growth and degradation, commonly several in a single grain. Low Y in portions of the cores of inherited zircons and monazites and in monazites and outer portions of zircons from the xenoliths appear to indicate growth in equilibrium with abundant garnet.

KEY WORDS: granite, xenolith, accessory minerals, zircon, monazite, anatexis, Mojave Desert, California.

1. Introduction

1.1. Granites as probes of the crust and its history: potential and limitations

Granitic rocks are essential components of the continental crust, and in fact their relatively low density and resulting buoyancy account for its preservation. Most contain products of crustal anatexis, the partial melting of pre-existing crust, and thus they may preserve information about ancient and remote crustal regions and the conditions that resulted in melting. A full understanding of anatexis is a major goal of petrologists, tectonicists, and Earth historians in general, because it promises to elucidate the nature and history of inaccessible portions of the crust and to clarify major dynamic processes of tectonic and geochemical reorganisation. Such an understanding demands answers to the following questions:

(1) *What were the conditions and materials involved in magma generation*? Specifically: (a) At what depth and temperature did melting take place? (b) Was melting restricted entirely to the crust, or was there a contribution of mass from the mantle? (c) What was (were) the protolith(s) of the crustal sources? (d) How heterogeneous was the source material? (e) What were the melt fractions involved? (f) What was the tectonic setting during melting?

(2) *By what mechanism was crustal anatexis induced*? (a) Through mantle heat input? By increased heat flow (conductive heat transfer) or advection via emplacement of mantle-derived plutons? (b) Through thermal blanketing? By sedimentary of tectonic burial? Involving materials with unusually high heat productivity? (c) Through rapid unroofing, resulting in decompressional melting? (d) Through an influx of hot aqueous fluids, introducing heat and/or inducing wet melting?

(3) *What was the previous history of the crust from which the magmas were derived*? (a) How old was the crust? (b) What events (igneous, metamorphic, sedimentary, deformational) had it undergone?

Although they are the essential permanent component of the crust and thus have the potential for long-term tracking of Earth history, continuing production of granites in fact commonly obscures the interpretation of that history. Anatexis yields magmas that homogenise much of the variability that could be inherited from source regions and erases much of the direct evidence for what that region was like. Most restite is altered and re-equilibrated during ascent and crystallisation to the extent that it is often at best a shadowy hint of the origins of the magma. What is commonly delivered to shallower, more accessible regions is a seemingly bland rock that has extracted mass but relatively little information from its source region.

1.2 Purpose

The purpose of this paper is to discuss ways of enhancing information retrieval from granites by overcoming some of the obstacles mentioned above. We use as an example the relatively well-studied Old Woman-Piute Batholith in southeastern California, which, like most granites, has resisted yielding definitive answers concerning its past. We focus here on two directions of study that are proving fruitful. The first involves the fortuitous, though not exceptional, preservation of lower crustal xenoliths in a younger dyke exposed near the batholith. These xenoliths provide direct information about the materials that were present beneath the batholith; they also record the effects of a major anatectic event, one that may in fact have yielded the exposed granites. The second study is of accessory minerals, especially zircons, in the batholith. It is well known that the most convincing restitic material in granitic rocks is zircon with demonstrable inheritance. We emphasise that investigations of geochemistry and zonation patterns of accessory minerals can greatly enhance our understanding of the nature and history of the crust from whence they came. Combining these studies with conventional and *in situ* U-Pb studies can provide a powerful method for unravelling petrogenesis and crustal evolution.

2. The Old Woman-Piute Batholith

2.1. Introduction

The late Cretaceous Old Woman-Piute Batholith, a medium-sized (400 km^2) composite batholith in the eastern Mojave Desert of southeastern California, includes several discrete plutons of metaluminous, biotite ± hornblende granodiorite and peraluminous, two-mica monzogranite. Both intruded a metamorphic complex comprising diverse rocks of both Proterozoic and Palaeozoic age.

We have studied this batholith using a variety of methods, including major and trace element and isotope geochemistry, phase chemistry and barometry, and U–Pb geochronology (both conventional and SHRIMP). Results are discussed in Mittlefehldt and Miller (1983), Foster *et al.* (1989), Miller *et al.* (1990), and Foster *et al.* (1992), and summarised below. We present the answers to the questions of Section 1.1 that have arisen from the previous studies, and enumerate the remaining puzzles. In the following sections we show how xenolith and accessory mineral studies are providing a more complete picture.

2.2. Regional setting

The oldest well-dated rocks in the eastern Mojave Desert are about 1·8 Ga, but Nd and Pb isotopic compositions and ages of detrital zircons indicate that at least part of the crust dates back to about 2·5 Ga (e.g. Silver *et al.* 1977; Bennett & DePaolo 1987; Wooden *et al.* 1988; Wooden & Miller 1990; Wooden unpub. data). The early crust preserved at the surface is dominantly granite with less abundant mafic igneous and supracrustal lithologies. An orogenic episode at 1·71 Ga, called the Ivanpah Orogeny by Wooden and Miller (1990), was accompanied by intrusion of abundant granites and metamorphism that attained granulite grade. Copious post-orogenic granitic plutonism lasted from about 1·68 to 1·62 Ga, and anorogenic granites and alkalic rocks were emplaced later, at 1·40–1·44 Ga (e.g. Silver *et al.* 1977; Gleason *et al.* 1988; Anderson & Bender 1989; Wooden & Miller 1990). A 1·1–1·2 Ga event is marked by abundant dolerite dykes (e.g. Hammond 1986). Cratonal sedimentation commenced at the beginning of the Palaeozoic and continued into the Triassic (Stone *et al.* 1983).

The craton here was intensely reactivated beginning in late Triassic or Jurassic time. Abundant granites were emplaced during the Jurassic and again during the Cretaceous. Major thrusting appears to have begun during the Jurassic, perhaps synchronous with mid-Jurassic plutonism (e.g. Foster *et al.* (1992); McCaffrey *et al.* 1991), and continued until the latter part of the Cretaceous (Fletcher & Karlstrom 1990). Denudational tectonism (Carl *et al.* 1991) immediately followed the latest stages of granite intrusion, during which the Old Woman-Piute Batholith was emplaced. It remains uncertain whether deformation that accompanied batholith emplacement represented the last stage of shortening or whether extension had already begun. We favour the latter interpretation.

The Mojave crust was reactivated again during the mid-Tertiary, when it underwent moderate to extreme extension (e.g. Davis & Lister 1988). The rocks exposed in the Old Woman and Piute Mountains were only mildly extended, but extension-related volcanism and dyking were widespread (Hileman *et al.* 1990; Miller & Miller 1991).

2.3. The granites

Metaluminous and peraluminous granites of the Old Woman-Piute Batholith crystallised at 72–73 Ma; ages of the two suites are indistinguishable (Fig. 1; cf. Foster *et al.* 1989; Foster *et al.* (1992); Miller *et al.* 1990). Zircon inheritance in both suites is large and diverse, expecially in the peraluminous monzogranites (Fig. 1, Table 1). The upper intercept of the granodiorite discordia based on conventional U–Pb methods is 1·61 Ga; the zircon fractions from the monzogranite are highly discordant and scatter about a line with an upper intercept of about 1·67 Ga. The well-defined granodiorite chord is shown by SHRIMP data to be an artifact of mixing of inherited zircons with ages representing both the 1·4 and 1·6–1·8 Ga periods. The monzogranite has the same age and inheritance pattern defined by SHRIMP, but is much richer in inherited zircons; all of the apparently scattered zircon fractions lie within the cone defined by the lower intercept, 1·4 Ga, and 1·8 Ga.

The batholith was emplaced in the middle crust, at a depth of about 15 km (estimated from hornblende and garnet-plagioclase-biotite-muscovite barometry of the granites as well as garnet-plagioclase-Al_2SiO_5 barometry from the country rocks; Foster *et al.* (1992). Temperatures of the monzogranite and granodiorite magmas are estimated from zircon saturation thermometry (Watson & Harrison 1983; Harrison & Watson 1983) to have been about 750 and 800°C, respectively. (Our estimates for zircon and monazite saturation take into account contribution of inherited (restitic) grains to Zr and LREE concentrations. Inherent imprecision in our estimates of melt compositions, though large, introduce only tens of degrees of error in T estimates.) A similar temperature is indicated for the monzogranites by the monazite saturation thermometer (Rapp *et al.* 1987).

The geochemical characteristics of the granites are discussed in Miller *et al.* (1990). Briefly, the granodiorites are less silicic than the monzogranites (57–69% vs 71–76%). Both have intermediate Na and K, with K_2O/Na_2O somewhat <1 in granodiorite and near 1 in monzogranite. No elements are unusually high or low in abundance for subduction-related continental arc granites.

The monzogranites are isotopically similar to the Proterozoic eastern Mojave crust that they intrude, with high $^{87}Sr/^{86}Sr$ and low $^{143}Nd/^{144}Nd$ ratios (Fig. 2a; cf.

Bennett & DePaolo 1987). Granodiorites have ratios of $^{87}Sr/^{86}Sr$ and $^{143}Nd/^{144}Nd$ that are somewhat lower and higher, respectively, than the monzogranites but still far outside the range of primitive, mantle-derived magmas. Lead isotope ratios of both monzogranites and granodiorites fall within the distinctive eastern Mojave Proterozoic array of Wooden and Miller (1990), lying very close to a 1·7 Ga Pb/Pb isochron with elevated initial ratios and having high model Th/U ratios (Fig. 2b). Oxygen isotope ratios for both suites are intermediate between those of primitive magmas and sedimentary rocks ($\delta^{18}O$ 7·2–9·3‰).

2.4. Interpretation of granites: implications for anatexis

Here we attempt to answer, or at least constrain, the questions that were posed in Section 1.1 for this particular batholith, based upon both our petrological data and on regional knowledge. Those conclusions marked with "*" are justified more fully in Miller *et al.* (1990).

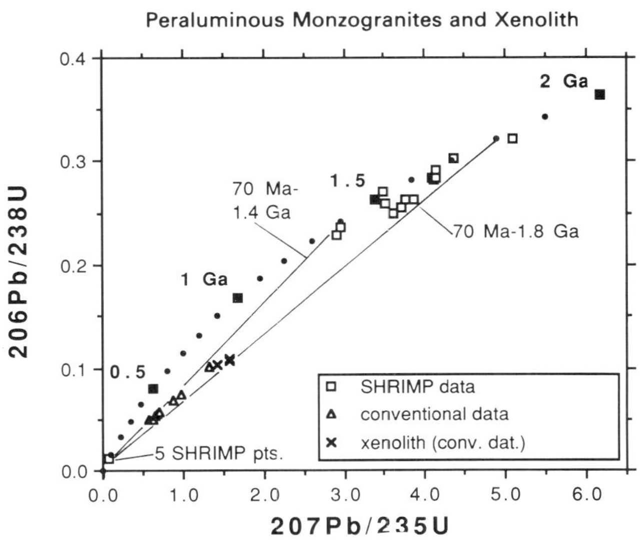

(a)

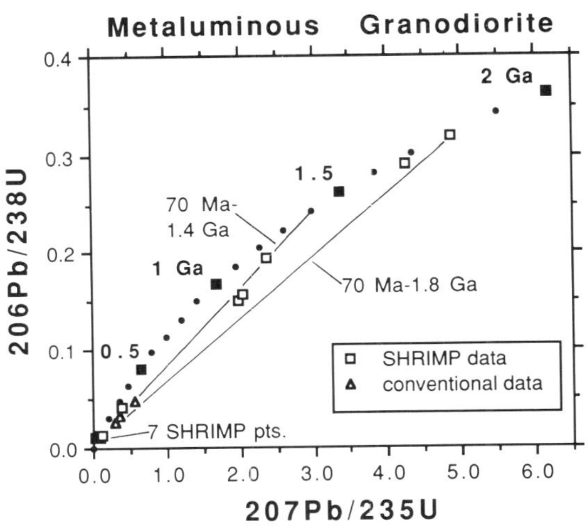

(b)

Figure 1 Conventional (Miller *et al.* 1990; Wooden, unpub. data) and SHRIMP (Foster *et al.* 1989) U-Pb data from Old Woman-Piute granites and lower crustal xenolith; reference discordia lines connect age of granite (*c.* 70 Ma) with 1·4 and 1·8 Ga upper intercepts: (a) peraluminous monzogranite (samples SW1, W496 and BGMG) and quartzose lower crustal xenolith PT8A; (b) metaluminous granodiorite, sample ND-12.

2.4.1. Conditions and materials involved in magma generation. (a) Temperatures in the source region must have been at least 750–800°C, as indicated by the calculated magma temperatures. There are no constraints on depth, except that it was greater than 15 km (the depth of emplacement). (b)* The monzogranite was probably entirely the product of crustal melting. The granodiorite is equivocal: it unquestionably had a large crustal component, but it could have had an important mantle contribution as well. (c)* Source materials for the monzogranite were broadly quartzofeldspathic and dominantly of early Proterozoic age; the granodiorite was derived from intermediate, or perhaps mafic, igneous crust with a possible mantle contribution as well. No sedimentary input is required in either case, though an immature sedimentary source (e.g. greywacke) or a relatively small mature sedimentary component (e.g. shale) is possible for the monzogranite. (d)* Isotopic variability requires some heterogeneity in the source regions of each magma, but otherwise there are no constraints on variability. (e)* Melt fractions are difficult to assess, but were probably larger (>50%?) for the granodiorite than for the monzogranite (probably <50%). (f) The upper crust was either in the final stages of contraction, or in a period of post-contraction extension.

2.4.2. Mechanism of anatexis. Circumstantial evidence can be used to support any of the four mechanisms in Section 1.1. (a) Barton (1990) has proposed a model for the Great Basin to the N whereby progressive heating of the crust during protracted Jurassic through to Cretaceous subduction ultimately resulted in almost pure crustal melting; the same model could apply here. Although there is no direct evidence for copious Cretaceous intrusion of the lower crust by subduction zone magmas, a substantial subcrustal input into the granodiorite is possible. Abundant mafic magma, ponded at depth and/or mixed with crustal melts, would be consistent with observations in similar continent margin settings (e.g. Foster & Hyndman 1990). (b) Jurassic crustal thickening could have provided sufficient thermal blanketing to induce anatexis by the late Cretaceous (cf. Patiño Douce *et al.* 1990; Wright & Wooden 1991). (c) Late Cretaceous unroofing may have been sufficient to induce decompression melting in an already warm lower crust. (d) There is some evidence that abundant hot fluids passed through the eastern Mojave crust during the late Cretaceous, potentially triggering melting (e.g. Hoisch & Hamilton 1990).

2.4.3. Prior history of the crustal source. (a) The SHRIMP data demonstrate that the crust involved in the melting event (and/or the source area of metasedimentary rocks in this crust) was at least 1·8 Ga, and Nd and Pb isotope data suggest that it was several hundred million years older, in other words, that its age was similar or identical to that of the crust that is presently exposed at the surface. (b) The distribution of SHRIMP ages further reveals that this crust had a complex history from 1·8–1·4 Ga, a history that again is indistinguishable from that of present-day upper crust. The nature of the Proterozoic events that occurred in the source region is not established with certainty (i.e. whether the ages represent magmatic crystallisation or heating of pre-existing crust).

The granites themselves, together with surface geology, thus provide important constraints on the melting process and crustal history, but many questions remain unanswered. In the following sections we pursue these questions using different approaches.

3. Lower crustal xenoliths: implications for anatexis

3.1. Introduction

We have discovered garnet-rich, lower crustal xenoliths in dykes of mid-Tertiary age (*c.* 20 Ma) in the Old Woman and Piute Mountains (Staude & Miller 1987; Hanchar 1990; Hanchar, Miller, Wooden, & Staude in prep.). One dyke in the northern Piute Mountains is particularly rich in xenoliths and is the subject of detailed study. These xenoliths are especially important in deciphering anatexis in this region because, firstly, they are the only known samples of ancient lower crust within 200 km of the eastern Mojave Desert and, secondly, because they appear to record directly a late Phanerozoic melting event. In this section we briefly describe the xenoliths, cite the evidence that they are residues of anatexis, evaluate the possibility that they are in fact the residue from the Old Woman-Piute Batholith-forming event, and comment on their significance for crustal melting in general.

3.2. Nature of the xenoliths

All of the xenoliths are characterised by the presence of garnet, often in great abundance (to 70%). We categorise them on the basis of mineral assemblage (*not* abundance), dividing them into the following categories: metaluminous mafic (hornblende + plagioclase ± quartz ± clinopyroxene), aluminous (kyanite ± sillimanite + quartz + plagioclase + biotite ± K-feldspar) and quartzofeldspathic (plagioclase + quartz ± biotite). We further distinguish a quartzose category on the basis of its extremely high quartz content (>60 modal %). The quartzofeldspathic xenoliths are most abundant (>50%); each of the other three types accounts for ≥10%. Rutile, apatite, and zircon are all nearly ubiquitous, with rutile particularly abundant (to >2 modal %). Sphene is present in several samples, and monazite in most aluminous and quartzofeldspathic (including quartzose) xenoliths. Plagioclase is primarily andesine, ranging in composition from about An_{25}–An_{50}. All xenoliths have moderately coarse, crystalloblastic textures typical of high-grade metamorphic rocks, except for one with a well-defined sequential texture (euhedral grains of garnet, hornblende, biotite, rutile, apatite, zircon, and REE-rich epidote, all enclosed poikilitically within coarse plagioclase grains; quartz is interstitial) that we distinguish as igneous-textured.

Both thermobarometry and mineral assemblages require that the xenoliths (igneous-textured as well as metamorphic) equilibrated at pressures in excess of about 1 GPa and temperatures ≥750°C. Mean calculated P and T are 1·1 GPa and 760°C. The absence of orthopyroxene, presence of kyanite + quartz + K-feldspar without muscovite, and coexistence of garnet and rutile and absence of ilmenite are all consistent with these estimates, althogh we cannot rule out the possibility that the thermobarometers measure points on the retrograde path. Presence of biotite and hornblende limits the peak T to <950–1000°C.

The diverse protoliths included mafic to intermediate igneous rocks (metaluminous mafic xenoliths), pelites and

Table 1 Ion probe U-Pb data for zircons

	U (ppm)	Th (ppm)	Pb* (ppm)	^{204}Pb (ppb)	^{206}Pb/^{238}U (Ma ± 1σ)	^{207}Pb/^{235}U (Ma ± 1σ)	^{207}Pb/^{206}Pb (±1σ)
SW-1 Sweetwater Wash monzogranite							
1·1	575	160	140	49	1438 ± 16	1554 ± 14	0·10197(103)
1·2	549	136	139	40	1464 ± 16	1574 ± 14	0·10235(111)
2·1	270	151	75	55	1539 ± 18	1523 ± 20	0·09126(181)
3·1	459	73	5	17	72 ± 2	55 ± 8	0·03557(547)
4·1	1110	170	297	36	1606 ± 18	1660 ± 12	0·10247(063)
6·1	284	93	80	23	1608 ± 18	1654 ± 17	0·10159(149)
7·1	3730	29	38	21	75 ± 1	75 ± 2	0·04783(119)
7·2	1180	39	11	29	68 ± 1	79 ± 5	0·05523(334)
8·1	290	109	66	46	1328 ± 15	1386 ± 17	0·09001(157)
8·2	635	152	164	39	1506 ± 17	1586 ± 14	0·09891(106)
9·1	519	61	5	20	73 ± 1	84 ± 8	0·04153(566)
9·2	218	115	59	34	1505 ± 18	1604 ± 20	0·10293(193)
10·1	491	149	157	36	1802 ± 20	1836 ± 14	0·11063(105)
11·1	348	131	98	40	1600 ± 18	1655 ± 17	0·10217(147)
12·1	2000	77	21	27	75 ± 1	75 ± 3	0·04716(209)
13·1	262	68	74	23	1650 ± 18	1664 ± 11	0·09982(050)
13·2	1650	183	398	40	1490 ± 16	1531 ± 11	0·09523(055)
13·3	2600	228	565	40	1366 ± 15	1399 ± 11	0·08873(046)
14·1	1490	177	501	7	1702 ± 18	1704 ± 12	0·10114(052)
ND-12 Old Woman granodiorite							
1·1	869	30	8	22	70 ± 1	72 ± 6	0·04867(379)
5·1	1060	257	12	30	79 ± 1	90 ± 6	0·05447(353)
10·1	282	155	3	19	70 ± 1	69 ± 15	0·04602(1019)
12·1	215	44	8	36	262 ± 5	327 ± 22	0·06648(512)
12·2	1150	60	158	21	903 ± 15	1112 ± 14	0·09389(090)
12·3	383	127	62	90	948 ± 16	1127 ± 20	0·08767(190)
14·1	541	23	5	24	73 ± 1	90 ± 9	0·06145(624)
15·1	289	215	4	29	75 ± 2	22 ± 19	0·01421(1174)
16·1	296	141	4	31	78 ± 2	121 ± 20	0·07415(1282)
17·1	809	374	266	59	1791 ± 18	1796 ± 18	0·10577(089)
18·1	272	119	80	35	1648 ± 26	1683 ± 21	0·10221(158)
18·2	160	57	31	53	1143 ± 19	1232 ± 26	0·08643(256)
19·1	419	401	6	26	78 ± 2	86 ± 15	0·05378(947)

Note: Methods described in Foster *et al.* 1989; these data were presented in Figure 3 of that paper but not tabulated.

both clay- and quartz-rich sandstones (aluminous and quartzose xenoliths), and probably chemically immature sandstones and relatively felsic volcanic and plutonic rocks (quartzofeldspathic xenoliths). The bulk chemistry of the xenoliths, however, does not match expected unmodified protolith compositions: in particular, many are poor in K and Rb, rich in Al and Ti, and have high Fe/Mg, Al/Ca and heavy REE relative to "normal" igneous and sedimentary rocks with the same SiO_2 concentrations (e.g. Fig. 3). A simple, closed-system interpretation of the protoliths is thus precluded.

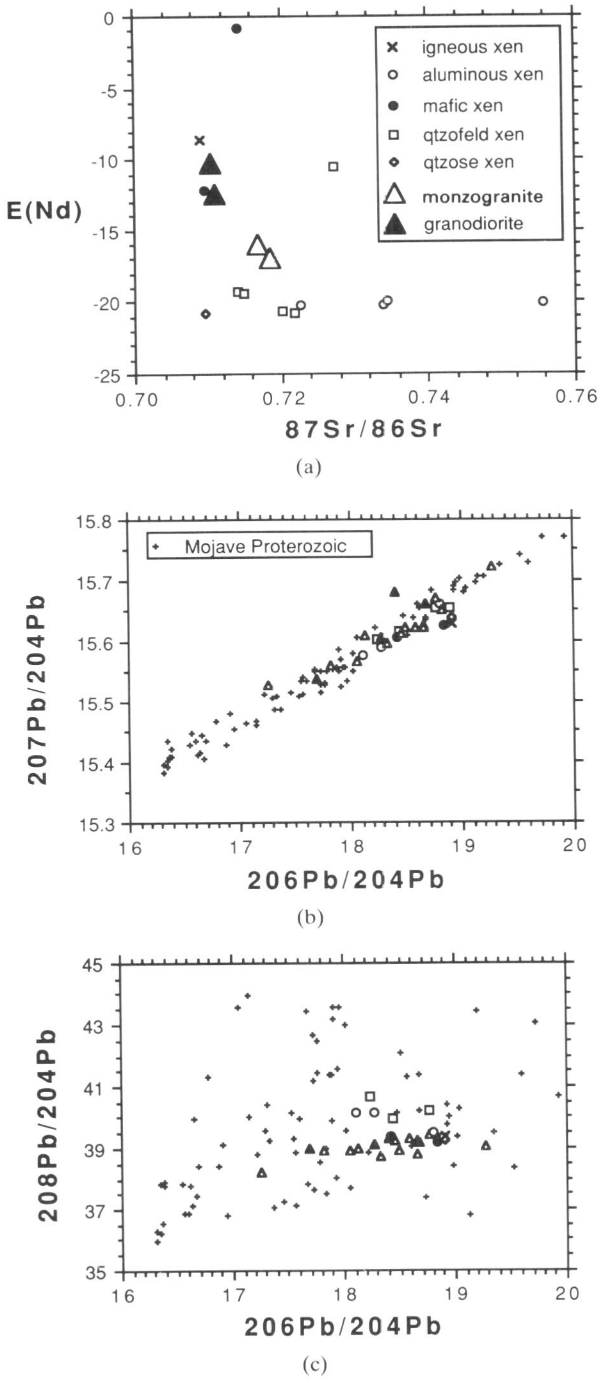

Figure 2 Isotope data for granites and lower crustal xenoliths: (a) ε_{Nd} vs $^{87}Sr/^{86}Sr$ at 70 Ma; (b), (c) present-day Pb isotope ratios; Proterozoic rocks from the Old Woman and Piute Mountains (Wooden and C. Cook, unpub. data) and elsewhere in the eastern Mojave Desert (Wooden & Miller 1990, and unpub. data) are shown for comparison; eight granite analyses are of feldspars and therefore represent approximately initial ratios; all other points represent whole rock analyses (relatively low U/Pb and Th/Pb indicate that changes in ratios over the last 70 Ma are small).

Isotopically, the xenoliths appear clearly to be part of the Mojave early Proterozoic crust (cf. Wooden & Miller 1990). A majority cluster near the lower end of the array of $^{143}Nd/^{144}Nd$ ratios reported for this region by Bennett and DePaolo (1987) (ε_{Nd} *c.* -21), with others ranging upwards to -1 (Fig. 2). $^{87}Sr/^{86}Sr$ ratios range from 0·710–0·756, again consistent with an early Proterozoic age. Lead isotope ratios are in the middle of the eastern Mojave array (Fig. 2; Wooden *et al.* 1988). However, although isotopic ratios very clearly ally these rocks with exposed crust, model ages for all three systems are entirely unrealistic, in general much too old but in some cases too young (Nd: generally 2–4 Ga; Sr: most 4–10 Ga). Three highly discordant, tightly clustered zircon fractions from a quartzose xenolith strongly suggest that the zircon initially crystallised in the early Proterozoic and experienced a major late Phanerozoic thermal event (Fig. 1).

3.3. Residues of anatexis

We attribute the modal and geochemical idiosyncracies of the xenoliths to anatexis: their nature is in all respects consistent with residues of partial melting in the deep crust. Their compositional peculiarities most strikingly correlate with abundance of modal garnet and reflect as well paucity of K-feldspar and abundance of rutile. Subtraction of a granitic melt fraction from diverse crustal rocks at depth could yield precisely the type of residue that we observe. Experimental studies show that melting of shales, sandstones, and intermediate igneous rocks at 1–1·5 GPa

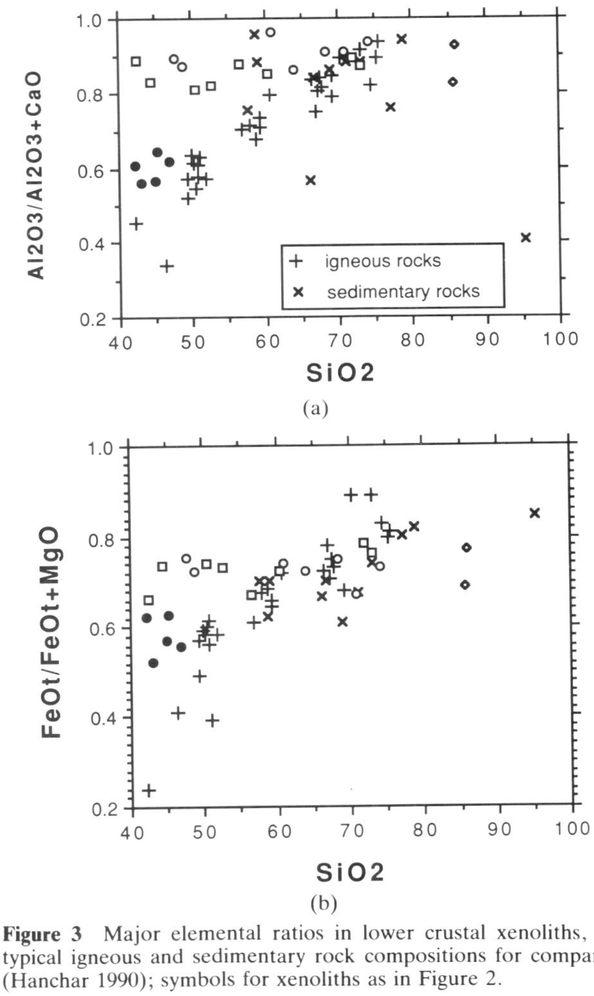

Figure 3 Major elemental ratios in lower crustal xenoliths, with typical igneous and sedimentary rock compositions for comparison (Hanchar 1990); symbols for xenoliths as in Figure 2.

and temperatures of 850–950°C yields granitic melt + garnet-rich residue that contains rutile ± quartz ± clinopyroxene ± hornblende ± biotite (e.g. Vielzeuf & Holloway 1988; Carroll & Wyllie 1990; Patiño Douce & Johnston 1991). This garnet-rich residue would be relatively enriched in Fe and Ti; it would be depleted in K-feldspar (hence in K), and depending on its initial composition compared to granitic melt either enriched or improverished in quartz (and Si) and plagioclase. The broad range in SiO_2 concentration and in other elemental abundances is thus a function of diverse protoliths and varying fractions of melt extracted. Major element mass balance modelling verifies that addition of typical granitic melt to our xenoliths can yield reasonable protolith compositions, and suggests highly variable, and some cases very large, melt fractions ($F < 10\%$–$>70\%$; Hanchar 1990; Hanchar *et al.* (in prep.)). Calculated melts were highly diverse in trace element concentrations, but "protoliths" composed of these "melts" + xenoliths have reasonable trace element concentrations (Hanchar 1990; Hanchar *et al.* (in prep.)).

Isotopic data verify that the xenoliths underwent a chemical reorganisation in the relatively recent past (late Phanerozoic; Wooden *et al.* 1990). Nd and Sr isotope ratios begin to diverge from reasonable values in a short time, given present-day Sm/Nd and Rb/Sr. This disturbance was the result of preferential loss of Nd and Rb, or in a few cases Sm or Sr, by extraction in melt. Melts in equilibrium with residues with xenolith mineralogies would in general have appropriate ratios to explain the depletion (Hanchar 1990; Hanchar *et al.* (in prep.)). Rb/Sr ratios are higher in melts than in xenoliths in almost all cases, but for biotite-rich samples, which have low rather than high Rb–Sr model ages, calculated melts hav lower Rb/Sr. Similarly, abundant garnet would generally result in lower Sm/Nd in coexisting melt, but monazite counterbalances this effect in some samples. Estimates of melt fraction lost based upon reasonable assumed age, initial $^{87}Sr/^{86}Sr$ and Rb and Sr concentrations in coexisting melts are consistent with those based on major elements. Both methods suggest that the mean amount of melt extracted was roughly 50%.

3.4. Are these xenoliths the residue of the melting event that produced the Old Woman-Piute Batholith?

3.4.1. Tests of the hypothesis. The hypothesis that the xenolith material is the residue of Old Woman-Piute anatexis implies that (a) temperature and depth of emplacement of the granites should be consistent with *P* and *T* of peak metamorphism; (b) major and trace element chemistry of the granites should be consistent with that of melts equilibrated with xenolith mineralogy; (c) isotopic compositions of granites should be compatible with those of the xenoliths; and (d) the age of peak metamorphism recorded in the xenoliths should be essentially the same as that of the batholith. We shall address these issues in sequence.

3.4.2. Conditions of melting and granite emplacement. Emplacement depth of the granites (*c.* 15 km) is consistent with the apparent xenolith metamorphic depth of about 40 km, requiring 25 km of ascent. Zircon and monazite saturation thermometry suggesting a *T* of about 800°C for the granodioritic magma and 750°C for the monzogranite magma (Section 2.3) is also consistent with conditions calculated for the xenoliths. However, a case can be made to suggest that the temperature of anatexis of the xenoliths was higher, perhaps 850–950°C, and that *T* from thermometry reflects re-equilibration. First, extensive melting should not occur in crustal rocks at temperatures lower than about 850°C if the only available water is structurally bound in micas and amphiboles (e.g. Clemens & Vielzeuf 1987; Vielzeuf & Holloway 1988; Patiño Douce & Johnston 1991; Rapp 1990). If the melts were very wet, it is unlikely that they would have equilibrated with clinopyroxene at relatively low *T* and ascent would have been impeded. Furthermore, assemblages in the aluminous and quartzofeldspathic xenoliths suggest that the reaction biotite → garnet + rutile + melt was important during melting. According to Patiño Douce and Johnston (1991), this reaction takes place at temperatures near 900°C. Thus, consistency of magma temperatures and xenolith metamorphism remain equivocal.

3.4.3. Elemental chemistry. The xenoliths, with their ubiquitous moderately sodic plagioclase and almost ubiquitous quartz, would have been in equilibrium with quite felsic melts that were poor in mafic constituents, even if temperatures were as high as 900–950°C. (e.g. Patiño Douce & Johnston 1991; Rapp 1990; Miller *et al.* 1986; Carroll & Wyllie 1990). Those derived from the metaluminous sources would have been metaluminous or weakly peraluminous, whereas those derived from more aluminous sources would be peraluminous, especially in the case of the Al_2SiO_5-bearing material (e.g. Patiño Douce & Johnston 1991; Miller *et al.* 1986; Conrad *et al.* 1988). It is easy to reconcile the compositions of the granites with such starting materials, especially if the more mafic granodiorites were either cumulate-rich or contained abundant re-equilibrated restite, which seems required by the fact that their solidi would be well in excess of 1,000°C.

Calculated trace element concentrations in the xenolith melts are extremely variable (Fig. 4). The absolute abundances are governed to a considerable extent by choice of Kd's, which are rarely known with confidence to within a factor of 2. Granite melts derived from such diverse source materials either must be extraordinarily heterogeneous (we know of no examples of such heterogeneity), or they are partially homogenised during coalescence and ascent (in our view more likely). The calculated trace element concentrations are generally consistent with the Old Woman-Piute granite abundances; that is, the granites could be mixes of melts from various appropriate xenoliths (Fig. 4), though melts from some xenoliths would have to have contributed much larger mass fractions that others. There are apparent discrepancies in heavy REE and Eu anomalies. Heavy REE content is lower in the calculated melt than it is in the granites, and Eu anomalies are less negative, or even positive. Incomplete equilibration of garnet with melt as melting and garnet production progressed could minimise HREE depletion and relative Eu enrichment in the melt (cf. Qin 1991).

3.4.4. Isotopic compositions. As with trace element compositions, Nd and Sr isotopic ratios of the melt fractions derived from the xenoliths must have been highly diverse and, to produce the semblance of homogeneity that the plutons possess, there must have been partial isotopic homogenisation of the melt fractions during coalescence and ascent (Fig. 5). Viewed in this way, the xenoliths are appropriate source materials for the Old Woman-Piute granites: the metaluminous xenoliths are the best matches for the granodiorite, and a combination of diverse xenoliths can readily match the monzogranite. However, we emphasise again that the isotopic "match" is by no means unique, owing to the wide range of ratios among the xenoliths, and that certain xenolith types must have contributed disproportionately to the melts to achieve the observed granite isotopic compositions (cf. Figs 2, 5).

A puzzling aspect of the isotopic data is the relative homogeneity in Pb composition (Fig. 2). In contrast to Nd and Sr, they are more restricted in Pb isotope ratios than the granites, though they fall in the same area along the same $^{206}Pb/^{204}Pb$ vs $^{207}Pb/^{204}Pb$ trend, on average at slightly higher values. The xenoliths also have a somewhat higher average $^{208}Pb/^{204}Pb$ ratio. The more radiogenic Pb in the xenoliths could be a result of retention of radiogenic Pb in zircon and perhaps monazite in the (xenolith) residue (cf. Hogan & Sinha 1991; Miller *et al.* 1988), but this does not explain the tighter clustering of xenolith Pb, or the similar spread in ratios of both monzogranites and granodiorites.

3.4.5. Age of metamorphism and of granites. We do not have good constraints on the age of metamorphism of the xenoliths. Both the U–Pb data in Figure 1 and the disturbed Rb–Sr, Sm–Nd, and U–Pb systems point to a late Phanerozoic event, but they are no more specific than that. Any time between the Jurassic and the end of the Cretaceous is geologically reasonable, as is the mid-Tertiary. Thus, the late Cretaceous age of the granites is perfectly plausible but not required as the time of xenolith metamorphism and anatexis.

We plan to test the timing by dating rims of zircons and cores and rims of monazites from the xenoliths by high-resolution ion microprobe, which should clarify the age of peak metamorphism.

3.4.6. Significance of the igneous-textured xenolith. The igneous-textured xenolith has the same type of high-pressure assemblage as the other xenoliths and thermobarometry yields the same estimates of P and T. We interpret it to be a cumulate that formed at depth from a magma that was present during metamorphism. Its mineralogy and elemental and isotopic composition suggest that it could have been comagmatic with the granodiorite of the Old Woman-Piute Batholith. It has a similar mineral assemblage as the granodiorite (plagioclase + biotite + hornblende + quartz + apatite + zircon + REE-rich epidote are shared by both). Garnet and rutile, present in the xenolith but not in the granodiorite, would become unstable during ascent from 40 to 15 km. K-feldspar, which crystallised late from the granodiorite magma, is absent in the xenolith; this may indicate the stage at which melt was extracted from the cumulate. Sphene and magnetite, present in the granodiorite but absent in the xenolith, may reflect changes in f_{O_2} during ascent and emplacement. Calculated abundances of Rb, Sr, and heavy REE in melt coexisting with the xenolith are essentially identical to those of the mean granodiorite (Fig. 4). Ba is higher in calculated melts by a factor of two, which is probably not unreasonable, given uncertainties in Kd's. Isotopic compositions are also very similar (Fig. 2).

3.4.7. Summary. An iron-clad case cannot be made that the xenoliths are residues of the anatectic event that

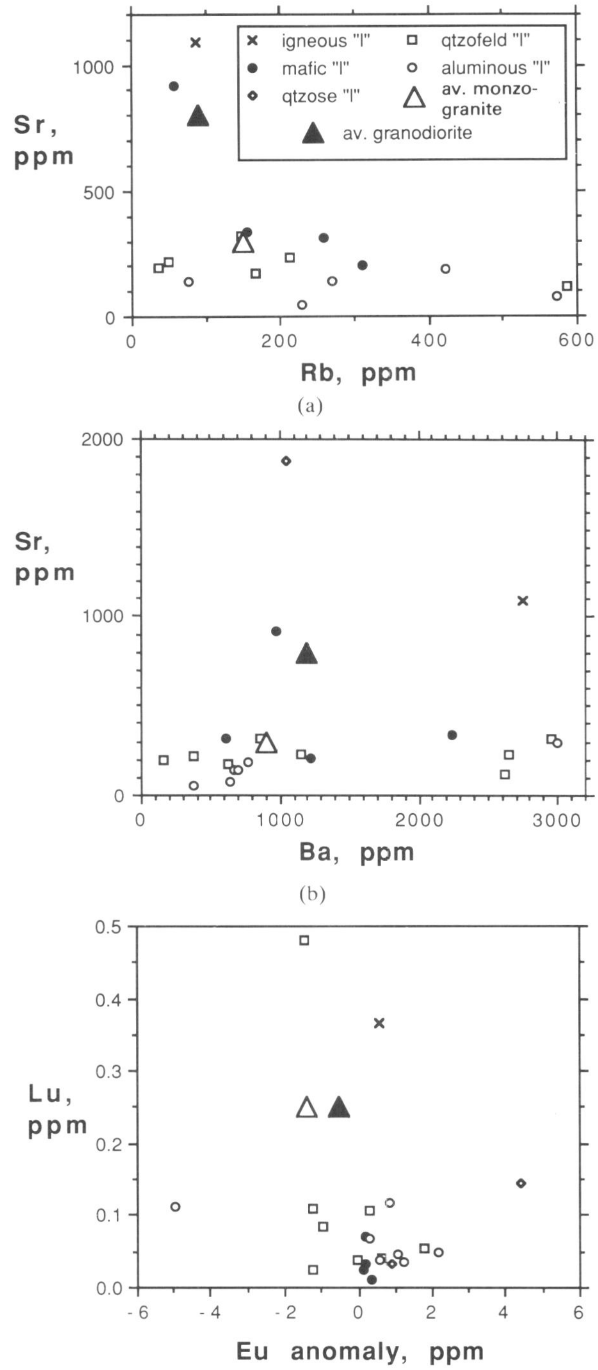

Figure 4 Calculated abundances of some trace elements in granitic melts coexisting with xenoliths (Hanchar 1990), with mean Old Woman-Piute granite compositions for comparison.

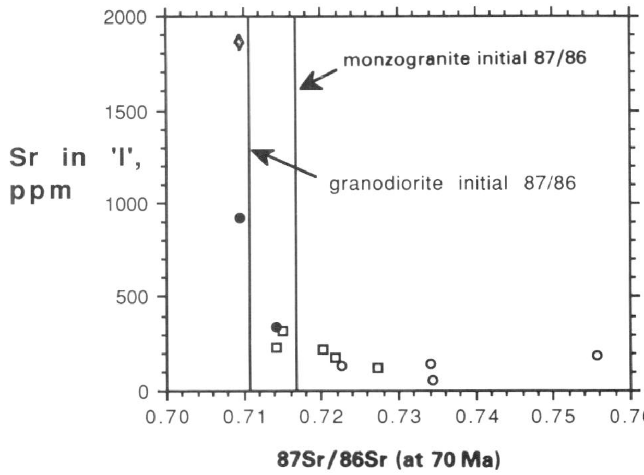

Figure 5 Calculated Sr concentrations of melts coexisting with xenoliths vs $^{87}Sr/^{86}Sr$ of the xenoliths (presumably that of the extracted melts, neglecting the minimal increase in the xenolith ratios since the melting event); the figure suggests ways in which melt increments could be combined to yield the observed ratios in the granites; symbols as in Figure 2.

produced the Old Woman-Piute Batholith, and in fact there are some problems with this interpretation. However, their characteristics are in many ways consistent with such a relationship. Perhaps the best support for this hypothesis lies in the facts that the Old Woman-Piute granites are the most abundant moderately felsic late Phanerozoic igneous rocks in the area, and that they are the only known igneous rocks that are at all appropriate products of melting of the xenoliths (cf. Jurassic rocks: Gerber *et al.* 1991; earlier Cretaceous rocks: Karlstrom, Miller, Kingsbury, & Wooden (in prep.); Tertiary rocks: Miller & Miller, 1991). The latter argument, that the Old Woman-Piute granites are the only appropriate products, loses some impact given the possibility that xenolith-derived melts were only a contaminant of more mafic magmas. Dating the Phanerozoic zircon and monazite growth should be an almost definitive test.

3.5. Interpretation of the xenoliths: implications for anatexis *assuming that they are Old Woman-Piute residue*

Assuming that the xenoliths represent the residue of melting that produced the Old Woman-Piute Batholith, we attempt below to answer some of the questions posed in Section 1.1.

3.5.1. Conditions and materials involved in magma generation. (a) Melting occurred at a depth of about 40 km and a *T* of at least 750–800°C, and perhaps near 900°C. (b) Melts could have been entirely derived from the crust represented by the xenoliths. (c) The source materials of the monzogranite probably included predominantly immature sedimentary and perhaps fairly felsic igneous rocks, with important contributions from more mature sediments and intermediate igneous rocks. The granodiorite was most probably derived primarily from intermediate igneous rocks with some contribution from more felsic material. (d) The source was extremely heterogeneous. (e) Melt fractions were highly variable, as diverse rocks provided melt increments ranging from near zero to more than 70% of their initial mass. The *mean* melt fraction appears to have been in the order of roughly 50%, a percentage that may have had physical significance on a large scale but did not on a centimetre scale. (f) The tectonic setting is uncertain. However, the conditions (750–950°C at 40 km) imply a thickened crust and a moderately, but not strongly, elevated geotherm. The dominant assemblages and textures show no evidence for decompression (no late, low *P* assemblages). Ion probe data for monazites and zircon rims may enable us to date the duration of metamorphism, which will constrain the tectonic history.

3.5.2. Mechanism of anatexis. (a) Enhanced mantle heat input is possible but by no means required to explain the *P-T* conditions. No mantle-derived magma input is needed. (b) Earlier Mesozoic crustal thickening and subsequent thermal relaxation is consistent with the apparent moderately elevated temperatures at great depth. (c) Unroofing was not the mechanism for inducing melting. (d) There is no evidence for major fluid influx.

4.5.3. Previous history of the crustal source. (a) Planned ion probe studies may help to resolve the age of crustal formation in this region, but we cannot answer this question yet. (b) The xenoliths record in their protoliths igneous and sedimentary events. Furthermore, the presence at great depth of supracrustal material documents very large-scale underthrusting, probably during one or both of the known orogenic episodes that affected this region (early Proterozoic, Mesozoic).

4. U-Th-bearing accessory minerals

4.1. Introduction

U- and Th-bearing accessory minerals have extraordinary potential for elucidating crustal histories. In addition to being the principal tools for U-Th-Pb geochronology, they tend to be stable and to recrystallise very sluggishly throughout a wide range of crustal conditions and compositions, and they are in general highly insoluble in melts and fluids. Therefore, they retain information about their heritage through multiple cycles of crustal processes. Individual minerals commonly record new events without losing their record of previous events, and each mineral has unique responses to different crustal processes (e.g. diffusivity of Pb, rate of recrystallisation, solubility).

Growth and degradation (i.e. abrasion, fracturing, resorption) histories of accessory minerals are recorded in almost ubiquitous compositional zonation patterns. These patterns are commonly difficult or impossible to see using standard transmitted or reflected light microscopy, but we have found that both backscattered electron (BSE) and cathodoluminescence (CL) methods image the patterns in spectacular detail (see also Patterson *et al.* 1989; Vavra 1990). BSE patterns reveal variation in mean atomic number (e.g. in zircon, Hf concentration is the main control of intensity). Colour and brightness in CL are controlled by poorly known trace elemental variations, but we have found that those accessory minerals that luminesce (notably zircon) have patterns that are essentially the same as for BSE, though commonly subtler details appear with CL.

We assume that specific types of growth or degradation must be limited to certain environments. Thus, geometry of zonation patterns, together with chemistry of the zones, can reveal the sequence of environments in which a mineral has grown or been degraded. By combining studies of zircon patterns in various rock types with standard petrographical observations, descriptions from the literature and reasoning about the dynamics of textural modification, a model characterising growth and degradation histories as recorded in zonation patterns can be constructed (Miller & Hanchar 1991). We apply these results to the interpretation of Old Woman-Piute granite and xenolith history below, and present some tentative interpretations of zonation patterns observed in other accessory minerals.

4.2. Zircons from Old Woman-Piute granites and xenoliths

4.2.1. Geochronology of the granites. SHRIMP studies demonstrate that both monzogranite and granodiorite zircons have common and diverse inheritance (Fig. 1, Table 1). In both rock units, analyses with clear inheritance range from concordant to strongly discordant. Points are mostly slightly to moderately discordant, with ages determined by assuming a lower intercept of 70 Ma spanning the entire 1·4–1·8 Ga common range of Mojave basement (including the 1·44–1·62 Ga period for which no ages are reported). Figure 6 reveals the likely source of much of the discordance, and suggests how selection of SHRIMP points may be enhanced by detailed initial study of zonation patterns. The zircons in our study rarely had zoning that was obvious in transmitted or reflected light, and unfortunately we did not undertake CL or BSE studies until after the grains had been analysed by SHRIMP. The discordant points generally cross fundamental zone boundaries, suggesting that with small enough primary ion beams concordant ages could be obtained for each episode of zircon growth; Pb loss may have been negligible in the events that these zircons experienced after initial crystallisation. Magmatic rims are evident in most grains, but widths vary dramatically and we clearly missed small inherited cores and narrow magmatic rims in many cases (we suspect that this is typical of analyses made without benefit of CL or BSE studies). (We use the term "magmatic" to refer to new growth from the melts of the Old Woman-Piute Batholith.)

4.2.2. Zircon histories. Figure 6 reveals the remarkable variety of zonation patterns in zircons from the Old Woman-Piute Batholith. The more pronounced inheritance in monzogranite zircons is shown to be as much a function of relative sizes of cores and rims as of differences in how many grains are inherited: cores are far larger and magmatic rims smaller in the coarse monzogranite zircons than in the coarse granodiorite zircons that were studied. The only grains that in general appear to represent entirely magmatic growth in the monzogranite are tiny, acicular, euhedral grains observed in thin-sections. These grains are typically $10–30 \times 2–6\ \mu m$, too small to be separated in routine mineral separation or to be analysed by SHRIMP. Apparently, magmatic zircons in the granodiorite are more abundant and attain larger sizes.

The monzogranite zircons in particular show spectacular diversity. We have encountered almost all of the zonation features in sample SW1 that we have found in a reconnaissance study of zircons from a wide range of rock types (Miller & Hanchar 1991). Some examples, with our interpretations of their histories, are shown in Figure 6. These zircons, taken in sum, display a long and complex igneous, metamorphic, and probably sedimentary Proterozoic history for the source region.

The granodiorite zircons do not show as great diversity as the monzogranite zircons. This may reflect a less heterogeneous source region, but it probably also results at least in part from the smaller size of the cores: the smaller the core, the less opportunity for it to reveal complex multiple zones. The lower mass fraction of inherited zircon in the granodiorite is a function of the higher solubility of

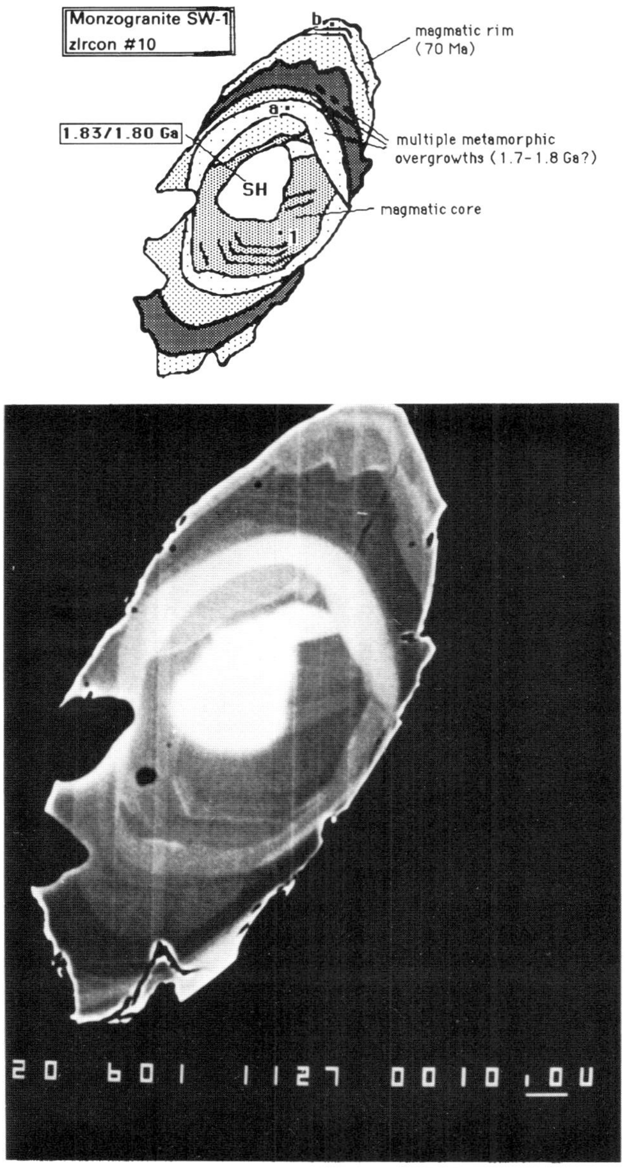

(a)

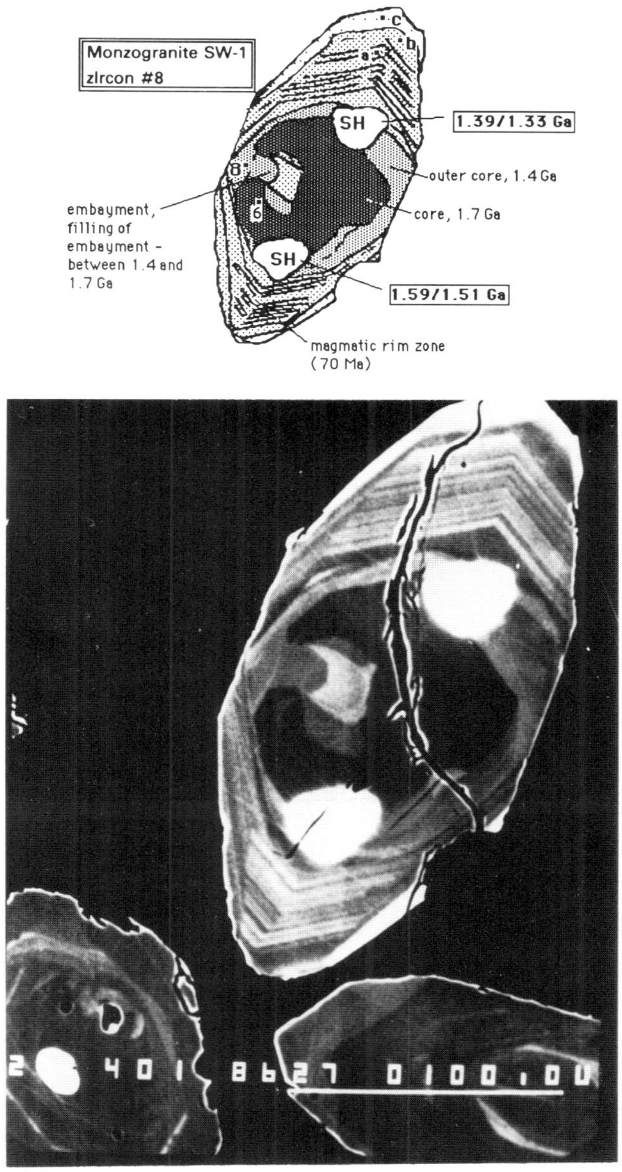

(b)

Figure 6 Backscattered electron micrographs of granite zircons (a)–(d) and xenotime (e): white spots labelled SH are points analysed by SHRIMP (brightness indicates charging where the carbon coat was damaged); otherwise, brightness in zircons is largely determined by Hf concentration (higher = brighter); interpretations, SHRIMP point ages ($^{207}Pb/^{235}U$, $^{206}Pb/^{238}U$), and locations of analytical points in Table 2 are given in sketches.

zircon in this hotter, metaluminous magma as compared with the monzogranite magma (Watson & Harrison 1983; Harrison & Watson 1983).

4.2.3. Chemistry of the zones. Compositions of representative points from monzogranite and granodiorite zircons are shown in Table 2. The main substituting element is Hf; HfO_2 concentration ranges from *c.* 1–2 wt% (rarely to 3 wt%) and is the dominant control of BSE intensity variations. Most magmatic portions of grains have HfO_2 concentration between 1·5 and 2 wt%. Other elements are highly variable, with perhaps the most noteworthy being Y. Y_2O_3 concentration ranges from less than our detection limit (*c.* 0·02 wt%) to 0·2 wt% in inherited cores in granodiorite zircons, and from below detection to 0·4 wt% in cores in monzogranites; concentrations in magmatic portions of grains are toward the high ends of these ranges. We suspect that much of the variation in Y is attributable to the modal assemblage with which the zircon zone was in equilibrium during its growth, in particular to the abundance of garnet (see Section 4.3 below). Partitioning of Y between zircon and the remainder of a rock or magma can vary by more than an order of magnitude depending upon whether garnet is absent or abundant.

4.3. Zircons from the xenoliths

Three zircon fractions from a large quartzose xenolith yielded extremely discordant U-Pb data points that cluster closely about one-third of the way from 72 Ma on concordia to *c.* 1·8 Ga (Section 3·2). A large fraction of the U-Pb system of these zircons must be late Phanerozoic in age. Zonation in these zircons suggests that they have not experienced Pb loss, but rather have ancient cores and thick, young metamorphic rims. The geometry of the cores and rims is consistent with about two-thirds of the mass being rim material, which, if U contents are similar and the age of the rims is Mesozoic or younger, would yield the observed discordance. With the exception of those in the igneous-textured xenolith, which are euhedral with concentric euhedral or sector zoning, all observed zircons have thick rims that appear to be of metamorphic origin.

The xenolith zircons have 1·2–2 wt% HfO_2 (Table 3). Y_2O_3 is extremely low in the rims (generally $\leq$0·05 wt%), which suggests that they grew in equilibrium with the abundant garnet that is now present. Cores are more variable in Y content (below detection to 0·8 wt%), perhaps

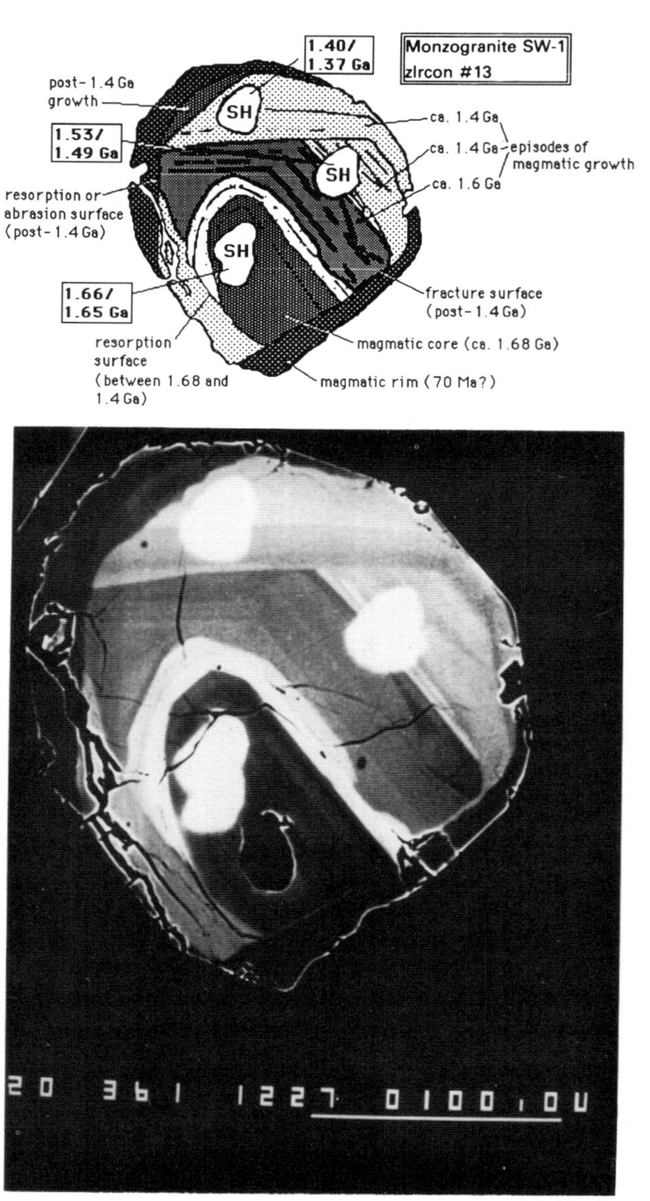

(c)

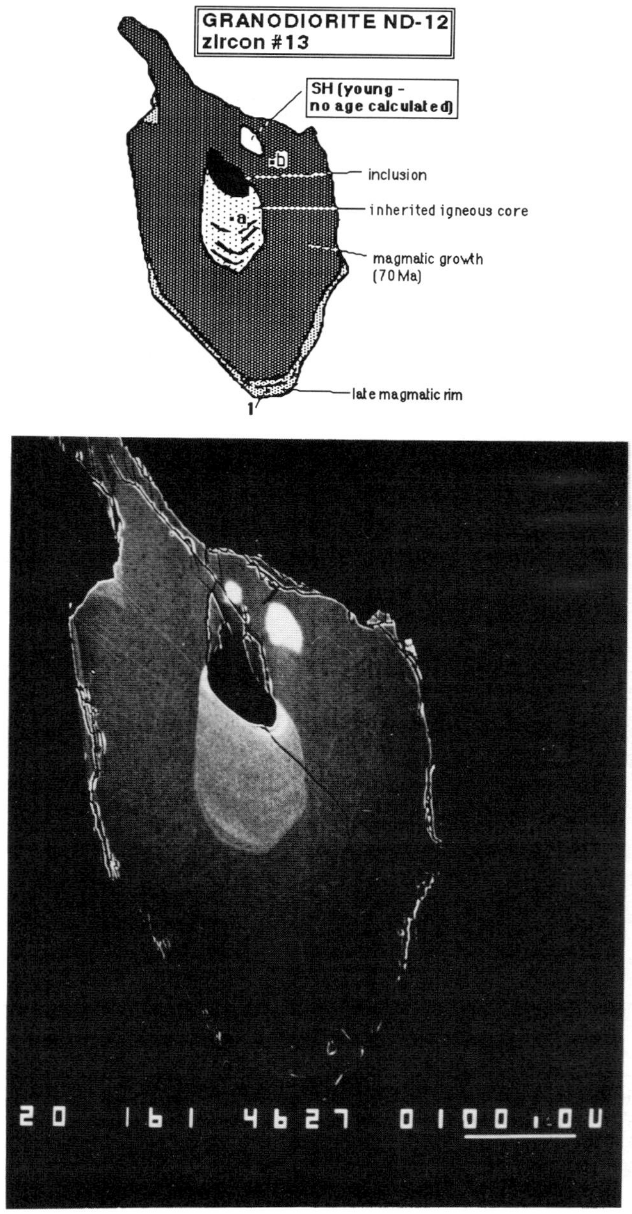

(d)

Figure 6 (*Continued*).

reflecting variable garnet content during earlier stages of zircon growth.

4.4. Other U-Th accessory minerals

Several other U- and Th-bearing accessory minerals that are widespread in crustal rocks are present in the Old Woman-Piute granites and in the xenoliths. These include apatite (ubiquitous), sphene (ubiquitous in granodiorite; in some xenoliths), monazite (ubiquitous in monzogranite; in many xenoliths), xenotime (sparse but probably ubiquitous in monzogranite; present in some xenoliths), and allanite (in most or all granodiorite samples; a light REE-rich epidote with strong sector and concentric euhedral zoning is present in the igneous-textured xenolith). All of these generally have low solubility in melts and thus have the potential to be carried as inherited, restitic grains with some record of crustal history from a monzogranite source region (e.g. Rapp *et al.* 1987; Green & Pearson 1987; Watson & Harrison 1983; Harrison & Watson 1983). We have emphasised monazite and to a lesser extent xenotime in our studies so far because they both can preserve U-Pb inheritance and because both show distinctive zonation patterns in our samples.

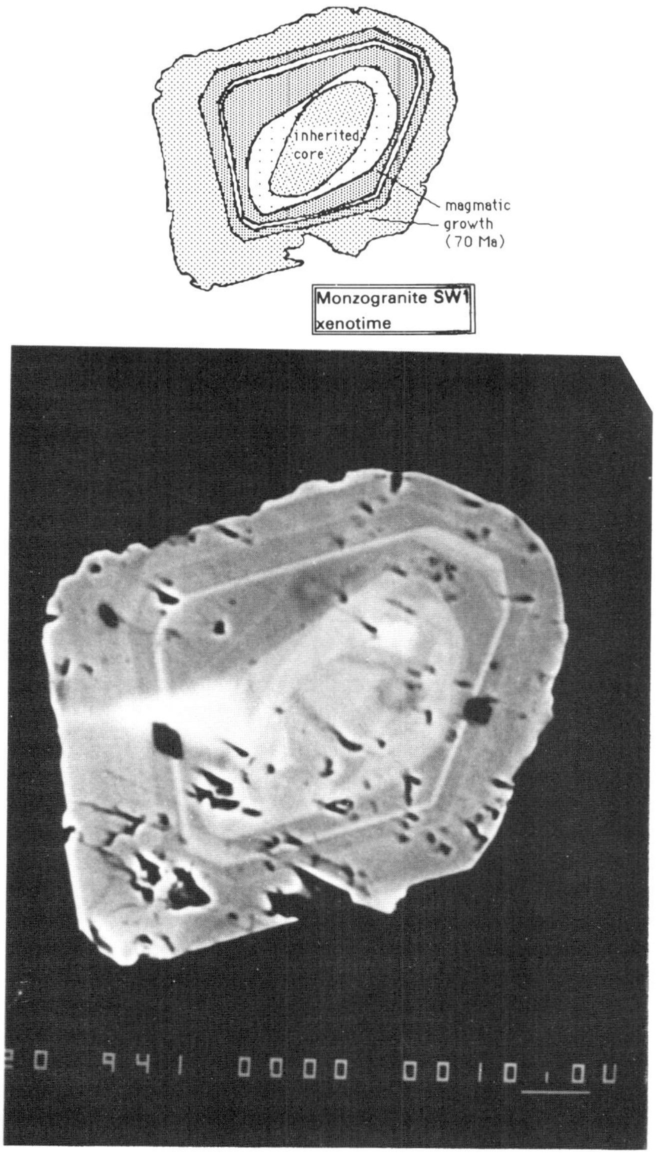

(e)

Figure 6 (*Continued*).

We do not yet have any geochronological data from xenotimes. We have, however, noted complex patterns from grains in the monzogranite that appear to demonstrate inherited cores and magmatic rims (Fig. 6(f)). Representative analyses are shown in Table 4.

Conventional U-Pb analyses of monazites from the monzogranite demonstrate strong inheritance, though not so pronounced as that of zircons (Kingsbury *et al.* 1990; Kingsbury 1990; cf. Copeland *et al.* 1988), with an upper intercept of about 1·6 Ga. Monazites are highly variable in composition in both the monzogranite and the xenoliths, with a large range in Th content and La varying inversely with Y and Sm (Table 4); geometry of zones (visible in BSE but not CL) is complex. Magmatic rims of grains from the monzogranite are commonly richer in Y than what we take to be inherited cores. Monazite in xenoliths shows an enormous range in Th content (0·6–11 wt% ThO_2), but Y is uniformly very low (only one analysed point, a core with 1·31 wt%, has >0·11 wt% Y_2O_3). The Y concentrations suggest that monazites grew in equilibrium with the abundant garnet that is now in the xenoliths (i.e. they represent peak metamorphism), and that the cores of monzogranite monazites also grew in a garnet-rich environment.

4.5. Interpretation of accessory minerals: implications for anatexis

The following can be inferred from studies of the accessories in the granites.

4.5.1. Conditions and materials involved in magma generation. No information on (a), (b) or (e), conditions, presence or absence of mantle component, and melt fractions. (c) There clearly were igneous- and probably sedimentary-protolith rocks in the source. At least some of these rocks had undergone one or more metamorphic episodes before anatexis. (d) The monzogranite had an extremely heterogeneous source; the granodiorite may have had a somewhat more homogeneous source, but the origins of the smaller zircon cores are harder to diagnose. (f) The accessory minerals have not yet provided us with information about tectonic setting. However, ion probe U-Pb data for monazites and zircon rims from xenoliths and for monazite cores and outer zones of zircons from the granites may shed light on the Phanerozoic metamorphic history of the source, and therefore on tectonic history.

4.5.2. Mechanism of anatexis. No direct information.

4.5.3. Previous history of the crustal source. (a) The crust that yielded the granites was at least 1·8 Ga old. Further dating may reveal earlier components. (b) This crust had a complex and protracted history from early to middle Proterozoic time. It experienced repeated igneous, metamorphic and probably sedimentary events during this interval.

5. Discussion

5.1. Anatexis in the eastern Mojave crust

Assuming that the xenoliths are the residues of the late Cretaceous anatectic event and combining their constraints with those provided by accessory minerals, regional geology, and the monzogranites themselves, we propose that the following can be inferred about the melting event and the nature and history of the Old Woman-Piute source region:

5.1.1. Conditions and materials involved in magma generation. Melting took place at a depth of about 40 km and T between 750 and 900°C. The source materials were

Table 2 Analyses of representative zircons from granites

*Monzogranite SW*1[1]

grain-	8[2]					10[2]			A		B[3]
	6	*8*	*a*	*b*	*c*	*1*	*a*	*b*	*core*	*rim*	
ZrO_2	63·54	63·59	63·94	64·50	63·52	65·41	64·73	64·52	65·14	64·43	64·46
SiO_2	34·78	34·32	32·87	33·07	32·89	33·22	33·33	32·87	31·81	31·90	32·50
HfO_2	1·33	1·41	3·04	2·39	3·68	1·73	2·27	2·70	1·78	2·06	1·25
TiO_2	0·05	0·04	0·08	0·07	0·11	0·06	0·06	0·07	nd	nd	nd
U_3O_8	0·03	0·01	0·19	0·15	0·78	0·11	0·31	0·14	0·19	0·06	0·02
Ce_2O_3	0·01	0·01	0·01	0·00	0·00	0·03	0·01	0·01	nd	nd	nd
Y_2O_3	0·01	0·10	0·37	0·15	0·22	0·05	0·01	0·17	0·04	0·46	0·31
Total	99·75	99·48	100·50	100·33	101·20	100·61	100·74	100·48	100·30	100·52	98·72

*Granodiorite ND*12

grain-	13[2]			5	
	1	*a*	*b*	*core*	*rim*
ZrO_2	65·85	66·25	64·84	64·52	64·81
SiO_2	32·89	33·57	33·21	33·20	34·29
HfO_2	1·65	2·26	1·62	1·51	1·54
TiO_2	0·04	0·07	nd	nd	nd
Ce_2O_3	0·02	0·01	nd	nd	nd
Y_2O_3	0·11	0·00	0·01	0·21	0·02
Total	100·60	102·22	99·73	99·72	100·74

Note: [1]labelled OW1 in Foster *et al.* 1989; Miller *et al.* 1990; [2]see Figure 16 for locations of analyses; [3]small, acicular, euhedral (magmatic).

Table 3 Representative analyses of zircons from xenoliths

Sample	XJ111, #4		XJ150, #3		PT8A, #2[1]		XJ169A, #3
	core	rim	core	rim	core	rim	core
ZrO_2	64·65	66·18	65·81	66·92	68·37	66·54	65·37
SiO_2	30·91	31·51	31·34	31·24	30·53	31·47	31·41
HfO_2	1·55	1·78	1·94	1·36	1·55	0·119	1·11
TiO_2	0·06	0·08	0·21	0·41	0·05	0·04	0·04
U_3O_8	0·03	0·04	0·06	0·08	0·03	0·00	0·07
Ce_2O_3	0·01	0·00	0·02	0·01	0·01	0·00	0·01
Y_2O_3	0·63	0·07	0·01	0·00	0·04	0·03	0·40
Total	97·87	99·62	99·43	99·97	100·63	99·31	98·53

highly diverse metaigneous and metasedimentary rocks of early and middle Proterozoic age (1·4, 1·6–1·8 Ga). The granodiorite had on average a more mafic source (intermediate igneous) than the monzogranite (diverse clastic sediment + igneous), but both had a range of contributing compositions. Melting was probably entirely intracrustal. Melt fractions were highly variable in the source region (probably 0 to >70%), and melt increments were extremely variable in trace element and isotopic composition. To form plutons with only moderate variability, these increments must have coalesced and homogenised in the source region or during ascent. Decompression probably did not occur, or at least was not an important factor, during peak metamorphism and anatexis, and little if any heat input was required from intruding mantle magmas.

5.1.2. Mechanism of anatexis. The most plausible *dominant* factor in inducing melting appears to be thermal blanketing resulting from mid-Mesozoic crustal thickening.

5.1.3. Previous history of the crustal source. The source region comprised multiple early to middle Proterozoic intrusions and early Proterozoic clastic sedimentary rocks.

Table 4 Representative analyses of monazites and xenotimes from granites and xenoliths

	Monazites											*Xenotime*[3]	
Sample	SW1[1]			SW3[1]		XJ150[2]				PT8A[2]		SW1[1]	
grain	#5			#1	#2	#1		#2		#1		#1	
	core	mid	rim	in bio	in Ksp	core	rim	core	rim	core	rim	core	rim
La_2O_3	18·41	15·71	16·50	17·67	12·76	15·56	14·56	11·92	13·56	11·38	13·29	bd	bd
Ce_2O_3	33·71	32·32	32·90	30·83	28·60	30·59	29·18	27·22	29·29	27·39	27·68	0·02	0·09
Nd_2O_3	10·78	11·69	11·91	10·47	12·28	17·39	16·47	15·26	16·48	15·59	15·72	0·08	0·38
Sm_2O_3	1·12	1·82	1·59	1·08	2·51	2·14	1·78	1·46	1·70	1·93	1·98	0·32	0·76
Gd_2O_3	nd	nd	nd	0·44	1·66	0·53	0·44	0·31	0·39	0·98	0·78	1·06	2·39
Ho_2O_3	nd	nd	nd	0·01	0·08	nd	nd	nd	nd	nd	nd	1·57	1·48
Yb_2O_3	nd	nd	nd	nd	nd	0·00	0·00	0·00	0·00	0·00	0·00	4·28	3·04
Y_2O_3	0·43	1·81	1·11	0·44	1·85	0·05	0·07	0·04	0·04	0·29	0·10	38·80	39·45
ThO_2	4·56	3·67	3·65	3·31	3·22	0·58	1·22	9·88	4·25	7·49	5·61	0·59	0·15
U_3O_8	nd	nd	nd	0·12	0·41	0·13	0·14	0·19	0·14	0·28	0·17	2·18	0·47
$P2O_5$	30·38	30·81	30·45	29·65	30·17	30·69	33·24	28·86	30·18	29·42	29·63	31·84	33·51
SiO_2	0·54	0·16	0·22	0·31	0·26	0·09	0·45	0·91	0·35	0·70	0·62	0·40	0·03
CaO	0·56	0·67	0·67	0·52	0·64	0·45	0·61	1·50	0·90	1·47	1·49	0·01	0·04
FeO	nd	nd	nd	nd	nd	0·03	0·63	0·00	0·01	0·00	0·02	nd	nd
Total	100·49	98·66	98·97	94·84	94·44	98·29	98·79	97·61	97·34	96·98	96·98	81·13	81·78

Notes: [1]granite; [2]xenolith; [3]low totals reflect in part heavy REE not analysed; inclusion of oxides of Tb, Dy, Er, Tm, and Lu in smooth chondrite-normalised abundances adds 11–12 wt%.

Many of these rocks underwent one Proterozoic metamorphic event, and some underwent several. During the Proterozoic, or during the Mesozoic, or both, some or all of these rocks were underthrust from supracrustal to deep crustal positions. A Mesozoic underthrusting event probably set the stage for late Cretaceous anatexis.

5.2. Implications for crustal melting

The heterogeneity of the source region and of melt increments that characterised the Old Woman-Piute melting event and the great variation in local melt fraction are to be expected in crustal melting (e.g. Miller *et al.* 1988). Mixing processes must be effective to yield moderately homogeneous plutons, whether they are wholly or only partially derived from the crust. This anatectic event was also typical in that it involved H_2O-undersaturated melting and a subsequent long distance ascent, and that the only *recognisable* restitic material in the granites is inherited accessory minerals (though the restite fraction may have been appreciable).

5.3. Applying xenolith and accessory mineral studies to understanding granite generation and crustal histories

Lower crustal xenoliths are not available everywhere that there are granites, but where they are, they are an invaluable source of information, at least about the available materials at depth and crustal history, and at best about the melting process itself. Accessory minerals, on the other hand, are ubiquitous in granites, and many, probably most, granites contain demonstrably inherited zircons and possibly monazites and xenotimes. Studies that combine zone geometry and chemistry with ion probe U-Pb geochronology can be extremely powerful in eliciting information about the shadowy source regions of granite magmas.

6. Acknowledgements

Comments from Bruce Patterson and an anonymous reviewer helped us to clarify the manuscript. Jim Kingsbury's studies of monazite contributed to our understanding of accessory mineral behaviour. This research was supported by NSF grants EAR-8609801 and 890432.

References

Anderson, J. L. & Bender, E. E. 1989. Nature and origin of Proterozoic A-type granitic magmatism in the southwestern United States of America. LITHOS **23,** 19–52.

Barton, M. D. 1990. Cretaceous magmatism, metamorphism, and metallogeny in the east-central Great Basin. *In* Anderson, J. L. (ed.) *The nature and origin of Cordilleran Magmatism.* GEOL SOC AM MEM **174,** 283–302.

Bennett, V. C. & DePaolo, D. J. 1987. Proterozoic crustal history of the western United States as determined by neodymium isotopic mapping. GEOL SOC AM BULL **99,** 674–85.

Brouand, M., Banzet, G. & Barbey, P. 1990. Zircon behaviour during crustal anatexis. Evidence from the Tibetan slab migmatites. J VOLCANOL GEOTHERM RES **44,** 143–62.

Carl, B. S., Miller, C. F. & Foster, D. A. 1991. Western Old Woman Mountains shear zone: evidence for latest Cretaceous ductile extension in the Cordilleran orogenic belt. GEOLOGY **19,** 893–6.

Carroll, M. R. & Wyllie, P. J. 1990. The system tonalite-H_2O at 15 kbar and the genesis of calc-alkaline magmas. AM MINERAL **75,** 345–57.

Clemens, J. D. & Vielzeuf, D. 1987. Constraints on melting and magma production in the crust. EARTH PLANET SCI LETT **86,** 287–306.

Conrad, W. K., Nicholls, I. A. & Wall, V. J. 1988. Water-saturated and undersaturated melting of metaluminous and peraluminous crustal compositions at 10 kb; evidence for the origin of silicic magmas in the Taupo volcanic zone, New Zealand, and other occurrences. J PETROL **29,** 765–803.

Copeland, P., Parrish, R. R. & Harrison, T. M. 1988. Identification of inherited Pb in monazite and its implications for U-Pb systematics. NATURE **333,** 760–3.

Davis, G. A. & Lister, G. S. 1988. Detachment faulting in continental extension, perspectives from the southwestern U.S. Cordillera. GEOL SOC AM SPEC PAP **218,** 133–59.

Ellis, D. J. & Green, D. H. 1979. An experimental study of the effect of Ca upon garnet-clinopyroxene Fe-Mg exchange equilibria. CONTRIB MINERAL PETROL **71,** 13–22.

Ferry, J. M. & Spear, F. S. 1978. Experimental calibration of the partitioning of Fe and Mg between biotite and garnet. CONTRIB MINERAL PETROL **66,** 113–17.

Fletcher, J. M. & Karlstrom, K. E. 1990. Late Cretaceous ductile deformation, metamorphism, and plutonism in the Piute Mountains, eastern Mojave Desert. J GEOPHYS RES **95,** 487–500.

Foster, D. A. & Hyndman, D. W. 1990. Large-scale anatexis: the importance of subcrustal magma intrusion. EOS **71,** 299–300.

Foster, D. A, Harrison, T. M. & Miller, C. F. 1989. Age, inheritance, and uplift of the Old Woman-Piute Batholith, California, and implications for K-feldspar age spectra. J GEOL **97,** 232–43.

Foster, D. A., Miller, C. F. & Harrison, T. M. 1992. The timing and character of crustal thickening, metamorphism, plutonism, and unroofing in the Old Woman Mountains area, California: evidence from $^{40}Ar/^{39}Ar$ thermochronology and thermobarometry. GEOL SOC AM BULL **104,** 176–91.

Gerber, M. E., Miller, C. F. Wooden, J. L. & Foster, D. A. 1991. Plutonism at the eastern edge of the Cordilleran Jurassic magmatic belt, Mojave Desert, California. GEOL SOC AM ABSTR PROG **23,** A249.

Gleason, J. D., Miller, C. F. & Wooden, J. L. 1988. Barrel Spring alkalic complex: 1·4 Ga anorogenic plutonism in the Old Woman-Piute Range, eastern Mojave Desert, California. GEOL SOC AM ABSTR PROG **20,** 164.

Graham, C. M. & Powell, R. 1984. A garnet-hornblende geothermometer: calibration, testing, and application to the Pelona Schist, southern California. J METAMORPHIC GEOL **2,** 13–31.

Green, T. H. & Pearson, N. J. 1987. An experimental study of the Nb and Ta partitioning between Ti-rich minerals and silicate liquids at high pressure and temperature. GEOCHIM COSMOCHIM ACTA **51,** 55–62.

Hammond, J. G. 1986. Geochemistry and petrogenesis of Proterozoic diabase in the southern Death Valley region of California. CONTRIB MINERAL PETROL **93,** 321–21.

Hanchar, J. M. 1990. *Lower crustal xenoliths in a mid-Tertiary dike, Piute Mountains, southeastern California.* Unpublished Master's Thesis, Vanderbilt University, Nashville, Tennessee, U.S.A.

Harrison, T. M. & Watson, E. B. 1983. Kinetics of zircon dissolution and Zr diffusion in granitic melts of variable water content. CONTRIB MINERAL PETROL **84,** 66–72.

Hileman, G. E., Miller, C. F. & Knoll, M. A. 1990. Mid-Tertiary structural evolution of the Old Woman Mountains area: implications for crustal extension across southeastern California. J GEOPHYS RES **95,** 581–614.

Hogan, J. P. & Sinha, A. K. 1991. The effect of accessory minerals on the redistribution of lead isotopes during crustal anatexis: a model. GEOCHIM COSMOCHIM ACTA **55,** 335–48.

Hoisch, T. D. & Hamilton, W. B. 1990. Granite generation by fluid-induced anatexis. EOS **71,** 694.

Kingsbury, J. A. 1990. *Utility of monazite in geochronology: examples from the Old Woman-Piute Mountains, southeastern California.* Unpublished Master's Thesis, Vanderbilt University, Nashville, Tennessee, U.S.A.

Kingsbury, J. A., Miller, C. F., Wooden, J. L. & Harrison, T. M. 1990. Utility of monazite in geochronology: examples from the old Woman-Piute Mountains, southeastern California. EOS **71,** 654.

Kohn, M. J. & Spear, F. S. 1990. Two new geobarometers for garnet amphibolites, with applications to southeastern Vermont. AM MINERAL **75,** 89–96.

McCaffrey, K. J. W., Howard, K. A. & Bailey, C. M. 1991. Jurassic syntectonic pluton emplacement in the Clipper Mountains, eastern Mojave Desert, California. GEOL SOC AM ABSTR PROG **23,** A250.

Miller, C. F. & Hanchar, J. M. 1991. Zircon zonation patterns and interpretation of crustal histories. GEOL SOC AM ABSTR PROG **23,** A46.

Miller, J. S. & Miller, C. F. 1991. Tertiary extension-related volcanism, Old Woman Mountains area, eastern Mojave Desert, California. J GEOPHYS RES **96,** 13629–44.

Miller, C. F., Rapp, R. P. & Watson, E. B. 1986. AFM mineral-felsic liquid phase relations; potential for elucidation of the origin and evolution of felsic magmas. GEOL SOC AM ABSTR PROG **18,** 695.

Miller, C. F., Watson, E. B. & Harrison, T. M. 1988. Perspectives on the source, segregation, and transport of granitoid magmas. TRANS R SOC EDINBURGH EARTH SCI **79,** 135–56.

Miller, C. F., Wooden, J. L., Bennett, V. C., Wright, J. E., Solomon, G. C. & Hurst, R. W. 1990. Petrogenesis of the composite peraluminous-metaluminous Old Woman-Piute Range Batholith, southeastern California: isotopic constraints. *In* Anderson, J. L. (ed.) *The Nature and Origin of Cordilleran Magmatism.* GEOL SOC AM MEM **174,** 99–109.

Mittlefeheldt, D. W. & Miller, C. F. 1983. Geochemistry of Sweetwater Wash pluton, California: implications for 'anomalous' trace element behavior during differentiation of felsic magmas. GEOCHIM COSMOCHIM ACTA **47,** 109–24.

Newton, R. C. & Haselton, H. T. 1981. Thermodynamics of the garnet-plagioclase-Al_2SiO_5-quartz geobarometer. *In* Newton, R. C., Navrotsky, A. & Wood, B. J. (eds) *Thermodynamics of minerals and melts,* 129–45. New York: Springer.

Newton, R. C. & Perkins, D. III, 1981. Thermodynamic calibration of geobarometers based on the assemblages garnet-plagioclase-orthopyroxene (clinopyroxene)-quartz. AM MINERAL **67,** 203–22.

Patiño Douce, A. E., Humphreys, E. D. & Johnston, A. D. 1990. Anatexis and metamorphism in the tectonically thickened continental crust exemplified by the Sevier hinterland, western North America. EARTH PLANET SCI LETT **97,** 290–315.

Patiño Douce, A. E. & Johnston, A. D. 1991. Phase equilibria and melt productivity in the pelitic system: implications for the origins of peraluminous granitoids and aluminous granulites. CONTRIB MINERAL PETROL **107,** 202–18.

Patterson, B. A. Stephens, W. E. & Herd, D. A. 1989. Zoning in granitoid accessory minerals as revealed by backscattered electron imagery. MINERAL MAG **53,** 55–62.

Qin, Z. 1991. Disequilibrium partial melting model and fractionation of highly incompatible trace elements during partial melting. GEOL SOC AM ABSTR PROG **23,** A45.

Rapp, R. P. 1990. *Vapor absent partial melting of amphibolite/eclogite at 8–32 kbar, implications for the origin and growth of continental crust.* Ph.D. Dissertation, Rensselaer Polytechnic Institute, Troy, NY, U.S.A.

Rapp, R. P., Ryerson, F. J. & Miller, C. F. 1987. Experimental evidence bearing on the stability of monazite during crustal anatexis. GEOPHYS RES LETT **14,** 307–10.

Silver, L. T., Anderson, C. A., Crittenden, M. & Robertson, J. M. 1977. Chronostratigraphic elements of the Precambrian rocks of the southwestern and far western United States. GEOL SOC AM ABSTR PROG **9,** 1176.

Staude, J. M. G. & Miller, C. F. 1987. Lower crustal xenoliths from a Tertiary composite dike, Piute Mountains, S.E. California. GEOL SOC AM ABSTR PROG **19,** 454.

Stone, P., Howard, K. A. & Hamilton, W. 1983. Correlation of metamorphosed Paleozoic strata of the southeastern Mojave Desert, California and Arizona. GEOL SOC AM BULL **99,** 1135–47.

Vavra, G., 1990. On the kinematics of zircon growth and its petrogenetic significance: a cathodoluminescence study. CONTRIB MINERAL PETROL **106,** 90–9.

Vielzeuf, D. & Holloway, J. R. 1988. Experimental determination of fluid-absent melting relations in the pelitic system. Consequences for crustal differentiation. CONTRIB MINERAL PETROL **98,** 257–76.

Watson, E. B. & Harrison, T. M. 1983. Zircon saturation revisited: temperature and compositional effects in a variety of crustal magma types. EARTH PLANET SCI LETT **64,** 295–304.

Wooden, J. L. & Miller, D. M. 1990. Chronologic and isotopic framework for Early Proterozoic crustal evolution in the eastern Mojave Desert region, southeastern California. J GEOPHYS RES **95,** 20133–46.

Wooden, J. L., Bennett, V. C., Hanchar, J. M. & Miller, C. F. 1990. Lower crustal xenoliths in Tertiary dykes, Old Woman Mountains area, California: II. Pb, Nd and Sr isotopic data. GEOL SOC AM ABSTR PROG **22,** 95.

Wooden, J. L., Stacey, J. S. & Howard, K. A. 1988. Pb isotopic evidence for the formation of Proterozoic crust in the southwestern United States. *In* Ernst, W. G. (ed.) *Metamorphism and crustal evolution of the western United States,* 69–86. Englewood Cliffs, N.J.: Prentice-Hall.

Wright, J. E. & Wooden, J. L. 1991. New Sr, Nd, and Pb isotopic data from plutons in the northern Great Basin: implications for crustal structure and granite petrogenesis in the hinterland of the Sevier thrust belt. GEOLOGY **19,** 457–60.

CALVIN F. MILLER, Department of Geology, Vanderbilt University, Nashville, TN 37235, U.S.A.
JOHN M. HANCHAR, Department of Geology, Vanderbilt University, Nashville, TN 37235, U.S.A.; present address: Department of Earth and Environmental Sciences, Rensselaer Polytechnic Institute, Troy, NY 12180-3590, U.S.A.
JOSEPH L. WOODEN, U.S. Geological Survey, Menlo Park, CA 94025, U.S.A.
VICTORIA C. BENNETT, Research School of Earth Sciences, Australian National University, Carberra, ACT 2601, Australia.
T. MARK HARRISON, Department of Earth and Space Sciences, University of California, Los Angeles, CA 90024, U.S.A.
DAVID A. WARK, Department of Earth and Environmental Sciences, Rensselaer Polytechnic Institute, Troy, NY 12180-3590, U.S.A.
DAVID A. FOSTER, Department of Geological Sciences, State University of New York, Albany, NY 12222, U.S.A.; present address: Victorian Institute of Earth and Planetary Sciences and Department of Geology, LaTrobe University, Bundoora, Victoria 3083, Australia.

Transactions of the Royal Society of Edinburgh: Earth Sciences, **83,** 63–69, 1992

Evidence for ascent of differentiated liquids in a silicic magma chamber found in a granitic pluton

Gail A. Mahood and Paula C. Cornejo

ABSTRACT: Fluid dynamic modelling of crystallising calc-alkalic magma bodies has predicted that differentiated liquids will ascend as boundary layers and that accumulation of these buoyant liquids near chamber roofs will result in compositionally stratified magma chambers. This paper reports physical features in La Gloria Pluton that can be interpreted as trapped ascending differentiated liquids. Leucogranitic layers decimetres thick, which are locally stratified, are trapped beneath overhanging wall contacts. The same felsic magmas were also preserved where they were injected into the wall rocks as dykes and as large sill complexes. These rocks do not represent differentiated magmas produced by crystallisation along the exposed walls because the felsic layers occur *at* the wall rock contact, not inboard of it. Rather, we speculate that evolved felsic liquids are generated by crystallisation all across the deep levels of chambers and that initial melt segregation occurs by flowage of melt into tension fractures. Melt bodies so formed may be large enough to have significant ascent velocities as diapirs and/or dykes. The other way in which the leucogranite occurrence is at variance with the convective fractionation model is that the ascending liquids did not feed a highly differentiated cap to the chamber, as the composition at the roof, although the most felsic in this vertically and concentrically zoned pluton, is considerably more mafic than the trapped leucogranitic liquids. This suggests that these evolved liquids were usually mixed back into the main body of the chamber. Backmixing may be general in continental-margin calc-alkalic magmatic systems, which, in contrast to those in intracontinental settings, rarely produce volcanic rocks more silicic than rhyodacite. That the highly differentiated liquids are preserved at all at La Gloria is a result of the unusual stepped nature of the contact and the entirely passive mode of emplacement of the pluton, which, in contrast to ballooning in place, does not result in wall zones being "scoured".

KEY WORDS: convective fractionation, differentiation mechanisms, gash veins, leucogranite, melt segregation, rhyolites, zoned plutons.

Models for differentiation within felsic to intermediate magma chambers have been strongly influenced by recent studies applying theoretical and experimental fluid dynamics to the effects of crystallisation on convection and the development of compositional gradients within magma chambers. These studies have echoed Bowen's assertion (1928) that crystal settling may not be the dominant way in which fractional crystallisation occurs (e.g. Shaw 1965; Campbell 1978; McBirney & Noyes 1979; Hildreth 1981). It is the magma at the lower levels of chambers that is likely to be crystallising rather than magma near the roof because the temperature along the adiabat for convecting magma decreases less with decreasing pressure than does the temperature of the liquidus (e.g. McBirney & Noyes 1979). Development of H_2O-rich zones near the chamber margins (e.g. Shaw 1974) may suppress crystallisation there. These suggestions are consistent with the observation that most zoned ignimbrites record chambers that had greater concentrations of crystals at deeper levels (Hildreth 1981). Thus one expects the earliest stages of crystallisation of a granitic chamber to occur not at the roof or upper walls but at deeper levels (although somewhat above the zone in direct physical contact with new magma inputs from lower crustal melting and homogenisation zones).

Several groups of researchers doing theoretical calculations and "tank experiments", in which aqueous solutions are cooled externally to produce crystallisation along tank walls (e.g. McBirney & Noyes 1979; Chen & Turner 1980; Turner & Gustafson 1981; McBirney *et al.* 1985; Huppert *et al.* 1986; Turner & Campbell 1986), developed a general model for the differentiation of felsic to intermediate magma within crustal magma chambers (e.g. Sparks *et al.* 1985; Nilson *et al.* 1985; Turner & Campbell 1986). The resulting convective fractionation model can be summarised as follows: crystallisation along the walls of the magma chamber removes dense components from the liquid, leaving a residual liquid less dense than either the crystallising mush at the wall or the adjacent unaffected magma. This differentiated liquid rises inboard of the mush and, when it reaches the roof zone, displaces denser magma there. Successive ascending liquids enter the progressively zoned column at the level appropriate to their density. Given the right thermal gradients, convection in the zoned column can occur in discrete layers, resulting in a chamber with step-function compositional gradients.

The tank experiments are fascinating, and they have had great heuristic value in stimulating our imagination about magma chamber processes. The quantitative application of experiments on simple aqueous solutions to magmas, however, is not yet possible, and it can be difficult to draw even qualitative conclusions due to the problems of dynamical similitude (see discussions in McBirney 1985;

Spera *et al.* 1986; Sparks & Huppert 1987). Application to felsic magmas is even more problematic than it is for basaltic systems due to their greater viscosity, crystal content, and non-Newtonian behaviour. Moreover, Spera *et al.* (1984) have argued that upward marginal flow is unlikely in silica-rich magmas due to their large Lewis number (i.e. the dominance of thermal over chemical diffusion in the boundary layer) that prevents a chemical boundary layer countercurrent to the thermally driven downwelling. Because computer calculations and experiments in tanks can be produced much more quickly than detailed studies of zoned plutons or ignimbrites, both the theoretical calculations and experiments, as well as the models based on them, have far outrun our observational data base. As a result, these models are only weakly constrained by field evidence and remain largely untested in silicic and intermediate systems. Some corroborative field data demonstrating step-function zoning has been offered in support of the convective fractionation model (e.g. Fridrich & Mahood 1987; Boden 1989), but direct physical evidence for the processes predicted by the convective fractionation model has not been reported for silicic intrusions.

This paper addresses the following questions stimulated by the convective fractionation model: (1) Is there evidence for boundary layers of felsic liquid ascending along the margins of magma chambers, either as preserved boundary layers or as flow layering and compaction features that can be inferred to have formed in accumulating crystals due to their passage? (2) If such boundary layers can be identified, can it be demonstrated that they channelled differentiated magma to the roofs of chambers? We present relevant field and petrological data on occurrences of leucogranite in La Gloria Pluton, a spectacularly exposed zoned intrusion with more than 2·5 km of vertical relief in the southern Andes. Although confident that these leucogranites represent highly differentiated liquids that were ascending along the margins of the chamber due to their buoyancy, we discuss field and petrological relationships that are not entirely consistent with the convective fractionation model as presently formulated.

1. Regional geological setting of La Gloria Pluton

La Gloria Pluton is located in the foothills of the southern Andes about 40 km E of Santiago, Chile (Fig. 1), within a belt of mid-Miocene granodiorite and quartz monzonite stocks and plutons (Drake *et al.* 1982). Outcrops of La Gloria Pluton define an elongate exposure 20 km long and 5 km wide that is parallel to the N–S structural grain of a thick sequence of Mesozoic to Cainozoic volcanic arc and associated sedimentary formations folded and faulted in association with Tertiary uplift of the Andes (Thiele 1980). The host rock to La Gloria Pluton is the Abanico Formation, which, in the immediate vicinity of the pluton, consists of a massive section dominated by intermediate composition lavas, sills, and breccias, with minor interbedded volcaniclastic sandstones and siltstones.

2. Form of intrusion

Deep fluvial and glacial erosion has carved a rugged topography exposing the apical roof zone, sloping walls, sill complexes emanating from the pluton, and more than 2·5 km of vertical relief on the pluton over a short horizontal distance. The Rio Colorado slices E–W across the pluton,

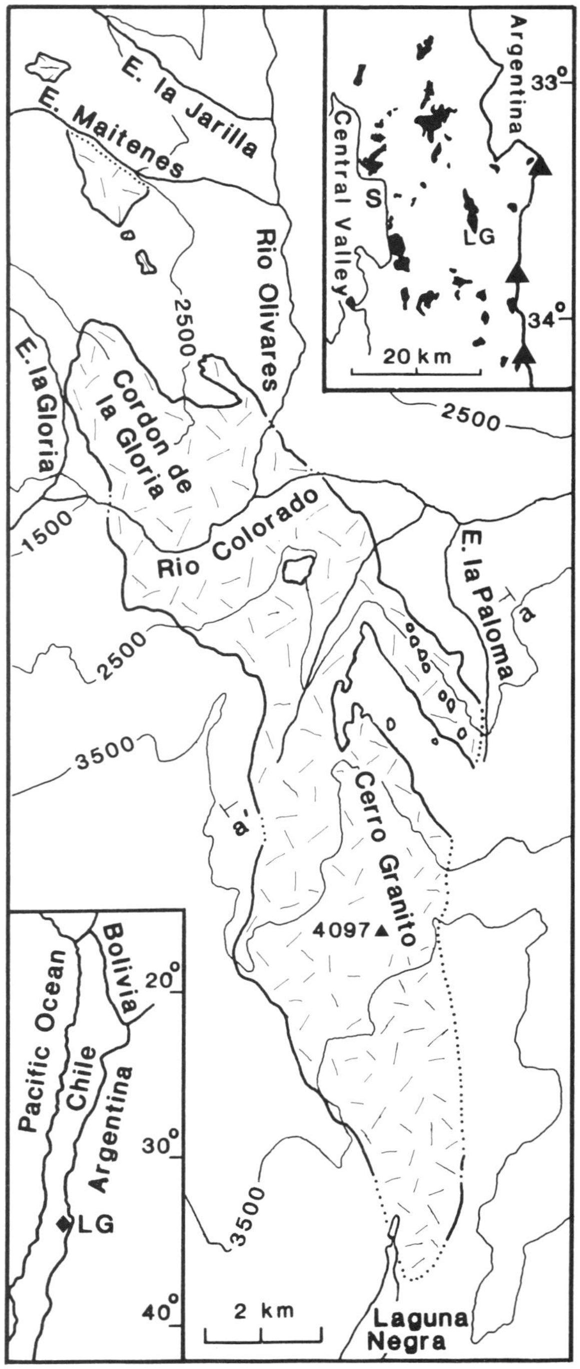

Figure 1 Location map of La Gloria Pluton: upper right inset shows Miocene plutons (solid black), including La Gloria Pluton (LG), relative to the central valley of Chile, the city of Santiago (S), and the crest of the Andes, marked by the political border and, in this segment, by three active stratovolcanoes (triangles); main map shows major drainages (E = Estero), topographic contours in metres, and outcrop pattern of the pluton (hachured); contacts are dotted where buried by Quaternary deposits, which have been left off the map for the sake of clarity but consist of river deposits, alluvium, landslides, debris flows, and till; wall rock (unpatterned) to the pluton consists of a massive section of Cretaceous-Tertiary arc rocks dominated by intermediate composition lavas, sills, and breccias, with minor interbedded volcaniclastic sandstones and siltstones and redbeds; intersection of sloping topography with a pluton-wall rock contact that is stepped on the scale of centimetres to hundreds of metres results in the irregular contact on the eastern side of the pluton on the slope W of Estero La Paloma, including tiny windows of pluton exposed through the large wall-rock step and scraps of wall rock remaining on the granite slope below.

providing a near-horizontal traverse at an elevation of about 1,500 m (Fig. 1). Wall contacts exposed from there to elevations of about 3,500 m define an overall shape of the pluton similar to an overturned canoe (Figs 1, 2). Projection of wall contacts suggests that the highest portion of the apical roof of the pluton lay no more than 100 m above the highest exposed granite on Cerro Granito ridge (elevation 4,097 m, Fig. 1). The apical roof dips away from this high point, diving beneath the surface at an elevation of about 3,200 m at both the northern and southern ends of exposure (Fig. 1).

The shallow depth of emplacement and the massive and rectilinear jointed character of the wall rocks were both important in controlling the style of intrusion of the pluton. The pluton was emplaced entirely passively (at least at the exposed level) as an isolated body at a shallow depth, about 4 km as estimated from the reconstructed thickness of the stratigraphic section (Aguirre 1960). The pluton made room for itself at the exposed level by stoping, and excellent exposures that catch this process "in the act" demonstrate how it proceeded. Magma, nearly always leucogranitic in composition, injected laterally along shallowly dipping bedding planes, and then short dyke sections formed along joint surfaces, releasing stope blocks into the magma body. Blocks in all stages of dislodgement are preserved, including free-floating blocks centimetres to decimetres in diameter that are angular near the contact but become progressively rounded within only metres from the contact. Stoping of jointed wall-rock blocks resulted in the stepped character of the walls and roof of the pluton. These steps occur on all scales, ranging in size from centimetres to hundreds of metres. Intersection of sloping topography with one especially large step results in the irregular contact on the eastern side of the pluton on the slope W of Estero la Paloma (Figs 1, 2), and smaller steps result in small windows of pluton exposed through the wall-rock step and scraps of wall rock remaining on the granite slope below (Fig. 1).

Because the Abanico Formation is gently folded, the bedding-plane "step" and joint-surface "risers" are not necessarily horizontal and vertical, respectively. On the eastern side of the pluton S of the Rio Colorado, where the steps are most easily observed due to less formidable terrain, the Abanico Formation dips about 50° NW. As a result, the giant step in the eastern contact of the pluton dips toward the body, forming overhanging contacts.

The presence of the steps on both margins of the pluton and the occurrence of steps down to the centimetre scale argue that the stepped nature of the contact is intrinsic to the intrusion process and is not a result of post-solidification faulting. Additional evidence that the pluton margins have

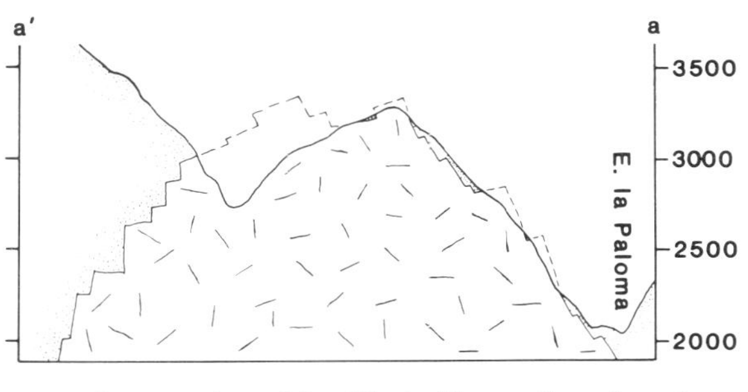

Figure 2 Cross section of La Gloria Pluton: line of section a-a′ shown on Figure 1; extrapolation of contacts above and below ground is based on additional sections to the N and S; general shape of the pluton (hachured) is similar to an over-turned canoe, but stoping of massively bedded and jointed volcaniclastic wall rocks (dotted) has led to a stepped contact in detail.

not been faulted is that sills and dykes can be traced continuously from the pluton into the wall rocks without displacement.

3. Compositional and mineralogical zoning of the pluton

La Gloria Pluton is zoned vertically and horizontally in bulk composition, mineral modes (Table 1), and textures. The core zone exposed at the lowest level in the Rio Colorado is an equigranular, medium-grained, hornblende-biotite granodiorite/quartz monzodiorite that contains abundant mafic enclaves and locally exhibits plagioclase-clinopyroxene cumulate textures. There is a systematic decrease in the abundance of mafic silicates as a function of elevation, as well as an increase in the K-feldspar/plagioclase ratio and in Fe-Ti oxides and amphibole as proportions of the mafic phases. It grades through an inequigranular, coarse-grained, quartz monzodiorite at the middle levels of the pluton to a quartz monzonite with large euhedral crystals of K-feldspar and hornblende at the highest exposures on Cerro Granito ridge. The borders of the pluton are inequigranular at all levels. The euhedral nature and larger size of the mafic phases at the borders and near the roof suggest that magmatic water was concentrated there. The f_{HF}, f_{HCl}, and f_{HF}/f_{HCl}, as determined by analysis of biotite and amphibole, also increased toward the roof zone (Cornejo & Mahood (submitted)).

4. Leucogranites at stepped contacts

The most felsic rocks in the pluton are leucogranites, which are commonly found at the margin of the pluton. They range from aplitic rocks with just a trace of biotite to medium-grained, equigranular rocks with sparse biotite, amphibole, and sphene. Locally, biotite- and K-feldspar-rich pegmatite a few centimetres thick is present within a few tens of centimetres of the wall-rock contacts in association with leucogranite.

Leucogranite occurs as centimetre-scale selvages along upright contacts and as thicker layers beneath steps, as well

Table 1 Average compositions and modes of La Gloria Pluton.

	Level of pluton		Leucogranites	
	Lower	Upper	Transitional	Evolved
SiO_2 (wt. %)	65·3	66·5	77·3	77·5
TiO_2	0·57	0·56	0·20	0·14
Al_2O_3	16·3	16·1	12·9	12·3
Fe_2O_3t	4·41	4·38	0·74	0·65
MnO	0·07	0·07	0·01	0·01
MgO	1·85	1·25	0·13	0·09
CaO	3·80	2·47	1·58	0·70
Na_2O	4·20	4·12	3·51	2·59
K_2O	3·28	4·51	3·61	5·95
P_2O_5	0·15	0·14	0·04	0·03
Rb (ppm)	86	147	91	155
Sr	503	332	411	139
Zr	135	274	106	83
Quartz (model %)	15·6	17·8	33·8	34·7
K-fsp	16·4	35·6	48·7	47·7
Plagioclase	52·0	37·7	15·3	15·5
Amphibole	10·0	6·1	0·0	0·0
Biotite	3·0	0·4	1·0	1·5
Titanite	0·4	0·1	0·0	0·0
Mt + Ilm	1·4	1·8	0·8	0·3
Zr + Ap	0·6	0·3	0·1	0·0

Note: Major-element analyses normalised to 100%, volatile-free.

as in sill and dyke complexes that extend from these occurrences out into the wallrocks. Leucogranite is injected into the country rocks as sills following bedding planes and less commonly as dykes that extend from the "risers" of the steps and parallel the main margin of the pluton (Fig. 3). These sills and dykes range from centimetres to metres thick and, although they typically extend into the wall rocks only a few metres, a few large sills extend for hundreds of metres N of the Rio Colorado on the eastern side of the pluton. Leucogranite dykes cut the pluton locally within tens of metres of its borders but are very rare in the core of the pluton.

It is the occurrence of leucogranite layers beneath overhanging steps in the contacts that is the main focus of this paper. Although present at a number of localities along the contact, the best examples are in Estero La Gloria and especially W of Estero La Paloma (Fig. 1). In Estero La Gloria, the gently folded Abanico Formation dips eastward, and near the head of Estero La Paloma it dips westward; thus in both localities, steps in the contacts produced by the stoping of massive, hornfelsed wall rocks have led to steps that dip toward the centre of the pluton, producing overhanging contacts. In these two localities, nearly every observed jog or step in the wall contact is underlain by 10–70 cm of leucogranite. These layers of leucogranite always have approximately horizontal contacts with the body of quartz monzodiorite below. As a result, many of the leucogranite occurrences are as isolated layers ponded beneath steps in the contacts (Figs 3, 4). Leucogranite is only rarely present along the "risers" of the steps, where it is invariably thinner than beneath adjacent steps. Usually, along the more steeply inclined parts of the stepped contact and along subvertical walls there is no evidence of such felsic layers; rather, grain size and modal proportions of the mafic minerals commonly increase toward the walls.

Locally beneath the steps, the leucogranite itself is layered and modally and compositionally zoned (Fig. 3). In one occurrence just W of the head of Estero La Paloma, a nearly mafic-free aplite occurs as a 20-cm layer beneath a step in the contact, and a 40-cm leucogranite layer lies directly beneath, in sharp (but not cross-cutting) contact with both the layer above and the granite below (Fig. 4). The granite beneath these layers is more leucocratic than usual and grades into the normal quartz monzodiorite over two vertical metres. As the contact is approached, the proportion of K-feldspar increases and the proportion of mafics, especially amphibole, decreases drastically.

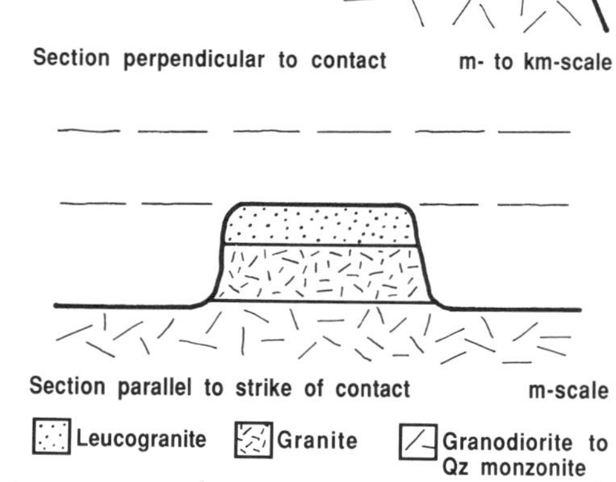

Figure 3 Schematic diagrams illustrating the occurrences of leucogranitic layers.

5. Interpretation of occurrence of leucogranites

The most striking thing about the association of the fine-grained leucogranite with the wall contacts is that there is strong evidence that leucogranitic liquid rose along the walls of the chamber in a fashion analogous to that predicted for fractionated boundary layers. The evidence for this is the disposition of the leucogranite along the stepped wall-rock contact. Along "risers" it is usually absent; medium-grained quartz monzodiorite is in contact with the wall rock. It is also absent where, as a consequence of the sense of dip of the wall rocks, the steps dip in the same direction as the contact overall. Leucogranite is nearly always present wherever a step dips toward the centre of the pluton; the exceptions are very small steps (less than a few decimetres). These relationships are interpreted as indicating that ascending layers of low-density, differentiated magma were trapped beneath overhangs in the chamber wall, but where the walls were vertical or where the steps dipped in the same direction as the overall contact, they rose unimpeded. Once ponded beneath an overhang, low flow rates in this stagnant zone would tend to preserve the differentiated liquids there.

Additional evidence that these liquids were trapped beneath overhanging contacts is the presence of accumulations of small, angular, stoped blocks and of more rounded enclaves (which may have originated as stoped wall rock) in these "backwater" locales. Evidence that the evolved melts were emplaced buoyantly is that, locally, the leucogranite itself is layered internally, with the most felsic material at the top and with near-horizontal contacts between layers in slight discordance to the inward-sloping wall-rock step (Fig. 3). The observation that contacts are sharp, but not cross-cutting, between leucogranite layers and between them and the main body of the granite is important, because

Figure 4 Photograph of internally stratified felsic layer trapped beneath wall-rock overhang, from a locality just W of the head of Estero La Paloma: wall-rock step dips gently back and to the right; uppermost, nearly mafic-free true granite is 20 cm thick and is in abrupt but not cross-cutting contact with the underlying layer of mafic-poor granite; beneath these two layers the granite is more leucocratic than usual and over a vertical distance of 2 m grades continuously to normal quartz monzodiorite.

it indicates that differentiated liquid was rising while the adjacent parts of the chamber still contained a significant fraction of melt. This means that the leucocratic layers are not simply very late-stage, cross-cutting features; rather, the chemistry and occurrence of the leucogranites has some bearing on general differentiation processes within the chamber.

The observation at one locality W of Estero La Paloma that the volcaniclastic sediments of the wall rock had been partially melted near a stepped contact (although the melt had not segregated), led to consideration of an alternative explanation for the leucogranite layers: that the leucogranites were melts of stoped blocks of wall rock, which subsequently rose through the chamber. This possibility was tested as part of a larger isotopic study of the pluton (Mahood & Halliday, unpublished data) by analysing the Sr and Nd isotopic compositions of samples from the body of the pluton, from the wall rocks, and from the most evolved of the leucogranite layers. The $^{143}Nd/^{144}Nd$ ratio of five samples of the body of the pluton are essentially within analytical error of 0·51281 (0·512799–0·512817) and are distinctly different from values of 0·512865 and 0·512871 for wall-rock samples, one each on the western and eastern sides of the pluton, the latter collected from near the leucogranite sample. The leucogranite layer, at 0·512807, matches the pluton. This isotopic evidence indicates that the evolved liquids originate by differentiation of the La Gloria magma body and not by melting of wall rocks.

6. Comparison with theoretical and experimental models

The field relations at La Gloria Pluton can be used to place crude constraints on the size of the rising layers, in order to compare them with the size of boundary layers predicted by the convective fractionation model. Scaling from simple experiments to predict the size of boundary layers in natural systems is difficult, but models suggest that they would be centimetres to tens of centimetres wide (Shaw 1974; Brandeis *et al.* 1984; Nilson *et al.* 1985). The observation that at La Gloria leucogranite is not present under short steps, which have only a few centimetres of vertical relief, suggests that the ascending layers were at least decimetres in width, as one would expect layers only a centimetre or so thick to be brought to a stop beneath overhangs centimetres high. Because the rise of the leucomelts is driven by buoyancy, the rate of ascent should be greater for thicker layers. As a result, layers that greatly exceed in thickness the local relief on the walls may rise sufficiently rapidly for liquid not to be trapped. The decimetre to metre sizes of sills and dykes of leucogranite that extend from the leucogranitic layers into the wall rock also suggest that the ascending layers were at least decimetres in thickness. The reasoning here is that homogeneous dykes and sills could not be produced by the successive propagation of small batches of magma that had accumulated beneath the overhangs; they must have been supplied by larger batches of magma that were rising along the wall at the time of sill/dyke emplacement. On the other hand, it seems unlikely that the ascending currents were much larger than metre-sized because, if they had been tens to hundreds of metres, they could have entirely filled huge steps in the contact (e.g. W of Estero La Paloma), and, by virtue of their large size and buoyancy, such layers would not readily have been stirred back into the main body of the chamber. Instead, what is observed is leucogranite beneath the smaller-scale steps that decorate the large steps like that W of Estero La Paloma, where it has been protected from being mixed back into the body of the chamber.

Accepting that the leucogranites represent trapped, ascending, evolved liquids, the next question to ask is whether they have a spatial distribution and compositional relationship to the rest of the pluton as predicted by the convective fractionation model. The first way in which the La Gloria occurrence does not match the convective fractionation model is in the relationship of the composition of the leucogranites to that of the upper levels of the pluton. In the convective fractionation model, the uppermost level of a chamber contains evolved magma that accumulated from ascent of differentiated boundary layers, and it grades downward to progressively more-mafic compositions produced by boundary layers formed during earlier stages of fractionation of the chamber. If this model applies, one would expect the upper reaches of the pluton to resemble the leucogranites trapped beneath overhangs. At La Gloria, however, the roof-zone rock is not nearly this felsic, and there is a large gap in composition between it and the leucogranites (i.e. from 66 to 77% SiO_2, Table 1). This suggests that if evolved liquids were produced in any volume, they were mixed back into the body of the chamber and did not remain discrete liquids capable of accumulating in large volumes near the roof.

The second way in which the La Gloria leucogranite occurrences are at variance with the model is the location of the leucogranites with respect to the chamber walls. If the leucogranitic liquids were produced as boundary layers inboard of crystals growing inward from cool walls, as the model predicts, then at La Gloria this crystallisation must have occurred below exposed levels because the leucogranite is always *at* the wall contacts, rather than inboard of them. This suggests that at exposed levels the pluton did not crystallise inward from its margin, analogous to "border groups" in mafic intrusions.

In contrast to mafic intrusions, granitic plutons lack chilled margins; on the contrary, local coarsening of grain size toward pluton margins is a feature that has been noted generally for mesozonal plutons. This suggests that, as opposed to the convective fractionation model, silicic magma chambers do *not* crystallise from the walls inward. In large chambers, any chill zone that might have formed on initial intrusion can be wholly or partially resorbed due to the thermal energy of continuing magma inputs and is facilitated by the relatively low heats of fusion of granitic minerals. Concentration of H_2O and other volatiles at the walls (e.g. Cornejo & Mahood (submitted)) may suppress crystallisation there and, by increasing diffusivities, enhance crystal growth rates and result in larger crystals.

7. Speculation on the segregation and preservation of high-silica melts

If leucomelts are not produced as boundary layers adjacent to crystallising walls, where do they originate? We speculate that evolved felsic liquids are generated not at chamber margins but by crystallisation all across a chamber at its deep levels. Crystallisation would occur at depth first (McCarthy & Fripp 1978) because the adiabat for convecting magma is steeper than liquidi (e.g. McBirney & Noyes 1979). The establishment of roofward and marginward enrichments in volatiles would enhance this effect.

Numerous workers have emphasised the difficulty of segregating silicic magmas from crystal-liquid mushes and have postulated that liquids can segregate and coalesce only when the volume of melt is so large that continuity of the crystalline mush breaks down at about 20–40% melt (see Wickham (1987) for a recent discussion). This is based on the fact that gravity-driven compaction of crystals is too slow a process to effectively liberate viscous, felsic melt from low-melt-fraction mush over the lifespans of typical magmatic systems. This conclusion, however, presents a problem for the generation of the leucogranites observed at La Gloria, as they are much too low in Fe and Mg to be products of only moderate amounts of fractionation. Moreover, the conclusion that small melt fractions cannot be extracted seems to be at variance with the existence of large volumes of high-silica rhyolite. This seems especially so for cases such as Yellowstone, where thousands of cubic kilometres are produced, or rhyolite dome fields like Glass Mountain, California, in which high-silica rhyolites with <1 ppm Sr require extreme fractionation of hundreds or thousands of cubic kilometres of felsic magma (Mahood 1990). It would seem that these rhyolites are *prima facie* evidence that segregation of small melt fractions occurs. So, too, are the leucogranitic dykes that commonly cut mesozonal granite plutons.

It is the occurrence of these leucocratic dykes and related gash veins filled with aplite that suggests a solution to the apparent dilemma of extraction of small melt fractions. Perhaps the operative process for initial melt segregation is not compaction of crystals, but formation of tension gashes into which highly evolved melt is expressed. Because dilatant fractures can form in a crystal-liquid mush only when there is a contiguous crystal network that can sustain brittle fracture, i.e. at crystallinites >70–80% (van der Molen & Paterson 1979), this separation process *favours* production of small melt fractions, in contrast to crystal compaction, which is biased toward separation of large melt fractions. Dilatant fractures would be expected to occur under many conditions, because, for a magma at magmastatic equilibrium, any imposed stress is likely to produce deviatoric stresses that somewhere in the body will produce extensional strains. Processes that might result in these strains are eruptive drawdown, stoping, a fresh input of magma, faulting, vapour saturation (in shallow-level plutons), or even the volume decrease on crystallisation; in other words, a wide range of phenomena that would provide multiple opportunities over the lifespan of a magmatic system for the segregation of evolved melts.

Crystal-liquid mushes might be expected to behave brittlely on short time-scales, but deform ductilely over the larger life of a chamber. This property would facilitate separation of evolved liquids. Melts that flow into low-pressure zones of tension fractures produced by movements of the mush may form bodies large enough to have significant ascent velocities as diapirs (or dykes, depending on the viscosity contrast between mush and melt) during the intervening periods when the host magma deforms ductilely. Such diapirs could be expected to survive the low-intensity convection that would characterise silicic magmas (H. Huppert, pers. comm. 1991). The high viscosities of the magmas would suppress, though certainly not entirely prevent, mixing between the diapiric evolved melt and the host magma in its path. Silicic diapirs that encountered the chamber margins before losing their identity through mixing would be constrained to rise along the sloping walls, where evidence of their ascent could be preserved beneath steps in the wall contacts. It would be preserved because the more mafic and therefore denser magma of the body of the chamber could not displace it beneath overhangs. Only a more evolved melt could displace trapped differentiated liquid, by rising buoyantly to the highest point, displacing the resident liquid downward.

It was noted above that at La Gloria ascending evolved melts did not collect at the roof zone to form a highly differentiated cap to the chamber, as would be predicted by the convective fractionation model; the composition at the roof is considerably more mafic than the trapped leucogranites (Table 1). It is unlikely that evolved melts that might have accumulated near the roof were lost due to eruption, as the grainsize of the pluton actually increases slightly roofward, thereby showing no evidence of pressure quenching of the magma that now occupies the former chamber as solid rock (although not precluding venting of the chamber at some earlier stage of its history). This suggests that the extremely differentiated liquids represented by the leucogranites were usually mixed back into the main body of the chamber. This may be the fate of most of the ascending diapirs: progressive mixing with the granitic magma in its path until it becomes so like the surrounding magma in density that its buoyant rise ceases. Only those diapirs that ascend just a short distance before encountering the sloping walls of the chamber are likely to retain their highly evolved compositions intact.

This backmixing process may be general in continental-margin calc-alkalic magmatic systems, which rarely produce volcanic rocks (i.e. crystal-poor magmas) more silicic than rhyodacite. That the highly differentiated liquids are preserved at all at La Gloria is a result of (1) the unusual stepped nature of the pluton-wall rock contact providing overhanging havens for buoyant highly evolved melts, (2) the entirely passive mode of emplacement of the pluton, which, in contrast to a pluton that balloons in place, does not result in the wall zones being "scoured", and (3) the pluton being a simple, isolated body that has not had its margins disturbed by the emplacement of multiple magma pulses or other nearby plutons.

In contrast to continental-margin arcs, magmatic systems in intracontinental extensional settings commonly produce significant volumes of rhyolite and high-silica rhyolite. Perhaps they are more common here because the extensional environment provides more opportunities for formation of dilatant cracks in crystal-liquid mushes and consequent separation of highly evolved melts, as well as more spaces into which these liquids can escape, accumulate, and be preserved as discrete melts.

8. Acknowledgements

Supported by NSF Grants EAR-8805074 and INT-8822573. GAM thanks Wes Hildreth for recognising La Gloria Pluton (while hiking through it to reach the active volcanoes) as a place where it might be possible to scamper around on pluton wall and roof contacts, and dedicates this paper to Bob Drake for saving her from a potentially fatal fall on La Gloria when, besotted with hypoxia and fancy exposures, she did something stupid. We appreciate a helpful review by C. R. Bacon.

9. References

Aguirre, L. 1960. Geologia de los Andes de Chile Central, provincia de Aconcagua. INST INVEST GEOL CHILE **9**.

Boden, D. R. 1989. Evidence for step-function zoning of magma and eruptive dynamics, Toquima caldera complex, Nevada. J VOLCANOL GEOTHERM RES **37**, 39–57.

Bowen, N. L. 1928. *The evolution of the igneous rocks* New York: Dover.

Brandeis, G., Jaupart, C. & Allegre, C. J. 1984. Nucleation, crystal growth and the thermal regime of cooling magmas. J GEOPHYS RES **89,** 10, 161–77.

Campbell, I. H. 1978. Some problems with cumulus theory. LITHOS **11,** 311–23.

Chen, C. F. & Turner, J. S. 1980. Crystallization in a double-diffusive system. J GEOPHYS RES **85,** 2573–93.

Cornejo, P. C. 1990. *Geology, mineral compositions, and magmatic gradients of a zoned pluton: La Gloria pluton, central Chilean Andes.* M.S. thesis, Stanford University, Stanford.

Cornejo, P. C. & Mahood, G. A. (submitted). Seeing past the effects of re-equilibration to reconstruct magmatic gradients in plutons: La Gloria pluton, central Chilean Andes. CONTRIB MINERAL PETROL (submitted).

Drake, R. E., Vergara, M., Munizaga, F. & Vicente, J. C. 1982. Geochronology of Mesozoic-Cenozoic magmatism in Central Chile, Lat. 31°–36° S. EARTH SCI REV **18,** 353–63.

Fridrich, C. J. & Mahood, G. A. 1987. Compositional layers in the zoned magma chamber of the Grizzly Peak Tuff. GEOLOGY **15,** 299–303.

Helz, R. T. 1987. Differentiation behavior of Kilauea Iki lava lake, Kilauea Volcano, Hawaii: An overview of past and current work. *In* Mysen, B. O. (ed.) *Magmatic processes: physicochemical principles.* GEOCHEM SOC SPEC PUB **1,** 241–58.

Hibbard, M. J. & Watters, R. J. 1985. Fracturing and diking in incompletely crystallized granitic plutons. LITHOS **18,** 1–12.

Hildreth, W. 1981. Gradients in silicic magma chambers: implications for lithospheric magmatism. J GEOPHYS RES **86,** 10,153–92.

Huppert, H. E., Sparks, R. S. J., Wilson, J. R. & Hallworth, M. A. 1986. Cooling and crystallization at an inclined plane. EARTH PLANET SCILETT **79,** 319–28.

Mahood, G. A. 1990. A second reply to "Comments by R. S. J. Sparks, H. E. Huppert, and C. J. N. Wilson, on 'Evidence for long residence times of rhyolitic magma in the Long Valley magmatic system: the isotopic record in the precaldera lavas of Glass Mountain:". EARTH PLANET SCI LETT **99,** 395–9.

McBirney, A. R. 1985. Further considerations of double-diffusive stratification and layering in the Skaergaard intrusion. J PETROL **26,** 993–1001.

McBirney, A. R., Baker, B. H. & Nilson, R. H. 1985. Liquid fractionation. Part I: Basic principles and experimental simulations. J VOLCANOL GEOTHERM RES **24,** 1–24.

McBirney, A. R. & Noyes, R. M. 1979. Crystallization and layering of the Skaergaard intrusion. J PETROL **20,** 487–554.

McCarthy, T. S. & Fripp, R. E. P. 1978. The crystallization history of a granitic magma, as revealed by trace element abundances. J GEOL **88,** 211–4.

Nilson, R. B., McBirney, A. R. & Baker, B. H. 1985. Liquid fractionation. Part II: Fluid dynamics and quantitative implications for magmatic systems. J VOLCANOL GEOTHERM RES **24,** 25–54.

Shaw, H. R. 1965. Comments on viscosity, crystal settling and convection in granitic magmas. AMER J SCI **264,** 120–52.

Shaw, H. R. 1974. Diffusion of H_2O in granitic liquids: Part I. Experimental data; Part II. Mass transfer in magma chambers. *In* Hofmann, A. W., Giletti, B. J., Yoder, H. S., Jr. & Yund, R. A. (eds.) *Geochemical Transport and Kinetics,* **139–70.** Washington: Carnegie Institute.

Sparks, R. S. J., Huppert, H. E. & Turner, J. S. 1985. The fluid dynamics of evolving magma chambers. PHILOS TRANS R SOC LONDON **A310,** 511–34.

Sparks, R. S. J. & Huppert, H. E. 1987. Laboratory experiments with aqueous solutions modelling magma chamber processes. I. Discussion of their validity and gelogical application. *In* Parsons, I. (ed.) *Origins of Igneous Layering,* 527–38. Dordrecht: D. Reidel.

Spera, F. J., Yuen, D. A. & Kemp, D. V. 1984. Mass transfer rates along vertical walls in magma chambers and marginal upwelling. NATURE **310,** 764–7.

Spera, F. J., Yuen, D. A., Clark, S. & Hong, H. J. 1986. Double-diffusive convection in magma chambers: single or multiple layers. GEOPHYS RES LETT **13,** 153–6.

Thiele, R. 1980. Hoja Santiago, Region Metropolitana. INST INVEST GEOL CHILE 1:250,000 CARTA **39.**

Turner, J. S. & Campbell, I. H. 1986. Convection and mixing in magma chambers. EARTH SCI REV **23,** 255–52.

Turner, J. S. & Gustafson, L. B. 1981. Fluid motions and compositional gradients produced by crystallization or melting at vertical boundaries. J VOLCANOL GEOTHERM RES **11,** 93–125.

van der Molen, I. & Paterson, M. S. 1979. Experimental deformation of partially melted granite. CONTRIB MINERAL PETROL **70,** 299–318.

Wickham, S. M. 1987. The segregation and emplacement of granitic magmas. J GEOL SOC LONDON **A144,** 281–97.

GAIL A. MAHOOD and PAULA C. CORNEJO, Department of Geology, Stanford University, Stanford, CA 94305, U.S.A.

Transactions of the Royal Society of Edinburgh: Earth Sciences, **83,** 71–81, 1992

Granites and rhyolites from the northwestern U.S.A.: temporal variation in magmatic processes and relations to tectonic setting

Marc D. Norman, William P. Leeman and Stanley A. Mertzman

ABSTRACT: Cretaceous and Cainozoic granites and rhyolites in the northwestern U.S.A. provide a record of silicic magmatism related to diverse tectonic settings and large-scale variations in crustal structure. The Late Cretaceous Idaho Batholith is a tonalitic to granitic Cordilleran batholith that was produced during plate convergence. Rocks of the batholith tend to be sodic ($Na_2O > K_2O$), with fractionated HREE, negligible Eu anomalies, and high Sr contents, suggesting their generation from relatively mafic sources at a depth sufficient to stabilise garnet. In contrast, Neogene rhyolites of the Snake River Plain, which erupted in an extensional environment, are potassic ($K_2O > Na_2O$), with unfractionated HREE patterns, negative Eu anomalies, and low Sr contents, suggesting a shallower, more feldspathic source with abundant plagioclase. Eocene age volcanic and plutonic rocks have compositions transitional between those of the Cretaceous batholith and the Neogene rhyolites. These data are consistent with a progressively shallowing locus of silicic magma generation as the tectonic regime changed from convergence to extension.

KEY WORDS: granite, rhyolite, Idaho Batholith, Snake River Plain, Cretaceous, Cainozoic.

Silicic magmatism in the Pacific Northwest provides a record of crustal evolution spanning the last ~120 Ma. During this time, igneous activity was not continuous, but occurred in three major episodes (Fig. 1) that correspond to fundamentally different tectonic settings. Magmatism also spanned major crustal discontinuities, including boundaries between cratonal and accreted terranes. Silicic magmatism in this region therefore provides an opportunity to investigate the relations between magma compositions, igneous processes, tectonic settings, and crustal structures. This paper presents a comparison of the compositions of silicic magmatism associated with each tectonomagmatic episode, focusing on how the processes of magma generation within the crust changed through time, and possible relations to the tectonic environment.

The regional tectonic setting of the northwestern U.S.A. during Cretaceous and Cainozoic time has changed from predominantly related to plate convergence, to predominantly related to continental extension and rifting (Christiansen & Lipman 1972; Christiansen & McKee 1978). Late Palaeozoic and Mesozoic convergence along the western margin of N America produced an extensive Andean-style continental volcanic arc, the roots of which are exposed as a belt of tonalitic to granitic Cordilleran batholiths from southern Mexico to Alaska (Coney 1978; Hamilton 1988). The Late Cretaceous Idaho Batholith is one of these Cordilleran batholiths.

Latest Cretaceous to middle Tertiary time encompasses a transitional tectonic setting that is not well understood, but is certainly related to an evolving continental margin. A magmatic quiescence occurred during latest Cretaceous and Paleocene time as the Laramide orogeny built the Rocky Mountains E of the formerly active volcanic arc. There is a close correspondence in time between the Laramide contraction and a doubling in the rate of convergence between N America and the oceanic Farallon plate to the W (Rea & Duncan 1986). Rapid convergence may have caused a shallowing of the subducted plate which shut down the volcanic arc (Dickinson *et al.* 1988).

The Laramide orogeny was followed by an outbreak of magmatism across the region. Widespread igneous activity ~55–40 Ma emplaced mafic, intermediate and silicic volcanic and plutonic rocks in a broad belt across the Pacific Northwest and Canadian Cordillera (Armstrong & Ward 1991). Of particular interest here is the Challis magmatic field, which consists of Eocene volcanic and plutonic rocks exposed on the flanks of the Idaho Batholith (McIntyre *et al.* 1982; Lewis & Kiilsgaard 1991; Norman & Mertzman 1991). Normal and strike-slip faulting coincident with Eocene magmatism demonstrates an incipient extensional environment at this time, possibly related to a slowing of the rate of plate convergence and plate reorganisation (Bennett 1986; Norman & Leeman 1989; Carlson *et al.* 1991). The degree to which active subduction-related magma systems contributed to the Eocene activity is unclear (Dudas 1991; O'Brien *et al.* 1991; Norman & Mertzman 1991; Suayah & Rogers 1991).

Magmatism again declined across most of the northwestern U.S.A. during late Eocene to middle Miocene time as the volcanism migrated westward and southward. The principal activity in the northwestern U.S.A. during this time was the establishment of the Cascades volcanic arc along the western margin of N America (Armstrong & Ward 1991). Small volumes of basalt and andesite erupted in Montana, Idaho and Oregon, but silicic activity was rare. In contrast, large volumes of ignimbrite erupted at this time in the Great Basin and farther to the SW. Deformation during this time was principally extension along low-angle detachment faults, manifest in several metamorphic core complexes (Miller 1983; Wust, 1986a, b).

Intense volcanism in the northwestern U.S.A. resumed in the middle Miocene as Late Cainozoic Basin and Range-style extension and block faulting became established. Magmatism was "fundamentally basaltic", but included a large component of rhyolitic lavas and ash flow tuffs in a bimodal association (Christiansen & Lipman 1972; Christiansen & McKee 1978). Eruption of tholeiitic flood basalts from fissures in Washington, Oregon, and western Idaho began ~17–18 Ma (Hooper *et al.* 1984; Carlson & Hart 1987). Shortly thereafter, large silicic eruptive centres became established in southern Oregon, Idaho and northern Nevada, producing large volumes of rhyolite ash-flow tuffs and lava (Ekren *et al.* 1984; Bonnichsen & Kaufman 1987; Rytuba & McKee 1984). Similar eruptive centres propagated across the region along two major axes: one which trends NE–SW across southern Idaho (the Snake River Plain–Yellowstone Plateau) and a second more diffuse NW–SE trend across eastern Oregon.

A major crustal boundary in the region juxtaposes accreted terranes of Palaeozoic and Mesozoic oceanic sediments and volcanic arc assemblages to the W, against Precambrian metasedimentary and high-grade cratonic rocks to the E (Vallier & Brooks 1987). Assembly of these terranes with N America appears to have occurred during early Cretaceous time, before the main phase of Idaho Batholith intrusive activity. The location of this boundary between the craton and accreted terranes is reflected in regional compositional variations of Cretaceous and Cainozoic mafic and silicic rocks (Zartman 1974; Armstrong *et al.* 1977; Leeman 1982; Farmer & DePaolo 1983; Hart 1985; Fleck & Criss 1985; Menzies 1989; Fleck 1990). For example, igneous rocks W of this boundary, i.e. those erupted through the accreted terranes, have isotopic compositions similar to young oceanic or volcanic arc basalts, consistent with their derivation either from asthenospheric, oceanic-type mantle or relatively juvenile crustal material. To the E of this boundary, igneous rocks have compositions suggesting source regions primarily in old, evolved continental lithosphere. This is true both for basalts and for silicic rocks and is especially well illustrated along the western margin of the Idaho Batholith which intrudes the boundary (discussed in more detail below).

Thus, study of silicic rocks in the northwestern U.S.A. provides an opportunity to compare magmatism from diverse tectonic settings and crustal sources. In this paper, we shall concentrate on silicic rocks of central and southwestern Idaho, focusing on temporal changes in the compositions of magmas produced from a relatively restricted parcel of cratonic crust but in a variety of tectonic settings.

1. Lithologic descriptions

1.1. Idaho Batholith (Kg)

The Idaho Batholith is a composite mass of tonalite, granodiorite, and granite of Late Cretaceous age (Armstrong *et al.* 1977; Hyndman 1983, 1984; Kiilsgaard & Lewis 1985; Criss & Fleck 1987). The lithological zonation of the batholith, from tonalite and diorite at its margins through a granodioritic main phase to a granitic core, appears to represent the temporal evolution of batholith magmatism as younger ages are preserved toward the core. Orogenic plutons of Permian–Triassic and Jurassic–Early Cretaceous age occur in the accreted terranes, but the main phase of intrusive activity occurred ~72–95 Ma, after the docking of the accreted terranes with N America (Vallier & Brooks 1987). Palaeocene and Eocene ages for silicic plutonic rocks in the area are common (Criss & Fleck 1987); however, these younger ages are now thought to be either cooling ages or related to the Challis magmatic episode (next section). Proterozoic and Palaeozoic metasedimentary rocks host the batholith, but the area is often depicted as part of the Wyoming craton, and Archaean rocks may be present locally at depth.

The Idaho Batholith intrudes the boundary between the

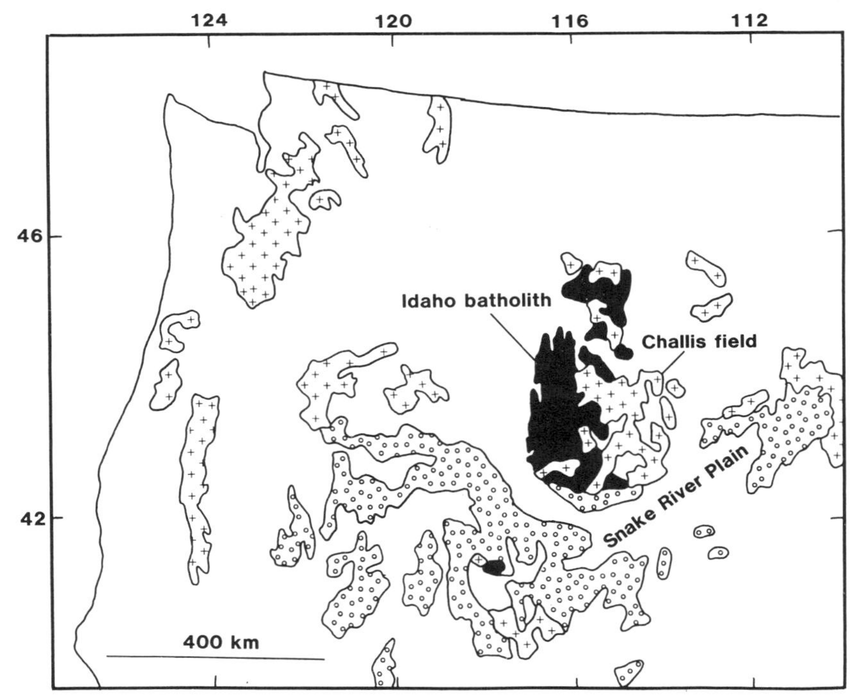

Figure 1 A generalised map of the northwestern U.S.A. illustrating the distribution of silicic rocks of Cretaceous (black), Eocene (+), and Miocene (○) ages; locations of the Idaho Batholith, the Challis magmatic field, and the Snake River Plain are shown.

craton and the accreted terranes. Regional variations in the composition of the batholith appear to reflect variable contributions of primary arc magma and several components from the continental lithosphere (Armstrong *et al.* 1977; Fleck & Criss 1985; Fleck 1990). W of the craton/accreted terrane boundary, compositions of batholith plutons are consistent with their derivation from accreted arc terranes and/or active arc magmatism (e.g. $^{87}Sr/^{86}Sr \leq 0{\cdot}704$). Toward the E, batholith compositions change rapidly across a narrow (5–15 km) transition zone to values suggesting major contributions from the older continental lithosphere (e.g. $^{87}Sr/^{86}Sr \geq 0{\cdot}706$).

Volcanic rocks coeval with the Idaho Batholith generally are absent, having been eroded during the unroofing of the batholith during late Cretaceous and early Tertiary uplift. An exceptional occurrence is the Elkhorn Mountains volcanism of western Montana. These volcanic rocks are genetically associated with the Boulder Batholith, a relatively small, Late Cretaceous epizonal pluton related only indirectly to the Idaho Batholith (Tilling 1973; Hamilton & Myers 1974).

1.2. Challis magmatism (Tc)

Challis volcanic rocks comprise 51–40 Ma mafic, intermediate, and silicic lavas and tuffs erupted from fissures and calderas in central and southwestern Idaho (McIntyre *et al.* 1982; Fisher *et al.* 1983; Hardyman 1988). Challis activity began with a lower sequence of mafic to intermediate lava flows which erupted from numerous small fissure vents. An upper sequence of more silicic ash-flow tuffs and lavas associated with cauldron complexes became established ~48 Ma. The locations of these cauldron complexes and associated mineralisation are controlled largely by the NE–SW-trending trans-Challis fault system which parallels regional structures (Bennett & Knowles 1985; Bennett 1986). Two suites of Eocene plutonic rocks, one dioritic and the other granitic, and cogenetic hypabyssal and volcanic equivalents of these suites have been recognised (Bennett & Knowles 1985; Hardyman & Fisher 1985; Lewis & Kiilsgaard 1991). The Challis volcanic field appears to be restricted to areas underlain by the craton, and compositions of Challis volcanic and plutonic rocks suggest they were produced predominantly from sources within the continental lithosphere (Norman & Leeman 1989; Norman & Mertzman 1991; Lewis & Kiilsgaard 1991).

1.3. Snake River Plain rhyolites (Tr)

The Snake River Plain–Yellowstone Plateau volcanic province has been the principal focus of Neogene magmatism in the region of the northwestern U.S.A. underlain by the craton. Although Late Cainozoic basalts are widely exposed, rhyolitic extrusive rocks are volumetrically dominant. The general trend of younger ages to the E (Armstrong *et al.* 1975) has been ascribed to the westward movement of the N American plate over a mantle hot spot currently located beneath the Yellowstone Plateau (Morgan 1972). However, the age progression of volcanism along the Plain is not perfect and its northeasterly trend is sub-parallel to regional structures such as the trans-Challis fault system, casting some doubt on the hot spot hypothesis. There is little doubt that initiation of major Miocene activity in this region corresponded with the onset of Basin and Range-style extension, and the Snake River Plain must be a zone of accommodation between the extended Basin and Range province to the S, and the less attenuated Idaho Batholith region to the N.

Neogene rhyolites (Tr) of the Snake River Plain and Yellowstone Plateau comprise large volume ash-flow tuffs and lava flows erupted from major volcanic centres (Armstrong *et al.* 1975; Doe *et al.* 1982; Ekren *et al.* 1984; Rytuba & McKee 1984; Bonnichsen & Kaufman 1987; Honjo & Leeman 1987; Hildreth *et al.* 1991). Poorly-developed calderas, and the unusually dry and fairly hot (≥900°C) compositions of rhyolites from some eruptive centres in the western and central Snake River Plain suggest that the rhyolites may have erupted more or less directly from middle to lower crustal depths (Ekren *et al.* 1984; Bonnichsen & Kaufman 1987; Honjo & Leeman 1987). No Miocene plutonic rocks have been recognised in this area.

The discussion here presents data for rhyolites from the Owyhee region of southwestern Idaho and from the Magic Reservoir eruptive centre of central Idaho. The Owyhee area is near the western edge of the craton and may be a transitional zone between the craton to the E and accreted terranes to the W. The Magic Reservoir eruptive centre is a major silicic centre in the north–central Snake River Plain, near the margin of the Snake River Plain and the Idaho Batholith (Honjo & Leeman 1987). Compositions of rhyolites from these two areas are representative of Miocene and Pliocene rhyolites from the region, although a comprehensive review of the geochemistry of Snake River Plain rhyolites is not yet available.

2. Major element geochemistry

Representative major and trace element analyses of granodiorites from the Idaho Batholith, a Challis ash-flow tuff from southwestern Idaho and Snake River Plain rhyolites from southwestern Idaho are presented in Table 1. These data represent new analyses for all three suites of rocks obtained by X-ray fluorescence (S. Mertzman, analyst, Franklin and Marshall) and instrumental neutron activation (B. Chappell, analyst, ANU). Additional data from Lewis & Kiilsgaard (1991), Honjo & Leeman (1987), Leeman & Honjo (unpublished), and Leeman, Hooper, Manduca, Fleck, Clarke, and Lewis (unpublished) are also included in this discussion.

Rocks of the Idaho Batholith tend to have the highest contents of Al, Ca, and Na, and the lowest contents of Fe, Ti, and K for a given silica value (Fig. 2). MgO contents of all three suites are similar, although some Idaho Batholith rocks and especially the more mafic Challis lavas extend to fairly magnesian compositions. Challis samples form a continuous compositional series, from basalt through rhyolite with 78% SiO_2. We shall concentrate our comparisons on the more silicic Challis samples, those with >60% SiO_2. The Idaho Batholith also displays a fairly broad compositional range, although some of the most mafic samples may be at least partial cumulates. By comparison, the Miocene rhyolites of the Snake River Plain are more restricted in composition, ranging in silica content only from 68–76%. For the "plagiophile" elements (Al, Ca, Na, K), Challis compositions are bracketed by those of the Idaho Batholith granitic rocks and the Snake River Plain rhyolites. This seems to be a common feature of these three suites, and may be an important guide to changes in source compositions.

The Idaho Batholith is relatively sodic, with Na_2O/K_2O ratios usually ≥1 and extending to quite high values (Fig. 2). Miocene rhyolites are more potassic, with Na_2O/K_2O ratios ≤1. Challis rocks have intermediate Na_2O/K_2O, scattering around 1 and not extending to values as high as those of the

Table 1 Representative major and trace element analyses of rocks from the Cretaceous Idaho Batholith (Kg), Eocene Challis volcanics (Tc), and Miocene rhyolites from the Snake River Plain (Tr).

	Kg			Tc	Tr			
Sample:	87-23	87-24	87-25	Tcv-4	Tmc-1	Tjl-2	84-58	84-59
SiO_2	69·80	69·96	70·56	65·16	69·17	73·85	76·84	73·6
TiO_2	0·52	0·40	0·34	0·56	0·55	0·4	0·15	0·21
Al_2O_3	15·04	16·05	15·41	15·46	13·71	12·97	12·26	12·27
Fe_2O_3	3·39	2·38	2·18	4·58	3·39	2·77	1·76	2·08
MnO	0·06	0·03	0·03	0·07	0·06	0·03	0·02	0·03
MgO	0·96	0·62	0·52	1·53	0·37	0·13	0·01	0·02
CaO	3·03	2·65	2·34	3·43	1·59	0·92	0·38	0·66
Na_2O	3·56	4·27	4·01	3·30	2·85	3·17	3·05	2·71
K_2O	2·83	3·00	3·42	3·74	5·65	5·29	5·36	5·95
P_2O_5	0·16	0·10	0·08	0·18	0·08	0·05	0·01	0·02
LOI	0·68	0·83	0·76	2·06	2·87	0·59	0·57	2·5
Sum	100·03	100·29	99·65	100·07	100·29	100·17	100·21	100·05
Rb	104	75	88	111	174	193	257	206
Sr	632	953	835	451	107	65	6	26
Ba	946	1142	1100	1227	1506	1088	354	933
Y	13	5	5	12	62	35	44	59
Zr	181	199	174	198	562	408	270	307
V	43	29	24	72	20	15	4	5
Ni	6	5	5	5	4	5	3	4
Cr	13	24	20	40	6	10	4	10
Sc	4·4	2·9	2·9	7·3	6·6	4·3	1·7	2·2
Co	4·9	2·2	1·8	10	2·8	1·9	0·5	0·6
Cs	4·6	3·1	2·1	1·63	4	2·3	4·1	3·9
La	52·9	50·1	45·1	45·1	72·2	64·8	84·2	92·1
Ce	108	104	92	85·1	147	114	155	189
Nd	42·9	42·2	37·9		71	50·4	68·5	75·9
Sm	6·77	5·97	5·67	5·9	13·6	8·67	11·6	13·2
Eu	1·74	1·55	1·47	1·3	2·02	1·14	0·44	1
Gd	4·7	3·5	3·5		11·6	6·93	8·5	11·4
Tb	1·0	0·6	0·7	0·74	2·11	1·5	2	2·2
Ho	1·0	0·5	0·6		2·28	1·47	2·1	2·5
Yb	1·82	0·78	0·99	1·82	6·79	4	4·61	6·09
Lu	0·26	0·10	0·13	0·30	0·94	0·56	0·6	0·82
Hf	5·4	5·8	5·6	5·1	13·8	11·6	9	9·4
Ta	4·6	2·0	2·5	0·98	4·07	4·45	6·7	5·5
Th	18	9	12	20	20·6	29·4	43	34
U	6·5	2	2·5	4·9	5·3	6·1	9	6·5

Idaho Batholith. The more sodic compositions of the batholith and Challis samples are similar to those of Lachlan I-type granites (Fig. 3). Compositions of the more potassic Snake River Plain rhyolites are transitional between Lachlan I-type and S-type granites, with the least-fractionated sample showing I-type affinities (Fig. 3). (The term "fractionated" is used loosely here, without specific reference to a particular physical mechanism such as removal of crystals from a melt. It refers more to compositional characteristics, with those samples of a given series having, for example, higher Si and lower Fe and Ca contents being the "more fractionated" samples.) The alumina saturation indices (ASI = molar $Al_2O_3/CaO + Na_2O + K_2O$) of all suites overlap and increase systematically in the more evolved samples (Fig. 4). In each suite, ASI is <1 in the least fractionated samples, and increases in the more fractionated samples. Virtually all samples have ASI $\leq 1{\cdot}1$, which is considered typical of I-type sources (Chappell & White 1984).

Because we are comparing granites and rhyolites, a fair question is: how well do plutonic and volcanic compositions compare? For the three suites considered here, this can be addressed directly only for the Challis co-genetic volcanic and plutonic rocks. For these samples, there is a nearly complete compositional overlap for the Challis plutonic and volcanic samples (Fig. 5). A similar result has been found for co-genetic volcanic and plutonic rocks from the Andes Mountains (Atherton *et al.* 1979; Thorpe & Francis 1979), the Lachlan Fold Belt (Wyborn *et al.* 1981), the southwestern U.S.A. (Lipman 1988), and the Late Cretaceous Elkhorn Mountains volcanism and Boulder Batholith of western Montana (Tilling 1973; Hamilton & Myers 1974). A generalised model relating the compositions of co-genetic volcanic and plutonic rocks has been presented by Macdonald and Smith (1988). We shall assume that the compositions of the volcanic and plutonic suites carry similar information on the nature of their sources and can be meaningfully compared.

3. Trace element geochemistry

3.1. Rare-earth elements

All three suites are LREE-enriched with chondrite-normalised La values ~100–300 and similar slopes between La and Sm (Fig. 6). LREE contents of the Snake River Plain rhyolites extend to somewhat higher values than those of the Idaho Batholith and the Challis rocks. Important differences between the three suites occur in their HREE contents and the size of their Eu anomalies. Rocks of the Idaho Batholith tend to be relatively depleted in HREE, with steeply sloping patterns and variable but generally high La/Yb (Figs 6, 7; see also Shuster & Bickford 1985; Foster & Hyndman 1990). The more fractionated samples (i.e.

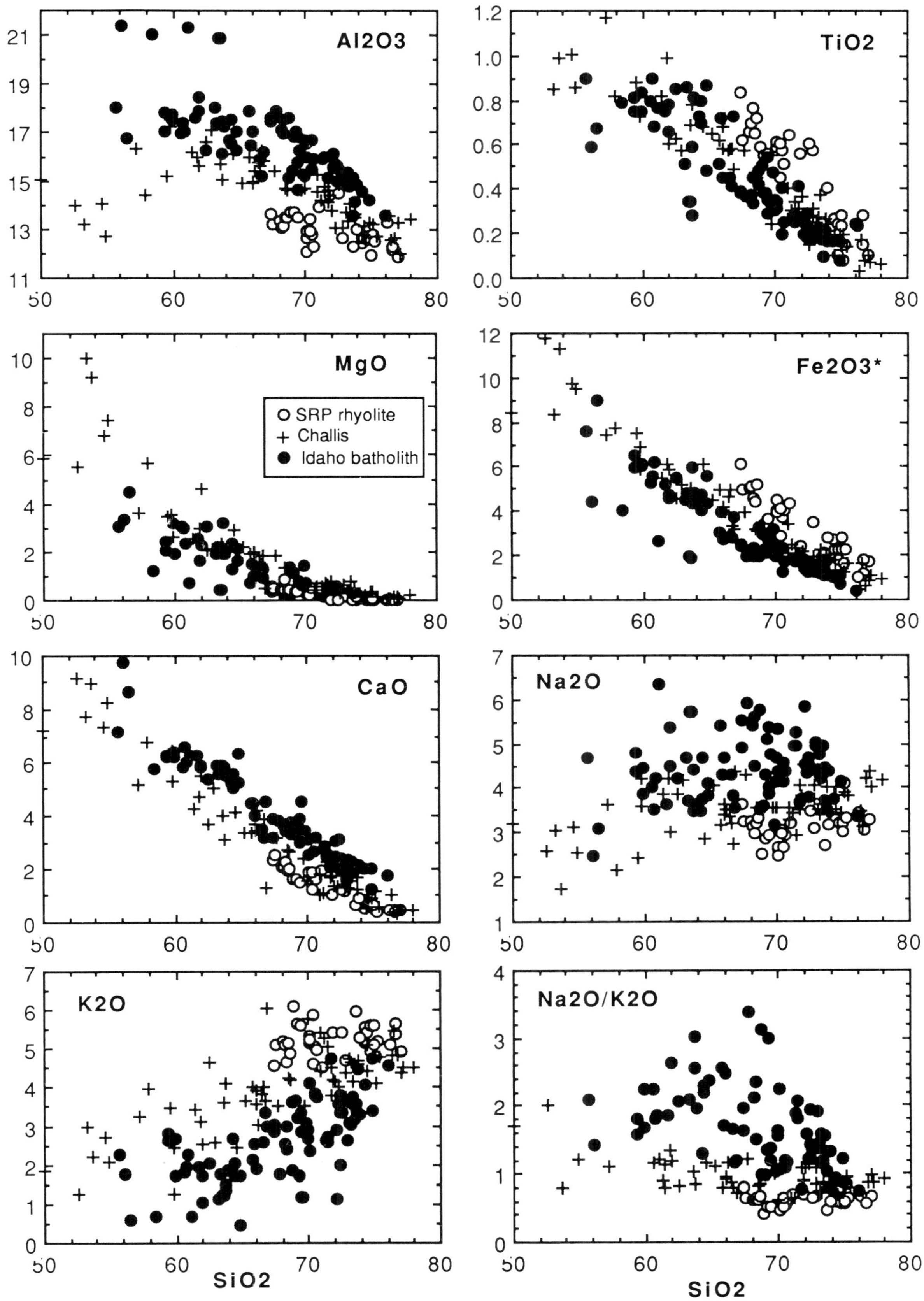

Figure 2 Major element variation diagrams for rocks of the Idaho Batholith (filled circles), Challis volcanic and plutonic rocks (crosses), and rhyolites of the Snake River Plain (open circles); rocks of the Idaho Batholith have high contents of Al_2O_3, CaO, Na_2O, and Na_2O/K_2O, and low contents of TiO_2, Fe_2O_3* (total Fe), and K_2O compared to Challis and Snake River Plain rocks with similar SiO_2 values; the Challis suite typically has compositions intermediate between those of the Idaho Batholith and the Snake River Plain samples.

lowest Fe content) tend to have the highest La/Yb. Minor positive and negative Eu anomalies are found, but rocks of the Idaho Batholith typically lack significant Eu anomalies. Challis volcanic rocks have higher HREE contents than is typical for the Idaho Batholith (lower normalised La/Yb) and have either negligible or negative Eu anomalies (Figs 6, 7). Positive Eu anomalies are rare in Challis plutonic and volcanic rocks. Fractionated Challis granites develop deep negative Eu anomalies and relatively flat HREE patterns. Neogene rhyolites have REE characteristics similar to fractionated Challis granites in that they have ubiquitous negative Eu anomalies and consistently flat HREE patterns (low normalised La/Yb) (Figs 6, 7).

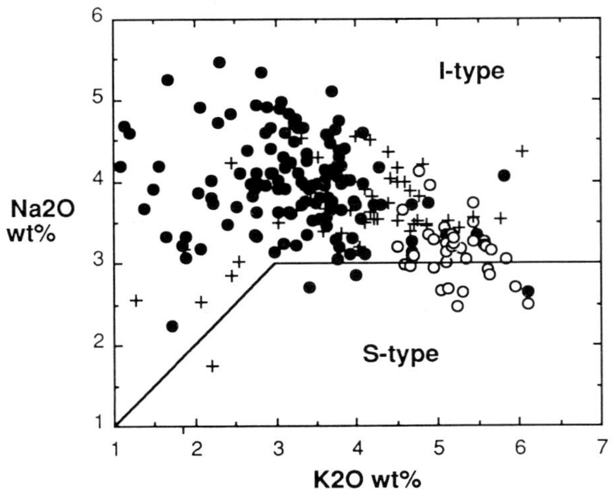

Figure 3 Variation of Na_2O and K_2O contents in the Idaho Batholith (filled circles), Challis volcanic and plutonic rocks (crosses), and rhyolites of the Snake River Plain (open circles), relative to the fields for Lachlan Fold Belt I-type and S-type granites (after Chappell & White 1984); S-type field encompasses virtually all Lachlan S-type granites, but some I-type samples overlap the boundary; Idaho Batholith and Challis samples have compositions very similar to Lachlan I-type granites; Snake River Plain rhyolites have compositions transitional between Lachlan I-type and S-type granites; the least evolved rhyolites tend to have more I-type characteristics.

3.2. Sr and Rb contents

Sr and Rb contents also highlight important compositional differences among the three suites (Fig. 8). Typically, the Idaho Batholith contains relatively low Rb and high Sr contents, and rhyolites of the Snake River Plain have higher Rb and lower Sr contents. Sr concentrations in the Idaho Batholith can be quite high, reaching up to ~2500 ppm; more typical Sr contents would be 600–1000 ppm. Rb concentrations in the Idaho Batholith typically are 10–100 ppm, with a few higher values in the more evolved samples. Sr contents of the Snake River Plain rhyolites range from <10 to 250 ppm; Rb contents range from 80 to 440 ppm. Challis samples have a broad range of Sr and Rb contents that extend from values similar to those of the Idaho Batholith in the less evolved samples, to values more like those of the Snake River Plain rhyolites in the more fractionated samples. Rb/Sr does not change much with fractionation in the Idaho Batholith (Fig. 8). Rb/Sr values of Snake River Plain rhyolites are higher than those of the Idaho Batholith, and increase rapidly with fractionation. Challis samples again span the range observed for the other two suites, showing a well-defined increase in Rb/Sr with decreasing Fe content (i.e. increasing fractionation).

Sr contents of Idaho Batholith and Challis samples are poorly correlated with the size of their Eu anomaly (Fig. 9). This is particularly true for the less evolved Challis rocks, which have ubiquitous negative Eu anomalies and a considerable range of Sr contents, but the size of these Eu anomalies is largely independent of Sr concentration. The more evolved Challis granitic rocks do develop deep negative Eu anomalies and correspondingly low Sr contents (Fig. 9). Snake River Plain rhyolites, on the other hand, clearly show the effects of plagioclase fractionation with increasingly large negative Eu anomalies corresponding to lower Sr contents.

4. Discussion

Compositions of Cretaceous and Cainozoic silicic rocks in the northwestern Pacific provide constraints on the processes of magma generation and evolution, the nature of

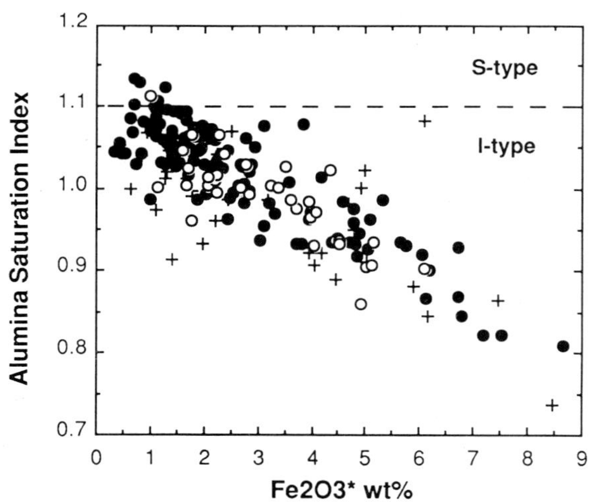

Figure 4 Fe_2O_3* (total Fe content) vs Alumina Saturation Index (ASI = $Al_2O_3/CaO + Na_2O + K_2O$ atomic) for samples of the Idaho Batholith, Challis volcanic and plutonic rocks, and Snake River Plain rhyolites; symbols are the same as on previous diagrams; virtually all samples have ASI < 1.1, consistent with an I-type classification (Chappell & White 1984).

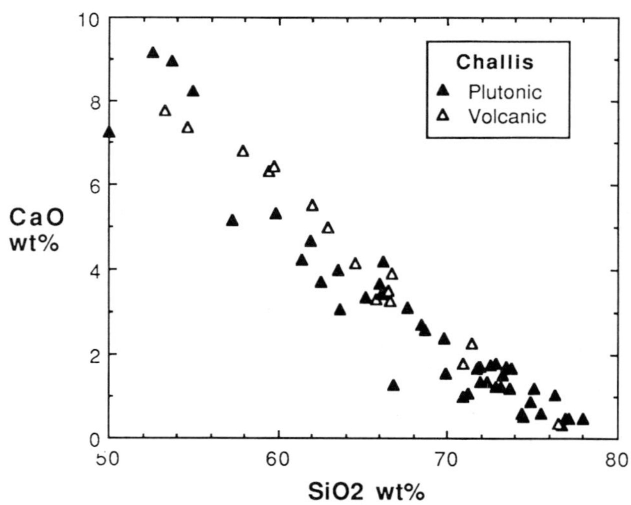

Figure 5 Variation of SiO_2 vs CaO content for Challis plutonic (filled triangles) and volcanic (open triangles) samples; a nearly complete compositional overlap between volcanic and plutonic Challis samples provides a justification for comparing the plutonic Idaho Batholith and volcanic Snake River Plain rhyolite compositions.

their source rocks, and the links between magmatism and tectonic setting. Significant compositional differences between the three suites discussed here suggest a temporal evolution of magmatic compositions which can be related to changes in the nature of their source regions and conditions of magma generation.

Granitic rocks of the Idaho Batholith and the Snake River Plain rhyolites represent endmembers for the temporal sequence. In many respects, Challis samples have compositions intermediate between those of the other two suites. Considering the three suites discussed here as time steps, observed temporal changes in silicic magma composition include: (1) decreasing Na/K: the Idaho Batholith typically has $Na_2O > K_2O$, Challis samples have $Na_2O \approx K_2O$, and rhyolites of the Snake River Plain have $Na_2O < K_2O$; (2) decreasing La/Yb, which is controlled by HREE concentration: rocks of the Idaho Batholith have steeply dipping HREE patterns, Snake River plain rhyolites have flat HREE patterns that are enriched relative to those of the batholith, and Challis rocks have REE patterns steeper than those of the Miocene rhyolites and comparable to those of the least evolved samples from the Idaho Batholith; (3) increasing magnitude of Eu anomalies: small positive and negative Eu anomalies that scatter around $Eu/Eu^* = 1$ (i.e. no Eu anomaly) are typical of the Idaho Batholith, Challis rocks have negligible or small negative Eu anomalies, and Snake River Plain rhyolites have ubiquitous negative Eu anomalies; (4) decreasing Sr concentrations: high Sr concentrations are typical of the Idaho Batholith, whereas Snake River Plain rhyolites are notably poor in Sr and Challis samples have intermediate compositions. These trends are summarised in Figure 10. We interpret these geochemical characteristics as indicating a change from

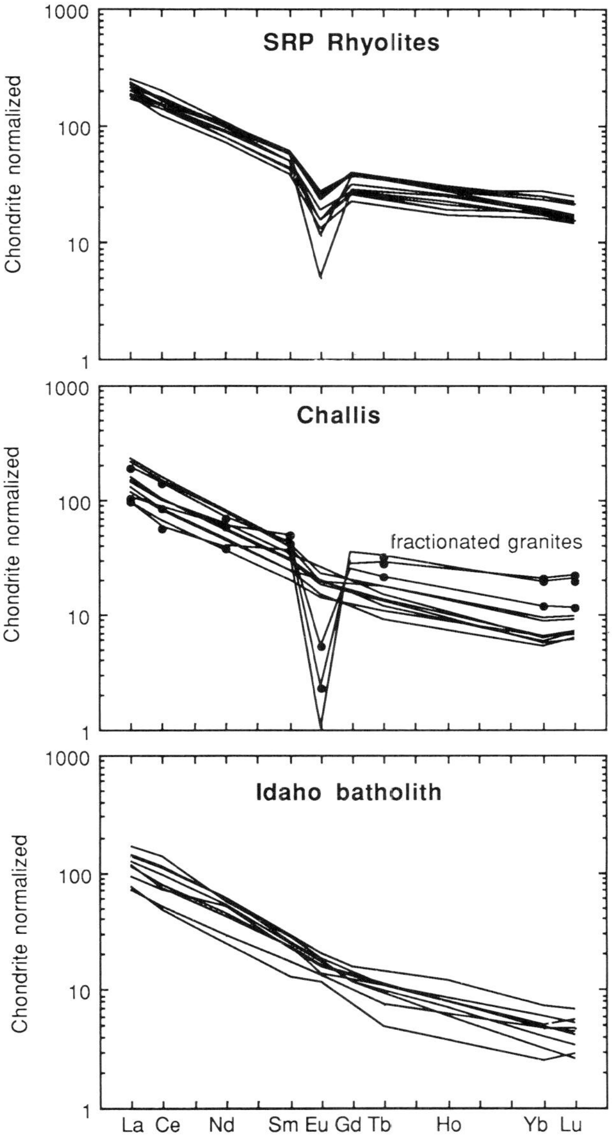

Figure 6 Chondrite-normalised rare-earth element (REE) patterns of representative samples from the Idaho Batholith, Challis volcanic and plutonic rocks, and Snake River Plain rhyolites; all samples are enriched in the light REE; rocks of the Idaho Batholith have fractionated heavy REE (HREE) patterns and lack significant Eu anomalies; Snake River Plain rhyolites have nearly flat HREE patterns and negative Eu anomalies; Challis samples typically have compositions intermediate between those of the Idaho Batholith and Snake River Plain rhyolites, with either negligible or slight negative Eu anomalies; fractionated Challis samples develop REE patterns somewhat similar to those of the Snake River Plain rhyolites; the REE patterns of Snake River Plain rhyolites are similar to that of the average upper crust (Taylor & McLennan 1985).

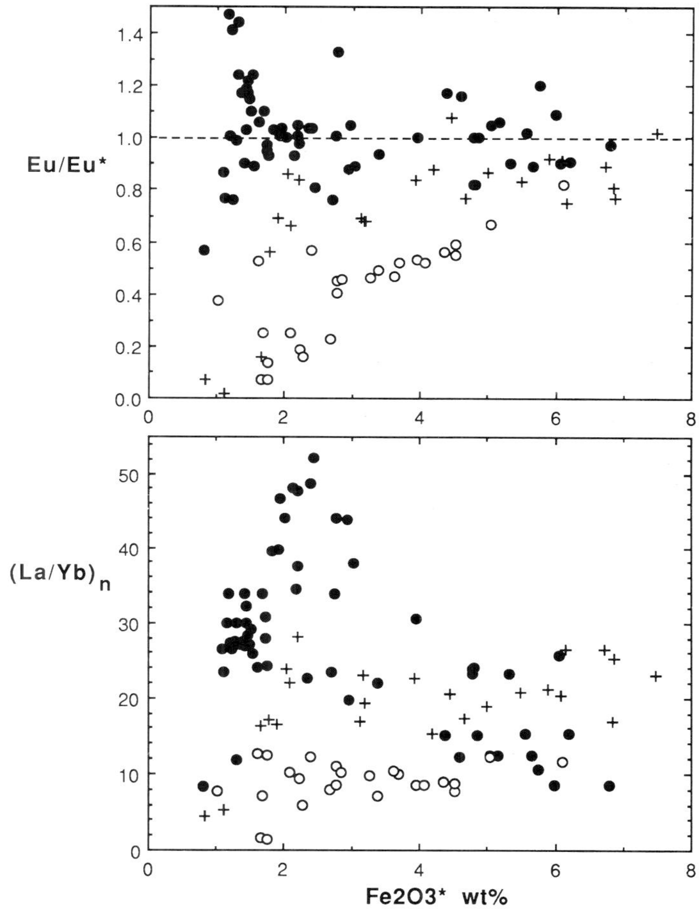

Figure 7 Variation of chondrite-normalised La/Yb ratios and magnitude of the Eu anomalies (Eu/Eu^*) with total Fe content ($Fe_2O_3^*$) for the Idaho Batholith, Challis samples, and Snake River Plain rhyolites; symbols are the same as on previous diagrams; $Eu/Eu^* = 1$ indicates no Eu anomaly; $Eu/Eu^* > 1$ and <1 indicates positive and negative Eu anomalies, respectively; the Idaho Batholith has variable $(La/Yb)_n$, including some highly fractionated compositions, but Eu/Eu^* which scatter around 1; Idaho Batholith samples tend to develop more fractionated REE patterns (higher La/Yb) with increasing fractionation (lower $Fe_2O_3^*$); Snake River Plain rhyolites have relatively constant $(La/Yb)_n$, and consistently negative Eu anomalies that become increasingly pronounced (smaller Eu/Eu^*) with fractionation; Challis samples typically have negative Eu anomalies and fairly constant $(La/Yb)_n$.

garnet-bearing sources for the Idaho Batholith, to plagioclase-dominated sources for rhyolites of the Snake River Plain. Such a change in source mineralogy suggests a general trend of decreasing depth of magma generation through time.

Granitic rocks of the Cretaceous Idaho Batholith have trace element compositions consistent with magma generation at lower crustal or upper mantle depths where garnet is a stable phase. REE patterns in the Idaho Batholith are similar to those of tonalitic rocks from the western margin of the Peninsular Range Batholith (Gromet & Silver 1987) and to those of Archaean tonalites (see summaries in Taylor & McLennan 1985, and Rudnick & Taylor 1986). Such a pattern apparently requires residual garnet in the magmatic source region; amphibolite sources without garnet do not seem capable of producing magmas with such a pattern (Rudnick & Taylor 1986; Gromet & Silver 1987). The lack of significant Eu anomalies, the relatively high Sr contents and the lack of a correlation between Sr content and indicators of fractionation (e.g. Fe content) seem to preclude significant residual plagioclase in the source of the Idaho Batholith. Likewise, plagioclase fractionation is unlikely to have been the predominant process controlling the compositional variations observed in the batholith. Although some crystal accumulation is to be expected during crystallisation of the batholith, the effects of this process appear to be relatively minor. Mafic to intermediate composition, garnet-bearing lithologies seem the most likely source material for the magmas of the Idaho Batholith.

The low Sr contents, negative Eu anomalies, and flat HREE patterns of the Neogene rhyolites contrast sharply with the compositional characteristics of the Idaho Batholith and suggest generation of the Snake River Plain rhyolitic magmas from felsic crust at shallower depths within the crust where garnet is absent and residual plagioclase is more abundant. Plagioclase fractionation was also a common process during the evolution of these magmas, as shown by the decreasing Sr contents and increasing magnitude of the Eu anomaly in the more evolved rocks. REE patterns of the Snake River Plain rhyolites are very similar to those of the average upper continental crust, for which a similar model involving intracrustal melting leaving plagioclase as a residual phase has been proposed (Taylor & McLennan 1985).

Source characteristics of the Eocene Challis magmatism are more problematic, but the compositional characteristics of Challis rocks generally are transitional between those of the Idaho Batholith and the Snake River Plain rhyolites, suggesting intermediate depths of magma generation. The more fractionated Challis samples clearly show the effects of

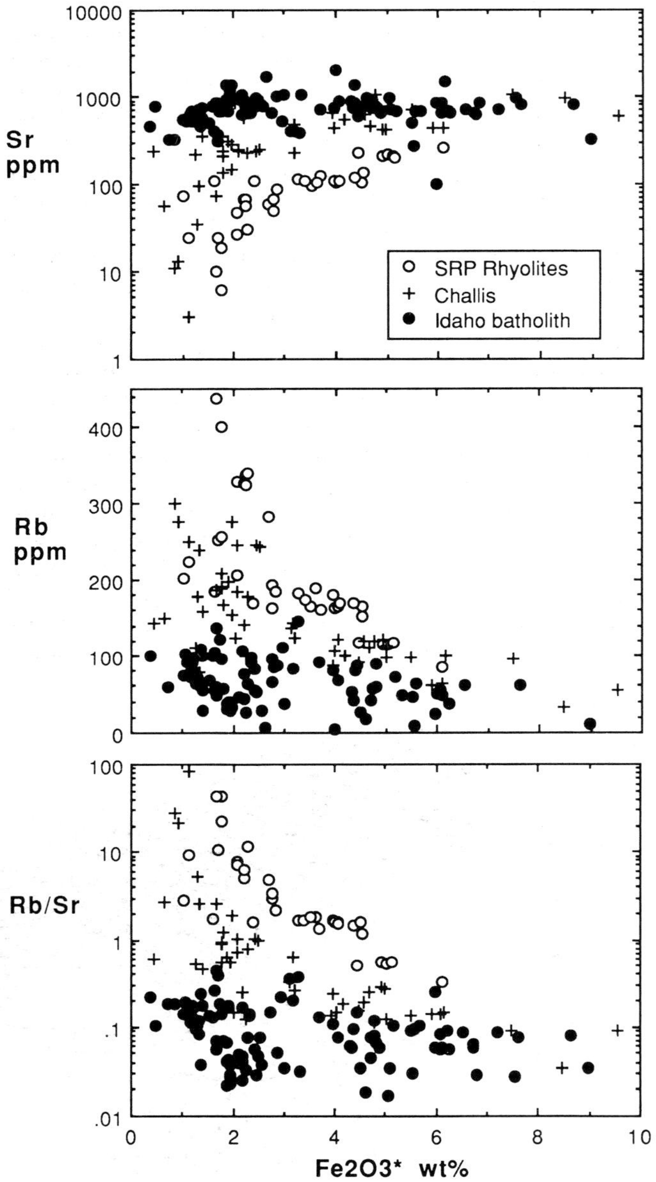

Figure 8 Variation of Sr, Rb, and Rb/Sr contents with Fe_2O_3* for the Idaho Batholith, Challis rocks, and Snake River Plain rhyolites; symbols are the same as on previous diagrams; Idaho Batholith samples have relatively constant Sr and Rb contents; Snake River Plain rhyolites have lower Sr and higher Rb contents, which decrease and increase, respectively, with decreasing Fe_2O_3* content; Challis samples have Sr and Rb contents comparable to the Idaho Batholith in the least fractionated samples, which evolve to compositions more similar to those of the Snake River Plain rhyolites in the more fractionated samples.

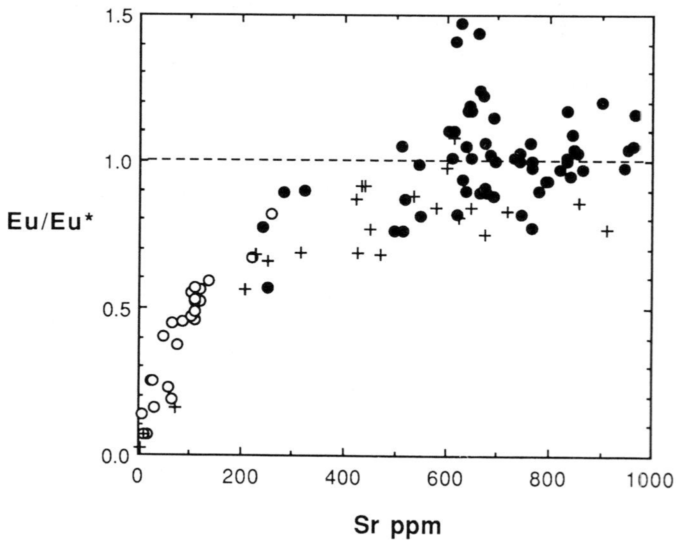

Figure 9 Variation of magnitude of the Eu anomaly (Eu/Eu*; see Fig. 7) with Sr content for the Idaho Batholith, Challis samples, and rhyolites of the Snake River Plain; symbols are the same as on previous diagrams; the Idaho Batholith has relatively high Sr contents, and Eu/Eu* which are variable but scatter around 1 and are not well-correlated with Sr content; Snake River Plain rhyolites have relatively low Sr contents, ubiquitous negative anomalies, and a good correlation between the Eu/Eu* and Sr content; Challis samples typically have negative Eu anomalies that are largely independent of Sr content in the least fractionated samples, and lower Sr contents and larger Eu anomalies in the more fractionated samples.

plagioclase fractionation, and the formation of co-genetic volcanic-plutonic caldera complexes attests to the shallow level history of at least some Challis magmas. Mafic and intermediate Challis lavas have compositions suggesting mixed crustal and mantle source characteristics, leading to suggestions they may have formed near the crust-mantle boundary (Norman & Mertzman 1991). This would be consistent with the depth estimates developed here.

The seemingly progressive change from garnet-bearing sources for the Idaho batholith to plagioclase-dominant sources for the rhyolitic magmas of the Snake River Plain would be consistent with a general trend of decreasing depth of magma generation through time. Experimental determinations and petrological studies of crustal xenoliths suggest that pressures of at least 10 kbar (1 kbar = 100 MPa) (35 km depth) would be implied by a garnet-bearing source of the Idaho Batholith. For moderate degrees of melting (10–40%) of a mafic to intermediate source, garnet coexists with melts of tonalitic to trondhjemitic compositions at pressures greater than 8 kbar (Rapp *et al.* 1991). Garnet phenocrysts and garnet-bearing xenoliths in I-type high-silica andesites and dacites from New Zealand apparently crystallised at pressures of 10–15 kbar (Green 1991), and garnet-bearing crustal xenoliths from the Mojave Desert are thought to represent residues after granite extraction at approximately 10 kbar (Miller *et al.* 1991). Residual garnet in the source region of the Idaho Batholith would seem to imply generation of these magmas at depths of greater than 35 km if they formed by equilibration and segregation of these magmas from an I-type source region. Considerably greater depths of magma generation are possible; for example, Rapp *et al.* (1991) find that maximum production of tonalitic magmas from an eclogitic source occurs at ~22 kbar (~77 km). The clear influence of plagioclase in Snake River Plain rhyolites indicates these magmas formed and evolved within the continental crust, probably at pressures less than 10 kbar (<35 km depth).

Sodic Archaean tonalites with compositional characteristics similar to the Idaho Batholith may have formed by melting of shallowly subducted slabs (McCulloch 1991; McCulloch & Bennett 1991). However, isotopic data indicate that the Idaho Batholith probably was not produced by direct melting of a subducted oceanic slab, but predominantly from within the continental lithosphere (Fleck & Criss 1985; Fleck 1990). The presence of crustal rocks at depths greater than 35 km would imply considerable crustal thickening associated with Mesozoic plate convergence and subduction. The Proterozoic country rocks of the Idaho Batholith are low and medium grade metasedimentary rocks, so this model would also require significant transport of batholithic magmas through the continental lithosphere.

Alternatively, the Idaho Batholith may have formed by remobilisation of Precambrian tonalitic gneisses. In this case, batholith magmas would carry the image of their sources (Chappell *et al.* 1987), and no constraints on the depths of magma generation could be inferred. Precambrian orthogneisses with the general required compositional characteristics are found as xenoliths in some Snake River Plain basaltic lavas (Leeman *et al.* 1985) and as basement outcrops farther E (Wooden & Mueller 1988). However, restite enclaves and relict mineral cores are sparse in the Idaho Batholith (Hyndman 1983, 1984; Kiilsgaard & Lewis 1985), and although more detailed work is needed, at this time we favour an origin for the batholith by segregation of magma from a garnet-bearing source at depth.

The inferred depths of silicic magma generation seem to correlate with the tectonic environment, with the locus of melting apparently migrating to shallower levels within the lithosphere as convergence gave way to extension. Cretaceous magmatism was associated with plate convergence, compressional deformation and crustal thickening. Neogene magmatism accompanied crustal thinning associated with rifting and extension. Eocene activity is more problematic but appears to represent an intermediate depth of magma evolution associated with a transitional tectonic environment. Decreasing depths of silicic magma generation may be related to increasing crustal attenuation, increasing density of the lower crust following a prolonged history of basalt intrusion and melt extraction, or the ease of magma transport through the crust aided by an extensional tectonic setting.

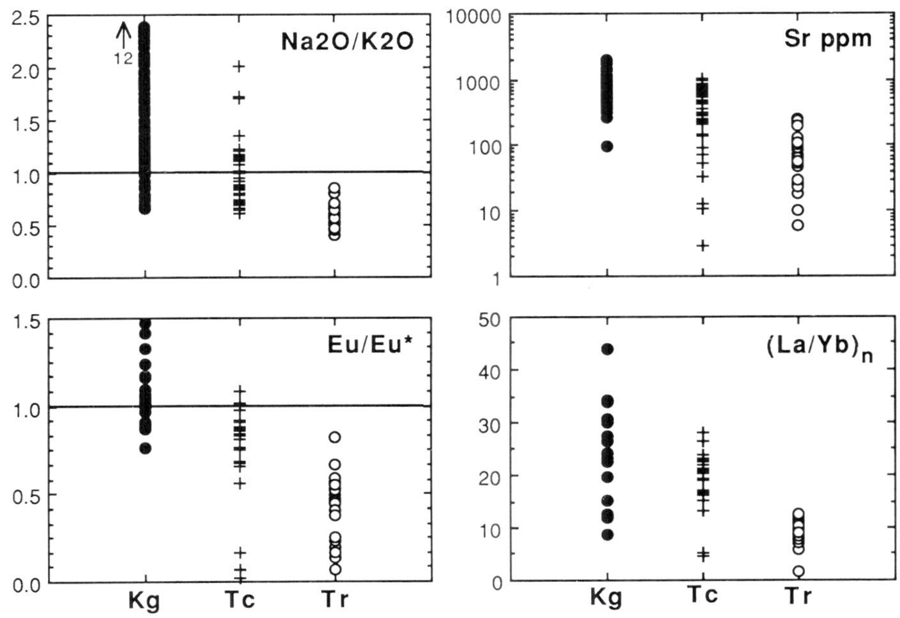

Figure 10 A summary of compositional parameters for the Idaho Batholith (Kg), Challis samples (Tc), and rhyolites of the Snake River Plain (Tr), illustrating the temporal evolution of magma compositions.

5. Conclusions

(1) The compositions of Cretaceous and Cainozoic granites and rhyolites in the northwestern U.S.A. provide a record of intracrustal magmatism in convergent and extensional tectonic settings.

(2) Cretaceous granitic rocks of the Idaho Batholith have sodic, HREE-depleted compositions, typically lacking significant Eu anomalies. These characteristics are consistent with their generation from garnet-bearing, mafic to intermediate source rocks. Neogene rhyolites of the Snake River Plain have more potassic compositions with higher HREE abundances and ubiquitous negative Eu anomalies. These magmas appear to have been generated from more felsic, plagioclase-dominated sources. Eocene volcanic and plutonic rocks have compositions transitional between the Cretaceous granitic rocks and the Neogene rhyolites.

(3) The change from Cretaceous to Neogene silicic magma compositions suggests melting at progressively shallower depths associated with the transition from plate convergence to extension.

6. Acknowledgements

INAA data provided by Bruce Chappell are gratefully acknowledged. We thank Eric Christiansen for a useful review. MDN is supported by NASA grant NAG9-454 to K. Keil. Portions of this research were funded by grants from the U.S. National Science Foundation (EAR83-20358 and EAR90-18437) to WPL. This is Planetary Geosciences publication number 669, and SOEST contribution number 2646.

7. References

Armstrong, R. L. & Ward, P. 1991. Evolving geographic patterns of Cenozoic magmatism in the North American Cordillera: the temporal and spatial association of magmatism and metamorphic core complexes. J GEOPHYS RES **96,** 13201–24.

Armstrong, R. L., Leeman, W. P. & Malde, H. E. 1975. K–Ar dating, Quaternary and Neogene volcanic rocks of the Snake River Plain, Idaho. AM J SCI **275,** 225–51.

Armstrong, R. L., Taubeneck, W. H. & Hales, P. O. 1977. Rb–Sr and K–Ar geochronometry of Mesozoic granitic rocks and their Sr isotopic composition, Oregon, Washington, and Idaho. GEOL SOC AM BULL **88,** 397–411.

Atherton, M. P., McCourt, W. J., Sanderson, L. M. & Taylor, W. P. 1979. The geochemical character of the segmented Peruvian Coastal batholith and associated volcanics. *In* Atherton, M. P. & Tarney, J. (eds) *Origin of granite batholiths,* 45–64. Kent, UK: Shiva.

Bennett, E. H. 1986. Relationship of the Trans-Challis fault system in central Idaho to Eocene and Basin and Range extensions. GEOLOGY **14,** 481–4.

Bennett, E. H. & Knowles, C. R. 1985. Tertiary plutons and related rocks in central Idaho. US GEOL SURV BULL **1658A-S,** McIntyre D. H. (ed.) 81–95.

Bonnichsen, B. & Kaufman, D. F. 1987. Physical features of rhyolite lava flows in the Snake River Plain volcanic province, southwestern Idaho. GEOL SOC AM SPEC PAP **212,** 119–45.

Carlson, D. H., Fleck, R., Moye, F. J. & Fox, K. F. 1991. Geology, geochemistry, and isotopic character of the Colville Igneous Complex, northeastern Washington. J GEOPHYS RES **96,** 13313–33.

Carlson, R. W. & Hart, W. K. 1987. Crustal genesis on the Oregon Plateau. J GEOPHYS RES **92,** 6191–206.

Chappell, B. W. & White, A. J. R. 1984. I- and S-type granites in the Lachlan Fold Belt, southeastern Australia. *In* Keqin, X. & Guanchi, T. (eds), *Geology of granites and their metallogenic relations,* 87–101. Beijing: Science Press.

Chappell, B. W., White, A. J. R. & Wyborn, D. 1987. The importance of residual source material (restite) in granite petrogenesis. J PETROL **28,** 1111–38.

Christiansen, R. L. & Lipman, P. W. 1972. Cenozoic volcanism and plate tectonic evolution of the western United States II. Late Cenozoic. PHILOS TRANS R SOC LONDON **A271,** 249–84..

Christiansen, R. L. & McKee, E. H. 1978. Late Cenozoic volcanic and tectonic evolution of the Great Basin and Columbia intermontane regions. *In* Smith, R. B., & Eaton, G.P. (eds). *Cenozoic Tectonics and Regional Geophysics of the Western Cordillera.* GEOL SOC AM MEM **152,** 283–311.

Coney, P. J. 1978. Mesozoic-Cenozoic Cordilleran plate tectonics. *In* Smith, R. B. & Eaton, G. P. (eds). *Cenozoic tectonics and regional geophysics of the Western Cordillera.* GEOL SOC AM MEM **152,** 33–50.

Criss, R. E. & Fleck, R. J. 1987. Petrogenesis, geochronology, and hydrothermal systems of the northern Idaho batholith and adjacent areas based on $^{18}O/^{16}O$, D/H, $^{87}Sr/^{86}Sr$, K–Ar, and $^{40}Ar/^{39}Ar$ studies. US GEOL SURV PROF PAPER **1436,** 95–137.

Dickinson, W. R., Klute, M. A., Hayes, M. J., Janecke, S. U., Lundin, E. R., McKittrick, M. A. & Olivares, M. A. 1988. Paleogeographic and paleotectonic setting of Laramide sedimentary basins in the central Rocky Mountains region. GEOL SOC AM BULL **100,** 1023–39.

Doe, B. R., Leeman, W. P., Christiansen, R. L. & Hedge, C. E. 1982. Lead and strontium isotopes and related trace elements as genetic tracers in the upper Cenozoic rhyolite–basalt association of the Yellowstone Plateau volcanic field. J GEOPHYS RES **87,** 4785–806.

Dudas, F. O. 1991. Geochemistry of igneous rocks from the Crazy Mountains, Montana, and tectonic models for the Montana alkalic province. J GEOPHYS RES **96,** 13261–77.

Ekren, E. B., McIntyre, D. H. & Bennett, E. H. 1984. High-temperature, large-volume, lavalike ash-flow tuffs without calderas in southwestern Idaho. US GEOL SURV PROF PAP **1272**.

Farmer, G. L. & DePaolo, D. J. 1983. Origin of Mesozoic and Tertiary granite in the western United States and implications for pre-Mesozoic crustal structure 1. Nd and Sr isotopic studies in the geocline of the northern Great Basin. J GEOPHYS RES **88,** 3379–401.

Fisher, F. S., McIntyre, D. H. & Johnson, K. M. 1983. Geologic map of the Challis 1° × 2° quadrangle, Idah. US GEOL SURV OPEN FILE REP **83-523**.

Fleck, R. J. 1990. Neodymium, strontium, and trace element evidence of crustal anatexis and magma mixing in the Idaho batholith. *In* Anderson, J. L. (ed.) *The nature and origin of Cordilleran Magmatism* 359–73. GEOL SOC AM MEM **174**.

Fleck, R. J. & Criss, R. E. 1985. Strontium and oxygen isotopic variations in Mesozoic and Tertiary plutons of central Idaho. CONTRIB MINERAL PETROL **90,** 291–308.

Foster, D. A. & Hyndman, D. W. 1990. Magma mixing and mingling between synplutonic mafic dikes and granite in the Idaho-Bitterroot batholith. *In* Anderson, J. L. (ed.). *The Nature and Origin of Cordilleran magmatism,* 347–58 GEOL SOC AM MEM **174**.

Green, T. H. 1991. Significance of garnet-bearing I-type volcanics and high-level intrusives from Northland, New Zealand. *In* Chappell, B. W. (ed) *Abstr Second Hutton Symposium Granites and Related Rocks,* 40. Canberra: Bureau of Mineral Resources Record 1991/25.

Gromet, L. P. & Silver, L. T. 1987. REE variations across the Peninsular Ranges batholith: Implications for batholithic petrogenesis and crustal growth in magmatic arcs. J PETROL **28,** 75–125.

Hamilton, W. B. 1988. Tectonic setting and variations with depth of some Cretaceous and Cenozoic structural and magmatic systems of the western United States. *In* Ernst, W. G. (ed.), *Metamorphism and Crustal Evolution of the Western United States* 1–40.

Hamilton, W. & Myers, W. B. 1974. Nature of the Boulder batholith of Montana. GEOL SOC AM BULL **85,** 365–78.

Hardyman, R. F. 1988. Eocene magmatism, Challis volcanic field, central Idaho: an overview. GEOL SOC AM ABSTR PROG **20,** 419.

Hardyman, R. F. & Fisher, F. S. 1985. Rhyolite intrusions and associated mineral deposits in the Challis volcanic field, Challis quadrangle. US GEOL SURV BULL **1658A-S,** McIntyre, D. H. (ed.), 167–79.

Hart, W. K. 1985. Chemical and isotopic evidence for mixing between depleted and enriched mantle, northwestern U.S.A. GEOCHIM COSMOCHIM ACTA **49,** 131–44.

Hildreth, W., Halliday, A. N. & Christiansen, R. L. 1991. Isotopic and chemical evidence concerning the genesis and contamination of basaltic and rhyolitic magma beneath the Yellowstone Plateau volcanic field. J PETROL **32,** 63–138.

Hill, R. I., Chappell, B. W. & Campbell, I. H. 1991. Late Archean granitoids of the southeastern Yilgarn Block Western Australia: age, geochemistry, and origin. *In* Chappell B. W. (ed.) *Abstracts of the Second Hutton Symposium on Granites and Related Rocks,* 46. Canberra: Bureau of Mineral Resources Record 1991/25.

Honjo, N., McElwee, K. R., Duncan, R. A. & Leeman, W. P. 1986. K–Ar ages of volcanic rocks from the Magic Reservoir eruptive center, Snake River Plain, Idaho. ISOCHRON/WEST **46,** 15–19.

Honjo, N. & Leeman, W. P. 1987. Origin of hybrid ferrolatite lavas from Magic Reservoir eruptive center, Snake River Plain, Idaho. CONTRIB MINERAL PETROL **96,** 163–77.

Hooper, P. R., Kleck, W. D., Knowles, C. R., Reidel, S. P. & Theiessen, R. L. 1984. Imnaha basalt, Columbia River Basalt Group. J PETROL **25,** 473–500.

Hyndman, D. W. 1983. The Idaho batholith and associated plutons, Idaho and western Montana. GEOL SOC AM MEM **159,** 213–40.

Hyndman, D. W. 1984. A petrographic and chemical section through the northern Idaho batholith. J GEOL **92,** 83–102.

Kiilsgaard, T. H. & Lewis, R. S. 1985. Plutonic rocks of Cretaceous age and faults in the Atlanta lobe of the Idaho batholith, Challis quadrangle. US GEOL SURV BULL **1658A-S,** McIntyre, D. H. (ed.), 29–42.

Leeman, W. P. 1982. Tectonic and magmatic significance of strontium isotopic variations in Cenozoic volcanic rocks from the western United States. GEOL SOC AM BULL **93,** 487–503.

Leeman, W. P., Menzies, M. A., Matty, D. J. & Embree, G. F. 1985. Strontium, neodymium, and lead isotopic compositions of deep crustal xenoliths from the Snake River Plain: evidence for Archean basement. EARTH PLANET SCI LETT **75,** 354–68

Lewis, R. S. & Kiilsgaard, T. H. 1991. Eocene plutonic rocks in south central Idaho. J GEOPHYS RES **96,** 13295–311.

Lipman, P. W. 1988. Evolution of silicic magma in the upper crust: the mid-Tertiary Latir volcanic firld and its cogenetic granitic batholith, northern New Mexico, U.S.A. TRANS R SOC EDINBURGH EARTH SCI **79,** 265–88.

Macdonald, R. & Smith, R. L. 1988. Relationships between silicic plutonism and volcanism: geochemical evidence. TRANS R SOC EDINBURGH EARTH SCI **79,** 257–63.

McCulloch, M. T. 1991. The role of subducted slabs in an evolving earth. EARTH PLANET SCI LETT (in press).

McCulloch, M. T. & Bennett, V. C. 1991. Progressive growth of the Earth's continental crust and depleted mantle: geochemical and geodynamical constraints. GEOCHIM COSMOCHIM ACTA (submitted).

McIntyre, D. H., Ekren, E. B. & Hardyman, R. F. 1982. Stratigraphic and structural framework of the Challis volcanics in the eastern half of the Challis 1° × 2° quadrangle. *In* Bonichsen, B. & Breckenridge, R. M. (eds) *Cenozoic Geology of Idaho.* BULL IDAHO BUR MINES GEOL **26,** 3–22.

Menzies, M. A. 1989. Cratonic, circumcratonic, and oceanic mantle domains beneath the western United States. J GEOPHYS RES **94,** 7899–915.

Miller, C. F., Hanchar, J. M., Wooden, J. L., Bennett, V. C., Harrison, T. M., Wark, D. A. & Foster, D. A. 1991. Source regions of granitoid plutons: evidence from lower crustal xenoliths and inherited accessory minerals. *In* Chappell, B. W. (ed.) *Abstracts of the Second Hutton Symposium on Granites and Related Rocks,* 68. Canberra: Bureau of Mineral Resources Record 1991/25.

Miller, D. M. 1983. Strain on a gneiss dome in the Albion Mountains metamorphic core complex, Idaho. AM J SCI **283,** 605–32.

Morgan, J. P. 1972. Plate motions and deep mantle convection. *In* Shagam, R. (ed.) GEOL SOC AM MEM **132,** 7–22.

Morgan, L. A., Doherty, D. A. & Leeman, W. P. 1984. Ignimbrites of the eastern Snake River Plain: evidence for major caldera-forming eruptions. J GEOPHYS RES **89,** 8665–78.

Norman, M. D. & Leeman, W. P. 1989. Geochemical evolution of Cretaceous–Cenozoic magmatism and its relation to tectonic setting, southwestern Idaho. EARTH PLANET SCI LETT **94,** 78–96.

Norman, M. D. & Mertzman, S. A. 1991. Petrogenesis of Challis volcanics from central and southwestern Idaho: trace element and Pb isotopic evidence. J GEOPHYS RES **96,** 13729–93.

O'Brien, H. E., Irving, A. J. & McCallum, I. S. 1991. Eocene potassic magmatism in the Highwood Mountains, Montana: petrology, geochemistry, and tectonic implications. J GEOPHYS RES **96,** 13237–60.

Rapp, R. P., Watson, E. B. & Miller, C. F. 1991. Partial melting of amphibolite/eclogite and the origin of Archean trondhjemites and tonalites. PRECAMBRIAN RES **51,** 1–25.

Rea, D. K. & Duncan, R. A. 1986. North Pacific plate convergence: a quantitative record of the past 140 m.y. GEOLOGY **14,** 373–6.

Rudnick, R. L. & Taylor, S. R. 1986. Geochemical constraints on the origin of Archaean tonalitic-trondhjemitic rocks and implications for lower crustal compositions. *In* Dawson, J. B., Carsell, D. A., Hall, J. & Wedpohl K. H. (eds) *The Nature of the Lower Continental Crust* GEOL SOC SPEC PUB **24,** 179–91.

Rytuba, J. J. & McKee, E. H. 1984. Peralkaline ash flow tuffs and calderas of the McDermitt volcanic field, southeast Oregon and north central Nevada. J GEOPHYS RES **89,** 8616–28.

Schuster, R. D. & Bickford, M. E. 1985. Chemical and isotopic evidence for the petrogenesis of the northeastern Idaho Batholith. J GEOL **93,** 727–42.

Suayah, I. B. & Rogers, J. J. W. 1991. Petrology of the Lower Tertiary Clarno formation in north central Oregon: the importance of magma mixing. J GEOPHYS RES **96,** 13357–71.

Taylor, S. R. & McLennan, S. M. 1985. *The continental crust: its composition and evolution.* Oxford: Blackwell Scientific Publishing.

Tilling, R. I. 1973. Boulder batholith, Montana: a product of two contemporaneous but chemically distinct magma series. GEOL SOC AM BULL **84,** 3879–900.

Thorpe, R. S. & Francis, P. W. 1979. Petrogenetic relationships of volcanic and intrusive rocks of the Andes. *In* Atherton, M. P. & Tarney, J. (eds), *Origin of granite batholiths: geochemical evidence,* 65–75. Kent, UK: Shiva.

Vallier, T. L. & Brooks, H. C. 1987. *The geology of the Blue Mountains.* US GEOL SUR PROF PAP **1436**

Wooden, J. L. & Mueller, P. A. 1988. Pb, Sr, and Nd isotopic compositions of a suite of Late Archean, igneous rocks, eastern Beartooth Mountains: implications for crust-mantle evolution. EARTH PLANET SCI LETT **87,** 59–72.

Wust, S. L. 1986a. Extensional deformation with northwest vergence, Pioneer core complex, central Idaho. GEOLOGY **14,** 712–4.

Wust, S. L. 1986b. Regional correlation of extension directions in Cordilleran metamorphic core complexes. GEOLOGY **14,** 828–30.

Wyborn, D., Chappell, B. W. & Johnston, R. M. 1981. Three S-type volcanic suites from the Lachlan Fold Belt, southeast Australia. J GEOPHYS RES **86,** 10335–48.

Zartman, R. E. 1974. Lead provinces in the Cordillera of the western United States and their geologic significance. ECON GEOL **69,** 792–805.

MARC D. NORMAN, Planetary Geosciences, Department of Geology and Geophysics, School of Ocean and Earth Science and Technology, University of Hawaii at Manoa, Honolulu, HI 96822, U.S.A.

WILLIAM P. LEEMAN, Keith-Wiess Geological Laboratories, Rice University, Houston, TX 77251, U.S.A.

STANLEY A. MERTZMAN, Department of Geology, Franklin and Marshall College, Lancaster, PA 17604, U.S.A.

Transactions of the Royal Society of Edinburgh: Earth Sciences, **83**, 83–93, 1992

Granite genesis and the mechanics of convergent orogenic belts with application to the southern Adelaide Fold Belt

Mike Sandiford, John Foden, Shaohua Zhou and Simon Turner

ABSTRACT: Two models for the heating responsible for granite generation during convergent deformation may be distinguished on the basis of the length- and time-scales associated with the thermal perturbation, namely: (1) long-lived, lithospheric-scale heating as a conductive response to the deformation, and (2) transient, localised heating as a response to advective heat sources such as mantle-derived melts. The strong temperature dependence of lithospheric rheology implies that the heat advected within rising granites may affect the distribution and rates of deformation within the developing orogen in a way that reflects the thermal regime attendant on granite formation; this contention is supported by numerical models of lithospheric deformation based on the thin-sheet approximation. The model results are compared with geological and isotopic constraints on granite genesis in the southern Adelaide Fold Belt where intrusion spans a 25 Ma convergent deformation cycle, from about 516 to 490 Ma, resulting in crustal thickening to 50–55 km. High-T metamorphism in this belt is spatially restricted to an axis of magmatic activity where the intensity and complexity of deformation is significantly greater, and may have started earlier, than in adjacent low-grade areas. The implication is that granite generation and emplacement is a causative factor in localising deformation, and on the basis of the results of the mechanical models suggests that granite formation occurred in response to localised, transient crustal heating by mantle melts. This is consistent with the Nd- and Sr-isotopic composition of the granites which seems to reflect mixed sources with components derived both from the depleted contemporary mantle and the older crust.

KEY WORDS: granite, mechanics, deformation, metamorphism.

The advection of heat within rising granitic bodies clearly plays an important role in the thermal evolution of many orogenic belts, particularly in orogenic belts that develop, and eventually preserve, high-T, low-P metamorphic belts. An important question relating to the geodynamic evolution of such belts is whether the heating responsible for granite formation represents the conductive response to the crustal thickening and associated mantle deformation, or, alternatively, the response to other transient processes such as the emplacement of mantle-derived magmas into the deforming lithosphere. The strong temperature dependence of lithospheric rheology (e.g. Brace & Kholstedt 1980) implies that the advection of heat within magmatic bodies ascending through the lithosphere may have considerable impact on the mechanical evolution of the deforming lithosphere, and indeed such a relationship has been proposed by a number of workers (e.g. Miller & Gans 1989; Sandiford & Powell 1991; Collins & Vernon 1991). Numerical simulations reported by Sandford *et al.* (1991) show that in orogenic belts appropriate to the development and eventual preservation of high-T, low-P metamorphism during convergence, the mechanical consequences of granite formation may vary considerably with the mechanism of granite formation. Thus we may expect that the observed spatial and temporal relationships between granite formation, deformation and metamorphism are strongly influenced by the mode of granite formation. However, such calculations are strongly dependent on the assumed rheology about which there is considerable uncertainty, and it is therefore useful to examine the results in the light of documented case histories. In this paper we briefly review some theoretical thermal constraints on, and mechanical consequences of, granite formation in orogenic belts and discuss the results in the light of geological and isotopic constraints on the evolution of granitic magmatism in the Early Phanerozoic southern Adelaide Fold Belt in S Australia.

1. Thermal constraints on granite genesis

Melting of crustal rocks may begin at temperatures as low as 600–650°C in rocks of appropriate composition; however, it is likely that significantly higher temperatures are required to produce the melt proportions and scale of melting appropriate to segregation of granitic melts. We term the temperature for granite formation by crustal melting T_{crit} and consider a likely minimum for T_{crit} of 800°C. Such temperatures are well in excess of any steady-state geotherm appropriate to normal continental crust, and therefore granite formation involving melting of crustal rocks may only occur in response to significant thermal perturbations. The common association of granitic rocks with deformation and metamorphism in ancient and modern fold belts testifies to the fact that orogenesis is frequently accompanied by extremely perturbed thermal regimes. We can distinguish between a number of the processes contributing to the perturbation of the temperature field in orogenic belts on the basis of the characteristic length- and time-scales of the perturbation. In this respect there are two distinct classes of process relevant to granite formation, namely (1) lithospheric-scale heating as a conductive response to the deformation, and (2) localised heating as a response to advective, transient heat sources.

Lithospheric-scale thermal perturbations must be produced by large-scale convergent deformation in orogenic belts. The displacement of material with respect to the Earth's surface results in the advection of heat and also changes the vertical distribution of heat sources within the lithosphere, with the thermal evolution of the deforming lithosphere depending on the inherited thermal structure, the way deformation is partitioned between crust and mantle lithosphere and the rate of deformation. The increased radiogenic heat production of thickened crust may lead to significant heating, but, as shown by Sandiford and Powell (1991) for typical heat production distributions in continental lithosphere, the amount of heating depends to a large extent on the way the mantle lithosphere responds to the deformation (Fig. 1a). During deformation the evolving temperature field is sensitive to the rate of deformation. For lithosphere of the order of 100 km thick vertical strain rates ($\dot{\varepsilon}_z$) must be less than a few times $10^{-16}\,s^{-1}$, for the potential thermal structure to be attained during deformation. At $\dot{\varepsilon}_z$ greater than about $10^{-15}\,s^{-1}$, the deformation will be essentially isothermal. The characteristic time-scale associated with a perturbation due to lithospheric deformation is of the order of the thermal time constant of the lithosphere, that is 10^7–10^8 years, and is therefore longlived with respect to the time-scales of active deformation in orogenic belts. As shown by Thompson (1989), granite generation may occur much earlier in the convergent deformation cycle if the thermal structure is significantly perturbed before convergence, for example by extension, and in the models described in the following sections we assume an initially perturbed thermal regime characterised by an abnormal initial surface heatflow of 85 mW m^{-2} (Fig. 1b).

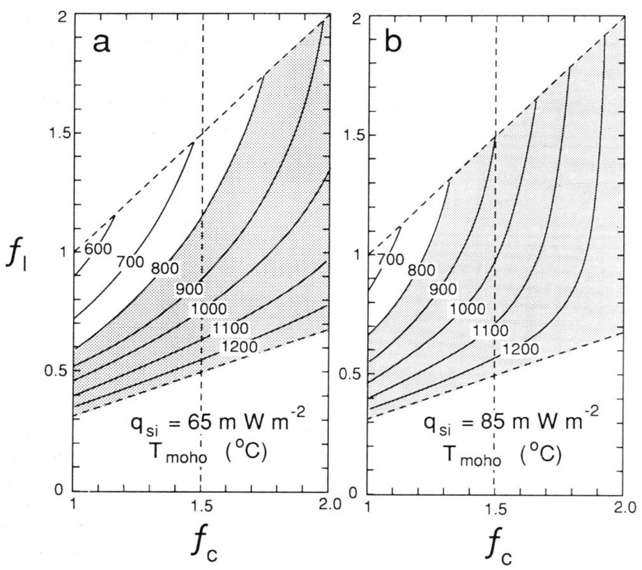

Figure 1 The effects of variable crustal and mantle thickening on the potential conductive response to deformation can be illustrated by showing the steady state or *potential* temperature as a function of f_c and f_l, where f_c and f_l are the ratio of deformed to initial crustal and lithospheric (crust and mantle) thicknesses, respectively (Sandiford & Powell 1990). This figure shows the potential Moho temperature for two different initial thermal states characterised by the initial surface heat flow, q_{si}: (a) q_{si} is 65 mW m^{-2} with a corresponding T_{Moho} of 560°C; (b) q_{si} is 85 mW m^{-2} with a corresponding T_{Moho} of 650°C. In (a) and (b), the heat source is assumed to be homogeneously distributed in the crust and absent in the mantle, the conductivity, of magnitude 3 W m^{-1} K^{-1}, is assumed to be temperature independent, the crust is initially 34 km thick and the mantle lithosphere initially 66 km thick.

Much more localised transient heating may accompany the emplacement of high temperature mantle-derived basaltic melts into the lithosphere. As shown by Huppert and Sparks (1988) this mechanism may lead to the generation of significant volumes of granitic melt, with the volume of melt produced depending upon a number of parameters, particularly the temperature of the country rocks at the time of emplacement. However, even for temperatures appropriate to the Moho for *normal* continental lithosphere, this mechanism can efficiently generate granitic melt. For example, Huppert and Sparks (1988) estimate that for country rock temperatures of 500°C, roof-rock melting above a basaltic sill may produce granitic melt volumes equivalent to that of the basaltic sill. The time-scale for thermal perturbations associated with mantle magmatism is of the order of 10^4–10^6 years (Huppert & Sparks 1988) and is therefore shortlived with respect to the time-scales of active deformation in orogenic belts.

2. Mechanical consequences of magmatic heat advection in a deforming lithosphere

In this section we consider the mechanical response of the lithosphere subject to compression during granite generation and advection in response to the two different scales of heating described above. In order to do this, we adopt a simple description in which the lithosphere is considered as a thin viscous sheet overlying an inviscid asthenosphere. This is the so-called thin sheet approximation which treats only vertical averages of stresses and strain rates (e.g. Bird & Piper 1980; Sonder & England 1986), and which allows the orogenic force balance to be expressed as:

$$F_d + F_b = F_l, \tag{1}$$

where F_d is the tectonic driving force, F_l is the strength of the lithosphere and F_b is a measure of the potential energy stored in the orogen and may be considered as the buoyancy force exerted by the deformed lithosphere on its surroundings (e.g. Molnar & Lyon-Caen 1988; Sandiford & Powell 1990). Forces are considered to be positive when directed towards the orogen. Thus in the discussion below F_d is positive and F_b is negative.

With the thin sheet approximation the strength of the lithosphere in compression is given by:

$$F_l = \int_{surface}^{z_l} (\sigma_{xx} - \sigma_{zz})\, dz, \tag{2}$$

where z_l is the depth of the base of the lithosphere. We adopt a lithospheric rheology in which failure is given by a combination of frictional sliding and power-law or Dorn-law creep, and assume that crust and mantle failure is governed by quartz and olivine respectively (see Appendix for details of the assumed rheology). An important feature of this rheology is that the lithosphere has distinct strength maxima, one corresponding to the transition from frictional sliding to ductile creep in the crust and the other to the transition from quartz- to olivine-dominated rheology at the Moho. In the discussion that follows, we consider the evolution of an orogen that is built through time in response to the application of a driving force of constant magnitude. We then briefly consider the evolution of such an orogen as the driving forces are relaxed.

2.1. Evolution with constant driving force

From equation (1), the deformation of a lithosphere subject to an externally-applied driving force of magnitude F_d can be understood as a balance between the effective driving force, $F_{ed} = F_d + F_b$, and the strength of the lithosphere, F_l. Assuming constant F_d, then F_{ed} will, in general, decrease as the deformation advances, with the rate of decrease dependent on the rate at which potential energy is stored in the orogen (which in turn depends primarily on the evolving density structure of the lithosphere, see Sandiford & Powell 1990). The theoretical limit to convergent deformation applies when F_b is equal in magnitude, but opposite in sign, to F_d, although in practice the limit will apply for $|F_b|$ less than $|F_d|$ by an amount related to the limiting strength of the lithosphere at the thermal regime appropriate to the deformed lithosphere. The limiting strength, F_{lim}, can be defined as the lithospheric strength at the limit of geologically significant strain rates. In the following discussion we consider this limiting strain rate to be $\dot{\varepsilon}_z = 3 \times 10^{-17}\ s^{-1}$. Figure 2a, which shows the dependence of strength of our model lithosphere in compression as a function of Moho temperature and strain rate, indicates that, for a Moho temperature of 800°C, F_{lim} is of the order of $1{\cdot}2 \times 10^{12} Nm^{-1}$.

Just as F_{ed} changes as a consequence of the evolving potential energy of the orogen, F_l changes as the thermal structure of the orogen evolves in response to the deformation and the advection of heat within granites and other magmas. As is appropriate to the viscous rheology outlined in the Appendix, F_l is only defined uniquely for a specified strain rate. When coupled with an equation relating deformation to the evolving thermal structure, equation 1 and the constitutive relations outlined in the Appendix can be solved for $\dot{\varepsilon}_z$ as illustrated in Figure 3, which shows the temporal evolution of crustal thickening rate $\dot{\varepsilon}_z$, crustal thickening, defined by the crustal thickening factor f_c, and buoyancy forces expressed as the ratio $F'_b = -(F_b/F_d)$, for a lithosphere subject to a driving force of $5 \times 10^{12}\ Nm^{-1}$ and inheriting a perturbed thermal regime characterised by an initial heatflow of $85\ mWm^{-2}$ (the full details of the mechanical models presented here are given in Sandiford *et al.* (1991). Note that the limit to significant strains occurs at F'_b approximately 0·75 since for the lower crustal temperatures achieved in these models, T_{Moho} approximately 900°C, the ratio F_{lim}/F_d is of the order of 0·2.

Figure 4 shows the same evolution as in Figure 3 but with granitic melt segregated when T_{Moho} reaches $T_{crit} = 800°C$ and emplaced as a 2·3 km-thick sill at 10 km depth. This depth corresponds approximately to a strength maxima associated with the transition from frictional sliding in ductile creep and consequently the heating associated with granitic ascent induces a transient weakening reflected in an increase in $\dot{\varepsilon}_z$ of up to about $5 \times 10^{-14}\ s^{-1}$, and a cumulative increase in f_c of about 5%. The weakening is short-lived because the dissipation of heat associated with the ascent of the granite is short-lived. Also, the rapid increase in the buoyancy forces associated with the transient high strain-rate deformation implies a concomitant reduction in F_{ed}. Note that the crustal thickening at the end of the transient high strain-rate deformation is greater by a few per cent than the limiting crustal thickening in the absence of heat advection by granites (Fig. 3). This highlights the important point that advective heat transfer processes operating in an orogen subject to a constant driving force are always capable of inducing a finite-strain increment. The magnitude of this strain increment is clearly dependent on the amount of heating and particularly its disposition with respect to the strength controlling regions in the lithosphere (for further details see Sandiford *et al.* 1991), as well as the bulk strength of the lithosphere at the time of the advective heating. For the model lithosphere considered here, the magnitude of the strain increment is severely limited because by the time the advective process is initiated ($T_{Moho} > 800°C$), the strength of the lithosphere has been significantly reduced by conductive heating with the consequence that F_b approaches $(F_d - F_{lim})$.

Figure 5 shows the same evolution as in Figure 3 but with granitic melt segregated at arbitrary times related to input of a transient localised heat source appropriate to a basaltic melt. The basalt is modelled as a 3 km-thick sill emplaced at the Moho with the granite modelled as a 2 km-thick sill at 800°C emplaced at 10 km depth. For this geometry there are two significant components to the weakening, one associated with thermal effects of granite emplacement which may affect the strength maxima in the upper crust and a potentially more important weakening associated with basalt

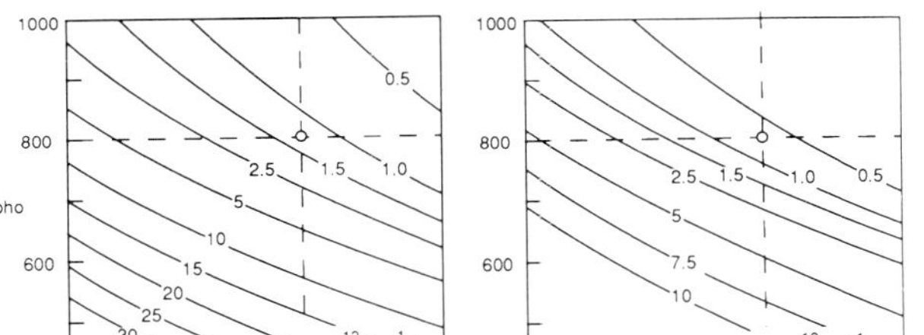

Figure 2 The strength of the model lithosphere, F_l, expressed as a function of T_{Moho} and the rate of crustal thickening, $\dot{\varepsilon}_z$: (a) lithospheric strength in compression; (b) lithospheric strength in tension; the difference occurs because of the different expressions governing frictional sliding in these two regimes (see equations 4 and 5 in Appendix). At the limiting rate for the accumulation of geologically significant strains ($\dot{\varepsilon}_z$ about $3 \times 10^{-17}\ s^{-1}$) and at $T_{Moho} = 800°C$, appropriate to the generation of granites in the deep crust, the strength of our lithosphere in compression is approximately $1{\cdot}2 \times 10^{12}\ Nm^{-1}$.

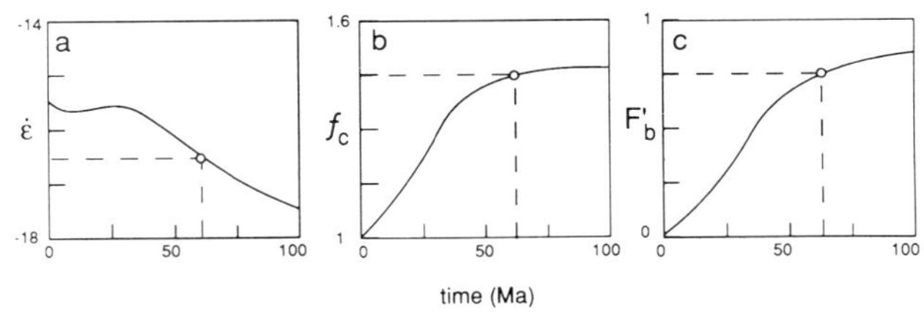

Figure 3 Model results presented in terms of the temporal evolution of: (a) $\log_{10}$ ($\dot{\varepsilon}_z$) (units of s^{-1}), (b) crustal thickening strain, f_c, (c) F'_b, the ratio of buoyancy forces to initial driving forces; results are for a total model run time of 100 Ma, with $F_d \times 5 \times 10^{12}\ N\ m^{-1}$, and with no granite generation; dashed lines show the co-ordinates at the limit of accumulation of geologically significant strains.

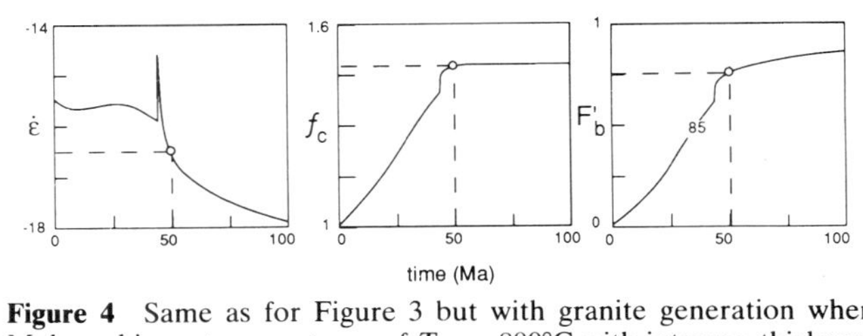

Figure 4 Same as for Figure 3 but with granite generation when Moho achieves temperatures of $T_{crit} = 800°C$ with intruson thickness of 2·3 km emplaced at 10 km depth.

emplacement which may temporarily destroy the strength maxima at the Moho. Moreover, since there is no linkage between granitic segregation and the overall thermal and hence mechanical state of the lithosphere, it is possible that the advective heat transfer associated with magmatism may occur very early in the deformation, and as shown result in a significant amount of weakening.

In the experiments described above, an initially perturbed thermal regime has been assumed in order to allow deformation at geologically significant rates for a convergent driving force ($F_d \times 5 \times 10^{12}$ Nm^{-1}) that would naturally limit the crustal thickening to $f_c = 1{\cdot}5$ (see Sandiford & Powell 1991, for a discussion of the forces appropriate to orogens in which low-P, high-T metamorphic terranes may be formed and preserved). For this driving force, lithosphere with a more typical thermal regime will undergo no significant deformation. This is illustrated in Figure 6, which shows the response of our model lithosphere with an initial heatflow of ($q_{si} = 65$ mW m^{-2}), with model parameters otherwise similar to Figure 5. In this model insignificant deformation occurs until the emplacement of basalt and the associated granite generation. Interestingly, once deformation is localised by the process of magmatic heat advection, and the thermal effects of the advective process have decayed, the deformation continues at a significant rate (in this case with $\dot{\varepsilon}_z$ approximately 2×10^{-16} s^{-1}). This is because the lithosphere remains weakened through the conductive response to the initial deformation and illustrates the possibility that mantle magmatism may be an effective trigger localising the initial deformation in a convergent orogen.

2.2. Evolution following relaxation of driving forces

In the experiments described in the preceding sections, the convergent deformation effectively ceased when F_b' approached 0·8. This implies that with relaxation of the force driving convergent deformation, the orogen will be

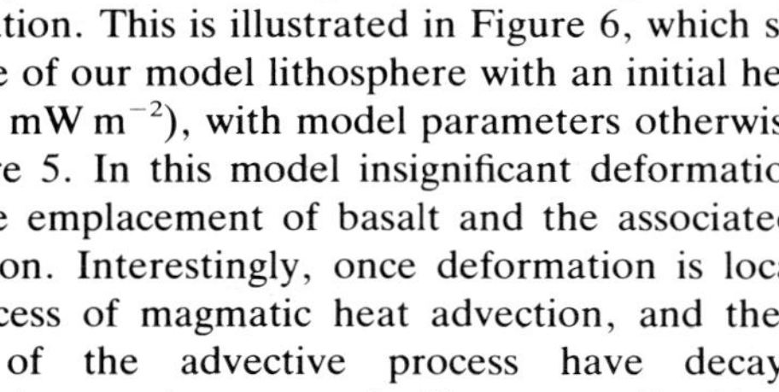

Figure 5 Same as for Figure 3 but for granite genesis due to intrusion of mafic magma sill ($z = 3$ km) at the Moho at times 15 Ma, 30 Ma and 45 Ma; total model time of 60 Ma, $F_d = 4{\cdot}5 \times 10^{12}$ N m^{-1}; thickness of the basaltic sill emplaced at the Moho is 3 km and the granite sill is 2 km emplaced at 10 km depth.

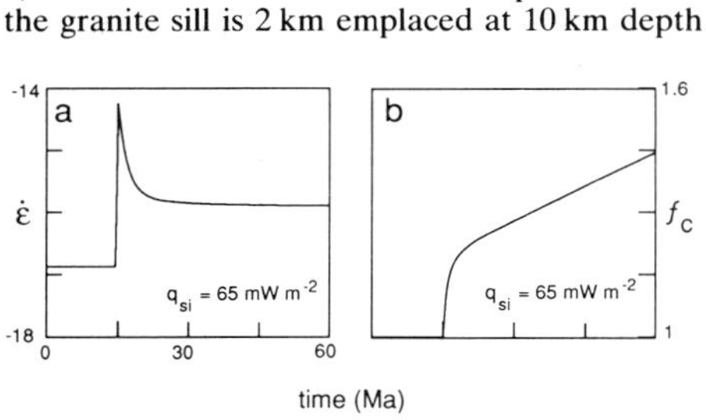

Figure 6 Evolution of a lithosphere with $q_{si} = 65$ mW m^{-2} subject to $F_d = 7 \times 10^{12}$ N m^{-1} and intruded by mafic magma with associated granite genesis at time 15 Ma (with identical intrusion geometry as in Fig. 5); absence of significant deformation before the magmatic episode suggests that mafic magmatism may provide sufficient thermal weakening to localise deformation in convergent orogenic belts.

subject to a significant extensional buoyancy force. Since the lithosphere is weaker in extension than in compression (compare Fig. 2b with 2a) this raises the possibility that convergent deformation is followed by significant extensional deformation. As shown by Sonder *et al.* (1987), England (1987) and Zhou and Sandiford (1991), whether or not extension occurs is dependent to a great extent on the thermal state of the lithosphere which is usefully characterised by the Moho temperatures (Fig. 2b). For example, Figure 2b shows that, at the limit of geologically significant strain rates, our model lithosphere is incapable of supporting extensional buoyancy forces of more than about $0{\cdot}5$–$1{\cdot}0 \times 10^{12}$ Nm^{-1} if T_{Moho} is greater than about 800°C. If the bulk strength and temperature-dependence of our model lithosphere is appropriate to the continental lithosphere (see discussion below) then an important implication is that, for orogens in which granite genesis occurs as a conductive response to a lithospheric-scale thermal perturbation with a characteristic timescale of 10^7–10^8 Ma, we may expect that convergent deformation is followed by significant extensional deformation. In contrast, where granites are generated by localised transient heating episodes with time scales of the order of 10^4–10^6 Ma, there is no necessary requirement for elevated Moho temperatures during relaxation of the forces driving convergence and therefore such terranes will not necessarily undergo late extension.

2.3. Discussion

In-as-much as we have based our calculations on laboratory-determined rheological parameters (e.g. Patterson 1987) and we have assumed a very simple lithospheric compositional structure, there are a number of important uncertainties that require comment. From the point of view of the role of magmas in the mechanics of lithospheric deformation, the main concerns are with the bulk strength of the lithosphere, the strength-depth profile with particular reference to the existence of distinct strength maxima in the mid-upper crust and upper mantle lithosphere, and the temperature and stress-dependence of the constitutive relations appropriate to the lithosphere. These concerns are briefly discussed below.

Any consideration of the strength of the continental lithosphere is complicated by its sensitivity to thermal structure. However, it is possible to limit to the range of acceptable lithospheric strength on the basis of a number of observations concerning the energetics of active mountain belts. For example, the excess potential energy associated with the Tibetan Plateau over normal continental lithosphere (and the mid-ocean ridge systems) is approximately 1×10^{13} Nm^{-1} (Zhou & Sandiford 1991), and we may assume that this plateau represents the condition where F_{ed} approaches F_{lim} (Molnar & Lyon-Caen 1988). T_{Moho} is probably significantly elevated beneath the Tibetan Plateau and therefore F_{lim} will be of the order of a few times 10^{12} Nm^{-1} or less (Fig. 2a) and thus the force driving convergence between India and Asia is likely to be of the order of 1–2×10^{13} Nm^{-1}. Any continental lithosphere appropriate to the formation of an orogen must therefore be able to deform at geologically significant rates when subject to driving forces of this magnitude. Figure 2a shows that for the rheology outlined in Appendix 1, and for a bulk strength of 10^{13} Nm^{-1}, our model lithosphere is capable of deforming at significant compressional strain rates providing Moho temperatures are of the order of 550–600°C, which is probably appropriate to a mildy perturbed lithospheric thermal structure.

The existence of two strength-controlling zones in our models results from the assumption that the strength of the crust and the mantle are controlled by quartz and olivine, respectively. Since olivine is a significant component in almost all mantle rocks, this assumption is likely to be valid for the mantle. In contrast, the compositional structure of the continental crust is complex, and it is not at all self-evident that quartz will always be strength-limiting, especially in the lower parts of the crust, and an alternative model for crustal strength involves a number of strength-controlling regions appropriate to a complex compositional structure (e.g. Ord & Hobbs 1989). However, there is some geological support for the notion that much of the lower crust represents a weak zone during orogenesis, at least at elevated temperatures. For example, Gans (1987) has suggested that the very flat Moho topography in the Basin and Range Province reflects the annihilation of Moho gradients generated during extension by lower-crustal channel flow. While not constraining the detailed strength-depth profile, such behaviour implies that the lower crust is significantly weaker than either the upper mantle or the mid-upper crust (see also Bird 1991), and thus provides some support for the general form of the strength-depth profile adopted here.

England and co-workers have shown that the length scales of deformation in thin-sheet models of continental deformation are sensitive to the parameters governing the averaged rheology of the thin sheet (see England & Jackson 1989, for an accessible review). Therefore, comparison of the thin-sheet predictions with the observed distribution of deformation within active orogenic belts provides one important test of continental rheology, and suggests an averaged power-law relation with a stress exponent of at least 3, as employed in the models presented here.

While these observations can be taken as providing some support for the general form of the rheology assumed in the models described here, they in no way validate the precise details of this rheology, and one of our significant concerns is whether we can use geological observations to further constrain our understanding of the rheology of the continental lithosphere. Indeed, it is our hope that we may be able to integrate the results obtained from the types of modelling strategies outlined here with the observed spatial and temporal distribution of deformation and magmatism in orogenic belts to begin to assess rheological parameters. To this end, we reiterate the main implications of the model results described above:

(1) Granite generation as a conductive response to lithospheric deformation must occur late in the deformation cycle. This is not principally because of the necessary time for this mechanism to generate lower crustal temperatures appropriate to granite formation (although this is certainly a contributing factor), but rather it is because for these thermal regimes the lithosphere is very weak and at any given driving force the great proportion of the potential deformation will already have occurred (of course, it should be noted that it may be possible to produce rather different temporal and spatial relationships between deformation and magmatism by varying the driving forces in an *ad hoc* fashion).

(2) Granite generation associated with a localised transient thermal perturbation due to basalt intrusion in the deep crust and/or upper mantle may have a potentially dramatic effect on the deformation if the perturbation occurs while there is still significant lithospheric strength, that is before T_{Moho} reaches about 650–700°C. Moreover, for orogens forming in response to the relatively small driving forces that seem appropriate for the formation and eventual preservation of low-P, high-T metamorphic belts (Sandiford & Powell 1991) it is possible that the advective heating may provide the trigger that initially localises deformation within the orogen, in which case we may expect granites to accompany the first increments of deformation.

(3) A further important difference between the two mechanisms for generating granites relates to the stability of the lithosphere to extensional deformation following relaxation of the forces driving convergence. If the thermal regime responsible for granite genesis is representative of the lithospheric thermal regime, then it seems unlikely that the lithosphere could escape significant late-stage extension, since for a given thermal regime the lithosphere is necessarily weaker in tension than compression.

In the following sections, we describe a natural example of granite magmatism from the southern Adelaide Fold Belt, where geological and isotopic data together with insights gained from the models described above can be used to constrain the dynamics of that particular belt.

3. Role of granites in the southern Adelaide Fold belt

3.1. Geological background

The Adelaide Fold Belt is part of a Late Proterozoic to Early Palaeozoic continental marginal orogen which extended over 5000 km along the south-eastern edge of the early Gondwana supercontinent. Within the S Australian sector, this belt was deformed, metamorphosed and intruded by granites during the 520–480 Ma Delamerian Orogeny. From about 900 Ma, thick, dominantly shallow marine sedimentary sequences, comprising the Adelaidean system, were deposited in a series of basins developed within and perhaps on the margins of the Archaean to mid Proterozoic basement of the Gawler Craton (Preiss 1987). In the early stages, subsidence was associated with the extrusion of voluminous mafic continental tholeiitic volcanics (Crawford & Hilyard 1990) in rift-related grabens. This suite encompasses the Wooltana, Depot Creek and Beda Volcanics, and the Gairdner dyke swarm, loosely referred to here as the Early Adelaidean mafic suites. A renewed phase of sedimentation in the southern part of the belt began at about 526 Ma, with the deposition of the Normanville and Kanmantoo Groups (Jenkins 1990; Cooper *et al.* 1990). The Normanville Group contains mafic lavas (the Truro volcanics) that include very alkaline compositions and yet have Nd- and Sr-isotopic compositions approaching that of the Cambrian depleted mantle. Emergence of a sediment provenance during the deposition of the Kanmantoo Group is witnessed by the dramatic influx of clastics that are, at least in part, turbiditic. Mafic dykes, now metamorphosed and deformed, suggest magmatism may have continued during this interval, or resumed shortly after deposition, with a much less alkaline signature than the Truro Volcanics. Deposition had apparently ceased before the 516 Ma emplacement of the earliest of the granites (the Rathjen and Tanunda Creek granites, Foden *et al.* in prep.) associated with the Delamerian Orogeny.

Granite emplacement in the southern Adelaide Fold Belt spans a period of about 30 Ma (Milnes *et al.* 1977; Foden *et al.* in prep.) with two distinct chemical, spatial and temporal, associations developed (Foden *et al.* 1990; Turner & Foden 1990):

(1) an *older suite* of dominantly I-type granites (encompassing a spectrum through to S-type) characterised

by (variably developed) tectonic fabrics, with zircon ion-microprobe ages in the range 516–490 Ma, and found within presently-exposed Kanmantoo Group sediments; and

(2) a *younger suite* of undeformed A-type granites associated with further mafic magmas, with intrusion ages of 490–480 Ma, and now found as isolated monadnocks within Tertiary sediments to the E of the main exposures of the Adelaide Fold Belt (Turner *et al.* 1992).

The following discussion emphasises that the very significant differences in the geochemical character of the two granite suites reflects important changes in the process of granite magma formation as the available source rock compositions are likely to have been much the same throughout this interval.

3.2. Deformation, metamorphism and timing of granite magmatism

The older suite of granites are widely distributed within the exposed Kanmantoo Group sediments, being most abundant in a NNW-trending belt to the E of Adelaide (Fig. 7). In this region, the granites occupy the core of a regional metamorphic culmination characterised by the progressive development of biotite, staurolite-andalusite, sillimanite and migmatite zones (Offler & Fleming 1968; Mancktelow (1990; Dymoke & Sandiford 1992). Lateral gradients in T_{max} reach $10°C\,km^{-1}$ (Dymoke & Sandiford 1992). The dominant structures in this region are N-trending, upright, open to tight, folds which cut obliquely across the NNW-trending metamorphic gradients, thus affording a unique opportunity to observe the development of, and timing of mineral growth within, single structures as they pass up-grade. In the biotite zone, away from the main granite belt, the structure is relatively simple, being dominated by open upright folds, typically with a single axial planar foliation.

Porphyroblastic rocks of the andalusite-staurolite and sillimanite zones preserve evidence of a more complicated structural history as complex inclusion trails in porphyroblasts and crenulated fabrics. However, the dominant fabrics and the highest temperature mineral assemblages are associated with the same upright structures as at lower grades. In contrast, in the migmatite zone the dominant foliation is crenulated by fabrics associated with the upright folds. The majority of the migmatitic segregations are contained in the early foliation, although some occur as axial fabrics to the upright folds (Fleming & White 1984). The early foliation is concordant with, and contained within, the oldest, sheet-like, granitic bodies such as the Rathjen gneiss (a tabular body with currently-exposed dimensions approximately $10 \times 3 \times 0{\cdot}4$ kms). Numerous mesoscopic isoclinal folds are observed associated with this early foliation; however, macroscopic folds have not been demonstrated convincingly, in part, no doubt, because the Kanmantoo succession is a monotonous sequence in which stratigraphic markers are difficult to recognise.

These observations strongly support the notion that heat focused in the highest-grade regions diffused laterally into the lower-grade regions, producing progressively lower maximum temperatures achieved at progressively later times, and therefore strongly implicate the currently-exposed magmatic bodies as being representative of the heat source for the low-P, high-T metamorphism (we note that, in order to explain the total heat budget of the belt, it is necessary to appeal to significantly greater volumes of melt than present in the current erosion surface). The emplacement of the oldest of the granites appears to have predated, or, more probably, was associated with, the earlier increments of strain within this part of the belt. The Palmer Granite, and all other granites younger than about

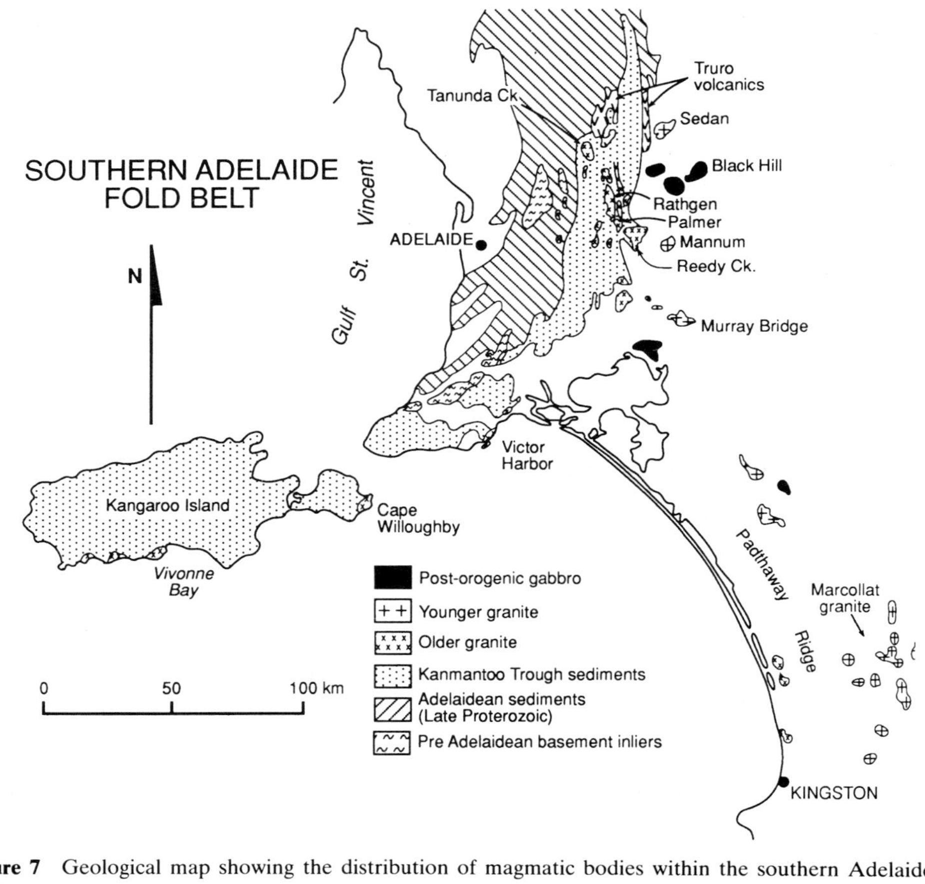

Figure 7 Geological map showing the distribution of magmatic bodies within the southern Adelaide Fold Belt.

506 ± 6 Ma (the recalculated age of the Palmer Granite, Foden *et al.* in prep.), preserve only evidence of the upright foliation. The youngest demonstrated zircon ion-microprobe age for a foliated intrusive is from the Reedy Creek Granodiorite at 490 Ma. While these data apparently constrain the age of the older foliation, they allow two interpretations for the timing of upright structures, namely (1) the upright structures developed progressively, or episodically, over a significant period of time (from 516–487 Ma), or (2) the upright structures postdate the youngest members of the older granite suite (from 490–487 Ma, with the younger constraint provided by the observation that the foliated Reedy Creek Granodiorite is cross cut by an undeformed diorite dyke with ion-microprobe zircon age of 487 ± 2 Ma). While we cannot discriminate between these possibilities on the basis of the available data and field observations, an important constraint on the kinematic significance of these structures is provided by the timing of the main crustal thickening. Layer-parallel quartz veins in the andalusite-staurolite and sillimanite zones, formed during the earlier of the two main fabric-forming episodes, contain kyanite overprinted by andalusite which in turn is partially replaced by fibrous sillimanite (Sandiford *et al.* 1990), suggesting that the heating attendant on attainment of peak temperatures (coincident with the second, upright structures in these zones) was essentially isobaric at pressures of about 400 MPa. The implication is that tectonic burial to depths of approximately 14 km was achieved early in the structural history. Note that since the current crustal thickness is of the order of 35–40 km (Greenhalgh *et al.* 1991), and there is no evidence of significant, subsequent extension in this part of the belt, the crustal thickness during peak metamorphism must have been of the order of 50–55 km.

3.3. Geochemistry of the granites and potential source regions

The older suite of granites are mainly I-types ranging from tonalite to true granite, but do include a continuous variation to granites with S-type character. The granites are typically biotite and/or hornblende-bearing, are relatively calcic, with high Ba, Sr contents and low Nb, Ti, Y, F and Ga contents. The isotopic compositions range from those, like the Reedy Creek granodiorite, having very low initial $^{87}Sr/^{86}Sr$ ratios (0·7060) and high initial ε_{Nd} (−2·5 – −3·1), through to those, like the Vivonne Bay Granite, with high

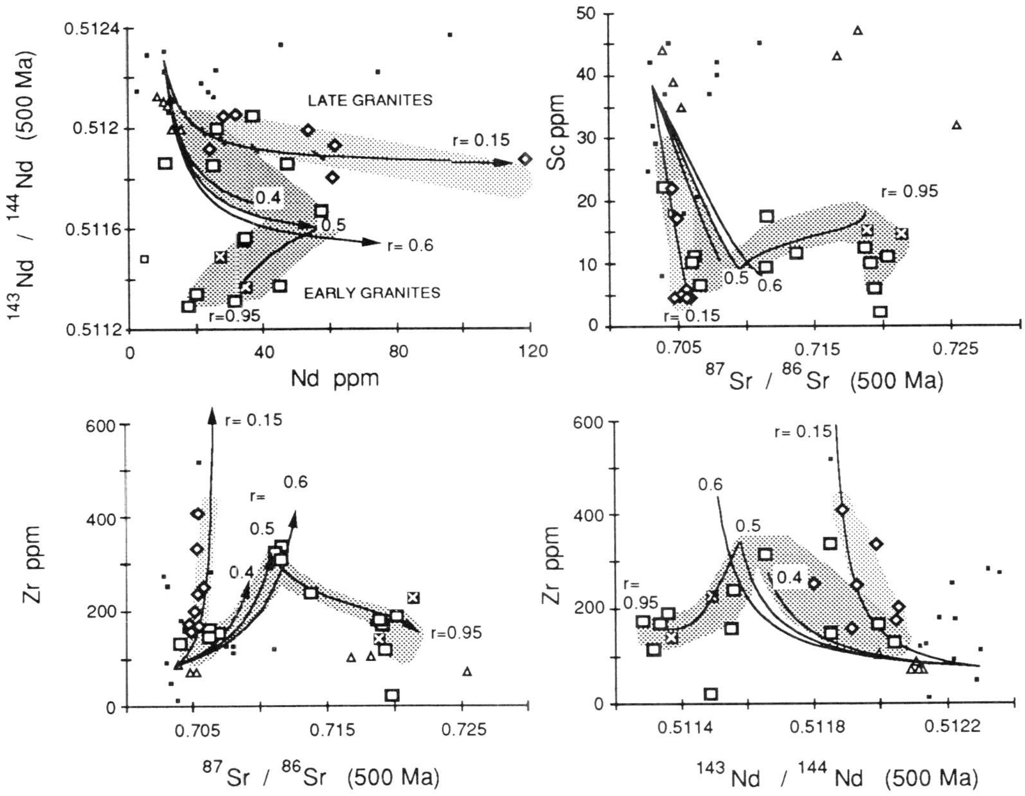

Figure 8 Trace element vs isotopic composition variations at 500 Ma for rocks of the Adelaide Fold Belt: filled squares—Delamerian basalts; open triangles—Adelaidean basalts, open squares—younger Delamerian granite suite, crossed squares—average compositions of Adelaidean and Kanmantoo sediments; trend of the older granite suite (516–490 Ma) shown in darker shading, trend of the younger granite suite shown in the lighter shading; curves represent AFC trends involving a 500 Ma basaltic end-member and an average crustal composition calculated using the equation of DePaolo (1981) with r = assimilation rate/rate of crystallisation. The parameters of the modelled curves are as follows: basaltic end-member, Nd = 5·8 ppm, $^{143}Nd/^{144}Nd$ (500 Ma) = 0·51230, Sr = 120 ppm, $^{87}Sr/^{86}Sr$ (500 Ma) = 0·7028, Zr = 50 ppm, Sc = 38 ppm, Sm = 1·5 ppm; crustal endmember: Nd = 30 ppm, $^{143}Nd/^{144}Nd$ (500 Ma) = 0·511135, Sr = 150 ppm, $^{87}Sr/^{86}Sr$ (500 Ma) = 0·7119, Zr = 180 ppm, Sc = 15 ppm, Sm = 7 ppm. AFC curves r = 0·4, 0·5, 0·6 terminate at 55% crystallisation, the curve r = 0·95 represents a further 50% crystallisation. Bulk distribution coefficients used are as follows: D_{Nd} = 0·0001 and 1·1, D_{Sr} = 0·005 and 2·5, D_{Zr} = 0·005 and 1·2, D_{Sc} = 2 and 1·8, where the second values listed are those used in the last assimilation-dominated (r = 0·95) stage. Note that the isotopic data utilised in Figures 8, 9 and 10 was largely determined at the Department of Geology, University of Adelaide, using a Finnigan-Mat 261 mass-spectrometer; data and methods are described in Foden *et al.* (in prep.); the major and trace element compositions were also determined at the University of Adelaide using a Phillips 1480, 100 kV XRF; in most cases each datapoint represents a single pluton.

initial $^{87}Sr/^{86}Sr$ (0·719) ratios and lower ε_{Nd} (−13·6). The group shows continuous correlated variation of Nd- and Sr-isotopic compositions with trace-element concentration (Fig. 8). Nd-model ages range between 1·1 and 2·4 Ga (Fig. 10) and the oldest Nd-model ages and highest Sr-isotopic ratios are the same as values in the Kanmantoo and Adelaidean sedimentary country rocks (see next section).

The younger suite of granites is associated with some mafic intrusions and thus forms part of a bimodal suite (Turner & Foden 1990; Turner *et al.* 1992). They are characteristically single-feldspar A-types, with high F, LREE, Y, Nb and Ga contents, high Rb/Sr and Nb/Y and Nb/Zr ratios, low Sr and Ba contents, and very low Sc, Ni and Cr contents. Isotopic systems are remarkably primitive, overlapping the values for contemporary mafic rocks (Fig. 9, 10). Initial $^{143}Nd/^{144}Nd$ ratios are high with ε_{Nd} values (at 487 Ma) from −4 to +2 and initial $^{87}Sr/^{86}Sr$ ratios are in the range 0·7035–0·7068 (Fig. 8). Nd-model ages cluster very strongly about 1·1 Ga (Fig. 10). Although initial Nd- and Sr-isotopic variation is correlated with trace-element concentration variation, total isotopic change is very slight compared to the very large changes in trace-element content (Fig. 8). This is in contrast with the older granite suite where isotopic variation is dominant.

Potential source regions available to contribute to the Delamerian magmas are as follows. The mid to upper crust into which the magmas were emplaced consists of Late Proterozoic Adelaidean metasediments (at 500 Ma average $\varepsilon_{Nd} = -10$ and $^{87}Sr/^{86}Sr = 0{\cdot}7286$) and metasediments of the Cambrian Kanmantoo Group (at 500 Ma average $\varepsilon_{Nd} = -12$ and $^{87}Sr/^{86}Sr = 0{\cdot}7201$). Early Proterozoic basement slivers within the thrust belt that forms the western margin of the southern Adelaide Fold Belt suggest that much of the mid-crust beneath the fold belt may be similar to the Early Proterozoic and Archaean gneissic rocks now widely exposed on the Gawler Craton. On the craton, lower Proterozoic felsic rocks consist of orthopyroxene-bearing granites and alkali granites with a typical range (at 500 Ma) of ε_{Nd} from −12 to −17 and $^{87}Sr/^{86}Sr$ from 0·7718 to >1·0, whereas the mafic lithologies, including gabbros and basaltic dykes, show ε_{Nd} in the range −17–−21 and $^{87}Sr/^{86}Sr$ in the range 0·7090–0·7300. At 500 Ma, the Archaean Cape Carnot felsic gneisses have ε_{Nd} around −27 and $^{87}Sr/^{86}Sr$ between 0·7369 and 0·7810, whereas associated mafic granulites have ε_{Nd} between −11 and −15 and $^{87}Sr/^{86}Sr$ from 0·71194–0·7276. It is probable that the composition of the Adelaidean and Kanmantoo shaley sediments represents some average of all these exposed older crustal components. Lower-most crust and upper mantle xenoliths from kimberlites in S Australia are granulitic and eclogitic (McCulloch *et al.* 1982; Arculus *et al.* 1990), with ε_{Nd} (500 Ma) from −12 to −21, and $^{87}Sr/^{86}Sr$ (500 Ma) from 0·7066 to 0·71196. The deeper subcontinental lithospheric mantle may be similar to mantle xenoliths sampled by Recent volcanics in S Australia (Mt Gambier) and western Victoria where the mantle lithosphere has been enriched for long enough to allow isotopic divergence from the depleted mantle (McDonough *et al.* 1985; McDonough & McCulloch 1987; O'Reilly & Griffin 1988; Griffin *et al.* 1988). At 500 Ma, ε_{Nd} ranges from +6 to −1·1 while $^{87}Sr/^{86}Sr$ is

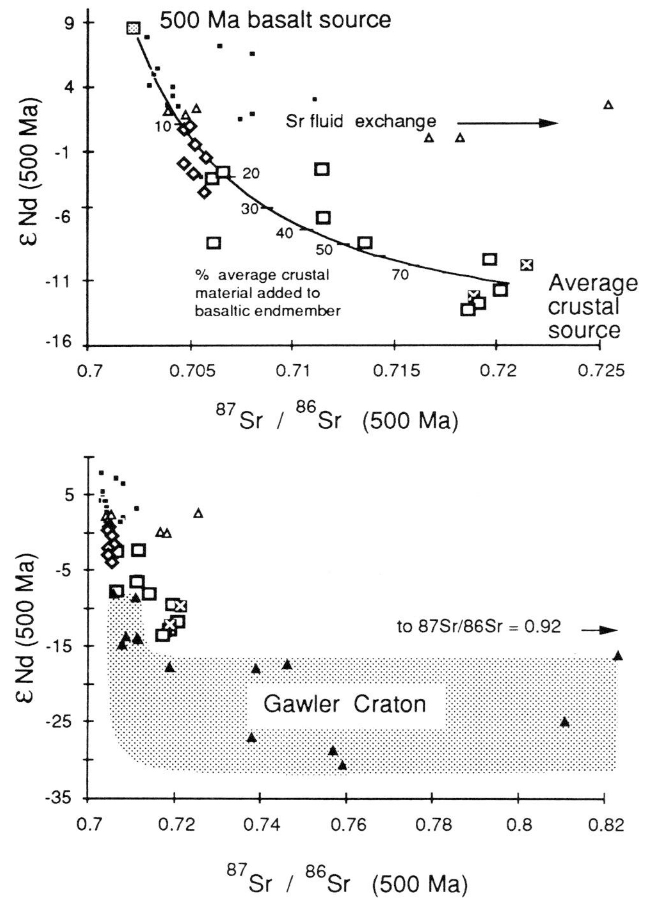

Figure 9 ε_{Nd} (500 Ma) and $^{87}Sr/^{86}Sr$ (500 Ma) variations for the same samples shown with the symbols as in Figure 8: part B shows the field of the Gawler Craton earlier Proterozoic and Archaean samples; curve shows the simple average crust-basalt mixing relationship using the same data as listed in Figure 8.

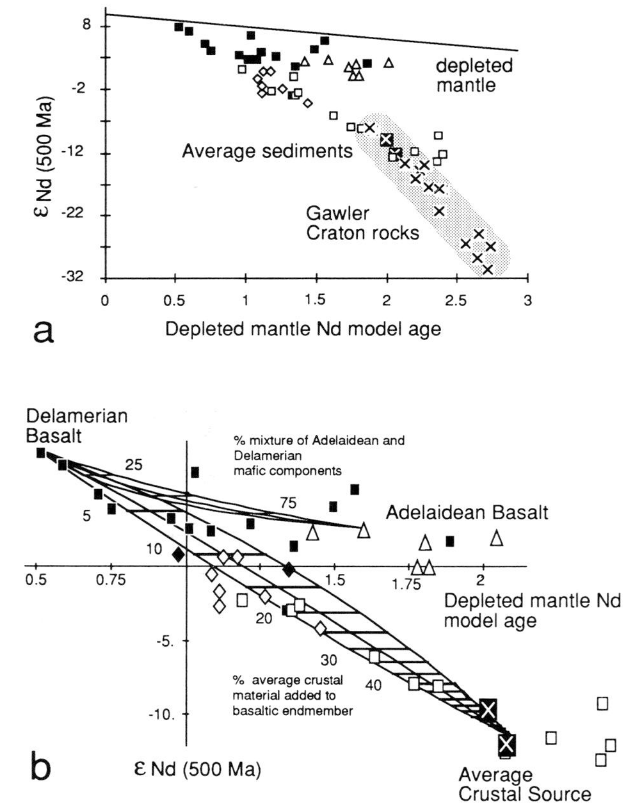

Figure 10 (a) ε_{Nd} (500 Ma) vs depleted mantle model age for samples illustrated in Figure 8 plus Gawler Craton data; (b) modelled two-end-member mixing between 500 Ma depleted mantle-derived basalt and the average crust and Adelaidean basalt. The separate mixing curves are for Sm = 1, 1·25 and 1·5 ppm, respectively.

0·7035–0·7069. Finally, melts from the contemporary depleted mantle would have ε_{Nd}(500 Ma) = +7·5, $^{87}Sr/^{86}Sr$ (500 Ma) = 0·70248.

3.4. Role of contemporary mantle in granite genesis

The isotopic compositions of the Delamerian and Early Adelaidean mafic rocks, the two Delamerian granite suites, the average composition of Kanmantoo and Adelaidean sedimentary rocks (assumed to represent the average mid-upper crustal compositions) and the field of Archaean to mid Proterozoic basement lithologies of the Gawler Craton are shown in ε_{Nd}-$^{87}Sr/^{86}Sr$ (at 500 Ma) space in Figure 9. Note that the splaying of the compositions of some of the mafic rocks to higher $^{87}Sr/^{86}Sr$ values reflects alteration effects during metamorphism, a feature to which the REE group (and Nd isotopes) seems to have been largely immune (Crawford & Hilyard 1991). Figure 9 shows that the older Delamerian granite suite forms a continuous trend from the field of Delamerian and Early Adelaidean mafic rocks to the field of the average mid-upper crust. This trend is suggestive of mixing since, on this type of diagram of initial isotope values, fractional crystallisation by itself will not produce any variation. If, as is more probable, contamination (mixing) and fractional crystallisation occur simultaneously, then the shape of the curve on this diagram may be modified, due to changing Sr/Nd ratios. Figure 9a illustrates a mixture of the average crustal composition and the most primitive Delamerian basalt, which reproduces the observed range in older granite compositions, implying that they mostly represent primitive mafic magmas contaminated by between 15% and 40% mid-upper crustal material, extending to approximately 100% fusion of crustal material in the case of the *S-type granites*. Note that the younger Delamerian granite suite defines a distinct trend with much less total Nd- and Sr-isotopic variation than the older granites.

Figure 10 illustrates the variation of Nd-model age (depleted mantle) versus ε_{Nd} (at 500 Ma). On this diagram, most of the Delamerian data (both mafic and felsic) form a linear trend between the field including average Kanmantoo and Adelaidean sediment compositions and S-type granites and the most primitive Delamerian basalts intersecting the depleted mantle at 500 Ma. The implication is that the granites are crust-mantle mixtures. Note that on this diagram Early Adelaidean mafic rocks do not lie on this trend, implying that these, or some deep crustal equivalents, cannot be a major source for the Delamerian granites. Figure 10a illustrates how the mixing relationships vary with different initial Sm/Nd ratios in the mafic end member, with the observed data falling within the mixing field for a mafic end member, and provides support for the notion that most of the I-type granites of the older suite are contemporary mantle derivatives contaminated by between 15% and 40% crustal material.

Figure 8 illustrates the variation of selected trace elements as a function of isotopic composition and emphasises the difference between the older and younger granite suites. The former show very large-scale isotopic range and limited trace element variation, while the latter show the opposite. The older granites all lie on curving trends which connect the Delamerian basalt field with the average crustal compositions. In converging on the average-sediment compositions, elements such as Nd and Zr show first enrichment and then depletion. Since these elements are strongly incompatible in basaltic magmas, but may become compatible in lower temperature felsic magmas, this trend suggests a two-stage combined assimilation and fractional crystallisation history (e.g. DePaolo 1981). In the first stage, the ratio, r, of *assimilation rate* to *fractionation rate* is moderate ($r = 0{\cdot}5$) and K_d values are low. In the second stage there is more extensive assimilation ($r = 0{\cdot}95$) amounting to direct mixing. As illustrated in Figure 8, strongly compatible elements such as Sc also follow this trend, with concentrations rising slightly in the late assimilation-controlled stage (not a trend explicible by any melting, fractionation, or restite unmixing scenario). By contrast, the younger granites are the potential products of the same end members (basalt and crust), but the process is dominated by fractionation with very minor assimilation ($r = 0{\cdot}15$).

4. Discussion

Granite magmatism has clearly played an important role in the thermal evolution of the southern Adelaide Fold Belt, and any attempt to understand the geodynamic evolution of this belt, and many other similar belts, requires an evaluation of the extent to which magmatism has been focused by deformation or, alternatively, deformation has been localised by magmatism. The apparent generation of granites throughout the convergent deformation history, the lack of significant late extensional deformation, and a characteristic isotopic signature reflecting significant involvement of mantle melts would seem to lend support for a primary role for magmas; this contention is also supported by numerical models based on the thin-sheet approximation. In this scenario, the evolution of the orogen may be seen in terms of the progressive localisation of convergent deformation, primarily as a consequence of weakening attendant on advection of heat in mantle and, to a lesser extent, crustal magmas, in a lithosphere subject to an externally-applied compressive stress field. Since each increment of crustal thickening strain results in an increase in the potential energy stored in the orogen, the potential for further deformation is reduced, with the kinematics reflecting an interaction between the forces driving convergence and the potential energy stored in the orogen. Since this interaction is mediated by the mechanical strength of the lithosphere, it is sensitive to the evolving thermal regime.

In some ways the mechanism of granite formation implied by our analysis resembles the "MASH" (melting, assimilation, storage, homogenisation) model of Hildreth and Moorbath (1988) who point out that contamination of basaltic magmas will be promoted by factors that facilitate deep crustal stalling or storage. In the southern Adelaide Fold Belt, the transition from the older highly-contaminated suite to the younger highly-fractionated suite reflects a dramatic change in the nature of the petrogenetic processes, which we believe are coupled to the mechanical state of the lithosphere.

One important, and general, insight provided by the discussion above concerns the distinction between so-called *syn-tectonic* granites and *post-tectonic* granites commonly made on the basis of whether or not the granites are strained, and which is often taken to reflect an important episode in the evolution of an orogenic belt such as a reduction in the forces driving convergence. As we have shown, the kinematics vary with the evolution of potential energy in an orogenic belt. Thus without independent evidence concerning the magnitude of the potential energy stored in the orogen, evidence for a change in the kinematics is not sufficient to infer changes in the driving forces.

Finding observations that reflect the way in which potential energy has evolved in ancient orogens represents a formidable, and generally unsolved, challenge. However, it may be that information regarding the relative magnitude of buoyancy forces and driving forces is recorded in the way in which magmas interact with the lithosphere during their ascent. For example, in the southern Adelaide Fold Belt there is a clear difference in the geochemical evolution of the early and late granite suites which we attribute to very different degrees of crustal assimilation (see Fig. 8). One reason for this variation may be that the extent of assimilation is governed by the averaged state of stress within the lithosphere. The field evidence suggests that this transition in geochemical behaviour corresponds to the transition from the condition $F_{ed} > F_{lim}$ to $F_{ed} < F_{lim}$. There are three possibilities for such a transition: (1) an increase in the buoyancy forces, F_b, as a direct consequence of crustal thickening strain at constant driving force, F_d; (2) an increase in F_b as a consequence of changes in the density structure of the lithosphere not directly related to crustal thickening strain such as convective thinning of the lower thermal boundary layer of the mantle lithosphere (e.g. Houseman *et al.* 1981; Sandiford & Powell 1990), at constant F_d; (3) a decrease in F_d. In the first case, there can be no qualitative difference in the bulk deviatoric stress field which must remain compressional, at a level dictated by F_{lim}. However, the remaining two possibilities allow a dramatic change in the stress field with the potential of developing extensional deviatoric stresses (Sandiford & Powell 1990). We suggest that the dramatic change in the geochemical transition from early to late granites reflects a qualitative change in stress field produced by one of these two mechanisms.

5. Acknowledgements

This work represents the outgrowth of continuing projects concerned with the thermal and mechanical evolution of the southern Adelaide Fold Belt, funded in part by the ARC and Adelaide University URG grant schemes, and supported by the Department of Geology. Our current understanding of this belt reflects the important contributions of a number of others involved in the projects, and we especially acknowledge the contributions of Nick Martin, Geoff Fraser, Simon Mitchell and Peter Dymoke and the expert technical assistance of David Bruce and John Stanley in the fields of mass spectrometry and XRF analysis.

6. Appendix

It has become common to model failure of the continental lithosphere as a combination of frictional sliding and strongly temperature-dependent ductile-creep processes. We follow this approach and assume a very simple compositional structure for the lithosphere, consisting of a quartz-dominated crust and an olivine-dominated mantle that behave according to the constitutive relations outlined below.

At low temperature and high strain rate, the failure mechanism in the upper crust and mantle is modelled as frictional sliding according to Byrlee's law (Byrlee 1968; Brace & Kholstedt 1980);

$$\tau_s = \sigma_0 + \mu\sigma_n(1-\lambda) \quad (3)$$

where τ_s, σ_0, μ, σ_n and λ are, respectively, the shear stress, the cohesion, the coefficient of friction, normal stress on the failure plane and the ratio of fluid pore pressure to the normal stress. Assuming λ is constant, then for biaxial horizontal compression the maximum and minimum principle stresses are related by (e.g. Sonder & England 1986):

$$\sigma_1 - \sigma_3 = \frac{2(\sigma_0 - \mu\sigma_z(1-\lambda))}{\sqrt{\mu^2+1}-\mu}, \quad (4)$$

and for biaxial horizontal tension:

$$\sigma_1 - \sigma_3 = \frac{2(\sigma_0 - \mu\sigma_z(1-\lambda))}{\sqrt{\mu^2+1}+\mu}. \quad (5)$$

At high temperature the failure mechanism is modelled as:

(1) power-law creep in the crust for all $(\sigma_1 - \sigma_3)$ and in the mantle for $(\sigma_1 - \sigma_3) < 200$ MPa (e.g. Brace & Kholstedt 1980):

$$\sigma_1 - \sigma_3 = \left(\frac{\dot{\varepsilon}}{A_p}\right)^{1/n} \exp\left(\frac{Q_p}{nRT}\right), \quad (6)$$

where A_p is a material constant, n is the power law exponent, Q_p is the activation energy, R is the gas constant, $\dot{\varepsilon}$ is the strain rate and T is material temperature; or

(2) Dorn-law creep in the mantle for $(\sigma_1 - \sigma_3) > 200$ MPa (e.g. Brace & Kholstedt 1980):

$$\sigma_1 - \sigma_3 = \sigma_d\left[1 - \sqrt{\frac{RT}{Q_d}\ln\left(\frac{\dot{\varepsilon}_d}{\dot{\varepsilon}}\right)}\right], \quad (7)$$

where Q_d is the activation energy, σ_d is the threshold stress and $\dot{\varepsilon}_d$ is the pre-exponential constant for Dorn-law creep.

We have used the parameter range outlined by Brace and Kholstedt (1980) for the rheological constants with the exception that the activation energy for power-law creep in quartz used here is $1{\cdot}5 \times 10^5$ J mol^{-1} (Jaoul *et al.* 1984) and, following England (1987) and others, the pre-exponential constant for power-law creep in quartz, A_p, has been corrected to 5×10^{-6} s^{-1} MPa^{-3}. We take the value of λ for the crust to be 0·4 and for the mantle to be 0. Further details of the modelling are given in Sandford *et al.* (1991).

7. References

Arculus, R. J., Ferguson, J., Chapell, B. W., Smith, D., McCulloch, M. T., Jackson, I., Hensel, H. D., Taylor, S. R., Knutson, J. & Gust, D. A. 1990. Eclogites and granultes in the lower continental crust: examples from eastern Australia and southwestern USA. *In* Smith, D. C. (ed) *Eclogites and related rocks,* 335–80. Amsterdam: Elsevier.

Bird, P. 1991. Lateral extrusion of lower crust from under high topography in the isostatic limit. J GEOPHYS RES **96,** 10,275–86.

Bird, P. & Piper, K. 1980. Plane stress finite element modes of tectonic flow in southern California. PHYS EARTH PLANET INT **21,** 158–75.

Brace, W. F. & Kholstedt, D. L. 1980. Limits on lithospheric stress imposed by laboratory experiments J GEOPHYS RES **85,** 6248–52.

Byrlee, J. D. 1968. Brittle-ductile transition in rocks J GEOPHYS RES **73,** 4741–50.

Collins, W. J. & Vernon, R. H. 1991. Orogeny associated with anticlockwise P-T-t paths:evidence from low-P,high-T terranes in the Arunta Inlier, central Australia. GEOLOGY (in press).

Cooper, J. A., Jenkins, R. J. F., Compston, W. & Williams, I. S. 1990. Ion microprobe U-Pb zircon dating within the lower Cambrian of South Australia. GEOL SOC AUST ABSTR **27,** 96.

Crawford, A. J. & Hilyard, D. 1990. Geochemistry of Late Proterozoic tholeiitic flood basalts, Adelaide Geosyncline, South Australia. GEOL SOC AUST SPEC PUBL **16,** 49–67.

DePaolo, D. 1981. Trace element and isotopic effects of combined

wallrock assimilation and fractional crystallisation. EARTH PLANET SCI LETT **53,** 189–202.
Dymoke, P. & Sandiford, M. 1992. Phase relationships in Buchan Facies Series pelitic assemblages: Calculations with application to andalusite-staurolite parageneses in the Mount Lofty Ranges, South Australia. CONTRIB MINERAL PETROL (in press).
England, P. C. 1987. Diffuse deformation: length scales, rates and metamorphic evolution. PHILOS TRANS R SOC LONDON **A321,** 3–22.
England, P. C. & Jackson, J. 1989. Active deformation of the continents. ANN REV EARTH PLANET SCI **17,** 197–226.
Fleming, P. D. & White, A. J. R. 1984. Relationships between deformation and partial melting in the Palmer migmatites. AUST J EARTH SCI **31,** 351–60.
Foden, J. D., Turner, S. P. & Morrison, R. S. 1990. The tectonic implications of Delamerian magmatism in South Australia and western Victoria. GEOL SOC AUST SPEC PUBL **16,** 465–82.
Foden, J. D., Williams, I., Compston, W., Michard, A. & Turner, S. P. in prep. Some constraints on the timing of the Delamerian Orogeny in the Southern Adelaide Fold Belt: Implications for the evolution of the eastern margin of Gondwanaland (in prep.).
Gans, P. B. 1987. An open-system, two-layer crustal stretching model for the eastern Great basin. TECTONICS **6,** 1–12.
Greenhalgh, S. A., Tapley, D. & von der Borch, C. C. 1991. Explosion seismic determination of crustal structure beneath the Adelaide Geosyncline, South Australia. PHYS EARTH PLANET INT **58,** 323–43.
Griffin, W. L., O'Reilly, S. Y. & Stabel, A. 1988. Mantle metasomatism beneath western Victoria, Australia: II. Isotopic geochemistry of Cr-diopside lherzolites and Al-augite pyroxenites. GEOCHIM COSMOCHIM ACTA **52,** 449–59.
Hildreth, W. & Moorbath, S. 1988. Crustal contributions to arc magmatism in the Andes of Central Chile. CONTRIB MINERAL PETROL **98,** 455–89.
Houseman, G. A., McKenzie, D. P. & Molnar, P. 1981. Convective instability of a thickened boundary layer and its relevance for the thermal evolution of continental convergent belts. J GEOPHYS RES **86,** 6115–32.
Huppert, H. E. & Sparks, R. S. J. 1988. The generation of granitic magmas by intrusion of basalt into continental crust. J PETROL **29,** 599–624.
Jaoul, O., Tullis, J. & Kronenberg, A. 1984. The effect of varying water contents on the creep behaviour of Heavitree Quartzite. J GEOPHYS RES **89,** 4298–312.
Jenkins, R. J. F. 1990. The Adelaide Foldbelt: tectonic reappraisal. GEOL SOC AUST SPEC PUBL **16,** 396–420.
Mancktelow, N. S. 1990. The structure of the southern Adelaide Fold Belt, south Australia. GEOL SOC AUST SPEC PUBL **16,** 369–95.
McCulloch, M. T., Arculus, R. J., Chappell, B. W. & Ferguson, J. 1982. Isotopic and geochemical studies of nodules in kimberlite have implications for the lower continental crust. NATURE **300,** 166–9.
McDonough, W. F., McCulloch, M. T. & Sun, S. S. 1985. Isotopic and geochemical systematics in Tertiary-Recent basalts from southeastern Australia and implications for the evolution of the subcontinental lithosphere. GEOCHIM COSMOCHIM ACTA **49,** 2051–67.
McDonough, W. F. & McCulloch, M. T. 1987. The southeast Australian lithospheric mantle: isotopic and geochemical constraints on its growth and evolution. EARTH PLANET SCI LETT **86,** 327–40.
Miller, E. L. & Gans, P. B. 1989. Cretaceous crustal structure and metamorphism in the hinterland of the Sevier thrust belt. GEOLOGY **17,** 59–62.
Milnes, A. R., Compston, W. & Daily, B. 1977. Pre- to syn-tectonic emplacement of early Palaeozoic granites in southeastern South Australia. J GEOL SOC AUST **24,** 87–106.
Molnar, P. & Lyon-Caen, H. 1988. Some simple physical aspects of the support, structure, and evolution of mountain belts. GEOL SOC AM SPEC PAP **218,** 179–207.
Offler, R. & Fleming, P. D. 1968. A synthesis of folding and metamorphism in the Mt Lofty Ranges, South Australia. J GEOL SOC AUST **15,** 245–66.
Ord, A. & Hobbs, B. E. 1989. The strength of the continental crust. TECTONOPHYSICS **158,** 269–89.
O'Reilly, S. Y. & Griffin, W. L. 1988. Mantle metasomatism beneath western Victoria, Australia: I. Metasomatic processes in Cr-diopside lherzolites. GEOCHIM COSMOCHIM ACTA **52,** 443–47.
Patterson, M. S. 1987. Problems in extrapolation of laboratory rheological data. TECTONOPHYSICS **133,** 33–43.
Preiss, W. V. 1987. The Adelaide Geosyncline: Late Proterozoic stratigraphy, sedimentation, palaeontology and tectonics. BULL GEOL SURV S AUST **53**.
Sandiford, M., Oliver, R. L., Mills, K. J. & Allen, R. V. 1990. A cordierite-staurolite-muscovite association, east of Springton, Mt Lofty Ranges; implications for the metamorphic evolution of the Kanmantoo Group. GEOL SOC AUST SPEC PUBL **16,** 483–95.
Sandiford, M., Martin, N., Zhou, S. and Fraser, G. 1991. Mechanical consequences of granite emplacement during high-T, low-P metamorphism and the origin of 'anticlockwise' PT paths. EARTH PLANET SCI LETT **107,** 164–72.
Sandiford, M. & Powell, R. 1990. Some isostatic and thermal consequences of the vertical strain geometry in convergent orogens. EARTH PLANET SCI LETT **98,** 154–65.
Sandiford, M. & Powell, R. 1991. Isostatic and thermal constraints on the evolution of high temperature low pressure metamorphic terrains in convergent orogens. J METAMORPH GEOL **9,** 333–40.
Sonder, L. J., England, P. C., Wernicke, B. P. & Christiansen, R. L. 1987. A physical model for Cenozoic extension of western North America. *In* Coward, M. P., Dewey, J. F. & Hancock, P. L. (eds) *Continental Extensional Tectonics*. GEOL SOC LONDON SPEC PUBL **28,** 187–201.
Sonder, L. J. & England, P. 1986. Vertical averages of rheology of the continental lithosphere: relation to thin sheet parameters. EARTH PLANET SCI LETT **77,** 81–90.
Thompson, P. 1989. Moderate over-thickening of thinned sialic crust and the origin of granitic magmatism and regional metamorphism in low-P-high-T terranes. GEOLOGY **17,** 520–23.
Turner, S. P., Foden, J. D. & Morrison, R. 1992. Derivation of A-type magma by fractionation of basaltic magma and an example from the Padthaway Ridge, South Australia. LITHOS (in press).
Turner, S. P. & Foden, J. D. 1990. The nature of mafic magmatism through the evolution of the Adelaide Foldbelt and subsequent Delamerian Orogeny. *In* Parker, J. (ed) *Mafic dykes and emplacement mechanisms*. Rotterdam: Balkema, 431–5.
Zhou, S. & Sandiford, M. 1991. On the stability of isostatically compensated mountain belts. J GEOPHYS RES (submitted).

MIKE SANDIFORD, JOHN FODEN, SHAOHUA ZHOU and SIMON TURNER, Department of Geology and Geophysics, University of Adelaide, G.P.O. Box 498, Adelaide, South Australia 5001, Australia.

Transactions of the Royal Society of Edinburgh: Earth Sciences, **83**, 95–106, 1992

Migmatite and melt segregation at Cooma, New South Wales

D. J. Ellis and M. Obata

ABSTRACT: The Cooma Complex of southeastern New South Wales comprises an andalusite-bearing S-type granodiorite surrounded by migmatites and low-pressure metamorphosed pelitic and psammitic sediments. The migmatite formed by the melting reaction:

$$\text{Biotite} + \text{Andalusite} + \text{K-feldspar} + \text{Quartz} + \text{V} = \text{Cordierite} + \text{Liquid}$$

at about 350–400 MPa P_{H_2O}, 670–730°C.

The melanosome consists of biotite + cordierite + andalusite + K-feldspar + plagioclase + quartz + ilmenite, whereas the leucosome consists of cordierite + K-feldspar + quartz with extremely rare biotite and plagioclase. In a closed system, freezing of the leucosome melt patches should have resulted in cordierite back-reaction with melt to produce biotite and andalusite. The virtually anhydrous mineralogy of the leucosome patches, lack of cordierite reaction and the absence of biotite selvedges at the leucosome–melanosome contacts, indicates that the melt did not completely solidify *in situ*. These observations can be explained by an initial peritectic melting reaction in the migmatite being arrested from back-reaction upon cooling because of the removal of *hydrous* melt, enabling leucosome cordierite to escape back-reaction. We propose that the melanosome is the residue of partial melting but that the leucosome patches do not represent frozen melt segregations but rather the liquidus minerals (cumulates) which precipitated from the melt.

In the restite-rich granodiorite from the core of the Cooma Complex, cordierite of similar composition to that in the migmatite has reaction rims of biotite and andalusite and there are coexisting biotite and andalusite in the matrix. The granodiorite consisted of about 50 wt% melt together with resite biotite, quartz and plagioclase, which can possibly be identified in the surrounding migmatite. Previous work suggested that the Cooma Granodiorite can be derived from a mixture of the surrounding metasediments which are of similar composition in the high and low-grade areas surrounding the granodiorite. Re-examination of those data shows that the high-grade metasediments are more An-rich than the low-grade rocks. The Cooma Granodiorite is very similar to the high-grade rocks in terms of Or–Ab–An ratio. This suggests derivation of the Cooma Granodiorite from the high-grade rocks and not from the relatively An-poor low-grade rocks which are typical of exposed sediments in the Lachlan Fold Belt. It is most likely that the granodiorite and envelope of high-grade rocks have been emplaced into the compositionally different lower grade rocks from slightly greater depths.

KEY WORDS: cordierite, granodiorite, leucosome, liquidus phase diagram, melanosome, migmatite, restite

The study of intrusive granites provides information on the products of partial melting, with the proposal that many felsic compositions are a combination of melt plus entrained restite mineralogy (see White & Chappell 1988 and references therein). Most high-level felsic igneous rocks with intrusive relations have risen kilometres from their source regions. Thus the extent to which the restite mineralogy faithfully preserves that of the source is difficult to ascertain because of the possibility of subsequent modification (White & Chappell 1977, p. 18).

The potential modification of source restite during ascent has been discussed by Chappell *et al.* (1987) who showed that compositionally identical plutonic and volcanic rocks contain different minerals. This led to the concept of primary and secondary restite. As the Cooma Granodiorite is known to be very similar in major and trace element and isotopic composition to the surrounding sedimentary envelope (Chappell 1984), it provides an excellent example of high-grade metamorphism and granite formation in which primary restite can be directly related to the surrounding metamorphic rocks.

The Cooma Complex is a good example of low-pressure partial melting and production of an S-type granodiorite. It has been suggested on field, petrographic, isotopic and geochemical grounds that the granodiorite was formed virtually *in situ* by partial melting of metasediments similar to those surrounding the pluton (Joplin 1962; Pidgeon & Compston 1965; White *et al.* 1974). It therefore provides a petrographical and mineralogical test of the resite model, as a progression can be seen in the field from unmelted sediments through migmatites to the granodiorite. It is therefore possible to constrain the likely partial melting reactions which produced the migmatite and test the consistency from migmatite through to granodiorite in terms of liquidus and restitic minerals.

Abbreviations used in this paper:

Als	andalusite or sillimanite
And	andalusite
Kf	alkali feldspar
Bio	biotite
L	liquid
Pl	plagioclase feldspar
Crd	cordierite
Dum	dumortierite
V	vapour
Grt	garnet
Ilm	ilmenite
Mus	muscovite
Qtz	quartz
Mg	Mg/Mg + Fe × 100 (molecular)

1. Geology

The Cooma region (Fig. 1) is a classic area in geology because it was one of the first localities from which andalusite–sillimanite type regional metamorphism was described (Joplin 1942). A succinct description of the Cooma Complex was given by Flood and Vernon (1978). The Cooma Granodiorite is a "regional aureole" type of granite (White *et al.* 1974, p. 161) occurring in the Lachlan Fold Belt of southeastern Australia. It is a small, elliptical pluton, about 8 km in maximum dimension, surrounded by metapelitic and metapsammitic gneisses and schists (Joplin 1942, 1943; Hopwood 1976) in a terrane of low-pressure regional metamorphism (Vallance 1967, 1969). The granodiorite occurs in the highest grade rocks, namely migmatitic K-feldspar–cordierite–andalusite–(sillimanite) gneisses, locally enclosing small patches that resemble the Cooma Granodiorite proper (Vallance 1969, p. 182). The metamorphic grade decreases to chlorite zone away from the granodiorite (see Flood & Vernon 1978, fig. 1). The granodiorite varies from gneissic to massive and contains abundant metasedimentary xenoliths. It consists of quartz, oligoclase, K-feldspar and biotite, with andalusite, sillimanite, cordierite and secondary muscovite. These minerals also occur in the adjacent high-grade metamorphic rocks (Joplin 1942; Hopwood 1976; Chappell & White 1976). Pidgeon and Compston (1965) determined a Sr isotope age of 460 Ma for the lower grade schists (Binjura Beds) and 415 Ma for the Cooma Granodiorite and high-grade metamorphic zones.

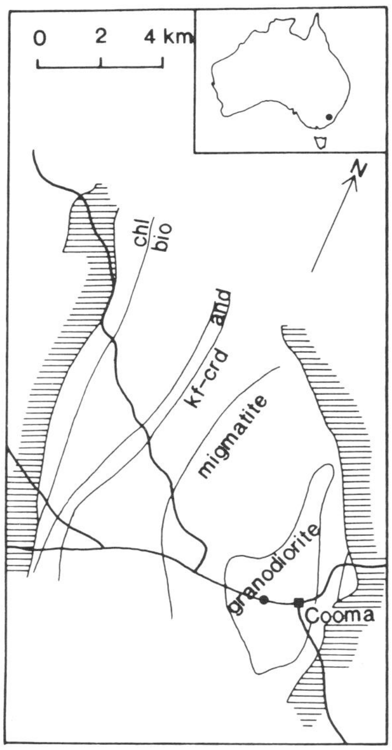

Figure 1 Geological map of the Cooma region, southern New South Wales (simplified after Munksgaard 1988). The sample location of the granodiorite and migmatite are shown by the square and circle respectively. The granodiorite sample is from Nanny Goat Hill in the town of Cooma. The shaded area represents basalt cover rocks.

Munksgaard (1988) carried out a detailed bulk-rock chemical and O and Sr isotopic study of the granodiorite and compared it with the rocks of the surrounding metasedimentary envelope. He supported the previous suggestions that the high-grade metapelitic and metapsammitic gneisses immediately surrounding the granodiorite are the geochemical equivalents of the more distant, low-grade metasediments. He proposed that the granodiorite can be modelled as a subequal mixture of pelite and psammite and confirmed the suggestion of Chappell and White (1976) and White and Chappell (1988) that it represents a partial melt from which a solid residue (restite) has not significantly separated. Based on slight but consistently lower $\delta^{18}O$ values in the granodiorite (average +12·2‰) compared with the probable mixed pelitic-psammitic source (average +13·5‰), he suggested that H_2O-rich fluid flux was an important factor in the formation of the Cooma Granodiorite.

Flood and Vernon (1978) proposed that although the granodiorite probably formed by partial melting of the adjacent high-grade migmatitic rocks, this high-grade envelope may have been dragged up to higher crustal levels by the intruding granitic diapir. Thus the metamorphic zones around Cooma may be a greatly compressed section of the metamorphic grade changes between the depths of magma generation and magma emplacement.

The migmatite zone grades into the Cooma Granodiorite proper (Joplin 1942). In places the granodiorite has a strong foliation but normally it is a massive, medium- to coarse-grained rock rich in biotite and with xenoliths exclusively of metasedimentary origin. Most xenoliths are schistose or gneissic biotite-rich types, sometimes with sillimanite, and there are fragments of milky quartz derived from quartz veins in the metamoprhic rocks. The high silica content of the granodiorite is also noted in the geochemical plots of White and Chappell (1988, fig. 2). Cordierite and andalusite are also present in the granodiorite, reflecting its aluminous composition (Chappell & White 1976). The cordierite is invariably partly or completely altered to dull green pinite. Large single grains of sillimanite (up to 5 mm) are also rarely found (W. E. Cameron, *pers. comm.* 1989).

One reason why so little mineralogical data have been published on the Cooma rocks is the extensive weathering. However recent road excavations unearthed fresh migmatite on the Cooma–Berridale road (Chappell & White, 1976, fig. 7) approximately 3 km W of the centre of Cooma (Fig. 1). In outcrop, the migmatite has irregular layering (stromatic) in which there was a moderate degree of melting (terminology after Ashworth 1985, table 1.1). The leucosome is subordinate in volume to melanosome. Leucosome layers are generally parallel to the foliation of the melanosome but some are discordant. In hand specimen the melanosome is a medium-grained biotite gneiss. Cordierite and andalusite are hardly recognised with the naked eye. In contrast the leucosome veins, which are up to several centimetres thick, contain spectacular pale-blue cordierites, up to 10 mm across, set in a slightly finer-grained quartz–feldspar matrix. The leucosome is notable for the extreme rarity of biotite and a complete absence of any biotite-rich selvedges. In addition the cordierite is not altered to secondary pinite as is typical in the lower-grade rocks and the granodiorite.

2. Petrography

2.1. Melanosome

In thin-section the melanosome consists of foliated brown biotite, poikiloblastic cordierite and orthoclase, quartz, ilmenite, andalusite, rare plagioclase and accessory tourmaline and zircon (Fig. 2). The cordierite and orthoclase are crowded with rounded inclusions of biotite, feldspar and quartz. Extremely thin needles of fibrolitic sillimanite are present as inclusions in the cores of some melanosome cordierites. There is no evidence for disequilibrium amongst these phases and the texture is of uniform character. In addition very thin wisps of fibrolite are concentrated at the end of, and parallel to, the cleavage trace of biotite. The andalusite is much coarser grained and abundant throughout the rock.

2.2 Leucosome

Melanosome–leucosome contacts are marked by an abrupt change in grain-size and mode, with biotite being extremely rare in the leucosome. The large euhedral to rounded cordierites are free of inclusions, in marked contrast to their poikiloblastic habit in the melanosome (Fig. 2). The rims of the leucosome cordierites that we have examined show virtually no evidence of reaction, although one grain was observed with a flake of biotite with andalusite at the rim. In rare cases, the cordierite is replaced by pale green Ti-free biotite of the same *Mg* as the cordierite, a feature ascribed to a later, subsolidus replacement of the cordierite. Some cordierites also show partial replacement along cracks by clay mineral. Orthoclase contains rare inclusions of euhedral plagioclase. The orthoclase is euhedral where it is in contact with interstitial quartz. Plagioclase does not occur in the groundmass of the leucosome. Pale-blue dumortierite is a distinctive minor accessory in the leucosome, as is rare subhedral apatite. Only one grain of ilmenite was observed in thin-sections of the leucosome. One or two flakes of brown biotite were observed in the matrix of the rock in hand specimen.

2.3. Granodiorite

The granodiorite is distinct from the migmatite in the abundance of euhedral to subhedral plagioclase phenocrysts (Fig. 2), characterised by large cores of calcic andesine overgrown by a distinct euhedral 1–2 mm rim of strongly zoned and more sodic oligoclase. Orthoclase occurs as separate grains throughout the rock. The granodiorite rarely contains fresh cordierite. Cordierite phenocrysts have rare brown biotite inclusions as do those in the melanosome, whereas cordierite in the leucosome is inclusion-free. It has a distinct reaction rim of brown biotite and andalusite (Fig. 2C). Subhedral to granular aggregates of andalusite as well as biotite are common throughout the matrix of the granodiorite. Ilmenite is a common accessory in both the granodiorite and the melanosome.

3. Mineral chemistry

Mineral compositions were determined with either the TPD Si(Li) detector electron probe or a CAMECA crystal spectrometer probe at the Research School of Earth

Figure 2A Thin-section photograph of the melanosome. Poikiloblastic cordierite and orthoclase crystals with inclusions of quartz, biotite and rare plagioclase are set in a foliated matrix of biotite, feldspar, andalusite, ilmenite and secondary fibrolite. Cross-polarised light.

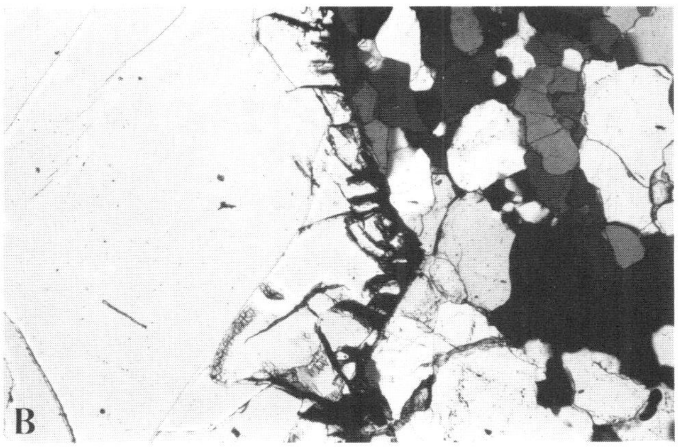

Figure 2B Thin-section photograph of the leucosome. The rock consists of euhedral to subhedral crystals of cordierite together with orthoclase, rare plagioclase, subhedral apatite and extremely rare biotite. The dark alteration around the edge of the euhedral cordierite is clay. Cross-polarised light.

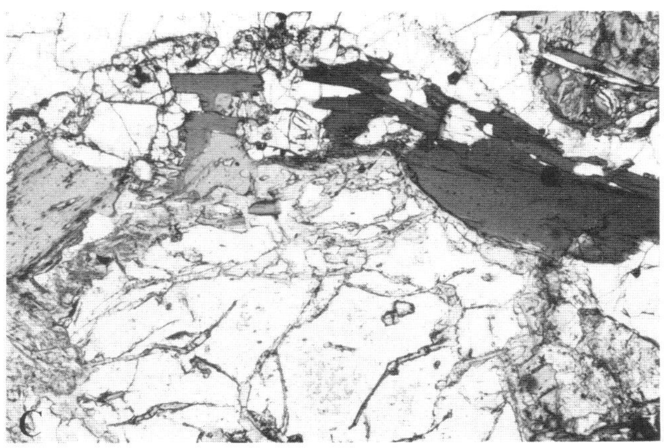

Figure 2C Thin-section photograph of the Cooma Granodiorite. Cordierite phenocrysts (lower half of figure) have inclusions of quartz and a reaction rim of biotite and andalusite. The matrix consists of orthoclase, plagioclase, biotite, andalusite and quartz. Plane-polarised light. The width of each figure is 2·5 mm.

Table 1 Compositions of the minerals in the migmatite and Cooma Granodiorite

	1. Crd	2. Crd	3. Bio	4. Bio	5. Pl	6. Kf	7. Ilm	8. Crd	9. Bio	10. Pl	11. Kf	12. Crd	13. Bio	14. And	15. Bio	16. Bio	17. Ilm	18. Pl	19. Pl	20. Kf
SiO_2	50·03	50·53	35·41	36·16	60·31	64·66	—	48·55	36·56	61·01	64·38	48·64	34·87	36·30	35·59	34·30	—	57·17	62·30	65·65
TiO_2	—	—	3·18	1·15	—	—	53·58	—	1·11	—	—	—	2·50	—	3·24	3·91	54·87	—	—	—
Al_2O_3	33·35	33·76	18·78	21·14	25·63	18·49	—	34·38	20·74	25·17	18·51	34·07	20·61	63·17	20·20	20·38	—	27·57	24·90	19·92
FeO	10·48	8·98	19·93	19·43	0·13	—	44·33	10·02	18·55	—	—	9·50	18·89	0·48	19·29	21·28	42·23	—	—	—
MnO	0·19	0·41	—	—	—	—	1·60	0·42	—	—	—	0·43	0·31	—	0·27	0·40	3·29	0·03	—	—
MgO	6·99	7·44	7·94	9·64	—	—	0·49	7·13	9·47	—	—	7·01	8·89	0·03	8·63	6·86	0·19	—	—	—
CaO	—	—	—	0·11	6·85	—	—	—	—	6·55	—	—	—	—	0·03	0·03	—	8·36	5·22	0·05
Na_2O	—	0·20	—	—	7·62	1·37	—	0·14	—	7·66	1·53	0·30	0·17	—	0·09	0·14	—	6·49	8·35	2·05
K_2O	—	—	9·53	9·62	0·27	14·86	—	0·01	9·90	0·31	14·10	—	8·78	—	8·92	9·28	—	0·11	0·17	12·94
	18(O)	18(O)	22(O)	22(O)	8(O)	8(O)	3(O)	18(O)	22(O)	8(O)	8(O)	18(O)	22(O)	5(O)	22(O)	22(O)	3(O)	8(O)	8(O)	8(O)
Si	5·049	5·043	5·434	5·363	2·666	2·993	—	4·923	5·453	2·694	2·995	4·959	5·290	0·982	5·315	5·205	—	2·563	2·732	2·971
Ti	—	—	0·367	0·128	—	—	1·009		0·117	—	—	—	0·285	—	0·364	0·446	1·024	—	—	—
Al	3·967	3·972	3·396	3·697	1·336	1·009	—	4·111	3·648	1·310	1·015	4·093	3·685	2·015	3·608	3·646	—	1·456	1·287	1·062
Fe	0·884	0·749	2·557	2·410	0·005	—	0·923	0·851	2·315	—	—	0·810	2·397	0·011	2·409	2·701	0·876	—	—	—
Mn	0·016	0·035	—	—	—	—	0·034	0·036	—	—	—	0·037	0·040	—	0·034	0·052	0·069	0·001	—	—
Mg	1·052	1·152	1·816	2·133	—	—	0·013	1·079	2·107	—	—	1·065	0·011	0·001	1·920	1·552	0·007	—	—	—
Ca	—	—	—	0·018	0·325	—	—	—	—	0·310	—	—	—	—	0·005	0·006	—	0·402	0·245	0·002
Na	—	0·039	—	—	0·653	0·123	—	0·028	—	0·656	0·138	0·058	0·049	—	0·025	0·043	—	0·565	0·710	0·180
K	—	—	1·816	1·820	0·015	0·878	—	0·001	1·885	0·018	0·837	—	1·699	—	1·700	1·797	—	0·006	0·009	0·747
Sum	10·968	10·990	15·386	15·569	5·000	5·003	1·979	11·029	15·525	4·998	4·985	11·022	15·456	3·009	15·380	15·448	1·976	4·993	4·983	4·962
Mg	54·3	60·6	41·5	46·9				55·9	47·6			56·8	45·6		44·3	36·5				

Notes: 1. Melanosome Crd core. 2. Melanosome Crd rim at Bio. 3. Melanosome Fe-rich Bio. 4. Melanosome Mg-rich Bio. 5. Melanosome Plag inclusion in Kf. 6. Melanosome Kf host to above Plag. Overall Kf range Or/Or + Ab 83–91. 7. Melanosome ilm. 8. Leucosome Crd core. Overall range Mg_{56-53}. 9. Leucosome Bio (rare). 10. Leucosome Plag inclusion in Kf. 11. Leucosome Kf host to above Plag. Overall range Or/Or + Ab 82–92. 12. Granodiorite Crd. 13. Granodiorite Bio reaction rim to above Crd. 14. Granodiorite andalusite reaction rim to above Crd. 15. Granodiorite Bio next to Ilm in groundmass. 16. Granodiorite Bio inclusion in Plag. 17. Granodiorite Ilm. 18. Granodiorite Plag core. 19. Granodiorite Plag rim overgrowing above core. 20. Granodiorite Kfeld (Na-rich).

Sciences at ANU. Representative analyses are presented in Table 1.

3.1. Cordierite

There is an overlap in the compositions of cordierites from the three rock types (Fig. 3), with the overall range in *Mg* from 53 to 60. Melanosome cordierite tends to have slight reverse zoning in some grains, with cores Mg_{54} to rims of $Mg_{60\cdot5}$. In the leucosome, cordierite cores are Mg_{56} and most grains are of uniform composition although one rim was normally zoned to a slightly more Fe-rich composition (Mg_{53}). Cordierite from the granodiorite is Mg_{56}, similar to that of the leucosome.

3.2. Biotite

Biotite from the melanosome is in the range Mg_{40-47}. There is no obvious change in Al content with *Mg*-number, but the melanosome biotites tend to have a lower Al content than those of the granodiorite (Fig. 3). Ti is within the range 0·1–0·4 atoms per formula unit, similar to that of the migmatite biotites. There is a negative correlation between the Ti content and the *Mg* number of the biotites, with an overlap in compositions from the migmatite and granodiorite. The single grain of biotite in the leucosome has Mg_{45-48}, which is at the Mg-rich end of the range in melanosome biotite compositions. Biotite in the granodiorite is always more aluminous than the migmatite biotites (Fig. 3) and varies in composition with location. The biotite from the reaction rim to cordierite is the most Mg-rich (Mg_{46}) whereas that in feldspar and the matrix extends to a more Fe-rich (Mg_{36}) and Ti-rich composition than in the migmatite.

Both the melanosome and granodiorite have coexisting Crd–Bio–Als whereas the leucosome has only Crd–(Bio) (Fig. 3). The different cordierite zoning patterns and tie-line

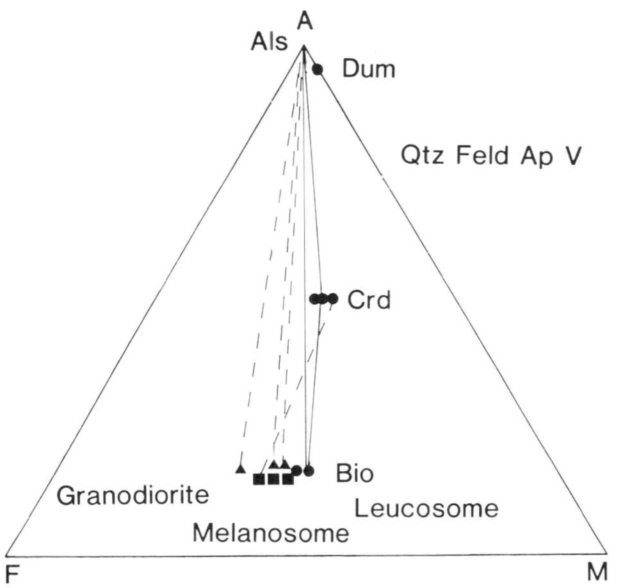

Figure 3 AFM projection showing the compositions of coexisting phases in the different rock types. Biotite is shown by triangles for the granodiorite, squares for the melanosome and circles for the leucosome. The small overlapping range in cordierite compositions for all three rock types is shown by circles. The solid tie line shows the three-phase triangle Crd–Bio–Als at the condition of melt separation from the leucosome. The dashed lines show the compositions of coexisting Crd–Bio in the melanosome and Bio–And in the granodiorite due to subsolidus Fe–Mg exchange and crystal–liquid re-equilibration respectively. (A = Al_2O_3–Na_2O–K_2O, F = total iron as FeO, M = MgO molecular proportions in this and later diagrams.)

rotation from the leucosome to the melanosome are consistent with a small amount of Fe–Mg subsolidus exchange between Crd and Bio in the melanosome (Fe Crd + Mg Bio = Mg Crd + Fe Bio proceeding from left to right with decreasing temperature). Thus the cordierite in the melanosome ranges from the composition of the leucosome (Mg_{56}) up to a maximum of Mg_{63} at the rim near biotite.

3.3. Feldspars

Plagioclase is of very minor abundance in the migmatite. Plagioclase inclusions in orthoclase of the melanosome are typically calcic (An_{33}) but one grain has a core of An_{15}. Plagioclase in the leucosome shows a range of compositions due to zoning (Fig. 4). The most calcic composition (An_{31}) is from a euhedral plagioclase inclusion in orthoclase which is very similar to that of the melanosome. The cores of the leucosome plagioclases tend to concentrate around this value. There is continuous but also some reverse zoning down to An_{16} ($Or_1Ab_{73}An_{16}$). In contrast, the granodiorite contains abundant plagioclase consisting of large, optically uniform cores overgrown by more sodic rims of euhedral, zoned plagioclase. The cores are distinctly more calcic (An_{33}–An_{41}) than those of the migmatite (Fig. 4). However the zoned plagioclase overgrowths (An_{21}–An_{25}) are similar to the zoned plagioclase from the leucosome.

Orthoclase in the melanosome is of similar composition to orthoclase cores from the leucosome, but the latter has orthoclase rims which extend to more sodic compositions (Fig. 4). Orthoclase from the granodiorite is similar to that of the migmatite and, unlike plagioclase, does not have distinct cores differing from that of the migmatite (Fig. 4). The melanosome orthoclase ranges from Or/Or + Ab = 86–91. Orthoclase of the leucosome ranges from Or/Or + Ab = 82–92, with one rim of 79. Orthoclase of the granodiorite does not extend from such potassic compositions and is in the range Or/Or + Ab = 76–80. Oscillatory zoning is absent from these feldspars.

3.4. Ilmenite

Ilmenite is absent from the leucosome, but ilmenite from the melanosome and granodiorite are very similar in

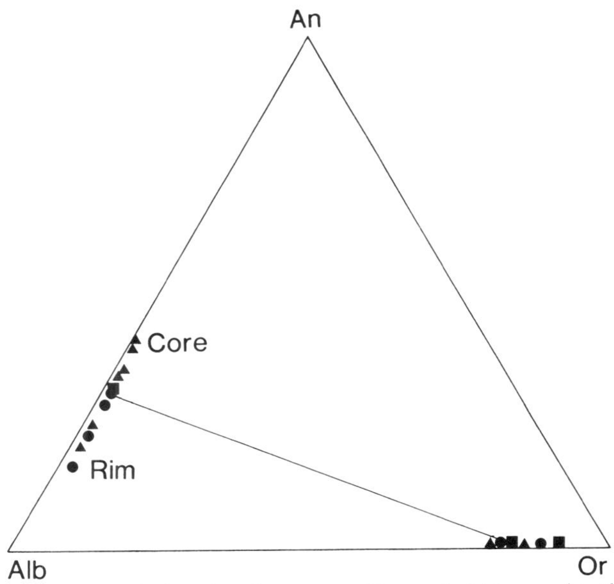

Figure 4 Variation in the composition of feldspars in the granodiorite (triangles), melanosome (squares) and leucosome (circles). Note the more calcic composition of the cores to granodiorite plagioclases. The tie line connects a plagioclase inclusion in its host orthoclase from the leucosome.

composition (Table 1). Structural formula calculations indicate an absence of ferric iron in both cases, with a small amount of Mn and even less Mg being present.

4. Discussion

4.1. Melting reactions

The liquidus mineralogy in any partial melting reaction, as identified in a leucosome, should ideally correlate with the residual mineralogy in the melanosome (as defined by Mehnert 1968 and summarised in Ashworth 1985, table 1.1). The very nature of partial melting is to produce a quartz–feldspar-rich melt and a residue rich in ferromagnesian minerals. However, the extent to which the leucosome and melanosome in a particular migmatite faithfully record this ideal, in terms of mineral compositions and rock compositions, is fraught with complications.

Local melt segregation and subsequent freezing could, under fortuitous circumstances, produce a leucosome identical in composition to the melt. Dehydration melting reactions involving muscovite, biotite or amphibole involve peritectic reactions (see Ellis 1986). In a closed system the melt formed by a peritectic reaction would not freeze of its own composition but would be modified by reaction with the minerals present. The occurrence of biotite selvedges separating leucosome from melanosome in some migmatites has been cited as evidence for this process (Le Breton & Thompson 1988). A consequence of this back-reaction is that the leucosome may not necessarily be of the same composition as the original melt.

Discussions of partial melting using simple system analogues (such as Mus + Qtz in the KASH system) involve only discontinuous reactions. In more complex systems and certainly in natural rocks with a higher variance, the cooling of a melt involves continuous reactions in which the remaining melt composition progressively changes with decreasing temperature, irrespective of whether the beginning of melting was due to a eutectic or peritectic reaction. Therefore an unambiguous definition and convincing demonstration of melt and restite compositions in terms of leucosome and melanosome (or 'minimum-temperature melt' and restite in felsic suites) requires careful examination. The fact that many leucosome compositions 'fail the test' of being respectable minimum melts (see Olsen 1985 and references therein) may perhaps be viewed with the above comments.

A further problem, which is evident in the Cooma example and elsewhere, lies in the fact that many leucosomes of presumed melt origin are virtually anhydrous in their mineralogy, even though the associated melanosome is rich in hydrous minerals. Any felsic melt produced by dehydration melting must have a high H_2O content (e.g. Burnham 1969; Burnham & Nekvasil 1986; Nekvasil & Burnham 1987), which implies that cooling in a closed system would result in the leucosome melt crystallising new hydrous minerals at moderate pressures.

The mineral compositions for the leucosome, melanosome and granodiorite have striking similarities, though differing significantly in modal proportions and textures. We propose that the migmatite formed by the melting reaction:

$$\mathrm{Bio + Als + Kf + Qtz + V = Crd + L} \qquad (1)$$

and that the melanosome represents a residue from partial melting and that the leucosome represents the initial site of segregation and partial solidification of the melt fraction. However this melt was largely removed from the migmatite, leaving behind the early precipitating crystalline (anhydrous) phases. Thus the bulk composition of the leucosome is not a low-pressure anatectic melt but an anhydrous cumulate, and the cordierite in the leucosome generally does not show evidence of back-reaction as would be expected upon freezing of an isochemical closed system. In contrast, the granodiorite has cordierite which underwent back-reaction to biotite and andalusite, and the groundmass has then crystallised coexisting biotite and andalusite from the remaining liquid.

Primary muscovite appears to be absent from the migmatite and the granodiorite probably due to its subsolidus breakdown at lower grades. Vernon (1978) showed that andalusite porphyroblasts developed at the expense of muscovite as a result of the reaction:

$$\mathrm{Crd + Musc = Bio + And + Qtz} \qquad (2)$$

Thus the partial melting reactions in the migmatite are more likely to have occurred at shallower depths than those required for the intersection of the muscovite-quartz curve with the solidus for pelitic rocks (Fig. 5). However, the errors in P–T estimates cannot entirely rule out the possibility of muscovite being involved in melting, although there is no petrographic evidence for this.

The absence of coexisting garnet + cordierite and the stable coexistence of Bio + Als + Crd means that the metamorphic grade must have been below that of the reaction

$$\mathrm{Bio + Als + Qtz = Crd + Grt + Kf + H_2O} \qquad (3)$$

(see Thompson 1976, fig. 7; Abbott & Clarke 1979, fig. 3). The stability of Mus + Qtz and Bio + Als and the minimum melting curves for quartz–feldspar systems, places the Cooma rocks at approximately 350–400 MPa and 670–730°C. This lies within the two boundaries for the andalusite–sillimanite reaction of Holdaway (1971) and Richardson *et al.* (1969).

The compositions of coexisting Crd–Bio–(Als–Kf–Qtz) may further constrain the P, T, P_{H_2O} conditions (Fig. 5) using the experimentally determined isopleth data of Holdaway and Lee (1977) and the calculated isopleth data and petrogenetic grids of Thompson (1976).

In the presence of excess aluminosilicate at low pressures, the melting of common pelitic rocks can be modelled by water excess and dehydration melting reactions involving biotite. In the limiting pure Fe or Mg systems KFASH or KMASH the dehydration reaction

$$\mathrm{Bio + Als + Qtz = Crd + Kf + V} \qquad (4)$$

terminates at the H_2O saturated solidus (see e.g. Grant 1985, fig. 3.6). The presence of MgO in the Cooma rocks (cordierite Mg_{56} biotite Mg_{42}) means that the beginning of melting and phase equilibria can be more closely modelled by the KFMASH chemical system where the reaction

$$\mathrm{Bio + Als + Kf + Qtz + V = Crd + L} \qquad (1)$$

occurs (see Grant 1985, fig. 3.14; Ellis 1986, fig. 6; Vielzeuf & Holloway 1988, fig. 5).

The discontinuous univariant melting reactions illustrated in Figure 5 can be discussed in more detail by the use of isobaric, pseudobinary T–X (Fe–Mg) sections to illustrate the partial melting of a specific rock composition (see Ellis 1986, fig. 8 for details). There are two possible melting reactions which produced the migmatite at Cooma, depending on the mineralogy just at the beginning of melting. If the initial rock consisted of Bio–Crd–Als–Kf–Qtz(–V) then the beginning of melting would be defined by the discontinuous peritectic reaction (1) above. This is represented by the univariant reaction (1) in the P–T diagram of Figure 5 and by the horizontal line in the isobaric T–X section of Figure 6. In this case the original gneiss would produce melt patches (leucosome) and a residue (melanosome) of Bio + Crd + Kf + Qtz.

This sort of discontinuous melting reaction is capable of producing an abrupt appearance of migmatite in the field, as it affects a wide range (Fe–Mg) of rocks at the one temperature. In contrast, if at the same P_{H_2O}–T conditions the initial Cooma gneiss (A/AFM <50) passed from a divariant Crd + Bio + Als assemblage to a (trivariant) Crd + Bio assemblage by the divariant (KFMASH) reaction

$$\text{Bio} + \text{Als} + \text{Qtz} = \text{Crd} + \text{Kf} + \text{V} \qquad (5)$$

(given by a divariant loop in Fig. 6), then this new assemblage would commence melting at a slightly higher temperature by the continuous reaction

$$\text{Crd} + \text{Kf} + \text{Qtz} + \text{V} = \text{Bio} + \text{L}. \qquad (6)$$

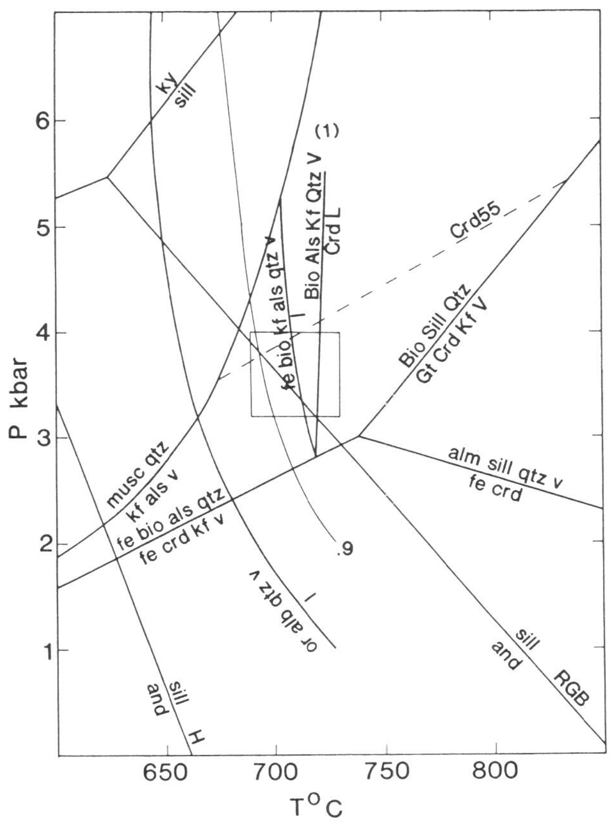

Figure 5 P–T diagram ($P_{H_2O} = P_{total}$) giving the estimated conditions of formation of the Cooma rocks defined by the square. Limiting end-member systems are given in lower case, reactions in the Fe–Mg KFMASH system have minerals and liquid starting in upper case. The aluminosilicate phase boundaries are of Richardson *et al.* (1969)—RGB, and Holdaway (1971)—H. The water-saturated solidus for the Qtz–Or–Als-water system is from Johannes (1985, fig. 2.2). The curve labelled by 0·9 represents the solidus for the Qtz–Or–Als-water system for an activity of water = 0·9, calculated by Nekvasil (1988, table 1). The muscovite-out curve is from Thompson (1976, fig. 7). The melting reactions for biotite with excess aluminosilicate, K-feldspar, quartz and vapour for the pure Fe system and the Fe–Mg system are from Grant, (1985, fig. 3.15). Above the temperature of melting for these two biotite reactions the biotite and muscovite dehydration reactions are actually metastable. The dashed line shows the P–T location of Mg_{55} cordierite in the assemblage Bio–Crd–Als–Kf–Qtz–V calculated by Holdaway and Lee (1977). An activity of water in the melt less than 0·9 would prohibit Mg_{55} cordierite phenocrysts in the granodiorite from reacting to biotite with andalusite as the aluminosilicate polymorph using the RGB boundary.

At temperatures below the discontinuous Fe–Mg reaction (1) only relatively iron-rich rocks would begin melting by the continuous reaction

$$\text{Bio} + \text{Als} + \text{Kf} + \text{Qtz} + \text{V} = \text{L}. \qquad (7)$$

The large cordierites in the leucosome are unlikely to be the product of the melting reaction (1). The fact that the compositions of the phases in the leucosome and melanosome are so similar suggests that the two were in equilibrium. In this case the melting would have been followed by segregation of melt into patches, some of which are discordant to the melanosome foliation. It is difficult to envisage how a liquid could segregate from the residue and selectively carry with it 10 mm sized, euhedral cordierites, yet exclude all the biotite and the smaller poikiloblastic cordierites of the same composition. Thus we interpret the large euhedral cordierites as having crystallised directly from the melt after the melt segregated as leucosome patches.

4.2. Origin of the leucosome

Notable features of the leucosome are the anhydrous nature of the mineralogy, absence of biotite inclusions in cordierite (in contrast to those of the melanosome and granodiorite), the fact that cordierite has almost entirely escaped back-reaction to Bio + Als or was later joined by biotite, and the absence of biotite selvedges separating leucosome from melanosome. Other leucosome patches in the Cooma area are reported to have biotite-rich selvedges (Chappell & White 1976) which we interpret as possible examples of melt that did not escape from the migmatite.

We interpret these observations as evidence that a significant proportion of the original melt in the leucosome has escaped from the migmatite, leaving behind the early crystallising liquidus phases (Crd + Kf + Qtz + (Plag)) as cumulates. Our arguments for this can be illustrated from Figures 3 and 7 using the crystal–liquid Fe–Mg equilibria between cordierite and coexisting melt (Ellis 1986, table 4). The melanosome, or residue of partial melting contains Crd (Mg_{56}), Bio (Mg_{42-45}) and ilmenite as the ferromagnesian phases. In contrast the leucosome has virtually only cordierite of Mg_{56} and extremely rare biotite (Mg_{42}). Thus the bulk rock composition of the leucosome has a higher *Mg* number (*ca* 56) than the residue (*ca* 45). In terms of crystal–liquid equilibria, this is impossible as at low temperatures the relative partitioning of Fe–Mg is *Mg* Crd > Bio > Liq > Grt > Ilm (Grant 1985; Ellis 1986; Le Breton & Thompson 1988). An approximate *Mg* value for the liquid can be calculated from the experimental data of Ellis (1986), see Table 2. Liquid *Mg* values in the range 28–35 are less than those of the biotite of the melanosome and considerably less than the leucosome with an *Mg* value of *ca* 56.

The granodiorite grades into the migmatite without sharply defined intrusive contacts. Here the mineral assumblage is Crd–Bio–And–(Feld–Qtz–Ilm). The cordierite phenocrysts have a reaction rim of Bio–And which is consistent with the earlier proposed back-reaction which should have occurred in the migmatite if it was a closed system:

$$\text{Crd} + \text{L} = \text{Bio} + \text{Als} + \text{Kf} + \text{Qtz} + \text{V}. \qquad (7)$$

It is noteworthy that the granodiorite biotite extends to much more Fe-rich compositions (Mg_{37}) than in the migmatite and in the reaction rim around the cordierite. These features can be illustrated in a series of ternary liquidus phase diagrams relating the univariant melting reactions of Figure 5 and the T–X sections of Figure 6 to

Table 2 Relative compositions of coexisting cordierite and liquid as a function of temperature using the experimental data of Ellis (1986, table 4)

T°C	*Mg* Crd	*Mg* Liquid
650	57	35·6
700	57	32·8
750	57	30
800	57	28

the order of crystallisation of the non-feldspar phases. An initial metamorphic assemblage of Bio–Crd–Als–Kf–Qtz would begin melting by peritectic reaction (1) to produce liquid A (Fig. 7). A further increase in temperature would produce a less aluminous liquid such as indicated by point B, lying on the Crd–Bio boundary curve of Figure 7. Thus the residual mineralogy of the melanosome would include Crd Mg_{57}, Bio Mg_{45} and the liquid would have the composition Mg_{32}.

The fact that the melanosome contains both Bio and Crd and the leucosome only Crd is difficult to reconcile at the one P–T. However, Flood and Vernon (1978) suggested that the Cooma Granodiorite dragged up its migmatic surrounds from greater depth. Any slight movement to shallower levels after melt segregation into the leucosome patches would result in an expansion of the liquidus field of cordierite relative to biotite. Thus a small temperature interval in which cordierite crystallised without biotite would ensue until the Bio–Crd cotectic was again reached (Fig. 7 points B to C). This explains the absence of biotite inclusions in leucosome cordierites, but their presence in cordierite from the melanosome and granodiorite. It is proposed that at that stage the remaining melt was expelled from the migmatite and possibly formed pegmatites. When the liquid reached the biotite-cordierite liquidus boundary at the slightly lower pressure P2 (point C of Fig. 7) the peritectic reaction Crd + L = Bio then occurred. The liquid then evolved to slightly more aluminous compositions again and at point D the initial melting reaction (1) was reversed:

$$\text{Crd} + \text{L} = \text{Bio} + \text{Als} + \text{Kf} + \text{Qtz} + \text{V}. \qquad (7)$$

The granodiorite cordierites have reaction rims of biotite and andalusite.

Further cooling from D to E resulted in continued cotectic crystallisation of biotite–andalusite–(feldspar–quartz) from the remaining melt as it varied to more Fe-rich compositions from D to E until the magma completely solidified (Figs 6, 7). The granodiorite contains abundant groundmass biotite and andalusite, together with feldspar and quartz. It was noted earlier that the biotites in the granodiorite extend to more Fe-rich compositions than in the migmatite, consistent with this explanation.

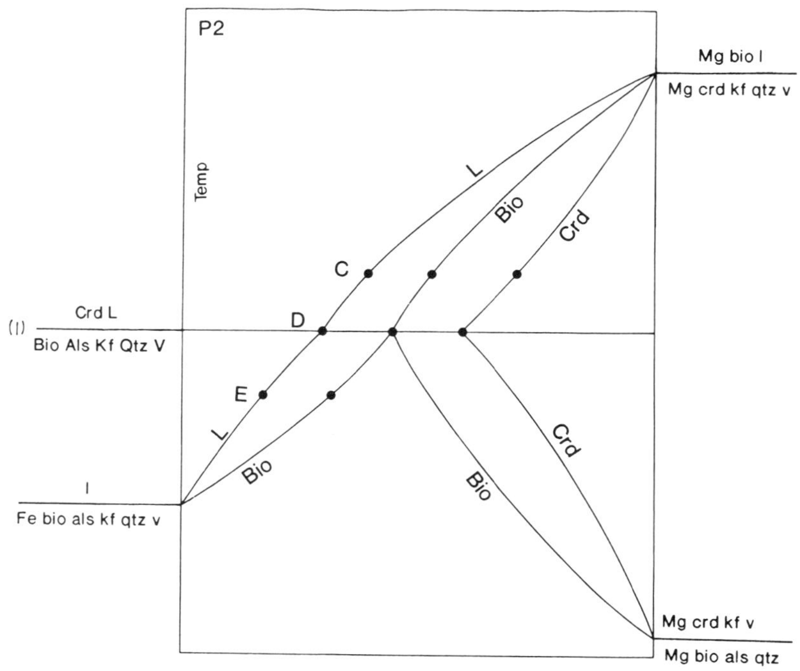

Figure 6 T–X (Fe–Mg) schematic, pseudobinary phase diagram, 'projected' from feldspar, quartz and water at pressure P2 of Figure 7. Below the temperature (D) of the discontinuous Fe–Mg melting reaction Bio + Als + Kf + Qtz + V = Crd + L, Fe-rich rocks with excess Bio + Als melt via the continuous reaction Bio + Als + Kf + Qtz + V = L whereby biotite and melt become more Mg-rich with increasing temperature. The reaction terminates at low temperature at the discontinuous limiting pure-Fe eutectic melting reaction Fe Bio + Als + Kf + Qtz + V = L. More Mg-rich rocks consist of the assemblage Crd + Bio + Als which becomes more Fe-rich with increasing temperature due to the continuous Fe–Mg dehydration reaction Bio + Als + Qtz = Crd + Kf + V. This reaction terminates at low temperature at the discontiuous pure Mg reaction Mg Bio + Als + Qtz = Mg Crd + Kf + V. Above the temperature of the discontinuous Fe–Mg melting reaction, cordierite, biotite and melt all become more Mg-rich with increasing temperature due to the continuous Fe–Mg reaction Bio + Kf + Qtz + V = Crd + L. This continuous reaction terminates at higher temperature at the discontinuous limiting pure-Mg peritectic melting reaction Crd + Kf + Qtz + V = Bio + L. (The discontinuous pure-Mg eutectic melting reaction is with excess aluminosilicate—Crd + Als + Kf + Qtz + V = L and is at slightly lower temperatures than the above reaction. For clarity it is not shown on this diagram.) The points C, D and E are also shown in Figure 7 at pressure P2.

4.3. The liquid component of the Cooma Granodiorite

The composition of the Cooma Granodiorite is similar in major and trace element composition to a calculated mixture of the surrounding Ordovician metapelites and metapsammites (see Munksgaard 1988, fig. 2; White & Chappell 1988, fig. 3). This confirms earlier suggestions that the granodiorite was derived by melting of the surrounding rocks (Joplin 1962; Pidgeon & Compston 1965; White *et al.* 1974). In addition, the fact that the granodiorite has normative quartz–feldspar values which lie quite distant from the cotectic in the granite system (Munksgaard 1988, fig. 3; White & Chappell 1988, fig. 1) supports the belief that the granodiorite represents a partial melt from which a solid residue has not escaped to a significant extent. Our calculations (Table 3) suggest that the granodiorite originally consisted of about 50 wt% melt together with biotite, plagioclase and quartz. Our justification for this is presented below.

The former presence of this melt in the granodiorite is indicated by the petrographic evidence for reaction (1). The composition of this melt is given by point A of Figure 7. The previously cited experimental data show that a minimum-temperature melt saturated with either muscovite at higher pressure or andalusite at lower pressure would be slightly more aluminous than cordierite when projected onto an AFM tenary diagram from feldspar, quartz and water vapour (Fig. 7). In addition, independent experimental data require that a melt be more Fe-rich ($\sim Mg_{32}$) than coexisting cordierite and biotite (*Mg* Crd > Bio > Melt).

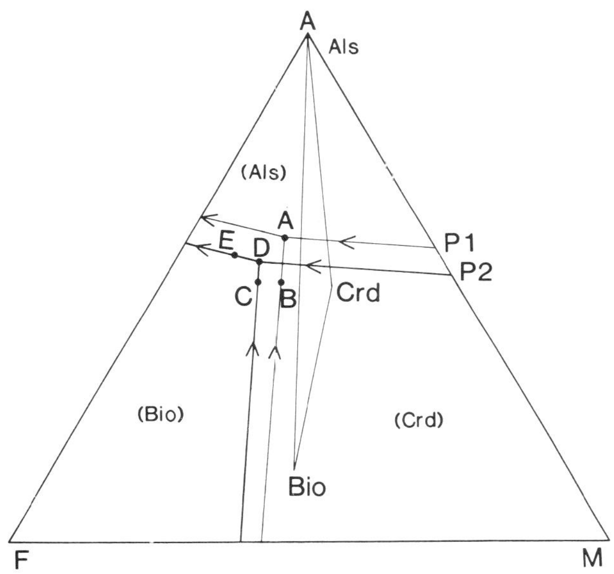

Figure 7 Schematic ternary AFM liquidus diagram projected from feldspar, quartz and excess water for the conditions of initial partial melting at pressure P1 of the migmatite, and the slightly lower pressure, P2 of solidification of the granodiorite. The liquidus phase for each field is shown in brackets. Arrows point in the direction of down temperature on the liquidus boundaries. The three-phase assemblage Crd–Bio–Als–(Feld-Qtz–V) commences melting to produce liquid A via the reacion Bio + Als + Feld + Qtz + V = Crd + L. Further increase in temperature changes the composition of the liquid from A to B after the aluminosilicate is reacted out. At the temperature B a slight decrease in pressure to P2 expands the liquidus field of Crd so that the melt then has a small temperature interval during which only Crd crystallises from the melt. At the lower temperature when the liquid reaches the peritectic line at C, Crd reacts with liquid to form biotite and the liquid evolves from C to D. At D Crd reacts with liquid to form Bio and And. Cordierite which is now completely mantled by Bio–And in the granodiorite is protected from further reaction and the remaining liquid now evolves to more Fe-rich compositions by further crystallisation of groundmass biotite and andalusite along the Bio–And cotectic from D to E, where the last melt freezes.

Table 3 Composition of the Cooma Granodiorite, estimated melt and restite component (total iron as FeO and readjusted anhydrous for calculations)

	1	2	3	4
SiO_2	73·18	70·60	37·95	57·34
TiO_2	0·55	0·78	1·15	
Al_2O_3	13·94	17·72	21·53	27·64
FeO	4·08	2·48	19·26	
MnO	0·06	0·13		
MgO	1·79	0·65	9·83	
CaO	0·97	0·60		8·39
Na_2O	1·51	2·07		6·51
K_2O	3·79	4·68	10·28	0·11
P_2O_5	0·13	0·27		

Notes: Granodiorite = 0·48 Melt + 0·15 Bio + 0·08 Plag + 0·29 Qtz (wt. fractions). 1. Cooma Granodiorite C1 (White & Chappell 1988, table 1). 2. Calculated composition of the melt in the Cooma Granodiorite. 3. Resitite biotite (composition of leucosome biotite). 4. Restite plagioclase (core composition).

Published analyses of the Cooma Granodiorite (Munksgaard 1988; White & Chappell 1988) show a small range in compositions when plotted on an AFM diagram (Fig. 8). It has an *Mg* of 43·9 (data of White & Chappell 1988; table 1) that is virtually the same as the biotite in the rock (Fig. 8). The granodiorite is also much less aluminous than cordierite on an AFM diagram. The bulk chemical composition of the Cooma Granodiorite does not satisfy the criteria for being entirely a liquid composition, as such a melt must very early in the crystallisation history have been similar to that of point A in Figure 7. Although removal of the analysed biotite from the Cooma Granodiorite will produce a more aluminous melt in terms of A/AFM ratios (Fig. 8), the coincidence in *Mg* value of the granodiorite biotite to the bulk rock does not satisfy the other (independent) constraint that this melt should also be more Fe-rich. We believe that the biotite in the granodiorite (restite and that which crystallised from the melt) has re-equilibrated in Mg–Fe composition during solidification of the granodiorite.

In contrast, the rare biotite in the leucosome is more Mg-rich (Fig. 8) and should have been very similar in composition to that which was initially in the granodiorite. The leucosome biotite, which it is argued represents the liquidus composition, escaped re-equilibration because the melt subsequently left the migmatite. Removal of this biotite from the bulk composition of the granodiorite extrapolates to a liquid that is both more aluminous than cordierite and more Fe-rich (Fig. 8, Table 3). Thus the two independent criteria for the composition of the melt at the time of reaction (1) indicated by point D in Figures 7 and 8 are satisfied and coincide fairly well by the removal of leucosome rather than re-equilibrated granodiorite biotite. Simple mass-balance calculations require that some 15 wt% of the biotite present in the granodiorite did not actually crystallise from the melt.

The high SiO_2 value for the calculated remaining rock is in accord with earlier geochemical arguments (White & Chappell 1988, fig. 2) that the granodiorite is rich in quartz which did not crystallise directly from the melt. It is possible to use the quartz–feldspar–liquid–vapour calculations of Nekvasil (1988) to estimate the amount of quartz and feldspar which are also restitic in the Cooma Granodiorite.

We estimate that approximately 20 wt% quartz was restitic or xenocrystal. This estimate is based on the location of the quartz–alkali feldspar cotectic at 500 MPa, H_2O-saturated (Nekvasil 1988, table 1) projecting the composition of the Cooma Granodiorite into the quartz–feldspar system from biotite ($K_2(MgFe)_6Al_2Si_6O_{20}(OH)_4$).

Unlike biotite which re-equilibrates rapidly with melt, the plagioclase in the granodiorite is zoned to much more sodic compositions than the granodiorite bulk rock Ab/Ab + An ratio (Figs 4, 9). Indeed the bulk composition is less An/An + Ab rich than the most calcic core to the plagioclases. In contrast the rim and groundmass plagioclases show a very wide range in compositions, similar to those of the migmatite, and extend to more sodic values than the bulk composition of the granodiorite. This zoning is due to the very slow diffusion rates in plagioclase (see Johannes 1985, and references therein), which therefore enables the residual melt fraction to evolve to sodic compositions, in contrast to the *Mg* value which is buffered to a far narrower range because of the fast Fe–Mg diffusion in biotite.

Evidence to support a xenocrystal origin for the large plagioclase crystals in the granodiorite was presented 50 years ago by Joplin (1942, pp. 187–93). She noted that two types of granodiorite are found, which only differ in the presence of porphyritic plagioclase in addition to porphyritic orthoclase. The plagioclase-rich variety is present in the centre of the complex and is apparently surrounded by the other variety. Plagioclase occurs both as large tabular phenocrysts (xenocrysts) with euhedral igneous overgrowths and as small clear grains in the allotriomorphic groundmass. Some of these xenocrystal cores show strain.

The composition of the Cooma Granodiorite in terms of Or–Ab–An content, after removal of K_2O to satisfy biotite and CaO to satisfy apatite, is shown in Figure 9 for the 500 MPa projection of Nekvasil (1988, fig. 10b). The granodiorite lies well away from the alkali feldspar-saturated surface towards plagioclase and requires unacceptably high temperatures to be a liquid composition. In particular the presence of magmatic andalusite places limits on the temperature of this melt which must have been saturated in alkali feldspar. The Cooma Granodiorite lies on a tie line between melt located on the alkali feldspar–plagioclase feldspar cotectic surface, with a low anorthite content, and the compositions of the plagioclase cores (Fig. 9). This indicates the presence of approximately 8 wt% xenocrystal or restitic plagioclase in the granodiorite.

The compositions of biotite, cordierite, ilmenite, andalusite, alkali feldspar and the sodic rims and groundmass plagioclase in the granodiorite can all be matched extremely well in the migmatite sample that we have studied in detail here. In contrast, the more calcic plagioclase core compositions cannot be identified in this particular migmatite sample. Munksgaard (1988) found that on Harker Diagrams the metasediments of Cooma form well-defined linear arrays. He concluded that the granodiorite occupies a position intermediate between the metapelitic and meta-

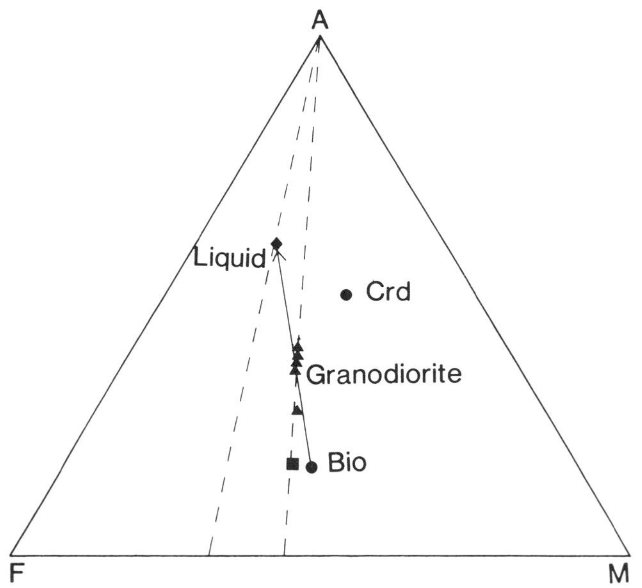

Figure 8 AFM plot of the composition of the Cooma Granodiorite (triangles), the biotite and cordierite in the leucosome (dots), the biotite from the granodiorite (square) and the calculated composition of the liquid in the restite-rich granodiorite. The dashed line from the granodiorite biotite composition to Als shows that removal of this composition biotite has no effect on the *Mg* number of the extrapolated, more aluminous residual melt composition. In contrast removal of leucosome biotite composition from the granodiorite extrapolates along the arrow to both a more aluminous and lower *Mg* number melt composition. The liquid composition was determined by the intersection of the biotite-granodiorite extrapolation line with the *Mg* value of the liquid (shown by the left-hand dashed line) constrained by the independent Crd–liq Fe–Mg experimental data of Ellis (1986). the Cooma Grandiorite rock compositions are from White and Chappell (1988) and Munksgaard (1988).

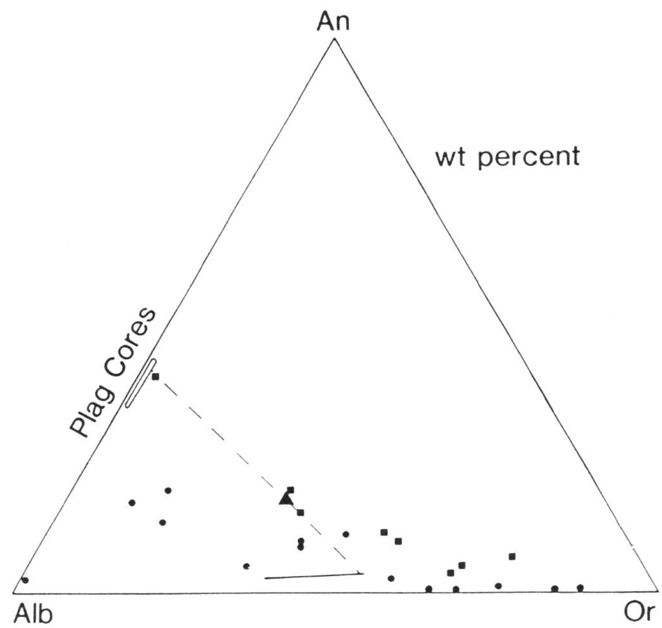

Figure 9 An–Or–Ab (wt%) projection of the compositions of the low-grade metasediments (circles) and the high-grade metasediments (squares) from the data of Munksgaard (1988); the Cooma Granodiorite (triangle) from the data of White and Chappell (1988); together with the range in composition of the plagioclase cores in the granodiorite (Table 1). The straight line represents the water, quartz, plagioclase and alkali feldspar saturated five-phase cotectic for the projected compositions of low temperature liquids at 500 MPa calculated by Nekvasil (1988). The dashed line connects the plagioclase cores to the bulk rock composition of the granodiorite and the inferred composition of the melt. Note that the high-grade metasediments are more anorthite rich than the low-grade metasediments, and trend to the same plagioclase composition as the restitic plagioclase cores in the granodiorite. The granodiorite falls on the trend of the high-grade metasediments and not that of the low-grade metasediments. Data were projected after removal of apatite and all Fe and Mg as biotite ($K_2(Mg, Fe)_6Al_2Si_6O_{20}(OH)_4$). Note that no Ti was included in this biotite calculation.

psammitic rocks of its surrounding aureole, and with few exceptions the granodiorite samples lie on a crudely linear Harker Diagram array defined by the metasediments. His calculations showed that the best fit to the average Cooma Granodiorite composition is obtained for a pelitic/psammitic mix in the proportions 40:60. He notes that the main misfits were CaO, Na_2O, Sr, Pb and Mn and suggested the granodiorite may have formed from sediments slightly richer in feldspar than its surrounding metasedimentary envelope. In view of Munksgaard's (1988) observation that there is an important misfit in some elements, the low-grade and high-grade metasediments are also plotted in terms of the relative proportions of Or–Ab–An, projected from biotite, quartz, apatite, in Figure 9. This clearly shows that the high-grade gneisses are not identical to the low-grade rocks and are relatively enriched in anorthite. The low-grade metasediments show a trend defined by a mixture of clay and moderately sodic detrital plagioclase. Wyborn and Chappell (1983) found that the Ordovician and Silurian metasediments of the Snowy Mountains region (W of Cooma) are essentially a mixture of quartz and clay minerals together with a small but variable detrital feldspar component. In contrast, the high-grade metasediments show a trend to significantly higher anorthite contents. Previous work in the Lachlan Fold Belt has shown that the exposed low-grade sediments are not Ca-rich enough to be a suitable source for the LFB granites, except for the Cooma Supersuite of granites (Chappell 1984, fig. 4). The Cooma Granodiorite has a chemical composition that is distinctly different from most of the S-type granites in the Lachlan Fold Belt (LFB), containing significantly less Ca, Na and Sr (Chappell 1984). The S-type Bullenbalong Suite of the LFB has even higher CaO contents than the Cooma Granodiorite (White & Chappell 1988) and certainly requires a less mature sedimentary basement source than the exposed Ordovician and Silurian sediments (see also Wyborn & Chappell 1983). We suggest that in the specific case of the Cooma Granodiorite, which has a slightly higher CaO content than the surrounding low-grade metasediments (Munksgaard 1988; White & Chappell 1988) this may be explained by derivation from a Ca-rich source similar to the high-grade rocks and not to the low-grade rocks. Based on structural and bulk rock chemical considerations, Steele *et al.* (1991) have made a similar suggestion that the Cooma Supersuite granites were derived from protoliths similar to the Ca-rich, migmatitic Gundowring Terrane of Northeast Victoria.

In view of the suggestion of Flood and Vernon (1978) that the Cooma body has dragged up with it the surrounding high-grade metamorphic envelope it seems possible that the relatively An-rich high-grade metasediments may also provide the necessary source for the Ca-rich plagioclases in the granodiorite. However our observation that the partial melting process produces a more calcic residue may negate the need to invoke a more calcic, less mature sedimentary protolith than the exposed Ordovician sediments for the specific case of the Cooma Granodiorite. The absence of such calcic plagioclase compositions from the particular migmatite studied here reflects its more potassic sedimentary composition compared to the overall range in the metasedimentary rocks of Cooma (Fig. 9).

5. Acknowledgements

This work has benefited from discussions with Drs B. W. Chappell and W. E. Cameron. In addition we thank Dr W. E. Cameron for kindly supplying us with the sample of Cooma Granodiorite containing fresh cordierite. We thank Professor A. J. R. White, Dr R. A. Binns and Mr D. A. Steele for comments on various drafts of this work. This work was commenced while Masaaki Obata was on study leave at the Department of Geology, The Australian National University and was supported by the Australian Academy of Science and the Japanese Society for the Promotion of Science (JSPS). David Ellis acknowledges the financial assistance of the Australian Research Council, and an ANU Faculties Research Fund. The authors wish to express their sincere gratitude to Professor P. Brown, St. Andrews, Dr Mike Rickard and Maree Coldrick, ANU for their assistance in editorial matters.

6. References

Abbott, R. N. & Clarke, D. B. 1979. Hypothetical liquidus relationships in the subsystem Al_2O_3–FeO–MgO projected from quartz, alkali feldspar and plagioclase for $a(H_2O) \leq 1$. CAN MINERAL **17,** 549–60.

Ashworth, J. R. 1985. Introduction. *In* Ashworth J. R. (ed.) *Migmatites,* 1–35. Glasgow: Blackie.

Burnham, C. W. 1969. The importance of volatile consituents. *In* Yoder, H. S. (ed.) *The evolution of the igneous rocks (Fiftieth anniversary perspectives),* 439–82. Princeton: Princeton University Press.

Burnham, C. W. & Nekvasil, H. 1986. Equilibrium properties of granite pegmatite magmas. AMER MINERAL **71,** 239–63.

Chappell, B. W. 1984. Source rocks of I- and S-type granites in the Lachlan Fold Belt, southeastern Australia. PHIL TRANS R SOC LONDON **A310,** 693–707.

Chappell, B. W. & White, A. J. R. 1976. Plutonic rocks of the Lachlan Mobile Zone. 25th Int Geol Congr., Sydney, Excursion Guide No. 13C.

Chappell, B. W., White, A. J. R. & Wyborn, D. 1987. The importance of residual source material (restite) in granite petrogenesis. J PETROL **28,** 1111–38.

Ellis, D. J. 1986. Garnet-Liquid Fe–Mg equilibria and implications for the beginning of melting in the crust and subduction zones. AM J SCI **286,** 765–91.

Flood, R. H. & Vernon, R. H. 1978. The Cooma Granodiorite, Australia: an example of in situ crustal anatexis? GEOLOGY **6,** 81–84.

Grant, J. A. 1985. Phase equilibria in partial melting of pelitic rocks. *In* Ashworth J. R. (ed.) *Migmatites,* 86–144. Glasgow: Blackie.

Holdaway, M. J. 1971. Stability of andalusite and the aluminium-silicate phase diagram. AM J SCI **271,** 97–131.

Holdaway, M. J. & Lee, S. M. 1977. Fe–Mg cordierite stability in high grade pelitic rocks based on experimental, theoretical and natural observations. CONTRIB MINER PETROL **63,** 175–98.

Hopwood, T. P. 1976. Stratigraphy and structural summary of the Cooma metamorphic complex. J GEOL SOC AUST **23,** 345–60.

Johannes, W. 1985. The significance of experimental studies for the formation of migmatites. *In* Ashworth, J. R. (ed.). *Migmatites* 36–85. Glasgow: Blackie.

Joplin, G. A. 1942. Petrologic studies in the Ordovician of New South Wales I: The Cooma Complex. PROC LINN SOC NSW **67,** 156–96.

Joplin, G. A. 1943. Petrologic studies in the Ordovician of New South Wales II: The northern extension of the Cooma Complex. PROC LINN SOC NSW **68,** 159–83.

Joplin, G. A. 1962. An apparent magmatic cycle in the Tasman geosyncline. J GEOL SOC AUST **9,** 51–69.

Le Breton, N. & Thompson, A. B. 1988. Fluid-absent (dehydration) melting of biotite in metapelites in the early stages of crustal anatexis. CONTRIB MINER PETROL **99,** 226–37.

Menhert, K. R. 1968. *Migmatites and the origin of granites.* Amsterdam: Elsevier.

Munskgaard, N. C. 1988. Source of the Cooma Granodiorite, New South Wales—a possible role of fluid-rock interactions. AUST J EARTH SCI **35,** 363–77.

Nekvasil, H. 1988. Calculated effect of anorthite component on the crystallisation paths of H_2O-undersaturated haplogranitic melts. AM MINERAL **73,** 966–81.

Nekvasil, H. & Burnham. C. W. 1987. The individual effects of pressure and water content on phase equilibria in the "granite" system. In Magmatic Processes: Physicochemical principles. GEOCHEM SOC SPEC PUBL **1,** 433–45.

Olsen, S. N. 1985. Mass balance in migmatites. *In* Ashworth J. R. (ed.). *Migmatites,* pp. 145–79. Glasgow: Blackie.

Pidgeon, R. T. & Compston, W. 1965. The age and origin of the Cooma granite and its associated metamorphic zones, NSW J PETROL **6,** 193–222.

Richardson, S. W., Gilbert, M. C. & Bell, P. M. 1969. Experimental determination of kyanite-andalusite and andalusite-sillimanite equilibria: the aluminium silicate triple point. AM J SCI **267,** 259–72.

Steele, D. A., Price, R. C., Fleming, P. D. & Gray, C. M. 1991. The origin of Cooma Supersuite granites: source protoliths and early magmatic processes. *In* Chappell, B. W. (Ed.) *Second Hutton Symposium on granites and related rocks, Canberra 1991* (*Abstracts*). REC BMR GEOL GEOPHYS 1991/25, 100.

Thompson, A. B. 1976. Mineral reactions in pelitic rocks. II. Calculations of some P–T–X(Fe–Mg) phase relations. AM J SCI **276,** 425–54.

Vallance, T. G. 1967. Palaeozoic low-pressure regional metamorphism in Australia. DANSK GEOL FOREN MEDD **17,** 494–503.

Vallance, T. G. 1969. Southern and Central Highlands fold belt: plutonic and metamorphic rocks. J GEOL SOC AUST **16,** 180–200.

Vernon, R. H. 1978. Pseudomorphous replacement of cordierite by symplectic intergrowths of andalusite, biotite and quartz. LITHOS **11,** 283–9.

Vielzeuf, D. & Holloway, J. R. 1988. Experimental determination of the fluid-absent melting relations in the pelitic system. CONTRIB MINER PETROL **98,** 257–276.

White, A. J. R. & Chappell, B. W. 1977. Ultrametamorphism and granitoid genesis. TECTONOPHYSICS **43,** 7–22.

White, A. J. R. & Chappell, B. W. 1988. Some supracrustal (S-type) granites of the Lachlan Fold Belt. TRANS R SOC EDINBURGH EARTH SCI **79,** 169–181.

White, A. J. R., Chappell, B. W. & Cleary, J. R. 1974. Geological setting and emplacement of some Australian Palaeozoic batholiths and implications for intrusive mechanisms. PAC GEOL **8,** 159–71.

Wyborn, L. A. & Chappell, B. W. 1983. Chemistry of the Ordovician and Silurian greywackes of the Snowy Mountains, Southeastern Australia: an example of chemical evolution of sediments with time. CHEM GEOL **39,** 81–92.

D. J. ELLIS, Department of Geology, The Australian National University, GPO Box 4 Canberra City ACT 2601 Australia.

M. OBATA, Department of Geology, Faculty of Science, Kumamoto University, Kumamoto 860, Japan.

Transactions of the Royal Society of Edinburgh: Earth Sciences, **83**, 107–114, 1992

Using granite to image the thermal state of the source terrane

E-an Zen

ABSTRACT: It should be possible to infer the thermal state of the source terrane for granitic bodies, provided we have independent means to establish the chemical nature of this terrane. The chemical nature of the granitic rocks, including their degree of hydration, implies the solidus temperature. The concentration of the heat-producing radioactive elements in the granite (K, U, and Th) probably provides an upper estimate of their concentration in the source rock, which is an important thermal parameter. The depth and ambient temperature of the country rock into which the granite magma intruded provide useful boundary conditions for the thermal regime at the crustal level of anatexis. These constraints in turn form the bases for estimating the subcrustal thermal flux as well as the effective thermal interface for enhanced heat flow from below that resulted in anatexis. These inferences, in combination with other *field-based* parameters such as uplift rates and permissible time lapses for the geological events, permit realistic thermal modelling for the formation of granitic batholiths. The procedure is applied to the Late Cretaceous Pioneer and Boulder batholiths in southwestern Montana, U.S.A. The modelling results suggest that mantle upwelling, not subduction or thrust loading, caused anatexis. The isotopic chemistry of the granitic rocks rules out direct mixing of mantle magma, and field relations rule out crustal thinning as causes for partial melting.

KEY WORDS: anatexis, heat productivity, heat source, thermal regime.

Granitic rocks could potentially instruct us about the thermal state of their source terranes during anatexis, provided we have some independent information about the nature of the source. The nature of the source rocks thus forms an essential part of the constraints on the thermal regime of anatexis. Inquiry into the thermal regimes and constraints during anatexis is boosted by the application of computer modelling, which in recent years has received considerable attention (for example, Hanson & Barton 1989; Zen 1988a; DeYoreo *et al.* 1989).

I have previously described (Zen 1988a) a particular procedure for computing the thermal configuration of a thrust-thickened crust and the amount of melt developed through stepwise anatexis, for given initial and boundary conditions, using a flexible one-dimensional thermal modelling program, 1DT (Haugerud 1986). Because the 1DT program allows interruption of the computations at arbitrary moments, when the boundary conditions may be changed, it readily permits changes in the mantle heat flux (approximating subduction or upwelling), crustal thickness (approximating thrusting or tectonic denudation), the temperature profile within the crust (approximating melting for specified values of the latent heat), or the uplift rate (approximating steady denudation, i.e. uplift precisely balanced by erosion).

1. The Late Cretaceous batholiths of southwestern Montana

In this paper, I shall combine field-based tectonic and petrological data and this modelling approach to delimit conditions for the genesis of the magmas of two Late Cretaceous (Laramide) calcalkaline batholiths in southwestern Montana, U.S.A.—the Boulder and Pioneer batholiths (Fig. 1).

The Boulder Batholith, nearly 5000 km^2 in area, contains plutons that range in rock type from syenogabbro to granite. Granodiorite is the principal rock type; the Butte Quartz Monzonite, which in the Streckeisen classification (Streckeisen 1973) is a granodiorite, occupies about two-thirds of the area of this batholith. The rocks predominantly carry hornblende and biotite as the mafic phases; two-mica granite is confined to a single small satellite body (the Moose Creek Granite; Zen, unpublished data). The Boulder Batholith ranges in age from about 79 to 68 Ma (Campanian to Maastrichtian). The petrogenesis and chemistry of this batholith were summarised by Tilling (1973); age data are from Tilling *et al.* (1968), and a map showing distribution of the plutons was compiled by Smedes *et al.* (1988).

The Pioneer Batholith (Fig. 2) is considerably smaller than the Boulder Batholith, though it, too, consists of discrete plutons that range from quartz diorite and tonalite, to the dominant granodiorite and granite. Hornblende and biotite are the mafic phases; two-mica granite is minor (Zen 1988b, in prep.; Arth *et al.* 1986; Snee 1982). In age, the plutons range from 80–65 Ma (Campanian to Maastrichtian, possibly earliest Palaeocene), the same as the ages of the Boulder Batholith especially when the dating uncertainties are included. In petrography, bulk chemistry, age, and tectonic history, and, as we shall see, in Pb isotope data, the two batholiths are nearly identical twins.

Despite these similarities, the two batholiths have distinctly different sizes and initial strontium ratios (iSr). The Pioneer Batholith is characterised by iSr values of 0·7113–0·7160 (Arth *et al.* 1986), whereas the Boulder

Batholith is characterised by iSr values of 0·7055–0·7092 (Doe *et al.* 1968). These two batholiths, so discriminated, are neatly separated on the ground by a system of high-angle faults known as the Trusty Gulch-Sawmill Gulch Fault System (Zen 1988b), as is shown on Figure 2.

2. Estimating melting properties based on nature of the source rocks

2.1. Nature of the source rocks

Can we constrain the nature of the source rock? This knowledge is important because it could significantly affect the realistic modelling of melting properties. Some Pb-isotope data for both batholiths are available (Doe *et al.* 1968, 1986); they are consistent with a model source age of 1·9 ± 0·2 Ga for the Pioneer Batholith, and 2·1 ± 0·3 Ga for the Boulder Batholith (Zen, in prep.). For our purposes these numbers are indistinguishable. The high iSr values for the Pioneer Batholith also show that the magmas had old crustal sources, consistent with the Pb-isotope data. No published study of the Sm/Nd systematics is yet available for either batholith.

2.2. Melting properties of the source rocks

As already stated, hornblende-biotite granodiorites dominate both batholiths, and neither contains strongly peraluminous granite. In addition, the basement rocks were metamorphosed in the upper amphibolite to granulite facies during the Archaean or the Proterozoic eras (Lowell 1965; Sahinen 1939; James & Hedge 1980, and references therein). Taken together, these facts indicate that the source rock was already partially dehydrated and probably deprived of some of its low-melting material. Because the plutons solidified at very shallow levels, mostly less than 3 km in the eastern lobe of the Pioneer Batholith for which the modelling results and field data apply (Zen 1988b, and in prep.; Snee 1982), the magmas must also have been undersaturated in H_2O, consistent with anatexis in a vapour-absent regime; any H_2O-rich vapour phase could have evolved only toward the end of solidification at shallow crustal levels. For these reasons, I infer that the solidus temperature was closer to 800°C than to 650°C; the lower temperature would be more appropriate for a source rich in readily-fusible hydrous pelitic material (Wyllie 1977).

3. Estimating thermal productivity of source rocks

3.1. Granite as images of thermal productivity of source rocks

Thermal modelling requires knowing the heat productivity of the source material. In the paragraphs following, I argue that the granitic rocks could indeed provide useful clues. In granulite-grade source rocks, K occurs mainly in feldspar

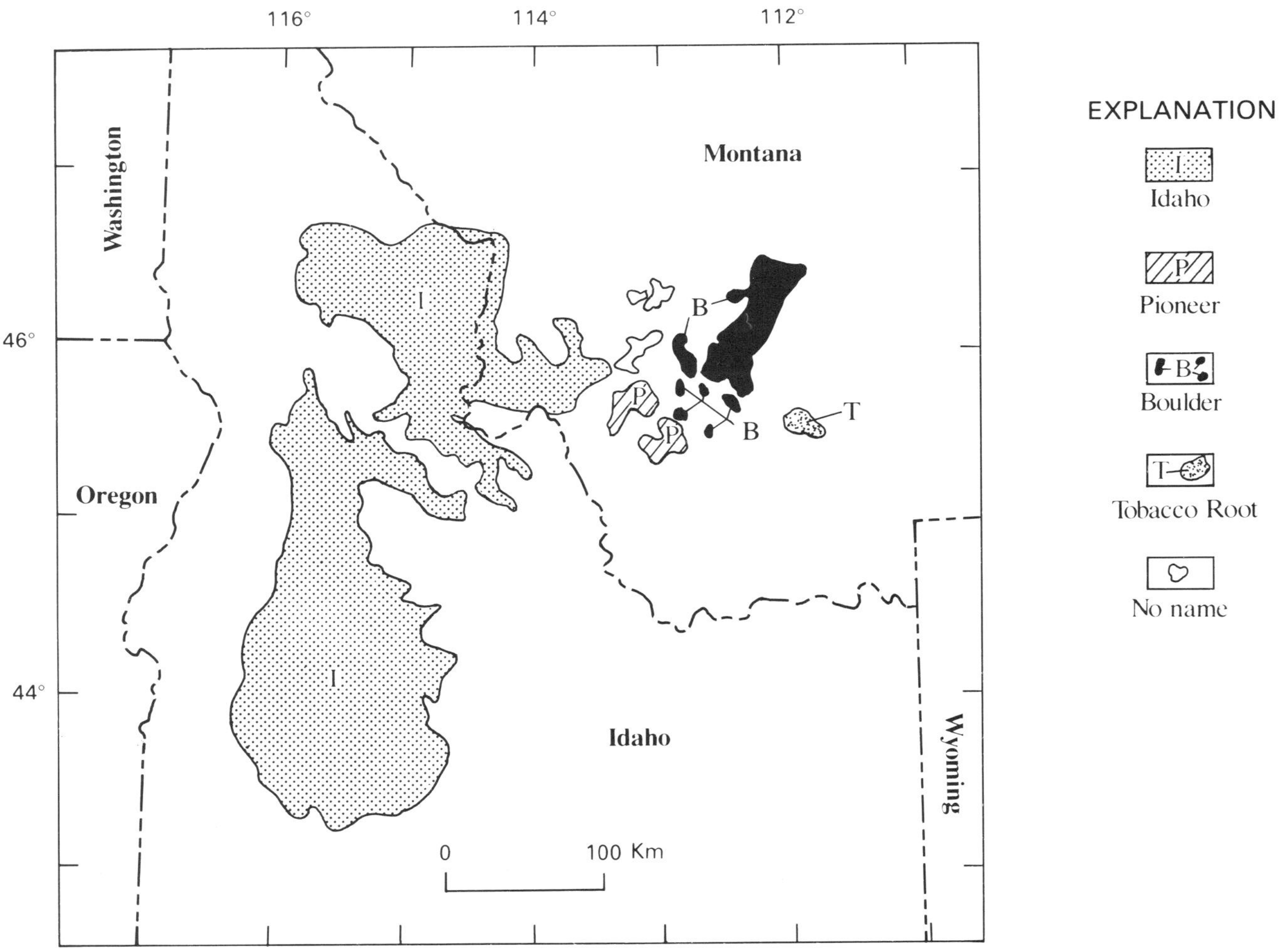

Figure 1 Index map showing major batholiths based on King and Beikman 1974 and modified by data from Ruppel *et al.* (1983), Snee (1982), and my own data.

and biotite, precisely the minerals expected to contribute substantially to the first melt. Indeed, the breakdown of biotite probably initiated melting (Patiño-Douce *et al.* 1990; Clemens & Vielzeuf 1987; Vielzeuf & Holloway 1988). U and Th occur mainly in accessory phases such as zircon, allanite, apatite, sphene, and monazite. These minerals may dissolve in the melt, but voluminous petrographic reports document that the more refractory phases such as zircon and allanite are common as restite minerals. It is impossible, however, to accurately estimate the proportion of U and Th carried in the magma, either dissolved or entrained, versus what is left behind during melt separation. The difficulty is particularly aggravated by our common inability to demonstrate the petrogenetic linkages between a specific source rock and a specific granite; even more rarely can both rocks be sampled.

3.2. Data from migmatites

Study of the distribution of U and Th between leucosomes and melanosomes of migmatites, where the source rock-melt linkage is plausible, could help (see, for instance, Sawyer 1987, Weber *et al.* 1985), even though the possible effect of entrainment of restite might not be recorded because of the short distances of melt migration. Table 1 shows a paired data set for equilibrium partial fusion of the Quetico metasedimentary rocks for southwestern Ontario, Canada (Sawyer 1987), demonstrating that, under some circumstances, U and Th are strongly partitioned into the

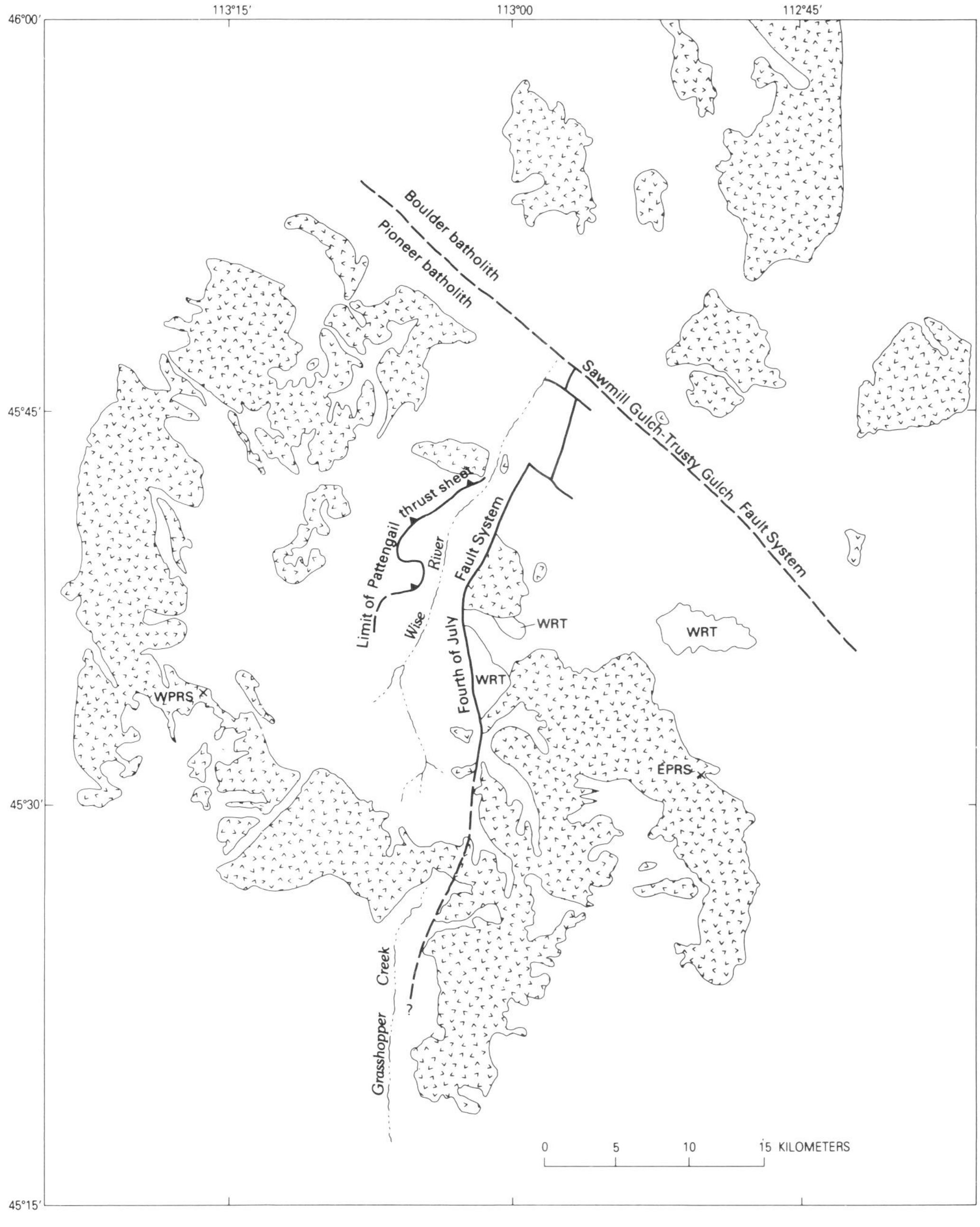

Figure 2 Map of the Pioneer Batholith and some satellite bodies of the Boulder Batholith; distribution of plutons and the Wise River thrust sheet (WRT) based on Zen (1988b, in preparation), and Snee (1982); EPRS, the eastern Pioneer reference site for modelling, discussed in this paper; WPRS, the western Pioneer reference site, not discussed here.

magma (here, the samples might include some entrained restite minerals).

3.3. Data from the Cooma Granodiorite

Data from the Cooma Granodiorite of the Lachlan Fold Belt, Australia, are also of interest because, ever since the early studies by Joplin (1942, 1943), the Cooma Granodiorite has been regarded by many geologists as an example of *in situ* anatexis (White & Chappell 1977; Munksgaard 1988). (This hypothesis has been challenged (Flood & Vernon 1978), but I shall provisionally accept it in this discussion.) If the Cooma Granodiorite is a local anatectic product, its chemical composition can be directly compared with that of the supposed source rocks, which are the Ordovician sedimentary rocks surrounding the pluton (Joplin 1942). Munksgaard (1988) compiled data on the Th and U contents of the Cooma Granodiorite, and of the surrounding source sedimentary rocks. Excerpts of Munksgaard's tabulated data, given in Table 1, clearly show that, if restite formed during the formation of the Cooma Granodiorite, it did not concentrate Th and U.

3.4. Data summary

Table 1 also gives a compilation of the concentrations of Th and U in the Boulder Batholith (Tilling & Gottfried 1969) and in the Pioneer Batholith (Zen, in prep.), and compares these values with those of other rocks in addition to the Cooma Complex and the leucosome-melanosome pair of the Quetico metasedimentary rocks. These rocks include "average" diorite, quartz diorite, and granodiorite (all from Clark *et al.* 1966); rocks of the Canadian shield and the "average crust" (however defined; both from Tilling & Gottfried 1969); Archaean metamorphic rocks from the Beartooth Mountains (Wooden *et al.* 1988); and some Archaean granites of Wyoming (Stuckless *et al.* 1981). Data from the last group may be biased toward high values because these granites are exceptionally rich in radioactive elements and many are associated with economical U resources.

It is probably reckless to generalise from so few data, but I do suggest that, based on what we know, and when we include consideration of K, the heat productivity of the granites reasonably gives upper limits for those of the sources. This conclusion is consistent with the measured values for the Pioneer Batholith, averaging about 2 $\mu W/m^3$ (Zen, in prep.) and for the Boulder Batholith, averaging about 3 $\mu W/m^3$ (Tilling & Gottfried 1969), when compared to the typical Archaean cratonal value for N America, about 1 $\mu W/m^3$ (Lachenbruch & Sass 1977). In the models I have used the last value for the source rocks because it may be more representative than the "upper limits" given by the batholiths.

4. Distribution of heat productivity with depth

Another factor that must be considered for purposes of thermal modelling is the vertical distribution of heat productivity within the crust. Figure 3 depicts four distribution schemes, all decreasing monotonically with depth: (1), uniform value of heat productivity, A, of 1 $\mu W/m^3$, from the surface to 15 km; $A = 0$ below that level; (2) The *same total integrated productivity*, uniformly distributed to 30 km depth, with implied halving of the value of A; (3) A decreases linearly with depth, from a surface value of 1 $\mu W/m^3$ to zero at 30 km; (4) A decreases logarithmically with depth (Lachenbruch 1970) from a surface value of 1 $\mu W/m^3$, so as to preserve the integrated productivity between the surface and 30 km.

Most of the model runs used scheme (1). Runs 625A and 625B (Table 2) respectively tested schemes (2) and (3). Both of these latter schemes agree better than does (1) with the

Table 1 U and Th contents of southwestern Montana granitic rocks and some comparisons (ppm)

	Mean U	Mean Th
Boulder Batholith ($n = 60$)	3·9	15·4
Pioneer Batholith ($n = 18$)	3·2	14·0
Quetico migmatite, Canada		
Source metasediments ($n = 1$)	2·6	6·7
Equilibrium partial melt ($n = 3$)	3·3	15·3
Cooma area, Australia		
Cooma granodiorite	3·5 ($n = 4$)	18·8 ($n = 5$)
High grade psammite	3·9 ($n = 7$)	18·7 ($n = 7$)
Low grade psammite	4·6 ($n = 5$)	17·3 ($n = 7$)
High grade pelite	5·0 ($n = 2$)	20·7 ($n = 3$)
Low grade pelite	4·3 ($n = 6$)	18·7 ($n = 6$)
Archaean metamorphic rocks,		
Beartooth Mountains ($n = 35$)	1·7	9·1
Details		
Meta-andesite ($n = 9$)	2·2	7·3
Meta-pelite ($n = 2$)	2·0	8·0
Meta-dacite ($n = 18$)	2·5	10·0
Meta-rhyolite ($n = 3$)	0·7	8·0
Meta-greywacke ($n = 3$)	1·7	24·3
Archaean mafic granite,		
Wind River Range ($n = 46$)	3·5	30·5
Average diorite & quartz diorite	2·0 ($n = 69$)	8·5 ($n = 47$)
Average granodiorite	2·6 ($n = 156$)	9·3 ($n = 92$)
Canadian Shield Rocks ($n = 330$)	2·5	10·3
Average continental crust	2·8	10·0

Note: Sources of data: Boulder Batholith, Canadian shield rocks, and average crust, Tilling and Gottfried (1969). Pioneer batholith, Zen (in preparation). Archaean rocks of Beartooth Mountains, Wooden *et al.* (1988). Archaean granites of Wyoming, Stuckless *et al.* (1981). Rocks of the Cooma Complex, Australia, Munksgaard (1988). Migmatite of the Quetico Metasedimentary Belt, Canada, Sawyer (1987). Other rocks, Clark *et al.* (1966).

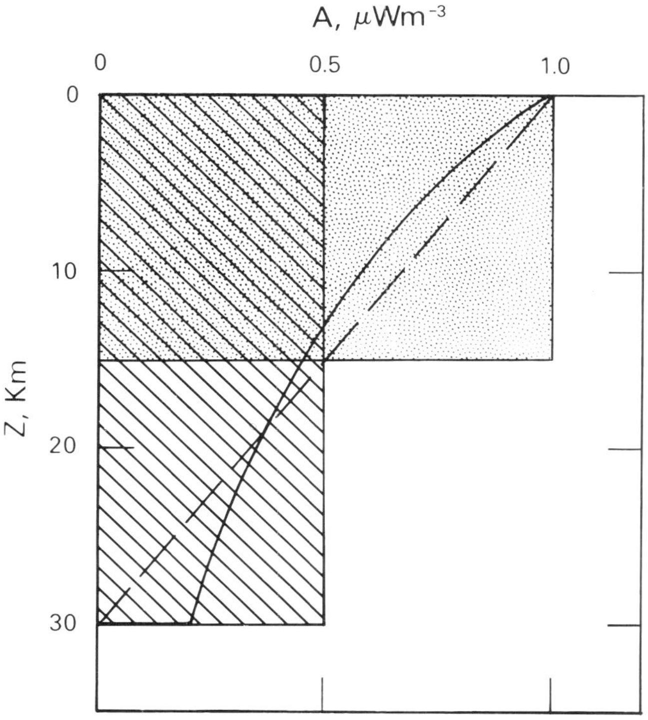

Figure 3 Four possible schemes of vertical distribution of heat productivity, A, within the crust: (1, stipple pattern): uniform distribution of A at $1\ \mu Wm^{-3}$, but only within the upper 15 km; (2, hached pattern): uniform distribution of $A = 0.5\ \mu Wm^{-3}$ within the entire 30 km of the crust so that the integrated crustal productivity is same as above; (3 dashed line): A decreases with depth from $1\ \mu Wm^{-3}$ at the surface to 0 at 30 km, so that the integrated crustal productivity is same as above; and (4, solid line): A decreases logarithmically, from a value of $1\ \mu Wm^{-3}$ at the surface, such that the integrated crustal value remains same as above for the 30-km-thick crust.

known entrainment of minerals that are Th and U carriers; yet they predict only about 20–30% more melt production within the time period for uplift that is limited by geological relations (this time limit, based on field relations augmented by palynological and isotopical chronology, permits a maximum of about 10 m.y. between the arrival of thrust sheets terminating sedimentation and the intrusion of the major plutons). Although scheme (4) cannot be modelled by 1DT, its effect of melt production clearly must be close to that of (3).

5. Constraints on mode of energy transport to site of anatexis

What was the mode of energy transport *into* the zone of anatexis? For the Montana batholiths at least, I rule out aqueous fluid influx to the site of melting as a source of energy. This is because the melt must have been undersaturated in H_2O during anatexis, yet must have contained enough H_2O to stabilise hornblende and biotite. For such magmas, presumably intermediate between tonalite and granite in bulk composition, no more than 10% H_2O can be in the melt (Wyllie 1977), and much of this H_2O must have been derived from breakdown of hydrous minerals within the source rocks. If we allow 3% of the 10% H_2O to be introduced from a region of higher pressure and temperature, we can roughly calculate the quantity of heat delivered on the basis of the data of Burnham *et al.* (1969). At most, about 10% of the total rock might be melted by this mechanism—significant, but far from adequate to even overcome the critical melting fraction and to cause melt mobilisation.

The possibility that energy was carried directly by mixing mantle-derived mafic magma (Huppert & Sparks 1988), can

Table 2 Parameters used in thermal modelling, and principal results

Run No.	Incubation period m.y. ip_1	ip_2	Depth of thermal interface	Melting interval °C	Maximum ambient temperature °C (a)	Total Fusion (b) End of ip_2	5 m.y. later	10 m.y. later
N5A	35	2	36	700–1060	181	4·61	5·88	7·26
N1A	35	2	36	740–1100	182	3·73	4·88	6·21
N6A	35	2	36	800–1160	183	2·59	3·56	4·71
N3A	35	2	46	700–1060	164	2·17	3·15	4·31
N2A	35	2	46	740–1100	165	1·61	2·43	3·44
N4A	35	2	46	800–1160	166	0·81	1·42	2·24
625A (c)	35	2	46	800–1160	161	1·25	2·00	2·99
625B (d)	35	2	46	800–1160	162	0·96	1·60	2·44
10B	35	2	60	700–1060	143	0·31	0·71	1·34
10C	35	2	60	740–1100	143	0·09	0·35	0·82
N7A	35	2	60	800–1160	143	0·00	0·05	0·29

Notes: (a) At original land surface.
(b) Figures in these three columns represent equivalent thickness of residuum-free melt generated in the crustal column, in km. See Zen (1988a). "Later" refers to time lapse since end of ip_2.
(c) Same as N4A except that the heat production, A, is $0{\cdot}5\ \mu Wm^{-3}$ throughout the 30 km of the crust, so that the integrated heat production is the same as for N4A.
(d) Same as N4A except that the heat production, A, is $1{\cdot}0\ \mu Wm^{-3}$ at $Z = 0$ km and decreases linearly to 0 at $Z = 30$ km, so that the integrated heat production is the same as for N4A.
Other parameters used in modelling (see Zen 1988a for definitions of terms):
Dimensions of model cell: 1 km
Thermal conductivity: $2\ W\ .\ m^{-1}\ K^{-1}$
Heat production, A (upper 15 km only): $1\ \mu W\ .\ m^{-3}$.
Heat capacity, $0{\cdot}9\ kJ\ .\ kg^{-1}\ .\ K^{-1}$.
Bulk density: $2780\ kg\ .\ m^{-3}$.
Thermal diffusivity, $8 \times 10^{-7}\ m^2\ .\ s^{-1}$.
Mantle flux for the original continental geotherm: $30\ mW\ .\ M^{-2}$.
Enhanced mantle flux below thermal interface: $75\ mW\ .\ m^{-2}$.
Fusion Step: 40°C.
Parametric heat of fusion, Δh_f (Zen 1988a): 300 K
Time interval for adjusting thermal profile: 5 m.y.
Thrust sheet is 6 km thick.
Decompression (uplift balanced by erosion) rate is 0·1 km/m.y.

be ruled out as a substantial process for the Pioneer batholith because of its high initial Sr ratios.

Thus I infer that conduction was the chief mode of energy transfer for anatexis for this batholith. This is significant because conductive heating is slow on a crustal scale (Zen 1988a); there is a time lag of a few tens of millions of years between the geological trigger than led to the heating and the production of large volumes of magma. In my modelling, I start with a typical continental geotherm. At a specific moment into the model time, the basal heat flux is increased at some particular monitoring depth to initiate a new thermal regime. This depth is referred to as the thermal interface, geologically corresponding, for example, to the top of mantle upwelling or of the subducted slab.

6. Application of model results to real rocks

6.1. Relating model constraints to regional geology

Because southwestern Montana was part of the craton during the Cretaceous compressional orogenies, crustal thinning is not a likely cause for major heating of the lower crust. The mapped Laramide thrust sheets in the area (Zen 1988a, in prep.; Snee 1978; Pearson & Zen 1985) contain little-metamorphosed sedimentary rocks, and the thrusting was near the land surface, involving a few kilometres of load. This amount of thickening could make only a minor contribution to the volume of melt. Moreover, the timing is all wrong because the first plutons of the batholith were intruded even before the Cretaceous sedimentation was cut off by the arrival of the thrust sheets (Zen 1988a).

For the Pioneer Batholith, the autochthonous country rocks at the level of granite emplacement but away from contact aureoles are unmetamorphosed sediments that still retain smectite (Zen in prep.). So an ambient temperature of about 160°C seems an appropriate upper limit for the country rock at the time and depth of intrusion, and model results that led to higher ambient temperatures were discarded. Then, using the other thermal information already discussed, we can constrain the minimum requisite subcrustal heat flux as well as the maximum depth of the thermal interface for appreciable melting within reasonable geological time intervals to give a batholith, say, a couple of kilometres thick (the constraint on the depth of the thermal interface rules out advection of heat by mafic magmas at shallow depths; advection at depths below the interface is, in contrast, implied by the need to maintain the heat flux over millions of years).

The permissible time interval, based on field, palynological, and geochronological data, is 5 ± 5 m.y. between the start of uplift and the event of intrusion. This time limit is used in combination with a field-based denudation rate (assumed to be the rate of uplift precisely balanced by erosion) of 0·1 km/m.y. Denudation promotes cooling, and is the dominant way by which the magmatic system is eventually quenched. Detailed arguments for the selection of these model values are given elsewhere (Zen in prep.).

6.2. Geological interpretation of results of thermal modelling

Table 2 summarises the principal results of my modelling, with volume of melt given in cubic km per square km of area. Clearly, not all of the melt generated can be expected to be pooled and to intrude, so the model volumes may be optimistic. In compensation, however, nor do the volumes include entrained restitic material.

For reasons already discussed, both the 700°C and the 740°C solidi are too cold. The placement of the thermal interface at 36 km—equal to a 30 km crust plus a 6-km thrust sheet—would cause the ambient temperature to exceed smectite thermal stability, so it, too, is unacceptable. However, a depth of 46 km for the thermal interface and a solidus temperature of 800°C give just under 2 km^3 of melt per square km. Thermal interfaces deeper than 50 km do not yield much crustal melting within reasonable time intervals because energy cannot be delivered fast enough to overwhelm its conductive dissipation so as to raise the temperature and cause melting. Varying the duration of heating by 5 m.y., which is all that the geological record could allow, changes the volume by no more than 50% and so does not affect the basic picture.

A melt thickness of 2–3 km forms a rather thin batholith, but perhaps this is as it should be. The Pioneer Batholith is located at the eastern limit of the main Laramide-age magmatic belt in this part of the northern Rocky Mountains; there may be a good thermal reason for this limit. Farther N, the closely similar, coeval, and spatially aligned Boulder Batholith has initial Sr ratios of 0·7055–0·7092, and a similar model source rock age; it could therefore tolerate incorporation of some mantle material. Combined with the greater heat productivity of the Boulder Batholith rocks (Zen in prep.), this factor might help to account for its larger size as well. E of this alignment of the two batholiths, that is, toward the craton, Late Cretaceous plutonism drops off precipitously. The major exception is the miniscule Tobacco Root Batholith, which has a single published initial Sr ratio of 0·704 (Shuster *et al.* 1989). On this account alone, this batholith might owe its presence to the mixing of considerable mantle material and more efficient energy transfer.

7. Conclusion: subduction versus mantle upwelling as energy source

In conclusion, we can speculate on the relative importance of shallow subduction versus mantle upwelling as energy sources. Shallow subduction has been often suggested as a way to generate the voluminous Tertiary magmas in the northern Rockies (for example, Lipman *et al.* 1972; Bird 1988). As we have seen, a heat source placed deeper than 50 km below the initial land surface is thermally impotent. But because the nearest Late Cretaceous continental margin was then at least 300 km away, beyond Idaho (see, for example, Armstrong *et al.* 1977; Fleck & Criss 1985; Lund & Snee 1988), a subduction slab no more than 50 km deep would seem both a mechanical and a geological monstrosity. This difficulty is accentuated by the thermal need for the slab to maintain a heat flux of about 75 mW/m^2 for 20–30 million years, thermally like a 15 m.y.-old oceanic crust. So I infer that mantle upwelling, perhaps a hot spot, may have been the major thermal cause for these batholiths.

8. Acknowledgements

I thank Fred Barker, Mike Brown, Bruce Doe, Jane Hammarstrom, Susan Gunn, Zell Peterman, Lee Silver, Larry Snee, and Bob Tilling for helpful discussions and for their sharing of data. I am grateful to Edward Sawyer for sharing data and insights on U and Th partitioning in leucosome-melanosome pairs. John Clemens, Jane Hammarstrom, Pete Toulmin and Allan White reviewed the manuscript and I thank them for their sagacious comments. This contribution is dedicated to the late Richard L. Armstrong, a pioneer thinker of the tectonics of the northern Rocky Mountains area.

References

Armstrong, R. L., Taubeneck, W. H. & Hales, P. O. 1977. Rb-Sr and K-Ar geochronometry of Mesozoic granitic rocks and their Sr isotopic composition, Oregon, Washington, and Idaho. GEOL SOC AM BULL **88,** 397–411.

Arth, J. G., Zen, E-an, Sellers, George & Hammarstrom, Jane 1986. High initial Sr isotopic ratios and evidence for magma mixing in the Pioneer batholith of southwest Montana: J GEOL **94,** 419–30.

Bird, Peter 1988. Formation of the Rocky Mountains, western United States: A continuum computer model: SCIENCE **239,** 1501–07.

Burnham, C. W., Holloway, J. R. & Davis, N. F. 1969. Thermodynamic properties of water to 1000°C and 10,000 bars. GEOL SOC AM SPEC PAP **132**.

Clark, S. P., Jr., Peterman, Z. E. & Heier, K. S. 1966. Abundances of uranium, thorium, and potassium. *In* S. P. Clark, Jr. (ed.) *Handbook of physical constants,* revised edn, 521–41. GEOL SOC AM MEM **97**.

Clemens, J. D. & Vielzeuf, D. 1987. Constraints on melting and magma production in the crust. EARTH PLANET SCI LETT **86,** 287–306.

DeYoreo, J. J., Lux, D. R. & Guidotti, C. V. 1989. The role of crustal anatexis and magma migration in the thermal evolution of regions of thickened continental crust. *In* Daly, J. S., Cliff, R. A. & Yardley, B. W. D. (eds) *Evolution of metamorphic belts.* GEOL SOC LONDON SPEC PUB **43,** 187–202.

Doe, B. R., Tilling, R. I., Hedge, C. E. & Klepper, M. R. 1968. Lead and strontium isotope studies of the Boulder batholith, southwestern Montana. ECON GEOL **63,** 884–906.

Doe, B. R., Berger, B. R. & Elliott, J. E. 1986. Lead-isotope evaluation of selected ores and mineral prospects in the Butte and Dillon 1° × 2° quadrangles, Montana. US GEOL SURV OPEN-FILE REP **86–111**.

Fleck, R. J. & Criss, R. E. 1985. Strontium and oxygen isotopic variations in Mesozoic and Tertiary plutons of central Idaho. CONTRIB MINERAL PETROL **90,** 291–308.

Flood, R. H. & Vernon, R. H. 1978. The Cooma granodiorite, Australia: an example of in situ crustal anatexis? GEOLOGY **6,** 81–4.

Hanson, R. B. & Barton, M. D. 1989. Thermal development of low-pressure metamorphic belt: results from two-dimensional numerical models. J GEOPHYS RES **94,** 10,363–77.

Haugerud, R. A. 1986. 1DT—an interactive, screen-oriented microcomputer program for simulation of 1-dimensional geothermal histories. US GEOL SURV OPEN-FILE REP **86–511**.

Huppert, H. E. & Sparks, R. S. J. 1988. The generation of granitic magmas by intrusion of basalt into continental crust. J PETROL **29,** 599–624.

James, H. L. & Hedge, C. E. 1980. Age of the basement rocks of southwest Montana. GEOL SOC AM BULL Pt. I, **91,** 11–5.

Joplin, G. A. 1942. Petrological studies in the Ordovician of New South Wales, I. The Cooma Complex. PROC LINN SOC NSW **67,** 156–96.

Joplin, G. A. 1943. Petrological studies in the Ordovician of New South Wales, II. The northern extension of the Cooma Complex. PROC LINN SOC NSW **68,** 159–83.

King, P. B. & Beikman, H. M. 1974. *Geologic Map of the United States,* scale 1:2,500,000: US. Geological Survey.

Lachenbruch, A. H. 1970. Crustal temperature and heat production: implications of the linear heat-flow relations. J GEOPHYS RES **75,** 3291–300.

Lachenbruch, A. H. & Sass, J. H. 1977. Heat flow in the United States and the thermal regime of the crust. *In* J. G. Heacock (ed.) *The earth's crust, its nature and physical properties.* AM GEOPHYS UNION, GEOPHYS MONOGR **20,** 626–75.

Lipman, P. W., Prostka, H. J. & Christiansen, R. L. 1972. Cenozoic volcanism and plate-tectonic evolution of the western United States. I. Early and Middle Cenozoic. PHILOS TRANS R SOC LONDON **A271,** 217–48.

Lowell, W. R. 1965. *Geologic map of the Bannack-Grayling area, Beaverhead County, Montana,* scale 1:31,680: U.S. Geological Survey, Miscellaneous Geologic Investigations Map I–433.

Lund, Karen & Snee, L. W. 1988. Tectonic development of the Salmon River suture zone, western Idaho: Conclusions and questions. GEOL SOC AM ABSTR PROG **20,** 429.

Munksgaard, N. C. 1988. Source of the Cooma Granodiorite, New South Wales—a possible role of fluid-rock interactions. AUST J EARTH SCI **35,** 363–77.

Patiño-Douce, A. E., Humphreys, E. D. & Johnston, A. D. 1990. Anatexis and metamorphism in tectonically thickened continental crust exemplified by the Sevier hinterland, western North America. EARTH PLANET SCI LETT **97,** 290–315.

Pearson, R. C. & Zen, E-an 1985. *Geologic map of the east Pioneer Mountains, Beaverhead County, Montana,* scale 1:50,000, with text: U.S. Geological Survey Miscellaneous Field Studies Map MF-1806-A.

Ruppel, E. T., O'Neill, J. M. & Lopez, D. A. 1983. *Preliminary geologic map of the Dillon 1° × 2° quadrangle, Montana,* scale 1:250,000: US GEOL SURV OPEN-FILE REP **83–168,** 2 sheets.

Sahinen, U. M. 1939. Geology and ore deposits of the Rochester and adjacent mining districts, Madison County, Montana. MONTANA BUR MINES GEOL MEM **19**.

Sawyer, E. W. 1987. The role of partial melting and fractional crystallization in determining discordant migmatite leucosome compositions: J PETROL **28,** 445–73.

Shuster, R. D., Mueller, P. A., D'Arcy, K. & Heatherington, A. 1989. Nd evidence of the age and nature of source regions of the northeastern Idaho batholith and Tobacco Root batholith, Montana. GEOL SOC AM ABSTR PROG **21,** 325.

Smedes, H. W., Klepper, M. R. & Tilling, R. I. 1988. *Preliminary map of plutonic units of the Boulder batholith, southwestern Montana,* scale 1:200,000: US GEOL SURV OPEN-FILE REP **88–238,** 1 sheet with text.

Snee, L. W. 1978. *Petrography, K/Ar ages, and field relations of the igneous rocks of part of the Pioneer batholith, southwestern Montana* (M.A. thesis, Ohio State University, Columbus, Ohio).

Snee, L. W. 1982. *Emplacement and cooling of the Pioneer batholith, southwestern Montana* (Ph.D. thesis, Ohio State University, Columbus, Ohio).

Streckeisen, A. L. 1973. Plutonic rocks: classification and nomenclature recommended by the IUGS Subcommittee on the systematics of igneous rocks. GEOTIMES **18** (10), 26–30.

Stuckless, J. S., Bunker, C. M., Bush, C. A. & VanTrump, George, Jr. 1981. Radioelement concentrations in Archean granites of central Wyoming; US GEOL SURV OPEN-FILE REP **81–948**.

Tilling, R. I. 1973. Boulder batholith, Montana—a product of two contemporaneous but chemically distinct magma series. GEOL SOC AM BULL **84,** 3879–900.

Tilling, R. I., Klepper, M. R. & Obradovich, J. D. 1968. K-Ar ages and time span of emplacement of the Boulder batholith, Montana. AM J SCI **266,** 671–89.

Tilling, R. I. & Gottfried, David 1969. Distribution of thorium, uranium, and potassium in igneous rocks of the Boulder batholith region, Montana, and its bearing on radiogenic heat production and heat flow. US GEOL SURV PROF PAP **614-E**.

Vielzeuf, D. & Holloway, J. R. 1988. Experimental determination of the fluid-absent melting relations in the pelitic system. CONTRIB MINERAL PETROL **98,** 257–76.

Weber, C., Barbey, P., Cuney, M. & Martin, H. 1985. Trace element behaviour during migmatization. Evidence for a complex melt-residuum-fluid interaction in the St. Malo migmatitic dome (France). CONTRIB MINERAL PETROL **90,** 52–62.

White, A. J. R. & Chappell, B. W. 1977. Ultrametamorphism and granitoid genesis. TECTONOPHYSICS **43,** 7–22.

Wooden, J. L., Mueller, P. A., Mogk, D. A. & Bowes, D. R. 1988. A review of the geochemistry and geochronology of Archean rocks of the Beartooth Mountains, Montana and Wyoming. *In* Lewis, S. E. & Berg, R. B. (eds) *Precambrian and Mesozoic plate margins.* MONTANA BUR MINES GEOL SPEC PUBL **96,** 23–42.

Wyllie, P. J. 1977. Crustal anatexis: an experimental review. TECTONOPHYSICS **43,** 41–71.

Zen, E-an 1988a. Thermal modelling of stepwise anatexis in a thrust-thickened sialic crust. TRANS R SOC EDINBURGH EARTH SCI **79,** 223–35.

Zen, E-an 1988b. Bedrock geology of the Vipond Park 15-minute, Stine Mountain 7–1/2 minute, and Maurice Mountain 7–1/2 minute quadrangles, Pioneer Mountains, Beaverhead County, Montana. US GEOL SURV BULL **1625**.

Zen, E-an (in prep.) Generation of magmas of the Pioneer batholith: a geologically constrained thermal model. GEOL SOC AM SPEC PAP.

E-AN ZEN, Department of Geology, University of Maryland, College Park, Maryland 20742, U.S.A.

Transactions of the Royal Society of Edinburgh: Earth Sciences, **83**, 115–126, 1992

Petrogenesis of felsic I-type granites: an example from northern Queensland

David C. Champion and Bruce W. Chappell

ABSTRACT: Felsic I-type granites and associated volcanic rocks of Carboniferous age are extensively developed over an area of 15,000 km^2 in northern Queensland. These granites have been subdivided into four supersuites: Almaden, Claret Creek, Ootann and O'Briens Creek.

Granites of the Almaden Supersuite are intermediate to felsic (56–72% SiO_2) and are characterised by high K_2O, K/K(K + Na), Rb, Rb/Sr, Th, U and relatively low Ba and Sr. The Claret Creek Supersuite granites are a little more felsic (65–77% SiO_2), and are chemically distinctive, having higher Al_2O_3, CaO, Na_2O and Sr, and lower K_2O, Rb, Th and U than granites of the Almaden Supersuite.

Granites of the Ootann and O'Briens Creek supersuites all contain more than 70% SiO_2 and these comprise more than 90% of the total area of granites. These two supersuites are characterised by low Sr, Sr/Y and large negative Eu/Eu*, with the more evolved rocks becoming strongly depleted in TiO_2, FeO* MgO, CaO, Ba, Sr, Sc, V, Cr, Ni, Eu, Ce_N/Y_N and K/Rb, and enriched in Rb, Pb, Th, U and Rb/Sr. Granites belonging to the O'Briens Creek Supersuite contain significantly higher abundances of HFSE, HREE and F (0·2–0·5 wt%) than those of the Ootann Supersuite, and as such have developed some characteristics of A-type granites.

Geochemical and isotopic properties suggest that all granites are of crustal derivation. The granites of all supersuites have very similar initial $^{87}Sr/^{86}Sr$ and ε_{Nd} of 0·710 and −7·0–−8·0, respectively, except where they outcrop within Proterozoic country rocks, when they have more evolved ε_{Nd} (−8·0–−11·0). Depleted-mantle model ages cluster around 1·5 Ga. The isotope systematics and geochemistry indicate that these granites were not derived from the equivalents of any exposed country rocks.

Models for the petrogenesis of these granites all appear to require the involvement of a long-lived and isotopically homogeneous crustal protolith, that most probably underplated the crust in the Proterozoic. Granites of the two more felsic supersuites were either derived by varying degrees of partial melting from this protolith of andesitic to dacitic composition, and/or were produced by a two-stage process by remelting of intermediate rocks similar in composition to the mafic end-members of the Almaden Supersuite. The resulting primary partial melts for the Ootann and O'Briens Creek supersuites underwent extensive, high-level, feldspar-dominated, crystal fractionation.

KEY WORDS: A-type characteristics, crustal melting, crystal fractionation, fractionated granite, isotopically homogeneous, protolith, underplating.

The origin of very felsic (>70% SiO_2) granites and rhyolites, particularly when they occur in large volumes, has long been controversial. Debate is even greater for those felsic granites/rhyolites in which there is extreme enrichment and depletion of elements. Petrogenetic models for these granites are varied, both with reference to their possible source rocks, and also to the processes that led to their felsic character. The majority of proposed models are typically coupled with extensive crystal fractionation (e.g. Groves & McCarthy 1978; Christiansen *et al.* 1984). Proposed models for "special" source rocks for felsic granites include residual or granulitic crust (Christiansen *et al.* 1986; Manning & Pichavant 1988); tonalitic crust (Creaser *et al.* 1991); mantle-derived material underplated onto the lower crust (Wyborn *et al.* 1988) or hybridised with crustal material (Hildreth *et al.* 1991); mixtures of crustal and mantle-derived melts (Leat *et al.* 1986); and even lithospheric mantle, invariably metasomatised (Plant *et al.* 1985).

An additional problem in deciphering the prehistory of strongly fractionated felsic granites is that a great many of these granites display textural evidence that suggests interaction with a fluid phase (e.g. Martin & Bowden 1981; Taylor *et al.* 1981; Kleeman & Twist 1989). Many workers have invoked this mineralogical/geochemical overprinting of the original assemblage and its inherent geochemistry, by various types of fluids, e.g. late stage residual fluids, hydrothermal fluids, halogen-enriched fluids, to produce or help produce the characteristic enriched/depleted geochemical signatures of strongly fractionated felsic granites (e.g. Martin & Bowden 1981; Charoy 1986; Kleeman & Twist 1989).

Extensive felsic granites (and volcanics) occur in northern Queensland (Fig. 1). The Carboniferous I-type granites (CIG) of the Mount Surprise–Herberton Group comprise perhaps one of the best examples, worldwide, of felsic and highly evolved I-type granites developed on a large scale (greater than 8,000 km^2, averaging 74% SiO_2) making them ideally suited for the study of the petrogenesis of felsic granites. Further, the CIG have been the focus of fairly extensive study, with over 700 whole-rock granite analyses available for discussion (from Bailey 1969; Sheraton 1974; Richards 1981; Johnston 1984; Witt 1985; Pollard 1988; Blevin 1990; Clarke 1990; Champion 1991).

1. Regional setting

The oldest exposed rocks in northern Queensland are the Proterozoic rocks of the Georgetown-Coen Province comprising the Georgetown, Dargalong, Yambo and Coen inliers, which outcrop in a N-S belt over a distance of 900 km and cover an are of over 70,000 km^2. Although only the Georgetown Inlier (e.g. Withnall *et al.* 1980, 1988) has been studied in any detail, it is evident that the inliers share many similarities and are comprised of variably metamorphosed and multiply deformed supracrustal sequences (e.g. de Keyser & Lucas 1968; Willmott *et al.* 1973). The depositional ages of these rocks are not known but are constrained by the presence of widespread I- and S-type granites to be pre-1550 Ma.

To the E of the Georgetown-Coen Province, the Hodgkinson Province comprises Ordovician to Devonian supracrustal rocks outcropping over 75,000 km^2 (Fig. 1). The province is dominated by deep-water, quartzofeldspathic, mainly turbiditic rocks, whose mineralogy, geochemistry and isotope chemistry are consistent with derivation from rocks similar to the Proterozoic supracrustal rocks exposed to the W (de Keyser & Lucas 1968; Bultitude *et al.* 1990). No older basement rocks have been identified for either the Georgetown-Coen or Hodgkinson provinces. The two provinces are separated by the Palmerville Fault which extends over 400 km (Fig. 1) and has been suggested to represent a long-lived, fundamental zone of weakness in the upper crust (Shaw *et al.* 1987).

2. Field relations and mineralogy

The Carboniferous I-type granites (CIG) outcrop over 8,000 km^2 in the Mount Surprise-Herberton area where they intrude both the Georgetown-Coen and Hodgkinson provinces (Fig. 1). The CIG constitute part of the extensive (>30,000 km^2) Upper Palaeozoic Coastal Ranges Igneous Province of northern Queensland and form part of a terrane that experienced widespread recurrent felsic igneous activity from the Middle Proterozoic to early Triassic.

Previous workers on the granites (Bailey 1977; Richards 1981; Johnston 1984; Witt 1985; Pollard 1988), and Champion (1991) have grouped individual granite plutons, on the basis of petrography and geochemistry, into granite suites (27 major suites are now recognised). These suites were grouped into three major supersuites and one minor supersuite, namely the Almaden (600 km^2), Ootann (5,000 km^2), O'Briens Creek (2,500 km^2) and Claret Creek (250 km^2) supersuites (Fig. 1).

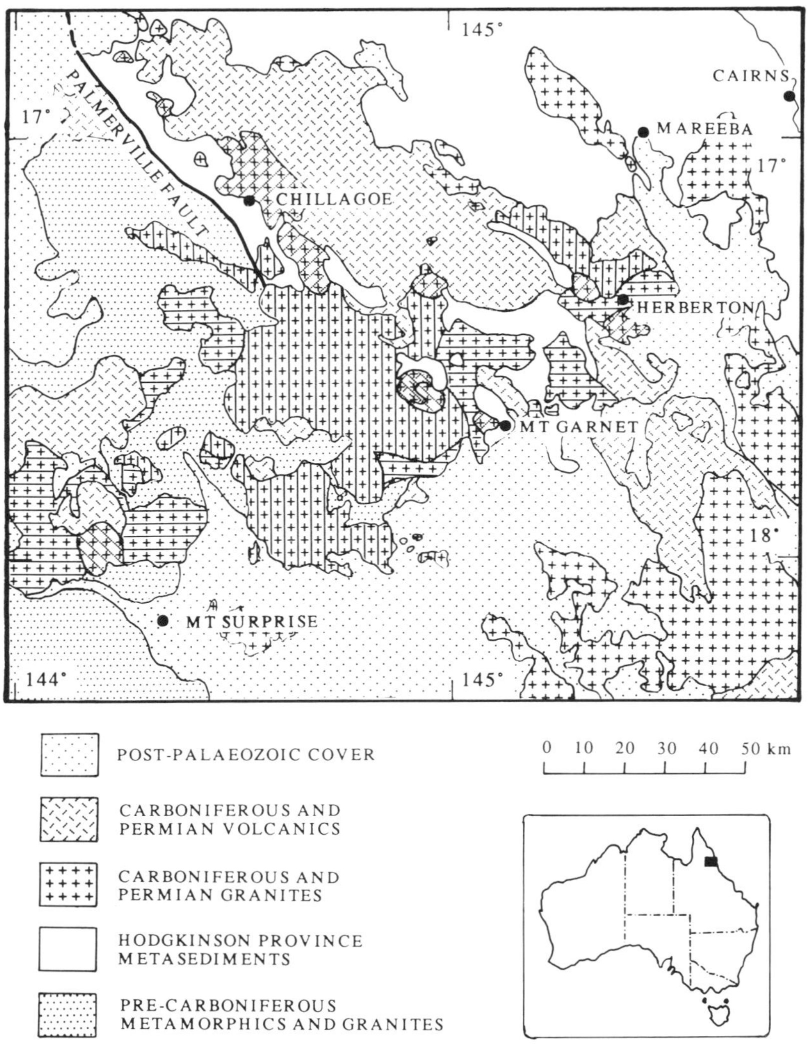

Figure 1 Generalised distribution of the granites of the O'Briens Creek (horizontal lines), Ootann (vertical lines), Almaden (NE lines) and Claret Creek (NW lines) supersuites.

The CIG vary from biotite-hornblende granodiorites and rarer pyroxene-biotite-hornblende quartz diorites to biotite leucogranites. Accessory minerals include apatite, zircon, ilmenite, magnetite, allanite, fluorite and less common xenotime, monazite and thorite. Individual plutons range from over 200 km^2 to 10 km^2, down to less than 1 km^2 for mirogranite units, the latter mostly in the O'Briens Creek Supersuite. The association with contemporaneous volcanic rocks and the presence of miarolitic cavities show that the granites are all high level. The granites are post-tectonic and appear to have been emplaced in a tensional environment (Oversby *et al.* 1980). Detailed petrography of the granites has been provided by numerous workers (e.g. Sheraton & Labonne 1978; Pollard *et al.* 1983).

The O'Briens Creek granites and more felsic members of the Ootann Supersuite contain widespread evidence of textural re-equilibration, manifested as alteration of plagioclase, alkali feldspar and biotite and growth of minerals such as chlorite, secondary biotite, muscovite, albite and fluorite (e.g. Pollard *et al.* 1983; Pollard 1988; Taylor & Pollard 1988). These changes were considered by Pollard *et al.* (1983) and Taylor and Pollard (1988) to represent *in situ* grain-boundary-controlled interaction with residual late-magmatic or post-magmatic fluids. Blevin (1990) and Champion (1991) suggested that this "selected pervasive" background alteration essentially represented a closed system on the sample-size scale, i.e. was largely isochemical consistent with the geochemical evidence (see below).

Most of the CIG have not been accurately dated. Available data indicate that the CIG record a general easterly decrease in crystallisation ages that may span some 50 Ma, from around 330 Ma in the W (Black & Holmes unpublished data, in Black & McCulloch 1990) to 280 Ma in the Herberton area (Black *et al.* 1978). Superimposed on this easterly age decrease is the evidence that the O'Briens Creek granites are either older than, or of similar age to, the Almaden, Claret Creek and Ootann granites. This relationship has been convincingly demonstrated for the granites E of the Palmerville Fault, where dating has yielded dominant ages of around 300 Ma for the Almaden, Ootann and Claret Creek granites and 315 Ma for the O'Briens Creek granites (Black 1978; Black *et al.* 1978; Richards 1981; Johnston & Black 1986).

3. Geochemical and isotopic systematics

The CIG share many common geochemical features (Table 1). In all supersuites, TiO_2, FeO*, MgO, CaO, Sr, Eu, V, Cr, Ni, Co, Cu, Zn decrease with increasing fractionation, down to very low levels in the most fractionated granites of the Ootann and O'Briens Creek supersuites (Fig. 2). The granites vary from metaluminous to slightly peraluminous, with the Aluminium Saturation Index (ASI) increasing from 0·85 in the Almaden Supersuite to a near-constant value slightly greater than 1·0 for the O'Briens Creek Supersuite. The granites of the Almaden, Ootann and O'Briens Creek supersuites all posses high K_20, K/(K + Na), Rb, Pb, Th and U (Table 1) and fall into the high-K field of Peccerillo and Taylor (1976) (Fig. 3).

3.1. Ootann and O'Briens Creek supersuites

3.1.1. Major element geochemistry. Granites of the Ootann (70–77% SiO_2) and O'Briens Creek (73–77% SiO_2) supersuites are very felsic and comprise over 90% of the outcropping CIG, with a surface area of greater than 7,000 km^2 (Fig. 1). Major element compositions of granites of the two supersuites are very similar (Table 1). The granites have high Differentiation Indices (averages of 89 and 95, respectively) and tightly cluster around the 100–200 MPa H_2O-saturated granite minima of Tuttle and Bowen (1958). Major element trends with increasing fractionation are typified by decreasing TiO_2, FeO*, MgO, CaO, Al_2O_3 and P_2O_5, and increasing SiO_2 and K_2O.

In the most fractionated O'Briens Creek granites, SiO_2, Na_2O and K_2O, and to a lesser extent Al_2O_3, exhibit opposite behaviour with SiO_2 and K_2O decreasing and Al_2O_3 and Na_2O increasing with increasing fractionation. This is evident on the normative Q-Ab-Or triangular plots, where the most fractionated granites trend towards the Ab apex. Such geochemical trends are found elsewhere, e.g. the topaz rhyolites of the western U.S.A. (Chritiansen *et al.* 1984, 1986), and the topaz granites of Finland (Haapala 1988), and have been produced experimentally in H_2O-saturated haplogranite systems by the addition of F and/or B (e.g. Pichavant 1981; Manning 1982). A similar process may very well have been operative for the O'Briens Creek granites, where F increases from 0·15 wt% to greater than 0·6 wt% with increasing fractionation. In the Ootann Supersuite granites, where F is less abundant (<0·25 wt%) and exhibits only minor increases with increasing fractionation, there is no tendency for the granites to become more sodic.

3.1.2. Trace element and rare-earth element geochemistry. The granites of the Ootann and O'Briens Creek supersuites are marked by significant variations in their trace element contents, particularly in the more fractionated granites (Fig. 2, Table 1). With increasing fractionation Ba, Sr, Eu, Sc, V, Cr, Ni, Mn and Zn strongly decrease (up to ten-fold variation), and Cs, Rb, Th and U increase. The degree of fractionation is well represented by the K/Rb ratio which decreases from 220 to 90 in the Ootann granites and from 130 to less than 50 in the O'Briens Creek granites (Fig. 4). Similarly, Rb/Sr ratios increase from 1 to 11 in the Ootann granites, to very large ratios of 100 and more in the O'Briens Creek granites. Rb, Th and U in the O'Briens Creek granites are present at high concentrations, reaching levels of 800 ppm, 80 ppm and 30 ppm, respectively (Figs 2, 7). Relative to the average low-Ca granite, the Ootann and O'Briens Creek granites have higher Rb and Th, and lower Ba, Zr, Nb, La, Ce and V.

The Ootann Supersuite granites are moderately LREE-enriched (50 to 150 times chondrites) with Ce_N/Y_N ratios of 7 to 2. Eu/Eu* decreases from 0·8 to 0·12 with increasing fractionation. The O'Briens Creek granites range from moderately LREE-enriched to HREE-enriched, with Ce_N/Y_N ratios decreasing from 5 to less than 1. In the latter granites, Y and the HREE increase with fractionation to moderate to high levels (up to 100 times chondrites). Eu levels are very low in the O'Briens Creek granites with Eu/Eu* decreasing from 0·2 to less than 0·01. The coupling of incompatible HREE behaviour and more compatible LREE behaviour in the O'Briens Creek granites has resulted in a pronounced flattening of REE patterns, similar to that documented elsewhere, e.g. the topaz rhyolites of the U.S.A. (Christiansen *et al.* 1986, Fig. 40) and the Bishop Tuff (Hildreth 1979).

3.1.3. Fractionation trends of high field strength elements. The Ootann and O'Briens Creek granites are very similar, particularly for the less fractionated granites of both supersuites, where discrimination between the supersuites is difficult. The major chemical differences are in

Table 1 Selected geochemical analyses of Carboniferous I-type granites of the Claret Creek, Almaden, Ootann and O'Briens Creek supersuites; 'AL' samples from Richards (1981), the rest from Champion (1991).

	CLARET CK	ALMADEN		OOTANN			O'BRIENS CREEK		
Sample	D179	AL001	AL038	AL035	D130	AL112	D167a	D162	D186
SiO_2	72·43	64·76	66·04	73·51	75·54	76·80	75·19	75·87	75·58
TiO_2	0·28	0·57	0·53	0·28	0·12	0·06	0·16	0·06	0·03
Al_2O_3	13·94	15·78	15·41	14·40	12·46	13·20	12·09	12·32	12·50
Fe_2O_3	1·07	1·00	1·51	0·42	0·50	0·10	0·61	0·48	0·46
FeO	1·27	3·54	2·68	1·46	0·51	0·53	0·86	0·72	0·47
MnO	0·04	0·08	0·08	0·07	0·05	0·03	0·02	0·06	0·02
MgO	0·46	2·30	2·06	0·29	0·23	0·08	0·16	0·07	0·05
CaO	1·63	4·77	4·62	1·63	0·95	0·41	0·80	0·46	0·55
Na_2O	4·28	2·00	3·04	3·85	3·43	3·38	3·07	3·25	3·95
K_2O	3·33	2·76	2·94	4·19	4·63	5·25	4·99	4·91	4·50
P_2O_5	0·06	0·10	0·08	0·05	0·02	0·02	0·03	0·01	$<$0·01
Trace elements (ppm)									
F	1392				1117		2250	3841	6796
Cs	2·7				5·6		5·5	12·9	35·9
Ba	945	350	350	560	140	155	180	24	4
Rb	133	127	151	202	349	328	296	484	745
Sr	186	202	174	121	45	23	32	7	2·5
Pb	16	17	16	17	26	42	31	92	51
Th	14·4	17·2	19·0	29·0	31·5	43·0	40·4	41·8	44·2
U	3·1	2·4	2·6	4·4	11·6	3·2	10·2	14·6	5·9
Zr	194	107	97	171	90	72	174	98	142
Hf	5·9				4·0		5·4	4·2	8·6
Nb	9·0	6·5	7·0	10·5	18·0	9·5	9·0	17·0	39·0
Ta	2·9				7·4		6·3	7·7	10·3
Y	23	19	23	39	52	28	57	102	165
Sc	3·6				3·9		3·6	4·8	3·7
V	12	89	84	8	6	$<$1	4	$<$1	$<$1
Cr	$<$1	22	19	2	$<$1	$<$1	$<$1	$<$1	$<$1
Ni	3	4	4	$<$1	2	$<$1	2	$<$1	$<$1
Cu	2	7	4	1	2	5	1	8	$<$1
Zn	39	56	45	25	19	7	23	88	15
Ga	12·4	16·0	15·6	16·2	11·0	14·2	13·2	14·6	18·6
Rare-earth elements (ppm)									
La	43·6	22·0	25·0	45·0	18·8	25·0	55·3	35·8	17·7
Ce	90·9	48·0	52·0	102·0	46·6	62·0	127·9	93·0	49·9
Pr		6·4	5·9	9·9		9·2			
Nd	35·2	23·8	20·7	36·6	22·8	32·9	56·0	48·0	25·7
Sm	5·66	4·70	4·18	6·94	5·43	6·26	10·8	12·6	8·79
Eu	1·0	0·96	0·77	0·94	0·33	0·24	0·54	0·13	0·04
Gd	4·20	3·69	3·39	5·98	4·64	5·75	9·61	13·06	9·47
Tb	0·86	0·62	0·56	1·0	1·44	1·0	2·04	2·66	2·96
Dy		3·49	3·70	6·27		6·40			
Ho	0·88	0·79	0·83	1·42	1·90	1·41	2·42	3·84	5·0
Er		2·29	2·43	4·24		4·43			
Tm	0·45				0·94		0·73	1·10	1·30
Yb	2·53	2·33	2·82	4·22	6·45	4·99	5·38	10·4	22·0
Lu	0·36				0·99		0·72	1·44	3·16

the abundances and behaviour of the HREE, Y, Nb, Ta, Zr, Th, Ga, F and Sn, which all occur at higher levels in the O'Briens Creek granites relative to the Ootann granites (Figs 2, 5). The strong incompatible behaviour of Nb, Ga, Y and the HREE in the first group of granites is very characteristic, with the relative magnitude in variation of these elements being strongly suite-specific (Fig. 6). The high Ga, Nb and Y in many of the granites of the O'Briens Creek Supersuite make them superficially similar to A-type granites (Collins *et al.* 1982). The widespread O'Briens Creek Supersuite granites are, therefore, a prime example of felsic I-type granites that have, by extensive fractionation, developed many of the characteristics of A-type granites. It is noteworthy, however, that in the O'Briens Creek granites, unlike A-type granites, Ga, Nb and Y (and Sn and F) increase with fractionation from baseline levels more typical of I-type granites, i.e. similar to those in the Ootann Supersuite granites.

3.1.4. Evidence for hydrothermal alteration. The suite-specific nature of Nb, Ga and Y behaviour (Fig. 6), the good correlation of these elements with more traditional indicators of fractionation (e.g. K/Rb, Rb/Sr), and the F-alkali relationships (as predicted by experimental work) in the suites within the O'Briens Creek and Ootann supersuites, strongly support the contention that the geochemical signature in the CIG is largely magmatic. Similar conclusions were reached for comparably fractionated felsic granites by Christiansen *et al.* (1984, 1986), Mackenzie *et al.* (1988), and Kontak (1990). There are, however, exceptions e.g. the granites of the Go Sam Suite (Johnston & Black 1986), of the O'Briens Creek Supersuite, contain evidence for hydrothermal alteration resulting in element transfer. These alteration-induced chemical changes in the CIG are, however, erratic and readily recognisable. They do not appear to reinforce magmatic trends, e.g. Pollard (1988, pp. 50–51) failed to find any consistent relationship between geochemistry and degree of alteration. It is possible that the low CaO and anorthite component in the felsic CIG are due to alteration and loss of Ca. However, CaO (and normative An) strongly decrease with fractionation, until an approximately constant value of 0·5% is reached (Fig. 4). It is notable in this regard that the fractionation trends of the granites of the Ootann and O'Briens Creek supersuites on the An-Ab-Or plot are very similar to those predicted by Nekvasil (1988, fig. 10) on theoretical grounds.

3.2. Almaden Supersuite

The granites of the Almaden Supersuite are the most mafic of the CIG (Table 1), with SiO_2 from 57–72% (average of 67·5% SiO_2). Relative to the average high-Ca granite (Turekian & Wedepohl 1961) the Almaden granites contains lower TiO_2, Al_2O_3, Na_2O and P_2O_5, and greater MgO, CaO and K_2O. The Almaden granites are characterised by moderate to high K_2O and $K/(K+Na)$ (averages of 3·3 and 0·42, respectively) and fall into the high-K field of Peccerillo and Taylor (1976) (Fig. 3). Both Na_2O and K_2O increase with increasing SiO_2. Like the granites of the Ootann and O'Briens Creek supersuites, the Almaden granites contain moderate to high Rb, Pb, Th and U (averages of 169 ppm, 16 ppm, 23·5 ppm and 4·4 ppm, respectively; Figs 2, 7; Tables 1, 2). Despite the moderate to high K_2O, the granites of the Almaden Supersuite have low K/Rb ratios (average 159) compared to similar granites elsewhere (Figs 4, 7). The most mafic Almaden granite, the Retire Monzodiorite, has, at SiO_2 values of 58–62%, low K/Rb ratios of 130–160 (Fig. 7). Contrary to the other large ion lithophile elements Ba (<500 ppm) and Sr (<250 ppm) are present at moderate to

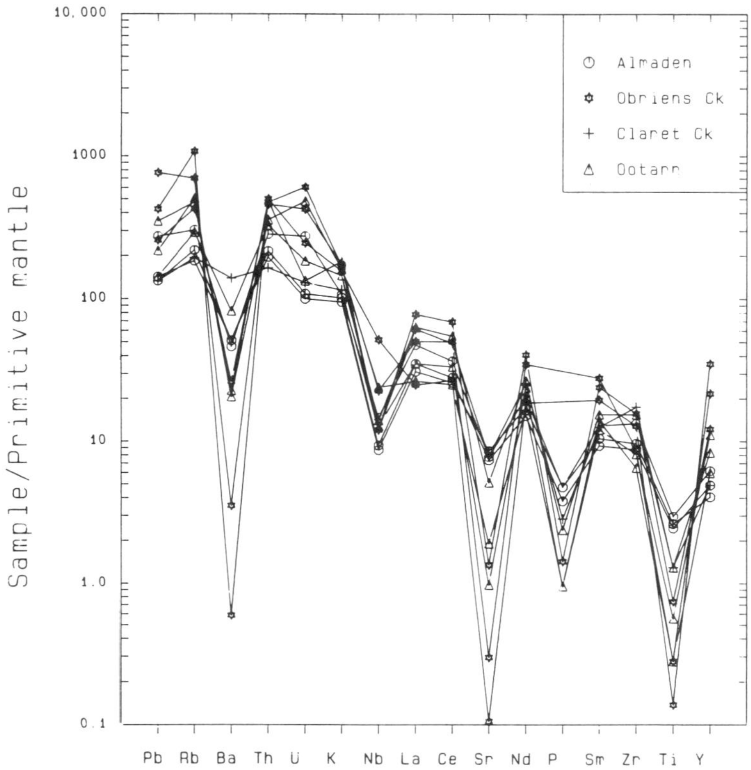

Figure 2 Multi-element diagram for selected granites of the O'Briens Creek, Ootann, Almaden and Claret Creek supersuites; analyses listed in Table 1.

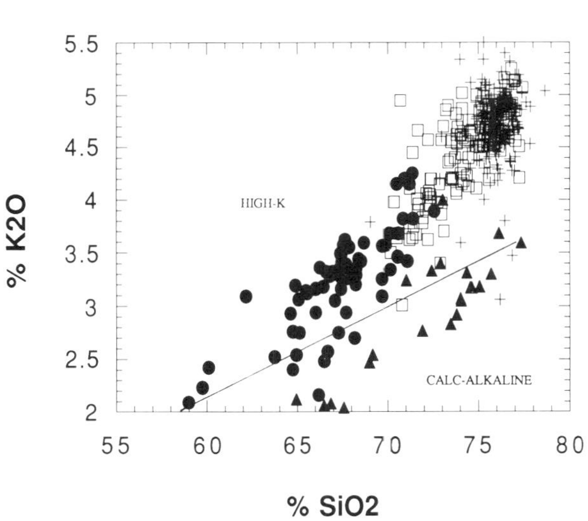

Figure 3 Comparison of wt% K_2O levels in the granites of the O'Briens Creek (crosses), Ootann (open squares), Almaden (full circles) and Claret Creek (full triangles) supersuites, with the calc-alkaline and high-K fields of Peccerillo and Taylor (1976).

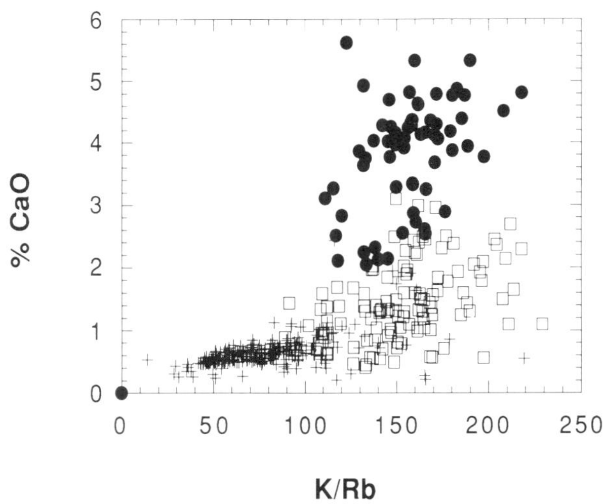

Figure 4 Variation of CaO with increasing fractionation (decreasing K/Rb ratio) in the granites of the O'Briens Creek (crosses), Ootann (open squares) and Almaden (full circles) supersuites; note the limited change in CaO in the O'Briens Creek Supersuite, despite the change in degree of fractionation.

low levels (Fig. 2; Tables 1, 2), with correspondingly high Ca/Sr, Rb/Sr and Rb/Ba ratios (averages of 181, 1·0 and 0·5, respectively). Moderate negative Eu anomalies characterise all the Almaden granites, mainly in the range 0·5–0·7. The granites are LREE enriched (40–120 times chondrites), with Ce_N/Y_N of 3 to 8. The HREE (7–12 times chondrites) are not fractionated, with relatively flat chondrite-normalised patterns.

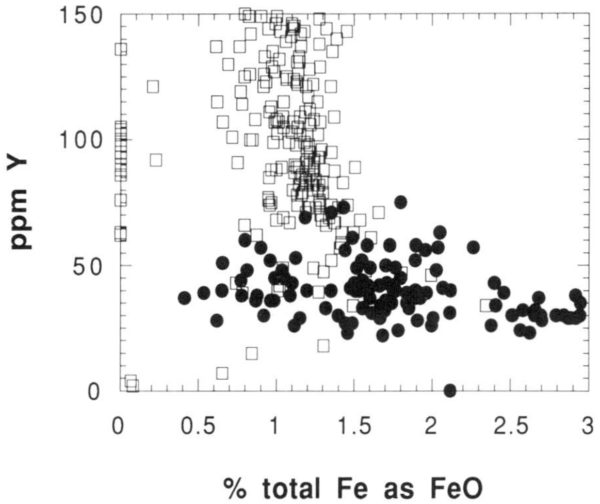

Figure 5 Y variation in the Ootann (full circles) and O'Briens Creek (open squares) supersuites, showing the large increase in Y with increasing fractionation in the latter granites.

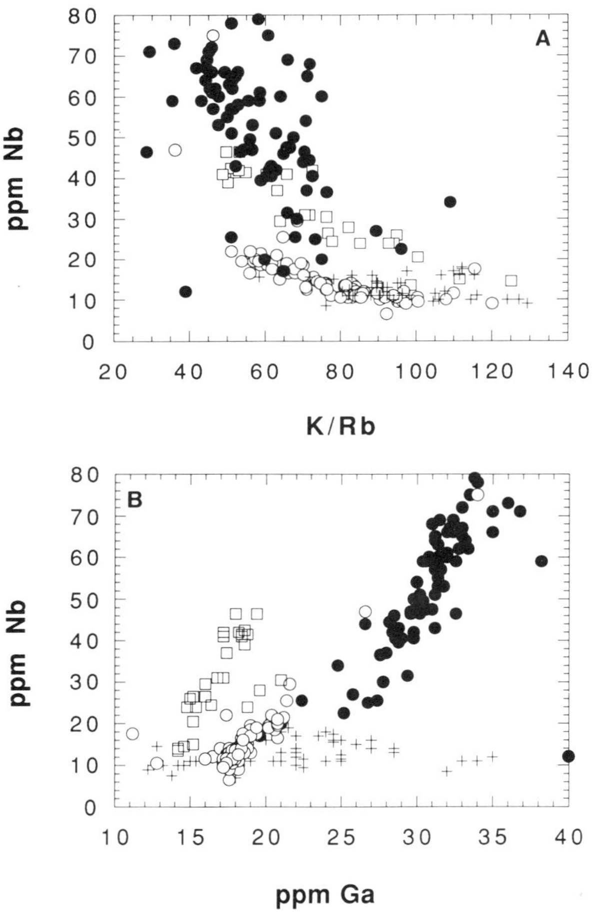

Figure 6 (A) Variation of Nb with increasing fractionation (K/Rb) and (B) Nb and Ga, in four suites of the O'Briens Creek Supersuite, showing the suite-specific behaviour of Nb and Ga consistent with a magmatic origin.

Although more mafic, the supersuite is otherwise geochemically similar to the Ootann Supersuite granites. Minor differences include slightly lower TiO_2, P_2O_5 and LREE in the Almaden granites. The major geochemical distinctions between the two supersuites are in the elements CaO, K_2O, Rb, Ba and Sr, particularly the relationships between these elements. The geochemical characteristics of the Almaden granites described above (i.e. high Rb and Ca/Sr, low Ba, Sr and K/Rb) are not evident in the Ootann granites and clearly discriminate the two supersuites (Figs 4, 7).

3.3. Claret Creek Supersuite

The Claret Creek Supersuite granites are the most geochemically distinctive of the CIG. They exhibit a large compositional range (66–76% SiO_2), they are sodic in nature (Bailey 1977) and fall in the medium-K (calc-alkaline) class of Peccerillo and Taylor (1976) (Fig. 3). Major differences between the Claret Creek granites and the granites of the Almaden, Ootann and O'Briens Creek supersuites include higher Al_2O_3, Na_2O, CaO and Sr, and distinctly lower K_2O, Rb, Th, U and Y in the former (Figs 2, 3; Tables 1, 2). REE data for the Claret Creek Supersuite show a similar abundance pattern to the Almaden granites with La_N/Sm_N of 5·0, Eu/Eu* of 0·60 and unfractionated HREE.

3.4. Origin of geochemical variations

The major and trace element variations in the CIG appear to be largely due to crystal fractionation processes. Major and trace element modelling of the CIG suggest that the major element chemical variation in these suites can be satisfactorily modelled by 20% to more than 70% crystal fractionation with prevalent involvement of plagioclase. Major differences in the fractionating mineral assemblages involved the behaviour of the ferromagnesian phases which, based on petrographical observations, changed from pyroxene to hornblende to biotite, and the increase in the feldspar/ferromagnesian mineral ratio, with increasing SiO_2

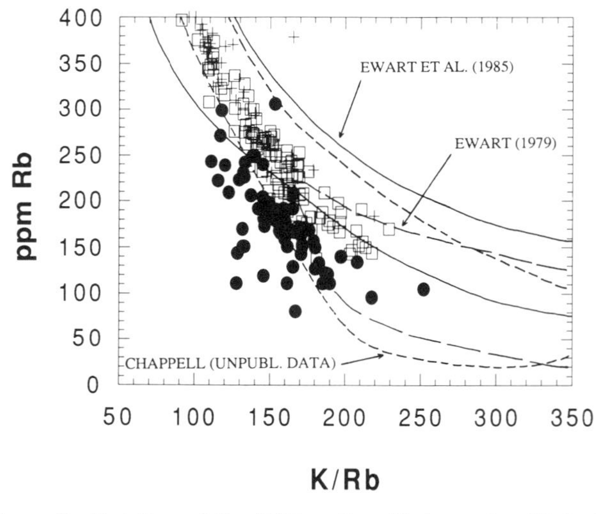

Figure 7 Variation of the K/Rb ratio with increasing Rb in the granites of the O'Briens Creek (crosses), Ootann (open squares) and Almaden (full circles) supersuites, compared with similar felsic rocks from elsewhere; note the low K/Rb ratios of the granites of the Almaden Supersuite relative to the O'Briens Creek and Ootannn supersuites and granites from elsewhere; fields shown are those for the I-type granites of the Lachlan Fold Belt (Chappell unpublished data), anorogenic felsic volcanics from eastern Australia (Ewart *et al.* 1985) and for high-K orogenic volcanics (Ewart 1979).

content, from the Almaden to the Ootann to the O'Briens Creek supersuites. Additionally, alkali feldspar and quartz were present as fractionating phases in the more felsic granites in the O'Briens Creek Supersuite. The differing mineral assemblages can readily explain aspects of the variable geochemical behaviour of the supersuites, such as the varying trends of F, Ba and Ca/Sr ratios with increasing fractionation.

There can be little doubt that the granites of the O'Briens Creek and Ootann supersuites are crustal in origin, as suggested by their strongly fractionated felsic nature and their Nd and Sr isotope systematics (see below). Similarly, the geochemical nature of the Almaden, in particular the moderate to high K/(K + Na), K/Rb, Ca/Sr, Rb/Sr and Rb/Ba ratios and moderate to low Eu/Eu*, Ba/La and La/Th, coupled with evolved but constant Sr and Nd isotope systematics strongly suggest a predominantly crustal origin (e.g. Hildreth & Moorbath 1988).

3.5. Isotopic variation

Sr and Nd isotope studies have demonstrated two contrasting groups within the CIG. E of the Palmerville Fault, granites representative of all supersuites have an initial $^{87}Sr/^{86}Sr$ ratio of approximately 0·710 and ε_{Nd} values of −7 to −8 (e.g. Black 1978, 1980; Johnston & Black 1986; Black & McCulloch 1990; Champion 1991) (Fig. 8). The limited data for those CIG W of the fault suggest that the situation there is more complicated. The western granites have more evolved isotopic signatures, i.e. initial $^{87}Sr/^{86}Sr$ ratios of 0·710 and higher, and ε_{Nd} from −8·0 to −11·0 (Black 1978; Black & McCulloch 1990; Champion 1991) (Fig. 8) requiring at least two additional end-members. There are a number of possible mechanisms capable of explaining the isotopic variation of the western granites, the most obvious

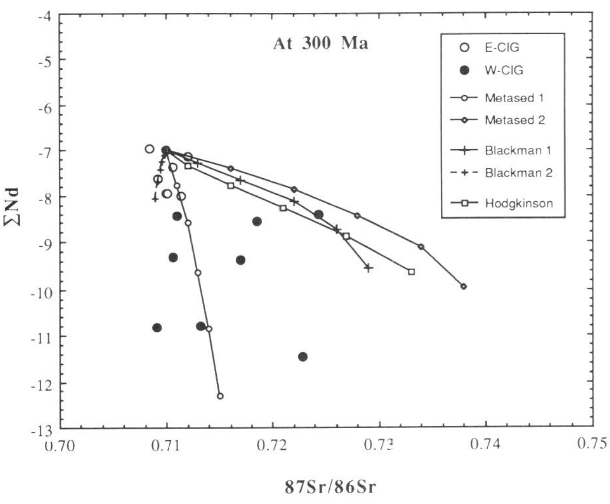

Figure 8 Nd-Sr isotope diagram, showing the isotopic variation between the granites W (W-CIG) and E (E-CIG) of the Palmerville Fault; also shown are combined assimilation and fractional crystallisation (AFC) curves for Proterozoic metasediments: Metased 1—$\varepsilon_{Nd} = -22{\cdot}0$, Nd = 45 ppm, $^{87}Sr/^{86}Sr = 0{\cdot}72$, Sr = 250 ppm; Metased 2—$\varepsilon_{Nd} = -16{\cdot}0$, Nd = 40 ppm, $^{87}Sr/^{86}Sr = 0{\cdot}75$, Sr = 250 ppm; Proterozoic granites: Blackman 1—$\varepsilon_{Nd} = -20{\cdot}0$, Nd = 20 ppm, $^{87}Sr/^{86}Sr = 0{\cdot}74$, Sr = 200 ppm; Blackman 2—$\varepsilon_{Nd} = -15.2$, Nd = 12 ppm, $^{87}Sr/^{86}Sr = 0{\cdot}709$, Sr = 685 ppm; and Hodgkinson Province sediments: Hodgkinson—$\varepsilon_{Nd} = -15{\cdot}0$, Nd = 40 ppm, $^{87}Sr/^{86}Sr = 0{\cdot}75$, Sr = 100 ppm. The model Carboniferous I-type granite (for the Ootann and O'Briens Creek supersuites) has: $\varepsilon_{Nd} = -7{\cdot}0$, Nd = 25 ppm, $^{87}Sr/^{86}Sr = 0{\cdot}71$, Sr = 50 ppm; AFC curves were calculated from the equation of DePaolo (1981a), using an r value (mass of assimilant/mass of fractionate) of 0·1 and bulk partition coefficients of 2 and 3 for Nd and Sr, respectively, and are shown for 0–50% fractionation.

being the incorporation (by means of assimilation or contamination or at the source) of the more evolved Proterozoic crust. Combined assimilation and fractional crystallisation (AFC) models (DePaolo 1981a) (Fig. 8) for a variety of local intermediate to felsic assimilants show that variable assimilation of local country rocks, combined with fractionation, can account for most of the isotopic variation in the CIG, in particular the wide variation in ε_{Nd}.

Uncorrected Nd-depleted mantle model ages range from 1·5–2·5 Ga. When corrected for fractionation of Sm/Nd (using an average value of 0·12 for $^{147}Sm/^{144}Nd$ calculated from over 100 CIG samples), all except two of the model ages fall in the range 1·5–1·7 Ga. This range overlaps the ages of the first two orogenic events in the Georgetown Inlier (Black *et al.* 1979) and the Middle Proterozoic granites (Black & McCulloch 1990), strongly suggesting that the source rocks for the CIG were younger than the Proterozoic basement of the Georgetown and Coen inliers into which they were emplaced. The CIG, therefore, appear to be, as first suggested by Chappell and Stephens (1988), a good example of I-type granites derived from an underplated source. Such a process would also appear to be the most feasible in producing the voluminous source rocks (see below) required to produce the CIG.

The combined geochemical and isotopic data indicate that no local exposed rocks could have been the sole source rocks for the CIG. On the basis of a similar Nd isotopic signature, Black and McCulloch (1990) suggested that the Dido Tonalite was a possible source for the CIG. The Dido Tonalite, however, with intermediate to felsic compositions (65–72% SiO_2), high Sr (700–850 ppm), and low Rb (<100), Th (<10 ppm), U (<2 ppm), and Rb/Sr ratios (<0·2) (Sheraton & Labonne 1978; Champion 1991) could not have been the precursor for the CIG. Mafic lower crustal xenoliths have similar ε_{Nd} values to the CIG (Rudnick 1990). It is difficult to evaluate the possible role of these xenoliths as suitable protoliths for the CIG as they have lost an unknown amount of Rb, Th, U and K (Rudnick & Taylor 1987).

4. Petrogenesis

4.1. Isotopic and geochemical constraints

The combined petrographic, geochemical and isotopic information available for the CIG provides considerable constraints on their petrogenesis. A petrogenetic model for the CIG must take in to account the following points:

(1) The very similar initial $^{143}Nd/^{144}Nd$ and $^{87}Sr/^{86}Sr$ isotope compositions require that members of all supersuites, even the sodic Claret Creek Supersuite, were derived, ultimately, from isotopically uniform sources. There is little or no evidence for isotopic equilibration in the lower crust (e.g. Rudnick 1990). None of the exposed country rocks are feasible source rocks for the CIG. Furthermore, Sr and Nd isotope data indicate a long-lived crustal protolith, with model ages younger than the Proterozoic country rocks. This indicates that the protolith was underplated onto or was emplaced into the lower crust for this area.

(2) The distinctive geochemistry of the granite supersuites indicates that, although the protolith was isotopically homogeneous, it was not chemically homogeneous. The isotopic constraints, in particular identical initial $^{87}Sr/^{86}Sr$ ratios, suggest that this chemical heterogeneity must have been relatively short lived, most probably less than 50 Ma. The geochemistry of the granites of the Almaden Supersuite suggest that the protolith for those granites was intermediate

to mafic (<60% SiO_2) with moderate K_2O, Th, U, Rb contents.

(3) The highly fractionated granites of the O'Briens Creek Supersuite are at least as old as the granites of the Ootann and Almaden supersuites, suggesting that the strongly fractionated O'Briens Creek granites were not produced from a residual source (cf. Collins *et al.* 1982; Manning & Pichavant 1988), or at least not from a protolith that melted to produce the Almaden and Ootann granites.

(4) The Carboniferous granites were all infracrustal in origin. The preserved geochemical signatures of the granites are dominantly magmatic. Geochemical variation within the granites is consistent with crystal fractionation models.

(5) The granites outcrop over 8,000 km^2 (as also separately do the volcanic rocks). Assuming that these rocks result from 30% partial melting, then a protolith of both large volume and area is required. The processes that produced these granites operated episodically or continuously over at least 30 Ma.

4.2. Petrogenetic models

It is apparent from the above that, in the absence of wholesale crustal Nd, Sr and Pb isotopic re-equilibration (either before or during melting), the possible candidates for source rocks for the CIG are limited. The most plausible petrogenetic models for the CIG would appear to be those involving a large-scale isotopically-homogeneous protolith that was underplated to the lower crust of northern Queensland, most probably some time in the Proterozoic. Two end-member models can be invoked: (1) decreasing degrees of partial melting of a predominantly intermediate protolith to produce the Almaden, Ootann and O'Briens Creek granites, respectively; or, (2) partial melting of a basaltic to intermediate protolith to produce the Almaden (and Claret Creek) granites, which have themselves been remelted to produce the granites of the Ootann and O'Briens Creek supersuites. This melting may have been accompanied, by, or preceded by, enrichment and depletion of K_2O, Rb, Th, U, etc., evidence of which may be present in the lower crustal xenoliths (Rudnick & Taylor 1987) and granites of the Claret Creek Supersuite (see below).

It is difficult to decide between models 1 or 2 (above). The preponderance of Ootann and O'Briens Creek granites and the age relationships would seem to indicate that the felsic granites were not derived from an intermediate product (such as the Almaden Supersuite) but were derived directly from the protolith. There are, however, some aspects in favour of the second model; in particular, it is a more feasible mechanism for producing both the isotopic homogeneity and geochemical variability of the granites (e.g. suites and supersuites) over such a large compositional range (58–77% SiO_2) and of satisfying rheological arguments regarding the percentage of partial melting (e.g. Van der Molen & Paterson 1979). It is quite feasible that more extensive mafic granites exist at depth given the subvolcanic nature of the presently exposed felsic granites. It is also possible that a combination of both mechanisms (models 1 and 2) was operative, with the granite protoliths consisting of mixtures of the original protolith and derivative partial melts.

One other possible model for the CIG is the hybrid hypothesis of Hildreth *et al.* (1991), who proposed that the voluminous high silica rhyolites of the Yellowstone Plateau volcanic field were derived from a hybrid source that represented a large-scale, isotopically-mixed combination of contemporaneous basalts and crustal rocks. This model is capable of producing the overall isotopic homogeneity of the majority of the CIG and, paradoxically, also the isotopic variability, with the more radiogenic isotopic signature of the western granites due to greater contribution of the crustal source and/or less effective hybridisation. A number of difficulties are evident, however. Firstly, if the Almaden Supersuite granites represent differentiates of hybridised mantle melts, then this hybridisation must have been similarly efficient for all suites of the Almaden Supersuite, given the virtually identical ε_{Nd} of the granites despite the geochemical variability evident in these granites, particularly in the LILE. Secondly, it is difficult to imagine how a consistent isotopic signature could have been produced episodically over such a long period of time (50 Ma) and a large areal extent; the Yellowstone rhyolites, although spanning only 2·2 Ma, exhibit pronounced isotopic variations related to caldera collapse and source, i.e. intracaldera or extracaldera (Hildreth *et al.* 1991, Fig. 6).

4.3. Possible source rocks

Regardless of which model is proposed for the generation of the Almaden Supersuite (and Claret Creek Supersuite) granites, it is more-or-less implicit that the Ootann and O'Briens Creek granites were derived from partial melting of intermediate source rocks that were possibly similar in composition to the more mafic granites of the Almaden Supersuite. This hypothesis has been modelled utilising the Retire Monzodiorite, the most mafic unit (57–63% SiO_2) of the Almaden Supersuite. The Retire Monzodiorite is characterised by moderate to high K_2O, Rb, Th and U content and is a fertile source rock with abundant modal quartz, alkali feldspar and biotite (Table 2).

From this modelling, it is evident that the Ootann Supersuite granites could have been derived from a protolith compositionally similar to the Retire Monzodiorite with the proviso either that the protolith was richer in P_2O_5, Sr and Zr than the Retire Monzodiorite, or that the more mafic end-members of the Ootann granites are partly cumulative. Calculated partition coefficients for 30–50% partial melting are consistent with published partition coefficients, with Ba, Pb, Th, U, Rb and the REE behaving incompatibly, and V, Cr, Ni, Cu, Sc, Zn, Sr and Eu behaving compatibly. Such a protolith would, after partial melting, produce a residue dominated by plagioclase, pyroxenes and hornblende. Calculated residue compositions are similar in many respects to those reported for lower crustal xenoliths from the area (Table 3).

Difficulties were encountered with modelling of the O'Briens Creek Supersuite, chiefly with F and Ba, with results suggesting that either the source rocks for the supersuite were lower in Ba and/or the supersuite contains more mafic end-members not yet recognised. The latter is not unexpected given the geochemical convergence of the less differentiated O'Briens Creek and Ootann granites. A more mafic protolith (than the Retire Monzodiorite) is not tenable given the very low levels of V and Cr in the O'Briens Creek Supersuite. The very similar geochemistry of the Ootann and O'Briens Creek supersuites strongly suggests that these two supersuites had a very similar, if not identical, source. The most feasible solution for the genesis of granites of the O'Briens Creek Supersuite would appear to be one involving smaller degrees of partial melting of a similar protolith to that for the Ootann Supersuite coupled with more extensive fractionation than the latter, consistent with the low levels of compatible trace elements (e.g. Halliday *et al.* 1991). It is noted that partial melting of such

Table 2 Representative major and trace element chemistry and modal data for the Retire Monzodiorite of the Retire Suite; note the amounts of modal quartz, alkali feldspar and biotite, and the moderate to high LILE-content; data from Bailey (1969) and Blevin (unpublished data 1990); no geochemical data are available for B59.

	B73 Bailey	B52 Bailey	B59 Bailey	BLE969 Blevin	BLE857 Blevin	BLE975 Blevin
SiO_2	62·16	59·76	—	60·58	62·28	58·51
TiO_2	0·64	0·68	—	0·79	0·65	0·79
Al_2O_3	15·55	16·46	—	16·46	15·70	16·93
Fe_2O_3	1·35	1·12	—	1·93	1·93	2·13
FeO	4·11	4·88	—	4·44	3·47	4·70
MnO	0·11	0·11	—	0·14	0·10	0·15
MgO	2·97	4·14	—	2·59	2·94	2·95
CaO	5·62	6·90	—	6·15	5·58	6·85
Na_2O	2·63	2·74	—	2·93	2·77	3·05
K_2O	3·09	2·23	—	2·33	2·85	1·88
P_2O_5	0·10	0·10	—	0·14	0·10	0·14
LOI	0·14	0·27	—	1·16	0·98	1·00
Trace elements (ppm)						
F			—	900		900
Ba	317	280	—	327	302	281
Rb	209	144	—	138	193	113
Sr	184	212	—	236	200	237
Pb	33	22	—	7	20	12
Th	25·0	18·0	—	18	22	10
U	6·6	4·9	—	3	3	1
Zr	148	105	—	135	149	108
Nb	11·0	10·0	—			
Y	29	27	—	29	31	29
Sc	20	23	—	20		
V	127	145	—	150	128	170
Cr	33	52	—	14	31	13
Ni	30	15	—	6	11	7
Ga	15·0	16·0				
Modes %						
Quartz	20·2	14·7	19·0			
Plagioclase	42·3	48·6	43·8			
Alkali feldspar	14·5	9·2	20·3			
Biotite	14·4	11·8	9·3			
Hornblende	5·1	5·2	5·5			
Pyroxene	2·5	9·6	1·6			
Accessories	0·1					
Opaques	0·9	0·8	0·5			

rocks as the Retire Monzodiorite would result in a residue with little or no hydrous phases (see discussion by Creaser *et al.* 1991), overcoming the difficulties of incorporating F into the melt (e.g. Manning & Pichavant 1988).

It is apparent that although the granites of the Claret Creek Supersuite are geochemically distinct from the other granites of the CIG, the isotopic data suggest a fundamental relationship between all the CIG. The differing chemistry of the Claret Creek granites relative to the granites of the Almaden Supersuite could not have been produced by greater degrees of partial melting or via reworking of a residual source, as these processes could not have produced

Table 3 Calculated residual compositions after extraction of 30–45% partial melt from the Retire Monzodiorite; compositions of local lower crustal xenoliths from Rudnick & Taylor (1987) and Stolz & Davies (1989) with similar Sr and Nd isotopic signature to the Carboniferous granites are shown for comparison.

	RESIDUAL COMPOSITIONS						LOWER CRUSTAL XENOLITHS				
	30% melting		40% melting		45% melting		Rudnick & Taylor			Stolz & Davies	
	min	max	min	max	min	max	85–108	85–120	85–100	32–32	32–22
SiO_2	50·70	58·39	47·26	56·23	45·07	54·85	48·90	52·39	51·70	51·52	50·32
TiO_2	0·75	0·98	0·81	1·07	0·84	1·14	1·09	0·57	0·88	0·69	2·61
Al_2O_3	16·06	18·49	16·35	19·18	16·53	19·62	15·16	16·50	11·26	21·29	20·80
Fe_2O_3	1·16	2·67	1·18	2·95	1·19	3·12				3·30	3·54
FeO	4·17	7·07	4·56	7·95	4·81	8·50				3·14	9·52
FeO*	6·28	8·88	6·88	9·91	7·26	10·57	12·42	9·45	13·30	6·11	12·71
MnO	0·12	0·19	0·13	0·21	0·13	0·22	0·22	0·15	0·20	0·13	0·26
MgO	3·37	6·96	3·81	7·99	4·09	8·65	8·48	9·17	14·50	5·38	4·85
CaO	6·94	10·90	7·70	12·32	8·18	13·22	12·70	9·30	6·82	10·85	4·89
Na_2O	1·71	3·09	1·50	3·10	1·36	3·11	0·89	2·34	1·06	2·94	2·36
K_2O	0·65	2·81	0·13	2·65	<0·20	2·55	0·16	0·19	0·16	0·24	0·19
P_2O_5	0·09	0·16	0·09	0·17	0·09	0·18	0·32	0·28	0·19	0·04	<0·02

Notes: Minimum (min) and maximum (max) refer to partial melting of the most mafic (58% SiO_2) and most felsic (62% SiO_2) Retire Monzodiorite; all lower crustal xenoliths are from the Hill 32 locality, McBride basalt province, northern Queensland (cf. Rudnick & Taylor 1987).

similar concentrations of the LREE and the ferromagnesian elements, and the similar mg numbers in the two supersuites. Importantly, the combination of low K_2O and high Ba suggest that the geochemical characteristics of the Claret Creek Supersuite are inherited from the source. The Claret Creek Supersuite appears to best fit a model involving derivation of the granites at greater depths from a protolith originally similar to that of the other CIG but which has experienced depletion of the more incompatible LILE. In this regard, it is significant that lower crustal xenoliths from the area are strongly depleted in K_2O, Rb, Th and U (Rudnick & Taylor 1987) but not in Ba, Sr and Pb, which Rudnick and Taylor suggested was due to mobilisation during high-grade metamorphism. This depletion event most probably occurred at *c.* 300 Ma which was a time of widespread granulite formation (Rudnick & Williams 1987). It may also be speculated that the protoliths for the Almaden, Ootann and O'Briens Creek supersuites were enriched in these elements at that time.

5. Conclusions and implications

The voluminous Carboniferous granites of northern Queensland, especially the granites of the Ootann and O'Briens Creek supersuites, are excellent examples of strongly fractionated felsic I-type granites and, therefore, provide important information on such granites and the geochemical trends that develop with fractionation in such rocks.

Two models are proposed for the origin of the felsic granites of the Ootann and O'Briens Creek supersuites, each essentially involving a protolith of andesitic to dacitic composition that is moderately enriched in LILE. As documented by many workers, e.g. Ewart (1979), Hildreth & Moorbath (1988), such andesitic and dacitic rocks are not uncommon. A multi-stage process is required for the production of felsic granites such as those of the Ootann and O'Briens Creek supersuites, commencing with initial emplacement and/or underplating of mantle-derived melts, i.e. basalts into the crust. It is considered unlikely that felsic granites, particularly strongly fractionated granites such as those in northern Queensland, can be derived by crystal fractionation from basaltic rocks, even when accompanied by assimilation (DePaolo 1981a, b). Additionally, it is doubtful whether such rocks alone are sufficiently enriched in quartz and K and other LILE to be capable of producing true granitic melts; most experimental evidence suggests that partial melting of basaltic and other low-K rocks will produce predominantly sodic melts such as tonalities, trondhjemites and sodic granodiorites (e.g. Rutter & Wyllie 1988; Rushmer 1990). A variety of processes, which may have operated separately or in tandem, appear to be required to render these mantle-derived melts more fertile. These include such processes as: accompanying fractionation (e.g. Wyborn *et al.* 1988), accompanying crustal contamination or hybridisation (e.g. DePaolo 1981b; Hildreth & Moorbath 1988), and/or by reworking of this protolith (as proposed for the present work; also Chappell & Stephens 1988; Hildreth *et al.* 1991). Felsic granites, such as the granites of the Ootann and O'Briens Creek supersuites, can then be derived by partial melting of this modified protolith. As evidenced by the granites of the Ootann and O'Briens Creek supersuites, and indicated by many workers (e.g. Mahood & Halliday 1988; Halliday *et al.* 1991), the subsequent felsic partial melts are further fractionated by extensive, feldspar-dominated, crystal fractionation, producing granites that are strongly depleted in TiO_2, FeO*, MgO, CaO, Ba, Sr, Sc, V, Cr, Ni, Eu, Ce_N/Y_N and K/Rb, and enriched in Rb, Pb, Th, U and RB/Sr.

A number of important aspects are evident from the results of the present work. Firstly, it is apparent that felsic I-type granites, e.g. the O'Briens Creek Supersuite, can, with extensive crystal fractionation, develop many of the characteristics of A-type granites, i.e. high abundances of HFSE, HREE and F. This shows that caution must be exercised in automatically assigning strongly fractionated felsic granites as A-type granites (see also Whalen *et al.* 1987). Secondly, there is little evidence to substantiate that a protolith more felsic than andesitic to dacitic is required for production of fractionated felsic granites. Thirdly, there is the possibility that rapid reworking may be more common than previously recognised, e.g. Hildreth *et al.* (1991), Atherton & Petford (1991). This raises a number of questions regarding petrogenesis of felsic granites elsewhere. For example, the topaz rhyolites of the western U.S.A. may indeed have been derived from reworking of intermediate melts produced from a more typical Proterozoic protolith, rather than from small degrees of partial melting of a low-LILE felsic granulitic source as proposed by Christiansen *et al.* (1984, 1986), as may also be the case for the high silica rhyolites of the Sierra La Primavera, Mexico (Mahood & Halliday 1988). Petrogenetic models for felsic igneous rocks should take into consideration the possibility of multi-stage processes occurring over relatively short time-scales.

6. Acknowledgements

This work was supported through a Australian Postgraduate Research Award to Champion and Australian Mineral Industry Research Association grants to Chappell. We thank R. J. Bultitude, P. L. Blevin, A. J. R. White, J. Kilpatrick, D. J. Ellis, C. Johnston, D. N. G. Richards, G. Clarke, D. Mackenzie and D. Wyborn for assistance and continued discussion throughout this study.

7. References

Atherton, M. P. & Petford, N. 1991. The origin and geotectonic setting of Andean granitoids. REC BMR GEOL GEOPHYS **1991**/25, 4.

Bailey, J. C. 1969. *Geochemistry of Georgetown Granites.* Unpublished Ph.D. Thesis, The Australian National University, Canberra.

Bailey, J. C. 1977. Petrochemistry of the Claret Creek Ring Complex, northeast Queensland. J GEOL SOC AUST **24,** 1–14.

Black, L. P. 1978. Isotopic ages of rocks from the Georgetown-Mount Garnet-Herberton area, north Queensland. REP BMR GEOL GEOPHYS **200**.

Black, L. P. 1980. Rb-Sr systematics of the Claret Creek ring complex and their bearing on the origin of Upper Palaeozoic igneous rocks in northeast Queensland. J GEOL SOC AUST **27,** 157–66.

Black, L. P. & McCulloch, M. T. 1990. Isotopic evidence for the dependence of recurrent felsic magmatism on new crust formation: an example from the Georgetown region of Northeastern Australia. GEOCHIM COSMOCHIM ACTA **54,** 183–96.

Black, L. P., Blake, D. H. & Olatunji, J. A. 1978. Ages of granites and associated mineralisation in the Herberton tinfield of northeast Queensland. J BMR GEOL GEOPHYS **3,** 173–80.

Black, L. P., Bell, T. H., Rubenach, M. J. & Withnall, I. W. 1979. Geochronology of discrete structural-metamorphic events in a multiply deformed Precambrian terrian. TECTONOPHYSICS **54,** 103–37.

Blevin, P. L. 1990. *The tungsten-molybdenum-bismuth hydrothermal mineralising system at Bamford Hill, northeast Queensland, Australia.* Unpublished Ph.D. Thesis, James Cook University of North Queensland.

Bultitude, R. J., Donchak, P. J., Domagala, J., Fordham, B. G. & Champion, D. C. 1990. Geology and tectonics of the

Hodgkinson Province, north Queensland. PROC PAC RIM CONGRESS 1990 **3,** 75–81.

Champion, D. C. 1991. *Felsic granites of far north Queensland.* Unpublished Ph.D. Thesis, The Australian National University, Canberra.

Chappell, B. W. & Stephens, W. E. 1988. Origin of infracrustal (I-type) granite magmas. TRANS R SOC EDINBURGH EARTH SCI **79,** 71–86.

Charoy, B. 1986. The genesis of the Cornubian Batholith (south-west England): the example of the Carnmenellis pluton. J PETROL **27,** 571–604.

Christiansen, E. H., Bikum, J. V., Sheridan, M. F. & Burt, D. M. 1984. Geochemical evolution of topaz rhyolites from the Thomas Range and Spor Mountain, Utah. AM MINERAL **69,** 223–36.

Christiansen, E. H., Sheridan, M. F. & Burt, D. M. 1986. The geology and geochemistry of Cenozoic topaz rhyolites from the western United States. GEOL SOC AM SPEC PAP **205**.

Clarke, G. C. 1990. *The role of fluids in the evolution of granites and associated Mo, W and Sn deposits, Herberton District, North Queensland.* Unpublished Ph.D. Thesis, James Cook University of North Queensland.

Collins, W. J., Beams, S. D., White, A. J. R. & Chappell B. W. 1982. Nature and origin of A-type granites with particular reference to southeast Australia. CONTRIB MINERAL PETROL **80,** 189–200.

Creaser, R. A., Price, R. C. & Wormald, R. J. 1991. A-type granites revisited: assessment of a residual-source model. GEOLOGY **19,** 163–6.

de Keyser, K. & Lucas, K. G. 1968. Geology of the Hodgkinson and Laura Basins, north Queensland. BULL BMR GEOL GEOPHYS **84**.

DePaolo, D. J. 1981a. Trace element and isotopic effects of combined wallrock assimilation and fractional crystallization. EARTH PLANET SCI LETT **53,** 189–202.

DePaolo, D. J. 1981b. A neodymium and strontium isotope study of the Mesozoic calc-alkaline granitic batholiths of the Sierra Nevada and Peninsular Ranges, California. J GEOPHYS RES **86,** 10470–501.

Ewart, A. 1979. A review of the mineralogy and chemistry of Tertiary-Recent dacitic, latitic, rhyolitic, and related salic volcanic rocks. *In* Barker, F. (ed.) *Trondhjemites, dacites and related rocks,* 13–21. Amsterdam: Elsevier.

Ewart, A., Chappell, B. W. & Le Maitre, R. W. 1985. Aspects of the mineralogy and chemistry of the intermediate-silicic Cainozoic volcanic rocks of eastern Australia. Part 1: introduction and geochemistry. AUST J EARTH SCI **32,** 359–82.

Groves, D. I. & McCarthy, T. S. 1978. Fractional crystallization and the orign of tin deposits in granitoids. MINERAL DEPOSITA **13,** 11–26.

Haapala, I. 1988. Metallogeny of the Proterozoic rapakivi granites of Finland. *In* Taylor, R. P. & Strong, D. F. (eds) *Recent advances in the geology of granite-related mineral deposits,* 124–32. Quebec: Canadian Institute Mining and Metallurgy.

Halliday, A. N., Davidson, J. P., Hildreth, W. & Holden, P. 1991. Modelling the petrogenesis of high Rb/Sr magmas. CHEM GEOL **92,** 107–14.

Hildreth, E. W. 1979. The Bishop Tuff: evidence for the origin of compositional zonation in silicic magma chambers. GEOL SOC AM SPEC PAP **180,** 43–75.

Hildreth, E. W., Halliday, A. N. & Christiansen, R. L. 1991. Isotopic and chemical evidence concerning the genesis and contamination of basaltic and rhyolitic magma beneath the Yellowstone Plateau Volcanic Field. J PETROL **32,** 63–138.

Hildreth, E. W. & Moorbath, S. 1988. Crustal contributions to arc magmatism in the Andes of Central Chile. CONTRIB MINERAL PETROL **98,** 455–89.

Johnston, C. J. 1984. *Granitoids of the Coolgarra Batholith, Mount Garnet, north Queensland.* Unpublished Ph.D. Thesis, James Cook University of North Queensland.

Johnston, C. J. & Black, L. P. 1986. Rb-Sr systematics of the Coolgarra Batholith, north Queensland. AUST J EARTH SCI **33,** 309–24.

Kleeman, G. & Twist, D. 1989. The compositionally-zoned sheet-like granite pluton of the Bushveld Complex: evidence bearing on the nature of A-type magmatism. J PETROL **30,** 1383–414.

Kontak, D. J. 1990. The East Kemptville topaz-muscovite leucogranite, Nova Scotia. I. geological setting and whole-rock geochemistry. CAN MINERAL **27,** 818–25.

Leat, P. T., Thompson, R. N., Morrison, M. A., Hendry, G. L. & Trayhorn, S. C. 1986. Geodynamic significance of post-Variscan intrusive and extrusive potassic magmatism in SW England. TRANS R SOC EDINBURGH EARTH SCI **77,** 349–60.

Mackenzie, D. E., Black, L. P. & Sun, S-S. 1988. Origin of alkali-feldspar granites: An example from the Poimena granite, northeast Tasmania, Australia. GEOCHIM COSMOCHIM ACTA **52,** 2507–24.

Mahood, G. A. & Halliday, A. N. 1988. Generation of high-silica rhyolite: A Nd, Sr, and O isotopic study of Sierra La Primavera, Mexican Neovolcanic Belt. CONTRIB MINERAL PETROL **100,** 183–91.

Manning, D. A. C. 1982. An experimental study of the effects of fluorine on the crystallisation of granitic melts. *In* Evans, A. M. (ed.) *Metallisation associated with acid magmatism,* 191–203. Devon: Wiley and Sons.

Manning, D. A. C. & Pichavant, M. 1988. Volatiles and their bearing on the behaviour of metals in granitic systems. *In* Taylor, R. P. & Strong, D. F. (eds) *Recent advances in the geology of granite-related mineral deposits,* 13–24. Quebec: Canadian Institute Mining and Metallurgy.

Martin, R. F. & Bowden, P. 1981. Peraluminous granites produced by rock-fluid interaction in the Ririwai nonorogenic ring-complex, Nigeria: mineralogical evidence. CAN MINERAL **19,** 65–82.

Nekvasil, H. 1988. Calculated effect of anorthite component on the crystallisation paths of H_2O-undersaturated haplogranitic melts. AM MINERAL **73,** 966–81.

Oversby, B. S., Black, L. P. & Sheraton, J. W. 1980. Late Palaeozoic continental volcanism in northeastern Australia. *In* Henderson, R. A. & Stephenson, P. J. (eds) *The geology and geophysics of Northeastern Australia.* 247–68. Brisbane: Geological Society of Australia, Queensland Division.

Peccerillo, A. & Taylor, S. R. 1976. Geochemistry of Eocene calc-alkaline volcanic rocks from the Kastamonu area, northern Turkey. CONTRIB MINERAL PETROL **58,** 63–81.

Pichavant, M. 1981. An experimental study of the effect of boron on a water saturated haplogranite at 1 kbar vapour pressure. CONTRIB MINERAL PETROL **76,** 430–9.

Plant, J. A., O'Brien, C. O., Tarney, J. & Hurdley, J. 1985. Geochemical criteria for the recognition of high heat production granites. In *High heat production (HHP) granites, hydrothermal circulation and ore genesis,* 263–86. London: The Institution of Mining and Metallurgy.

Pollard, P. J. 1988. Petrogenesis of tin-bearing granites of the Emuford district, Herberton tinfield, Australia. AUST J EARTH SCI **35,** 39–57.

Pollard, P. J., Millburn, D., Taylor R. G. & Cuff, C. 1983. Mineralogical and textural modifications in granites associated with tin mineralization, Herberton-Mt Garnet Tinfield, Queensland. *In Permian Geology of Queensland,* 413–30. Brisbane: Geological Society of Australia, Queensland Division.

Richards, D. N. G. 1981. *Granitoids of the northern Tate Batholith, Chillagoe, northern Queensland.* Unpublished Ph.D. Thesis, James Cook University of North Queensland.

Rudnick, R. L. 1990. Nd and Sr isotopic compositions of lower crustal xenoliths from north Queensland, Australia: implications for Nd model ages and crustal growth processes. CHEM GEOL **83,** 195–208.

Rudnick, R. L. & Taylor, S. R. 1987. The composition and petrogenesis of the lower crust: a xenolith study. J GEOPHYS RES **92,** 13,981–4005.

Rudnick, R. L. & Williams, I. S. 1987. Dating the lower crust by ion microprobe. EARTH PLANET SCI LETT **85,** 145–61.

Rushmer, T. 1990. Partial melting of two amphibolites: contrasting experimental results under fluid-absent conditions. CONTRIB MINERAL PETROL **107,** 41–59.

Rutter, M. J. & Wyllie, P. J. 1988. Melting of vapour-absent tonalite at 10 kbar to simulate dehydration-melting in the deep crust. NATURE **331,** 159–60.

Shaw, R. D., Fawckner, J. F. & Bultitude, R. J. 1987. The Palmerville Fault system: A major imbricate thrust system in the northern Tasmanides, north Queensland. AUST J EARTH SCI **34,** 69–93.

Sheraton, J. W. 1974. Chemical analyses of acid igneous rocks of northeast Queensland. REC BMR GEOL GEOPHYS **1974**/62.

Sheraton, J. W. & Labonne, B. 1978. Petrology and geochemistry of acid igneous rocks of northeast Queensland. BULL BMR GEOL GOEPHYS **169**.

Stolz, A. J. & Davies, G. R. 1989. Metasomatised lower crustal and upper mantle xenoliths from north Queensland: Chemical and isotopic evidence bearing on the composition and source of fluid phase. GEOCHIM COSMOCHIM ACTA **53,** 649–60.

Taylor, R. P., Strong, D. F. & Fryer, B. J. 1981. Volatile control of contrasting trace element distributions in peralkaline granitic and volcanic rocks. CONTRIB MINERAL PETROL **77,** 267–71.

Taylor, R. G. & Pollard, P. J. 1988. Pervasive hydrothermal alteration in tin-bearing granites and implications for the evolution of ore-bearing magmatic fluids. *In* Taylor, R. P. & Strong, D. F. (eds) *Recent advances in the geology of granite-related mineral deposits,* 86–95. Quebec: Canadian Institute Mining and Metallurgy.

Turekian, K. K. & Wedepohl, K. H. 1961. Distribution of the elements in some major units of the earth's crust. BULL GEOL SOC AM **72,** 175–92.

Tuttle, O. F. & Bowen, N. L. 1958. Origin of granite in the light of experimental studies in the system $NaAlSi_3O_8$-$KAlSi_3O_8$-SiO_2-H_2O. MEM GEOL SOC AM **74**.

Van der Molen, I. & Paterson, M. S. 1979. Experimental deformation of partially-melted granite. CONTRIB MINERAL PETROL **70,** 299–318.

Whalen, J. B., Currie, K. L. & Chappell, B. W. 1987. A-type granites: geochemical characteristics, discrimination and petrogenesis. CONTRIB MINERAL PETROL **95,** 407–19.

Willmott, W. F., Whitaker, W. G., Palfreyman, W. D. & Trail, D. S. 1973. Igneous and metamorphic rocks of Cape York Peninsular and Torres Strait. BULL BMR GEOL GEOPHYS **135**.

Withnall, I. W., Bain, J. H. C. & Rubenach, M. J. 1980. The Precambrian geology of northeastern Queensland. *In* Henderson, R. A. & Stephenson, P. J. (eds) *The geology and geophysics of Northeastern Australia,* 109–28. Brisbane: Geological Society of Australia, Queensland Division.

Withnall, I. W., Bain, J. H. C., Draper, J. J., Mackenzie, D. E. & Oversby, B. S. 1988. Proterozoic stratigraphy and tectonic history of the Georgetown Inlier, northeastern Australia. PRECAMBRIAN RES **40/41,** 429–46.

Witt, W. K. 1985. *Diffuse (background), and fracture-controlled feldspathic alteration in tin-mineralised granites of the Irvinebank-Emuford area, northeast Queensland.* Unpublished Ph.D. Thesis, James Cook University of North Queensland.

Wyborn, L. A. I., Page, R. W. & McCulloch, M. T. 1988. Petrology, geochronology and isotope geochemistry of the post-1,820 Ma granites of the Mount Isa Inlier: mechamisms for the generation of Proterozoic anorogenic granites. PRECAMBRIAN RES **40/41,** 509–41.

D. C. CHAMPION and B. W. CHAPPELL, Department of Geology, The Australian National University, GPO Box 4, Canberra City, ACT 2601, Australia.

Transactions of the Royal Society of Edinburgh: Earth Sciences, **83**, 127–138, 1992

The Layos Granite, Hercynian Complex of Toledo (Spain): an example of parautochthonous restite-rich granite in a granulitic area

L. Barbero and C. Villaseca

ABSTRACT: The Layos Granite forms elongated massifs within the Toledo Complex of central Spain. It is late-tectonic with respect to the F2 regional phase and simultaneous with the metamorphic peak of the region, which reached a maximum temperature of 800–850°C and pressures of 400–600 MPa. Field studies indicate that this intrusion belongs to the "regional migmatite terrane granite" type. This granite is typically interlayered with sill-like veins and elongated bodies of cordierite/garnet-bearing leucogranites. Enclaves are widespread and comprise restitic types (quartz lumps, biotite, cordierite and sillimanite-rich enclaves) and refractory metamorphic country-rocks including orthogneisses, amphibolites, quartzites, conglomerates and calc-silicate rocks.

These granites vary from quartz-rich tonalites to melamonzogranites and define a S-type trend on a QAP plot. Cordierite and biotite are the mafic phases of the rocks. The particularly high percentage of cordierite (10%–30%) varies inversely with the silica content. Sillimanite is a common accessory mineral, always included in cordierite, suggesting a restitic origin. The mineral chemistry of the Layos Granite is similar to that of the leucogranites and country-rock peraluminous granulites (kinzigites), indicating a close approach to equilibrium. The uniform composition of plagioclase (An_{25}), the high albitic content of the K-feldspar, the continuous variation in the Fe/Mg ratios of the mafic minerals, and the high Ti content of the biotites (2·5–6·5%) suggest a genetic relationship.

Geochemically, the Layos Granite is strongly peraluminous. Normative corundum lies between 4% and 10% and varies inversely with increase in SiO_2. The CaO content is typically low (<1·25%) and shows little variation; similarly the LILE show a limited range. On many variation diagrams, linear trends from peraluminous granulites to the Layos Granite and associated leucogranite can be observed. The chemical characteristics argue against an igneous fractionation or fusion mechanism for the diversification of the Layos Granite. A restite unmixing model between a granulitic pole (represented by the granulites of the Toledo Complex) and a minimum melt (leucogranites) could explain the main chemical variation of the Layos Granite. Melting of a pelitic protolith under anhydrous conditions (biotite dehydration melting) could lead to minimum-temperature melt compositions and a strongly peraluminous residuum.

For the most mafic granites (61–63% SiO_2), it is estimated that the trapped restite component must have been around 65%. This high proportion of restite is close to the estimated rheological critical melt fraction, but field evidence suggests that this critical value has been exceeded. This high restite component implies high viscosity of the melt which, together with the anhydrous assemblage of the Layos Granite and the associated leucogranites, indicates H_2O-undersaturated melting conditions. Under such conditions, the high viscosity magma (crystal-liquid mush) had a restricted movement capacity, leading to the development of parautochthonous plutonic bodies.

KEY WORDS: cordierite-rich granites, dry-melting, Iberian Variscan belt, restite-unmixing, source materials, strongly peraluminous.

The Complejo Plutono-Migmatítico de Toledo, hereafter termed the Toledo Complex, forms an E–W elongated block of high-grade metamorphic and plutonic rocks to the S of Toledo in central Spain. It is part of the Centro–Iberian zone of the axial zone of the Hercynian Belt (Julivert *et al.* 1974). To the N, the Toledo Complex is overlain by post-Mesozoic sedimentary deposits and to the S it is bounded by a major listric fault which brings it into contact with low-grade Palaeozoic sediments in which was emplaced a post-tectonic epizonal batholith (Fig. 1).

The Toledo Complex comprises a metamorphic assemblage of presumed pre-Ordovician age, consisting mainly of metapelites with some orthogneisses and minor metapsammites and marbles. The metamorphism has granulitic facies characteristics and is assumed to be Hercynian in age. The granulite terrane exhumation is explained by extensional thinning developed by great listric faults giving rise to a core-complex geometry (Casquet *et al.* 1988). Emplaced in the metamorphic rocks is a syntectonic felsic to mafic suite of igneous rocks (Barbero *et al.* 1990)

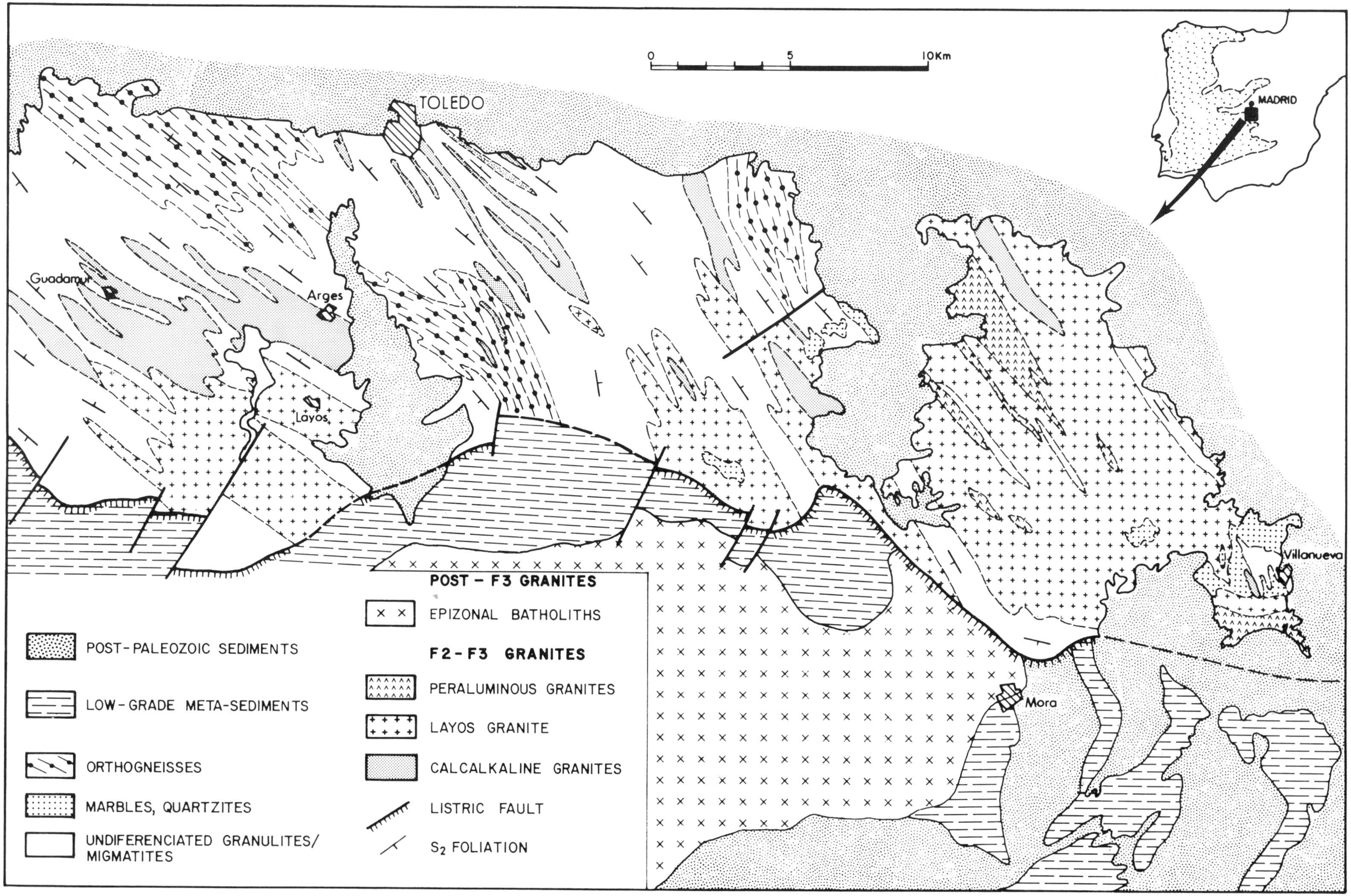

Figure 1 Geological map of the Toledo Complex; inset map indicates the location in the Hercynian Iberian Belt.

which was contemporaneous with or slightly earlier than bodies of heterogeneous strongly peraluminous parautochthonous granites.

This paper deals with petrographical, mineralogical, geochemical and genetic aspects of one of the most peraluminous of these granites, the Layos Granite.

1. Geological setting

The main structure of the Toledo Complex is a strain-slip foliation (S2) with 110°–170° regional strike, usually dipping to the E. This foliation is seen in the metamorphic rocks, mainly in the nonpelitic types (orthogneisses, psammitic paragneisses). In the pelitic varieties, the migmatisation event has erased the S2 foliation and has generated granoblastic non-foliated rocks with a marked banded structure coherent with the S2 foliation of the orthogneisses and metapsammites. In general, the F1 tectonic phase is almost completely obliterated and only some sporadic folds surrounded by S2 and axial F1 surfaces affected by S2 can be observed (Martín Escorza & López Martínez 1978). The F3 tectonic phase causes large- to small-scale asymmetric folds with axial surfaces near-vertical and parallel to the S2 foliation. A weak mineral elongation lineation is sporadically developed, but S3 axial surface foliations are not common. There are two later tectonic phases that cause rotation and folding of the pre-existent structures (Martín Escorza & López Martínez 1978).

The metamorphic rocks of the Toledo Complex are of both sedimentary and igneous origin. The main metaigneous types are leucocratic gneisses and augen gneisses. These rocks are similar to those of the Spanish Central System, probably of late-Cambrian or early-Ordovician age (Vialette *et al.* 1987; Wildberg *et al.* 1989). These metaigneous rocks are exposed mainly in the northern part of the complex (Fig. 1). They retain a foliated structure and because of their low mica content are poorly migmatised.

The metasedimentary types include peraluminous granulites (kinzigitic in affinity), biotitic gneisses, quartzites, amphibolites, marbles and conglomerates, the first being the most abundant type. The granulites exhibit a strong migmatitic layering which defines a stromatic structure characterised by a cordierite-rich mesosome (with Qtz + Kfs + Pl + Crd + Bt + Grt + Sp + Sil ± Ilm) which alternates with narrow leucosomes (Fig. 2(A)), varying from garnet- to cordierite-, and to a lesser extent biotite-, bearing leucogranites. Especially noteworthy is the presence of large rounded garnet crystals in the leucosomes that mainly consist of Qtz + Kfs ± Pl (Fig. 2(A)). In general, it is difficult to find melanosomes or mafic rims around the

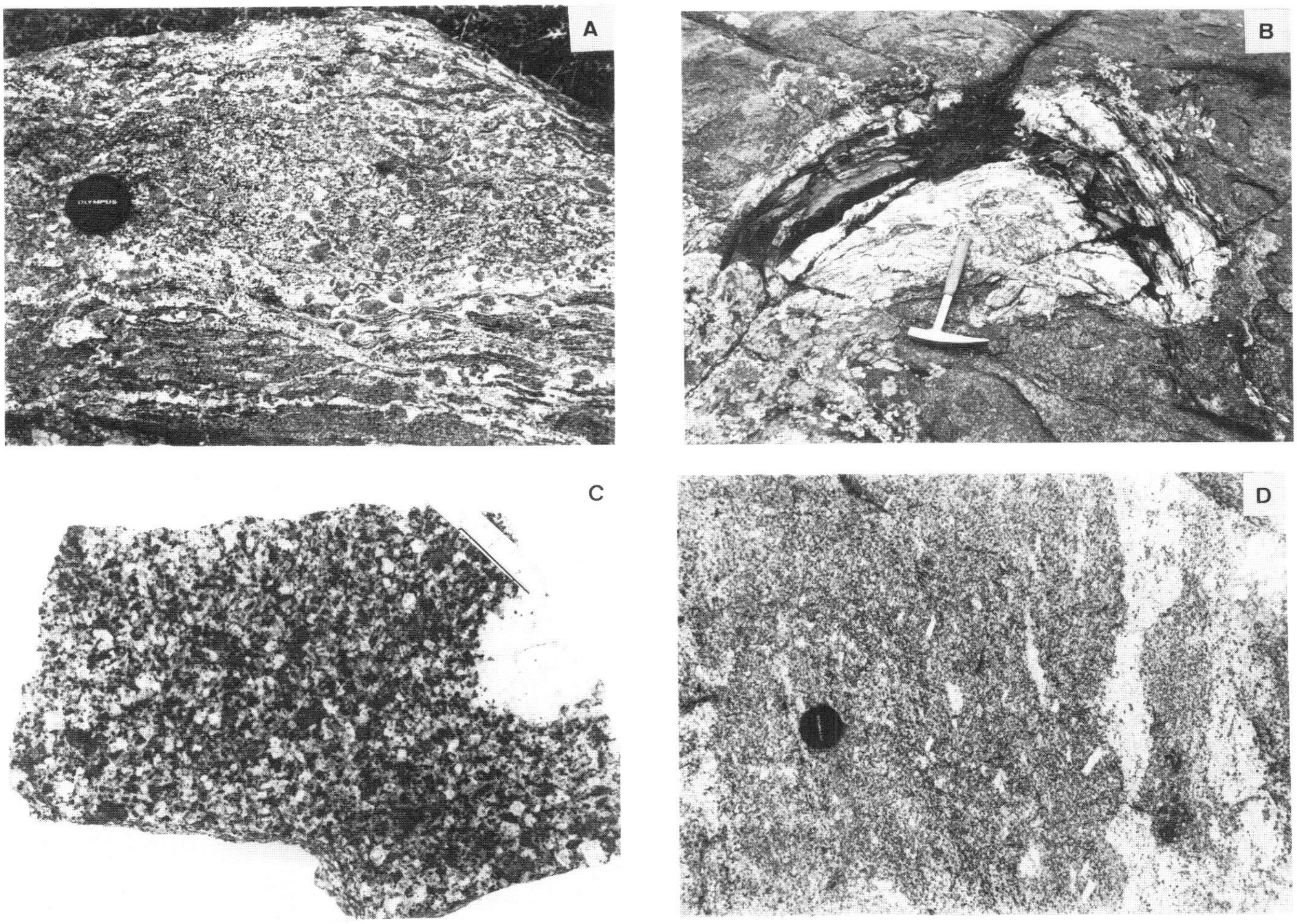

Figure 2 (A) General aspect of the peraluminous migmatitic granulites, showing the stromatic structure marked by the garnet-bearing leucosomes and the cordierite-rich dark mesosome. (B) Partially migmatised enclave of folded metamorphic country rock enclosed in the Layos Granite. (C) Hand-sample aspect of the cordierite-rich Layos Granite; a K-feldspar lump is also shown. (D) Field aspect of the Layos Granite in which it is possible to observe some K-feldspar phenocrysts and also a vein-like associated peraluminous leucogranite.

leucosomes. Sometimes the stromatic structure of the rock is disrupted, leading to diatexitic rocks with similar appearance to the Layos Granite.

Marbles, calc-silicate rocks and quartzites are abundant in the eastern part of the Toledo Complex (Fig. 1). Elsewhere these rocks appear as layers boudinaged and occasionally disrupted in the migmatitic pelites.

Two groups of igneous rocks may be distinguished, namely calcalkine granites and associated mafic rocks, and peraluminous granites (Fig. 1). Both were emplaced relatively late with respect to the second deformational phase; they sometimes show intrusive features and locally cut the S2 foliation. In places it is possible to observe enclaves with early F3 folds enclosed in the granites (Fig. 2(B)). Nevertheless, all the granites show local development of foliation and phenocryst alignment with the S2 direction, and in some places these structures are folded by the third phase (F3). Thus the age of emplacement of these granites must be close to the second tectonic phase, probably between F2 and F3.

The calcalkaline granites and associated mafic rocks comprise three different suites which have been studied by Barbero *et al.* (1990). The first consists of olivine–pyroxene gabbros that outcrop as isolated sills, enclaves and small massifs within the calcalkaline granites. The second consists of amphibole gabbros that are always associated with the intermediate and felsic granites; locally it is possible to observe mingling processes between them. The third suite comprises intermediate–felsic granites, with variable porphyritic texture, in which are typically found microgranular tonalitic enclaves (Barbero *et al.* 1990).

The peraluminous granites form two groups based on the content of normative corundum. Moderately peraluminous granites (normative corundum $\leq 4\%$) consist of the Moncloa type, a K-feldspar megacryst-bearing porphyritic granodiorite; the Villanueva type, a garnet-bearing microporphyritic granite; and the Fuente-Topino type, a biotitic leucogranite with cordierite aggregates. Strongly peraluminous granites (normative corundum $>4\%$) are structurally and lithologically heterogeneous and rich in xenoliths and restitic components. To this group belongs the cordierite-bearing Layos Granite which is commonly associated with peraluminous leucogranites, resembling the vein-like types of migmatite leucosomes.

2. Metamorphic grade

The coexistence of the assemblage Grt + Crd + Sil + Bt + Sp + Kfs in pelitic rocks, together with the presence of orthopyroxene in metamorphic coronas around olivine in the gabbros (Barbero & Villaseca 1988) are indicative of granulite facies. The extensive migmatisation and anatexis observed in the pelitic rocks is also consistent with high-grade metamorphic conditions. The coexistence of garnet, biotite and cordierite indicates low-pressure conditions (Green & Ringwood 1967); this is consistent with the absence of garnet in the coronitic gabbros.

The results of several calculations using different mineral equilibria in the peraluminous granulites, biotitic gneisses and metabasites are summarised in Table 1. A maximum pressure limit has been estimated with the GRIPS barometer (Bohlen & Liotta 1986; Ghent & Stout 1984) of around 700 MPa for temperatures near 800°C. The garnet–biotite, cordierite–biotite and spinel–cordierite thermometers yield temperatures between 720 and 870°C. To avoid closure problems with the Fe–Mg interchange thermometers (Frost & Chacko 1989) we have estimated the cross-over of the GASP barometer with other *P–T*-dependent equilibria (e.g. spinel–cordierite). This cross-over indicates temperatures between 770 and 820°C, in agreement with the other temperature estimates, and pressures of 500–600 MPa, clearly under the GRIPS limit.

The formation of cordierite by garnet destabilisation (coronas of cordierite around garnet are frequent in these granulites), gives pressures 50–150 MPa lower than the estimated metamorphic peak conditions, suggesting an isothermal decompression process in the area. This is very frequent in migmatitic Variscan terranes (Ibarguchi & Martínez 1982; Jones & Brown 1990). Temperature estimates in the metamorphic assemblages of the metabasites, based on Opx–Cpx and Opx–Bt thermometers (Table 1), give values between 770–950°C, confirming the granulitic character of the area. In summary, the metamorphic peak conditions of the Toledo Complex are within the range of 800 ± 50°C and 500 ± 100 MPa, clearly in the granulite facies. Under these conditions, biotite breakdown and subsequent dehydration melting reactions can occur; thus the development of granulite facies paragenesis and partial melting are concomitant processes. This coincidence is not fortuitous as in regions which are inferred to have yielded significant quantities of granitic melts by fluid-absent partial fusion, temperature estimates are always restricted to this range of 800–850°C, indicating either a buffering effect and/or the inability of geothermometers to record higher temperatures (Vielzeuf *et al.* 1990).

3. Petrography and mineralogy of the Layos Granite

The Layos Granite has a distinctive dark colour (Fig. 2(C)) due to the high proportion of cordierite, reaching 30% in

Table 1 Peak metamorphic temperatures (°C) and pressure (MPa) estimates from various calibrations

		Temperature							Pressure		
		Grt–Bt		Grt–Crd		Sp–Crd	Opx–Cpx	Opx–Bt	GASP	Sp–Crd	Grt–Crd
Sample	Type	T1	T2	T3	T4						
87085	Granulite	826	841	827	797	868					
T-356	Granulite	807	821	793	762	751			540	610	400
90961	Granulite	774	785	829	799	839			560	590	420
T-361	Gneiss	743	757	722	690	768			620	560	540
89083	Gabbro						903	951			
89081	Gabbro						837				
87075	Gabbro							772			

Notes: All temperature and pressure estimates are based on averaged garnet core, and biotite or cordierite from matrix. Garnet–biotite thermometers are: T1 = Ferry & Spear (1978), T2 = Hodges & Spear (1982); garnet–cordierite thermometers are: T3 = Holdaway & Lee (1977), T4 = Lavrenteva & Perchuck (1981); spinel–cordierite thermometry after Vielzeuf (1983); orthopyroxene–clinopyroxene thermometry after Wells (1977); orthopyroxene–biotite thermometry after Sengupta *et al.* (1990); GASP barometer is calculated at 800°C following the method of Ganguly & Saxena (1984); spinel–cordierite barometry after Vielzeuf (1983); garnet–cordierite barometer of Holdaway & Lee (1977).

the most mafic rocks, with a mean value of 20%. Petrographically it varies from quartz-rich tonalites to melamonzogranites, with granodiorites being the most abundant type. On a QAP plot (Fig. 3) it defines a typical S-type distribution. In places it is possible to observe porphyritic varieties characterised by K-feldspar phenocrysts aligned in 120°–140° directions; locally a gradual transition between porphyritic and non-porphyritic types can be observed. The Layos Granite is little deformed and only a weak mineral elongation or compositional banding is locally observed.

This granite is closely related to peraluminous leucogranites that occur as veins or dyke-like bodies within it, and locally define a banded aspect to the granite (Fig. 2(D)). It is also possible to find synplutonic emplacement processes with other strongly peraluminous varieties, which are indicated by the occurrence of imbrication zones following F3 directions. In these zones the interdigitation processes can produce mingled facies between the Layos Granite and the coeval peraluminous granite.

A major characteristic of the Layos Granite is the abundance of enclaves, which comprise three groups. The first group consists of refractory metamorphic country rock enclaves corresponding to the types already described: orthogneisses, amphibolites, quartzitic conglomerates, calc-silicate rocks and marbles. These enclaves are the most abundant and can reach a maximum size of 2–3 m, although usually they are not greater than 20–25 cm in size (Fig. 2(B)).

The second group consists of restitic enclaves which are, in decreasing order of abundance:

(a) Quartz lumps ranging in size from 1–2 cm up to 50 cm. In some of the larger lumps there is a gneissic partial rim around the quartz, suggesting a derivation from vein quartz in the original metamorphic sequence.

(b) Biotite-bearing enclaves usually 1–2 cm in size and rounded in form. They commonly present biotite-bearing layers alternating with cordierite-sillimanite- and minor spinel-bearing layers, resembling the regional peraluminous granulites.

(c) Cordierite-rich enclaves (1–2 cm) generally formed with intergrowths of cordierite and quartz. Minor biotite and sillimanite are locally present.

(d) Sillimanite-rich enclaves of rounded shape and up to 25 cm in diameter. The aluminosilicate is sometimes included in cordierite or plagioclase crystals with minor biotite, K-feldspar, spinel and quartz.

(e) K-feldspar lumps showing corroded borders and perthitic textures that are locally associated with quartz, forming micro-pegmatitic textures. It is possible to find inclusions of quartz, plagioclase, and minor biotite and sillimanite within these crystals. Exceptionally, they can reach up to 25 cm in diameter, but generally they are 3–4 cm in size. They are occasionally surrounded by a biotite rim (Fig. 2(C)).

The third group consists of mafic igneous enclaves. These enclaves are very scarce and can be found in places where mafic rocks outcrop (Villanueva region, see Fig. 1) and it seems likely that they were incorporated at the emplacement level. Amphibole-biotite-bearing quartz diorites and gabbros are the main types. Scarce microgranular enclaves with quartz, plagioclase, biotite and accessory garnet are also present.

The Layos Granite is a medium-grained rock with a hypidiomorphic texture. Occasionally it shows a porphyritic texture defined by euhedral K-feldspar phenocrysts. The major minerals are quartz, plagioclase, K-feldspar, cordierite and biotite. Accessory minerals include apatite, zircon, monazite, tourmaline, sillimanite, garnet, ilmenite (Ilm_{98}), rutile and iron sulphides. Selected mineral compositions are given in Table 2.

Plagioclase composition ranges between An_{25}–An_{23}. In some samples there is a narrow rim with more albitic composition (An_{12}–An_5). Particularly remarkable is the absence of zoning (variations greater than 3% in An are not observed). Thus, neither calcic cores nor more complex and marked zoning are found, in contrast to the other peraluminous or calc-alkaline allochthonous granites (e.g. Rottura *et al.* 1990; Chappell *et al.* 1987). It is noteworthy that the plagioclase has the same composition over the range from tonalites to monzogranites and is also similar to that of the peraluminous granulites and leucogranites (Fig. 4(A)).

K-feldspar compositions vary between Or_{92} and Or_{71}. Occasionally the K-feldspar crystals display some perthitic veins. Sporadically, sillimanite needles, biotite, plagioclase and quartz are included in these crystals. Like the plagioclase, the K-feldspar of the Layos Granites and those of the peraluminous granulites and leucogranites have similar compositions. The relatively high albitic component of the K-feldspar resembles that obtained for minimum-temperature melts and partially fused pelitic xenoliths (Grapes 1985).

The most abundant ferromagnesian phase is cordierite. This mineral usually occurs as euhedral or subhedral grains, with frequent inclusions of acicular sillimanite, biotite and quartz, suggesting a restitic origin for this mineral (Chappell *et al.* 1987). The Fe/(Fe + Mg) ratios vary from 0·38–0·48; they are close to the ratios displayed by the cordierites of the peraluminous granulites (0·38–0·53) and leucogranites (0·44–0·62). The very low Na + K + Ca contents of these cordierites are characteristic of granulite facies cordierites (Vry *et al.* 1990). According to those authors, the low Na + K + Ca values imply relatively high X_{CO_2} in the cordierite (0·3–0·7). Figure 4 is a Mn vs Na + K + Ca plot on which it is clear that the cordierites of the Layos Granite are Mn-, Ca- and alkali-poor compared to those of the typical peraluminous Hercynian Iberian batholiths (Casillas 1989; Rottura *et al.* 1989; Andonaegui 1990).

Biotite occurs as subhedral crystals with a typical interstitial character (Andonaegui 1990). The relatively high Ti and low Al^{VI} contents are typical of granulite facies biotites (Dymek 1983). These high Ti and Al^{VI} values

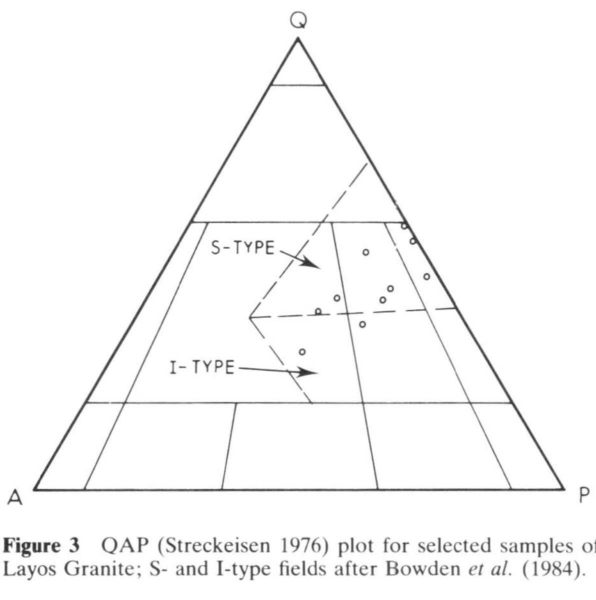

Figure 3 QAP (Streckeisen 1976) plot for selected samples of the Layos Granite; S- and I-type fields after Bowden *et al.* (1984).

Table 2 Representative microprobe analyses of mineral phases of the Layos Granite

	Pl	Kfs	Bt-1	Bt-2	Crd-1	Crd-2	Grt	Ilm
SiO_2	62·51	65·15	34·48	34·93	48·69	48·08	37·41	0·01
TiO_2	0·00	0·01	5·12	3·11	0·00	0·00	0·05	54·19
Al_2O_3	23·18	18·88	16·84	18·18	31·49	31·41	20·28	0·04
FeO*	0·10		19·58	19·82	9·13	10·32	32·37	45·78
Cr_2O_3			0·12	0·00	0·00	0·01	0·04	0·01
MnO				0·08	0·27	0·23	2·50	1·80
MgO			7·84	8·15	6·74	6·68	4·22	0·18
CaO	4·94	0·05	0·04	0·02	0·01	0·00	0·73	0·00
Na_2O	8·86	1·96	0·17	0·08	0·18	0·20		
K_2O	0·26	13·69	9·08	8·99	0·02	0·02		
Total	99·85	99·74	93·27	93·36	96·53	96·95	97·60	102·01
Oxygen	32	32	22	22	18	18	12	3
Si	11·11	11·95	5·39	5·44	5·12	5·07	6·11	0·00
Al(IV)	4·85	4·08	2·61	2·56			0·00	
Al(VI)			0·50	0·78	3·90	3·90	3·90	
Fe	0·01	0·00	2·56	2·58	0·80	0·91	4·42	0·96
Ti			0·40	0·36	0·00	0·00	0·01	1·01
Mn					0·02	0·02	0·35	0·04
Mg			1·83	1·89	1·06	1·05	1·03	0·01
Ca	0·94	0·01	0·01	0·00	0·00	0·00	0·13	
Na	3·05	0·70	0·05	0·02	0·04	0·04		
K	0·06	3·21	1·81	1·78	0·00	0·00		

Notes: Pl = plagioclase, Kfs = K-feldspar; Bt = biotite; Crd = cordierite; Grt = garnet; Ilm = ilmenite.

contrast with the lower values displayed by the biotites of other Variscan peraluminous granites (Fig. 4(C)) (cordierite-bearing granites; Casillas 1989; Andonaegui 1990; Rottura *et al.* 1989). The biotites of the Layos Granite have Fe/(Fe + Mg) ratios very similar to those of the peraluminous granulites, although there is a narrow variation range. This fact suggests a possible restitic origin for this mineral.

Finally, the accessory garnet (Alm_{75} Py_{17} Sp_6 Gr_2) is very low in the spessartine molecule compared to other peraluminous monzogranite–granodiorite plutons (Villaseca & Barbero in press). Similarly, ilmenites are also Mn-poor (≤2·3%) and contrast with the main granitic plutons of the area (Andonaegui 1990; Brandebourger 1984; Casillas 1989), or with the S-type granites (Whalen & Chappell 1988).

In summary, the feldspar and ferromagnesian phases are compositionally different from those of the main peraluminous Hercynian batholiths of the area and from other cordierite-bearing granites (Flood & Shaw 1975; Speer 1981), and define an exotic granitic mineralogical assemblage.

4. Geochemistry of the Layos Granite

Major elements were determined by ICP spectrometry. Trace elements were analysed by XRF spectrometry using the method described by Brändle and Cerqueira (1972). Rare-earth elements were analysed by CNRS (Nancy, France) by ICP spectrometry. Seventeen samples of the Layos Granite from quartz-rich tonalites to monzogranites, six peraluminous granulites and five leucocratic rocks have been analysed. A further 13 analyses of Layos Granite and associated leucogranites are available from Andonaegui (1990). From these 41 analyses we have made a selection of the most significant rock compositions as given in Table 3.

The Layos Granite ranges in SiO_2 content from 61–72%. Most of the samples are clearly richer in modal quartz than the normal granitic series (Fig. 3). The granite is strongly peraluminous with normative corundum values ranging from 4–10%, the mean value being 7%. Normative corundum varies inversely with the silica content. This strongly peraluminous character is also shown on the A–B diagram (Debon & Le Fort 1983), where the Layos Granite defines a strongly positive trend with high A values (Fig. 5); this feature is not very common in other granitic series, and only occurs in some S-type granites (White & Chappell 1987) and some restite-rich types (Clarke & Lyons 1986). On the A–B diagram the most mafic varieties of the Layos Granite plot close to the peraluminous granulites and in the field of pelitic sediment composition (Fig. 5). The felsic members of the Layos Granite trend towards the leucogranitic field. The strongly peraluminous character is in marked contrast with the less peraluminous Iberian Hercynian cordierite-bearing batholiths (normative corundum ≤3%). Furthermore, these allochthonous and post-tectonic plutons show an opposite trend of increase in the peraluminous index with silica content.

The major elements of the Layos Granite show two noteworthy characteristics: firstly, the high Fe, Mg and Ti contents (in some cases $FeO_t + MgO + TiO_2 > 10\%$) reflect the high modal abundance of cordierite and biotite; secondly, the very low variation in the CaO content which precludes igneous fractionation processes involving plagioclase. This limited CaO variation in the Layos Granite is in the same range as in the peraluminous granulites and leucogranites (Fig. 6), and is in accord with the similar plagioclase composition in the three lithological groups.

Similar limited variation, as silica increases, is shown by other major elements and especially the alkalis (Fig. 6). Nevertheless, on Harker Diagrams, the Layos Granite displays linear trends which suggest either mixing between contrasting end-members, or unmixing of some intermediate types into more mafic and felsic end-members. Trace element data display more complex trends (Fig. 6) with generally greater scatter. Nevertheless, some of them show more or less defined ranges of variation. For example, Ba, Rb, Sr and Cr tend to decrease with increasing silica content. The high Cr–Ni content of these granites is controlled by the different proportion of ferromagnesian phases rich in these elements (i.e. biotite, cordierite).

REE patterns of the Layos Granite are characterised by a negative Eu anomaly (Fig. 7); the different samples have similar LREE and more variable HREE patterns ($Gd_N/Lu_N = 0{\cdot}8$–36) than those of the peraluminous granulites. The associated leucogranites have either as much REE as the metasediments or less total REE contents (particularly HREE), with different patterns. The sample with the lowest total REE and the most marked positive Eu anomaly is characteristic of some minimum-temperature granite compositions typical of disequilibrium melting processes in relatively anhydrous environments (Moller & Muecke 1984; Barbey *et al.* 1989) or possibly is the consequence of some crystal accumulation of feldspars (Sawyer 1987).

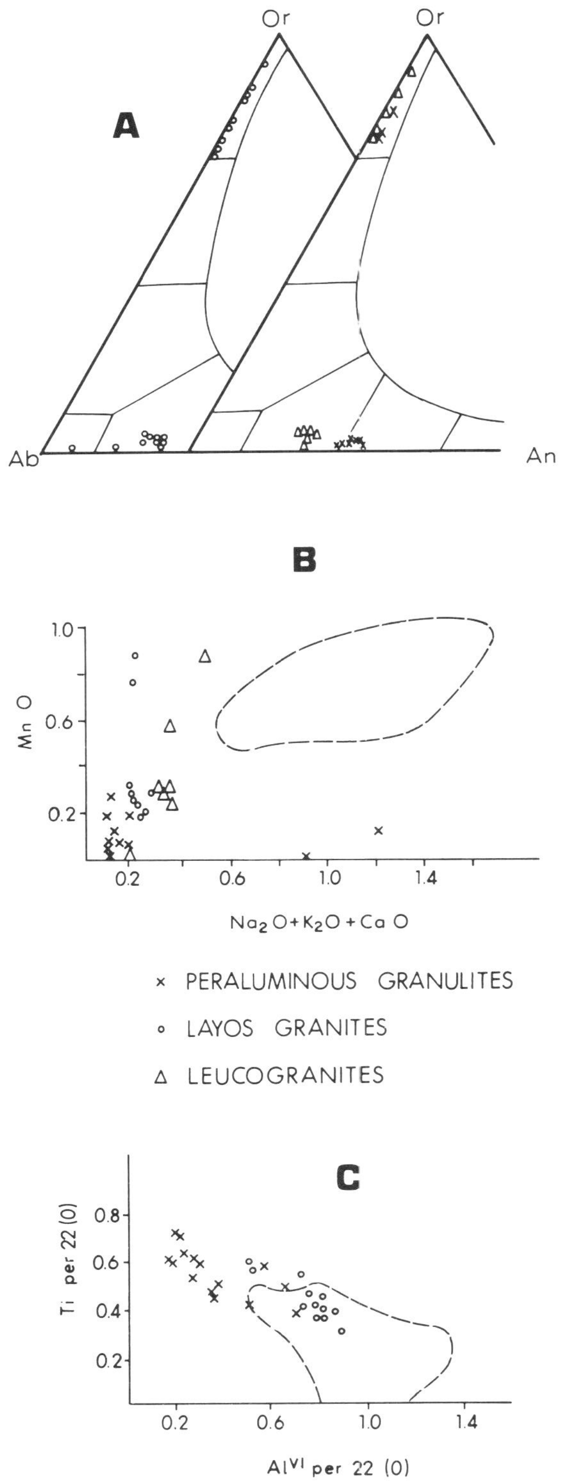

Figure 4 (A) Feldspar compositions on the Or–Ab–An molecular diagram. (B) Cordierite composition on a MnO vs $Na_2O + K_2O + CaO$ plot; dashed field corresponds to cordierites of typical peraluminous Hercynian batholiths. (C) Biotite composition in terms of Ti and Al^{VI} contents; dashed field corresponds to biotites of typical peraluminous Hercynian batholiths (see text for explanation).

5. Granite petrogenesis

5.1. Main evolutionary mechanism

The Layos Granite is a catazonal pluton and a regional-migmatite terrane granite. Its anatectic origin is confirmed by: its parautochthonous character indicated by the absence of a contact aureole in the adjacent metamorphic rocks; the thermodynamical equilibrium between its mineral assemblages and those of the country rock granulites; and the close relationships between the age of emplacement and the migmatisation event.

On many Harker Diagrams the variation trend of the Layos Granite is clearly linear and lies between the compositional fields of granulites and leucogranites (Figs 5, 6). A crystal fractionation process seems to be unlikely

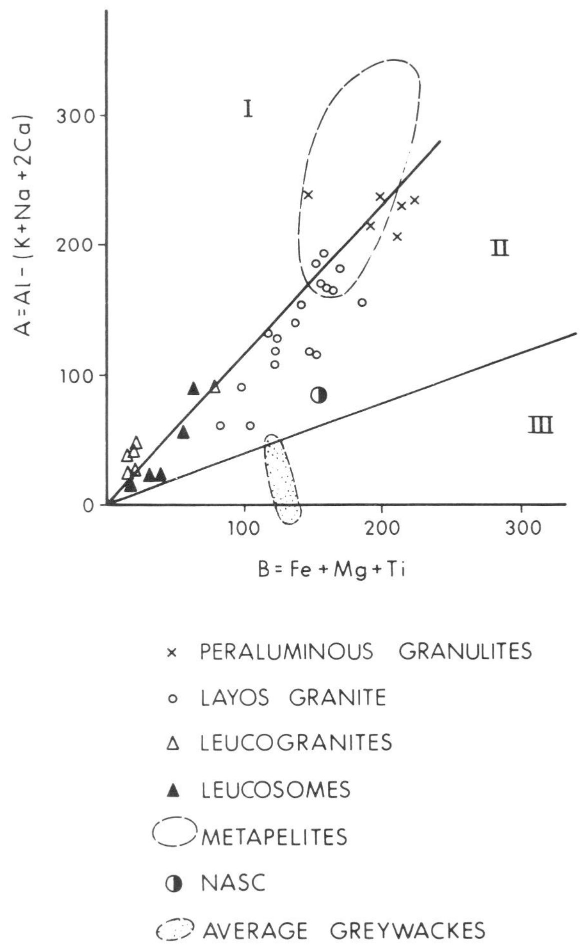

Figure 5 Plot of the peraluminous granulites, Layos Granite and leucogranites in the A–B diagram after Debon & Le Fort (1983); composition fields of metapelites of the Spanish Central System (Villaseca 1983), average greywackes (McRae & Nesbitt 1980) and North American Shale Composite (NASC of Gromet *et al.* 1984) are also reported (see text for explanation); Field I = muscovite > biotite granites; Field II = muscovite ± biotite granites; Field III = muscovite < biotite granites.

Table 3 Major, trace element and corundum normative contents of representative pelitic granulites (1 to 3), Layos Granites (4 to 11) and associated leucogranites (12 to 15); sample 15 is from Andonaegui (1990); sample 12 corresponds to a leucosome band

	1	2	3	4	5	6	7	8	9	10	11	12	13	14	15
Sample	90961	T-383	T-388	89103	90537	91085	90538	90966	90964	90968	89861	T-382	90959	T-400	81925
SiO_2	56·96	57·15	58·19	61·37	63·20	64·61	66·07	68·78	69·82	71·14	72·52	69·50	72·50	74·18	74·36
TiO_2	1·12	1·13	1·12	0·61	0·70	0·82	0·81	0·96	0·53	0·64	0·66	0·52	0·26	0·13	0·06
Al_2O_3	20·10	20·16	19·43	18·45	17·88	15·52	17·03	13·63	14·49	13·44	12·30	14·96	14·81	13·78	14·40
Fe_2O_3	3·11	*9·39	*8·89	1·01	0·19	1·74	0·64	*6·20	1·26	*5·40	0·00	*2·56	0·15	*0·81	0·37
FeO	6·55			5·49	5·79	3·59	5·26		2·47		5·12		0·88		0·75
MgO	3·66	3·39	3·40	2·67	2·38	1·90	2·17	1·70	1·28	1·32	0·29	0·25	0·40		
MnO	0·08	0·08	0·08	0·25	0·07	0·08	0·07	0·06	0·06	0·01	0·00	0·01	0·05		
CaO	1·45	1·54	1·52	0·81	1·22	1·25	1·09	0·76	0·99	1·02	1·01	1·25	1·13		
Na_2O	1·75	1·79	1·82	2·01	2·01	2·61	2·32	2·27	2·71	2·66	3·65	3·25	3·21		
K_2O	2·63	3·90	3·65	4·03	4·27	3·29	3·81	3·65	4·72	2·77	2·77	7·00	5·22	5·34	4·21
P_2O_5	0·20	0·13	0·14	0·23	0·52	0·35	0·23	0·16	0·33	0·19	0·36	0·34	0·28	0·16	0·22
H_2O	1·31	1·06	1·42	2·20	1·42	2·35	1·27	1·51	1·69	1·46	1·10	1·16	0·57	0·64	0·82
Total	98·92	99·72	99·66	99·13	99·65	98·11	100·77	99·68	100·35	99·71	99·75	99·64	99·75	99·79	99·98
Ba	547	762	898	604	839	697	596	663	699	570	278	1174	848	573	614
Cr	541	145	137	364	418	267	427	268	193	390	281	25	20	5	138
Ni	80	57	54	65	45	43	45	38	23	50	37	10	5	5	41
Rb	130	172	136	161	149	127	155	165	169	115	161	263	135	209	79
Sr	187	243	256	173	202	193	192	130	235	151	114	261	197	155	208
Th	16	19	23	16	12	21	7	15	13	9	8	22	12	12	1
Y	35	27	25	23	38	48	25	23	28	18	10	8	9	5	11
Zr	173	218	219	123	195	296	190	309	207	213	198	166	133	71	87
La	61·6	69·4	67·4	33·5			37·6	42·9	32·2	34·0	29·9	50·0	30·8	25·2	
Ce	112·6	135·7	131·8	64·5			71·4	94·2	62·9	74·7	59·8	114·6	60·9	56·0	
Nd	50·9	54·8	54·2	29·1			33·9	42·3	30·0	33·3	28·2	51·3	30·1	20·3	
Sm	10·3	11·0	10·98	6·5			7·5	9·4	7·1	7·6	6·2	11·9	6·3	4·0	
Eu	1·76	1·93	2·09	1·25			1·36	1·40	1·37	1·48	0·90	2·06	1·62	1·01	
Gd	8·23	9·03	8·24	5·43			6·40	7·81	6·64	6·77	4·99	9·65	4·68	3·32	
Dy	6·95	7·07	6·37	4·54			5·55	5·82	5·11	4·96	2·94	3·95	3·95	1·44	
Er	3·34	3·70	3·37	3·05			3·13	3·55	2·76	2·47	1·26	1·14	1·14	0·55	
Yb	3·14	3·30	3·03	4·25			3·22	3·50	3·29	1·99	0·97	0·51	0·51	0·19	
Lu	0·52	0·59	0·58	0·82			0·54	0·63	0·58	0·36	0·17	0·13	0·13	0·07	
C	11·97	11·37	12·41	9·86	8·98	6·23	7·66	5·94	3·91	6·43	3·93	4·28	1·99	0·93	3·03

Notes: * = total Fe as Fe_2O_3.

because of the absence of CaO variation and because an igneous granitic series could not evolve without plagioclase fractionation. However, in the most evolved rocks some porphyritic varieties locally display a gradual transition towards the much more abundant non-porphyritic types, possibly due to limited crystal fractionation. Other processes such as varying degrees of partial melting seem also to be unlikely. In such a fusion process a strong enrichment in incompatible elements and little variation in the compatible ones is to be expected (Minster & Allègre 1977). On incompatible versus compatible element diagrams, the Layos Granite shows little variation for both types of elements, suggesting that a fusion mechanism cannot explain their chemical variability.

Some trace element concentrations in the Layos Granite are quite variable. Zr, P and Th contents show the same variations as in the pelitic granulites, and are always higher than predicted from their solubility in granitic melts at temperatures in the range of this anatectic terrane (Watson 1987). This oversaturation in some HFS elements is better explained by inheritance of zircon, monazite and apatite in the major refractory phases and in the incongruent solids produced in the dehydration melting (biotite, cordierite, garnet, . . .).

The REE contents of the Layos Granite show an inverse variation with silica increase, that is, the more restite-rich a granite, the higher the REE content. This fact again suggests that the accessory phases, which tend to be REE-rich, are mainly of restitic origin or are included in the solid melt residuum (Barbey *et al.* 1990). In general, the Layos Granite has total REE contents intermediate between those of the metapelites and leucogranites. The increase in the La/Lu ratio of the leucogranites with respect to that of the peraluminous granulites and metapelites can be explained by the presence of residual garnet in these rocks, although some of that mineral was consumed in the cordierite-producing reactions.

In view of these observations, we strongly prefer a restite-unmixing process for the origin and diversification of the Layos Granite. In this restite-unmixing model the restitic pole will be represented by the peraluminous granulites, while the minimum-temperature melts correspond to the associated leucogranites.

5.2. Nature of the source

The most likely composition for the source rock would be that of a pelitic sediment. The normative corundum content (>4%), increasing in the less felsic rocks, the low Na2O (≤3%) and CaO contents (≤1·5%), and the LILE content of the Layos Granite (Rb ≈ 100–170 ppm, Sr ≈ 150–250 ppm and Ba ≈ 300–850 ppm), indicate a pelitic parent (Miller 1985). In several geochemical plots, the peraluminous granulites fall in the compositional fields of pelites. In Figure 5 we have plotted the variation field of pelitic metasediments of the Toledo Complex and the Spanish Central System, together with N American Shale Composite (NASC) (Gromet *et al.* 1984), and the mean values of some greywacke sediments (McRae & Nesbitt 1980). It is clear that the peraluminous granulites plot in the pelitic field and far from the other sedimentary rock compositions. Furthermore, on chondrite-normalised spidergrams the Layos Granite and the granulites show similar patterns, with marked negative anomalies in Nb, Sr, P and Ti similar to other S-type granites and pelitic sediments (Thompson *et al.* 1984).

The other pole of the unmixing line corresponds to the leucogranites. We have sampled large leucogranitic massifs and also some leucosomes and little bands. There are some geochemical differences between the leucogranites of the large massifs and those of leucosomes, especially in Na_2O, K_2O, Ba, Rb and Sr (Fig. 6). These differences are also evident in the Qtz–Ab–Or diagram in which the massif-type leucogranites plot close to the theoretical minimum-temperature melt composition in anhydrous conditions, being richer in the Or component than H_2O-saturated melts (Ebadi & Johannes 1991). Some leucosomes are shifted from those thermal minimum compositions. Furthermore, the composition of the leucogranites is similar in major elements to that obtained by experimental melting of metapelitic protoliths (Wickham 1987a; Vielzeuf & Holloway 1988; Patiño & Johnston 1991).

The limited isotopic data on the Layos granodiorite and associated leucogranites (Andonaegui 1990, samples 81942, 81925 and 81926 of table III-6), clearly suggest a crustal source ($^{87}Sr/^{86}Sr \geq 0{\cdot}713$, Fig. 8). The anatectic origin is more evident in the Layos Granite than in other peraluminous Hercynian granites (Del Moro 1987; Andonaegui 1990; Barbero *et al.* 1990; Rottura *et al.* 1990).

5.3. Constraints on melt extraction and emplacement

The available petrographical, mineralogical and geochemical data strongly suggest a restite-unmixing model for the formation of the Layos Granite. An estimate based on the linear patterns on variation diagrams indicates a maximum restitic component of about 65% for the most mafic rocks (SiO_2 61–63%), and about 25–30% for rocks with 67–68% SiO_2. This high proportion of restite in the less evolved types is in good agreement with the observed modal proportion of restitic minerals. Thus, the most mafic rocks have modal cordierite contents of around 30% and about 70% of the quartz may be restitic, although not easily identifiable as such (Chappell *et al.* 1987).

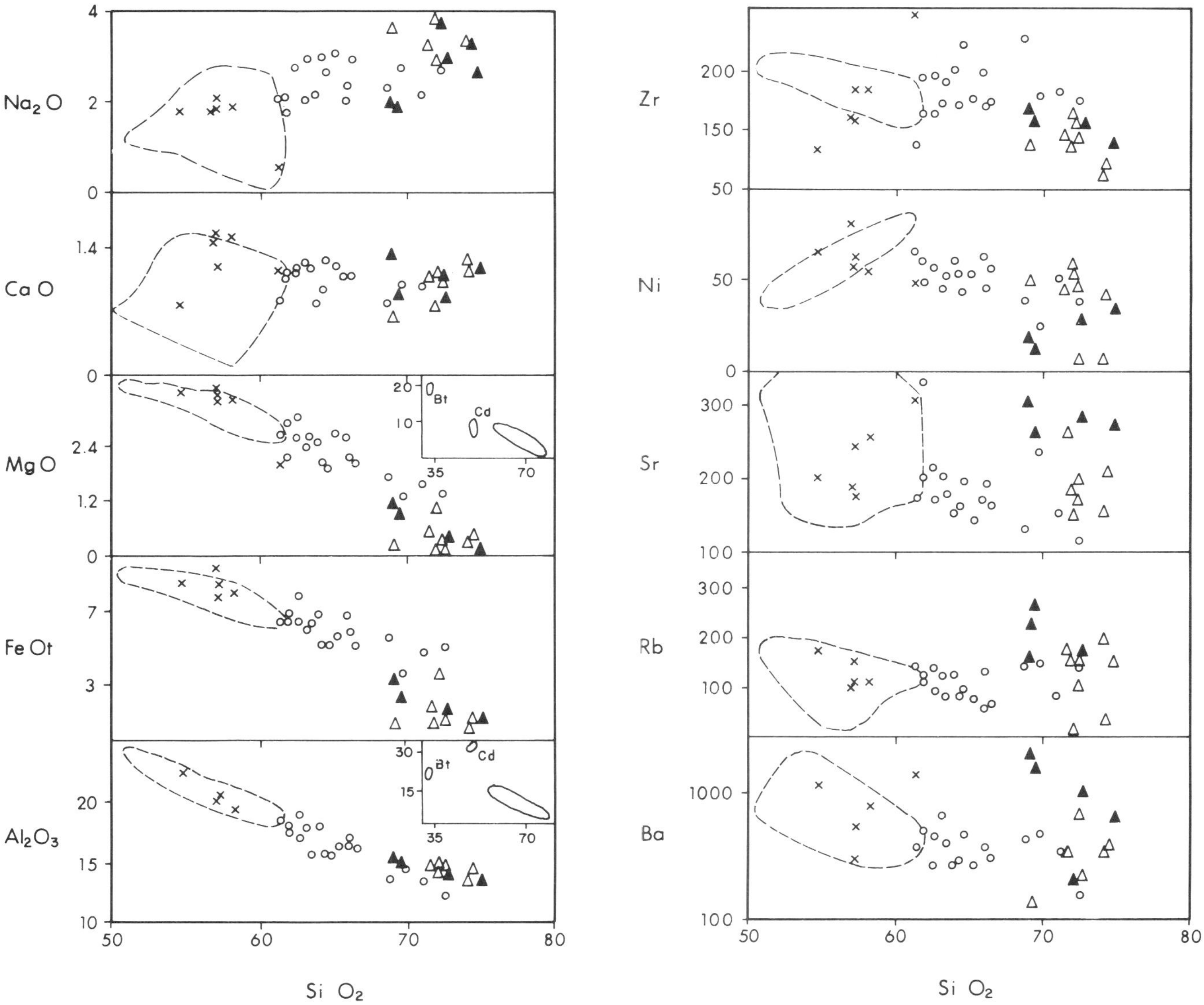

Figure 6 Plots of granulites, Layos Granite and leucogranites in Harker Diagrams; in all diagrams dashed fields represent the compositional variation of the Sierra de Guadarrama metapelites; insets in SiO_2 vs Al_2O_3 and SiO_2 vs MgO diagrams represent the composition of the biotites and cordierites and the variation trend of the Layos Granite (see text for explanation); symbols as in Figure 5.

All these data suggest a melt proportion of between 30% and 60%, always higher than the rheological critical melt fraction. In high strain state conditions, as could be envisaged in the tectonometamorphic climax of the Toledo Complex, segregation would be possible even with as little as 25% high-viscosity H_2O-undersaturated felsic melt (Arzi 1978; Van der Molen & Paterson 1979; Wickham 1987b; Miller *et al.* 1988).

Assuming that the metapelites contain about 25% biotite with 4% H_2O in the biotite and $X_w = 0{\cdot}41$ in the melt, at 800°C and 500 MPa (see Le Breton & Thompson 1988), the amount of melt produced could be around 35% in volume. This value is above the critical melt fraction, and would allow the segregation and accumulation of granitic magma. However, it seems likely that the Layos Granite was never fully melted and that restitic materials (biotite, cordierite, quartz, etc.) were redistributed by mechanical disaggregation due to convective overturn of large (cubic kilometre-sized) partially melted systems (Wickham 1987b). As a crystal-liquid mush, this anhydrous high-viscosity magma had a restricted movement capacity, giving rise to the aforementioned parautochthonous character (Fig. 9).

Other peraluminous magmas generated in this anatectic event contain a lower melt fraction, as represented by near minimum-temperature melt composition (Wickham 1987b). They could be synplutonically segregated as veins or dyke-like bodies within the Layos Granite, and generate a local complex-banded structure (Figs 2, 9).

6. Conclusions

The Toledo Complex consists of high-grade metamorphic rocks which were syntectonically intruded by a mafic–felsic calcalkaline association and by heterogeneous peraluminous granites to which the Layos Granite belongs.

The strongly peraluminous Layos Granite is characterised by a high modal proportion of cordierite and other restitic minerals. Its synchronous emplacement during the regional

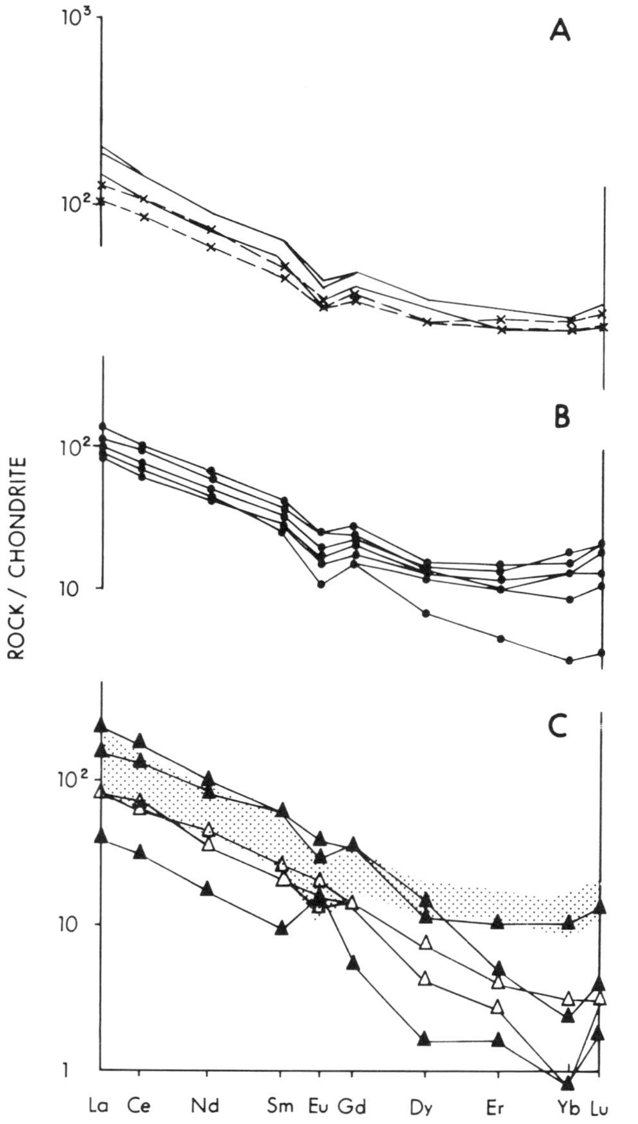

Figure 7 Chondrite-normalised (Masuda *et al.* 1973) REE patterns: (A) Peraluminous granulites (solid line) and metapelites (dashed line); (B) Layos Granite; (C) Leucogranites and leucosomes (solid triangles). The pelitic–granitic variation field (shaded area) is also shown.

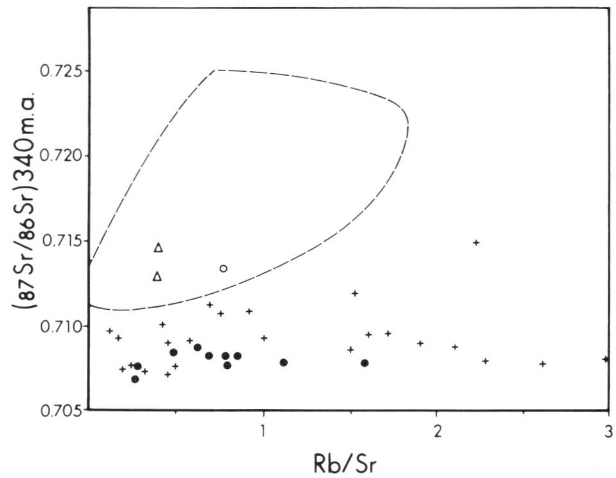

Figure 8 Rb/Sr vs $(^{87}Sr/^{86}Sr)_{340\,M.a.}$ plot showing the leucogranites ($n = 2$) and Layos Granite ($n = 1$); dashed field represents the variation of metapelites from several zones of the Hercynian belt (Ivrea, Strona and Calabria, Del Moro 1987); symbols: circle = Layos granite; triangle = leucogranites; dots = Toledo Complex calcalkaline granites (Barbero *et al.* 1990); crosses = cordierite-bearing Hercynian batholiths (Rottura *et al.* 1990; Andonaegui 1990); see text for explanation.

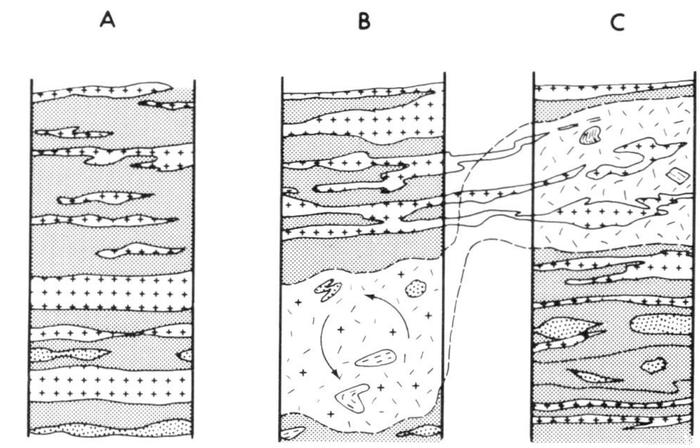

Figure 9 Schematic petrogenetic model for the generation and emplacement of the Layos Granite and associated leucogranites. (A) Development of wide migmatite zones in which pods of leucogranitic melts segregate. (B) The rheological critical melt fraction is exceeded and convective homogenisation occurs. (C) The resultant large volume of crystal-liquid mush segregates and with limited ascent within the crust develops syntectonic intrusions accompanied by segregation of leucogranitic bodies.

metamorphic climax and its restite-rich character lead to unusual mineralogical features: plagioclase is almost unzoned and does not vary in composition from the mafic to the more felsic members of the series; K-feldspar has relatively high albitic content; cordierite has low alkali and Mn content and usually contains sillimanite inclusions; biotite has high Ti and low Al^{VI} contents.

Geochemically the most remarkable characteristics are the high normative corundum values (as much as 10%) increasing in the more mafic types and the very limited variation in CaO content in the whole series.

In the light of these features, the chemical variations observed in the Layos Granite can be best explained by a restite-unmixing process between an acid pole represented by the associated minimum-temperature melt leucogranites and a restite pole constituted by the peraluminous granulites that are spatially, chemically and mineralogically associated with the Layos Granite. The source material of the Layos Granite has the chemical and isotopic characteristics of a pelitic metasediment. The high proportion of restitic material in the Layos Granite confers a relatively restricted movement capacity, resulting in a parautochthonous emplacement of this crystal-liquid mush.

The Layos Granite has contrasted petrographical and geochemical features compared to the main cordierite-bearing batholiths of the Iberian Variscan Belt. These monzogranitic batholiths usually define Gueret-type trends on an A–B plot, being more peraluminous in the more felsic varieties. We consider that this voluminous plutonism, late to post-tectonically emplaced in upper levels, has experienced considerable mobility within the crust and has to be derived from different source materials, probably more granodioritic in composition.

The scarcity of restite-rich granites such as the Layos Granite implies that dry melting of pelitic protoliths is not the usual mechanism of melt generation in this part of the Iberian Hercynian Belt.

7. Acknowledgements

We are especially grateful to Norman Snelling for his constructive critical review and for the revision of the English version of the text. We wish also to acknowledge the contributions of Alexandro Rottura, Bruce Chappell and two anonymous reviewers which greatly improved this manuscript. Microprobe analyses were performed at Oviedo University (Spain) kindly provided by L. G. Corretge. We thank Olga Fernández, Mayte Larrea and Maribel Sevillano for their collaboration in sample preparation and in ICP spectrometry. This work is financially supported by a CICYT (PB88-0117) project of the Ministerio de Educación y Ciencia and by a FPI-MEC grant to one of the authors (LB).

8. References

Andonaegui, P. 1990. *Geoquímica y geocronología de los granitoides del sur de Toledo.* Tesis doctoral, Universidad Complutense de Madrid, Spain.

Arzi, A. A. 1978. Critical phenomena in the rheology of partially melted rocks. TECTONOPHYSICS **44,** 173–89.

Barbero, L. & Villaseca, C. 1988. Gabros coroníticos en el macizo cristalino de Toledo. GEOGACETA **5,** 67–9.

Barbero, L., Villaseca, C. & Andonaegui, P. 1990. On the origin of the gabbro–tonalite–monzogranite association from Toledo area (Hercynian Iberian belt). SCHWEIZ MINERAL PETROGR MITT **7,** 209–21.

Barbey, P., Bertrand, J. M., Angoua, S. & Dautei, D. 1989. Petrology and U/Pb chronology of the Telohat migmatites (Aleksod, Central Hoggar). CONTRIB MINERAL PETROL **101,** 207–19.

Barbey, P., Macaudiere, J. & Uzenti, J. P. 1990. High-pressure dehydration melting of metapelites: evidence from migmatites of Yaoundí (Cameroon). J PETROL **31,** 401–28.

Bohlen, J. R. & Liotta, J. J. (1986). A barometer for garnet amphibolites and garnet granulites. J PETROL **27,** 1024–34.

Bowden, P., Batchelor, R. A., Chappell, B. W., Didier, J. & Lameyre, J. 1984. Petrological, geochemical and source criteria for the classification of granitic rocks: a discussion. PHYS EARTH PLANET INTERIORS **35,** 1–11.

Brandebourger, E. 1984. *Les granitoides hercyniens tardifs de la Sierra de Guadarrama (Système Central, Espagne).* Pétrographie et geochimie, These 3eme Cycle, CNRS, France.

Brändle, J. L. & Cerqueira, I. 1972. Determinación de elementos menores en rocas silicatadas por fluorescencia de rayos X. ESTUD GEOL **28,** 445–51.

Casillas, R. 1989. *Las asociaciones plutónicas tardihercínicas del sector occidental de la Sierra de Guadarrama—Sistema Central Español, (Las Navas del Marqués–San Martín de Valdeiglesias). Petrología, geoquímica, génesis y evolución.* Tesis Doctoral, Universidad Complutense de Madrid, Spain.

Casquet, C., Fuster, J. M., González Casado, J. M., Peinado, M. & Villaseca, C. 1988. Extensional tectonics and granite emplacement in the Spanish Central System. A discussion. EUROPEAN SCI FOUND (SPEC VOL), PROC 5th WORK, 65–76.

Chappell, B. W., White, A. J. R. & Wyborn, D. 1987. The importance of residual source material (restite) in granite petrogenesis. J PETROL **28,** 1111–38.

Clarke, R. G. & Lyons, J. B. 1986. Petrogenesis of the Kingsman intrusive suite: peraluminous granitoids of Western New Hampshire. J PETROL **27,** 1365–93.

Debon, F. & Le Fort, P. 1983. A chemical-mineralogical classification of common plutonic rock and associations. TRANS R SOC EDINBURGH: EARTH SCI **73,** 135–49.

Del Moro, A. 1987. Sistematica Rb/Sr di alcune magmatiti tardo-erciniche dell'area italiana. RIC SCI EDUC PERM **52,** 107–32.

Dymek, F. R. 1983. Titanium, aluminium and interlayered cation substitution in biotite from high-grade gneisses, west Greenland. AM MINERAL **68,** 880–99.

Ebadi, A. & Johannes, W. 1991. Beginning of melting and composition of first melts in the system Qz–Ab–Or–H_2O–CO_2. CONTRIB MINERAL PETROL **106,** 286–95.

Ferry, J. M. & Spear, F. S. 1978. Experimental calibration of the partitioning of Fe and Mg between garnet and biotite. CONTRIB MINERAL PETROL **66,** 113–7.

Flood, R. H. & Shaw, S. E. 1975. A cordierite-bearing granite suite from the New England batholith. N.S.W. Australia. CONTRIB MINERAL PETROL **52,** 157–64.

Frost, B. R. & Chacko, T. 1989. The granulite uncertainty principle: limitations on thermobarometry in granulites. J GEOL **97,** 435–50.

Ganguly, J. & Saxena, S. K. 1984. Mixing properties of aluminosilicate garnets: constrains for natural and experimental data and application to geothermo-barometry. AM MINERAL **69,** 88–97.

Ghent, E. D. & Stout, M. Z. 1984. TiO_2 activity in metamorphosed pelitic and basic rocks: principles and applications to metamorphism in southeastern Canadian Cordillera. CONTRIB MINERAL PETROL **86,** 248–55.

Grapes, R. H. 1985. Melting and thermal reconstruction of pelitic xenolith, Wehr volcano, east Eifel, west Germany. J PETROL **27,** 343–96.

Green, D. & Ringwood, A. 1967. An experimental investigation of the gabbro to eclogite transformation and its petrological applications. GEOCHIM COSMOCHIM ACTA **31,** 767–833.

Gromet, L. P., Dymek, R. F., Haskin, L. A. & Korotev, R. L. 1984. The "North American Shale composite": its compilation, major and trace element characteristics. GEOCHIM COSMOCHIM ACTA **48,** 2469–82.

Hodges, K. V. & Spear, F. S. 1982. Geothermometry, geobarometry and the Al_2O_3 triple point at Mt Moosilauke, New Hampshire. AM MINERAL **67,** 1118–34.

Holdaway, M. J. & Lee, S. N. 1977. Fe–Mg cordierite stability in high-grade pelitic rocks based on experimental, theoretical and natural observations. CONTRIB MINERAL PETROL **63,** 175–98.

Ibarguchi, J. I. G. & Martínez, F. J. 1982. Petrology of garnet–cordierite–sillimanite gneisses from the El Tormes thermal Dome, Iberian Hercynian foldbelt (W Spain). CONTRIB MINERAL PETROL **80,** 14–24.

Jones, K. A. & Brown, M. 1990. High-temperature 'clockwise' P–T paths and melting in the development of regional migmatites: an example from southern Brittany, France. J MET GEOL **8,** 551–78.

Julivert, M., Fontboté, J. M., Ribeiro, A. & Conde, L. E. 1974. *Memoria explicativa del mapa tectónico de la Península Ibérica y Baleares. Escala 1:100,000.* I.G.M.E., Madrid, Spain.

Lavrenteva, E. V. & Perchuck, L. L. 1981. Cordierite–garnet thermometer. A collection of theses. ACAD SCI USSR **259,** 607–700.

Le Breton, N. & Thompson, A. B. 1988. Fluid-absent (dehydration) melting of biotite in metapelites in the early stages of crustal anatexis. CONTRIB MINERAL PETROL **99,** 226–37.

Martín Escorza, C. & López Martínez, J. 1978. Análisis mesoestructural en la Unidad Migmatítica de Toledo. ESTUD GEOL **34,** 34–43.

Masuda, A., Nakamura, N. & Tanaka, T. 1973. Fine structures of mutual normalized rare earth patterns of chondrites. GEOCHIM COSMOCHIM ACTA **37,** 239–48.

McRae, N. D. & Nesbitt, H. W. 1980. Partial melting of common metasedimentary rocks: a mass balance approach. CONTRIB MINERAL PETROL **75,** 21–6.

Miller, C. F. 1985. Are strongly peraluminous magmas derived from pelitic sedimentary sources? J GEOL **93,** 673–89.

Miller, C. F., Watson, E. B. & Harrison, T. M. 1988. Perspectives on the source, segregation and transport of granitoid magmas. TRANS R SOC EDINBURGH EARTH SCI **79,** 135–56.

Minster, F. J. & Allègre, C. J. 1977. Systematic use of trace elements in igneous processes. Part I: Fractional crystallization processes in volcanic suites. CONTRIB MINERAL PETROL **60,** 57–75.

Moller, P. & Muecke, G. K. 1984. Significance of europium anomalies in silicate melts and crystal melt equilibria: a re-evaluation. CONTRIB MINERAL PETROL **87,** 242–50.

Patiño, A. & Johnston, A. 1991. Phase equilibria and melt productivity in the pelitic system: implications for the origin of peraluminous granitoids and aluminous granulites. CONTRIB MINERAL PETROL **107,** 202–18.

Rottura, A., Bargossi, G. M., Caironi, V., D'Amico, C. & Maccarrone, E. 1989. Petrology and geochemistry of late-Hercynian granites from the Western Central System of the Iberian Massif. EUR J MINER **1,** 667–83.

Rottura, A., Bargossi, G. M., Caironi, V., Del Moro, A., Maccarrone, E., Macera, P., Paglionico, A. Petrini, R. Piccarreta, G. & Poli, G. 1990. Petrogenesis of contrasting Hercynian granitoids from the Calabrian Arc, southern Italy. LITHOS **24,** 97–119.

Sawyer, E. W. 1987. The role of partial melting and fractional crystallisation in determining discordant migmatite leucosome compositions. J PETROL **28,** 445–73.

Sengupta, P., Dasgupta, S. K., Bhattacharya, P. K. & Mukherjee, M. 1990. An orthopyroxene–biotite geothermometer and its application in crustal granulites and mantle derived rocks. J MET GEOL **8,** 191–8.

Speer, J. A. 1981. Petrology of cordierite- and almandine-bearing granitoid plutons of the southern Appalachian piedmont, U.S.A. CAN MINERAL **19,** 35–46.

Streckeisen, A. 1976. To each plutonic rock its proper name. EARTH SCI REV **12,** 1–13.

Thompson, R. N., Morrison, M. A., Hendry, G. L. & Parry, S. J. 1984. An assessment on the relative roles of crust and mantle magma genesis: an elemental approach. PHILOS TRANS R SOC LONDON **A310,** 549–90.

Van der Molen, I. & Paterson, M. S. 1979. Experimental deformation of partially-melted granite. CONTRIB MINERAL PETROL **70,** 299–318.

Vialette, Y., Casquet, C., Fúster, J. M., Ibarrola, E., Navidad, M., Peinado, M. & Villaseca, C. 1987. Geochronological study of orthogneisses from the Sierra de Guadarrama (Spanish Central System). N JARB MINER MH JG **H10,** 465–79.

Vielzeuf, D. 1983. The spinel and quartz associations in high grade xenoliths from Tallante (SE Spain) and their potential use in geothermometry and barometry. CONTRIB MINERAL PETROL **82,** 301–11.

Vielzeuf, D., Clemens, J. D., Pin, C. & Moinet, E. 1990. Granites, granulites and crustal evolution. *In* Vielzeuf, D. & Vidal, Ph. (eds) *Granulites and crustal evolution.* NATO ASI SCI SER C **311,** 59–85.

Vielzeuf, D. & Holloway, J. R. 1988. Experimental determination of the fluid-absent melting relations in the pelitic system. Consequences for crustal differentiation. CONTRIB MINERAL PETROL **98,** 257–76.

Villaseca, C. 1983. *Evolución metamórfica del sector centro-septentrional de la Sierra de Guadarrama.* Tesis Doctoral 216/84, Universidad Complutense de Madrid, Spain.

Villaseca, C. & Barbero, L. (in press). Los granates de rocas metapelíticas de la región central española: implicaciones en el origen de granates en granitoides. GEOGACETA (in press).

Vry, J. K., Brown, E. B. & Valley, J. W. 1990. Cordierite volatile content and the role of CO_2 in high-grade metamorphism. AM MINERAL **75,** 71–88.

Watson, E. B. 1987. The role of accessory minerals in granitoid geochemistry. 1ST HUTTON SYMP ABSTR, 19–21.

Wells, P. R. A. 1977. Pyroxene thermometry in simple and complex systems. CONTRIB MINERAL PETROL **62,** 129–39.

Whalen, J. B. & Chappell, B. W. 1988. Opaque mineralogy and mafic mineral chemistry of I- and S-type granites of the Lachlan fold belt, southern Australia. AM MINERAL **73,** 281–96.

White, A. J. R. & Chappell, B. W. 1987. Some supracrustal (S-type) granites of the Lachlan Fold Belt. TRANS R SOC EDINBURGH EARTH SCI **79,** 169–81.

Wickham, S. M. 1987a. Crustal anatexis and granite petrogenesis during low pressure regional metamorphism in the Trois Seigneurs massif, Pyrenees, France. J PETROL **28,** 127–69.

Wickham, S. M. 1987b. The segregation and emplacement of granitic magmas. J GEOL SOC LONDON **144,** 281–97.

Wildberg, H. G. H., Bischoff, L. & Baumann, A. 1989. U–Pb ages of zircons from meta-igneous and metasedimentary rocks of the Sierra de Guadarrama: implications for the Central Iberian crustal evolution. CONTRIB MINERAL PETROL **103,** 253–62.

L. BARBERO and C. VILLASECA, Dpto. Petrología y Geoquímica, Fac. C.C. Geológicas, Universidad Complutense, 28040 Madrid, Spain.

Transactions of the Royal Society of Edinburgh: Earth Sciences, **83**, 139–144, 1992

Restite-melt and mafic–felsic magma mixing and mingling in an S-type dacite, Cerro del Hoyazo, southeastern Spain

H. P. Zeck

ABSTRACT: Approximately 10–15 vol% of the Neogene Hoyazo dacite consists of Al-rich restite rock inclusions ($Al_2O_3 = 20–45\%$) and monocrystal inclusions derived therefrom. Restite material and dacitic melt were formed syngenetically from a (semi-)pelitic rock sequence by means of anatexis. Restite rock fragments and dacite show similar high $\delta^{18}O$ values (13–16‰) corresponding to those found for sedimentary material. Striking monocrystal restite inclusions in the dacite rock are graphite crystals measuring a few hundred μm, 0·5–10 mm blue cordierite crystals and 2–10 mm ruby red crystals of almandine-rich garnet ($1{\cdot}1 \pm 0{\cdot}2$ vol%). Although the almandine crystals are perfectly euhedral, they are identical in every respect to the crystals found in the Al-rich restite rock inclusions and cannot be crystallisation products of the magmatic melt. The dacite also contains many inclusions of quartz gabbroic and basaltoid material which contains inclusions identical to the restite material found in the dacitic glass base. Many basaltoid inclusions show well-developed chilled borders. These inclusions may represent a more mafic magma of deeper origin which mixed with some dacite magma before mingling into it.

KEY WORDS: almandine-bearing volcanics, Alpine belt, Betic Cordilleras, cordierite-bearing volcanics, Neogene volcanism in southeastern Spain.

Mixing of restite material and anatectic melt in varying proportions and mixing/mingling of independent magmas have become recognised models to explain variation within magmatic complexes (e.g. White & Chappell 1977; Scambos *et al.* 1986; Chappell *et al.* 1987). Much of the evidence comes from textural and mineralogical observations. However, in deep-seated rocks the plutonic crystallisation history tends to blur the original magmatic relations. Corresponding volcanic rocks may show these relations in more pristine form.

The *c.* 1 km^2 Hoyazo dacite exposure in SE Spain represents an example of an erupted anatectic melt containing abundant restite material. Another striking feature of the Hoyazo dacite is the occurrence of more mafic magmatic rock inclusions with well-developed chilled borders, representing intermingled magmatic material. The rock is well exposed and reasonably, at places exceptionally, fresh. This allows one to demonstrate convincingly the petrogenetically critical textural and mineralogical relations. This paper summarises the pertinent petrology of this occurrence, which was given in full in Zeck (1968, 1970), and relates it to recent developments.

1. Hoyazo dacite: geology, lithology, petrography, isotope data

The Hoyazo dacite is part of the 13–7 Ma-old (Di Battistini *et al.* 1987) volcanic province in southeastern Spain, postdating the Alpine nappe emplacement in the Betic Cordilleras (Fig. 1). Cerro del Hoyazo forms a shallow hill rising 100–200 m above the Campo de Níjar at the southern foot of the Sierra Alhamilla. It consists of an erosional remnant of volcanic rock, a pipe and surrounding block lava, which at *c.* 6 Ma ago was covered by reef deposits (Dabrio *et al.* 1981). Subsequent erosion excavated a large depression through the overlying carbonate cover, within the softer volcanic rock. This resulted in a deceptive crater-like morphology.

The fresh rock consists of a glass base containing crystallisation products of the magmatic melt, and a range of rock inclusions and monocrystal inclusions derived therefrom. Glass forms *c.* 50 vol% and magmatic precipitation products include plagioclase (*c.* 10 vol%), cordierite (*c.* 9 vol%) and biotite (*c.* 8 vol%), plus small amounts of sillimanite and possibly apatite and zircon. These minerals form isolated, euhedral crystals enclosed within the glass (Fig. 2a). Plagioclase crystals show a rather simple zoning, An_{85}–An_{50}. Cordierite consistently forms transformation sector trillings (Zeck 1972).

The rock inclusions consist of: (1) cordierite-biotite quartz diorite which is thought to represent a plutonic crystallisation product of the dacite magma (cognate inclusions); (2) Al-rich restite material syngenetic with the dacitic melt; (3) quartz gabbroic and basaltoid material mixed/mingled into the dacite magma; and (4) accidental inclusions (schists, quartzites, amphibolites). The total content of inclusions measuring over 1 cm is $1{\cdot}4 \pm 0{\cdot}3$ vol% and was determined by point counting in the field, by repeatedly throwing out a fishing net with tape-marked nodes with a mutual distance of *c.* 20 cm, and counting the shadow points on the rock surface. The relative proportions of the various types of rock inclusion were determined in the detrital material: Al-rich restite material *c.* 60%, igneous inclusions *c.* 40%, and accidental inclusions amounting to 1% or less.

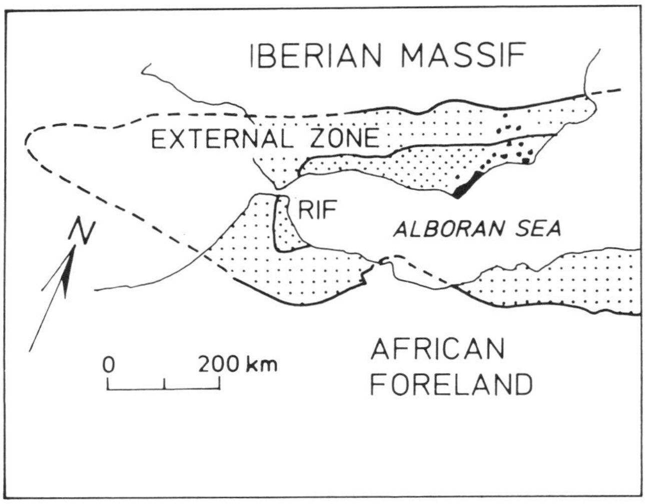

Figure 1 Location of the Hoyazo dacite within the Neogene volcanic province in southeastern Spain; volcanic rocks (black dots) postdate Alpine nappe emplacement, the pattern of which is indicated by the distribution of the Internal (dense stipple) and External Zones in the Betic-Rif-Maghrebian orocline.

The Al-rich restite rock inclusions ($Al_2O_3 = 20–45\%$) consist of: (1) almandine-biotite-fibrolite-plagioclase rock with a well-developed metamorphic foliation (back cover, Zeck 1968), and without quartz or K-feldspar; (2) cordierite–quartz rock without plagioclase or K-feldspar; and (3) spinel–cordierite hornfels. The latter represents deep-seated recrystallisation products of the foliated almandine–biotite–fibrolite–plagioclase rock, the major restite material. Striking monocrystal restite inclusions in the dacitic rock are a few hundred μm-sized graphite crystals (Fig. 2d), 0·5–10 mm cordierite crystals and 2–10 mm, ruby red garnet crystals, the latter forming $1{\cdot}1 \pm 0{\cdot}2$ vol% of the dacitic rock. The garnet crystals are usually perfectly euhedral (Fig. 2e) and have therefore been interpreted by some authors (López Ruiz *et al.* 1977; López Ruiz & Badiola 1980) as crystallisation products of the melt. However, this interpretation is untenable as these crystals are in every respect identical to the crystals found in the Al-rich restite rock inclusions (Fig. 2g); size, colour, inclusions (quartz, biotite, fibrolite felt) and refringence are identical, and microprobe studies by Munksgaard (1985) indicate closely similar chemical composition and zoning ($Alm_{70\text{-}82}Py_{7\text{-}13}Gr_{15\text{-}2}Sp_{10\text{-}1}$, core to rim). Cordierite restite crystals are anhedral and have inclusions of fibrolite (Fig. 3a). The larger examples can be seen in hand specimen as clear, blue crystals (Cerro del Hoyazo is the "type locality" for cordierite; Zeck 1968).

Other monocrystal restite inclusions are quartz, plagioclase, spinel and zircon. Quartz occurs as up to 10 cm-sized fragments, probably representing very coarse-grained quartz-rich veins or streaks within the melted rock complex. Smaller crystals appear clearly corroded. Plagioclase restite crystals have andesine composition and are anhedral (Fig. 2c). They correspond to crystals found in restite almandine–biotite–fibrolite–plagioclase rock inclusions. Spinel crystals are rare, usually 50–300 μm in size and dark lilac or green in colour (Fig. 3c). They are hercynitic in composition and were derived from spinel–cordierite hornfels material. Zircon crystals are usually anhedral (Fig. 2h). All in all, about 10–15 vol% of the Hoyazo dacite consists of restite material, predominantly Al-rich.

Practically all restite material has "clean" contacts towards the glass (Figs 2c, d, e, f; 3a, c). Notable exceptions are spinel–cordierite hornfels fragments which may show a reaction rim consisting of plagioclase An_{85}–An_{50}. In some cases, restite cordierite crystals show some overgrowth of magmatic cordierite, and in a very few cases a thin reaction rim of plagioclase crystals is present (Fig. 3b).

The igneous rock inclusions have a rounded form (Fig. 3e), vary in size from *c.* 1 mm to *c.* 60 cm and consist of cordierite–biotite quartz diorite, quartz gabbroic and basaltoid material. Plagioclase in the cordierite–biotite quartz diorite inclusions is An_{55}–An_{35} and preliminary analytical results suggest a variable SiO_2 content (*c.* 58–69%) which on average corresponds to that of the bulk dacite (*c.* 64% SiO_2); CaO and FeO seem similar, MgO slightly higher. These rocks are thought to represent deep-seated crystallisation products of the dacite magma (cognate inclusions). The quartz–gabbroic inclusions are characterised by more calcic plagioclase (An_{90}–An_{50}) and, further, contain quartz, hornblende, biotite and pyroxene. There is *c.* 5–10 vol% (altered) vesicular, interstitial glass, representing quenched residual melt. Preliminary analytical results indicate SiO_2 contents of 55–59% for this type of inclusion; CaO seems to be 5–9% (vs. 3–4% in dacite), FeO ~ 6% (vs. ~ 4% in dacite) and MgO 5–8% (vs. 3–5% in dacite). Euhedral, monocrystal plagioclase (Fig. 2b), hornblende (Fig. 3d) and orthopyroxene inclusions derived from these rocks inclusions are present in the dacitic glass base. The basaltoid inclusions are more fine-grained, consisting mainly of plagioclase (An_{90}–An_{50}), hornblende, orthopyroxene, clinopyroxene, biotite and considerable amounts (up to *c.* 30 vol%) of glass. Preliminary analytical results indicate SiO_2 contents of 51–53% and CaO, FeO and MgO contents similar to those found for the quartz gabbroic inclusions. Many igneous rock inclusions contain some inclusions of restite material identical to that found within the dacite glass base (Fig. 3e, f). The Al-rich fragments are invariably surrounded by a reaction rim of interlocking crystals of calcic plagioclase which may enclose some crystals of spinel. The basaltoid inclusions contain embayed quartz crystals, up to about 1 cm in size, which are surrounded by a reaction rim consisting of vesicular glass containing some pyroxene or hornblende crystals. Many basaltoid rock inclusions show well-developed chilled borders (Fig. 3g, h), usually 2–3 mm in thickness, in which plagioclase grain size may range inwardly from 75–500 μm.

Oxygen isotope analyses by Munksgaard (1984) indicate $\delta^{18}O$ values for the dacite ranging from 13·1–15·6‰, similar

Figure 2 Petrography of Hoyazo dacite: bar on each photomicrograph represents 250 μm.

(a) Glass base containing crystallisation products of magmatic melt; plagioclase (An_{85}–An_{50}), cordierite (euhedral sector trilling, basal section) and biotite can be recognised.

(b) Large, broken plagioclase crystal (An_{90}–An_{50}) showing complex zoning with patchy zoning in centre (incongruent melt inclusions with complimentary anorthite-rich rims); intermixed monocrystal inclusion derived from quartz–gabbroic rock inclusion.

(c) Andesine crystal fragment derived from almandine–biotite–fibrolite–plagioclase rock inclusion, the major restite material.

(d) Graphite restite crystals, some bent around a plagioclase crystal which is similar to that in (b); derived from Al-rich rock material.

(e) Euhedral almandine crystal within glass base. Note the absence of any reaction or precipitation products on the crystal margin.

(f) Close-up of almandine crystal margin.

(g) Euhedral almandine crystal within fibrolite felt in almandine–biotite–fibrolite–plagioclase rock inclusion.

(h) Inherited, restite, zircon crystal with biotite.

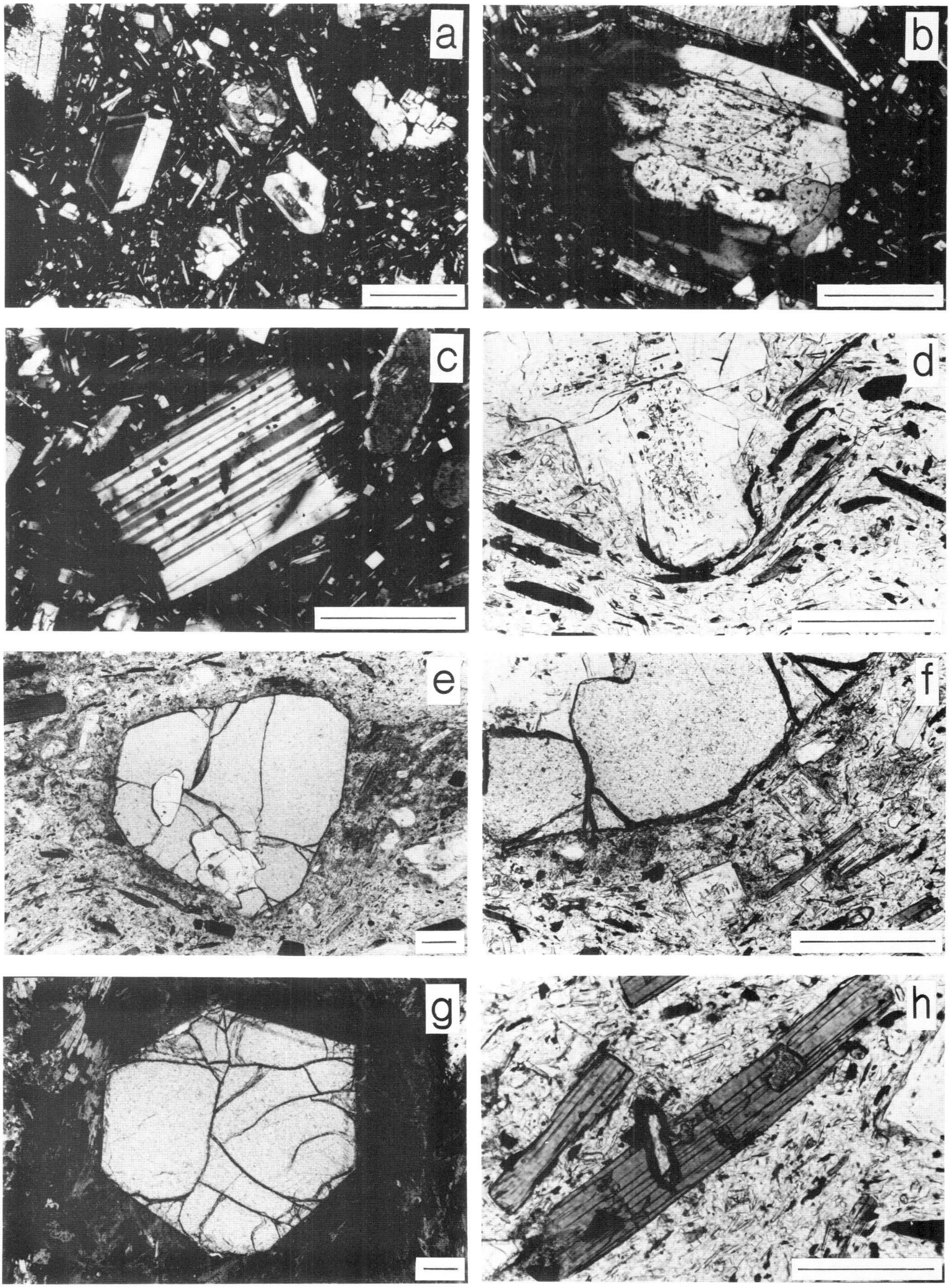
a
b
c
d
e
f
g
h

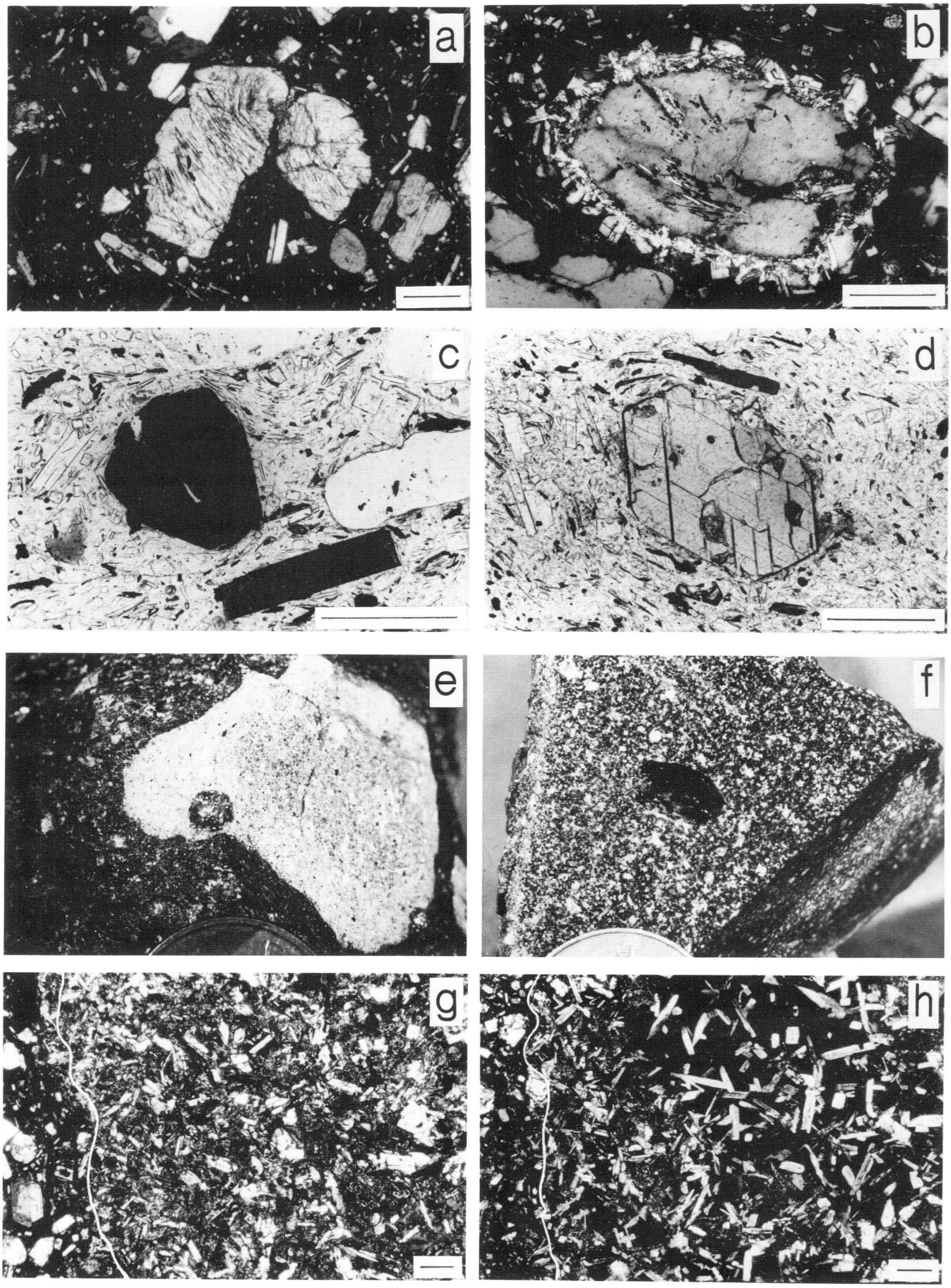
a
b
c
d
e
f
g
h

to values for the Al-rich restite rock inclusions (13·0–16·2‰) and the cordierite–biotite quartz diorite (cognate) inclusions (15‰). The quartz gabbroic and basaltoid inclusions have lower $\delta^{18}O$ values (11·1–13·1‰), though still very high in comparison with values found in mantle-derived basalts.

2. Conclusions

The available data indicate that the Al-rich inclusion material and the dacitic melt were derived syngenetically from a (semi-)pelitic rock sequence by means of anatexis. Critical features are: (1) texture, mineralogy and Al-rich composition ($Al_2O_3 \equiv c.$ 20–45%) of the primary restite rock inclusions which correspond to the material which makes up melanosomes in pelitic migmatites; (2) high restite content (*c.* 10–15 vol%) and its regular distribution throughout the dacite clearly favour an "internal" origin; (3) perfect cotectic split into (a) plagioclase-bearing restite without quartz or K-feldspar and (b) quartz-bearing restite without plagioclase or K-feldspar strongly favours an origin by partial melting; and (4) high and similar $\delta^{18}O$ (13–16‰) in the dacite and restite rock inclusions, corresponding to values given for sedimentary material (Margaritz *et al.* 1978), strongly suggest an S-type character for the dacite (cf. Chappell *et al.* 1987).

Cordierite-biotite quartz diorite inclusions have similar mineralogy, SiO_2, CaO and FeO contents and $\delta^{18}O$ values as the dacite itself and are thought to represent cognate material. The quartz gabbroic and basaltoid inclusions have different mineralogy and are lower in SiO_2 and $\delta^{18}O$ and higher in MgO, CaO and FeO in comparison with the dacite, suggesting that they represent magma(s) of more mafic character independent from the dacite. The $\delta^{18}O$ values (11–13‰) support interaction of these magma(s) with the dacitic magma and/or similar crustal material. This mixing process was not necessarily a simple one-stage process involving a directly mantle-derived basaltic magma. The quartz gabbroic material shows some characteristics of restite material as defined by Chappell *et al.* (1987), e.g. plagioclase composition and (patchy) zoning An_{90}–An_{50}. However, this material, in many cases with *c.* 30 vol% hornblende and CaO contents of 5–9%, can hardly qualify as restite material for the peraluminous Hoyazo dacite.

Figure 3 Petrography of Hoyazo dacite, continued; bar represents 250 μm.
(a) Anhedral cordierite restite fragment with inclusions of fibrolite; note the clean crystal margin.
(b) Same type of cordierite, but with a thin reaction rim consisting of plagioclase and some biotite; this is rare in the Hoyazo dacite.
(c) Euhedral hercynite restite crystal, derived from spinel–cordierite hornfels; note absence of reaction rim, and rounded quartz crystal *c.* 100 μm to the right.
(d) Euhedral hornblende crystal, intermixed monocrystal inclusion derived from quartz-rich gabbroic rock inclusion.
(e) Basaltoid rock inclusion with monocrystal inclusion of almandine, indicating intermixing of dacite magma into more basic magma before mingling; rim of 25 mm Ø coin for scale.
(f) Basaltoid rock inclusion, slightly more coarse-grained than in (e), showing foliated almandine–cordierite–biotite fragment derived from almandine–biotite–fibrolite–plagioclase rock restite, again suggesting mixing before mingling; same scale as (e).
(g) Chilled border of *c.* 10 cm sized basaltoid inclusion (right) towards dacitic glass (left); contact traced; plagioclase crystals range from ~ 75 μm at the margin, to ~ 400 μm in the centre.
(h) as (g).

The fine-grained basaltoid inclusions with *c.* 30 vol% of glass and showing chilled borders suggest that at a later stage a new influx from the same magma chamber or from another one into the dacitic magma at depth, was directly followed by eruption and quenching of the dacitic melt and its inclusions (Zeck 1970). This is supported by the wide variety of mineral phases in the dacite and the scarcity of reaction rims, indicating that equilibrium was not reached and hardly attempted.

The regular occurrence within the basaltoid inclusions of inclusions of garnet, cordierite, fibrolite, quartz, etc., which make up only 10–15 vol% in the dacite itself, clearly indicates that considerable mixing with the dacite magma took place *before* mingling. This conclusion disagrees with suggestions of other authors (e.g. Poli & Tommasini 1991; Barbarin & Didier 1992) suggesting mingling of mafic magma blebs into acid magmatic melt, *followed* by hybridisation and disintegration leading to a hybrid igneous rock. As felsic-mafic relations within the Hoyazo dacite seem in no way uncommon, it is suggested that early felsic–mafic magma mixing may be of more general occurrence.

A great deal of the variation observed within these types of inclusions, here and in other occurrences, such as degree of chemical difference with host rock, colour index and pyroxene/hornblende/biotite ratios, grain size and proportion of matrix/glass, content of intermixed feldspar crystals, etc., may be explained by how effective the felsic–mafic interaction has been. Major factors which decide the fate of mixing are the difference in liquidus–solidus temperatures of the two magmas, controlling to a large extent the viscosity difference *and the way contact was made* (Zeck 1970). Realistic modelling of felsic–mafic magma mixing/mingling should allow for several stages of (more) felsic–(more) mafic interaction with variable cooling-effective mass relations (Zeck 1970; Bacon 1986).

3. Acknowledgements

This work was supported by grants 11-7589 and 11-8311 from the Danish Research Council (SNF). Dr J. C. Bailey, Copenhagen, is thanked for comments on the manuscript.

4. References

Bacon, C. R. 1986. Magmatic inclusions in silicic and intermediate volcanic rocks. J GEOPHYS RES **91**, 6091–112.
Barbarin, B. & Didier, J. 1992. Genesis and evolution of mafic magmatic enclaves through various types of interactions between coeval felsic and mafic magmas. TRANS R SOC EDINBURGH EARTH SCI **83**, 145–53.
Chappell, B. W., White, A. J. R. & Wyborn, D. 1987. The importance of residual source material (restite) in granite petrogenesis. J PETROL **28**, 1111–38.
Dabrio, C. J., Esteban, M. & Martin, J. M. 1981. The coral reef of Níjar, Messinian (uppermost Miocene), Almería province, SE Spain. J SEDIMENT PETROL **51**, 521–39.
Di Battistini, G. Toscani, Laccarino, S. & Villa, I. M. 1987. K/Ar ages and the geological setting of calc-alkaline volcanic rocks from Sierra de Gata, SE Spain. NEUES JAHRB MINERAL, MONATSH, 1987, 369–83.
López-Ruiz, J., Badiola, E. R. & García Cacho, L. 1977. Origine des grenats des roches calco-alcalines du Sud-Est de l'Espagne. BULL VOLCANOL **40**, 141–52.
López-Ruiz, J. & Badiola, R. E. 1980. La región volcánica neogena del sureste de España. ESTUD GEOL **36**, 5–63.
Magaritz, M., Whitford, D. J. & James, D. E. 1978. Oxygen isotopes and the origin of high $^{87}Sr/^{86}Sr$ andesites. EARTH PLANET SCI LETT **40**, 220–30.
Munksgaard, N. C. 1984. High $\delta^{18}O$ and possible pre-eruptional

Rb–Sr isochrons in cordierite–bearing Neogene volcanics from SE Spain. CONTRIB MINERAL PETROL **87,** 3651–8.

Munksgaard, N. C. 1985. A non-magmatic origin for compositionally zoned euhedral garnets in silicic Neogene volcanics from SE Spain. NEUES JAHRB MINERAL MONATSH, 1985, 73–82.

Poli, G. & Tommasini, S. 1991. Model for the origin and significance of microgranular enclaves in calc-alkaline granitoids. J PETROL **32,** 657–66.

Scambos, T. A., Loiselle, M. C. & Wones, D. R. 1986. The Centre Pond pluton: the restite part of the story (phase separation and melt evolution in granitoid genesis). AM J SCI **286,** 241–80.

White, A. J. R. & Chappell, B. W. 1977. Ultrametamorphism and granitoid genesis. TECTONOPHYSICS **43,** 7–22.

Zeck, H. P. 1968. *Anatectic origin and further petrogenesis of almandine-bearing biotite–cordierite–labradorite–dacite with many inclusions of restite and basaltoid material, Cerro del Hoyazo, SE Spain.* Thesis, Amsterdam University.

Zeck, H. P. 1970. An erupted migmatite from Cerro del Hoyazo, SE Spain. CONTRIB MINERAL PETROL **26,** 225–46.

Zeck, H. P. 1972. Transformation trillings in cordierite. J PETROL **13,** 367–80.

H.P. ZECK, Geological Institute, Øster Voldgade 10, 1350 K Copenhagen, Denmark.

Transactions of the Royal Society of Edinburgh: Earth Sciences, **83,** 145–153, 1992

Genesis and evolution of mafic microgranular enclaves through various types of interaction between coexisting felsic and mafic magmas

Bernard Barbarin and Jean Didier

ABSTRACT: Thermal, mechanical and chemical exchange occurs between felsic and mafic magmas in dynamic magma systems. The occurrence and efficiency of such exchanges are constrained mainly by the intensive parameters, the compositions, and the mass fractions of the coexisting magmas. As these interacting parameters do not change simultaneously during the evolution of the granite systems, the exchanges appear sequentially, and affect magmatic systems at different structural levels, i.e. in magma chambers at depth, in the conduits, or after emplacement. Hybridisation processes are especially effective in the plutonic environment because contrasting magmas can interact over a long time-span before cooling. The different exchanges are complementary and tend to reduce the contrasts between the coexisting magmas. They can be extensive or limited in space and time; they are either combined into mixing processes which produce homogeneous rocks, or only into mingling processes which produce rocks with heterogeneities of various size-scales. Mafic microgranular enclaves represent the most common heterogeneities present in the granite plutons. The composite enclaves and the many types of mafic microgranular enclaves commonly associated in a single pluton, or in polygenic enclave swarms, are produced by the sequential occurrence of various exchanges between coexisting magmas with constantly changing intensive parameters and mass fractions. The complex succession and repetition of exchanges, and the resulting partial chemical and complete isotopic equilibration, mask the original identities of the initial components.

KEY WORDS: granite, mafic microgranular enclaves, magma mixing, mingling, hybridisation processes, thermal exchange, chemical exchange, mechanical exchange

Many models have been proposed for the genesis and evolution of mafic microgranular enclaves (MME) (see reviews in Didier 1973, 1987; Barbarin & Didier 1991). In recent years, four models have been frequently cited by petrologists: the restite model (e.g. White & Chappell 1977; Chappell *et al.* 1987; Chappell & White 1991), the closed-system differentiation model or autolith model (e.g. Pabst 1928; Fershtater & Borodina 1977, 1991; Dodge & Kistler 1990), the disrupted dyke model (e.g. Pitcher 1991), and the model of coexisting mantle-derived mafic magmas (e.g. Didier & Lameyre 1969; Huppert & Sparks 1988; Wiebe 1991). These models have been reviewed and discussed in detail by various authors in Didier and Barbarin (1991), and generally involve coexisting magmas of contrasting compositions which are in contact and interact. Even in the restite model, the "restite" enclaves do not represent solid rocks enclosed in the granite magmas, but mushes consisting of restitic crystals impregnated in anatectic liquids (Chappell *et al.* 1987).

The aim of this contribution is to review the different types of interaction between coexisting magmas of contrasting characteristics regardless of their origins and genetic relationships, to discuss the parameters which control the occurrence and efficiency of these interactions, and to consider the role of exchange processes in the genesis of MME and their host granites. The various types of interaction are induced by the thermal and compositional contrasts between the coexisting magmas, and are closely controlled by their relative physical properties and mass fractions. All these processes are complementary and tend to reduce or suppress the differences between these magmas. In plutonic bodies, there are many kinds of heterogeneities ranging in scale from less than one millimeter to some kilometers (Table 1).

1. Terminology-definitions

Thermal, chemical and mechanical interactions are the principal types of exchange between coexisting magmas (Fig. 1). The term "mingling", or "co-mingling", generally indicates mechanical interaction in which the original magmas partially retain their identities in the mixtures. The term "mixing" involves chemical interactions that complete the mechanical interactions in order to produce homogeneous "hybrid rocks" in which the identities of the original magmas are obscured. As the term "mixing" is commonly used "to include all scales and types of interactions between coexisting magmas" (Wiebe 1980, p. 197), it is then necessary to specify "mechanical mixing" when talking about "mingling" or mechanical interactions, and "thorough mixing" or "chemical mixing" when talking about "mixing" or mechanical plus chemical interactions.

2. Physical properties of coexisting magmas and interaction processes

In granite systems, interactions between coexisting felsic and mafic magmas are mainly constrained by their initial temperatures, compositions, H_2O content, relative volumes, crystallinities, absolute values of viscosity, and subsequently by their relative rheology. As these intensive and interacting parameters do not change simultaneously during the evolution of the systems, the various types of interactions may appear sequentially.

Table 1 Heterogeneities present at different scales in the granite plutons (from Oldenburg *et al.* 1989)

Heterogeneity	Scale, m
Crystal zoning	$<10^{-3}$
Xenocrysts Mafic clots	10^{-3}–10^{-2}
Enclaves Schlieren	10^{-2}–1
Enclave swarms Mafic dykes Composite dykes	1–10
Pluton zonation	10–10^4

Mafic magmas from which the MME crystallise are hotter and of basic composition (45–50 wt% SiO_2), whereas felsic magmas of the granite hosts are cooler and acid (65–75 wt% SiO_2). Because of these contrasts in temperature and composition, the coexisting magmas have very different physical properties. Depending on the chemical contrast, effective viscosities of mafic magmas at a given volume fraction of crystals are from one to four orders of magnitude lower than those of felsic magmas. Viscosity contrasts have an important influence on the dynamics of magma

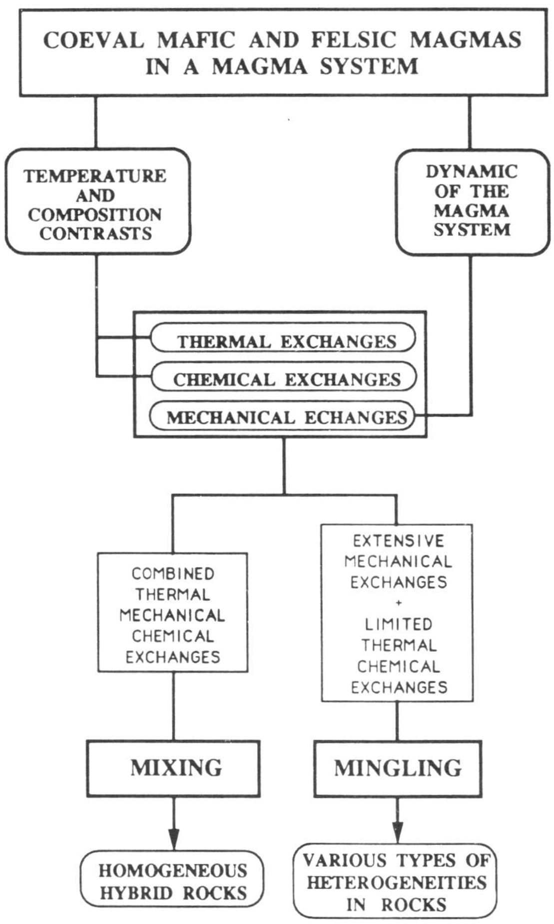

Figure 1 The different types of exchange occurring between two coexisting and contrasting magmas in a dynamic magma system: their combination into mixing or mingling processes.

hybridisation (Huppert *et al.* 1984): low viscosity contrasts between two magmas generally favour mixing, whereas high viscosity contrasts prevent mixing and only permit mingling. In any case, hybridisation processes are efficient only between magmas with low absolute viscosities.

In their rheological modelling, Sparks and Marshall (1986) consider that viscosity and crystal contents of mafic and felsic magmas also vary as a function of their mass fractions. Whether interacting magmas mix or mingle is a function of their relative volumes and temperatures: a large proportion of mafic magma and thermal equilibration favour mixing, whereas a large proportion of felsic magma and a large temperature contrast only permit mingling. In a composition-mixing diagram, Sparks and Marshall (1986, fig. 10, p. 116) define a domain in which physical properties and mass fractions of the coexisting magmas permit mixing. The extension of this domain is very dependent upon initial temperatures, chemical compositions, H_2O content, and the crystallinity of either or both of the end-members. The relative volumes, temperatures, and viscosities of the coexisting magmas not only play a critical role in determining the occurrence and efficiency of the various forms of interaction, but also constrain the spatial extent of the areas affected by these interactions.

3. Thermal exchange

Thermal exchange is much more rapid than mechanical or chemical exchange. Thermal diffusion is typically several orders of magnitude faster than chemical diffusion in silicate melts (Shaw *et al.* 1976; Hofmann 1980; Watson 1981; Sparks & Marshall 1986). The initially high thermal contrast between felsic and mafic magmas leads to thermal exchange (Sparks *et al.* 1977; Huppert & Sparks 1988; Blake & Koyaguchi 1991). Detailed study of the textures of MME and their host rocks provides constraints on the processes. Fine grain-size, acicular and hollow minerals, and occasional chilled margins of the MME indicate the rapid crystallisation and local quenching of mafic magma after rapid loss of heat to the enclosing granite magma.

The mass fractions of the coexisting magmas also have a significant effect on their thermal evolution. Mafic bodies of great size reach thermal equilibrium with the granite magma only after a relatively long time-span during which extensive chemical exchange can occur and modify their original identity, whereas small blobs of mafic magma reach thermal equilibrium very quickly and become sufficiently rigid to limit and even inhibit chemical exchange (Fernandez & Barbarin 1991). Small quenched enclaves retain most closely the characteristics of their parental magma.

4. Mechanical exchange

Two coexisting magmas can also exchange blobs of magmas or phenocrysts. Blobs of one magma enclosed in the other form enclaves, whereas phenocrysts of one magma introduced into the other form xenocrysts.

The size of enclaves is very variable, from clots of a few minerals on the millimetre scale to very large enclaves reaching several kilometres. The most common enclaves of granites range from 10–20 cm in size. The morphology of enclaves is also variable, from the common ovoid shapes to fairly rare lenticular or cigar-like shapes. The morphology and the relative orientation of enclaves compared with the flow foliation of granite hosts provide information on the relative rheology of coexisting and contrasting magmas during mechanical mixing (Fernandez & Barbarin 1991).

Phenocrysts can be transferred back and forth from one magma to another. Given that the compositional contrast between two magmas is relatively strong, phenocrysts crystallised in one magma and mechanically transferred will not be in equilibrium with the new host magma. Some phenocrysts recrystallise entirely, whereas others partially dissolve and become mantled with minerals in equilibrium with the surrounding magma. Xenocrysts occur as clots of mafic minerals and as ovoid or rounded crystals surrounded by reaction aureoles. They can also form cores of some strongly zoned crystals such as plagioclase crystals (e.g. Barbarin 1990). Generally, mafic minerals such as amphibole, pyroxene and olivine from the mafic magma occur as clots of imbricated mafic minerals, whereas quartz and K-feldspar phenocrysts from the felsic magma form xenocrysts with, respectively, ocellar and rapakivi textures in the mafic magma. Cores of complexly-zoned plagioclase originate in either the felsic or the mafic magma. Some large accessory minerals such as apatitie or zircon can also be transferred from one magma to another and form xenocrysts. Xenocrysts are generally associated with relatively homogeneous hybrid granites comprising large masses of mafic and felsic rocks (e.g. Barbarin 1988). Exceptionally, xenocrysts can be transferred back in the magma from which they initially crystallised (Gourgaud *et al.* 1981; Cantagrel *et al.* 1984).

In addition to enclaves and xenocrysts, another type of mafic/felsic association occurs where mafic magmas are abundant or dominant, i.e. at the contact zones between large volumes of felsic and mafic rocks. These associations form banded structures or net-veined complexes. Contacts between mafic and felsic rocks are generally cuspate or lobate, and more locally straight. They are commonly bordered by basalt-like fine-grained margins a few centimetres thick, which delimit the mafic masses. In these associations, both felsic and mafic rocks generally have a fine grain-size. The generally lobate and cuspate contacts and the textures of the associated rocks suggest that the two magmas had very low viscosities when they came in contact, and then crystallised very quickly because of high degrees of undercooling.

Banded structures or net-veined associations, enclaves and xenocrysts are the results of mechanical exchange between coexisting and contrasting magmas. In granite plutons, enclaves are generally common and may be locally abundant, xenocrysts are fairly discrete, and banded structures and net-veined associations are exceptional. The latter are generally restricted to subvolcanic plutons. Conversely, banded structures and net-veined associations are quite frequent in lavas and pyroclastic deposits.

Mechanical exchange is possible only where there is active stirring of the two magmas. Experiments on magma mixing by forced convective stirring (Kouchi & Sunagawa 1983, 1985) show that where two contrasting magmas are vigorously stirred, banded structures occur at the contact zone. These then divide into smaller and smaller blobs of magma that are pulled away from the contact zones and scattered in the host magma. Each blob of magma can be so reduced that it consists only of clots of some minerals, or of single minerals which react with their host and take on a typically xenocrystic aspect. Depending on the intensity of the stirring, mechanical exchange can be limited to banded structures, to occurrences of MME, or can produce xenocrysts. The scarcity of banded structures in plutons can result from the mingling or mechanical mixing of magmas at depth before the ascent of granites and their enclaves. Like synplutonic dykes or monogenic enclave swarms (Barbarin 1991a), the scarce net-veined associations present in the plutons are produced by late injections of large volumes of mafic magma into cooling granites and interaction at emplacement levels.

It is important to specify where mechanical exchange occurs, in other words, the location and causes of stirring. Stirring can occur at depth, within the magma chamber where two contrasting magmas come into contact, in the conduit during emplacement, and locally after emplacement. In magma chambers, stirring can be produced by thermal gradients which induce double-diffusive convection. In addition to convective stirring, injection of mafic magma into felsic magma as a fountain can also yield vigorous stirring, formation of double-diffusive convective layers and possibly overturning (Campbell & Turner 1985, 1986, 1989). In conduits, the relative ascent velocities of two magmas with different physical properties, particularly with different densities and viscosities, also results in turbulent stirring (Koyaguchi 1985, 1987; Koyaguchi & Blake 1991). At the emplacement level, magmatic flow can also produce limited mechanical exchange.

Where two magmas are in contact without the effect of stirring, enclaves can be produced by viscous fingering. The less viscous magma injects into discontinuities of the more viscous magma which has begun crystallising. These discontinuities could be formed by early fractures (Hibbard & Watters 1985) related to the loss in volume of crystallising magmas. Viscous fingering follows a fractal law (Bébien *et al.* 1987). The "fingers" of magma can then be detached from the main stock and form enclaves.

5. Chemical exchange

Although more discrete than mechanical exchange, chemical exchange complements the other hybridisation processes and plays a prominent role in the generation of hybrid rocks (see Debon 1991). Chemical elements diffuse between coexisting magmas, during the whole time-span from meeting of the magmas to their consolidation, as long as homogenisation is not attained. Mechanical transfer and assimilation of xenocrysts also alter the initial compositions of the contrasting magmas (e.g. Briot 1990). Chemical exchange generally acts after thermal equilibration. Experiments carried out on major element diffusion between melts of contrasting compositions have demonstrated a systematic and differential migration of all elements on both sides of the contact surface (e.g. Yoder 1973; Alibert 1980; Alibert & Carron 1980; Watson 1982; Johnston & Wyllie 1988). Experiments also indicate that this selective interdiffusion of elements across interfaces between the melts is significantly enhanced by the presence of fluid phases, and especially of H_2O (Alibert & Delbove 1980; Johnston & Wyllie 1988).

The rate of chemical transfer depends on the chemical contrast and on the physical state of both components. The rate is greatest for the fully-molten state, decreases progressively with the crystallisation of the mafic component and becomes very limited or nil when the mafic component reaches a fully-solid state. Thus, as the size of enclaves controls their cooling rate, so the magnitude of chemical transfer increases with enclave size. Consequently, the chemical differences between small MME and host granites are more important than those between large enclaves or mafic bodies and host granites (e.g.Barbarin 1988; Dorais *et al.* 1990), and large MME or mafic bodies are frequently isotopically equilibrated with the granite hosts whereas smaller enclaves are not (e.g. Pin *et al.* 1990; Pin 1991). In any case, the MME may partially or fully preserve the chemical and isotopic features of parental magmas above a certain threshold size, which is a function of parameters such as temperature, cooling rate, fluid composition and mobility, or type of element considered (Baker 1989, 1990; Lesher 1990).

Experiments indicate that chemical diffusivities vary widely for different elements and isotopes. Chemical diffusivity measurements on dacitic-rhyolitic melt systems indicate that isotopic equilibration proceeds faster than chemical equilibration, and that isotopic equilibration is faster for Sr than for Nd (Baker 1989, 1990; Lesher 1990). Enclaves and the granite hosts are isotopically equilibrated for Rb–Sr in most cases (e.g. Barbarin *et al.* 1985; Dodge & Kistler 1990; Pin *et al.* 1990; Pin 1991; Fourcade & Javoy 1991). Isotopic equilibration studies for Nd–Sm are equivocal (e.g. Holden *et al.* 1987; Pin *et al.* 1990). Although oxygen isotopes are also affected by equilibration processes, isotopic equilibration is not achieved (Fourcade & Javoy 1991).

Major and trace elements are distinct in enclaves and their hosts. However, they show parallel variations. For instance, REE are generally more abundant and less fractionated in enclaves than in their hosts. Nevertheless, the REE content is higher in enclaves enclosed in REE-rich granites, and lower in enclaves enclosed in REE-poor granites. Similarly, granites with slightly fractionated REE enclose enclaves with slightly fractionated REE (e.g. Barbarin 1989). MgO and FeO_t contents of enclaves and their hosts remain fairly constant for several enclave-host pairs from different localities of the same pluton or from contiguous granite plutons. The Mg/Fe + Mg ratio is also rather similar for enclaves and their hosts (e.g. Cocirta & Orsini 1986; Debon 1991). These parallel variations of the major and trace elements of enclaves and the host granites suggest partial chemical equilibration and also reflect some similarity in their mineral compositions. For instance, K-feldspar-rich granites contain K-feldspar-bearing enclaves, and where sphene is present in granites it is also present in the enclaves. Where the chemical compositions of minerals vary from one exposure to another in the same pluton, they vary in both the enclaves and the host granites of the pairs (Barbarin *et al.* 1989; Allen 1991). Enclaves and the host granites generally contain the same minerals with the same chemical compositions, but in different proportions. Bowen (1922) suggests that this same mineralogy may be produced through reactions of mafic enclaves with the enclosing magmas.

Scarce zoned enclaves consisting of a mafic core surrounded by generally one and occasionally two or more irregular hybrid zones, separated by sharp to diffuse contacts (Debon 1975; Alibert 1980; Bacon 1986; Cramer & Kwak 1988), may also be produced by partial chemical equilibration. Continuous felsic haloes formed in the host granite in contact with the MME also result from local mass transfer of elements by diffusion. Some authors also suggest that the fine-grained margins of enclaves, which are generally enriched in mafic minerals and poor in felsic minerals, are also produced through chemical diffusion (Belin 1988; Bussy 1990). Detailed studies of petrographic, mineralogical and chemical features of the cores of enclaves, through the contact zones to the host granites (Wiebe 1973; Gamble 1979; Belin 1988; Cramer & Kwak 1988; Barbarin 1989; Bussy 1990) indicate the species and quantity of elements that are transferred either from the enclaves to the host granites or *vice versa*. During selective interdiffusion, CaO, MgO, and FeO_t from the mafic magmas are introduced in the felsic magmas and result in the crystallisation of mafic silicates, whereas SiO_2, K_2O, (and Na_2O) from the felsic magmas are introduced in the mafic magmas and result in the crystallisation of poikilitic quartz and K-feldspar. Introduction of H_2O into the mafic magmas also induces the destabilisation of anhydrous mafic minerals and their transformation into H_2O-bearing mafic minerals. Olivine or pyroxene phenocrysts are generally replaced by clots of amphibole or biotite. Exceptionally, olivine and pyroxene relics can be observed in the cores of some clots of mafic minerals in MME enclosed in tonalites. Once again, these types of phenocryst relics are more frequent in the magmatic enclaves of volcanic rocks, where they are better preserved because of faster cooling and chilling of the mineral associations.

Detailed studies of chemical variations through mafic/felsic interfaces also indicate some gradients distinct from gradients observed experimentally (e.g. Yoder 1973; Alibert 1980; Johnston & Wyllie 1988). These chemical transfers against the concentration gradients ("uphill" diffusion) are related to the crystallisation of some minerals which maintain the concentration of the elements at a low value in the melt. "Uphill" diffusion of K due to crystallisation of biotite is described in detail by Johnston and Wyllie (1988). Diffusion of trace elements is also controlled by the crystallisation of some accessory minerals (e.g. Bussy 1990).

Diffusive processes such as steady state, which are common in granite plutons after emplacement and before consolidation, are generally limited in the space for most elements. Chemical exchange is generally enhanced by vigorous stirring and by mechanical fragmentation of the mafic magma, which reduces the size of the enclaves and increases the contact surfaces between the coexisting magmas. Stirring also favours chemical exchange because it continually renews the felsic magma in contact with the enclaves and maintains the chemical difference between the magmas. Chemical, like mechanical, exchange is obstructed by quenched fine-grained margins which form shells around the enclaves. These shells prohibit mass and crystal transfer, and probably inhibit chemical exchange.

Because isotopes have the highest chemical diffusivities and isotopic exchange proceeds much faster than chemical exchange (Baker 1989, 1990; Lesher 1990), isotopic equilibration is generally completed over the entire pluton, whereas chemical equilibration is partial and local, and commonly concerns individual MME and the adjacent host granite. Chemical similarities and isotopic equilibration are differently interpreted. For instance, Dodge and Kistler (1990) consider that enclaves and the granite hosts have the same isotopic features because they are cogenetic and not because of isotopic equilibration. Isotopic investigation of the various components of composite and synplutonic dykes in some granites of the Sierra Nevada batholith (Barbarin *et al.* 1989) indicates that the initial mafic and felsic components of these dykes have different isotopic features and that the more hybrid components have isotopically equilibrated (Barbarin 1989).

6. Relationships between various hybridisation processes

Exchange processes do not act independently; they are generally associated and frequently combined. The various processes generally succeed and interact during magma mixing events (Fig. 2). When hot mafic magma and felsic magma are in contact, the mafic magma quenches and is disrupted into small blobs. This mechanical exchange increases the interface areas between the magmas and promotes chemical exchange. Chemical exchange reduces composition contrast and degrades sharp interfaces between the components into graded hybrid zones. The newly-formed hybrid magmas may play the role of an additional component in the mixing system and react with the mafic magma, the felsic magma, or with both. Hybridisation processes are favoured by the presence of hybrid magmas because of closer physical properties. Complete homogenisation can be obtained only after many successive stages of hybridisation with formation of more and more homogeneous hybrid magmas.

All these processes start at the contact areas, and progressively affect larger volumes of magmas. The hybridisation processes are believed to be relatively fast compared with the time required for crystallisation. They can be completed at any moment as soon as homogenisation is reached or crystallinities of the magmas become too high due to cooling. This model of hybridisation allows partial to complete homogenisation of initially contrasting magmas, and is at variance with the results of numerical modelling, which conclude that mixing between two components with contrasting temperatures, compositions and physical features is limited or very difficult (e.g. Sparks & Marshall 1986).

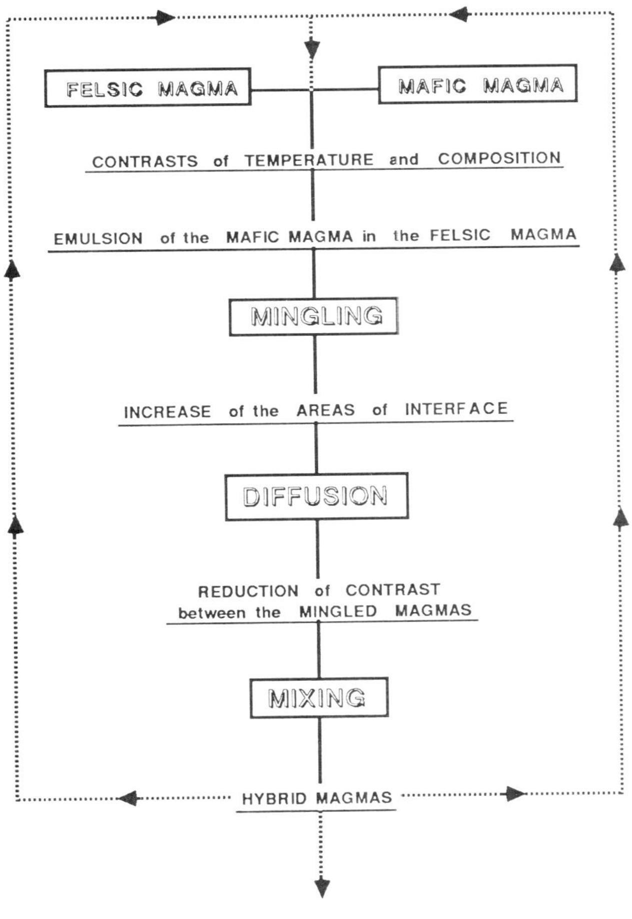

Figure 2 Succession and combination of the different types of interaction at contacts between two coexisting magmas; the hybrid magmas produced remain in the contact zone and interact with either the mafic or the felsic magmas.

Succession and repetition of the various exchange processes and constant evolution of the end-members affected by these processes produce the composite enclaves which frequently consist of hybrid rocks rich in xenocrysts (e.g. Barbarin 1988), and the many types of MME that are associated in a single pluton or in polygenic swarms. Each type of enclave is generated by limited mingling between two small volumes of already hybrid magma. In addition to the variety in composition, the mass fractions of the components also play a large part in the production of the many types of enclave.

7. Location of exchange processes

Interaction processes act as soon as two contrasting magmas come into contact, and for as long as thermal and compositional gradients exist, until consolidation. The constant evolution of physical properties of the coexisting magmas and of magma system dynamics will be successively more or less favourable to the various interactions.

In the case of calc-alkaline plutonism, and with the assumption that the mafic magmas come from the upper mantle (e.g. Didier & Lameyre 1969; Huppert & Sparks 1988; Wiebe 1991), the following succession of exchange interactions is proposed (Table 2). Mantle-derived magmas are injected into the lower crust, where they either pond or form dense dyke swarms. Thermal transfer initiates melting of crustal rocks and the production of crustal melt. This crustal melt mixes with hot mantle-derived magmas, then with more and more hybrid mafic magmas, through extensive and combined mechanical and chemical exchange. The mass fraction of the hot mafic end-member remains greater than that of the crustal anatectic melt for a fairly long time, during which thorough mixing produces homogenised hybrid magmas; this is favoured by the dominance of the high-temperature magma. When injection of mafic magmas stops, the crustal rocks continue to melt and to produce anatectic magmas for a quite long period (Huppert & Sparks 1988). The felsic component, i.e. the crustal melt, becomes more and more abundant and its mass fraction increases to more than that of the mafic component. Mixing becomes less likely and thorough mixing is even inhibited where the volume of felsic magma is much larger than that of the mafic component. Convective stirring, which continues, permits only extensive mechanical exchange, and mingling produces only MME. Where mafic magma is injected as a fountain into a magma chamber filled with felsic crustal or slightly hybrid magma, the greater dominance of the felsic end-member over the mafic end-member also permits only mingling, i.e. scattering of blobs of mafic magmas within the felsic magmas to form MME. Thermal exchange results in the chilling of the mafic blobs.

Interactions also occur when magmas are evacuated from their chambers and move upward via a dense swarm of dykes. Forceful injection in relatively narrow conduits of magmas of contrasting densities is accompanied by turbulent stirring (Koyaguchi 1985, 1986; Blake & Campbell 1986). Mechanical exchange dominates over chemical exchange within such conduits. Furthermore, during this short time-span, mechanical and limited chemical exchange take place on a very local scale. Hybridisation within a given conduit differs from hybridisation within other conduits because differently effective interaction processes occur and evolve end-members with different compositions and mass fractions.

Mingling in conduits produces the various types of MME. Conversely, because the volume of granite magmas involved in the mingling is much larger than the volume of enclaves, compositions of the granite magmas are only slightly affected and no difference occurs between the granite magmas affected by distinct and local mingling processes.

At the emplacement level, magmatic flow concentrates MME along viscosity heterogenities to form polygenic swarms in which they are associated with other relatively rigid elements present in the magmas, including accessory minerals, amphibole and biotite phenocrysts, and xenoliths (Barbarin 1991a). After emplacement, when the magma is no longer mobile, extensive chemical exchange is active until consolidation of the pluton. This results in partial chemical equilibrium and sometimes complete isotopic equilibrium between enclaves and host. Limited *in-situ* chemical exchange results in the development of generally porphyritic hybrid zones, or fine-grained margins, to discrete zoned enclaves (e.g. Cramer & Kwak 1988) and felsic haloes in the host granite around the enclaves. The frequently discontinuous and variably thick hybrid zones are separated by sharp to diffuse contacts. They also have diffuse contacts with both the cores of enclaves and the host granites. These hybrid zones which form through *in situ* chemical exchange at the emplacement level differ from the hybrid zones produced within conduits. The latter display sharp contacts with the host granite, and look like composite enclaves consisting of mafic enclave enclosed in the hybrid host.
host.

The porphyritic zones should not be confused with another type of mafic and porphyritic rock that also surrounds some enclaves. Such rocks, which are rich in accessory minerals, large amphibole and biotite phenocrysts, are not produced by mixing but represent high-level cumulates disrupted by late shearing induced by magma flowage and preserved around the relatively rigid enclaves. These rocks generally form the schlieren-layerings and the host of MME in some polygenic swarms.

8. Relative rheology of coexisting magmas and types of interaction

In most plutons, mafic magmas are not introduced into felsic magmas solely at the beginning of the plutonic history, but are generally injected into the granite systems at different stages of the crystallisation of felsic magmas. The different types of interaction that successively occur depend on the crystallinity and rheology of the felsic magmas at the time of mafic magma injection (Fig. 3).

(1) Thorough mixing, involving thermal, mechanical and chemical exchange between mafic magma and the crustal melt, results in the genesis of homogeneous hybrid magmas which might produce calc-alkaline granites. Phenocrysts which are already present in one of the components, or in both, can be transferred and altered into xenocrysts. Xenocrysts are especially abundant in the hybrid rocks.

(2) If mafic magmas are introduced slightly later, when the effective viscosity of the felsic magma is still significantly low, the viscosity contrast of the coexisting magmas is sufficiently large to permit extensive mechanical exchange only. Blobs of mafic magma form MME in the felsic magma. The viscosity in the host felsic magma generally remains low enough to allow either scattering of enclaves throughout the pluton by convection and other dispersal forces, or the local concentration of enclaves into polygenic swarms by segregation processes.

(3) If mafic magmas are introduced when the felsic magmas are partially crystallised, they are channelled into the early fractures of the nearly-solid granite rocks, and mechanically interact with the felsic components only locally to form fragmented or composite synplutonic dykes. With increasing viscosity contrast, the shapes of mafic parcels vary from rounded blobs with crenulate margins to angular fragments with breccia structures. In synplutonic or composite dykes, chemical exchange is limited, such that there is no isotopic equilibration between enclaves and their host (Barbarin *et al.* 1989; Barbarin 1989).

Table 2 Role of the various types of exchange at different structural levels, their causes and their results

Place of exchanges	Causes	Nature of exchanges	Results
EMPLACEMENT LEVEL (upper crust)	Late injections of mafic magmas	LOCAL MECHANICAL AND CHEMICAL EXCHANGE	Enclaves in monogenic swarms Synplutonic dykes (Net-veined complexes)
	Compositional contrasts between enclaves and host	EXTENSIVE CHEMICAL EXCHANGE	Partial chemical equilibrium and complete isotopic equilibrium between enclaves and host
		LIMITED CHEMICAL EXCHANGE	Fine-grained margins, felsic haloes, and scarce zoned enclaves with discontinuous hybrid zones at enclave-host contacts
MAGMA CONDUITS (dyke swarms)	Relative movement of magmas of contrasted densities	EXTENSIVE MECHANICAL EXCHANGE	Various types of enclave Composite enclaves Xenocrysts
		LIMITED CHEMICAL EXCHANGE	Hybrid rocks surrounding enclaves
MAGMA CHAMBER (lower crust)	Temperature contrasts	THERMAL EXCHANGE	Melting of crustal rocks and production of anatectic melt Chilling of mafic magma (fine grain-size, chilled margins, acicular and hollow crystals)
	Convective stirring Magma fountains	EXTENSIVE MECHANICAL EXCHANGE	(Banded structures) Enclaves Xenocrysts
	Compositional contrasts between coeval magmas	EXTENSIVE CHEMICAL EXCHANGE (INTERDIFFUSION)	Hybrid magmas (granites?)

(4) Late injection of mafic magma into quite solid granitic rock results in the formation of continuous mafic dykes. Rheology contrasts between the two components are so high that most exchange is inhibited.

The various types of interactions described above can occur within the same pluton (e.g. the Dinkey Creek and Bass Lake plutons, Sierra Nevada Batholith, California: Barbarin 1989, 1991b) or in the same limited area (e.g. the HDB quarry in the Pulau Ubin island, Malay Peninsula: Kumar 1988). The model proposed (Fig. 3) indicates that the two hypotheses which explain MME formation by mingling of mafic and felsic magmas (see Barbarin & Didier (1991), for review), or by disruption of once-continuous synplutonic or synmagmatic dykes (see Pitcher (1991), for review), respectively reflect distinct rheology contrasts. Exceptional occurrences of synplutonic mafic dykes passing into enclave trains in some plutons (Roddick & Armstrong 1959; Cobbing & Pitcher 1972; Barnes 1983; Furman & Spera 1985; Allen 1991) reflect local variations of effective viscosity within the enclosing granites.

9. Conclusions

The efficiency and spatial extent of thermal, mechanical and chemical exchange between coexisting magmas depend on their thermal and compositional contrasts, physical properties, mass fractions, and the dynamics of the magma system. Hybridisation processes occur at different stages in the evolution of the magma system and are generally combined in order to reduce contrasts between the end-members as long as the system has sufficient energy.

The sequence and efficiency of the various hybridisation processes are largely controlled by the relative rheology of the coexisting and contrasting magmas. The rheology of the participating magmas is complex because it depends on various factors, especially temperature, crystallinity and other physical and chemical parameters, which constantly vary during crystallisation. In the common situation where mafic magmas are injected at different stages of crystallisation of felsic magmas, the styles and efficiency of these interaction processes are extremely dependent upon the rheological properties of the host at the time of introduction of the mafic magma into the granite system.

Because exchange processes are evident only where they are limited, mingling is easier to prove than mixing. Geochemical and isotopic data, however, permit the tracing of magma mixing in some plutons. MME represent only 1–2% of the volume of the calc-alkaline granite plutons, although the volume of mafic magma involved in the genesis of these granites can be up to 70% or more according to one model of the isotope data (e.g. DePaolo 1981). Enclaves then represent tiny portions of the mafic magma that are resistant to the various types of interaction and succeed in partially retaining their original identity. The intermediate compositions of most enclaves are distinct from that of the mafic magma from which they originate. Even if they can be easily distinguished from their host, they are, however, strongly affected by the interaction processes. MME are hybrid rocks like the host granites and are produced by mechanical and chemical exchange between the same end-members as those that produce the relatively homogeneous granites through more extensive hybridisation.

10. Acknowledgements

We are grateful to Bruce Chappell for his invitation to the Second Hutton Symposium in Canberra, which incited us to write this paper. Careful reviews by Ed Stephens and an anonymous reviewer largely improved the earlier version of the manuscript.

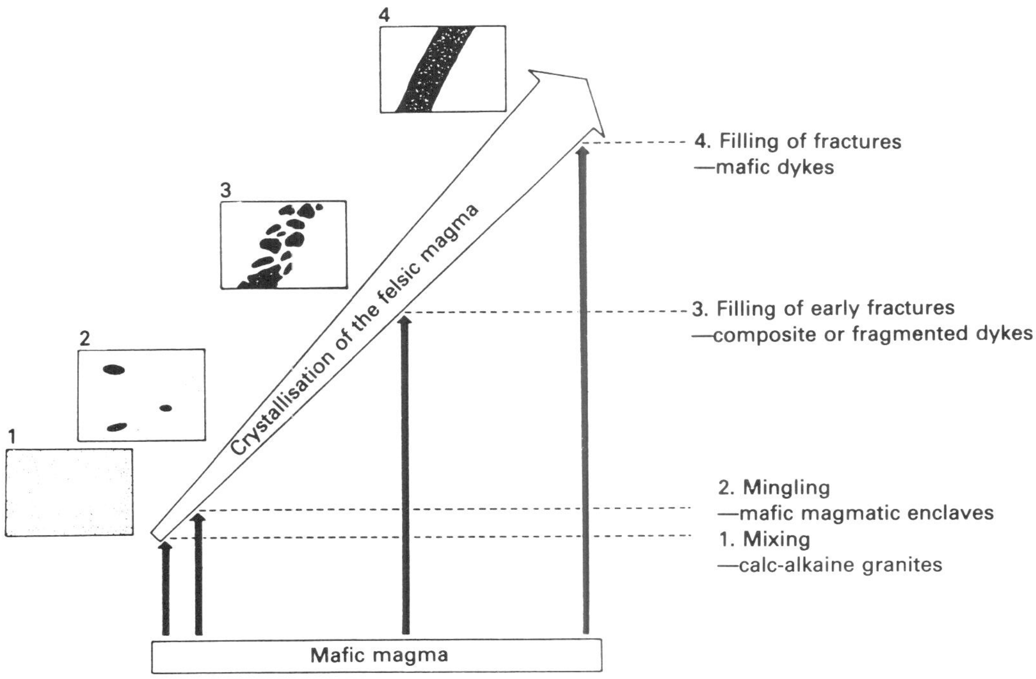

Figure 3 Sketch showing the different types of hybridisation obtained by injection of mafic magma into a granite system at different stages of crystallisation of the felsic magma; increase of the crystallinity and subsequently of the bulk viscosity of the felsic magma progressively reduces exchanges between the coexisting magmas. (from Barbarin 1989).

11. References

Alibert, C. 1980. *Etude expérimentale de la diffusion chimique des éléments majeurs entre verres et liquides de composition rhyolitique, basaltique et phonolitique.* Unpublished thesis, Université P. et M. Curie, Paris VII, France.

Alibert, C. & Carron, J. P. 1980. Données expérimentales sur la diffusion des éléments majeurs entre verres ou liquides de compositions basaltique, rhyolitique et phonolitique, entre 900°C et 1300°C, à pression ordinaire. EARTH PLANET SCI LETT **47,** 294–306.

Alibert, C. & Delbove, F. 1980. Données préliminaires sur le rôle de l'eau dans la diffusion chimique entre rhyolite et phonolite fondues à 900°C sous une pression d'eau de 4 kbar. C R ACAD SCI PARIS **291,** 789–92.

Allen, C. M. 1991. Local equilibrium of mafic enclaves and granitoids of the Turtle pluton, southeast California: Mineral, chemical, and isotopic evidence. AM MINERAL **76,** 574–88.

Bacon, C. R. 1986. Magmatic inclusions in silicic and intermediate volcanic rocks. J GEOPHYS RES **91B,** 6091–112.

Baker, D. R. 1989. Tracer versus trace element diffusion: diffusional decoupling of Sr concentration from Sr isotope composition. GEOCHIM COSMOCHIM ACTA **53,** 3015–23.

Baker, D. R. 1990. Chemical interdiffusion of dacite and rhyolite: anhydrous measurements at 1 atm and 10 Kbar, application of transition state theory, and diffusion in zoned magma chambers. CONTRIB MINERAL PETROL **104,** 407–23.

Barbarin, B. 1988. Field evidence for successive mixing and mingling between the Piolard Diorite and the Saint Julien-la-Vêtre Monzogranite (Nord Forez, Massif Central, France). CAN J EARTH SCI **25,** 49–59.

Barbarin, B. 1989. Importance des différents processus d'hybridation dans les plutons granitiques du batholite de la Sierra Nevada, Californie. SCHWEIZ MINERAL PETROGR MITT **69,** 303–15.

Barbarin, B. 1990. Plagioclase xenocrysts and mafic magmatic enclaves in some granitoids of the Sierra Nevada Batholith, California. J GEOPHYS RES **95,** 17747–56.

Barbarin, B. 1991a. Contrasted origins for the "polygenic" and "monogenic" enclave swarms in some granitoids of the Sierra Nevada batholith, California. TERRA ABSTR **2,** 32.

Barbarin, B. 1991b. Enclaves of the Mesozoic calc-alkaline granitoids of the Sierra Nevada batholith, California. *In* Didier, J. & Barbarin, B. (eds) *Enclaves and granite petrology,* DEV PETROL **13,** 135–53. Amsterdam: Elsevier.

Barbarin, B., Dodge, F. C. W. & Kistler, R. W. 1985. REE contents and Rb–Sr systematics of mafic enclaves and other associated mafic rocks, central Sierra Nevada, California. EOS **66,** 1150.

Barbarin, B., Dodge, F. C. W., Kistler, R. W. & Bateman, P. C. 1989. Mafic inclusions and associated aggregates and dikes in granitoid rocks, central Sierra Nevada batholith. Analytic data. U S GEOL SURV BULL **1899,** 1–28.

Barbarin, B. & Didier, J. 1991. Review of the main hypotheses proposed for the genesis and evolution of mafic microgranular enclaves. *In* Didier, J. & Barbarin, B. (eds) *Enclaves and granite petrology,* DEV PETROL **13,** 367–73. Amsterdam: Elsevier.

Barnes, C. G. 1983. Petrology and upward zonation of the Wooley Creek Batholith, Klamath Mountains, California. J PETROL **24,** 495–537.

Bateman, P. C. & Chappell, B. W. 1979. Crystallization, fractionation, and solidification of the Tuolumne Intrusive Series, Yosemite National Park, California. GEOL SOC AM BULL **90,** 465–82.

Bébien, J., Gagny, C. & Soussi Tanani, S. 1987. Les associations de magmas acides et basiques: des objets fractals? C R ACAD SCI PARIS **305,** 277–80.

Belin, J. M. 1988. Evolution des enclaves basiques et de leur matrice dans un granite. L'exemple du granite porphyroïde de Saint-Gervais d'Auvergne (Massif Central français). C R ACAD SCI PARIS **307,** 387–93.

Blake, S. & Campbell, I. H. 1986. The dynamics of magma-mixing during flow in volcanic conduits. CONTRIB MINERAL PETROL **94,** 72–81.

Blake, S. & Koyaguchi, T. 1991. Insights on the magma mixing model from volcanic rocks. *In* Didier, J. & Barbarin, B. (eds) *Enclaves and granite petrology,* DEV PETROL **13,** 403–13. Amsterdam: Elsevier.

Bowen, N. L. 1922. The behavior of inclusions in igneous magmas. J PETROL **30,** 513–70.

Briot, D. 1990. Magma mixing versus xenocryst assimilation: The genesis of trachyandesites in Sancy volcano, Massif Central, France. LITHOS **25,** 227–41.

Bussy, F. 1990. *Pétrogenèse des enclaves microgrenues associées aux granitoïdes calcoalcalins: exemple des massifs varisque du Mont-Blanc (Alpes occidentales) et miocène du Monte Capanne (Ile d'Elbe, Italie).* MÉM GÉOL **7,** Université de Lausanne.

Campbell, I. H. & Turner, J. S. 1985. Turbulent mixing between fluids with different viscosities. NATURE **313,** 39–42.

Campbell, I. H. & Turner, J. S. 1986. The influence of viscosity on fountains in magma chambers. J PETROL **27,** 1–30.

Campbell, I. H. & Turner, J. S. 1989. Fountain in magma chambers. J PETROL **30,** 885–923.

Cantagrel, J. M., Didier, J. & Gourgaud, A. 1984. Magma mixing: origin of intermediate rocks and "enclaves" from volcanism to plutonism. PHYS EARTH PLANET SCI **35,** 63–76.

Chappell, B. W., White, A. J. R. & Wyborn, D. 1987. The importance of residual source material (restite) in granite petrogenesis. J PETROL **28,** 1111–38.

Chappell, B. W. & White, A. J. R. 1991. Restite enclaves and the restite model. *In* Didier, J. & Barbarin, B. (eds) *Enclaves and granite petrology,* DEV PETROL **13,** 375–81. Amsterdam: Elsevier.

Cobbing, E. J. & Pitcher, W. S. 1972. The Coastal Batholith of Central Peru. J GEOL SOC LONDON **128,** 421–60.

Cocirta, C. & Orsini, J. B. 1986. Signification de la diversité de composition des enclaves "microgrenues" sombres en contexte plutonique. L'exemple des plutons calcocalcalins de Bono et Buddusò (Sardaigne septentrionale). C R ACAD SCI PARIS **302,** 331–6.

Cramer, J. J. & Kwak, T. A. P. 1988. A geochemical study of zoned inclusions in granitic rocks. AM J SCI **288,** 827–71.

Debon, F. 1975. *Les massifs granitoïdes à structure concentrique de Cauterets-Panticosa (Pyrénées occidentales) et leurs enclaves. Une étude pétrographique et géochimique.* MEM SCI TERRE **33.**

Debon, F. 1991. Comparative major element chemistry in various "microgranular enclave-plutonic host" pairs. *In* Didier, J. & Barbarin, B. (eds) *Enclaves and granite petrology,* DEV PETROL **13,** 293–312. Amsterdam: Elsevier.

DePaolo, D. J. 1981. A neodymium and strontium isotopic study of the Mesozoic calcalkaline granitic batholiths of the Sierra Nevada and Peninsular Ranges, California. J GEOPHYS RES **86,** 10470–88.

Didier, D. J. 1973. *Granites and their enclaves. The bearing of enclaves on the origin of granites.* DEV PETROL **3** Amsterdam: Elsevier.

Didier, J. 1987. Contribution of enclave studies to the understanding of origin and evolution of granitic magmas. GEOL RUNDSCHAU **76,** 41–50.

Didier, J. & Barbarin, B. (eds.) 1991. *Enclaves and granite petrology.* DEV PETROL **13**. Amsterdam: Elsevier.

Didier, J. & Lameyre, J. 1969. Les granites du Massif Central français: étude comparée des leucogranites et granodiorites. CONTRIB MINERAL PETROL **24,** 219–38.

Dodge, F. C. W. & Kistler, R. W. 1990. Some additional observations on inclusions in the granitic rocks of the Sierra Nevada. J GEOPHYS RES **95,** 17841–8.

Dorais, M. J., Whitney, J. A. & Roden, M. F. 1990. Origin of mafic enclaves in the Dinkey Creek pluton, central Sierra Nevada Batholith, California. J PETROL **31,** 853–81.

Fernandez, A. N. & Barbarin, B. 1991. Relative rheology of coexisting mafic and felsic magmas: Nature of resulting interaction processes. Shape and mineral fabrics of mafic microgranular enclaves. *In* Didier, J. & Barbarin, B. (eds) *Enclaves and granite petrology,* DEV PETROL **13,** 263–75. Amsterdam: Elsevier.

Fershtater, G. B. & Borodina, N. S. 1977. Petrology of autoliths in granitic rocks. INT GEOL REV **19,** 458–68.

Fershtater, G. B. & Borodina, N. S. 1991. Enclaves in the Hercynian granitoids of the Ural Mountains, U.S.S.R.. *In* Didier, J. & Barbarin, B. (eds) *Enclaves and granite petrology,* DEV PETROL **13,** 83–94. Amsterdam: Elsevier.

Fourcade, S. & Javoy, M. 1991. Sr–Nd–O isotopic features of mafic microgranular enclaves and host granitoids from the Pyrenees, France: Evidence for their hybrid nature and inference on their origin. *In* Didier, J. & Barbarin, B. (eds) *Enclaves and granite petrology,* DEV PETROL **13,** 345–64. Amsterdam: Elsevier.

Furman, T. & Spera, F. J. 1985. Co-mingling of acid and basic magma with implications for the origin of mafic I-type xenoliths: field and petrochemical relations of an usual dike complex at Eagle Peak Lake, Sequoia National Park, California, U.S.A. J VOLCANOL GEOTHERM RES **24,** 151–78.

Gamble, J. A. 1979. Some relationships between coexisting granitic and basaltic magmas and the genesis of hybrid rocks in the Tertiary central complex of Slieve Gullion, northeast Ireland. J VOLCANOL GEOTHERM RES **5,** 297–316.

Gourgaud, A., Cantagrel, J. M. & Vincent, P. M. 1981. Mélange de magmas et pétrogenèse des trachyandésites du Mont-Dore (Massif Central français). C R ACAD SCI PARIS **293,** 711–6.

Hibbard, M. J. & Watters, R. J. 1985. Fracturing and diking in incompletely crystallized granitic plutons. LITHOS **18,** 1–12.

Hofmann, A. W. 1980. Diffusion in natural silicate melts: a critical review. *In* Hargraves, R. B. (ed.) *Physics of magmatic processes,* 385–417. Princeton, New Jersey: Princeton University Press.

Holden, P., Halliday, A. N. & Stephens, W. E. 1987. Neodymium and strontium isotope content of microdiorite enclaves points to mantle input into granitoid production. NATURE **330,** 53–6.

Huppert, H. E., Sparks, R. S. J. & Turner. J. S. 1984. Some effects of viscosity on the dynamics of replenished magma chamber. J GEOPHYS RES **89,** 6857–77.

Huppert, H. E. & Sparks, R. S. J. 1988. The generation of granitic magmas by intrusion of basalt into the crust. J PETROL **29,** 599–624.

Johnston, A. D. & Wyllie, P. J. 1988. Interaction of granitic and basic magmas: experimental observations on contamination processes at 10 kbar with H_2O. CONTRIB MINERAL PETROL **98,** 352–62.

Kouchi, A. & Sunagawa, I. 1983. Mixing basaltic and dacitic magmas by forced convection. NATURE **304,** 527–8.

Kouchi, A. & Sunagawa, I. 1985. A model for mixing basaltic and dacitic magmas as deduced from experimental data. CONTRIB MINERAL PETROL **89,** 17–23.

Koyaguchi, T. 1985. Magma mixing in a conduit. J VOLCANOL GEOTHERM RES **25,** 365–9.

Koyaguchi, T. 1986. Evidence for two-stage mixing in magmatic inclusions and rhyolitic lava domes on Niijima Island, Japan. J VOLCANOL GEOTHERM RES **29,** 71–98.

Koyaguchi, T. 1987. Magma mixing in a squeezed conduit. EARTH PLANET SCI LETT **84,** 339–44.

Koyaguchi, T. & Blake, S. 1991. Origin of mafic enclaves: Constraints on the magma mixing model from fluid dynamic experiments. *In* Didier, J. & Barbarin, B. (eds) *Enclaves and granite petrology,* DEV PETROL **13,** 415–429. Amsterdam: Elsevier.

Kumar, S. C. 1988. Microgranular enclaves in granitoids: agents of magma mixing. J SOUTHEAST ASIAN EARTH SCI **2,** 109–21.

Lesher, C. E. 1990. Decoupling of chemical and isotopic exchange during magma mixing. NATURE **344,** 235–7.

Oldenburg, C. L. M., Spera, F. J., Yuen, D. A. & Sewell, L. G. 1989. Dynamic mixing in magma bodies: theory, simulations and implications. J GEOPHYS RES **94,** 9215–36.

Pabst, A. 1928. Observations on inclusions in the granitic rocks of the Sierra Nevada. UNIV CALIFORNIA PUB DEPT GEOL SCI **17,** 325–86.

Pin, C. 1991. Sr–Nd isotopic study of igneous and metasedimentary enclaves in some Hercynian granitoids from the Massif Central, France. *In* Didier, J. & Barbarin, B. (eds) *Enclaves and granite petrology,* DEV PETROL **13,** 333–43. Amsterdam: Elsevier.

Pin, C., Binon, M., Belin, J. M., Barbarin, B. & Clemens, J. D. 1990. Origin of microgranular enclaves in granitoids: equivocal Sr–Nd evidence from Hercynian rocks in the Massif Central (France). J GEOPHYS RES **95,** 17821–8.

Pitcher, W. S. 1991. Synplutonic dykes and mafic enclaves. *In* Didier, J. & Barbarin, B. (eds) *Enclaves and granite petrology,* DEV PETROL **13,** 383–91. Amsterdam: Elsevier.

Roddick, J. A. & Armstrong, J. E. 1959. Relict dikes in the Coast Mountains near Vancouver, B. C., J GEOL **67,** 603–13.

Shaw, H. R., Smith, R. L & Hildreth, W. 1976. Thermogravitational mechanisms for chemical variations in zoned magma chambers. GEOL SOC AM ABSTR PROG **8,** 1102.

Sparks, R. S. J., Sigurdsson, H. & Wilson, L. 1977. Magma mixing: a mechanism for triggering acid explosive eruptions. NATURE **267,** 315–8.

Sparks, R. S. J. & Marshall, L. 1986. Thermal and mechanical constraints on mixing between mafic and silicic magmas. J VOLCANOL GEOTHERM RES **29,** 99–124.

Watson, E. B. 1981. Diffusion in magmas at depth in the Earth: the effects of pressure and dissolved H_2O. EARTH PLANET SCI LETT **52,** 291–301.

Watson, E. B. 1982. Basalt contamination by continental crust: some experiments and models. CONTRIB MINERAL PETROL **80,** 73–87.

White, A. J. R. & Chappell, B. W. 1977. Ultrametamorphism and granitoid genesis. TECTONOPHYSICS **43,** 7–22.

Wiebe, R. A. 1973. Relations between coexisting basaltic and granitic magmas in a composite dike. AM J SCI **273,** 130–51.

Wiebe, R. A. 1980. Commingling of contrasted magmas in the plutonic environment: examples from the Nain anorthositic complex. J GEOL **88,** 197–209.

Wiebe, R. A. 1991. Commingling of contrasted magmas and generation of mafic enclaves in granitic rocks. *In* Didier, J. & Barbarin, B. (eds) *Enclaves and granite petrology,* DEV PETROL **13,** 393–402. Amsterdam: Elsevier.

Yoder, H. S. Jr 1973. Contemporaneous basaltic and rhyolitic magmas. AM MINERAL **58,** 153–71.

BERNARD BARBARIN, Laboratoire de Pétrographie-Volcanologie, Université de Paris-Sud, Bât. 504, F-91405 Orsay Cedex, France.
JEAN DIDIER, Département des Sciences de la Terre, Université Blaise Pascal et URA 10 C.N.R.S., 5, rue Kessler, F-63038 Clermont-Ferrand Cedex, France.

Transactions of the Royal Society of Edinburgh: Earth Sciences, **83,** 155–164, 1992

C-type magmas: igneous charnockites and their extrusive equivalents

Jonathan A. Kilpatrick and David J. Ellis

ABSTRACT: Igneous charnockites are characterised by distinctively high abundances of K_2O, TiO_2, P_2O_5 and LIL elements and low CaO at a given SiO_2 level compared to metamorphic charnockites, and I-, S- and A-type granites. They form a distinctive type of intrusive igneous rocks, the Charnockite Magma Type (CMT or C-type), which generally lack hornblende and consist of pyroxene, alkali feldspar, plagioclase, quartz, biotite, apatite, ilmenite and titanomagnetite. Although this mineral assemblage superficially resembles that of metamorphic charnockites, magmatic charnockites are characterised by inverted pigeonite, exceptionally calcic alkali feldspar, potassic plagioclase, and coexisting opaque oxides, all with crystallisation temperatures of 950–1050°C. Apatite is a ubiquitous phase which, together with the very high concentrations of Zr and TiO_2 over a wide silica range, is consistent with the derivation of the Charnockite Magma Type by very high temperature partial melting and fractionation.

The credibility of intrusive charnockites as a magmatic type has historically foundered because of their apparent restriction to granulite belts and the absence of any reported extrusive equivalents. We report examples of volcanic rocks, of various ages, with the same distinctive major and trace element compositions, mineral assemblages and high temperatures of crystallisation as intrusive charnockites.

The Charnockite Magma Type is considered to be derived by melting of a hornblende-free or poor, LILE-enriched fertile granulite source which had not been geochemically depleted by a previous partial melting event but which was dehydrated in an earlier metamorphism. Whereas H_2O-saturated melting produces migmatites or "failed" granites, and vapour-absent melting of an amphibolite can produce I-type granites, according to this model the vapour-absent melting of a hornblende-free or hornblende-poor granulite at even higher temperatures produces charnockites.

KEY WORDS: C-types, granulites, magmatic, metamorphic, volcanics

A dichotomy has developed in the literature as to the maximum temperatures which can reasonably be attained during crustal metamorphism, and by implication, crustal anatexis and granite formation. Several recent reviews of granulites (Bohlen & Boettcher 1981; Newton & Hansen 1983) convey the impression of average granulite facies metamorphism occurring between 700–850°C. Indeed, a recent model of granulite formation implies that many granulites do not require higher temperatures than many amphibolites, but may simply form by the dehydration of amphibolites due to the reduced activity of H_2O caused by CO_2 flushing on a regional scale (Friend 1981; Newton & Hansen 1983). In contrast, there are now well-documented examples of regional granulite terranes which equilibrated at temperatures in excess of 950–1000°C (see Ellis 1980; Ellis *et al.* 1980; Harley 1989; Bertrand *et al.* 1991, and references therein).

That such temperatures are realised in crustal rocks has considerable implications for the range of crustal melts which can be produced. Not only is temperature an important factor, but the composition and presence or absence of a free fluid phase has a profound influence on the formation and compositions of granites (see Nekvasil & Burnham 1987, and references therein). At one extreme, it is generally agreed that H_2O-excess melting at low temperatures (670–700°C) produces H_2O-saturated melts. Field evidence for such a process is provided by migmatites, which have been termed "failed" granites (Le Breton & Thompson 1988; Clemens 1991) because of the inability of H_2O-saturated melts to rise to shallower levels without freezing. There is now a common view that most granites form at higher temperatures (*c.* 800–900°C) through fluid-absent melting reactions involving quartz and feldspar together with hornblende in di-normative (mafic-intermediate) and biotite (and possibly muscovite) in C-normative (pelitic) source rocks. The I- and S-type classification scheme of White and Chappell (1977) reflects these two constrasting source compositions.

The general reticence to accept high temperatures for crustal metamorphism as a relatively common occurrence is also reflected in the igneous rock literature. Thus, a common perception is that in the absence of hornblende in di-normative rocks, crustal melting under very reduced activities of H_2O is unlikely in the crust (see e.g. Wyllie 1977; Clemens 1991). Although there are regional granulite terranes equilibrated at temperatures in excess of 1000°C, there has never been a serious appraisal of the possible granitic rocks that could be produced under such conditions. We here report distinctive geochemical and mineralogical evidence which identifies the Charnockite Magma Type as a separate igneous association (C-type), distinct from I-, S- and A-type granites of various ages, island arc magmas (M-type), and the volumetrically minor felsic differentiates of tholeiitic intrusions. We suggest that the C-type charnockite intrusions represent the products of very high temperature crustal melting of "dry", but geochemically fertile, granulites occurring at temperatures well in excess of those needed for normal I- and S-type granites.

1. Charnockites: historical review

Holland (1900) first applied the term charnockite to a hypersthene-bearing granite from Madras in Southern India, to which Howie (1954) later ascribed a magmatic origin. From that time charnockites have been described from most continents, and it has been observed that charnockites are usually restricted to intermediate and high pressure granulite terranes, with typically a granulitic mineralogy. Charnockites have been the focus of much debate over an igneous versus metamorphic origin (see Cooray 1969; de Waard 1969; Ravich 1972; Sheraton 1982; Newton & Hansen 1983; Young & Ellis 1990). Many recent studies have suggested that charnockitisation was caused by CO_2 infiltration into pre-existing amphibolites and granulites, particularly the classic charnockite areas of Sri Lanka and southern India (Janardhan *et al.* 1979; 1982; Friend 1981; Condie *et al.* 1982, 1986; Condie & Allen 1984; Hansen *et al.* 1984, 1987; Allen *et al.* 1985). In these areas, precursor amphibolites are said to have been converted to orthopyroxene–biotite–feldspar–quartz charnockitic gneisses at 700–800°C and 500–570 MPa (5–7·5 kbar) (Condie & Allen 1984; Hansen *et al.* 1984, 1987). The large scale over which this CO_2 streaming must have occurred lead Newton and Hansen (1983) to suggest that almost all charnockites that have intrusive field relations were once normal hornblende-bearing granites which have undergone a later subsolidus charnockitisation (dehydration).

In this paper we show that intrusive charnockites exist in a number of continents, and that they define a previously unrecognised and distinctive magma type, the Charnockite Magma Type (CMT or C-type). Field, geochemical and petrographic features of these igneous charnockites distinguish them from metamorphic charnockites.

Previously, igneous charnockites have been thought to be restricted in occurrence to plutonic rocks in granulite terranes. This is shown by the fact that the IUGS classification of igneous rocks (Le Maitre 1988) classifies charnockitic plutonic rocks but makes no mention of extrusive equivalents. Indeed, an examination of standard petrological textbooks reveals a lack of interest in charnockites as possible igneous rocks: earlier textbooks (e.g. Turner & Verhoogen 1960; Barth 1967) devote a few pages to the subject, whereas some more recent textbooks (Cox *et al.* 1979; Wilson 1989) make no mention of charnockites. We have identified extrusive equivalents of intrusive charnockites for the first time. These volcanics have the same distinctive geochemical and mineralogical characteristics as the C-type plutonic rocks.

2. Petrography and mineralogy of the Charnockite Magma Type

A characteristic feature of most intrusive charnockites is that they have been deformed and recrystallised to varying degrees following emplacement. This results in much of the original igneous texture and mineralogy being modified to lower temperature, although still maintaining charnockitic assemblages (e.g. Young & Ellis 1990, 1991; Young & Black 1991). However, where deformation was not completely pervasive, distinctive high temperature textural and mineralogical features remain.

The CMT primarily consists of magmas that crystallised to give orthopyroxene ± clinopyroxene + biotite + feldspar + quartz (with accessory ilmenite and apatite). Charnockites (*sensu stricto*) should have a quartz:alkali-feldspar:plagioclase ratio that is granitic, as defined by Le Maitre (1988); however, within the CMT we have also included those magmas (generally regarded as differentiates) which can be considered cogenetic to charnockites (*sensu stricto*) and which may not necessarily contain the true charnockite petrography, as defined above, i.e. may contain hornblende and little or no orthopyroxene. These differentiates retain the characteristic geochemical signature of magmatic charnockites, as defined below, inherited from their parental magmas. The crystallisation of an almost anhydrous assemblage causes true igneous charnockites to fractionate to more hydrous differentiated magmas which precipitate a higher proportion of hornblende and lesser pyroxene than the precursor magmas; these differentiated magmas depart from the characteristic charnockite mineral assemblage but are still intrinsically related, through magmatic processes, to the origin of igneous charnockites. Thus, we prefer to use the term Charnockite Magma Type not only to include those magmas which contain the charnockite assemblage and have a distinctive geochemical signature, but also to include and recognise the differentiation of these magmas. The recognition of the distinct geochemical signature of the CMT in these differentiates is fundamental in classifying them as C-type magmas.

The petrography and order of crystallisation of igneous charnockites is very similar to that deduced by Barth and contrasts with that of common hornblende-bearing granites (Barth 1967, p. 223, Figures III–V). An addition to the scheme of Barth is the additional presence of early crystallising pigeonite, which we believe to have been a liquidus phase. Unlike many I-type granites and calc-alkaline suites, in which hornblende crystallises before biotite (see also Wones & Dodge 1977), in charnockites biotite is the hydrous phase, and minor hornblende, if present at all, is either a very late stage crystallisation product from the higher SiO_2 variants, or else a secondary subsolidus phase. Unlike many I-type granites and island arc magmas, euhedral hornblende is absent from the early crystallisation sequence in the CMT and never occurs as inclusions in the other minerals.

The Ardery Charnockitic Intrusions (ACI) of the Windmill Islands, Antarctica (Blight & Oliver 1977; Kilpatrick *et al.* 1990; Kilpatrick & Ellis 1991) provide an excellent example of the CMT, having undergone little deformation and still preserving the original igneous textures and assemblages which, to varying degrees, may also still be seen preserved in more deformed and recrystallised charnockite intrusions. The ACI are characterised by the essential assemblage pyroxenes–feldspars–quartz–biotite–apatite–opaque oxides. Coarse-grained minerals vary from euhedral to subhedral and display good textural evidence for sequential crystallisation, unlike the metamorphic charnockites of Sri Lanka and India. Many samples are porphyritic with prominent phenocrysts of alkali feldspar and lesser plagioclase. Unlike I- and S-type granites and island arc rocks, which characteristically have complex oscillatory zoned feldspars, the charnockite feldspars show simple, normal zoning. Alkali feldspar is commonly strongly perthitic and of composition Or_{80}–Or_{95}. Early crystallising alkali feldspar may contain in excess of 11 mol% An, attesting to high temperatures of crystallisation (Fuhrman & Lindsley 1988). Plagioclase feldspar generally ranges from An_{55}–An_{30} throughout and is often strongly mesoperthitic. The feldspars commonly contain inclusions of euhedral pyroxene, apatite and opaque oxides. In some instances the earliest formed pyroxene is pigeonite with $m^* = 40–45$ ($m^* = 100^*Mg/Mg + Fe^{2+}_{tot}$), overgrown by orthopyroxene (Fig. 1). At lower temperatures, pigeonite is replaced by orthopyroxene and clinopyroxene in the crystallisation sequence, all of which can develop complex subsolidus exsolutions attributable to the original high temperatures of crystallisation. Apatite is a ubiquitous phase with an apparently wide crystallising temperature interval, and occurs as squat euhedral to subhedral crystals within pigeonite (Fig. 1) and other early crystallising minerals, subhedral-anhedral inclusions in opaques and biotite, and as acicular crystals. Biotite is the characteristic hydrous phase, joining the above minerals well below the liquidus. Euhedral crystals of titanomagnetite and ilmenite occur within feldspars and pyroxenes, while being typically anhedral in the groundmass. Opaque oxide pairs isolated within early crystallising phases have compositions indicating equilibration temperatures of 950–1050°C at an oxygen fugacity equivalent to QFM. Quartz occurs as subhedral inclusions concentrated in zones towards the rims of some feldspars and in the groundmass. With increasing SiO_2 content, there is a decrease in the modal proportion of ferromagnesian minerals and apatite, and an increase in quartz and feldspar. Inverted pigeonite is apparently absent from rocks with greater than 70% SiO_2, but its place is not taken by fayalite because of the relatively Mg-rich composition of these magmas.

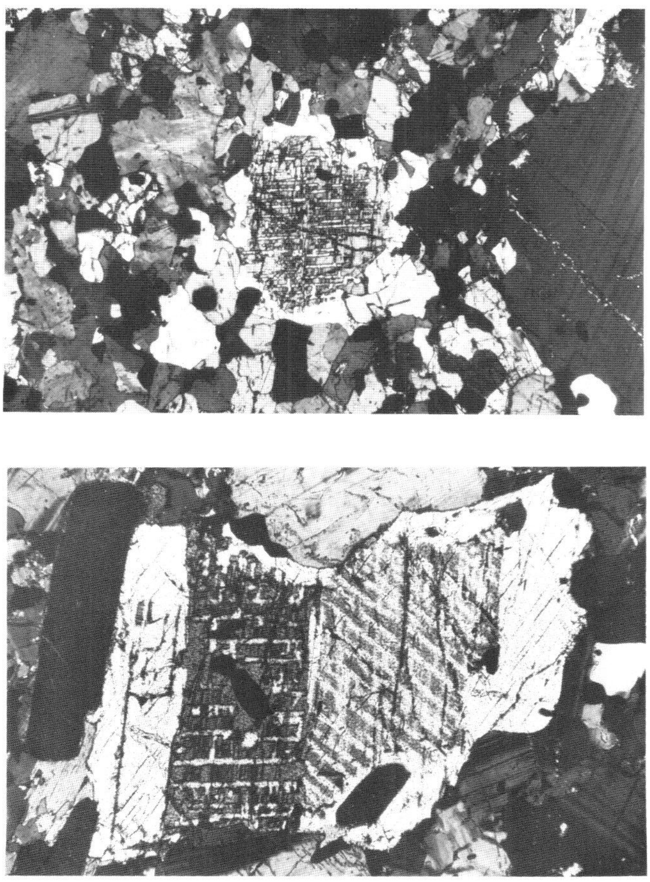

Figure 1(A), (B) Inverted pigeonite surrounded by orthopyroxene; note the presence of large euhedral apatite, both as inclusions in pigeonite and in the groundmass of (b); cross-polars, width of pictures 2·5 mm.

3. Geochemistry of intrusive charnockites

We have compiled a database of chemical analyses of magmatic charnockites and associated cumulates and derivative rocks. This is based on our own data for Antarctic and Central Australian charnockites and published analyses from elsewhere for which a clear indication of a magmatic origin (i.e. magmatic layering, intrusive contacts, enclave swarms, high crystallisation temperatures) has been reported. The Ardery Charnockitic Intrusions, Antarctica (Blight & Oliver 1977; Kilpatrick *et al.* 1990) are an excellent example of the Charnockite Magma Type. They are a Proterozoic series of related plutons which show field evidence for intrusion and subsequent differentiation, with abundant evidence of the related cumulates. The initial magma was itself a charnockite with about 62% SiO_2, but there is an overall range from 45–78% SiO_2 with the lower SiO_2 levels representing cumulates.

Over the broad SiO_2 range of 45–78% (Table 1), the CMT can be distinguished and characterised by their very high TiO_2, P_2O_5, and K_2O, lower CaO, and lower m* for a given SiO_2 content, relative to metamorphic charnockites (Fig. 2). The CMT's features also distinguish them from other magma types. It can be seen from Figure 2 that the CMT represents a significant departure, to varying degrees, from most crustal magmas as represented, for example, by the Lachlan Fold Belt granites. The extreme, and coupled, abundances of TiO_2, P_2O_5 and K_2O, as well as lower CaO, are unique among igneous rocks. The high TiO_2 and P_2O_5 are reflected in the high modal abundances of opaque oxide and apatite throughout. The distinctive nature of the magmatic charnockites, associated cumulates and derivative magmas justifies the definition of a previously unrecognised magma type: the Charnockite Magma Type or C-type.

The intrusive CMT members occur dominantly within late Archaean–Proterozoic granulite terranes from Australia (Musgrave Ranges), Antarctica (Windmill Islands, MacRobertson Land, Prince Charles Mountains, and the Bunger Hills), Africa (Limpopo Belt), North America (Adirondacks), and Scandinavia. The same distinctive chemical features are found in all these locatities. While other common magma types, such as modern island arc magmas (Ewart 1982), can show very high K_2O contents at a given SiO_2 level, they are characterised by low TiO_2 abundances. In addition, there is a general broad correlation between increasing H_2O content and the early appearance of hornblende at the expense of pyroxene in high-K island arc series (Ewart 1982). In contrast, the CMT is distinct in its coupled enrichment in TiO_2, P_2O_5, and K_2O and inferred low H_2O contents and general lack of hornblende. The relatively low CaO is illustrated in Figure 3(A). I-type granites and island arc calc-alkaline series show a trend from mafic towards higher relative Al compositions and away from amphibole (Fig. 3(A)). This is consistent with the control of feldspar-amphibole either along a cotectic or through resite unmixing (see Ellis & Thompson 1986). In contrast, the CMT is significantly offset from this trend to a low Ca compositional range, consistent with the scarcity of amphibole. There is also a striking relatively higher FeO + MgO at a given SiO_2 level in the CMT compared to other common mafic-felsic di-normative suites (Fig. 3).

By comparison, an average I-type granite of the Lachlan

Table 1 Representative analyses of the Charnockite Magma Suite

	1	2	3	4	5	6	7	8	9	10	11	12
	Ardery Charnockitic Intrusions					Bunger Hills Charn.	Mawson Charn.	Prince Charles Mountains Charn.	Limpopo Belt Charn.	Musgraves Ranges Charn.	Norway Charn.	Sweden Charn.
SiO_2	56·23	62·02	66·37	70·75	77·34	60·98	60·96	63·15	60·76	61·50	60·21	61·16
TiO_2	2·55	1·63	1·16	0·75	0·23	1·83	2·09	1·33	1·40	1·44	1·53	1·30
Al_2O_3	14·39	14·38	13·85	13·28	11·80	13·57	13·97	14·27	14·15	15·40	13·91	14·9
FeO_{tot}	11·01	7·99	6·47	4·25	1·27	9·90	9·24	7·08	7·76	7·14	9·97	7·09
MnO	0·20	0·16	0·09	0·09	0·02	0·19	0·22	0·14	0·17	0·14	0·11	0·14
MgO	2·28	1·89	1·44	0·82	0·19	2·07	2·53	2·57	3·84	1·41	1·63	1·88
CaO	6·55	4·89	3·67	2·46	0·69	5·24	5·01	4·78	5·43	4·77	3·17	4·37
Na_2O	2·86	2·90	2·51	2·85	1·86	2·27	1·82	2·78	3·08	3·68	4·12	4·02
K_2O	2·98	3·66	4·08	4·51	6·53	3·46	3·62	3·26	2·72	3·88	4·59	4·63
P_2O_5	0·94	0·48	0·35	0·25	0·06	0·49	0·53	0·64	0·68	0·65	0·66	0·52
m*	26·99	29·61	28·43	25·61	21·18	27·1	32·76	39·27	46·89	25·98	22·56	32·04
Ba	1450	1180	1140	915	335	1517	1140		1699			
Rb	70	111	153	182	185	91	114		89			
Sr	267	245	186	139	81	284	169		854			
Pb	19	24	24	31	28	21	17					
Th	3	6	4·5	26	13	1	4					
U	<0·5	1	1	1	1	<0·5	1					
Zr	303	317	305	281	114	340	460		287			
Nb	21·5	20	10	15	1·5	17	28		17			
Y	59	48	37	45	6	55	46		47			
La	55	42	41	50	13	61	65					
Ce	134	102	95	114	20	130	135					
V	162	96	90	37	8	139	93		131			
Cr	4	10	11	5	<1	16	26		188			
Ni	7	13	15	4	1	4	9		64			
Cu	24	19	19	8	2	14	17		23			
Zn	118	100	75	71	14	117	95		126			
Ga	21	19·5	18	16·5	12·5	19	21					

Notes: $m^* = 100^*Mg/(Mg + Fe^{+2}_{tot})$; Major elements normalised to 100% volatile free; 1–5, 7, 10: Unpublished data; 6: Sheraton *et al.* (in press); 8: Ravich *et al.* (1985); 9: van Reenen & Roering (1990); 11: Duchesne *et al.* (1987); 12: Hubbard & Whitley (1979).

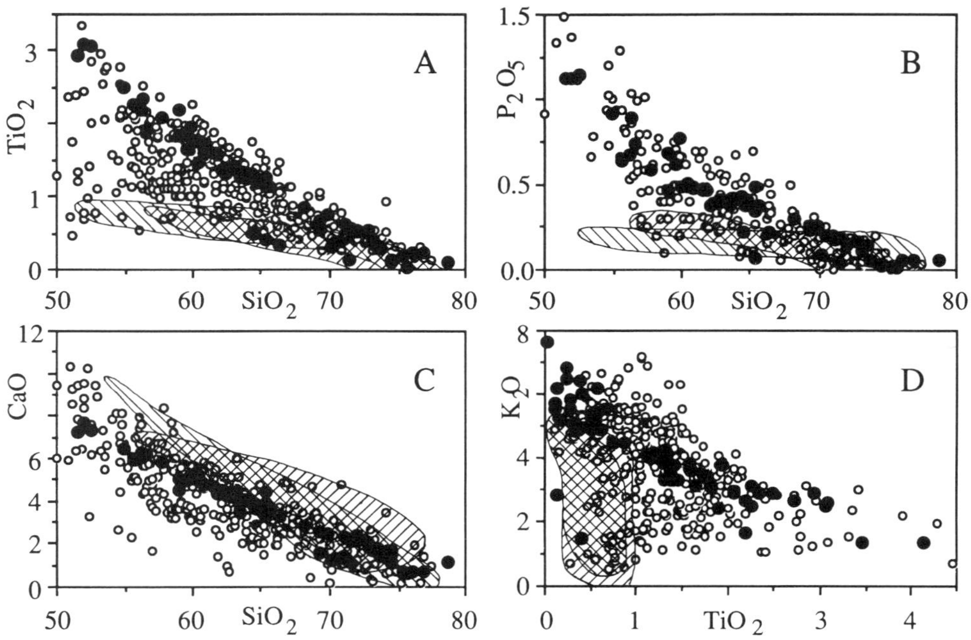

Figure 2 Geochemical variation diagrams showing members of the Charnockite Magma Type (○), Ardery Charnockitic Intrusions (●), and the fields for metamorphic charnockites (\\\\\\\\\\) and Lachlan Fold Belt granites (/////) (comprising 1958 analyses); references for the data, where published, are available from the authors.

Fold Belt (White & Chappell 1983) shows considerably higher CaO and lower K_2O, TiO_2 and P_2O_5 relative to typical CMT analyses (Fig. 3). Also the CMT is clearly different from the C-normative S-type granites derived by the partial melting of greywackes (Table 2).

There are marked geochemical and mineralogical differences between A-type granites (Loiselle & Wones 1979; Collins *et al.* 1982; Whalen *et al.* 1987) and the CMT. A-types are commonly high SiO_2 rocks, whereas the CMT is characterised by a wider range of SiO_2 contents, with parental compositions having about 62% SiO_2. Although A-types are characteristically enriched in HFSE and certain incompatible elements (Table 2), an important characteristic is the extremely high ratio of Fe to Mg. Thus A-type magmas generally have m* of less than 10 and commonly less than 5, whereas unfractionated I-types and calc-alkaline island arc suites have m* of ≈45. The CMT has m* ≈ 25–40 over a very wide SiO_2 range (Table 1). This is illustrated in Figure 3, where it can be seen that the CMT defines a trend towards relative alkali enrichment with increasing SiO_2 which is quite distinct from that of the more Fe-rich A-types and the more Mg-rich I-types. Also shown are the volumetrically minor granophyric products of differentiated continental tholeiitic basalt intrusions. This shows that direct fractionation of a tholeiitic parent magma produces "granites" which are extremely Fe-rich compared to charnockites. Indeed there is a closer resemblance between A-types and continental basalt fractionation products (Turner *et al.* 1991) than there is between A-type granites and charnockites, even though enriched trace element abundances in all three might superficially suggest a similar origin.

These differences in geochemistry are also reflected in the contrasting mineralogies and temperatures of formation. A notable feature of many A-type granites (and basalt-derived granophyres) is the presence of fayalite.

Early (or restite) phases are most commonly clinopyroxene and orthopyroxene in I-type granites but, because of reaction during cooling, the final crystallisation products are hornblende and biotite. These pyroxenes are still present in volcanic rocks of equivalent composition, but the temperatures are too low for the crystallisation of pigeonite. It is generally agreed that I-type magmas are produced by vapour-absent melting of hornblende at temperatures around 850–900°C. In fractionated I-types in which m* reaches about 10, fayalite may be an early crystallisation product and it is subsequently made over to an aggregate of annitic biotite, but this reaction with the magma is probably earlier in the crystallisation sequence than for the more anhydrous A-types in which the biotite is very late and interstitial.

In contrast, charnockitic magmas crystallise early pigeonite, commonly preserved in the volcanic rocks. In intrusive charnockites this may be seen as inverted pigeonite. Because of the low H_2O and high temperature of the charnockite magma, pyroxenes are stable down to the

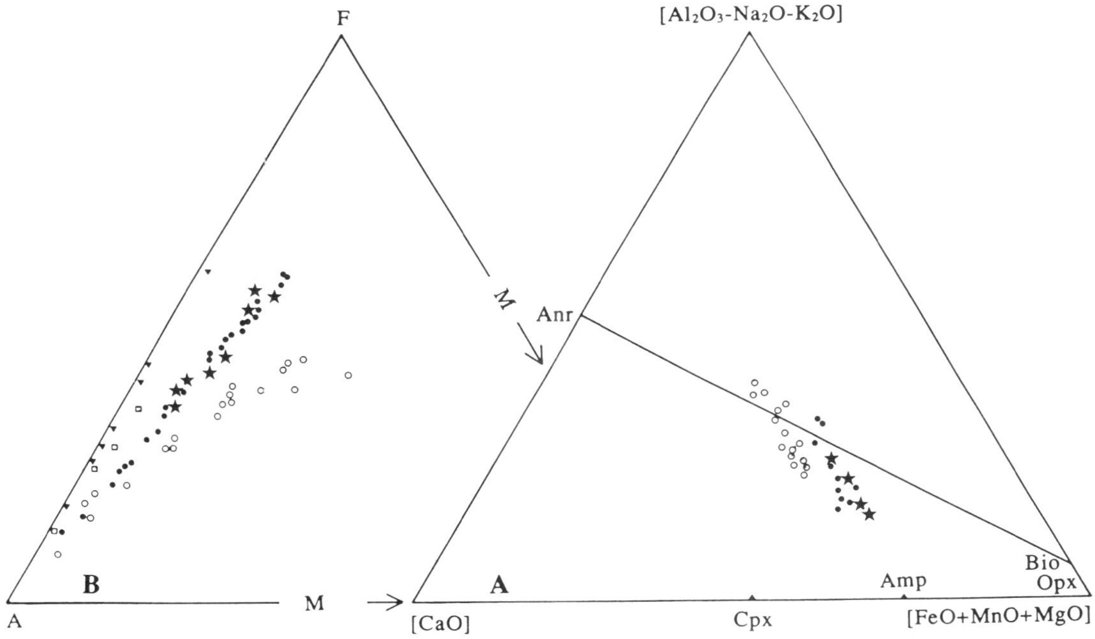

Figure 3(A) ACF (molecular %) plot (projected from SiO_2 and H_2O) of the relative compositions of representative I-type granites (open circles) from the Lachlan Fold Belt and the Tuolumne Intrusive Series, California, and extrusive equivalents of the Charnockite Magma Type (stars), compared to the charnockite intrusives discussed here (closed circles); (Chappell *et al.* 1991, Tables 2, 12; White & Chappell 1977, Table 1; Bateman & Chappell 1979, Table 2).

Figure 3(B) AFM (wt%, total iron as FeO) plot of I-type granites and calc-alkaline island arc suites (open circles) compared to the Charnockite Magma Type (closed circles), the more Fe-rich A-type granites (open squares), differentiates of tholeiitic basalts (inverted triangles), and the extrusive equivalents of the Charnockite Magma Type (stars); although all very high SiO_2 rocks merge towards the alkali-rich end of the spectrum, the Charnockite Magma Type is clearly different from these other rock suites; (data sources: I-types: see Fig. 3 and Turner & Verhoogen 1960, Table 28 for Cascades volcanics; A-types: Collins *et al.* 1982; Whalen *et al.* 1987; differentiated basalt intrusion granophyres, granites and rhyolite: Tasmanian granophyre, Hergt 1987; Skaergaard granophyre, Naslund 1989; Bushveld Complex granites, Kleeman & Twist 1989, Table 5, Iceland rhyolite, MacDonald *et al.* 1990).

Table 2 Selected analyses of contrasting magma types

	1 average I-type $n = 991$	2 average S-type $n = 578$	3 average A-type $n = 148$	4 average Mumbulla Suite $n = 8$	5 average Gabo Suite $n = 9$	6 Bushveld Granite	7 Skaergaard granophyre	8 Iceland rhyolite	9 Tasmanian granophyre
SiO_2	69·17	70·27	73·81	77·21	73·04	72·70	71·00	73·48	67·29
TiO_2	0·43	0·48	0·26	0·13	0·37	0·26	0·45	0·21	0·83
Al_2O_3	14·33	14·10	12·40	11·79	12·62	12·16	11·60	12·58	11·09
FeO_{tot}	3·23	3·37	2·70	1·17	2·98	3·74	4·94	2·99	8·48
MnO	0·07	0·06	0·06	0·03	0·08	0·06	0·09	0·08	0·14
MgO	1·42	1·42	0·20	0·04	0·33	0·05	0·11	0·03	0·20
CaO	3·20	2·03	0·75	0·39	0·96	1·13	1·72	0·38	3·60
Na_2O	3·13	2·41	4·07	3·08	3·70	3·76	4·44	5·23	2·89
K_2O	3·4	3·96	4·65	5·00	4·11	4·88	3·43	4·16	3·08
P_2O_5	0·11	0·15	0·04	0·02	0·08	0·02	0·05	0·01	0·19
m*	43·9	42·9	11·7	5·7	16·5	2·3	3·8	1·8	4·0
Ba	538	468	352	575	767	1137	840	180	635
Rb	151	217	169	242	167	170	75	125	121
Sr	247	120	48	43	148	84	130	15	101
Pb	19	27	24	37	28	20			17
Th	18	18	23	26	21	22			14
U	4	4	5	6	5	5			2
Zr	151	165	528	170	490	396	1080	1045	334
Nb	11	12	37	19	25	21	90	173	16
Y	28	32	75	90	83	57	170	128	59
La				64	62	76	103	127	39
Ce	64	64	137	150	153	170	222	249	88
V	60	56	6	2	6		9		1
Cr				<1	2		25		1
Ni	7	13	<1	<1	1	9	8		1
Cu	9	11	2	9	4	3			19
Zn	49	62	120	122	133				111
Ga	16	17	24·6	20	21	19			18

Notes: $m^* = 100^*Mg/(Mg + Fe^{+2}_{tot})$; 1, 2, & 3: Whalen *et al.* (1987); 4 & 5: Collins *et al.* (1982); 6: Kleeman & Twist (1989); 7: Naslund (1989); 8: Macdonald *et al.* (1990); 9: Hergt (1987).

solidus. Because of the high K_2O content, biotite crystallises early and coexists with orthopyroxene (see Barth 1967, p. 223), whereas hornblende, when present, is late.

4. The chemical distinction between the charnockite magma type and metamorphic charnockites

The metamorphic charnockite data illustrated in Figure 2 consist of gneisses containing the diagnostic assemblage of two feldspars + orthopyroxene + biotite + quartz for which the mineral compositions generally indicate re-equilibration at ≈700°C, and for which there may be field evidence for CO_2 streaming, but no evidence for a magmatic origin. Analyses are from southern India, Sri Lanka, and Norway. Metamorphic charnockites have much lower TiO_2, P_2O_5 and K_2O, and higher CaO than the CMT (Fig. 2), and thus the two groups can be readily distinguished.

There is general coincidence between the compositional field of metamorphic charnockites and that of the Lachlan Fold Belt granites (Fig. 2), which lends support to the contention that metamorphic charnockites are derived from pre-existing "normal" granites and gneisses possibly through CO_2 streaming. Condie and Allen (1984) suggest that the range in rock compositions of metamorphic charnockites was due to hornblende fractionation in the precursor igneous rocks which themselves had normal crustal magma compositions.

5. Volcanic equivalents of the charnockite magma type

We have surveyed the literature of felsic volcanic rocks in order to evaluate whether extrusive equivalents of the CMT exist, as it is well known that chemically-related plutonic and volcanic rock units, with the same phenocryst assemblages, occur within other granite suites (e.g. Wyborn & Chappell 1986). Many large continental basalt provinces are in fact bimodal with a silica gap between basalts and high silica extrusives. In some cases, these felsic extrusive magmas match members of the intrusive CMT in major and trace element geochemistry and mineralogy (Table 3). A detailed discussion of the relations between basalt and the CMT is presented elsewhere (Kilpatrick & Ellis in prep.), and discussion here is restricted to felsic volcanics. The compositionally homogeneous Cretaceous Etendeka quartz latites and the Lebombo rhyolites of the Karoo Province, together with the Lower Cretaceous felsic volcanics of the Parana Province, show petrographic, mineralogical, and geochemical similarities to the plutonic members of the CMT. Petrographic similarities are evident in the absence of primary hydrous phases, and within the Etendeka quartz latites both orthopyroxene and pigeonite occur within a generally glassy groundmass (Erlank *et al.* 1984). These volcanics represent relatively high temperature (>1000°C), dry, crust-derived magmas (Cleverly *et al.* 1984; Duncan *et al.* 1984; Erlank *et al.* 1984; Bellieni *et al.* 1986) which are not derived from the voluminous continental flood basalts

Table 3 Representative analyses of proposed volcanic equivalents to the CMS

	A Parana Quartz Latite BRA-12	B Karoo Quartz Latite KL-20	C Yardea Dacite
SiO_2	67·10	67·12	67·08
TiO_2	0·88	0·94	0·74
Al_2O_3	12·86	12·92	14·24
FeO_{tot}	5·21	5·74	4·64
MnO	0·10	0·10	0·10
MgO	1·33	1·39	0·84
CaO	2·99	3·27	2·47
Na_2O	2·77	2·89	3·43
K_2O	3·82	3·87	4·75
P_2O_5	0·27	0·28	0·20
Ba	632	585	1090
Rb	170	161	189
Sr	136	139	203
Pb	36·6	24·1	30
Th	14·7	12·1	29·5
U	5·5	3·9	5
Zr	259	254	406
Nb	19·5	22·3	18
Y	35	34·1	49
La	42·9	32·3	71
Ce	95·4	76	145
Nd	39·4	34·5	
Sc		19	13
V	62	75	26
Cr	7·3	13·4	3
Ni	2·74	2·3	4
Zn	67	76	80
Ga	16·8	16	20·2

Notes: A & B: Duncan *et al.* 1984; C: Creaser & White 1991.

(CFB) with which they are spacially associated. Figure 4 compares the geochemical signature of two representative CMT magmas with 60·9 and 65·3% SiO_2 from the ACI (Table 1; analyses 2 and 3 respectively) with examples of possible felsic extrusive equivalents (Table 3). Sample BRA-12 (Erlank 1984) and sample KL-20 (Erlank 1984) represent quartz latites both having 67·1% SiO_2 that are associated with the continental flood basalt provinces, the Parana and Karoo respectively (Table 3). Although the more felsic quartz latites have slightly lower abundances of Ti and P, as is expected from Figure 2(A), overall they are nearly identical with the CMT magmas. The only major departures in the patterns are the higher Th and U, and lower Ba and Eu of the quartz latites, possibly consistent with derivation from different provinces, and minor upper crustal assimilation and feldspar fractionation by these felsic high temperature volcanics. The REE patterns of the CMT magmas and the quartz latites are remarkably similar with La/Yb ratios of 10·9 and 10·4, respectively.

The Middle Proterozoic Yardea Dacite, of the Gawler Range Volcanics (GRV) of South Australia, shares the CMT signature and has an essentially identical pattern to the felsic volcanics cited above (Table 3; Fig. 4). The GRV is an extensive (>25,000 km^3) bimodal volcanic province (Giles 1988; Stewart & Foden 1990; Creaser & White 1991), in which many of the felsic volcanics have high TiO_2 and P_2O_5. They show similarities to the CMT, particularly in the mineralogy; for example, the GRV rarely contain hornblende, but contain phenocrysts of quartz, K-feldspar, plagioclase, clinopyroxene, orthopyroxene, biotite, and magnetite in a quartz + K-feldspar groundmass. The Yardea Dacite (1592 ± 3 Ma; Fanning *et al.* 1988) in these volcanics contains pigeonite phenocrysts indicating high temperatures of crystallisation (≈900–1000°C; Creaser & White 1991) and also has an unusual fabric for a felsic volcanic rock most likely indicative of eruption of an unusually dry magma. Stewart and Foden (1990) found the entire GRV to have high K_2O with samples also high in Ba, Th, Zr, Y, K, Rb and LREE.

The felsic lavas represented in Figure 4 show significant departures from the crustal signature of the average I-type granite of White and Chappell (1983), with the latter having considerably lower Pb, Sr, Nd, P, Zr, Ti and Y, and similar Ba, Th, U, K, La and Ce. These volcanics also show a similar low CaO relative to I-type granites and lie entirely within the separate field defined by the CMT (Fig. 3). This

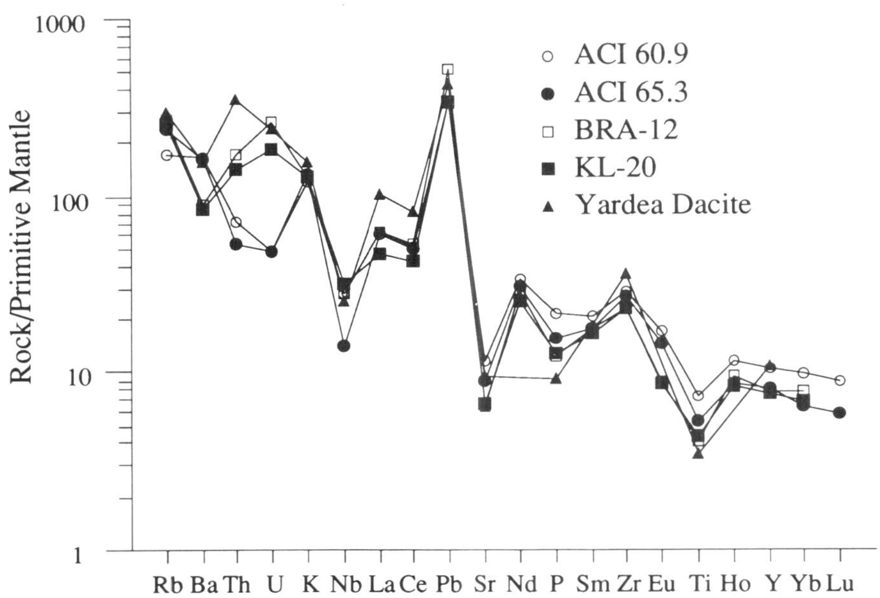

Figure 4 Primitive mantle-normalised diagram of two representative Ardery Charnockitic Intrusions (ACI) magmas and possible C-type volcanics; data from Tables 1 (anal 2, 3) and 3; normalising values from Sun & McDonough (1989).

is also evident in terms of the relative Fe-enrichment of the various suites (Fig. 3), where once again these high temperature volcanics show a clear resemblance to charnockite intrusives both in terms of relative Mg/Fe and of the range in relative alkali (and SiO_2) values.

We therefore suggest that these examples with their petrographic and geochemical similarity to the plutonic members of the CMT are actually C-type volcanics. This examination for possible C-type extrusives was far from exhaustive, but it is hoped that this work will encourage others to reassess their data in light of these findings.

It is not considered that the association of C-type volcanics with coeval continental flood basalts is evidence that the CMT is itself the direct fractionation product of these basalts. In fact the possibility of such a link between the felsic volcanics and the CFB has been dismissed by many workers (see Erlank 1984; Bellieni *et al.* 1986; Hawkesworth *et al.* 1988). Rather, it is considered that these felsic volcanics originated either through crustal anatexis (Cox 1972; Erlank *et al.* 1984), or partial melting of underplated basalt (Cleverly *et al.* 1984).

While the fractionation of a high P_2O_5 and TiO_2 (HPT) basalt could conceivably produce intermediate magmas with a HPT signature superficially similar to the CMT, the eruption of extremely large volumes of uniform composition felsic magma is at variance with the relative volumes and range in rock types produced by basalt fractionation (see MacDonald *et al.* 1990, as an example of the products of basalt fractionation). Therefore such evidence as SiO_2 gaps and large volumes (as shown by the Parana and Karoo CFB terranes), high temperatures, and the geochemical and mineralogical criteria discussed above, are required to identify C-type volcanic rocks.

6. The origin of the charnockite magma type

Our preferred model is that the Charnockite Magma Type results from the partial melting of a previously dehydrated mafic to intermediate di-normative source rock, which is, or is almost, hornblende-free due to a previous dehydration occurring well below the solidus, possibly caused by a metamorphic episode involving a mixed, but low-H_2O fluid phase.

Alternative models such as AFC and fractional crystallisation processes involving a HPT parental basalt are not consistent with REE data for the specific case of the Ardery Charnockitic Intrusions (unpublished data). At present there are insufficient isotopic data, for many CMT members, to adequately and realistically control such petrogenetic modelling. More importantly, there is a lack of contemporaneous HPT basaltic dykes associated with CMT intrusions to support such modelling. Furthermore, CMT intrusions have parental magmas with $\approx$62% SiO_2 and have nowhere been reported as showing direct evidence of a parental HPT basalt.

A granulitic source which lacks hornblende contrasts with the amphibolite source required for I-type granites and it is known to significantly affect the composition of the magmas produced by partial melting. Examination of the melting reactions of amphibolite in H_2O-excess conditions and dehydration (Ellis & Thompson 1986) shows that any magma produced by melting from such a source will crystallise hornblende with freezing. Pyroxene-bearing felsic magmas are typically produced by higher temperatures of melting of the same source (Ellis & Thompson 1986); with cooling they display the classical Bowen reaction of pyroxene back-reacting with the magma to produce amphibole. This back-reaction is generally absent in intrusive charnockites and instead, biotite co-exists with pyroxene, a feature recognised by Barth (1967, p. 223). Many initially H_2O-undersaturated calc-alkaline explosive igneous rocks contain pyroxene + biotite + feldspar phenocrysts (Ewart 1982) and may represent the high temperature dehydration melting of amphibolite to produce magmas crystallising pyroxene, which subsequently ascend rapidly to pressures lower than hornblende stability (Ellis & Thompson 1986). These volcanic rocks have chemically equivalent intrusive rocks which contain primary hornblende (White & Chappell 1983; Wyborn & Chappell 1986). These extrusive pyroxene-bearing magmas, however, are not C-type magmas.

We suggest that a necessary prerequisite for charnockite formation is the earlier, subsolidus dehydration of hornblende to leave a fertile granulite which would then melt at much higher temperatures under fluid-absent conditions. The generally undepleted nature of incompatible elements in the CMT indicates a geochemically fertile granulite source, whereas the lower CaO of the CMT, relative to hornblende-bearing I-type granites, argues against the involvement of hornblende in the source rock. Experimental studies (Naney 1983) indicate the extremely low H_2O content needed to eliminate hornblende from the liquidus. The production of a two-pyroxene + two-feldspar + biotite + quartz granulite, where biotite is more stable relative to hornblende, requires the presence (and excess) of K-feldspar (Wones & Dodge 1977). Thus an intermediate high-K rather than basaltic source rock is required for the CMT.

Much of the discussion of the mineralogy of metamorphic charnockites has involved the reaction Bi + Qtz = Opx + Kf + V for metamorphic rocks and the reaction Bi + Qtz = Opx + Kf + L during partial melting (Newton & Hansen 1983; Bhattacharya & Sen 1986; Hansen *et al.* 1987; Peterson & Newton 1990). We contend that these are not the key equilibria in the generation of magmatic charnockites. The above reactions also do not constrain the liquidus temperatures, liquidus mineralogy, or ultimately the melting reactions which we consider produced the CMT. In this regard the low CaO, the general lack of hornblende, and the presence of early pigeonite are key features for understanding the origin of magmatic charnockites.

In a review of crustal melting, Clemens (1991) acknowledged that there are regional granulites which equilibrated at $\approx$1000°C (see Ellis 1980; Ellis *et al.* 1980) but states that there was no evidence of anhydrous granite magmas generated in the deep crust and that such temperature ranges are not normally accessible. Some recent reviews of granulite facies metamorphism (e.g. Bohlen & Boettcher 1981) have restricted granulites worldwide to a conservative 700–800°C. Ellis *et al.* (1980) and Ellis (1980) showed that there are in fact large Precambrian granulite belts which have equilibrated at ~1000°C (see also Harley 1989; Sandiford & Powell 1986; Ellis 1987). We contend that the geochemically and mineralogically distinctive plutonic and volcanic Charnockite Magma Type, recognised from at least five continents, represents the consequence of such metamorphism. Such temperatures may be realised in a number of different geological settings. They can originate through the input of large volumes of basalt into stabilised, but fertile, lower crustal granulites to produce charnockites. Equally, we contend that they may form within the basal levels of thickened crust, with the thickening due to Himalayan-type plate margin activity, although in this scenario there is probably still a need for basaltic input to the heat supply. Recent modelling in the thermal regime of a doubly thickened crust (England & Richardson 1977; Thompson & England 1984) has shown that extreme temperatures of metamorphism can be generated. Distinctive crustal melts may be one result of such a process (Ellis 1987).

The upward migration of HPT basaltic magmas to underplate and/or intrude the lowest crust is a possible mechanism to account for the geochemical signature of C-type magmas. Partial melting of a mixture between two significantly different lower crustal source rocks could explain much of the variation of C-types. These two source materials were, firstly, crust-derived dry granulites which were not geochemically depleted as a result of a previous melting event, and secondly, mafic-intermediate dry granulites derived from HPT basaltic magmas intruded into the lower crust, or a mixture of such granulites and basalts.

7. Acknowledgements

We gratefully acknowledge the generous logistical support of the Antarctic Division, Hobart in enabling one of us (Kilpatrick) to undertake field work in Antarctica, and the field support of the Casey 1988 wintering expeditioners. This research was supported through an Australian Postgraduate Research Award to Kilpatrick and Antarctic Science Advisory Council and Australian Research Council grants to Ellis. We thank D. N. Young, D. C. Champion, B. W. Chappell, A. Ewart, S-s. Sun and A. J. R. White for stimulating discussions during this study.

8. References

Allen P., Condie, K. C. & Narayana, B. L. 1985. The geochemistry of prograde and retrograde charnockite-gneiss reactions in sourthern India. GEOCHIM COSMOCHIM ACTA **49,** 323–36.

Barth, T. F. W. 1967. *Theoretical petrology,* 2nd edn New York: John Wiley and Sons.

Bateman, P. C. & Chappell, B. W 1979. Crystallization, fractionation, and solidification of the Tuolumne Intrusive Series, Yosemite National Park, California. GEOL SOC AM BULL **90,** 465–82.

Bellieni, G. M., Comin-Chiaramonti, P., Marques, L. S., Melfi, A. J., Nardy, A. J. R., Papatrechas, C., Piccirillo, E. M., Roisenberg, A. & Stolfa, D. 1986. Petrogenetic aspects of acid and basaltic lavas from the Parana Plateau (Brazil): geological, mineralogical and petrological relationships. J PETROL **27,** 915–44.

Bertrand, P., Ellis, D. J. & Green, D. H. 1991. The stability of sapphirine–quartz and hypersthene–sillimanite–quartz assemblages: an experimental investigation in the system FeO—MgO—Al_2O_3—SiO_2 under H_2O and CO_2 conditions. CONTRIB MINERAL PETROL **108,** 55–71.

Bhattacharya, A. Sen, S. K. 1986. Granulite metamorphism, fluid buffering, and dehydration melting in the Madras charnockites and metapelites. J. PETROL **27,** 1119–41.

Blight, D. F. & Oliver, R. L. 1977. The metamorphic geology of the Windmill Islands, Antarctica: a preliminary account. J GEOL SOC AUST **24,** 239–62.

Bohlen, S. R. & Boettcher, A. L. 1981. Experimental investigations and geological applications of orthopyroxene geobarometry. AM MINERAL **66,** 951–64.

Chappell, B. W., White, A. J. R. & Williams, I. S. 1991. Excursion guide: A transverse section through granites of the Lachlan Fold Belt. BUR MINER RES GEOL GEOPHYS AUST, RECORD 1991/22.

Clemens, J. D. 1991. The Granulite-granite connexion. *In* Vielzeuf, D. & Vidal, Ph. (eds) *Granulites and crustal evolution,* 25–36. Netherlands: Kluwer Academic Publishers.

Cleverly, R. W., Betton, P. J. & Bristow, J. W. 1984. Geochemistry and petrogenesis of the Lebombo rhyolites. GEOL SOC SOUTH AFRICA, SPEC PUBL **13,** 171–94.

Collins, W. J., Beams, S. D., White, A. J. R. & Chappell, B. W. 1982. Nature and origin of A-Type granites with particular reference to southeastern Australia. CONTRIB MINERAL PETROL **80,** 189–200.

Condie, K. C. & Allen, P. 1984. Origin of Archaean charnockites from southern India. *In* Kroner, A. (ed.) *Archaean geochemistry,* 182–203. Berlin: Springer.

Condie, K. C., Allen, P. & Narayana, B. L. 1982. Geochemistry of the Archean low- to high-grade transition zone, southern India. CONTRIB MINERAL PETROL **81,** 157–67.

Condie, K. C., Bowling, G. P. & Allen, P. 1986. Origin of granites in an Archean high-grade terrane, southern India. CONTRIB MINERAL PETROL **92,** 93–103.

Cooray, P. G. 1969. Charnockites as metamorphic rocks. AM J SCI **267,** 969–82.

Cox, K. G. 1972. The Karroo volcanic cycle. J GEOL SOC LONDON **128,** 311–36.

Cox, K. G., Bell, J. D. & Pankhurst, R. J. 1979. *The interpretation of igneous rocks.* London: Allen and Unwin.

Creaser, R. A. & White, A. J. R. 1991. Yardea Dacite—Large-volume, high-temperature felsic volcanism from the middle Proterozoic of South Australia. GEOLOGY **19,** 48–51.

de Waard, D. 1969. The occurrence of charnockite in the Adirondacks: a note on the origin and definition of charnockite. AM J SCI **267,** 983–87.

Duchesne, J. C., Caruba, R. & Iacconi, P. 1987. Zircon in charnockitic rocks from Rogaland (southwest Norway); petrogenetic implications. LITHOS **20,** 357–68.

Duncan, A. R., Erlank, A. J. & Marsh, J. S. 1984. Regional geochemistry of the Karoo Igneous Province. GEOL SOC SOUTH AFRICA, SPEC PUBL **13,** 355–88.

Ellis, D. J. 1980. Osumilite–sapphirine–quartz granulites from Enderby Land, Antarctica: P-T conditions of metamorphism, implications for garnet–cordierite equilibria and the evolution of the deep crust. CONTRIB MINERAL PETROL **74,** 201–10.

Ellis, D. J. 1987. Origin and evolution of granulites in normal and thickened crusts. GEOLOGY **15,** 167–70.

Ellis, D. J. & Thompson, A. B. 1986. Subsolidus and partial melting reactions in the quartz-excess $CaO + MgO + Al_2O_2 + SiO_2 + H_2O$ system under water-excess and water-deficient conditions to 10 kb: some implications for the origin of peraluminous melts from mafic rocks. J PETROL **27,** 91–121.

Ellis, D. J., Sheraton, J. W., England, R. N. & Dallwitz, W. B. 1980. Osumilite–sapphirine–quartz granulites from Enderby Land, Antarctica—mineral assemblages and reactions. CONTRIB MINERAL PETROL **72,** 123–43.

England, P. C. & Richardson, S. W. 1977. The influence of erosion upon the mineral facies of rocks from different metamorphic environments. J GEOL SOC London **134,** 201–13.

Erlank, A. J. (ed) 1984. Petrogenesis of the volcanic rocks of the Karoo Province. GEOL SOC SOUTH AFRICA, SPEC PUBL **13**.

Erlank, A. J., Marsh, J. S., Duncan, A. R., Miller R. McG., Hawkesworth, C. J., Betton, P. J. & Rex, D. C. 1984. Geochemistry and petrogenesis of the Etendeka Volcanic rocks from SWA/Namibia. GEOL SOC SOUTH AFRICA, SPEC PUBL **13,** 195–245.

Ewart, A. 1982. The mineralogy and petrology of Tertiary–Recent orogenic volcanic rocks: with special reference to the andesitic-basaltic compositional range, *In* Thorpe, R. S. (ed.) *Andesites,* 25–95. New York: John Wiley & Sons.

Fanning, C. M., Flint, R. B., Parker, A. J., Ludwig, K. R. & Blisset, A. H. 1988. Redefined Proterozoic tectonic evolution of the Gawler Craton, South Australia, through U–Pb zircon geochronology. PRECAMBRIAN RES **40/41,** 363–86.

Friend, C. R. L. 1981. Charnockite and granite formation and influx of CO_2 at Kabbaldurga. NATURE **294,** 550–52.

Fuhrman, M. L. & Lindsley, D. H. 1988. Ternary-feldspar modeling and thermometry. AM Mineral **73,** 201–15.

Giles, C. W. 1988. Petrogenesis of the Proterozoic Gawler Range Volcanics, South Australia. PRECAMBRIAN RES **40/41,** 407–27.

Hansen, E. C., Newton, R. C. & Janardhan, A. S. 1984. Fluid inclusions in rocks from the amphibolite-facies gneiss to charnockite progression in southern Karnataka, India: direct evidence concerning the fluids of granulite metamorphism. J METAMORPHIC GEOL **2,** 249–64.

Hansen, E. C., Janardhan, A. S., Newton, R. C., Prame, W. K. B.

N. & Ravindra Kumar, G. R. 1987. Arrested charnockite formation in southern India and Sri Lanka. CONTRIB MINERAL PETROL **96,** 255–44.

Harley, S. L. 1989. The origins of granulites: a metamorphic perspective. GEOL MAG **126,** 215–47.

Hawkesworth, C., Mantovani, M. & Peate, D. 1988. Lithospheric remobilisation during Parana CFB magmatism. J PETROL SPEC LITHOSPHERE ISSUE, 205–23.

Hergt, J. M. 1987. *The origin and evolution of the Tasmanian dolerites.* Unpublished Ph.D. thesis, The Australian National University, Canberra.

Holland, T. H. 1900. The charnockite series, a group of Archaean hypersthenic rocks in Peninsular India. GEOL SURV INDIA MEM **28,** 119–249.

Howie, R. A. 1954. The geochemistry of the charnockite series of Madras, India. TRANS R SOC EDINBURGH **62,** 725–68.

Hubbard, F. H. & Whitley, J. E. 1979. REE in charnockite and associated rocks, southwest Sweden. LITHOS **12,** 1–11.

Janardhan, A. S., Newton, R. C. & Smith, J. V. 1979. Ancient crustal metamorphism at low pH_2O: charnockite formation at Kabbaldurga, south India. NATURE **278,** 511–14.

Janardhan, A. S., Newton, R. C. & Hansen, E. C. 1982. The transformation of amphibolite facies gneiss to charnockite in southern Karnataka and northern Tamil Nadu, India. CONTRIB MINERAL PETROL **79,** 130–49.

Kilpatrick, J. A. & Ellis, D. J. 1991. The Charnockite Magma Suite: A new and geochemically distinct magma suite, its characterisation and petrogenetic implications. *In* Chappell, B. W. (ed.) *Second Hutton Symposium on Granites and Related Rocks. Abstracts,* 57. Canberra: BMR, GGA, R 1991/25.

Kilpatrick, J. A., Ellis, D. J. & Young, D. N. 1990. Field aspects of the Ardery Charnockitic Intrusions, Windmill Islands, Antarctica. A dynamic magma chamber. GEOL SOC AUST ABSTR **25,** 290.

Kleeman, G. J. Twist, D. 1989. The Compositionally-zoned sheet-like granite pluton of the Bushveld Complex: evidence bearing on the nature of A-type magmatism. J. PETROL **30,** 1383–414.

Le Breton, N. & Thompson, A. B. 1988. Fluid-absent (dehydration) melting of biotite in metapelites in the early stages of crustal anatexis. CONTRIB MINERAL PETROL **99,** 226–37.

Le Maitre, R. W. 1988. *A classification of igneous rocks and glossary of terms.* Oxford: Blackwell Scientific Publications.

Loiselle, M. C. & Wones, D. R. 1979. Characteristics and origin of anorgenic granites. GEOL SOC AM PROG ABSTR **11,** 468.

MacDonald, R., McGarvie, D. W., Pinkerton, H., Smith, R. L. & Palacz, Z. A. 1990. Petrogenetic evolution of the Torfajokull Volcanic Complex, Iceland I. Relationship between the magma types J PETROL **31,** 429–59.

Naney, M. T. 1983. Phase equilibria of rock-forming ferromagnesian silicates in granitic systems. AM J SCI **283,** 993–1033.

Naslund, H. R. 1989. Petrology of the Basistoppen Sill, East Greenland: A calculated magma differentiation trend. J PETROL **30,** 299–319.

Nekvasil, H. & Burnham, C. W. 1987. The individual effects of pressure and water content on phase equilibria in the "granite" system. GEOCHEM SOC, SPEC PUBL **1,** 433–45.

Newton, R. C. & Hansen, E. C. 1983. The origin of Proterozic and late Archean charnockites-evidence from field relations and experimental petrology. GEOL SOC AM MEM **161,** 167–78.

Peterson, J. W. & Newton, R. C. 1990. Experimental biotite-quartz melting in the KMASH-CO_2 system and the role of CO_2 in the petrogenesis of granites and related rocks. AM MINERAL **75,** 1029–42.

Ravich, M. G. 1972. The charnockite problem. *In* Adie, R. J. (ed.) *Antarctic geology and geophysics,* 523–26. Oslo: Scandinavian University Books.

Ravich, M. G., Solov'ev, D. S. & Fedorov, L. V. 1985. *Geological structure of MacRobertson Land (East Antarctica).* Rotterdam: A. A. Balkema.

Sandiford, M. & Powell, R. 1986. Pyroxene exsolution in granulites from Fyfe Hills, Enderby Land, Antarctica: Evidence for 1000°C metamorphic temperatures in Archean continental crust. AM MINERAL **71,** 946–54.

Sheraton, J. W. 1982. Origin of charnockitic rocks of MacRobertson Land. *In* Craddock, C. (ed.) *Antarctic geoscience,* 489–97. Madison: University of Wisconsin Press.

Sheraton, J. W., Black, L. P. & Tindle, A. G. 1992. Petrogenesis of plutonic rocks in a Proterozoic granulite-facies terrane—The Bunger Hills, East Antarctica. CHEM GEOL (in press).

Stewart, K. & Foden, J. 1990. The fundamental role of mantle derived magma in the production of a large volume felsic volcanic province, Gawler Ranges, South Australia. GEOL SOC AUST ABSTR **27,** 96.

Sun, S.-s. & McDonough, W. F. 1989. Chemical and isotopic systematics of oceanic basalts: implications for mantle composition and processes. *In* Saunders, A. D. & Norry, M. J. (eds) *Magmatism in the ocean basins.* GEOL SOC SPEC PUBL **42,** 313–45.

Thompson, A. B. & England, P. C. 1984. Pressure-temperature-time paths of regional metamorphism II. Their inference and interpretation using mineral assemblages in metamorphic rocks. J PETROL **25,** 929–55.

Turner, F. J. & Verhoogen, J. 1960. *Igneous and metamorphic petrology,* 2nd edn. New York: McGraw-Hill.

Turner, S. P., Foden, J. D. & Morrison, R. S. 1991. Derivation of A-type magma by fractionation of basaltic magma with an example from the Padthaway Ridge, South Australia. *In* Chappell, B. W. (ed.) *Second Hutton Symposium on Granites and Related Rocks. Abstracts.,* 107. Canbera: BMR, GGA, R 1991/25.

van Reenen, D. D. & Roering, C. (eds) 1990. *The Limpopo Belt: A field workshop on granulites and deep crustal tectonics.*: Department of Geology, Rand Afrikaans University, Johannesburg, South Africa.

Whalen, J. B., Currie, K. L. & Chappell, B. W. 1987. A-type granites: geochemical characteristics, discrimination and petrogenesis. CONTRIB MINERAL PETROL **95,** 407–19.

White, A. J. R. & Chappell, B. W. 1977. Ultrametamorphism and granitoid genesis. TECTONOPHYSICS **43,** 7-22.

White, A. J. R. & Chappell, B. W. 1983. Granitoid types and their distribution in the Lachlan Fold Belt, southeastern Australia. GEOL SOC AM MEM **159,** 21–34.

Wilson, M. 1989. *Igneous petrogenesis. A global tectonic approach.* London: Unwin Hyman Ltd.

Wones, D. R. & Dodge, F. C. W. 1977. The stability of phlogopite in the presence of quartz and diopside. *In* Fraser, D. G. (ed.) *Thermodynamics in geology.* Amsterdam: D. Reidel Pub. Co.

Wyborn, D. & Chappell, B. W. 1986. The petrogenetic significance of chemically related plutonic and volcanic rock units. GEOL MAG **123,** 619–28.

Wyllie, P. J. 1977. Crustal anatexis: an experimental review. TECTONOPHYSICS **43,** 41–71.

Young, D. N. & Black, L. P. 1991. U-Pb zircon dating of Proterozoic igneous charnockites from the Mawson coast, East Antarctica. ANTARCT SCI **3,** 205–16.

Young, D. N. & Ellis, D. J. 1990. Petrology of Proterozoic igneous charnockites from Mawson, Antarctica: high-temperature synorogenic granites produced by anatexis in a thickened crust. GEOL SOC AUST ABSTR **25,** 261–62.

Young, D. N. & Ellis, D. J. 1991. The intrusive Mawson charnockites: evidence for a compressional plate margin setting of the Proterozoic mobile belt of East Antarctica. *In* Thomson, M. R. A., Crame, J. A. & Thomson, J. W. (eds) *Geological evolution of Antarctica,* 25–31. Cambridge: Cambridge University Press.

J. A. KILPATRICK and D. J. ELLIS, Department of Geology, The Australian National University, GPO Box 4, Canberra City ACT 2601, Australia.

Transactions of the Royal Society of Edinburgh: Earth Sciences, **83**, 165–171, 1992

Tectonic setting and origin of the Proterozoic rapakivi granites of southeastern Fennoscandia

Ilmari Haapala and O. Tapani Rämö

ABSTRACT: The 1·65–1·54 Ga rapakivi granites of southeastern Fennoscandia represent the silicic members of a bimodal magmatic association in which the mafic members are tholeiitic diabase dykes and minor gabbroic-anorthositic bodies. They are metaluminous to slightly peraluminous A-type granites and occur as high-level batholiths and stocks in an E–W-trending belt extending from Soviet Karelia to southwestern Finland. The Soviet Karelian granites were emplaced into the contact zone between Archaean craton and Svecofennian juvenile 1·9 Ga-old crust, while the Finnish granites were intruded into the Svecofennian crust. Deep seismic soundings show that the rapakivi granites and the contemporaneous, mainly WNW or NW-trending diabase dyke swarms are situated in a zone of relatively thin crust. Below the Wiborg Batholith there exists a domal structure in the lithosphere in which a transitional zone (mafic underplate) occurs between the crust and the mantle.

The Nd isotopic evolution of the rapakivi granites ($\varepsilon_{Nd}(T)$ −3·1– −0·2) corresponds to the evolution of the 1·9 Ga-old Svecofennian crust, as do their Pb isotopic compositions. This implies that the Finnish granites represent anatectic melts of the Svecofennian crust. In contrast, the Soviet Karelian granites show isotopic composition indicative of substantial incorporation of Archaean lower crust material. Petrochemical modelling of one of the Finnish batholiths shows that its parental magma could have been generated by *c*. 20% melting of a granodioritic source and that fractional crystallisation was important during the subsequent evolution of this magma.

The rapakivi granites are redefined as A-type granites that show the rapakivi texture at least in larger batholiths. The field, geochemical, and seismic data indicate that the classical Finnish rapakivi granites were generated in an anorogenic extensional regime by partial melting of the lower/middle crust. The melting, and possibly also the extensional tectonics, were related to upwellings of hot mantle material which led to intrusion of mafic magmas at the base and into the crust.

KEY WORDS: anorogenic, A-type, bimodal, diabase, extensional, Finland, isotopes (Nd, Pb), mantle upwelling, partial melting.

The terms "rapakivi granite" and "rapakivi texture" were introduced to international geological literature one hundred years ago by J. J. Sederholm (1891) in his classical paper "Ueber die finnländischen Rapakiwigesteine". Since then, similar granites have been described from a number of Precambrian cratonic areas on different continents, but southern Finland is regarded as the type area of the rapakivi granites. Sederholm (1891, 1928) observed that the Finnish rapakivi granites form a special group of young (Subjotnian) massive potassium-rich granites which are usually porphyritic and contain two generations of quartz and alkali feldspar, the early alkali feldspar often occurring as ovoidal megacrysts surrounded by an oligoclase-andesine shell (the rapakivi texture). The traditional, still commonly-used terminology distinguishes several petrographical varieties of rapakivi granites: wiborgite (alkali feldspar ovoids generally mantled by plagioclase), pyterlite (most of the alkali feldspar ovoids unmantled), and a number of other porphyritic or even-grained granites, many of which do not contain any ovoidal alkali feldspar megacrysts (Wahl 1925; Vorma 1976). In fact, only the largest southeastern Fennoscandian batholiths (Salmi, Wiborg, Laitila, Vehmaa, Åland) contain varieties which can be called wiborgite. Recent studies have shown that the chemical and mineralogical peculiarities as well as the mode of occurrence and magmatic association of the rapakivi granites meet the characteristics of subalkaline A-type granites (Haapala & Rämö 1990; Emslie 1991; Rämö 1991). Accordingly, the rapakivi granites may be redefined as A-type granites characterised by the presence, at least in the larger batholiths, of granite varieties showing the rapakivi texture.

1. Tectonic setting of the rapakivi granites

The major tectonic units of southeastern Fennoscandia are: (1) the Archaean granite-greenstone belt complex partly covered by early Proterozoic supracrustal rocks; (2) the early Proterozoic (1·9–1·8 Ga) Svecofennian orogenic domain accreted to the eastern Archaean craton; and (3) the anorogenic 1·65–1·54 Ga-old rapakivi granite complexes (Fig. 1). Substantial amounts of new continental crust were formed during the early (synorogenic) Svecofennian stage (1·9–1·87 Ga) by extrusion and intrusion of tholeiitic and calc-alkaline volcanic and plutonic rocks (I-type tonalites, granodiorites and granites, lesser gabbros) (Huhma 1986; Front & Nurmi 1987). During subsequent late-orogenic and post-orogenic (post-collisional) stages, remelting of the tectonically thickened crust produced 1·83 Ga-old potassium-rich S-type granites and small 1·80–1·79 Ga-old granite-monzonite plutons (Nurmi &

Haapala 1986; Huhma 1986; Hubbard & Branigan 1987; Lindberg & Eklund 1988).

The rapakivi granites of southeastern Fennoscandia occur as high-level multiple batholiths and stocks along a roughly E–W-trending belt extending from the eastern side of Lake Ladoga, U.S.S.R., to the archipelago of southwestern Finland (Fig. 1). Geophysical studies indicate that the rapakivi batholiths are subhorizontal sheet-like bodies not more than 10 km in thickness (Laurén 1970; Korja & Elo 1990; Larin *et al.* 1991). Silicic dyke rocks (quartz-feldspar porphyries) and minor gabbroic and anorthositic bodies are commonly associated with the rapakivi batholiths. The highest intrusion ages (1·65–1·62 Ga) are recorded for the Wiborg Batholith and its satellite plutons in southeastern Finland, whereas the rapakivi granites of southwestern Finland and Soviet Karelia show ages from 1·59–1·54 Ga (Vaasjoki *et al.* 1991; Neymark *et al.* 1991; Suominen 1991; see also Vaasjoki 1977). The rapakivi granites of southeastern Fennoscandia are thus markedly younger than the 1·77–1·75 Ga-old rapakivi granites of the Ukrainian shield (Esipchuk 1991).

Swarms of tholeiitic diabase dykes, running generally in a WNW or NW direction (in the archipelago of southwestern Finland mainly in a NE direction) transect the Svecofennian crust in the surroundings of the rapakivi complexes (Fig. 1). The isotopic ages of the diabase dykes conform to the ages of the rapakivi granites: 1665–1635 Ma in southeastern Finland, about 1580 Ma in southwestern Finland (Vaasjoki *et al.* 1991; Suominen 1991). The extensive subparallel diabase dyke swarms are good evidence for an extensional tectonic setting during the emplacement of the rapakivi granites. The simultaneous existence of mafic and silicic magmas has been verified in several localities in composite dykes and also in the plutonic association, and locally mingling and mixing of the two magmas has led to formation of intermediate rocks (Eklund 1991; Rämö 1991; Salonsaari & Haapala 1991).

After the emplacement of the voluminous 1·65–1·54 Ga-old rapakivi granites and related mafic rocks, no metamorphic events have affected the bedrock of southeastern Fennoscandia. Minor later igneous activity is represented by intrusion and extrusion of Jotnian tholeiitic basalts that produced the 1·26 Ga-old diabase dykes associated with the Jotnian sandstone-filled graben in southwestern Finland as well as the diabase dykes and the porphyry interlayers in the Jotnian sandstone (Salmi suite) on the northeastern shore of Lake Ladoga in U.S.S.R. (Suominen 1991; Larin *et al.* 1991).

Recent deep seismic soundings (Luosto *et al.* 1990; Korja 1991; Luosto 1991) have shown that the rapakivi granites and the diabase dyke swarms are situated in a broad E–W-trending belt of relatively low crustal thickness (Fig. 2). Below the Wiborg Batholith, the mantle forms a dome-like structure bulging into the crust. In this dome structure two "Moho boundaries" (M_1 and M_2) are observed, at depths of 40 km and 50 km, respectively (Fig.

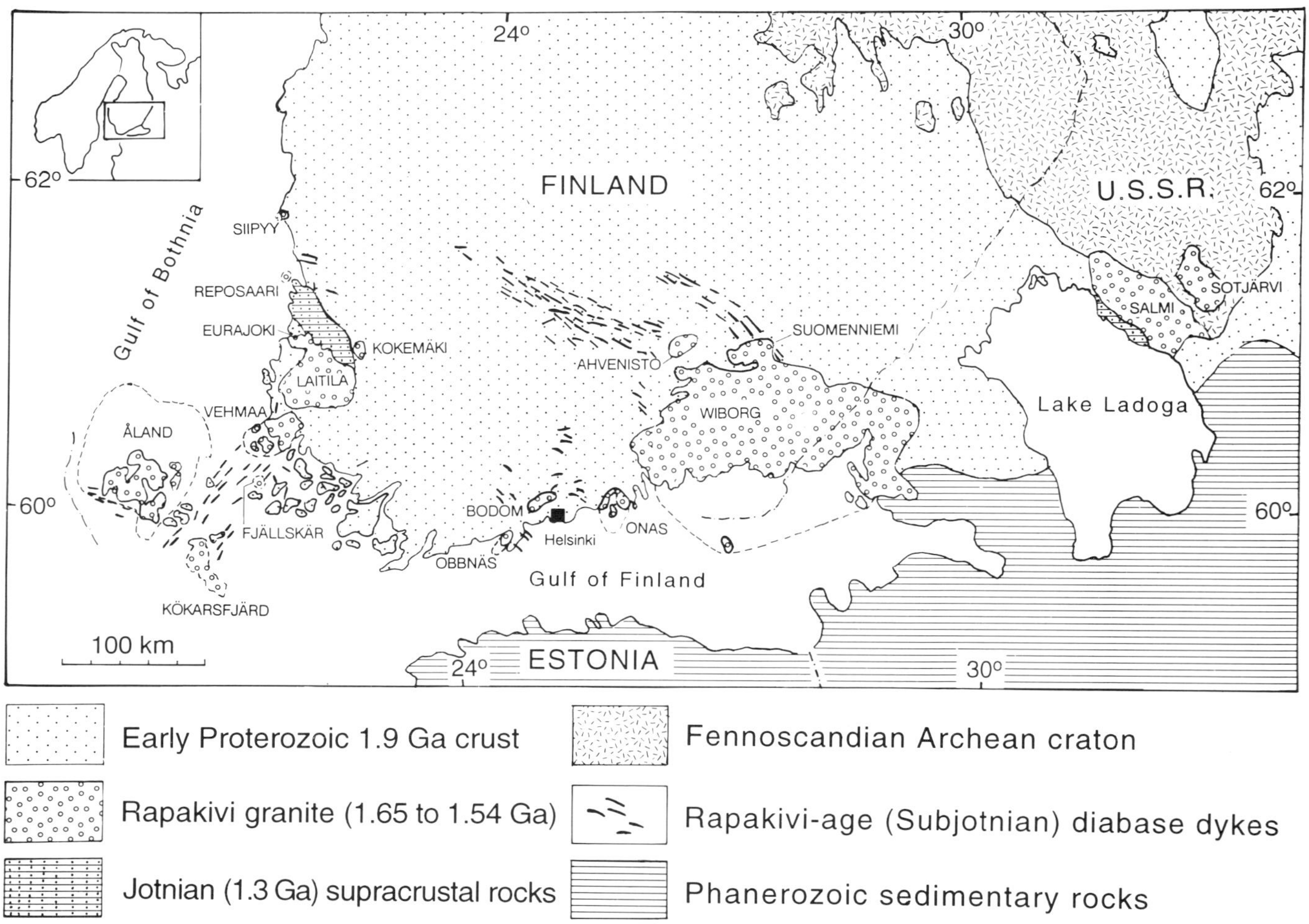

Figure 1 Map showing the major tectonic units, rapakivi granite batholiths and stocks, and contemporaneous diabase dykes of southern Finland and adjacent Soviet Karelia; dashed lines indicate assumed areal extents of the rapakivi granite intrusions; modified from Rämö (1991, fig. 1).

3). The transitional zone between M_1 and M_2 may represent a mafic underplate related to diapiric uprise of mantle magmas, in agreement with the models of Haapala (1989), Rämö and Haapala (1991), and Korja (1991). The seismic soundings also show that below the rapakivi granites there is, at a depth of 10–14 km, a high-velocity (6·4–6·5 km/s) body which is interpreted to represent anorthosite and gabbro (Korja 1991).

It is possible that later block movements, the Jotnian basaltic magmatism and isostatic changes have affected the mantle-crust boundary. Because signs of later orogenic activity are, however, lacking in southeastern Fennoscandia and because the Jotnian magmatism was local and volumetrically minor, the present mantle topography probably still reflects the geotectonic situation 1·6 Ga ago.

2. Chemical composition of the rapakivi granites

The rapakivi granites of southeastern Fennoscandia have several compositional features that make them distinct from other granite suites of the region. These include high Si, K, F, Rb, Ga, Zr, Hf, Th, U, Zn, and REE (except Eu) abundances as well as high K/Na and very high Fe/Mg and Ga/Al ratios, and low abundances of Ca, Mg, Al, P, and Sr (Sahama 1945; Vorma 1976; Haapala 1977; Nurmi & Haapala 1986; Rämö 1991). The youngest intrusive phases are often strongly anomalous topaz-bearing granites showing lowered K/Na ratios, extreme enrichments in F (up to 1·5 wt%), Ga (up to 60 ppm), Rb (up to 1,000 ppm), elevated Sn abundances (up to 100 ppm), strong depletions

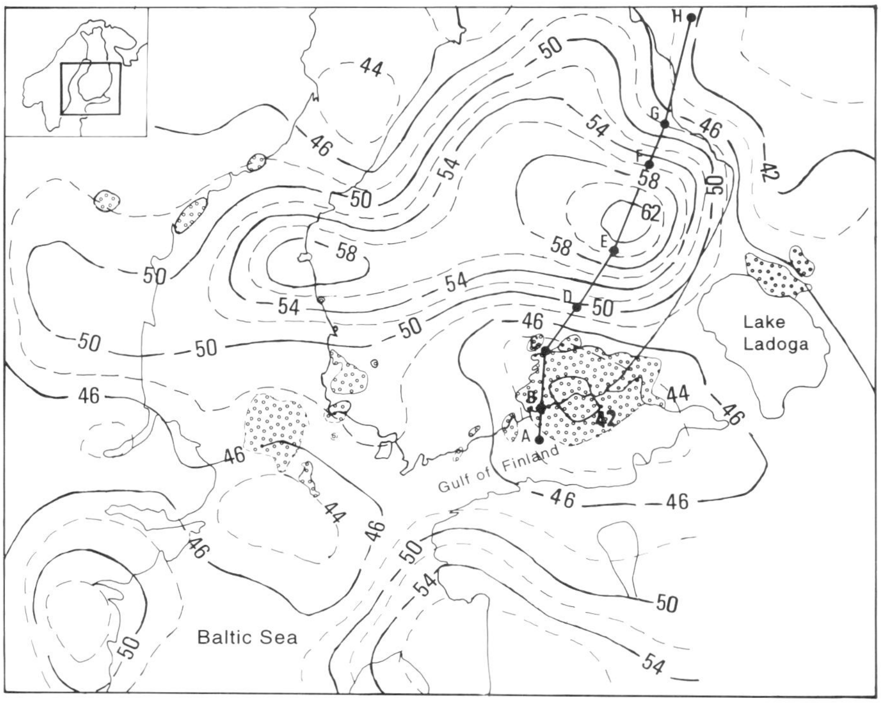

Figure 2 Map showing Moho depth contours (km) of a part of the Fennoscandian shield (after Luosto 1991 and Babel Working Group 1991); exposed rapakivi granites and the location of the deep seismic Baltic profile (A to H) are indicated (see Fig. 3).

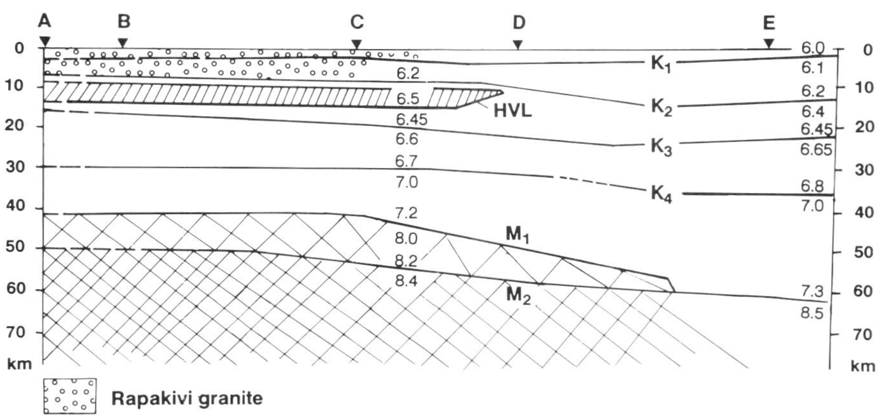

Figure 3 Crustal cross section for a part (A to E) of the Baltic profile (see Fig. 2): K_1, K_2, K_3, and K_4 are P-wave velocity boundaries within the crust, HVL a high-velocity layer beneath rapakivi granite, M_1 the Moho discontinuity, and M_2 another discontinuity in the upper mantle; width of the cross-section corresponds to 300 km; modified from Luosto *et al.* (1990, fig. 16).

in Ba and Sr (down to <100 and <10 ppm, respectively), and flat REE-patterns (as opposed to the LREE-enriched patterns normally encountered in the rapakivi granites; Vorma 1976; Haapala 1977; Rämö 1991).

Regarding alumina saturation, the rapakivi granites are generally metaluminous or, especially the late-stage granites, slightly peraluminous (Fig. 4(a)). They also show alkaline affinities (e.g. relatively high K and Na contents, Fig. 4(b)), yet no peralkaline granites have been found in the southeastern Fennoscandian intrusions. In the Suomenniemi Batholith (Fig. 1), however, a minor peralkaline hypersolvus intrusive phase (aegirine-augite alkali-feldspar syenite) has recently been found (Rämö 1991). In the tectonomagmatic diagrams of Pearce *et al.* (1984) the southeastern Fennoscandian rapakivi granites, associated silicic dyke rocks, and alkali-feldspar syenites plot almost exclusively in the within-plate granite field (Haapala & Rämö 1990; Rämö 1991). Due to their distinctly elevated Ga/Al ratios (10,000* Ga/Al is generally in the order of 3 to 4) they identify themselves as A-type granites in the discrimination diagrams of Whalen *et al.* (1987).

The wiborgitic rapakivi granite varieties are biotite-hornblende granites with 65–70 wt% SiO_2 and 1·7–3·1 wt% CaO, whereas the typical pyterlites are generally biotite granites with 75–77 wt% SiO_2 and 0·8–1·0 wt% CaO. However, many porphyritic and even-grained granites that have neither mantled nor unmantled alkali feldspar ovoids are chemically akin to wiborgites or pyterlites. Thus, the formation of the rapakivi texture (mantled and unmantled ovoidal alkali feldspar megacrysts, two generations of quartz and alkali feldspar) is controlled by the chemical composition as well as the physical conditions of the crystallising granite magma. In the light of recent theoretical and petrological studies, decompression of the ascending and partly crystallised granite magma (Whitney 1975; Nekvasil 1991) and mixing of the granite magma with a mafic magma (Hibbard 1981; Wark & Stimac 1991) are the main mechanisms to generate the stability changes needed for the formation of the rapakivi texture.

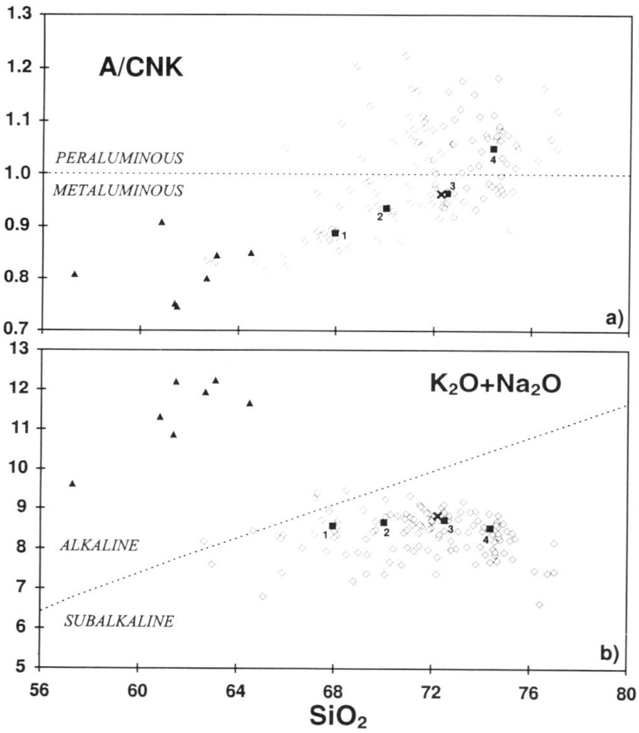

Figure 4 Analyses of the rapakivi granites and associated silicic dyke rocks (diamonds) and alkali-feldspar syenites (closed triangles) plotted in (a) A/CNK vs SiO_2 and (b) $K_2O + Na_2O$ vs SiO_2 variation diagrams; also shown are the average analyses (closed squares) of the main rock types in the Suomenniemi rapakivi granite batholith (1—hornblende granite; 2—biotite-hornblende granite; 3—biotite granite; 4—topaz-bearing granite) and the composition of its hypothetical parental magma (cross; see Rämö 1991); A/CNK in (a) denotes molecular $Al_2O_3/(CaO + K_2O + Na_2O)$ ratio; line separating the alkaline and subalkaline fields in (b) is from Irvine and Baragar (1971); data from Vorma (1976), Haapala (1977), Bergman (1981), Törnroos (1984), and Rämö (1991).

3. Origin of the rapakivi granites

Recently, detailed isotopic and general geochemical studies have been carried out on the rapakivi granites and related mafic rocks of southeastern Fennoscandia (Haapala & Rämö 1990; Rämö 1990; Beljaev *et al.* 1991; Neymark & Amelin 1991; Rämö & Haapala 1991; Rämö 1991). The main objective of these studies has been to assess the origin of the granites: do they represent anatectic melts of pre-existing continental crust or is there a substantial mantle component involved? Items of interest have also been the source characteristics and evolution of the contemporaneous mantle-derived mafic rocks (diabase dykes, gabbros, and anorthosites). In the following, Nd and Pb isotopic data acquired for the Finnish and Soviet Karelian rapakivi granite batholiths are summarised to evaluate the origin of these granites.

3.1. Nd isotopic constraints

Altogether 29 whole-rock Nd isotopic analyses of the southeastern Fennoscandian rapakivi granites have been reported by Haapala and Rämö (1990), Rämö and Haapala (1991), and Rämö (1991). The Finnish granites, which were emplaced within the Svecofennian 1·9 Ga crust, all show very similar Nd isotopic characteristics. They are strongly enriched in LREE ($^{147}Sm/^{144}Nd$ ratios range from 0·1–0·12) and show a narrow range of slightly negative $\varepsilon_{Nd}(T)$ values (−3·1–−0·2). Their T_{DM} model ages (DePaolo 1981) vary from 2·0–2·1 Ga (excluding the topaz-bearing late-stage granites that show Sm/Nd ratios affected by fractionation of allanite and/or monazite).

The Soviet Karelian granites that are found in the contact zone between the Svecofennian and Archaean domains show distinctly less radiogenic Nd isotopic compositions than the Finnish granites. Their $\varepsilon_{Nd}(T)$ values range from −8·1 to −5·7 and their T_{DM} model ages are in the order of 2·3–2·5 Ga (excluding the topaz-bearing granites). Similar unradiogenic Nd isotopic compositions have been reported for one of the Soviet Karelian batholiths (Salmi) also by Neymark and Amelin (1991).

The Nd isotopic compositions of the Finnish and Soviet Karelian rapakivi granites are shown in an ε_{Nd} versus age diagram in Figure 5. Also shown are the compositional space and time-integrated evolution of the juvenile Svecofennian 1·9 Ga crust (Huhma 1986; Patchett & Kouvo 1986) and average evolution of Archaean continental crust. The time-integrated evolution of the Finnish rapakivi granites conforms closely to the evolution of the Svecofennian crust (the average T_{DM} model ages are 2·06 ± 0·03 (1 S.D.) and 2·12 ± 0·14 (1σ) Ga, respectively) which suggests that the Finnish granites actually derive from the juvenile crust. The clearly more negative $\varepsilon_{Nd}(T)$ values of the Soviet Karelian rapakivi granites imply that there was a significant contribution of older crustal material to these granites. Because in southeastern Fennoscandia there is no indisputable evidence for an early Proterozoic crustal domain older than the 1·9 Ga crust (cf. Huhma *et al.* 1991), the source for the unradiogenic Nd was most probably Archaean in age. On the basis of the Nd data, Rämö (1991, pp. 101–103) proposed a mixing model according to which the Soviet Karelian granites consist of 30–45% Archaean crustal material, the remainder being reworked early Proterozoic crust.

3.2. Pb isotopic constraints

Rämö (1991) reported Pb isotopic analyses of 29 whole rocks and 18 alkali feldspar fractions of the Finnish and Soviet Karelian granites. The data are summarised in a $^{207}Pb/^{204}Pb$ vs $^{206}Pb/^{204}Pb$ diagram in Figure 6. In accordance with the Nd isotopic data, the Finnish and Soviet Karelian granites are distinct also in terms of their Pb isotopic composition. The Finnish samples plot along the average growth curve of the Svecofennian crust, while the Soviet Karelian data show very low Pb isotopic ratios that plot distinctly below the evolution of the juvenile crust. In accordance with the interpretation of the Nd isotopic data, the Pb isotopic compositions also imply that the Finnish granites represent anatectic melts of the early Proterozoic Svecofennian crust and that the Soviet Karelian granites are mixtures of early Proterozoic and Archaean crustal sources. Modelling the Pb isotopic data, Rämö (1991, pp. 103–104) concluded that the Nd mixing model can be refined and suggested that the Soviet Karelian granites are about one-to-one mixtures of early Proterozoic and Archaean lower crust, the latter of which shows retarded Pb isotopic evolution having an apparent Stacey and Kramers (1975) μ_2 value of about 8 (Fig. 6).

In modelling the magmatic evolution of the granites of one of the Finnish batholiths (Suomenniemi), Rämö (1991) concluded that the parental magma for that batholith was possibly generated by about 20% melting of an intermediate to silicic (granodioritic) protolith, presuming that melting was started by anhydrous breakdown of biotite and amphibole. Fusion of these minerals is the obvious reason for the elevated contents of F, and possibly also for those of Rb, Ga, Zn, Sn, and Nb in the rapakivi granites. Petrochemical modelling (Rämö 1991) also suggests that the

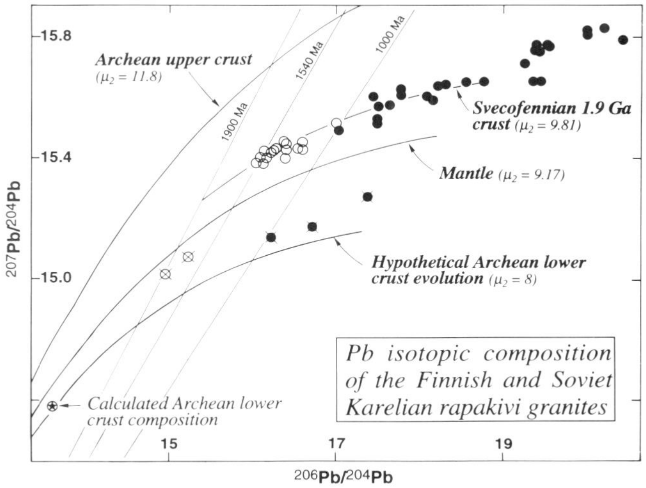

Figure 6 Pb isotopic data of the Finnish and Soviet Karelian rapakivi granites plotted in $^{207}Pb/^{204}Pb$ vs $^{206}Pb/^{204}Pb$ diagram, showing the growth curves of major sources for Pb in the Fennoscandian shield; the latter comprise Archaean upper crust ($\mu_2 = 11{\cdot}8$), mantle ($\mu_2 = 9{\cdot}17$), juvenile 1·9 Ga-old Svecofennian crust ($\mu_2 = 9{\cdot}81$) (all approximated on the basis of the galena Pb data of Vaasjoki 1981), and hypothetical Archaean lower crust (apparent $\mu_2 = 8$); Stacey and Kramers (1975) model isochrons for 1000, 1540, and 1900 Ma are also shown; rapakivi samples: open circles—Finnish alkali feldspars; closed circles—Finnish whole rocks; crossed open circles—Soviet Karelian alkali feldspars; crossed closed circles—Soviet Karelian whole rocks; a hypothetical Archaean lower crust composition (Rämö, 1991) is also shown; modified from Rämö, (1991, fig. 48).

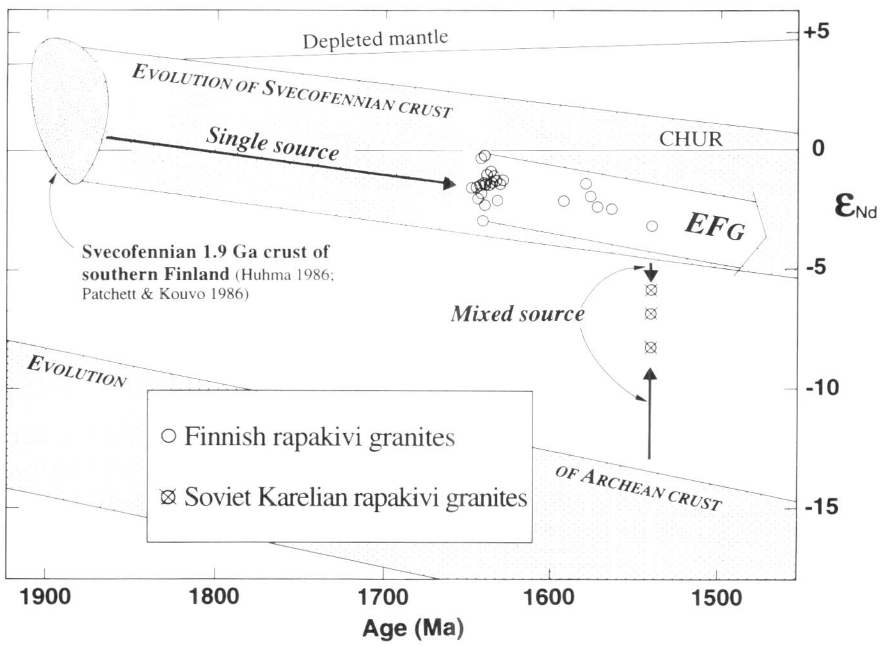

Figure 5 Nd isotopic data (ε_{Nd} versus age plot) of the rapakivi granites of southeastern Fennoscandia, illustrating their single source (Finnish granites) and mixed source (Soviet Karelian granites) origin; EFg denotes the overall evolution path of the Finnish rapakivi granites; Nd isotopic composition and time-integrated evolution of the Svecofennian 1·9 Ga-old juvenile crust and the evolution path of average Archaean crust are shown; marked in the figure are also the evolution lines of undifferentiated Earth (CHUR; DePaolo & Wasserburg 1976) and depleted mantle (DePaolo 1981).

subsequent evolution of the rapakivi granite magma in the Suomenniemi Batholith was largely controlled by fractionation of an alkali feldspar, quartz and clinopyroxene-dominated mineral assemblage, resulting in a peraluminous trend for the residual liquids (cf. Fig. 4a).

4. Summary and conclusions

The rapakivi granites of southeastern Fennoscandia are 200–350 Ma younger than the Svecofennian orogenic (magmatic arc) igneous activity, and 150–250 Ma younger than the voluminously minor postorogenic (post-collisional) granitic magmatism. A direct connection of the rapakivi granites to the Svecofennian orogeny is thus improbable. The rapakivi magmatism was clearly bimodal, the mafic members being represented by tholeiitic diabase dykes, gabbros, and anorthosites, the silicic members by various granites and quartz-feldspar porphyry dykes. The simultaneous existence of mafic and silicic magmas can be verified at several localities in composite dykes and in plutonic association.

The southeastern Fennoscandian rapakivi granites are metaluminous to slightly peraluminous A-type granites. Nd and Pb isotopic studies show that the source of the magmas was the juvenile 1·9 Ga-old Svecofennian crust for the Finnish granites, but that the rapakivi granites of Soviet Karelia, located at the border between the Archaean and early Proterozoic crustal domains, show a major Archaean source component. The parallel diabase dyke swarms indicate an extensional tectonic setting during the emplacement of the granites. The location of the rapakivi granites in a roughly E–W-trending belt of low crustal thickness supports this explanation. The mantle dome below the Wiborg Batholith can be interpreted as a relict of a mantle diapir. Partial melting of the lower/middle crust to yield the rapakivi granite magmas was probably caused by upwelling hot mantle material (mafic underplating).

It is worth noting that the granites of the Wiborg rapakivi Batholith and its satellite intrusions are older (1·65–1·62 Ga) than the rapakivi granites of southwestern Finland and Soviet Karelia (1·59–1·54 Ga). This excludes continental plate overriding a single hot mantle plume as a possible explanation (cf. Pesonen *et al.* 1989). A different model for the partial melting of the crust is presented by Windley (1988, 1991; see also Vorma 1976). According to this model, the rapakivi granites of Fennoscandia were generated in a post-collisional setting by partial melting of a tectonically thickened (but subsequently stretched and thinned) crust, the necessary heat for the fusion being produced internally within the thick crust. However, the bimodal magmatic association, the simultaneous existence of mafic and silicic magmas, the extensive diabase dyke swarms, the domed mantle below the Wiborg Batholith, and the large age gap between the orogenic-postorogenic granites and the rapakivi granites can more plausibly be explained by the model involving uprise of hot mantle material, mafic underplating, and partial melting of the lower/middle crust caused by the extra heat brought into the system by mafic magmas.

Although the eastern Fennoscandian rapakivi granites cannot be regarded as direct products of orogenic processes, an overall causal connection between the Svecofennian orogeny and the generation of the rapakivi granites exists. It was the Svecofennian orogeny that generated the crust which acted as the protolith for the Finnish rapakivi granites, and Archaean crust-forming processes produced the source which contributed to the generation of the Soviet Karelian granites. Furthermore, it is possible that the Svecofennian orogeny resulted in unstable domains in the subcontinental mantle (e.g. relicts of subducted oceanic crust), which may eventually have contributed to the onset of the partial melting in the upper mantle 1·6 Ga ago.

5. Acknowledgements

Riitta Fagerström drafted most of the figures. This paper is IGCP Project 315 (Rapakivi Granites and Related Rocks) Publication No. 8.

6. References

Babel Working Group 1991. Integrated seismic studies of the Baltic Shield using data in the Gulf of Bothnia region. GEOPHYS J INT (in press).

Beljaev, A., Larin, A. & Amelin, Yu. 1991. Geochemistry. *In* Haapala, I., Rämö, O. T. & Salonsaari, P. T. (eds) *Salmi batholith and Pitkäranta ore field in Soviet Karelia,* IGCP Project 315 Symposium Rapakivi Granites and Related Rocks, Excursion Guide. GEOL SURV FINLAND GUIDE **33,** 11–9.

Bergman, L. 1981. *Berggrunden inom Signilskär, Mariehamn och Geta kartblad*—Signilskärin, Maarianhaminan ja Getan karttaalueiden kallioperä (in Swedish); English summary: *Pre-Quaternary rocks of the Signilskär, Mariehamn and Geta mapsheet areas.* Geological map of Finland 1:100,000, Explanation to the map of rocks, 0034 + 0043 Signilskär, 1012 Maarianhamina and 1021 Geta, Geological Survey of Finland.

DePaolo, D. J. 1981. Neodymium isotopes in the Colorado Front range and crust-mantle evolution in the Proterozoic. NATURE **291,** 193–6.

DePaolo, D. J. & Wasserburg, G. J. 1976. Nd isotopic variations and petrogenetic models. GEOPHYS RES LETT **3,** 249–52.

Eklund, O. 1991. *Betydelsen av magmamixing som magmatisk evolutionsprocess i tolkningen av post- och anorogena bergarters uppkomst, Åland SV Finland.* Unpublished thesis, Åbo Akademi University, Institute for Geology and Mineralogy (partly in English).

Emslie, R. F. 1991. Granitoids of rapakivi granite-anorthosite and related associations. *In* Haapala, I. & Condie, K. C. (eds) Precambrian granitoids—petrogenesis, geochemistry and metallogeny. PRECAMBRIAN RES **51,** 173–92.

Esipchuk, K. Ye. 1991. Rapakivi granites of the Ukrainian shield: phases and facies. *In* Haapala, I. & Rämö, O. T. (eds) *Symposium on Rapakivi Granites and Related Rocks Abstract Volume.* GEOL SURV FINLAND GUIDE **34,** 16.

Front, K. & Nurmi, P. A. 1987. Characteristics and geological setting of synkinematic Svecokarelian granitoids in southern Finland. PRECAMBRIAN RES **35,** 207–24.

Haapala, I. 1977. Petrography and geochemistry of the Eurajoki stock, a rapakivi-granite complex with greisen-type mineralization in southwestern Finland. GEOL SURV FINLAND BULL **286**.

Haapala, I. 1989. Suomen rapakivigraniiteista, Rapakivi granites of Finland. ACAD SCI FENNICA YEARB **1988–1989** (in Finnish with an English summary).

Haapala, I. & Rämö, O. T. 1990. Petrogenesis of the Proterozoic rapakivi granites of Finland. *In* Stein, H. J. & Hannah, J. L. (eds) *Ore-bearing granite systems*; *Petrogenesis and mineralizing processes.* GEOL SOC AM SPEC PAP **246,** 275–86.

Hibbard, M. J. 1981. The magma mixing origin of mantled feldspars. CONTRIB MINERAL PETROL **76,** 158–70.

Hubbard, F. & Branigan, N. 1987. Late Svecofennian magmatism and tectonism, Åland, Southwest Finland. PRECAMBRIAN RES **35,** 241–56.

Huhma, H. 1986. Sm–Nd, U–Pb and Pb–Pb isotopic evidence for the origin of the Early Proterozoic Svecokarelian crust in Finland. GEOL SURV FINLAND BULL **337**.

Huhma, H., Claesson, S., Kinny, P. D. & Williams, I. S. 1991. The growth of early Proterozoic crust: new evidence from Svecofennian detrital zircons. TERRA NOVA **3,** 175–8.

Irvine, T. N. & Baragar, W. R. A. 1971. A guide to the chemical classification of the common volcanic rocks. CANAD J EARTH SCI **8,** 523–48.

Korja, A. 1991. *Crustal and upper mantle structure of the Wiborg batholith, SE Finland.* Institute of Seismology, University of Helsinki, Report S-25, 97–101.

Korja, A. & Elo, S. 1990. Crustal and upper mantle structure of the Wiborg batholith, SE Finland. *Abstracts of the second symposium on the Baltic shield, Lund, Sweden, June 5–7, 1990,* 57.

Larin, A., Beljaev, A. & Stepanov, K. 1991. Geological setting of the Salmi batholith. *In* Haapala, I., Rämö, O. T. & Salonsaari, P. T. (eds) *Salmi batholith and Pitkäranta ore field in Soviet Karelia,* IGCP Project 315 Symposium Rapakivi Granites and Related Rocks, Excursion Guide. GEOL SURV FINLAND GUIDE **33,** 6–7.

Laurén, L. 1970. An interpretation of the negative gravity anomalies associated with the rapakivi granites and the Jotnian sandstone in southern Finland. GEOL FÖREN STOCKHOLM FÖRH **92,** 21–34.

Lindberg, L. & Eklund, O. 1988. Interactions between basaltic and granitic magmas in a Svecofennian postorogenic granitoid intrusion, Åland, southwest Finland. LITHOS **22,** 13–23.

Luosto, U. 1991. *Moho depth maps of the Fennoscandian shield based on seismic refraction data.* Institute of Seismology, University of Helsinki, Report S-25, 43-9.

Luosto, U., Tiira, T., Korhonen, H., Azbel, I., Burmin, V., Buyanov, A., Kosminskaya, I., Ionkis, V. & Sharov, N. 1990. Crust and upper mantle structure along the DSS Baltic profile in SE Finland. GEOPHY J INT **101,** 89–110.

Nekvasil, H. 1991. Ascent of felsic magma and formation of rapakivi. AM MINERAL **76,** 1279–90.

Neymark, L. & Amelin, Yu. 1991. Isotope geochemistry. *In* Haapala, I., Rämö, O. T. & Salonsaari, P. T. (eds) *Salmi batholith and Pitkäranta ore field in Soviet Karelia,* IGCP Project 315 Symposium Rapakivi Granites and Related Rocks, Excursion Guide. GEOL SURV FINLAND GUIDE **33,** 39–40.

Neymark, L., Amelin, Yu. & Larin, A. 1991. Geochronology of the rocks of the Salmi batholith. *In* Haapala, I., Rämö, O. T. & Salonsaari, P. T. (eds) *Salmi batholith and Pitkäranta ore field in Soviet Karelia,* IGCP Project 315 Symposium Rapakivi Granites and Related Rocks, Excursion Guide. GEOL SURV FINLAND GUIDE **33,** 34–6.

Nurmi, P. A. & Haapala, I. 1986. The Proterozoic granitoids of Finland: granite types, metallogeny and relation to crustal evolution. BULL GEOL SOC FINLAND **58,** 203–33.

Patchett, J. & Kouvo, O. 1986. Origin of continental crust of 1·9–1·7 Ga age. Nd isotopes and U–Pb zircon ages in the Svecokarelian terrain of South Finland. CONTRIB MINERAL PETROL **92,** 1–12.

Pearce, J. A., Harris, N. B. W. & Tindle, A. G. 1984. Trace element discrimination diagrams for the tectonic interpretation of granitic rocks. J PETROL **25,** 956–83.

Pesonen, L. J., Torsvik, T. H., Elming, S.-Å. & Bylund, G. 1989. Crustal evolution of Fennoscandia–palaeomagnetic constraints. TECTONOPHYSICS **162,** 27–49.

Rämö, O. T. 1990. Diabase dyke swarms and silicic magmatism–evidence from the Proterozoic of Finland. *In* Parker, A. J., Rickwood, P. C. & Tucker, D. H. (eds) *Mafic dykes and emplacement mechanisms,* 185–99. Rotterdam: A.A. Balkema.

Rämö, O. T. 1991. Petrogenesis of the Proterozoic rapakivi granites and related basic rocks of southeastern Fennoscandia: Nd and Pb isotopic and general geochemical constraints. GEOL SURV FINLAND BULL **355**.

Rämö, O. T. & Haapala, I. 1991. The rapakivi granites of eastern Fennoscandia: a review with insights into their origin in the light of new Sm–Nd isotopic data. *In* Gower, C. F., Rivers, T. & Ryan, B. (eds) *Mid-Proterozoic Laurentia-Baltica.* GEOL ASSOC CANAD SPEC PAP **38,** 401–15.

Sahama, Th. G. 1945. On the chemistry of the east Fennoscandian rapakivi granites. BULL COMM GÉOL FINLANDE **136**.

Salonsaari, P. T. & Haapala, I. 1991. The Jaala-Iitti rapakivi complex: an example of bimodal magmatism and hybridization in the Wiborg rapakivi batholith, Finland. *In* Haapala, I. & Rämö, O. T. (eds) *Symposium on Rapakivi Granites and Related Rocks Abstract Volume.* GEOL SURV FINLAND GUIDE **34,** 45.

Sederholm, J. J. 1891. Ueber die finnländischen Rapakiwigesteine. TSCHERMAK'S MINERAL PETROGR MITT **12,** 1–31.

Sederholm, J. J. 1928. On orbicular granites, spotted and nodular granites etc. and on the rapakivi texture. BULL COMM GÉOL FINLANDE **83**.

Stacey, J. S. & Kramers, J. D. 1975. Approximation of terrestrial lead isotope evolution by a two-stage model. EARTH PLANET SCI LETT **26,** 207–21.

Suominen, V. 1991. The chronostratigraphy of SW Finland with special reference to the Postjotnian and Subjotnian diabases. GEOL SURV FINLAND BULL **356**.

Törnroos, R. 1984. Petrography, mineral chemistry and petrochemistry of granite porphyry dykes from Sibbo, southern Finland. GEOL SURV FINLAND BULL **326**.

Vaasjoki, M. 1977. Rapakivi granites and other postorogenic rocks in Finland; Their age and the lead isotopic composition of certain associated galena mineralizations. GEOL SURV FINLAND BULL **294**.

Vaasjoki, M. 1981. The lead isotopic composition of some Finnish galenas. GEOL SURV FINLAND BULL **316**.

Vaasjoki, M., Rämö, O. T. & Sakko, M. 1991. New U–Pb ages from the Wiborg rapakivi area: constraints on the temporal evolution of the rapakivi granite-anorthosite-diabase dyke association of southeastern Finland. *In* Haapala, I. & Condie, K. C. (eds) *Precambrian granitoids–petrogenesis, geochemistry and metallogeny.* PRECAMBRIAN RES **51,** 227–43.

Vorma, A. 1976. On the petrochemistry of rapakivi granites with special reference to the Laitila massif, southwestern Finland. GEOL SURV FINLAND BULL **285**.

Wahl, W. 1925. Die Gesteine des wiborger Rapakiwigebietes. FENNIA **45** (20).

Wark, D. A. & Stimac, J. A. 1991. Experimental evidence for the origin of rapakivi texture by a dissolution- and diffusion-controlled mechanism. *In* Haapala, I. & Rämö, O. T. (eds) *Symposium on Rapakivi Granites and Related Rocks Abstract Volume.* GEOL SURV FINLAND GUIDE **34,** 63.

Whalen, J. B., Currie, K. L. & Chappell, B. W. 1987. A-type granites; Geochemical characteristics, discrimination, and petrogenesis. CONTRIB MINERAL PETROL **95,** 407–19.

Whitney, J. A. 1975. The effects of pressure, temperature, and X_{H_2O} on phase assemblages in four synthetic rock compositions. J GEOL **83,** 1–31.

Windley, B. F. 1988. Mid-Proterozoic "anorogenic" magmatism: its pre-orogenic and post-orogenic environments in the Grenville, Ketilidian and Svecofennian collisional belts. *Geological Association of Canada, Mineralogical Association of Canada, Canadian Society of Petroleum Geologists Joint Annual Meeting Program with Abstracts* 13, A136.

Windley, B. F. 1991. Two groups of rapakivi granites and related "anorogenic" rocks: their evolution in a plate tectonic framework during the formation and during the breakup of a supercontinent. *In* Haapala, I. & Rämö, O. T. (eds) *Symposium on Rapakivi Granites and Related Rocks Abstract Volume.* GEOL SURV FINLAND GUIDE **34,** 63.

ILMARI HAAPALA and O. TAPANI RÄMÖ, Department of Geology, University of Helsinki, Box 115, SF-00171 Helsinki, Finland.

Transactions of the Royal Society of Edinburgh: Earth Sciences, **83**, 173–178, 1992

The rapakivi granites of S Greenland—crustal melting in response to extensional tectonics and magmatic underplating

P. E. Brown, T. J. Dempster, T. N. Harrison and D. H. W. Hutton

ABSTRACT: Early Proterozoic rapakivi intrusions in S Greenland occur as thick sheets which have ramp–flat geometry and were intruded along the median planes of active ductile extensional shear zones. These shear zones and their intrusions were linked via transfer zones in a major three-dimensional framework. At high structural levels (*c*. 6 km) the rapakivi intrusions developed thermal aureoles which overprint the regional assemblages, whereas at deeper levels in the regional structure they are contemporaneous with regional metamorphism. Thermobarometry on the regional and contact assemblages indicates low pressure granulite facies conditions (200–400 MPa, 650°–800°C) suggesting very high thermal gradients. The rapakivi suite and associated norites have low initial $^{87}Sr/^{86}Sr$ together with positive ε_{Nd} values, indicating the involvement of predominantly young crust and/or mantle component in the generation of the igneous suite. It is considered that the voluminous norites are closely related to the mafic melts which underplated the juvenile crust to trigger the generation of the monzonitic rapakivi suite. Taken together, the data are consistent with a model of Proterozoic lithospheric extension, thinning of relatively juvenile continental crust and compression of mantle isotherms, resulting in high crustal heat flow, mafic underplating, and crustal melting with emplacement of magmas along a linked network of extensional shear zones.

KEY WORDS: Bimodal magmatism, extension, granulites

Large volumes of rapakivi granites and associated norites, intruded about 1740 Ma (Allart 1976; Gulson & Krogh 1975; Dempster *et al.* 1991a,b), invade the early Proterozoic Ketilidian mobile belt in Greenland. In some localities exceptionally good exposures enable observations to be made on the three-dimensional shape of the intrusions. In this paper we review the role of a strong extensional tectonic regime on the mechanism of intrusion of the granites, the genesis of which is considered in relation to an exceptionally high geothermal gradient and the strongly bimodal nature of the magmatism.

1. Intrusion mechanism

Over the greater part of the area invaded by the rapakivi intrusions, the Ketilidian country rocks consist of regionally flat-lying granulite facies migmatites and schists which, prior to the emplacement of the rapakivi bodies, were intruded by an extensive, concordant sheet of microcline granite which attained a thickness of approximately 1 km (Bridgwater *et al.* 1974).

Key observations on the relationships between the rapakivi intrusions and the regional flat-lying structures (Hutton *et al.* 1990) have been made in the area of Qernertoq (Fig. 1) and are depicted in Figures 2, 3, and 4. To the N (Fig. 2, locality b, and Fig. 3b) a sheet of rapakivi up to 400 m thick was intruded along the sub-horizontal boundary between underlying migmatites and overlying microcline granite. To the S (Fig. 2, locality a, and Fig. 3a) the rapakivi sheet turns down until at sea level the contact is about vertical or steeply inclined southwards. Regional foliation in the schists of the footwall of the intrusion also turns down to the S, but not so steeply, so that it is obliquely cut by the igneous contact. This deflection of the footwall banding was caused by extensional shear with top to the S sense of movement (Hutton *et al.* 1990). A new cleavage, formed during this extensional shear, intensifies upwards through the footwall towards the base of the rapakivi sheet, with which it is concordant. A strong S–C fabric and N–S stretching lineation were developed in the shear zone where all small-scale shear sense indicators show top to the S movement. Similar observations and deductions can be made where the microcline granite forms the hanging wall of the structure and the evidence is in accord with the existence of an extensional shear zone around and concordant with the rapakivi sheet.

That deformation and intrusion were contemporaneous is indicated by textures in the thermal aureole of the intrusion, which show overprinting of shear fabrics followed by further deformation. Also, in the rapakivi sheet itself there are fabrics formed by pre-full-crystallisation deformation and crystal plastic strain (Hutton 1988) and localised solid state shear. Intrusion, formation of the thermal aureole and shearing were thus coeval and the rapakivi sheet was emplaced along an active, ductile, extensional shear zone.

Hutton *et al.* (1990) have made a comparison of the ramp and flat geometry of the rapakivi intrusion with structures found in extensional sedimentary basins and on a wider scale have demonstrated the existence of rapakivi ramps and flats connected by sinistral transfer faults analogous to such faults found in brittle systems at higher crustal levels.

2. Metamorphism

Granulite facies gneisses and migmatites are widely developed throughout the southern part of the Ketilidian belt (Dawes 1970; Bridgwater *et al.* 1974), where small-scale extensional shears are frequently occupied by hyperthene-garnet pegmatites. Typcial metasedimentary schists contain quartz, K-feldspar, plagioclase, cordierite, usually garnet and/or orthopyroxene and there is a general lack of reaction textures, suggesting a close approach to equilibrium (Dempster *et al.* 1991a). Biotite is uncommon in the highest grade rocks of the Augpilagtoq area (Fig. 1) but is conspicuous in the region of Prins Christian Sund and Qernertoq where orthopyroxene is less common, suggesting a somewhat lower metamorphic grade in the E than the W. Rapakivi intrusions in the Augpilagtoq area were evidently emplaced close to the peak of regional metamorphism and lack thermal aureoles. Further to the E in the Qernertoq area and at higher crustal levels, the thermal aureole is marked by the development of biotite at the expense of garnet and, where present, orthopyroxene, this being followed in the inner aureole by renewed growth of cordierite and garnet.

A number of thermometers and barometers applied to the metamorphic rocks (Dempster *et al.* 1991a) indicate temperatures of 650–800°C and pressures of 200–400 MPa. The results of the application of the techniques of Powell and Holland (1988) and Holland and Powell (1990) are depicted in Figure 5, which suggests a somewhat lower pressure for the Qernertoq aureole as compared to the Augpilagtoq region. Exceptionally high regional geothermal gradients are indicated, in excess of 60°C/km.

3. Petrology

The rapakivi intrusions are of extremely coarse grain size, with the long axis of feldspar ovoids frequently attaining 10 cm and sometimes 15 cm. This makes the determination of both modal and chemical compositions difficult and, to alleviate this, average analyses have been used to determine the composition (Table 1) and norms (Fig. 6) of the Prins Christian Sund, Qernertoq and Tasermiut intrusions. The rocks are quartz monzonites in which plagioclase mantles around K-feldspar ovoids are usually well developed. Dempster *et al.* (1991b) have used Sr, O and H isotopes to suggest an origin for the mantles through an extended history of high and low temperature exsolution. Plagioclase as phenocrysts and in the groundmass has compositions in the range An_{25}–An_{40}, with rare phenocryst cores up to An_{60}.

Both a black and a white facies of the granites have been recognised, the colour relating essentially to that of the feldspars, there being no major difference in bulk chemistry or radiogenic isotopes between the two facies. Contacts between the black and white facies appear in some cases to be secondary and due to infiltration of fluids into the black granite, but more commonly the contact appears to be intrusive and the differences attributable to the original fluid content of the magmas. The white facies has an entirely hydrous mafic mineralogy, whereas that of the black facies is largely anhydrous.

Olivine with composition Fa_{93}–Fa_{96} is common in the black granite of Prins Christian Sund (Harrison *et al.* 1990a) where it occurs with clinopyroxene ($En_{20}Fs_{78}Wo_2$) and quartz. Orthopyroxene and clinopyroxene (ferroaugite) are found predominantly in the black granites, although rare clinopyroxene is found in the white facies on Qernertoq. Amphibole, usually ferroedenite or ferrohornblende, is ubiquitous and in the black granites appears on textural evidence to be exclusively subsolidus. Fe-rich biotite is more abundant in the white than the black facies and in the latter is mainly subsolidus, forming extensive fringes around other magmatic mafic silicates. Ilmenite close to end-member composition is the only opaque phase, zircon and apatite are ubiquitous, and allanite is present in the Prins Christian Sund and Qernertoq areas but appears to be absent further W. Fluorite is uncommon throughout.

4. Discussion

4.1. Granite type

Although of monzonitic rather than granitic composition, the Greenland rapakivi intrusions plot in discrimination

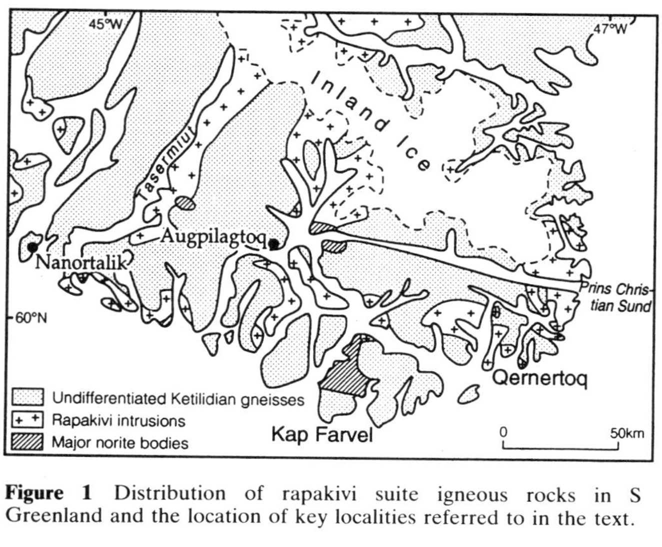

Figure 1 Distribution of rapakivi suite igneous rocks in S Greenland and the location of key localities referred to in the text.

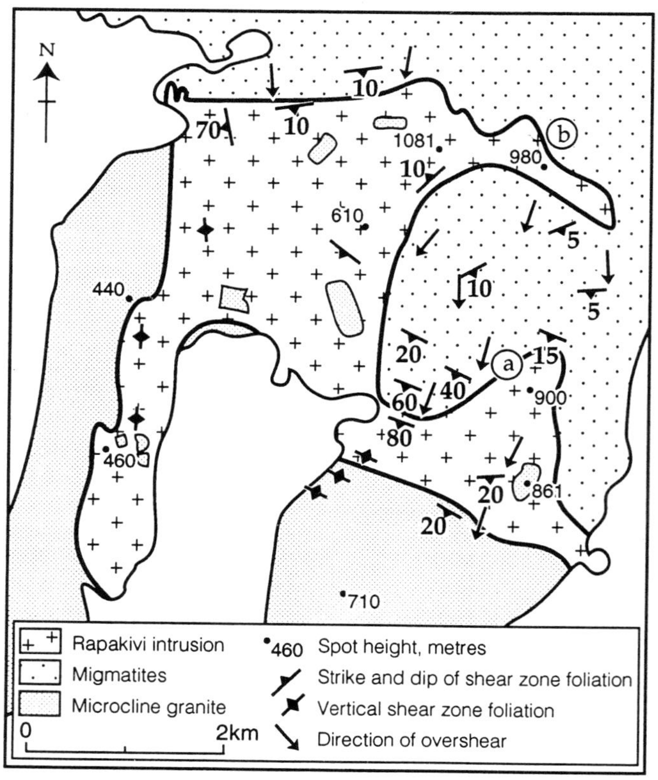

Figure 2 Geological map of part of Qernertoq; localities marked a and b refer to photographs (a) and (b) in Figure 3.

Figure 3 (a) View from the W to summit 990 m (Fig. 2) showing the down-turn of foliation in the metamorphic rocks and ramp structure in the rapakivi intrusion; Figure 4 shows the main geological units seen in this view; note the microcline granite on right-hand summit.
(b) Flat-lying rapakivi sheet overlying migmatities (Fig. 2, locality b).
(c) Subhorizontal contact of rapakivi sheet; locality on N side of summit 1081 (Fig. 2).
(d) Vertical contact of rapakivi intrusion against microcline granite on W side of Qernertoq; view looking W beneath summit 440 m (Fig. 2).
(e) Stoped blocks with rapakivi ovoids conspicuous in matrix; Augpilagtoq area.

diagrams utilising concentrations of high field strength elements, in the field of A-type granites (Fig. 7). They have the A-type characteristics of being formed from high temperature, anhydrous, anorogenic magmas.

Emslie (1991) has summarised the mineralogical and chemical features of the granitic members of the rapakivi granite—anorthosite association and suggested that the rapakivi granites form a distinctive subset of the A-type category. Particularly characteristic are high FeO/FeO + MgO, high $Na_2O + K_2O$, low Al_2O_3, ilmenite rather than magnetite, and bimodal association with mafic magmas. The Greenland monzonitic rapakivis exhibit these features, but have somewhat high Al_2O_3. The bimodal association with noritic mafic rocks is very marked (Harrison *et al.* 1990b).

4.2. Cumulate features

In Figure 8 the norms of average analyses are plotted in the system Q–Qr–Ab under water deficient conditions (Johannes & Holtz 1990). In this system the shifts of "minimum" melt composition with increasing water deficiency and rising pressure are pertinent to the monzonitic composition of the rapakivi intrusions. An increase of An in the system under conditions of increasing pressure and decreasing water content will also produce melts enriched in Or and poor in Q (Nekvasil 1988). Although voluminous, water-deficient granite magmas are unlikely to be "minimum" melts, the comparison with the experimental results suggests that the monzonitic rapakivis might approach such compositions.

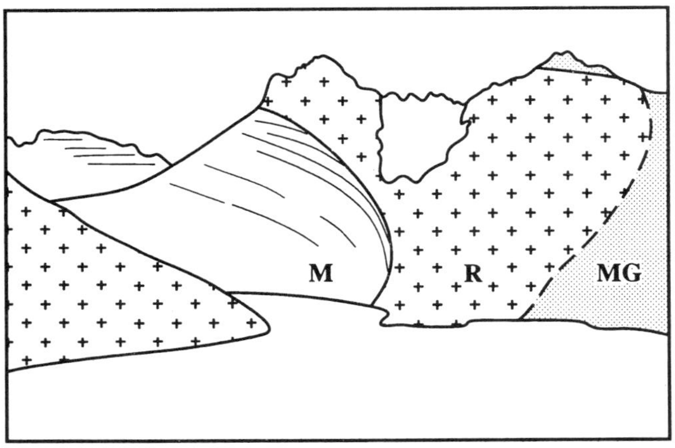

Figure 4 Refers to Figure 3a; M, metamorphics, R, rapakivi intrusion, MG, microcline granite.

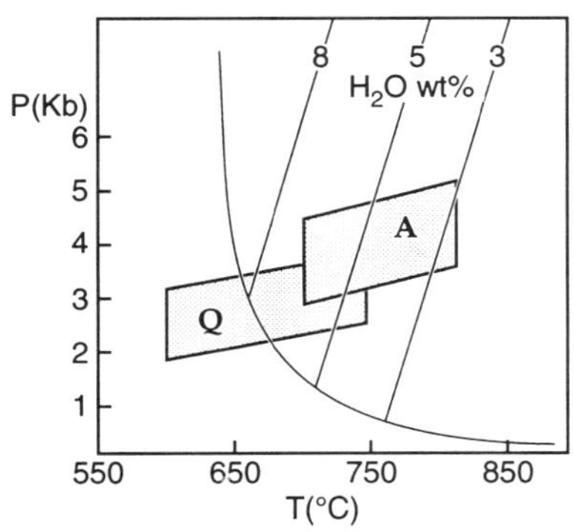

Figure 5 Estimated metamorphic conditions using Powell & Holland (1988) and Holland & Powell (1990); Q, Qernertoq, A, Augpilagtoq; water-deficient melting curves in the granite system from Johannes & Holtz (1990).

Table 1 Mean standard deviations for ten analyses from the Qernertoq, seven analyses from Prins Christian Sund and fifteen analyses from Tasermiut.

	Tasermiut (10)		Prins Christian Sund (7)		Qernertoq (15)	
	x	*σ*	*x*	*σ*	*x*	*σ*
SiO_2	63·44	2·59	63·93	1·68	61·10	2·33
TiO_2	0·87	0·28	0·76	0·13	0·87	0·20
Al_2O_3	16·19	0·88	16·23	1·02	18·06	1·06
Fe_2O_3	1·25	0·34	2·02	0·44	6·21*	1·44
FeO	4·83	1·82	4·50	1·10	—	—
MnO	0·06	0·03	0·06	0·03	0·07	0·03
MgO	1·78	0·75	0·99	0·34	1·75	0·98
CaO	2·79	0·88	3·46	0·58	4·05	0·95
Na_2O	2·73	0·38	2·64	0·37	2·73	0·32
K_2O	4·81	1·18	4·24	0·41	3·94	1·06
P_2O_5	0·26	0·12	0·27	0·06	0·32	0·12
Trace elements ppm						
Rb	142	43	122	27	103	25
Sr	284	62	329	49	426	88
Ba	996	191	1040	470	1320	393
Y	35	10	47	14	28	6
Nb	20	4	27	2	17	3
Zr	298	94	480	111	271	143

On the other hand, the rocks are dominated by large alkali feldspar ovoids which can be shown to have a thermal and movement history suggestive of cumulate processes, so that it is improbable that any of the material analysed is actually representative of original liquid compositions. It is common to find, together in the same exposure, feldspars which have undergone partial dissolution to form the

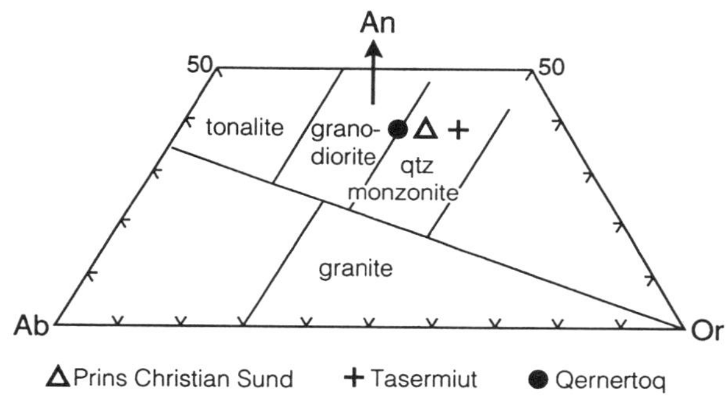

Figure 6 Classification of average analyses (Table 1) in the normative feldspar triangle (after O'Connor 1965).

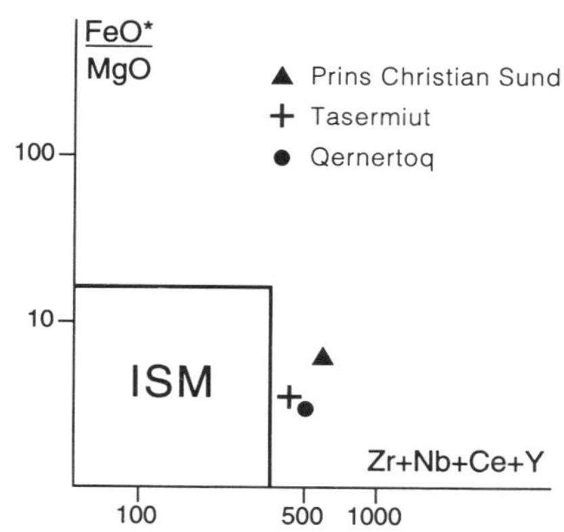

Figure 7 Discrimination diagram FeO/MgO vs Zr + Nb + Ce + Y, showing the field I-, S- and M-type granites plus fractionated granites (Whalen *et al.* 1987); Greenland rapakivi averages fall in the area of A-type granites.

characteristic ovoid shape, mantled ovoids which have lost the original K-feldspar core through dissolution, euhedral rectangular alkali feldspars and feldspars, sometimes mantled, which have been fractured and broken in transport. Cumulate layering, though uncommon, has also been described (Finch *et al.* 1990). Huppert and Sparks (1988) in discussing the melting of crustal rocks by mafic magma, suggested that the site of melting will also be one of crystallisation and in this light some of the rapakivi feldspars may have been transported as phenocrysts from the deep source region to the level of consolidation. Alkali-feldspar accumulation is thus a feature of the Greenland intrusions. Positive Eu anomalies which would be indicative of plagioclase accumulation have not been found and a weak negative anomaly is present in specimens from Prins Christian Sund (Figs 9, 10). The overall REE abundances in the Greenland rocks are similar to those of rapakivi granites elsewhere.

4.3. Crustal melting and tectonic setting

The rapakivi intrusions are the culmination of evolution of the Ketilidian crust; they were preceded by regional migmatisation and the voluminous microcline granite sheet which intrudes the flat-lying metamorphic complex. This extended metamorphic and magmatic history suggests that water-depleted crustal rocks may have been the source of the dry, high temperature rapakivi magmas. On the other hand, it is unlikely that the very large volume of K-rich rapakivi magma had a source in rocks severely depleted by previous melt extraction, a point made by Emslie (1991) with respect to the rapakivi granite association in general. In this respect Creaser *et al.* (1991) have criticised the suggestion by Collins *et al.* (1982) that the source material of A-type granites is a dry granulite-type residue depleted by the prior extraction of an I-type melt, a potentially more fertile tonalite to granodiorite source being favoured.

From a Nd isotope study, which included samples of rapakivi granites, Patchett and Bridgwater (1984) demonstrated that the Ketilidian crust consisted of 85–95% material newly derived from the mantle. The remaining 5–15% was concluded to be Archaean material, probably sedimentary, which was mixed with the mantle-derived component to provide the Ketilidian crust with a young, overall homogeneous, isotopic signature. An island arc origin for the mantle-derived material was proposed for which Nd concentrations suggest a basalt to andesite composition, though this is not closely constrained.

This newly formed crust was the source of the rapakivi magmas. The low pressure, high temperature metamorphism which led to temperatures of up to 800°C at 300 MPa took place in a regime of strong extensional tectonics which controlled the rise and emplacement of the rapakivi intrusions. Such high temperatures require that the extended and thinned crust was underplated by mafic magma and evidence for this is seen in the noritic gabbros associated with the rapakivi intrusions. Evidence of mixing and hybridisation has not been seen in the rapakivi granites, but cannot yet be ruled out. At the present level of exposure the mafic and rapakivi magmas mingle but do not mix (Harrison *et al.* 1990b).

Juvenile crust, extensional tectonics and magmatic underplating suggest that a present-day analogy for the Ketilidian flat-lying complex may be in a back-arc environment. Much of the central area of the Ketilidian belt, to the N and W of the flat-lying complex, is occupied by the calc-alkaline Julianehaab batholith which has been likened to an arc-type of intrusive complex (Watterson 1978).

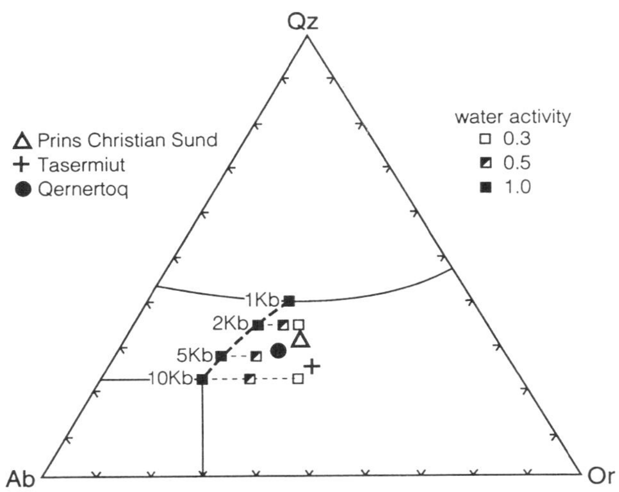

Figure 8 The water-deficient system Q–Or–Ab at pressures up to 10 kbar (Johannes & Holtz 1990) and the position of Greenland rapakivi averages.

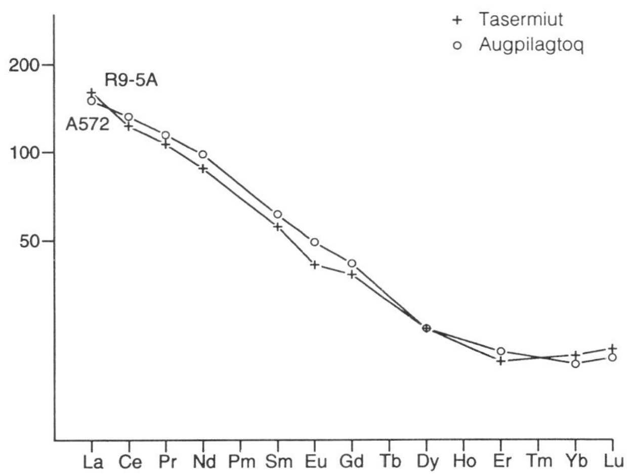

Figure 9 Rare-earth element patterns for rapakivi granite specimens from the Tasermiut and Augpilagtoq area.

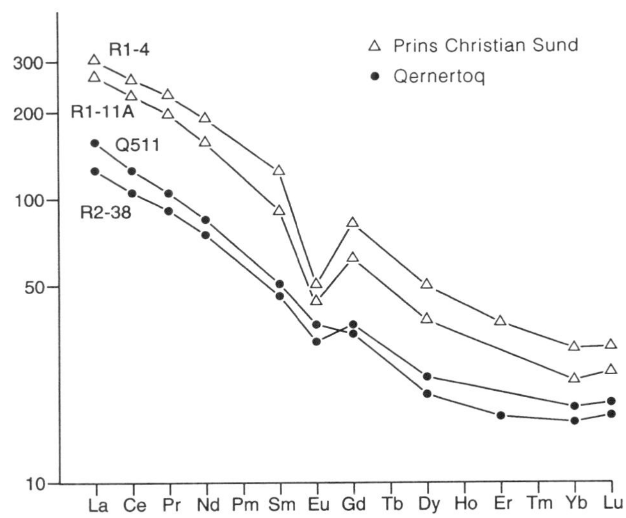

Figure 10 Rare-earth element patterns for rapakivi granite specimens from Prins Christian Sund and Qernertoq.

An alternative tectonic scenario involving Himalayan-style thickening of the crust followed by extensional collapse has been proposed by Windley (1991). This model seems largely based on statements that the structure of this area is "characterised by open to tight folds" in addition to "large-scale, recumbent isoclines and nappes". From our own extensive field observations in the high grade gneisses, we know of only one large fold structure (cf. Allart 1976) and indeed the gneisses are characterised by an almost complete lack of fold structure. The igneous and metamorphic history of the area also casts considerable doubt on the model of Windley (1991). Thus not only is the magmatism obviously bimodal but there is also a complete absence of high-pressure metamorphic mineral assemblages either within the lower grade schists or preserved as relicts in the gneisses.

5. Acknowledgements

In particular we thank Sue Becker for contributing to the project both in the field and laboratory. Others who have been involved include I. Parsons, W. Brown, A. Finch and J. Reavy. The work was supported by NERC grant GR/6336 to PEB, who also acknowledges support from the University of Aberdeen. D. Bridgwater is thanked for advice, GGU for interest and encouragement and W. S. Pitcher for commenting on the manuscript. This is IGCP-315 contribution no. 7.

6. References

Allart, J. H. 1976. The Ketilidian mobile belt in South Greenland. *In* Escher, A. & Watt, W. S. (eds) *Geology of Greenland*, 121–51. GRONL GEOL UNDERS COPENHAGEN.

Bridgwater, D., Sutton, J. & Watterson, J. 1974. Crustal downfolding associated with igneous activity. TECTONOPHYSICS **21,** 57–77.

Collins, W. J., Beams, S. D., White, A. J. R. & Chappell, B. W. 1982. Nature and origin of A-type granites with particular reference to south-eastern Australia. CONTRIB MINERAL PETROL **80,** 189–200.

Creaser, R. A., Price, R. C. & Wormald, R. J. 1991. A-type granites revisted: Assessment of a residual-source model. GEOLOGY **19,** 163–6.

Dawes, P. R. 1970. The plutonic history of the Tasiussaq area, South Greenland, with special reference to a high-grade gneiss complex. BULL GRONL GEOL UNDERS **88**.

Dempster, P. J., Hutton, D. H. W., Harrison, T. N., Brown, P. E. & Jenkin, G. R. T. 1991a. Textural evolution of rapakivi granites, south Greenland—Sr, O and H isotopic investigations. CONTRIB MINERAL PETROL **107,** 459–71.

Dempster, T. J., Harrison, T. N., Brown, P. E. & Hutton, D. H. W. 1991b. Low-pressure granulites from the Ketilidian mobile belt of south Greenland. J PETROL **32,** 979–1004.

Emslie, R. F. 1991. Granitoids of rapakivi granite—anorthosite and related associations. PRECAMBRIAN RES **51,** 173–92.

Finch, A. A., Reavy, R. J., Harrison, T. N. & Brown, P. E. 1990. Rapakivi granites, south Greenland: hydrothermal alteration of igneous layering. J GEOL SOC **147,** 739–42.

Gulson, B. L. & Krogh, T. E. 1975. Evidence of multiple intrusion, possible resetting of U-Pb ages, and new crystallisation of zircons in the post-tectonic intrusions (Rapakivi granites) and gneisses from south Greenland. GEOCHIM COSMOCHIM ACTA **39,** 65–82.

Harrison, T. N., Parsons, I. & Brown, P. E. 1990a. Fayalite-bearing rapakivi granites from the Prins Christian Sund pluton, south Greenland. MINERAL MAG **54,** 57–64.

Harrison, T. N., Reavy, R. J., Finch, A. A. & Brown, P. E. 1990b. Co-existing mafic and felsic magmas in the early Proterozoic rapakivi granite suite of Southern Greenland. BULL GEOL SOC DENMARK **38,** 53–8.

Holland, T. J. B. & Powell, R. 1990. An enlarged and updated internally consistent thermodynamic data set with uncertainties and correlations: the system K_2O–Na_2O–CaO–MgO–MnO–FeO–Fe_2O_3–Al_2O_3–TiO_2–SiO_2–C–H_2–O_2. J METAM GEOL **8,** 89–124.

Huppert, H. E. & Sparks, R. S. 1988. The generation of granitic magmas by intrusion of basalt into continental crust. J PETROL **29,** 599–624.

Hutton, D. H. W. 1988. Granite emplacement mechanisms and tectonic controls: inferences from deformation studies. TRANS R SOC EDINBURGH EARTH SCI **79,** 245–55.

Hutton, D. H. W., Dempster, T. J., Brown, P. E. & Becker, S. M. 1990. A new mechanism of granite emplacement: rapakivi intrusions in active extensional shear zones. NATURE **343,** 452–4.

Johannes, W. & Holtz, F. 1990. Formation and composition of H_2O–undersaturated granitic melts. *In* Ashworth, J. R. & Brown, M. (eds) *High temperature metamorphism and crustal anatexis,* 87–104. London: Unwin Hyman.

Nekvasil, H. 1988. Calculated effect of anorthite component on the crystallisation paths of H_2O-undersaturated haplogranitic melts. AM MINERAL **73,** 966–81.

O'Connor, J. T. 1965. A classification for quartz-rich igneous rocks based on feldspar ratios. GEOL SURV AM Prof Pap **525-B,** 79–84.

Patchett, P. J. & Bridgwater, D. 1984. Origin of continental crust of 1·9–1·7 Ga age defined by Nd isotopes in the Ketilidian terrain of South Greenland. CONTRIB MINERAL PETROL **87,** 311–8.

Pearce, J. A., Harris, N. B. W. & Tindle, A. G. 1984. Trace element discrimination diagrams for the tectonic interpretation of granitic rocks. J PETROL **25,** 956–83.

Powell, R. & Holland, T. J. S. 1988. An internally consistent thermodynamic data set with uncertainties and correlations: 3 Applications to Geobarometry, worked examples and a computer programme. J METAM GEOL 6, 173–204.

Watterson, J. 1978. Proterozoic intraplate deformation in the light of South-east Asian neotectonics. NATURE **273,** 636–40.

Whalen, J. B., Currie, K. L. & Chappell, B. W. 1987. A-type granites: geochemical characteristics, discrimination and petrogenesis. CONTRIB MINERAL PETROL **95,** 407–19.

Windley, B. F. 1991. Early Proterozoic collision tectonics, and rapakivi granites in an extensional thrust-thickened crust: the Ketilidian orogen, South Greenland. TECTONOPHYSICS **195,** 1–10.

P. E. BROWN, Department of Geology, University of St Andrews, St Andrews, KY16 9ST, Scotland.
T. J. DEMPSTER, Department of Geology and Applied Geology, University of Glasgow, Glasgow G12 8QQ, Scotland.
T. N. HARRISON, c/o Department of Geology and Petroleum Geology, University of Aberdeen, Aberdeen, AB9 2UE, Scotland.
D. H. W. HUTTON, Department of Earth Sciences, University of Durham, South Road, Durham, DH1 3LA, England.

Transactions of the Royal Society of Edinburgh: Earth Sciences, **83,** 179–190, 1992

A nested diapir model for the reversely zoned Turtle Pluton, southeastern California

Charlotte M. Allen

ABSTRACT: Most zoned plutons described in the geological literature have mafic rims and felsic cores and are referred to as "normally zoned", whereas relatively few "reversely zoned" intrusions (felsic rims and mafic cores) have been described. That unusual zonation pattern has been variously attributed to *in situ* processes or to the reordering of an underlying, vertically stratified, magma chamber either by intrusion through an orifice or by emplacement of composite diapirs. The Turtle Pluton is an early Cretaceous, reversely zoned, intrusion that is divided into four facies: a Rim Sequence that is graditionally zoned from bt + ilm + musc monzogranite to hb + bt + mt + sph granodiorite; a Core Facies of more homogeneous hb + bt + mt + sph granodiorite to quartz monzodiorite; between these two facies, a structural discontinuity termed the Schlieren Zone; and an Eastern Facies of monzogranite to granodiorite. Field relationships, distribution of strain, and geochemical and isotopic studies (including a range of initial ^{87}Sr/^{86}Sr from 0·7085–0·7065) suggest that the reverse zonation of the Turtle Pluton is the result of sequential emplacement of two diapirs each derived from the same underlying, vertically stratified, magma chamber, and that the Rim Sequence zonation is chiefly the result of mixing of intermediate and felsic magmas from distinct sources accompanied by minor fractional crystallisation.

KEY WORDS: early Cretaceous, mafic enclaves, magma mixing, Mojave Desert, source rocks

Of the compositionally zoned plutons described in the literature, relatively few are composed of a more felsic rim and a more mafic core (Ermanovics 1970; Ayuso 1984; Fridrich & Mahood 1984; Nabelek *et al.* 1986; Bourne & Danis 1987; Wernicke 1987; Nironen 1989); these are termed reversely zoned in contrast to the "normal" case of mafic rims and felsic cores. Whereas *in situ* fractional crystallisation through sidewall crystallisation and solidification is commonly invoked as the principal mechanism that produces normally zoned plutons, this model cannot be applied to reversely zoned plutons particularly as the minerals with the highest crystallisation temperatures reside in the core.

Descriptions of reversely zoned plutons and emplacement models have been summarised by Ayuso (1984), Fridrich and Mahood (1984) and Bourne and Danis (1987). The generalised models are: (1) *in situ* crystallisation from wall rocks inwards and concentration of mafic minerals (cumulate and/or restite) toward the core by flow segregation; (2) *in situ* crystallisation from the core outwards due to volatile enrichment and depression of liquidus temperatures in magmas close to the walls; (3) rearrangement of a horizontally stratified magma chamber by intrusion of stratified magmas through an orifice to a higher level; and (4) sequential emplacement of one diapir into another, both generated from the same evolving magma chamber. The first two models require that compositional variation was generated at or near the level of exposure, and the latter two models imply the existence of a layered magma chamber at depth that was reorganised through a process of intrusion to a higher level.

The Turtle Pluton is a reversely zoned intrusion that is exposed over 20 km^2 in the southern Turtle Mountains of southeastern California, and it may be directly related to, or continuous in the subsurface with, other plutons in the area (Figs 1, 2). The field, strain, geochemical, and isotopic data presented here suggest that the pluton is the product of nested emplacement of two consanguineous diapirs and that one diapir was compositionally zoned prior to intrusion. The compositional zonation is thought to be the product of magma mixing of granitic and intermediate magmas (not basaltic enclaves) and concomitant fractional crystallisation.

1. Geological setting

The Turtle Pluton is one of several plutons of known or inferred Cretaceous age that occurs in the study area (Fig. 1) which lies W and S of a belt of core complexes in an extensional terrane (Howard & John 1988). The area is on the craton inboard of the eastern margin of Jurassic igneous activity at this latitude.

The southern Turtle Mountains are composed of Proterozoic igneous and metamorphic rocks intruded by largely undeformed Cretaceous plutons, all of which are overlain by shallowly dipping Tertiary volcanic and lesser sedimentary rocks. The subhorizontal dips of the 20–21 Ma lava flows indicate that the range has not been tilted greatly since that time and is approximately autochthonous as compared to nearby greatly rotated core complexes (Howard *et al.* 1982; Howard & John 1988). These observations and lack of known major late Mesozoic structures suggest the Turtle Pluton has not been significantly rotated since intrusion.

The pre-Mesozoic terrane of the southern Turtle Mountains is primarily composed of granite gneiss of the Virginia May Mine (Howard *et al.* 1982), a biotite quartzofeldspathic gneiss associated with leucocratic granitic dykes. Subsidiary to this rock type are: mafic augen gneiss, granite gneiss, amphibolite, minor ophitic dolerite, and minor, generally undeformed, hornblende diorite or gabbro (Howard *et al.* 1988). No significant contact metamorphic effects have been observed around the Turtle Pluton.

2. The Turtle Pluton

The Turtle Pluton concordantly intrudes steeply foliated Proterozoic gneiss and the Patton Granite of unknown age (Fig. 1) and it is itself intruded by the Fortification Granodiorite, the Target Granite, and garnet aplite dykes.

The Turtle Pluton can be divided into four facies (Fig. 2). The Rim Sequence (RS) composes the periphery of the intrusion on three sides and is gradationally zoned with respect to mineral modes (Figs 3, 4) and composition, grainsize, and whole-rock composition, from medium-grained biotite monzogranite at the rim, to porphyritic and coarse-grained biotite hornblende granodiorite, through coarse-grained biotite hornblende granodiorite, towards the core. With the exception of unfoliated monzogranite adjacent to the country rock, the RS has a foliation that parallels the country rock contact. The foliation pattern suggests that the pluton is a steeply SSE-plunging cylinder near the present exposure level (Fig. 2). Tabular minerals commonly define a subvertical lineation suggestive of magmatic flow.

The Schlieren Zone (SZ), at the contact between the RS and Core Facies (CF) and concentric with the outline of the pluton, is a zone tens of metres wide of highly strained, coarse-grained granodiorite characterised by a well-developed, steep foliation defined by alignment of mafic minerals, schlieren, mafic enclaves, and elongate, subvertical screens of wall rock gneiss (Fig. 5). A steep lineation defined by prismatic minerals is locally present.

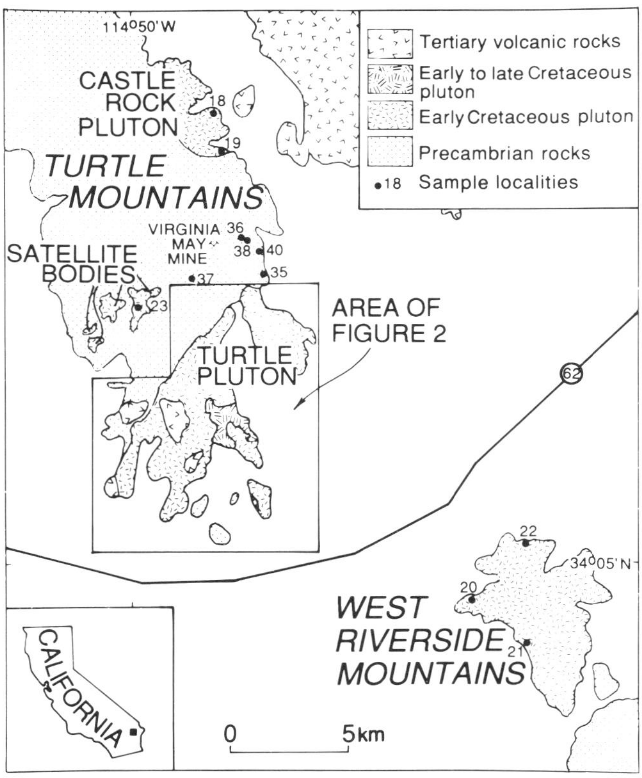

Figure 1 Simplified geological map of the southern Turtle Mountains area.

The gradational changes of rock type, mineralogy, and mineral chemistry across the RS and into the CF are represented by a N–S traverse (A–A′, Fig. 2). The modal changes along the traverse are given in Figures 3 and 4. The salient features are the change of rock type from monzogranite near the country rock contact to granodiorite toward the core of the pluton, the change of opaque mineralogy from ilmenite to magnetite (+ sphene), the occurrence of allanite in hornblende-free and hornblende-bearing granodiorite, and the change of the dominant ferromagnesian mineral from biotite to hornblende progressively into the pluton.

The CF varies in composition in a general way from coarse-grained granodiorite similar to the RS, to coarse-grained quartz monzodiorite towards the centre of the pluton adjacent the Target Granite (Fig. 3). On either side of the SZ, the CF and RS are generally indistinguishable except along their southern contact where the RS contains K-feldspar phenocrysts and is more felsic than the CF. The foliation in the CF is generally less well developed than that in the RS, although locally a strong foliation may be present. The CF is younger than the RS, as demonstrated by a dyke of CF that crosscuts porphyritic granodiorite of the RS.

The Eastern Facies of the Turtle Pluton is composed of porphyritic and nonporphyritic monzogranite and granodiorite. The porphyritic portion contains white oikocrysts of K-feldspar not observed elsewhere in the pluton. Although this facies has a country rock contact apparently continuous with that of the RS, and it has geochemical similarities with both the RS and CF, the rocks of the Eastern Facies lack the orderly zonation observed in the RS, and the intrusive relationship of the Eastern Facies with the other facies is unknown.

Fine-grained mafic enclaves of basaltic to andesitic composition are common in all facies of the Turtle Pluton and are more voluminous toward the core of the pluton (Waugh 1985). Partially disaggregated mafic dykes observed in the periphery of the pluton suggest that the dismemberment of dykes may be the source of some mafic enclaves; moreover, some enclaves are texturally indistinguishable from the dykes (Allen 1991). Coarse-grained, dioritic enclaves are also present locally in all facies of the pluton and the larger mappable bodies are shown in Figure 2. Other than in the SZ, xenoliths of country rock are found only within metres of exposed wall rock contacts or covered, inferred contacts.

2.1. Geochronological data

The Turtle Pluton is thought to be early Cretaceous in age (110 Ma) based on $^{235}U/^{207}Pb$ ages on sphene from three samples (manuscript *in prep.*), including multiple fractions from one. U–Pb zircon geochronology on a sample from the crosscutting Target Granite, and a well-fitted Rb–Sr isochron on garnet aplite dykes both yield about 100 Ma and indicate this is the minimum age of the Turtle pluton.

Inaccuracies in age determinations of up to 25% do not affect the petrogenetic models discussed here.

2.2. Emplacement conditions

The Turtle Pluton crystallised at pressures between 200 and 400 MPa as determined from hornblende geobarometry based on total Al contents of hornblende from granodioritic rocks containing the assemblage pl + qtz + ksp + bt + hb + sph + mt (Allen 1991). In hornblende-free monzogranite, the presence of sparse melt-precipitated muscovite crystals also suggests pressures of at least 200–400 MPa (Zen 1988). Such muscovite in the younger, cross-cutting garnet aplites indicates similar emplacement depths.

Crystallisation temperatures are less well constrained, but can be inferred by comparison with crystallisation experiments on monzogranite and granodiorite at 200 MPa (Naney 1983). The absence of pyroxene in the Turtle Pluton suggests crystallisation temperatures of less than 780–790°C for granodiorite and 730–750°C for monzogranite, depending on H_2O content; however, total reaction of pyroxene with melt and, therefore, higher liquidus temperatures cannot be entirely discounted.

The most unusual feature of crystallisation sequences deduced from petrographic observations is the presence of biotite before the crystallisation of hornblende. Biotite could be either a restite phase or an early crystallising mineral in Turtle Pluton. Lack of other evidence of major restite phases, clinopyroxene and calcic plagioclase cores

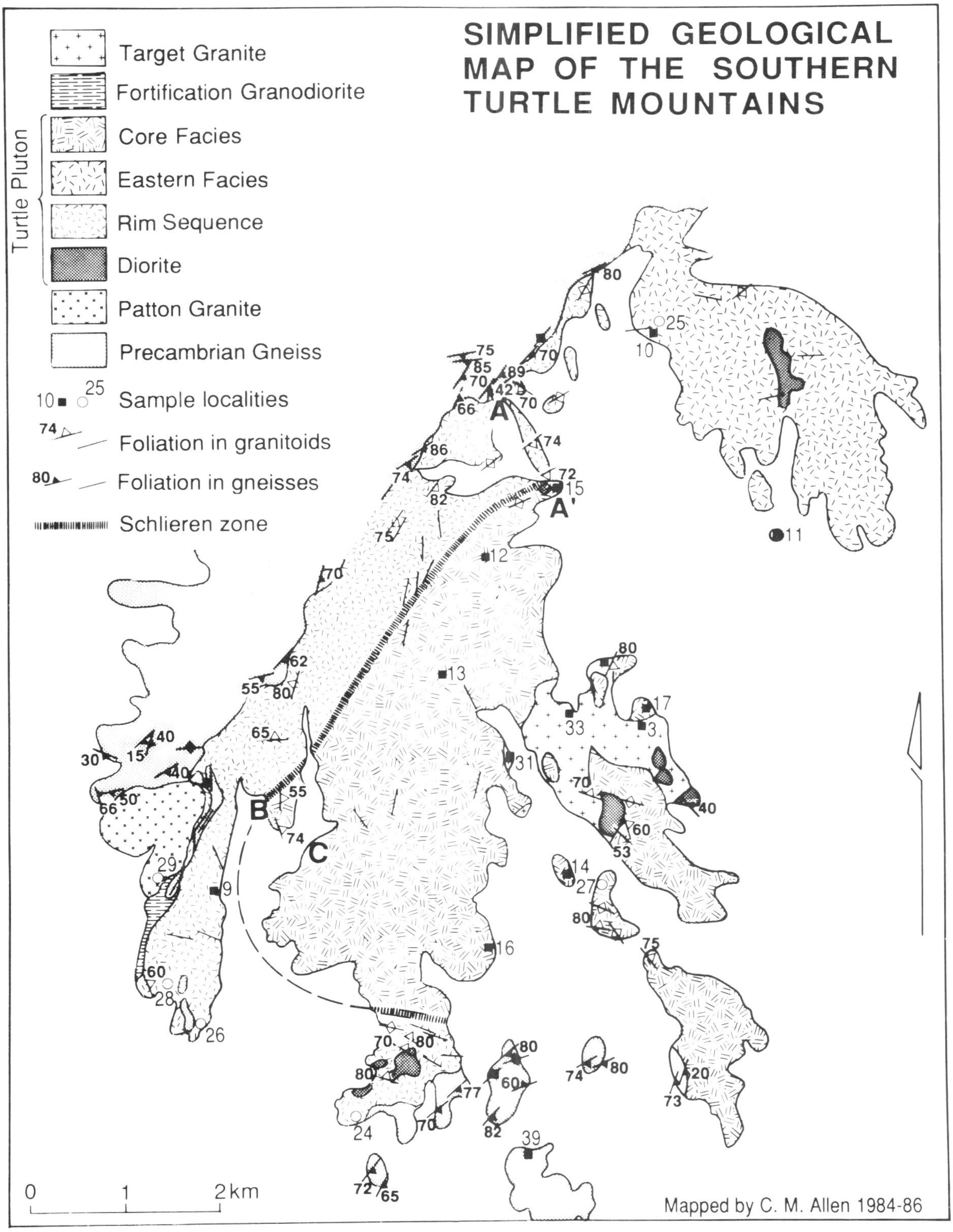

Figure 2 Geological map of the Turtle Pluton and environs.

that might be expected to be restite in metaluminous magma (Chappell *et al.* 1987), suggests that a restite origin is unlikely. The textural evidence of biotite preceding hornblende and its reaction to hornblende is observed in all of the hornblende-bearing granodiorites of the Turtle Pluton (62–68% SiO_2). The same sequence of biotite before hornblende has also been reported in the granitic rocks of the Huntington Lake Quadrangle, Sierra Nevada (Bateman & Wones 1972) and the Yerrington Batholith (Dilles 1987), which contrasts with the sequence, hornblende followed by biotite, predicted by Bowen's Reaction Series (Bowen 1922) observed in many other calcalkaline plutons (e.g. Bateman & Chappell 1979; Speer 1987). The relative stability of both mineral groups has been discussed by Wones and Gilbert (1982) and Hewitt and Wones (1984). Given that the mineral and whole-rock compositions in the Turtle Pluton are not unusual, the sequence could be explained by an initially low H_2O activity (<3·5% H_2O, Naney 1983) which increased during fractional crystallisation.

The change of opaque mineral assemblage (ilmenite to magnetite) and the concomitant appearance of sphene in rocks with distance from the wall-rock contact and into the pluton suggest the reaction

$$3FeTiO_3 + 3SiO_2 + 3CaO + 0{\cdot}5O_2 = Fe_3O_4 + 3CaTiSiO_5$$

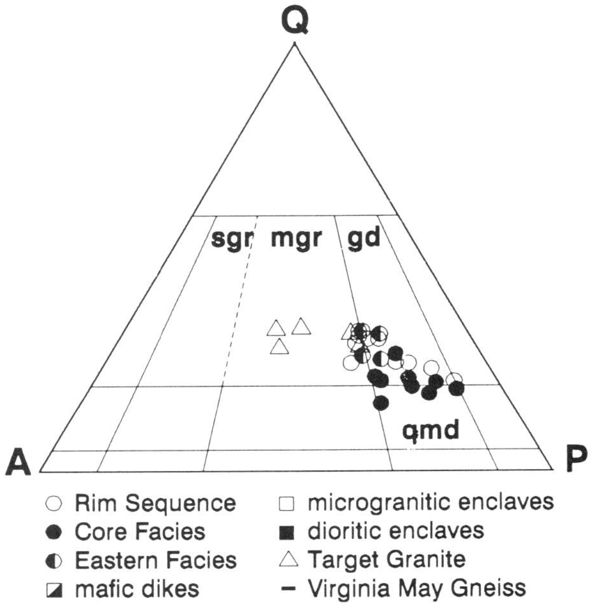

Figure 3 Modes of granitic rocks after Le Bas and Streckeisen (1991); sgr = syenogranite, mgr = monzogranite, gd = granodiorite, qmd = quartz monzodiorite; symbols are consistent for all figures.

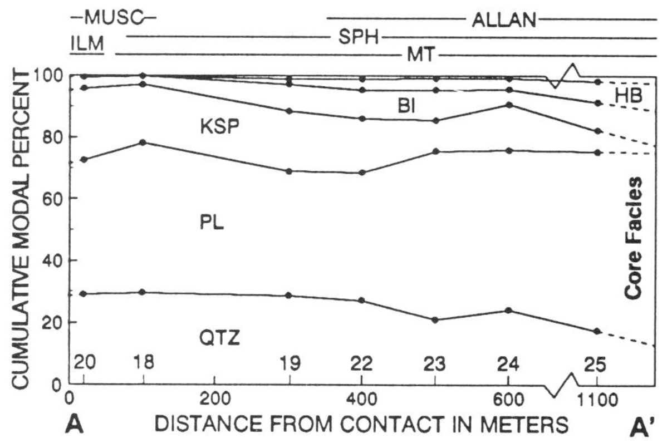

Figure 4 Modal changes across the Rim Sequence, traverse A–A′ of Figure 2.

which could be reflecting an increase of any of f_{O_2}, a_{Ca}, and a_{Si} (Zen 1988).

An understanding of the physical state of the magma during the emplacement process can be obtained from estimates of strain across the Turtle Pluton. These estimates of strain can be derived from measurement of axial lengths of enclaves (Ramsay 1989; Hutton 1982, 1988). Although this method includes some tentative assumptions such as all enclaves initially had the same spherical shape, the method appears to be useful in a broad range of cases (Ramsay 1989; Hutton 1988, and references therein). In the Turtle Pluton, axial lengths of mafic enclaves were measured within metre-square grids at 28 locations on subvertical and subhorizontal surfaces along a two-segment traverse of the RS, SZ and into the CF (Fig. 2). Each horizontal grid sampling is accompanied by a nearby vertical one. As far as was possible, these samplings were randomly selected from the available outcrops. The averages of the axial ratios for each grid (n = 7 to 31 enclaves) are presented in Figure 6, which represents a total of 445 enclave measurements along a composite 2·5 km traverse (Waugh 1985; Allen 1989). Although the magnitude of the standard deviation is subequal to the mean in many cases, in most X/Z is subequal to Y/Z. These measurements suggest that strain was flattening strain, that the highest cumulative strain occurred in the SZ, that strain was variable in the RS, and that strain in the CF was similar to the lowest strain measured in the RS. These results are in apparent conflict with the presence of subvertical mineral lineations in many

Figure 5 Tabular screen of Proterozoic gneiss in the Schlieren Zone; arrows indicate the extent of the screen in the background; pen for scale.

parts of the RS and SZ. One possible explanation of this apparent contradiction is that the aligned phenocrysts resulted from magmatic flow and the discoid enclaves from late-stage flattening.

Many zoned plutons can be successfully modelled by a balloon inflation model (Ramsay 1989). Strain in these plutons is essentially pure flattening strain, and the greatest strain occurs at the margins and decreases towards the interior of the pluton. In such cases, the measured strains are interpreted to result from the crystallisation front moving progressively into the pluton as material was added to the core. Thus, rocks at the margin of balloon-inflation type plutons record the greatest total strain history. Clearly, the Turtle Pluton does not fit this model, although the pluton apparently cooled from the exterior to the interior given the fine-grained margin of the pluton (chilled margin) and the dyke of CF that intrudes the RS. Instead, field observations and strain data indicate that the highest strain in the Turtle Pluton occurred at the effective wall of the intruding CF, i.e. in a zone prone to record high strain, which in this case would have been the mostly crystalline portion of the RS adjacent to the mostly non-crystalline CF. These results are very similar to those from a study of the normally zoned Criffell Pluton, where it was found that highest strains occurred along an internal contact of monzogranite and granodiorite, not at the margin of the pluton (Courrioux 1987). This two-stage emplacement model for the Turtle Pluton will be examined geochemically.

2.3. Geochemistry

RS samples (Table 1) define an orderly, linear array on Harker Diagrams (Fig. 7) over the silica range of 65·0–73·3%. Magnitudes of correlation coefficients (r) are greater than 0·957, except MnO and Na_2O. CF samples plot along the RS regression line for all oxides other than Al_2O_3, Na_2O, K_2O, and P_2O_5. With the exception of the most mafic of the CF samples (17), thought to be a cumulate based on trace element data, the CF samples have a relatively small compositional range for most oxides ($SiO_2 = 61{\cdot}7$–$64{\cdot}8\%$) except for alkalis and alumina. CF samples can be subdivided into K-rich and K-poor groups. The samples collected closer to the younger Target Granite are relatively enriched in alkalis for their silica contents as compared to the other CF samples and to the RS trend. The K-poor samples lie close to the RS regression line. In addition, two samples from the Eastern Facies ($SiO_2 = 67{\cdot}4$ and $68{\cdot}2\%$) generally fall along the RS trend.

The three types of mafic rock found within the Turtle Pluton (fine- and coarse-grained mafic enclaves, and mafic dykes) have a broad range of major oxide contents over a limited silica range of 50·3–54·8% (Allen 1991). Except for TiO_2, FeO, and CaO, there is essentially no overlap of the mafic rock field and the RS regression line on Harker Diagrams.

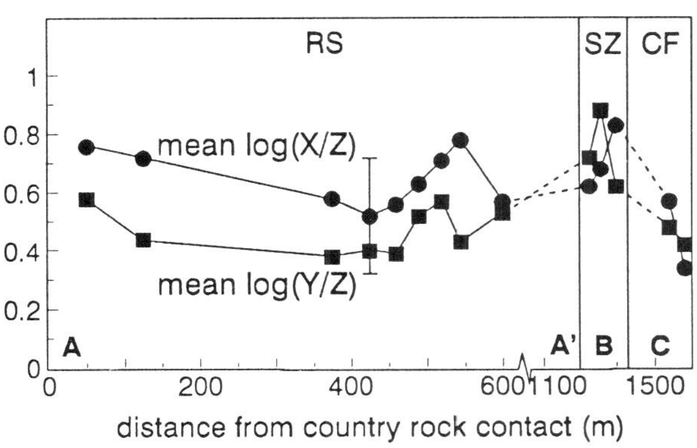

Figure 6 Mean axial lengths of mafic enclaves along most of traverse A–A′, and in areas B and C (see Fig. 2); X/Z represents measurements on vertical surfaces radial to the pluton contact; Y/Z represents measurements on horizontal surfaces; one standard deviation bar given is representative of that for both X/Z and Y/Z.

Country rocks sampled from the southern Turtle Mountains are Proterozoic gneisses. With the exception of one mafic augen gneiss, these rocks have silica contents of 72·2–76·3%, and the RS regression line intersects the field defined by these samples for all oxides except Al_2O_3, K_2O, and Na_2O (Howard *et al.* 1982; C. M. Allen, S. E. Shaw & K. A. Howard *unpub. data*).

High-valence trace elements are plotted against TiO_2 (Fig. 8). For RS data there is a strong positive correlation of TiO_2 with Sc and V, a moderate one with Cr, and none with Y. CF and Eastern Facies samples are roughly colinear with the RS trend, except for a positive correlation of Y and TiO_2 for CF samples. Most mafic rocks have high Sc, V, Y, and Cr contents for a given TiO_2 content relative to Turtle Pluton samples.

Three large ion lithophile trace elements (Rb, Sr, Ba) are plotted against K_2O (Fig. 9). RS samples define a linear trend of positive slope ($r = 0{\cdot}897$) for Rb versus K_2O and most other granitic and granodioritic samples also lie close to the regression line. Some mafic rocks occur on this trend, but the mafic enclaves lie along a different, steeper positive slope. The relationship among the rock types is similar for Sr versus K_2O (including a high correlation for RS data, $r = -0{\cdot}990$), except that the trends have the opposite slope. In contrast, Ba–K_2O correlation for RS data is poor. Granitic samples from that unit have Ba concentrations greater than 1100 ppm, whereas RS granodioritic rocks and CF samples contain Ba concentrations in the range of 600–900 ppm. Mafic rocks have Ba concentrations from about 100–600 ppm at low K_2O contents.

Correlation of Zr (130–155 ppm) with SiO_2 and other trace elements from the RS is poor. Given that this element occurs essentially in just one phase, zircon, the low correlation may be partly a function of sample size (>4 kg).

These major and trace element data suggest that neither mafic rocks nor country rocks are directly involved in the processes resulting in linear major and trace element trends displayed by the RS and some CF samples, that there is no significant geochemical discontinuity at the SZ, that the Eastern Facies is similar to the RS, and that perhaps the K-rich nature of CF samples proximal to the Target Granite is due to contact metamorphic effects. Moreover, the curvilinear trend displayed by Ba data indicates that the RS cannot be modelled by a two end-member mixing process alone.

2.4. Mineral compositions

Mineral compositions suggest that the principal break in zonation across the Turtle Pluton occurs at the monzogranite/granodiorite interface within the RS and not at the SZ. The granitic rocks not only contain ilmenite rather than magnetite and sphene, but also contain biotite of composition distinct from that in granodioritic rocks of the RS and CF (Fig. 10). Biotites in the granitic rocks contain higher concentratons of tetrahedral Al, Mn, and F, and have higher Fe/(Fe + Mg). All biotites in the granodioritic rocks appear to have similar compositions and may change Mn composition systematically across the RS and into the CF. Furthermore, plagioclases in the granitic rocks are normally zoned, whereas plagioclase is reversely zoned in all granodioritic rocks of the RS and CF (Fig. 11). There are at

Table 1 Major and trace element analyses[a].

Sample#	CA855	BW8420	BW8418	BW8418	BW8422	BW8423	BW8424	BW8425	CA85120	CA854	CA8450	CA8513
ID#	1	2	3	4	5	6	7	8	9	10	11	12
Type[b]	RSgr	RSgr	RSgd	RSgd	RSgd	RSgd	RSgd	RSgd	RSgd	EFgd	EFgd	CFgd
SiO_2	72·98	73·32	70·45	68·26	67·34	66·90	66·49	64·96	71·10	68·18	67·41	64·81
TiO_2	0·21	0·20	0·28	0·40	0·43	0·45	0·47	0·53	0·31	0·36	0·47	0·65
Al_2O_3	14·50	14·55	15·23	15·65	16·25	16·31	16·14	16·56	15·63	15·41	16·78	16·38
FeO_{TOT}	1·89	1·49	2·20	3·21	3·21	3·43	3·61	4·41	2·92	2·26	4·45	6·06
MnO	0·10	0·08	0·11	0·11	0·10	0·10	0·11	0·10	0·06	0·09	0·13	0·11
MgO	0·53	0·46	0·88	1·29	1·26	1·77	1·56	1·95	0·77	1·25	1·64	2·40
CaO	1·82	1·87	2·61	3·52	3·93	4·12	4·18	5·22	2·78	3·35	4·81	5·44
Na_2O	3·56	3·92	3·52	3·33	3·60	3·16	3·39	2·85	3·05	3·73	3·00	2·56
K_2O	3·53	3·47	3·49	3·07	2·79	2·82	2·80	2·19	3·71	3·72	2·88	2·87
P_2O_5	0·07	0·09	0·12	0·16	0·18	0·19	0·20	0·23	0·14	0·15	0·23	0·21
LOI	0·28	0·30	0·44	0·37	0·76	0·84	0·67	0·51	0·28	0·63	0·59	0·63
Total	99·47	99·75	99·33	99·37	99·85	100·09	99·62	99·51	100·75	99·49	102·39	102·12
ASI[c]	1·12	1·07	1·06	1·03	1·01	1·03	0·99	1·00	1·11	0·95	1·00	0·95
Rb	93	102	103	95	88	90	83	55	100	112	75	92
Sr	287	269	311	382	430	440	429	558	336	357	609	427
Ba	1570	1230	820	830	630	780	730	860		560	1130	430
V	14	15	25		62	65	69	85		51		104
Sc	3	3	5		8	8	9	8		6		12
Cr	2	4	5		14	16	16	7		13		12
Zr	130	135	152		139	142	124	155		113		141
Y	23	24	25		25	25	26	22		21		25

Sample#	CA84143	CA84147	CA85122	CA8614	CA8566	H79123A	H8493	CA8596	H8492	H79118	H79124
ID#	13	14	15	16	34	35	36	37	38	39	40
Type[b]	CFqmd	CFqmd	CFqmd	CFqmd	Xvm	Xvm	Xvm	Xvm	Xvm	Xgg	Xag
SiO_2	63·03	62·59	61·68	58·68	75·72	75·10	74·40	73·56	72·20	73·70	68·90
TiO_2	0·59	0·64	0·63	0·84	0·25	0·32	0·02	0·20	0·57	0·24	0·75
Al_2O_3	16·27	16·71	17·18	18·11	11·44	11·30	13·70	13·18	11·30	13·50	13·40
FeO_{TOT}	5·54	5·45	5·47	7·83	3·01	4·29	0·63	1·99	5·89	2·20	5·15
MnO	0·13	0·11	0·16	0·14	0·10	0·03	0·02	0·10	0·13	0·03	0·08
MgO	2·39	2·35	2·32	3·16	0·98	0·75	0·12	1·07	1·08	2·32	1·09
CaO	5·27	5·59	5·98	7·45	0·75	0·69	1·18	1·38	1·26	1·30	2·33
Na_2O	2·33	2·48	2·73	2·62	2·36	2·48	2·34	2·50	1·55	2·65	2·67
K_2O	3·35	3·03	1·91	1·73	4·40	4·07	6·49	4·40	4·71	5·41	4·59
P_2O_5	0·25	0·25	0·23	0·30	0·02	0·05	0·05	0·03	0·13	0·06	0·29
LOI	0·62	0·81	0·46	0·62	0·52	0·40	0·53	0·34	0·61	0·51	0·47
Total	99·77	100·01	98·75	101·48	99·55	99·48	99·48	98·75	99·43	101·92	99·72
ASI[c]	0·95	0·95	0·98	0·92	1·14	1·16	1·05	1·16	1·14	1·07	0·98
Rb	102	90	58	58	138		154	133	128	317	135
Sr	426	448	597	515	58		127	310	100	107	176
Ba	760	770	760	450							
V	95	103	98	142							
Sc	12	12	12	20							
Cr	16	16	18	16							
Zr	166	174	149	219							
Y	29	33	25	45							

Notes: (a) Major elements in wt% and trace elements in ppm by weight. All major elements, Rb and Sr by XRF at Virginia Tech; Ba and other trace elements by ICP–OES at Texas Tech unless otherwise noted. All samples beginning with H were provided by K. A. Howard and S. E. Shaw; (b) gr = monzogranite, gd = granodiorite, qmd = quartz monzodiorite, vm = Virginia May Gneiss, gg = granite gneiss, ag = augen gneiss, RS = Rim Sequence, CF = Core Facies, EF = Eastern Facies, X = Proterozoic; (c) ASI is the aluminium saturation index = mol $Al_2O_3/(Na_2O + K_2O + CaO)$

least two hypotheses to explain the observed plagioclase zonation patterns, and, as an aside, the change of crystal size observed in the RS. Loomis (1982) and Loomis and Welber (1982) interpreted the pattern of normally zoned, smaller plagioclase at the perimeter of a pluton and reversely zoned, larger ones in a pluton's core to result from undercooling, fractional crystallisation, and volatile loss (Loomis 1982; Loomis & Welber 1982). Hibbard (1981) attributed reversely zoned plagioclase to the addition of hotter, more mafic magma to a cooler more felsic one. If the second hypothesis is correct as suggested by Sr isotope data, then the granitic rocks apparently did not undergo significant magma mixing.

2.5. Strontium isotope data

Whole-rock Sr isotope data are presented in Table 2, where initial $^{87}Sr/^{86}Sr$ ratios are referred to as SRI. All data are corrected to 110 Ma except data from the Target Granite (99 Ma). A linear regression of the same eight RS samples regressed for major elements results in a false age of 310 Ma, or a slope equivalent to 200 Ma on this time-corrected plot (Fig. 12A). The intercept for this regression is 0·7057. Such false ages result from the mixing of two isotopically distinct reservoirs (end-member compositions) that lie along the regression, or mixing, line. Scatter along the regression can be interpreted as shifts of Rb and Sr concentrations due fractional crystallisation. Note that some samples from the CF (one collected along the trend of the RS traverse), fall along the postulated mixing line, and that isotopic data, like the major and trace element data, indicate that the SZ is primarily a structural but not a geochemical discontinuity. In addition, data from the Eastern Facies fall near the mixing line defined by the regression of RS samples. The Eastern Facies, plus the two most mafic samples from the RS and some CF samples (low K) have a similar SRI of 0·7067 ± 0·0002. This SRI is similar to that of other analysed early Cretaceous plutons in the study area (Allen 1989). Three CF samples proximal to the Target Granite have isotopic compositions intermediate between more distant CF samples and the cross-cutting, relatively non-radiogenic Target Granite, suggesting the CF may have been altered by intrusion of the younger monzogranite.

The relationship of mafic enclaves and dykes to the Turtle Pluton has been discussed in detail (Allen 1991). Mafic enclaves have Sr isotopic compositions that generally overlap the RS data. Mafic dykes and dioritic enclaves are distinctly less radiogenic (Fig. 12A) and most do not lie on the mixing line. Based on these data and major and trace element compositions, the similarity of mafic enclave and host granite isotopic compositions is thought to result from local interaction of enclaves with host granite during their magmatic and post-magmatic histories, and is not thought to be indicative of a consanguineous relationship (Allen 1991). Dykes have suffered some interaction, but they preserve more primitive isotopic ratios, and dioritic enclaves are interpreted as accidental xenoliths of compositions unlike presently exposed country rock.

3. Petrogenetic model

Although linear data arrays on major and trace element variation diagrams can be interpreted as the result of several

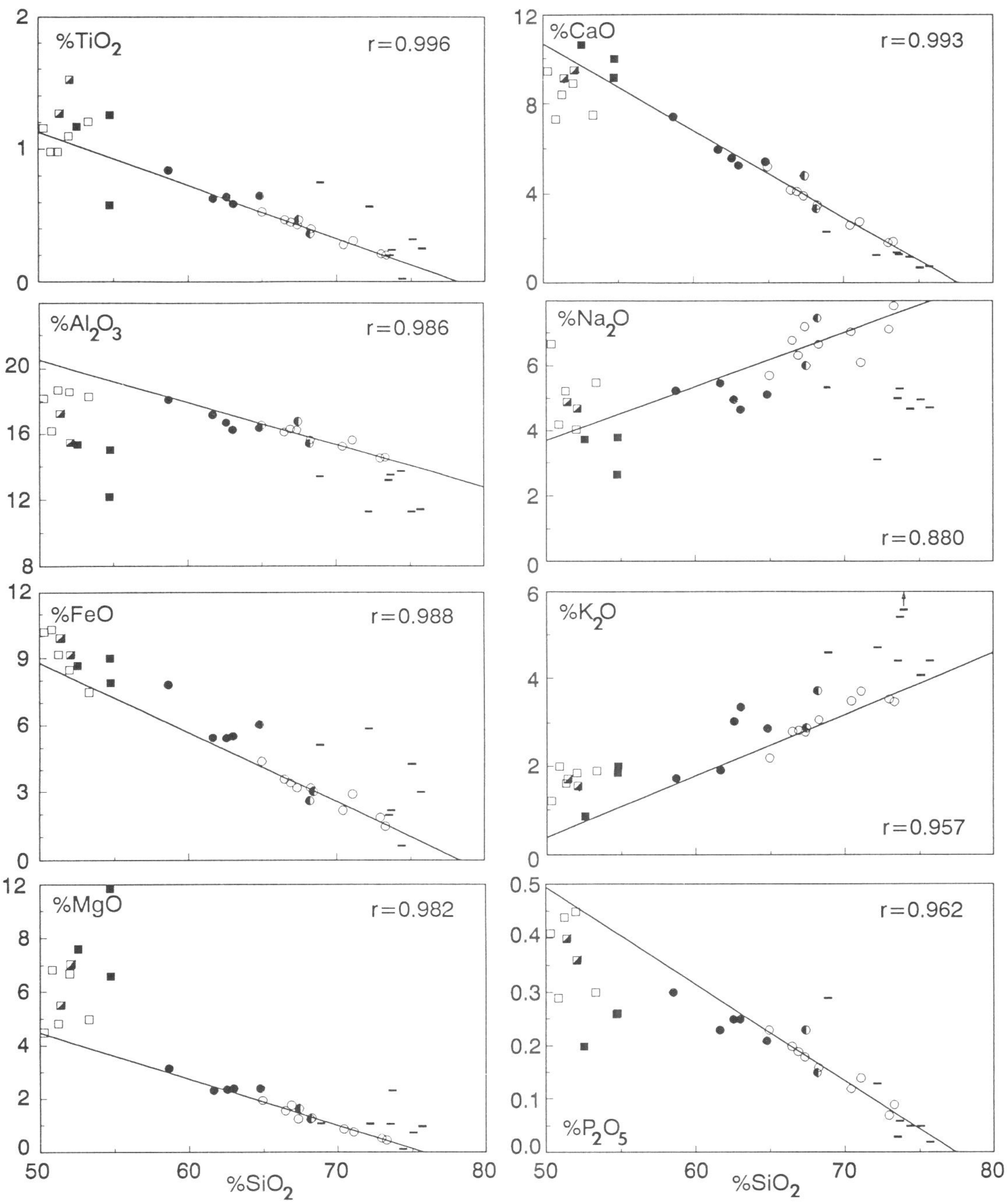

Figure 7 Harker Diagrams: regression lines and correlation coefficients for RS data are given; note that the data for the mafic rocks and the Proterozoic country rocks tend not to fall along the RS regression lines; symbols as in Figure 3.

different processes, false ages or pseudochrons (third decimal place variation in SRI in this case) can only result from the mixing of two isotopically distinct reservoirs. Fractional crystallisation alone cannot account for the isotopic variability. Restite unmixing or progressive partial melting of a single source cannot account for the isotopic data from the Turtle Pluton, as there is no evidence of retention of Sr-controlling phases, notably plagioclase, from the source. Bulk assimilation of country rock cannot account for the compositional variation across the RS, given that the country rock compositions are not colinear with the RS compositions on most major and trace element plots. Only magma mixing between two reservoirs (including assimilation of partial melts from country rock by a more mafic magma) can account for the systematic changes of Sr isotopic composition observed across the RS and into the CF.

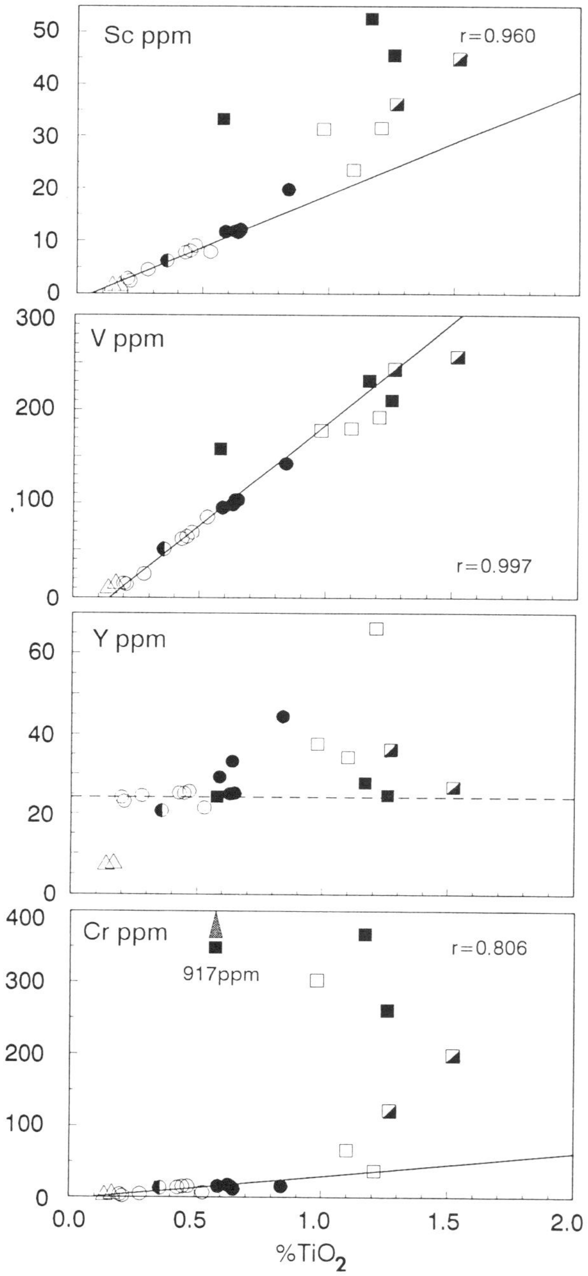

Figure 8 Concentrations of high field strength elements plotted against TiO_2; regression lines and correlation coefficients for RS data are given, except for Y where there is no correlation with TiO_2.

A test of magma mixing is the preservation of sample order from an isochron diagram to a reciprocal or companion plot (Langmuir *et al.* 1977; Fig. 12A, B). Among the RS samples, only sample 1 shifts position. Samples 1, 2 and 3, the hornblende-free granites, all have the same SRI within error (0·7085 ± 0·0001) and could be interpreted as cumulates and fractionates of the same magma, the most evolved magma represented in the pluton. Note, however, that the granitic samples 1 and 2 have some mineral and whole-rock compositions that make them distinct from the granodioritic samples including sample 3 (Section 2.4) and therefore the granitic samples are distinct from the remainder of the pluton. The shifts in sample order among those with SRI of about 0·7067 are probably the result of fractional crystallisation.

The two magmatic end-members which mixed to form the pseudochron defined by the RS data must bracket the data and lie along the regression or mixing line. The mafic end-member is constrained to have SRI between 0·7057 (the intercept of the mixing line) and 0·7065 (the composition of the most primitive RS samples and several other granodioritic samples). These two potential mafic end-member compositions correspond to Sr concentrations of about 1000 and 600 ppm, respectively (Fig. 12B). The end-member composition characterised by SRI of 0·7065 and 600 ppm Sr is the preferred model end-member based on: (a) the preponderance of rocks with SRI of 0·7067 ± 0·0002 that occur in the area, (b) the general

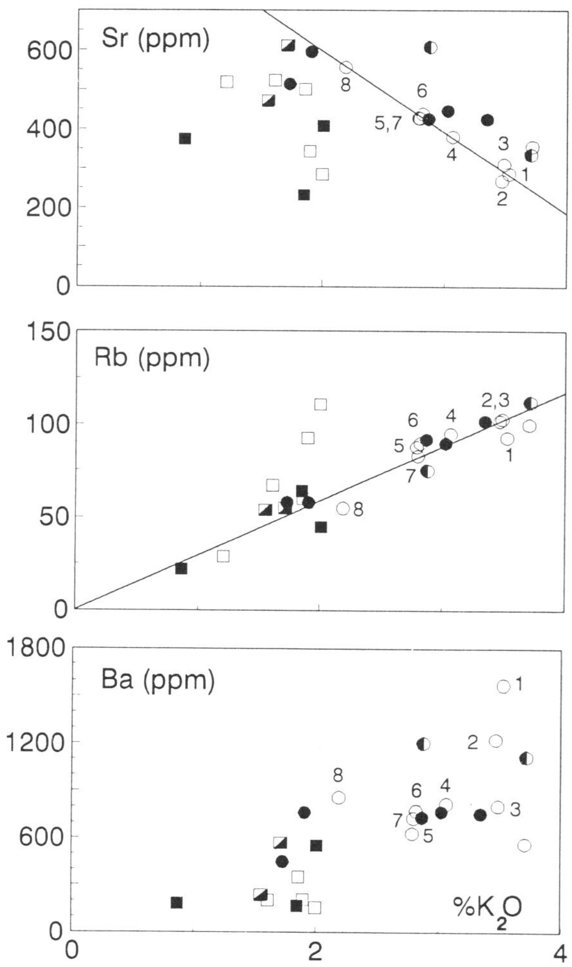

Figure 9 Concentrations of Sr, Rb, and Ba plotted against K_2O; regression lines for RS data are given.

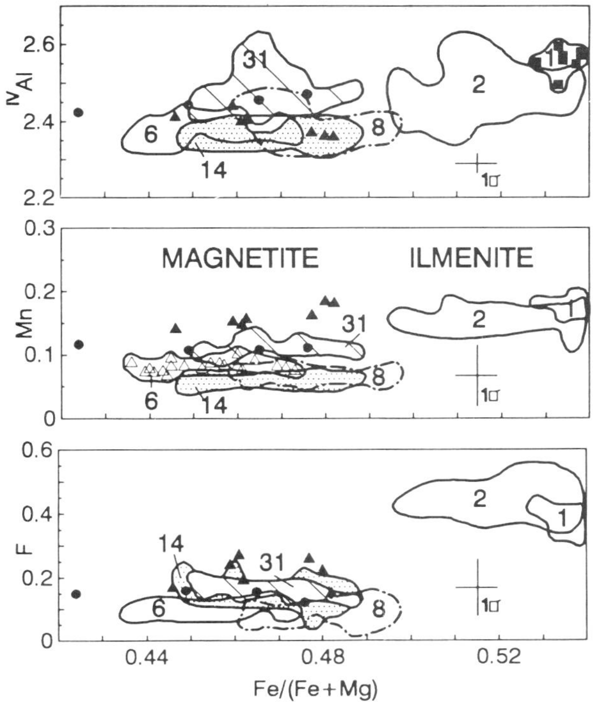

Figure 10 Biotite compositions of granites; oxide assemblages are labelled MAGNETITE or ILMENITE; solid triangles = sample 3, dots = sample 4, solid squares in field of sample 1 show data density represented by all envelopes.

argument that the difference of physico-chemical properties between smaller volumes of basaltic magma and larger volumes of granitic magma tends to inhibit mixing (Sparks & Marshall 1986; Frost & Mahood 1987), and (c) the interpretation of the role of basaltic to andesitic enclaves and dykes in the generation of the Turtle Pluton as given by Allen (1991). Sample 5 ($SiO_2 = 61{\cdot}7\%$) is thought to represent the primitive end-member as it is the most mafic rock of CF that is not a cumulate. One more mafic sample (16) contains 219 ppm Zr, which suggests significant mineral accumulation relative to other CF samples. Selecting a silica content of 61·5%, and assuming that the linear regressions of major and trace elements and Sr isotopic ratios adequately represent the RS and contiguous CF, the primitive end-member has Sr = 597 ppm, Rb = 56 ppm, and SRI = 0·7065 (Table 3).

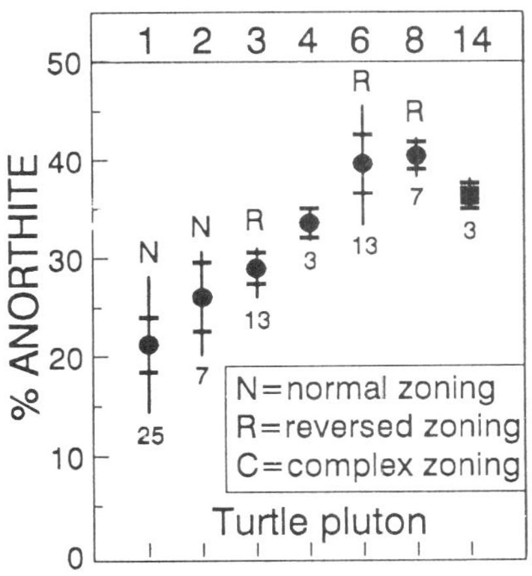

Figure 11 Plagioclase compositions from the Turtle area; symbols represent means, lines give ranges, and cross bars give 1 standard deviation; dot = RS, square = CF; numbers below bars are numbers of analyses; note the increase of An with progress into the pluton, and the change from normally to reversely zoned plagioclase near the periphery of the pluton.

Table 2 Rb–Sr Isotope Data.

Sample#	ID#	$^{87}Rb/^{86}Sr^{a}$	$^{87}Sr/^{86}Sr^{b}$	SRI[c]
Rim Sequence (110 Ma)				
CA85-5	1	0·92	0·71007	0·7086
BW84-20	2	1·08	0·71021	0·7085
BW84-18	3	0·97	0·70991	0·7084
BW84-19	4	0·70	0·70908	0·7080
BW84-22	5	0·59	0·70829	0·7074
BW84-23	6	0·60	0·70823	0·7073
BW84-24	7	0·56	0·70782	0·7069
BW84-25	8	0·28	0·70698	0·7065
CA85-120	9	0·86	0·70954	0·7082
Eastern Facies (110 Ma)				
CA85-4	10	0·90	0·70816	0·7068
CA84-50	11	0·36	0·70699	0·7064
Core Facies (110 Ma)				
CA85-13	12	0·58	0·70778	0·7069
CA84-143	13	0·68	0·70680	0·7057
CA84-147	14	0·57	0·70663	0·7057
CA84-147		0·57	0·70653	0·7056
CA85-122	15	0·28	0·70701	0·7066
CA86-14	16	0·30	0·70715	0·7067
CA85-25	17	0·60	0·70685	0·7059
Target Granite (100 Ma)				
CA85-18	40	1·87	0·70766	0·7050
CA85-111	41	1·39	0·70701	0·7050
CA85-101	42	0·62	0·70584	0·7050
CA85-15	43	0·60	0·70540	0·7045

Notes: (a) Error in ratio less than 0·94%; (b) Standard error less than 0·0001; (c) SRI = $^{87}Sr/^{86}Sr$ calculated at given age

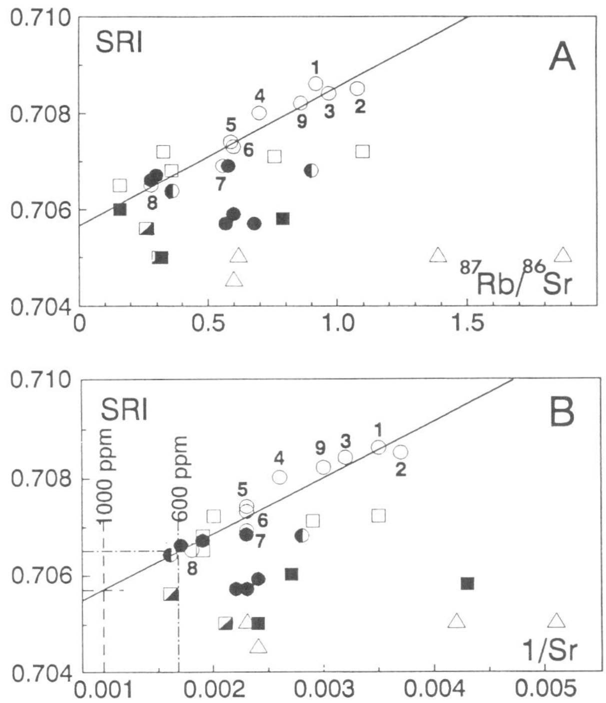

Figure 12 Initial $^{87}Sr/^{86}Sr$ plotted against A $^{87}Rb/^{86}Sr$, and B 1/Sr (ppm); symbols as in Figure 9; regressions of Rim Sequence data are given; note the positive slope on A equivalent to 200 Ma and an intercept of 0·7057; order of data for Rim Sequence is maintained from plot A to B, except for samples 1, 2, and 3, all of which lack hornblende; the maintenance of sample order suggests magma mixing is a viable mechanism to yield the pseudochron, but other mechanisms are necessary to explain samples 1, 2 and 3; note the number of granodiorites with SRI of about 0·7067 and the low SRI of the Target Granite; end-member Sr concentrations are taken as either 1000 ppm or 600 ppm for the mafic end-member.

Table 3 Calculated end-member compositions.

	Mafic	Felsic
Major elements (wt%)		
SiO_2	61·50	76·00
TiO_2	0·65	0·09
Al_2O_3	17·40	13·91
FeO	5·37	0·59
MnO	0·11	0·10
MgO	2·42	0·01
CaO	6·15	0·71
Na_2O	2·76	4·04
K_2O	1·94	4·06
P_2O_5	0·26	0·04
TOTAL	98·56	99·55
Trace elements (ppm)		
Sr	597	184
Rb	58	116
Ba	*700	*900
Sc	12	1
V	106	0
Cr	19	1
Y	25	24
Zr	148	132
SRI	0·7065	0·7108
$^{87}Rb/^{86}Sr$	0·28	1·79

Note: * Values are selected, not calculated

The felsic end-member must be at least as radiogenic as the RS monzogranites (SRI = 0·7085) but it is otherwise unconstrained by isotopic data. The mineral data given above (Section 4) suggest that the Turtle Pluton monzogranite samples (about 73% SiO_2, 285 ppm Sr, and 100 ppm Rb) have suffered little magma mixing and, therefore, that they may approach the felsic end-member composition. However, bulk composition variability at near constant SRI suggests significant fractionation has affected these samples. Assuming linear regressions adequately represent the RS, a theoretical upper limit of the felsic end-member composition can be estimated by calculating the silica composition at which the first major element is depleted. In the case of the RS, MgO is the first oxide depleted at 76·1% SiO_2. (One trace element, V, is depleted at a lower SiO_2 composition, 74·8%.) An end-member of SiO_2 = 76·0% corresponds to 185 ppm Sr and 120 ppm Rb (Table 3) and approaches minimum-temperature melt compositions (White & Chappell 1977). Referring to the Sr-companion plot (Fig. 12B), the calculated 76% SiO_2 (minimum-temperature melt) end-member indicates a maximum possible SRI of 0·7108, a ratio less than any sampled Proterozoic rock in the southern Turtle Mountains (SRI at 110 Ma of 0·7313–0·9120, n = 8; Allen 1989). If this end-member were derived solely from Proterozoic crust of 1·8–1·4 Ga age with an initial isotopic composition like the bulk Earth model of DePaolo and Wasserburg (1976) as indicated by a Rb–Sr isochron from six Proterozoic gneiss samples (Allen 1989), it must have been a relatively high Rb/Sr magma derived from rocks with a long-term, low $^{87}Rb/^{86}Sr$ of 0·36–0·43 (Rb/Sr = 0·13–0·16). Alternatively, the minimum-temperature melt could have been derived from younger rocks that are not exposed at the surface.

Inverting the argument, if these are reasonable mafic and felsic end-member compositions (Table 3) and only magma mixing between two end-members occurred to form the RS, then a concave-up curve should obtain. The curvature of the mixing hyperbola is determined by the ratio of the Y-axis denominators (^{86}Sr concentrations) of the end-members which is essentially 0·31 (185/600) for all plots. This ratio and the end-member compositions determine the position of the mixing hyperbola (see Langmuir *et al.* 1977) and should accurately describe major and trace element data when plotted against SRI (Fig. 13). It is an adequate, first order model for most elements, more so for Sr and TiO_2 than other elements, but mixing cannot account for the curvilinear trends defined by Ba on this and other variation diagrams (Fig. 9). A second obvious mechanism that could account for the scatter of data away from mixing curve is fractional crystallisation, suggesting that a model of combined assimilation and fractional crystallisation (AFC) similar to that of DePaolo (1981) might be appropriate except that "assimilation" should be regarded as addition of a second magma. The scatter of data to both sides of the simple mixing curve indicates that some rocks are cumulates and others are fractionates, and that no single AFC curve can account for all of the data.

The more striking aspect of these data from the Turtle Pluton, however, is not the degree of scatter but the linear coherence with respect to many major and trace elements, and SRI. The gradational changes of composition suggest an orderly process in a magma chamber. The Turtle Pluton probably began as a volume of granitic magma that was a partial melt of a low Rb/Sr source rock (perhaps intermediate to mafic in composition). The granitic magma was mixed by convection with a batch of intermediate magma, and fractionated during side-wall crystallisation, mixed with more intermediate magma, and so on, until a layered, density-stratified chamber was formed at the roof above a more homogeneous intermediate magma (after the model of McBirney *et al.* 1985). Such a layered igneous stratigraphy could give rise to the gradational series of mixed magmas sampled across the RS. The CF, in this model, would be the reservoir of intermediate magma derived from a partial melt of fairly mafic, old crust, or a basalt contaminated by the assimilation of crustal rocks with or without fractional crystallisation. The mechanism by which this underlying reservoir came to reside in the core of

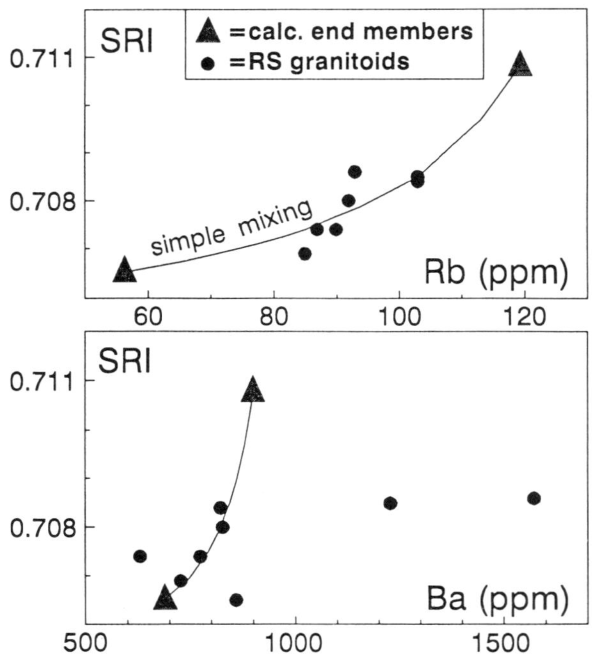

Figure 13 Plots of Rb and Ba against initial $^{87}Sr/^{86}Sr$ showing that magma mixing alone between the two calculated end-members (triangles) cannot account for the Rim Sequence data, and that a family of AFC curves is necessary to explain the data given that data lie both above and below the model mixing line.

the layered magma chamber is in part answered by the occurrence of the SZ complete with screens of country rock. A possible mechanism to explain the presence of the SZ, the chemical coherence across the SZ, and the presence of screens very similar to exposed Virginia May Gneiss in the SZ, is that the layered portion of the chamber was buoyant enough to rise as a diapir, only later to be injected by a second diapir of consanguineous intermediate magma capped by a discontinuous carapace of Virginia May gneiss at about 110 Ma. Less is known about the Eastern Facies, but Sr isotopic data and major and trace element analyses suggest these rocks are fractionates of magma similar to the CF.

4. Comparison with other reversely-zoned plutons

This study of the Turtle Pluton supports the models presented by Bourne and Danis (1987) and Fridrich and Mahood (1984), who suggested that reversely zoned plutons result from intrusion of a layered magma chamber to a higher level and inversion of the igneous layering during intrusion. Estimated emplacement pressures of 200–400 MPa (and perhaps greater) indicate, however, that the Turtle Pluton is not subvolcanic. In the Turtle Pluton, the zonation is primarily the result of magma mixing, and fractional crystallisation produced second-order effects. The Turtle Pluton appears to comprise two nested and related diapirs derived from a single magma chamber; the first diapir to be emplaced was the gradationally zoned RS with or without the fractionated magmas now represented by the Eastern Facies, and the second diapir was the more homogeneous CF. The Turtle Pluton crystallised at about 110 Ma and was sufficiently solid to be dyked by aplites and pegmatites, including garnet-bearing aplites at about 100 Ma, and to be stoped by the compositionally unrelated Target Granite at about the same time. This study of the Turtle Pluton suggests that multiple diapirs arising from the same evolving magma chamber can produce a reversely zoned intrusion.

5. Acknowledgements

This work is part of a Ph.D. dissertation completed at Virginia Polytechnic Institute and State University which was supervised by A. K. Sinha, and was partially underwritten by the Graduate Intern Program of the U.S. Geological Survey. Unpublished analyses of some Proterozoic samples were kindly provided by S. E. Shaw and K. A. Howard. R. A. Tosdal very generously provided U–Pb analyses of sphene. C. G. Barnes and coworkers supplied ICP analyses of trace elements. Constructive reviews by G. A. Mahood, W. E. Stephens and B. W. Chappell greatly helped in focusing the manuscript.

6. References

Allen, C. M. 1989. *Petrogenesis of the reversely zoned Turtle pluton, southeastern California.* Unpublished Ph.D. Thesis, Virginia Polytechnic Institute and State University.

Allen, C. M. 1991. Local equilibrium of mafic enclaves and granitoids of the Turtle pluton, S.E. California: mineral chemical and isotopic evidence. AM MINERAL **76,** 574–88.

Ayuso, R. A. 1984. Field relations, crystallization, and petrography of reversely zoned granitic pluton in the Bottle Lake Complex, Maine. PROF PAP US GEOL SURV **1320**.

Bateman, P. C. & Chappell, B. W. 1979. Crystallization, fractionation, and solidification of the Tuolume Intrusive Series, Yosemite National Park, California. BULL GEOL SOC AM **90** 465–82.

Bateman, P. C. & Wones, D. R. 1972. Geologic map of the Huntington Lake quadrangle, central Sierra Nevada, California. US GEOL SURV QUADRANGLE MAP **GQ 987**.

Bourne, J. & Danis, D. 1987. A proposed model for the formation of reversely zoned plutons based on a study of the Lacorne Complex, Superior Province, Quebec, CAN J EARTH SCI **24,** 2506–20.

Bowen, N. L. 1922. The reaction principal in petrogenesis. J. GEOL **30,** 177–98.

Chappell, B. W., White, A. J. R. & Wyborn, D. 1987. The importance of residual source material (restite) in granite petrogenesis. J PETROL **28,** 1111–38.

Courrioux, G. 1987. Oblique diapirism: the Criffell granodiorite/granite zoned pluton (southwest Scotland). J STRUCTURAL GEOL **9,** 313–30.

DePaolo, D. J. 1981. Trace element and isotopic effects of combined wallrock assimilation and fractional crystallization. EARTH PLANET SCI LETT **53,** 189–202.

DePaolo, D. J. & Wasserburg, G. L. 1976. Inferences about magma sources and mantle structure from variations of $^{143}Nd/^{144}Nd$. GEOPHYS RES LETT **7,** 743–6.

Dilles, J. H. 1987. Petrology of the Yerrington Batholith, Nevada: evidence for evolution of porphyry copper ore deposits. ECON GEOL **82,** 1750–89.

Ermanovics, I. F. 1970. Zonal structure of the Perth Road monzonite, Grenville Province, Ontario. CAN J EARTH SCI **7,** 414–34.

Fridrich, C. J. & Mahood, G. A. 1984. Reverse zoning in the resurgent intrusions of the Grizzly Peak cauldron, Sawatch Range, Colorado. BULL GEOL SOC AM **95,** 779–87.

Frost, T. P. & Mahood, G. A. 1987. Field, chemical, and physical constraints on mafic-felsic magma interaction in the Lamark Granodiorite, Sierra Nevada, California. BULL GEOL SOC AM **99,** 272–91.

Hewitt, D. A. & Wones, D. R. 1984. Experimental phase relations of the micas. *In* Bailey, S. W. (ed.) *Micas* MINERAL SOC AM, REV MINERAL **13,** 201–56.

Hibbard, M. J. 1981. The magma mixing origin of mantled feldspars. CONTRIB MINERAL PETROL **76,** 158–70.

Howard, K. A., Stone, P., Pernokas, M. A. & Marvin, R. F. 1982. Geologic and geochronologic reconnaissance of the Turtle Mountains area, California: West border of the Whipple detachment terrane. *In* Frost, E. G. & Martin, D. L. (eds) *Mesozoic–Cenozoic Tectonic Evolution of the Colorado River Region, California, Arizona, and Nevada,* 314–54. San Diego: Cordilleran Publishers.

Howard, K. A., Nielson, J. E., Simpson, R. W., Hazlett, R. W., Alminas, H. V., Nakata, J. K. & McDonnell, J. R. Jr., 1988. Mineral Resources of the Turtle Mountains Wilderness Study Area, San Bernardino County, California. BULL US GEOL SURV **1717-B**.

Howard, K. A. & John, B. E. 1988. Crustal extension along a rooted system of imbricate low-angle faults: Colorado River extensional corridor, California and Arizona. *In* Dewey, J. F. & Hancock, P. L. (eds) *Continental Extensional Tectonics.* GEOL SOC LONDON SPEC PAP **28,** 299–311.

Hutton, D. H. W. 1982. A method for the determination of initial shape of deformed xenoliths. TECTONOPHYSICS **85,** 45–50.

Hutton, D. H. W. 1988. Granite emplacement mechanisms and tectonic controls: inferences from deformation studies TRANS R SOC EDINBURGH EARTH SCI **79,** 245–55.

Langmuir, C. H., Vocke, R. D. & Hanson, G. N. 1977. A general mixing equation with applications to Icelandic basalts. EARTH PLANET SCI LETT **37,** 380–92.

Le Bas, M. J. & Streckeisen, A. L. 1991. THE IUGS systematics of igneous rocks. J. GEOL SOC LONGON **148,** 825–33.

Loomis, T. P. 1982. Numerical simulations of crystallization processes of plagioclase in complex melts: the origin of irregular and oscillatory zoning in plagioclase. CONTRIB MINERAL PETROL **81,** 219–29.

Loomis, T. P. & Welber, P. W. 1982. Crystallization processes in the Rocky Hill Granodiorite pluton, California: an interpretation based on compositional zoning of plagioclase. CONTRIB MINERAL PETROL **81,** 230–9.

McBirney, A. R., Baker, B. H. & Nilson, R. H. 1985. Liquid factionation. Part I: basic principles and experimental simulations. J. VOLCANOL GEOTHERM RES **24,** 1–24.

Nabelek, P. I., Papike, J. J. & Laul, J. C. 1986. The Notch Peak Granite Stock, Utah: Origin of reverse zoning and petrogenesis. J PETROL **27,** 1035–9.

Naney, M. T. 1983. Phase equilibria of rock-forming ferromagnesian silicates in granitic systems. AM J SCI **283,** 993–1033.

Nironen, M. 1989. Emplacement and structural setting of granitoids in the Early Proterozoic Tampere and Savo Melts, Finland—implications for contrasting crustal evolution. BULL GEOL SURV FINLAND **346**.

Ramsay, J. C. 1989. Emplacement kinematics of a granite diapir: the Chindamore batholith, Zimbabwe, J. STRUCT GEOL **11,** 191–209.

Sparks, R. S. J. & Marshall, L. A. 1986. Thermal and mechanical constraints on mixing between mafic and silicic magmas. J VOLCANOL GEOTHERM RES **29,** 99–124.

Speer, J. A. 1987. Evolution of magmatic AFM mineral assemblages in granitoid rocks: the hornblende + melt = biotite reaction in the Liberty Hill pluton, South Carolina. AM MINERAL **72,** 863–78.

Waugh, B. J. 1985. *The origin of mafic enclaves within a granodioritic pluton of the Turtle Mountains, California,* Unpublished B.Sc. Thesis, Carleton College,

Wernicke, R. S. 1987. *Geology petrography and geochemistry of the El Topo Pluton, Baja, California Norte, Mexico.* Unpublished MS Thesis, San Diego State University,

White, A. J. R. & Chappell, B. W. 1977. Ultrametamorphism and granitoid genesis. TECTONOPHYSICS **43,** 7–22.

Wones, D. R. & Gilbert, M. C. 1982. Amphiboles in the igneous environment. In Veblen, D. R. and Ribbe, P. H. (eds) *Amphiboles: petrology and experimental phase relations.* MINERAL SOC AM, REV MINERAL **9B,** 355–90.

Zen, E-an. 1988. Phase relations of peraluminous granitic rocks and their petrogenetic implications. ANN REV EARTH PLANET SCI **16,** 21–51.

C. M. ALLEN, Department of Geology, The Australian National University, GPO Box 4, Canberra City, ACT 2601, Australia.

Transactions of the Royal Society of Edinburgh: Earth Sciences, **83**, 191–199, 1992

Spatial, compositional and rheological constraints on the origin of zoning in the Criffell pluton, Scotland

W. E. Stephens

ABSTRACT: The Criffell pluton in southwestern Scotland (397 Ma, a Newer Granite of late Caledonian age) is concentrically zoned with outer granodiorites of typically I-type aspect passing into inner granite with more evolved characteristics. The zonation is examined in terms of the compositional surfaces of bulk parameters such as SiO_2 and Rb/Sr and compositional variation is best modelled as multi-pulse, there being greater variation in bulk composition between pulses than within pulse. Published variations in Sr, Nd and O isotopes reflect the derivation of the pulses from separate and isotopically distinct sources. Other evidence for open-system behaviour includes mingling with mafic magmas to form enclaves, whereas closed-system behaviour is indicated by restite separation in the early granodiorites, and fractional crystallisation in the late granites. A dominant infracrustal I-type magma formed the first pulse followed by magma derived from more evolved crustal rocks (mainly metasediments of varying ages and maturities). Experimental fluid-absent melting of amphibolite and metapelite at about 900°C has shown that significant quantities of melt can be generated, respectively with I-type and S-type characteristics. Despite having similar bulk compositions, these melts have very different viscosities and densities for the same H_2O contents ($\eta_{\text{S-type}} > \eta_{\text{I-type}}$ and $\rho_{\text{S-type}} \leq \rho_{\text{I-type}}$). It is argued that the rheological controls on magma escape from the source region along complex and tortuous pathways favour the more fluid I-type melts over the more viscous (and only slightly less dense) S-type melts. This constraint could have the effect of reversing the expected buoyancy-driven emplacement sequence, and may represent an alternative rheological differentiation mechanism for the formation of some zoned plutons.

KEY WORDS: granite, compositional zoning, Caledonian

Zoned plutons, along with inverted ash flow sequences, are often taken to reflect transient states of silicic magma chambers, zoned plutons as plan views and tuff sequences representing inverted sections (although a few plutons offer a partial view of the third dimension, e.g. the Puscao pluton in the Andes described by Taylor 1976). Very many petrogenetic studies of zoned plutons have emphasised the global variations in chemical characteristics (e.g. Bateman & Chappell 1979; Tindle & Pearce 1981) but spatial variations are only rarely considered, except perhaps by structural geologists (e.g. Holder 1979). If zoned plutons are to be taken as representatives of magma chambers, then it is essential to consider compositional variation within a spatial context and to appreciate that any particular composition represents the time-integrated effects of magmatic processes up to the moment when the consolidation front passed its final position within the pluton.

The first aim of this study is to examine the form of spatial variation in major and trace element compositions and their correlations with published isotopic data within a single normally-zoned pluton. The data are interpreted in terms of two key parameters, namely (a) whether the magmas evolved in *closed* or *open* systems, and (b) whether the rocks fall on a positively evolving trend or a negatively evolving igneous trend. These parameters can be used as the basis of a descriptive scheme to chart the development of a zoned pluton. The Criffel pluton is a good example of compositional zonation in the Caledonian orogen of Britain and Ireland, and the first part of this paper is a case study of the application of these concepts.

A characteristic feature of most zoned plutons is the fact that they are *normally* zoned from a more mafic margin to a more felsic core. This sequence is consistent with *in situ* fractional crystallisation inwards from the walls of a single parental magma (e.g. Vance 1961; Tindle & Pearce 1981). However, isotopic studies over the last decade or so have revealed that most zoned plutons could not have formed solely by simple closed-system processes (e.g. Kistler *et al.* 1986). The second aim of this paper is to consider the implications for granite magma source regions of this petrogenetic requirement for open systems. As a consequence of the contrasting rheological properties that are predicted for melts derived from mixed I-type/S-type sources, a model is developed in which many of the features of zoned granite plutons are inherited directly from the source region, without the requirement to involve subsequent contamination. The regular geometrical zonation of the pluton is attributed by the model to differences in the rates at which magmas can escape from their source, as governed by contrasts in their rheological properties.

1. Zonation of the Criffell pluton

Zoning in the Criffell pluton takes the form of concentric shells trending from an outer rim of unambiguously I-type granodiorite into a core of two-mica granite which has some of the features typical of S-type granites (Chappell & White 1974), and was most probably derived from immature metasediments. The pluton is situated in the SW of Scotland

(Fig. 1) and is a "Newer Granite" emplaced during the waning stages of the Caledonian orogen (Stephens 1988). Most contemporary plutons in Scotland are I-types, whereas S-types occur only in those plutons associated with the earlier peak of Barrovian and Buchan metamorphism, or were emplaced close to the Scotland–northern England terrane boundary after collision (Stephens 1988). The Criffell pluton lies very close to this boundary and was emplaced very late in the Caledonian sequence (397 ± 2 Ma, Halliday *et al.* 1980) and the I-type and mild S-type characteristics are undoubtedly, in part, a reflection of its complex tectonic setting. The aim of this paper is to address the petrogenesis of plutons which have a zonation pattern from outer I-types to inner S-types, most of this sequence being present at Criffell, although the strongly peraluminous compositions of Miller (1985) are not represented.

Figure 1 shows the distribution of petrological zones in the Criffel pluton. The outer rim is a clinopyroxene-bearing hornblende biotite granodiorite (CBH) which passes into a hornblende biotite granodiorite (HB) without obvious contact in the field, although there is geochemical evidence to suggest that these members may comprise two discrete pulses (but not necessarily bounded at the petrological boundaries). The biotite granite (B) is locally porphyritic, and an internal contact between the biotite granite and the granodiorites can be mapped through this zone in some parts of the pluton. The inner two zones are muscovite biotite granite (MB) and biotite muscovite granite (BM), the latter containing large discrete flakes of apparently primary muscovite.

Sampling of the pluton on a grid yielded 180 rocks (Fig. 1) which have been analysed for major and trace elements by X-ray fluorescence (see Stephens *et al.* 1985 for details). A subset was analysed for Sr and O isotopes (Halliday *et al.* 1980) and a further subset for Nd isotopes. Holden *et al.* (1987) and Stephens *et al.* (1991) studied enclave-host granite pairs for their Sr and Nd isotope compositions.

The petrogenesis of this pluton was studied by Phillips (1956) and Phillips *et al.* (1981), who regarded the zonation as having developed by *in situ* crystal-liquid processes driven

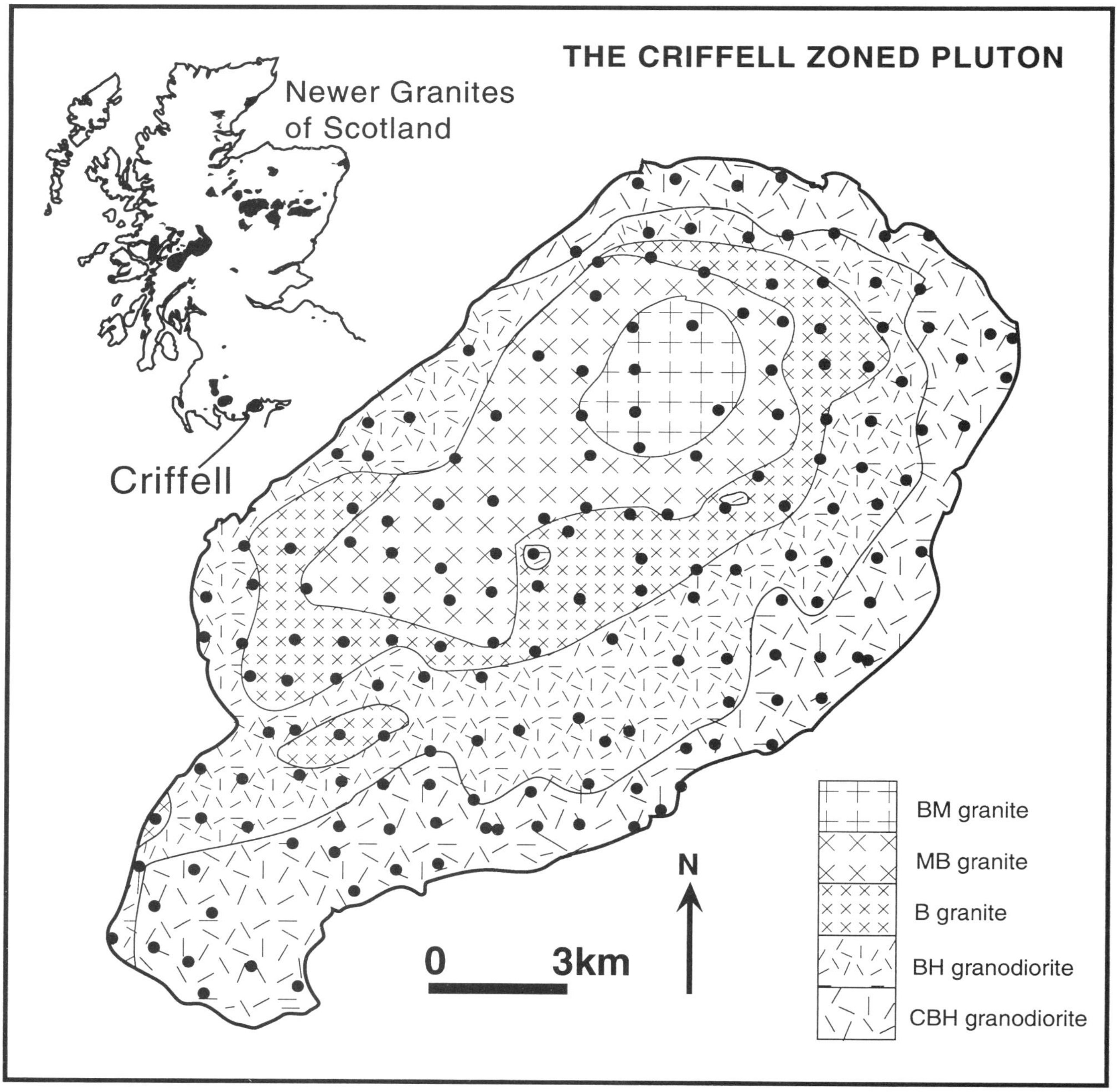

Figure 1 The Criffell pluton, showing its location in Scotland, petrological zones and sample localities (black dots).

by convective motion, and Halliday *et al.* (1980), Stephens and Halliday (1980), Stephens *et al.* (1985), and Holden *et al.* (1987), who used isotopes to demonstrate that the observed variations in composition could not be accounted for entirely by closed-system processes. Specifically, those authors proposed that the pluton was generated from a deep crustal source, that enclaves have a mantle origin and had partially equilibrated with the host granodiorite, and that the central core of the pluton involved an evolved source (most probably dominated by immature metasediments) either by assimilation-fractional crystallisation (AFC) or by the incorporation of anatectic melts.

2. Compositional zonation surfaces

The spatial structure of compositional variation is very important to any consideration of *in situ* processes, yet it is rarely considered. This is at least partly due to the fact that a very dense sampling grid is required which requires good exposures and a major analytical effort. In this section the spatial variations in SiO_2 (taken to reflect bulk composition) and Rb/Sr (taken as an indicator of feldspar fractionation) are considered.

The surface of SiO_2 concentration over the Criffell pluton in geographic space is shown in Figure 2(a)–(d). Contour map Figure 2(a) shows a marked concentric structure with SiO_2 varying from around 60–62% at the outer contact in the CBH granodiorites, to a maximum of about 72% in the BM granite. This map was generated by using computer-contouring algorithms, but does not differ significantly from the results of hand contouring. Either method generates contour lines whose course is influenced by the distribution of values in the region of interest. However, in the case of a composite pluton the information required is not the spatial distribution over the whole pluton, but the variation within individual pulses. For this reason the data have been recontoured by using an internal boundary as a barrier to influence in the contouring routine, as if it were a fault. In that mode, the data outside the boundary have no influence on the contour values inside, and the inner and outer parts of the pluton were contoured separately either side of the discontinuity and then juxtaposed. The result is Figure 2(c) which shows a very different compositional surface. In essence, the effect of the internal contact on contouring is to amplify the "step" between the outer and inner pulse and to suppress the apparent within-pulse variation. These contrasts are illustrated in three-dimensions by isometric projections (Fig. 2(b) and (d)), viewing the compositional surfaces from the southeastern corner of the equivalent maps downwards at an angle of 15° from the horizontal.

Essentially, the outer granodiorite pulse is shown by the above method to vary around 64 ± 2 % SiO_2 in a very noisy pattern with crude concentric structure, whereas the inner granite pulse varies more systematically around 70 ± 2 % SiO_2 towards a peak in the BM granite. The magnitude of the compositional step varies from 2–6 % SiO_2, a feature not apparent in the model presented in Figure 2(a). Harker and other variation diagrams are deceptive in this respect: by disregarding spatial dimensions the "step" is taken to be the difference between the most evolved primitive member and the most primitive evolved member regardless of whether or not they are spatially contiguous.

Spatial variation in Rb/Sr is rather different (Fig. 2(e) and (f)). There is no significant variation in the ratio within the granodiorite, whereas the inner granite shows a fine zonation structure with Rb/Sr increasing to a peak in the BM granite precisely at the maximum in SiO_2. The granite pulse clearly shows evidence for a systematic spatial evolution, presumably through the fractional crystallisation of plagioclase and/or alkali feldspar, and the regularity of its form suggests the action of processes indigenous to the magma chamber.

In summary, objective contouring demonstrates that the granodiorite member is poorly zoned, with a "noisy" structure, following a crude concentric form, whereas the granite member has a slightly more organised, "less-noisy" structure with clear concentric form. The apparent compositional step between these pulses is typically as great as the whole within-pulse variation. In contrast, the incompatible trace element Rb shows no significant variation in the outer parts of the pluton, but increases exponentially towards the pluton core.

The map analysis described in this section enables constraints to be placed on the three-dimensional form of compositional zonation within the pluton, and to place some constraints on how it might have evolved. In order to gain a clearer picture of the processes which contributed to the development of this structure, it is now necessary to examine the interplay of open and closed system processes.

3. Open and closed system processes

The relative importance of fractional crystallisation, restite separation, assimilation, magma mixing and anatexis in generating compositional zonation in plutons is a key problem in granite petrogenesis. Most studies conclude that one or perhaps two of these processes can account for the zonation observed in most plutons. The thesis presented here is that most, if not all, of these processes may be operating at different times and in different parts of the same pluton. If this thesis, is correct, then the challenge is to quantify the contributions of each to the petrogenesis of rocks from different parts of the zoned pluton, underlining the need for detailed, spatially oriented studies. These processes are discussed below under the headings *open system* and *closed system,* and it is proposed that an evolving pluton may have periods when it is open to influxes of magma from different sources or to contamination, whereas before and/or afterwards it may evolve normally by crystal-liquid processes (mainly restite separation and/or fractional crystallisation) or not evolve at all.

Open system processes operating on a typical infracrustal granite are normally easily distinguished using Nd and Sr isotope ratios (Fig. 3(a)). The *initial magma* of relatively primitive crustal composition (slightly positive ε_{Sr} and slightly negative ε_{Nd}) can be modified *progressively* (i.e. changing to more evolved magmatic compositions) by the assimilation of wall rocks or by the incorporation of more silicic anatectic melts. Their progress will normally be directed towards the bottom and/or right of the ε_{Sr}–ε_{Nd} diagram where the end-member contaminants or anatectic sources would tend to fall (Fig. 3(a)). Alternatively, the open system might develop compositions counter to the normal evolutionary trend, through interaction of the *initial magma* with batches of more primitive mafic melts. The typical product of such interaction is the igneous enclave; studies of isotopic equilibration within enclaves (Stephens *et al.* 1991) suggest that the enclaves will equilibrate through the remaining melt phase and, if the enclave is still permeable to melt, then this equilibration will be considerable. Isotopically we have predicted this to have a sigmoidal form on Figure 3(a), with enclaves equilibrating extensively (and more rapidly for Sr than Nd), whereas the effect on the host granite may be small (Stephens *et al.* 1991). This is consistent with the experimental results of interdiffusion in rhyolite dacite systems obtained by Lesher (1990).

Closed system effects would not normally be perceptible in the ε_{Sr}–ε_{Nd} diagram except for the special case where the granite has a high initial Rb/Sr ratio and undergoes protracted cooling; then a significant increase in the ε_{Sr} might be observed but there would be no effect on the ε_{Nd} (Fig. 3(a)). Possible development trends include *closed systems* in which the *melt* evolves *either* to more silicic compositions by fractional crystallisation and/or restite separation, *or* to less silicic compositions through blending with mafic magmas. *Cumulates* will be either mafic or felsic.

Zoned granite plutons clearly have complex dynamic histories and it is important to known the dominant

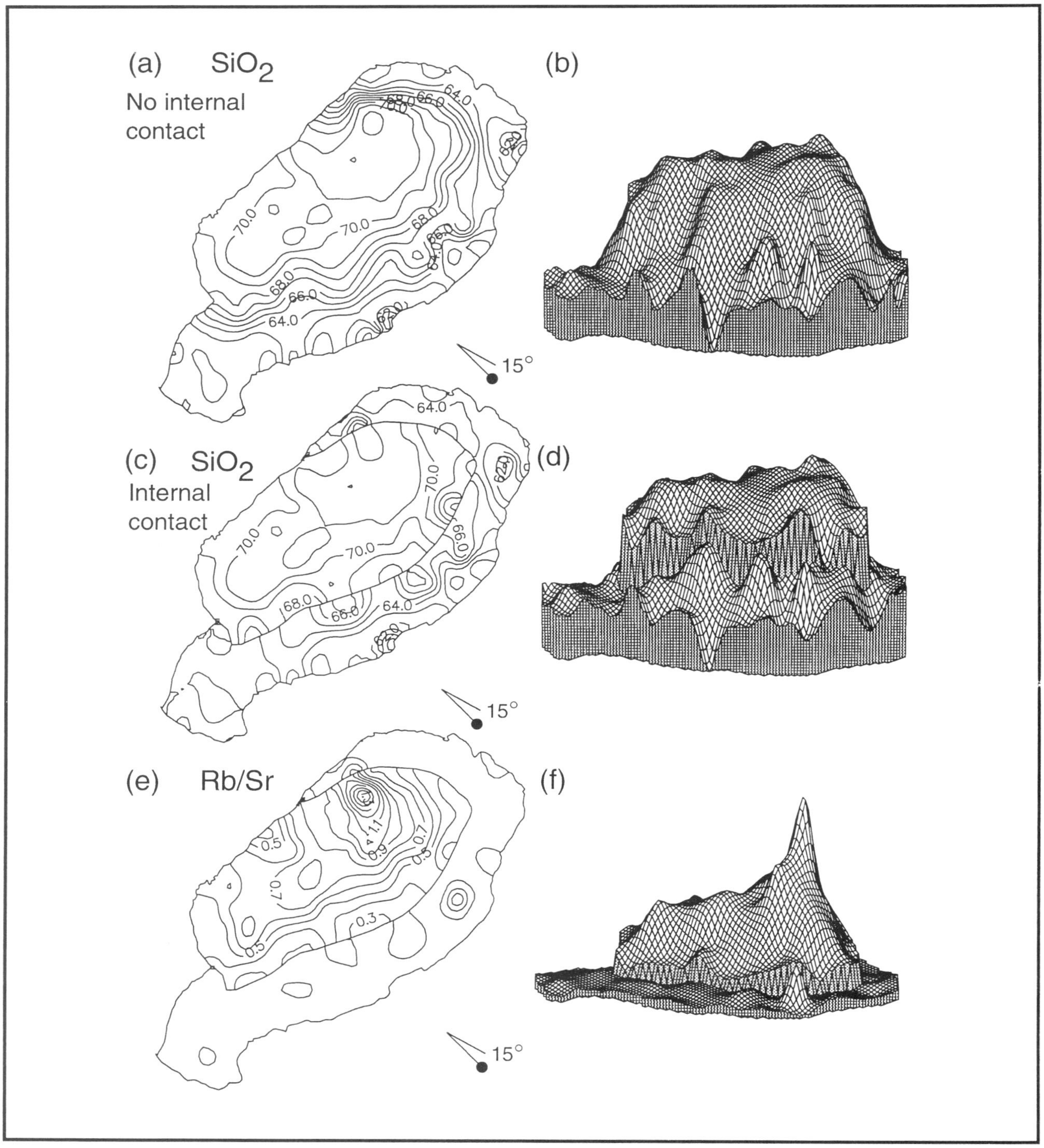

Figure 2 Maps of whole-rock variations within the Criffell pluton:
(a) Contour model for SiO_2 over the whole pluton.
(b) Isometric view of the surface of (a), taken from southeastern corner at an angle of 15° from the horizontal, as indicated.
(c) Two-pulse contour model, with pulses separated by the internal contact which isolates each pulse in the contouring routine.
(d) Isometric view of the surface of (c), taken from southeastern corner at an angle of 15° from the horizontal, as indicated.
(e) Rb/Sr map for 2-pulse model.
(f) Isometric view of the surface of (e), taken from southeastern corner at an angle of 15° from the horizontal, as indicated.

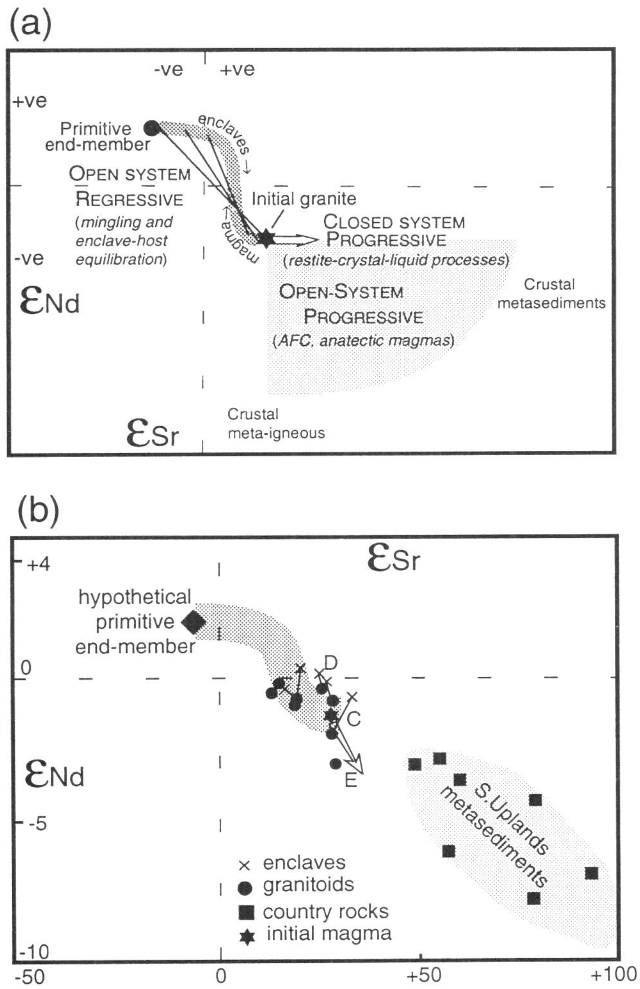

Figure 3 (a) Schematic models of open and closed system environments within Nd–Sr space (see text for details).
(b) Suggested evolution of the Criffell pluton in Nd–Sr space (data from Halliday *et al.* 1980; O'Nions *et al.* 1983; Holden *et al.* 1987).

process(es) operating at various stages in that history. A summary of the compositional development may be charted using a figure constructed on the basis of open/closed system versus compositional trends. The development of the Criffell pluton is described in this form in Figure 4(a).

The principal difference within the Criffell pluton is between the outer granodiorite and the inner granite pulses, and the contrasts in bulk compositions are 64 vs 70% for mean SiO_2 content, metaluminous vs peraluminous, 0·7055 vs 0·7070 for initial ($^{87}Sr/^{86}Sr$), −1 vs −3 for ε_{Nd}, and 9 vs 11‰ for $\delta^{18}O$. These differences require a dominantly open-system explanation, and Stephens *et al.* (1985) concluded that assimilation-fractional crystallisation (AFC, DePaolo 1981) processes could adequately account for most of the observed features. An alternative, now considered equally probable, is that the magmas (C and E on Fig. 4(a) and (b)) were derived from distinct sources. The granite magma pulse has radiogenic isotope characteristics consistent with its entire derivation from a deep crustal source similar to some of the surrounding metasediments (Fig. 3(b)). In either case, the data require that the developing pluton went through a major, open-system event during the middle stages of plutonic evolution in which the magma became progressively more evolved.

In finer detail it is possible to discern the effects of other processes, perhaps less important contributors to the overall zonation, but locally significant. Mafic enclaves with "igneous" textures are very abundant in the outer part of the pluton. They have ε_{Nd} values some 1–2 units more positive than their hosts and appear to have equilibrated significantly from a postulated primitive parental magma (see Fig. 4(a) and (b)) and in so doing they will have modified the composition of the host granite. However, even if equilibration with mafic magma does require there to have been a consequent change in the composition of the host, the reservoir represented by the host magma can be considered infinite compared with the mass of enclaves, and the bulk effect of this open system regressive stage is almost certainly small on the pluton scale (Fig. 3(b)). Restite separation is evident in the main granodiorite stage. Amphibole-rich clots have been studied within the Strontian pluton using backscattered electron imaging and electron probe microanalysis (Stephens & Castro 1991), and the data indicate a probable restitic source for these clots. Identical material (though rather less abundant) is common within the outer granodiorites of the Criffell pluton. Some of the bulk variation within the Criffell granodiorites is almost certainly due to restite separation during a closed system progressive stage. Finally, trace element data, including the Rb/Sr maps of Figure 2(e), require that the most central granites evolved by the fractional removal of plagioclase and/or alkali feldspar. This marks the final stage in the evolution of the pluton.

The Criffell pluton thus seems to have undergone a complex sequence of magmatic evolutionary events, and this is schematically illustrated in Figure 4(a) in terms of open/closed systems and evolutionary trends. The thick solid line of Figure 4(a) represents the dominant process which gave rise to the rock from the initial magma pulse through consolidation of the pluton, probably, but not necessarily, inwards from the margin. The line is schematic and has no dimensions of space or time. Similarly, the four quadrants are nominal and not scaled; thus the early excursion into the open system stage (C to D on Fig. 4(a)) is shown as a large loop, whereas if the diagram was scaled to reflect importance then this loop might be too small to represent on the diagram. Between the end of the first granodiorite pulse and beginning of the fresh granite pulse, there is the suggestion in the intermediate Sr isotopic initial ratios that some hybrid interaction may have occurred. Again this was probably very limited in its contribution to overall zonation. A cartoon illustrating a conceptual model for the development of the pluton in crustal section is presented in Figure 4(b).

4. Characteristics of melts in heterogeneous source regions

Appealing to open-system processes has consequences for the thermal structure of source regions and ascent paths. AFC models derive heat from the heat of crystallisation of minerals and this will severely limit the efficacy of the process. For the purpose of this paper, a closed system is regarded as one in which the magma pulse within the pluton has not been modified by interaction with any external matter since leaving its source, i.e. a true image of its source as proposed by Chappell (1979). Whereas end-members in open system processes can often be identified by isotopic techniques, the processes operating on the system are less easy to identify. For open system processes to have a major effect on a pluton-wide scale, the energy requirements may be considerable. As an alternative, it is worth considering whether or not sources are heterogeneous and whether apparent open system geochemical effects can actually be derived from within the source region. If a pluton is 1000 km^3 in volume and was derived by 30% melting, then there must have been ≈3000 km^3 of source rock. The deep crust, the likely source region for most I-type granites, is probably heterogeneous on various scales (Patchett 1991; Miller *et al.* 1987) and it is conceivable that most source regions are heterogeneous on this scale. Some of the consequences of melting in a heterogeneous source region are considered below.

The isotopic characteristics of the members of the Criffell

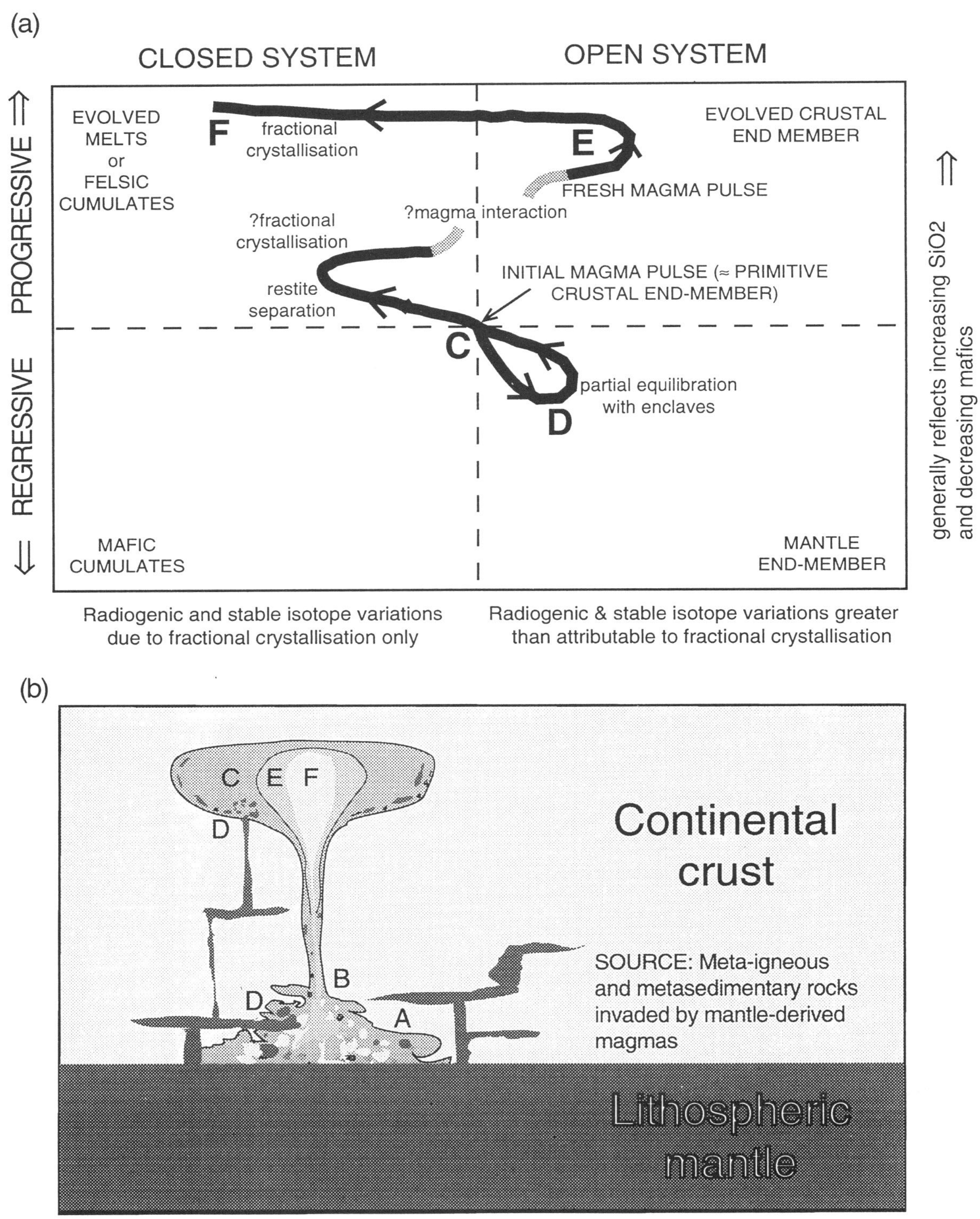

Figure 4 (a) Summary of petrogenetic evolution of the Criffell pluton in terms of open and closed systems, and evolution of the magma composition. This granodiorite magma (C) evolved to more primitive compositions by mingling with mafic magma in an open system (D), until the system closed. The magma then evolved to more silicic compositions in a closed system through restite separation and possibly fractional crystallisation. A pulse of new magma from a different and more evolved source (open system) was emplaced into the centre of the pluton (E) and finally evolved by fractional crystallisation in a closed system (F).

(b) Cartoon to illustrate the likely development of the Criffel pluton from magma genesis in a heterogeneous I-type/S-type source (A). Rates of melt segregation controlled by rheological contrasts (see text for details) result in more mafic magma (granodiorite) being first to escape the source (B). Further development through open and closed system plutonic processes (C to F) are the same as for Figure 4(a).

pluton (and most of those in southern Scotland) indicate an infracrustal (Chappell & Stephens 1988) origin from a relatively young and unevolved source. Analogous (if not actual) sources occur within the host Lower Palaeozoic sequence of the region, with volcanic rocks and immature greywackes being common in the lower (and presumably deeper) part of the sequence, and potentially representing I-type and immature S-type sources, respectively, reflecting the types of sources of the magmas comprising this pluton.

In order to examine some possible consequences of melting heterogeneous I-type/S-type sources, experimental studies of melting I-type and S-type rocks have been collated in Figure 5. Vapour absent melting of metabasalt at 0·3 GPa (Beard & Lofgren 1989) and garnet tonalite (Rutter & Wyllie 1988) at 1 GPa are taken to represent the I-types, whereas vapour absent melting of metapelite (Patiño Douce & Johnston 1991) at 1 GPa is taken to represent S-types.

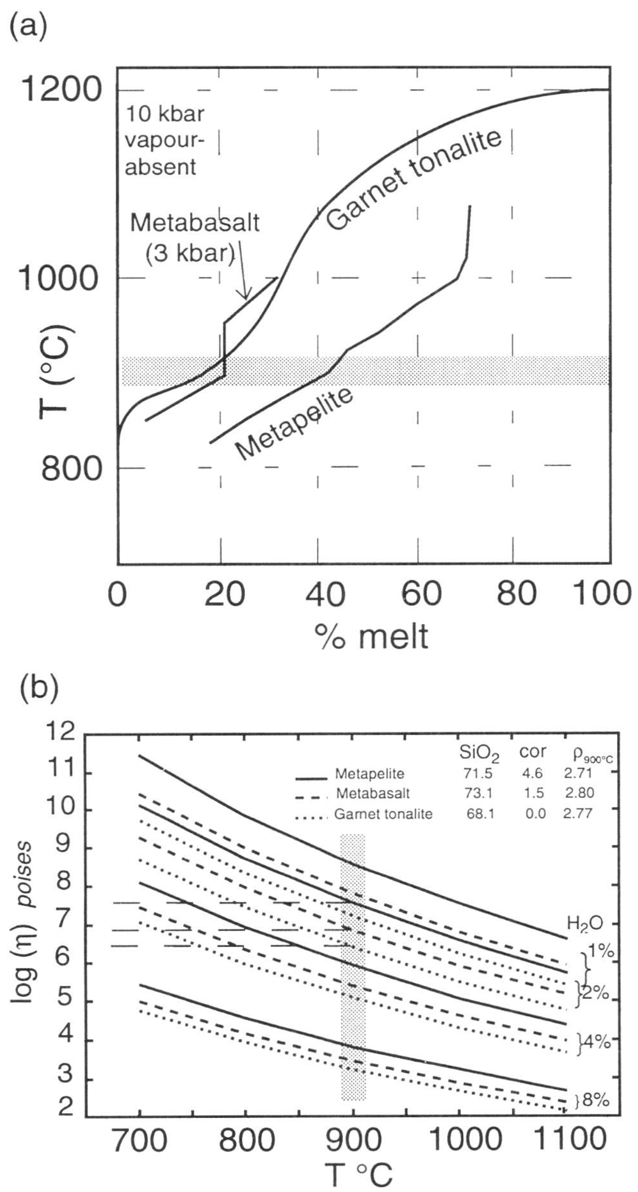

Figure 5 (a) Melt fraction–T plots for metabasalt (Beard & Lofgren 1989), garnet tonalite (Rutter & Wyllie 1988), and metapelite (Patiño Douce & Johnston 1991); cor = % CIPW normative corundum, and ρ = density in k gm^{-3}.
(b) Calculated viscosity of the three melt compositions plotted in Figure 5(a) at various conditions of T and H_2O contents; viscosity model of Shaw (1972) used assuming no suspended crystals.

Reasonable quantities of melt are generated from both sources at 900°C, about 20% for the I-types and 40% for the S-type (Fig. 5(a)). Glass compositions for melts at this temperature vary from 68·1–73·1% SiO_2 for the I-types and an intermediate value of 71·5% for the S-type. However, the S-type has 4·6% normative corundum, whereas the I-types have 0–1·5% of that normative component. Melt viscosities have been calculated from these bulk melt compositions using the method of Shaw (1972) at various likely temperatures and H_2O contents (Fig. 5(b)). (Anhydrous densities at 900°C calculated using the method of Bottinga and Weill (1970), are also presented on this figure.) The important result is that whereas the maximum density contrast between the two I-types and the single S-type is only about 3%, the maximum viscosity contrast at reasonable H_2O contents is typically an order to magnitude. Despite the I-type metabasalt melt having higher SiO_2, it has a lower calculated viscosity than the S-type metapelite (at the same temperature and H_2O content), which is presumably a function of relatively low Fe, Mg and Ca of the S-type melt.

We can conclude that isothermal melting of a heterogeneous I-type/S-type source will generate abundant melt at 900°C and that increasing the pelite content will significantly increase the viscosity of the melt, but not the SiO_2 content (at the same H_2O level). A greywacke will probably behave in some intermediate way. The important question is whether these melts will homogenise before ascent, presumably only to differentiate again at a high level. It is argued later that by keeping these melts largely isolated there is no requirement for a later major differentiation stage and that the geometry of zoned plutons may have been rheologically controlled.

5. Rheological segregation as a model for zoned plutons

Carrigan and Eichelberger (1990) proposed that co-existing magmas of contrasting density could be emplaced in reverse order to that expected on the basis of density alone, the controlling factor being a process they termed *viscosity segregation*. They suggested that magmas of different viscosities will respond differentially to shear stresses while flowing up a conduit, and will segregate in order to attain an equilibrium Newtonian velocity flow profile. They attributed the origin of some silicic-mafic zonation in lava flows to this process. Intuitively, however, it seems likely that the diameters of any conduits required to transport perhaps 1000 km^3 of magma would be too great to generate the velocity gradients for effective viscosity segregation, and anyway the process does not apply at low flow rates. A similar argument was used against the Bagnold effect (flow differentiation) for porphyritic pluton interiors by Barrière (1976).

The case has been made above that co-existing I-type magmas of similar bulk compositions (70 ± 3% SiO_2) have only 3% contrast in density but an order of magnitude difference in viscosity (assuming the same H_2O content). Viscosity is clearly an important parameter in the segregation of magmas from their sources, and an order of magnitude contrast in viscosity is bound to have a significant effect. Nevertheless, most recent fluid dynamic studies of magma behaviour are based on the effects of thermal and compositional density gradients in melts resulting in mass transfer. The density and viscosity contrasts of the experimental I-type and S-type melts described above are

$\eta_{S\text{-type}} \gg \eta_{I\text{-type}}$, but $\rho_{S\text{-type}} \leq \rho_{I\text{-type}}$, and whereas even a small density contrast will be very significant for co-existing melts it is likely that viscosity contrasts will be important in the early stages of melt segregation and accumulation from source.

Other rheological properties should also be considered as influencing the rate of melt release at the melt-source segregation stage. The Bingham yield stress must be overcome before a melt can be mobilised, and Hulme (1974) and McBirney and Murase (1984) demonstrated a strong linear correlation between finite yield strength and SiO_2 content of a range of silicate melts. The magnitude of this yield stress might be expected to affect the rate and efficacy of melt segregation. In the early stages of melting, when melt is in the form of a thin film in contact with a large surface area of solid in the source, surface tension will be an important control on the efficiency of segregation. The importance of surface tension contrasts between different melt structures is not known as few experimental determinations exist; the only published data on rock melts are those for andesites and more basic melts by Murase and McBirney (1973) and Walker and Mullins (1981), and for diopside-anorthite mixtures by Taniguchi (1988). These findings indicate that most intermediate to basic melts have similar isothermal surface tension coefficients, and the potentially important observation that the surface tension of silicate melts increases with temperature (most common substances decrease). No surface tension measurements have yet been made for rhyolitic melts.

The relative importance of these three parameters in determining rates of melt segregation is very poorly understood for siliceous magma systems, but might be very important. In the situation where a source of heterogeneous I-type/S-type composition melts isothermally, the rate of melt escape is likely to be influenced significantly by melt rheology. Finite yield stress and viscosity will tend to act together, namely to inhibit melt segragation in the more siliceous melts, but the action of surface tension is not known. It is envisaged that a complex, porous source region presents a labyrinth of tortuous paths for melt to escape its source, and an ideal environment in which to segregate melts of contrasting rheological properties, either by flow or by surface effects. As acknowledged, the relative importance of viscosity, Bingham yield stress and surface tension are not known and may vary, thus the process postulated above is termed *rheological segregation* in order to distinguish it from the *viscosity segregation* defined by Carrigan and Eichelberger (1990) in which viscosity is identified as the primary control, and flow is defined as operating in a conduit.

A very simple analogue of rheological segregation can be constructed with engine oil and water in which $\eta_{oil} \gg \eta_{water}$, and $\rho_{oil} < \rho_{water}$ at room temperature. An agitated mixture of these two fluids very rapidly reverts to a stable stratification in which the viscous oil overlies the water. However, passing a plug of steel wool down through a column of the same agitated oil-water mixture before stratification has the effect of efficiently transferring the oil to the bottom of the column. This reversed stratification is presumably due to high surface tension between the oil and steel wool causing retention of the oil in the porous medium at the base of the column, despite the very strong inverse density gradient. This experiment is not an accurate or scale model for magma source regions, but is intended only to demonstrate that contrasts in rheological properties may have marked effects on a mixture of two fluids in transfer through a porous solid medium.

A simple consequence of rheological segregation is that the release of melt from its source may be compositionally-controlled, more siliceous melts being released more slowly than less siliceous melts. If these melts ascend at similar rates (given their small density contrasts) then in a zoned pluton the granodiorite/tonalite magmas will be emplaced first and the monzogranites/granites will follow later. Thus concentric zonation could be controlled at source by rheological segregation in a heterogeneous source, and this would also account for the bulk chemical and isotopic variations while not requiring any unusual thermal regime outside the source.

Whereas crystal-liquid processes have dominated thinking about the origin of compositional variation in zoned plutons for several decades (e.g. Nockolds 1940), and buoyancy-driven forces have dominated thinking about plutonic emplacement since the days of Grout's experiments with syrups and oils (Grout 1945), it seems timely now to review whether indeed these are the dominant controls. This contribution offers some alternative mechanisms based on the close correlations between bulk magma composition and rheological properties.

6. Conclusions

Compositional variation in the Criffell zoned pluton is the aggregate effect of between-pulse and within-pulse variations. Between-pulse variations are generally of a greater magnitude than any within-pulse variation.

The spatial aspect of zonation is often lost in variation diagrams in which adjacent points on a trend need not be spatially contiguous. Interpretations of *in situ* process based on variation diagrams may be flawed.

The Criffel pluton evolved by a complex interplay of open and closed system processes. The initial magma first followed an open system path through equilibration with mafic magmas and the formation of mafic enclaves, but then became closed and evolved dominantly by restite separation. With the arrival of a fresh impulse of magma from a distinct source, the pluton once more evolved as an open system. Finally, behaving essentially as a closed system, the magma underwent fractional crystallisation until the pluton consolidated.

Melting of heterogeneous I-type/S-type source regions produces abundant melts of both types at temperatures around 900°C. Co-existence of such magmas in a large source volume is possible, and compositional heterogeneity in the pluton may simply reflect compositional heterogeneity in the source region.

Such melting of heterogeneous I-type/S-type source regions produces melts in which the contrast in viscosity is far greater (an order of magnitude) than the contrast in density (a few %) for similar bulk compositions. Viscosity, and potentially yield stress and surface tension contrasts, may act individually or in concert to impede the release of siliceous magmas from a heterogeneous source containing both siliceous and intermediate magmas. Such rheological segregation may be a significant differentiation mechanism in controlling the timing of emplacement and thus the geometry of zoned plutons.

8. References

Barrière, M. 1976. Flowage differentiation; limitation of the "Bagnold effect" to narrow intrusion. CONTRIB MINERAL PETROL **55**, 139–45.

Bateman, P. C. & Chappell, B. W. 1979. Crystallisation,

fractionation and solidification of the Tuolumne Intrusive Series, Yosemite National Park, California. BULL GEOL SOC AM **90,** 465–82.

Beard, J. S. & Lofgren, G. E. 1989. Effects of water on the composition and partial melts of greenstone and amphibolite. SCIENCE **244,** 195–7.

Bottinga, J. & Weill, D. F. 1970. Density of liquid silicate systems calculated from partial molar volumes of oxide components. AM J SCI **269,** 169–82.

Bottinga, J. & Weill, D. F. 1972. The viscosity of magmatic silicate liquids: a model for calculation. AM J SCI **272,** 438–75.

Carrigan, C. R. & Eichelberger, J. C. 1990. Zoning of magmas by viscosity in volcanic conduits. NATURE **343,** 248–51.

Chappell, B. W. 1979. Granites as images of their source rocks. GEOL SOC AM PROG ABS TR **11,** 400.

Chappell, B. W. & Stephens, W. E. 1988. Origin of infracrustal (I-type) granite magmas. TRANS R SOC EDINBURGH EARTH SCI **79,** 71–86.

Chappell, B. W. & White, A. J. R. 1974. Two contrasting magma types. PAC GEOL **8,** 173–4.

DePaolo, D. J. 1981. Trace element and isotopic effects of combined wallrock assimilation and fractional crystallisation. EARTH PLANET SCI LETT **53,** 189–202.

Grout, F. F. 1945. Scale models of structures related to batholiths. AM J SCI **243A,** 260.

Halliday, A. N., Stephens, W. E. & Harmon, R. S. 1980. Rb–Sr and O isotopic relationships in 3 zoned Caledonian granitic plutons, Southern Uplands, Scotland: evidence for varied sources and hybridization of magmas. J GEOL SOC LONDON **137,** 329–48.

Holden, P., Halliday, A. N. & Stephens, W. E. 1987. Neodymium and strontium isotope content of microdiorite enclaves points to mantle input to granitoid production. NATURE **330,** 51–6.

Holder, M. T. 1979. An emplacement mechanism for post-tectonic granites and its implications for their geochemical features. *In* Atherton, M. P. & Tarney, J. (eds) *Origin of granite batholiths: geochemical evidence,* 116–28. Orpington: Shiva Press.

Hulme, G. 1974. The interpretation of lava flow morphology. GEOPHYS J ROY ASTR SOC **39,** 361–83.

Kistler, R. W., Chappell, B. W., Peck, D. L. & Bateman, P. C. 1986. Isotopic variation in the Tuolumne Intrusive Suite, central Sierra Nevada, California. CONTRIB MINERAL PETROL **94,** 205–20.

Lesher, C. E. 1990. Decoupling of chemical and isotopic exchange during magma mixing. NATURE **334,** 235–7.

McBirney, A. R. & Murase, T. 1984. Rheological properties of magmas. ANN REV EARTH PLANET SCI **12,** 337–57.

Miller, C. F. 1985. Are strongly peraluminous magmas derived from pelitic sedimentary sources? J GEOL **93,** 673–89.

Miller, C. F., Watson, E. B. & Harrison, T. M. 1987. Perspectives on source, segregation and transport of granitoid magmas. TRANS R SOC EDINBURGH EARTH SCI **79,** 135–56.

Murase, T. & McBirney, A. R. 1973. Properties of some common igneous rocks and their melts at high temperature. GEOL SOC AM BULL **84,** 3563–92.

Nockolds, S. R. 1940. The Garabal Hill-Glen Fyne igneous complex. J GEOL SOC LONDON **96,** 451–511.

O'Nions, R. K., Hamilton, P. J. & Hooker, P. J. 1983. A Nd isotope investigation of sediments related to crustal development in the British Isles. EARTH PLANET SCI LETT **63,** 229–40.

Patchett, P. J. 1991. Isotope geochemistry and chemical evolution of the mantle and crust. *Rev Geophys for 1991, Supplement.* AM GEOPHYS UNION, 457–70.

Patiño Douce, A. E. & Johnston, A. D. 1991. Phase equilibria and melt productivity in the pelitic system: Implications for the origin of peraluminous granitoids and aluminous granulites. CONTRIB MINERAL PETROL **107,** 202–18.

Phillips, W. J. 1956. The Criffell-Dalbeattie granodiorite complex. J GEOL SOC LONDON **112,** 221–39.

Phillips, W. J., Fuge, R. & Phillips, N. 1981. Convection and crystallisation in the Criffell-Dalbeattie pluton. J GEOL SOC LONDON **138,** 351–66.

Rutter, M. J. & Wyllie, P. J. 1988. Melting of vapour-absent tonalite at 10 kbar to simulate dehydration-melting in the deep crust. NATURE **331,** 159–60.

Shaw, H. R. 1972. Viscosities of magmatic silicate liquids: an empirical method of prediction. AM J SCI **266,** 255–64.

Stephens, W. E. 1988. Granitoid plutonism in the Caledonian orogen of Europe. *In* Harris, A. L. *et al.* (eds) *The Caledonian-Appalachian orogen.* GEOL SOC LONDON SPEC PUBL **38,** 389–403.

Stephens, W. E. & Castro, A. 1991. Mafic clots in the Strontian pluton: restite or the debris of magma mingling? TERRA ABSTR **4,** 31–2.

Stephens, W. E. & Halliday, A. N. 1980. Discontinuities in the composition surface of a zoned pluton, Criffell, Scotland. GEOL SOC AM BULL **91,** 165–70.

Stephens, W. E. & Halliday, A. N. 1984. Geochemical contrasts between late Caledonian granitoid plutons of northern, central and southern Scotland. TRANS R SOC EDINBURGH EARTH SCI **75,** 259–73.

Stephens, W. E., Whitley, J. E., Thirlwall, M. F. & Halliday, A. N. 1985. The Criffell zoned pluton: correlated behaviour of rare-earth element abundances with isotopic systems. CONTRIB MINERAL PETROL **89,** 226–38.

Stephens, W. E., Holden, P. & Henney, P. J. 1991. Microdioritic enclaves within the Scottish Caledonian granitoids and their significance for crustal magmatism. *In* Didier, J. & Barbarin, B. (eds) *Granites and their enclaves,* 2nd edn, 125–134. Amsterdam: Elsevier.

Taniguchi, H. 1988. Surface tension of melts in the system $CaMgSi_2O_6$–$CaAl_2Si_2O_8$ and its structural significance. CONTRIB MINERAL PETROL **100,** 484–9.

Taylor, W. P. 1976. Intrusion and differentiation of granitic magma at a high level in the crust: the Puscao pluton, Lima Province, Peru. J PETROL **17,** 194–218.

Tindle, A. G. & Pearce, J. A. 1981. Petrogenetic modelling of in situ fractional crystallisation in the zoned Loch Doon pluton, Scotland. CONTRIB MINERAL PETROL **78,** 196–207.

Vance, J. A. 1961. Zoned granitic intrusions–an alternative hypothesis of origin. GEOL SOC AM BULL **72,** 1723–8.

Walker, D. & Mullins, O. 1981. Surface tension of natural silicate melts from 1,200°–1,500°C and implications for melt structure. CONTRIB MINERAL PETROL **76,** 455–62.

W. E. STEPHENS, Department of Geography & Geology, University of St Andrews, St Andrews, Fife KY16 9ST, Scotland.

Transactions of the Royal Society of Edinburgh: Earth Sciences, **83**, 201–209, 1992

Proterozoic granite types in Australia: implications for lower crust composition, structure and evolution

L. A. I. Wyborn, D. Wyborn, R. G. Warren and B. J. Drummond

ABSTRACT: Granites and their associated comagmatic felsic volcanic rocks occur in most Proterozoic provinces of Australia. Using multi-element, primordial-mantle-normalised abundance diagrams and various petrological characteristics, Australian Proterozoic granites can be subdivided into five groups: (i) I-type, Sr-depleted, Y-undepleted, restite-dominated, (ii) I-type, Sr-depleted, Y-undepleted, fractionated, low in incompatible elements, (iii) I-type Sr-depleted, Y-undepleted, enriched in incompatible elements (anorogenic granites), (iv) I-type, Sr-undepleted, Y-depleted, (v) S-type, Sr-depleted, Y-undepleted. The four Sr-depleted groups dominate, and group (iv) is of very limited extent. A comparison of these Proterozoic granites with Australian and Papua New Guinean granites of other time periods shows that these characteristic Sr-depleted Y-undepleted patterns are also dominant in early Palaeozoic granites. They are significantly different from those of granites in modern island arcs associated with subduction, and with most granites from Archaean terranes, where the multi-element diagrams are dominated by Sr-undepleted, Y-depleted patterns.

The Sr-depleted, Y-undepleted patterns are thought to indicate source regions that contained plagioclase but not garnet, whilst the Sr-undepleted, Y-depleted patterns are taken to correspond with the presence of garnet, but not plagioclase, in the source rocks. The Sr-depleted, Y-undepleted patterns also only occur in regions where the lower crustal structure is dominated by an underplated mafic layer with a P-wave velocity of 7·2–7·4 km/s. In contrast, in regions where the granites are dominated by Sr-undepleted, Y-depleted patterns, such as in the Archaean and in Cainozoic island arcs, this intermediate velocity layer is not present, and the crust-mantle boundary is very sharp.

Two other distinctive compositional changes have been noted among the I-type granites of different age. Firstly, Na is highest in Archaean and Cainozoic granites, and lowest in early Proterozoic granites; Palaeozoic and Mesozoic granites have intermediate values. Secondly, late Archaean and Proterozoic granites are the most enriched in K, Th and U, while the Cainozoic and early Archaean tonalites are the most depleted; Palaeozoic and Mesozoic granites again contain intermediate amounts of those elements.

KEY WORDS: granite sources, multi-element diagrams, Sr-depletion, underplating, Y-depletion.

Granitic rocks and their associated comagmatic felsic volcanic rocks occur in most Proterozoic domains of Australia (Fig. 1) and, in many, constitute almost 30% of the exposed rocks. Within these domains, felsic igneous rocks occur at specific time intervals and are not distributed evenly through time (Page 1988); the chemical compositions also appear to systematically change with time (L. Wyborn *et al.* 1987, 1988a). This paper seeks to characterise Australian Proterozoic granite and felsic volcanic compositions and to compare them with their Archaean and Phanerozoic analogues. The compositions will be discussed in relation to lower crustal structure, using data from seismic refraction surveys. The term granite is used in a broad sense and will incorporate syenogranite, monzogranite, granodiorite, diorite, and tonalite, as well as their comagmatic volcanic rocks. Most of the data in this paper, unless otherwise referenced, are synthesised from the whole-rock geochemical database ROCKCHEM held by the Bureau of Mineral Resources, Geology and Geophysics, Australia (BMR) which contains 24,000 whole-rock analyses from Australia and Papua New Guinea. Likewise, unless individually referenced, all geochronological age determinations are from Page (1988) and L. Wyborn *et al.* (1987).

1. Classification of Proterozoic felsic magmas

In comparing felsic magma compositions in this paper, several different chemical classifications are used in order to highlight the differences among the various granite types. Although some granites may have a direct mantle derivation, the majority of granites are interpreted to be derived by partial melting of lower crustal sources (e.g. Chappell & Stephens 1988). Thus, the chemical composition of any individual granite is likely to be the end product of the combination of many geological events, and the observed chemical signature may result from the source or sources and/or the pressure-temperature conditions at the time of formation.

1.1. Multi-element variation diagram

One of the most useful diagrams for comparing granites is a modification of the multi-element, primordial-mantle-normalised abundance diagrams published by Sun and McDonough (1989), in which the elements are ordered in increasing compatibility with respect to the mantle source from left to right. In this paper, the main modifications to the diagram are that, first, La, Ce and Nd are the only rare earths used and, second, Na is added at the right-hand end of the element axis (Fig. 2). As the primordial mantle is the *ultimate* (terrestrial) source of all crustal magmas, all elements are normalised to the primordial mantle so as to assess the degree of chemical fractionation of the granites from this original common source. The diagrams, therefore, represent the integrated effects of all the episodes of melting and other processes of fractionation that have contributed to the production of the granites. As pointed out by Tarney *et al.* (1987), rather simple and consistent patterns emerge using this technique.

Using this diagram, Tarney *et al.* (1987) and Wyborn *et al.* (1988b) showed that the granites containing less than 70% SiO_2 can be divided into two groups: Sr-undepleted and Y-depleted, or Sr-depleted and Y-undepleted. These two groups were interpreted by Tarney *et al.* (1987) and Wyborn *et al.* (1988b) to represent a difference between granites which at some stage of their generation had garnet stable in their source regions (Y-depleted, Sr-undepleted), and those granites which had plagioclase-bearing source regions (Y-undepleted, Sr-depleted), i.e. the source regions, at the time of partial melting, contained either plagioclase or garnet, but rarely both. Both groups are, in all cases, Nb-depleted.

In granites containing more than 70% SiO_2, fractionation can affect these multi-element patterns, and this may lead to erroneous conclusions. For example, plagioclase separation, either by fractional crystallisation or restite separation, will result in Sr-depletion in more felsic granites, while Y may concentrate substantially with fractionation. In Figure 2, the range of SiO_2 compositions is also given for each granite suite as an indication of whether or not fractionation might have affected the multi-element diagram.

1.2. S-type versus I-type

The Proterozoic granites under discussion will be subdivided on the S(edimentary)- and I(gneous)-type classification of Chappell and White (1974), which was later modified into S(upracrustal)- and I(infracrustal)-types by Chappell and White (1984, 1992) depending on whether the source had or had not been exposed to supracrustal processes. In the Proterozoic rocks, the more mafic S-types contain garnet and/or cordierite and/or muscovite, while the more mafic I-types contain hornblende and rarely clinopyroxene. In S-type granites of any age, the alumina saturation index (ASI; Zen 1986) (mol $Al_2O_3/(CaO + K_2O + Na_2O)$) is $>1{\cdot}1$ at levels of $<72\%$ SiO_2. The cut-off value of 72% is to distinguish between those I-type metaluminous granites which may fractionate to ASI values $\sim 1{\cdot}1$ at SiO_2 values $>72\%$ SiO_2.

1.3. Range of SiO_2 within a granite/felsic volcanic system

The total range of SiO_2 content of any coeval igneous system can be quite variable. The granites may be unimodal, with either a range in SiO_2 contents from 48–70% and a peak at around 54–56%, or else a range in SiO_2 contents from 58–78% and a peak of 72–75% SiO_2. In the classification of I-type granites of Chappell and Stephens (1988), the former group would represent of the I-(tonalitic) type while the latter group would be the I-(granodioritic) type. Alternatively, felsic magmas can be part of a bimodal

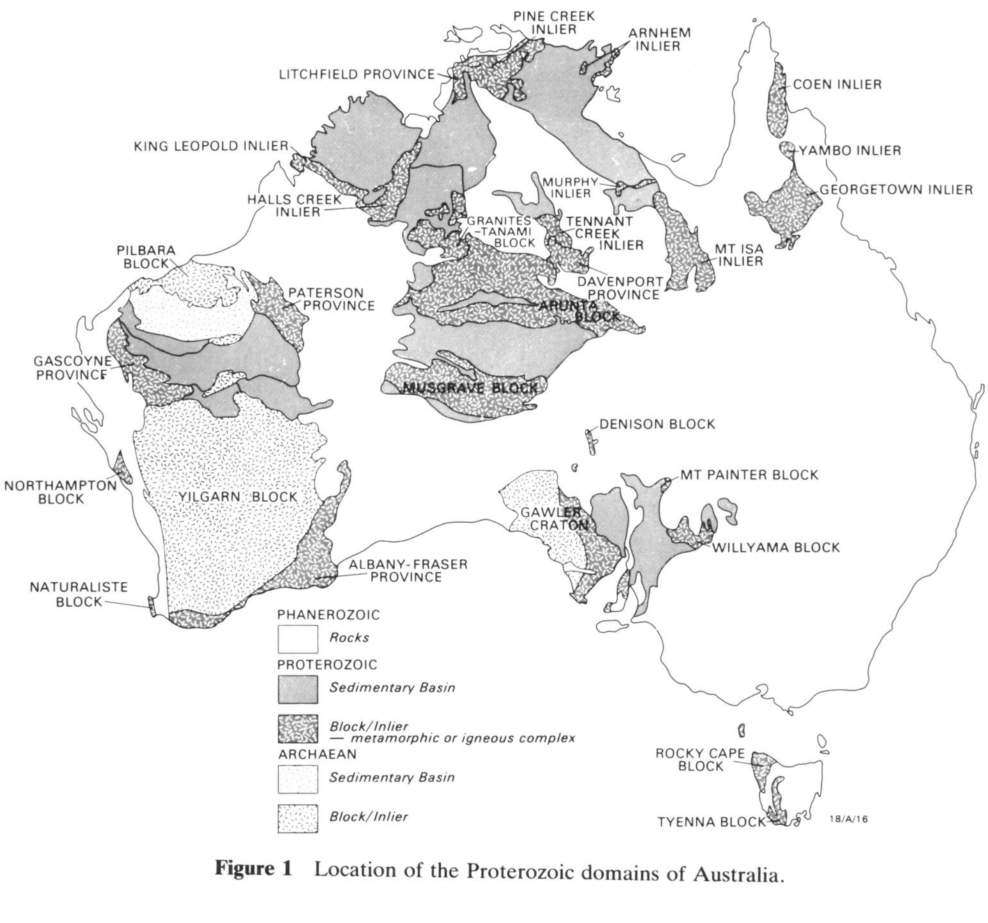

Figure 1 Location of the Proterozoic domains of Australia.

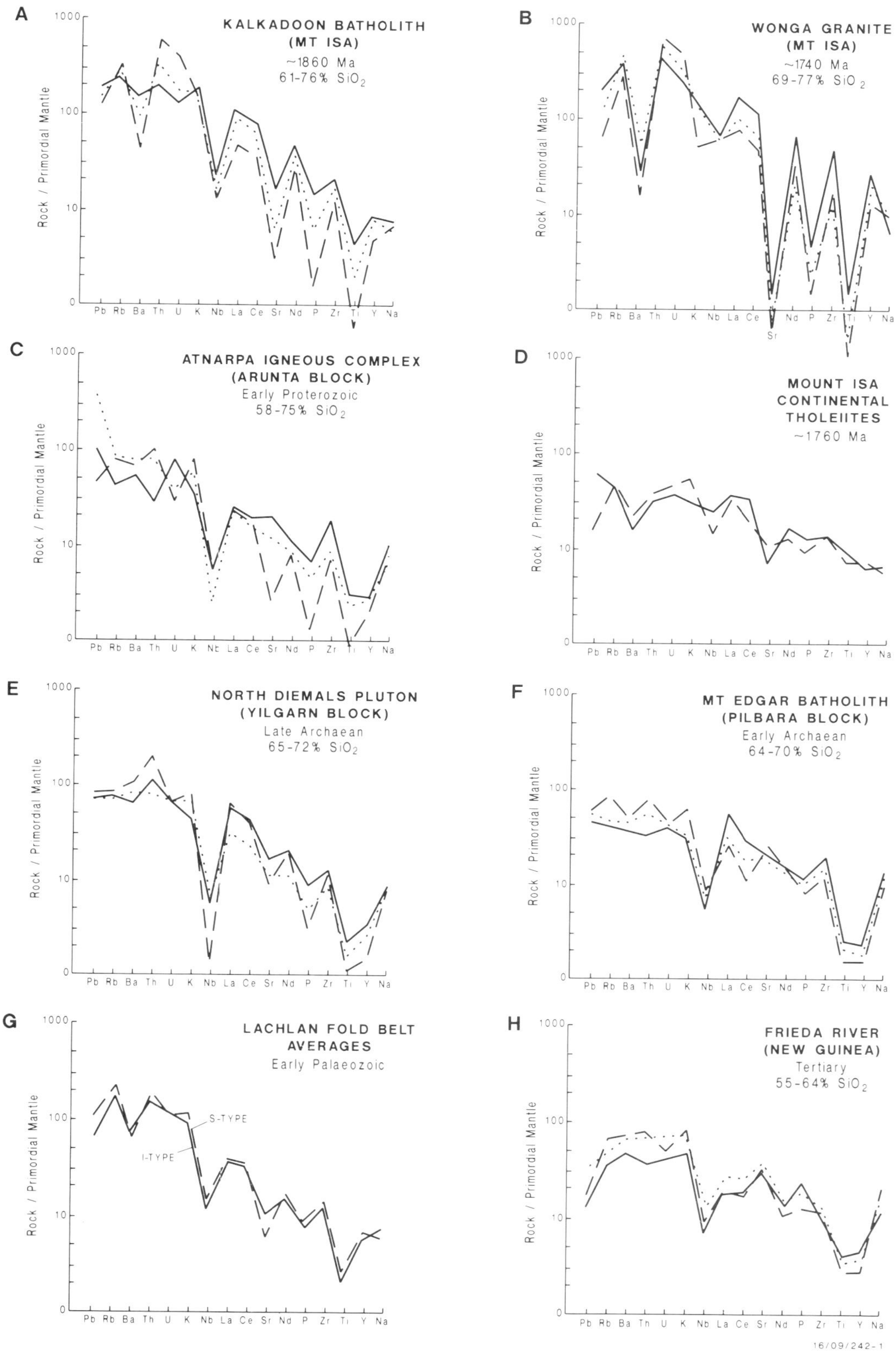

Figure 2 Multi-element primordial-mantle-normalised abundance diagrams for representative granites from the Australian continent; normalising values are from Sun & McDonough (1989); in (D) the dashed line and solid line are two individual samples; in (G) the dashed line is an average S-type and the solid line an average I-type granite of the Lachlan Fold Belt; in all other figures the solid line is the lowest SiO_2 content, the dotted line an intermediate SiO_2 content and the dashed line the highest SiO_2 content.

igneous event in which there are coeval mafic igneous rocks present; in these bimodal suites there is a general absence of rocks with SiO_2 between 52% and 58%.

1.4. Identification of different crystallisation regimes

Most granites are generated by partial melting. This process can produce varying amounts of melt depending on the *P–T*-fluid conditions at, and the composition of, the source region. In addition, granites may contain abundant individual crystals of restite (White & Chappell 1977; Chappell *et al.* 1987) and are, therefore, two-component mixes of melt and restite. The proportion of the two end-members can vary from those granites that are pure liquids to those that contain abundant restite.

If a granite is restite-rich, then the crystallisation regime is dominated by freezing of the interstitial liquid; comagmatic volcanic rocks consist of the same crystal-liquid mix (Wyborn & Chappell 1986). Restite-dominated plutons can be distinguished from melt-dominated ones by near-constant K/Rb ratios over the range of SiO_2, and by comagmatic volcanic suites that have compositions identical to the intrusive granites.

Unlike restite-rich granites, ones that intrude as liquids can undergo crystal-liquid fractionation. Each individual pluton will have a distinctive composition, and the comagmatic volcanic rocks, in turn, will have compositions distinct from, and generally more siliceous than, their cogenetic intrusions (Wyborn & Chappell 1986; D. Wyborn *et al.* 1987). These individual plutons can consist of mappable bodies of leucogranite, small zoned plutons and essentially homogeneous bodies of granodiorite to monzogranite composition. Chemically, the granites of this melt-rich type are distinguished by decreasing K/Rb with increasing SiO_2, exponentially increasing values of incompatible elements such as Rb and U at values of >72% SiO_2, and by general lack of chemical cohesion between individual members of the suite, both intrusive and extrusive.

2. Major granite types in the Australian Proterozoic

Based on the above criteria, the major Proterozoic felsic magma systems within the Australian continent can be divided into five main groups: (i) I-type, Sr-depleted, Y-undepleted, restite-dominated; (ii) I-type, Sr-depleted, Y-undepleted, fractionated, low in incompatible elements; (iii) I-type, Sr-depleted, Y-undepleted, enriched in incompatible elements (anorogenic); (iv) I-type, Sr-undepleted, Y-depleted; (v) S-type, Sr-depleted, Y-undepleted. Granites of a particular group appear to dominate during a certain time interval, but are not necessarily exclusive to it.

2.1. I-type, Sr-depleted, Y-undepleted, restite-dominated granites

2.1.1. Distribution, age and geological setting. Although felsic magmas of this group are geographically fairly widely distributed, their ages appear to be primarily 1870–1850 Ma. The group includes the Kalkadoon Granodiorite and Leichhardt Volcanics (Mount Isa Inlier), the Nimbuwah Complex (Pine Creek Inlier), the Whitewater Volcanics and comagmatic granites of the King Leopold Inlier (Griffin 1991) and the Halls Creek Inlier, the Ali Curung and Alarinjela Complexes (Arunta Block) and the Tennant Creek and Cabbage Gum granites (Tennant Creek Inlier). On geological evidence, the felsic magmas were emplaced immediately following a low-*P*, high-*T* metamorphic and deformational event. Most of the comagmatic volcanic rocks are subaerial. There is no known mineralisation associated with any member of this group.

2.1.2. Petrological and chemical characteristics. The granites in this group are dominated by granodiorite to monzogranite compositions and, where analysed, felsic volcanic rocks are chemically indistinguishable from their comagmatic intrusives. These rocks are considered I(granodiorite)-type because biotite, hornblende and rarely clinopyroxene are the mafic silicate minerals present and the majority of the granites have ASI values of <1·1. The SiO_2 values range from 57–77%, with a dominance in the range 68–72%; mafic igneous rocks are rarely associated with this group. Harker diagrams are characterised by linear trends, and the granites have near-constant K/Rb ratios with increasing SiO_2 (Wyborn & Page 1983; Wyborn 1988). The constant K/Rb ratios, combined with the great similarity between the intrusive and extrusive rock compositions, implies that fractionation occurred by restite unmixing. Multi-element patterns have a characteristic Sr-depleted, Y-undepleted trend throughout the total range of SiO_2 values (Fig. 2(A)), indicating that garnet had never been a major stable phase in the sources of these granites.

2.2 I-type, Sr-depleted, Y-undepleted, fractionated granites, low in incompatible elements

2.2.1. Distribution, age and geological setting. Granites of this type are known from two time periods. Most were emplaced during the period 1850–1800 Ma and include the following; the Cullen Batholith (Pine Creek Inlier), parts of the Nicholson Granite and the Cliffdale Volcanics (Murphy Inlier), the Ewen Batholith (Mount Isa Inlier), possibly the Harverson Suite (Arunta Block), unnamed intrusive porphyries and Bernborough Formation (Tennant Creek Block), the Lewis Granite and Winnecke Granophyre (Granites Tanami Block), the Minnie Creek Batholith (Gascoyne Province) and the Donnington Suite (Gawler Craton; Mortimer *et al.* 1988). A second group was emplaced around 700 Ma and includes the Mount Crofton Granite Complex (Paterson Province; Goellnicht *et al.* 1991). These granites form complex outcrop patterns and many distinct phases are mapped within them. In addition, in any province where the restite-dominated granites also occur, this fractionated group is always younger. This group is the most metallogenically significant among Australian Proterozoic granitic rocks. Some granites contain vein Sn, W, Cu and U deposits, while Au deposits are located near the contact aureoles (e.g. Telfer, Goellnicht *et al.* 1991; Cullen Mineral Field, Stuart-Smith & Needham 1984).

2.2.2. Petrological and chemical characteristics. These granites have a much wider SiO_2 range than the restite-controlled group and can include cumulate quartz gabbros and diorites as the more mafic end-members with SiO_2 as low as 52%. However, the dominant composition is granodiorite to monzogranite, containing between 72% and 76% SiO_2. The felsic magmas have I-(granodioritic) type affinities. For rocks in which $SiO_2 < 70\%$, biotite and hornblende are the mafic silicate minerals. The more mafic granites have ASI values of 0·9, but the more highly siliceous, fractionated rocks have ASI values of up to 1·2 and the chemical variation within these highly siliceous granites is controlled by fractional crystallisation processes. Rb, Y, Li, and U increase rapidly with increasing SiO_2, and the contents of these elements at equivalent SiO_2 values are often double those of the restite-controlled suites; K/Rb also decreases with increasing SiO_2 in this fractionated group. The multi-element pattern is similar to the I-type restite-controlled granites in that they are Sr-depleted and Y-undepleted.

2.3. I-type, Sr-depleted, Y-undepleted, enriched in incompatible elements (anorogenic granites)

2.3.1. Distribution, age and geological setting. Proterozoic granites and volcanic rocks of this type are widely distributed throughout Australia in both space and time, although they do appear to be more voluminous during the period from 1800–1500 Ma. While the granites of this group are termed anorogenic, they are always coeval with major extensional events, although they are laterally displaced from the maximum area of surface rift development (Wyborn *et al.* 1988a). In any province where these igneous rocks enriched in incompatible elements occur, they are always preceded by the less enriched I-type granites. Wyborn *et al.* (1988a) argued that in the Mount Isa Inlier the source of these enriched granites was relatively anhydrous, mantle-derived, lower crust that was underplated during earlier Proterozoic extensional or compressional events.

2.3.2. Petrology and geochemistry. At least three distinct but related subgroups occur. The first subgroup has very high F, Zr, Nb, and Y contents relative to the older I-type granites in the same region, and is invariably associated with mafic intrusions. This subgroup consists predominantly of extrusive rocks and ranges in SiO_2 content from 63–77% with a peak in the 69–72% interval. This subgroup is predominantly 1800–1780 Ma in age and includes the Argylla Volcanics and Bottletree Formation (Mount Isa Inlier), Myola Volcanics (Gawler Craton), and the Tidnamurkana Volcanics (Denison Block). The 1060 Ma Tollu Volcanics (Musgrave Block; Giles 1981) are also chemically similar. In the Mount Isa Inlier, this subgroup was interpreted by Bultitude and Wyborn (1982) to have formed by remelting of the anhydrous residue left behind after the removal of the I-type magma and was considered to consist of true A-types, as defined by Collins *et al.* (1982); this subgroup contains the only Proterozoic igneous rocks in Australia that possibly fit this definition.

The second subgroup also has high F values, but the levels of the incompatible elements such as Y, Zr, and Nb are variable, being high in some suites and low in others. There is evidence of fractionation in these suites, with distinctly high silica plutons that are the most enriched in elements such as K_2O, F, Th, U, Rb; K/Rb decreases with increasing SiO_2. This subgroup is also associated with coeval mafic intrusives but, unlike the preceding subgroup, is not spatially related to earlier I-types. This group includes the Malone Creek Granite (Pine Creek Inlier), the Wonga and Sybella granites (Mount Isa Inlier), the Mount Swan, Mount Ida and Ennugan Mountains granites and unnamed orthogneisses (Arunta Block), the McGregor Volcanics, Moonta Porphyry and Moody Suite granites (Gawler Craton).

The third subgroup is not as highly enriched in F, and is characterised by high $Fe_2O_3/(FeO + Fe_2O_3)$ relative to other Proterozoic granites. The granites are around 1600–1450 Ma in age, and the occurrence of extensive areas of granite breccias and hematite-rich zones is characteristic; late K- or Na-rich alteration is also common. Granites of this subgroup have undergone fractional crystallisation, and there is a considerable range in SiO_2 values from 52–78%, with a peak SiO_2 content of 69–74%. These granites have higher Na_2O contents than other Proterozoic Sr-depleted granites and the most fractionated granites are extremely high in U, with values up to 20 ppm being quite common. The two best examples of this subgroup are the Williams Batholith (Mount Isa Inlier) and the Gawler Range Volcanics, Hiltaba and Charleston granites, Olympic Dam Suite (central Gawler Craton and adjacent Stuart Shelf; Creaser 1989; Oreskes & Einaudi 1990). This latter suite is associated with the significant Olympic Dam Cu-U-Au-Ag deposit.

The multi-element patterns for all subgroups are relatively similar and are characterised by extreme enrichments of Rb, Th, and U and depletions of Sr, P and Ti (Fig. 2B) relative to the previous two groups.

2.4. I-type, Sr-undepleted, Y-depleted

2.4.1. Distribution, age and geological setting. These granites are extremely restricted in distribution and occur as relatively small suites in only three provinces. They include the Marramungee Granite (Mount Isa Inlier), the Dougalls, Ord River and Mabel Downs Tonalites (Halls Creek Inlier; Ogasawara 1988) and the Huckitta Tonalite (Foden *et al.* 1988), the Alice Springs Granite and the Atnarpa Igneous Complex (Arunta Block). In any of these areas, this Y-depleted group appears to be the earliest granite intruded.

2.4.2. Petrological and chemical characteristics. Compositionally, these granites range from tonalites to trondhjemites and leucogranites. The range in SiO_2 is from 55–75% with a peak in SiO_2 content at around 65%; they are classified as I-(granodioritic) type. The granites are typically low in incompatible elements and are characterised by Na $\gg$ K. This group has very distinctive multi-element patterns (Fig. 2(C)), with strong depletion of Y in all samples. Sr is undepleted in the mafic rocks, but becomes depleted in the related felsic rocks as a result of plagioclase fractionation. Moreover, the Y-depletion indicates that garnet was stable in the source region of these granites.

2.5. S-type granites, Sr-depleted, Y-undepleted

2.5.1. Distribution, age and geological setting. Proterozoic S-type granites may be more difficult to identify positively than younger S-types, as their sources tend to be predominantly feldspathic sediments which were not as chemically evolved as the sources of Palaeozoic S-types (Wyborn *et al.* 1988a). S-type granites are found to be 1880–1500 Ma in age and include granites in the Litchfield Province (Pietsch & Edgoose 1988), the Bradshaw Granite (Arnhem Block), the Big Toby Granite and numerous small, unnamed, high-K, muscovite microgranites (Mount Isa Inlier), the Forsyth and Esmeralda granites and Croydon Volcanics (Georgetown Inlier), the Devils Marbles and Elkedra granites (Davenport Province), the Jinka and Barrow Creek granites (Arunta Block) and the Potosi Gneiss (Broken Hill Block).

2.5.2. Petrology and geochemistry. All granites of this group contain biotite, with varying amounts of garnet, cordierite and muscovite. Rocks containing less than 72% SiO_2 have ASI values from 1·1–1·3. Some of the Proterozoic S-type granites are closely associated with migmatites and high-grade regional metamorphic rocks and are typically restite-controlled granites; others intrude into low metamorphic grades and some of these granites are strongly fractionated (e.g. Barrow Creek Granite), with decreasing K/Rb and increasing Li and Rb with increasing fractionation. The SiO_2 range is from 65–77% with a dominance in the range of 74–77%. The multi-element patterns are typically Sr-depleted, Y-undepleted, with strong depletions also in P, Ti and Ba.

3. Comparison of Proterozoic granites with granites of other ages

3.1. Comparison between Proterozoic and Archaean granites

A distinct contrast between the Archaean and Proterozoic granites of Australia has already been noted by Groves *et al.* (1987). Using multi-element diagrams, the majority of late Archaean I-type granites of the Yilgarn Block (data from Bettenay 1988) can be divided into three chemically distinct groups. In contrast to Proterozoic granites, the majority of the late Archaean granites are dominated by Y-depleted patterns. The first Archaean group identified comprises the early gneisses (SiO_2 range 65–75% SiO_2) which are characterised by Y-depletion and no Sr depletion, even at high SiO_2 values. The second group, which intrudes these gneisses, has a SiO_2 range from 50–75% with a peak of 68–70%; it is also characterised by Y-depletion but, as with the Proterozoic Y-depleted granites, in this group the degree of Sr-depleted changes from being low at low SiO_2 to high at high SiO_2 (Fig. 2(E)). The third Archaean group is typical of some of the youngest granites: the granites in this group have distinctive Sr-depleted, Y-undepleted patterns and are very similar to the majority of Proterozoic granites, in particular to the anorogenic suites. However, spatially this third group is minor. Late Archaean and Proterozoic granites are similar in that both are enriched in K_2O, Th and U (Fig. 2) but, although the K_2O values are similar in the two groups, the Na_2O values, at equivalent K_2O contents, are about 1% higher for the late Archaean rocks (Wyborn 1988).

Multi-element diagrams for the early Archaean granites from the Pilbara region show the same three types of patterns as found in granites of the late Archaean Yilgarn Craton, with again the Y-depleted types dominating. The early Archaean tonalites are characteristically Sr-undepleted and Y-depleted (Fig. 2(F), data from Collins 1984). Other granites show patterns which vary from Sr-undepletion, Y-depletion for more mafic compositions to Sr- and Y-depletion for the more felsic rocks. Late, high-SiO_2 granites, which are of small areal extent, show patterns which are Sr-depleted, Y-undepleted, similar to Proterozoic anorogenic granites. An important distinction of the early Archaean granites is that they are not as enriched in K_2O, Th or U as the late Archaean to Proterozoic ones (Fig. 2), although the Na_2O values in the early and late Archaean groups are similar.

3.2. Comparison of Proterozoic granites with Phanerozoic granites

3.2.1. Early Palaeozoic. Both S- and I-type early Palaeozoic granites from the Lachlan Fold Belt are characterised by Sr-depleted, Y-undepleted patterns and therefore resemble the typical Proterozoic granites (Fig. 2(G), averages plotted from Chappell & White 1984). The Palaeozoic granites also have similar SiO_2 ranges, with most granites being more felsic than 60% SiO_2, but they are not as enriched in K, Th and U as the Proterozoic or late Archaean granites (Fig. 2). A conspicuous exception is the Palaeozoic Moruya Batholith (Griffin *et al.* 1978) which is Sr-undepleted, Y-depleted at more mafic compositions, but which becomes Sr-depleted with increasing SiO_2, and so is similar to the Proterozoic Sr-undepleted, Y-depleted type (as in Fig. 2(C)). Palaeozoic A-types have multi-element patterns similar to the anorogenic suites in the Proterozoic, but are volumetrically less significant.

3.2.2. Late Palaeozoic-Mesozoic-Cainozoic. The patterns of younger granites are more diverse than those of any other age, presumably a reflection of the greater diversity of lower crustal compositions. The Y-depleted type becomes increasingly abundant in Australia, in the late Palaeozoic New England Fold Belt. On multi-element diagrams, M-type granites related to contemporaneous subduction in a young arc system, such as the Frieda River Complex of Papua-New Guinea (Whalen *et al.* 1982) are characterised by Sr-undepleted, Y-depleted granite throughout the complete silica range (Fig. 2(H)). These granites are more mafic than most Precambrian and Palaeozoic suites, with a range from 55–64% SiO_2. These granites have low K_2O, Th and U abundances like those of the early Archaean.

4. Changes in Australian granite compositions through time

From the above data, there appear to be consistent patterns of granite composition that change with time throughout the Australia continent. The noteworthy geochemical aspects that change systematically with time are summarised below:

Proterozoic and early Palaeozoic granites in Australia are dominated by Sr-depleted, Y-undepleted signatures. I-type granites of those ages are dominantly I-(granodioritic) type. In contrast, the Archaean and the late Palaeozoic to Cainozoic granites are characterised by Y-depleted signatures, and in the more mafic rocks there is no Sr depletion. In the late Archaean and Palaeozoic, these granites, as a result of plagioclase fractionation, may become Sr-depleted with increasing SiO_2.

Anorogenic or A-type granites of any age are characterised by Sr-depleted, Y-undepleted patterns. This type of granite is abundant in the Proterozoic, but is less common in both older and younger terranes.

The Na_2O/K_2O ratio of granites changes with time; the Archaean and late Palaeozoic to Mesozoic granites contain higher Na than their Proterozoic and early Palaeozoic counterparts. Late Archaean and Proterozoic granites characteristically have the highest K_2O, Th and U values, with Palaeozoic granites having the next highest values; early Archaean and Cainozoic granites have the lowest values.

5. Derivation of Australian Proterozoic granites

The majority of Australian Proterozoic granites are Sr-depleted, Y-undepleted and characteristically have high levels of K, Th and U. In any area, the granites that are low in incompatible elements usually precede the later anorogenic granites which are enriched in incompatible elements.

The source of Australian Proterozoic granites is enigmatic. As is shown in Figure 2, most Archaean granites and gneisses are Y-depleted, and hence cannot be the source of the voluminous Proterozoic granites. This is also supported by available isotopic data which indicate that, although the majority of Australian Proterozoic granites have a recognisable crustal prehistory, all sources were formed in the Proterozoic (McCulloch 1987). Thus, even the S-types cannot be derived from Archaean sediments or Archaean-derived sediments. Moreover, the chemical composition of the granites precludes a direct derivation from the mantle (L. Wyborn *et al.* 1987). The multi-element patterns of the majority of Australian Proterozoic granites are also unlike those from modern subduction zones. Therefore, subduction, at least as we see it today, apparently did not play a major role in the generation of these Sr-depleted, Y-undepleted granites.

Etheridge *et al.* (1987) and L. Wyborn *et al.* (1987) suggested that most Proterozoic I-type granites were derived from fractionated mantle-derived underplates that were added successively to the base of the crust during major Proterozoic tectonothermal events. In the Mount Isa Inlier, Wyborn *et al.* (1988a) suggested that the source for one of the major anorogenic I-type granites were underplated during an earlier magmatic event during which enriched continental tholeiites were emplaced. The late I-type granite only occurs where these enriched tholeiites occur (Wyborn *et al.* 1988a). These enriched tholeiites also have the Sr-depleted, Y-undepleted pattern (Fig. 2(D)), that is characteristic of the majority of Proterozoic granites (Fig. 2(A), (B)), and hence some of the I-type granites could be derived by partial melting of similar sources.

Another attraction of the underplating hypothesis is that, as L. Wyborn *et al.* (1987) have noted, the composition of mafic igneous rocks evolved with time during the Proterozoic, with the degree and depth of melting changing consistently from early, high-temperature, depleted tholeiites, to enriched tholeiites, to later alkalic rocks. Thus, the change in compositions of I-type granites with time could simply reflect the changing composition of the underplated material. A problem of this model is that to derive the granites from these sources would require vast volumes of such underplated material.

6. Relationship of granite compositions to lower crustal structure as determined for seismic refraction surveys

Seismic refraction profiles, which provide information on the seismic velocity structure of the crust, show consistent changes with the age of the Australian crust, and provide evidence for the underplating model proposed for the Proterozoic of Australia (Drummond *et al.* 1988). The seismic refraction data clearly show a high-velocity layer characterised by seismic velocities of 7·4–7·6 km/s underneath Proterozoic provinces, which has been interpreted by Drummond and Collins (1986) to be largely underplated mafic material in the granulite stability field. Wyborn *et al.* (1988b) assumed this layer to be the source region for the majority of Proterozoic I-type granites. It is about 15 km thick in Proterozoic and Palaeozoic provinces and thinner in younger provinces (Fig. 3). It is made up of at least 45% mafic granulite and is only apparent in provinces where the granites are dominated by plagioclase-stable source regions (Sr-depleted, Y-undepleted). This layer is believed to have formed by progressive underplating of the lower crust during major tectonothermal events, particularly extensional events, as these often involve the intrusion of major volumes of mafic mantle-derived material.

In contrast, this high-velocity layer (7·4–7·6 km/s) is not found beneath island arc or continental margin environments (Gill 1981), nor is it present in Archaean terranes (Drummond 1988), where I-type granites characteristically have a Sr-undepleted, Y-depleted multi-element pattern (Fig. 3). This has been taken to imply that at some stage in the development of these granites, garnet was stable in their source region. However, seismic refraction profiles cannot delineate a substantial source region within the present-day lower crust where garnet could be stable. Therefore the initial source region for these granites was probably the upper mantle.

Enriched, so-called anorogenic, I-type granites are also more prominent in the Proterozoic than at any other period, and are also thought to have been derived by partial melting of this high-velocity underplated layer. Although the mafic lower crustal layer is present in both Proterozoic and Palaeozoic terranes, major crustal rift basins containing substantial amounts of mantle-derived continental tholeiite are typical of the Proterozoic (L. Wyborn *et al.* 1987; Wyborn 1988) and appear less common in the Palaeozoic. Hence, their suggested lesser abundance in the Palaeozoic could imply that there was less suitable material added to the lower crust and/or that there were insufficiently high temperature crustal events to melt this layer. In contrast, anorogenic granites are not a major granite type in Archaean areas, nor are they abundant in Papua-New Guinea; that is, in areas which were not extensively underplated.

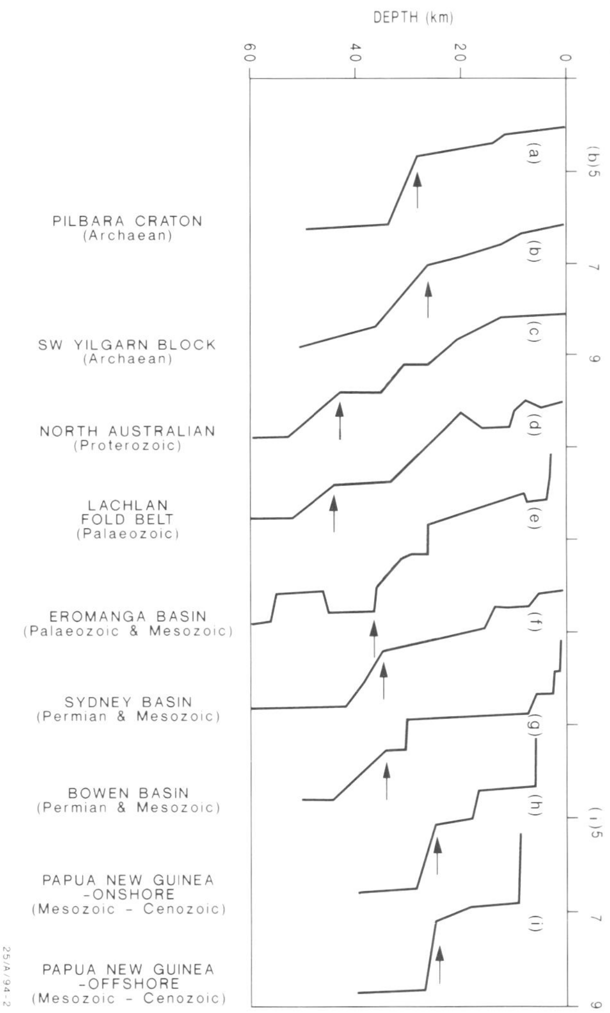

Figure 3 Representative P-wave velocity versus depth models for nine regions in order of decreasing age (after Drummond & Collins 1986): velocities have all been adjusted to 1 GPa and 25°C; shaded areas show regions in the crust where the velocity is greater than 7·0 km/s and the arrows indicate the interpreted base of the crust where the velocity is greater than 8·0 km/s; velocity scale 5–7–9 (km/s) steps across the diagram for successive profiles, left to right, as shown by (a) referring to profile (a), (b) to profile (b) and so on.

7. Rationalising the changes in Australian granite compositions

This review of granite compositions has shown that there are three major systematic changes with time in Australian granite compositions. The first is a change from Y-depleted granites in the Archaean to Y-undepleted granites in the Proterozoic and early Palaeozoic, back to Y-depleted granites in the late Palaeozoic to Cainozoic. This change probably reflects a transition from granites whose prehistory involved, at some stage, melting in the garnet stability field (Y-depleted) to sources that were dominated by plagioclase (Sr-depleted).

The two distinct patterns, dominated by either garnet or plagioclase in the source, are obviously related to depth of partial melting, with the plagioclase-bearing sources occurring at shallower levels than the garnet-bearing sources. The latter would have resided at depths greater than 45 km, and probably greater than 60 km (Gromet & Silver 1987). In Australia, the plagioclase-dominated sources fit well with the underplating models proposed for the Proterozoic (e.g. Wyborn 1988). The presence of garnet in the source region does not necessarily imply contemporaneous subduction; rather it implies melting at a particular depth, which may or may not have been generated by subduction. In the Proterozoic terranes, the garnet-sourced granites are a minority, and are generally the oldest granites in a particular area. Rather than suggesting that subduction was directly involved in the generation of these granites, it is plausible that initially the thermal anomaly that caused the major crustal melting event was at a greater depth than for the later and more voluminous granites which came from shallower, plagioclase-dominated, source regions. Further, the presence of Y-depleted granite does not necessarily mean that the melting event which produced the granites took place at a depth of more than 45 km. The depletion could have been generated in an earlier partial melting episode which underplated appropriate source material in the lower crust. Any granite generated from that material would have the Y-depleted signature unless significant fractionation occurred during crystallisation. Such a scenario of partial melting and granite formation reproducing the composition of the source rocks (remagmatisation) has been proposed for the generation of the Palaeozoic Moruya Batholith (Chappell & Stephens 1988).

The second systematic change with time in Australian granite compositions is that of the Na content of the large I-type batholiths, with Na being most abundant in the Archaean and Cainozoic granites, least in the early Proterozoic, and intermediate for granites of other ages. This transition in Na content is probably coupled with the change from garnet to plagioclase dominance in the source regions of the granites, with Na being lower in amount in those granites that were derived from plagioclase-bearing source regions.

The third systematic change is the variation in K, Th, and U contents in I-type granites from depletion in the early Archaean, to strong enrichment in the late Archaean and Proterozoic, followed by progressive depletion to levels in the Cainozoic which are equivalent to those in the early Archaean. This change is enigmatic. We speculate that it is possible that these elements were held in stable phases in the early Archaean mantle, began to be transferred to the crust in the late Archaean, and have been progressively depleted in the mantle thereafter.

8. Acknowledgements

This paper has benefited greatly from detailed comments by John Sheraton and Charlotte Allan. The paper is published with permission of the Executive Director, Bureau of Mineral Resources, Geology and Geophysics, Australia.

9. References

Bettenay, L. 1988. Granitoid batholiths of the Eastern Goldfields Province, Yilgarn Block: characteristics and significance to gold mineralisation. *In* Ho. S. E. Groves, D. I. (eds) *Advances in understanding Precambrian Gold Deposits*, Vol 2. GEOL DEPT & UNIV EXTENSION, UNIV WEST AUST PUBL **12,** 227–37.

Bultitude, R. J. & Wyborn, L. A. I. 1982. Distribution and geochemistry of volcanic rocks in the Duchess-Urandangi region, Queensland. BMR J AUST GEOL GEOPHYS **7,** 99–112.

Chappell, B. W. & Stephens, W. E. 1988. Origin of infracrustal (I-type) granite magmas. TRANS R SOC EDINBURGH EARTH SCI **79,** 71–86.

Chappell, B. W. & White, A. J. R. 1974. Two contrasting granite types. PAC GEOL **8,** 173–4.

Chappell, B. W. & White, A. J. R. 1984. I-type and S-type granites in the Lachlan Fold Belt, southeastern Australia. *In* Xu Kequin & Tu Guangchi (eds) *Geology of granites and their metallogenic relations,* Proceedings of the International Symposium, Nanjing University, Nanjing, China, 87–101. Beijing: Science Press.

Chappell, B. W. & White, A. J. R. 1992. I-type and S-type granites in the Lachlan Fold Belt. TRANS R SOC EDINBURGH EARTH SCI **83,** 1–26.

Chappell, B. W., White, A. J. R. & Wyborn, D. 1987. The importance of residual source material (restite) in granite petrogenesis. J PETROL **28,** 1111–38.

Collins, W. J. 1984. *Geological Evolution of an Archaean Batholith.* Unpublished Ph.D. Thesis, La Trobe University.

Collins, W. J., Beams, S. D., White, A. J. R. & Chappell, B. W. 1982. Nature and origin of A-type granites with particular reference to southeastern Australia. CONTRIB MINERAL PETROL **80,** 189–200.

Creaser, R. A. 1989. *The geology and petrology of middle Proterozoic felsic magmatism of the Stuart Shelf, South Australia.* Unpublished Ph.D. thesis, La Trobe University.

Drummond, B. J. 1982. Seismic constraints on the chemical composition of the crust of the Pilbara Craton, Northwest Australia. REV BRASIL GEOSCI **12,** 113–20.

Drummond, B. J. & Collins, C. D. N. 1986. Seismic evidence for underplating of the lower continental crust of Australia. EARTH PLANET SCI LETT **79,** 361–72.

Drummond, B. J., Wyborn, L. A. I., Wyborn, D. & Tarney, J.F., 1988. Temporal changes in continental growth patterns revealed by seismic crustal structure and granite geochemistry. GEOL SOC AUST ABSTR **21,** 118–19.

Etheridge, M. A., Rutland, R. W. R. & Wyborn, L. A. I. 1987. Orogenesis and tectonic process in the early to middle Proterozoic of northern Australia. AM GEOPHYS UNION, GEODYNAMIC SER **17,** 131–47.

Foden, J. D., Buick, I. S. & Mortimer, G. E. 1988. The petrology and geochemistry of granitic gneisses from the east Arunta Inlier, Central Australia: implications for Proterozoic crustal development. PRECAMBRIAN RES **40/41,** 233–59.

Giles, X. X. 1981. *A comparative study of Archaean and Proterozoic felsic volcanic associations in Southern Australia.* Unpublished Ph.D. thesis, University of Adelaide.

Gill, J. B. 1981. *Orogenic andesites and plate tectonics.* Berlin: Springer.

Goellnicht, N. M., Groves, D. I. & McNaughton, N. J. 1991. Late Proterozoic fractionated granitoids of the mineralised Telfer area, Paterson Province, Western Australia. PRECAMBRIAN RES **51,** 375–91.

Griffin, T. J. 1991. Early Proterozoic calcalkaline magmatism in the Hooper Complex, west Kimberley, Western Australia. Abstracts for the Second Hutton Symposium on granites and related rocks, Canberra, 1991. REC BMR GEOL GEOPHYS **1991/25,** 41.

Griffin, T. J., White, A. J. R. & Chappell, B. W. 1978. The Moruya batholith and geochemical contrasts between the

Moruya and Jindabyne Suites. J GEOL SOC AUST **25,** 235–47.

Gromet, L. P. & Silver, L. T., 1987. REE variations across the Peninsular Ranges Batholith: implications for batholithic petrogenesis and crustal growth in magmatic arcs. J PETROL **28,** 75–125.

Groves, D. I., Wyborn, L. A. I., Rutland, R. W. R., Drummond, B. J., Etheridge, M. A. & Sun, S. S. 1987. Precambrian granites and crustal evolution in Australia. *Abstracts, international symposium on granites and associated mineralizations,* Salvador, Brazil.

McCulloch, M. T. 1987. Sm–Nd isotopic constraints on the evolution of Precambrian crust in the Australian continent. AM GEOPHYS UNION, GEODYNAMIC SER **17,** 115–30.

Mortimer, G. E., Cooper, J. A. & Oliver, R. L. 1988. The geochemical evolution of Proterozoic granitoids near Port Lincoln in the Gawler Orogenic Domain of South Australia. PRECAMBRIAN RES **40/41,** 387–406.

Ogasawara, M. 1988. Geochemistry of the early Proterozoic granitoids in the Halls Creek orogenic subprovince, Northern Australia. PRECAMBRIAN RES **40/41,** 469–86.

Oreskes, N. & Einaudi, M. 1990. Origin of rare earth element-enriched hematite breccias at the Olympic Dam Cu-U-Au-Ag deposit, Roxby Downs, South Australia. ECON GEOL **85,** 1–28.

Page, R. W. 1988. Geochronology of early to middle Proterozoic fold belts in Northern Australia: a review. PRECAMBRIAN RES **40/41,** 1–19.

Pietsch & Edgoose, 1988. The stratigraphy, metamorphism and tectonics of the early Proterozoic Litchfield Province and western Pine Creek Geosyncline, Northern Territory. PRECAMBRIAN RES **40/41,** 565–88.

Stuart-Smith, P. G. & Needham, R. S. 1984. Hydrothermal mineral deposits and their association with granitoids in the Cullen Mineral Field, Northern Territory, *Darwin Conference, Australasian Institute of Mining and Metallurgy,* 329–38.

Sun, S. S. & McDonough, W. F. 1989. Chemical and isotopic systematics of oceanic basalts: implications for mantle composition and processes. GEOL SOC LONDON SPEC PUBL **42,** 313–45.

Tarney, J., Wyborn, L. A. I., Sheraton, J. W. & Wyborn, D. 1987. Trace element differences between Archaean, Proterozoic and Phanerozoic crustal components—implications for crustal growth processes. *In* Ashwal, L. D. (ed.) *Workshop on the growth of continental crust.* LUNAR PLANET INST TECH REP **88·02,** 139–40.

Whalen, J. B., Britten, R. M. & McDougall, I. 1982. Geochronology and geochemistry of the Frieda River Prospect area, Papua New Guinea. ECON GEOL **77,** 592–616.

White, A. J. R. & Chappell, B. W. 1977. Ultrametamorphism and granitoid genesis. TECTONOPHYSICS **43,** 7–22.

Wyborn, D. & Chappell, B. W. 1986. The petrogenetic significance of chemically related plutonic and volcanic rock units. GEOL MAG **123,** 619–28.

Wyborn, D., Turner, B. S. & Chappell, B. W. 1987. The Boggy Plain Supersuite: a distinctive belt of I-type igneous rocks of potential economic significance in the Lachlan Fold Belt. AUST J EARTH SCI **34,** 21–43.

Wyborn, L. A. I. 1988. Petrology, geochemistry and age of a major Australian 1880–1840 Ma felsic volcano-plutonic suite: a model for intracontinental felsic magma generation. PRECAMBRIAN RES **40/41,** 37–60.

Wyborn, L. A. I. & Page, R. W. 1983. The Proterozoic Kalkadoon and Ewen Batholiths, Mount Isa Inlier, Queensland: source, chemistry, age and metamorphism. BMR J AUST GEOL GEOPHYS **8,** 53–69.

Wyborn, L. A. I., Page, R. W. & Parker, A. J. 1987. Geochemical and geochronological signatures in Australian Proterozoic igneous rocks. GEOL SOC LONDON SPEC PUBL 33, 377–94.

Wyborn, L. A. I., Page, R. W. & McCulloch, M. T. 1988a. Petrology, geochronology, and isotope geochemistry of the post-1820 Ma granites of the Mount Isa Inlier: mechanisms for the generation of Proterozoic anorogenic granites. PRECAMBRIAN RES **40/41,** 509–41.

Wyborn, L. A. I., Wyborn, D., Chappell, B. W., Sheraton, J., Tarney, J. F., Collins, W. J. & Drummond, B. J. 1988b. Geological evolution of granite compositions with time in the Australian continent–implications for tectonic and mantle process. GEOL SOC AUST ABSTR **21,** 434–5.

Zen, E-an, 1986. Aluminum enrichment in silicate melts by fractional crystallization: some mineralogic and petrographic constraints. J PETROL **27,** 1095–117.

L. A. I. WYBORN, D. WYBORN, R. G. WARREN and B. J. DRUMMOND, Bureau of Mineral Resources, Geology and Geophysics, GPO Box 378, Canberra City, ACT 2601, Australia.

Transactions of the Royal Society of Edinburgh: Earth Sciences, **83,** 211–226, 1992

Late Archaean granites of the southeastern Yilgarn Block, Western Australia: age, geochemistry, and origin

R. I. Hill, B. W. Chappell and I. H. Campbell

ABSTRACT: Late Archaean granitic rocks from the southern Yilgarn Craton of Western Australia have a close temporal relationship to the basaltic and komatiitic volcanism which occurs within spatially associated greenstone belts. Greenstone volcanism apparently began ~2715 Ma ago, whereas voluminous felsic magmatism (both extrusive and intrusive) began about 2690 Ma ago. A brief but voluminous episode of crust-derived magmatism ~2690–2685 Ma ago resulted in the emplacement of a diverse assemblage of plutons having granodioritic, monzogranitic and tonalitic compositions. This early felsic episode was followed immediately by the emplacement of mafic sills, and, after a further time delay, by a second episode of voluminous crust-derived magmatism dominated by monzogranite but containing plutons covering a wide compositional range, including diorite, granodiorite and tonalite. The products of this 2665–2660 Ma magmatic episode now form a significant fraction of the exposed southern Yilgarn Craton. Later magmatism, which continued to at least 2600 Ma ago, appears largely restricted to rocks having unusually fractionated compositions.

The magmatic sequence basalt–voluminous crust-derived magmatism–later diverse magmatism, is interpreted in terms of a dynamically-based model for the ascent of the head of a new mantle plume. In this model basalts and komatiites are derived by decompression melting of rising plume material, and the crust-derived magmas result after conductive transport of heat from the top of the plume head into overlying continental crust. This type of magmatic evolution, the fundamentally bimodal nature of the magmatism, the presence of high-Mg volcanics (komatiites), and the areal extent of the late Archaean magmatic event, are all suggested to be characteristic of crustal reworking above mantle plumes rather than resulting from other processes, such as those related to subduction.

KEY WORDS: crust-derived melts, mantle plumes, heat, conduction, basalt, komatiite, granite, granodiorite, monzogranite, greenstone, geochronology.

The development of a detailed understanding of the origin of the granitic rocks that comprise a significant proportion of the exposed area of the Archaean shields has been hampered by a lack of precise geochronological control both on the intrusive rocks themselves, as well as on spatially associated supracrustal rocks such as the greenstone belts which also comprise an important constituent of Archaean crust. The Yilgarn Craton of southwestern Australia is no exception to this generalisation. Rocks of broadly granitic composition, including gneissic as well as unfoliated, clearly intrusive granities, are estimated to constitute approximately 70% of the basement rocks of the Yilgarn Craton of southwestern Australia (e.g. Gee *et al.* 1981). The ages of most are known only in the most general terms. Early geochronological studies using the Rb–Sr and Pb–Pb methods indicated that the bulk of these rocks had probable emplacement ages of 2600–2700 Ma, although some were considerably older (Turek 1966; Arriens 1971; Oversby 1975). More recent zircon U–Pb studies provide confirmation of the earlier work and reinforce previous conclusions as to the areal significance of the late Archaean magmatic event; rocks having zircon U–Pb ages in the range 2600–2700 Ma are now known from across the craton (Fig. 1).

With the generally higher precision obtainable using ion microprobe zircon U–Pb techniques compared to the Rb–Sr and Pb–Pb methods (commonly ±5 to ±8 Ma versus ±30 Ma or more; errors quoted in this paper are the 95% confidence limits unless stated otherwise) it has been possible to demonstrate that at least one group of late Archaean granites was contemporaneous with felsic volcanism, that initiation of felsic magmatism in the southeastern Yilgarn about 2690 Ma followed closely after eruption of thick sequences of basaltic rocks, that there are two main felsic magmatic episodes separated in time by about 30 Ma, and that rocks of essentially identical composition formed at different times (Campbell & Hill 1988; Hill & Campbell 1992). Also, with the SHRIMP ion microprobe it has been possible to demonstrate the presence of much older (to 3450 Ma) xenocrystic zircon in samples from a range of the late Archaean rock types, including granodiorites, granites and tonalites, felsic volcanic rocks, and basalts (Compston *et al.* 1986; Claoué-Long *et al.* 1988; Campbell & Hill 1988; Hill *et al.* 1989; Hill & Campbell 1992). The presence of this old, probably inherited, zircon has been used to argue that the late Archaean felsic rocks were derived through partial melting of pre-existing continental crust (Campbell & Hill 1988; Hill *et al.* 1989). Although a similar conclusion had been arrived at earlier by Oversby (1975) on the basis of Pb isotope studies, her model was controversial, with a number of workers tending to discount the importance of crustal reworking in the genesis of the Late Archean granites (e.g. McCulloch 1987).

This paper provides a summary of our studies of the geochronology and geochemistry of the late Archaean intrusive rocks of the southern Yilgarn Craton. Most of our samples are from the Norseman-Wiluna Belt, a greenstone-dominated terrain that comprises much of the easternmost portion of the exposed Yilgarn (Gee 1979; Fig. 1). The geochronology shows that emplacement of the granites is an integral part of the greenstone-forming thermal event that dominates the supracrustal rocks of the southeastern Yilgarn; accordingly, a brief description of the greenstones is provided to give context for the data on the granites. Finally, we summarise a previously-described model in which the late Archaean evolution of this region is suggested to be the consequence of crustal reworking above the head of a mantle plume. Both this model and the approach on which it is based may have wider application.

1. Late Archaean geology of the southern part of the Norseman–Wiluna Belt, Western Australia

The oldest known rocks in the area are components within the so-called greenstone belts, sequences of altered basaltic ("greenstone"), felsic volcanic, and volcanogenic sedimentary rocks (e.g. Doepel 1973; Gemutz & Theron, 1975; Swager & Griffin 1990). The most complete section through the greenstone stratigraphy occurs in the vicinity of Norseman (Fig. 2), where from bottom to top the sequence is comprised of a lower unit (the Penneshaw Formation) which contains felsic volcanic rocks with an age of 2938 ± 10 Ma; interbedded banded iron formation, felsic volcanic and volcanogenic rocks (Noganyer Formation); an 8 km-thick sequence of pillowed basalts and mafic sills which includes a minor komatiitic component (Woolyeenyer Formation); and an upper sequence of felsic volcanic and associated sedimentary rocks (Mt Kirk Formation: age 2688 ± 8 Ma). An upper limit for the time of initiation of basalt eruption is given by the age (2714 ± 5 Ma) of an internally differentiated gabbroic sill interpreted to represent a feeder to overlying basaltic flows and which cuts the lower part of the Woolyeenyer Formation. The Mt Kirk Formation is cut by a number of sills of broadly gabbroic composition, and all units are cut by a variety of plutons, dykes and sills of broadly granitic composition.

The greenstone sequences of the Norseman area have been correlated with similar sequences to the N, at Kambalda, Kalgoorlie, and in the Ora Banda region (Swager & Griffin 1990). Although the lower two units (Penneshaw felsic volcanics and Noganyer banded iron formation and felsic volcanics) are recognised only at Norseman, the rest of the Norseman sequence of lower basalt–komatiite–upper basalt–felsic volcanics–mafic sills can be correlated over much of the southern part of the Norseman-Wiluna Belt, possibly as far N as Menzies and Mt Ida, respectively 300 and 400 km away (Swager & Griffin 1990; R.E.T. Hill *et al.* 1988). Available geochronological data (Campbell & Hill 1988; Claoué-Long *et al.* 1988) provide support for this correlation.

The supracrustal rocks have been folded, faulted, and

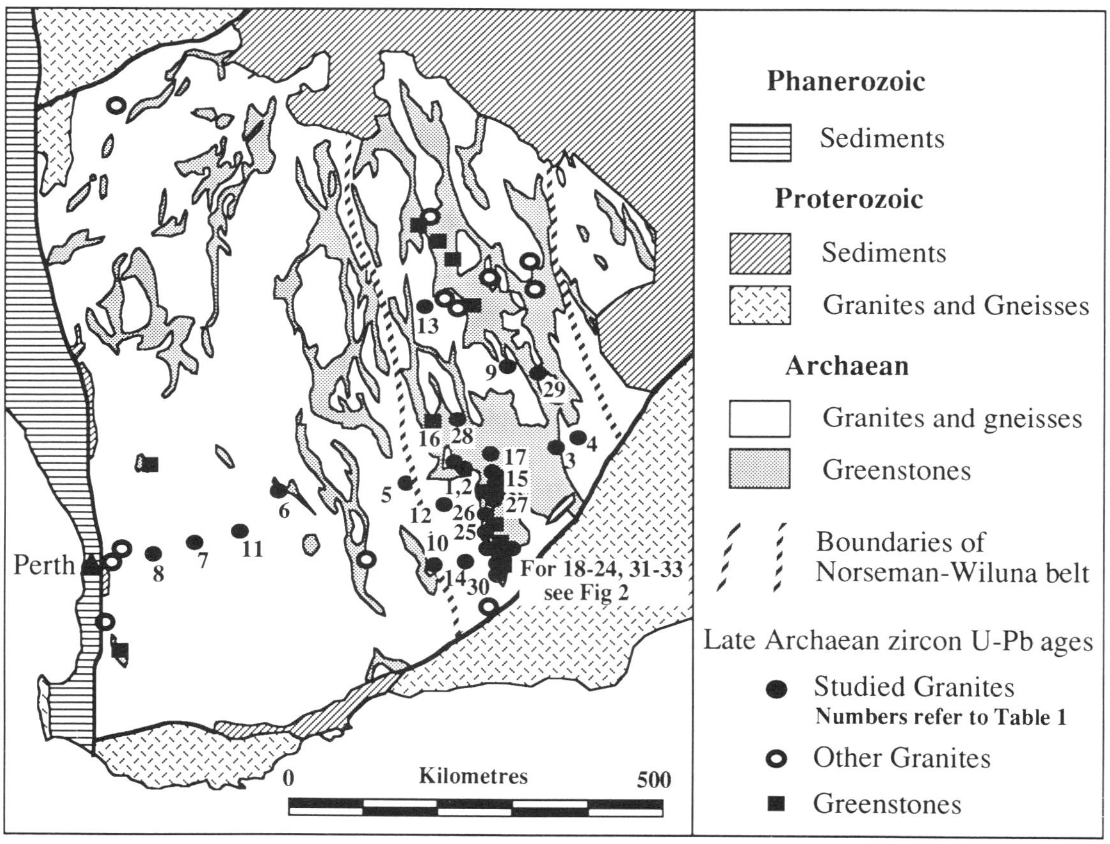

Figure 1 Sketch geological map of the Yilgarn Craton of southwestern Australia, showing location of studied granite samples (solid circles; numbers refer to Table 1), and other granites (open circles) and greenstone samples (solid squares) for which late Archaean zircon U–Pb ages are available; data from Table 1, Wilde & Pidgeon (1986), Claoué-Long *et al.* (1988), Compston *et al.* (1987), Kinny *et al.* (1990), Pidgeon *et al.* (1990), and our unpublished data; boundaries of Norseman-Wiluna Belt after Gee (1979).

intruded by compositionally diverse granites. Individual intrusions range in size from dykes and sills less than 1 m thick to large plutons having dimensions of tens of kilometres. In the vicinity of Norseman, the felsic volcanic rocks which comprise the upper part of the greenstone sequence were folded and faulted prior to intrusion of a large pluton of granodiorite (Fig. 2). Because the age of the felsic volcanic rocks (2688 ± 8 Ma) is not resolvably different from the age of that pluton (2686 ± 6 Ma), at Norseman important folding and faulting must have occurred essentially simultaneously with felsic volcanic eruption and granodiorite emplacement. These new age data thus provide support for the hypothesis of Campbell and Hill (1988) that a substantial part of the tectonism within this region may have occurred in response to the rise of huge volumes of felsic magma within the crust 2685–2690 Ma ago. This is the process that was termed "granite tectonics" by Chappell *et al.* (1987).

2. The late Archaean granites

An extensive summary of previous work on Yilgarn granites is given by Cassidy *et al.* (1991). Because of the paucity of geochronological and, to a lesser extent, compositional data for the granites, most classification schemes employed so far have relied at least in part on inferred structural setting. Sofoulis (1963) introduced the terms *internal* for plutons that intrude the greenstone sequences and *external* for plutons that comprise the interiors of the large belts of predominantly granitic rock that separate individual greenstone belts. Bettenay (1977) subdivided the external granites into syn-kinematic and post-kinematic types. Syn-kinematic granites occur as marginal screens to, or elongate plutons within, the greenstone belts, and should probably be grouped with the internal granites (Cassidy *et al.* 1991). Post-kinematic granites are generally leucocratic biotite monzogranites characterised by a limited geochemical range (Bettenay 1977; Cassidy *et al.* 1991). Although lack of outcrop remains a major problem, particularly with respect to proper characterisation of the external granites, available data suggest that the internal granites are compositionally much more diverse than the external granites. Plutons comprised of hornblende granodiorite, hornblende and/or biotite monzogranite, and biotite syenogranite dominate the internal granites, with lesser amounts of quartz diorite, tonalite, trondhjemite, quartz syenite, syenite, and alkali-feldspar granite present as well (e.g. Hallberg 1985; Cassidy *et al.* 1991). By way of contrast, exposures of the external granites are dominated by biotite monzogranite (e.g. Bettenay 1977; Libby 1978). The greenstone sequences are cut also by large numbers of dykes and small intrusions of compositionally diverse porphyritic granites (e.g. Perring *et al.* 1991).

Despite poor exposure, all attempts at estimating the relative abundance of the different granite types have concluded that the late Archaean granites are dominated by, and have an average composition of, monzogranite with a K_2O content of about 3·7% (Lambert & Heier 1968; Bettenay 1977; Libby 1978; Hallberg 1985). Recent work (e.g. Hallberg 1985; Swager & Griffin 1990), utilising aeromagnetic data which often allow individual plutons to be delineated even when largely obscured by a thick mantle of Cainozoic sediment, apparently provides support for the conclusions of the earlier studies which assumed that outcrop is random and not biased in favour of particular rock types.

Most recent models for the origin of these late Archaean granites involve derivation through partial melting of older, complex, continental crust (e.g. Oversby 1975; Bettenay 1977; Hill *et al.* 1989; Cassidy *et al.* 1991). However,

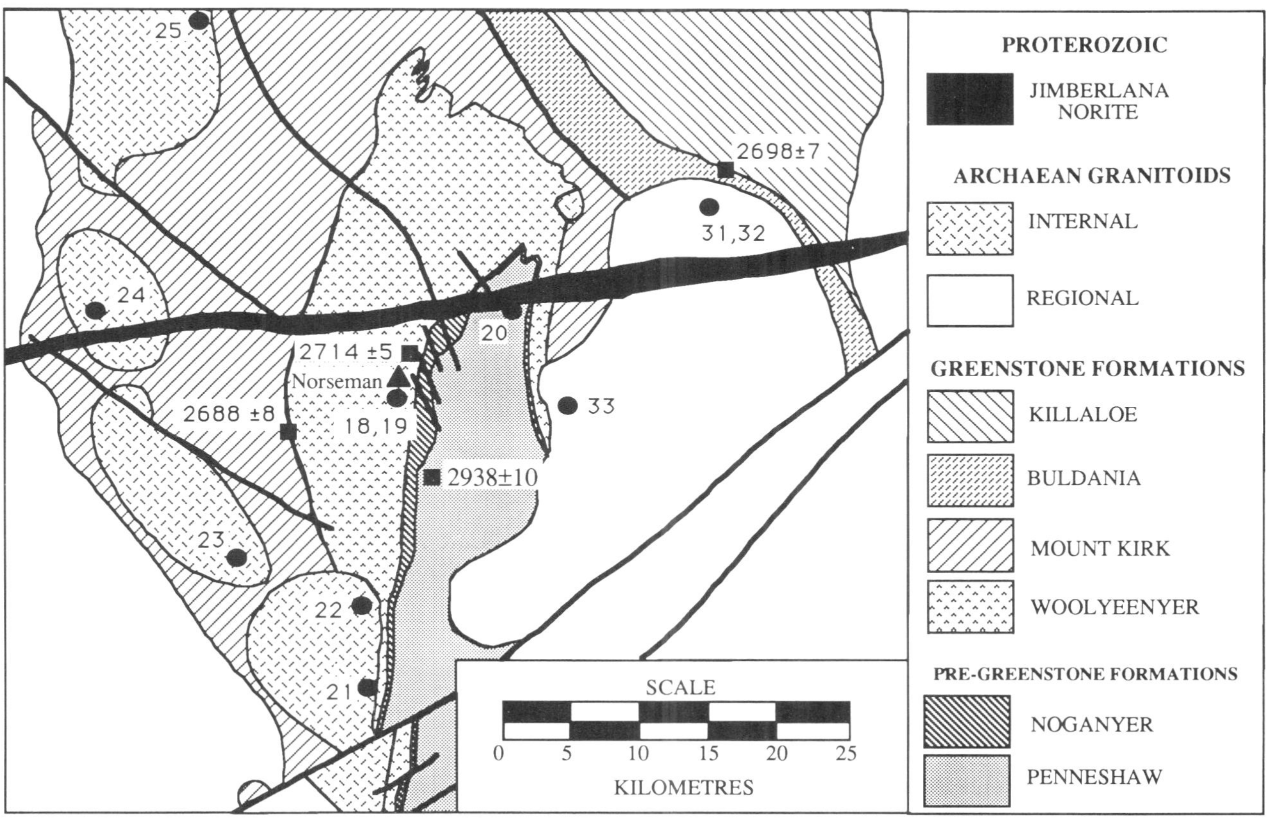

Figure 2 Sketch geological map of the area around Norseman, showing location of studied granite samples (numbered dots), and available ages for the greenstone sequences (solid squares).

Table 1 Location, selected compositional data, and age (where available) for some southern Yilgarn granites

Sample Number	Rock type	Name	Location	SiO_2 wt %	Na_2O wt %	K_2O wt %	Sample age ±95% C.L.	Pluton age* ±95% C.L.
"Late" granites and syenites								
1. 71–738	Biotite granite	Mungari	30°53·1′S, 121°15·1′E	73·9	3·8	4·8	2603 ± 19	2602 ± 11
2. 88–589	Biotite granite	Mungari	30°53·1′S, 121°15·1′E	74·4	3·7	4·9	2600 ± 12	2602 ± 11
3. 71–742	Biotite granite	Karonie	30°55·8′S, 122°33·0′E	71·2	6·2	4·4	2638 ± 26	
4. 86–326	Hb biotite syenite	Dingo Rock	30°57·8′S, 122°37·9′E	69·3	5·2	5·0		
Late highly fractionated granites								
5. 86–301	Biotite granite	Woolgangie	31°10·8′S, 120°29·0′E	66·9	3·9	4·3	2646 ± 6	
6. 86–305	Porph bi granite dyke	Westonia	31°17·6′S, 118°41·8′E	74·9	3·4	5·1	2637 ± 7	
7. 86–307	Biotite granite	Doodlakine	31°36·9′S, 117°51·1′E	68·1	3·8	4·3	2645 ± 7	
8. 86–330	Biotite granite	Meckering	31°36·9′S, 117°1·5′E	73·9	3·6	4·8	2628 ± 8	
Coarse-grained granites containing K-feldspar megacrysts								
9. 72–859	Biotite granite	Menangina Rocks	29°50·3′S, 121°56·4′E	71·7	4·5	3·9	~2665	
10. 72–864	Biotite granite	Wheeler Rocks	32°11·2′S, 120°53·5′E	70·9	4·0	4·5	2661 ± 8[2]	
11. 86–306	Biotite granite	Merredin	31°27·0′S, 118°17·9′E	71·0	3·7	3·5	2665 ± 4	
12. 88–512	Biotite granite	Queen Victoria Rocks	31°17·5′S, 120°55·3′E	71·0	4·4	3·4	2687 ± 6	
13. 88–469	Biotite granite	Union Jack Well	28°43·2′S, 120°46·7′E	72·6	4·1	3·9	2664 ± 8	
14. 89–64	Biotite granite	Bronzite Ridge	32°10·6′S, 121, 17·9′E	71·3	4·3	3·4	2665 ± 7[3]	
Tonalites								
15. 71–1079	Bi-hb tonalite	Kambalda	~31°11′S, ~121°41′E	70·8	5·3	2·3	2662 ± 6[1]	
16. 86–338	Bi-hb tonalite	Bonnie Vale	30°51·9′S, 121°9·6′E	71·2	5·0	1·7	2680 ± 5	
17. 88–585	Bi-hb tonalite dyke	New Celebration	31°1·8′S, 121°36·1′S	70·5	6·4	2·0	2686 ± 4	2665 ± 4
18. 89–54	Bi-hb tonalite	Norseman	32°13·2′S. 121°46·8′E	68·7	4·3	1·4	2664 ± 4	2665 ± 4
19. 89–57	Bi-hb tonalite	Norseman	32°13·2′S, 121°46·8′E	69·0	4·3	1·4	2666 ± 6	
20. 89–63	Bi-hb tonalite	Lake Brazier	32°8·7′S, 121°52·2′E	64·5	4·7	1·5	2664 ± 3[3]	
Intrusive centres of "gneiss domes"								
21. 86–319	Biotite granite	Dundas Rocks	32°23·3′S, 121°46·0′E	74·1	3·7	4·9		
22. 88–557	Biotite granite	Dundas pluton	32°20·7′S, 121°45·6′E	74·2	3·8	4·9		
23. 88–552	Biotite granite	Goodia dome	32°19·6′S, 121°38·2′E	73·7	4·5	4·3		
24. 88–554	Biotite granite	Theatre Rocks	32°8·4′S, 121°33·3′E	72·0	4·5	3·6	2672 ± 7[3]	? ~ 2665
25. 88–515	Biotite tonalite	Pioneer dome	31°52·0′S, 121°37·8′E	71·6	5·0	2·6		
26. 89–52	Biotite tonalite	Widgiemooltha dome	31°36·2′S, 121°36·0′E	71·6	5·4	2·6		
27. 88–514	Biotite granite	Yallari Well	31°7·5′S, 121′25·8′E	68·9	5·3	4·3		
Porphyritic granodiorites								
28. 86–337	Porph bi-hb grd	Lady Bountiful	30°30·2′S, 121°12·4′E	66·3	4·7	2·8		? ~ 2665
29. 88–600	Porph bi-hb grd	Porphyry	29°4·6′S, 122°16·8′E	66·8	5·5	2·9	2667 ± 4	
Early granodiorites								
30. 71–905	Bi-hb granodiorite	Stennet Rocks	32°35·7′S, 121°3·6′E	71·8	4·8	2·7	2691 ± 8[2]	
31. 71–908	Bi-hb granodiorite	Buldania Rocks	32°4·7′S, 122°2·0′E	69·5	4·8	2·3	2689 ± 22[2]	2686 ± 6
32. 71–909	Bi granite	Buldania Rocks	32°4·7′S, 122°2·0′E	75·8	3·2	5·6		2686 ± 6
33. 88–553	Bi-hb granodiorite	Buldania pluton	32°12·6′S, 121°55·0′E	65·0	4·5	3·0	2686 ± 6[3]	

Notes: * If different from the age assignment for an individual sample. Abbreviations: Bi (biotite), hb (hornblende); grd (granodiorite); porph (porphyritic). Rocks listed as biotite granites and biotite tonalites contain at most a trace of hornblende. References to previously published data:[1] Hill and Compston (1986);[2] Hill *et al.* (1989);[3] Hill and Campbell (1992).

substantial disagreement still exists as to the tectonic environment that prevailed during greenstone evolution and subsequent granite emplacement (summary in Cassidy *et al.* 1991).

Table 1 and Figures 3 and 4 summarise the age and geochemistry of granites from the southern Yilgarn Block. In the table, which lists the 25 samples for which we have good age control, we use the SiO_2, Na_2O and K_2O contents to indicate the compositional range of late Archaean southern Yilgarn granites. Although most SiO_2 values are in the relatively restricted range 66–74%, Na_2O and K_2O abundances vary considerably (3·2–6·4% and 1·4–5·1%, respectively). We have divided the analyses in Table 1 into eight groups largely on the basis of petrography and chemical composition; because a number of these groupings have age significance as well, the geochronological data have been used in assigning labels. Although two groups (those associated with 'gneiss domes') are delineated on the basis of structural setting, largely to facilitate comparison with previous work, we suggest that they are genetically related, and together can be related to a third compositional grouping, the tonalites.

The abundances of selected elements are shown diagrammatically on primitive mantle normalised diagrams for Pb, Rb, Ba, Th, U, K, Nb, La, Ce, Sr, Nd, P, Zr, Ti, Y, and Na (Fig. 3), whereas rare-earth element (REE) abundances are illustrated on standard chondrite-normalised plots (Fig. 4). We have adopted the primitive mantle diagrams of Wyborn *et al.* (1987, 1992), in part to facilitate comparison of the late Archaean granites with those of the Proterozoic, studied by these workers. It should be kept in mind that because of our restricted number of samples, the discussion below is intended to give an idea of the range of compositions and the age relationships of the late Archaean granites, and is not intended to be either a comprehensive review of the geochemistry of the late Archaean granites nor a rigid classification scheme.

Perhaps the most striking feature of these data is that all analyses show marked depletion of Nb and Ti, a feature often taken as indicative of island arc or island arc-like processes; if any of these rocks formed through partial melting of hydrated basaltic rocks within the greenstones (e.g. Glikson 1972), then this process must have left similar residual phases to those produced during island arc petrogenesis.

2.1. Granodiorites

Biotite-hornblende-sphene granodiorite underlies considerable areas to the E and W of the Norseman greenstones. Although these rocks are generally relatively K-poor, having <15% K-feldspar and <3·0% K_2O, K-feldspar may still occur as large prominent phenocrysts. Sample 88-553, for instance, from near the margin of a large pluton E of Norseman, contains prominent aligned K-feldspar phenocrysts up to several centimetres in length.

The four samples listed in Table 1 come from two plutons having ages of 2691 ± 8 Ma and 2686 ± 6 Ma; these rocks are thus contemporaneous with the felsic volcanic rocks within the greenstone sequences, eruption of which began 2688 ± 8 Ma ago. Three samples from one pluton (88–553, 71–908, 71–909) give an indication of fractionation trends within a single body; as SiO_2 increases from 65·0 to 75·8%, abundances of the REE, Nb, Sr, P, Zr, Ti, and Y decrease markedly, whereas Rb and Th increase. Other elements shown on the diagrams do not show a consistent increase or decrease. Abundances of elements such as Mg, Fe, Mn, Cr, V and Ni, which are not shown on the diagrams, also decrease markedly. These patterns are consistent with early and continued precipitation of plagioclase, hornblende, apatite, zircon, sphene, and possibly biotite; that is, they are consistent with fractionation involving the observed early crystallising phases.

Relatively steep REE patterns such as those observed for the granodiorite group (with chondrite-normalised La/Lu ratios $La_N/Lu_N = 25–35$ in (Fig. 4)) are often interpreted as evidence for residual garnet during the partial melting event. Small negative Eu anomalies for even the most mafic members of the group could result from the presence of residual plagioclase during the granodiorite-forming event, from small amounts of feldspar fractionation, or be a feature inherited from the source materials. As mentioned above, we infer that the large depletions of Nb and Ti are source features.

2.2. Porphyritic granodiorites

Two samples of relatively mafic seriate-porphyritic granodiorite are from small plutons intrusive into the greenstones and that host economically significant gold mineralisation, at Lady Bountiful (86–337) and Porphyry (88–600). The Liberty Granodiorite, which hosts the Lady Bountiful deposit, has been described by Witt and Swager (1989); our analysis is very similar to analyses of the Liberty Granodiorite reported by these workers. The Porphyry pluton has an age of 2667 ± 4 Ma but the age of the Liberty Granodiorite is controversial. Hill and Campbell (1989) reported an ion probe $^{207}Pb/^{206}Pb$ age of 2593 ± 10 Ma for zircons from 86–337 which they inferred to date emplacement. This interpretation was challenged by McNaughton and Cassidy (1990) on the basis of Pb–Pb data which yielded an apparent age of 2675 ± 23 Ma. Subsequent ion probe analysis of zircons from samples provided by McNaughton and Cassidy document extensive disturbance of the zircon U–Pb systems for three samples from the southern margin of the Liberty Granodiorite, and indicate a minimum age of ~2665 Ma for this pluton. We infer that the Liberty Granodiorite is at least 2665 Ma old.

Our sample 86–337 was obtained from a locality adjacent to the major Lady Bountiful gold deposit, and one interpretation of the combined isotopic data is that the zircons from this sample were extensively disturbed in a major ore-forming hydrothermal system. However, this sample is chemically similar to analyses of samples distant from known ore-forming systems reported by Witt and Swager (1989); this similarity may indicate that chemical modification of rocks at the margin of this hydrothermal system has been minor.

The two samples listed in Table 1 are quite similar, having ~66·5% SiO_2, high Na_2O (4·7, 5·5%), and low K_2O (2·8, 2·9%). Analyses of samples from the Porphyry pluton summarised by Cassidy *et al.* (1991) have higher K_2O and lower Na_2O than does our sample; unlike the Liberty case, these differences may indicate considerable hydrothermal alteration of rocks adjacent to the gold deposit. Na_2O/K_2O ratios of 1·7 and 1·9 derived from the data in Table 1 are similar to those from the "normal" granodiorites (1·5–2·0); differences between the two groups are largely textural (but probably also include age). There are some small differences also in trace element characteristics, including steeper REE patterns [$La_N/Lu_N \sim 65$] and generally higher Sr and Ba; however, these apparent differences may result in large part because of comparison of the more mafic porphyritic

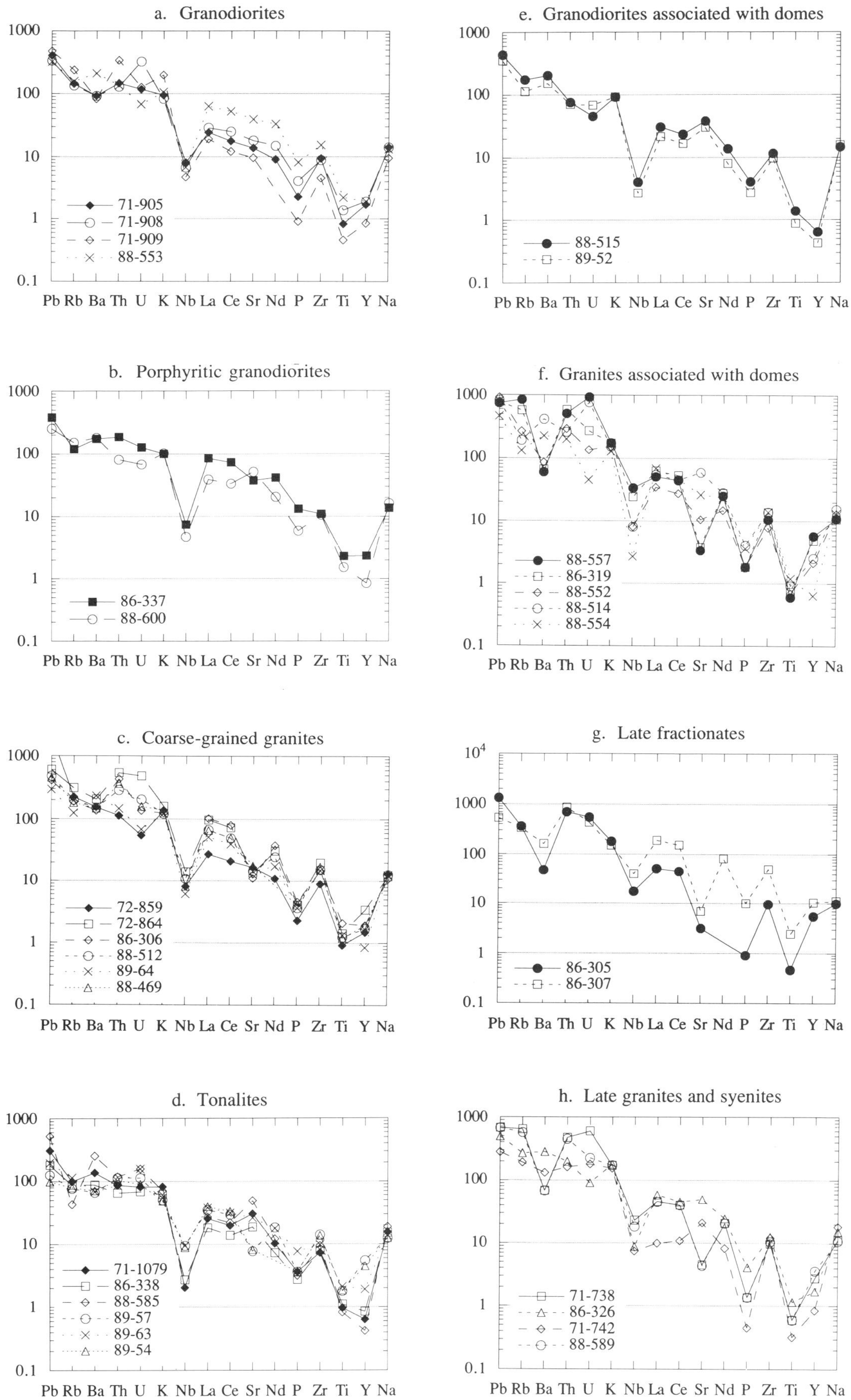

Figure 3 Primitive-mantle-normalised diagrams for southern Yilgarn granites; details of samples (locations, ages, etc.) are given in Table 1; normalising values are from Sun & McDonough (1989) and McDonough & Frey (1990).

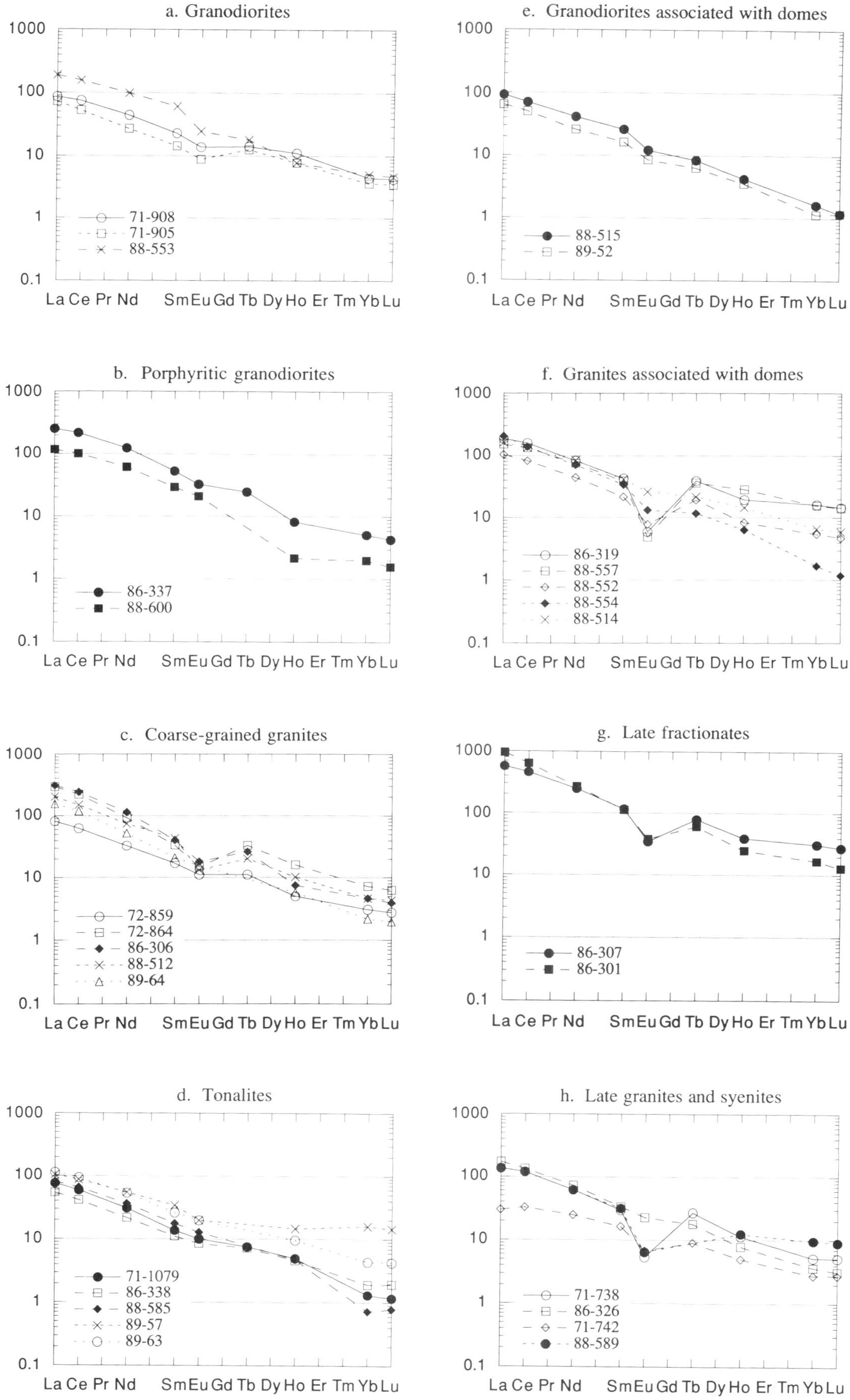

Figure 4 Chondrite-normalised rare earth element plots for southern Yilgarn granites; details of samples (locations, ages, etc) are given in Table 1; normalising values are from McDonough & Frey (1990).

granodiorites with more silicic and more fractionated samples from the "normal" granodiorites. Abundances of the most incompatible elements (Pb through to K on Fig. 3) are consistently about 80–200 times primitive mantle, and Sr shows either small positive or negative anomalies relative to neighbouring elements. One feature that does distinguish clearly these rocks from the "normal" granodiorites is the much less marked negative P anomaly of the porphyritic group (Fig. 3).

The overall compositional similarities suggest that these rocks may have had a similar origin to that inferred for the other granodiorites, that is, derivation through partial melting of pre-existing crust within the field of garnet stability. However, the presence of relatively young ($2718 \pm 13\ (1\sigma)$ Ma, $2741 \pm 15\ (1\sigma)$ Ma) xenocrystic zircon within the Porphyry body points to the involvement of much younger crust than the 3100–3400 Ma material thought to dominate the source of the older granodiorites, all of which are from well to the S of the inferred northeastern margin of ancient crust which runs approximately through Kalgoorlie (Oversby 1975; Hill *et al.* 1989).

2.3. Coarse-grained granites

A common rock of the southeastern Yilgarn is a coarse-grained biotite monzogranite containing prominent phenocrysts of pink K-feldspar. Although the six samples listed on Table 1 are from widely dispersed localities (for example, samples 86–306 and 88–469 are respectively 300 km W and 400 km N of sample 89–64), geochemical characteristics are internally consistent, and, overall, similar to those of the granodiorites. The differences are that SiO_2, Na_2O and K_2O abundances are higher (ranges of 70·9–72·6%, 3·7–4·5%, and 3·4–4·5% for the granites, respectively), abundances of the most incompatible elements (Pb, Rb, Ba, Th, U and K) are generally higher but are variable, and there are negative Sr and Eu anomalies of variable magnitude. Five samples have ages within the restricted range 2661–2665 Ma; the sixth (88–512) is older, 2687 ± 6 Ma. To date, volcanic rocks with ages of 2661–2665 Ma are not known in the Yilgarn.

Because of the overall compositional homogeneity, the simple mineralogy dominated by two feldspars plus quartz, and the lack of abundant more mafic rocks within the younger (2661–2665 Ma) group, we speculate that these samples closely mimic original liquid compositions, and that fractionation was unimportant in their genesis. Given such a petrogenesis, Sr depletions could result from retention of plagioclase within the source during the melting event; steep REE patterns, with $La_N/Lu_N = 40–100$, also suggest the presence of residual garnet. The magnitudes of the Sr and Eu anomalies are correlated, consistent with retention in the source of variable amounts of plagioclase feldspar. Although this is our preferred interpretation, more detailed work on individual plutons is needed to rule out the possibility that the trace element patterns result from consistent patterns of crystal fractionation.

Our interpretation thus is that whereas the granodiorites separated from a source containing residual garnet but little or no residual plagioclase, the coarse-grained granites separated from a garnet-bearing residue that contained variable amounts of plagioclase. This could have resulted because either the granodiorites formed at higher pressures and/or because they represent sufficiently high degrees of melting to deplete substantially the amount of residual plagioclase in the source. Available isotope data (Oversby 1975; McCulloch 1987) can be interpreted as indicating that the granodiorites and the granites were derived from melting of isotopically similar sources.

2.4. Tonalites

Small bodies of biotite-hornblende tonalite and biotite tonalite that cut the folded greenstone sequences are volumetrically insignificant when compared to the granites (Bettenay 1977). The new geochronological data which show that many of these masses are younger than the large granodiorite plutons and contemporaneous (2690–2685 Ma) felsic volcanic rocks invalidates earlier models which suggested that they form part of a magmatic sequence evolving towards more K-rich rocks with time (e.g. Glikson 1972). Most of the tonalites are quite leucocratic, being composed largely of sodic plagioclase and quartz with relatively small amounts (total of 10–20%) of the mafic phases hornblende and biotite; some would be trondhjemites by the IUGS classification (Le Maitre 1989). The high plagioclase plus quartz abundances are reflected in high SiO_2 contents, generally in the range 68–72%.

Element abundance patterns are very consistent, with most samples having moderate concentrations of the highly incompatible elements (50–150 times primitive mantle), marked negative Nb and Ti anomalies, and moderate negative P anomalies. REE patterns are highly variable, with $La_N/Lu_N = 7–100$. Sr abundances vary considerably, with some samples having relative enrichments and others relative depletions. Three samples (86–338, 71–1079, and 88–585, with 0·02–0·05 ppm Lu) are unusually depleted in the heavy rare-earth elements (HREE), having Lu abundances comparable to the average value for chondritic meteorites of 0·0254 ppm (McDonough & Frey 1990). Sample 89–57 from Norseman is distinct from the other tonalites in having much higher abundances of HREE (Fig. 4).

Composition is independent of age: sample 86–338 from Bonnie Vale N of Coolgardie (age 2680 ± 5 Ma) has a very similar element abundance pattern to sample 71–1079 from the 2662 ± 6 Ma Red Hill Granite Complex at Kambalda. Samples 89–63 and 88–585 contain scarce xenocrystic zircon with ages of $2793 \pm 5\ (1\sigma)$ Ma and $2723 \pm 2\ (1\sigma)$ Ma, respectively, whereas 89–54 and 89–57 (which are from the same pluton) contain abundant inherited zircon (including old cores surrounded by younger magmatic rims) with an age of 2685 ± 4 Ma.

2.5. Granites and granodiorites associated with "gneiss domes"

Six roughly ovoid bodies of very felsic, variably fine- to medium-grained biotite granite, monzogranite, or Na-rich granodiorite intrude the western margin of the Kalgoorlie-Norseman greenstone sequences from near Lake Dundas S of Norseman, northwards to near Coolgardie. Because these rocks commonly are foliated, these bodies are often referred to as "gneiss domes" (e.g. Archibald *et al.* 1981) or "domes" (Swager & Griffin 1990). From S to N they are Dundas Rocks, Goodia, Theatre Rocks, Pioneer, Widgiemooltha and Horse Rocks (nomenclature from Swager & Griffin 1990, with the exception of Dundas Rocks). A combination of zircon scarcity, high levels of metamictisation with resultant easy isotopic disturbance, and common zircon inheritance, have led to difficulties in dating these bodies. Hill and Campbell (1992) report an apparent age of 2672 ± 7 Ma for the Theatre Rocks mass, but note that is not possible to resolve two probable populations having ages ~2660 and ~2680 Ma. This sample contains xenocrystic zircon with an age ~2800 Ma, as does the Widgiemooltha Monzogranite to the N; this is equal to the inferred ages of banded gneisses spatially associated with the Pioneer and Horse Rocks domes obtained by the Rb–Sr and Sm–Nd total-rock methods (McCulloch *et al.* 1983).

These structurally similar bodies have highly variable chemistry. Granite samples from the Pioneer (88–515) and Widgiemooltha (89–52) domes are K-poor and generally Na-rich relative to samples from the other four domes; they are biotite granodiorites rather than biotite monzogranites. In fact, these two samples (88–515 and 89–52) are compositionally very similar to the tonalites, to which Hill and Campbell (1992) suggest they are related. Samples from the other four domes appear to be related more by position than by any other parameter. Two samples from the southernmost dome (Dundas Rocks) are from very close to its roof (sample 88–557 was obtained from a drillhole which first penetrated a considerable thickness of basaltic rock), and are very enriched in the most incompatible elements (Rb, Th and U abundances are 300–1000 times primitive mantle) and extremely depleted in Sr and Eu. The next sample moving N (88–552 from the Goodia Dome) has these characteristics but not at anything like the same level; the Dundas Rocks samples could be related to the Goodia sample by extensive feldspar fractionation.

Moving northwards again, the Theatre Rocks sample (88–554) is intermediate between the Goodia sample on the one hand, and those from Pioneer and Widgiemooltha on the other. Theatre Rocks has a small negative Sr anomaly, and a moderate negative Eu anomaly, compared to positive Sr anomalies and small negative Eu anomalies for the two latter samples. Our interpretation is that all of the "gneiss dome" plutons are related by feldspar fractionation from an initial tonalitic composition, and that the "gneiss dome" plutons have the same origin as the small tonalite bodies elsewhere in the greenstone belt. The presence of relatively young (~2800 Ma or less) inherited zircon suggests an origin through partial melting of relatively young pre-existing crust, possibly such as that exposed in the gneisses marginal to the Pioneer and Horse Rocks dome (Hill & Campbell 1992).

2.6. Late, fractionated granites

Three samples of generally fine- to medium-grained biotite granite from a traverse between Coolgardie and Perth are characterised by young (2628 ± 8 Ma to 2646 ± 6 Ma) ages, high incompatible element (200–1000 times primitive mantle) and light rare-earth element (130–1000 times chondrites for La) abundances, and large relative Sr and P depletions (Figs 3, 4). A sample from a dyke of porphyritic biotite granite from Westonia (86–305; age 2637 ± 7 Ma) has similar characteristics. These four samples cover a substantial range in SiO_2 content of from 66·9–74·9%; as SiO_2 increases, Ba, Th, REE, Nb, Sr, P, Zr and Ti abundances decrease markedly, by factors of 7 to 10 for La and P, for example; Pb, Rb and U abundances, however, increase. No attempt has been made as yet to model the origin of these variations.

2.7. Late granites and syenites

Structurally late plutons such as the Mungari Granite have long been recognised as having a distinctive composition (e.g. Glikson 1972; Oversby 1975). Data for three such plutons are given in Table 1, and illustrated in Figures 3 and 4. The Mungari Granite (71–738, 88–589), a medium-grained fluorite-bearing biotite granite, has very high Rb, Th and U abundances, and marked depletions in the trace elements incorporated into feldspars (Ba, Sr and Eu). The Eu depletion is superimposed on a fairly steep REE pattern, with $La_N/Lu_N = 35$. Mungari, in fact, is chemically quite similar to the granites that comprise the "gneiss domes" S of Norseman, and may have a similar origin, thought to have involved feldspar fractionation from an original tonalitic parent.

The other two samples are compositionally different. Sample 86–326 is of a hornblende-biotite syenite from Dingo Rock (130 km E of Kalgoorlie), whereas 71–742 is a fine-grained granite from near Karonie, 110 km E of Kalgoorlie. The syenite has high Na_2O and K_2O (5·2 and 5·0%), slightly elevated (200–300 times primitive mantle) Pb, Rb, Ba and Th compared to most southeastern Yilgarn granites, and a REE pattern with a steep slope ($La_N/Lu_N = 54$) and no Eu anomaly. The Karonie Granite has generally lower abundance levels of the incompatible elements, much lower abundances of the LREE, and a flatter REE pattern ($La_N/Lu_N = 11$) than most southeastern Yilgarn granites.

2.8. Other types of intrusives

As well as the compositional groups summarised above, for which we have both chemical and age data, there is a volumetrically minor and compositionally diverse group of more mafic intrusives which so far remain undated. Compositions include biotite-rich diorite ($SiO_2 = 46{\cdot}2\%$, $Na_2O = 4{\cdot}7\%$, $K_2O = 2{\cdot}3\%$), quartz diorite and very mafic tonalite, and quartz monzonite (e.g. $SiO_2 = 57{\cdot}7\%$, $Na_2O = 5{\cdot}3\%$, $K_2O = 4{\cdot}1\%$). Cassidy *et al.* (1991) also note the presence of such rocks, and provide geochemical data for quartz diorites from the Lawlers Tonalite composite pluton. The relationship of these small intrusives to the much more abundant more felsic rocks is at present unclear.

2.9. Conclusions from the combined geochemical and geochronological data

Two conclusions derive directly from the data presented in Table 1 and summarised on Figures 3 and 4; these are that the range of observed geochemical characteristics and ages probably precludes derivation of all rocks by identical processes from a uniform source, and the rocks of a particular composition can form at different times. Taking the second conclusion first, examples of granites, tonalites and mafic granodiorites from the earlier (~2690–2685 Ma) magmatic episode have virtually identical major element compositions to rocks from the later (2665–2660 Ma) plutons.

However, the variation in the major and trace element compositions of these three main groups (granodiorites, granites, and tonalites) would appear to require significant differences in their petrogenesis. For instance, felsic differentiates from the granodiorite group have much smaller Eu anomalies and no Sr anomalies compared to coarse-grained granites having similar major SiO_2 contents; these differences in principle can result from different source residual mineralogy, from variations in the nature and proportions of fractionating phases, or from both. However, within the one granodiorite pluton for which we have multiple analyses, variations in the Eu and Sr anomalies are small (Figs 3a, 4a), a feature which causes us to suggest that the observed variations are more likely to be related to source rather than later magmatic processes. Again, the more felsic members of the tonalite group have quite different compositions (particularly Na_2O and K_2O contents) when compared to the other two groups. It is possible that these variations could result from one of at least three main causes, that is, differences in source composition, or removal of large quantities of K-rich phases from liquids parental to the tonalites, but not from the parental liquids giving the other compositions, or from retention of a greater proportion of K-rich phases in the tonalite source. However, we suggest that the observed differences are most simply interpreted in terms of compositional differences between the respective sources so that K-rich phases were far less important in the sources which yielded the tonalites than in those which gave rise to the granodiorite and granite groups. This basic concept, that granites, particularly the more mafic ones, reflect the compositions of their source rocks, has been developed in detail elsewhere (e.g. White & Chappell 1977, 1988; Chappell *et al.* 1987; Chappell & Stephens 1988; Chappell & White 1992).

Finally, the often incompatible-element-enriched late fractionates, late granites and syenites appear to have had a different origin from the much more voluminous older rocks. Some of these have been compared with so-called A-types from younger provinces by Cassidy *et al.* (1991).

3. Models for the origin of the granitic rocks of the Yilgarn

The evidence for substantial involvement of pre-existing continental crustal material as source material for the genesis of the bulk of the late Archaean granitic rocks of the eastern Yilgarn now seems to be accepted by most workers (e.g. Hill *et al.* 1989; Witt & Swager 1989; Cassidy *et al.* 1991). The evidence for this includes isotopic data for both granites and ores (Oversby 1975; Browning *et al.* 1987) and the difficulty of producing large volumes of K-rich granite directly from the mantle (e.g. Bettenay 1977; Hill *et al.* 1989), as well as the documentation of the presence of old xenocrystic zircons in a wide variety of rock types (Compston *et al.* 1986; Campbell & Hill 1988; Claoué-Long *et al.* 1988; Hill & Campbell 1992). The presence of xenocrystic zircon with ages as old as 3400 Ma in komatiitic basalts that predate initiation of felsic magmatism has been used to argue that the feeders to the basaltic pile, at least at Kambalda, passed through much older, complex pre-existing continental crust (Compston *et al.* 1986). Evidence for an ensialic environment for the development of the Norseman-Wiluna Belt also comes from the presence of gneiss remnants with inferred pre-greenstone fabrics in some intrusive granites (e.g. Archibald *et al.* 1981). The abundance of monzogranites having approximately minimum-temperature melt compositions (for crustal P and T), combined with a relative paucity of mafic rocks of the same age, would also appear to require a crustal origin for the late Archaean granitic rocks.

Arguments are now centred around elucidation of the tectonic environment in which these late Archaean granites were produced. The principle candidates are a subduction-related continental margin or marginal basin situation (e.g. Gee *et al.* 1981; Hallberg 1986; Barley *et al.* 1989; Barley & Groves 1990; Cassidy *et al.* 1991), and a mantle plume initiated continental reworking, which can take place entirely within a continent or may involve reworking of an older continental margin (Campbell & Hill 1988; Hill *et al.* 1992). Proponents of the two different models have used two quite different approaches in their attempts to elucidate the origin of the late Archaean granites of the Yilgarn Craton. The first has been to use comparison with possible modern analogues in an attempt to find a compositional and tectonic "match". The second has been to use the available data in an attempt to understand the processes involved in the origin of key rock types, and from that to infer the tectonic environment.

3.1. The plate margin model

Evidence in favour of a plate margin ("subduction") environment has been reviewed by Barley *et al.* (1989), Barley and Groves (1990), and Cassidy *et al.* (1991). Barley and Groves (1990, p. 5) provide a succinct statement of the "comparative" approach to elucidating tectonic environment: "To discriminate between possible tectonic settings it is necessary to describe lithostratigraphic associations and the evolution of structural domains in sufficient detail to allow rigorous comparison with modern tectonic environments."

Barley *et al.* (1989) and Barley and Groves (1990) recognise two main tectonostratigraphic terranes within the Norseman-Wiluna Belt, a western association of tholeiites, komatiites and sulfidic sediments, and an eastern association of tholeiites, calc-alkaline volcanics, and volcanogenic sediments. The eastern association includes a number of large, partly subaerial, dominantly calc-alkaline centres (such as those at Welcome Well and Mt Ida), and although a complete compositional spectrum from basalts through to rhyolites is observed, andesitic and dacitic compositions are dominant. Pyroclastic and feldspathic sedimentary rocks believed to be derived from the eastern terrane overlie lithologies of the western komatiite-containing terrane. These two tectonostratigraphic (eastern calc-alkaline, western komatiite-containing) terranes are compared, respectively, with modern volcanic arcs and the oceanic crust and volcano-sedimentary cover of back-arc basins. Because of the evidence for older crust beneath the southern part of the belt, it is inferred that the Norseman-Wiluna Belt formed at an obliquely convergent craton or proto-craton margin and may represent the remains of a volcanic arc and back-arc basin complex developed in response to westward-dipping subduction.

This interpretation appears to be in conflict with the mapping of Hallberg (1985), who inferred that the oldest rocks in the area of the eastern association of Barley *et al.* (1989) are a thick sequence of komatiites, basalts, and chemical sediments (banded iron formation, and chert) and that these are overlain, locally at least, by conglomeratic sedimentary rocks, and by further basalts. This fundamentally mafic sequence is overlain by volcanic and volcaniclastic rocks derived from a number of emergent andesitic volcanic centres (including those at Welcome Well and Mt Ida), the products of which are intruded and overlain by more rocks having basaltic compositions. The youngest rocks described by Hallberg (1985) are basalts and high-Mg basalts and their intrusive equivalents. At least in the area around Leonora, there appears to be little difference in the overall nature (both are fundamentally mafic) of the lower parts of the two associations of Barley *et al.* (1989). However, both Hallberg (1985) and Barley *et al.* (1989) agree that the andesitic volcanic centres predate at least some of the mafic magmatism.

Cassidy *et al.* (1991), in a study of the geochemistry of the plutonic rocks of the Norseman-Wiluna belt, similarly use the comparative approach, and argue that because of a general compositional similarity between the "average" Norseman-Wiluna Belt granite and some Phanerozoic I-type suites, particularly that of the central and eastern domains of the Peninsular Ranges Batholith of southern and Baja California, the Norseman-Wiluna Belt granites have a similar origin. Table 2 compares average analyses of Yilgarn granites with granites from the 380–430 Ma Lachlan Fold Belt of eastern Australia, the 80–125 Ma Peninsular Ranges Batholith, Quaternary volcanic rocks from the Bismark Arc, and an estimated average composition of the upper crust. Although there is a tendency for simple averages to obscure petrogenetically-important trends within groups of data, there are broad similarities in the average compositions of the granitic rocks from the Archaean and Phanerozoic terranes.

Use of the comparative approach in this manner at least implicitly carries the assumption that all of the rocks used in the calculation of the average are related, and have formed from similar source compositions through similar petrological processes. While this may be the case for the Peninsular Ranges granites (e.g. Silver *et al.* 1979; Hill *et al.* 1986; Silver & Chappell 1988), it is unlikely to be the case for the Norseman-Wiluna Belt area, where we suggest that at least four combinations of source composition and petrogenesis were involved.

3.1.1. Problems with the arc model. The Peninsular Ranges Batholith, the origin of which is clearly related in some way to the subduction of oceanic crust beneath the North American continental margin (e.g. Silver *et al.* 1979; Hill *et al.* 1986; Gromet & Silver 1987; Silver & Chappell 1988), does share many chemical characteristics with the Norseman-Wiluna granites. REE abundance patterns are quite similar, as noted by Cassidy *et al.* (1991), as are average compositions. This should not be surprising, as there are broad similarities in the bulk compositions of all three "granitic" groups listed in Table 2 (1 + 2, 3 and 4), while the Bismark Arc rocks, as shown in Table 2, are generally more mafic. However, when considered in detail, there are significant compositional differences, particularly in CaO and K_2O contents, the former of which is characteristically higher and the latter lower (at similar SiO_2) in those rocks clearly associated with subduction (Peninsular Ranges Batholith, Bismark Arc) than in those where a relationship to subduction is more tenuous, or even unlikely (Lachlan Fold Belt, Chappell & Stephens 1988).

Such differences may result from the relative importance of older continental crustal materials (which are characteristically enriched in K relative to the mantle, and to oceanic crust) in the generation of those rocks, and thus may provide some information regarding proximity to older pre-existing continental crust. For instance, the source for the Lachlan Fold Belt granites has been inferred to consist almost entirely of older continental crustal materials (e.g. White & Chappell 1977, 1988; Chappell *et al.* 1987; Chappell & Stephens 1988; Chappell & White, 1992), and that of the Peninsular Ranges Batholith to contain from 0–35% pre-existing crust (Hill *et al.* 1986), whereas incorporation of old crust is likely to be unimportant in the generation of the Bismark Arc (Whalen 1985). These estimates are consistent with the inferred tectonic setting of the three areas: the Lachlan magmatism appears to have taken place largely within a continent, the Peninsular Ranges Batholith developed along a continental margin, and the Bismark Arc in an intra-oceanic setting.

However, a more detailed look at the overall composition of magmatism within different environments points to significant differences between the Norseman-Wiluna Belt rocks and the Phanerozoic analogue preferred by Cassidy *et al.* (1991), the Peninsular Ranges Batholith. Figure 5 uses the SiO_2 contents of plutonic and volcanic rocks to compare more directly the relative distributions of the different rock types within these two areas. For the Norseman-Wiluna Belt, the average SiO_2 content of the internal granites is lower than that of the mode, and although the internal granites show a much wider range of compositions, overall

Table 2 Comparison of average compositions of Yilgarn granites with other compositions

	1	2		3	4	5	6
	mean ± S.D.	mean ± S.D.	range	mean	mean	mean	mean
SiO_2	72·63 ± 1·86	69·74 ± 4·42	54·8–75·8	68·98	67·12	60·13	66·0
TiO_2	0·24 ± 0·09	0·36 ± 0·29	0·01–1·76	0·43	0·63	0·98	0·5
Al_2O_3	14·18 ± 0·72	15·00 ± 1·02	11·7–18·7	14·32	15·98	16·42	15·2
Fe_2O_3	0·82 ± 0·32	2·69 ± 1·56	0·50–10·24	1·03	1·10	1·91	
FeO	0·98 ± 0·36			2·44	2·92	3·73	4·5
MnO	0·03 ± 0·01	0·05 ± 0·05	0·01–0·39	0·07	0·07	0·11	
MgO	0·45 ± 0·20	1·00 ± 0·98	0·06–5·75	1·56	1·75	3·23	2·2
CaO	1·71 ± 0·51	2·36 ± 1·53	0·29–8·32	3·35	4·55	5·85	
Na_2O	4·07 ± 0·44	4·56 ± 0·74	2·90–7·92	3·08	4·15	4·05	3·9
K_2O	4·01 ± 0·81	3·03 ± 1·27	0·27–5·89	3·30	2·13	1·40	4·4
P_2O_5	0·09 ± 0·04	0·14 ± 0·11	0·01–0·68				
Ba	1058 ± 403	1117 ± 512	29–3040				
Rb	162 ± 64	118 ± 92	5–465				
Sr	239 ± 127	596 ± 369	20–1655				
Pb	46 ± 25	31 ± 17	5–98				
Th	32 ± 18	18 ± 19	2–88				
Zr	164 ± 70	154 ± 85	5–817				
Nb	9 ± 6	6 ± 3	1–17				
Y	14 ± 9	10 ± 6	1–36				
La	56 ± 36	46 ± 31	3–154				
Ce	96 ± 60	94 ± 56	8–218				
V	15 ± 9	51 ± 42	1–177				
Cr	5 ± 3	30 ± 34	2–246				
Ni	3 ± 2	17 ± 18	1–114				
Zn	42 ± 22	66 ± 24	22–137				

Notes: 1. Average of 87 late Archaean post-kinematic, external granites from the southern Yilgarn (Cassidy *et al.* 1991).
2. Average (and total range) of 230 (for major elements only) late Archaean granites from the Norseman-Wiluna Belt (Cassidy *et al.* 1991).
3. Average of 948 analyses of I-type granites from the Lachlan Fold Belt (Chappell & Stephens 1988).
4. Average of 297 analyses of granitic rocks from the Peninsular Ranges Batholith having $SiO_2 > 53\%$ (Chappell & Stephens 1988).
5. Average of 64 analyses of volcanic rocks from zone E of the Bismark volcanic arc (data from Johnson & Chappell 1979).
6. Estimated chemical composition of the upper crust (Taylor & McLennan 1985).

there is little difference in the composition of the bulk of the internal and external granites, and of the granites and the felsic volcanic rocks. However, the shape of the Norseman-Wiluna distribution is fundamentally different from that of the Peninsular Ranges Batholith. Although these histograms may to some extent reflect sampling patterns rather than areal distributions (for instance, because many of the Peninsular Ranges samples were collected for zircon geochronology, mafic lithologies are probably under-represented in the total sample), they show that there are significant differences in the compositional spectrum between the eastern Yilgarn rocks and the suggested modern analogue.

Considered overall, the Norseman-Wiluna Belt has a strongly bimodal distribution of igneous rock types (a feature best shown by the data for the volcanic rocks), whereas the Peninsular Ranges Batholith shows a broad compositional range with a mode at compositions equivalent to felsic andesite or silica-poor dacite. We note, however, that Figure 5 does not contain any data from the andesitic volcanic centres that stratigraphically appear to underlie at least part of the mafic sequence in the Leonora-Laverton area. With this caveat, the Norseman-Wiluna pattern is quite unlike that for orogenic andesites figured by Gill (1981), which shows a very broad maximum at 53–61% SiO_2. Mafic rocks are uncommon within the areas underlain by the external granites, a fact usually taken as resulting from different exposure levels (e.g. Archibald *et al.* 1981).

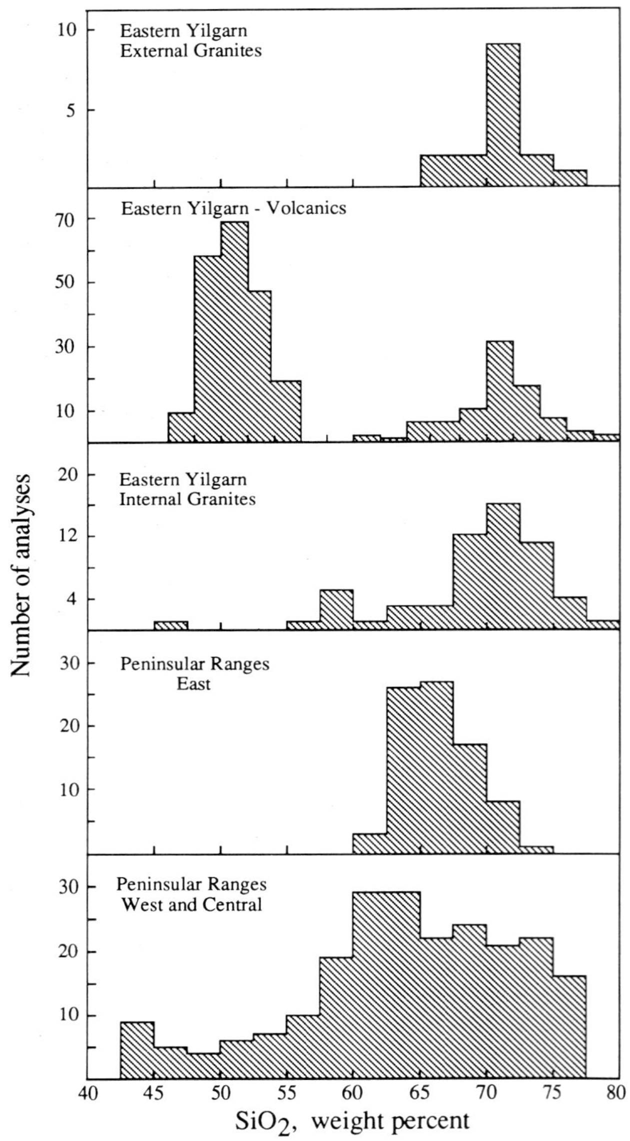

Figure 5 Histograms comparing SiO_2 contents of southern Yilgarn and Peninsular Ranges intrusive and extrusive rocks; the strongly bimodal character of the Yilgarn data set contrasts with the much broader compositional spectrum observed for the Peninsular Ranges samples; data sources: Eastern Yilgarn external granites (our unpublished data); Eastern Yilgarn volcanics (Hallberg 1972); Eastern Yilgarn internal granites (our unpublished data, and Table 3 of Cassidy *et al.* 1991); Peninsular Ranges E (Hill *et al.* 1988); Peninsular Ranges W and central (Silver & Chappell 1988).

The marked isotopic, age, and trace element compositional zoning within the Peninsular Ranges Batholith is not found in the Norseman-Wiluna Belt rocks. This zoning is very systematic and significant and it apparently relates to distance from a subduction zone (Silver *et al.* 1979; Gromet & Silver 1987; Silver & Chappell 1988); similar variations have been observed for modern subduction-related systems such as the Sunda Arc (Whitford *et al.* 1979; see also Gill 1981). Systematic geographical variations such as these have not yet been observed for Yilgarn rocks. Another characteristic of the Peninsular Ranges Batholith which differentiates it from the Norseman-Wiluna belt is the presence within the former of continuous magmatism over a period of at least 45 Ma (80 Ma if the extension of the magmatic arc across the Gulf of California is considered). This contrasts with the magmatic history of the Norseman-Wiluna Belt which is characterised by short periods of intense activity separated by periods of relative quiescence.

Finally, the bulk of the southern Yilgarn magmatism is represented not by the internal granites but by the huge areas of synchronous and compositionally similar external granites. On an areally-weighted basis, the average composition of the late Archaean magmatic activity is much closer to that of the average post-tectonic granite of Table 2 than to the Norseman-Wiluna average discussed by Cassidy *et al.* (1991). As the external and internal granites formed simultaneously, any model for the origin of the Norseman-Wiluna Belt granites must also provide a satisfactory solution to the problem of the origin of the volumetrically-dominant monzogranites that comprise the external granite terranes. The shift to more felsic compositions that results from inclusion of the external granites in the averages (but which is implicit also in the difference between the mean and the mode in the internal granites) tends to increase significantly the compositional differences between the Yilgarn granites and the Peninsular Ranges Batholith average. Consideration of the external granites points to another fundamental difference between the Yilgarn and Peninsular Ranges provinces; whereas, in the Peninsular Ranges, magmatism at any one time was restricted to an elongate belt thousands of kilometres long but at most several tens of kilometres wide (e.g. Silver *et al.* 1979; Silver & Chappell 1988), the late Archaean event apparently affected almost all of the Yilgarn Craton, an equant area about 800 km across, more or less simultaneously (Table 1; Nieuwland & Compston 1981; Pidgeon *et al.* 1990; Kinny *et al.* 1990).

On the basis of these characteristics (compositionally limited range for the granites, average felsic composition, magmatic episodicity, and areal extent and geographic form of the magmatic event), we suggest that a more appropriate Phanerozoic analogue for the southern Yilgarn is the Lachlan Fold Belt of eastern Australia. Even here, though, there are differences, which are inferred to result, at least in part, from differences in composition of the source regions for the granites.

3.2. Identification of processes operating in the generation of granitic terranes

Identification of the processes involved in the generation of the magmatic rocks found within a particular area may have considerable utility in attempts to unravel the origin and tectonic setting of granite production. In particular, we suggest that two important questions to address are: how was heat transported from the mantle into the crust, and what evidence is there which may be used to infer the type of melting within the mantle contemporaneous with felsic magmatism?

The first question, broadly taken, has two answers: by conduction, or by advection. Heat transport by conduction will usually involve a distinct time gap (of a few Ma to several tens of Ma) between mafic and felsic magmatism, as heat is conducted from the hot mantle source into the overlying crust. In contrast, heat transfer by advection will usually result in the near-simultaneous eruption of mafic and felsic magmas.

The second question regarding the type of mantle melting associated with felsic magmatism also has two answers: melting can result from adiabatic ascent of either normal or anomalously hot mantle ("decompression melting"), or can occur when a flux (usually H_2O) is added to the mantle ("flux melting"). Decompression melting (e.g. McKenzie & Bickle 1988) is characteristic of constructive plate margins (particularly mid-ocean ridges) and hotspots, and usually results in the production of magmas with low volatile contents. Although the basaltic rocks resulting from decompression melting vary widely in composition, usually only the more alkali-rich variants are volatile-rich. Flux melting, on the other hand, is characteristically associated with subduction zones, and commonly produces magmas relatively enriched in volatile components, particularly H_2O (e.g. Burnham 1979). Basaltic products might include high-Al basalts and boninites. More generally, mantle-derived rocks associated with subduction tend to be more SiO_2-rich, having compositions in the basaltic andesite or andesitic fields (e.g. Gill 1981). This characteristic can result both from the melting of hydrated basalt (e.g. Burnham 1979), or from injection of high-SiO_2 continental materials into the source regions via subduction. Clues to the origin of the mantle-derived rocks that constitute a common, if often minor, component of granitic terranes (e.g. Chappell & Stephens 1988; Silver & Chappell 1988; Stephens & Halliday 1988) might be obtained from both extrusive and intrusive components, as well as from study of the mafic inclusions or enclaves that are common in many granites.

Clear answers to both questions provide important information regarding the tectonic setting of magma generation. The presence of mafic magmas resulting from flux melting strongly implicates subduction, whereas evidence for the products of decompression melting points to either a constructive margin or within-plate origin.

The process-oriented approach has been applied to the problem of elucidation of the tectonic environment of magma genesis applicable to the late Archaean of the Yilgarn by Campbell and Hill (1988), who argue that the inferred time interval, ~15 Ma, between the eruption of the first basalt and the beginning of voluminous crust-derived felsic magmatism was consistent with a model in which partial melting of the lower crust was the result of conduction of heat upwards from the top of the hot basalt source region. This approach differs from most attempts at inferring the origin of crust-derived magmatism in that it explicitly uses an observation (a time interval) which may have direct physical significance in order to distinguish between plausible physical processes (heat transport through conduction versus heat transport by advection) involved in melt generation. It includes the explicit assumption that there is a direct relationship between the two magmatic events; the appropriateness and some of the implications of this assumption have been explored by Hill (1991) and Hill *et al.* (1992).

Although more data have shown the time interval to be a little longer (~25 Ma) than was first indicated, our original conclusion still holds: the time gap between the start of mantle-derived mafic magmatism and crust-derived felsic magmatism is most satisfactorily explained as being the time taken to conduct sufficient heat from the basalt source into the crust in order to initiate melting.

A second characteristic of the Norseman-Wiluna Belt, that we infer has important implications for understanding the tectonic environment, is the presence within the greenstone sequences of komatiite, a variety of basalt inferred to have formed at unusually high temperatures (e.g. Green *et al.* 1975; Jarvis & Campbell 1983; McKenzie 1984; McKenzie & Bickle 1988; Campbell *et al.* 1989). The modern analogue of komatiite (picrite) is characteristic of "hotspots", and has been inferred to form by partial melting within narrow columns of anomalously hot material rising within the mantle, that is, through melting within mantle plumes (e.g. Schilling 1973; Campbell & Griffiths 1990). The recent integration of results from experimental petrology with a dynamical model for mantle plumes (Campbell *et al.* 1989; Campbell & Griffiths 1990) points to the presence of high-Mg magmas such as picrites or komatiites as indicating the involvement of a mantle plume. Consequently, the presence of komatiites within the Yilgarn sequences may be taken as direct evidence for the involvement of a mantle plume in the late Archaean geological evolution of this region.

In summary, these two observations (the presence of a time gap between initiation of basaltic volcanism and voluminous felsic magmatism, and the presence of picrites) have been used to infer the operation of two important processes (conduction of heat from a basalt source region to the crust, and presence of a mantle plume). We suggest that this process-oriented approach may complement the comparative approach in studies of ancient magmatic provinces.

3.3. The mantle plume model

Finally, there is one more question that may be answerable: does the area of interest show characteristics expected to be associated with one of the two principal modes (Davies 1988a, b; Davies & Richards 1992) of mantle convection? Magmatic provinces clearly related to the plate-scale mode tend to be elongate, often having lengths of many thousands of kilometres and widths of only one to two hundred kilometres, whereas those associated with the starting phase of mantle plumes (plume heads: see Griffiths & Campbell 1990; Campbell & Griffiths 1990) are predicted to be more equant and have diameters ~2000 km. Other characteristics of plume head provinces include: the occasional presence of high-T mantle-derived melts such as picrites or komatiites; a magmatic evolution from early basalt followed, after a time gap, by crust-derived magmatism, in turn followed by a prolonged magmatic history; an overall time-scale for magmatism of as much as 100 Ma; the fundamentally bimodal nature of volcanism (although voluminous basalt is not always present); early uplift that precedes basaltic magmatism; and the presence during the latter (particularly the felsic) part of the magmatic record of contemporaneous extension (Hill *et al.* 1992).

For the Yilgarn, we come to three conclusions: the observed time gap implies heat transport by conduction, the compositions of the mafic rocks imply an origin through decompression melting of rising mantle, and the presence of komatiites indicates the presence of a mantle plume. We have further developed the Yilgarn model in terms of theoretical work on the origin of mantle plumes (Campbell *et al.* 1989), and infer that the late Archaean granites of the Yilgarn developed in response to the rise of a new mantle plume ~2715 Ma ago. In this model, decompression melting within the rising head of a new mantle plume gave rise to the voluminous basalts of the greenstone belts, whereas melting of hot source material ascending within the axial conduit yielded komatiites. The greenstone sequences are thus the late Archaean equivalent of a continental flood basalt province (see Griffiths & Campbell 1990; Campbell & Griffiths 1990). Conduction of heat upwards from the top of the plume head into overlying primitive but complex continental crust initiated, after a time lag ~25 Ma, partial melting of the lower crust. We infer that initial melting took place largely within the field of garnet stability. The developing partial melt layer blocked the rise of mantle-derived melts, which were trapped near the base of the crust. Rise of the crust-derived magmas gave rise to the voluminous felsic volcanic materials within the greenstone belts, and to the early granodiorites and granites. Late intrusion of mafic magmas followed immediately after the first main crustal magmatic event.

A second main crustal magmatic episode followed after a second time gap; this time the bulk of the material produced was of granitic composition. Geochemical characteristics, particularly negative Sr and Eu anomalies, are interpreted as indicating a lower-pressure origin (field of plagioclase plus garnet stability) than that inferred for the earlier granodiorite-dominated anatectic event (garnet predominant). Available isotopic data (Oversby 1975) suggest similar source compositions were involved during both episodes, at least for the area S of Kalgoorlie. Volumetrically less significant tonalites, and late fractionated granites and syenites, appear to have a different origin, the former possibly involving partial melting of a source containing a component of younger (~2700–2800 Ma from the ages of inherited zircons) supracrustal material. Note that while the compositions of the granitic rocks may be used to make inferences about the pressures and temperatures pertaining during partial melting, and provide information about the compositions of their source regions, they cannot be used to infer the nature of the heat source that resulted in crustal anatexis.

We infer that the distinguishing characteristics of plume-related magmatism as compared to subduction-related magmatism are the presence within the former of distinct magmatic episodes rather than a magmatic continuum, the fundamentally bimodal nature of the magmatic products of plume provinces, and the presence within plume provinces of essentially contemporaneous magmatism over a large (up to ~2000 km) equant area rather than the restriction of the locus of magmatism to well-defined linear belts that is a characteristic of many modern subduction-related arcs.

Finally, we note that geochronological work is showing that the close temporal relationship between granites and greenstones observed for the Norseman-Wiluna Belt is not unusual (e.g. Abitibi Subprovince of the Superior Province of Canada: Corfu *et al.* 1989; Wong *et al.* 1991; Pilbara Craton of Western Australia: Williams & Collins 1990). If these other terranes are shown to have a similar magmatic evolution to that of the southeastern Yilgarn, then the model developed for the southeastern Yilgarn may have wider application.

4. Conclusions

New geochemical data for granitic rocks from the southern Yilgarn Block of Western Australia further document the overall compositional heterogeneity of the late Archaean magmatism of this region. However, much of the documented compositional variation comes from analyses of rock types insignificant in area, as shown by published studies on the average composition of Yilgarn crust. The dominant rock type of the southern Yilgarn appears to be coarse-grained 2660–2665 Ma monzogranite characterised by negative Sr and Eu anomalies and a REE pattern with a steep slope. Other important rock types are equigranular and porphyritic granodiorites, a variety of rocks of tonalitic composition, and more fractionated rocks apparently derived from an original tonalitic parent.

The areally-extensive granite plutonism forms an integral part of a late Archaean thermal event that began with the eruption of thick sequences of basaltic rocks ~2715 Ma ago and which continued until at least the emplacement of small plutons of biotite granite at ~2600 Ma. However, voluminous felsic magmatism within the eastern part of the Yilgarn (the Norseman-Wiluna belt) is largely restricted to two brief episodes 2665–2660 Ma and 2690–2685 Ma ago.

The mantle plume model developed for the origin of the crust-derived rocks that form an important part of the late Archaean thermal and structural reworking of the Yilgarn Craton may have applications to other areas of areally extensive continental intraplate magmatism.

5. Acknowledgements

Our geochronological work in the Yilgarn has received substantial support from Aztec Exploration, Australian Consolidated Minerals, BP Minerals, Central Norseman Gold Corporation, Hill Minerals, Pancontinental, Placer Exploration, West Coast Holdings and Western Mining Corporation. Access to, and assistance with, the SHRIMP ion microprobe has been generously provided by Bill Compston and Ian Williams; Jan Bitmead, Elizabeth Jagodzinski and Richard Rudowski are thanked for assistance with sample preparation and data collection. Our ideas on the application of the mantle plume model to the late Archaean geology of the Yilgarn Craton have benefited considerably from discussions with Ross Griffiths and Geoff Davies. We thank two journal reviewers for their constructive comments.

6. References

Archibald, N. J., Bettenay, L. F., Bickle, M. J. & Groves, D. I. 1981. Evolution of Archaean crust in the Eastern Goldfields province of the Yilgarn Block, Western Australia. GEOL SOC AUST SPEC PUBL **7,** 491–504.

Arriens, P. A. 1971. The Archaean geochronology of Australia. GEOL SOC AUST SPEC PUBL **3,** 11–23.

Barley, M. E., Eisenlohr, B. N., Groves, D. I., Perring, C. S. & Vearncombe, J. R. 1989. Late Archean convergent margin tectonics and gold mineralization: A new look at the Norseman-Wiluna belt, Western Australia. GEOLOGY, **17,** 826–9.

Barley, M. E. & Groves, D. I. 1990. Deciphering the tectonic evolution of Archaean greenstone belts: the importance of contrasting histories to the distribution of mineralization in the Yilgarn Craton, Western Australia. PRECAMB RES **46,** 3–20.

Bettenay, L. F. 1977. *Regional geology and petrogenesis of Archaean granitoids in the Southeastern Yilgarn Block, Western Australia.* Unpublished Ph.D. Thesis, University of Western Australia, Perth.

Browning, P., Groves, D. I., Blockley, J. G. & Rosman, K. J. R. 1987. Lead isotope constraints on the age and source of gold mineralization in the Archean Yilgarn Block, Western Australia. ECON GEOL **82,** 971–86.

Burnham, C. W. 1979. Magmas and hydrothermal fluids. *In* Barnes H. L. (ed.) *Geochemistry of hydrothermal ore deposits,* 2nd edn, 71–136. New York: Wiley.

Campbell, I. H., Griffiths, R. W. & Hill, R. I. 1989. Melting in an Archaean mantle plume: heads it's basalts, tails it's komatiites. NATURE **339,** 697–9.

Campbell, I. H. & Griffiths, R. W. 1990. Implications of mantle plume structure for the origin of flood basalts. EARTH PLANET SCI LETT **99,** 79–93.

Campbell, I. H. & Hill, R. I. 1988. A two-stage model for the formation of the granite-greenstone terrains of the Kalgoorlie-Norseman area, Western Australia. EARTH PLANET SCI LETT **90,** 11–25.

Cassidy, K. F., Barley, M. E., Groves, D. I., Perring, C. S. & Hallberg, J. A. 1991. An overview of the nature, distribution and inferred tectonic setting of granitoids in the late-Archaean Norseman-Wiluna Belt. PRECAMB RES **51,** 51–83.

Chappell, B. W. & Stephens, W. E. 1988. Origin of infracrustal (I-type) magmas. TRANS R SOC EDINBURGH EARTH SCI **79,** 71–86.

Chappell, B. W. & White, A. J. R. 1992. I-type and S-type granites in the Lachlan Fold Belt. TRANS R SOC EDINBURGH EARTH SCI **83,** 1–26.

Chappell, B. W., White, A. J. R. & Wyborn, D. 1987. The importance of residual source material (restite) in granite petrogenesis. J PETROL **28,** 1111–38.

Claoué-Long, J. C., Compston, W. & Cowden, A. 1988. The age of the Kambalda greenstones resolved by ion-microprobe: Implications for Archaean dating methods. EARTH PLANET SCI LETT **89,** 239–59.

Compston, W., Williams, I. S., Campbell, I. H. & Gresham, J. J. 1986. Zircon xenocrysts from the Kambalda volcanics: Age constraints and direct evidence for older continental crust below the Kambalda-Norseman greenstones. EARTH PLANET SCI LETT **76,** 299–311.

Compston, W., Williams, I. S. & McCulloch, M. T. 1987. Contrasting zircon U–Pb and model Sm–Nd ages for the Archaean Logue Brook granite. AUST J EARTH SCI **33,** 193–200.

Corfu, F., Krogh, T. E., Kwok, Y. Y. & Jensen, L. S. 1989. U–Pb geochronology in the south-western Abitibi greenstone belt, Superior Province. CAN J EARTH SCI **26,** 1747–63.

Davies, G. F. 1988a. Role of the lithosphere in mantle convection. J GEOPHYS RES **93,** 10451–66.

Davies, G. F. 1988b. Ocean bathymetry and mantle convection, 1. Large-scale flow and hot-spots. J GEOPHYS RES **93,** 10467–80.

Davies, G. F. & Richards, M. A. 1992. Mantle convection. J GEOL (in press).

Doepel, J. J. G. 1973. Norseman. WESTERN AUSTRALIA GEOL SURV 1:250 000 SERIES EXPLANATORY NOTES.

Gee, R. D. 1979. Structure and tectonic style of the Western Australian Shield. TECTONOPHYSICS **58,** 327–69.

Gee, R. D., Baxter, J. L., Wilde, S. A. & Williams, I. R. 1981. Crustal development in the Archaean Yilgarn Block, Western Australia. GEOL SOC AUST SPEC PUBL **7,** 43–56.

Gemutz, I. & Theron, A. 1975. The Archaean between Coolgardie and Norseman—stratigraphy and mineralisation. *In* Knight, C. L. (ed.) *Economic geology of Australia and New Guinea, I. Metals*, 66–74. AUST INST MIN METALL MONOGRAPH **5**.

Gill, J. 1981. *Orogenic andesites and plate tectonics.* Berlin: Springer.

Glikson, A. Y. 1972. Early Precambrian evidence of a primitive ocean crust and island nuclei of sodic granite. GEOL SOC AM BULL **83,** 3323–44.

Green, D. H., Nicholls, I. A., Viljoen, M. & Viljoen, R. 1975. Experimental demonstration of the existence of peridotitic liquids in earliest Archean magmatism. GEOLOGY **3,** 11–14.

Griffiths, R. W. & Campbell, I. H. 1990. Stirring and structure in mantle plumes. EARTH PLANET SCI LETT **99,** 66–78.

Gromet, L. P. & Silver, L. T. 1987. REE variations across the Peninsular Ranges batholith: implications for batholithic petrogenesis and crustal growth in magmatic arcs. J PETROL **28,** 75–125.

Hallberg, J. A. 1972. Geochemistry of Archaean volcanic belts in the Eastern Goldfields region of Western Australia. J PETROL **13,** 45–56.

Hallberg, J. A. 1985. *Geology and mineral deposits of the Leonora-Laverton area, Northeastern Yilgarn Block, Western Australia.* Perth: Hesperian Press.

Hallberg, J. A. 1986. Archaean basin development and crustal extension in the northeastern Yilgarn Block, Western Australia. PRECAMB RES **31,** 133–56.

Hill, R. E. T., Gole, M. J. & Barnes, S. J. 1988. *Physical volcanology of komatiites: A field guide to the komatiites between Kalgoorlie and Wiluna, Eastern Goldfields Province, Yilgarn Block, Western Australia.* Perth: Geol Soc Australia, Western Australia Division.

Hill, R. I. 1988. San Jacinto intrusive complex, 1. Geology and mineral chemistry, and a model for intermittent recharge of tonalitic magma chambers. J GEOPHYS RES **93,** 10325–48.

Hill, R. I. 1991. Starting plumes and continental breakup. EARTH PLANET SCI LETT **104,** 398–416.

Hill, R. I. & Campbell, I. H. 1989. A post-metamorphic age for gold mineralisation at Lady Bountiful, Yilgarn Block, Western Australia. AUST J EARTH SCI **36,** 313–6.

Hill, R. I. & Campbell, I. H. 1992. Age and origin of granites from the Norseman region of Western Australia. AUST J EARTH SCI (in press).

Hill, R. I. & Compston, W. 1986. Age of granite emplacement, southeastern Yilgarn Block, Western Australia. AUST NAT UNIV RES SCH EARTH SCI ANN REP, 70–71.

Hill, R. I., Silver, L. T. & Taylor, H. P., Jr. 1986. Coupled Sr–O isotope variations as an indicator of source heterogeneity for the northern Peninsular Ranges batholith. CONTRIB MINERAL PETROL **92,** 351–61.

Hill, R. I., Chappell, B. W. & Silver, L. T. 1988. San Jacinto intrusive complex, 2. geochemistry. J GEOPHYS RES **93,** 10349–72.

Hill, R. I., Campbell, I. H. & Compston, W. 1989. Age and origin of granitic rocks in the Kalgoorlie-Norseman region of Western Australia: Implications for the origin of Archaean crust. GEOCHIM COSMOCHIM ACTA 53, 1259–75.

Hill, R. I., Campbell, I. H., Davies, G. F. & Griffiths, R. W. 1992 Mantle plumes and continental tectonics. SCIENCE (in press).

Jarvis, G. T. & Campbell, I. H. 1983. Archean komatiites and geotherms: solution to an apparent contradiction. GEOPHYS RES LETT **10,** 1133–6.

Johnson, R. W. & Chappell, B. W. 1979. Chemical analyses of rocks from the Late Cainozoic volcanoes of north-central New Britain and the Witu Islands, Papua New Guinea. REP BMR GEOL GEOPHYS **209**.

Kinny, P. D., Wijbrans, J. R., Froude, D. O., Williams, I. S. & Compston, W. 1990. Age constraints on the geological evolution of the Narryer Gneiss Complex, Western Australia. AUST J EARTH SCI **37,** 51–69.

Lambert, I. B. & Heier, K. S. 1968. Geochemical investigations of deep-seated rocks in the Australian shield. LITHOS **1,** 30–53.

Le Maitre, R. W. 1989. *A classification of igneous rocks and glossary of terms.* Oxford: Blackwell.

Libby, W. G. 1978. Regional variation in granitic rocks. GEOL SURV WESTERN AUST REP **9,** 53–109.

McCulloch, M. T. 1987. Sm–Nd isotopic constraints on the evolution of Precambrian crust in the Australian continent. *In* Kröner A. (ed.) *Proterozoic lithospheric evolution.* AMER GEOPHYS UNION GEODYNAMICS SER **17,** 115–30.

McCulloch, M. T., Compston, W. & Froude, D. 1983. Sm–Nd and Rb–Sr dating of Archaean gneisses, eastern Yilgarn Block, Western Australia. GEOL SOC AUST **30,** 149–53.

McDonough, W. F. & Frey, F. A. 1990. Geochemistry and mineralogy of rare earth elements. *In* Lipin, B. R. & McKay, G. A. (eds.) *Rare earth elements.* MIN SOC AM REV MINERAL **21,** 99–145.

McKenzie, D. 1984. The generation and compaction of partially molten rock. J PETROL **25,** 713–65.

McKenzie, D. & Bickle, M. J. 1988. The volume and composition of melt generated by extension of the lithosphere. J PETROL **29,** 625–79.

McNaughton, N. J. & Cassidy, K. F. 1990. A reassessment of the age of the Liberty Granodiorite: implications for a model of synchronous mesothermal gold mineralisation within the Norseman-Wiluna belt, Western Australia. AUST J EARTH SCI **37,** 373–6.

Nieuwland, D. A. & Compston, W. 1981. Crustal evolution in the Yilgarn Block near Perth, Western Australia. GEOL SOC AUST SPEC PUBL **7,** 159–71.

Oversby, V. M. 1975. Lead isotope systematics and ages of Archaean acid intrusives in the Kalgoorlie-Norseman area, Western Australia. GEOCHIM COSMOCHIM ACTA **39,** 1107–25.

Perring, C. S., Groves, D. I., Shellabear, J. N. & Hallberg, J. A. 1991. The "porphyry-gold" association in the Norseman-Wiluna belt of Western Australia: implications for models of Archaean gold metallogeny. PRECAMB RES **51,** 85–113.

Pidgeon, R. T., Wilde, S. A., Compston, W. & Shield, M. W. 1990. Archaean evolution of the Wongan Hills Greenstone Belt, Yilgarn Craton, Western Australia. AUST J EARTH SCI **37,** 279–92.

Schilling. J.-G. 1973. Iceland mantle plume: geochemical study of Reykjanes Ridge. NATURE **242,** 565–72.

Silver, L. T., Taylor, H. P., Jr., & Chappell, B. W. 1979. Some petrological, geochemical and geochronological observations of the Peninsular Ranges batholith near the international border of the U.S.A. and Mexico, *In* Abbott, P. L. & Todd, V. R. (eds) GEOL SOC AM 1979 GUIDEB 83–110.

Silver, L. T. & Chappell, B. W. 1988. The Peninsular Ranges batholith: an insight into the evolution of the Cordilleran batholiths of southwestern North America. TRANS R SOC EDINBURGH EARTH SCI **79,** 105–21.

Sofoulis, J. 1963. Boorabbin. WESTERN AUST GEOL SURV 1:250 000 SER EXPLANATORY NOTES.

Stephens, W. E. & Halliday, A. N. 1988. Geochemical contrasts between late Caledonian granitoid plutons of northern, central and southern Scotland. TRANS R SOC EDINBURGH EARTH SCI **79,** 259–73.

Sun, S.-S. & McDonough, W. F. 1989. Chemical and isotopic systematics of oceanic basalts: implications for mantle composition and processes. *In* Saunders, A. D. & Norry, M. J. (eds) *Magmatism in the ocean basins.* GEOL SOC LONDON SPEC PUBL **42,** 313–45.

Swager, C. & Griffin, T. J. 1990. *Geology of the Archaean Kalgoorlie Terrane.* Perth: Geological Survey of Western Australia.

Taylor, S. R. & McLennan, S. M. 1985. *The continental crust: its composition and evolution.* Oxford: Blackwell.

Turek, A. 1966. *Rubidium-strontium isotopic studies in the Kalgoorlie-Norseman area, Western Australia.* Unpublished Ph.D. Thesis, The Australian National University, Canberra.

Whalen, J. B. 1985. Geochemistry of an island-arc plutonic suite: the Uasilau-Yau Yau Intrusive Complex, New Britain, P.N.G. J PETROL **26,** 603–32.

White, A. J. R. & Chappell, B. W. 1977. Ultrametamorphism and granitoid genesis. TECTONOPHYSICS **43,** 7–22.

White, A. J. R. & Chappell, B. W. 1988. Some supracrustal (S-type) granites of the Lachlan Fold Belt. TRANS R SOC EDINBURGH EARTH SCI **79,** 169–81.

Whitford, D. J., Nicholls, I. A. & Taylor, S. R. 1979. Spatial variations in the geochemistry of Quaternary lavas across the Sunda arc in Java and Bali. CONTRIB MINERAL PETROL **70,** 341–56.

Wilde, S. A. & Pidgeon, R. T. 1986. Geology and geochronology of the Saddleback Greenstone Belt in the Archaean Yilgarn Block, southwestern Australia. AUST J EARTH SCI **33,** 491–501.

Williams, I. S. & Collins, W. J. 1990. Granite-greenstone terranes in the Pilbara Block, Australia, as coeval volcano-plutonic complexes; evidence from U–Pb zircon dating of the Mount Edgar Batholith. EARTH PLANET SCI LETT **97,** 41–53.

Witt, W. K. & Swager, C. P. 1989. Structural setting and geochemistry of the Archaean I-type granitoids in the Bardoc-Coolgardie area of the Eastern Goldfields Province, Western Australia. PRECAMBRIAN RES **44,** 323–51.

Wong, L., Davis, D. W., Krogh, T. E. & Robert, F. 1991. U–Pb zircon and rutile chronology of Archean greenstone formation and gold mineralization in the Val d'Or region, Quebec. EARTH PLANET SCI LETT **104,** 325–36.

Wyborn, L. A. I., Page, R. W. & Parker, A. J. 1987. Geochemical and geochronological signatures in Australian Proterozoic igneous rocks. *In* Pharaoh, T. C., Beckinsale, R. D. & Rickard, D. (eds). *Geochemistry and mineralization of proterozoic volcanic suites.* 377–394. GEOL SOC LONDON SPEC PUBL **33,** 377–94.

Wyborn, L. A. I., Wyborn, D., Warren, R. G. & Drummond, B. J. 1992. Proterozoic granite types in Australia: implications for lower crustal structure and evolution. TRANS R SOC EDINBURGH EARTH SCI **83,** 201–09.

R. I. HILL, Research School of Earth Sciences, The Australian National University, GPO Box 4, Canberra City, ACT 2601, Australia.
B. W. CHAPPELL, Department of Geology, The Australian National University, GPO Box 4, Canberra City, ACT 2601, Australia.
I. H. CAMPBELL, Research School of Earth Sciences, The Australian National University, GPO Box 4, Canberra City, ACT 2601, Australia.

Transactions of the Royal Society of Edinburgh: Earth Sciences, **83,** 227–233, 1992

Anorogenic granite magma genesis: new isotopic data for the southern sector of the British Tertiary Igneous Province

I. G. Meighan, A. E. Fallick and A. G. McCormick

ABSTRACT: It is now generally accepted that British Tertiary granites contain crustal and mantle components. Genesis principally by differentiation of crustally contaminated basaltic magmas is widely held and silicic melts with some remarkable trace element similarities were generated within *different* upper crust along the St Kilda/Skye – Carlingford zone.

New whole-rock (and mineral) O isotope data for the southern sector of the province (N Arran, Ailsa Craig, Mourne Mountains, Slieve Gullion, etc) reveal that $\delta^{18}O$ lies in the range $+5{\cdot}1$ to $+9{\cdot}7$‰ for most of the analysed granites, meteoric water-rock interaction having been in general less intensive than at Skye and Mull. Nevertheless, highly ^{18}O-depleted country rocks (with $\delta^{18}O < 0$) exist adjacent to the N Arran and Mourne Mountains granite plutons. There is as yet no evidence for the existence of low $-{}^{18}O$ granitic melts in this southern sector where magmatic $\delta^{18}O$ compositions (up to *c.* $+9{\cdot}5$‰) can be inferred for some of the intrusions.

New Nd (and Sr) isotope data indicate that although there is some similarity in initial $^{87}Sr/^{86}Sr$ ratios between the northern and southern sector granites, in northeastern Ireland initial ε_{Nd} values for the analysed Tertiary acid major intrusions range from $-3{\cdot}9$ to $-4{\cdot}5$. This is in marked contrast to the Skye granites, some of which have values below -20, reflecting the involvement of different lithosphere.

KEY WORDS: Arran, Mourne Mountains, Ailsa Craig plutons, A-types; basaltic differentiation; crustal contamination; meteoric water-rock interaction; Sr, Nd, O isotopes; trace elements.

The anorogenic and (?) plume-associated (White & McKenzie 1989) British Tertiary Igneous Province (BTIP) straddles the major structural elements of the British Isles, having developed across a region of highly varied upper crustal lithologies *and* differing basement (Fig. 1, Table 1) just before the opening of the NE Atlantic Ocean. It is thus an excellent field laboratory for studying essentially coeval (55–61 Ma) peraluminous-peralkaline granites and the contamination of contemporaneous basic melts (including MORB-like compositions), most of which belong to a relatively short-lived magmatic event of <8 Ma (e.g. Mussett *et al.* 1988).

Although relatively small, the BTIP has exerted a profound influence on petrological and geochemical thinking and indeed James Hutton made many fundamental observations on the Northern Granite Complex of Arran, including areas of different texture (grain size) within this pluton (G. Y. Craig, pers. comm. 1991).

The province has become renowned for the classic field mapping of J. E. Richey, A. Harker, E. B. Bailey and others. For example, Richey discovered four (now five) separate granite ring dykes in the two centres of the Mourne Mountains central complex (Richey 1928) and mapped this area of approximately 150 km^2 during only six weeks in the summer of 1925 (W. Q. Kennedy, pers. comm. 1972). In this study he also introduced the 'G' terminology now so familiar in enumerating the individual intrusions in a granite complex. Subsequently, through the work of S. R. Moorbath, H. P. Taylor Jr., R. W. Forester, A. P. Dickin and other geochemists, this province has featured in several pioneering isotopic studies on granite magma genesis, crustal contamination of basic melts and meteoric water-rock interaction. Thompson (1982) has published the most recent major review.

A full range of felsic rocks appears in the BTIP high level (volcanic-subvolcanic) "acid-basic" association: (1) major intrusions of granophyre ± felsite in such central intrusive complexes as Mull, Skye and Slieve Gullion: recently some of the felsites have been re-interpreted as welded tuffs (e.g. Sparks 1988; Bell & Emeleus 1988); (2) granites *sensu stricto* in such central intrusive complexes as the Mourne Mountains, Arran and Lundy; (3) acid minor intrusions associated with the central intrusive complexes; (4) rhyolitic (/obsidian) flows (/welded tuffs) within the basic lava sequences (which are never present regularly *throughout* these), e.g. County Antrim whose felsic lavas include some very highly fractionated high-silica rhyolite compositions (Meighan 1979; Meighan *et al.* 1984, 1988). Gravity studies (e.g. Bott & Tuson 1973) and in places field evidence indicate that overall the felsic rocks are considerably less abundant than basic/ultrabasic ones. Petrogenetically this has been taken to indicate sufficient basic magma to have generated the granitic melts by either crystal fractionation or (principally upper) crustal melting.

Normatively the felsic rocks range from peralkaline to peraluminous, but on a province-wide scale peralkaline compositions are subordinate and are absent from the major *granite* bodies (Mourne Mountains and N Arran where principally biotite granites, some highly fractionated, are encountered). Individual granitic bodies tend to exhibit relatively homogeneous chemistry (e.g. Brown & Rushton 1960) but where chemical variation exists within specific intrusions this can be attributed to magmatic pulses (e.g. Meighan *et al.* 1984) or acid-basic hybridisation: predictably these A-type intrusions (Eby 1990) lack any continuous concentric zoning (or internal fabric) and the differentiation of the acid melts dictated by geochemical evidence (Meighan 1979) must have occurred below the present exposure level.

In historical sequence the "majority view" of BTIP felsic magma genesis initially swung from basaltic differentiation to high-level crustal melting (Table 2), followed in the 1970s by a period of fierce debate between these two extremes. This has passed into a phase dominated by some degree of "petrogenetic convergence" in which mantle and crustal components are recognised (Meighan *et al.* 1984; Dickin *et al.* 1984) with seemingly no uncontaminated basaltic differentiate granites or "100% crustal melt" granites in the province. For Skye, the basic parent has been identified as the MORB-like Preshal Mhor basalt magma type (Dickin *et al.* 1984) and the BTIP granites in general may contain substantial asthenospheric mantle components (e.g. Thompson & Morrison 1988).

Despite the current recognition of mantle and crustal involvement, BTIP silicic magma genesis still poses many problems, especially the precise quantitative breakdown of the various components in each body. Furthermore, the relatively recent recognition that zoned magma chambers (ferrobasalt-rhyolite) existed at some of the central complexes (Marshall & Sparks 1984) indicates that magma mixing may have been an additional, complicating process for some intrusions. Also, it is significant that for Iceland, in a truly oceanic sector of the wider N Atlantic Tertiary Igneous Province, felsic magma genesis continues to be highly controversial, e.g. Macdonald *et al.* (1987, 1990) argue for fractional crystallisation (selective contamination) of basalt, whereas Thy *et al.* (1990) and Sigmarsson *et al.* (1991) envisage partial melting of the Icelandic oceanic crust.

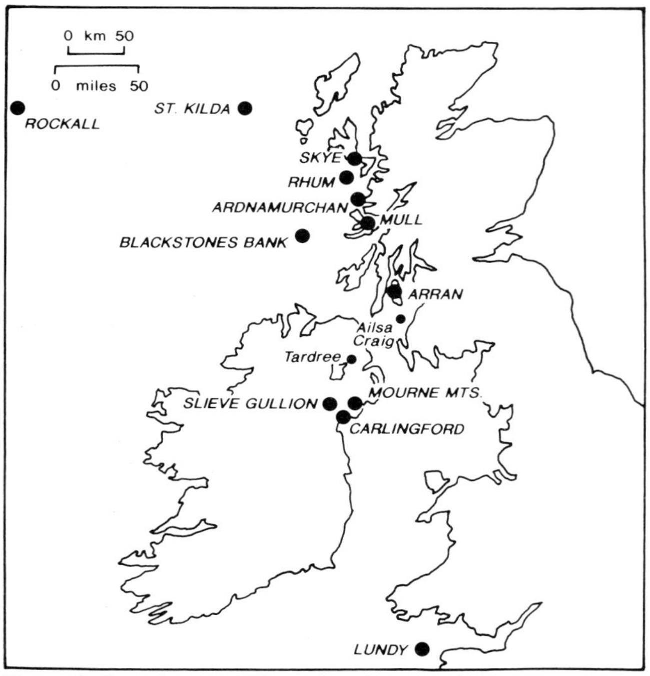

Figure 1 Location map illustrating the central complexes of the British Tertiary Igneous Province together with Ailsa Craig and Tardree, Co. Antrim.

Table 2 Petrogenetic history of BTIP granites.

1924, etc.	Basaltic differentiation
(1950s	Some granitisation)
1960s	Upper crustal melting
Early 1970s	Still argument whether "*pure*" crustal melts or *closed system* basaltic differentiates
Early 1980s	Petrogenetic convergence: differentiation of crustally contaminated basic melts
	No "pure" crustal melts (or closed system basaltic differentiates)?

Several approaches have been utilised to investigate BTIP granite magma genesis, including experimental petrology, trace element geochemistry/multielement (spidergram) plots and isotopic studies. Combined isotopic work (e.g. Sr–Nd–Pb) has been particularly revealing, but unfortunately there is a gross geographic imbalance here: for Skye (Table 3) there is a profusion of data (Sr, Pb, Nd, O, H, C and recently (Dickin *et al.* 1987) Ce isotopic ratios) for the

Table 3 Published isotopic coverage for some BTIP central complexes and the isotopic ranges for its granites.

Skye	Sr, Pb, Nd, O, H, C, Ce
Mull	Sr, (Pb), (Nd), O (H)
Arran	Sr, Pb (O)
St Kilda	Sr
Carlingford	Sr

Initial $^{87}Sr/^{86}Sr$: 0·7028 ± 9 (Ailsa Craig), 0·70379 ± 5 (St Kilda)—*c.* 0·732 (Coire Uaigneich Granophyre, Skye): many intrusions are in the 0·708–0·712 range

Initial ε_{Nd}: −0·9 (Lundy) to below −20 (Skye)

$\delta^{18}O$ (‰ SMOW): <0 to *c.* +9·5‰ (none >+10‰)

There are no published U-Pb zircon studies

Data sources: Harrison *et al.* (1987) for Ailsa Craig, Harding *et al.* (1984) for St Kilda, Thorpe *et al.* (1990) for Lundy, this study and numerous others cited in the references.

Table 1 BTIP central complexes on Skye–Lundy transect: oldest basement, structural setting and crustal lithologies.

Central complex	Oldest basement	Structural setting	Relevant crustal lithologies
Skye	Archaean	N of H.B.F. & I.S.	Lewisian gneisses (granulite and amphibolite facies), Torridonian arkoses, etc.
Mull	Archaean	N of H.B.F. & I.S., on G.G.F.	Lewisian and Moinian basement rocks, etc.
Arran	Proterozoic	N & S of H.B.F., N of I.S.	Dalradian metasediments, Proterozoic basement rocks, etc.
Mourne Mts	Proterozoic?	S of S.U.F., N of (but very near) I.S.	Lower Palaeozoic turbidites, Caledonian granites, Proterozoic basement rocks?
Lundy	Proterozoic/Palaeozoic	S of I.S.	Palaeozoic metasediments, granites.

Notes: H.B.F. = Highland Boundary Fault, I.S. = Iapetus Suture, G.G.F. = Great Glen Fault, S.U.F. = Southern Uplands Fault.

acid and basic rocks. For Mull, Rhum, Arran and Lundy there is rather less complete coverage, and for other central complexes (e.g. St Kilda, Carlingford) the only published isotopic data are for Sr.

This paper reviews certain aspects of the last two decades of research on BTIP granite magmatism and summarises new isotopic results for the southern sector of the province (N Arran, Mourne Mountains, etc). For some of the central complexes involved in this study, e.g. the Mourne Mountains, there is a large new isotopic database and the full, individual results will be published elsewhere. These were determined by the standard techniques of the Scottish Universities Research and Reactor Centre, East Kilbride.

1. BTIP granite research since 1970

The last two decades have seen many important contributions on BTIP granitic rocks principally in the fields of trace element and isotope geochemistry. Some of these are summarised below (in no particular order of priority).

(1) Taylor and Forester (1971) demonstrated that for Skye and Mull, two of the central complexes with major exposures of felsic rocks, in general the silicic and basic rocks have suffered substantial high-temperature meteoric water-rock interaction with whole-rock $\partial^{18}O$ values now <0 in some cases. Later studies, e.g. Pankhurst *et al.* 1978, Walsh *et al.* (1979), have suggested that for the most part this hydrothermal activity produced remarkably little change in the pristine chemistry of the granitic rocks (but cf. Ferry 1985) and thus magma chemistries can be deduced provided the major fluid channelways are avoided. Unfortunately, full stable isotope studies have not been extended to all the other BTIP central complexes, for some of which, e.g. St Kilda and Carlingford, there are no published determinations (Table 3).

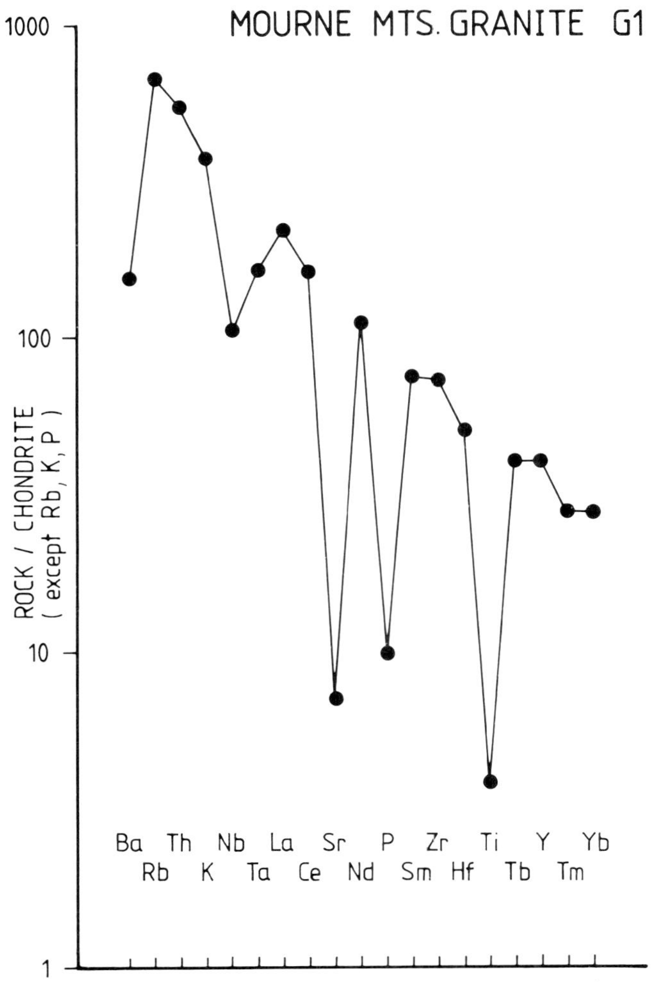

Figure 2 Normalised trace-element pattern (spidergram) for Mourne Mountains granite G1. Normalisation as in Thompson (1982) for sample G1/17 (Meighan *et al.* 1984).

(2) Meighan (1979) and others have presented trace element data which indicate that some of the intrusive and extrusive/peralkaline and subalkaline felsic melts are the products of extreme (feldspar-dominated) crystal fractionation. Differentiation of the BTIP *granitic* magmas is now generally accepted (Thompson 1982; Gamble *et al.* 1992) and is the only reasonable explanation for the acid rocks in the province with <5 ppm Sr, Eu/Eu* < 0·1 and Rb/Sr > 100. Clearly, this conclusion dictates that many of the felsic rocks cannot be considered as potential primary magma compositions formed by melting of "normal crustal sources", whereas the more primitive granitic rocks must be evaluated in this context. Significantly, spidergrams suggest that many of the latter also solidified from melts which had undergone crystal fractionation, e.g. Mourne Mountains granite G1 (Fig. 2) with *c.* 70% SiO_2 shows Ba, Sr, P and Ti relative depletions suggestive of fractionation which included feldspars and accessory minerals.

(3) The mantle component in BTIP felsic magmas can be explained by:

(i) mixing of crustal melt and *closed system* basaltic differentiate granites, with or without subsequent refusion and evolution of such hybrid systems (Thompson 1982), this magma mixing hypothesis being possible in, for example, a zoned magma chamber situation (Marshall & Sparks 1984);

(ii) differentiation of crustally contaminated basic melts.

The second model is now the consensus view, and it is apparent (principally from research on Skye and Mull) that individual magmas may have been subjected to (a) different *styles of contamination,* and (b) different *contaminants* (Table 4). For example, the crustal component may have been introduced as a partial melt formed adjacent to a plexus of minor intrusions in the lower crust *and* subsequently during combined assimilation–fractional crystallisation (AFC) in upper crustal magma reservoirs. Two styles of selective contamination have been advocated for Skye. One involves actual magma derived from, for example, the selective melting of leucogranite sheets in lower crustal intermediate gneisses (Dickin *et al.* 1987), whereas the other is contamination at elemental/isotopic level. The latter has been applied especially in explaining the massive crustal Pb isotopic component in some of the Skye basic rocks and may also be relevant to the Sr isotopic compositions of certain granites, e.g. the Mourne Mountains as discussed below. Finally, it is now generally accepted that the crustal component in BTIP granites was not introduced during meteoric water-rock interaction (cf. Taylor &

Table 4 BTIP granites: possible modes of incorporation for the crustal component.

1. Lower crustal melt contaminates basic magmas before fractional crystallisation: melting controlled by basic minor intrusions and may have been selective in terms of rock type (e.g. leucogranites within intermediate gneisses).
2. AFC in high-level basic magma reservoirs.
3. Mixing of basaltic differentiate and upper crustal melt granites, e.g. where zoned magma chambers.
4. Crystal fractionation of basic melts plus selective (*elemental/isotopic*) contamination.

Forester 1971) but at the magmatic stage. Indeed some of the new data presented below provide further support for this in that there are granites in the Mourne Mountains and North Arran with "normal" whole-rock $\delta^{18}O$ values up to *c.* 9–9·5‰ and *elevated* initial Sr isotope ratios. Also, for certain individual bodies (e.g. Mourne Mountains biotite granite G3, Meighan *et al.* 1988), excellent quality Rb–Sr whole-rock isochrons with low MSWD values testify to the introduction of the crustal component before the final differentiation of the acid magmas.

(4) The crustal component in most Skye granites is believed to be *c.* 10% (Dickin *et al.* 1984) but it seems to be very much higher (approximately 67%) in the relatively early Coire Uaigneich granophyre (Table 3; Dickin & Exley 1981). Indeed, there is a tendency throughout the province for the earliest felsic melts at each central complex to have had the highest $^{87}Sr/^{86}Sr$ ratios (e.g. Skye, Mull, Arran, Mourne Mountains). Unfortunately, there are insufficient available Nd isotopic data to allow comment in this context, but for the Mourne Mountains (Table 7) it is interesting that the elevated initial Sr isotope ratio for the G1 granite (Meighan *et al.* 1988) is *not* accompanied by initial ε_{Nd} values significantly different from the later granites (G2–G5 inclusive).

The presence of a crustal component in BTIP granites seems irrefutable, but it is interesting to speculate that the Tertiary lithospheric mantle beneath Skye and Mull could have had radiogenic isotopic compositions very similar to parts of the Lewisian crust (Menzies *et al.* 1987; Thompson & Morrison 1988). Clearly this point is very important in studying the crustal contamination of BTIP basic magmas and indicates that the crustal component in some of the granites could be substantially lower than current estimates suggest. Nevertheless, in the southern sector of the province, where magmatic $\partial^{18}O$ compositions can be inferred more readily for the granites (see Section 3.1), values up to *c.* 9·5‰ (in excess of the normal basaltic differentiate range) point to measurable crustal involvement.

From (1)–(4) above it is apparent that petrogenetic studies on BTIP granites must now evaluate a formidable array of processes: (1) meteoric water-rock interaction (and/or autometasomatism) which may have modified their pristine chemistry; (2) crystal fractionation (feldspar(s), ferromagnesian minerals, accessory minerals) in the *felsic* melts and also mixing of these with other felsic (or more basic) magmas; (3) crustal melting (in general obvious potential restitic material (Chappell *et al.* 1987) does not occur in these granites so the problem of restite-melt separation is suppressed); (4) both upper and lower crustal contamination of the basic/more evolved magmas (bulk and/or selective); (5) mantle melting which resulted in both lithospheric and asthenospheric components in the basic melts, the former with radiogenic isotopic signatures similar to parts of the crust.

2. BTIP felsic magmas: a geochemical comparison across the province

As noted above, the BTIP developed over a region of highly varied upper crustal lithologies (Table 1). In certain areas, these include some high Sr (>400 ppm) compositions with $Ce_N/Yb_N > 10$, e.g. Lewisian gneisses in northwestern Scotland and Caledonian (Newry complex) granodiorites in northeastern Ireland. Nevertheless, the Tertiary felsic rocks exhibit some important general geochemical similarities such as *low Sr* (<100 ppm), relatively *low* Ce_N/Yb_N ratios (<8) and obvious *negative Eu anomalies.* Table 5 illustrates this point for specific examples with relatively similar SiO_2/Thornton-Tuttle differentiation index values in order to reduce the variable of differing degrees of fractional crystallisation in the acid melts. The remarkable similarity from St Kilda/Skye to Carlingford is what would be expected among silicic magmas generated principally by basaltic differentiation. Of the variation that does exist, clearly some (Sr, Eu/Eu*) can be attributed to the rocks not exhibiting identical degrees of fractionation, and the remainder (e.g. Ce_N/Yb_N) is within the known range for Tertiary basic melts (e.g. Thompson *et al.* 1980, Lyle 1985).

3. New isotopic data for the S sector of the BTIP

Overall, the granites of the southern sector (Arran, Ailsa Craig, northeastern Ireland, Lundy) are believed to have similar petrogenesis to their northern counterparts (e.g. Meighan 1979, Meighan *et al.* 1984, Dickin *et al.* 1981, Thorpe *et al.* 1990), although it must be noted that Stone (1988, 1990) assigns general S-type status to the Lundy intrusions. The paucity of published isotopic data factor (Table 3) is particularly serious for parts of this sector, especially for O. Recognising this and the importance of information for several isotopic systems, new data for Arran (O) and northeastern Ireland (O, Nd, Sr) are summarised and discussed in relation to the northern sector.

Table 5 Selective geochemical data for certain felsic rocks from northeastern Ireland, Arran, Mull, Skye and St Kilda (southern to northern sector of BTIP).

Location	Carlingford	Slieve Gullion	Mourne Mts	County Antrim	Arran	North Arran	Mull Centre 3	Skye W Red Hills Centre	St Kilda
Lithology (Q.U.B. sample number)	Granophyre, W side of Slieve Foye (Q.U.B. 45)	Ring Complex Granophyre (Q.U.B. 42)	Granite G1 (Q.U.B. 121)	Porphyritic Obsidian, Sandy Braes (Q.U.B. 4)	Corriegills Pitchstone (Q.U.B. 72)	Coarse Granite (Q.U.B. 6)	Toll Doire Granophyre (Q.U.B. 70)	Loch Ainort Epigranite (Q.U.B. 68)	Glen Bay Granophyre (Q.U.B. 73)
SiO_2 (wt.%)	72·2	69·6	70·7	75·5	74·0	76·1	75·9	72·7	72·9
D.I.	87·7	88·2	88·6	92·3	93·4	93·5	94·6	90·9	93·7
Sr (p.p.m.)	70	80	83	43	34	57	40	55	44
Ce_N/Yb_N	4·00	4·60	5·74	5·52	3·10	2·83	3·04	2·39	3·74
Eu/Eu*	0·46	0·62	0·49	0·32	0·41	0·39	0·48	0·48	0·47
Crustal setting	'Southern Uplands'	'Southern Uplands'	'Southern Uplands'	'Midland Valley'	Midland Valley	on HBF	Scottish Highlands	Scottish Highlands	'Scottish Highlands'

Notes: Major elements and Sr were determined by X-ray fluorescence analysis, Ce, Yb and Eu by neutron activation analysis (as in Meighan *et al.* 1984); D.I. = Thornton—Tuttle differentiation index; HBF = Highland Boundary Fault; Q.U.B. = Queen's University, Belfast.

3.1. O isotopic data: N Arran, Ailsa Craig and northeastern Ireland

Table 6 summarises new O isotopic data for the granitic rocks of N Arran (the coarse biotite granite and later, more fractionated, 'fine' biotite granite) and the Mourne Mountains (amphibole–biotite granites G1 and G5 together with the very fractionated biotite granites G2, G3 and G4: McCormick *et al.* under review). Data for the N Arran (Dalradian) and Mourne Mountains (Silurian metasedimentary) country rocks are included, as are results for the Ailsa Craig peralkaline microgranite and other non-peralkaline rocks from northeastern Ireland (the Tardree rhyolite of County Antrim and granophyres from Slieve Gullion and Carlingford). Accepting that the database for Slieve Gullion/Carlingford is meagre (three/one samples, respectively), the following conclusions can be drawn:

(1) Excluding four samples (two from N Arran, one from Slieve Gullion and the Carlingford granophyre), all the Tertiary rocks lie in the range +5·1–+9·7‰. However, the less extensive mineral data indicate that these cannot all be accepted as *magmatic* compositions. $\Delta_{QTZ\text{-}A.F.}$ is clearly in excess of two (i.e. outside the magmatic equilibrium range) in some of the N Arran, Mourne Mountains and Slieve Gullion rocks. Nevertheless, it is apparent that the majority of the analysed samples from the southern sector of the BTIP suffered at the most slight meteoric water-rock interaction, this phenomenon having been less intensive than at Skye and Mull where negative whole-rock $\partial^{18}O$ values are encountered. However, the Carlingford central complex, which yielded a whole-rock value of *c.* 0, clearly merits fuller investigation.

(2) The $\partial^{18}O_{QTZ}$ values determined on some of the samples all exceed +7‰ and in combination with the whole-rock data indicate that in contrast to Skye (Forester & Taylor 1977) there is as yet no evidence for the existence of low $-^{18}O$ granitic magmas in the southern sector. However, in the absence of mineral analyses the very depleted Carlingford sample cannot be assessed in this context.

(3) Accepting that there has been at the most relatively minor meteoric water-rock interaction for most of the granitic rocks under discussion, it follows that *magmatic* $\partial^{18}O$ values can be inferred more readily than in the northern sector. For the eastern Mourne Mountains, these range from *c.* +7·5‰ (G1 granite) to about +9·5‰ (G2), and for N Arran again a maximum value of about +9·5‰ is suggested. The latter figure is significantly higher than the maximum on the normal basaltic differentiate range (+7·5‰) and is taken to reflect the crustal component in these granites. By contrast, on its general chemistry and low initial Sr isotope ratio (Table 3) Ailsa Craig appears as a potential basaltic differentiate with a minimal crustal component. It is unlikely that this small, isolated, peralkaline microgranite mass was subjected to intensive meteoric water-rock interaction, and, interestingly, the whole-rock O isotope analyses do lie within the basaltic differentiate range. However, $\partial^{18}O$ mineral determinations are required before the data reported here can be accepted as magmatic values.

(4) In both N Arran and the Mourne Mountains the immediate aureole rocks are severely depleted in ^{18}O (with $\delta^{18}O < 0$) and clearly the meteoric fluids wrought more spectacular changes than in the plutons themselves. In N Arran, ^{18}O-depleted Dalradian country rocks exist until approximately 1 km out from the pluton margin with in general the most depleted values near this.

Finally, it must be noted that at the Mourne Mountains and N Arran there is no evidence that Tertiary lava sequences were erupted. This is in marked contrast to Skye and Mull, and it seems that no large-scale, meteoric-hydrothermal convection systems had been established in the upper crust while granite genesis occurred beneath the Mourne and N Arran regions. Such systems might have accelerated the cooling of the granitic melts, preventing extreme fractionation, and in this context it is significant that the Mourne Mountains and N Arran Tertiary centres contain some of the most highly fractionated silicic compositions in the BTIP. This is evidenced by the "fine" biotite granite of N Arran and Mourne Mountains biotite granites G2, G3 and G4, all of which are enriched in Rb and severely depleted in Sr with Rb/Sr ≫ 10 and Eu/Eu* < 0·15 (Meighan *et al.* 1984).

3.2. Nd and Sr isotopic data for the southern sector of the BTIP

Table 7 summarises the existing and new Nd and Sr isotopic data for the southern sector of the province and allows the following conclusions to be drawn:

(1) Excepting Ailsa Craig (for which no Nd data are available), none of the analysed compositions can be a

Table 6 O isotope data (‰ SMOW) for the felsic rocks of N Arran, Ailsa Craig and northeastern Ireland and the N Arran/Mourne Mountains country rocks.

	Coarse biotite granite, N Arran	Fine biotite granite, N Arran	Ailsa Craig peralkaline microgranite	Mourne Mts granites G1–G5 inclusive	Slieve Gullion granophyres/ microgranite	Carlingford granophyre	Tardree rhyolite, Co. Antrim
$\delta^{18}O_{W.R.}$	+1·3–+8·9 (12) +5·3–+8·9 (11)	+4·7,† +5·1 (2)	+6·6, +7·4* (1)	+6·1–+9·7 (36)	+8·3, +5·7,† +4·2 (3)	0·0, +0·5* (1)	+8·4 (1)
$\delta^{18}O_{QTZ}$	+8·8–+9·9 (4)	+8·3† (1)		+7·2–+9·6 (20)	+8·7† (1)		
$\Delta_{QTZ\text{-}A.F.}$	2·3–4·2 (4)	4·7† (1)		0·9–3·9 (15)	4·6† (1)		

Whole-rock values for country rocks: Dalradian metapelites, North Arran −3·3–+10·5 (13); Silurian metasediments, Mourne Mts −4·6–−0·5 (4) (Lower Palaeozoic metasediments of Southern Uplands +10·8–+13·9).

The individual Mourne granite ranges are: G1 6·1–8·1, G2 6·1–9·7, G3 6·9–8·7, G4 6·7–7·9, G5 6·9–8·5 (all values positive). An aplite in the coarse granite, N Arran, gave duplicate results of +9·2, +9·0.

Notes: Number of samples in brackets; W.R. = whole rock, QTZ = quartz, A.F. = alkali feldspar, * = duplicate determination, † = same sample; Slieve Gullion data are from Gamble *et al.* (1992); Carlingford sample is as in Table 5; Arran data are from an unpublished study with J. D. Hudson and A. P. Dickin.

Table 7 Initial ε_{Nd} and $^{87}Sr/^{86}Sr$ data for the felsic rocks of the southern sector of the BTIP.

	Initial ε_{Nd}	$(^{87}Sr/^{86}Sr)_o$
Arran		0·7071–0·7186
Ailsa Craig		0·7028
Mourne Mts (G1–G5 inclusive)	−3·9–−4·5 (8)	0·7085–0·7147
Slieve Gullion	−4·0–−4·1 (3)	0·7097–0·7104
Carlingford		0·7099
Tardree rhyolite, Co. Antrim	−3·4 (1)	0·7073
Lundy	−0·9–−1·9 (5)	0·7146

Data sources: this study, Dickin *et al.* (1981) (Arran), Harrison *et al.* (1987) (Ailsa Craig), Gamble *et al.* (1992) (Slieve Gullion), Thorpe *et al.* (1990) (Lundy); Nd result for Tardree was determined at Royal Holloway and Bedford New College as in Gamble *et al.* (1992); number of samples in brackets; Carlingford (Q.U.B. 45) and Tardree (Q.U.B. 361) samples are as in Table 6.

typical, closed-system basaltic differentiate: initial $^{87}Sr/^{86}Sr$ is elevated and variable, but in many cases similar to the northern sector values (Table 3), and initial ε_{Nd} values are in the narrow range of −3·9–−4·5 for the Irish major intrusive rocks. Significantly, and in marked contrast to Skye (Table 3), no strongly negative ε_{Nd} values have been obtained so far for felsic rocks from this southern sector where there is thus no evidence for the involvement of Lewisian (as opposed to younger) basement in Tertiary granite genesis.

(2) The Mourne Mountains granites have initial ε_{Nd} and $\partial^{18}O$ *magmatic* values between those for the upper crust in this region (low-grade Lower Palaeozoic metasediments) and asthenospheric mantle. In general, the initial Sr and Nd isotopic results (and O where magmatic $\partial^{18}O$ values may be inferred) can be taken to support the presence of both crustal and mantle components in the southern sector granitic melts, but trace element evidence (e.g. Meighan *et al.* 1984) can also be very important in revealing the mantle component. By contrast, the available isotopic data for Ailsa Craig (Sr, O) do not reveal an obvious crustal component in this peralkaline microgranite.

(3) In the Mourne Mountains, the first intrusion (G1) is the least siliceous and most Sr-rich granite of this central complex and has a higher initial $^{87}Sr/^{86}Sr$ ratio (0·7129 ± 2) than the more fractionated G2–G4 inclusive (0·7085 ± 17–0·7114 ± 3). However, the initial ε_{Nd} values for the two analysed samples of G1 (−3·9, −4·0) are not significantly different from those of the later granites (−4·1–−4·5). In explaining this apparent decoupling of the Sm–Nd and Rb–Sr isotopic systems, it is advocated that selective (elemental/isotopic) contamination of the granitic melts (involving Sr but not Nd) must be given serious consideration.

4. Conclusions

(1) It is now generally accepted that BTIP granite magmas contained crustal and mantle components and it is widely held that most were differentiates of crustally contaminated basaltic melts. The geochemical similarity of certain silicic rocks over the St Kilda/Skye-Carlingford zone is readily explained by the granitic magmas having been generated principally by basaltic differentiation.

(2) New oxygen isotope data (for whole rocks and minerals) indicate that in general meteoric water-rock interaction was less severe in the southern sector of the BTIP than at the Skye and Mull central complexes. Most of the analysed granites have whole-rock $\partial^{18}O$ values in excess of +5‰ and for some magmatic values can be inferred (whose upper limit of *c.* +9.5‰ (N Arran, Mourne Mountains) is taken to reflect their crustal component). Nevertheless, highly ^{18}O-depleted country rocks (with whole-rock $\partial^{18}O$ values < 0) occur adjacent to the Mourne Mountains and N Arran granite plutons. There is as yet no evidence for the existence of low-^{18}O granite magmas in the southern sector.

(3) It is noted that where intensive, large-scale meteoric water-rock interaction did not occur as it did at Skye and Mull, highly fractionated granitic melts developed (e.g. Mourne Mountains, N Arran).

(4) New Nd (and Sr) isotope data indicate that although there is some similarity in initial $^{87}Sr/^{86}Sr$ ratios between the northern and southern sector granites, in northeastern Ireland initial ε_{Nd} values for the Tertiary acid major intrusions range from −3·9–−4·5. This is in marked contrast to the Skye granites, some of which have values below −20 reflecting the involvement of different lithosphere. The Mourne Mountains granites display an appreciable range in initial $^{87}Sr/^{86}Sr$ ratios, yet initial ε_{Nd} values are relatively constant at *c.* −4.

5. Acknowledgements

IGM gratefully acknowledges financial assistance from The Queen's University of Belfast which enabled him to attend the Second Hutton Symposium in Canberra. AGM acknowledges a postgraduate studentship awarded by the Department of Education for Northern Ireland. The Scottish Universities Research and Reactor Centre is supported by N.E.R.C. and Scottish universities. We are grateful to the following for their assistance and patience in the production of this paper: Peter Brown (University of St Andrews); Emer Carlin and Kathleen Irvine (Q.U.B.).

6. References

Bell, B. R. & Emeleus, C. H. 1988. A review of silicic pyroclastic rocks of the British Tertiary Volcanic Province. *In* Morton, A. C. & Parson, L. M. (eds) *Early Tertiary Volcanism and the Opening of the NE Atlantic*, 365–79. SPEC PUBL GEOL SOC LONDON **39**.

Bott, M. H. P. & Tuson, J. 1973. Deep Structure beneath the Tertiary Volcanic Regions of Skye, Mull and Ardnamurchan, North-west Scotland. NATURE PHYS SCI **242,** 114–6.

Brown, P. E. & Rushton, B. J. 1960. Some chemical data on the Mourne Mountain granite G2. GEOCHIM COSMOCHIM ACTA **18,** 193–9.

Chappell, B. W., White, A. J. R. & Wyborn, D. 1987. The Importance of Residual Source Material (Restite) in Granite Petrogenesis. J PETROL **28,** 1111–38.

Dickin, A. P. & Exley, R. A. 1981. Isotopic and Geochemical Evidence for Magma Mixing in the Petrogenesis of the Coire Uaigneich Granophyre, Isle of Skye, N.W. Scotland. CONTRIB MINERAL PETROL **76,** 98–108.

Dickin, A. P., Moorbath, S. & Welke, H. J. 1981. Isotope, trace element and major element geochemistry of Tertiary igneous rocks, Isle of Arran, Scotland. TRANS R SOC EDINBURGH EARTH SCI **72,** 159–70.

Dickin, A. P., Brown, J. L., Thompson, R. N., Halliday, A. N. & Morrison, M. A. 1984. Crustal contamination and the granite problem in the British Tertiary Volcanic Province. PHILOS TRANS R SOC LONDON **310A,** 755–80.

Dickin, A. P., Jones, N. W., Thirlwall, M. F. & Thompson, R. N. 1987. A Ce/Nd isotope study of crustal contamination processes affecting Palaeocene magmas in Skye, Northwest Scotland. CONTRIB MINERAL PETROL **96,** 455–64.

Eby, G. N. 1990. The A-type granitoids: A review of their occurrence and chemical characteristics and speculations on their petrogenesis. LITHOS **26,** 115–34.

Ferry, J. M. 1985. Hydrothermal alteration of Tertiary igneous rocks from the Isle of Skye, northwest Scotland II. Granites. CONTRIB MINERAL PETROL **91,** 283–304.

Forester, R. W. & Taylor, H. P. Jr., 1977. $^{18}O/^{16}O$, D/H and $^{13}C/^{12}C$ studies of the Tertiary igneous complex of Skye, Scotland. AM J SCI **277,** 136–77.

Gamble, J. A., Meighan, I. G. & McCormick, A. G. 1992. The petrogenesis of Tertiary microgranites and granophyres from the Slieve Gullion Central Complex, NE Ireland. J GEOL SOC LONDON, **149,** 93–106.

Harding, R. R., Merriman, R. J. & Nancarrow, P. H. A. 1984. St Kilda: an illustrated account of the geology. REP BR GEOL SURV **16**(7).

Harrison, R. K., Stone, P., Cameron, I. B., Elliot, R. W. & Harding, R. R. 1987. Geology, petrology and geochemistry of Ailsa Craig, Ayrshire. REP BR GEOL SURV **16**(9).

Lyle, P. 1985. The petrogenesis of the Tertiary basaltic and intermediate lavas of northeast Ireland. SCOTT J GEOL **21,** 71–84.

McCormick, A. G., Fallick, A. E., Harmon, R. S., Meighan, I. G. & Gibson, D. (under review). Oxygen and hydrogen isotope geochemistry of the Mourne Mountains Tertiary granites, Northern Ireland. J PETROL (under review).

Macdonald, R., Sparks, R. S. J., Sigurdsson, H., Mattey, D. P., McGarvie, D. W. & Smith, R. L. 1987. The 1875 eruption of Askja volcano, Iceland: combined fractional crystallization and selective contamination in the generation of rhyolitic magma. MINERAL MAG **51,** 183–202.

Macdonald, R., McGarvie, D. W., Pinkerton, H., Smith, R. L. & Palacz, Z. A. 1990. Petrogenetic Evolution of the Torfajökull Volcanic Complex, Iceland I. Relationship Between the Magma Types. J PETROL **31,** 429–59.

Marshall, L. A. & Sparks, R. S. J. 1984. Origin of some mixed-magma and net-veined ring intrusions. J GEOL SOC LONDON **141,** 171–82.

Meighan, I. G. 1979. The acid igneous rocks of the British Tertiary Province. BULL GEOL SURV GB **70,** 10–22.

Meighan, I. G., Gibson, D. & Hood, D. N. 1984. Some aspects of Tertiary acid magmatism in NE Ireland. MINERAL MAG **48,** 351–63.

Meighan, I. G., McCormick, A. G., Gibson, D., Gamble, J. A. & Graham, I. J. 1988. Rb–Sr isotopic determinations and the timing of Tertiary central complex magmatism in NE Ireland. *In* Morton, A. C. & Parson, L. M. (eds) *Early Tertiary Volcanism and the Opening of the NE Atlantic,* 349–60. SPEC PUBL GEOL SOC LONDON **39**.

Menzies, M. A., Halliday, A. N., Palacz, Z., Hunter, R. H., Upton, B. G. J., Aspen, P. & Hawkesworth, C. J. 1987. Evidence from mantle xenoliths for an enriched lithospheric keel under the Outer Hebrides. NATURE **325,** 44–7.

Mussett, A. E., Dagley, P. & Skelhorn, R. R. 1988. Time and duration of activity in the British Tertiary Igneous Province. *In* Morton, A. C. & Parson, L. M. (eds) *Early Tertiary Volcanism and the Opening of the NE Atlantic,* 337–48. SPEC PUBL GEOL SOC LONDON **39**.

Pankhurst, R. J., Walsh, J. N., Beckinsale, R. D. & Skelhorn, R. R. 1978. Isotopic and other geochemical evidence for the origin of the Loch Uisg Granophyre, Isle of Mull, Scotland. EARTH PLANET SCI LETT **38,** 355–63.

Richey, J. E. 1928. The structural relations of the Mourne granites (Northern Ireland). Q J GEOL SOC LONDON **83,** 653–88.

Sigmarsson, O., Hémond, C., Condomines, M., Fourcade, S. & Oskarsson, N. 1991. Origin of silicic magma in Iceland revealed by Th isotopes. GEOLOGY **19,** 621–4.

Sparks, R. S. J. 1988. Petrology and geochemistry of the Loch Ba ring dyke, Mull (N.W. Scotland): an example of extreme differentiation of tholeiitic magmas. CONTRIB MINERAL PETROL **100,** 446–61.

Stone, M. 1988. The significance of almandine garnets in the Lundy and Dartmoor granites. MINERAL MAG **52,** 651–8.

Stone, M. 1990. The Lundy granite: a geochemical and petrogenetic comparison with Hercynian and Tertiary granites. MINERAL MAG **54,** 431–46.

Taylor, H. P. Jr. & Forester, R. W. 1971. Low-O^{18} Igneous Rocks from the Intrusive Complexes of Skye, Mull and Ardnamurchan, Western Scotland. J PETROL **12,** 465–97.

Thompson, R. N. 1982. Magmatism of the British Tertiary Volcanic Province. SCOTT J GEOL **18,** 49–107.

Thompson, R. N., Gibson, I. L., Marriner, G. F., Mattey, D. P. & Morrison, M. A. 1980. Trace-Element Evidence of Multistage Mantle Fusion and Polybaric Fractional Crystallization in the Palaeocene Lavas of Skye, NW Scotland. J PETROL **21,** 265–93.

Thompson, R. N. & Morrison, M. A. 1988. Asthenospheric and lower-lithospheric mantle contributions to continental extensional magmatism: An example from the British Tertiary Province. CHEM GEOL **68,** 1–15.

Thorpe, R. S., Tindle, A. G. & Gledhill, A. 1990. The Petrology and Origin of the Tertiary Lundy Granite (Bristol Channel, UK). J PETROL **31,** 1379–406.

Thy, P., Beard, J. S. & Lofgren, G. E. 1990. Experimental constraints on the origin of Icelandic rhyolites. J GEOL **98,** 417–21.

Walsh, J. N., Beckinsale, R. D., Skelhorn, R. R. & Thorpe, R. S. 1979. Geochemistry and Petrogenesis of Tertiary Granitic Rocks from the Island of Mull, Northwest Scotland. CONTRIB MINERAL PETROL **71,** 99–116.

White, R. S. & McKenzie, D. P. 1989. Magmatism at rift zones: the generation of volcanic continental margins and flood basalts. J GEOPHYS RES **94,** 7685–729.

I. G. MEIGHAN and A. G. McCORMICK, Department of Geology, School of Geosciences, The Queen's University of Belfast, Belfast BT7 1NN, Northern Ireland.

A. E. FALLICK, Isotope Geology Unit, Scottish Universities Research and Reactor Centre, East Kilbride, Glasgow G75 0QU, Scotland.

Transactions of the Royal Society of Edinburgh: Earth Sciences, **83,** 235–245, 1992

The relationship between crustal magmatic underplating and granite genesis: an example from the Velay granite complex, Massif Central, France

B. J. Williamson, H. Downes and M. F. Thirlwall

ABSTRACT: The Velay granite pluton (Massif Central, France) is the youngest (304 ± 5 Ma) and largest (~6,900 km^2) of the major Massif Central monzogranites/granodiorites and was formed nearly 50 Ma after the cessation of Hercynian continental collision (Pin & Duthou 1990). It is a highly heterogeneous pluton consisting of I-type, high-Sr granites (Sr = 500–900 ppm) with low $\varepsilon_{Sr_{(304)}}$ (+35 to +41) and high $\varepsilon_{Nd_{(304)}}$ (−3 to −5), at its centre, grading into S-type and mixed I–S-type heterogeneous granites of more normal Sr content (100–420 ppm) and higher $\varepsilon_{Sr_{(304)}}$ (+40 to +210) and lower $\varepsilon_{Nd_{(304)}}$ (−3·8 to −7.3) at its margins.

The metasedimentary lower crust of the Massif Central was underplated/intruded by mafic mantle-derived magmas between 360 Ma and 300 Ma. From 300–280 Ma (Downes *et al.* 1991) underplating led to partial melting and granulite facies metamorphism of the underplated material (represented by felsic and mafic meta-igneous lower crustal xenoliths, $\varepsilon_{Sr_{(304)}}$ = −11 to +112, $\varepsilon_{Nd_{(304)}}$ = +2·2 to 8·2, Downes *et al* 1990). The partial melts assimilated mainly schist but also felsic gneiss and older granite country rock material ($\varepsilon_{Sr_{(304)}}$ = +100 to +300, $\varepsilon_{Nd_{(304)}}$ = −5 to −9) to produce the heterogeneous granites. Plagioclase and biotite were accumulated at the base of the intrusion which was intruded to high levels to form the high-Sr granites.

KEY WORDS: underplating, I-type, S-type, French Massif Central, Hercynian.

The crustal or mantle origin of granites is controversial (e.g. Allègre & Ben Othman 1980; Duthou *et al.* 1984; Pin & Duthou 1990). An extreme view is held by Chappell and White (1974) and White and Chappell (1977, 1983), for example, who consider that post-metamorphic granites are entirely crustally derived and those with mantle-type isotopic signatures have inherited these by melting of former igneous rocks in the crust. In contrast, most authors (e.g. DePaolo 1981; Didier *et al.* 1982; Pitcher 1982) consider that granites are derived partly from the crust and partly from the mantle.

Post-metamorphic Hercynian granites form around 50% of exposure in the Massif Central and this is therefore an excellent area in which to study their origin. The Velay granite is of particular interest as the nature of the lower crust directly beneath the pluton is well constrained. This is because Bournac, near the centre of the Velay complex, is one of the few localities in the Massif Central where granulite facies xenoliths of lower crustal origin have been brought to the surface by Tertiary volcanism (e.g. Downes *et al.* 1990).

Most Massif Central granites are considered to be crustal in origin (e.g. Downes & Duthou 1988; Turpin *et al.* 1990; Shaw 1991), although the precise nature of their crustal source is still debated. Turpin *et al.* (1990), for example, consider that the Guéret and Millevaches granites were produced by partial melting of upper crustal rocks, whereas Shaw (1991) considers that they had a source in the lower crust.

The Velay granite is unique among Massif Central monzogranites/granodiorites because of its high ε_{Sr_0} (Downes & Duthou 1988), its highly heterogeneous nature and the presence of an apparent ghost stratigraphy within its xenoliths, and is generally thought to have formed by *in situ* melting of local metamorphic country rocks (Didier 1973; Downes & Duthou 1988). The main question concerning the Velay complex is why such a large volume of granite was formed long (~50 Ma) after Hercynian continental collision (Pin & Duthou 1990) during a time of uplift, erosion and development of sedimentary troughs and basins (Pin & Duthou 1990) in a mainly extensional tectonic environment (Ledru *et al.* 1991). The production of such a large volume of granite must have required a substantial thermal input, possibly from the mantle. Mantle involvement in the formation of the Velay granite pluton is suggested by the presence of rounded dioritic enclaves. Furthermore, the metasedimentary lower crust of the Massif Central is considered to have been underplated by mantle-derived basic magmas around 300 Ma (Downes *et al.* 1990, 1991), contemporaneous with the intrusion of the Velay granite pluton. Involvement of the mantle in the petrogenesis of the Velay granite pluton would be in agreement with the model of Pin and Duthou (1990) that Massif Central granites, which they consider show increasing ε_{Nd_0} with decreasing age of emplacement, had progressively larger components from the mantle or were formed by partial melting of increasingly refractory crustal sources with time.

In this paper, trace elements, rare-earth elements (REE) and Sr and Nd isotopes are used to investigate the origin of the Velay granite which forms the bulk (~90%) of the granite in the Velay complex. The Velay granite is the equivalent of the "main anatectic granitoid" of Didier and Dupraz (1985) and is quite different from the younger, cross-cutting, "two-mica" and "tardi-migmatitique" ("post anatectic") granites and the older, deformed ("pre-anatectic") granites. Apart from age determinations (Rb–Sr whole-rock isochron, Caen-Vachette *et al.* 1982) and one Sr, Nd isotopic analysis (Downes & Duthou 1988), there are no published isotope data for the Velay granite.

1. Crustal evolution in the area of the Velay Complex

1.1. Age and nature of the present-day upper crust

The crust of the Massif Central is extremely heterogeneous as a result of a complex evolution involving two nappe formation episodes (420–320 Ma, Pin & Duthou 1990) causing southward overthrusting of an unmetamorphosed series, mainly pelites (now schists), by allochthonous high-medium-grade amphibolite and granulite facies gneisses, including the Leptyo-Amphibolite Group (480–500 Ma, Pin & Lancelot 1982). Locally, this overthrusting formed an inverted metamorphic zonation (Burg *et al.* 1984), which was intruded by granites during and after the thrust tectonic events (545–274 Ma, Duthou *et al.* 1984; Caen-Vachette *et al.* 1982). The Velay granite outcrops below the thrust plane in the low-grade schists (Fig. 1), although high-pressure kyanite-bearing rocks have been found in the south of the Velay, probably representing down-faulted blocks from the allochthonous zone. Ages of crustal material in close proximity to the Velay granite range from 545 ± 14 Ma (Caen–Vachette 1979) for meta-rhyolites in the Pilat low-pressure series, to 313 ± 8 Ma for the Margeride granite (Shaw 1991).

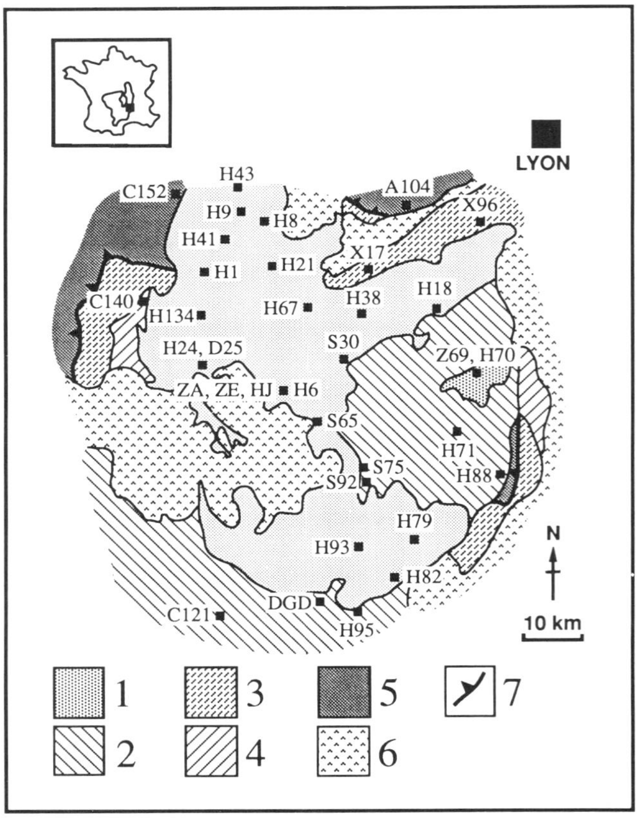

Figure 1 Sketch map of the Velay granite complex showing the sample locations of the heterogeneous (H) and high-Sr granites (S), country rocks (X, C, A) and xenoliths (D, G, D, Z); 1 = heterogeneous and high-Sr granites; 2 = gneisses, migmatites and some granites; 3 = low pressure schists; 4 = gneisses and older granites; 5 = medium-high pressure gneisses; 6 = Tertiary volcanics and Palaeozoic, Mesozoic and Tertiary sediments; 7 = major Hercynian thrust; inset shows an outline map of France and the Massif Central.

1.2. Age and nature of the lower crust

Granulite facies lower crustal xenoliths have been brought to the surface by Tertiary alkaline volcanics at Bournac, central Velay (Dupuy *et al.* 1977; Dostal *et al.* 1980; Leyreloup *et al.* 1977; Downes & Leyreloup 1986; Downes *et al.* 1990, 1991). The lower crust represented by these xenoliths is heterogeneous and consists of three petrologically distinct lithologies which are (a) felsic meta-igneous (b) mafic meta-igneous, and (c) metasedimentary in origin (Leyreloup *et al.* 1977; Downes & Leyreloup 1986; Downes *et al.* 1990). The oldest rocks are thought to be the metasediments and possibly some of the felsic meta-igneous rocks, which were intruded/underplated by mafic magmas, represented by the mafic meta-igneous xenoliths. The mafic magmas are believed to have formed layered intrusive gabbroic bodies or fossil magma chambers beneath much of the Massif Central (Dostal *et al.* 1980). The origin of the felsic meta-igneous xenoliths is more poorly understood. It is thought that they consist of two groups: (1) ancient meta-granites and (2) melts of the metasediments or extreme contaminants of the mafic magmas (Downes *et al.* 1991).

The ages of the Bournac lower crustal xenoliths are controversial (Downes *et al.* 1990, 1991). A majority of zircon lower intercept ages for metasedimentary lower crustal xenoliths range from 280–300 Ma, which are thought to be due to granulite facies metamorphism of the lower crust (Downes *et al.* 1990). The age of intrusion/underplating of the metasedimentary lower crust by mafic magmas (represented by the mafic meta-igneous xenoliths) is more poorly constrained as there are no zircon ages available. Whole-rock Sm–Nd data from the Ivrea zone (Voshage *et al.* 1990), which is thought to represent a section across the upper mantle and lower crust of the Hercynian basement of northern Italy, indicated that previously recognised emplacement ages (~600 Ma) of mafic magmas into the lower crust cannot be accepted and that an age of about 270 Ma is more appropriate. This age is within the range of ages (250–290 Ma) obtained by U-Pb determinations on zircons and monazites from the same rocks (Voshage *et al.* 1990). It is probable that the underplating event observed in the Ivrea zone occurred at the same time as that in the Massif Central. The Guéret granite (363 ± 13 Ma), however, is considered to have been partly derived from a mafic lower crustal source (Shaw 1991). It is therefore probable that the crust of the Massif Central was underplated and intruded between at least 360 Ma and *c.* 270 Ma and was accompanied by granulite facies metamorphism of the entire lower crust *c.* 300 Ma.

1.3. Nature of Velay granites

Velay granites are extremely varied, ranging from relatively homogeneous to banded and porphyritic with variable grain size and modal abundance of mafic minerals (see Section 9). Their colour index ranges from 5–25, they contain essential quartz and two feldspars, with modal compositions ranging from granite (s.s.) to granodiorite. The main mafic minerals are biotite and/or cordierite. The cordierite is either nodular or prismatic in habit. Nodular cordierite occurs up to 10 cm in diameter and has a characteristic grey/green colour if fresh, or brown if altered. Prismatic cordierite, found as 0·5–1 cm-long euhedral crystals, is rarer in the Velay granites but is often found in large aggregates, occasionally giving the rock a porphyritic texture. Accessory minerals in the Velay granite include zircon and apatite and more rarely monazite, andalusite, sillimanite and almandine (see Section 9).

Velay granites contain numerous xenoliths which form between 5% and 10% of the area of outcrop. These are mainly (70–80%) schists and biotite-rich schlieren, although there are also foliated meta-igneous xenoliths and "pillow"-shaped dioritic enclaves. The "pillow"-shaped dioritic enclaves are believed to have been mantle-derived and intruded contemporaneously with the main Velay pluton (Didier 1987).

2. Sampling and analytical notes

Twenty-four Sr and 13 Nd whole-rock isotope analyses have been obtained on the Velay granites and 11 Sr and five Nd isotope analyses for country rock and xenolith samples (Table 1, localities shown in Fig. 1). The granites were chosen to be representative of the wide petrographical variation of the Velay granites (see Section 9). The main types of country rocks collected were schists, gneisses and older granites which form the bulk of the Velay country rocks. Six samples chosen for isotope analysis included ortho- and para-gneisses, schists and amphibolites. A wide variety of xenoliths was collected and representative samples of each of the two main groups, dioritic enclave and dark biotite-rich gneisses, were chosen for isotope analysis (Table 1, Section 9).

Major and trace element analyses were carried out on an automated Philips PW1400 X-ray fluorescence spectrometer (XRF) at Royal Holloway and Bedford New College (RHBNC), using techniques described by Williamson (1991). An automated VG 354 multiple collector mass spectrometer was used for isotope analysis at the University of London Radiogenic Isotope Facility, RHBNC. A laboratory reference Nd standard analysed during the period of data acquisition yielded $^{143}Nd/^{144}Nd = 0{\cdot}511420 \pm 9$ (2σ n = 23) equivalent to a La Jolla reference standard of $0{\cdot}511855 \pm 4$ (2σ, n = 6), and N.B.S. SRM987 standards yielded $^{87}Sr/^{86}Sr = 0{\cdot}710247 \pm 20$ (σ, n = 33). Details of the technique employed are reported elsewhere (Thirlwall & Burnard 1990; Williamson 1991; Thirlwall 1992). Rare-earth elements (REE) were analysed by isotope dilution following the method of Thirwall (1982), adapted for multicollector analysis.

3. Geochemistry of Velay granites

All the Velay granites are peraluminous (mol. Al/(Na + K + 2Ca) = 1·0–1·8), indicating that they are either S-type or fractionated I-type granites (classification of Chappell & White 1974). They vary in their major element chemistry with SiO_2 = 60–78 wt% and Al_2O_3 = 12–19 wt%. On the trace element discrimination diagram of Pearce *et al.* (1984), they fall almost exclusively in the field of volcanic arc granites. There is however, good evidence that the Velay granites were formed 50 Ma after continent–continent collision and 60 Ma after the closure of an ocean domain in the Massif Central (Pin 1989).

Sr contents of Velay granites vary systematically with position within the pluton. The central region, around 8% of the area of the Velay granite, consists of granites with unusually high Sr (500–900 ppm, Table 1), which will be referred to as the high-Sr granites, grading into granites of more normal Sr content (100–420 ppm) at the margins, which will be referred to as the heterogeneous granites. Sr abundance in other Massif Central granites varies from 20–411 ppm (Downes & Duthou 1988; Turpin *et al.* 1990; Shaw 1991), although Sr contents of up to 900 ppm in granites are not unknown, for example in the peraluminous Strontian adamellite of Scotland (Stephens & Halliday 1984). The range of major and trace element compositions of the high-Sr granites has been modelled by 10% accumulation of 90% plagioclase and 10% biotite (Williamson 1991).

REE data for heterogeneous and high-Sr granites are given in Table 2 and are shown on chondrite-normalised diagrams in Figure 2. Heterogeneous granites have highly

Table 1 Sr and Nd isotope data for heterogeneous (H) and high-Sr granites (S), schist country rocks (X), gneiss and older granite country rocks (C), amphibolite country rock (A), dioritic enclaves (D) and biotite-rich meta-igneous xenoliths (Z); $^{143}Nd/^{144}Nd$ was normalised to $^{142}Nd/^{144}Nd = 1{\cdot}14187$ and $^{87}Sr/^{86}Sr$ was normalised to $^{88}Sr/^{86}Sr = 0{\cdot}1194$ (errors = 2σ); all ε-values were calculated at 304 Ma (see Text) using $^{87}Rb/^{86}Sr_{CHUR} = 0{\cdot}0839$ and $^{147}Sm/^{144}Nd_{CHUR} = 0{\cdot}1966$.

Sample	Rb ppm	Sr ppm	$^{87}Sr/^{86}Sr$	$^{87}Sr/^{86}Sr_0$	ε_{Sr_0}
H1	136	160	0·72737 ± 1	0·7167	179
H6	250	161	0·72946 ± 1	0·7100	84
H8	146	172	0·72386 ± 1	0·7133	129
H9	253	194	0·72710 ± 1	0·7108	95
H18	298	108	0·74624 ± 1	0·7117	107
H21	156	162	0·72992 ± 2	0·7179	195
H24	206	152	0·72892 ± 1	0·7120	111
H38	221	235	0·72176 ± 2	0·7100	83
H41	301	113	0·74383 ± 1	0·7105	90
H43	152	392	0·71326 ± 1	0·7084	61
H67	180	245	0·71922 ± 2	0·7100	84
H70	165	192	0·72427 ± 1	0·7135	133
H71	170	243	0·71699 ± 1	0·7083	58
H79	198	212	0·72513 ± 2	0·7135	132
H82	243	109	0·74222 ± 1	0·7143	145
H88	218	301	0·71694 ± 2	0·7079	53
H93	153	190	0·72385 ± 2	0·7138	137
H95	151	110	0·73590 ± 2	0·7187	207
H134	118	404	0·71068 ± 8	0·7070	41
HJ	225	218	0·72278 ± 2	0·7099	82
S30	161	807	0·70915 ± 1	0·7067	36
S65	142	896	0·70903 ± 1	0·7071	41
S75	142	645	0·70935 ± 4	0·7066	35
S92	158	534	0·71074 ± 1	0·7070	41
X17	199	183	0·73186 ± 2	0·7183	201
X96	150	72	0·74695 ± 2	0·7208	236
C121	202	86	0·75461 ± 5	0·7252	299
C140	145	212	0·72160 ± 1	0·7131	127
C152	159	213	0·72132 ± 1	0·7120	112
A104	4·1	186	0·70512 ± 2	0·7048	10
D25	188	565	0·71004 ± 1	0·7059	25
DGD	102	493	0·70881 ± 2	0·7062	30
Z69	206	749	0·71279 ± 4	0·7094	74
ZA	228	496	0·71357 ± 2	0·7078	52
ZE	199	268	0·71837 ± 2	0·7091	70
Sample	**Nd ppm**	**Sm ppm**	$^{143}Nd/^{144}Nd$	$^{143}Nd/^{144}Nd_0$	ε_{Nd_0}
H1	5·06	1·204	0·512262 ± 6	0·51197	−5·3
H6	28·35	7·041	0·512233 ± 7	0·51194	−6·1
H8	24·98	4·941	0·512085 ± 5	0·51185	−7·8
H18	78·67	14·520	0·512098 ± 4	0·51188	−7·2
H21	8·63	1·684	0·512029 ± 6	0·51179	−8·9
H41	29·92	4·93	0·512174 ± 5	0·51197	−5·3
H43	37·64	6·792	0·512196 ± 8	0·51198	−5·2
H82	32·87	8·744	0·512158 ± 4	0·51184	−8·0
H95	17·58	4·274	0·512334 ± 7	0·51204	−4·0
HJ	35·91	6·408	0·512147 ± 7	0·51193	−6·1
S30	30·26	5·169	0·512189 ± 5	0·51198	−5·1
S65	36·43	6·218	0·512269 ± 5	0·51206	−3·6
S92	22·82	3·777	0·512186 ± 5	0·51199	−5·1
X17	56·52	10·338	0·511997 ± 5	0·51178	−9·2
C121	4·48	1·224	0·512309 ± 4	0·51199	−5·2
D25	63·01	12·307	0·512339 ± 3	0·51210	−2·8
ZA	72·05	14·798	0·512270 ± 4	0·51202	−4·4
ZE	40·99	6·883	0·512213 ± 4	0·51201	−4·6

Table 2 Rare-earth element analyses obtained by isotope dilution (using the method of Thirlwall (1982), modified for a multiple collector mass spectrometer, Williamson 1991); samples names (H, S, X, etc.) as in Table 1.

	La	Ce	Nd	Sm	Eu	Gd	Dy	Er	Yb	Lu	La/Yb	Eu/Eu*
Heterogeneous granites												
H1	5·3	11·0	5·06	1·204	0·964	1·24	1·40	0·758	0·704	0·098	7·4	2·42
H6	29·9	61·5	28·36	7·042	0·838	2·79	1·29	0·398	0·353	0·068	85·4	0·58
H18	—	152·2	77·81	14·360	0·744	8·43	—	2·457	1·360	—	38·6	0·21
H21	11·1	21·6	8·63	1·685	0·993	1·50	1·84	1·326	1·455	0·216	7·6	1·92
H82	45·0	89·4	37·64	6·792	1·403	4·95	1·49	0·679	0·595	—	76·3	0·74
H95	32·8	69·5	32·87	8·744	0·668	4·44	3·05	1·411	1·210	—	27·1	0·33
HJ	19·2	41·3	17·58	4·274	0·485	4·39	6·28	3·940	3·720	0·494	5·2	0·34
	39·6	84·5	35·91	6·408	0·951	4·28	2·79	1·281	1·177	0·162	33·6	0·56
High-Sr granites												
	La	Ce	Nd	Sm	Eu	Gd	Dy	Er	Yb	Lu	La/Yb	Eu/Eu*
S92	30·4	62·5	22·85	3·777	0·819	2·26	2·61	0·881	0·746	—	40·5	0·86
S30	37·6	72·9	30·26	5·169	1·229	3·26	1·81	0·792	0·691	0·097	54·5	0·92
S65	44·8	87·5	36·43	6·218	1·367	3·93	2·19	0·948	0·798	—	56·1	0·85
Xenoliths and enclaves												
	La	Ce	Nd	Sm	Eu	Gd	Dy	Er	Yb	Lu	La/Yb	Eu/Eu*
D25	66·9	135·3	63·01	12·307	2·463	8·49	6·42	3·384	2·990	0·414	22·4	0·74
ZA	70·4	147·3	72·05	14·798	2·027	9·49	6·47	2·962	2·350	—	30·6	0·53
ZE	53·1	104·1	40·99	6·883	0·894	4·73	3·25	1·565	1·402	0·198	37·9	0·48
Country rocks												
	La	Ce	Nd	Sm	Eu	Gd	Dy	Er	Yb	Lu	La/Yb	Eu/Eu*
X17	64·5	124·3	54·52	10·338	2·028	3·12	4·21	4·285	3·946	—	16·5	1·10
C121	3·4	17·2	4·48	1·224	0·172	1·04	1·80	1·035	0·832	—	4·1	0·46

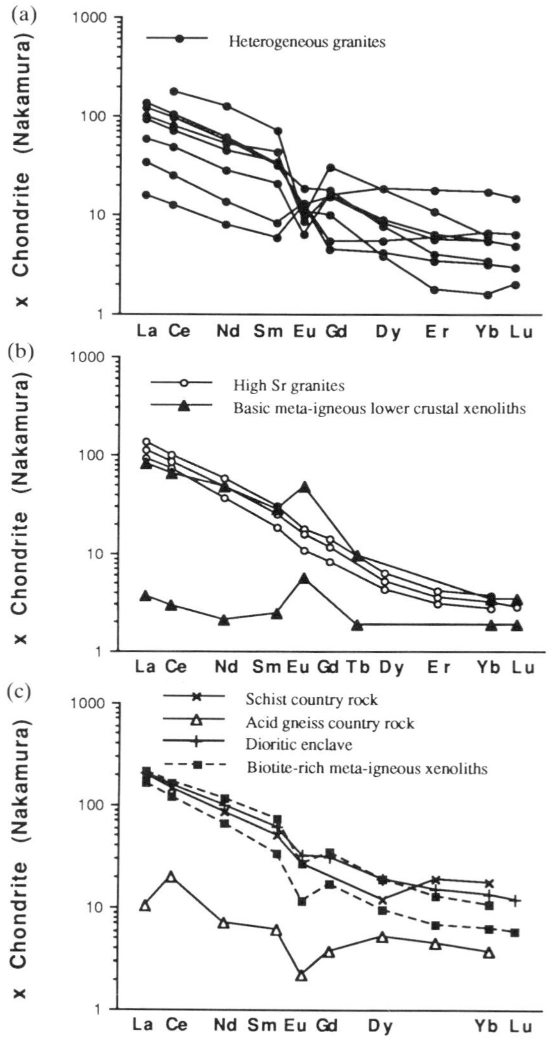

Figure 2 Chondrite-normalised REE patterns for: (a) the heterogeneous granites; (b) the high-Sr granites and mafic meta-igneous lower crustal xenoliths (Downes *et al.* 1990); (c) country rocks, enclaves and xenoliths; normalising coefficients from Nakamura (1974).

variable chondrite-normalised REE patterns with La/Yb = 5·2–76·3 and Eu/Eu* = 0·21–2·42 (Table 2, Fig. 2(a)). They generally have concave upward patterns in their heavy REE which may be the result of minerals such as garnet, clinopyroxene or amphibole having remained residual in their source. More unusual heterogeneous granite REE patterns have positive Eu anomalies (Eu/Eu* = 2·42). This may indicate that this sample has accumulated plagioclase or that it represents a plagioclase-rich residue. Another sample has a concave downward pattern in its heavy REE. This sample may have accumulated minerals such as garnet, clinopyroxene or amphibole. The wide variations in La/Yb ratios and the shapes of heterogeneous granite REE probably indicate that they had a highly heterogeneous source or were derived from a number of sources.

High-Sr granites have generally steeper, highly fractionated (La/Yb = 40·7–56·1), REE patterns compared with the heterogeneous granites (Table 2). They have smaller negative Eu anomalies (Eu/Eu* = 0.85–0.92) and have concave upward patterns in their heavy REE (Fig. 2(b)). The high-Sr granites lie within the range of, but are generally more light REE enriched than, typical mafic meta-igneous lower crustal xenoliths from Bournac (Fig. 2(b)), Downes *et al.* 1990).

Country rocks and meta-igneous xenoliths are also highly variable in their REE patterns, with La/Yb = 16·1–37.9 and Eu/Eu* = 0·46–0·74 (Fig. 2(c), Table 2). They have lower La/Yb, larger negative Eu anomalies and higher light REE contents than the high-Sr granites and have a similar range in REE patterns to the heterogeneous granites. Dioritic enclaves have REE patterns which are similar in shape to the high-Sr granites, although they have lower La/Yb and a higher total REE, probably because they are less fractionated.

4. Age determinations for the Velay granites

No Rb–Sr isochron was obtained for the Velay granites, probably because they contain varying amounts of biotite-rich restite which has retained its original isotopic signature. Even on a small, 22 cm, scale there is only partial

Table 3 Sr isotope data for biotites from Velay granites, enclaves and xenoliths and age determinations by Rb-Sr biotite-whole-rock isotopic dating (errors = 2σ); sample names (H, S, X, etc.) as in Table 1.

	Age	Rb ppm	Sr ppm	$^{87}Sr/^{86}Sr$
Heterogeneous granites				
H8	273 ± 6	667·0	6·6	1·9998 ± 2
H9	289 ± 5	940·4	7·1	2·5690 ± 5
H9	282 ± 5	972·3	7·1	2·5723 ± 7
H79	300 ± 8	786·3	7·1	2·21 ± 4
High-Sr granites				
S75	305 ± 5	672·8	10·1	1·613 ± 2
Dioritic enclaves				
D25	300 ± 5	393·9	32·8	0·8561 ± 3
DGD	305 ± 5	252·7	26·6	0·8263 ± 5
Biotite-rich xenolith				
Z69	307 ± 5	485·1	58·1	0·8175 ± 2

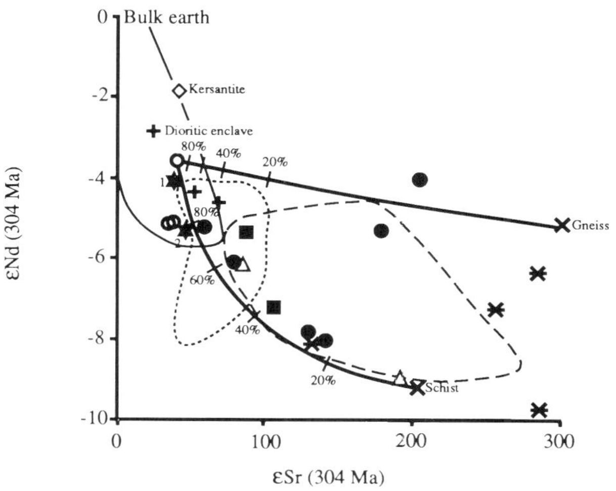

Symbols used in figures 3, 5 and 6

- ● Heterogeneous granites
- ○ High Sr granites
- ✚ Xenoliths and enclaves
- ✕ Country rocks
- ✱ Massif Central country rocks (Turpin *et al.*, 1990)
- ◇ Kersantite (Turpin *et al.*, 1988)

Fields for lower crustal xenoliths (recalculated to 304 Ma, Downes *et al.*, 1990) used in Figures 3, 5 and 6

Solid field	Basic meta-igneous lower crustal xenoliths
Dotted field	Acid meta-igneous lower crustal xenoliths
Dashed field	Metasedimentary lower crustal xenoliths

Figure 3 Initial ε_{Sr} vs ε_{Nd} for the heterogeneous and high-Sr granites, country rocks (including data from Turpin *et al.* 1990), xenoliths and enclaves: (1) = weighted mean for mafic meta-igneous lower crustal xenoliths, (2) = weighted mean for felsic and mafic meta-igneous lower crustal xenoliths (Downes *et al.* 1990); data for kersantites from Turpin *et al.* (1988); error bars are smaller than symbols and are therefore not shown; hyperbolae are divided into intervals of 20% mixing of the high-Sr granite component with melts of country rock compositions; see box for symbols.

equilibration of Sr isotopes between a typical sample of Velay granite and a partial melt from a xenolith (Williamson 1991). The Velay granites are therefore strongly variable in their initial Sr isotope ratios.

Rb–Sr biotite–whole-rock ages of between 300 ± 8 Ma and 273 ± 6 Ma have been obtained for the heterogeneous granites (Table 3). Two repeated analyses of one sample (H9) gave ages within error. The high-Sr granites yielded an age of 305 ± 5 Ma and a biotite-rich xenolith an age of 307 ± 5 Ma. It is therefore probable that the intrusion of the Velay granites occurred at around 304 Ma. This age is within error of a Rb–Sr whole-rock age of 298 ± 8 Ma determined by Caen-Vachette *et al.* (1982). "Pillow"-shaped dioritic enclaves yielded Rb–Sr biotite–whole-rock ages (Table 3) of 300 ± 5 Ma and 305 ± 5 Ma. These ages are within error of the intrusion age of the Velay granites and it is therefore likely that magmas represented by dioritic enclaves were intruded contemporaneously with Velay granites.

5. Sr and Nd isotope analyses

All $(^{87}Sr/^{86}Sr)_0$ and $(^{143}Nd/^{144}Nd)_0$ isotope ratios and ε_{Sr_0} and ε_{Nd_0} have been calculated using an age of 304 Ma (Table 1). The heterogeneous granites have a wide range of isotope compositions (ε_{Sr} = +40 to +210, ε_{Nd} − 3·8 to −7.3, Figs 4, 5, Table 1) and show no systematic relationship between this and their petrology. Both cordierite and biotite-bearing facies have a similar range in ε_{Sr} and ε_{Nd}, confirming that the heterogeneous granites are a single suite and probably not composed of two separate cordierite and biotite-bearing magmas. The heterogeneous granites have a wider variation in ε_{Sr} and ε_{Nd} than, and overlap with, the range of ε_{Sr} and ε_{Nd} for both I- and S-type granites from the Lachlan Fold Belt (I-types, ε_{Sr} = 35·5–180, ε_{Nd} = −0·9 to −9·2; S-types, ε_{Sr} = 154–254, ε_{Nd} = −6·6 to −11.1, after McCulloch & Chappell 1982).

High-Sr granites are fairly homogeneous in initial Sr and, to a lesser extent, in initial Nd isotope compositions (ε_{Sr} = +35 to 41, ε_{Nd} = −3 to −5, Fig. 3, Table 1). They have ε_{Sr} and ε_{Nd} which are closer to Bulk Earth values than other undeformed Massif Central granites (Duthou *et al.* 1984; Downes & Duthou 1988; Pin & Duthou 1990; Turpin

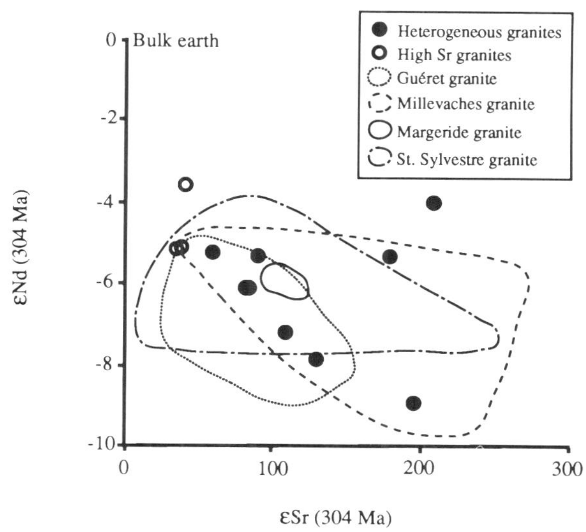

Figure 4 Initial ε_{Sr} vs ε_{Nd}; a comparison between the heterogeneous and high-Sr granites and other major Massif Central monzogranites/granodiorites including the Guéret and Millevaches (Turpin *et al.* 1990; Shaw 1991), the Margeride (Shaw 1991) and St Sylvestre granite (Turpin *et al.* 1990).

et al. 1990; Shaw 1991; Fig. 4) but lie within the range of I-type granites from the Lachlan Fold Belt (McCulloch & Chappell 1982).

Discussion

6.1. Origin of heterogeneous granites

Peraluminous granites are commonly thought to represent partial melts of upper crustal rocks during continental collision (e.g. Didier *et al.* 1982; Duthou *et al.* 1984), although possible sources for the heterogeneous Velay granites include: (1) schist country rocks; (2) gneiss and older granite country rocks; (3) the metasedimentary lower crust; (4) the meta-igneous lower crust, and (5) the mantle. Heterogeneous granites have highly variable chondrite-normalised REE patterns similar to those of their country rocks (Table 2) and overlap with upper crustal compositions on graphs of ε_{Sr} against 1/Sr (Fig. 5) and ε_{Sr} against ε_{Nd} (Fig. 3). It is therefore probable that the heterogeneous granites had a large source component in partial melts from the country rocks, although samples with lower ε_{Sr}, however, must have had an additional source. The heterogeneous granites are therefore unlikely to have been formed entirely by *in situ* melting as a result of anatexis, as proposed by Didier (1973). This supports evidence from field studies (Williamson 1991) that an alignment of xenoliths within the Velay granites is due to magmatic processes and is unlikely to represent ghost stratigraphy.

Dupraz (1983) showed that by virtue of its shape (a large dome with a flat summit and steep sides) the Velay granite was intruded from deeper levels in the crust than its present level of exposure. A possible alternative source for the heterogeneous granites with lower ε_{Sr} is the metasedimentary lower crust (Fig. 3). This has been suggested for the source of many other major Massif Central granites such as the Millevaches leucogranite (Downes & Duthou 1988; Shaw 1991).

On a graph of ε_{Sr} against ε_{Nd} (Fig. 3), the fields for the heterogeneous granites overlap with the field for metasedimentary lower crustal xenoliths (recalculated to 304 Ma, Downes *et al.* 1990). The metasedimentary lower crust of the Massif Central is, however, considered to have been the source for many older granites, such as the Arc de Fix orthogneiss and the Tournon, Margeride and la Chaise Dieu granites. It may therefore have been too refractory to have been a major source for the heterogeneous granites, although there is insufficient evidence to rule it out altogether.

Alternative low ε_{Sr} sources are the mantle and the felsic and mafic meta-igneous lower crust. From experimental studies and theoretical considerations (Wyllie 1983 a, b; Chappell & White 1974), it is very unlikely that peraluminous granites could be produced by partial melting of the mantle. It is also unlikely that the heterogeneous granites could have formed by mixing between mafic mantle-derived magmas and magmas with higher ε_{Sr}, such as melts of the country rocks, as a large component in mixing with low ε_{Sr} is needed. Dioritic enclaves (possibly representing mafic mantle-derived magmas, Didier 1987) are relatively rare in the Velay Complex and there is no geophysical evidence to suggest the presence of larger volumes of this type of material at deeper levels within the present-day upper crust.

A more likely low ε_{Sr} end member is the meta-igneous lower crust. In Figure 3, the heterogeneous granites overlap with the fields for felsic and mafic meta-igneous lower crustal xenoliths (Downes *et al.* 1990). On a graph of ε_{Sr} against Sr ppm (Fig. 6), a mixing hyperbola has been calculated between the weighted mean composition for country rocks (140 ppm Sr and $\varepsilon_{Sr} = 194{\cdot}0$, including data from Turpin *et al.* 1990, excluding amphibolites and mafic rocks) and the weighted mean composition for felsic and mafic meta-igneous lower crustal xenoliths (373 ppm Sr and $\varepsilon_{Sr} = 46{\cdot}9$, Downes *et al.* 1990, stars on Fig. 6). Mean compositions are weighted to account for variations in their contribution to the Sr isotopic budget by each sample depending on their Sr contents. The calculated mixing hyperbola is a good approximation to the trend of the heterogeneous granites and the variations in their compositions, and it is therefore possible that mixing between melts from two sources can account for the range in ε_{Sr} and ε_{Nd} of the heterogeneous granites. The two sources are: (1) country rocks, including both schists and gneisses

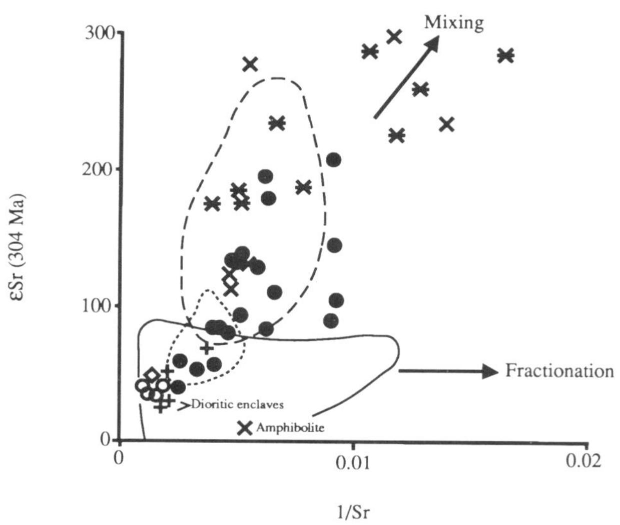

Figure 5 1/Sr vs initial ε_{Sr} for the heterogeneous and high-Sr Velay granites, country rocks (including data from Turpin *et al.* 1990) and xenoliths and enclaves; data for kersantites from Turpin *et al.* (1988) and data for lower crustal xenoliths from Downes *et al.* (1990); see box below Figure 3 for symbols.

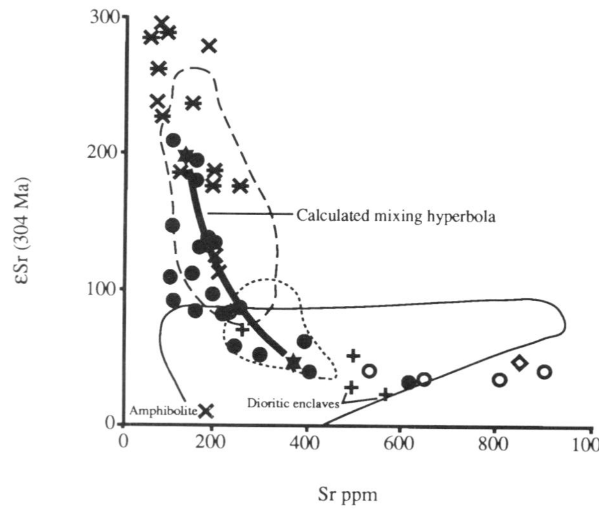

Figure 6 Sr vs ε_{Sr} for heterogeneous and high-Sr granites, country rocks (including data from Turpin *et al.* 1990), enclaves and xenoliths; also shown are weighted means (shown as stars) for country rocks (including data from Turpin *et al.* 1990) and felsic and mafic meta-igneous lower crustal xenoliths (Downes *et al.* 1990) and a calculated mixing hyperbola (see text) between these end members; data for kersantites from Turpin *et al.* (1988); see box below Figure 3 for symbols.

and older granites (with high ε_{Sr}), and (2) the felsic and mafic meta-igneous lower crust (with low ε_{Sr}), represented by meta-igneous lower crustal xenoliths (Downes *et al.* 1990).

6.2. Origin of high-Sr granites

High-Sr granites show little variation in their ε_{Sr} and, to a lesser extent, in their ε_{Nd} (Fig. 3). They clearly plot outside the field of the country rocks and the metasedimentary lower crust, and are therefore unlikely to have had a source in these lithologies. A possible source with low ε_{Sr} and high ε_{Nd} is the mantle (e.g. Didier *et al.* 1982; Ben Othman *et al.* 1984; Poli *et al.* 1989).

Spinel peridotite mantle xenoliths have been brought to the surface by alkaline volcanics in the Massif Central (e.g. Downes & Dupuy 1987). These contain two types, with unknown spatial distributions, which are: (1) LREE depleted with MORB-type ε_{Sr} and ε_{Nd} and (2) LREE enriched with higher ε_{Sr} and lower ε_{Nd}. They have ε_{Nd} less negative than $-2{\cdot}2$ which is generally much higher than that of the high-Sr granites $\varepsilon_{Nd} = -3{\cdot}6$ to $-5{\cdot}1$). Furthermore, Wyllie (1983 a, b), mainly on theoretical grounds, considered that granites cannot be primary magmas derived by partial melting of mantle peridotite or subducted gabbroic oceanic crust. The high-Sr granites are therefore unlikely to represent partial melts directly from the mantle.

The high-Sr granites may represent mantle-derived magmas which have undergone processes such as fractionation, mixing and AFC. Mantle-derived compositions which could represent less "mature" equivalents of the high-Sr granites include: (1) dioritic enclaves; (2) mafic meta-igneous lower crustal xenoliths, and (3) kersantites and minettes (Turpin *et al.* 1988), which will collectively be termed the DMK. Dioritic enclaves are thought to represent magmas which were intruded contemporaneously with the consolidation of the Velay granites (Didier 1973) and represent mafic mantle injections (Didier 1987). They have similar ε_{Sr} and ε_{Nd} values to mantle-derived mafic magmas which are thought to have underplated the lower crust at around 300 Ma (Downes *et al.* 1990, Fig. 3). It is therefore probable that there is a relationship between the dioritic enclaves and the mafic meta-igneous lower crust. In the Massif Central, late (290–315 Ma, Turpin *et al.* 1990) Hercynian minettes and kersantites are considered to have an origin in a mantle enriched by recycled crustal material (Turpin *et al.* 1988). DMK all have similar ε_{Sr} and ε_{Nd} to the high-Sr granites (Fig. 3).

Fractionation of shoshonitic magmas to granitic compositions has been modelled by fractionation of mainly hornblende and K-feldspar (Fowler 1988). The high-Sr granites and the dioritic enclaves, however, contain relatively little K-feldspar. Furthermore, where it occurs in the high-Sr granites it is considered to be a relatively late-formed mineral and is therefore unlikely to have fractionated. The compositions of the high-Sr granites cannot be modelled (using the modelling technique of Wright & Doherty 1970) by fractionation of hornblende from those of the dioritic enclaves, as the high-Sr granites have the same Al_2O_3 contents (~16·5 wt%) as the dioritic enclaves (Williamson 1991). Fractionation of hornblende (Al_2O_3 ~ 7 wt%) would have increased the Al_2O_3 contents of the high-Sr granites compared to those of the dioritic enclaves. Dioritic enclaves also have larger negative Eu anomalies (Eu/Eu* = 0·74) than the high-Sr granites (Eu/Eu* = 0·85–0·92, Fig. 2(b), Table 2). It is difficult to derive granites with smaller Eu anomalies from such dioritic magmas, as plagioclase is certain to be a fractionating phase. Fractionation from compositions similar to those of the kersantites and the mafic meta-igneous lower crustal xenoliths is equally unlikely, as they have major and trace element compositions similar to dioritic enclaves (Downes *et al.* 1990; Turpin *et al.* 1990).

The most convincing evidence that the high-Sr granites were not formed by fractionation of more mafic magmas is that the Sr contents of plagioclases contained within the high-Sr granites do not increase with increasing whole-rock Sr content and plagioclases do not have calcic or high-Sr cores. AFC of mafic magmas (represented by the DMK), with assimilation of country rock or metasedimentary lower crustal material, to form the high-Sr granites, is also unlikely as this would have significantly increased the ε_{Sr} of the high-Sr granites. Furthermore, mantle-derived magmas would probably not mix with a granite melt and are too rare in the Velay granites to provide a viable source composition in mixing which requires a large contribution from a high ε_{Nd} and low ε_{Sr} source.

Dioritic enclaves, meta-igneous lower crustal xenoliths and kersantites, therefore, probably do not represent a magmatic component in mixing to form the high-Sr granites. Their intrusion from the mantle, however, may have provided an important mechanism for heat transfer from the mantle, causing crustal melting and intrusion of the Velay granites.

As it is unlikely that the high-Sr granites were produced by partial melting of the mantle or by fractionation, mixing or AFC of mafic magmas, it is probable that they represent partial melts of the meta-igneous lower crust. On a graph of ε_{Sr} against ε_{Nd} (Fig. 3), the high-Sr granites plot between the weighted means for meta-igneous lower crustal xenoliths (felsic and mafic) and mafic meta-igneous lower crustal xenoliths alone. It is therefore probable that the source of the high-Sr granites contained a mixture of felsic and mafic materials which were probably both relatively fertile compared with the refractory metasedimentary lower crust. Because of the isotopic homogeneity of the high-Sr granites and the heterogeneous nature of meta-igneous lower crustal xenoliths, there must have been a high degree of homogenisation during the formation of the high-Sr granites.

6.3. Relationship between heterogeneous and high-Sr granites

On graphs of ε_{Sr} against Sr ppm (Fig. 6) and ε_{Sr} against ε_{Nd} (Fig. 3), the high-Sr granites lie at the low ε_{Sr} end of the trends of the heterogeneous granites. As there is no evidence that they form a separate intrusion, it is likely that the high-Sr granites represent an end member in mixing to produce the heterogeneous granites.

On the ε_{Sr} against ε_{Nd} diagram (Fig. 3), mixing hyperbolae have been calculated from the isotopically most primitive composition of the high-Sr granites (S65, Table 1) to each of the two main Velay country rock compositions: schists and older granites and gneisses. Most of the heterogeneous granites plot close to the high-Sr granite-schist mixing hyperbola and it is therefore probable that schists were the most dominant country rock source. Most of the heterogeneous granites lie between the two calculated mixing hyperbolae and, therefore, the compositions of the heterogeneous granites can be modelled by mixing between three end members: (1) partial melts of the meta-igneous lower crust (represented by the high-Sr granites); (2) partial melts from the schist country rocks; and (3) partial melts of the older granites and gneisses. H95 falls outside the two mixing hyperbolae, possibly as it contains garnet (see Section 9) which may have partially retained its original isotopic signature.

As the high-Sr granites were derived by partial melting of an originally igneous source, they are I-types granites (classification of Chappell & White 1974, 1982). Heterogeneous granites were probably either formed by partial melting of Velay country rocks (predominantly meta-pelitic schists) alone, and are therefore S-type granites, or by mixing between I-type granites (the high-Sr granites) and partial melts of country rocks, and are therefore mixed I-S-type granites. The Velay granite pluton is therefore zoned with I-type, high-Sr granites at its centre and heterogeneous, mainly S-type and mixed I-S-type granites at its margins.

6.4. Models for the emplacement of the Velay granites

Most large Massif Central monzogranites/grandodiorites were formed after the main period of Hercynian metamorphism (410–360 Ma, Duthou *et al.* 1984). This is curious given the usual close relationship between regional metamorphism and granite genesis (e.g. Stephansson 1975; Duthou *et al.* 1984). The formation of large volumes of granite long after the cessation of regional deformation and metamorphism is a major problem. As the formation of the Velay granite pluton cannot have been associated with the main period of Hercynian regional deformation and metamorphism associated with continental collision, an alternative heat source for melt generation is needed.

The probable sequence of events which led to the formation of the Velay granite pluton can be summarised as follows:

(1) Major tectonic episodes (420–320 Ma, Pin & Duthou 1990) associated with the Hercynian orogeny caused overthrusting by allochthonous, high-medium-grade amphibolite and granulite facies gneisses (including the Leptyno-Amphibolite Group) over a low-grade or unmetamorphosed authochthonous series, mainly pelites (now schists). The metasedimentary lower crust of the Massif Central was underplated/intruded by mafic, mantle-derived magmas (Downes *et al.* 1990). Underplating/intrusion is likely to have begun around 360 Ma as the Guéret granite (356 ± 10 Ma, Berthier *et al.* 1979) is considered to have had a source partly from mafic mantle-derived lower crust (Shaw 1991).

(2) Around 300 Ma (280–300 Ma), the entire lower crust of the Massif Central, including underplated and intruded mantle-derived mafic material, experienced granulite facies metamorphism (Downes *et al.* 1991). This is likely to have occurred as a result of further heat transfer from the mantle by continued underplating and intrusion of the lower crust by mantle-derived mafic magmas. Continued underplating and intrusion is demonstrated by the presence of dioritic enclaves (300 ± 5 Ma, Table 3) within the Velay granites which have very similar geochemical and isotopic compositions to mafic meta-igneous lower crustal xenoliths. Granulite facies metamorphism of the lower crust, continued underplating of the lower crust, and the intrusion of dioritic magmas to high levels in the crust, are likely to have caused partial melting of relative "fertile" Velay country rocks (schists and gneisses and older granites).

(3) Also around 300 Ma, granulite facies metamorphism within the lower crust and continued underplating by mantle-derived mafic magmas caused the lower crust to suffer partial melting. Underplated mafic mantle-derived crust provides a relatively "fertile" infracrustal source for I-type granites (Chappell & Stephens 1988). The metasedimentary lower crust is not likely to have yielded significant volumes of granite melt as it would have been relatively refractory following partial melting to produce many of the older granites in the Massif Central. Partial melts of the meta-igneous lower crust were emplaced to high levels in the crust. They intruded upwards and outwards and mixed with partial melts of the two main country rocks in the Velay complex (schists and gneisses and older granites) to produce the heterogeneous granites.

(4) Partial melts of the meta-igneous lower crust contained accumulated plagioclase at the base ("tail") of their intrusive body. The tail of the intrusion was then emplaced through the heterogeneous granites to high levels in the crust forming the high-Sr granites.

From studies of initial Sr and Nd isotopes, Pin and Duthou (1990) suggested that Massif Central monzogranites/granodiorites show increasing ε_{Nd}, and therefore an increasing mantle component and/or an increased component from mafic refractory crustal sources, with decreasing age of emplacement. The proposed petrogenesis of the Velay granite appears to be in good agreement with this model, expecially as the high-Sr granites have among the highest ε_{Nd_0} and are the youngest of the major Massif Central granites. It is therefore likely that the formation of the Velay granite pluton represents the climax of crustal growth at the end of the Hercynian orogeny in this region.

7. Conclusions

The Velay granite consists of high-Sr granites with low ε_{Sr} and high ε_{Nd} at its centre grading into heterogeneous granites with higher ε_{Sr} and lower ε_{Nd} nearer its margins. The range in initial isotopic compositions of the heterogeneous granites can be modelled by mixing between three end members: (1) partial melts of the meta-igneous lower crust (represented by the high-Sr granites); (2) partial melts of schist country rocks; and (3) partial melts of felsic gneisses and older granite country rocks. Most heterogeneous granites plot near the high-Sr granite-schist country rock mixing hyperbola. Whether this is due to sampling or because schists are more fusible is unclear.

The high-Sr granites are not considered to have had a source in the mantle, on the basis of their ε_{Sr} and ε_{Nd}, or the metasedimentary lower crust, as this is likely to have been too refractory following the formation of older granites in the Massif Central. Their ε_{Sr} and ε_{Nd} are similar to felsic and mafic meta-igneous lower crustal xenoliths and, because they are unlikely to represent less "mature" equivalents of mafic mantle-derived magmas (represented by mafic meta-igneous lower crustal xenoliths and dioritic enclaves), they were probably produced by partial melting of the meta-igneous lower crust.

The Velay granites are the youngest of the major monzogranites/granodiorites in the Massif Central. They were emplaced some 50 Ma after Hercynian continental collision and include facies (the high-Sr granites) with the lowest ε_{Sr} and the highest ε_{Nd} of all major Massif Central monzogranites/granodiorites. It is probable that late in the Hercynian orogeny (<360 Ma), mafic mantle-derived melts intruded and underplated the lower crust of the Massif Central (Downes *et al.* 1990). This material was then remelted due to further heat transfer from the mantle as a result of continued underplating and intrusion of the lower crust, as indicated by the presence of dioritic enclaves (~300 Ma) within the Velay granite pluton. The genesis of the high-Sr granites was therefore a two-stage process involving crustal underplating followed by remelting, a commonly proposed mechanism for the derivation of I-type granites (e.g. Atherton *et al.* 1979; Chappell & Stephens 1988). The high-Sr granites are therefore I-type (Chappell & White 1974, 1982; Chappell & Stephens 1988; Pitcher 1982), rather than M-type granites, which requires a source directly from the mantle. The Velay granite pluton is therefore zoned with I-type, high-Sr granites at its centre which grade into heterogeneous, mainly S-type and mixed I-S-type granites at its margins.

8. Acknowledgements

Ben Williamson acknowledges receipt of a NERC post-graduate research studentship at Birkbeck College, University of London. Isotope and XRF analyses were carried out at RHBNC laboratories which are University of London analytical services. Thanks are due to A. Long and A. Shaw for comments on the manuscript, G. Marriner and G. Ingram for technical help and G. Vitel (St. Etienne) for assistance in the field.

9. Appendix

Mineralogy of the heterogeneous and high Sr granites, country rocks and xenoliths: Q = quartz, Pl = plagioclase, Kf = K-feldspar, B = biotite, C = cordierite, A = apatite, Z = zircon, M = monazite, Hb = hornblende, G = garnet, Mu = muscovite, And = andalusite, S = sillimanite. Sample subsets are as in Table 1.

Sample	*Petrographic description*	*Mineralogy (in order of abundance)*
H1	Homogeneous granite with nodular cordierite	Q, Pl Kf, B, C, A, Z
H6	Homogeneous pale granite with nodular cordierite	Q, Kf, C, B, Z, A
H8	Banded granite with nodular cordierite	Q, Pl, Kf, C, B, A, Z
H9	Dark biotite granite with nodular cordierite	Kf, Pl, Q, B, C, A, Z
H18	Dark porphyritic granite with andalusite	Q, Kf, Pl, B, And, A, Z, M
H21	Homogeneous pale granite with nodular cordierite	Q, Pl, Kf, C, B, A, Z
H24	Dark porphyritic granite	Q, Pl, Kf, C, B, A, Z
H38	Homogeneous dark biotite-rich granite	Q, Kf, B, Pl, A, Z
H41	Dark biotite-rich granite	Q, Kf, Pl, B, A, Z
H43	Dark banded biotite granite	Q, Kf, Pl, B, A, Z, S
H67	Dark porphyritic granite with nodular cordierite	Q, B, Kf, Pl, C, A, Z
H70	Porphyritic granite with prismatic cordierite	Q, B, Kf, Pl, C, A, Z
H71	Dark granite with nodular cordierite	Q, Kf, B, C, Pl, A
H79	Pale porphyritic banded granite	Q, Kf, B, Pl, A, Z
H82	Dark granite with nodular cordierite	Q, Kf, B, Pl, A, Z
H88	Dark porphyritic and banded granite	Q, Pl, Kf, B, Z, A
H93	Dark porphyritic and banded granite	Kf, Q, Pl, B, A, Z, S
H95	Dark banded granite with nodular cordierite	Q, Kf, Pl, B, C, G, A, Z
H134	Pale granite with nodular cordierite	Q, Kf, Pl, B, A, Z
HJ	Dark biotite granite with nodular cordierite	Kf, Pl, Q, B, C, A, Z
High-Sr granites		
S30	Porphyritic granite with nodular cordierite	Kf, Pl, Q, B, C, A, Z
S65	Porphyritic granite with nodular cordierite	Kf, Q, Pl, B, C, A, Z
S75	Homogeneous granite with nodular cordierite	Pl, Kf, Q, B, C, Z, A
S92	Porphyritic granite with nodular cordierite	Q, Pl, Kf, B, C, A
Country rock		
X17	Mica schist	B, Qz, G, Pl, Mu, S, A, Z
X96	Mica schist	Qz, B, Pl, Mu, A
C121	Pale, banded ortho-gneiss	Q, Pl, B, Kf, A, Z
C140	Arc de Fix ortho-gneiss	Q, Pl, B, Kf, G, A, Z
C152	Sheared biotite meta-granite	Q, Pl, B, Kf, A, Z
A104	Amphibolite with relict igneous textures	Hb, Pl, Q, A, Z
Xenoliths and enclaves		
D25	Rounded dioritic enclave	Q, Pl, B, Hb, Z, A, M
Z69	Dark, banded biotite-rich meta-diorite	Q, B, Pl, Hb, A, Z
ZA	Dark, banded biotite-rich meta-diorite	Q, Pl, B, A, Z
ZE	Pale, banded meta-diorite	Q, Pl, B, A, Z

10. References

Allègre, C. J. & Ben Othman, D. 1980. Nd–Sr isotope relationship in granitoid rocks and continental crust development: a chemical approach to orogenesis. NATURE **286,** 335–41.

Atherton, M. P., McCourt, W. J., Sanderson, L. M. & Taylor, W. P. 1979. The geochemical character of the segmented Peruvian coastal batholith and associated volcanics. *In* Atherton, M. P. & Tarney, J. (eds) 45–64. Nantwich: Shiva.

Ben Othman, D., Fourcade, S. & Allègre, C. J. 1984. Recycling processes in granite-granodiorite complex genesis: the Querigut case studied by Nd–Sr isotope systematics. EARTH PLANET SCI LETT **69,** 290–300.

Berthier, F., Duthou, J. L. & Roques, M. 1979. Datation géochronologique Rb/Sr sur roches totales du granite de Guéret (Massif Central). Age fini-dévonien de mise en place de l'un de ses faciès types. BULL BRGM **2**(2) 59–72.

Burg, J. P., Leyreloup, A., Marchand, J. & Matte, Ph. 1984. Inverted metamorphic zonation and large-scale thrusting in the Variscan Belt: an example in the French Massif Central. *In* Hutton, D. W. H. & Sanderson, D. J., (eds), *Variscan tectonics of the North Atlantic Region.* GEOL SOC LONDON SPEC PUBL **14,** 47–61.

Caen-Vachette, M. 1979. Age Cambrien de rhyolites transformées en leptynites dans la série métamorphique du Pilat (Massif Central français). CR ACAD SCI PARIS **289,** 997–1000.

Caen-Vachette, M., Couturié, J.-P. & Didier, J. 1982. Ages radiométriques des granites anatechtique et tardimigmatique du Velay (Massif Central français). CR ACAD SCI PARIS **294,** 135–8.

Chappell, B. W. & Stephens, W. E. 1988. Origin of infracrustal (I-type) granite magmas. TRANS R SOC EDINBURGH EARTH SCI **79,** 71–86.

Chappell, B. W. & White, A. J. R. 1974. Two contrasting granite types. PAC GEOL **8,** 173–4.

Chappell, B. W. & White, A. J. R. 1982. I- and S-type granites in the Lachlan Fold Belt, southeastern Australia. *In* Zu Kepin & Tu Guangchi (eds), *Geology of granites and their metallogenic relationships* 87–101. Beijing: Science Press.

DePaolo, D. J. 1981. Neodymium isotopes in the Colorado Front Range and crust-mantle evolution in the Proterozoic. NATURE **291,** 193–6.

Didier, J. 1973. Granites and their enclaves. *The bearing of enclaves on the origin of granites,* 2, Developments in Petrology Series No. 3, 27–56. Amsterdam: Elsevier.

Didier, J. 1987. Contribution of enclave studies to the understanding of origin and evolution of granitic magmas. GEOL RDSCH **76,** 41–50.

Didier, J., Duthou, J.-L. & Lameyre, J. 1982. Mantle and crustal granites: genetic classification of orogenic granites and the nature of their enclaves. J VOLCANOL GEOTHERM RES **14** 125–32.

Didier, J. & Dupraz, J. 1985. Magmatic and metasomatic cordierites in the Velay Granitic Massif. *In The Crust, The Significance of Granites-Gneisses in the Lithosphere,* 35–77. Athens: Theophrastus.

Dostal, J., Dupuy, C. & Leyreloup, A. 1980. Geochemistry and petrology of meta-igneous granulitic xenoliths in Neogene volcanic rocks of the Massif Central, France-Implications for the lower crust. EARTH PLANET SCI LETT **50,** 31–40.

Downes, H. & Dupuy, C. 1987 Textural, isotopic and REE variations in spinel peridotite xenoliths, Massif Central, France. EARTH PLANET SCI LETT **82,** 121–35.

Downes, H. & Duthou, J.-L. 1988. Isotopic and trace-element arguments for the lower-crust origin of Hercynian granitoids and pre-Hercynian orthogneisses, Massif Central (France). CHEM GEOL **68,** 291–308.

Downes, H. & Leyreloup, A. 1986. Granulitic xenoliths from the French Massif Central: Petrology, Sr and Nd isotope systematics and model age estimates. *In* Dawson, J. B., Carswell, D. A., Hall, J. & Wedepohl, K. H. (eds) *Nature of the Lower Continental Crust.* GEOL SOC LONDON SPEC PUBL **24,** 319–330.

Downes, H., Dupuy, C. & Leyreloup, A. F. 1990. Crustal evolution of the Hercynian belt of Western Europe: Evidence from lower crustal xenoliths (French Massif Central). CHEM GEOL **83,** 209–31.

Downes, H., Kempton, P. D., Briot, D., Harmon, R. S. & Leyreloup, A. F. 1991. Pb- and O-isotope systematics in granulite facies xenoliths, French Massif Central: Implications for crustal processes. EARTH PLANET SCI LETT **102,** 342–57.

Dupraz, J. 1983. *Evolution du complexe anatectique du Velay: genèse de la cordiérite.* Thèse 3ème Cycle, Université de Clermont-Ferrand.

Dupuy, C., Leyreloup, A. & Verniers, J. 1977. The lower continental crust of the Massif Central (Bournac, France)- with special reference to REE, U and Th composition, evolution heat flow production. PHYS CHEM EARTH **11,** 401–16.

Duthou, J.-L., Cantagrel, J. M., Didier, J. & Vialette, Y. 1984. Palaeozoic granitoids from the French Massif Central: age and origin studied by 87Rb-87Sr system. PHYS EARTH PLANET INTERIORS **35,** 131–44.

Fowler, M. B. 1988. Ach'uaine hybrid appinite pipes: Evidence for mantle-derived shoshonitic parent magmas in Caledonian granite genesis. GEOLOGY **16,** 1026–30.

Ledru, P., Dallain, C., Lardeau, J. M. & Marignac, C. 1991. Tectonic evolution of the Velay granito-migmatitic dome, French Massif Central. TERRA ABST **3,** (1).

Leyreloup, A., Dupuy, C. & Andriambolona, R. 1977. Catazonal xenoliths in French Neogene volcanic rocks: composition of the lower crust, Part 2. Chemical composition and consequences of the evolution of the French Massif Central Precambrian crust. CONTRIBUT MINERAL PETROL **62,** 283–300.

McCulloch, M. T. & Chappell, B. W. 1982. Nd isotopic characteristics of S- and I-type granites. EARTH PLANET SCI LETT **58,** 51–64.

Nakamura, N. 1974 Determination of REE, Ba, Fe, Mg, Na and K in carbonaceous and ordinary chondrites. GEOCHIM COSMOCHIM ACTA **38,** 757–75.

Pearce, J. A., Harris, N. & Tindle, A. 1984. Trace element discrimination diagrams for the tectonic interpretation of granitic rocks. J PETROL **25,** 956–83.

Pin, C. 1989. *Essai sur la chronologie et l'évolution géodynamique de la chaîne hercynienne d'Europe. Dr d'Etat Thesis, Blaise Pascal University, Clermont-Ferrand.*

Pin, C. & Duthou, J.-L. 1990. Sources of Hercynian granitoids from the French Massif Central: Inferences from Nd isotopes and consequences for crustal evolution. CHEM GEOL **83,** 281–96.

Pin, C. & Lancelot, J. R. 1982. U-Pb dating of an early Palaeozoic bimodal magmatism in the French Massif Central and its further metamorphic evolution. CONTRIB MINERAL PETROL **79,** 1–12.

Pitcher, W. S. 1982. Granite type and tectonic environment. *In* Hsu, K. (ed) *Mountain building processes* 1–3, 19–40. London: Academic Press.

Poli, G., Ghezzo, C. & Conticelli, S. 1989. Geochemistry of granitic rocks from the Hercynian Sardinia-Corsica batholith: Implications for magma genesis. LITHOS **23,** 247–6.

Shaw, A. 1991. *The petrogenesis of Hercynian granites of the French Massif Central.* Ph.D. Thesis, Birkbeck College, University of London.

Stephansson, O. 1975. Polydiapirism of granitic rocks in the Svecofenrian of Central Sweden. PRECAMBRIAN RES **2,** 189–214.

Stephens, W. E. & Halliday, A. N. 1984. Geochemical contrasts between late Caledonian granitoid plutons of northern, central and southern Scotland. TRANS SOC EARTH SCI **75,** 259–73.

Thirlwall, M. F. 1982. A triple-filament method for rapid and precise analysis of rare-earth elements by isotope dilution. CHEM GEOL **35,** 155–66.

Thirlwall, M. F. 1992. Long-term reproducibility of multicollector Sr and Nd isotope ratio analysis. CHEM GEOL ISOTOPE GEOSCI **94,** 85–104.

Thirlwall, M. F. & Burnard, P. 1990. Pb–Sr–Nd isotope and chemical study of the origin of undersaturated and oversaturated shoshonitic magmas from the Borralan pluton, Assynt, NW Scotland. J GEOL SOC LOND **147** 259–69

Turpin, L., Velde, D. & Pinte, G. 1988. Geochemical comparisons between minettes and kersantites from the Western European Hercynian orogen: trace element and Pb–Sr–Nd isotope constraints on their origin. EARTH PLANET SCI LETT **87,** 73–86.

Turpin, L., Cuney, M., Friedrich, M., Bouchez, J.-L. & Auberin, M. 1990. Meta-igneous origin of Hercynian peraluminous granites in N.W. French Massif Central: implications for crustal history reconstructions. CONTRIB MINERAL PETROL **104,** 163–72.

Voshage, H., Hofmann, A. W., Mazzucchelli, M., Rivalenti, G., Sinigoi, S. Raczek, I. & Demarchi, G. 1990. Isotopic evidence from the Ivrea zone for a hybrid lower crust formed by magmatic underplating. NATURE **347,** 731–6.

White, A. J. R. & Chappel, B. W. 1977. Ultrametamorphism and granitoid genesis. TECTONOPHYSICS **43,** 7–22.

White, A. J. R. & Chappell, B. W. 1983. Granitoid types and their distribution in the Lachlan Fold belt, southeastern Australia. GEOL SOC AM MEM **159,** 21–33.

Williamson, B. J. 1991. *The petrology of the Velay granite complex, Massif Central, France.* Ph.D. Thesis, Birkbeck College, University of London.

Wright, T. L. & Doherty, P. C. 1970. A linear programming and least squares computer method of solving petrological mixing problems. BULL GEOL SOC AM **81,** 1995–2008.

Wyllie, P. S. 1983a. Experimental studies on biotite- and muscovite-granites and some crustal magmatic sources. *In* Atherton, M. P. & Gribble, C. D. eds) *Migmatites, melting and metamorphism,* 12–26. Orpington: Shiva.

Wyllie, P. S. 1983a. Experimental and thermal constraints on the deep-seated parentage of some granitoid magmas in subduction zones. *In* Atherton, M. P. & Gribble, C. (eds) *Migmatites, melting and metamorphism*, 37–51. Orpington: Shiva.

B. J. WILLIAMSON, Department of Geology, Birkbeck College, University of London, Malet Street, London WC1E 7HX, U.K.,; and Department of Geology, Royal Holloway and Bedford New College, University of London, Egham Hill, Egham, Surrey, TW20 0EX, U.K.
H. DOWNES, Department of Geology, Birkbeck College, University of London, Malet Street, London WC1E 7HX, U.K.
M. F. THIRLWALL, Department of Geology, Royal Holloway and Bedford New College, University of London, Egham Hill, Egham, Surrey TW20 0EX, U.K.

Transactions of the Royal Society of Edinburgh: Earth Sciences, **83**, 247–257, 1992

A stable isotopic ($\delta^{18}O$, δD) study of the late-Hercynian granites and their host-rocks in the Central Iberian Massif (Spain)

C. Recio, A. E. Fallick and J. M. Ugidos

ABSTRACT: Stable isotopic ratios (mainly $^{18}O/^{16}O$, but also D/H) have been measured for the three most important types of late-Hercynian granites, and their hosts, in the western area of the Central Iberian Massif (CIM), Spain. These granites are amphibole-bearing biotite granites, biotite granites and cordierite-bearing biotite granites. No intrusive relationships have been observed among them; the contact of each granite with the others is always gradational. Host-rocks are Precambrian/Cambrian metasediments, ranging from low-grade schists to migmatites (nebulites).

Whole-rock $\delta^{18}O_{SMOW}$ values are as follows: amphibole-bearing biotite granites $8{\cdot}9 \pm 0{\cdot}58‰$ (1σ, $n = 17$); biotite granites $9{\cdot}0 \pm 0{\cdot}35‰$ (1σ, $n = 11$); cordierite-bearing biotite granites $9{\cdot}6 \pm 0{\cdot}24‰$ (1σ, $n = 21$). $\delta^{18}O$ values for nebulites, into which some of these granites were emplaced, are significantly higher, at $11{\cdot}1 \pm 0{\cdot}58‰$ (1σ, $n = 13$). The Precambrian to Cambrian shales gave an average value of $\delta^{18}O = 11{\cdot}9 \pm 1{\cdot}23‰$ (1σ, $n = 5$). Whole-rock oxygen isotope ratios indicate that the origin of the granites was in neither purely sedimentary/metasedimentary rocks nor pristine mantle melts. $\delta^{18}O$ values close to $9{\cdot}0‰$ require a crustal protolith, having an important recycled component.

Oxygen isotope results are compatible with the cordierite-bearing granites being generated by assimilation of nebulite-like material by a biotite granite magma. However, $^{18}O/^{16}O$ of mineral separates obtained from the three different granites and the nebulite indicate that isotopic equilibrium, if ever reached, has not been preserved. The modified isotopic equilibrium is attributed to fluid activity, but mineral-pair δ–δ plots suggest that the granite system behaved as a closed system, and that the fluid was deuteric (magmatic) in origin. This implies that if assimilation did happen, it occurred at a temperature higher than the closure temperature of the different minerals to isotopic exchange. In a $\delta^{18}O$ vs δD plot, hornblende and biotite separates from the granites plot within the igneous field. A simple mesocrustal anatectic origin for the peraluminous late Hercynian granites of the western area of the CIM is difficult to sustain on the basis of the stable isotope data, consistent with other field, petrographic and geochemical evidence. Cordierite in the cordierite-bearing granites is not "restitic" from a deep source area, but rather is xenocrystic from the high-grade metamorphic country rock (nebulites).

KEY WORDS: Amphibole-bearing granites, assimilation, biotite granites, cordierite-bearing granites

In the western part of the Central Iberian Massif (CIM), Spain, three different late Hercynian granites are prevalent: cordierite-bearing biotite granites (CG), biotite granites (BG) and amphibole-bearing biotite granites (AG), suggesting different protoliths (at least for the extreme compositions). However, comparison of trace- and REE data for the three groups of granites, with each other and with the cordierite-rich host-rocks (nebulites), does not show up major differences. An origin for CG exclusively by anatexis of metasedimentary compositions similar to those of the country rocks is therefore difficult to sustain. These observations, together with petrographic data and the geological context in which the granites are found (see below), leads to a genetic model of the CG involving mixing of BG and host cordierite-rich nebulites (NB) (Ugidos 1990; Ugidos & Recio 1991). However, the nature of the original protoliths remains uncertain.

The use of stable isotopes as tracers of petrogenetic processes is now well established. The various types of rock that may take part in granite generation have been extensively characterised in the last few decades from the isotopic viewpoint (see reviews in Taylor 1968, 1977, 1980, 1986, 1988; Taylor & Sheppard 1986), as has been the effect of the processes involved on the fractionation of the light stable isotopes (see review in Valley *et al.* 1986). It has been found that whole-rock "oxygen isotopic ratios may be clearly diagnostic of (granite) source materials", and are ". . . . Valuable in discerning the character of a pluton if other criteria are inconclusive" (O'Neil & Chappell 1977; see also O'Neil *et al.* 1977).

The main focus of this work is therefore on the study of the $^{18}O/^{16}O$ ratios, both in whole-rock samples, in order to test the previous model and characterise the rocks and the possible protoliths, and in mineral separates, in order to investigate how the system behaved once the granite magmas were emplaced.

1. Geological setting and materials studied

The Central Iberian Massif (CIM) forms part of the so-called Central Iberian Zone (Lotze 1945; Julivert *et al.* 1974) and is characterised by a great variety of granites and high-grade metamorphic rocks, occupying approximately 10,000 km^2 (Fig. 1). The metamorphism in this region evolved from intermediate pressure conditions, during the first phases of Hercynian folding, to low-pressure, high-temperature conditions towards late orogenic stages, affecting mainly Upper Precambrian–Lower Cambrian pelites and greywackes (CXG).

Early Hercynian granites are generally autochthonous/subautochthonous sillimanite-bearing leucogranites and two-mica granites, associated with sillimanite/biotite-rich migmatites. Large volumes of late Hercynian granites intruded pre-heated country rocks (and earlier granites) during relatively rapid uplift. As a result of the heat input from the late granite magmas (coincident with a pressure drop), early metamorphic parageneses changed to those of low-pressure and high-temperature. The hosts of the late intrusions range from low-grade schists to hornfelses of the pyroxene-hornfels facies and very homogeneous cordierite-rich migmatites (here termed nebulites: NB). These high-grade metamorphic/anatectic rocks show relatively low-pressure mineral associations: cordierite-K-feldspar-sillimanite/andalusite, resulting from reactions involving biotite + muscovite and biotite + sillimanite developed during earlier metamorphic stages (Ugidos 1990, and references therein).

Each of AG, BG and CG ranges from medium-grained (sometimes porphyritic) granodiorite to monzogranite. AG have minor contents of amphibole (green hornblende as up to 1 cm prisms and/or irregular aggregates), allanite (zoned euhedral prisms; occasionally with an irregular inner zone) and sphene (euhedral/subhedral crystals). Plagioclase (andesine-oligoclase) shows complex zoning patterns and frequent synneusis textures. Subhedral biotite is the most abundant femic mineral. Alkali feldspars can reach 3–4 cm in porphyritic facies. BG are distinguished from AG in lacking amphibole and allanite, otherwise being quite similar. AG and BG have chemical affinities, and parallel or overlapping differentiation trends, suggesting two different granite magmas (Ugidos & Recio 1991). No intrusive relationships have been found between AG and BG: rather, gradual transitions are observed. CG show two extreme facies: an internal one (ICG), with less than one cordierite prism/m^2, found in gradual transition to BG; and a marginal one (MCG) with cordierite contents over 20 prisms/m^2, transitional to ICG. In both facies, cordierite is prismatic, very rarely not pinitised, and in size from a few mm up to 2–3 cm. Except for the presence of cordierite, BG and ICG are petrographically very similar. Chemical data for these granites are practically indistinguishable and show similar trends, although data for ICG are slightly more scattered than for BG. MCG have some muscovite, and occasionally sillimanite. Plagioclases (oligoclase; rarely andesine) have less complex zoning patterns than in BG or ICG, and synneusis textures are rare or absent. Comparison of chemical data between BG, ICG and MCG shows remarkable affinities of MCG with the other granites.

All three granite types have frequent microgranular

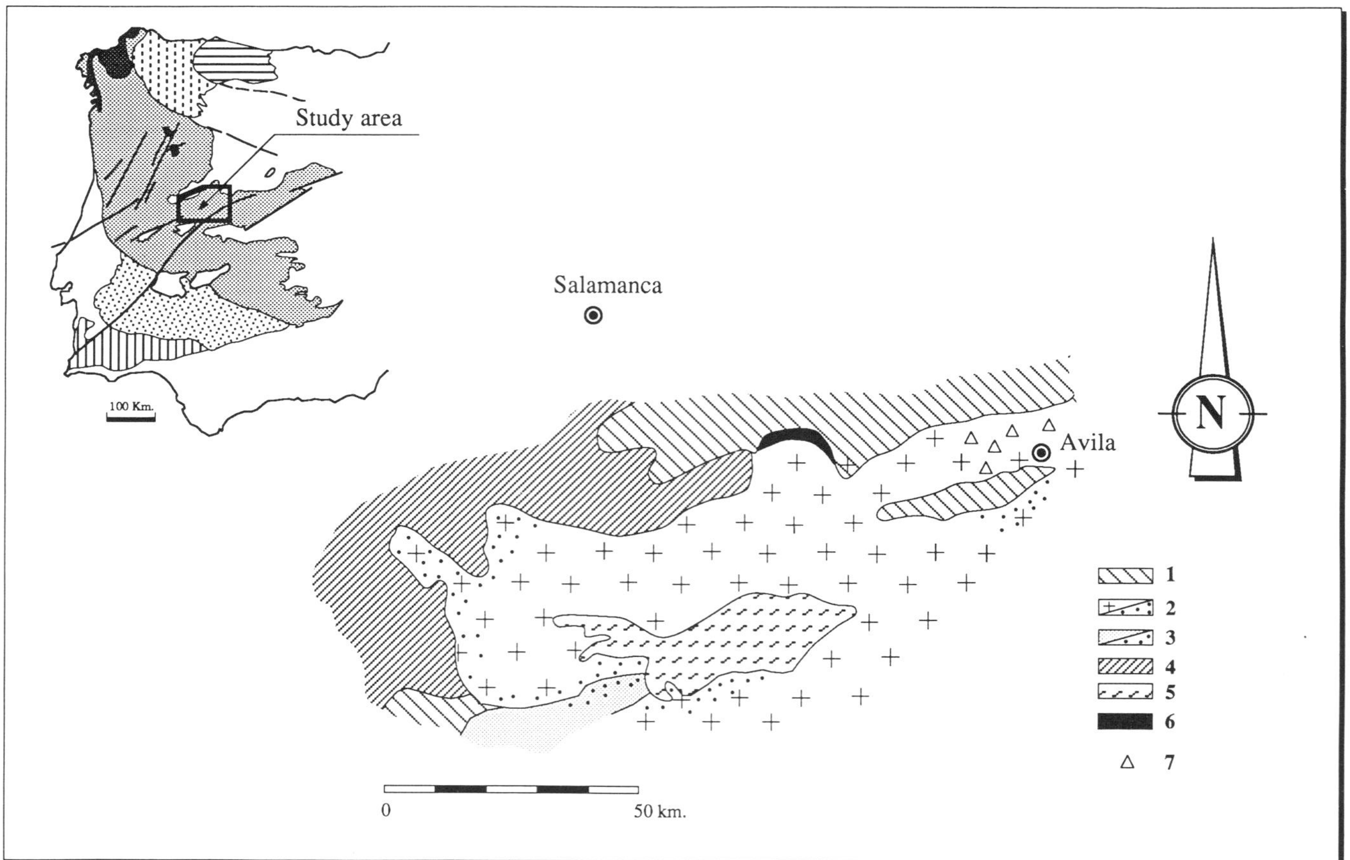

Figure 1 Geological map of the granites in the western part of the CIM, mainly based on Department of Petrology, University of Salamanca (1983): (1) post-palaeozoic sediments; (2) biotite granites ± cordierite (dots); (3) biotite ± sillimanite ± muscovite ± andalusite granites, ±cordierite (dots); (4) metamorphic rocks; (5) cordierite migmatites; (6) leucogranites and two-mica granites; (7) amphibole-bearing biotite granites.

enclaves (tonalites, quartzdiorites). Gabbro enclaves, although rare, can also be found. Metasedimentary enclaves are very scarce in AG and BG. An uncommon macroenclave of cordieritite has been found in BG. The only place at which cordierite has been found within BG is in the vicinity of the contact with this enclave (Ugidos 1988). ICG have some enclaves consisting of limestones, hornfelses and biotite/sillimanite-rich schists. Enclaves are more abundant in MCG, where they consist of hornfelses, limestones, schists, cordierite-rich migmatites and leucogranites. Apparent gradual transitions between CG and migmatites can also be observed. The migmatites are, in general, very homogeneous (NB), and have high cordierite contents (normally 10–20%, but higher contents are not uncommon). These cordierites are euhedral to subhedral, unaltered crystals, with or without sillimanite/biotite inclusions. NB are in gradual transition to less evolved migmatites. The major difference between NB and other migmatites is the mesosome/neosome ratio. Mesosome mineralogy consists mostly of sillimanite + biotite. Cordierite is the characteristic mineral of the neosome, together with quartz, oligoclase and alkali feldspar. Accessory andalusite can also occasionally be found in the migmatites. Chemical characteristics of NB and MCG strongly indicate that NB-type compositions are not probable source materials for MCG: CaO content in MCG is about twice that in NB, whereas Sr contents are similar in both rocks. Eu/Eu* values are lower in MCG for virtually identical REE patterns (except for Eu) in both MCG and NB (further details in Ugidos 1990; Recio 1990; Ugidos & Recio 1991).

2. Analytical methods

$^{18}O/^{16}O$ ratios have been determined in 72 whole rock powders and 85 mineral separates. Oxygen extraction for isotopic analysis used the technique of Clayton and Mayeda (1963), but employed a loading technique similar to that described by Friedman and Gleason (1973) and ClF_3 as reagent (Borthwick & Harmon 1982). Around 10 mg of finely powdered sample was loaded into nickel vessels and reacted for 15 hours at 650°C (quartz and feldspars) or 690°C (other minerals and whole-rock powders). The oxygen liberated was converted to CO_2 by means of a carbon-rod heated by a platinum wire. The yield of evolved CO_2 was manometrically measured. Isotope ratios were determined in a VG SIRA-10™ mass spectrometer. All samples were analysed at least in duplicate. During the period analyses took to complete, reference standards of known isotopic composition were run to check accuracy and precision. Results are reported in the usual notation, as $\delta^{18}O$ values relative to the V-SMOW reference standard. The value obtained for NBS-28 is 9·6‰. Typical overall reproducibility of duplicate runs is within ±0·2‰ (1σ). Whole rock oxygen isotope results are reported in Table 1.

A few samples representing each granite type and the enclaves were selected for mineral separation and isotopic analysis. Separates were prepared from the 165–215 μm fraction using a magnetic separator and heavy liquids. Purity of the separated phases was usually better than 95%, although occasionally hand-picking under a binocular microscope was found to be necessary. Fractions of quartz virtually 100% pure were obtained by reaction with fluorosilicic acid at ambient temperature for three days, and of biotite by crushing a preconcentrate in a TEMA™ mill filled with water (T. Dempster, pers. comm. 1989; see also Kelley & Bluck 1989). Samples were washed with reagent-grade acetone and twice-distilled water, and dried under a heat lamp before isotopic analysis. Checks for purity of the separate included optical examination of every sample, and analysis of a selected set of samples by XRD. Oxygen extractions and isotopic analyses were performed as for the whole rock samples. Mineral separate isotopic results are given in Table 2.

D/H ratios were measured on whole-rock powders and on hydrous mineral separates. Hydrogen extraction for isotopic analysis used the technique of Godfrey (1962), with modifications described by Jenkin (1988). Samples were loaded into degassed platinum crucibles and these were placed in quartz reaction tubes which were evacuated and heated under vacuum to 125°C overnight to remove any adsorbed H_2O. The degassed samples were melted by heating to 1,400°C using an induction furnace. Released water was reduced to hydrogen over uranium at 800°C, and the evolved gas yield was manometrically measured. D/H ratios were determined in a VG-602™ mass spectrometer, equipped with a modified mercury piston inlet. Isotopic ratios are reported as δD relative to V-SMOW. Reproducibility of duplicate runs is within ±2‰ (1σ). The value obtained for NBS-30 is −65‰. Hydrogen isotope results are reported in Table 3.

3. Whole-rock oxygen isotope data

Whole-rock $\delta^{18}O$ results are given in Table 1 and Figure 2. AG show a variation between $\delta^{18}O = +8{\cdot}0$‰ and $\delta^{18}O = +9{\cdot}8$‰, with an average of $\delta^{18}O = +8{\cdot}9 \pm 0{\cdot}58$‰ (1$\sigma$, for n = 17). The range of $\delta^{18}O$ values for BG is a bit wider, +7·7–9·8‰, with average +9·0 ± 0·35‰ (1σ; n = 11). Values for CG are between +8·6 and +10·5‰, with an average value for those in internal areas of the batholith (ICG) of $\delta^{18}O = +9{\cdot}6 \pm 0{\cdot}32$‰ (1$\sigma$; n = 9), and for those close to the margins of the batholith (MCG) of $\delta^{18}O = +9{\cdot}6 \pm 0{\cdot}21$‰ (1$\sigma$; n = 12). $\delta^{18}O$ values for NB are between +10·3 and 12·1‰, with average $\delta^{18}O = +11{\cdot}1 \pm 0{\cdot}58$‰ (1$\sigma$; n = 13). Some shale-type meta-sedimentary rocks of the CXG give values between +10·7 and 13·8‰, average $\delta^{18}O = +11{\cdot}9 \pm 1{\cdot}23$‰ (1$\sigma$; n = 5).

A few samples representing the enclaves encountered in the granites were analysed. Two microgranular (tonalitic) enclaves gave $\delta^{18}O$ of 6·6‰ and 8·6‰. A macroenclave of tonalitic-quarz dioritic affinities has $\delta^{18}O = 8{\cdot}0$‰. A decametric enclave of noritic gabbros has $\delta^{18}O = 4{\cdot}5$‰, and a decametric enclave of cordieritite has $\delta^{18}O = 12{\cdot}4$‰.

4. Discussion of whole-rock $\delta^{18}O$ data

No general geographical trend of $^{18}O/^{16}O$ variation has been observed. The fact that lower values tend to be concentrated in the northeastern corner of the study area is a consequence of AG distribution. This is also true for BG and CG: no systematic within-facies isotopic variation has been observed. Lithology, and not geography, controls the distribution of isotopic values.

Values for enclaves are variable, probably reflecting their different origin. It is notable, however, that $\delta^{18}O = 4{\cdot}5$‰ for UHR-28, the unweathered gabbro macro-enclave, must reflect a secondary process since it cannot be a mantle-derived value. The secondary character of this value is confirmed by hydrogen isotopes.

It is quite clear from Figure 2 and from the average values in Table 1 that the CG are more similar to the values of AG and BG than they are to those of NB. Thus, granites and nebulites can be distinguished as two distinct groups. A weighted average of $\delta^{18}O$ values for the granites is 9·2‰. This value, although virtually meaningless by itself, serves the purposes of the discussion below.

Typical values of mantle-derived materials are around 5·7–6·5‰ (Taylor 1980). The effects of fractional crystallisation have been discussed by, among others, Garlick (1966), Taylor (1968, 1977), and Taylor and Sheppard (1986), who concluded that at magmatic temperatures isotopic fractionations of more than 1‰ are unlikely. It is therefore unlikely that the granites of the CIM are purely fractionated materials after a mantle-derived magma.

Nebulites (NB) have $\delta^{18}O$ values that are close to the average value in shale-type metasediments. There is ample evidence in the western CIM (see, for example, Ugidos 1990) to suggest that the nebulites are the result of anatexis of metasediments which were originally similar to the shales analysed for this study. It is possible to follow the evolution of the shales in the field through higher-grade metamorphic rocks to migmatites (NB), so that the NB are truly rooted anatectic granites.

The presence of cordierites both in the NB and in the CG could be used to support a single-stage anatectic origin for CG. However, the marked differences in $\delta^{18}O$ between, on the one hand, the CG and, on the other hand, the NB and sediments of the CXG precludes this hypothesis. There is only one mechanism that can account for a lowering of several ‰ in the $^{18}O/^{16}O$ ratio (the difference between average shale and weighted average granite is 2·7‰): interaction with large quantities of low ^{18}O meteoric waters *at the magmatic stage,* a process for which there is no evidence in the geotectonic and regional metamorphic environment in which the granites of the CIM were emplaced. An origin for the granites different from that accepted for the nebulites needs to be invoked.

The average values of AG and BG (8·9 and 9·0‰, respectively) preclude their derivation directly from the mantle or pure sediments/metasediments. Rather, the $^{18}O/^{16}O$ ratios are intermediate between those extremes. Several models have been proposed for granites with similar $\delta^{18}O$ values. Interaction of lower crustal protoliths of varied nature with mantle-derived material has been proposed (Clayburn *et al.* 1983; Halliday *et al.* 1980; Harmon 1984; Harmon & Halliday 1980; Harmon *et al.* 1984; Ben-Othman *et al.* 1984, and references therein). It is also possible that granulite-facies metamorphic rocks present at depth were involved. Available oxygen isotope data for granulties are in the range $\delta^{18}O = 5–9‰$ (Wilson *et al.* 1970; Fourcade & Javoy 1973), thus allowing this type of material to be one of the contributors to the inferred deep-seated magma (if the problem of melting an essentially H_2O-free rock can be overcome; see also Michard-Vitrac *et al.* 1980). In any case, it seems clear that the protolith must be crustal and with I-type affinities.

Table 1 Whole rock $\delta^{18}O$ values in samples of the CIM; variation of the mean represents 1σ values.

AG		BG		ICG		MCG	
MIRA-1	8·3	MIRA-5	7·7	XSA-4	10·1	XSA-6	9·9
MIRA-2	8·4	VAA-1	9·4	XSA-5	10·3	XSA-9	10·5
MIRA-3	8·0	VAA-4	9·5	XSA-7	9·3	XSA-10	10·2
MIRA-4	8·3	VAA-5	8·6	XSA-8	9·6	XSA-11	9·9
CARA-3	8·6	VAA-6	9·2	BEU-14	9·2	MC-9	9·3
CARA-4	9·3	CARA-11	9·1	BEU-15	10·3	MC-10	9·3
CARA-8	9·3	CARA-15	8·7	BEU-16	8·6	MC-11	9·0
CARA-9	8·4	BEU-13	8·4	BEU-17	9·3	MC-12	9·3
CARA-6	8·9	PHU-1	9·8	GU-12	9·6	MC-13	9·7
CARA-1	9·8	BEU-12	9·2	Av.	9·58 ± 0·32	SEU-5	9·1
CARA-2	9·5	CARA-5	9·3			SEU-6	9·6
CARA-7	9·5	Av.	9·0 ± 0·35			SEU-7	9·3
VAA-11	9·8					Av.	9·59 ± 0·21
CARA-10	8·8						
CARA-12	8·8						
CARA-13	9·3						
CARA-14	8·2						
Av.	8·9 ± 0·58						

NB		Enclaves[1]		Metasediments[2]		Waters[3]	
BOA-6	10·8	BEU-18A	8·0	HR-41	12·2	Los Baños	−8·3
PHU-4	11·6	GU-12.A	8·6	HR-44	13·8	Cagalona	−8·4
PHU-2	10·8	GU-13.A	6·6	GU-21	10·7	Retortillo	−7·7
ABE-2	11·0	UHR-28	4·5	GU-23	11·5	Gamo River	−8·9
ABE-3	11·1	Ctta.	12·4	GU-24	11·3		
PHU-5	10·7			Av.	11·9 ± 1·23		
I-25	11·1						
I-30	11·9						
XSA-1	11·6						
XSA-2	10·3						
JAM-2	11·8						
XSA-3	10·5						
ABE-4	12·1						
Av.	11·05 ± 0·58						

Notes: (1) The samples under the heading "Enclaves" are a tonalite-quartz diorite macroenclave (BEU-18.A) within BG; two microgranular (tonalitic) enclaves (GU-12.A & 13.A) within CG; a noritic gabbro (UHR-28) macroenclave within CG, and a cordieritite macroenclave (Ctta.) within BG.

(2) The samples under the heading "Metasediments" are shales from the CXG.

(3) Samples under the heading "Waters" were taken from two different thermal springs (Retortillo and Los Baños, at source, before any kind of treatment) and from a cold spring (Cagalona), all of them well inside granitic regions. The sample labelled Gamo River was collected in a stream draining a granitic area, which all its length upstream flows on top of granites.

Table 2 $\delta^{18}O$ results for 84 mineral separates obtained from a selected set of samples of granites from the CIM and their enclaves.

NEBULITES	Q	KFd	Plg	Biot	Cord	Ilm	
ABE-4	13·3	11·8	13·5	9·1	11·0	6·0	
PHU-2	12·0	10·7	10·7	6·7	8·9		
JAM-2	13·7	12·3	13·3	8·4			
XSA-3	9·9	10·2	9·5	7·1	9.7		
AMPHIBOLE-BEARING GRANITES	Q	KFd	Plg	Biot	Hornb		
MIRA-3	10·7	9·9	9·0	5·1			
CARA-12	11·3	9·3	9·8	5·0	6·8		
CARA-13	10·9	10·2	10·2	5·4			
CARA-1					7·3		
CARA-2					7·2		
CARA-4					7·1		
CARA-14					7·2		
CORDIERITE-BEARING GRANITES	Q	KFd	Plg	Biot	Cord	Pin	Tourm
XSA-10	11·3	9·9	10·6	4·1		5·7	
SEU-6	10·3	9·7	10·7	6·4			
SEU-8	10·5	9·7	10·5	5·5			
MC-12	10·9	10·6	10·3	6·3			
MC-9.bis	11·2	10·45	11·0	5·3	8·5		
(H_2O-free)					4·4		
PEGMAT.							8·9
Near Peg.	10.6						
BIOTITE GRANITES	Q	KFd	Plg	Biot			
VAA-4	10·9	9·9	9·6	5·3			
XSA-7 (S-7)	10·7	9·6	10·5	5·5			
ENCLAVES	Q	KFd	Plg	Biot	Hyp	Mgt	
UHR-28	10·5	6·3	3·6	0·7	6·3	4·7	
GU-12.A	11·95	10·2	11·7	5·3			
BEU-18.A	10·9	10·8	9·6	5·3			

Notes: Q = quartz; KFd = K-feldspar; Plg = plagioclase; Biot = biotite; Cord = cordierite; Ilm = ilmenite; Hornb = hornblende; Pin = pinite; Tourm = tourmaline; Hyp = hypersthene; Mgt = magnetite.

5. Mineral separate $\delta^{18}O$: results and discussion

Mineral separate $\delta^{18}O$ values are given in Table 2, and plotted in Figure 3. Should perfect isotopic equilibrium between all the minerals of each independent rock have been achieved or preserved, the lines that pass through the points corresponding to each different mineral would not intersect each other, since the *order* of ^{18}O-enrichment of the minerals in the various rocks should be the same, whatever the type of rock concerned and the temperature. The observed reversals among quartz and feldspars show that equilibrium has not been perserved in many samples. There are, however, several features of Figure 3 worth considering: (1) quartz and biotite are, respectively, the most ^{18}O-rich and ^{18}O-depleted of the minerals, consistent with isotopic equilibrium; (2) cordierite and hornblende also

Table 3 Measured δD values in samples from the western part of the CIM; H_2O (%) is the water content of the mineral or rock as determined during hydrogen extraction; identification of samples under the heading "ENCLAVES" and "CXG SAMPLES" as in Table 1; W.R. = whole rock. n.d. = not determined.

Sample	Min	H_2O %	δD	Sample	Min	H_2O %	δD
AMPHIBOLE-BEARING GRANITES				NEBULITES			
CARA-1	Hornb	2·05	−89·3	ABE-4	Bt	3·82	−74·8
CARA-2	Hornb	2·01	−86·8	ABE-4	Cord	3·39	−63·6
CARA-4	Hornb	2·16	−75·4	JAM-2	Bt	3·59	−74·8
CARA-12	Hornb	2·03	−92·2	JAM-2	Cord	1·36	−60·1
CARA-14	Hornb	1·96	−73·9	XSA-3	Bt	4·20	−69·1
CARA-13	Bt	3·51	−77·2	XSA-3	Chl/Cord	2·71	−57·6
MIRA-3	Bt	3·40	−83·1	XSA-3	Cord	4·51	−51·8
CARA-12	Bt	3·55	−92·7	PHU-2	Bt	3·77	−54·6
BIOTITE GRANITES				CXG SAMPLES			
VAA-4	Bt	3·64	−71·6	HR-41	W.R.	3·94	−43·3
XSA-7	Bt	3·72	−69·8	HR-44	W.R.	4·00	−43·4
				GU-21	W.R.	3·22	−34·7
CORDIERITE-BEARING GRANITES				GU-23	W.R.	3·74	−49·4
XSA-10	Chl/Cord	3·73	−49·1	GU-24	W.R.	3·59	−42·4
XSA-10	Bt	3·80	−57·8				
SEU-8	Bt	3·84	−78·6				
MC-12	Bt	5·30	−53·2	WATERS			
SEU-6	Bt	3·42	−78·7	Retortillo			−57·1
MC-9.bis	Cord	4·94	−67·8	Los Baños			−60·4
				Cagalona			−60·7
ENCLAVES				Gamo River			−59·9
UHR-28	Bt	3·86	−65·0				
BEU-18.A	Bt	3·01	−74·5				
GU-12.A	Bt	3·19	−82·4				
Ctta.	W.R.	n.d.	−33·3				

fall into their theoretically-predicted positions; i.e. at higher values than biotite and lower values than both quartz and feldspars; (3) most reversals are between alkali feldspar and plagioclase.

The isotopic reversals between alkali feldspar and plagioclase could be due to several causes:

(a) The population of mineral grains reflects a composite source. This is a real possibility if some of the minerals represent assimilated host material and if the conditions (mainly temperature and time) were such that complete re-equilibration was not accomplished. The implications of this possibility will be discussed later.

(b) Perhaps a more realistic possibility is that feldspars have interacted with fluids. O'Neil and Taylor (1967) demonstrated that feldspars readily exchange oxygen isotopes with fluids, particulary if alkali-chloride solutions are involved, and the resultant feldspar usually acquires a cloudy aspect. Feldspars with cloudy aspect in thin section are frequent in the samples of the CIM. Na and K are the only major elements that do not define differentiation trends when plotted versus SiO_2 for BG (Ugidos & Recio 1991). Occasionally, albitic overgrowths of some plagioclases, and even some K-feldspar rims around plagioclase crystals, can be found. Also, the cordierites found within CG are very rich in Na_2O relative to those in NB (Recio 1990). So there is evidence for the existence of some Na-rich residual phase towards late stages of crystallisation. This may be related to fluids circulating within the granite which participated in the alteration of cordierite, since for such alteration to occur fluid different from that present within the channels is required (Schreyer 1985).

Independent evidence for the presence of fluids is also suggested by the relatively frequent episyenites developed in the granites though not sampled in this petrogenic study. The original mineralogy changes to chlorite, epidote, albite and red alkali feldspar when changing from granite to episyenite, and SiO_2 content may drop by as much as 10% (unpublished data of the authors).

Model (b) thus can reasonably explain the isotopic behaviour of alkali feldspar and plagioclase in the granites, and can also explain why these variations are erratic, such that sometimes alkali feldspar is isotopically heavier, whereas in other cases plagioclase is heavier than predicted. It can also explain why both feldspars sometimes behave normally with respect to other minerals such as biotite or quartz, as well as the fact that the fractionation between

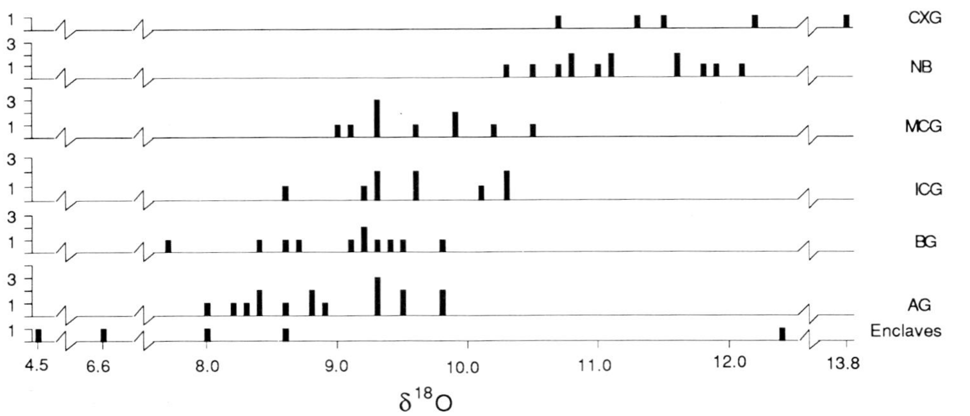

Figure 2 Histograms of $\delta^{18}O$ values, separated by rock type, in the samples of the CIM analysed for this study: CXG = "Complejo Esquisto-Grauváquico" samples; NB = Nebulites: MCG = cordierite-bearing biotite granites, marginal areas; ICG = cordierite-bearing biotite granites, internal areas; BG = biotite granites; AG = amphibole-bearing biotite granites.

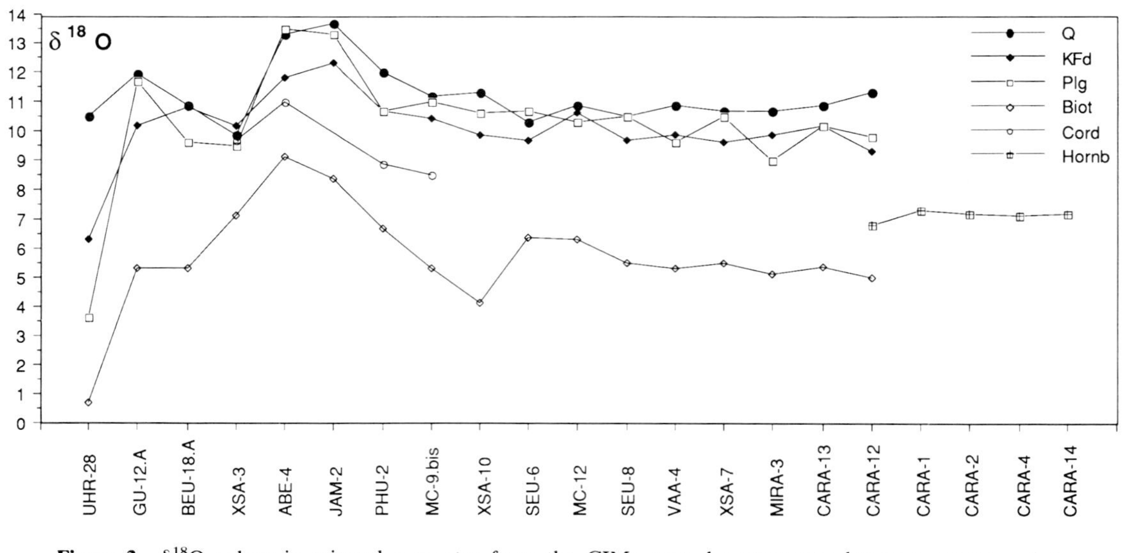

Figure 3 $\delta^{18}O$ values in mineral separates from the CIM; note the crossovers between quartz and the feldspars, indicating that the original isotopic equilibrium between mineral pairs in some rocks was not preserved.

alkali feldspar and plagioclase is relatively small. The fluid was magmatic, and had earlier been in equilibrium with the parent magma, its volume was relatively small compared with the volume of interactive rock, and interaction between feldspars and fluid took place at relatively high *T*, so that no major isotopic fractionation occurred. The granite system, as it will be argued below, was effectively a closed one.

Isotopic behaviour in NB is similar to that in the granites, in as much as it is mainly alkali feldspar and plagioclase that behave anomalously. Interaction with the same fluid that affected the granites is unlikely, since cordierites have generally not been modified. Feldspar disequilibrium could be an inherited feature, considering that the nebulite is really a (petrographically) homogenised migmatite, in which there are different proportions of "melted" (i.e. those from the neosome) and restitic minerals (i.e. unmelted mesosome material), and in which both "melted" and restitic plagioclase may be present. An extreme situation is encountered in sample XSA-3, in which the mesosome and neosome can still be recognised. Metamorphic minerals, such as sillimanite needles, are still present, and it is not surprising that isotopic fractionation between minerals in this rock is anomalous (Fig. 3). However, biotite behaves predictably while the other minerals (quartz, alkali feldspar, plagioclase, cordierite) all have $\delta^{18}O$ within 0·7‰.

It is reasonable to assume that although the granites did not preserve isotopic equilibrium for oxygen, they always behaved as an isotopically closed system (except when the fluid was lost from the system), since the fluid inferred to have acted on cordierite, alkali feldspar and plagioclase (see above) was probably internal to the system (deuteric) and not externally derived. The fact that the system was isotopically closed does not preclude it behaving as an open system with respect to mass *before reaching the isotopic closure T* of the first of the minerals. Arguments favouring this interpretation follow:

Gregory and Criss (1986) indicated that isotherms in δ-space are represented by straight lines, and that equilibrium attained at any given *T* will result in δ–δ plots which will be straight lines with slope unity. In Figure 4, δ–δ plots of different mineral pairs involving feldspars are presented. The thin line is the best fit to the sample points, and the thicker, dashed line is a slope of unity, for comparison. Should all the mineral pairs of the different samples have established equilibrium at a common *T*, sample points would have plotted on the slope unity line. Such being the case, the intercept of this line would have been Δ_{i-j}, the isotopic fractionation between the mineral pair considered, representative of the temperature at which the system established equilibrium. Although the sample points do plot close to a line the slope of which is never very different from unity, the intercept is far from the Δ_{i-j} expected for mineral pairs in each sample at the magmatic temperatures of the system.

Therefore, the minerals in the different samples *did not* equilibrate at a common temperature. This might have been expected from the discussion of Figure 3 and from the propensity of feldspars to continue oxygen isotope exchange to low temperature. Whether the system was closed (in the sense of consisting of mineral phases together with a coexisting fluid in constant proportions) remains a moot point. Gregory and Criss (1986) point out that "... by selecting the heaviest and lightest phases as the coordinate system to represent the rock ... all closed system multi-phase exchange trajectories will have *negative* slopes." It is clear from Table 2 and Figure 3 that the appropriate phases are quartz and biotite (the only common ubiquitous minerals other than the feldspars). In the δ–δ plot (Fig. 5(a)) for biotite and quartz, two very different trends are revealed. All of the granite samples plot close to a line with negative slope (highlighted in Fig. 5(b)), whereas nebulites and enclaves tend to follow the broken line with positive slope (there are a few exceptions). The fit in Figure 5(b) is acceptable (r = 0·675, significant at the 95% confidence level). In a review of this paper, Peter Nabelek has pointed out that the $\delta^{18}O$ values of feldspars (the other major mineral phase) are similar to those of quartz (see Figs 3, 4) so that there is an approximation to a two-phase system (biotite and quartz plus feldspar). One of the properties of δ-space, as discussed by Gregory and Criss (1986), is that straight lines in δ–δ plots transform into straight lines in Δ–δ plots. Figure 6 is a δ_{Bt} vs Δ_{Q-Bt} plot for the granites only; all data plot very close to the negative slope correlation line (r = 0·962 significant above the 99·9% confidence level). A characteristic of Δ–δ plots is that, subject to certain strict criteria (discussed by Gregory &

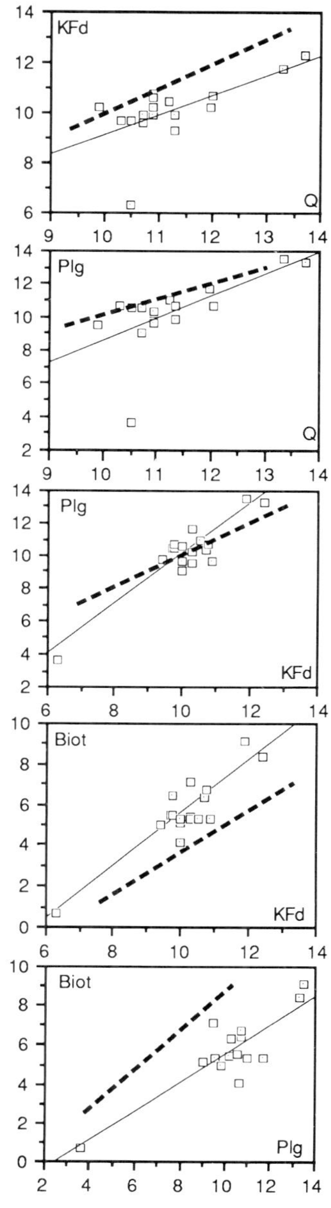

Figure 4 δ–δ plots of mineral pairs separated from samples of the CIM; thin line is best fit to data points; dashed line has slope unity; see text for discussion.

Criss 1986), the intercept ($\Delta = 0$) gives the system bulk $\delta^{18}O$. Although it has not been shown that these criteria have been fulfilled for the granites of this study, we note that the intercept of 9·16 is identical to the average granite whole rock value of 9·2‰.

In Figure 7 a δ–Δ plot, similar to that of Figure 6, is presented for all the samples. The trend of the granites (Fig. 6) is also approached by the enclaves. The NB, on the other hand, show complex behaviour: some samples plot along trend with the granites and enclaves, while others do not. This can be used to argue once more that granites and NB have different origins, and that the heat supplied by the granitic magma induced partial re-equilibration in the NB. The granites were the dominant component and Figure 7 indicates that at least some of the NB plot close to values typical of granites.

The oxygen isotope study of both whole rocks and mineral separates leads to the following conclusions: Granites and NB have very different protoliths, as indicated by their whole rock $\delta^{18}O$ values, typically around 9·2‰ for the granites and about 11·05‰ for the NB. The inferred protolith of the granites must have included an I-type component. The inferred protolith of the NB must have been in the upper crust, since only materials that have gone through low temperature weathering processes can have $\delta^{18}O$ values as high as those of the nebulites. Assimilation of an upper crustal protolith similar to nebulites can explain the isotopic variations seen in Figure 2.

Mineral separate data from the granite indicate that assimilation occurred at a temperature higher than the closure T of the first mineral to cease exchanging oxygen, so that after assimilation took place the system behaved as effectively *closed* for oxygen isotopes. Isotopic disequilibrium between mineral pairs could be caused by subsolidus exchange of feldspar with a deuteric fluid during the latest stages of consolidation of the magmas.

The oxygen isotopic ratios observed in the NB are consistent with their derivation by an anatectic process from materials whose oxygen isotopic composition is similar to the analysed metasediments of the CXG.

6. Hydrogen isotopes: results and discussion

The δD signatures of the 37 samples analysed are given in Table 3. The H_2O samples were collected in streams and thermal springs emanating from the granites. From Table 3, several observations can be made:

(1) coexisting biotite and hornblende in AG have very similar δD values

(2) biotite is always isotopically lighter than coexisting cordierite

(3) biotite separated from the enclaves is more similar to igneous than to sedimentary values, as represented by the CXG samples. The heaviest biotite of the enclaves ($\delta D = -65‰$) corresponds to a noritic gabbro (UHR-28) that, as already noted, is suspected of having suffered hydrothermal alteration.

(4) on the basis of hydrogen isotopes alone, it is not possible to clearly differentiate the biotites in the NB from those of the granites (but see Fig. 8 below).

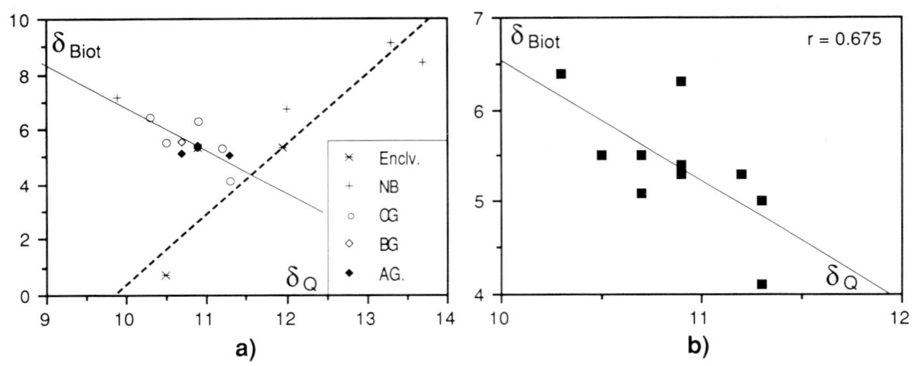

Figure 5 (a) δ_{Biot} vs δ_Q plot: all of the granite samples lie close to a trend of negative slope (solid line defined by 10 samples); nebulite and enclave samples scatter widely; for purposes of discussion, the dashed line of positive slope is fitted through five samples (three nebulites and two enclaves); Enclv = enclaves; NB = nebulites; CG = cordierite-bearing granites; BG = biotite granites; AG: amphibole-bearing biotite granites.
(b) Detail of δ-δ plot for biotite and quartz from the granite samples.

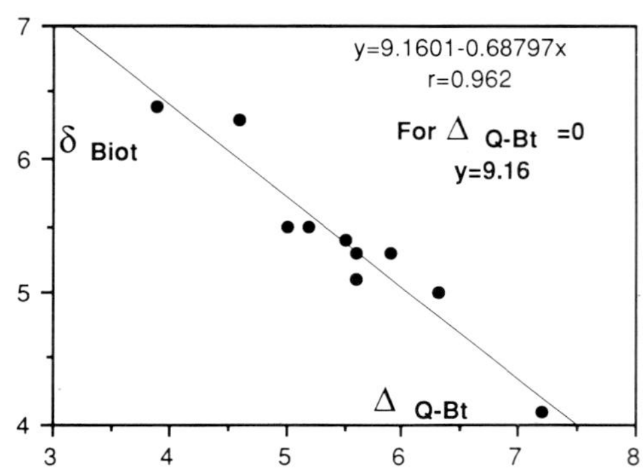

Figure 6 δ-Δ plot for biotite and quartz-biotite pairs from the granites of the CIM.

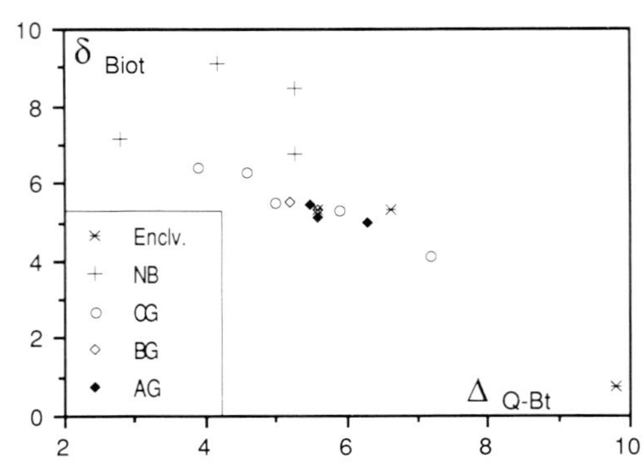

Figure 7 Δ_{Q-Bt} δ_{Bt} graph to show the relationship between granites, as plotted in Figure 6, and nebulites and enclaves; symbols as in Figure 5.

(5) variations attributable to mineral chemistry aside, there is a general tendency towards higher δD values from AG through BG, to CG and NB, with the samples of the CXG being the most D-rich (except for the cordieritite enclave which is the isotopically heaviest sample measured).

The combined δ^{18}O–δD results are plotted in Figure 8. The values plotted are the actual measured values, not the values of the calculated H_2O in equilibrium with the samples. Calculation of the isotopic composition of the H_2O in equilibrium with a given rock or mineral requires a series of assumptions regarding the temperature of interaction, and use of isotope fractionation factors. Because some fractionation factors are not known, the measured values are used in the discussion that follows. Note, however, that the field labelled "Cornb. Bath." (Fig. 8) corresponds to the isotopic values of H_2O in equilibrium with Cornubian Batholith rocks and minerals, at magmatic temperatures, as *calculated* by Sheppard (1977). Considering that H_2O concentrates D relative to hydrous minerals other than muscovite at magmatic temperatures (Suzuoki & Epstein 1976), and that the magnitude of $\alpha_{H_2O\text{-Melt}}$ may be limited by, or similar to $\alpha_{H_2O\text{-Mica}}$ when the H_2O content of the melt is low (Taylor, B. E. 1986), the field of the Cornubian Batholith rocks might be expected to be displaced towards lower δD values at similar δ^{18}O (see also Dobson *et al.* 1989).

Examination of Figure 8 reveals several points:

(1) Most of the samples plot in or very near the igneous field. The samples that deviate most from this field are:

(a) CXG samples: they have higher δD and δ^{18}O values, consistent with derivation from pre-Hercynian sediments.

(b) The δD value of the noritic gabbro enclave UHR-28 could well be a magmatic value (but also a typical meteoric water value; see Fig. 8). However, its δ^{18}O value is far too low, indicating postmagmatic modification. Its location midway towards values typical of present-day meteoric waters in the area makes it likely to have interacted with a low δ^{18}O fluid (almost certainly of meteoric derivation).

(c) The cordieritite enclave has the highest δD and the second highest δ^{18}O value of all the samples analysed, this being the most "sedimentary" signature found. The cordieritite has been completely pinitised and the δD and δ^{18}O signatures suggest that such pinitisation took place at high temperature, and consequently a deuteric fluid is inferred to have been involved. It is recalled that the cordieritite sample was taken from a macroenclave inside the BG.

(2) The δ^{18}O values of hornblendes are very constant (mean $\delta^{18}O = 7{\cdot}12 \pm 0{\cdot}19$‰, 1σ, $n = 5$), but higher than δ^{18}O values of biotites in the same rock type. The values suggest isotopic equilibrium between the two phases (see, for example, Taylor 1968). The similar δD values of biotites and hornblendes are to be expected. Hydrogen isotope fractionation for these minerals is controlled by the nature of the cations in octahedral coordination (see Suzuoki & Epstein 1976) and so the similar Fe/Mg (Graham *et al.* 1984) ratios for hornblendes and biotites from amphibole-bearing granites of the western CIM (Recio 1990) should lead to similar δD. The spread in δD values, however, may indicate interaction with a very small volume of fluid, sufficiently small not to modify the oxygen isotope ratios of the minerals, but large enough to affect their δD. Similar considerations apply to the other facies studied, as recorded by biotites and cordierites, as well as to the two micrograular (tonalitic) enclaves that have similar δ^{18}O, but different δD.

(3) It was stated earlier that in terms of δD, the biotites from the NB cannot be clearly distinguished from those separated from the granites. All biotites from the granites, as well as those from micrograular (tonalitic) enclaves, plot very close to the lower δ^{18}O limit of the igneous box, whereas biotites from the NB have distinctly higher δ^{18}O. However, the isotopic signature of cordierites discriminates between cordierites from the NB, falling in the field of Cornubian Batholith H_2O, and those from CG, having typical igneous values. This does not imply that the cordierites in the granites are of magmatic origin, but rather that they acquired such isotopic characteristics when they

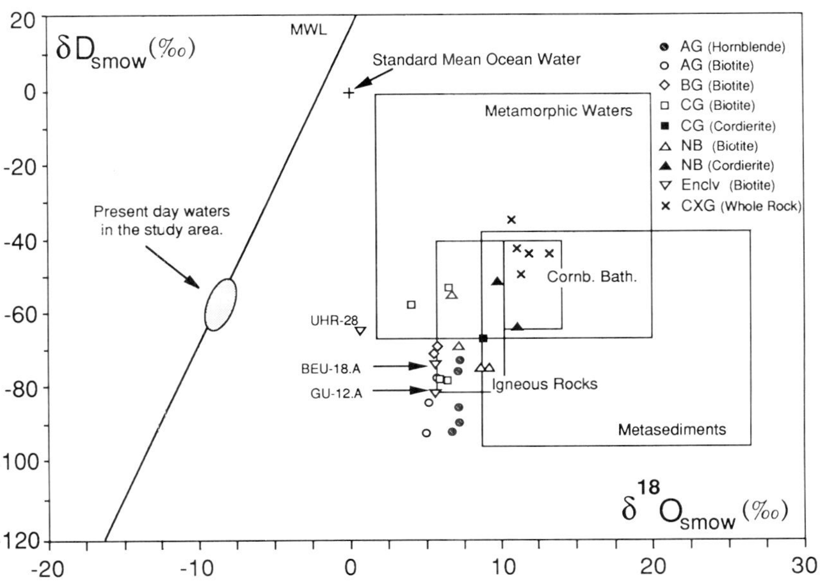

Figure 8 δ^{18}O–δD plot of the analysed samples from the western part of the CIM, separated by rock type (abbreviations as in Fig. 5) and phase analysed (within brackets); MWL = Meteoric Water Line of Craig (1961); field labelled "*Igneous rocks*" refers to the values of I-type granites; for explanation of the field labelled "*Cornb. Bath.*", see text; limits for the different fields shown are indicative only, after data of Sheppard (1986).

were incorporated into the granite as the result of assimilation. If the cordierite is originally from the NB and has been incorporated into the granite as a result of assimilation, interaction with a granitic deuteric fluid explains both the different δD values and the fact that the cordierite was pinitised. No single cordierite crystal in the granite is absolutely fresh.

(4) The CXG samples are distinctly different from any other rocks of this study, and have isotopic values corresponding to their presumed precursor. The CXG samples plot in an appropriate position (Fig. 8) to be a suitable hypothetical end member in a mixing process to give CG, involving a more primitive composition as the other end member.

In summary, therefore, it has been shown there is a correspondence between the hydrogen isotopic signature with that of oxygen, changing from lower (AG) towards higher (NB) values, with most of the granites having typically igneous signatures.

7. Summary and conclusions

The whole-rock $\delta^{18}O$ values of Figure 2 show that there is a marginal trend of increasing oxygen isotopic composition from the AG through BG to the CG. Significantly higher $\delta^{18}O$ are found in the NB compared to the entire granite population. The sediments of the CXG have $\delta^{18}O$ values close to the NB, though slightly higher on average. Since high-grade metamorphism can result in homogenisation and a slight lowering of $\delta^{18}O$ (Taylor & Sheppard 1986), the stable isotope data concur with field observations in suggesting that the NB were derived from the CXG sediments.

The $\delta^{18}O$ values corresponding to the granites, particularly those of AG and BG, cannot represent *solely* crustal, sediment-derived melts, nor purely mantle signatures. It is proposed that they represent a mixture of mantle and crustal materials. The $\delta^{18}O$ values observed, the type and $\delta^{18}O$ of enclaves included in the granites, as well as the thermal evolution of the region (Ugidos 1990), do not allow mantle contributions to be ruled out. Similar conclusions have been reached by Fourcade *et al.* (1989) for granitic rocks in Galicia; by Ben-Othman *et al.* (1984) for the Pyrenees (employing different isotope systems) and by many authors (see for example the review of Harmon *et al.* 1984) for the Caledonian of Scotland. When magma reaches relatively high crustal levels and encounters preheated metasedimentary rocks (see Ugidos 1990), assimilation proceeds readily, shifting the oxygen isotopic signatures towards $\delta^{18}O = 10‰$ and leading to generation of the CG. The CG was therefore the result of a second-stage process involving BG magmas plus NB, or similar rocks.

From mineral separate $\delta^{18}O$ values it is obvious that isotopic equilibrium between all minerals has not been preserved. This is to be expected if assimilation processes took place. Despite perfect equilibrium not having been preserved, the δ–Δ plot of Figure 6 suggests a common origin for the granites. Addition of NB to this plot (Fig. 7) causes loss of correlation, a clear indication that NB are external to the granite system, and consequently evidence that the granites cannot be derived primarily from the NB.

The lack of isotope equilibrium manifested by the $\delta^{18}O$ values of alkali feldspar and plagioclase as shown in Figure 3, at the same time as the whole rock system preserves magmatic values, is the result of deuteric fluid activity and subsolidus exchange.

All hydrogen isotope ratios, except for the gabbro macroenclave and the cordieritite, are primary. The isotopic signature of the gabbro is the result of hydrothermal alteration before incorporation into the granite, and the H_2O involved was most probably of meteoric derivation. The lack of re-equilibration back to magmatic values indicates a dry system at the time of incorporation. Other than the CXG samples, other granite samples plot in the igneous field, indicating unaltered granite values.

8. Acknowledgements

Financial support for this work was partly provided by the British Council (U.K.) and the Ministerio de Educación y Ciencia (Spain) through the "Acciones Integradas Hispano-Británicas" programme (Award SPN/705/3(89/9)128) and the EEC Twinning programme (Award PL89(4)0561). CR would like to thank the scientific and technical staff of the IGU of SURRC for their help, guidance and friendship during the tenure of a three-year grant awarded by the Spanish Ministry of Education and Science. We thank Peter Nabelek for a thoughtful and constructive review. Andrew Tait and Adrian Boyce are thanked for their help and W. E. Stephens for many contributions.

9. References

Ben-Othman, D., Fourcade, S. & Allègre, C. J. 1984. Recycling processes in granite-granodiorite complex genesis: the Querigut case studied by Nd–Sr isotope systematics EARTH PLANET SCI LETT **69,** 290–300.

Borthwick, J. & Harmon, R. S. 1982. A note regarding ClF_3 as an alternative to BrF_5 for oxygen isotope analysis. GEOCHIM COSMOCHIM ACTA **46,** 1665–8.

Clayburn, J. A. P., Harmon, R. S., Pankhurst, R. J. & Brown, J. F. 1983. Sr, O, and Pb isotope evidence for origin and evolution of Etive Igneous Complex, Scotland. NATURE **303,** 492–7.

Clayton, R. N. & Mayeda, T. K. 1963. The use of bromine pentafluroide in the extraction of oxygen from oxides and silicates for isotopic analysis. GEOCHIM COSMOCHIM ACTA **27,** 43–52.

Craig, H. 1961. Isotopic variations in meteoric waters. SCIENCE **133,** 1702–3.

Department of Petrology, University of Salamanca 1983. *Mapa 1:200000 de la síntesis geológica del basamento. Zona del cento-oeste español.* Salamanca: University of Salamanca.

Dobson, P. F., Epstein, S. & Stolper, E. M. 1989. Hydrogen isotope fractionation between coexisting vapor and silicate glasses and melts at low pressure. GEOCHIM COSMOCHIM ACTA **53,** 2723–30.

Fourcade, S., Peucat, J. J., Martineau, F., Cuesta, A., Corretgé, L. G. & Gil Ibarguchi, I. 1989. Análisis de isótopos de oxígeno y edad Rb–Sr del plutón zonado de Caldas de Reyes (Galicia, España). GEOGACETA **6,** 7–9.

Fourcade, S. & Javoy, M. 1973. Rapports $^{18}O/^{16}O$ dans les roches du vieux socle catazonal d'In Ouzzal (Sahara Algerien). CONTRIB MINERAL PETROL **42,** 235–44.

Friedman, I. & Gleason, J. D. 1973. Notes on the bromine pentafluoride technique of oxygen extraction. J RES US GEOL SURV **1**(6) 679–80.

Garlick, G. D. 1966. Oxygen isotope fractionation in igneous rocks. EARTH PLANET SCI LETT **1,** 361–8.

Godfrey, J. D. 1962. The deuterium content of hydrous minerals from the East–Central Sierra Nevada and Yosemite National Park. GEOCHIM COSMOCHIM ACTA **26,** 1215–45.

Graham, C. M., Harmon, R. S. & Sheppard, S. M. F. 1984. Experimental hydrogen isotope studies. Systematics of hydrogen isotope exchange between amphibole and water. AM MINERAL **69,** 128–38.

Gregory, R. T. & Criss, R. E. 1986. Isotopic exchange in open and closed systems". *In* Valley, J. W., Taylor, H. P. Jr & O'Neil, J. R. (eds) *Stable isotopes in high temperature geological processes.* MSA REV MINERAL **16,** 91–127.

Halliday, A. N., Stephens, W. E. & Harmon, R. S. 1980. Rb–Sr and O isotopic relationships in 3 zoned Caledonian granitic plutons, Southern Uplands, Scotland: Evidence for varied sources and hybridization of magma. J GEOL SOC LONDON **137,** 329–49.

Harmon, R. S. 1984. Stable isotope geochemistry of Caledonian granitoids from the British Isles and East Greenland. PHYS EARTH PLANET INT **35** (1–3), 105–20.

Harmon, R. S. & Halliday, A. N. 1980. Oxygen and strontium isotope relationships in the British late Caledonian granites. NATURE **283,** 21–5.

Harmon, R. S., Halliday, A. N., Clayburn, J. A. P. & Stephens, W. E. 1984. Chemical and isotopic systematics of the Caledonian intrusions of Scotland and Northern England: a guide to magma source region and magma-crust interaction. PHILOS TRANS R SOC LONDON **A310,** 709–42.

Jenkin, G. R. T. 1988. *Stable isotope studies in the caledonides of S. W. Connemara, Ireland.* Unpublished Ph.D. Thesis, University of Glasgow, Glasgow.

Julivert, M., Fontboté, J. M., Ribeiro, A. & Conde, L. E. 1974. *Memoria explicativa del Mapa Tectónico de la Peninsula Iberica y Baleares. A escala* 1:1,000.000. IGME. Madrid.

Kelley, S. & Bluck, B. J. 1989. Detrital mineral ages from the Southern Uplands using ^{40}Ar–^{39}Ar laser probe. J GEOL SOC LONDON **146,** 401–3.

Lotze, F. 1945. Zur Gliederung der Varisziden der Iberischen Meseta. Spanish translation by J. M. Rios: Observaciones respecto a la divisi6n de las Variscides de la Meseta Ibérica. PUB EXTRANJERAS SOBRE GEOL ESPAÑA **5,** 149–66 (1950).

Michard-Vitrac, A., Albarede, F., Dupuis, C. & Taylor, H. P., Jr. 1980. The genesis of Variscan (Hercynian) plutonic rocks: inferences from Sr, Pb and O studies on the Maladeta igneous complex, central Pyrenees (Spain). CONTRIB MINERAL PETROL **72,** 57–72.

O'Neil, J. R., Shaw, S. E. & Flood, R. H. 1977. Oxygen and hydrogen isotope compositions as indicators of granite genesis in the New England Batholith, Australia. CONTRIB MINERAL PETROL **62,** 313–28.

O'Neil, J. R. & Chappell, B. W. 1977. Oxygen and hydrogen isotope relations in the Berridale Batholith. J GEOL SOC LONDON **133,** 559–72.

O'Neil, J. R. & Taylor, H. P. Jr 1967. The oxygen isotope and cation exchange chemistry of feldspars. AM MINERAL **52,** 1414–37.

Recio, C. 1990. *The late Hercynian granitoids of the western area of the SCE: A stable (O, H, S) isotopic study.* Unpublished Tesis Doctoral, Universidad de Salamanca, Salamanca.

Schreyer, W. 1985. Experimental studies on cation substitutions and fluid incorporation in cordierite. BULL MINERAL **108,** 273–91.

Sheppard, S. M. F. 1977. The Cornubian Batholith, SW England: D/H and $^{18}O/^{16}O$ studies of kaolinite and other alteration minerals. J GEOL SOC LONDON **133,** 573–91.

Sheppard, S. M. F. 1986. Characterization and isotopic variations in natural waters. *In* Valley, J. W., Taylor, H. P., Jr. & O'Neil, J. R. (eds) *Stable isotopes in high temperature geological processes* MSA REV MINERAL **16,** 165–83.

Suzuoki, T. & Epstein, S. 1976. Hydrogen isotope fractionation between OH-bearing minerals and water. GEOCHIM COSMOCHIM ACTA **40,** 1229–40.

Taylor, B. E. 1986. Magmatic volatiles: Isotopic variation of C, H, and S. *In* Valley, J. W., Taylor, H. P., Jr. & O'Neil, J. R. (eds) *Stable isotopes in high temperature geological processes* MSA REV MINERAL **16,** 185–225.

Taylor, H. P., Jr. 1968. The oxygen isotope geochemistry of igneous rocks. CONTRIB MINERAL PETROL **19,** 1–71.

Taylor, H. P., Jr. 1977. Water/rock interactions and the origin of H_2O in granitic batholiths (Thirtieth William Smith Lecture). J GEOL SOC LONDON **133,** 509–58.

Taylor, H. P., Jr. 1980. The effects of assimilation of country rocks by magmas on $^{18}O/^{16}O$ and $^{87}Sr/^{86}Sr$ systematics in igneous rocks. EARTH PLANET SCI LETT **47,** 243–54.

Taylor, H. P., Jr. 1986. Igneous rocks: II. Isotopic case studies of circumpacific magmatism. *In* Valley, J. W., Taylor, H. P., Jr. & O'Neil, J. R. (eds) *Stable isotopes in high temperature geological processes* MSA REV MINERAL **16,** 273–317.

Taylor, H. P., Jr. 1988. Oxygen, hydrogen and strontium isotope constraints on the origin of granites. TRANS R SOC EDINBURGH EARTH SCI **79,** 317–38.

Taylor, H. P., Jr. & Sheppard, S. M. F. 1986. Igneous rocks: I. Processes of isotopic fractionation and isotope systematics. *In* Valley, J. W., Taylor, H. P. Jr. & O'Neil, J. R. (eds) *Stable isotopes in high temperature geological processes.* MSA REV MINERAL **16,** 227–71.

Ugidos, J. M. 1988. New aspects and considerations on the assimilation of cordierite-bearing rocks. REV SOC GEOL ESPAÑA **1,** 129–33.

Ugidos, J. M. 1990. Granites in the Central Iberian Massif as a Paradigm of Genetic Processes of Granitic Rocks: I-Types vs S-Types. *In* Dallmeyer, R. D. & Martínez García, E. (eds) *Pre-Mesozoic Geology of Iberia,* 189–206. Berlin: Springer.

Ugidos, J. M. & Recio, C. 1991. Geochemical characterization of late Hercynian granites from the Western Central Iberian Massif (CIM; Spain): Constraints on the origin of cordierite-bearing biotite granites. CHEM GEOL (submitted).

Valley, J. W., Taylor, H. P. Jr. & O'Neil, J. R. (eds) 1986. *Stable isotopes in high temperature geological processes,* MSA REV MINERAL **16**. Chelsea, Michigan: BookCrafters.

Wilson, A. F., Green, D. C. & Davidson, L. R. 1970. The use of oxygen isotope geothermometry on granulties and related intrusives, Musgrave Ranges, Central Australia. CONTRIB MINERAL PETROL **27,** 166–78.

C. RECIO, Servicio General de Analisis de Isótopos Estables, Facultad de Ciencias, Universidad de Salamanca, 37008–Salamanca, Spain.

A. E. FALLICK, Isotope Geology Unit, Scottish Universities Research & Reactor Centre, East Kilbride, Glasgow G75 0QU, Scotland, U.K.

J. M. UGIDOS, Departamento de Geología, Facultad de Ciencias, Universidad de Salamanca, 37008-Salamanca, Spain.

Transactions of the Royal Society of Edinburgh: Earth Sciences, **83**, 259–268, 1992

Genesis of the lower crustal garnet–orthopyroxene tonalites (S-type) of the Hidaka Metamorphic Belt, northern Japan

T. Shimura, M. Komatsu and J. T. Iiyama

ABSTRACT: The Hidaka Metamorphic Belt (HMB) in Hokkaido, northern Japan, consists of tilted metamorphic layers of an island-arc type crust from lower (granulite facies) to upper (very low-grade metasedimentary) horizons. Abundant granitic rocks, mainly S-type tonalites of crustal origin, intrude various metamorphic layers and are classified into four depth types, namely upper, middle, lower and basal. The basal orthopyroxene–garnet (S-type) tonalities were intruded into granulite facies country rocks. Textural and compositional evidence from minerals in the basal tonalite indicates that the crystallisation sequence is Grt–Pl–Opx–Bt–Qtz–Crd–Kfs, and that crystallisation took place at about 600 MPa and 900°C–700°C.

Some crystallisation experiments were carried out in an internally heated pressure vessel, using the basal tonalite, under the conditions of 300 and 600 MPa, 700–900°C, and with 0–20 wt% H_2O, respectively. The results show that the primary S-type tonalite magma was at a temperature above 900°C and contained 3–4 wt% H_2O at the beginning of crystallisation. In order to study the influence of normative orthoclase content on orthopyroxene crystallisation, some starting materials also included 15, 20 and 25% normative orthoclase, by adding $KAlSi_3O_8$ gel to the rock powder. Normative orthoclase content has an influence on the subliquidus crystallisation limit of orthopyroxene.

The changes in P–T conditions and chemical composition of the magma during ascent would generate the sequence from the basal to upper S-type granite. Opx-free S-type granitic magma can be generated from lower crustal Grt–Opx S-type granitic magma, by differentiation with falling magmatic temperature.

KEY WORDS: crystallisation, granulite, lower crust S-type tonalite.

Most S-type granitic and volcanic rocks are considered to have been derived through crustal anatexis of aluminous metasedimentary rocks (e.g. Chappell & White 1974; White & Chappell 1977). These S-type igneous bodies contain quartz, two feldspars and biotite, and also one or more aluminous phases such as muscovite, garnet, cordierite, aluminosilicates, tourmaline and spinel (Clarke 1981). In addition to these aluminous phases, orthopyroxene is another important mineral in some S-type granitic and related volcanic rocks.

Orthopyroxene-bearing S-type rocks have been reported from southeastern Australia (Birch & Gleadow 1974; Birch *et al.* 1977; Wyborn *et al.* 1981; Clemens & Wall 1981, 1984, 1988), New Zealand (Wood 1974; Barley 1987), southwestern Japan (Kawasaki 1980a, b; Murata 1982; Nakada 1983), and northern Japan (Komatsu 1983; Bamba 1985; Komatsu *et al.* 1986; Osanai *et al.* 1991; Maeda *et al.* 1991; Shimura 1992). A typical mineral assemblage for these rocks is Grt + Opx + Crd + Bt + Pl + Qtz + Kfs, although Grt, Crd and/or Kfs may be absent. They are poor in normative quartz and orthoclase compared with orthopyroxene-free S-type rocks (as shown in the an–ab–or–Q diagram, Fig. 1).

These Opx-bearing S-type suites are considered to have started to crystallise at relatively high pressure and temperature, that is, 500–700 MPa and above 800°C (Wyborn *et al.* 1981; Clemens & Wall 1981, 1984, 1988; Barley 1987; Maeda *et al.* 1991; Shimura 1991). These conditions are within those of the granulite facies and correspond to those of the lower crust of island arcs or continents only where enough heat is available. Almost all Opx-bearing rocks described previously, however, are volcanic rocks, and the crystallisation of S-type magmas of deep crustal origin has not yet been clarified.

The Grt–Opx tonalite masses in the Hidaka Metamorphic Belt (HMB), a tilted crustal section of an island arc (Komatsu *et al.* 1989; Osanai *et al.* 1991), are rare examples of S-type granites intruded into granulites and crystallised under granulite facies conditions. In addition, Opx-free S-type tonalite masses occur within amphibolite facies rocks of the belt. Using petrological and experimental studies, we discuss the crystallisation conditions, including compositional controls on Opx crystallisation, and the genesis of S-type granitic magmas.

1. The Hidaka Metamorphic Belt

1.1. Geology

The HMB is a high temperature–low pressure metamorphic belt of Tertiary age, occupying the axial zone of Hokkaido, northern Japan. It extends for about 140 km with a width of 10–20 km (Fig. 2). The Hidaka Main Thrust (HMT) defines the western extent of the belt and is in thrust contact with the western ophiolitic (Miyashita 1983) and accretionary complexes of upper Cretaceous age (Komatsu *et al.* 1983).

The HMB is considered to represent a tilted section of

island-arc type crust (Komatsu *et al.* 1983). The metamorphic grade systematically increases from very low grade in the E (top) to granulite facies in the W (base), and is divided into four metamorphic zones (I–IV) (Osanai 1985; Komatsu *et al.* 1989; Osanai *et al.* 1986, 1991). Temperatures estimated from the metamorphic rocks are *c.* 400°C (transition from I to II), *c.* 500°C (transition from II to III), 530–580°C (upper part of III), 600–670°C (lower part of III) and 720–840°C (IV) (Komatsu *et al.* 1989; Osanai *et al.* 1991). Pressures are 260–350 MPa in zone II, *c.* 500 MPa in III and 610–720 MPa in zone IV. Abundant gabbros, diorites and granites intrude the metamorphic zones and vary chemically and petrographically according to the zones which they intrude (e.g. Table 1, Fig. 5). Isotopic age determinations show that the main metamorphism and synchronous magmatism occurred at 50–40 Ma (Owada & Osanai 1988; Shimura *in prep.*). Tilting of the crust occurred at a later stage (Arita *et al.* 1986; Toyoshima 1990; Shimura 1992), *c.* 20 Ma (Osanai *et al.* 1991).

1.2. Granitic rocks

Granitic rocks are abundant in the HMB. They are mainly tonalitic in composition and can be classified into peraluminous and metaluminous types, corresponding, respectively, to the S-type and I-type granitic rocks of Chappell and White (1974) (see Komatsu *et al.* 1986; Owada 1989; Osanai *et al.* 1991; Shimura 1991). As described later (see Section 1.3), both types generally coexist within one pluton, but S-types greatly predominate. Initial Sr isotopic ratios of the S- and I-type granitic rocks range from 0·7048–0·7065, and from 0·7033–0·7048, respectively, and are nearly identical to the ranges of peltic metamorphic rocks (0·7043–0·7063) and of pyroxene amphibolites (0·7029–0·7049), respectively (Shibata & Ishihara 1979a, Okamoto 1983; Okamoto & Honma 1983; Ikeda 1984a, 1991; Owada & Osanai 1989: Shimura 1991). The S-type and I-type granitic rocks are considered to have been derived through the crustal anatexis of garnet–cordierite–biotite granulite and of pyroxene amphibolite in the

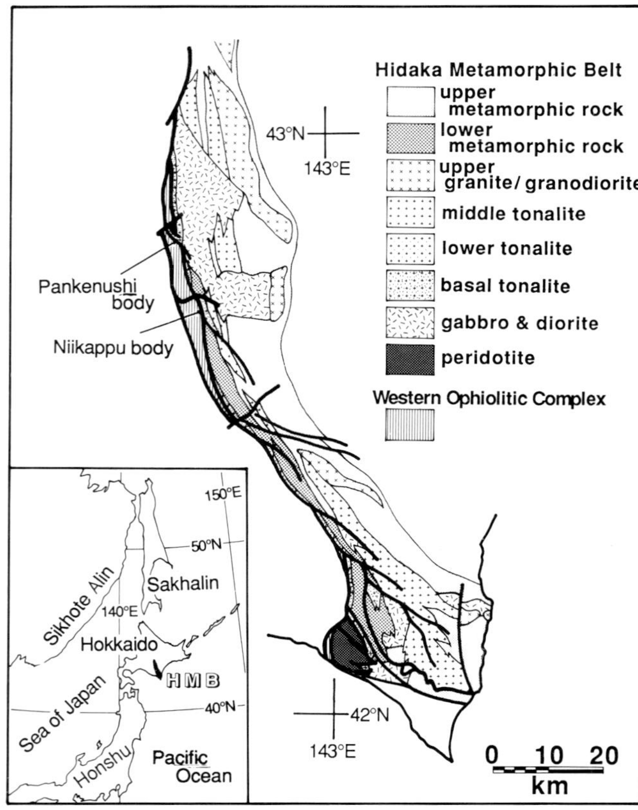

Figure 2 Simplified geological map of the Hidaka Metamorphic Belt, modified after Komatsu *et al.* (1986).

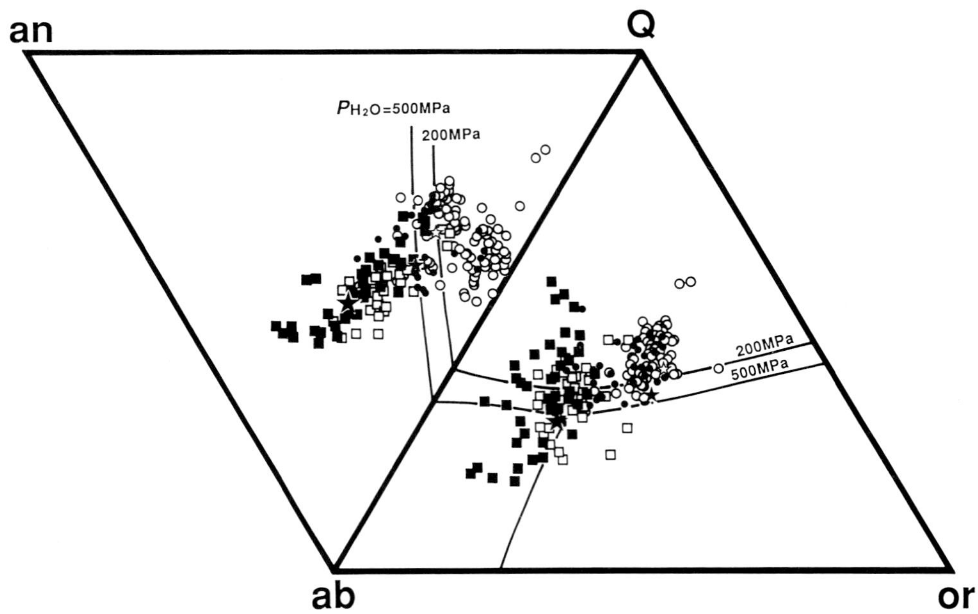

Figure 1 Composition of S-type igneous rocks on the normative an–ab–or–Q diagram; data are from Hines *et al.* (1978), Kawasaki (1980a), Wyborn *et al.* (1981), Clemens & Wall (1981, 1984, 1988), Murata (1982), Y. Takahashi (1986), Barley (1987), White & Chappell (1988), Owada (1989), Chappell *et al.* (1991), Wyborn *et al.* (1991), Shimura (1991); cotectic lines in the system an–ab–or–Q are also shown ($P_{H_2O} = P_{total}$) after Winkler (1979) and Nekvasil (1988); solid symbols indicate orthopyroxene-bearing rocks (or bodies); solid squares indicate basal and lower tonalite compositions of the Niikappu body, and open squares indicate middle tonalite composition in the HMB; small open and closed stars, and a large closed star indicate the starting material composition of experiments by Clemens and Wall (1981, 1988), and this study, respectively.

Table 1 Characteristic mineral assemblages of granitic rocks classified in terms of the S-type/I-type groups and intrusive levels.

Depth type	Intrusive level	Characteristic mineral assemblage I-type	S-type
Upper	very low grade metasediments	Bt + Hbl + Pl + Qtz	Bt + Pl + Qtz ± Kfs
Middle	zone I to II	Bt ± Hbl + Pl + Qtz	Bt ± Ms ± Crd + Pl + Qtz ± Kfs
Lower	zone III	Bt + Hbl + Opx ± Cpx + Pl + Qtz	Bt + Opx ± Crd + Pl + Qtz ± Kfs Bt + Grt ± Crd + Pl + Qtz ± Kfs
Basal	zone IV		Bt + Opx + Grt ± Crd + Pl + Qtz ± Kfs

Abbreviations are after Kretz (1983); I, II, III and IV refer to metamorphic zones of the Main Zone (after Osanai *et al.* 1991).

lowermost crust, respectively (Owada & Osanai 1989; Osanai *et al.* 1991).

The granitic rocks are intruded into the four metamorphic zones, and are grouped accordingly: (1) those intruded into zone I (very low grade metasedimentary rock), (2) those within zone I–II (greenschist facies to lower amphibolite facies), (3) those within zone III (upper amphibolite facies), and (4) those within zone IV (granulite facies). The granitic intrusions, except for those in zone I, generally have sheet-like forms and the granitic rocks have gneissose structures which are sub-parallel to those of country rocks. Metamorphic zones I to IV correspond to shallow through to deep crust, before tilting of the crust (Komatsu *et al.* 1983, 1989; Osanai *et al.* 1991), and the four granite types correspond to four depth of intrusion, that is: (1) upper granite-granodiorite, (2) middle tonalite-granodiorite, (3) lower tonalite, and (4) basal tonalite (Komatsu *et al.* 1986; Shimura 1992) (Fig. 2, Table 1). Each type has a characteristic mineral assemblage and chemical composition which changes gradually from lower to higher level tonalites. For example, the middle S-type tonalite is characterised by cordierite and/or muscovite, lower S-type tonalite by cordierite and orthopyroxene with or without garnet, and basal S-type tonalite by garnet and orthopyroxene with or without cordierite.

1.3. Basal tonalite

Basal tonalite is widely distributed along the HMT of the HMB, where the tonalite is strongly mylonitised (Osanai 1985; Bamba 1985; Osanai *et al.* 1986; Komatsu *et al.* 1986; Shimura 1992). Weakly mylonitised basal tonalite occurs in some places such as the Pankenushi River and the Niikappu River regions (Fig. 2). In the upper stream of the Niikappu River area, tonalitic rocks form a continuous large mass, intruding metamorphic rocks of the granulite facies to the lower amphibolite facies. The mass is composed of basal (W), lower and middle tonalites (E). Both S-type and I-type tonalities coexist throughout the body, although the former predominates. I-type tonalites occur as less than 20 m-wide bodies or in part as rounded forms less than 1 m in diameter within the S-type rocks, and show sharp contacts with S-type rocks. There are also small irregular-shaped patches of I-type tonalite, less than a few hundreds of millimetres across, within S-type tonalites, whose boundaries are vague, presumably due to mixing with the latter. The I-type tonalite appears to have intruded into the S-type rock before the complete consolidation of S-type magma.

Tonalites were injected into the country rocks to form agmatite along the margins of the intrusive mass. The tonalites carry abundant metamorphic blocks and fragments, giving a very heterogeneous character, even in the centre of the tonalite. The magmatic parts are referred to as tonalite matrix. This tonalite matrix has a characteristic mineral assemblage in each depth type (Table 1), but the mineral assemblage varies gradually or abruptly from the base to the top of the Niikappu River mass. The boundary between the basal and lower tonalities can be recognised by a sudden decrease in modal garnet, and the disappearance of large garnet grains (5–30 mm in diameter) around pelitic granulite enclaves, and finally by the disappearance of coexisting garnet and orthopyroxene. The boundary between the lower and middle tonalities is marked by the disappearance of orthopyroxene.

The country rocks to the basal tonalite of this mass are the granulite facies para- and ortho-gneisses which are also included in the tonalite as enclaves of various shapes and sizes. The enclaves are pelitic–psammitic granulite (Grt–Opx–Bt, Grt–Opx–Crd), two-pyroxene mafic granulite, two-pyroxene amphibolite, orthopyroxene–amphibolite, gabbroic granulite and metamorphosed ultramafic rocks, and show various extents of assimilation by tonalite magma.

The composition and crystallisation of magma free from the effects of assimilation are investigated here, using homogeneous rocks largely or totally free from pelitic enclaves (Fig. 3(A)). The minerals in the homogeneous tonalite matrix show compositional zoning, indicating that crystallisation took place during cooling of the magma. For example, garnet has $X_{Mg} = 0{\cdot}30$ in cores varying to $0{\cdot}14$ in rims, and plagioclase shows oscillatory zoning, becoming poorer in *An* towards the rim (An_{35}–An_{28}). The crystallising sequence of minerals from the tonalitic magma is inferred from the enclosing and reaction relations of the minerals; high X_{Mg} garnet grains are enclosed in plagioclase; high *An* plagioclase grains are enclosed in hypersthene with poikilitic texture (Fig. 3(B)). Biotite and quartz symplectites occur replacing orthopyroxene and as pseudomorphs after orthopyroxene (Fig. 3(C)). Cordierite and K-feldspar are never enclosed in other mineral grains and the latter occurs in interstices and is present in very small amounts. This textural evidence indicates that the crystallisation order is garnet–plagioclase–orthopyroxene–biotite–quartz–cordierite–K-feldspar (Fig. 4). X_{Mg} values of all ferromagnesian minerals have a sequence through crystallisation of Grt < Opx < Bt < Crd. The same sequence of X_{Mg} has been reported elsewhere (e.g. Grant 1973; Lonker 1980; Wyborn *et al.* 1981; Clemens & Wall 1981, 1988; Ellis 1986).

Mineral crystallisation temperatures, derived from coexisting mineral pairs (methods given in caption of Fig. 4), are shown in Figure 4. The temperature at which garnet begins to crystallise is thought to correspond approximately to the temperature at which the tonalite magma starts to crystallise. Therefore we have tentatively assigned as the earliest crystallisation temperature of the magma, the highest temperature estimated by the two-pyroxene geothermometer (e.g. Wood & Banno 1973; Wells 1977;

Kretz 1982; Lindsley 1983; Nickel & Brey 1984; Carlson & Lindsley 1988) from metamorphic enclaves of two-pyroxene mafic granulite, assuming that the metamorphic enclaves approximately preserve the highest temperature of the magma just before or after the crystallisation started. The calculated results using the rim composition for coexisting clinopyroxene ($X_{Mg} = 0{\cdot}69$–$0{\cdot}76$) and orthopyroxene ($X_{Mg} = 0{\cdot}56$–$0{\cdot}67$) range from 850–1,100°C. The range using the method by Wood and Banno (1973) and Wells (1977) is 880–950°C. We adopted 930°C as the highest temperature of the magma which formed the Niikappu tonalite body, which was confirmed from the melting-crystallisation experiment described in a later section.

The pressure at which the tonalite magma was emplaced was estimated from garnet mantle–cordierite core relationships using the Grt–Crd geobarometer (Hensen & Green 1973; Wells 1979), for garnet mantle, orthopyroxene core and plagioclase mantle using the Grt–Opx–Pl–Qtz geobarometer (Newton & Perkins 1982; Perkins & Chipera 1985), and for garnet mantle and orthopyroxene core using the Grt–Opx geobarometer (Harley & Green 1982). The estimated pressures range from 450–770 MPa, but are mostly concentrated near 600 MPa. According to Green (1976), garnet is stable in a granitic liquid at 700 MPa but not at 400 MPa. Clemens and Wall (1988) showed that garnet crystallises from a peraluminous granitic liquid at 100–500 MPa. They also reported that the garnet is stable at higher P–T and lower f_{H_2O} than cordierite. These constrain the minimum pressure of garnet crystallisation from granitic and tonalitic magmas. Garnet crystals in some orthopyroxene-bearing S-type volcanic rocks have reaction rims of cordierite-orthopyroxene (e.g. Wyborn *et al.* 1981; Clemens & Wall 1984). However, garnet of the basal tonalite of the HMB has no reaction rim and shows euhedral character (Fig. 3(A)). We have adopted 600 MPa as the pressure condition of the emplacement of the Niikappu tonalite; this is well within the range of pressures at which garnet crystallises from the tonalite magma.

The lower and middle tonalites might be derived from the same magma as that of the basal tonalite, considered from their continuity of occurrences (Niikappu body, Fig. 2) and chemical compositions (e.g. Figs 1 and 5; Table 2). Differences in the tonalites in different intrusive levels are thought to be controlled largely by the stability relations of minerals in P–T space, and in part by the crystallisation-differentiation of the minerals, especially garnet, orthopyroxene and plagioclase from the tonalite magma, as discussed in the next section.

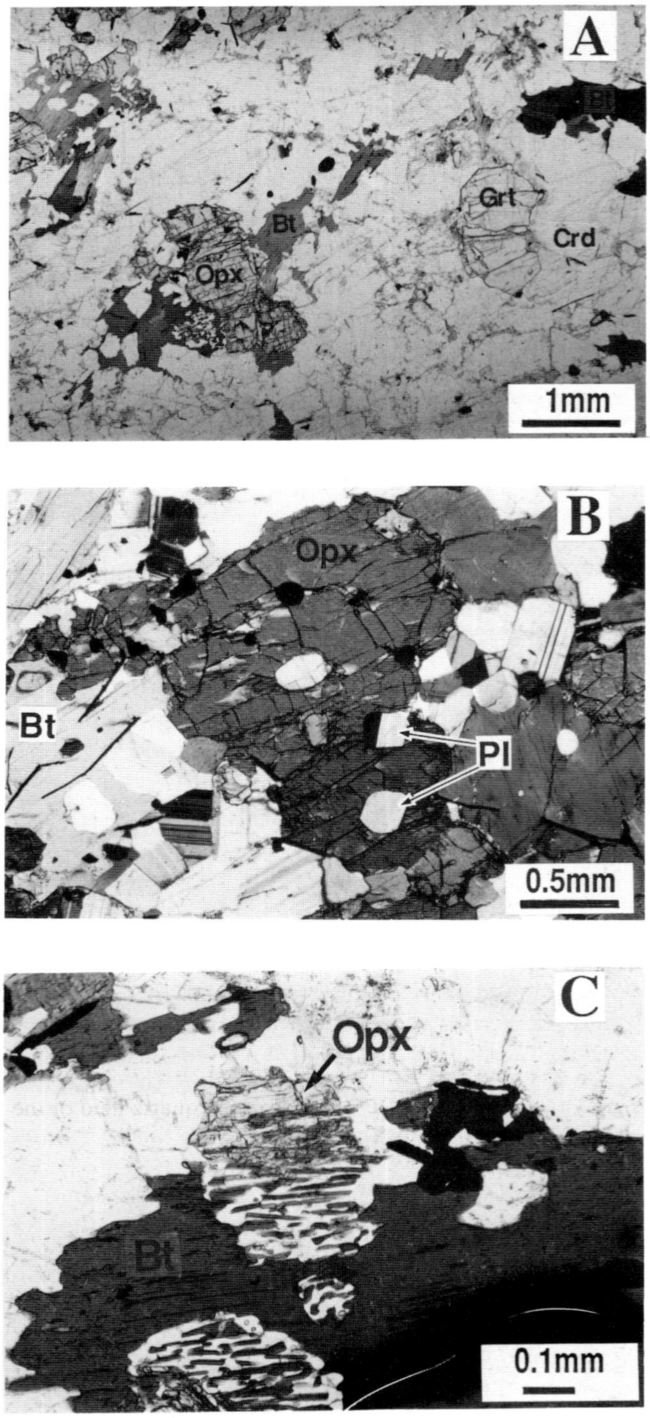

Figure 3 Photomicrographs of a typical homogeneous basal tonalite matrix, sample 71017: (A) with plane-polarised light, (B) plagioclase grains enclosed in orthopyroxene with poikilitic texture, crossed polars, and (C) biotite and quartz symplectite pseudomorphs of orthopyroxene, plane-polarised light.

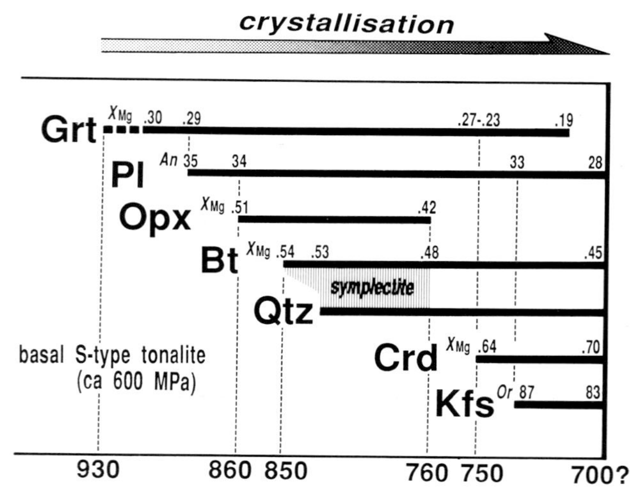

Figure 4 Crystallisation sequence of the basal S-type tonalite of the Niikappu body, determined from the textural evidence and the compositional zoning of each mineral; the temperature of 930°C was determined as explained in the text; other temperatures were roughly estimated, assuming that closely coexisting minerals formed at nearly the same stage; 860°C for garnet mantles and orthopyroxene cores by Lee & Ganguly (1988) (see Fig. 5 of Maeda *et al.* 1991), 850°C for the garnet mantles and biotite cores by Thompson (1976) and Indares and Martignole (1985), 850–760°C for orthopyroxene and symplecite biotite by Sengupta *et al.* (1990), and 750–700°C for garnet rims and cordierite cores to rims by Holdaway & Lee (1977) and Perchuk *et al.* (1985); abbreviations: $X_{Mg} = Mg/(Mg + Fe)$, $An = 100Ca/(Ca + Na + K)$, and $Or = 100K/(Ca + Na + K)$.

2. Experiments

The experimental study was carried out using natural basal tonalite, in order to clarify the chemical controls on the stability of orthopyroxene and the crystallisation sequence and the P–T–H_2O stability conditions of minerals in an S-type tonalitic magma.

2.1. Starting material and experimental methods

The basal tonalite (71017) used for the experiments has the mineral assemblage Grt + Opx + Crd + Bt + Pl + Qtz + Kfs. Its bulk chemical and mineral compositions are given in Tables 2 and 3. The amount of rock powder (<20 μm grain size) used was 0·2–0·3 g for each run, adding 0–20 wt% water, and about 2 mg anthracene ($C_{14}H_{10}$) to keep the iron species in ferrous form. In some runs, appropriate amounts of $KAlSi_3O_8$ gel were added to the rock powder to produce compositions with 15 (this is much the same as "71017"), 20 and 25% normative orthoclase. In some runs natural almandine garnet powder (<1 wt%) was added as seeds.

The starting material was sealed in Au tubes (35 mm long and 6 mm in diameter). The experiments were carried out using an internally heated pressure vessel, under Ar pressures of 300 and 600 MPa (±25 MPa) and 700–900°C (±5°C), for 24–219 hours duration, followed by quenching.

The experimental products were examined optically and by SEM and EDS, and analysed by EPMA. The stability field of a mineral was obtained by determining whether original grains of that mineral remained undissolved and/or whether new crystals of the same mineral had crystallised (Fig. 6). Generally, in a field where new crystals of a mineral form, original grains of the same mineral remained undissolved. However, this generalisation did not apply for garnet and cordierite (Tables 4 and 5); this may be due either to a low nucleation rate or to the difficulty in melting these minerals. The stability field of cordierite was determined by the conditions under which original grains of cordierite remained intact. The stability field of garnet was not clear at low temperatures ($T < 800$°C). Small amounts of cummingtonite crystallised in some low-temperature experiments; this may be metastable.

2.2. Experimental results at 600 MPa

The experimental results at 600 MPa and normative or =14·9% are given in Table 4 and in Figure 7 which shows the stability relations of the minerals in T-H_2O space.

The biotite breakdown temperature of about 850°C is very consistent with other experimental studies of peraluminous systems (e.g. Green 1976; Clemens & Wall 1981, 1988; Vielzeuf & Holloway 1988; Le Breton & Thompson 1988;

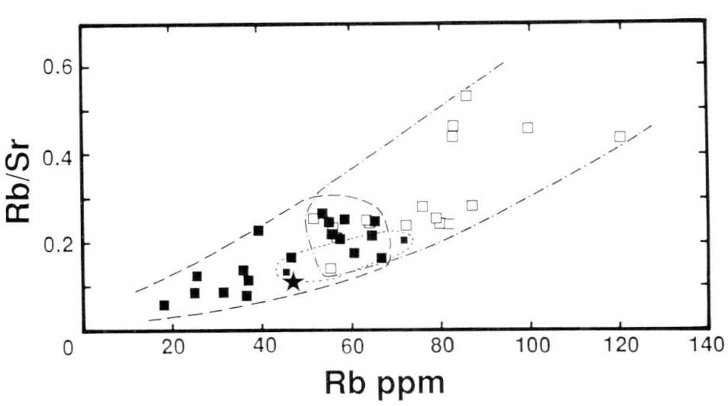

Figure 5 Bulk chemical composition of S-type granites of Niikappu Body on Rb/Sr–Rb (ppm) diagram; large solid square, small solid square, and open square indicate basal, lower, and middle S-type tonalite, respectively; star indicates basal S-type tonalite 71017.

Table 2 Bulk chemical compositions of a middle, lower, and basal S-type tonalite.

Sample Depth	71017 basal	0PP10 basal	0PP12 basal	0PP18 lower	0PP27 middle	1KB19A middle
SiO_2	65·00	69·99	67·98	65·61	65·34	70·78
TiO_2	0·72	0·59	0·73	0·65	0·86	0·44
Al_2O_3	16·25	15·19	14·23	15·37	14·52	14·91
Fe_2O_3	0·49	0·33	1·13	0·92	0·99	0·33
FeO	4·85	3·08	4·45	4·22	5·26	2·74
MnO	0·08	0·04	0·04	0·09	0·06	0·10
MgO	2·42	1·49	2·41	2·34	3·04	1·54
CaO	3·28	3·64	2·45	2·65	2·41	1·95
Na_2O	3·80	3·94	3·24	3·35	3·41	2·74
K_2O	2·53	1·43	2·81	3·23	2·70	3·35
P_2O_5	0·10	0·07	0·05	0·13	0·04	0·14
$H_2O(+)$	0·36	0·32	0·48	1·45	1·37	0·98
$H_2O(-)$	0·11	—	—	—	—	—
Total	100·00	100.00	100.00	100.00	100.00	100.00
CIPW NORM						
Q	19·40	29·37	26·62	22·41	22·34	34·19
C	1·53	0·61	1·50	1·88	1·71	3·59
or	14·89	8·33	16·43	19·27	16·07	19·86
ab	32·07	32·92	27·16	28·60	29·11	23·27
an	15·57	17·35	11·73	12·40	11·79	8·81
hy	13·42	8·09	12·01	12·01	15·20	8·09
mt	0·71	0·47	1·62	1·34	1·45	0·48
il	1·37	1·10	1·37	1·25	1·65	0·84
ap	0·23	0·16	0·12	0·30	0·09	0·32
Trace element						
Cr	89·0	55·9	95·7	84·3	119·0	53·7
Ni	33·9	26·2	32·3	34·6	40·1	19·4
Rb	46·5	36·2	64·0	71·4	99·3	71·9
Sr	432·3	454·9	299·7	354·1	218·4	305·7
V	166·4	130·9	168·4	175·2	224·9	91·4
Y	15·2	11·5	9·1	15·6	11·8	21·1
Zr	182·1	179·9	149·6	172·6	155·5	119·2
Ba	1147·2	1064·1	1311·5	1415·0	1098·9	1656·5
Pb	27·3	14·3	22·2	23·7	15·4	7·7
Th	4·2	—	10·0	4·7	—	—
Co	43·7	42·5	41·3	43·4	46·7	38·4
Nb	9·2	6·6	8·5	8·6	13·9	6·1
Cu	40·6	57·4	44·7	45·0	55·6	13·0
Zu	73·8	44·0	82·4	76·5	83·4	46·1

Sample 71017 is a starting material of the melting experiments discussed in this paper; the major element compositions are normalised to 100%.

Patiño Douce & Johnston 1991). There are, however, two differences with the results obtained by Clemens and Wall (1981, 1984, 1988); one is that plagioclase crystallises earlier than quartz, and the other is that orthopyroxene is stable down to the solidus, a point we discuss later.

The crystallisation order of plagioclase and quartz depends on the composition of the magma. The starting material composition of Clemens and Wall (1981, 1988) is slightly richer in normative Q and is in the quartz field of the system an–ab–or–Q (small stars in Fig. 1), while that of our study is closer to the an–ab side and in the plagioclase field (a large star in Fig. 1); hence, plagioclase crystallises first.

The experimental results at 600 MPa (Fig. 7) in which the T-H_2O conditions of crystallisation of the basal tonalite are controlled, can be compared with the natural crystallisation sequence shown in Figure 4. The crystallisation sequence from the experiment requires that H_2O contents of the magma were 3–4 wt% and that crystallisation began at a temperature above 900°C. This temperature appears reasonable, considering that the peraluminous magma was generated by crustal anatexis above the biotite breakdown temperature under granulite facies conditions. The inferred relatively low H_2O content of the magma may indicate that anatexis took place at H_2O-undersaturated conditions.

Table 3 Typical mineral compositions in a basal tonalite, sample 71017.

Mineral No.	Grt 765	Opx 1618	Bt 1760	Crd 1756	Pl 1728	Kfs 788
SiO_2	38·51	49·86	35·91	48·56	59·67	63·99
TiO_2	0·02	0·04	5·36	0·01	0·01	—
Al_2O_3	21·83	4·04	16·48	32·28	24·91	18·00
Cr_2O_3	—	0·02	0·05	0·01	0·12	—
FeO	30·11	29·76	17·96	6·89	—	—
MnO	2·51	1·16	0·11	0·16	—	—
MgO	6·70	15·94	10·69	8·92	—	—
CaO	1·13	0·12	—	0·02	6·98	0·09
Na_2O	—	0·01	0·11	0·06	7·46	1·44
K_2O	0·03	—	9·06	—	0·38	14·18
TOTAL	100·84	100·95	95·73	96·91	99·53	97·70
Formula						
Si	2·998	1·909	5·411	5·038	2·675	3·005
Ti	0·001	0·001	0·608	0·001	tr	—
Al	2·003	0·182	2·927	3·947	1·316	0·996
Cr	—	0·001	0·006	0·001	0.004	—
Fe	1·960	0·953	2·263	0·598	—	—
Mn	0·165	0·038	0·013	0·014	—	—
Mg	0·777	0·910	2·400	1·380	—	—
Ca	0·094	0·005	—	0·002	0·335	0·005
Na	—	0·001	0·033	0·012	0·648	0·131
K	0·003	—	1·742	—	0·021	0·850
TOTAL	8·001	3·999	15·403	10·993	5·000	4·987
O	12	6	22	18	8	8
X_{Mg}	0·284	0·488	0·515	0·698		
Ratio (%)						
Fe	65·40	51·03	48·53	30·23		
Mn	5·52					
Mg	25·94	48·72	51·47	69·77		
Ca	3·14	0·25			33·38	0·46
Na					64·49	13·31
K					2·13	86·23
Al^{IV}	0·002	0·091	2·589	0·962		
Al^{VI}	2·000	0·091	0·338	2·985		

3. Stability field of orthopyroxene

In the HMB, the middle and upper granitic rocks rarely contain orthopyroxene, while lower and basal tonalites have orthopyroxene and its pseudomorphs (consisting of biotite-quartz symplectite), as shown in Table 1. The stability of orthopyroxene is probably a function of magmatic composition and temperature. Clemens and Wall (1988) gave the following as an equilibrium relationship between biotite and orthopyroxene in S-type granitic rocks:

$$\underset{\text{orthopyroxene}}{3(Mg, Fe)SiO_3} + \underset{\text{liquid}}{KAlSi_3O_8} + \underset{\text{liquid or vapour}}{H_2O} = \underset{\text{biotite}}{K(Mg, Fe)_3AlSi_3O_{10}(OH)_2} + \underset{\text{quartz or liquid}}{3SiO_2}$$

This reaction has been studied by several authors (e.g. Hoffer & Grant 1980; Wendlandt 1981; Bohlen *et al.* 1983; Grant 1985, 1986; Peterson & Newton 1989, 1990). The reaction proceeds toward the right-hand side of the equation with decreasing temperature, and explains well the occurrence of biotite-quartz symplectite replacing orthopyroxene (Fig. 3(C)). It is inferred from the above reaction that the stability of orthopyroxene is controlled by $a_{KAlSi_3O_8}$ and X_{H_2O} in the magma. The experimental results in Figure 7 clearly show that, with increasing X_{H_2O}, biotite crystallises more abundantly than/or instead of orthopyroxene. In our case, however, as the H_2O content in the magma was relatively low, the role of $a_{KAlSi_3O_8}$ may be more important.

In order to examine the influence of normative orthoclase in the melt on the stability field of orthopyroxene, other experiments were carried out. Starting materials with variable normative orthoclase content were run under following conditions; $P = 300$ and 600 MPa, $T \leq 900°C$, and fixed H_2O: *c.* 4 wt% (Tables 4 and 5). In this experiment, the orthopyroxene stability field was determined by examination of whether or not new crystals of orthopyroxene had crystallised from the melt. Figure 8 shows the change of orthopyroxene stability field in $P–T$ space with various normative orthoclase contents. It is clear from this diagram that the orthopyroxene stability field systematically shifts to higher temperatures with increasing normative orthoclase. In the experiments with 15% normative orthoclase, orthopyroxene continuously crystallises down to the solidus (Figs 7, 8). In a magma with high normative orthoclase, orthopyroxene ceases to crystallise well above the solidus at high temperature, and biotite with or without quartz crystallises instead of orthopyroxene at low temperature.

The influence of the orthoclase component on orthopyroxene crystallisation thus explains the variation of mineral assemblages of tonalite, according to the intrusive levels, in the HMB (open squares in Fig. 1). The maximum

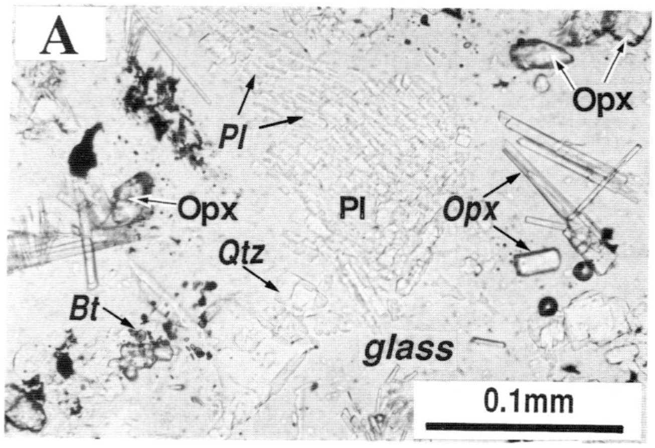

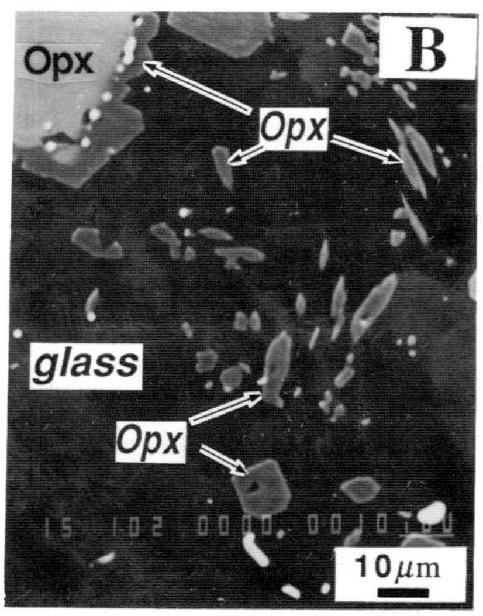

Figure 6 (A) photomicrographs of thin sections of an experimental product (run 8–3, 600 MPa and 847°C) taken with plane-polarised light; (B) BSE image photograph of a experimental product (run 11–3, 600 MPa and 900°C); italics indicate new products.

Table 4 Experimental results at 600 MPa for the basal S-type tonalite 71017 composition (normative or = 14·9).

Run No.	P (MPa)	T (°C)	H_2O (wt%)	Time (hours)	Remaining mineral	Run product
5P–1	580	700	4·0–4·8	123	Grt, Opx, Bt, Crd, Pl, Qtz, Kfs, Opaque	Opx, Bt, Pl, Qtz, Opaque, Glass
5P–2	580	700	5·5–5·9	123	Grt, Opx, Bt, Crd, Pl, Qtz, Opaque	Opx, Bt, Pl, Qtz, Cum?, Opaque, Glass, Vapor
5P–3	580	700	7·2–7·5	123	Grt, Opx, Bt, Crd, Pl, Qtz, Opaque	Opx, Bt, Pl, Qtz, Opaque, Glass, Vapor
5P–4	580	700	8·2–8·8	123	Grt, Opx, Bt, Crd, Pl, Qtz, Opaque	Opx, Bt, Pl, Qtz, Opaque, Glass, Vapor
5P–5	580	700	10·7–11·3	123	Grt, Opx, Bt, Crd, Pl, Qtz, Opaque	Opx, Bt, Pl, Qtz, Opaque, Glass, Vapor
5P–6	580	700	11·3–12·1	123	Grt, Opx, Bt, Crd, Pl, Qtz, Opaque	Opx, Bt, Pl, Qtz, Cum?, Opaque, Glass, Vapor
5P–7	580	700	14·9–15·3	123	Opx, Bt, Opaque	Opx, Bt, Pl, Cum?, Opaque, Glass, Vapor
3P–1	610	751	1·5–2·2	24	Grt, Opx, Bt, Crd, Pl, Qtz, Kfs, Opaque	Opaque, Glass
1–1	610	750	2·5–2·9	24	Grt, Opx, Bt, Crd, Pl, Qtz, Kfs, Opaque	Opaque, Glass
1–2	610	750	3·1–3·4	24	Grt, Opx, Bt, Crd, Pl, Qtz, Opaque	Opx, Bt, Pl, Qtz, Cum?, Opaque, Glass
3P–2	610	751	4·0–4·5	24	Grt, Opx, Bt, Pl, Qtz, Opaque	Opx, Bt, Pl, Qtz, Opaque, Glass
12–1	610	750	4·2–4·6	63	Grt, Opx, Bt, Pl, Qtz, Opaque	Opx, Bt, Pl, Qtz, Opaque, Glass
1–3	610	750	5·1–5·4	24	Grt, Opx, Bt, Pl, Qtz, Opaque	Opx, Bt, Pl, Qtz, Cum?, Opaque, Glass
3P–3	610	751	5·0–5·7	24	Grt, Opx, Bt, Pl, Qtz, Opaque	Opx, Bt, Pl, Qtz, Opaque, Glass
1–4	610	750	6·0–6·4	24	Grt, Opx, Bt, Pl, Opaque	Opx, Bt, Pl, Opaque, Glass
3P–4	610	751	6·6–7·0	24	Grt, Opx, Bt, Pl, Opaque	Opx, Bt, Pl, Opaque, Glass
1–5	610	750	7·0–7·4	24	Grt, Opx, Bt, Pl, Opaque	Opx, Bt, Pl, Opaque, Glass
3P–5	610	751	7·0–7·5	24	Opx, Bt, Pl, Opaque	Opx, Bt, Pl, Opaque, Glass
1–6	610	750	7·2–7·6	24	Grt, Opx, Bt, Pl, Opaque	Opx, Bt, Opaque, Glass
1–7	610	750	8·0–8·4	24	Grt, Opx, Bt, Opaque	Opx, Bt, Opaque, Glass, Vapor
1–8	610	750	8·2–8·6	24	Grt, Opx, Bt, Opaque	Opx, Bt, Opaque, Glass, Vapor
3P–6	610	751	9·4–9·9	24	Opx, Bt, Pl, Opaque	Opx, Bt, Opaque, Glass, Vapor
1–9	610	750	10·0–10·4	24	Opx, Bt, Opaque	Opx, Bt, Pl, Opaque, Glass Vapor
1–10	610	750	20·0–20·5	24	Opx, Bt, Opaque	Opx, Bt, Pl, Opaque, Glass, Vapor
2–1	610	800	0·36	29	Grt, Opx, Bt, Crd, Pl, Qtz, Kfs, Opaque	Opaque, Glass
2–2	610	800	1·0–1·4	29	Grt, Opx, Bt, Crd, Pl, Qtz, Opaque	Opx, Bt, Pl, Opaque, Glass
2–3	610	800	2·0–2·4	29	Grt, Opx, Bt, Crd?, Pl, Qtz, Opaque	Opx, Bt, Pl, Opaque, Glass
7–1	590	795	3·0–3·4	28	Grt, Opx, Bt, Pl, Qtz, Opaque	Opx, Bt, Pl, Qtz, Opaque, Glass
4P–1	600	801	3·3–3·8	39	Grt, Opx, Bt, Pl, Qtz, Opaque	Opx, Bt, Pl, Qtz, Opaque, Glass
2–4	610	800	4·1–4·4	29	Grt, Opx, Bt, Pl, Opaque	Opx, Bt, Pl, Opaque, Glass
2–5	610	800	5·1–5·4	29	Grt, Opx, Opaque	Opx, Bt, Pl, Opaque, Glass
7–2	590	795	5·2–5·5	28	Grt, Opx, Bt, Pl, Opaque	Grt, Opx, Bt, Pl, Opaque, Glass
4P–2	600	801	5·1–5·6	39	Grt, Opx, Bt, Pl, Opaque	Opx, Bt, Opaque, Glass
4P–3	600	801	5·7–6·2	39	Opx, Bt, Opaque	Opx, Bt, Opaque, Glass
2–6	610	800	6·0–6·3	29	Grt, Opx, Opaque	Opx, Bt, Opaque, Glass
2–7	610	800	6·2–6·6	29	Opx, Opaque	Opx, Bt, Opaque, Glass
2–8	610	800	7·0–7·4	29	Opx, Opaque	Opx, Bt, Opaque, Glass
2–9	610	800	9·1–9·5	29	Opx, Opaque	Opx, Bt, Opaque, Glass, Vapor
4P–4	600	801	10·0–10·3	39	Opx, Bt, Opaque	Opx, Bt, Opaque, Glass, Vapor
2–10	610	800	11·7–12·3	29	Opaque	Opx, Bt, Opaque, Glass, Vapor
10–1	610	853	0·7–0·9	80	Grt, Opx, Bt, Pl, Qtz, Opaque	Opx, Bt, Crd, Pl, Qtz, Opaque, Glass
8–1	600	847	1·7–2·2	29	Grt, Opx, Bt, Pl, Qtz, Opaque	Opx, Bt, Pl, Qtz, Opaque, Glass
8–2	600	847	2·5–2·8	29	Grt, Opx, Bt, Pl, Qtz, Opaque	Opx, Bt, Pl, Qtz, Opaque, Glass
8–3	600	847	2·9–3·1	29	Grt, Opx, Bt, Pl, Qtz, Opaque	Opx, Bt, Pl, Qtz, Opaque, Glass
8–4	600	847	3·6–3·9	29	Grt, Opx, Bt, Pl, Opaque	Grt, Opx, Bt, Pl, Opaque, Glass
10–2	610	853	4·4–4·7	80	Grt, Opx, Pl, Opaque	Grt, Opx, Pl, Opaque, Glass
8–5	600	847	4·7–5·1	29	Grt, Opx, Bt, Pl, Opaque	Grt, Opx, Bt, Pl, Opaque, Glass
10–3	610	853	5·0–5·3	80	Grt, Opx, Pl, Opaque	Grt, Opx, Opaque, Glass
10–4	610	853	7·2–7·4	80	Grt, Opx, Opaque	Grt, Opx, Bt, Opaque, Glass
10–5	610	853	8·8–9·3	80	Opaque	Opx, Bt, Pl, Opaque, Glass
11–4	600	900	1·7–1·9	71	Grt, Opx, Bt, Pl, Qtz, Opaque	Grt, Opx, Pl, Qtz, Opaque, Glass
11–1	600	900	3·1–3·4	71	Grt, Opx, Pl, Opaque	Grt, Opx, Pl, Opaque, Glass
11–2	600	900	4·4–4·6	71	Grt, Opx, Pl, Opaque	Grt, Opx, Opaque, Glass
11–3	600	900	4·6–4·8	71	Opx, Pl, Opaque	Opx, Opaque, Glass

Table 5 Experimental results at 300 MPa and with normative orthoclase = 15, 20 and 25%.

Run No.	P (MPA)	T (°C)	H_2O (wt%)	or (%)	Time (hours)	Remaining mineral	Run product
9–1	300	750	4·4–4·7	15	219	Grt, Opx, Bt, Crd, Pl, Qtz, Opaque	Opx, Bt, Pl, Qtz, Opaque, Glass
9–2	300	750	5·0–5·5	15	219	Grt, Opx, Bt, Crd, Pl, Qtz, Opaque	Opx, Bt, Pl, Qtz, Opaque, Glass
9–3	300	750	4·1–4·8	20	219	Grt, Opx, Bt, Crd, Pl, Qtz, Opaque	Opx, Bt, Pl, Qtz, Opaque, Glass
9–4	300	750	4·8–5·8	20	219	Grt, Opx, Bt, Crd, Pl, Qtz, Opaque	Opx, Bt, Pl, Qtz, Opaque, Glass
9–5	300	750	3·9–4·5	25	219	Grt, Opx, Bt, Crd, Pl, Qtz, Opaque	Bt, Pl, Qtz, Opaque, Glass
6–1	300	800	3·5–4·4	15	43	Grt, Opx, Bt, Crd, Pl, Qtz, Opaque	Opx, Bt, Pl, Qtz, Opaque, Glass
6–2	300	800	5·5–5·8	15	43	Grt, Opx, Bt, Crd, Pl, Qtz, Opaque	Opx, Bt, Pl, Opaque, Glass, Vapor
6–3	300	800	5·3–5·8	20	43	Opx, Bt, Pl, Qtz, Opaque	Opx, Bt, Pl, Qtz, Opaque, Glass
6–4	300	800	4·9–5·2	25	43	Opx, Bt, Pl, Qtz, Opaque	Opx, Bt, Pl, Opaque, Glass
12–2	600	750	3·8–4·9	20	63	Opx, Bt, Pl, Qtz, Opaque	Bt, Pl, Qtz, Opaque, Glass
12–3	600	750	3·8–4·2	25	63	Opx, Bt, Pl, Qtz, Opaque	Bt, Pl, Qtz, Opaque, Glass
7–3	590	795	3·1–3·5	20	28	Grt, Opx, Bt, Pl, Qtz, Opaque	Opx, Bt, Pl, Qtz, Opaque, Glass
7–4	590	795	4·6–4·9	20	28	Grt, Opx, Bt, Pl, Qtz, Opaque	Opx, Bt, Pl, Opaque, Glass
7–5	590	795	5·3–5·7	25	28	Grt, Opx, Bt, Pl, Qtz, Opaque	Bt, Pl, Opaque, Glass

normative orthoclase content of the basal tonalite is 16·4%, while that of the middle tonalite is 24·2%. Hence, the middle tonalite with high normative orthoclase lacks orthopyroxene but contains biotite and cordierite (and rare garnet) as mafic minerals.

4. Conclusions

The evolution and crystallisation of S-type tonalitic magmas in the HMB is summarised in Figure 9. The basal tonalite was intruded into granulite facies rocks and crystallised garnet, plagioclase, orthopyroxene, biotite, quartz, cordierite and rarely K-feldspar, in that order. Orthopyroxene was not completely replaced by biotite because of low $a_{KAlSi_3O_8}$ in the magma. The lower tonalite was derived from slightly differentiated magma and intruded into the upper amphibolite facies rocks, where it crystallised plagioclase, orthopyroxene, biotite, quartz, cordierite and K-feldspar. The middle tonalite was derived from an even more fractionated magma, and intruded into the lower amphibolite to greenschist facies rocks. Owing to an increase in normative orthoclase content in the magma, and a falling magmatic temperature, crystallisation might have started below the orthopyroxene crystallisation limit, and resulted in the crystallisation of biotite and cordierite as mafic minerals.

Voluminous S-type magmas are thought to be generated in the deep crust, above the biotite breakdown temperature (Vielzeuf & Holloway 1988) and under water-undersaturated conditions. Our study has shown that the stability of orthopyroxene in S-type magmas is controlled by $a_{KAlSi_3O_8}$ and/or a_{H_2O}, and that Opx crystallisation is largely controlled by $a_{KAlSi_3O_8}$ and X_{H_2O} in S-type magma of deep crustal origin.

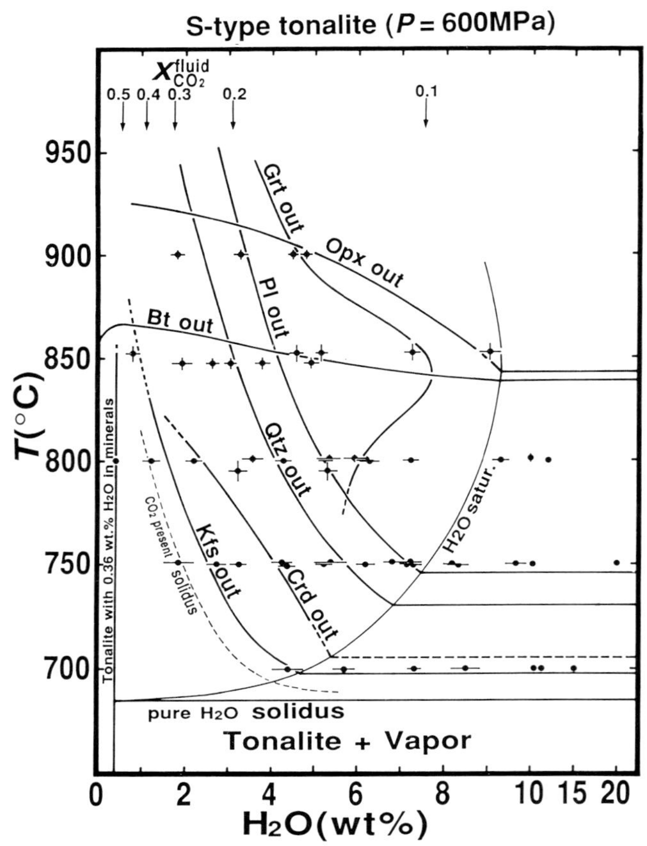

Figure 7 Experimental results plotted on H_2O (wt%) – T(°C) plane of the basal S-type tonalite composition (71017, see Tables 1 and 2) at 600 MPa; the uncertainty in H_2O is shown by error bars and is caused by increasing rate of the reaction of anthracene, $2C_{14}H_{10} + 33O_2 = 28CO_2 + 10H_2O$. X_{CO_2} in fluid was calculated by this reaction.

5. Acknowledgements

We thank K. Hirota and N. Furukawa for their help during the experiments. We wish to thank Y. Takahashi of the

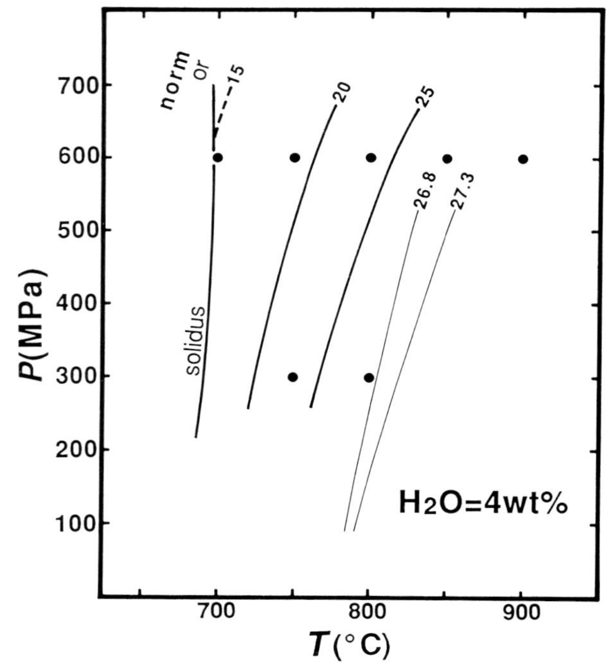

Figure 8 P–T diagram showing crystallisation field of orthopyroxene; thick lines and a thick broken line were obtained by this study, and thin lines taken from Clemens & Wall (1981, 1988); orthopyroxene crystallises on the right-hand side of each line; note that the crystallisation field of orthopyroxene is restricted to progressively higher temperatures with increasing orthoclase component in the system.

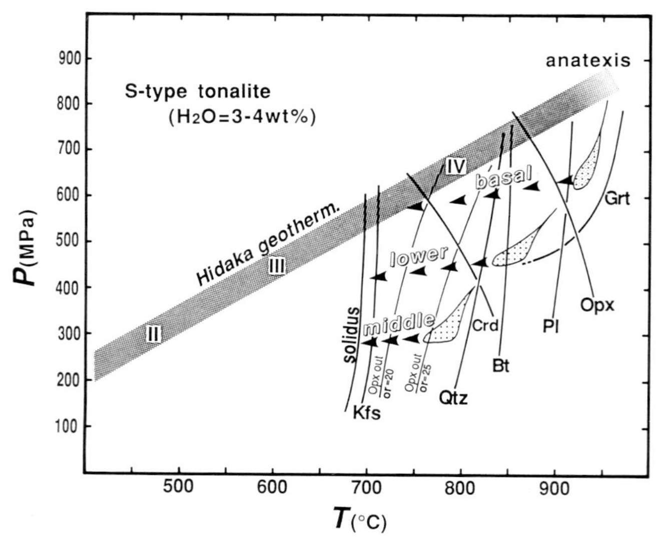

Figure 9 P–T diagram for the S-type tonalite composition with 3–4 wt% H_2O; parts of the garnet stability field and the dP/dT slope of each minerals are referred to Figure 1 of Green (1976) and Figure 4 of Clemens & Wall (1981); II, III, IV, and a shadowed wide line mark the metamorphic grades and a geothermal gradient of the HMB at the peak of metamorphism (after Osanai *et al.* 1991); diapirs show intrusion of buoyant magma, and arrows indicate the crystallisation at each intrusive depth.

Geological Survey of Japan, who kindly allowed us to use his unpublished data on the middle S-type tonalite. We are grateful to T. Watanabe, K. Arita, J. Maeda and Y. Osanai for discussions. This work was partly supported by a Grant-in-Aid for Scientific Research (No. 01540662, M. Komatsu; No. 02303008, H. Onuki) from the Ministry of Education, Science and Culture, Japan.

6. References

Arita, K., Toyoshima, T., Owada, M., Miyashita, S. & Jolivet, L. 1986. Tectonic movements of the Hidaka metamorphic belt, Hokkaido, Japan. MONOGR ASSOC GEOL COLLAB JAPAN **31,** 247–63.

Bamba, M. 1985. *Metamorphism of the Main Zone of the Hidaka metamorphic belt.* Unpublished Masters Thesis, Niigata University.

Barley, M. E. 1987. Origin and evolution of mid-Cretaceous, garnet-bearing, intermediate and silicic volcanics from Cantebury, New Zealand. J VOLCANOL GEOTHERM RES **32,** 247–67.

Birch, W. D. & Gleadow, J. W. 1974. The genesis of garnet and cordierite in acid volcanic rocks: evidence from the Cerberean cauldron, central Victoria, Australia. CONTRIB MINERAL PETROL **45,** 1–13.

Birch, W. D., Clemens, J. D. & Phillips, G. N. 1977. *Devonian acid igneous complexes in Central Victoria,* Guide to Excursion "C", Second Australian Geological Convention. Melbourne: National Museum of Victoria.

Bohlen, S. R., Boettcher, A.L., Wall, V. J. & Clemens, J. D. 1983. Stability of phlogopite–quartz and sanidine–quartz: a model for melting in the lower crust. CONTRIB MINERAL PETROL **83,** 270–7.

Carlson, W. D. & Lindsley, D. H. 1988. Thermochemistry of pyroxenes on the join $Mg_2Si_2O_6$–$CaMgSi_2O_6$. AM MINERAL **73,** 242–52.

Chappell, B. W. & White, A. J. R. 1974. Two contrasting granite types. PAC GEOL **8,** 173–4.

Chappell, B. W. & White, A. J. R. 1992. I- and S-type granites in the Lachlan Fold Belt. TRANS R SOC EDINBURGH EARTH SCI **83,** 1–26.

Chappell, B. W., White, A. J. R. & Williams, I.S. 1991. *A transverse section through granites of the Lachlan Fold Belt,* Second Hutton Symposium Field Guide. REC BMR GEOL GEOPHYS **1991**/22.

Clarke, D. B. 1981. The mineralogy of peraluminous granites: a review. CAN MINERAL **19,** 3–17.

Clemens, J. D. & Wall, V. J. 1981. Origin and crystallization of some peraluminous (S-type) grantitic magmas. CAN MINERAL **19,** 111–31.

Clemens, J. D. & Wall, V. J. 1984. Origin and evolution of a peraluminous silicic ignimbrite suite: the Violet Town Volcanics. CONTRIB MINERAL PETROL **88,** 354–71.

Clemens, J. D. & Wall, V. J. 1988. Controls on the mineralogy of S-type volcanic and plutonic rocks. LITHOS **21,** 56–66.

Ellis, D. J. 1986. Garnet-liquid Fe^{2+}-Mg equilibria and implications for the beginning of melting in the crust and subduction zones. AM J SCI **286,** 765–91.

Grant, J. A. 1973. Phase equilibria in high grade metamorphism and partial melting of pelitic rocks. AM J SCI **273,** 289–317.

Grant, J. A. 1985. Phase equilibria in partial melting of pelitic rocks. *In* Ashworth, J. R. (ed.) *Migmatites,* 86–144. Glasgow: Blackie.

Grant, J. A. 1986. Quartz–phlogopite–liquid equilibria and origins of charnockites. AM MINERAL **71,** 1071–5.

Green, T. H. 1976. Experimental generation of cordierite- or garnet-bearing grantitic liquids from a pelitic composition. GEOLOGY **4,** 85–8.

Harley, S. L. & Green, D. H. 1982. Garnet–Orthopyroxene barometry for granulites and peridotites. NATURE **300,** 697–701.

Hensen, B. J. & Green, D. H. 1973. Experimental study of the stability of cordierite and garnet in pelitic compositions at high pressures and temperatures. III. Synthesis of experimental data and geological applications. CONTRIB MINERAL PETROL **38,** 151–66.

Hine, R., Williams, I. S., Chappell, B. W. & White, A. J. R. 1978. Contrasts between I- and S-type granitoids of the Kosciusko Batholith. J GEOL SOC AUST **25,** 219–34.

Hoffer, E. & Grant, J. A. 1980. Experimental investigation of the formation of cordierite-orthopyroxene parageneses in pelitic rocks. CONTRIB MINERAL PETROL **73,** 15–22.

Holdaway, M. J. & Lee, S. M. 1977. Fe–Mg cordierite stability in high-grade pelitic rocks based on experimental, theoretical, and natural observations. CONTRIB MINERAL PETROL **63,** 175–98.

Ikeda, Y. 1984a. Petrological significance of granitic inclusions from Pliocene-early Pleistocene pyroclastic flow deposits in central Hokkaido, Japan, J JAPAN ASSOC MIN PETROL ECON GEOL **79,** 60–80.

Ikeda, Y. 1984b. Trace element abundances in some granites from the Hidaka Belt, Hokkaido. MAGMA **70,** 9–14.

Ikeda, Y. 1991. Geochemistry and magmatic evolution of Pliocene-early Pleistocene pyroclastic flow deposits in central Hokkaido, Japan. J GEOL SOC JAPAN **97,** 645–66.

Indares, A. & Martignole, J. 1985. Biotite–garnet geothermometry in the granulite facies: the influence of Ti and Al in biotite. AM MINERAL **70,** 272–8.

Kawasaki, M. 1980a. Omine acid rocks, Kii Peninsula—geology and major element chemistry. J JAPAN ASSOC MIN PETROL ECON GEOL **75,** 86–102.

Kawasaki, M. 1980b. Omine acid rocks, Kii Peninsula—mineralogy. J JAPAN ASSOC MIN PETROL ECON GEOL **75,** 146–59.

Komatsu, M. 1983. Hypersthene-garnet migmatite in the Pankenushi area, Hidaka Metamorphic Belt. *In* Komatsu, M. (ed.) *Petrological studies of deep crust and upper mantle rocks in island-arc,* Report of research project, grant in aid for scientific research, 1982, 26–9. Niigata: Niigata University.

Komatsu, M., Miyashita, S., Maeda, J., Osanai, Y. & Toyoshima, T. 1983. Disclosing of a deepest section of continental-type crust up-thrust as the final event of collision of arcs in Hokkaido, north Japan. *In* Hashimoto, M. & Uyeda, S. (eds) *Accretion tectonics in the Circum-Pacific regions,* 149–65. Tokyo: Terrapub.

Komatsu, J., Miyashita, S. & Arita, K. 1986. Composition and structure of the Hidaka metamophic belt, Hokkaido—historical review and present status. MONOGR ASSOC GEOL COLLAB JAPAN **31,** 189–203.

Komatsu, M., Osanai, Y., Toyoshima, T. & Miyashita, S. 1989. Evolution of the Hidaka metamorphic belt, northern Japan. GEOL SOC SPEC PUBL **43,** 487–93.

Kretz, R. 1982. Transfer and exchange equilibria in the portion of the pyroxene quadrilateral as deduced from natural and experimental data. GEOCHIM COSMOCHIM ACTA **46,** 411–21.

Kretz, R. 1983. Symbols for rock-forming minerals. AM MINERAL **68,** 277–9.

Le Breton, N. & Thompson, A. B. 1988. Fluid-absent (dehydration) melting of biotite in metapelites in the early stages of crustal anatexis. CONTRIB MINERAL PETROL **99,** 226–37.

Lee, H. Y. & Ganguly, J. 1988. Equilibrium compositions of coexisting garnet and orthopyroxene: experimental determinations in the system FeO–MgO–Al_2O_3–SiO_2, and applications. J PETROL **29,** 93–113.

Lindsley, D. H. 1983. Pyroxene thermometry. AM MINERAL **68,** 477–93.

Lonker, S. W. 1980. Conditions of metamorphism in high-grade pelitic gneisses from the Frontenac Axis, Ontario, Canada. CAN J EARTH SCI **17,** 1666–84.

Maeda, J., Shimura, S., Arita, K., Osanai, Y., Murata, M., Bamba, M. & Suetake, S. 1991. Chemical features of orthopyroxene in peraluminous igneous rocks. AM MINERAL **76,** 1676–84.

Miyashita, S. 1983. Reconstruction of the ophiolite succession in the western zone of the Hidaka Metamorphic Belt, Hokkaido. J GEOL SOC JAPAN **89,** 69–86.

Murata, M. 1982. S-type and I-type granitic rocks of the Ohmine district, central Kii peninsula. J JAPAN ASSOC MIN PETROL ECON GEOL **77,** 267–77.

Nakada, S. 1983. Zoned magma chamber of the Osuzuyama acid rocks, southwest Japan. J PETROL **24,** 471–94.

Nekvasil, H. 1988. Calculated effect of anorthite component on the crystallization paths of H_2O-undersaturated haplogranitic melts. AM MINERAL **73,** 966–81.

Newton, R. C. & Perkins, D. 1982. Thermodynamic calibration of geobarometers based on the assemblages garnet–plagioclase–orthopyroxene (clinopyroxene)–quartz. AM MINERAL **67,** 203–22.

Nickel, K. G. & Brey, G. 1984. Subsolidux orthopyroxene—clinopyroxene systematics in the system CaO–MgO–SiO_2 to

60 kb: a re-evaluation of the regular solution model. CONTRIB MINERAL PETROL **87,** 35–42.

Okamoto, Y. 1983. *Geochemical study of the granitic rocks in the Outer Zone, Southwest Japan and the Hidaka Belt, Hokkaido.* Unpublished Masters Thesis, Okayama University.

Okamoto, Y. & Honma, H. 1983. Oxygen and strontium isotope study of granitic and metamorphic rocks from the Hidaka belt, Hokkaido. MAGMA **67,** 151–5.

Osanai, Y. 1985. Geology and metamorphic zoning of the Main Zone of the Hidaka Metamorphic Belt in the Shizunai River region, Hokkaido J GEOL SOC JAPAN **91,** 259–78.

Osanai, Y., Arita, K. & Bamba, M. 1986. P–T conditions of granulite-facies rocks from the Hidaka metamorphic belt, Hokkaido, Japan. J GEOL SOC JAPAN **92,** 793–808.

Osanai, Y., Komatsu, M. & Owada, M. 1991. Metamorphism and granite genesis in the Hidaka Metamorphic Belt, Hokkaido, Japan. J METAMORPHIC GEOL **9,** 111–24.

Owada, M. 1989. Geology and chemical composition of granitic rocks in the southern part of the Hidaka metamorphic belt, with special reference to cordierite-bearing granitic rocks. J GEOL SOC JAPAN **95,** 227–40.

Owada, M. & Osanai, Y. 1988. Rb–Sr whole rock ages of homogeneous and heterogeneous granitic rocks from the Hidaka Metamorphic Belt, Hokkaido. *In* Komatsu, M. (ed.) *Tectonic belts in Hokkaido* 3, 38–44. Matsuyama: Ehime University.

Owada, M. & Osanai, Y. 1989. Genesis of granite in the Hidaka Metamorphic Belt. EARTH MON **11,** 252–7.

Patiño Douce, A. E. & Johnston, A. D. 1991. Phase equilibria and melt productivity in the pelitic system: implications for the origin of peraluminous granitoids and aluminous granulites. CONTRIB MINERAL PETROL **107,** 202–18.

Perchuk, L. L., Aranovich, L. Ya, Podlesskii, K. K. Lavrant'eva, I. V., Gerasimov, V. Yu, Fed'kin, V. V., Kitsul, V. I., Karasakov, L. P. & Berdnikov, N. V. 1985. Precambrian granulties of the Aldan shield, eastern Siberia, USSR. J METAMORPH GEOL **3,** 265–310.

Perkins, D. & Chipera, S. J. 1985. Garnet–orthopyroxene–plagioclase–quartz barometry: refinement and application to the English River subprovince and the Minnesota River valley. CONTRIB MINERAL PETROL **89,** 69–80.

Peterson, J. W. & Newton, R. 1989. Reversed experiments on biotite–quartz–feldspar melting in the system KMASH: implications for crustal anatexis. J GEOL **97,** 465–85.

Peterson, J. W. & Newton, R. 1990. Experimental biotite–quartz melting in the KMASH-CO_2 system and the role of CO_2 in the petrogenesis of granites and related rocks. AM MINERAL **75,** 1029–42.

Sengupta, P., Dasgupta, P., Bhattacharya, K. & Mukherjee, M. 1990. An orthopyroxene-biotite geothermometer and its application in crustal granulites and mantle-derived rocks. J METAMORPH GEOL **8,** 191–7.

Shibata, K. & Ishihara, S. 1979a. Rb–Sr whole-rock and K–Ar mineral ages of granitic rocks in Japan. GEOCHEM **13,** 113–9.

Shibata, K. & Ishihara, S. 1979b. Initial $^{87}Sr/^{86}Sr$ ratios of plutonic rocks from Japan. CONTRIB MINERAL PETROL **70,** 381–90.

Shimura, T. 1991. *Genesis and crystallization of the lower crustal S-type granitic magma.* Unpublished Doctoral Thesis, Hokkaido University.

Shimura, T. 1992. Intrusion of granitic magma and uplift tectonics of the Hidaka Metamorphic Belt, Hokkaido. J GEOL SOC JAPAN **98,** 1–20.

Takahashi, Y. 1986. *Petrological study of granitic rocks in the upper reaches of Satsunai river, Main Zone of the Hidaka Metamorphic Belt.* Unpublished Masters Thesis, Niigata University.

Thompson, A. B. 1976. Mineral reactions in pelitic rocks: II. Calculation of some P–T–X(Fe–Mg) phase reactions AM J SCI **276,** 425–54.

Toyoshima, T. 1990. Pseudotachylite from the Main Zone of the Hidaka metamorphic belt, Hokkaido, northern Japan. J METAMORPH GEOL **8,** 507–23.

Vielzeuf, D. & Holloway, R. 1988. Experimental determination of the fluid-absent melting relations in the pelitic system. Consequences for crustal differentiation. CONTRIB MINERAL PETROL **98,** 257–76.

Wells, P. R. A. 1977. Pyroxene thermometry in simple and complex systems. CONTRIB MINERAL PETROL **62,** 129–39.

Wells, P. R. A. 1979. Chemical and thermal evolution of archean sialic crust, southern west Greenland. J PETROL **20,** 187–226.

Wendlandt, R. F. 1981. Influence of CO_2 on melting of model granulite facies assemblages: a model for the genesis of charnockite. AM MINERAL **66,** 1164–74.

White, A. J. R. & Chappell, B. W. 1977. Ultrametamorphism and granitoid genesis. TECTONOPHYSICS **43,** 7–22.

White, A. J. R. & Chappell, B. W. 1988. Some supracrustal (S-type) granites of the Lachlan Fold Belt. TRANS R SOC EDINBURGH EARTH SCI **79,** 169–81.

Winkler, H. G. F. 1979. *Petrogenesis of metamorphic rocks,* 5th edn. New York: Springer.

Wood, B. J. & Banno, S. 1973. Garnet–orthopyroxene and orthopyroxene–clinopyroxene relationships in simple and complex systems. CONTRIB MINERAL PETROL **42,** 109–42.

Wood, C. P. 1974. Petrogenesis of garnet-bearing rhyolites from Canterbury, New Zealand. NEW ZEALAND J GEOL GEOPH **17,** 759–88.

Wyborn, D., Chappell, B. W. & Johnston, R. M. 1981. Three S-type volcanic suites from the Lachlan Fold Belt, southeast Australia. J GEOPHYS RES **86,** 10335–48.

Wyborn, D., White, A. J. R. & Chappell, B. W. 1991. *Enclaves in the S-type Cowra Granodiorite,* Second Hutton Symposium Field Guide. REC BMR GEOL GEOPHYS **1991**/24.

T. SHIMURA, Department of Geology and Mineralogy, Faculty of Science, Hokkaido University, Sapporo, 060, Japan.
M. KOMATSU, Department of Earth Science, Faculty of Science, Ehime University, Matsuyama, 790, Japan.
J. T. IIYAMA, Department of Earth Science, Faculty of Science, Chiba University, Chiba, 260, Japan.

Transactions of the Royal Society of Edinburgh: Earth Sciences, **83**, 269–280, 1992

Geochemistry and Rb–Sr geochronology of Mesozoic granites from Hong Kong

R. J. Sewell, D. P. F. Darbyshire, R. L. Langford and P. J. Strange

ABSTRACT: The granites of Hong Kong comprise a variety of assemblages dominated by chemically evolved compositions. They are divided into two suites based on petrographic, geochemical, and age criteria. The oldest and most primitive intrusive units are deformed biotite-hornblende granodiorites and monzogranites of the Lamma Suite. These rocks are characterised by high CaO (1·4–2·7%), and low Nb and Y contents. The Lion Rock Suite (LRS) is dominated by relatively undeformed monzogranite with subordinate quartz syenite and comprises three subgroups. Granites of subgroup I are separated into coarse- and fine- to medium-grained lithologies. The fine- to medium-grained granites are predominantly fluorite-bearing with silica contents ranging from 75·5–78%. They are characterised by high total REE, Ga, F, Rb, Nb, and Y contents and yield a Rb–Sr whole-rock isochron age of 155 ± 6 Ma with an initial $^{87}Sr/^{86}Sr$ ratio of 0·7101 ± 0·0060 (MSWD = 4·6). Granites of subgroup II comprise a diverse range of compositions (SiO_2 = 63–77%) and are characterised by highly variable trace element abundances. Coarse-grained granites yield an age of 148 ± 9 Ma with an initial $^{87}Sr/^{86}Sr$ ratio of 0·7060 ± 0·0006 (MSWD = 0·1). Granites of subgroup III are moderately to highly evolved (SiO_2 = 72·5–77·9%) and the silica-rich compositions are marked by enrichment in Y, Nb, Rb and depletion in Ba and Sr. Rb–Sr whole-rock isochron ages for individual plutons vary from 138 ± 1 to 136 ± 1 and corresponding initial $^{87}Sr/^{86}Sr$ ratios are 0·7080 ± 0·0002 (MSWD = 1·2) and 0·7092 ± 0·0006 (MSWD = 0·4). Granites of the Lamma Suite and coarse-grained granites of LRS subgroup I are interpreted as synorogenic I-types, whereas those of LRS subgroups II and III are interpreted as late-orogenic to postorogenic, fractionated I-types. Fine- to medium-grained granites of LRS subgroup I have distinctive A-type affinities and together with their association with quartz syenite indicate a transition from compressional to tensional tectonics.

KEY WORDS: Age dating, A-type, fluorite, fractionated, granodiorite, isotopes, I-type, petrology, S China, syenite, tectonic setting

Despite their relatively accessible location and excellent exposure, the Mesozoic granites of Hong Kong have not previously been studied in detail. The granites lie at the southern end of a broad (400 km) NE-trending belt of Mesozoic intrusives passing through the maritime provinces of southeastern China (Fujian and Guangdong, Fig. 1). Within this belt, granites are associated with voluminous intermediate to silicic volcanic deposits and are generally considered to have been emplaced in two main phases during the Yenshanian Orogeny (Jahn *et al.* 1976; Yang *et al.* 1986): an early phase ranging from 190–140 Ma comprising biotite monzogranite, granodiorite and diroite, and a late phase from 135–80 Ma including granodiorite, monzogranite and alkaline lithologies. Reconnaissance isotopic age dating of Hong Kong granites (Chandy & Snelling *in* Allen & Stephens 1971) has indicated K–Ar cooling ages in the range 143–117 Ma with a tendency to cluster about 135 Ma. However, sampling of intrusive units was based on a poorly understood sequence of emplacement and many lithologies were not analysed. This paper presents new Rb–Sr whole-rock age data and geochemistry for a wide range of Hong Kong granites and discusses aspects of their petrogenesis.

1. Regional geological setting

The Territory of Hong Kong is situated within the Lianhuashan Fault Zone, a broad (200 km) NE-trending fault-bounded depression located between the S China Fold System in the NW and the Maritime Fold System in the SE (Huang 1978, Fig. 1). Prominent NE-trending faults are generally thought to represent older zones of crustal weakness which have been periodically reactivated since the Proterozoic (Huan *et al.* 1982). Granite plutons (Fig. 2) cover approximately 35% of the exposed land area (1149 km^2) and intrude basement rocks consisting of deformed Devonian, Carboniferous, Permian, and Lower Jurassic metasedimentary rocks. The granites are associated with Upper Jurassic to Lower Cretaceous volcanic rocks comprising intermediate to silicic lavas, welded tuffs, and volcaniclastic sediments of the Repulse Bay Volcanic Group (Addison 1986). Calderas marked by ring faults are developed in eastern New Territories (Sai Kung and Long Harbour calderas) and on Lantau Island (Lantau caldera).

2. Analytical methods

Whole-rock geochemical analyses were determined by X-ray fluorescence spectrometry (University of Nottingham) using fusion beads for major elements and pressed-powder pellets for trace elements. REE analyses were obtained by inductively coupled plasma source emission spectroscopy at Royal Holloway and Bedford New College following the method of Walsh *et al.* (1981).

Concentrations of Rb and Sr as well as Rb/Sr atomic ratio for geochronology samples were determined by X-ray fluorescence spectrometry (Analytical Geochemical Group, British Geological Survey) on 20 g pellets pressed from -200 mesh whole rock powder. International reference standards were included with each batch of samples, and appropriate corrections were made for instrumental dead time, background and line interferences (Pankhurst & O'Nions 1973). After chemical extraction using ion exchange procedures, isotopic ratios were determined on an automated Micromass-354 mass spectrometer at the NERC Isotope Geosciences Laboratory.

Errors quoted throughout this paper are two standard deviations from measured values. The IUGS recommended value of $1{\cdot}42 \times 10^{-11}$ per year (Steiger & Jager 1977) was used for the decay constant of ^{87}Rb. Analytical uncertainties are estimated to be 0·02% for $^{87}Sr/^{86}Sr$ ratios and 1·0% for $^{87}Rb/^{86}Sr$ ratios. Replicate analyses of the NBS 987 Sr-isotopic standard yielded an average $^{87}Sr/^{86}Sr$ ratio of $0{\cdot}710204 \pm 0{\cdot}00023$ ($n = 46$). The data have been regressed according to the methods of York (1969) and Brooks *et al.* (1972).

3. Nomenclature, field characteristics and petrographic summary

Allen and Stephens (1971) recognised four main phases of Upper Jurassic syntectonic plutonic intrusion. Their subdivision was based on lithological appearance and mutual cross-cutting relationships, but no attempt was made to delineate individual plutons. Recent systematic field mapping of the Territory (Addison 1986; Strange & Shaw 1986; Langford *et al.* 1989; Strange *et al.* 1990) has involved detailed traverses of intrusive contacts enabling a clearly-defined pluton-based nomenclature for Hong Kong granites to be established (Strange *et al.* 1991).

The plutons consist predominantly of biotite monzogranites of normal to highly fractionated I-type (Chappell & White 1974) character. Two suites are distinguished on the basis of whole-rock major and trace element geochemistry, field and petrographic characteristics: the Lamma Suite comprising two relatively small batholithic bodies with poorly defined boundaries, and the younger Lion Rock Suite comprising several cross-cutting, granite plutons with well-defined boundaries (Table 1). Regional contact metamorphic features together with the close association of volcanic deposits suggest that the plutons were emplaced at high crustal levels.

3.1. Lamma Suite

The Lamma Suite includes the Tai Po Granodiorite and Sung Kong Granite which represent the oldest intrusive units. Isolated outcrops are exposed throughout much of the central and southern parts of the Territory (Fig. 2).

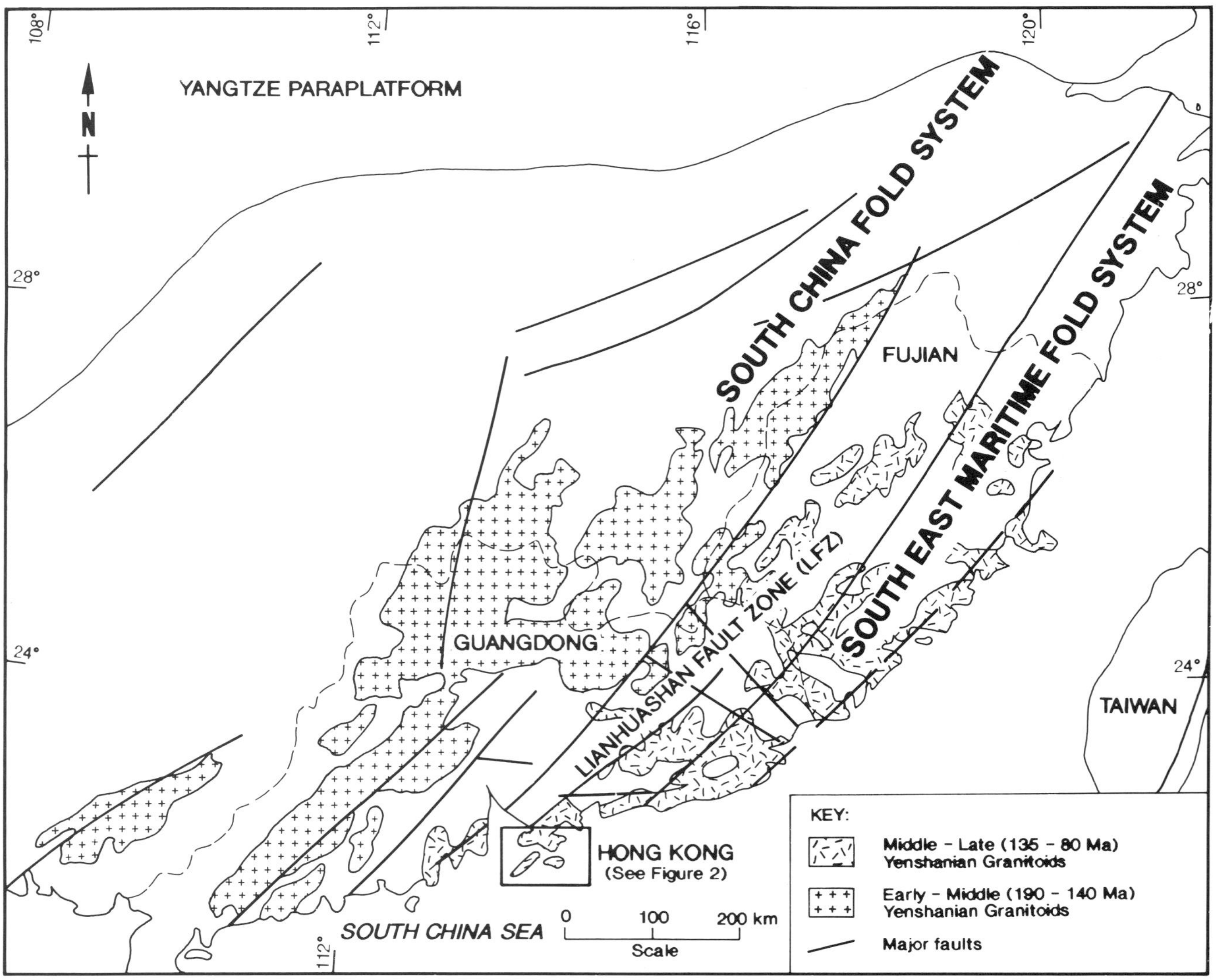

Figure 1 Principal tectonic elements and distribution of Mesozoic granites of southeastern China (after Jahn *et al.* 1976, Nanjing 1980).

The Tai Po Granodiorite, recognised as the oldest intrusive unit, is dominantly porphyritic and fine-grained, but may vary to medium-grained, non-porphyritic lithologies. Porphyritic granodiorite is characterised by large (up to 45 mm) euhedral alkali feldspar crystals set in a fine-grained groundmass and commonly contains abundant fine-grained mafic inclusions. The principal mafic minerals are euhedral dark brown biotite and subordinate subhedral dark green amphibole. Accessory minerals include magnetite, sphene, zircon, and apatite. Chlorite, epidote, and sericite are common subsolidus alteration minerals.

The Sung Kong Granite is recognised as being older than all other intrusive bodies except the Tai Po Granodiorite. It encompasses outcrops of coarse- and medium-grained porphyritic hornblende-biotite monzogranite in southern parts of the Territory. The main minerals are alkali feldspar, plagioclase, quartz, dark green biotite and amphibole, along with accessory magnetite, sphene, and zircon. Subsolidus alteration products include epidote, chlorite, sericite and green biotite.

3.2. Lion Rock Suite

Granites of the Lion Rock Suite (LRS) are divided into three subgroups based on field, petrographic, geochemical and age criteria (see Section 4).

The oldest subgroup is represented by the Tai Lam and Tsing Shan granites which form large (>20 km) circular to elliptical plutons in the northwestern New Territories (Fig. 2). The Tai Lam pluton is characterised by small relict bodies of coarse-grained monzogranite surrounded by a finer-grained, megacrystic leucogranite host. The chemical and petrographic evidence suggests that the Tai Lam pluton is a composite intrusive body: the fine- to medium-grained granite hosting an earlier coarse-grained central core. In the fine- to medium-grained lithologies, textural variation forms a continuum from primary texture granite (Pitfield *et al.* 1990) through transitional xenocrystic types, to equigranular microcrystic granite with an entirely derived texture. Such derivative porphyritic granites have been termed two-phase variants in Peninsular Malaysia (Cobbing *et al.* 1986) and are thought to represent a late, high-level pervasive emplacement of highly differentiated residual magma into a partly crystalline coarser-grained granite host. Greenish-brown biotite is the chief mafic mineral in the Tsing Shan and Tai Lam granites and is present as felted aggregates in the coarse-grained lithologies. Apatite, sphene, and zircon are the main accessory minerals of the coarse-grained granites and occur almost exclusively as inclusions in biotite. In the fine- to medium-grained granites biotite is much less abundant, and apatite, zircon, and sphene are accompanied by fluorite and allanite in the matrix. Garnet is present as rare xenocrysts. Subsolidus alteration products are chiefly chlorite and epidote.

Subgroup II comprises the Lantau and Sha Tin granite plutons, together with several bodies of quartz syenite (Fig. 2). The Sha Tin and Lantau granites exhibit marked textural zonation from coarse-grained inequigranular to fine-grained, alkali feldspar megacrystic lithologies similar to that displayed by fine- to medium-grained granites of subgroup I. In the New Territories, the Sha Tin granite forms a large (>20 km), elliptical multiphase pluton elongated subparallel to major NE-trending faults. It consists of a coarse-grained central core, surrounded by shells of medium- and fine-grained biotite monzogranite. The elliptical Lantau pluton in central Lantau Island comprises dominantly medium-grained monzogranite cut by fine-grained leucogranite. The pluton intrudes the Sung Kong Granite and is eclipsed in the S by the Lantau caldera (Fig. 2). A strongly

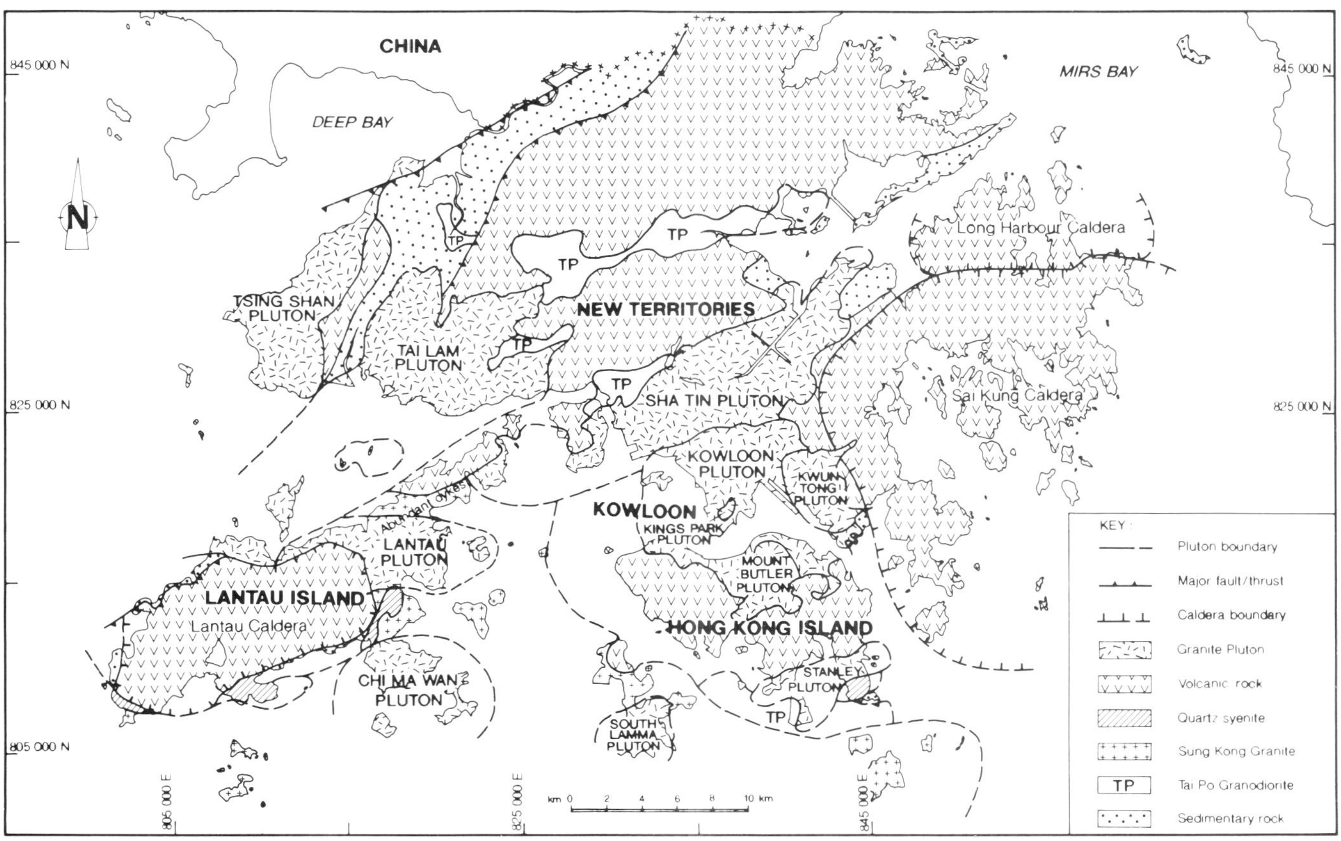

Figure 2 Simplified geological map of Hong Kong showing major structural features and distribution of granite plutons.

Table 1 Summary of nomenclature and field characteristics of Hong Kong granites.

	LAMMA SUITE	LION ROCK SUITE		
		Subgroup I	*Subgroup II*	*Subgroup III*
Colour	grey	light grey to white	light pink or grey	light grey or light pink
Rock type	hornblende-biotite granodiorite and monzogranite	biotite monzogranite to leucogranite	biotite monzogranite to leucogranite quartz syenite	biotite monzogranite to leucogranite
Alkali feldspar megacrysts	common	common	common	sparse
Pluton shape	irregular	elliptical	elliptical	circular
Area exposed (km^2)	65	120	100	130
Mineralisation	Pb, Zn, Mo (trace)	W, Mo, Sn, F	W, Mo, F (trace)	W, Mo, F (trace)
No. of plutons	2	2	4	7

E–W-oriented swarm of porphyritic microgranite and feldsparphyric rhyolite dykes on northeastern Lantau Island intrudes the Sha Tin and Lantau plutons. The main mafic mineral in granites of subgroup II is biotite which generally occurs as single flakes in fine- to medium-grained granites, or as felted aggregates in coarse-grained granites. The main accessory minerals are allanite, apatite, fluorite, sphene, zircon, and magnetite. Subsolidus alteration products include epidote, sericite, and chlorite.

Quartz syenite is present as several prominent bodies in southern parts of Hong Kong and Lantau islands as well as a host of minor intrusions in central and eastern parts of the Territory. The D'Aguilar Syenite in southeastern Hong Kong Island is porphyritic fine- and medium-grained. Small alkali feldspar phenocrysts are strongly oriented in a pilotaxitic groundmass comprising euhedral alkali feldspar and plagioclase with interstices filled with quartz. The main mafic minerals are brown biotite and green amphibole. Accessory minerals include allanite, sphene, zircon, and magnetite. The Lantau Syenite comprises three isolated plutonic bodies along the margin of the Lantau caldera on Lantau Island and is similar in texture and mineralogy to the D'Aguilar Syenite, although amphibole is much less abundant.

Sub-group III comprises smaller (4–15 km) plutons which are generally circular in outcrop and dominated by texturally uniform medium- or fine-grained biotite monzogranite and leucogranite. They include the Kowloon, Chi Ma Wan, Stanley, S Lamma, Mt Butler, and Kwun Tong plutons present in the southern and southeastern parts of the Territory (Fig. 2). A small body of fine-grained megacrystic monzogranite forming the King's Park intrusion in Kowloon is included in subgroup III, although age relationships are uncertain. Kowloon, Stanley, S Lamma, and Chi Ma Wan granites are equigranular with green to brown biotite as the chief mafic mineral. Zircon, apatite, sphene and magnetite are the accessory minerals. The fine-grained leucogranites of Mt Butler and Kwun Tong are commonly porphyritic comprising quartz, alkali feldspar, and sparse biotite phenocrysts set in an equigranular fine-grained matrix. Flakes of primary muscovite, pegmatite veins, and miarolitic cavities are evidence for post-intrusive hydrothermal activity. Accessory minerals include allanite, apatite, monazite, sphene, zircon, and magnetite.

4. Rb–Sr geochronology

Rb–Sr whole-rock isochron ages have been obtained for six plutons of the Lion Rock Suite. The data are presented in Table 2 and plotted on the isochron diagrams shown in

Table 2 Rb–Sr whole-rock analytical data for granites of Hong Kong; trace elements in ppm.

Pluton	No.	Rb	Sr	$^{87}Rb/^{86}Sr$	$^{87}Sr/^{86}Sr$	Pluton	No.	Rb	Sr	$^{87}Rb/^{86}Sr$	$^{87}Sr/^{86}Sr$
Subgroup I						*Subgroup III*					
Tai Lam	8747	524	15·0	103·5	0·936896	Kowloon	8699	261	26·8	28·36	0·763453
Tai Lam	8749	513	19·0	79·68	0·885908	Kowloon	8700	232	40·9	16·48	0·740535
Tai Lam	8750	494	19·5	74·70	0·878534	Kowloon	8701	231	30·4	22·09	0·751110
Tai Lam	8751	531	20·7	75·42	0·876532	Kowloon	8705	210	54·9	11·10	0·729919
Tai Lam	10403	451	25·8	51·21	0·823279	Kowloon	8706	236	35·2	19·47	0·746190
Tai Lam	10400	540	15·8	102·5	0·837618	Kowloon	8707	283	54·2	12·51	0·732425
						Kowloon	8708	197	108	5·258	0·718191
						Kowloon	8709	206	206	5·811	0·719491
						Stanley	8720	400	9·3	127·1	0·957725
Subgroup II						Stanley	8721	276	31·9	25·12	0·757346
Sha Tin	8694	231	143	4·681	0·715796	Stanley	8722	290	30·3	27·93	0·763043
Sha Tin	8695	230	133	5·002	0·716503	Mt Butler	8713	311	25·1	36·13	0·778365
Sha Tin	8696	224	155	4·188	0·714817	Mt Butler	8714	392	6·1	192·0	1·083850
Sha Tin	8697	228	108	6·138	0·718916	Mt Butler	8715	419	6·6	187·7	1·069672
Syenite	8678	163	201	2·350	0·711737	Mt Butler	8716	360	12·7	83·19	0·869777
Syenite	8679	166	190	2·519	0·712011	Mt Butler	8717	412	7·9	154·8	1·006707
Syenite	8680	154	175	2·551	0·712116	Kwun Tong	8723	353	28·8	35·70	0·778246
Syenite	8681	165	201	2·376	0·711782	Kwun Tong	8724	300	32·8	26·56	0·760447
Syenite	8682	165	190	2·521	0·712058	Kwun Tong	8725	348	26·1	38·85	0·784273
Syenite	8683	164	185	2·564	0·712064	Kwun Tong	8726	426	6·5	195·9	1·083587
Syenite	8685	168	196	2·488	0·711984	Kwun Tong	8727	454	7·3	238·7	1·169862
Syenite	8686	166	202	2·391	0·711817	Kwun Tong	8729	446	8·6	153·4	1·007870
Syenite	8687	159	192	2·401	0·711784	Kwun Tong	8730	436	8·4	187·2	1·070721
Syenite	8688	173	125	4·019	0·715195	Kwun Tong	8731	436	8·4	154·9	1·009654

Figure 3. Owing to problems in obtaining fresh material, the earlier plutonic bodies of the Lamma Suite have proved difficult to date by the Rb–Sr technique.

Six samples of the fine- to medium-grained granite of the Tai Lam pluton (subgroup I) were analysed, yielding an age of 155 ± 6 Ma and an initial $^{87}Sr/^{86}Sr$ ratio of 0·7101 ± 0·0060 (Fig. 3a). The high MSWD of 4·6 is solely due to one sample and omission of this data point from the regression results in a statistically good isochron (MSWD = 1·0) with the same parameters. However, there are no apparent reasons for excluding this sample.

Four samples from the coarse-grained core of the Sha Tin pluton (subgroup II) define a linear array indicating an age of 148 ± 9 Ma and an initial $^{87}Sr/^{86}Sr$ ratio of 0·7060 ± 0·0006 (MSWD = 0·1, Fig. 3b). However as these samples display some evidence for modification of the primary texture, this result should be considered a minimum estimate for the age of intrusion of the pluton.

Ten samples of D'Aguilar Syenite (subgroup II) yield an isochron age of 147 ± 8 Ma with an initial $^{87}Sr/^{86}Sr$ ratio of 0·7068 ± 0·0003 (Fig. 3c). Although the MSWD of 0·3 indicates an excellent fit of the data points to the isochron, a bimodal distribution is apparent with nine of the samples forming a single cluster and one sample with high Rb/Sr.

The best constrained ages are for the plutonic bodies in subgroup III of the LRS. Eight medium-grained granite samples of the Kowloon pluton define an excellent isochron (MSWD = 1·2) giving an age of 138 ± 1 Ma and initial $^{87}Sr/^{86}Sr$ ratio of 0·7080 ± 0·0002 (Fig. 3d). Only three samples from the Stanley pluton have been analysed, and one is considerably more enriched in ^{87}Sr than the others. The data therefore plot with a two-point distribution on an isochron yielding an age and initial ratio which is indistinguishable from that of the Kowloon pluton.

Five samples from the fine-grained granite of the Mt Butler pluton yield an isochron age of 136 ± 1 Ma with an initial $^{87}Sr/^{86}Sr$ ratio of 0·7084 ± 0·0010 (MSWD = 2·2, Fig. 3e). A similar age was obtained for the Kwun Tong pluton; analytical data for eight samples of the fine-grained granite define an excellent isochron (MSWD = 0·4, Fig. 3f) with initial $^{87}Sr/^{86}Sr$ ratio of 0·7092 ± 0·0006.

The age data confirm observed field relationships and indicate that emplacement of the plutons occurred between the two main Yenshanian intrusive phases reported from southern China (Jahn *et al.* 1976; Yang *et al.* 1986). A relatively continuous period of Early Jurassic to Late Cretaceous plutonism in southern China is therefore recorded.

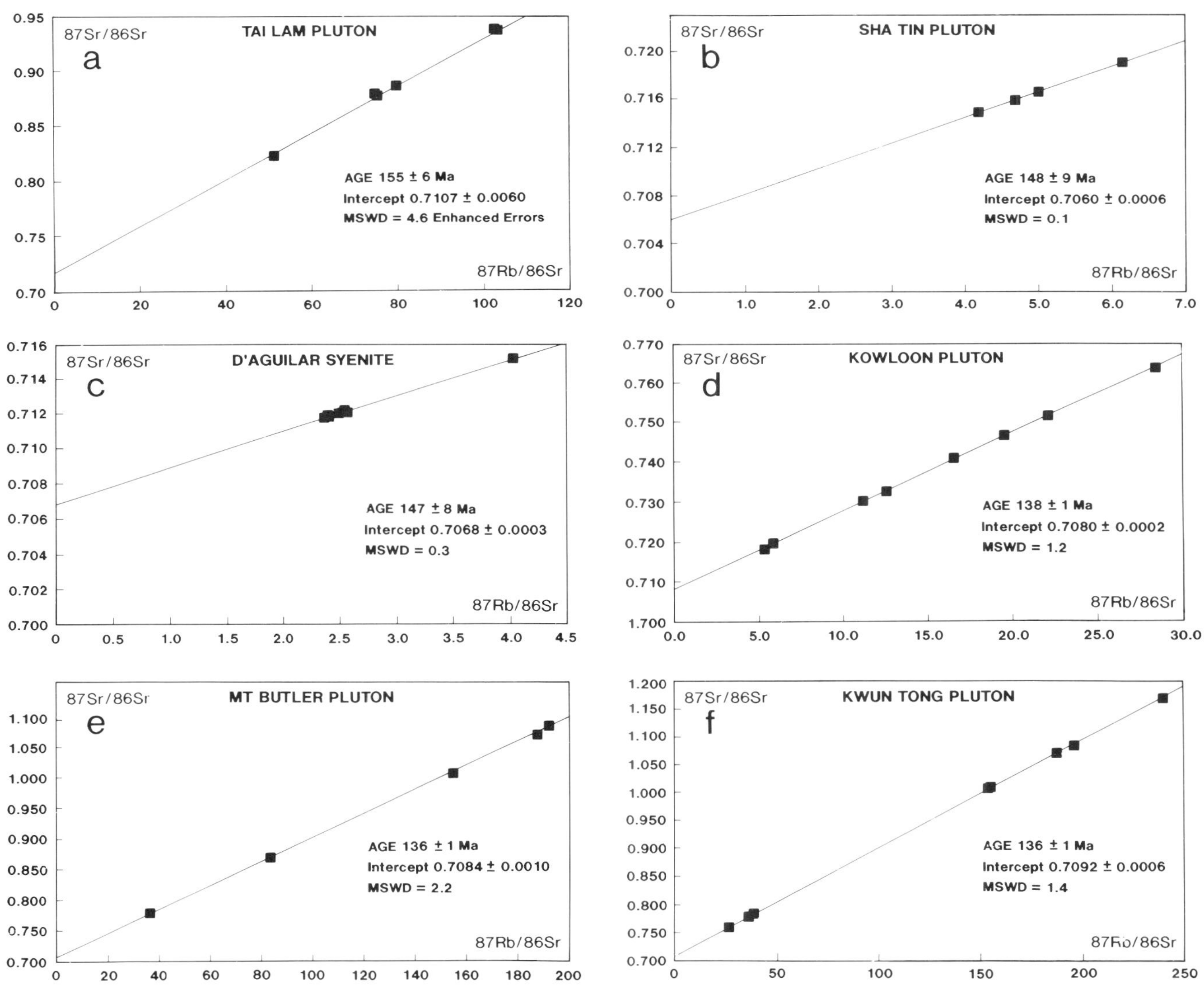

Figure 3a–f $^{87}Sr/^{86}Sr$ versus $^{87}Rb/^{86}Sr$ isochron diagrams for Hong Kong granites.

Table 3 Whole-rock major and trace element analytical data for representative Hong Kong granites.

	8743	8741	8511	8753	8750	9747	8697	8696	8732
wt%									
SiO_2	66·21	61·93	72·35	76·39	77·01	75·65	72·80	73·16	76·37
TiO_2	0·72	0·86	0·29	0·02	0·04	0·08	0·38	0·11	0·15
Al_2O_3	14·43	15·32	13·45	12·65	12·16	12·57	12·64	13·94	11·66
FeO_3t	5·29	6·36	2·70	0·99	1·08	1·38	3·29	1·09	1·56
MnO	0·10	0·07	0·04	0·06	0·05	0·04	0·09	0·02	0·02
MgO	1·59	3·23	0·36	—	—	0·04	0·39	0·08	0·03
CaO	3·23	4·13	1·99	0·68	0·73	0·99	1·51	0·87	0·87
Na_2O	3·10	2·83	2·56	3·86	3·50	3·12	3·53	3·38	3·04
K_2O	3·98	3·16	5·21	4·62	4·72	5·32	4·63	6·83	5·06
P_2O_5	0·18	0·16	0·06	—	—	0·01	0·08	0·02	0·01
LOI	0·93	1·53	0·63	0·58	0·43	0·48	0·33	0·25	1·23
TOTAL	99·76	99·58	99·64	99·85	99·72	99·68	99·67	99·75	100·00
ppm									
Ba	606	759	509	7	43	113	149	548	111
Nb	14	11	13	33	26	18	32	10	17
Rb	221	277	219	698	541	385	241	234	260
Sr	285	329	149	7	19	45	113	165	52
Y	41	36	38	156.	119	62	63	23	41
Zr	193	185	177	91	96	107	248	105	141
Ga	16	19	16	21	19	17	16	13	14
F	1137	4750	1286	5912	3766	1973	3206	712	1456
ppm									
La	49	46	78	11	21	29	168	64	51
Ce	96	89	150	29	49	63	304	114	98
Pr	11	10	16	4·1	6·4	7·7	31	11	11
Nd	42	38	58	20	26	30	104	38	37
Sm	8·0	7·2	10	8·3	9·1	7·8	17	6·0	6·8
Eu	1·54	1·39	0·97	0·10	0·19	0·39	0·81	1·06	0·48
Gd	7·3	6·6	7·6	13	12	8·3	13	4·2	5·8
Dy	7·1	6·4	6·5	21	17	9·6	12	3·6	6·0
Ho	1·33	1·19	1·16	4·1	3·2	1·80	2·14	0·68	1·15
Er	3·7	3·3	3·4	13	10	5·2	6·3	2·2	3·3
Yb	3·5	3·1	2·7	15	11	5·4	5·6	2·0	3·5
Lu	0·53	0·46	0·40	2·2	1·70	0·81	0·82	0·31	0·56

	8734	8683	8699	8707	8721	8710	8711	8717	8724
wt%									
SiO_2	70·79	67·40	77·47	76·92	77·42	75·92	75·83	76·75	75·68
TiO_2	0·40	0·33	0·08	0·09	0·06	0·16	0·17	0·02	0·06
Al_2O_3	13·47	15·52	11·74	12·01	11·81	12·03	11·97	12·23	12·94
Fe_2O_3t	3·00	3·31	1·06	1·28	0·99	1·59	1·64	0·84	1·02
MnO	0·08	0·12	0·08	0·06	0·08	0·04	0·05	0·04	0·05
MgO	0·39	0·43	—	0·07	—	0·15	0·01	—	—
CaO	1·64	1·59	0·63	0·70	0·72	0·90	0·94	0·49	0·71
Na_2O	3·40	3·74	3·60	3·40	3·61	3·37	3·49	4·05	3·98
K_2O	5·74	6·81	4·94	5·16	4·85	5·27	5·21	4·94	5·29
P_2O_5	0·09	0·06	—	—	—	0·02	0·03	—	—
LOI	0·73	0·40	0·20	0·30	0·48	0·45	0·35	0·45	0·35
TOTAL	99·73	99·71	99·80	99·99	100·02	99·90	99·69	99·81	100·08
ppm									
Ba	678	483	26	107	25	242	219	—	28
Nb	20	10	16	16	22	23	27	39	17
Rb	224	169	276	247	291	290	283	446	314
Sr	167	194	29	55	32	91	87	7	34
Y	41	30	52	40	56	49	37	137	46
Zr	246	389	95	123	89	141	139	88	105
Ga	16	16	15	15	15	15	16	17	16
F	1180	875	1357	953	1633	2972	2080	2377	1378
ppm									
La	74	172	22	36	23	56	64	13	23
Ce	140	289	48	75	50	105	121	32	47
Pr	15	29	5·9	9·2	6·3	11	13	4·7	6·0
Nd	53	96	24	35	25	41	45	23	23
Sm	9·2	13	6·3	8·1	6·0	8·3	8·1	8·4	5·8
Eu	1·37	2·0	0·35	0·52	0·42	0·56	0·56	0·13	0·45
Gd	7·8	8·3	6·4	7·1	7·4	8·0	6·6	12	6·1
Dy	7·2	6·1	7·6	7·1	8·8	8·5	6·9	17	7·2
Ho	1·36	1·13	1·45	1·31	1·68	1·58	1·30	3·4	1·39
Er	3·4	3·4	4·7	4·0	5·2	4·7	4·1	11	4·2
Yb	3·6	2·8	5·2	4·3	5·4	4·4	4·0	12	4·6
Lu	0·57	0·45	0·79	0·66	0·82	0·66	0·61	1·81	0·73

See Footnote p. 7.

5. Major and trace element geochemistry

Representative major and trace element data from 193 analysed rocks are given in Table 3 and selected element oxides and trace elements are plotted in Figures 4 and 5. The two suites generally define separate evolutionary trends on silica variation diagrams and, within the LRS, there are slight but consistent differences both among and within subgroups.

Samples of both suites are metaluminous to slightly peraluminous (mol. $Al_2O_3/(Na_2O+K_2O+CaO)<1{\cdot}1$). Mafic compositions predominate in the Lamma Suite and fractionation is indicated by a broad range of silica values from 63–75%. Silica contents in the LRS are in the range 70–78% with a predominance of silica-rich (>75%) compositions. The quartz syenites vary from 63–68% SiO_2. Colinear trends of decreasing TiO_2, CaO, and Al_2O_3 with increasing magmatic evolution are shown by both suites, indicating separation of plagioclase and a Ti-rich mineral. At a given silica content, CaO and K_2O concentrations in the Lamma Suite are generally slightly higher and lower, respectively, relative to the LRS, although coarse-grained granites from subgroup I have intermediate abundances. Quartz syenites generally plot about an extension of the trend toward less differentiated compositions for LRS subgroup II. Differences in the proportion of modal amphibole among these samples are indicated by rather variable TiO_2 and CaO contents. The large spread of data points in subgroup II reflects the broad lithological variation of constituent plutons (see Section 3) and on selected plots (e.g. TiO_2, CaO, Al_2O_3) separate evolutionary paths are apparent. Both subgroup II and subgroup III of the LRS are dominated by silica-rich ($SiO_2>75\%$) compositions.

Chemical differences between the two suites and among subgroups are magnified on plots of trace elements (Fig. 5a–d). In general, moderately to highly evolved samples of the Lamma Suite are distinguished from the LRS in having lower abundances of Y, Nb, and Ba at a given silica content, reflecting different magmatic sources. The quartz syenites generally plot about an extension of the main LRS trend at

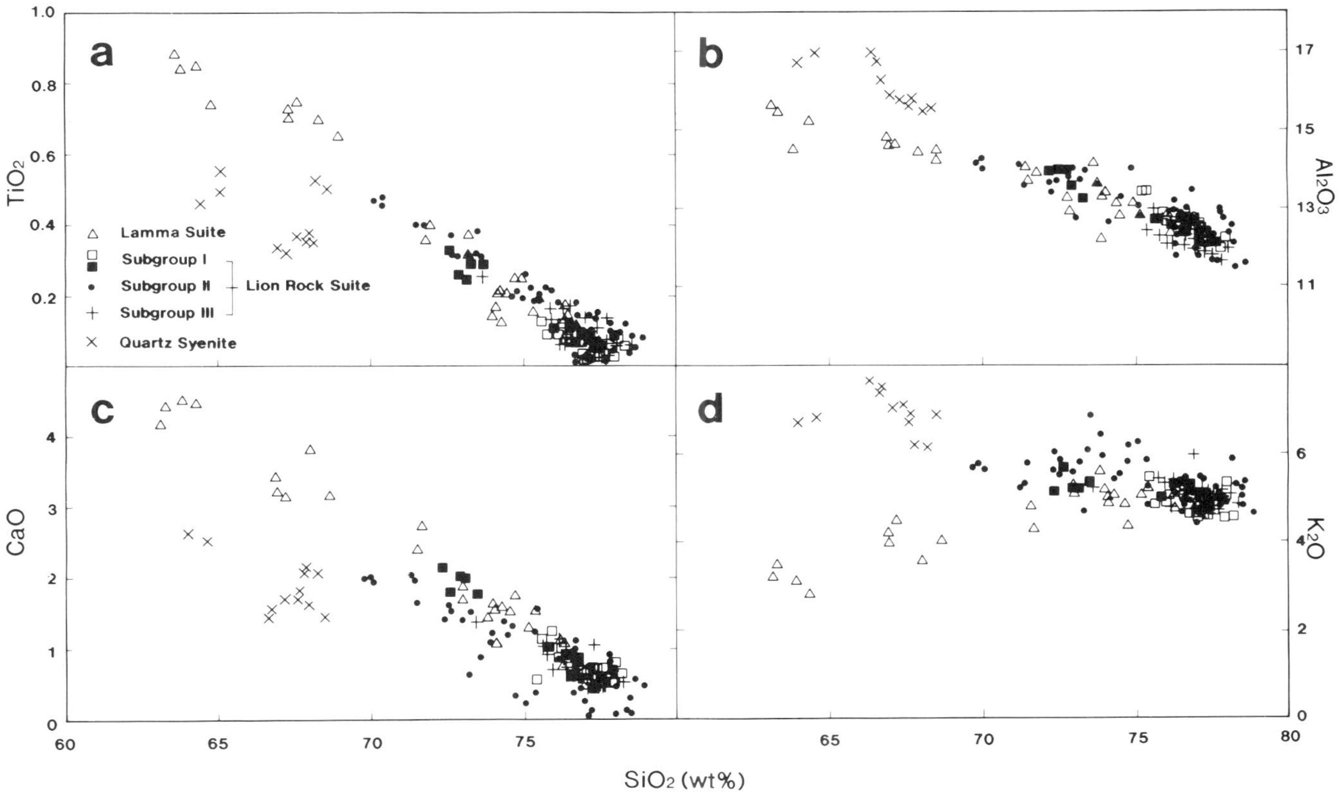

Figure 4 Selected major element silica variation diagrams for Hong Kong granites; numbers of samples comprising the main groups of plutons is as follows: Lamma Suite 25, LRS Subgroup I 37, LRS Subgroup II 89, LRS Subgroup III 42; filled squares = coarse-grained granites, open squares = fine- to medium-grained granites; samples of quartz syenite have been plotted separately in order to facilitate comparisons.

8743 Tai Po Granodiorite (828400E 824200N)
9747 Tsing Shan coarse-grained granite (810612E 829510N)
8741 Tai Po Granodiorite (824915E 827840N)
8697 Sha Tin coarse-grained granite (837650E 825910N)
8511 Tai Lam coarse-grained granite (816090E 827597N)
8696 Sha Tin coarse-grained granite (837400E 825680N)
8753 Tai Lam fine-grained granite (817900E 831300N)
8732 Lantau porphyritic dyke (822110E 820800N)
8750 Tai Lam fine-grained granite (818045E 832095N)
8734 Lantau porphyritic dyke (819370E 816540N)
8683 D'Aguilar Syenite (843630E 808320N)
8699 Kowloon medium-grained granite (839100E 822720N)
8707 Kowloon medium-grained granite (837840E 814940N)
8721 Stanley fine-grained granite (842470E 820360N)
8710 Chi Ma Wan medium-grained granite (818000E 808780N)
8711 Chi Ma Wan medium-grained granite (817990E 808800N)
8717 Mt Butler fine-grained granite (838920E 814440N)
8724 Kwun Tong fine-grained granite (842720E 820190N)

Major and trace elements by XRF analysis; Loss on Ignition (LOI) at 1000°C; Fe as total Fe_2O_3; REE by ICPES analysis; all grid references are based on the Hong Kong metric map series HM20C (1 : 20 000).

lower degrees of differentiation in accordance with major element trends, although Nb abundances of the amphibole-rich rocks are similar to those of the Lamma Suite. Ba contents in the amphibole-poor quartz syenite are distinctly higher than in the amphibole-rich quartz syenite, suggesting the latter may have been affected by late-stage hydrothermal alteration. Highly evolved ($SiO_2 > 75\%$) samples of LRS subgroup I have higher Y, Nb, and Rb contents than most other members of the suite, except for a small population of subgroup II samples which are among the most highly fractionated rocks in the Territory. There are consistent differences in trends and elemental abundances between the coarse-grained granites and fine- to medium-grained granites of subgroup I. Coarse-grained granites have slightly lower Rb, Y and Nb contents than the fine- to medium-grained rocks at similar SiO_2 content, and there is no continuous evolutionary trend between the two lithologies. Fine- to medium-grained granites also have distinctly higher concentrations of Ga and F (Table 3) and a narrower range of silica content relative to coarse-grained lithologies. Plots of Y, Nb, and Rb (Fig. 5a–c) abundances for LRS subgroup II reveal two discrete sample populations. One population is characterised by high Y, Nb, and Rb contents, whereas the other has distinctly lower concentrations of these elements at similar degrees of differentiation. Enrichment of Y, Nb, and Rb is a feature of silica-rich compositions in subgroup III.

On a Y/Nb versus SiO_2 plot (Fig. 6), the Lamma samples show a progressive decrease in Y/Nb ratio with increasing differentiation as a result of the crystallisation of amphibole. Coarse-grained granites of LRS subgroup I have distinctly higher Y/Nb ratios than Lamma samples at similar stages of magmatic evolution, but generally show little variation with increasing differentiation. This pattern is in sharp contrast to that of the fine- to medium-grained granites, which show a marked increase in Y/Nb ratio reaching high values (4–6) in the most evolved compositions. The quartz syenites generally have similar Y/Nb ratios to the Lamma samples, although slight variations are due to differing proportions of modal amphibole in constituent units. Differences in the Y/Nb ratios between suites and subgroups at similar stages of magmatic evolution probably reflect heterogeneous source materials, since the Y/Nb ratio is generally unaffected by crustal assimilation (Eby 1990). However, the lower and more variable Y/Nb ratios displayed by silica-rich compositions of LRS subgroups II and III may indicate a hydrothermal overprint in which accessory minerals have been removed by leaching.

The rather mafic I-type character of Lamma samples is illustrated on a Rb/Ba versus Zr + Ce + Y diagram (Fig. 7) where these rocks define a linear trend plotting below the fields for fractionated I-type granites from Tasmania (Housetop Batholith, Sawka *et al.* 1990) and Newfoundland (Ackley granite, Whalen 1983). The decrease in Zr + Ce + Y coupled with the increase in Rb/Ba ratio indicates the predominance of amphibole and plagioclase fractionation in these rocks. Coarse-grained granites of LRS subgroup I define a similar trend to the Lamma samples and generally plot in the field for fractionated I-types from Newfoundland. The decrease in Zr + Ce + Y in these rocks is due mainly to the separation of biotite and accessory minerals. The coarse-grained lithologies of LRS subgroup I differ markedly from the fine- to medium-grained granites, which show a sharp increase in Rb/Ba with slightly decreasing Zr + Ce + Y and overlap with the field for A-type granites from Newfoundland (Topsails Suite, Whalen *et al.* 1987). The increase in Rb/Ba in the finer-grained lithologies is indicative of more pronounced alkali feldspar fractionation. Samples of LRS subgroup II generally plot in the region of overlap between A-type granites and fractionated I-types

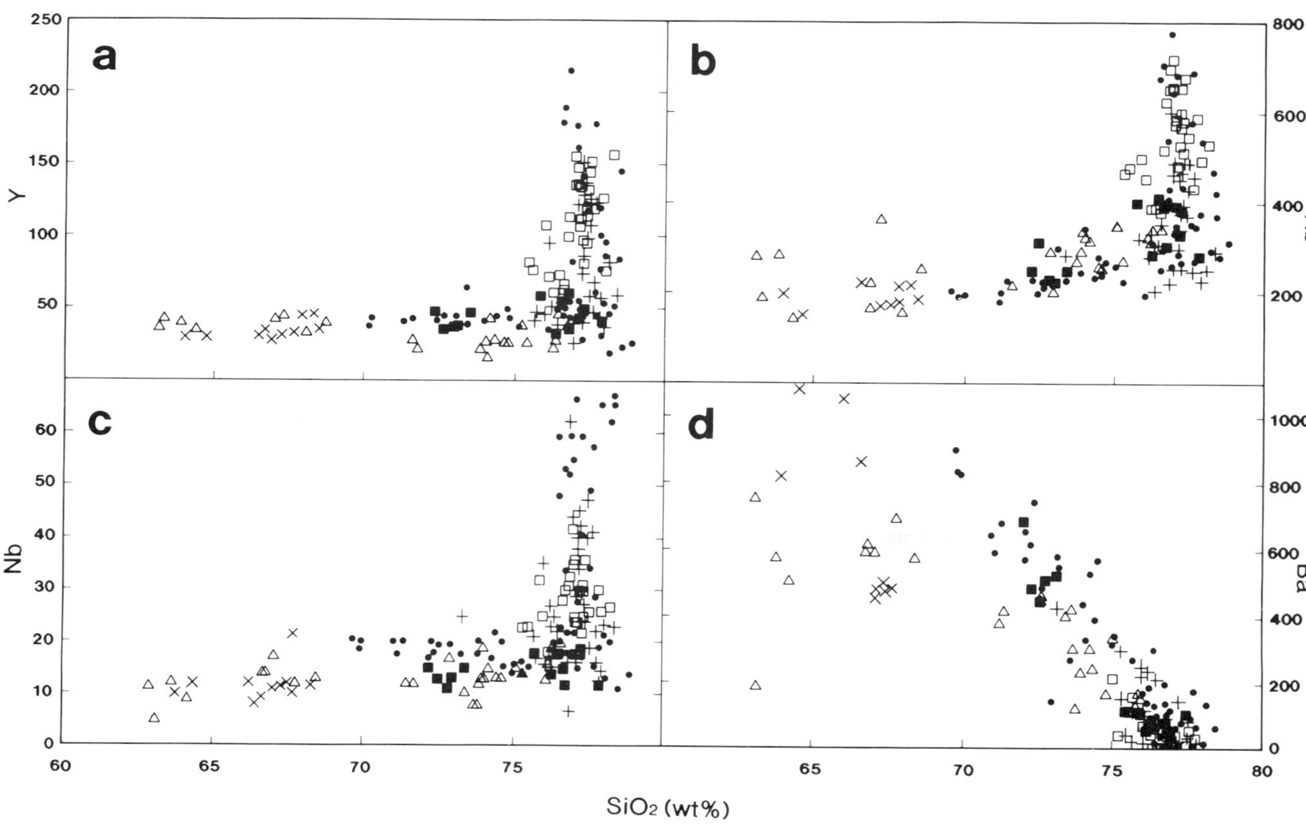

Figure 5 Selected trace element silica variation diagrams for Hong Kong granites (symbols as for Fig. 4).

from Tasmania and Newfoundland. A small population of these samples plots with highly evolved medium-to fine-grained granites of subgroup I. The least differentiated members of subgroup II plot in the field for A-type compositions along with a small population of quartz syenites. Samples of subgroup III plot outside the field for A-type granites and mostly in the field for fractionated I-types from Newfoundland.

6. REE geochemistry

Twenty whole-rock rare-earth element (REE) analyses were obtained for representative samples. Analytical data are given in Table 3 and compositions are depicted on chondrite-normalised diagrams in Fig. 8a–f.

Patterns for two samples from the Lamma Suite (Fig. 8a) are LREE-enriched with a slight negative Eu anomaly ($Eu/Eu^* = 0{\cdot}65$). Overall abundances of REE increase slightly with increasing SiO_2, a trend which has been reported in other mafic granites (Sawka *et al.* 1984). The shape of the patterns is similar to that of less differentiated LRS compositions, although the latter exhibit larger negative Eu anomalies.

With increasing magmatic evolution, REE patterns for two coarse-grained granites of LRS subgroup I (8511, 9477; Fig. 8b) show a decrease in LREE along with an increase in HREE abundance and deepening Eu anomaly. Such behaviour is indicative of crystallisation of plagioclase and a LREE-rich accessory mineral (Miller & Mittlefehldt 1982). Analogous REE variation is shown by the medium- to fine-grained granites, although at similar degrees of differentiation these rocks have notably higher HREE concentrations, deeper Eu anomalies, and flatter REE patterns approximating 50 times chondritic abundance. Such differences in REE concentrations between the two lithologies rules out a genetic relationship. The rather flat REE patterns in the fine- to medium-grained granites could have been produced by fractionation involving separation of biotite-bearing LREE-rich accessory minerals (e.g. allanite), although accessory minerals in these rocks are mostly confined to the matrix (see Section 3). Alternatively, the flat patterns could be attributed to hydrothermal processes in which leaching would deplete the granite of LREE-rich accessory minerals and albitisation would deepen the Eu anomaly. The presence of fluorite in these rocks is evidence for at least some metasomatic fluid interaction, although the lack of significant albitisation, greisenisation and kaolinitisation, along with the high overall REE concentration in these rocks, suggests this activity was minimal. The REE signatures of the fine- to medium-grained granites are thought largely to reflect magmatic concentrations and together with high Ga, F, Y, Nb, and Rb contents indicate similarities with A-type granites (Loiselle & Wones 1979).

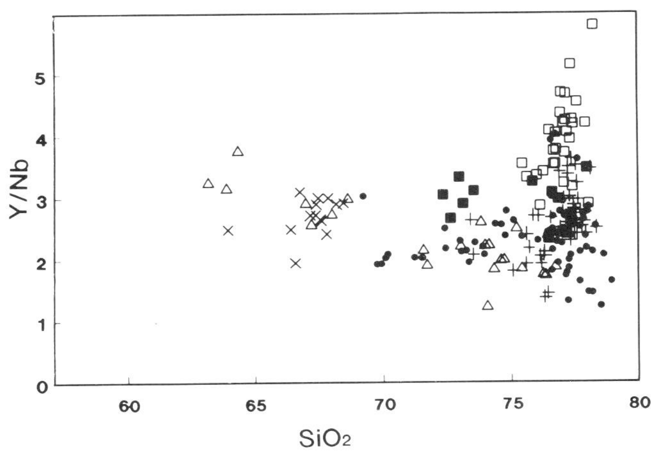

Figure 6 Plot of Y/Nb versus SiO_2 for Hong Kong granites (symbols as for Fig. 4).

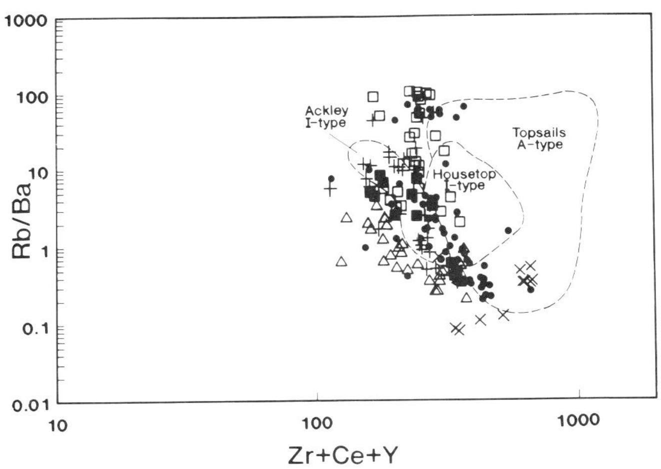

Figure 7 Rb/Ba versus Zr + Ce + Y (ppm) diagram for Hong Kong granites (symbols as for Fig. 4); data sources of fields for fractionated I-types (Ackley granite, Whalen 1983; Housetop Batholith, Sawka *et al.* 1990), and for A-types (Topsails suite, Whalen *et al.* 1987).

REE patterns for the rocks of LRS subgroup II (Fig. 8c) are LREE-enriched and exhibit a broad spectrum of compositions. The differing REE abundances for two samples at 73% SiO_2 indicate the presence of two discrete populations of granites as distinguished on major and trace elements plots (see Section 5). Crystallisation of plagioclase in all rocks of the subgroup is shown by increasing negative Eu anomalies, and a corresponding decrease in LREE abundance with increasing differentiation indicates separation of a LREE-rich mineral. The REE pattern for quartz syenite (D'Aguilar Syenite, Fig. 8d) is subparallel to that of subgroup II, although the slight convex downward curvature and marked negative Eu anomaly ($Eu/Eu^* = 0{\cdot}58$) indicates that amphibole as well as plagioclase was an important liquidus phase.

There is remarkable uniformity in the REE patterns for medium-grained granites of subgroup III (Fig. 8e). All samples show a similar magnitude of negative Eu anomaly, indicating plagioclase fractionation. HREE abundances are almost constant from weakly to strongly evolved compositions, whereas LREE contents increase only slightly with increasing differentiation. The fine-grained granites differ from the medium-grained lithologies in having almost flat REE patterns as well as large negative Eu anomalies (Fig. 8f). The more evolved Kwun Tong granite (8724) shows slightly lower LREE, and higher HREE contents than the Mt Butler granite (8717). Compared with the fine- to medium-grained granites of subgroup I, subgroup III fine-grained samples have slightly higher LREE, smaller negative Eu anomalies, and lower HREE, Ga, and F concentrations (Table 3) at similar degrees of differentiation. There is abundant physical evidence (see Section 3) in these rocks for post-emplacement hydrothermal activity. The REE signatures of the silica-rich fine-grained granites of subgroup III, together with the strong enrichment in Y, Nb and Rb, and depletion in Ba and Sr, are therefore thought to reflect a late-stage metasomatic overprint.

7. Discussion

Pervasive differences in major and trace element abundances between the two main suites, together with variation in initial $^{87}Sr/^{86}Sr$ ratios, major and trace element abundances and REE contents among subgroups of the LRS at similar degrees of differentiation, suggest that a range of source materials is involved in the genesis of Hong Kong granite magmas. Characterisation of these materials is beyond the scope of this paper, although inferences based on the available data can be made with regard to genetic classification and tectonic setting.

On a 10000*Ga/Al vs Zr + Nb + Ce + Y discrimination diagram (Fig. 9, after Eby 1990), the Lamma samples lie on the boundary between fields for unfractionated M-, I- and S-type granites and fractionated granites, and have only slightly lower Ga/Al ratios than coarse-grained granites of subgroup I. The fine- to medium-grained granites of subgroup I are distinguished from the coarse-grained lithologies in having higher Ga/Al ratios which plot outside the field for fractionated I-types from Tasmania and Newfoundland. Despite possessing high Nb, Ce, and Y abundances typical of A-type granites (see Section 5), the fine- to medium-grained granites have rather low Zr contents by comparison and this feature probably accounts for their positioning in the fractionated granite field. Granites of subgroup II have slightly higher total Zr + Nb + Ce + Y abundances than other members of the suite, but maintain similar Ga/Al ratios, while the quartz syenites have distinctly higher Zr + Nb + Ce + Y and slightly lower Ga/Al ratios. Overall, the chemistry of the quartz syenites is consistent with an A-type signature, although variation in major and trace element abundances suggests post-magmatic alteration may have affected the bulk rock chemistry. Granites of subgroup III have a broad range in total Zr + Nb + Ce + Y and Ga/Al ratios which overlap with

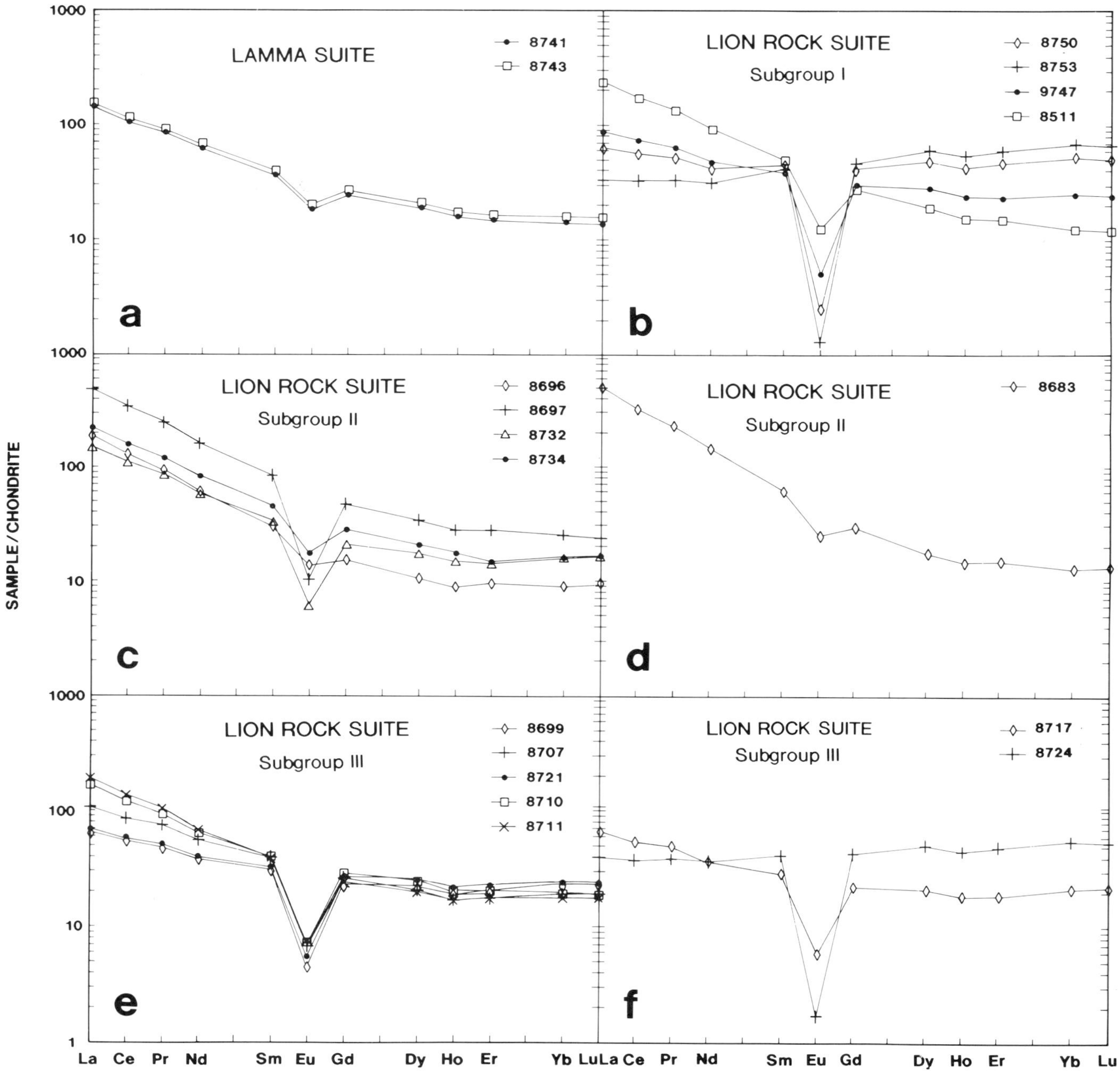

Figure 8a–f Chondrite-normalised REE patterns for Hong Kong granites (refer to Table 3 for sample descriptions).

fields for fractionated and unfractionated rocks. The high dispersion in total Zr + Nb + Ce + Y abundance reflects mobility of these elements during hydrothermal alteration (Harris *et al.* 1986). By contrast, the Ga/Al ratio appears to be only slightly affected by metasomatism.

On a Nb versus Y tectonic discrimination diagram (Pearce *et al.* 1984, Fig. 10), samples of the Lamma Suite mostly plot in the VAG + syn-COLG field which is consistent with the presence of mylonitised fabrics in many field outcrops. Deformation fabrics in younger granites are confined to coarse-grained lithologies of LRS subgroup I, which also plot in the VAG + syn-COLG field but overlap with the WPG field. The deformed rocks are best interpreted as having been emplaced in a synorogenic setting. By contrast, the fine- to medium-grained granites of subgroup I plot distinctly in the WPG field consistent with their A-type character. A-type granites are generally considered to have been emplaced in tensional (or non-compressive), continental rift zone, or oceanic basin environments (Loiselle & Wones 1979; Pearce *et al.* 1984; Whalen *et al.* 1987). Granites of subgroup II and III have slightly lower Y/Nb ratios than the granites of subgroup I and show linear differentiation trends extending from the VAG + syn-COLG field to the WPG field. Such behaviour is typical of fractionated I-type granites (e.g. Ackley granite, Whalen 1983) and suggest that these rocks were emplaced in a late-orogenic or postorogenic setting.

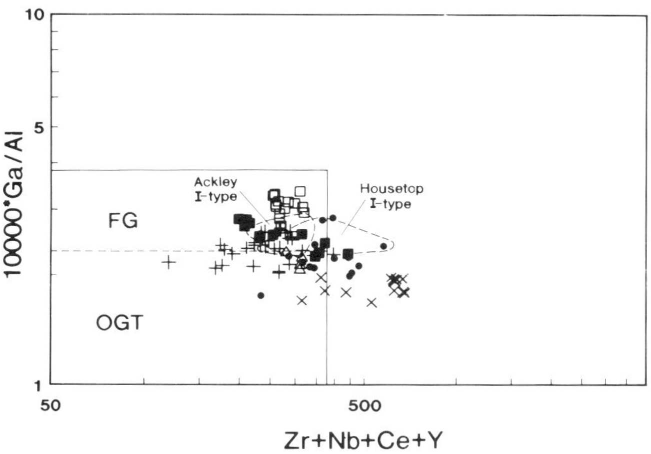

Figure 9 Granites of Hong Kong plotted in the 10000*Ga/Al versus Zr + Nb + Ce + Y (ppm) discriminant diagram for A-type granites after Eby (1990) (symbols as for Fig. 4, data sources for fractionated I-types as for Fig. 7); OGT = field for I-, S- and M-type granites; FG = field for fractionated I-type granites.

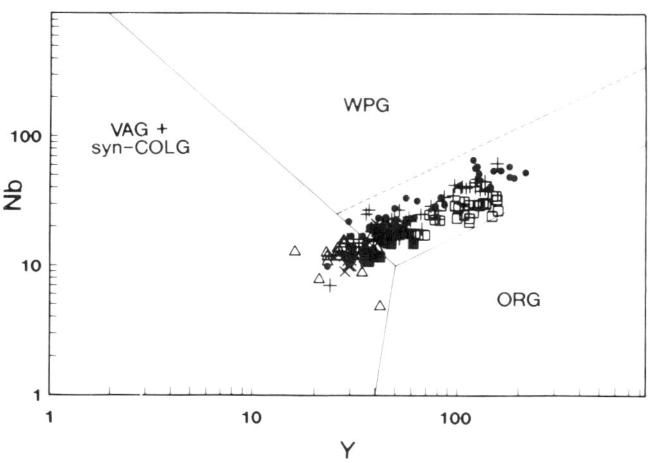

Figure 10 Data for Hong Kong granites plotted on a Y vs Nb tectonic discrimination diagram (after Pearce *et al.* 1984) (symbols as for Fig. 4); VAG = volcanic arc granites, syn-COLG = syn-collision granites, WPG = within-plate granites, ORG = ocean ridge granites.

Rb–Sr whole-rock age dating indicates that the granites of the LRS were emplaced over a period of at least 20 Ma and that each subgroup of plutons probably was intruded in pulses separated by several million years. Although absolute ages have not yet been determined for granites of the Lamma Suite and coarse-grained lithologies of LRS subgroup I, petrographic and chemical evidence suggests that these rocks were probably derived from primitive I-type sources. Granites with A-type affinity (LRS subgroup I) are intermediate in age between deformed, predominantly mafic I-type synorogenic granites (Lamma Suite) and undeformed, moderately to highly evolved late-orogenic or postorogenic I-type granites (LRS subgroups II and III). Together with the field association of quartz syenite and a linear dyke swarm (Lantau porphyritic granite and rhyolite dykes), such rocks appear to record a transition from mildly compressional to tensional tectonics. High initial $^{87}Sr/^{86}Sr$ ratios and F abundances in the A-type granites suggest they were derived from an isotopically evolved, halogen-rich, H_2O-poor source. Less radiogenic initial Sr isotopic signatures for the LRS subgroup II granites imply an increase in proportion of mantle-derived component in the source region. The more radiogenic initial Sr signatures exhibited by younger granite plutons (LRS subgroup III) probably reflect the development of active hydrothermal systems in these subvolcanic plutons as indicated by the presence of pegmatite veins and miarolitic cavities, and possibly more pronounced crustal contamination.

8. Acknowledgements

Rb–Sr data presented in this paper are recorded in NERC Isotope Geosciences Laboratory publication series No. 46. We are grateful to three anonymous reviewers for their helpful comments on improving earlier drafts. This paper is published with the approval of the Director of the Civil Engineering Deparment, Hong Kong Government.

9. References

Addison, R. 1986. *Geology of Sha Tin, 1:20 000 Sheet 7.* HONG KONG GEOL SURV MEM **1**.

Allen, P. M. & Stephens, E. A. 1971. *Report on the geological survey of Hong Kong.* Hong Kong: Government Press

Brooks, C., Hart, S. R. & Wendt, I. 1972. Realistic use of two error regression treatment as applied to rubidium-strontium data. REV GEOPHYS SPACE PHYS **10,** 551–57.

Chappell, B. W. & White, A. J. R. 1974. Two contrasting granite types. PAC GEOL **8,** 173–4.

Cobbing, E. J., Mallick, D. I. J., Pitfield, P. E. J. & Teoh, L. H. 1986. The granites of the Southeast Asian Tin Belt. J GEOL SOC LONDON **143,** 537–50.

Eby, G. N. 1990. The A-type granitoids: A review of their occurrence and chemical characteristics and speculations on their petrogenesis. *In* Woolley, A. R. & Ross, M. (eds) *Alkaline igneous rocks and carbonatites.* LITHOS **26,** 115–34.

Harris, N. B. W., Marzouki, F. M. H. & Ali, S. 1986. The Jabel Sayid Complex, Arabian Shield: geochemical constraints on the origin of peralkaline and related granites. J GEOL SOC LONDON **143,** 287–95.

Huan, W., Shi, Z. & Yan, J. 1982. Meso-Cenozoic tectonic evolution of eastern China and adjacent area and movement of the Pacific plate. (Abstr.) SCI GEOL SINICA **2,** 190.

Huang, C. C. 1978. An outline of the tectonic characteristics of China. ECLOGAE GEOL HELVETIAE **71,** 611–35.

Jahn, B. M., Chen, P. Y. & Yen, T. P. 1976. Rb–Sr ages of granitic rocks in southeastern China and their tectonic significance. BULL GEOL SOC AM **86,** 763–76.

Langford, R. L., Lai, K. W., Arthurton, R. S. & Shaw, R. 1989. *Geology of the Western New Territories, 1:20 000 sheets 2, 5 and 6.* HONG KONG GEOL SURV MEM **3**.

Loiselle, M. C. & Wones, D. R. 1979. Characteristics and origin of anorogenic granites. GEOL SOC AM, ABSTR PROG **11,** 468.

Miller, C. F. & Mittlefehldt, D. W. 1982. Depletion of light rare-earth elements in felsic magmas. GEOLOGY **10,** 129–33.

Pankhurst, R. J. & O'Nions, R. K. 1973. Determination of Rb/Sr and $^{87}Sr/^{86}Sr$ ratios of some standard rocks and evaluation of X-ray spectrometry in Rb–Sr geochemistry. CHEM GEOL **12,** 127–36.

Pearce, J. A., Harris, N. B. W. & Tindle, A. G. 1984. Trace element discrimination diagrams for the tectonic interpretation of granitic rocks. J PETROL **25,** 956–83.

Pitfield, P. E. J., Teoh, L. H. & Cobbing, E. J. 1990. Textural variation and tin mineralization in granites from the Main Range Province of the Southeast Asian Tin Belt. GEOL J **25,** 419–29.

Sawka, W. N., Chappell, B. W. & Norrish, K. 1984. Light rare-earth-element zoning in sphene and allanite during granitoid fractionation. GEOLOGY **12,** 131–34.

Sawka, W. N., Heizler, M. T., Kistler, R. W. & Chappell, B. W. 1990. Geochemistry of highly fractionated I- and S-type granites from the tin-tungsten province of western Tasmania. GEOL SOC AM SPEC PAP **246,** 161–79.

Steiger, R. H. & Jager, E. 1977. Subcommission on geochronology; convention on the use of decay constants in geo- and cosmo-chronology. EARTH PLANET SCI LETT **36,** 359–62.

Strange, P. J. & Shaw, R. 1986. *Geology of Hong Kong Island and Kowloon, 1:20 000 sheets 11 and 15.* HONG KONG GEOL SURV MEM **2**.

Strange, P. J., Shaw, R. & Addison, R. 1990. *Geology of Sai Kung and Clear Water Bay, 1:20 000 sheets 8, 12 and 16.* HONG KONG GEOL SURV MEM **4**.

Strange, P. J., Langford, R. L. & Sewell, R. J. 1991. Classification and nomenclature for Hong Kong granitoids. *In* Chappell, B. W. (ed.) *Second Hutton Symposium on Granites and Related Rocks, Canberra 1991.* BUR MIN RES, GEOL GEOPHYS REC 1991/25.

Walsh, J. N., Buckley, F. & Barker, J. 1981. The simultaneous determination of rare-earth elements in rocks using inductively coupled plasma source spectrometry. CHEM GEOL **33,** 141–53.

Whalen, J. B. 1983. The Ackley City Batholith, southeastern Newfoundland: evidence for crystal versus liquid-state fractionation. GEOCHIM COSMOCHIM ACTA **47,** 1443–57.

Whalen, J. B., Currie, K. L. & Chappell, B. W. 1987. A-type granites: geochemical characteristics, discrimination and petrogenesis. CONTRIB MINERAL PETROL **95,** 407–19.

Yang, Z., Cheng, Y. & Wang, 1986. *The geology of China,* Oxford Monographs on Geology and Geophysics 3. Oxford: Clarendon Press.

York, D. 1969. Least squares fitting of a straight line with correlated errors. EARTH PLANET SCI LETT **5,** 320–24.

R. J. SEWELL, Hong Kong Geological Survey, Geotechnical Engineering Office, 11/F Civil Engineering Building, 101 Princess Margaret Road, Homantin, Kowloon, Hong Kong.

D. P. F. DARBYSHIRE, NERC Isotope Geosciences Laboratory, Keyworth, Nottingham NG12 5GG. U.K.

R. L. LANGFORD, Hong Kong Geological Survey, Geotechnical Engineering Office, 11/F Civil Engineering Building, 101 Princess Margaret Road, Homantin, Kowloon, Hong Kong.

P. J. STRANGE, British Geological Survey, Keyworth, Nottingham NG12 5GG, U.K.

Transactions of the Royal Society of Edinburgh: Earth Sciences, **83,** 281–290, 1992

Granites of Edward VII Peninsula, Marie Byrd Land: anorogenic magmatism related to Antarctic–New Zealand rifting

S. D. Weaver, C. J. Adams, R. J. Pankhurst and I. L. Gibson

ABSTRACT: Syenogranites and monzogranites of Edward VII Peninsula, Marie Byrd Land, represent magmatism associated with continental rifting and the separation of New Zealand from W Antarctica in the mid-Cretaceous. These coarse-grained, leucocratic, subsolvus biotite granites occur as five small plutons cutting Lower Palaeozoic metasediments. Petrographic features include the predominance of microcline perthite over albite, bipyramidal smoky quartz, red-brown biotite and accessory ilmenite, zircon, apatite, monazite and fluorite. Enclaves are absent and miarolitic cavites are rare.

The granites are a weakly peraluminous, potassic, and highly fractionated suite with high concentrations of Rb, Nb, Y, HREE and F in the most evolved compositions. REE patterns vary from LREE-enriched ($Ce_N/Yb_N = 8{\cdot}4$), to flat REE patterns ($Ce_N/Yb_N = 1{\cdot}1$) with large negative Eu anomalies ($Eu/Eu^* = 0{\cdot}02$). Initial $^{87}Sr/^{86}Sr$ ratios are 0·7116–0·7206 and initial ε_{Nd} values are −5·5 to −7·7. Generalised fractionation trends for the suite are explicable in terms of the modal mineralogy. Monazite crystallisation exerted a predominant control on LREE concentrations.

The geochemistry of the Edward VII Peninsula granites suggests an infracrustal I-type source, and regionally available Devonian-Carboniferous I-type granodiorites and tonalites satisfy the isotopic constraints. The granites classify as A-type (preferred term A-subtype) and Within-Plate Granites on standard diagrams, but the least fractionated rocks clearly indicate the I-type, Volcanic Arc Granite geochemical signatures of their inferred crustal sources.

KEY WORDS: A-subtype, Cretaceous, fractionation trends, granodiorite-tonalite source, isotopes, I-type, monzogranite, petrogenesis, syenogranite.

The nature and origin of anorogenic granites has been the subject of considerable debate since Loiselle and Wones (1979) defined A-type granites, stressing their relatively anhydrous compositions and their occurrence in non-orogenic or postorogenic extensional tectonic settings. Collins *et al.* (1982) recognised A-type granites as a minor, late component of the Lachlan Fold Belt and suggested that they originated by partial melting of a residual felsic granulite that had previously generated I-type magmas. Anderson (1983) and Creaser *et al.* (1991) prefer a petrogenetic model in which A-type granites are produced by melting of I-type tonalite-granodiorite crust generated during a previous orogenic episode. A recent review by Eby (1990) emphasised the great diversity of geochemistry within the group of granites and related rocks classified as A-types. One of the major problems in identifying true A-types in the geological record is the difficulty of assessing independently the tectonic environment. It may be that some of the geochemical diversity ascribed previously to anorogenic granites is the result of erroneous conclusions regarding tectonic settings. The Edward VII Peninsula granites provide a useful case study of geologically young, undeformed rocks, the anorogenic environment of which can be inferred with some certainty.

1. Edward VII Peninsula granites

1.1. Region geology and field occurrence

Edward VII Peninsula is adjacent to the Ross embayment at the westernmost extremity of Marie Byrd Land (Fig. 1), one of several independent continental blocks that make up W Antarctica. On the southern and northern sides of the peninsula, scattered nunataks form the Rockefeller and Alexandra mountains. Geological knowledge of the area is summarised in the 1:250,000 map of Wade *et al.* (1977) which shows biotite-schists intruded by undeformed biotite leucogranites referred to as Byrd Coast Granite.

A NZ Antarctic Research Programme expedition worked on the peninsula in 1987/88 (Adams *et al.* 1989). Using K–Ar and Rb–Sr data, Adams and Weaver (1990) correlated the metasediments with the Swanson Formation of the Ford Ranges, Marie Byrd Land which, like similar quartzose turbiditic sequences on the western coast of the South Island, New Zealand, record a Late Ordovician low-grade metamorphism (Adams 1986). The newly recognised high-grade paragneiss-migmatite metamorphic complex of the Alexandra Mountains may represent a strong mid-Cretaceous thermal event (Adams & Weaver

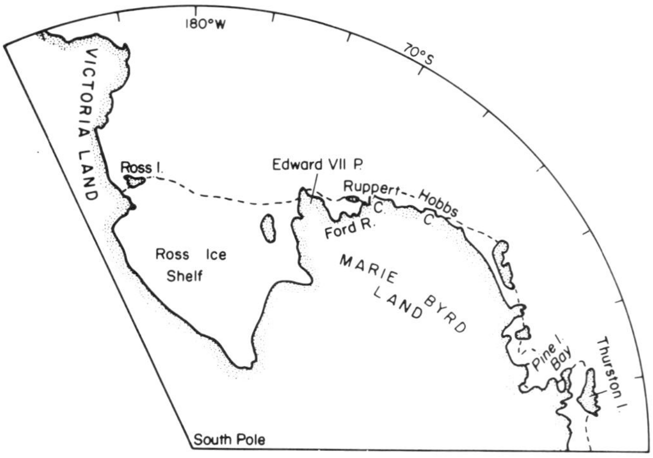

Figure 1 Location of Edward VII Peninsula and the Ford Ranges, Marie Byrd Land, West Antarctica.

1990) and is similar to the Fosdick Metamorphic Complex of the Ford Ranges, further to the E (Fig. 1). The Devonian-Carboniferous Ford Granodiorite of the Ford Ranges (Adams 1987) does not outcrop on Edward VII Peninsula but 300–360 Ma slate ages from nunataks in the central plateau area may suggest its occurrence at depth (Adams & Weaver 1990).

The granites of Edward VII Peninsula are coarse-grained, felsic, subsolvus syenogranites to monzogranites. They are undeformed, except for the development of flow foliations at their margins, and have sharp, generally steep, intrusive contacts with thermally metamorphosed Swanson Formation rocks. Geographically and petrologically, the granite nunataks fall into five groups (Bowman, Frazier, Paterson, Washington and Butler) apparently representing five separate plutons, each about 10–20 km in diameter (Fig. 2). Minor granite sheets, not obviously related to a particular pluton, occur at Mt Nilsen, Strider Rock and Breckinridge Peak which are largely composed of Swanson metasediments. Enclaves of magmatic rocks are absent but metasedimentary xenoliths occur near granite margins. Miarolitic cavities have only been observed at Mt Schlossbach. Apart from the minor development of aplite and pegmatite sheets, no other lithologies cross-cut or post-date the granites and there is a notable absence of mafic components from the whole peninsula.

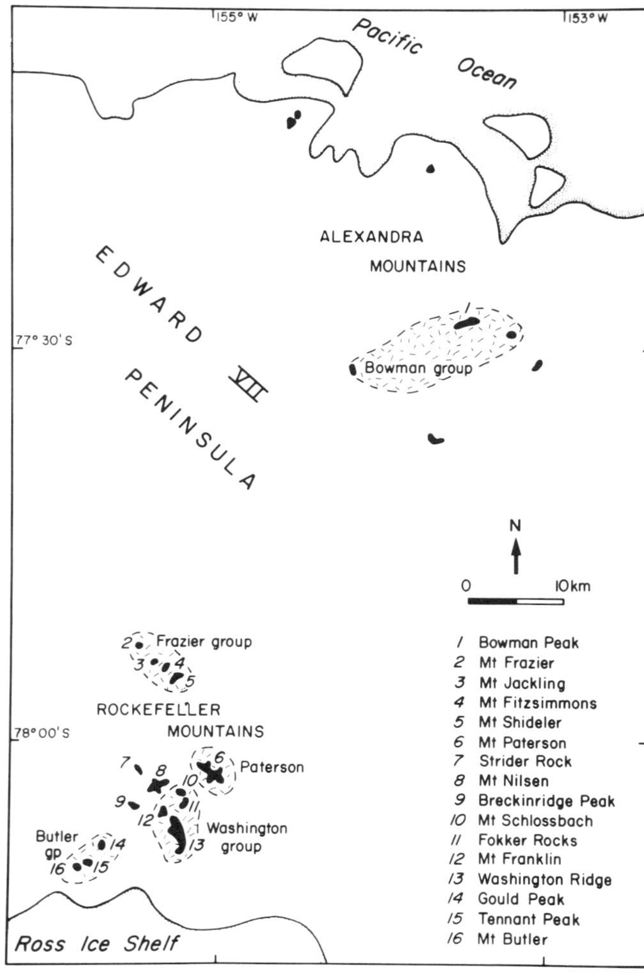

Figure 2 Distribution of granite plutons on Edward VII Peninsula; individual nunataks are numbered; these are grouped petrologically and comprise five distinct plutons.

1.2. Petrographic features

Granites of the Butler, Frazier and Paterson plutons are very coarsely crystalline grey-white monzogranites with euhedral white perthitic microcline crystals up to 50 mm in length, subordinate cream-coloured albite, smoky grey bipyramidal quartz and about 5 vol. % biotite. Granites of the Washington and Bowman plutons are coarse, equigranular, pink to buff syenogranites with salmon-pink perthitic microcline and minor cream-coloured albite. Grey bipyramidal quartz is particularly abundant (about 40%) and biotite makes up 1–3% in these very felsic rocks. There is no development of rapakivi texture. In the monzogranites, biotite is red-brown in colour which, together with the presence of accessory ilmenite, suggests relatively reducing magmatic conditions. Biotite in the syenogranites is brown to green-brown and, although ilmenite remains as the opaque phase, this and the pink colour of K-felspar suggests relatively more oxidising magmatic conditions. Other accessories include zircon, apatite, monazite and fluorite. Whole-rock compositions are weakly peraluminous with alumina saturation indices $<1{\cdot}1$.

The absence of material that could be interpreted as "restite" (mafic enclaves and calcic cores to plagioclase crystals) and the ubiquitous presence of quartz with bipyramidal terminations suggest that the granites were true melts which separated efficiently from their sources.

1.3. Geochronology and tectonic setting

Adams *et al.* (1989) reported K-Ar biotite ages for Edward VII Peninsula granites in the range 101–108 Ma. More recently, we have obtained Rb–Sr whole rock isochrons which define more precisely the crystallisation ages of the granites. These are summarised as follows (2σ error limits, MSWD $<2{\cdot}1$): Butler $97{\cdot}4 \pm 3{\cdot}2$ Ma, Frazier $96{\cdot}8 \pm 4{\cdot}1$ Ma, Washington $94{\cdot}6 \pm 2{\cdot}5$ Ma and Paterson $100{\cdot}6 \pm 3{\cdot}5$ Ma. Initial $^{87}Sr/^{86}Sr$ ratios are in the range 0·7117–0·7128. Byrd Coast Granite in the Ford Ranges has given older Rb–Sr ages which indicate two phases of intrusion, one at about 130–140 Ma and a second at about 110 Ma (Adams 1987). These rocks are also compositionally different from Edward VII Peninsula granites in having lower initial $^{87}Sr/^{86}Sr$ ratios of 0·7056–0·7079 (Weaver *et al.* 1991).

The emplacement age of the Edward VII Peninsula granites (95–100 Ma) is highly significant in terms of their tectonic setting. The New Zealand continental block separated from W Antarctica in the Late Cretaceous with

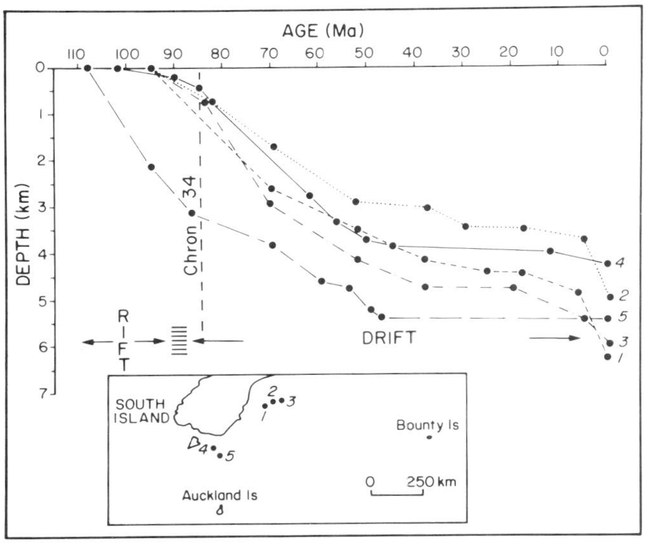

Figure 3 Compilation of hydrocarbon well subsidence curves for fault-controlled sedimentary basins on the SE New Zealand continental shelf (Field *et al.* 1989); these basins reach 4–6 km in thickness; continental rifting is thought to have coincided with the initiation of normal faulting; separation and the start of drifting occurred shortly before Chron 34.

the opening of the SW Pacific as the Campbell Plateau rifted away from Marie Byrd Land. The oldest oceanic magnetic anomaly (Chron 34) adjacent to the Campbell Plateau is dated at 84 Ma (Mayes *et al.* 1990). Clearly rifting began earlier than this. Mafic dykes emplaced parallel to the Ruppert and Hobbs coasts (Fig. 1) of Marie Byrd Land at about 90–110 Ma are thought to be rift-related (Grindley & Oliver 1983). The rift to drift history of the New Zealand continental block is one of continued crustal extension, subsidence and sedimentation. Hydrocarbon wells on he New Zealand continental shelf preserve this history (Fig. 3) and indicate that sedimentary basins began to form at about 95–110 Ma. Thus we conclude that Edward VII Peninsula magmatism coincided with the initiation of continental rifting along the Marie Byrd Land–New Zealand margin. It is also likely that the metamorphic rocks of the Fosdick and Alexandra mountains represent metamorphic core complexes related to the same extensional event.

2. Granite geochemistry and petrogenesis

2.1. Major and trace element variations and Sr and Nd isotopic ratios

Geochemical variations are illustrated by plots against SiO_2 for the 43 analysed rocks (Fig. 4) and 12 representative analyses are presented in Table 1. Analytical methods and estimates of precision for all elements are given by Weaver *et al.* (1990). Granite compositions are highly fractionated with $SiO_2 = 71–80\%$ and $FeO^* = 2{\cdot}9–0{\cdot}7\%$. The SiO_2 content of the Bowman samples is unusually high and this could possibly be accounted for by accumulation of bipyramidal quartz crystals. The minor intrusives are the least fractionated group. Throughout the suite $K_2O > Na_2O$. Concentrations of Rb, Nb, Y and HREE are high and strongly correlated with F content which reaches >6500 ppm in the most evolved compositions. These rocks have very low Ba and Sr concentrations, <20 and <5 ppm, respectively.

REE patterns (Fig. 5) vary from LREE-enriched in the least fractionated granites ($Ce_N/Yb_N = 8{\cdot}4$) with modest Eu anomalies ($Eu/Eu^* = 0{\cdot}76$), to essentially flat REE patterns with large negative Eu anomalies in the most evolved rocks ($Ce_N/Yb_N = 1{\cdot}1$, $Eu/Eu^* = 0{\cdot}02$). A feature of Figure 5 is a cross-over in the LREE patterns of the more evolved compositions. HREE, as judged by Yb, increase progressively but the LREE appear to peak in sample 10 and then decrease through sample 13 to sample 69. This suggests the late crystallisation of a phase which strongly concentrates LREE.

High precision isotopic data on four Edward VII Peninsula granites are given in Table 2. Initial $^{87}Sr/^{86}Sr$ isotopic ratios, assuming an age of 100 Ma, are in a range 0·7116–0·7206 which encompasses those ratios derived from Rb–Sr isochron measurements (see Section 1.3). This suggests continental crustal source materials with somewhat variable Rb/Sr ratios. Initial ε_{Nd} values, again calculated for a 100 Ma age of crystallisation, are strongly negative (−5·5 to −7·7) and consistent with a continental crustal source.

2.2. Generalised fractionation trends

In the Sr–Rb plot (Fig. 4), samples from individual plutons define approximately linear fields and often parallel or subparallel trends. The population does not define a single line of liquid evolution but the five plutons and the minor intrusives group do share a chemical coherence and comprise a granite suite (or arguably a supersuite). We can therefore assess only generalised fractionation trends in terms of the crystallising minerals.

Some of these trends are illustrated in Figure 6. The plot of Eu/Eu* versus Ba (Fig. 6A) illustrates the relative roles of alkali feldspar and plagioclase feldspar. Plagioclase crystallisation is dominant in the least evolved monozogranites, whereas alkali feldspar is almost entirely responsible for the trend in the more evolved monzogranites and syenogranites. Thus both plagioclase and alkali feldspar fractionation contribute to the generation of the significant negative Eu anomalies.

Crystallisation of biotite and ilmenite accounts for the reduction of Fe and Ti concentrations with increasing SiO_2 (Table 1, Fig. 4). Zircon is primarily responsible for the reduction of Zr with increasing SiO_2 (Fig. 4). Crystallisation vectors on Figure 6B indicate that zircon and apatite reduce the Y/Nb ratio, whereas ilmenite and biotite increase this ratio. The limited variation in the Y/Nb ratio suggests that the opposing effects of these minerals are approximately balanced. There is more than one order of magnitude change in the Ce/Nb ratio. The crystallisation vector for monazite (Fig. 6B) demonstrates that this mineral is responsible for this large variation. Monazite is present in trace proportions in all the monzogranites and syenogranites of Edward VII Peninsula. The cross-over of patterns in the LREE region of Figure 5 (Section 2.1) is therefore also ascribed to monazite fraction. Moreover, Th concentrations start to fall in the more evolved syenogranite compositions (Table 1), in accordance with the quoted very high partition coefficients of this element in monazite (Mittlefehldt & Miller 1983). Granite suites containing allanite in place of monazite should show similar reductions of LREE and Th concentrations in more fractionated compositions (Sawka 1988).

Another possible compositional control on trace element variations is the formation of halogen complexes. In the Edward VII Peninsula granites, Cl is low (up to 100 ppm) but F reaches 6700 ppm (Table 1). Fluoride complexes involving elements such as Ga, Nb, Zr, Zn, Y and REE

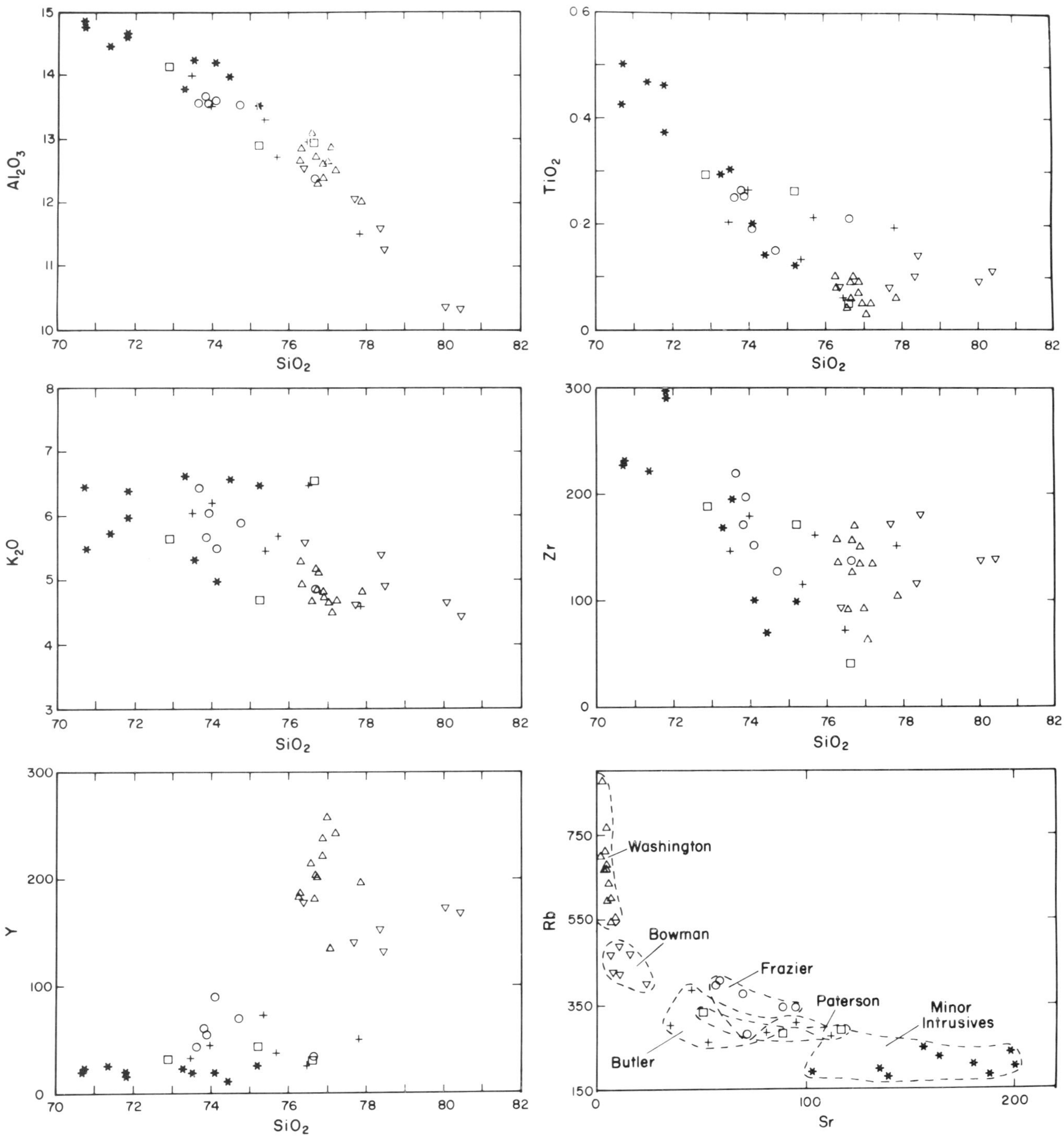

Figure 4 Selected plots to illustrate the geochemical variation of Edward VII granites; the compositional groupings of the different plutons are indicated by fields on the Rb vs Sr plot; major oxides are recalculated to 100% volatile-free.

(Collins *et al.* 1982) may have been stable in the melt and thus contributed to the geochemical variation. This possibility is difficult to evaluate but it seems not to have been a major compositional control. The elements Rb, Nb, Y and HREE have concentrations which rise in proportion to F, which simply indicates the relatively incompatible behaviour of this group. Zr decreases, and Ce increases then decreases, with F as expected if the concentrations of these elements are controlled essentially by the crystallisation of zircon and monazite + apatite, respectively. The increase in the Ga/Al ratio from monozogranites to syenogranites (10^3 Ga/Al increases from about 1 to 5) could be ascribed to the formation of fluoro-complexes of Ga that remain in the melt on crystallisation (Collins *et al.* 1982). However, a more plausible control is the crystallisation of calcic plagioclase which has a higher partition coefficient for Al than for Ga, irrespective of F activity.

In considering the fractionation trends described above, it must be emphasised that each Edward VII Peninsula pluton represents a separate magma batch that has experienced its own specific crystallisation history. The different magma batches must be related by different percentages of melting of a source which was compositionally variable at least in Rb/Sr and Sm/Nd ratios (Table 2). Therefore the percentage crystallisation calculations given on the crystallisation vectors of Figure 6 do not directly reflect the

Table 1 Representative analyses of Edward VII Peninsula (Marie Byrd Land) granites

	E7W58	E7W67	E7W69	E7W74	E7W10	E7W13	E7W100	E7W95	E7W87	E7W84	E7W90	E7W40
SiO_2	75·81	76·24	76·01	76·29	78·05	79·92	72·77	73·78	74·29	73·31	72·26	70·64
TiO_2	0·10	0·07	0·04	0·09	0·14	0·11	0·26	0·19	0·13	0·26	0·29	0·50
Al_2O_3	12·58	12·50	13·00	12·29	11·21	10·28	13·49	13·56	13·11	13·39	14·03	14·75
Fe_2O_3	1·48	1·42	1·07	1·75	1·65	1·49	2·06	2·15	1·64	1·98	2·20	2·86
MnO	<0·01	<0·01	<0·01	0·02	0·01	<0·01	0·03	0·04	0·02	0·04	<0·01	0·03
MgO	0·07	<0·05	<0·05	<0·05	0·10	0·11	0·32	0·21	0·13	0·33	0·29	0·69
CaO	0·84	0·70	0·58	0·80	0·72	0·56	1·20	1·08	0·91	1·25	1·53	1·91
Na_2O	3·25	3·42	3·88	3·25	2·70	2·47	2·79	3·03	2·94	2·33	2·85	2·83
K_2O	5·25	4·77	4·63	4·69	4·88	4·41	5·58	5·46	5·37	6·14	5·59	5·48
P_2O_5	0·02	0·01	0·01	0·02	0·03	0·02	0·08	0·07	0·05	0·08	0·09	0·19
LOI	0·59	0·49	0·36	0·32	0·05	0·78	0·86	1·04	0·68	0·66	0·89	0·14
TOTAL	100·00	99·68	99·64	99·57	99·54	100·16	99·44	100·61	99·27	99·77	100·03	100·02
ASI	1·00	1·04	1·04	1·03	1·02	1·04	1·05	1·05	1·07	1·05	1·04	1 ·05
F	6420	6750	6590	6620	4070	3630	1930	1920	1900	1410	1620	—
Cl	96	81	13	58	68	100	64	97	20	71	35	—
V	3	4	6	4	3	5	15	10	8	15	11	25
Cr	<3	<3	<3	<3	<3	<3	5	4	<3	5	4	5
Ni	<3	3	3	<3	<3	4	4	3	<3	4	4	6
Zn	57	62	51	105	47	54	37	51	36	33	29	53
Ga	31	34	36	31	24	23	20	23	22	20	22	22
Rb	552	709	875	634	399	422	341	394	386	277	288	206
Sr	9	4	3	6	24	11	95	57	45	112	117	200
Y	182	221	214	237	132	167	60	89	73	45	31	23
Zr	157	134	91	150	180	138	171	152	115	179	188	231
Nb	71	73	58	80	47	52	21	27	22	18	20	17
Ba	<20	<20	<20	<20	51	<20	408	171	140	446	446	667
La	69	58	50	67	106	77	60	46	42	58	32	52
Ce	*153·1*	*124·4*	*106·0*	*144·8*	*219·8*	*144·8*	*146·0*	*135·3*	*110·4*	*140·5*	*62·9*	*102*
Nd	*80·5*	*74·3*	*59·2*	*77·7*	*109·0*	*84·9*	*67·7*	*64·0*	*51·4*	*64·2*	*27·9*	—
Eu	*0·26*	*0·14*	*0·11*	*0·21*	*0·48*	*0·27*	*0·89*	*0·60*	*0·49*	*1·04*	*1·17*	—
Tb	*3·42*	*4·26*	*4·06*	*3·91*	*3·31*	*3·71*	*1·49*	*2·15*	*1·75*	*1·12*	*0·64*	—
Yb	*16·67*	*22·58*	*24·00*	*20·87*	*9·75*	*13·55*	*4·48*	*8·14*	*5·71*	*3·23*	*1·90*	—
Lu	*2·29*	*3·26*	*3·47*	*3·00*	*1·27*	*2·10*	*0·50*	*1·05*	*0·69*	*0·26*	—	—
Hf	*7·53*	*6·98*	*6·04*	*7·05*	*7·24*	*6·40*	*5·49*	*5·19*	*4·15*	*5·59*	*5·96*	—
Ta	*6·13*	*9·32*	*11·56*	*8·12*	*4·07*	*5·16*	*1·62*	*2·39*	*1·93*	*1·15*	*1·53*	—
Pb	57	43	53	43	40	33	38	38	42	36	33	34
Th	71·5	68·3	50·6	71·3	52·4	43·6	43·7	42·9	33·5	37·4	19·9	33

Explanation of Table 1: Major and trace elements by XRF analysis at University of Canterbury (Loss on Ignition (LOI) at 1000°C, Fe as total Fe_2O_3) except F by ion-specific electrode and italicised values by NAA (analyst I. L. Gibson, University of Waterloo). Samples as follows (prefix E7W omitted): 58 Mt Franklin, 67 Mt Schlossbach, 69 Fokker Rocks, 74 Washington Ridge, 10 & 13 Bowman Pk, 100 Mt Frazier, 95 Mt Shideler, 87 Gould Pk, 84 Mt Butler, 90 Mt Paterson, 40 Mt Nilsen. See Figure 2 for localities. ASI (Alumina Saturation Index) = mol $Al_2O_3/CaO + Na_2O + K_2O$.

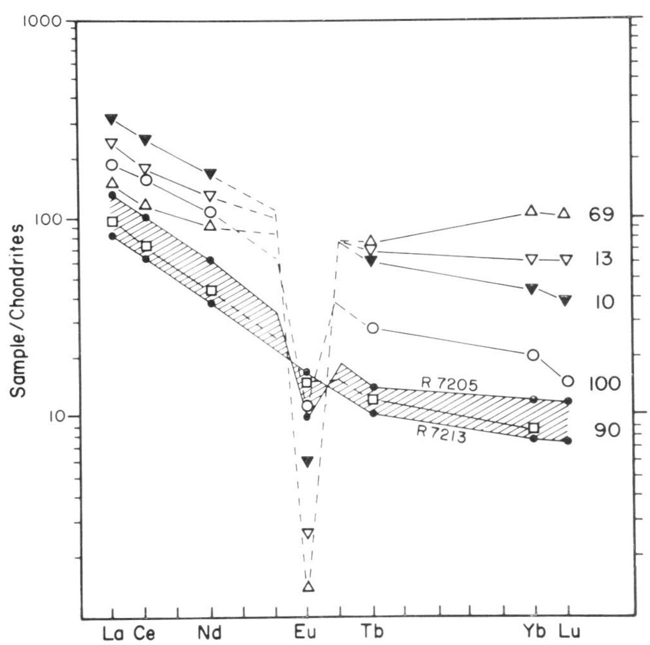

Figure 5 REE plot for selected Kind Edward VII Peninsula granites (see Table 1, symbols as in Fig. 4 except sample 10 which has closed symbol for clarity); shaded area represents range of REE concentrations in Ford Granodiorite (Table 2) defined by samples R7213 and R7205.

proportions of minerals crystallised from the melt, since they do not allow for differences in proportions of residual minerals during the melting process, or the effects of source compositional heterogeneity. The generalised fractionation trends do, however, allow the major mineral phases and their relative importance to be identified.

2.3. Source characteristics

A potential source material for Edward VII Pensinula granites which must be considered is Swanson Formation metasediment, although the weakly peraluminous compositions of the granites makes this unlikely. Adams (1986) quotes initial $^{87}Sr/^{86}Sr$ ratios for Swanson Formation slates of 0·712–0·718, using the determined cleavage age of 440 Ma. Recalculations of $^{87}Sr/^{86}Sr$ ratios to 100 Ma gives values >0·7600, greatly in excess of those observed in the Edward VII Peninsula granites. There are as yet no published Nd studies of Swanson Formation metasediments, but it seems that partial melting of these rocks cannot account for either the chemistry or isotope composition of the Edward VII Peninsula granites. Models involving contamination of more primitive magmas with Swanson Formation have not been investigated since there is a simple direct model that satisfies the present data.

The less fractionated part of the Edward VII Peninsula granite trend in Figure 6B overlaps the field denoting typical Volcanic Arc Granites (Pearce *et al.* 1984) and is close to

Table 2 Nd and Sr isotopic data on Edward VII Peninsula granites and Ford Granodiorite

Sample	Unit	Age	Sm	Nd	$^{147}Sm/^{144}Nd$	$^{143}Nd/^{144}Nd$	$^{143}Nd/^{144}Nd_i$	ε_{Nd}
E7W10	Ed VII	100	22·61	101·1	0·1352	0·512314	0·512226	−5·6
E7W67	Ed VII	100	20·96	66·82	0·1896	0·512325	0·512201	−6·1
E7W74	Ed VII	100	20·86	69·85	0·1805	0·512348	0·512236	−5·5
E7W87	Ed VII	100	12·23	51·66	0·1432	0·512233	0·512139	−7·3
E7W90	Ed VII	100	5·485	27·04	0·1226	0·512196	0·512116	−7·7
D1/3	Ford	350	6·097	30·45	0·1210	0·512336	0·512059	−2·5
		100					0·512257	−5·0
R7205	Ford	350	7·091	36·82	0·1164	0·512275	0·512008	−3·5
		100					0·512199	−6·1
R7460	Ford	350	5·492	31·78	0·1045	0·512282	0·512043	−2·9
		100					0·512214	−5·8

Sample	Unit	Age	Rb	Sr	Rb/Srwt	$^{87}Rb/^{86}Sr$	$^{87}Sr/^{86}Sr$	$^{87}Sr/^{86}Sr_i$
E7W10	Ed VII	100	391	25·3	15·47	45·10	0·784722	0·720634
E7W87	Ed VII	100	372	44·8	8·300	24·10	0·746910	0·712658
E7W90	Ed VII	100	279	112	2·494	7·22	0·721834	0·711568
D1/3	Ford	350	113	345	0·3256	·9423	0·710584	0·705889
		100						0·709245
R7205	Ford	350	279	132	2·116	6·138	0·736557	0·705974
		100						0·727834
R7460	Ford	350	120	287	0·4181	1·210	0·711827	0·705798
		100						0·710107

Note: $^{87}Sr/^{86}Sr$ and $^{143}Nd/^{144}Nd$ ratios were determined on Finnegan MAT 262 and VG 354 instruments at NIGL Keyworth, U.K., and are accurate to better than ±0·01% and ±0·005%, respectively.

the average crustal ratio of Taylor and McLennan (1985). An obvious potential crustal source, and one of typical VAG composition, is the Devonian-Carboniferous Ford Granodiorite (Weaver *et al.* 1991). This is a tonalite-granodiorite, biotite-rich, hornblende-bearing suite of 350–380 Ma age. As noted in Section 1.1, Ford Granodiorite does not outcrop on Edward VII Peninsula but may occur at depth, since there are extensive outcrops less than 200 km to the E in the Ford Ranges where a deeper crustal level is exposed. Granites of similar age and petrology also occur in northern Victoria Land (the Admiralty Intrusives) on the western side of the Ross Sea. Some representative analyses of Ford Granodiorite lithologies are given in Table 3. These are typically I-type granites, in many respects similar to those of the Lachlan and Caledonian fold belts (White & Chappell 1983; Chappell & Stephens 1988).

New Sr and Nd isotopic data for three Ford Granodiorite samples are presented in Table 2. Initial $^{87}Sr/^{86}Sr$ ratios are in the range 0·7058–0·7060, assuming an age of 350 Ma, and these agree well with the intercept ratios derived from the isochrons of Adams (1987). Initial ε_{Nd} values are in the range −2·5 to −3·5. These initial ratios and values have also been recalculated to 100 Ma, the age of Edward VII Peninsula magmatism (Table 2). The range of initial Sr isotopic ratios of Ford Granodiorite samples (at 100 Ma), 0·7092–0·7278, incorporates the range of Edward VII Peninsula granites (0·7116–0·7206). The variation in ε_{Nd} for the Ford Granodiorite samples at this time, −5·0 to −6·1, partly overlaps the Edward VII Peninsula granite range, −5·5 to −7·7. Isotopically, therefore, the Ford Granodiorite is a plausible source for the Edward VII Peninsula granites.

It is interesting to note that the envelope for REE patterns of Ford Granodiorite samples overlaps patterns of the least evolved Edward VII Peninsula granites (Fig. 5). Sample R7213, a granodiorite with $SiO_2 = 64{\cdot}8\%$, $Ce_N/Yb_N = 8{\cdot}21$ and $Eu/Eu^* = 0{\cdot}89$, has a very similar REE composition to sample 90, a Mt Paterson monzogranite with $SiO_2 = 72{\cdot}3\%$, $Ce_N/Yb_N = 8{\cdot}41$ and $Eu/Eu^* = 0{\cdot}76$. Sample R7213 is not considered to be parental to sample 90 (Fig. 5) because an unreasonably large degree of melting would be required to generate the REE concentrations of the latter. However, it is likely that 40–60% melting of a more primitive Devonian-Carboniferous tonalitic source (with similar Ce_N/Yb_N to R7213) could generate a Cretaceous monzogranite. This would satisfy the isotopic constraints and account for the similarities in REE patterns between source and derivative melt.

3. Classification of Edward VII Peninsula granites

3.1. A-types or I-types or both?

On the discrimination diagrams of Whalen *et al.* (1987), Edward VII Peninsula samples plot mostly within the fields for A-type granites. Two examples of these diagrams are illustrated in Figure 7. On the FeO*/MgO versus Zr + Nb + Ce + Y plot (Fig. 7A), some of the minor intrusions fall within the field of unfractionated granite types. Several monzogranites plot in the field denoting fractionated I- and S-types, whereas all syenogranites plot within the A-type field. In Figure 7B, the Edward VII Peninsula rocks form a linear array trending away from the M-, I- and S-type field. The average A-type of Whalen *et al.* (1987) falls within the Edward VII Peninsula array and the least fractionated samples plot close to average felsic I- and S-types.

The Edward VII Peninsula granites illustrate the difficulty of classifying highly fractionated biotite granites in terms of granite type. The weak peraluminous composition and the lack of modal Al-rich minerals (cordierite, Al_2SiO_5 polymorphs) preclude an S-type designation despite elevated initial $^{87}Sr/^{86}Sr$ ratios. The syenogranites would on their own be classed as A-types, but their association with monzogranites which trend back towards normal granite compositions suggests that the Edward VII Peninsula suite could also be regarded as a highly fractionated (felsic) I-type. Similar examples have been documented by Whalen *et al.* (1987) (Ackley Granite, Newfoundland) and Sawka *et al.* (1990) (Housetop and Meredith batholiths, Tasmania).

A basic principle of igneous petrology is that magmas (or melts) reflect the composition of their source materials. Edward VII Peninsula granites retain some of the compositional characteristics of their I-type sources (see

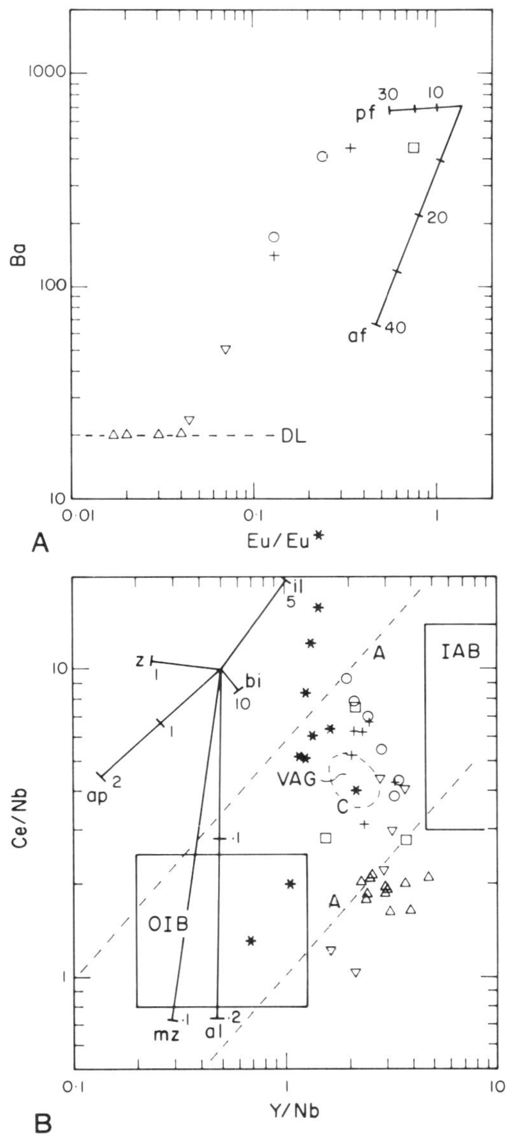

Figure 6 (A) Eu/Eu* versus Ba (ppm) plot showing alkali feldspar (af) and plagioclase feldspar (pf) fractionation trends calculated using the partition coefficients of Henderson (1982) and Mahood & Hildreth (1983); ticks mark percent crystallisation; DL = Ba detection limit.
(B) Y/Nb versus Ce/Nb plot after Eby (1990) for Edward VII Peninsula granites (symbols as in Fig. 4); OIB and IAB are fields of Oceanic Island Basalts and enriched Island Arc Basalts; A-type granites of Eby (1990) plot in a linear array between these fields; VAG is field of Volcanic Arc Granites constructed from analyses of Pearce *et al.* (1984) and after Eby (1990); C is average crustal composition from Taylor and McLennan (1985); fractionation trends showing percent crystallised have been calculated for biotite (bi), ilmenite (il), zircon (z), apatite (ap), monazite (mz) and allanite (al), using partition coefficients for Ce, Nb and Y as follows: biotite (2·8, 1·35, 0·35), ilmenite (1·2, 18, 0·4), zircon (2·4, 4·8, 85), apatite (42, 0·24, 71), monazite (2633, 5, 527) and allanite (1278, 3·2, 6·4); these were obtained from the data of Mahood & Hildreth (1983), Mittlefehldt & Miller (1983) and Sawka (1988); in some cases coefficients for Y and Nb were derived assuming that these elements behaved like a HREE and Ta respectively; note that allanite is not present in Edward VII Peninsula granites.

Section 2.3) but because of extensive fractionation the image is faint and distorted. Is this sufficient reason to give them a separate designation as A-types? Creaser *et al.* (1991) regard correctly A-types as a sub-group of I-type granites. M-types (Chappell & Stephens 1988) should also be regarded as an I-subtype, simply on the basis that their sources are igneous. Eby (1990), in his review of A-type

Table 3 Representative analyses of Ford Granodiorite, Ford Ranges, Marie Byrd Land

	D1/3	R7460	R7205	R7213
SiO_2	64·08	70·06	72·73	64·79
TiO_2	0·80	0·44	0·28	0·76
Al_2O_3	15·35	14·16	13·69	15·42
Fe_2O_3	4·80	2·88	1·85	4·71
MnO	0·04	0·55	0·04	0·08
MgO	2·85	1·13	0·31	2·17
CaO	4·46	2·58	1·35	4·10
Na_2O	3·46	3·92	3·37	3·78
K_2O	2·84	3·30	5·11	2·53
P_2O_5	0·02	0·14	0·10	0·26
LOI	0·43	0·56	0·36	0·86
TOTAL-	99·30	99·22	99·19	99·46
ASI	0·91	0·96	1·01	0·94
V	98	40	20	74
Cr	81	—	—	—
Ni	13	—	—	—
Zn	30	47	30	67
Ga	18	17	19	19
Rb	122	129	304	114
Sr	366	306	142	431
Y	28	22	31	20
Zr	153	145	158	162
Nb	8	15	20	17
Ba	433	533	395	464
La	37	44	46	29
Ce	*67·2*	*84·5*	91·6	*58·5*
Nd	*31·6*	*33·4*	40·2	*27·2*
Eu	*1·40*	*1·03*	0·84	*1·22*
Tb	*0·68*	*0·49*	0·68	*0·52*
Yb	*2·55*	*2·11*	3·00	*1·81*
Lu	*0·37*	*0·30*	0·41	*0·26*
Hf	*4·86*	*4·08*	4·53	*4·93*
Ta	*0·87*	*1·32*	2·26	*0·75*
Pb	11	22	28	14
Th	12·7	16·6	24·3	8·61

Notes: See Table 1 for analytical methods. Samples as follows: D1/3 Saunders Mt, R7460 Mt Warner, R7205 Mt Swan, R7213 Mt McCormick. See Weaver *et al.* (1991) for sample localities.

granites, identified a wide compositional diversity. In terms of ratios of incompatible and moderately incompatible trace elements, A-type granites plot in linear arrays between the fields of Ocean Islands Basalts and Island Arc Basalts (Fig. 6B). It seems that these highly fractionated crustal magmas still bear some geochemical imprint of their ultimate mantle sources. Edward VII Peninsula granites, like many A-types, have subduction-related geochemical signatures inherited from their immediate crustal source materials. Other A-types, including most peralkaline granites, have trace-element ratios similar to asthenospheric mantle sources. Despite the wide variety of A-type granite compositions identified, Eby (1990) related all to I-type sources. Thus the term *A-subtype* (rather than A-type) is more correct and in keeping with the fundamental division of granites into two types (I and S) according to the nature of their sources (White & Chappell 1983). The only unifying character of the A-subtypes is their anorogenic tectonic environment.

3.2. Tectonic discrimination diagrams

Most of the Edward VII Peninsula granites plot in the Within-Plate Granite (WPG) field on the widely used tectonic discrimination plots of Pearce *et al.* (1984) (Fig. 8). Assuming that Within-Plate Granites are synonymous with anorogenic granites, this designation is entirely compatible with the continental rift setting of the Edward VII Peninsula rocks.

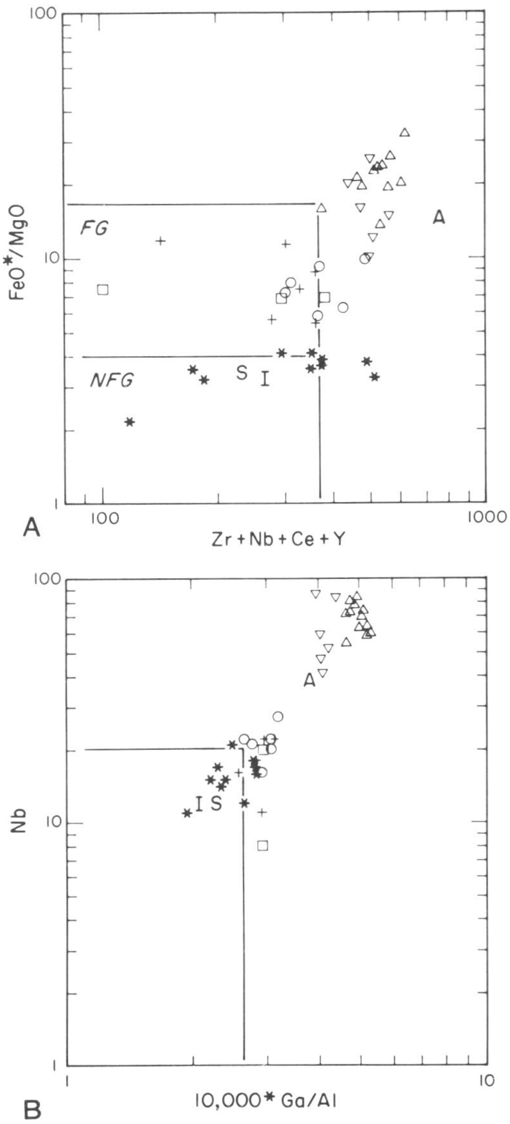

Figure 7 (A) Zr + Nb + Ce + Y (ppm) versus FeO*/MgO for Edward VII Peninsula granites (symbols as in Fig. 4) after Whalen *et al.* (1987); FG and NFG are fields for fractionated felsic granites and unfractionated M-, I- and S-type granites; S, I and A mark the average compositions of felsic S-type, felsic I-type and A-type granite.
(B) 10000 Ga/Al versus Nb (ppm) for Edward VII Peninsula granites; field is for M-, I- and S-type granites (after Whalen *et al.* (1987)).

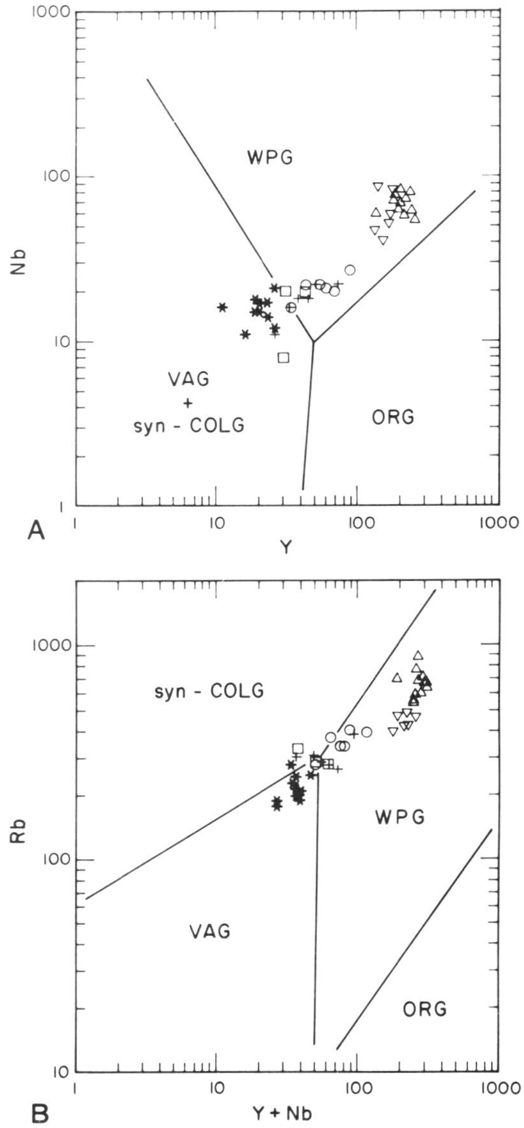

Figure 8 (A) Y (ppm) versus Nb (ppm) tectonic discrimination plot of Pearce *et al.* (1984): ORG—Ocean Ridge Granites, VAG—Volcanic Arc Granites, syn-COLG—syn-collision Granites, WPG—Within-Plate Granites; symbols as in Figure 4.
(B) Y + Nb (ppm) versus Rb (ppm) tectonic discrimination plot of Pearce *et al.* (1984).

The uniqueness of the association of tectonic setting and geochemical signature for mafic volcanic rocks has been questioned by many authors (e.g. Arculus 1987). This is a particular problem when mantle imprinted with the geochemical signature of subduction processes is reactivated during a subsequent and different tectonic regime. The problem is even more acute with granites which may be the products of a number of cycles of crustal magmatism. An inferred tectonic setting may be that of the source rather than the granites themselves (Chappell & Stephens 1988). Returning to Figure 8, although most of the Edward VII Peninsula granites suggest a WPG setting, the least fractionated samples, i.e. those closest to the composition of the crustal sources, plot in the Volcanic Arc Granite fields. This is compatible with the inferred Ford Granodiorite source (Section 2.3), samples of which plot logically in VAG fields (Weaver *et al.* 1991). In Figure 8 the Edward VII Peninsula data define distinct fractionation trends from the VAG to the WPG fields. This simply reflects the relatively incompatible nature of the elements concerned (Nb, Rb, Y) and the extreme degree of fractionation. The lesson here is that the position of the least fractionated samples represents the *source* and the trend to the most fractionated rocks represents the *process* (i.e. remelting of an evolved crustal source and extensive crystal fractionation). It is the process which correlates with the tectonic environment in this case.

The example of the Edward VII Peninsula granites illustrates the inherent dangers of making tectonic inferences from such discrimination diagrams. With no independent evidence concerning tectonic setting, and if the least fractionated granites only were represented, a logical but erroneous conclusion would be that the granites are of VAG affinity and directly related to subduction. If the most fractionated granites only were represented, the correct within-plate setting would be inferred. Given the complete set of samples as plotted in Figure 8, one might conclude, again erroneously, that they are an extremely fractionated suite of granites emplaced in an active margin.

4. Concluding remarks

The age and geological setting of Edward VII Peninsula granites lead to the conclusion that they represent anorogenic magmatism related to the initiation of continental rifting in the mid-Cretaceous. The geochemistry of the granites is consistent with an anorgenic, A-type designation. However, it is preferable to use the term *A-subtype* in recognition of the I-type nature of the sources of all anorgenic granites. Use of the term M-subtype rather than M-type is also advisable.

Edward VII Peninsula granites are similar to many high Y/Nb, A-subtype granites (Eby 1990) that have subduction-generated geochemical signatures inherited from their crustal sources. The probable source of the Cretaceous granites is I-type granodiorite-tonalite of Devonian–Carboniferous age. This model is in general accord with the suggestions of Anderson (1983) and Creaser *et al.* (1991) that anorogenic granites may be produced by melting of such crustal rocks. Creaser *et al.* (1991) calculate that 15–40% partial melting under vapour-absent conditions at about 900–950°C is required. This necessitates high geothermal gradients and the introduction of mantle-derived mafic melts into the crust. Mafic igneous rocks do not crop out on Edward VII Peninsula, but the degree of exposure is very limited. Mafic dykes of mid-Cretaceous age and layered alkaline gabbros (age unknown) are a prominent feature of the Ruppert and Hobbs coasts (Spörli & Craddock 1981) together with Cretaceous A-subtype granites and older granodiorite bodies that may be late Palaeozoic in age. New field observations (Bradshaw *et al.* 1991) of widespread mingling and mixing phenomena along the Ruppert and Hobbs coasts demonstrate the contemporaneity of mafic and granitic magmas, and give support to our petrogenetic model for the Edward VII Peninsula granites.

5. Acknowledgements

DSIR Antarctic and the pilots and crew of the VXE-6 Squadron, US Navy, are thanked for their logistic support for the 1987/88 NZARP expedition to Edward VII Peninsula. S.D.W. and C.J.A. thank Paul Broady and Peter Cleary for their assistance in the field. Darrin Gordon kindly performed the ISE determinations and Stephen Brown the XRF analyses. Ian Millar assisted with Sr–Nd analyses at the NERC Isotope Geosciences Laboratory, Keyworth: this paper constitutes NIGL publication No. 47. Discussions with John Bradshaw on New Zealand–Marie Byrd Land tectonics and with Nelson Eby on A-type granites have been particularly instructive.

6. References

Adams, C. J. 1986. Geochronological studies of the Swanson Formation of Marie Byrd Land, West Antarctica, and correlation with northern Victoria Land, East Antarctica, and South Island, New Zealand. NZ J GEOL GEOPHYS **29,** 345–58.

Adams, C. J. 1987. Geochronology of granite terranes in the Ford Ranges, Marie Byrd Land, West Antarctica. NZ J GEOL GEOPHYS **30,** 51–72.

Adams, C. J. & Weaver, S. D. 1990. Age and correlation of metamorphic basement in Edward VII Peninsula, Marie Byrd Land, West Antarctica, and correlation with Northern Victoria Land and Southern New Zealand. ZBL GEOL PALAONT TEIL I **1/2,** 75–86.

Adams, C. J., Broady, P., Cleary, P. J. & Weaver, S. D. 1989. Geological and biological expedition to Edward VII Peninsula, Marie Byrd Land, West Antarctica—1987/88: Field observations and initial results. NZ ANTARCTIC RECORD **9,** 5–33.

Anderson, J. L. 1983. Proterozoic anorogenic granite plutonism of North America. GEOL SOC AM MEM **161,** 133–54.

Arculus, R. J. 1987. The significance of source versus process in the tectonic controls of magma genesis. *In* Weaver, S. D. & Johnson, R. W. (eds) *Tectonic controls on magma chemistry,* 1–12. Amsterdam: Elsevier.

Bradshaw, J. D., Dalziel, I. W. D., Di Venere, V., Mukasa, S. B., Pankhurst, R. J., Storey, B. C. & Weaver, S. D. 1991. The southern rim of the Pacific: new work on the pre-Cenozoic rocks of Marie Byrd Land. ABSTR, 6TH INT SYMP ANTARCT EARTH SCI, 80–2. Tokyo: Natinal Institute of Polar Research.

Chappell, B. W. & Stephens, W. E. 1988. Origin of infracrustal (I-type) granite magmas. TRANS R SOC EDINBURGH EARTH SCI **79,** 71–86.

Collins, W. J., Beams, S. D., White, A. J. R. & Chappell, B. W. 1982. Nature and origin of A-type granites with particular reference to southeastern Australia. CONTRIB MINERAL PETROL **80,** 189–200.

Creaser, R. A., Price, R. C. & Wormald, R. J. 1991. A-type granites revisted: Assessment of a residual-source model. GEOLOGY **19,** 163–6.

Eby, G. N. 1990. The A-type granitoids: A review of their occurrence and chemical characteristics and speculations on their petrogenesis. LITHOS **26,** 115–34.

Field, B. D., Browne, G. H. & others. 1989. Cretaceous and Cenozoic sedimentary basins and geological evolution of the Canterbury Region, South Island, New Zealand. *New Zealand Geological Survey Basin Studies 2.* Lower Hutt: NZ Geological Survey.

Grindley, G. W. & Oliver, P. J. 1983. Palaeomagnetism of Cretaceous volcanic rocks from Marie Byrd Land, Antarctica. *In* Oliver, R. L., James, P. R. & Jago, J. D. (eds) *Antarctic Earth Science,* 133–9. Cambridge: Cambridge University Press.

Henderson, P. 1982. *Inorganic Geochemistry.* Oxford: Pergamon.

Loiselle, M. C. & Wones, D. R. 1979. Characteristics and origin of anorogenic granites. GEOL SOC AM ABSTR WITH PROG **11,** 468.

Mahood, G. & Hildreth, W. 1983. Large partition coefficients for trace elements in high-silica rhyolites. GEOCHIM COSMOCHIM ACTA **47,** 11–30.

Mayes, C. L., Lawver, L. A. & Sandwell, D. T. 1990. Tectonic history and new isochron chart of the South Pacific. J GEOPHYS RES **95,** 8543–67.

Mittlefehldt, D. W. & Miller, D. F. 1983. Geochemistry of the Sweetwater Wash Pluton, California: Implication for "anomalous" trace element behaviour during differentiation of felsic magmas. GEOCHIM COSMOCHIM ACTA **47,** 109–24.

Pearce, J. A., Harris, N. B. W. & Tindle, A. G. 1984. Trace element discrimination diagrams for the tectonic interpretation of granitic rocks. J PETROL **25,** 956–83.

Sawka, W. N. 1988. REE and trace element variations in accessory minerals and hornblende from the strongly zoned McMurry Meadows Pluton, California. TRANS R SOC EDINBURGH EARTH SCI **79,** 157–68.

Sawka, W. N., Heizler, M. T., Kistler, R. W. & Chappell, B. W. 1990. Geochemistry of highly fractionated I- and S-type granites from the Tin-Tungsten province of western Tasmania. *In* Stein, H. J. & Hannah, J. L. (eds) *Ore-bearing granite systems; Petrogenesis and mineralizing processes.* GEOL SOC AM SPEC PAP **246,** 161–179.

Spörli, K. B. & Craddock, D. 1981. Geology of the Ruppert Coast, Marie Byrd Land, Antarctica. *In* Cresswell, M. M. & Vella, P. (eds) *Gondwana V,* 243–50. Rotterdam: Balkema.

Taylor, S. R. & McLennan, S. M. 1985. *The continental crust: Its composition and evolution.* Oxford: Blackwell.

Wade, F. A., Cathey, C. A. & Oldham, J. B. 1977. Reconnaissance geologic map of the Alexandra Mountains quadrangle, Marie Byrd Land, Antarctica, 1:250,000. Antarctic Geologic Map A-5, US Antarctic Research Program.

Weaver, S. D., Gibson, I. L., Houghton, B. F. & Wilson, C. J. N. 1990. Mobility of rare earth and other elements during crystallization of peralkaline silicic lavas. J VOLCANOL GEOTHERM RES **43,** 57–70.

Weaver, S. D., Bradshaw, J. D. & Adams, C. J. 1991. Granitoids of the Ford Ranges, Marie Byrd Land, Antarctica. *In* Thomson, M. R. A., Crame, J. A. & Thomson, J. W. (eds) *Geological Evolution of Antarctica,* 345–51. Cambridge: Cambridge University Press.

Whalen, J. B., Currie,K. L. & Chappell, B. W. 1987. A-type granites: geochemical characteristics, discrimination and petrogenesis. CONTRIB MINERAL PETROL **95,** 407–19.

White, A. J. R. & Chappell, B. W. 1983. Granitoid types and their distribution in the Lachlan Fold Belt, southeastern Australia. GEOL SOC AM MEM **159,** 21–34.

S. D. WEAVER, Department of Geology, University of Canterbury, Christchurch, New Zealand.
C. J. ADAMS, Department of Scientific and Industrial Research, Physical Sciences, Lower Hutt, New Zealand.
R. J. PANKHURST, British Antarctic Survey, c/o Natural Environment Research Council, Isotope Geosciences Laboratory, Keyworth, Nottingham NG12 5GG, U.K.
I. L. GIBSON, Department of Earth Sciences, University of Waterloo, Waterloo, Ontario, N2L 3G1, Canada.

Transactions of the Royal Society of Edinburgh: Earth Sciences, **83**, 291–304, 1992

Triassic "Gondwana" granites of the Gastre district, North Patagonian Massif

C. W. Rapela, R. J. Pankhurst and S. M. Harrison

ABSTRACT: Granites of the Batholith of Central Patagonia were emplaced into late Precambrian metamorphic basement rocks and Palaeozoic orthogneisses (here dated by Rb–Sr whole-rock errorchrons at 346 ±35 Ma and 267 ±27 Ma), and are in fault-contact with younger volcanic rocks, mostly andesites. Two main suites of granites are recognised: both are much younger than their previously-supposed Late Palaeozoic age. The Gastre Suite is composed predominantly of hornblende-biotite granodiorite and monzogranite, often slightly foliated, and has yielded a Rb-Sr whole-rock isochron age of 220 ±3 Ma. The Lipetrén Suite includes biotite monzogranite and leucogranite, grading into quartz-feldspar porphyries and felsites, and has been dated at 208 ±1 Ma. A minor granodioritic unit yields an apparently Middle Jurassic age of 172 ±15 Ma. Textural evidence and hornblende geobarometry confirm that these are high-level sub-volcanic intrusions. Metaluminous compositions are common in the Gastre Suite, but are subordinate to highly siliceous (>70% SiO_2) and peraluminous varieties in the Lipetrén Suite. Despite this compositional bias, the granites are almost entirely calcalkaline and I-type, and have volcanic-arc rather than intraplate or collisional trace element characteristics. Initial $^{87}Sr/^{86}Sr$ ratios of 0·706 and ε_{Nd_t} values of −2·5 are also relatively primitive and are thought to indicate a juvenile crustal contribution (Nd "depleted mantle" model ages are less than 1,000 Ma).

The Triassic "Gondwana" magmatic episode is thus not an expression of Permo-Triassic collision of an allochthonous Patagonian terrane with the rest of southern S America, but may be related to the initial stages of supercontinent rifting.

KEY WORDS: Patagonia, batholith, I-type, calkalkaline, Triassic/Jurassic, rifting.

Pre-Andean magmatism in northern Patagonia is characterised by voluminous silicic plutonic and volcanic suites of Late Palaeozoic to Jurassic age, which constitute the Gondwana magmatic cycle of Patagonia (Llambías *et al.* 1984). The Jurassic silicic volcanic rocks (Tobífera or Chon-Aike group), which terminate this cycle, cover much of Patagonia and represent one of the largest rhyolitic provinces of the world; these were erupted just before the break-up of Gondwana (Dalziel *et al.* 1987; Uliana & Biddle 1987). To the N of the Río Colorado, the natural geographical limit of Patagonia, the Gondwana plutonic belts are parallel to the modern Andean chain (Pankhurst *et al.* 1988; Parada 1990) but, S of 39°S, discontinuous outcrops of the Gondwana magmatic belts diverge significantly southeastwards (see Fig. 1). South of the Río Colorado, two main Gondwana batholiths are recognised on the margins of the North Patagonian Massif (NPM)—the Somuncura Batholith and the Batholith of Central Patagonia (BCP). These older plutonic rocks of the Gondwana cycle intrude a pre-Mesozoic basement composed of schists, gneisses and granites (Cushamen and Mamil Choique formations) whose exact age is still uncertain: scattered K–Ar and model Rb–Sr ages indicate middle to late Palaeozoic ages, whereas whole-rock Rb–Sr isochrons of metamorphic rocks suggest the presence of basement rocks as old as Proterozoic (Linares *et al.* 1988; see also a recent review by Cingolani *et al.* 1991).

The best-exposed sector of the BCP is the Gastre area (Fig. 2) towards the NE, and SW outcrops are poorer and disappear beneath later volcanic sequences and sedimentary basins. Geochronological and geochemical evidence presented here (see Section 8 for analytical methods) show that the BCP is composed primarily of Late Triassic–Early Jurassic calcalkaline granitic suites that were emplaced as sub-volcanic intrusions into a local basement of Late Palaeozoic deformed granites and migmatites. Silicic ignimbrites interlayered with pyroclastic rocks carrying Late Triassic *Dicroidium* flora are most probably consanguineous with some of the BCP plutonic units (see Fig. 2). They are overlain by predominantly felsitic, ignimbritic and andesitic volcanic rocks of the Marifil and Taquetrén formations of the Chon-Aike series. Structural observations indicate that the BCP was emplaced along an active NW–SE-trending dextral shear-zone system up to 40 km wide (Rapela *et al.* 1991, see Fig. 2). Bands of mylonitic rocks and felsites are associated with the main shear zones. Emplacement and petrogenesis are discussed in relation to the earliest stage of rifting in the break-up of Gondwana.

1. Country rocks

Orthogneisses and foliated granites crop out to the S and W of Gastre, and extend into the southwestern part of the shear-zone system, although no direct contacts are seen. They are distinguished from the younger granites of the BCP by their petrographical characteristics and a more generally pervasive foliation or schistosity. In previous studies, all the supposed Palaeozoic outcrops were assigned to a single unit, the Mamil Choique Formation (Proserpio 1978). However, our observations and geochronological results suggest the existence of at least two separate units exposed around the dried-up bed of Laguna del Toro and in the hills of the Sierra del Medio, respectively (Fig. 2). Petrological and geochemical data for these Palaeozoic igneous rocks are compared with those of the later granites in Figures 4 and 6–10.

1.1. Laguna del Toro

These rocks have compositions ranging from tonalite to two-mica granite, with grey, white and pink varieties. They have an equigranular texture. Rb-Sr data for five samples of foliated biotite-granodiorite gneiss from the area SW of Gastre (Table 1) give a relatively poor fit to an isochron model (MSWD = 6·8). The excess scatter thus indicated was allowed for by multiplying error estimates by the square-root of MSWD, giving an age of 346 ±35 Ma and an initial $^{87}Sr/^{86}Sr$ ratio of 0·7067 ±0·0006 (Fig. 3(a)).

1.2. Sierra del Medio

This has been described previously by Llambías *et al.* (1984). The main rock types are tonalites and granites *sensu stricto,* mostly of equigranular texture but with minor development of porphyroids. Rb-Sr data for eight samples of the granodioritic gneiss of the Sierra del Medio give a good isochron fit (MSWD = 1.1), but with rather a limited range of Rb/Sr ratios (Fig. 3(b)). This results in only moderate definition of the age: 267 ±27 Ma with an initial $^{87}Sr/^{86}Sr$ ratio of 0.7068 ±0·0005. It is possible, but in our view highly unlikely, that the Sierra del Medio granitic magma was emplaced at an earlier time (with a lower initial $^{87}Sr/^{86}Sr$ ratio) and that the 267 ±27 Ma isochron represents perfect homogenisation during a metamorphic event at this time. However, a much simpler interpretation is that the isochron represents the true time of emplacement. Thus, although the initial $^{87}Sr/^{86}Sr$ ratio is comparable with that of the Laguna del Toro granite gneiss, the two data sets are incompatible and do not allow for cogenesis.

2. The Batholith of Central Patagonia

The granitic rocks of the BCP included in this study comprise two distinct superunits, together with a minor unit that apparently belongs to neither of these. The two main suites are distinguished partly on the basis of petrographical characteristics, but also on their chemical and isotopic features. The Rb-Sr data support the idea that the units constituting each of these suites were synchronous and comagmatic, and that the main period of emplacement was from late Triassic to earliest Jurassic times.

2.1. Gastre Suite

The Gastre Superunit comprises two units, composed respectively of foliated, grey hornblende-biotite granites and associated leucogranites (Fig. 4). The pale grey to pink leucogranites occur as minor bodies with sharply intrusive contacts against the foliated granodiorites. Modes overlap those of the Lipetrén Suite red granites (see below). Both units are notably richer in K-feldspar than the Palaeozoic basement granites, but still within the range of compositions found in the Andean Patagonian Batholith (Fig. 4).

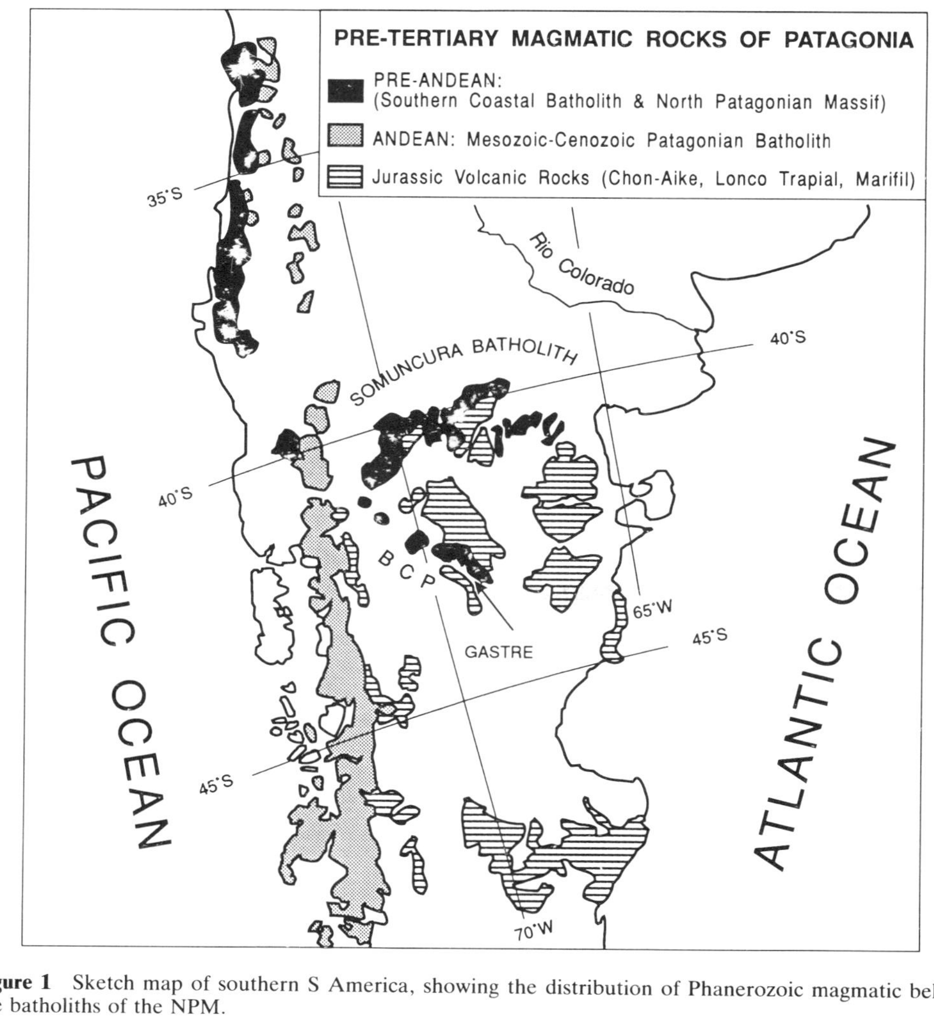

Figure 1 Sketch map of southern S America, showing the distribution of Phanerozoic magmatic belts and the batholiths of the NPM.

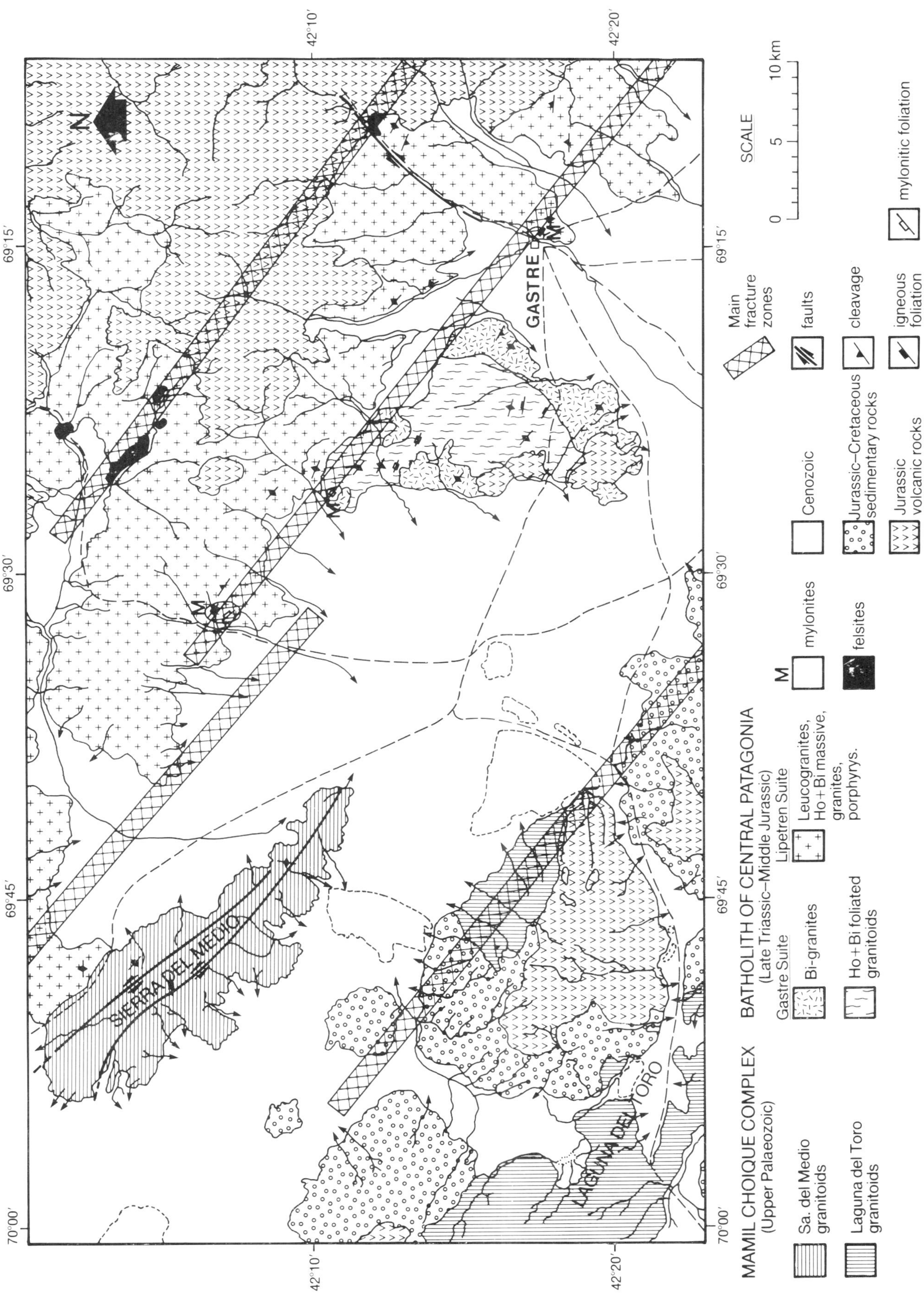

Figure 2 Geological sketch map of the Gastre area.

2.2. Lipetrén Suite

The Lipetrén Superunit contains four units: biotite-hornblende granites, leucogranophyres, porphyries and felsites.

2.2.1. Biotite-hornblende granites. These are fine- to medium-grained, porphyritic, grey/pink granites. In comparison with the other Mesozoic granites plotted in Figure 4, these rocks contain less modal quartz and K-feldspar, so that they fall close to the granodiorite field.

2.2.2. Granophyric leucogranite. This is the most extensive unit in outcrop and consists of medium-grained pink or red granite, with numerous aplite and felsite dykes and veins that often have diffuse contacts. Textures are subsolvus, with abundant miarolitic cavities (1–2 mm) and simplectic intergrowths between quartz, plagioclase and K-feldspar. There are also frequent granophyric textures, with albitisation and formation of replacement perthite. Modal compositions (Fig. 4) plot in a tight cluster coincident with those for granites of the Lipetrén Formation in its type area 150 km NW of Gastre (Rapela *et al.* 1991; Rapela & Alonso 1991).

2.2.3. Porphyries. These grey and pink rocks are characterised by the presence of phenocrysts of plagioclase accompanied by biotite and amphibole, in a fine-grained quartzo-feldspathic matrix: they include the most mafic rock-types of the suite. With reduction in the proportion of phenocrysts, textures grade into those of the felsites.

2.2.4. Felsites. These fine-grained rocks crop out as massive rocks in and around the structural lineaments. They are pink or brown in colour, similar to the leucogranites against which they have sharp contacts. In the area of La Horqueta they show flow-banding and contain disrupted auto-clasts 10–20 mm in size. They are composed of a microcrystalline matrix of quartz, feldspar and other minerals impossible to identify in thin-section.

Table 1 Rb–Sr analytical data for Gastre granites.

Sample	Type	Rb ppm	Sr ppm	$^{87}Rb/^{86}Sr$	$^{87}Sr/^{86}Sr$
1. *Laguna del Toro*					
GAS44	BGG	106	292	1·0191	0·711975
GAS46	BGG	108	201	1·5185	0·714077
GAS47	BGG	103	189	1·5584	0·714293
GAS48	BGG	86	353	0·6848	0·709884
GAS49	BGG	103	365	0·7911	0·710585
2. *Sierra del Medio*					
GAS25	2MGG	191	394	1·4295	0·712208
GAS26	2MGG	133	392	0·9891	0·710662
GAS27	2MGG	156	366	1·2426	0·711498
GAS28	2MGG	145	336	1·2499	0·711485
GAS29	2MGG	144	379	1·1189	0·710969
GAS30	2MGG	171	378	1·3295	0·711881
GAS31	2MGG	171	376	1·3231	0·711951
GAS32	2MGG	149	397	1·0695	0·710790
3. *Gastre Suite*					
GAS33	fol. HBG	148	430	0·9882	0·708618
GAS34	fol. HBG	139	397	1·0018	0·708681
GAS35	fol. HBG	122	469	0·7355	0·708166
GAS36	fol. HBG	167	297	1·6354	0·710617
GAS41	fol. HBG	105	483	0·6152	0·707582
GAS42	fol. HBG	109	473	0·6601	0·707646
GAS43	fol. HBG	94	516	0·5230	0·707195
GAS38	GS	191	180	3·1560	0·715590
GAS39	GS	194	191	3·0028	0·715017
GAS40	GS	187	189	2·9135	0·714693
4. *Lipetrén Suite*					
GAS20	BHG	158	308	1·5209	0·710171
GAS23	BHG	120	401	0·8623	0·708238
GAS24	BHG	140	393	1·0266	0·708638
GAS13	RBG	215	77	8·6765	0·731423
GAS14	RBG	214	68	9·9387	0·734808
GAS16	RBG	173	128	4·1104	0·717793
GAS17	RBG	153	201	2·2675	0·712516
GAS18	RBG	217	88	7·8003	0·728562
GAS19	RBG	203	47	13·7993	0·746344
GAS04	F	179	199	2·6943	0·714050
GAS10	F	209	26	25·8625	0·782771
GAS15	F	209	48	13·8122	0·746722
GAS02	MP	107	753	0·4111	0·707026
GAS03	MP	95	730	0·3844	0·706796
GAS05	MP	100	780	0·3696	0·706814
5. *Horqueta Granodiorite*					
GAS06	BHG	89	595	0·4386	0·708082
GAS07	BHG	100	534	0·5466	0·708207
GAS08	BHG	95	579	0·4816	0·708163
GAS11	BHG	159	371	1·2540	0·710033

Key: BGG = biotite granite-gneiss; 2MGG = 2-mica granodiorite-gneiss; fol. HBG = foliated hornblende-biotite granodiorite; GS = granite stock; BHG = biotite-hornblende granite; RBG = red biotite granite; F = felsite; MP = mafic porphyry.

2.3. Horqueta granodiorite

The rocks that compose this unit are mostly grey in colour, but occasionally appear pink with local concentration of K-feldspar. They are medium-grained and are cut by aplite dykes. They have high modal contents of quartz and plagioclase defining a transition to granitic compositions in QAP (Trend III in Fig. 4).

3. Geochronological results

The results of the Rb–Sr whole-rock analyses are presented in Table 1 and Sm-Nd data on selected samples in Table 2.

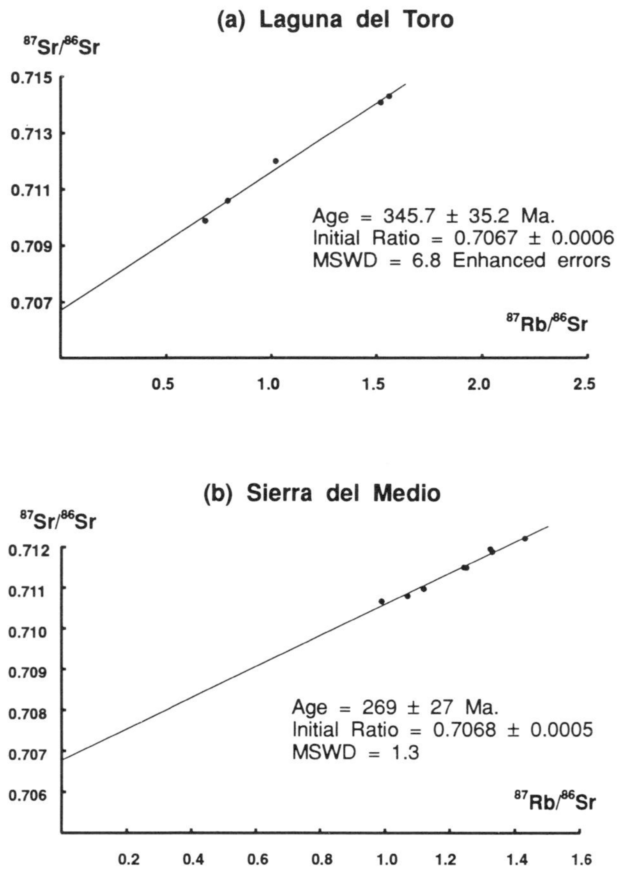

Figure 3 Rb–Sr whole-rock isochrons for the Palaeozoic basement granites of (a) Laguna del Toro, (b) Sierra del Medio.

3.1. Foliated granite of central Gastre (Gastre Suite)

Seven samples were analysed of the foliated hornblende-biotite granites W of Gastre, and three of an unfoliated red granite stock within the same area that was originally thought to be equivalent to the Lipetrén Suite. The Rb-Sr data for the foliated rocks define a reasonable isochron (MSWD = 3·0), giving an age of 210 ±11 Ma and an initial $^{87}Sr/^{86}Sr$ ratio of 0·7057 ±0·0001. The samples of the stock have higher but more uniform Rb/Sr ratios and do not define a useful independent isochron. They do, however, extend the isochron for the foliated granites, which they fit perfectly (reducing the MSWD to 2·4). In contrast, they do not fit the Lipetrén Suite isochron of Figure 5(b) well, resulting in a marked increase in MSWD from 2·7 to 6·4. It is concluded that the stock is not co-magmatic with the Lipetrén Suite but with the foliated suite within which it occurs. This is also borne out by the geochemical relationships of these rock types. The combined isochron gives an intrusion age of 220 ±3 Ma with an initial $^{87}Sr/^{86}Sr$ ratio of 0·7056 ±0·0001 (Fig. 5(a)).

These isochron parameters do not overlap within error with those for the Sierra del Medio foliated suite. Moreover, the lower age of the Gastre rocks is also associated with a lower initial $^{87}Sr/^{86}Sr$ ratio, so that partial closed-system resetting cannot be invoked as the reason for the differences. The data for these two rock units clearly establish that they belong to two independent intrusive episodes.

3.2. Lipetrén Suite

Only three samples of the biotite-hornblende granodiorite were analysed and two of these proved to have similar Rb/Sr ratios, so that an isochron age of 213 ±21 Ma obtained from this unit is not very precise despite the low MSWD (1·0). Interpretation of this rock type as coeval with the rest of the Lipetrén Suite allows a more useful assessment of its true age.

Six samples of the red granite unit also give a perfect isochron (MSWD = 1.0), this time with a good spread of data points, giving an age of 206 ±2 Ma and an initial $^{87}Sr/^{86}Sr$ ratio of 0·7058 ±0·0002 (Fig. 5(b)). Data for the aplitic body into which the red granite apparently grades northwards, and the pink felsite dykes cutting the other units, also lie on this isochron (nine points; MSWD = 3·4) without significantly altering the age or initial $^{87}Sr/^{86}Sr$ ratio. It is concluded that all these highly differentiated rocks were co-magmatic, as also suggested by the field relations and geochemistry.

Samples of the mafic porphyry unit have consistently low

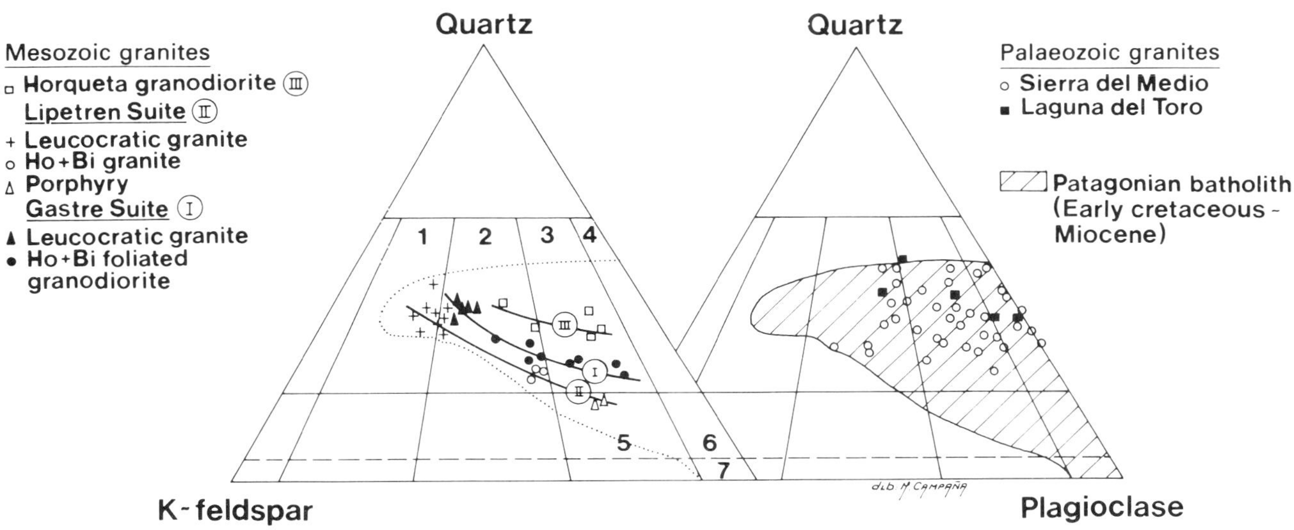

Figure 4 QAP diagram for Mesozoic granites of the Gastre area in comparison with the local Palaeozoic basement orthogneisses and the Cretaceous–Tertiary Patagonian Batholith.

Table 2 Sm–Nd analytical data for Gastre granites

Sample No.	Sm ppm	Nd ppm	$^{147}Sm/^{144}Nd$	$^{143}Nd/^{144}Nd$	$(^{143}Nd/^{144}Nd)_0$	ε_{Nd_t}	T_{chur}	T_{DM}	$(^{87}Sr/^{86}Sr)_0$*
Laguna del Toro									
GAS48	5·631	28·44	0·1197	0·512235	0·511961	−4·5	802	1251	0·706472
Sierra del Medio									
GAS25	2·333	10·41	0·1354	0·512242	0·512003	−5·7	990	1458	0·706717
GAS32	5·523	28·54	0·1170	0·512224	0·512017	−5·4	796	1235	0·706682
Gastre Suite									
GAS33	6·476	36·58	0·1070	0·512391	0·512237	−2·3	424	916	0·705526
GAS35	3·916	22·95	0·1032	0·512393	0·512244	−2·2	403	885	0·705865
GAS38	3·490	22·99	0·0918	0·512353	0·512221	−2·7	418	855	0·705715
GAS43	6·349	33·53	0·1145	0·512387	0·512222	−2·6	470	983	0·705559
Lipetrén Suite									
GAS03A	6·028	30·64	0·1189	0·512377	0·512215	−3·1	516	1038	0·705659
GAS17	3·092	20·08	0·0931	0·512375	0·512248	−2·4	391	838	0·705809
GAS20	2·980	10·08	0·0996	0·512390	0·512254	−2·3	393	863	0·705672

Notes: *Initial ratios calculated for time of crystallisation as in Figures 3 and 5; Rb-Sr data from Table 1. T_{chur}: model age for separation from uniform reservoir with chondritic REE composition. T_{DM}: model age for separation from depleted mantle generated 2,500 Ma ago.

Rb/Sr ratios and do not yield a useful independent age. However, these data points also lie well within error of the isochron defined by the red granite, a granitic enclave in the mafic porphyry and the hornblende-biotite granite. All 15 points together define a very precise isochron of 208 ±1 Ma; MSWD = 2·7; initial $^{87}Sr/^{86}Sr$ ratio = 0·7057 ±0·0001 (Fig. 5c), which we take as the time of emplacement of the Lipetrén Plutonic Suite NE of Gastre.

3.3. Horqueta granodiorite

Four samples of the grey biotite granodiorite of La Horqueta give an isochron relationship (MSWD = 1.1) corresponding to an age of 172 ±15 Ma and an initial $^{87}Sr/^{86}Sr$ ratio of 0·7070 (Fig. 5d). However, the distribution of the few data points is rather limited and in view of the remaining results this age is not considered to be statistically meaningful. In particular, since this unit is cut by felsitic dykes apparently emanating from the adjacent red granite-aplite (a sample of one of these, GAS-10, lies on the Lipetrén Suite isochron), it seems probable that the granodiorite may be older than is indicated by these few Rb-Sr determinations. Partial resetting of the latter during a Jurassic, or later, event seems a possible explanation.

4. Geochemistry of the Gastre granite suites

Geochemical data are presented in Table 3. The Gastre granites as a whole have moderate to high silica contents (the most mafic rock has 64% SiO_2). The AFM plot (Fig. 6) exhibits a typical calcalkaline trend. Harker diagrams for the BCP rocks show smooth progressive variations, whereas those for the Palaeozoic granites, especially the Sierra del Medio two-mica granodiorite, are dispersed (Fig. 7). With increasing differentiation there is a marked change from metaluminous granodiorites to strongly peraluminous granites (Fig. 8). In comparison with typical Andean granites at the same SiO_2 contents, K is slightly enriched and Ca depleted. The Peacock index of these suites is considerably lower than the range observed in the Patagonian Batholith and many other batholiths of the western American Cordillera. Nevertheless, there are no abnormal enrichments in LIL or HFS trace elements such as Ba, Ce, or Zr that would indicate an intraplate origin (Whalen *et al.* 1987). The Rb-Y-Nb discrimination diagram (Fig. 9) apparently confirms overall volcanic-arc affinities for these differentiated suites, and their relatively low Ga/Al ratios are more characteristic of normal I-type than A-type granites.

The Gastre and Lipetrén suites have essentially invariant ε_{Nd_t} values of −2·2 to −3·1 (Table 2), compatible with their indistinguishable initial $^{87}Sr/^{86}Sr$ ratios (Fig. 5). They are markedly less negative than the values of −4·5 to −5·7 shown by the samples of the Palaeozoic orthogneissic basement (which would have been slightly more negative still during the emplacement of the BCP). The Precambrian rocks of the NPM would have had considerably more radiogenic Sr (Linares *et al.* 1988), and therefore probably less radiogenic Nd, than the Palaeozoic granites analysed here. This indicates that the Triassic/Jurassic parent magmas were derived from more juvenile sources and incorporated little or no Nd from the country rocks. The chondrite-reservoir Nd model ages are all Palaeozoic and the depleted-mantle Nd model ages are 800 – 1,000 Ma (Table 2). On the assumption that these characteristics were inherited entirely from the source region, the last could be taken as a maximum age for the source rocks.

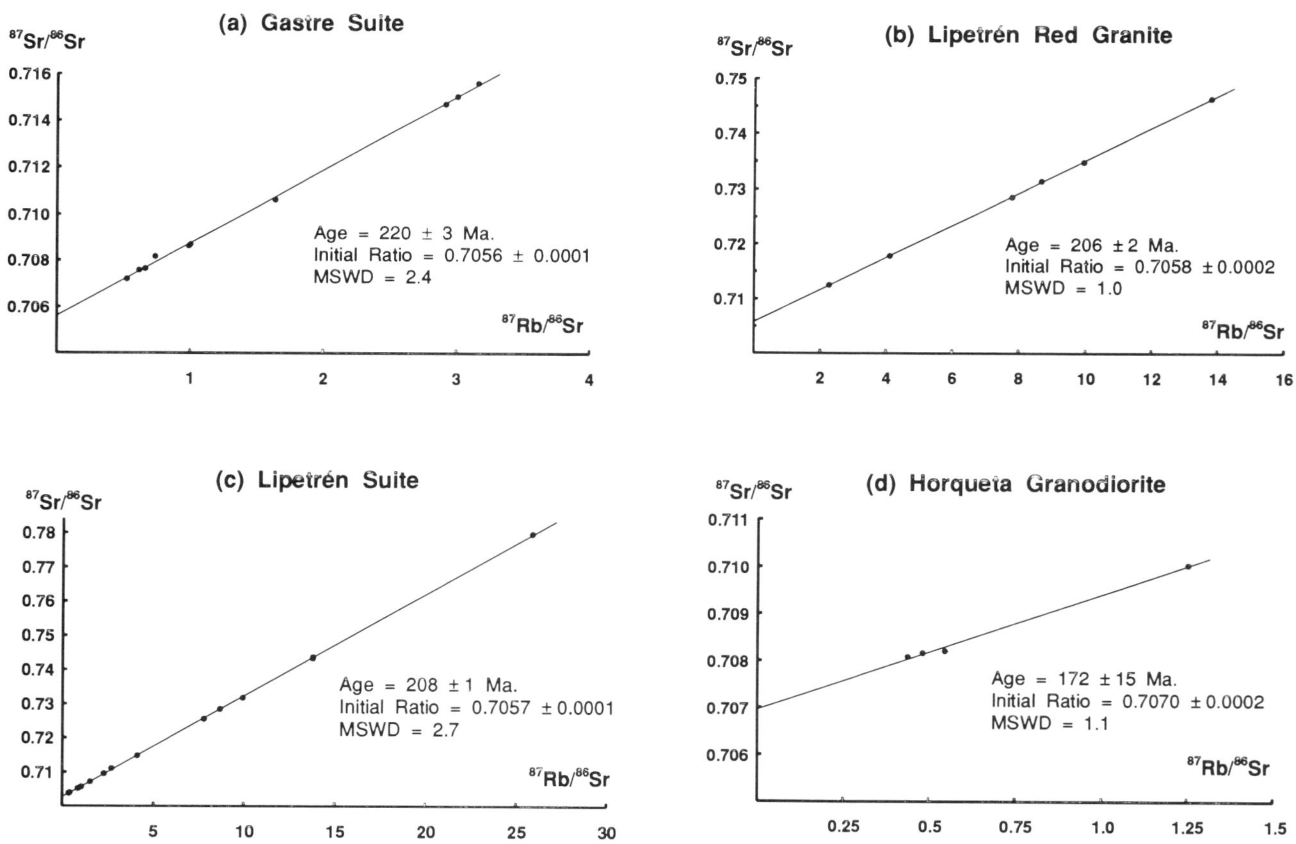

Figure 5 Rb-Sr whole-rock isochrons for Mesozoic granites of the BCP in the Gastre area: (a) Gastre Suite, (b) Lipetrén Red Granite, (c) Lipetrén Suite, (d) Horqueta granodiorite.

Table 3 Geochemical data

	PALAEOZOIC GRANITES									
	Sierra del Medio granites					*Laguna del Toro granites*				
	GAS 31	GAS 32	GAS 29	GAS 25	GAS 30	GAS 48	GAS 49	GAS 44	GAS 47	GAS 46
Major elements (%)										
SiO_2	66·68	68·98	70·72	71·55	72·33	65·85	69·88	71·95	74·69	76·10
TiO_2	0·61	0·53	0·33	0·30	0·39	0·76	0·41	0·37	0·12	0·12
Al_2O_3	16·00	15·45	15·20	14·96	13·90	15·45	13·98	13·95	13·25	13·00
Fe_2O_3	0·29	0·12	0·29	0·56	0·09	0·01	0·11	0·02	0·33	0·20
FeO	3·09	2·67	1·43	0·90	1·96	4·06	2·29	1·89	0·84	0·56
MnO	0·06	0·06	0·03	0·03	0·03	0·07	0·04	0·04	0·01	0·01
MgO	1·38	1·20	0·75	0·52	0·77	2·25	1·35	0·75	0·12	0·19
CaO	2·99	2·99	2·44	1·77	2·10	4·07	2·81	2·50	1·04	1·16
Na_2O	3·63	3·63	3·42	2·18	3·20	3·67	4·07	3·94	3·63	3·42
K_2O	3·82	2·76	4·07	5·99	5·05	2·53	3·82	3·41	5·05	4·07
P_2O_5	0·39	0·48	0·20	0·24	0·20	0·24	0·17	0·14	0·08	0·11
H_2O^+	0·60	0·63	0·65	0·15	0·46	0·46	0·55	0·48	0·47	0·60
H_2O^-	0·13	0·10	0·11	0·40	0·02	0·08	0·13	0·06	0·08	0·13
TOTAL	99·67	99·60	99·64	99·55	100·50	99·50	99·61	99·50	99·71	99·67
Trace elements (*ppm*)										
Rb	165	146	143	188	168	84	99	100	99	103
Sr	365	394	370	381	366	355	363	285	183	196
Ba	703	527	976	1498	1183	589	859	517	671	726
Ga	25	22	16	19	17	25	23	23	22	22
Y	32	26	61	16	17	16	8	8	5	8
Nb	14	13	9	8	10	8	6	8	7	9
Zr	136	130	121	88	94	147	135	139	47	57
La	—	25·2	—	15·2	—	29·1	—	—	—	19·9
Ce	—	45·3	—	23·3	—	57·6	—	—	—	35·3
Nd	—	22·3	—	11·0	—	25·0	—	—	—	14·9
Sm	—	4·0	—	2·2	—	2·8	—	—	—	3·0
Eu	—	0·9	—	0·9	—	1·1	—	—	—	0·4
Gd	—	3·9	—	2·1	—	4·0	—	—	—	2·0
Dy	—	3·5	—	2·8	—	3·3	—	—	—	1·3
Yb	—	1·1	—	0·8	—	1·0	—	—	—	0·1
Lu	—	0·1	—	0·1	—	0·2	—	—	—	0·1

	GASTRE SUITE									
	Foliated granodiorite						*Leucocratic granite*			*Mafic dyke*
	GAS 43	GAS 42	GAS 33	GAS 34	GAS 36	GAS 35	GAS 39	GAS 38	GAS 40	GAS 37
Major elements (%)										
SiO_2	63·11	63·92	66·02	67·08	69·64	70·85	73·35	75·10	75.10	58·60
TiO_2	0·74	0·72	0·57	0·69	0·48	0·69	0·38	0·26	0·28	0·93
Al_2O_3	14·98	15·80	14·13	14·50	13·40	14·20	14·20	12·30	13·02	14·81
Fe_2O_3	1·44	1·00	1·20	0·91	0·55	0·46	0·13	0·06	0·53	1·29
FeO	3·50	3·09	2·89	3·01	1·89	1·81	0·93	1·00	0·76	4·27
MnO	0·09	0·08	0·08	0·06	0·04	0·03	0·02	0·02	0·03	0·11
MgO	2·88	2·54	2·45	2·20	1·42	1·13	0·41	0·52	0·50	5·00
CaO	4·94	3·87	4·05	3·72	2·44	2·90	1·22	1·22	1·16	5·86
Na_2O	4·28	4·07	4·15	3·42	3·42	3·76	3·55	3·85	3·55	4·28
K_2O	2·76	3·41	4·05	3·82	5·46	3·41	4·64	5·46	4·88	3·41
P_2O_5	0·28	0·38	0·25	0·25	0·19	0·29	0·13	0·14	0·09	0·71
H_2O^+	0·50	0·58	0·50	0·60	0·58	0·35	0·43	0·19	0·38	0·50
H_2O^-	0·08	0·08	0·05	0·05	0·05	0·03	0·15	0·08	0·18	0·23
TOTAL	99·58	99·54	100·39	100·31	99·56	99·70	99·54	100·20	100·46	100·00
Trace elements (*ppm*)										
Rb	93	107	143	135	162	118	194	189	184	110
Sr	516	471	419	390	287	465	187	174	184	1041
Ba	860	1109	848	909	818	1368	809	777	800	1142
Ga	16	22	19	18	15	10	16	14	20	24
Y	25	19	22	20	23	9	13	15	17	26
Nb	7	8	9	8	9	4	7	9	8	6
Zr	133	209	174	259	196	135	128	132	128	180
La	36·3	—	32·6	—	—	27·3	—	45·1	—	69·4
Ce	70·8	—	67·4	—	—	40·0	—	74·2	—	155·7
Nd	33·6	—	32·2	—	—	15·9	—	21·9	—	84·2
Sm	6·8	—	4·3	—	—	2·7	—	3·3	—	15·8
Eu	1·1	—	1·0	—	—	0·7	—	0·4	—	3·8
Gd	5·0	—	4·5	—	—	2·1	—	2·7	—	10·0
Dy	4·4	—	3·8	—	—	1·8	—	2·4	—	5·4
Yb	2·4	—	2·1	—	—	0·8	—	1·4	—	1·7
Lu	0·3	—	0·3	—	—	0·1	—	0·2	—	0·3

continued . . .

Table 3—*contd*.

	Porphyry	Enclave	Felsite	LIPETRÉN SUITE Hornblende-biotite granodiorite			Red leucocratic granite				Horqueta granite	
	GAS 03A	GAS 02	GAS 04	GAS 24	GAS 23	GAS 20	GAS 17	GAS 13	GAS 15	GAS 14	GAS 06	GAS 11
Major elements (%)												
SiO_2	62·76	62·94	74·49	69·37	72·07	73·91	77·38	77·95	78·53	78·93	71·99	73·43
TiO_2	0·71	0·71	0·23	0·45	0·38	0·38	0·21	0·18	0·17	0·18	0·30	0·32
Al_2O_3	16·60	16·99	14·16	14·77	14·07	13·84	12·63	12·31	12·93	11·91	15·38	13·86
Fe_2O_3	2·53	2·45	0·67	1·98	1·16	1·05	0·67	0·78	0·02	0·55	1·02	1·44
FeO	2·42	2·18	0·80	1·66	1·62	1·03	0·39	0·16	0·62	0·28	1·17	1·15
MnO	0·10	0·10	0·02	0·07	0·07	0·05	0·04	0·05	0·02	0·02	0·04	0·05
MgO	1·99	1·92	0·42	1·44	1·10	0·70	0·25	0·18	0·12	0·12	0·77	0·63
CaO	4·28	3·52	1·13	3·21	2·75	1·83	1·10	0·46	0·30	0·30	3·06	2·05
Na_2O	3·29	3·10	2·62	2·74	2·92	2·55	2·47	2·44	2·47	2·47	2·92	2·47
K_2O	2·75	2·86	4·84	3·79	3·59	3·90	4·21	4·63	4·63	4·42	2·44	3·59
P_2O_5	0·30	0·28	0·05	0·04	0·16	0·11	0·05	0·07	0·05	0·04	0·13	0·11
H_2O^+	1·10	1·42	0·23	0·13	0·10	0·10	0·15	0·15	0·03	0·25	0·28	0·18
H_2O^-	0·70	1·48	0·23	0·38	0·50	0·18	0·10	0·40	0·18	0·25	0·12	0·25
TOTAL	99·53	99·95	99·89	100·03	100·49	99·63	99·65	99·76	100·07	99·69	99·62	99·53
Trace elements (ppm)												
Rb	98	106	178	134	117	155	153	213	210	213	89	154
Sr	740	747	191	378	394	295	195	71	44	62	586	356
Ba	1130	1052	840	886	952	829	615	465	195	359	1015	873
Ga	26	21	15	20	16	18	17	16	17	15	20	14
Y	20	20	15	11	11	11	13	25	19	24	9	14
Nb	7	7	8	8	7	7	8	14	11	13	5	8
Zr	252	264	132	138	134	112	85	92	72	93	127	156
La	35·4	—	34·9	—	—	24·8	28·4	—		—	—	—
Ce	71·1	—	63·8	—	—	45·3	50·6	—	—	—	—	—
Nd	33·0	—	22·1	—	—	17·4	17·4	—	—	—	—	—
Sm	6·6	—	4·1	—	—	3·2	2·2	—	—	—	—	—
Eu	1·4	—	0·5	—	—	0·6	0·3	—	—	—	—	—
Gd	4·5	—	2·8	—	—	2·1	2·0	—	—	—	—	—
Dy	4·5	—	2·6	—	—	2·0	4·1	—	—	—	—	—
Yb	1·7	—	1·5	—	—	1·1	1·5	—	—	—	—	—
Lu	0·2	—	0·2	—	—	0·2	0·2	—	—	—	—	—

When considered independently, the separate granitic suites of the BCP display certain systematic differences. The compositional range of the Gastre Suite is more restricted than that of the Lipetrén Suite, largely lacking the highly differentiated granophyric facies that is typical of the latter. Similarly the Al/Si ratio increases through the age sequence: Gastre Suite, Lipetrén Suite and Horqueta granodiorite, and strongly peraluminous rocks are restricted to the last two (Fig. 8). There is also a corresponding increase in Fe_2O_3/FeO ratio through this sequence, demonstrating increasing f_{O_2} conditions of crystallisation. Rare-earth element patterns for the BCP granites at Gastre (Fig. 10) are typically LREE-enriched, with chondrite-normalised La/Yb ratios ranging from 10 to 23 in the Gastre Suite but scarcely varying in the Lipetrén Suite (14 – 15). There is an increase in the magnitude of the negative Eu-anomaly with increasing SiO_2, in both suites, but with the maximum value of 0·5 – 0·4 occurring in the Lipetrén Suite leucogranites.

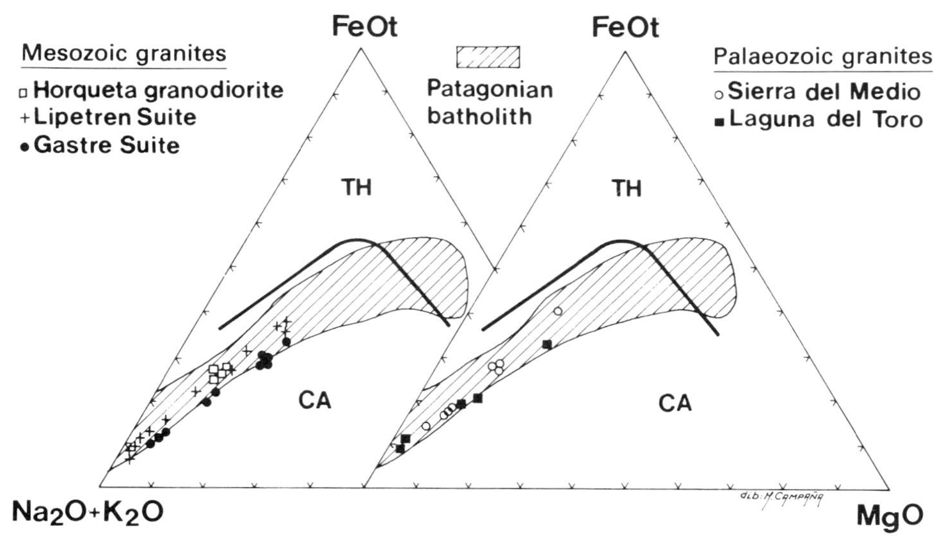

Figure 6 AFM plot for Mesozoic granites of the Gastre area in comparison with the local Palaeozoic basement orthogneisses.

The more differentiated rocks also often show relative depletion in the middle-range REE. Sr shows rather little variation in the Gastre Suite, but is highly fractionated within the Lipetrén Suite (800 – 15 ppm).

5. Petrogenesis

The overall similarities in geochemical and isotopic compositions for the Gastre and Lipetrén suites and the Horqueta granodiorite indicate a common model of origin. The geochronology and the somewhat different geochemical trends indicate that they each represent a separate pulse of magma.

The extreme variability of the Lipetrén Suite in particular, and features such as sharply falling Sr and middle REE, and increasingly pronounced Eu-anomaly, with increasing SiO_2, are strong arguments that the main cause of variation was fractionation of crystallising phases, notably plagioclase feldspar and hornblende. Petrographical textures and the close geographical and geochemical association between the Lipetrén Suite and the porphyry dykes and felsites, together with the development of high f_{O_2} conditions of crystallisation in the late stages, are interpreted as evidence for the emplacement of these rocks as a high-level sub-volcanic intrusion. This is confirmed by geobarometry based on the Al content of hornblendes, which implies crystallisation pressures very close to 1 kbar (1 kbar = 100 MPa) (Table 4).

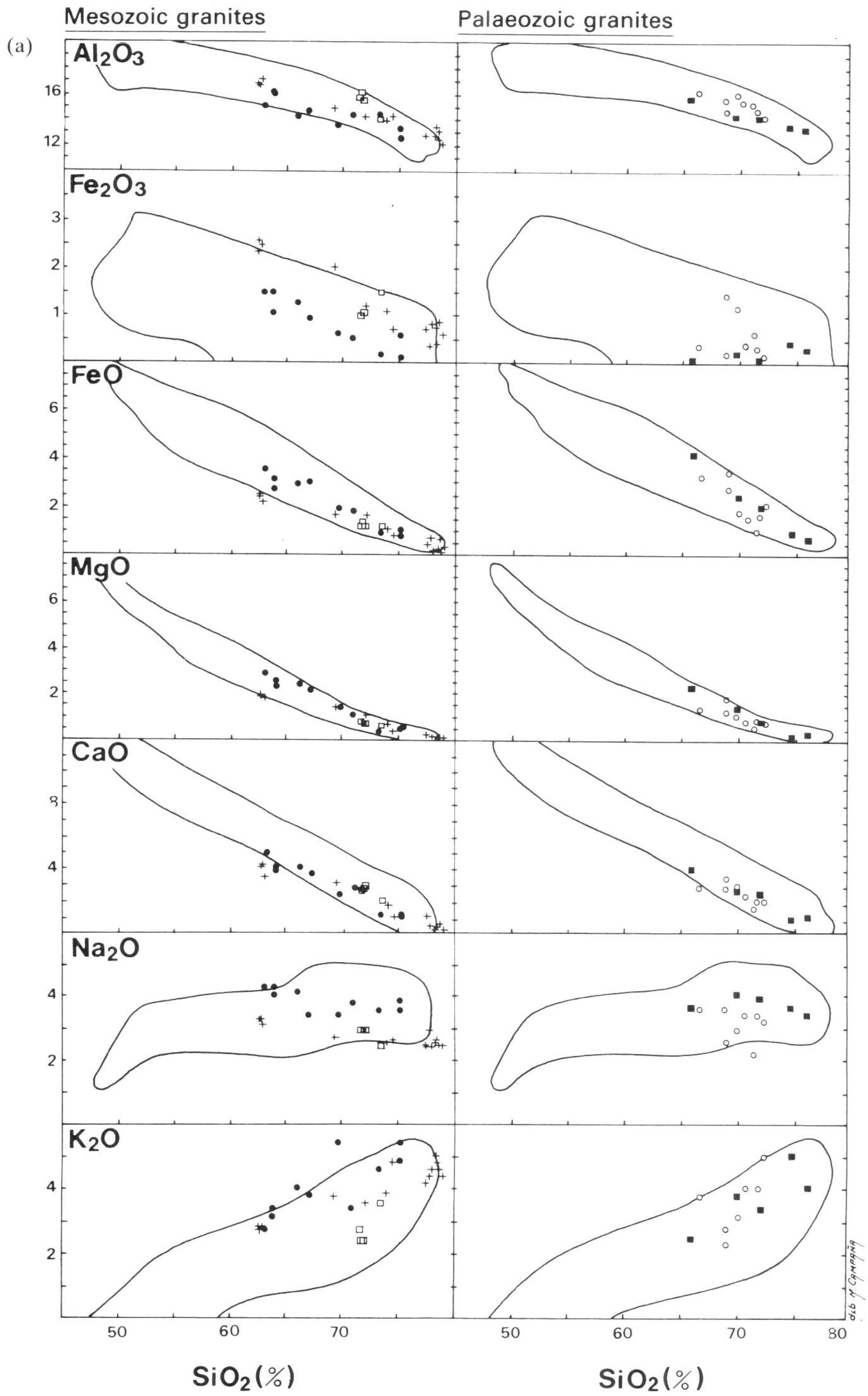

Figure 7 Harker plots for Mesozoic granites of the Gastre area and the local Palaeozoic basement orthogneisses: (a) major element variation.

The parent magmas were metaluminous and their source region was probably in the lower crust. Foliated granulite xenoliths found in Cainozoic alkalic complexes close to the Gastre area consist of calcic plagioclase, diopside, orthopyroxene magnetite and spinel (Corbella 1983) and may represent this Patagonian lower crust. In the northern sector of the NPM such granulitic xenoliths contain minor amounts of scapolite (1–5%) and amphibole (0–5%).

The isotopic and LIL-element characteristics of the parent magmas were rather more evolved than those of the Cretaceous–Tertiary magmas of the Andean Cordillera (e.g. the Patagonian Batholith) and they lack the mafic and intermediate representatives that dominate the latter rock associations. On the other hand, in both respects they match many of the Pre-Cretaceous granites of the Andes, which were termed "modified I-types" by Pankhurst (1990).

6. Plutonic events in the Gastre region and regional comparisons

The early history of the NPM remains obscure, although the Nd isotope data discussed above (Table 2) and the Rb-Sr

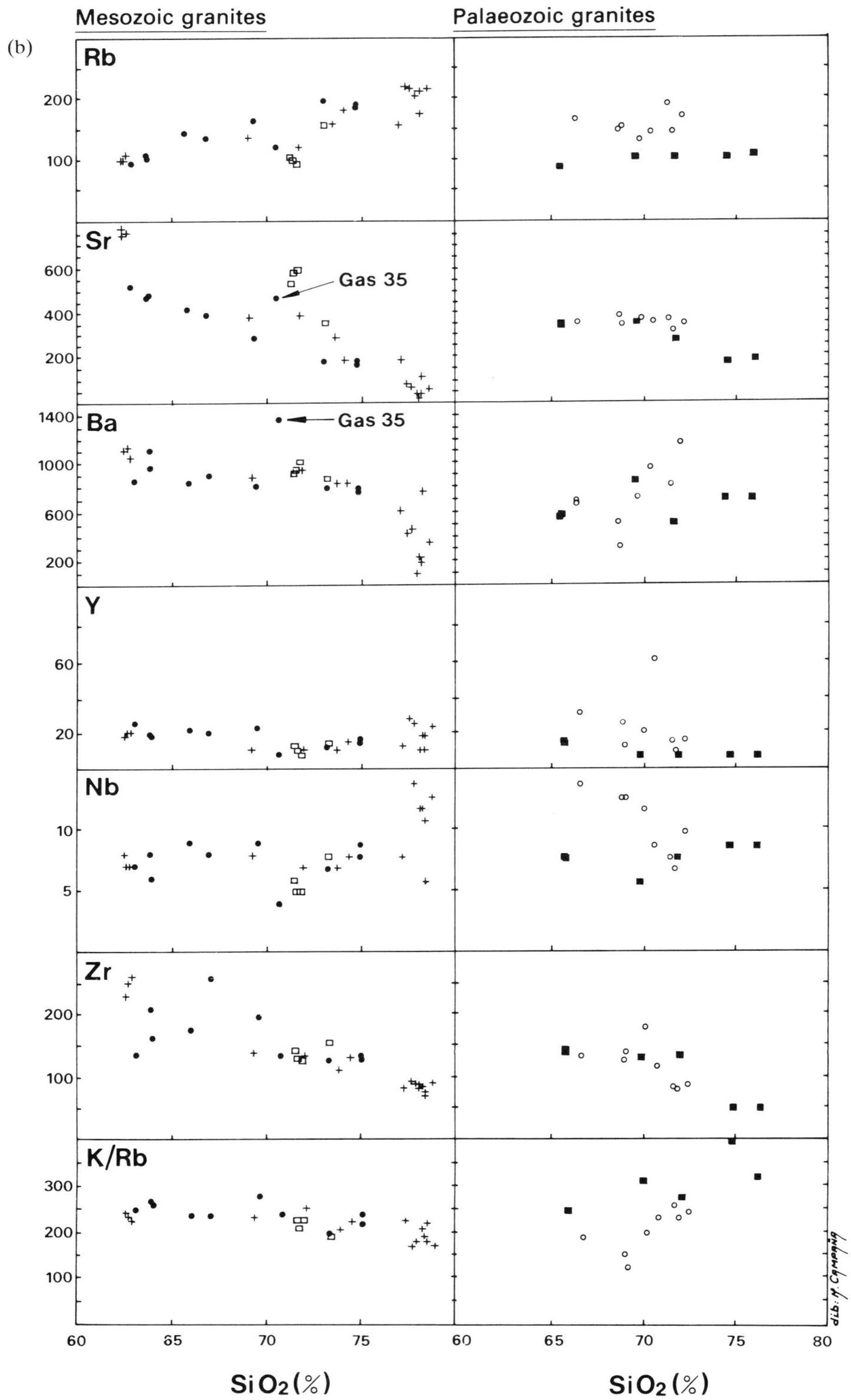

Figure 7 (b) Trace element variation; symbols as in Figure 6.

data of Linares *et al.* (1988) suggest that cratonisation was a mid- to late-Proterozoic event. The rather imprecise age of 346 ± 35 Ma for the Laguna del Toro orthogneiss is consistent, within error, with the pre-Late Carboniferous age previously assigned to the Mamil Choique formation in the Gastre region (Proserpio 1978). However, other parts of this formation are migmatitic in appearance and even include granitised metasediments (Caminos & Llambías 1984). Older Rb-Sr whole-rock isochrons have been reported for these migmatitic rocks in the area of Río Collon Cura: 480 ± 10 Ma (Linares *et al.* 1988; recalculated here as 480 ± 21 Ma after the exclusion of two analyses) and 489 ± 27 Ma. Together, these results suggest a rather complex Early Palaeozoic evolution of the basement.

The second intrusive event identified here is emplacement of the foliated peraluminous granodiorite of the Sierra del Medio at 267 ± 27 Ma (probably corresponding to Permian times). Permian plutonism has been widely assumed to be one of the main stages of evolution of the NPM (e.g. Stipanicic & Methol 1980; Llambías *et al.* 1984), based on K-Ar age determinations from widely spaced granitic outcrops. Linares *et al.* (1988) reported a Rb-Sr whole-rock isochron of 280 ± 5 Ma (1 σ error) and concordant K-Ar ages for granite samples from the Río Collon Cura, 100 km NW of Gastre. However, the age obtained for the Sierra del Medio granites is younger than previously supposed for these rocks, which were included in the Mamil Choique formation by Proserpio (1978).

An even greater revision of the presumed chronology is represented by the well-defined isochron ages of 220 ± 3 Ma and 208 ± 1 Ma for the most voluminous granitic rocks of the BCP, the Gastre and Lipetrén suites, respectively. These are clearly not Permian at all: the former is probably Late Triassic, whereas the latter is essentially coincident with the recommended Triassic/Jurassic boundary (e.g. Palmer 1983). These two ages are statistically distinct but are also sufficiently close to merit inclusion within a single magmatic episode. A growing body of evidence for the importance of granite plutonism of this age in northern Patagonia NW of Gastre now exists. Alonso (1987) reported a Rb-Sr isochron age of 195 ± 10 Ma for pre-Jurassic granites near Pilcaniyeu and further evidence for activity of this age in rocks overprinted by the Andean Cordillera was presented by Cingolani *et al.* (1992). Extensive rhyolitic and ignimbritic sequences, some interbedded with sediments carrying a *Dicroidium* flora, must be at least roughly coeval with the subvolcanic Lipetrén Suite, and may have been produced in a cauldron-type eruption.

The timing of these events in and around the BCP coincides with the stratigraphical definition of the Choiyoi group of felsic volcanics, first identified in the Mendoza region (Groeber 1946) but since recognised throughout northern Patagonia (e.g. Rapela & Kay 1988) and as far N as La Pampa province, where they have yielded several Rb-Sr and K-Ar ages of around 235 Ma (Linares *et al.* 1980). The uniformly felsic composition of the Gastre high-level granites and their close association with felsites suggest a genetic link with the Choyoi volcanic rocks not previously envisaged.

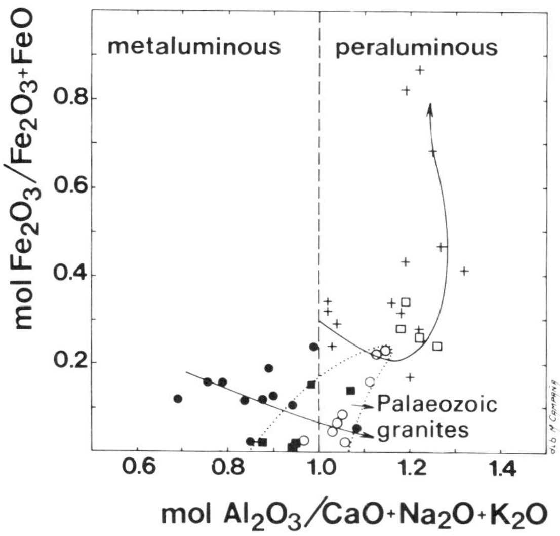

Figure 8 Alumina saturation index diagram for Mesozoic granites of the Gastre area and the local Palaeozoic basement orthogneisses; symbols as in Figure 6.

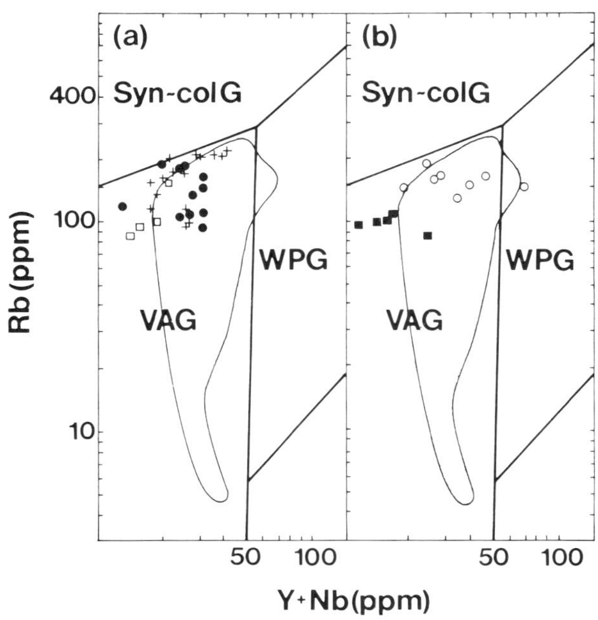

Figure 9 Rb-Y-Nb discrimination diagram for Mesozoic granites of the Gastre area and the local Palaeozoic basement orthogneisses; symbols as in Figure 6.

7. Tectonic implications

Partly on the basis of their presumed Permian age, Ramos (1984) proposed that the granites of the NPM were developed above a southward-dipping subduction zone during closure of a northern ocean basin, before Late Palaeozoic collision between an allochthonous "Patagonian terrane" and the rest of southern S America. It is clear from the now established Triassic/Jurassic age of the BCP granites that they cannot fulfil this role. Moreover, their I-type, calcalkaline, "volcanic arc" geochemical signatures appear to exclude the possibility that they were related to collisional tectonics in any way. Some evidence for Permian granitic magmatism in the area remains in the 269 ± 27 Ma isochron for the Sierra del Medio body reported here, the 280 ± 10 Ma isochron for the Río Collon Cura granite (Linares *et al.* 1988, with 2-σ error) and a recent revision of ages for the early stage of emplacement in the central part of the Somuncura Batholith (Pankhurst *et al.* in press). Nevertheless, the petrogenesis and tectonic significance of these granites have not yet been adequately assessed.

Basin development in southern S America during Middle Triassic to Early Jurassic times was controlled by a widespread extensional tectonics. Triassic extensional faulting produced a series of subparallel NW-SE-trending troughs indicative of pervasive rifting, but also with evidence of strike- slip faulting in places (Uliana *et al.* 1989). Extrusive centres in central Chile and Western Argentina (rhyolite-andesite lahars, ignimbrites and basalts) were structurally controlled and related to such faults (Charrier 1979; Ramos & Kay 1991).

Emplacement of the epizonal to subvolcanic granites of the BCP has also been inferred to be related to local pull-apart structures associated with NW-SE fracture zones (Rapela & Pankhurst 1992): mylonitic areas developed predominantly in the felsites and porphyries of the Lipetrén Suite appear to be a consequence of dextral strike-slip displacement and no direct evidence of regional extensional features (i.e. direct or detachment faults) has been observed. Since the igneous foliation of the older units (Gastre superunit) is parallel to the main fault zones and a fracture cleavage striking roughly N130° is widespread in all the granitic units, it is concluded that these faults were active during the whole history of batholith emplacement. Later reactivation of these faults has been also suggested by folding of overlying Cretaceous sedimentary sequences (Coira *et al.* 1975).

Thus there is ample evidence that the distribution and emplacement of the widespread *c.* 200 Ma silicic magmatism was controlled by regionally active NW–SE shear zones (cf. Hutton). The corollary, that the major Gastre shear zone system was related to the earliest stages of rifting of Gondwana, will be developed elsewhere (Rapela &

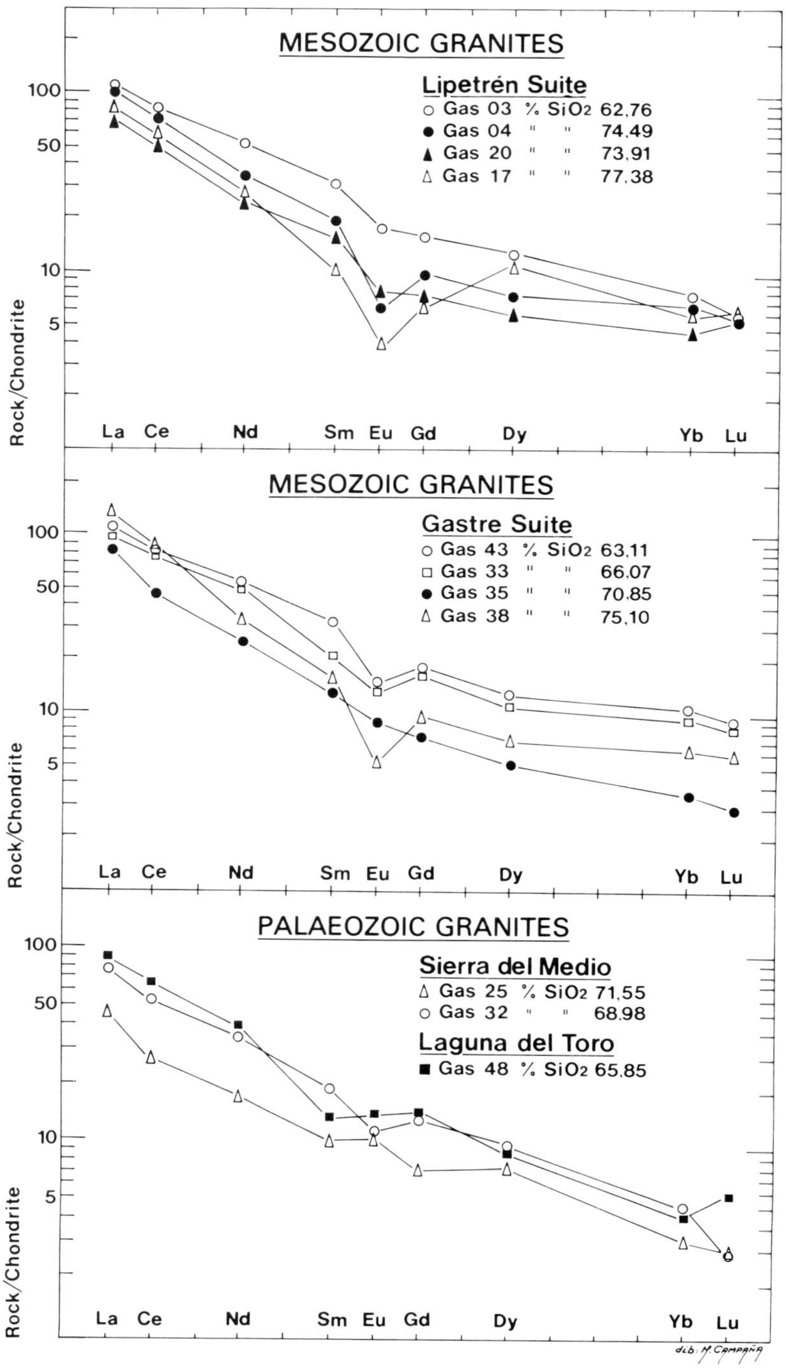

Figure 10 Chondrite-normalised REE patterns for Mesozoic granites of the Gastre area and the local Palaeozoic basement orthogneisses.

Table 4 Microprobe analyses of hornblende (Sample GAS20, Lipetrén Suite)

	Cation proportions					
	Rims				Cores	
Si	7·090	6·890	7·094	7·088	7·211	7·171
Al	1·043	1·213	1·040	1·086	0·930	0·959
K	0·159	0·161	0·151	0·152	0·094	0·119
Na	0·347	0·358	0·338	0·348	0·328	0·334
Ca	1·915	1·856	1·846	1·836	1·837	1·823
Fe	1·972	2·086	1·927	1·902	1·831	1·836
Mg	2·842	2·825	2·903	2·831	2·984	3·004
Mn	0·057	0·064	0·055	0·054	0·059	0·060
Cr	0·000	0·002	0·001	0·007	0·000	0·000
Ti	0·107	0·153	0·137	0·154	0·130	0·134
P(Kb)*	0·9	1·6	0·9	1·1	0·4	0·6

Notes: *Pressure of crystallisation calculated from calibration of Johnson & Rutherford (1988): $P = 4.28Al - 3{\cdot}54$. Modal composition of sample: plagioclase 45%, K-feldspar 27%, quartz 24%, biotite 3%, hornblende 1%, accessory minerals 0·2%.

Pankhurst 1992). At this stage, we simply note that the proposed deep crustal generation of these vast volumes of granitic and rhyolitic magmas fits with a scenario of underplating by ponded basalt and accompanying thinning of a weak, juvenile crust in an extensional regime.

8. Appendix: analytical methods

Major element and REE determinations were performed at the Centro de Investigaciones Geológicas, Universidad Nacional de La Plata. The determinations of Al, Fe_{total}, Ca, Mg, Mn, Na and K were carried out by atomic absorption, Si, P and Ti by colorimetry, and Fe by volumetry. The REE (La, Ce, Nd, Sm, Eu, Gd, Dy, Yb and Lu) were determined sequentially by inductively coupled plasma atomic emission spectrometry (ICP-AES). Samples were prepared using conventional rock dissolution (acid digestion) techniques, and the REE were separated by cation-exchange chromatography (Roenlandts & Michel 1986). The remaining trace elements (Ba, Zr, Nb, Y and Ga) were determined by X-ray fluorescence spectrometry in the Department of Geology, University of Keele, on the same samples analysed for major elements.

Whole-rock powders were analysed for Rb, Sr and Rb/Sr by X-ray fluorescence spectrometry at the British Geological Survey, London, using a method based on that of Pankhurst and O'Nions (1973). Sr, Sm and Nd were separated and isotopic compositions determined at the NERC Isotope Geology Centre, London, with a fully automated VG Isotopes 354 mass-spectrometer, using triple Faraday collectors in a combined peak-switching/simultaneous mode. During the period of these analyses, 10 measurements of the NBS 987 standard gave a mean $^{87}Sr/^{86}Sr$ ratio of 0·710235 ±0·000042 (2σ). The La Jolla standard (12 measurements) gave a $^{143}Nd/^{144}Nd$ ratio of 0·511825 ±0·000018. Isochron relationships were evaluated using the York-Williamson algorithm (Williamson 1968) and assuming a decay constant for ^{87}Rb of $1{\cdot}42 \times 10{-}11$ a-1. All errors quoted in the text are 2-σ, including those reported from previous literature.

9. Acknowledgements

We are grateful to our many colleagues in Centro de Investigaciones Geológicas and British Antarctic Survey who have either faciltiated or assisted with this work. M. Ingham and A. Robertson are thanked for their help with the XRF determination of Rb/Sr ratios, and G. Grissom (Stanford University) kindly provided the microprobe data in Table 4. This paper is a contribution to IGCP Projects Nos. 249 (Andean Magmatism and its Tectonic Setting) and 279 (Terranes in Latin America): it also constitutes NIGL Publication No. 49.

10. References

Alonso, G. B. 1987. Resultados geoquímicos y geocronológicos preliminares de los cuerpos graníticos de Pilcaniyeu, Provincia de Río Negro. ACTAS X CONGR GEOL ARGENTINO, TUCUMÁN, ARGENTINA **IV**, 27–9.

Caminos, R. & Llambías, E. J. 1984. El basamento cristalino. RELAT IX CONGR GEOL ARGENTINO, SAN CARLOS DE BARILOCHE **1** (2), 37–63.

Charrier, R. 1979. El Triásico de Chile y regiones adyacentes de Argentina: una reconstrucción paleogeográfica y paleoclimática. COMUNICACIONES **26**, 1–37.

Cingolani, C., Dalla Salda, L., Hervé, F., Munizaga, F., Pankhurst, R. J., Parada, M. A. & Rapela, C. W. 1991. The magmatic evolution of northern Patagonia: new impressions of pre-Andean and Andean tectonics. *In* Harmon, R. S. & Rapela, C. W. (eds) *Andean Magmatism and its Tectonic Setting*. GEOL SOC Am SPEC PUBL **265**, 29–44.

Coira, B., Nullo, F. E., Proserpio, C. & Ramos, V. A., 1975. Tectónica de basamento de la región occidental del Macizo Nordpatagónico (Provincias de Río Negro y Chubut), República Argentina. REV ASOC GEOL ARGENTINA **30** (4), 361–83.

Corbella, H. 1983. Hallazgo de rocas leucíticas en la Sierra de Los Chacays, Patagonia extrandina, Provincia de Chubut. REV ASOC ARGENTINA MINERAL PETROL SEDIMENTOL **14**(3–4), 88–96.

Dalziel, I. W. D., Storey, B. C., Garrett, S. W., Grunow, A. M., Herrod, L. B. D. & Pankhurst, R. J. 1987. Extensional tectonics and the fragmentation of Gondwanaland. *In* Coward, M. P., Dewey, J. F. & Hancock, P. L. (eds) *Continental Extension Tectonics*, SPEC PUBL GEOL SOC LONDON **28**, 431–41.

Groeber, P. 1946. Observaciones geológicas a lo largo del meridiano 70°. 1. Hoja Chos Malal. REV ASOC GEOL ARGENTINA **1**(3), 177–208.

Johnson, M. C. & Rutherford, M. J. 1988. Experimental calibration of an Aluminium-in-Hornblende geobarometer applicable to calc-alkaline rocks. EOS **69**, 1511.

Linares, E., Llambías, E. J. & Latorre, C. O. 1980. Geologías, E. J. & Latorre, C. O. 1980. Geología de la provincia de La Pampa, República Argentina, y geocronología de sus rocas metamórficas y eruptivas. REV ASOC GEOL ARGENTINA **35** (1), 87–146.

Linares, E., Cagnoni, M. C., Do Campo, M. & Ostera, H. A. 1988. Geochronology of metamorphic and eruptive rocks of southeastern Neuquen and northwestern Río Negro provinces, Argentine Republic. J S AMERICAN EARTH SCI **1** (1), 53–61.

Llambías, E. J., Caminos, R. & Rapela, C. W. 1984. Las plutonitas y vulcanitas del ciclo eruptivo gondwánico. RELAT IX CONGR GEOL ARGENTINO, SAN CARLOS DE BARILOCHE, ARGENTINA **1**(4), 85–117.

Palmer, A. R. 1983 (ed.). *The decade of North American geology* 1983 *geological time scale*. Boulder, Colorado: The Geological Society of America.

Pankhurst, R. J. 1990. The Paleozoic and Andean magmatic arcs of West Antarctica and southern South America. *In* Kay, S. M. & Rapela, C. W. (eds) *Plutonism from Antarctica to Alaska*. GEOL SOC AM SPEC PAP **241**, 1–7.

Pankhurst, R. J. & O'Nions, R. K. 1973. Determination of Rb/Sr and $^{87}Sr/^{86}Sr$ ratios of some standard rocks and evaluation of X-ray fluorescence spectrometry in Rb-Sr geochemistry. CHEM GEOL **12** (2), 127–36.

Pankhurst, R. J., Hole, M. J. & Brook, M. 1988. Isotope evidence for the origin of Andean granites. TRANS R SOC EDINBURGH EARTH SCI **79**, 123–33.

Pankhurst, R. J., Rapela, C. W., Caminos, R., Llambías, E. J. & Parica, C. (in press). A revised age for the granites of the central Somuncura batholith, North Patagonian Massif. J S AMERICAN EARTH SCI.

Parada, M. A. 1990. Granitoid plutonism in Central Chile and its geodynamic implications: a review. *In* Kay, S. M. & Rapela, C. W. (eds) *Plutonism from Antarctica to Alaska.* GEOL SOC AM SPEC PAP **241**.

Proserpio, C. A. 1978. Descripción geológica de la Hoja 42d, Gastre, Provincia del Chubut. SERV GEOL NAC BOL BUENOS AIRES ARGENTINA **161**.

Ramos, V. A. 1984. Patagonia—Un continente Paleozoico en la deriva? ACT CONGR GEOL ARGENTINO, SAN CARLOS DE BARILOCHE, ARGENTINA **II,** 311–25.

Ramos, V. A. & Kay, S. M. 1991. Triassic rift basalts of the Cuyo Basin, Central Argentina. *In* Harmon, R. S. & Rapela, C. W. (eds) *Andean Magmatism and its Tectonic Setting.* GEOL SOC AM SPEC PUB **265,** 79–91.

Rapela, C. W. & Kay, S. M. 1988. Late Paleozoic to Recent Magmatic Evolution of Northern Patagonia. EPISODES **11** (3), 176–82.

Rapela, C. W. & Pankhurst, R. J. 1992. The granites of northern Patagonia and the Gastre fault system in relation to the break-up of Gondwana. *In* Storey, B. C., Alabaster, A. & Pankhurst, R. J. (eds) *Magmatism and the causes of continental break-up.* GEOL SOC LONDON SPEC PUBL (in press).

Rapela, C. W., Días, G. F., Franzese, J. R., Alonso, G. & Benvenuto, A. R. 1991. El Batolito de la Patagonia Central: evidencias de un magmatismo Trásico–Jurásico asociado a fallas transcurrentes. REV GEOL CHILE **18**(2), 121–38.

Rapela, C. W. & Alonso, G. 1991. Composición litologica y geoquimica del batolito de la Patonia central. ACTAS VI CONGR GEOL CHILENO, VIÑA DEL MAR, CHILE **1,** 236–40.

Roenlandts, I & Michel, G. 1986. Sequential inductively coupled plasma determination of some rare-earth elements in five French geostandards. GEOSTANDARDS NEWSL **10**(2), 135–54.

Stipanicic, P. N. & Methol, E. J. 1980. Comarca Norpatagonica. 2° *Simposio de Geología Regional Argentina* II, 1,071–97. Córdoba: Academía Nacional de Ciencias.

Uliana, M. A., Biddle, K. T. & Cerdan, J. 1989. Mesozoic extension and the formation of Argentine sedimentary basins. *In* Tankard, A. J. & Balkwill, H. R. (eds) *Extensional Tectonics and Stratigraphy of the North Atlantic Margin* AM ASSOC PET GEOL MEM **46,** 599–613.

Uliana, M. A. & Biddle, K. T. 1987. Permian to Late Cenozoic evolution of northern Patagonia, main tectonic events, magmatic activity, and depositional trends. *In* McKenzie, G. D. (ed.) *Gondwana Six: Structure, Tectonics, and Geophysics.* AM GEOPHYS Union, Geophys Monogr **40,** 271–86.

Whalen, J. B., Currie, K. L. & Chappell, B. W. 1987. A-type granites: geochemical characteristics, discrimination and petrogenesis. CONTRIB MINERAL PETROL **95,** 407–19.

Williamson, J. H. 1968. Least-squares fitting of a straight line. CAN J PHYS **46,** 1845–7.

C. W. RAPELA, Centro de Investigaciones Geológicas, Universidad Nacional de La Plata, 644 Calle No. 1, 1900 La Plata, Argentina.

R. J. PANKHURST and S. M. HARRISON, British Antarctic Survey, c/o NERC Isotope Geosciences Laboratory, Kingsley Dunham Centre, Keyworth, Nottingham NG12 5GG, U.K.

Transactions of the Royal Society of Edinburgh: Earth Sciences, **83**, 305–316, 1992

The role of magma sources, oxidation states and fractionation in determining the granite metallogeny of eastern Australia

Phillip L. Blevin and Bruce W. Chappell

ABSTRACT: The ore-element associations of granite-related ore deposits in the eastern Australian Palaeozoic fold belts can be related to the inferred relative oxidation state, halogen content and degree of fractional crystallisation within the associated granite suites. Sn mineralisation is associated with both S- and I-type granites that are reduced and have undergone fractional crystallisation. Cu and Au are associated with magnetite- and/or sphene-bearing, oxidised, intermediate I-type suites. Mo is associated with similar granites that are more fractionated and oxidised. W is associated with a variety of granite types and shows little dependence on inferred magma redox state. The observed ore deposit-granite type distribution in eastern Australia, and the behaviour of ore elements during fractionation, is consistent with models of ore element sequestering by sulphides and Fe–Ti phases (e.g. pyrrhotite, ilmenite, sphene, magnetite) whose stability is nominally f_{O_2}-dependent. Fractional crystallisation acts to amplify this process through the progressive removal of compatible elements and the concentration of incompatible elements into decreasing melt volumes. The halogen content is also important. S-type granites are poorer in Cl than I-types. Cl decreases and F increases in both S- and I-type granites with fractional crystallisation. Low Cl contents combined with low magma f_{O_2} in themselves seem to provide an adequate explanation for the rarity of Mo, Cu, Pb and Zn type mineralisation with S-type granites. Although such properties of granite suites seem adequately to predict the associated ore-element assemblage to be expected in associated mineral deposits, additional factors determine whether or not there is associated economic mineralisation.

KEY WORDS: halogens, Lachlan Fold Belt, metallogenesis, mineral deposits, ore elements

The correlation of types of ore deposits (Cu, Mo, Sn, etc.) with specific classes of granites has long been a focus of attention in studies of mineral deposits. Broad associations are recognisable, such as those of porphyry Cu–Au deposits with oxidised intermediate magmas, and of Sn deposits with reduced and fractionated felsic magmas (Burnham & Ohmoto 1980; Lehmann 1990). That there are genetic links between those deposit types and their associated intrusives is no longer a controversial issue. Such associations suggest that processes operating within these magmatic systems were responsible for producing the ore element assemblages observed within ore deposits. This then invites questions regarding the genetic nature of such associations and the factors or processes, specific to those intrusive rocks, which were responsible for producing the ore element assemblage within the related ore deposits. Are those assemblages simple consequences of the metal concentrations inherited from the source region of the magma (e.g. Sn, Plimer 1980), or the result of subsequent magmatic processes (e.g. W, Newberry & Swanson 1986), or are they the result of the interplay of these factors in conjunction with the tectonic and stratigraphic setting?

In this discussion, we examine the importance of source protolith and the role of magmatic processes and intensive variables in producing the ore element assemblages observed within magmatic hydrothermal mineralisation, using the granites of the Lachlan Fold Belt (LFB) of southeastern Australia as an example.

1. Granites of the Lachlan Fold Belt

Aspects of the granite geology of the LFB have been discussed by Chappell (1984), Chappell and Stephens (1988), White and Chappell (1988) and Chappell and White (1992). Most of the following is an overview condensed from those studies and is repeated here because of its importance to the model that is being developed.

1.1. The I-S model: granites as images of their sources

All granites of the LFB are considered to have been derived by partial melting of the crust. S-type granites are inferred to have been sourced from sedimentary (supracrustal) protoliths. Those source rocks were depleted in Ca, Na and Sr due to the destruction of feldspars and the formation of clay minerals during earlier chemical weathering. I-type granites are inferred to have been derived from older igneous protoliths resident within the crust (infracrustal). They were not directly derived from the mantle or mantle wedge sources as in the case of arc-related M-type granites. A-type granites (Loiselle & Wones 1979; Collins *et al.* 1982) are also present within the LFB but are volumetrically insignificant (<1%) and they are not related to mineralisation in that region. They will not be discussed further here.

1.2. Granite suites, supersuites and the concept of basement terranes

An important feature of the LFB granites is that individual plutons can be grouped together into suites using petrographic, chemical and isotopic criteria (e.g. Hine *et al.* 1978). Granite suites are considered to have been derived from a narrow range of source-rock compositions. I-type suites are usually better defined than S-type suites and this probably reflects the more chemically homogeneous nature of their igneous protoliths. Suites with broadly similar characteristics, but differing in detail, can be grouped into supersuites (e.g. the Boggy Plain Supersuite of Wyborn *et al.* 1987).

Chappell *et al.* (1988) recognised regional petrographic and chemical patterns, or provincial characteristics, among granites in the LFB. Since the granites formed by melting of the deep crust, these patterns are thought to represent regional variations in the character and composition of that deep crust. On that basis, Chappell *et al.* (1988) recognised nine basement terranes within the LFB. These terranes do not generally correspond with those that have been proposed from near-surface structural features, or stratigraphic relationships within the rather monotonous cover of Ordovician flysch that extends throughout most of the LFB.

1.3. Fractionation mechanisms

Individual suites show distinctive trends when elements are plotted on variation diagrams against weight per cent SiO_2 or total Fe (FeO*). Some suites show the curved trends characteristic of fractional crystallisation processes, dominantly that of feldspar crystallisation, while plots for other suites produce straight line trends for elements even over a considerable range of composition (Fig. 1). The latter trends have been ascribed to varying degrees of fractionation between melt and the solid residual fraction (restite) carried up from the source (White & Chappell 1977; Chappell *et al.* 1987) and are the chemical expression of the degree of separation of those two fractions. An important implication of restite fractionation is that trace elements never become significantly enriched or depleted with respect to their original source concentrations. Fractional crystallisation may occur in the more felsic members of these suites that have rid themselves of their restite component; for example, the Koetong Supersuite (Chappell *et al.* in prep). Fractional crystallisation is a dominant process only in cases where a high-temperature melt-rich magma leaves its restite component (possibly at the source), and such behaviour was uncommon in producing the granites of the LFB.

A good measure of the presence and extent of fractional crystallisation within granite suites of the LFB is provided by Rb (Fig. 1). In the LFB, Rb concentrations above 250 ppm (in conjunction with appropriate sympathetic behaviour of Ba and Sr) can be taken as an indication that significant feldspar fractionation has occurred (Chappell *et al.* 1988). Rb (or K/Rb) is preferred to Ba or Sr as a discriminant in the LFB because marked differences occur in absolute Ba and Sr contents between different suites (as a result of the corresponding differences in source-rock compositions) and because Sr is low in S-type relative to I-type granites.

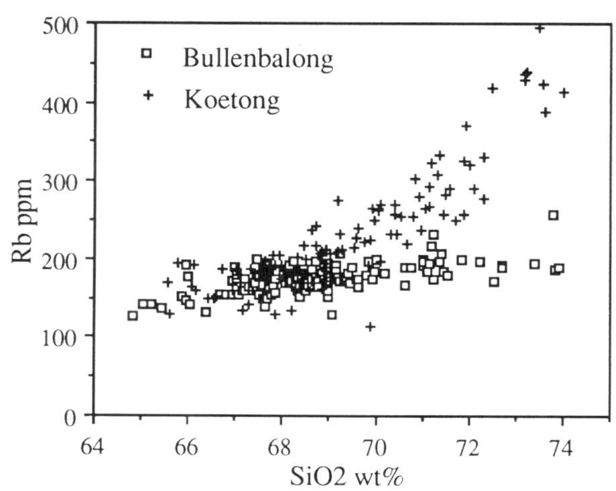

Figure 1 Comparison of Rb relative to SiO_2 in granites of the S-type Bullenbalong and Koetong supersuites.

1.4. Oxidation state of I- and S-type granites: a source-controlled variable

The petrographic, chemical and isotopic features that serve to discriminate between I- and S-type granites (e.g. Chappell & White 1984, 1992) are characteristics inherited from the protoliths of these rocks. One important difference between the two types for our present purposes is oxidation state, as expressed by Fe, and this will now be addressed in more detail.

Oxidation states of unaltered mafic to felsic magmas range over seven orders of magnitude of f_{O_2} (Carmichael 1991). Magmatic f_{O_2} is largely governed by the magmatic Fe^{3+}/Fe^{2+} ratio which is in turn dominantly controlled by the source region of the magma (Burnham & Ohmoto 1980; Carmichael 1991). S-type granites are typically more reduced than I-types (Chappell & White 1984) and are characterised by the presence of ilmenite, pyrrhotite and bone-coloured alkali-feldspar, whereas most I-types contain sphene, magnetite, pyrite and salmon-pink-coloured alkali feldspars. I-type granites are sometimes relatively reduced and they then contain ilmenite and bone-coloured alkali-feldspar in addition to hornblende (e.g. those of the Melbourne Basement Terrane; Chappell *et al.* 1988). Flood and Shaw (1975) ascribed the relatively reduced nature of S-type granites to the presence of graphite within the source-rocks. This acts to reduce S-type magmas to below the quartz-fayalite-magnetite buffer (QFM) towards the graphite-methane-(H_2O) buffer (GM). Reduced I-type granites do not possess that feature necessarily as a consequence of the involvement of reduced sedimentary material in their formation, but are more probably a reflection of the wide range of oxidation states present within infracrustal protoliths.

Comparison of weight per cent Fe_2O_3/FeO ratios for biotite and ilmenite from S- and I-type granites in eastern Australia (Fig. 2(a)) shows higher proportions of Fe^{3+} in I-type minerals (see also Whalen & Chappell 1988). This, coupled with the presence of magnetite and hornblende in the I-types translates to higher whole rock Fe_2O_3/FeO values for most I-type relative to S-type granites. A plot of three LFB supersuites (Fig. 2(b)) demonstrates that for each value of FeO* the ilmenite-bearing S-type Koetong and Strathbogie supersuites have lower Fe_2O_3/FeO ratios than the magnetite- and sphene-bearing I-type Boggy Plain Supersuite. Fe_2O_3/FeO provides a crude but (for the purposes of this paper) useful indicator of the *relative* oxidation state of granite magmas, provided that the chemical compositions of the granites being compared are similar, that the ratio reflects similar redox conditions to those inferred from the mineral assemblage and chemistry, and that the iron-bearing minerals have not been affected by later alteration. The clear discrimination of the three supersuites in Fe_2O_3/FeO–FeO^* space (Fig. 2(b)) is good evidence for source control of magma redox conditions in felsic granites. Burnham and Ohmoto (1980) regarded oxidised I-type and reduced S-type granites as analogues of magnetite-series and ilmenite-series granites, respectively (Ishihara 1981), placing the former f_{O_2}-T space at magmatic temperatures at $f_{O_2} >$ QFM, at or above Ni–NiO (NNO) buffer, and the latter at GM $< f_{O_2} <$ QFM, which is around the CO_2/CH_4 isofugacity curve at those temperatures. Limited unpublished mineralogical data for the LFB granites, and those of Whalen and Chappell (1988), support this conclusion for "typical" I- and S-type granites.

2. Granite–ore deposit relationships in eastern Australia

2.1. General granite metallogeny of eastern Australia

Granites in the Palaeozoic fold belts of eastern Australia are associated with characteristic ore deposit types discernable at a regional scale (Fig. 3). In the Hodgkinson Province of N Queensland, Sn and/or W are associated with both fractionated I- and S-type granites, and W + Mo ± Bi is associated only with oxidised and fractionated I-types. Cu, Pb and Ag occur with the less fractionated granodioritic I-type rocks of the Chillagoe region. Au is associated with I-type, and with unstudied granites, in the Thompson Fold Belt to the S, including the large Kidston deposit (61 t Au; Baker & Andrew 1991).

In the northern portion of the New England Fold Belt, Cu–Au and Mo deposits are common, though usually of low grade (Horton 1978; Murray 1990). The Mount Morgan deposit (370 kt Cu, 277 t Au; Arnold & Sillitoe 1989) is now regarded as magmatic in origin and it is associated with I-type tonalitic to granodioritic rocks. Other Cu–Au type deposits are commonly porphyry-like in style (Horton 1978) and are associated with I-type tonalitic, dioritic and granodioritic intrusions. The largest Mo deposit in eastern Australia at Anduramba (11·2 kt Mo; Murray 1990) is located in a porphyritic rhyolitic intrusive rock which is associated with a fractionated suite of oxidised I-type granites. In the southern section of the New England Fold Belt (Fig. 3), Sn ± W mineralisation is associated with fractionated granites that are probably I-type (possibly A-type) which have yielded ~200 kt of Sn, largely from alluvial workings (Markham & Basden 1974). Mo is also present and is associated with fractionated and oxidised I-type granites (Weber *et al.* 1978).

2.2. Granite type–mineral deposit type associations in the LFB

The Siluro-Devonian and Carboniferous LFB granites are associated with around one thousand mineral deposits which range in size from minor showings to world-class deposits of Sn and W. Au, Cu, Mo, Pb, Ag, Zn, Bi and F have also been mined. The general distribution of I- and S-type granites within the LFB is shown in Figure 4(a) and the

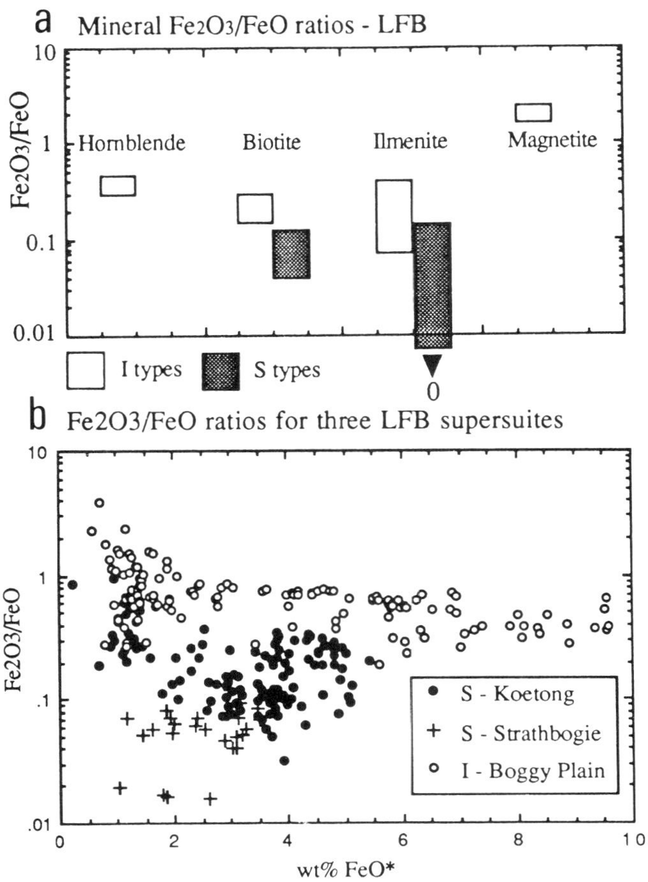

Figure 2 (a) Fe_2O_3/FeO ratios for minerals in I-type and S-type granites of LFB. (b) Comparison of Fe_2O_3/FeO relative to FeO^* for three LFB supersuites.

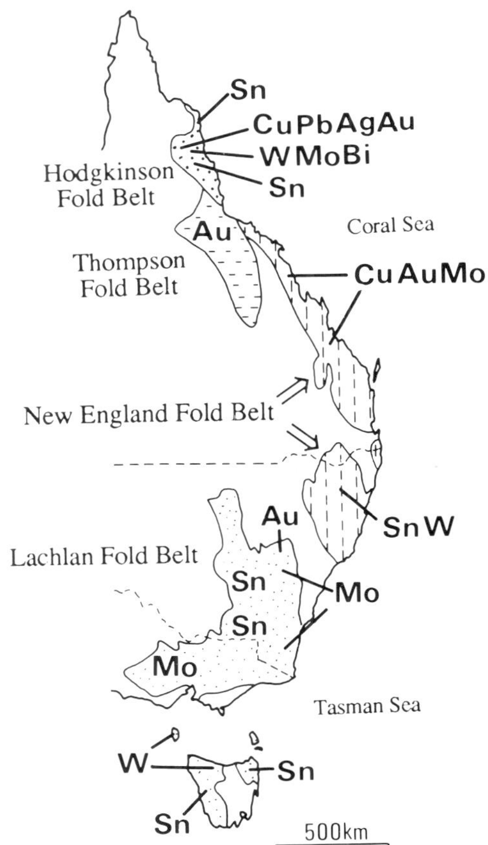

Figure 3 Distribution of important granite-metallogenic associations in the Palaeozoic fold belts of eastern Australia.

distribution of representative Sn, Mo and W deposits in Figure 4(b)–(d). The Koetong Supersuite comprises most S-type granites within the Wagga Basement Terrane (WBT; Chappell *et al.* 1988; Chappell *et al.* in prep.), while most S-type granites in the W and S of the Kosciusko Basement Terrane (KBT) are members of the Bullenbalong Supersuite. The I-type Boggy Plain Supersuite is present in the KBT, WBT, and Howqua basement Terrane (HBT).

2.2.1. Sn mineral deposits. Most Sn production within the LFB has come from deposits in Tasmania associated with fractionated and reduced S-type granites (Fig. 4(b)). The largest of these are carbonate replacement deposits at Renison, Mount Bischoff and Cleveland. Most mainland production (~60 kt) has come from the fractionated S-type Koetong Supersuite of the WBT. Production from the reduced and fractionated I-type Woolshed Valley granite in

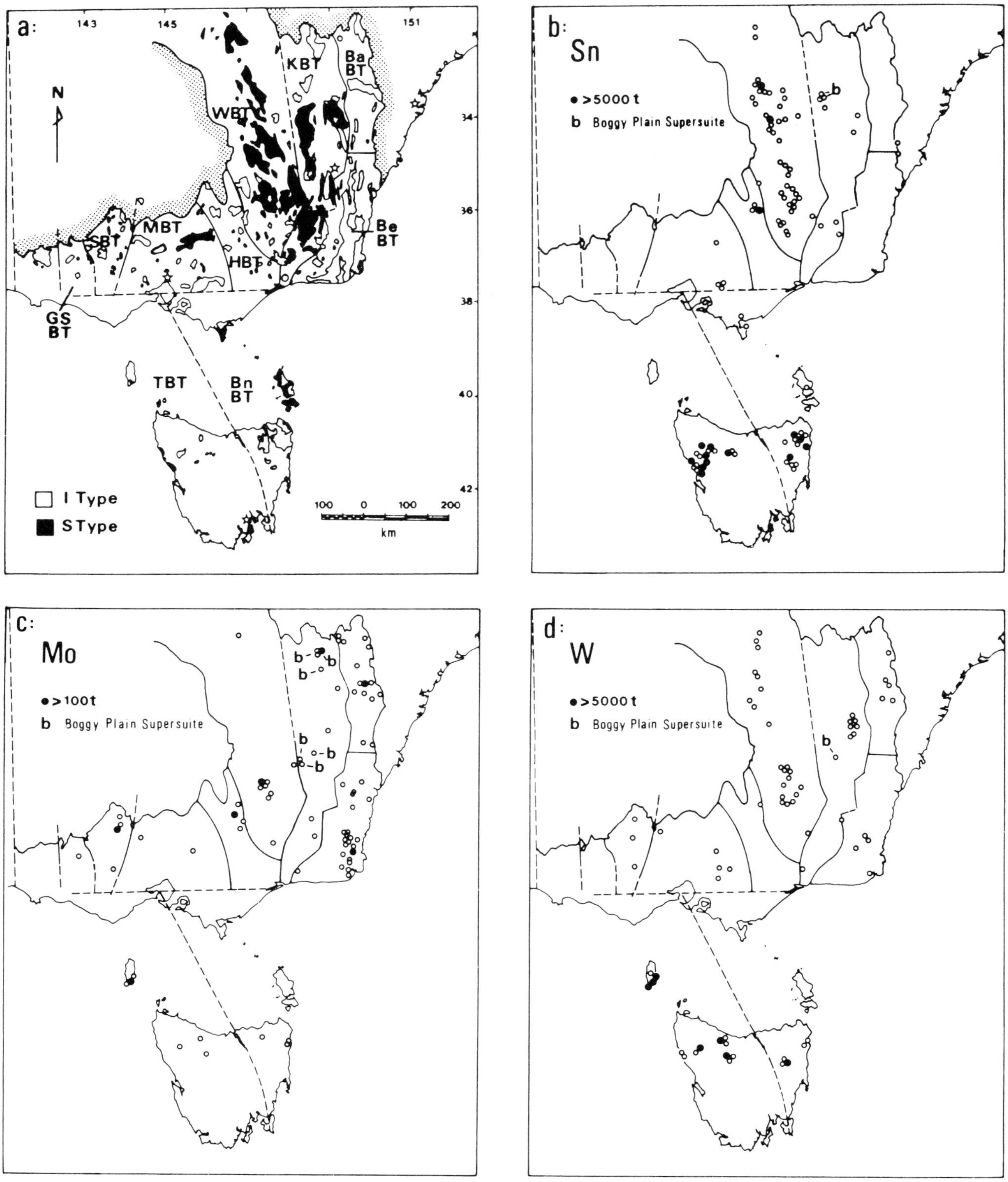

Figure 4 (a) Distribution of granite types in the LFB; basement terrane boundaries and abbreviations are from Chappell *et al.* (1988); distribution of representative granite-related ore deposits of Sn, Mo and W are shown in (b), (c) and (d), respectively.

the HBT (Fig. 4(a)) has all been alluvial (10 kt; Cochrane 1971). Sn production from elsewhere from the LFB has been minor (combined hard-rock production <50 t; total Sn production <300 t). The contrast between the Sn production associated with the restite-controlled Bullenbalong Supersuite of the KBT (<10 t) and the Koetong Supersuite of the WBT (~50 kt), is noteworthy.

2.2.2. Mo mineral deposits. The distribution of Mo deposits (Fig. 4(c)) emphasises the association of Mo deposits with oxidised I-type granites, in the Bathurst and Bega Basement terranes (BaBT and BeBT in Fig. 4(a)), with the Boggy Plain Supersuite, and elsewhere across the fold belt. Mo mineralisation usually occurs in close spatial association with fractionated felsic variants of these granites. It is noteworthy that the two largest Mo concentrations in the LFB are in the King Island W deposit (105 kt W vs 4·2 kt Mo; Williams *et al.* 1989; Wesolowski *et al.* 1988) and the Yeoval Cu prospect (85·1 kt Cu vs 2·6 kt Mo; Weber *et al.* 1978). The largest Mo-dominant deposit is the Mt Tennyson skarn (1·5 kt Mo) but even there, the Mo/W ratios approach unity in some parts of the deposit (Weber *et al.* 1978). Mt Moliagul in the Stavely Basement Terrane (SBT in Fig. 4(a)) comprises low- to moderate-grade disseminated Mo–Cu mineralisation in the felsic phases of an oxidised and somewhat fractionated granodiorite.

2.2.3. W mineral deposits. Numerous W occurrences are present in the LFB and the King Island scheelite skarn deposit (Fig. 4(d)) is world-class in size with a total resource of 105 kt W (Williams *et al.* 1989). Mo is also present as Mo-rich scheelite and molybdenite (Wesolowski *et al.* 1988). The deposit is associated with monzogranite to granodiorite with a range of SiO_2 from 58–69%. Late magmatic leucocratic apophyses and dykes are also present. The granites are biotite- and hornblende-bearing with early sphene, magnetite and traces of ilmenite, chalcopyrite and pyrrhotite (Wesolowski *et al.* 1988). The Housetop Suite, associated with the Kara W deposit, is oxidised, fractionated and contains hornblende, biotite and accessory magnetite and sphene.

Two W granite associations are present in the LFB: W with minor Mo associated with oxidised I-type granites, and W with greater or lesser amounts of Sn associated with fractionated and reduced I- and S-types (Fig. 4(d)). W is present as wolframite with Sn at W/Sn ratios of similar magnitude in the Aberfoyle, Moina and Storeys Creek deposits in Tasmania. In deposits where W and Mo occur together, the W/Mo ratio is highly variable, ranging from high (>10, e.g. King Island, Kara and Black Andrew), near unity (e.g. Mt Tennyson) to low (<10, e.g. Whipstick, Holbrook).

2.2.4. Cu and Au mineral deposits. Cu occurs in the Yeoval Cu prospect (85 kt Cu) along with several smaller deposits associated with granodioritic phases of the Yeoval Diorite Complex, a mafic member of the Boggy Plain Supersuite (Wyborn *et al.* 1987). Cu mineralisation is also associated with the Bannimboola Monzodiorite, also of that supersuite, and occurs as several small deposits associated with the oxidised Bete Belong Suite of eastern Victoria. Au is associated with the first two deposits and also has been produced in significant quantity from alluvial and hardrock workings in the Braidwood Granodiorite (~30 t Au). Gold is also currently produced from the Wonga mine located in the aureole of the Stawell pluton in western Victoria (Watchorn & Wilson 1989). Both the Braidwood and Stawell granites are oxidised. Numerous other hard-rock Au occurrences are present in the LFB but most are ambiguous in their genetic relationships to the associated granites.

2.3. Contrasts between mineralised granite suites

There is a consistent link between the ore-element assemblage in mineral deposits within the LFB, and the I-S, redox and fractionation characteristics of the associated granite suite. Variations between mineralised suites in terms of these parameters are shown in Figure 5. Suites associated with Sn and Cu–Au–Mo mineralisation plot as two distinct groups on both the $Fe_2O_3/FeO–FeO^*$ and $Fe_2O_3/FeO–Rb$ plots (Fig. 5). A continuum is present in the latter association from Cu (higher FeO*, lower Rb) through Au to Mo (lower FeO*, higher Rb). This trend is apparent even within single supersuites such as Boggy Plain, in which Cu mineralisation is associated with granodioritic and monzodioritic plutons and Mo and W mineralisation with felsic fractionated plutons.

W-mineralised suites show significant overlap with the Sn and Cu–Au–Mo associations in terms of Fe_2O_3/FeO at lower FeO* and higher Rb (Fig. 5). This is not unexpected, given the association of W with both Mo- and Sn-mineralisation. No continuum in ore-element assemblages appears to exist between W–Sn and W–Mo associations (i.e. W–Mo–Sn deposits). This suggests that the Sn and Cu–Au–Mo associations are distinct from each other in terms of these parameters and that W may be associated with the fractionated members of each. Granites associated with deposits having low W/Mo ratios tend to be more fractionated and more oxidised than those associated with high W/Mo ratio ore deposits. Granites associated with W–Sn mineralisation do not show consistent similar patterns (Section 4.3).

2.4. Contrasts between mineralised and unmineralised granite suites

Based on the observed distribution of ore deposits discussed above, significant Sn mineralisation is unlikely to occur with oxidised granite suites, and significant Mo, Cu and Au mineralisation is unlikely to be associated with granite suites that are reduced. On these grounds, S-type granites seem to be precluded from being prospective for Mo. In addition to the appropriate oxidation state, fractional crystallisation is also necessary. The absence of mineralisation associated with restite-fractionated suites, particularly S-types, is remarkable. The reduced but restite-fractionated S-type Bullenbalong Supersuite has a similar range of SiO_2 and FeO* to that of the Koetong Supersuite, but unpublished data show that Sn production associated with that extensive group of rocks has amounted to less than 50 t.

If fractional crystallisation is necessary for the production of substantial Sn and Mo mineralisation, other geochemical signatures that result from extended fractional crystallisation should also be evident. These would include high Rb, Nb, Ga and F, and low Sr, Ba, Ca and K/Rb. Restite-fractionated suites lack these trace element signatures and this emphasises the importance of using trace elements in this connection rather than major elements, which tend toward minimum-temperature melt compositions regardless of the fractionation process and the extent of its development.

Cu, Au and some W mineralisation are associated with granites of intermediate composition that do not show evidence of strong fractional crystallisation, but nevertheless are part of crystal fractionated suites (e.g. Au and Cu in the Bannimboola and Yeoval granodioritic to dioritic plutons of the Boggy Plain Supersuite). Apart from petrographic evidence for the presence of restite, it may be difficult to distinguish between the more mafic members of restite- and crystal-fractionated suites. Intermediate granites that are part of fractionally crystallised suites may be characterised by zoned ferromagnesian and feldspar minerals, lack distinct cores in plagioclase, and may be more likely to form zoned plutons (e.g. the Boggy Plain, Bannimboola and Yeoval plutons of the Boggy Plain Supersuite, Wyborn *et al.* 1987). Late-stage felsic variants may also be present. Such characteristics suggest that these plutons were initially melt-rich and had undergone fractional crystallisation.

3. Ore-element behaviour in felsic magmas: the role of magma oxidation state and fractionation mechanism

The above patterns of association between metallogenesis and granite type are largely empirical. In order to understand these relationships genetically, we need first to understand the behaviour of the ore-elements within felsic magmatic systems. The consistent metallogenic patterns shown by the LFB granites in terms of oxidation state and fractional crystallisation suggest that it is the combination of these features that determines the ore-element assemblages that are seen in related mineralisation.

Tacker and Candela (1987) recognised the importance of the redox-dependent melt-crystal partitioning behaviour of ore elements in the case of Mo. This concept can be extended to include W (Candela & Bouton 1990), Cu (Lynton *et al.* 1990), Sn (Lehmann 1990) and Au (by inference, based on its association with Mo and Cu). In magmatic-hydrothermal ore genesis, an ore element-bearing magmatic aqueous phase first evolves through exsolution from a crystallising melt. For this to occur, however, these ore elements must be available for partitioning into the aqueous phase, i.e. be present in the melt in sufficient abundance. For compatible elements, therefore, aqueous fluid exsolution needs to occur early rather than late during crystallisation (Candela & Holland 1986; Candela 1989).

The oxidation-state dependence on melt-crystal partitioning of ore elements can be tested by examining the behaviour of these elements within certain suites that represent a range of differing oxidation states and fractionation mechanisms. Such a test is doubly interesting because the metallogenic associations of these suites are also known.

3.1. Theoretical and experimental constraints on the behaviour of some ore elements

Sn speciation in melts is not fully understood, although its behaviour can be adequately modelled based on the redox pair Sn^{4+}/Sn^{2+} (Lehmann 1990). Octahedrally coordinated Sn^{4+} has an ionic radius similar to that of Ti^{4+} and Fe^{3+}, enabling it to substitute for these ions in minerals such as

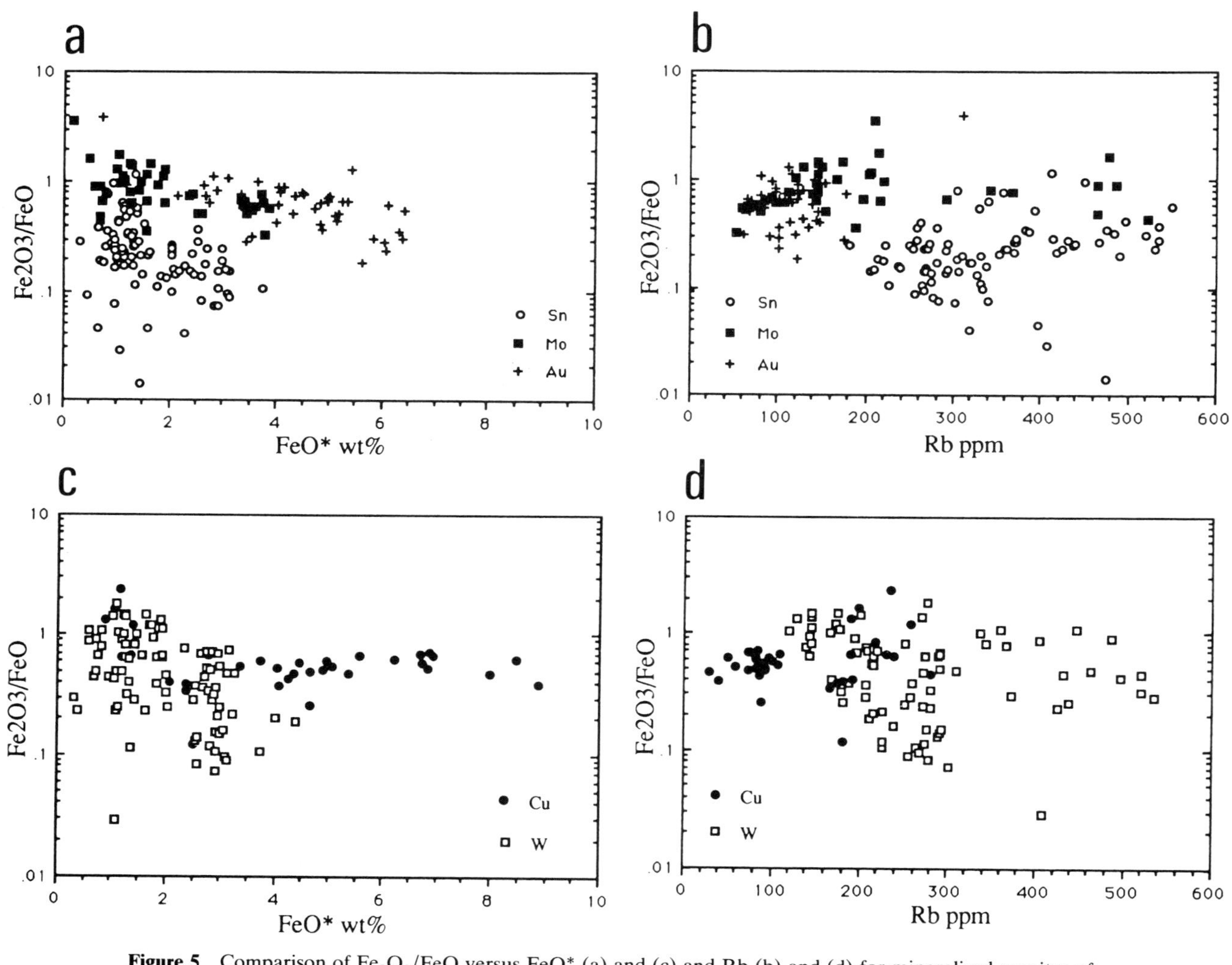

Figure 5 Comparison of Fe_2O_3/FeO versus FeO* (a) and (c) and Rb (b) and (d) for mineralised granites of the LFB; Cu, Mo and Au mineralised granites have distinctly higher Fe_2O_3/FeO ratios than those associated with Sn; Mo mineralised granites are distinguished from those related to Au and Cu by having generally lower FeO* and higher Rb (i.e. more felsic and fractionated); W mineralised granites show broad overlap with Mo and Sn mineralised granites in terms of both Fe_2O_3/FeO and Rb.

biotite, hornblende, sphene, ilmenite and magnetite. Sphene may accept substantial quantities of Sn and Ivanova and Butuzova (1968) have shown that sphene accepts more Sn than either Mo or W. In this scenario, the increase in the concentration of Sn in melts during fractional crystallisation would be enhanced by reduced conditions (see also Taylor & Wall 1990). The partitioning behaviour of Mo between magnetite, and of W and Mo between ilmenite and rhyolitic melts, has been studied respectively by Tacker and Candela (1987) and Candela and Bouton (1990). Those studies found that the partition coefficient for Mo between ilmenite and melt increased as f_{O_2} decreased from NNO to GM, in agreement with similar experiments using magnetite (Tacker & Candela 1987). In contrast, the partition coefficient for W decreased, although less markedly. W may also substitute for Ti in sphene (Simon 1972, cited in van Middelaar & Keith 1990) and biotite, although to a lesser extent than does Sn.

Cu, because of its chalcophile nature, preferentially partitions into sulfide phases. Lynton *et al.* (1990) showed this experimentally for Cu partitioning between pyrrhotite and high silica rhyolite for both f_{O_2} = NNO and GM. Cu should therefore act as a crystal-compatible element in those magmatic systems where pyrrhotite is stable ($f_{O_2} < \sim$NNO) and crystallises early. Au, Pb, Zn, Mo, Co and Ni would probably also be removed by early sulfide crystallisation.

3.2. Behaviour of S in reduced and oxidised melts

Whalen and Chappell (1988) demonstrated petrographically that sulfides (pyrrhotite > pyrite > chalcopyrite) are texturally early and relatively abundant in S-type granites, whereas in I-types, sulfides (pyrite > chalcopyrite > pyrrhotite) are texturally late and less abundant. The behaviour of S in magmatic systems has been studied experimentally by Katsura and Nagashima (1974), Haughton *et al.* (1974), and Carroll and Rutherford (1985). Pyrrhotite solubility is at a minimum near f_{O_2} = NNO. At lower f_{O_2}, S solubility is a function of the magma Fe content (all other factors held equal). At higher f_{O_2} pyrrhotite breaks down and anhydrite becomes stable in Ca-bearing magmas. This is due to an increase in the SO_2/H_2S ratio with increasing f_{O_2}. This is in agreement with our observation that the S (as sulfate) contents of apatites from I-type granites are higher than for S-types. The higher SO_2/H_2S ratio in relatively oxidised magmas encourages greater partitioning of S into the aqueous phase by virtue of the lower solubility of SO_2 compared to H_2S in hydrous magmas (Burnham & Ohmoto 1980).

3.3. Observed ore element behaviour in the LFB granites

Sn contents of the LFB granites are highest in suites that are both crystal-fractionated and reduced. The contrasting behaviour of Sn between the crystal-fractionated Koetong and the restite-fractionated Bullenbalong S-type supersuites (Fig. 6(a), (b)) emphasises the role of extended feldspar fractionation in the enrichment of Sn. Both supersuites are S-type, are reduced, have similar SiO_2 and FeO* ranges, and have similar Sn contents (~6 ppm) at their most mafic compositions. The increase in the Koetong Supersuite (Fig. 6(a)) contrasts with that of the fractionated but oxidised Boggy Plain Supersuite (Fig. 6(c)). Prolonged fractionation (here monitored by the rise and fall of Ba) does not result in such an increase in the abundance of Sn in the latter case. The only elevated values are associated with highly felsic members in which magnetite and sphene are minor or absent. Reduced, ilmenite-bearing I-type granites show similar behaviour to the S-types.

The difference in Mo behaviour between I and S-type granites (Fig. 7(a), (b)) indicates a lack of response by Mo when S-types undergo crystal fractionation and is consistent

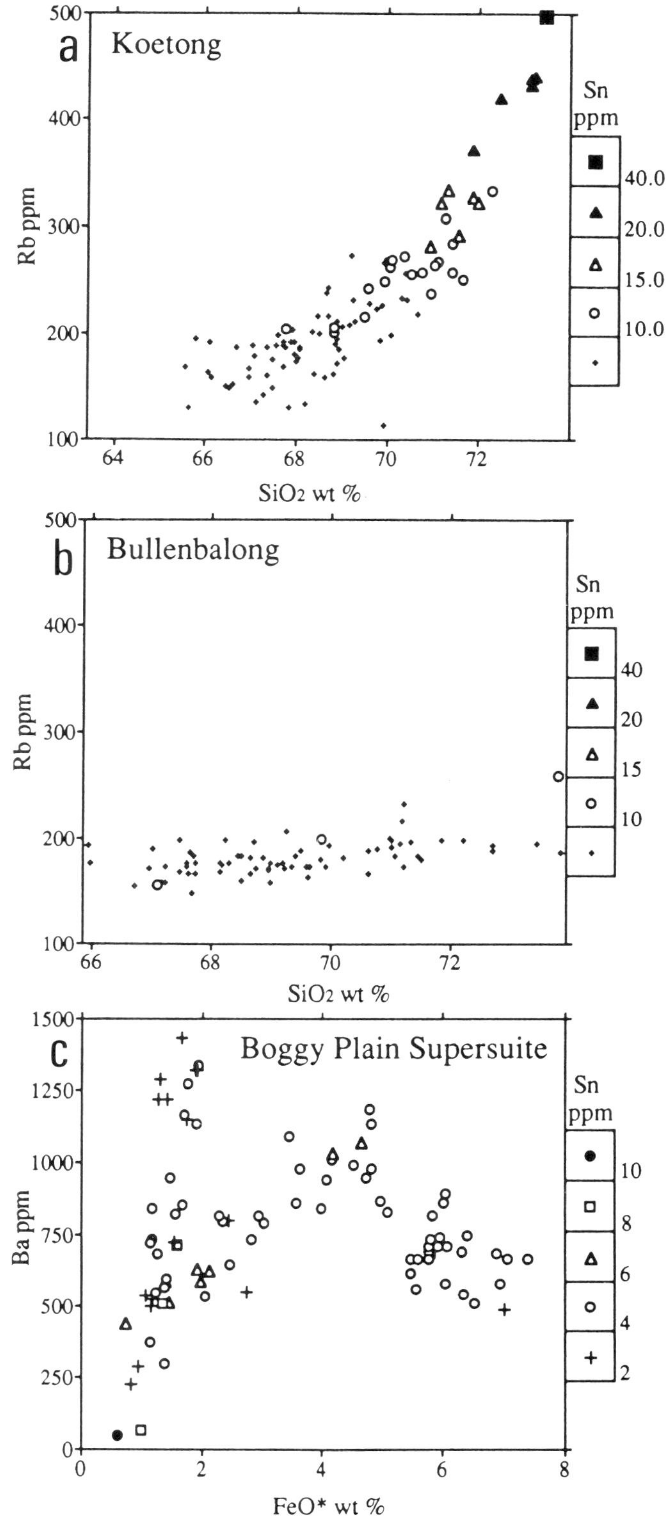

Figure 6 Behaviour of Sn within three LFB Supersuites; the S-type Koetong and Bullenbalong supersuites are inferred to have undergone dominant crystal- and restite-fractionation, respectively; a Ba versus FeO* plot for the oxidised I-type Boggy Plain Supersuite (c) shows the progressive build-up of Ba with decreasing FeO* until the incoming of biotite on the liquidus; only in the most felsic (<1% FeO) and fractionated (Ba <100 ppm) rocks does Sn increase in amount.

with the predicted behaviour for Mo in reduced melts (see above). Mo enrichment is most pronounced in oxidised I-type suites undergoing crystal fractionation and which are characterised by the presence of pink alkali-feldspars and sphene (Fig. 7(a)).

The distribution of W within S- and I-type granites is very similar (Fig. 8), despite the smaller data base, and shows that W increases in both types with crystal fractionation. The W content of the granite associated with the King Island scheelite skarn deposits (2–5 ppm, unpublished data; Wesolowski *et al.* 1988) is the same as that for geochemically similar but unmineralised granites.

Cu shows contrasts not only between I- and S-type granites but also within the I-type group itself (Fig. 9). This is despite the scatter inherent in Cu contents of crystalline plutonic rocks due to its mobile nature during subsolidus processes. Some suites such as those belonging to the Boggy Plain Supersuite are characterised by high Cu contents, particularly for the less felsic members. This suggests that these magmas may have acquired high initial Cu contents from their source protolith (a basaltic underplate; Wyborn *et al.* 1987) although this does not imply that the protolith itself was Cu-rich. Partial melting of hydrous S-bearing basaltic material may yield Cu-rich melts (Burnham 1979). Cu increases in concentration up to about $SiO_2 = 55\%$ and then falls steeply. This pattern of behaviour is similar to that found by Stanton (1990) for Cu in the porphyritic lavas of the Solomon Islands Younger Volcanic Suite (Fig. 10). Cu values above ~20 ppm are much less common in S-type granites (Fig. 9(b)) and the Cu content rapidly decreases with fractionation. Crystal fractionation of either I- or S-types diminishes the Cu content of the melt. The decline in the Cu content of I-type magmas, coupled with the increase in concentration of Mo during fractional crystallisation, results in a major change to the Cu/Mo ratio of the magma (Fig. 11).

3.4. Behaviour of Cl and F

The concentration of F and Cl in micas, amphiboles and apatite shows marked contrast between I- and S-type granites of both the LFB and the eastern Australian Palaeozoic region as a whole. F and Cl concentrations in biotites, expressed as intercept values (Munoz 1984), show that I-type granites are systematically richer in Cl than S-types (Fig. 12). F increases and Cl decreases with fractional crystallisation in both I- and S-type granites. This reflects the preferential partitioning of F into the melt phase. The decrease in Cl during fractionation may be a function of the preferential partitioning of Cl into crystallising OH-bearing minerals. Cl in silicate melts will partition strongly in favour of an aqueous phase (Kilinc & Burnham 1972). Halogen concentrations within micas largely reflect the halogen concentration within the aqueous phase with which the mica last equilibrated (Munoz 1984). If for most granites this phase is the magmatic aqueous phase

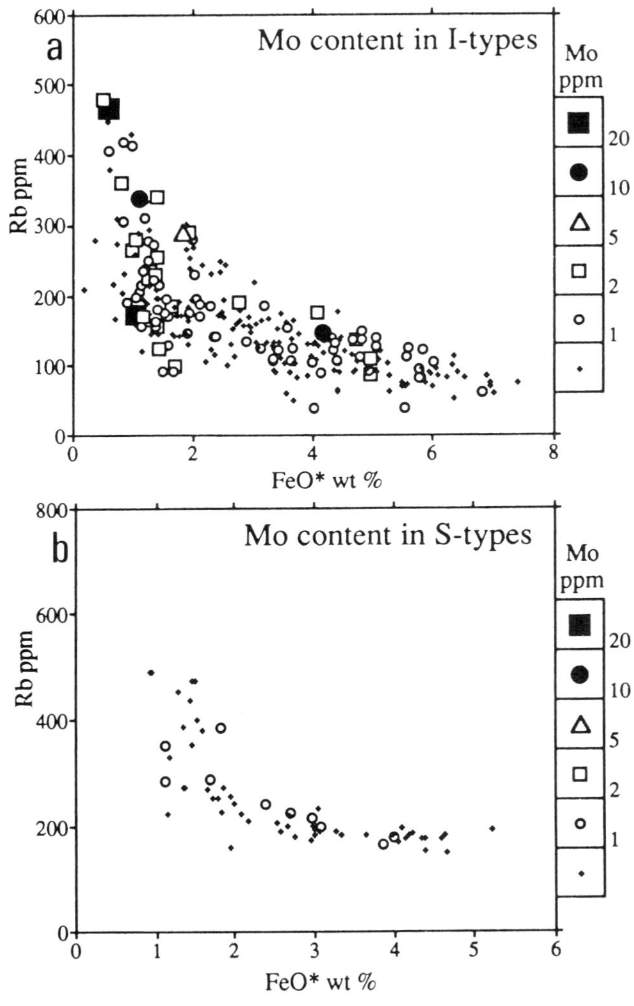

Figure 7 Behaviour of Mo in I- and S-type granites of the LFB; note the difference in abundance of Mo between I- and S-types; S-types show no increase in Mo with increasing Rb towards lower FeO*.

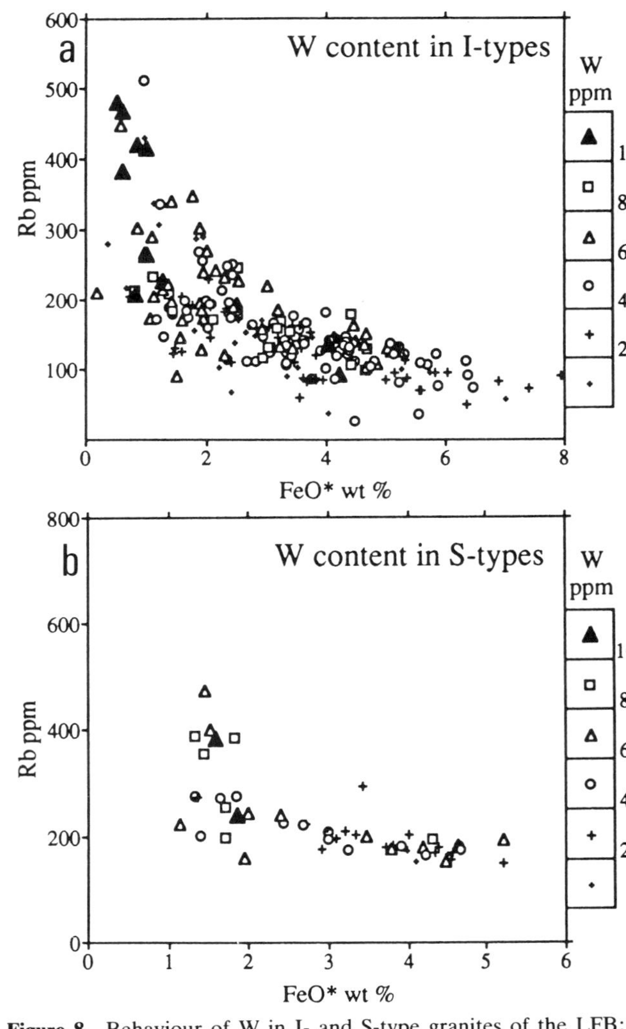

Figure 8 Behaviour of W in I- and S-type granites of the LFB; a general increase in W abundance with fractionation is evident in both I- and S-types.

exsolved during crystallisation, the Cl/F ratio of these fluids will be highest in less evolved granites (such as those associated with Cu mineralisation) and lowest in highly fractionated granites (such as those associated with Sn mineralisation).

4. Synthesis

4.1. Metallogenic models for I- and S-type granites

The foregoing shows that I- and S-type granites are fundamentally different in a variety of ways crucial from a metallogenic standpoint. These differences include source controls on magma redox and chemical compositions, contrasting fractionation mechanisms, redox-controlled crystal-melt partitioning of ore elements, and S and Cl concentrations. These differences can be integrated into metallogenic models for I- and S-type granites and, by inference, for granites in general. The two fundamental subdivisions are those based on oxidation state and the fractionation mechanism, as recognised above. Tentative models for the behaviour of these granite metallogenic associations are presented below. It should be emphasised that to some extent these represent working hypotheses and many details remain to be investigated and are thus subject to further interpretation and refinement.

4.1.1. Oxidised I-type granites undergoing fractional crystallisation. The behaviour of oxidised I-type magmas undergoing fractional crystallisation can be modelled from

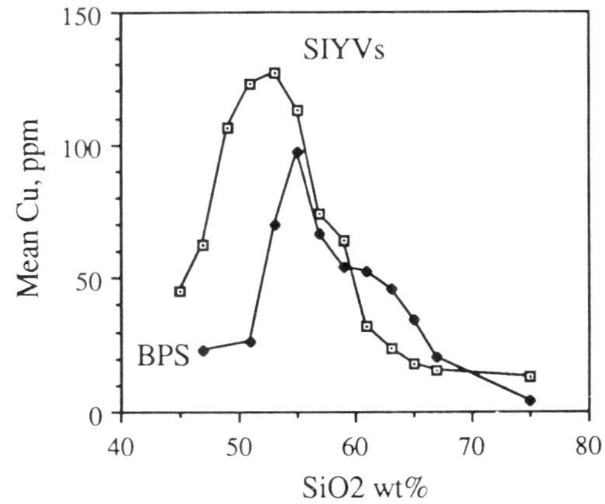

Figure 10 Comparison of Cu vs SiO_2 between the Solomon Islands Younger Volcanic Suite (SIYVs; Stanton 1990) and the Boggy Plain Supersuite (BPS); each point represents the mean Cu abundance (in ppm) calculated over 2% intervals of whole-rock SiO_2.

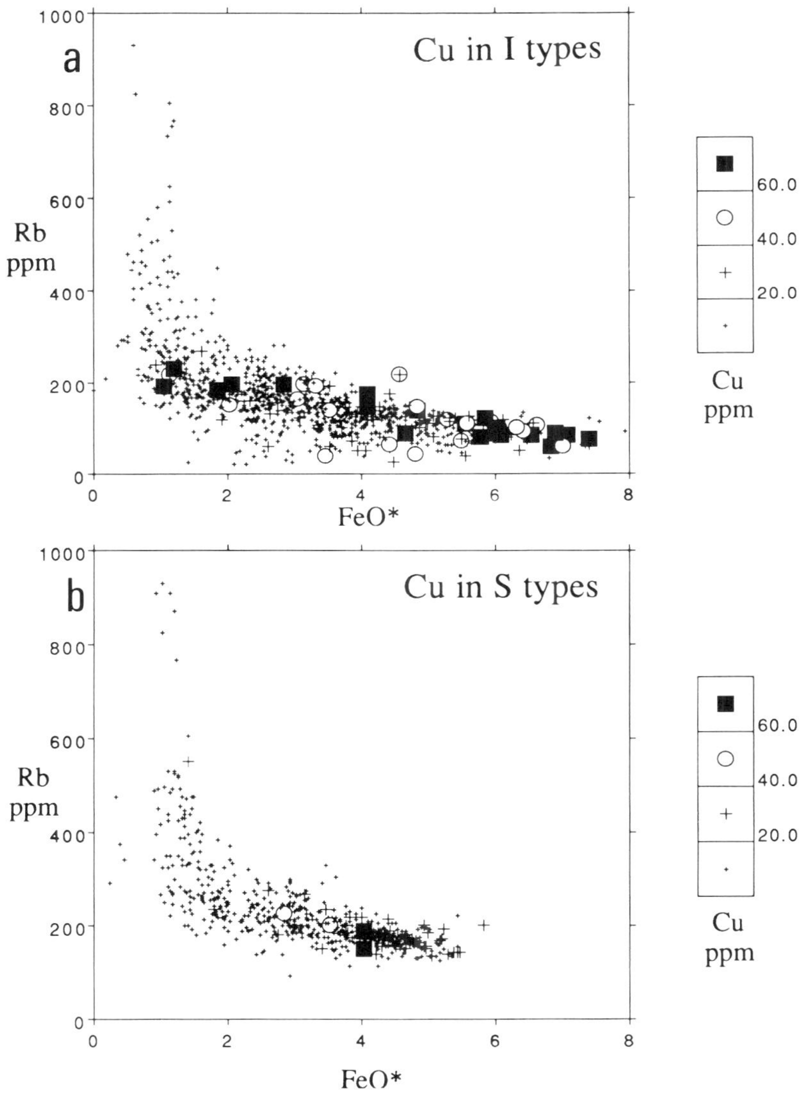

Figure 9 Behaviour of Cu in I- and S-type granites of the LFB; symbols refer to Cu abundance: small crosses <20, large crosses 20–40, circles 40–60, filled squares >60 ppm Cu; high Cu values at lower values of FeO* in the I-types belong to oxidised suites such as the Boggy Plain Supersuite; S-types have lower Cu values in general than do I-types, and S-type Cu values rapidly decrease towards lower FeO*.

the Boggy Plain Supersuite (Wyborn *et al.* 1987). Those authors concluded that the supersuite resulted from the high temperature melting of an underplated basaltic source within the crust. Oxidised magmas with $f_{O_2} > \sim$NNO would not have pyrrhotite stable as a crystalline phase, allowing Cu and other chalcophile elements to increase in concentration in the liquid during early fractional crystallisation. If crystallisation and early vapour exsolution were to occur at that stage, Cu, S and Cl and other incompatible elements (Au, Mo, Pb) would partition into the aqueous fluid phase (Candela & Holland 1986). Ni and Cr would also be removed from the melt by crystallising minerals such as (respectively) olivine and oxides. Continued fractional crystallisation would result in the progressive concentration of Mo and W. Sn would be sequestered by sphene and magnetite. The rapid decrease in Cu with further fractionation probably results from sulfide precipitation caused by decreasing S solubility as the melt becomes more felsic, fractionation of an Fe-bearing phase, or possibly through the loss of Cl and water during crystallisation. Plutons crystallising from these magmas are associated with W and/or Mo mineralisation that is characterised by high Mo/Cu ratios. It is only in highly fractionated, very felsic members of such a suite, where sphene, biotite and Fe–Ti oxides are either rare or absent, that increases in Sn contents occur. Any resulting Sn mineralisation at that stage is very minor.

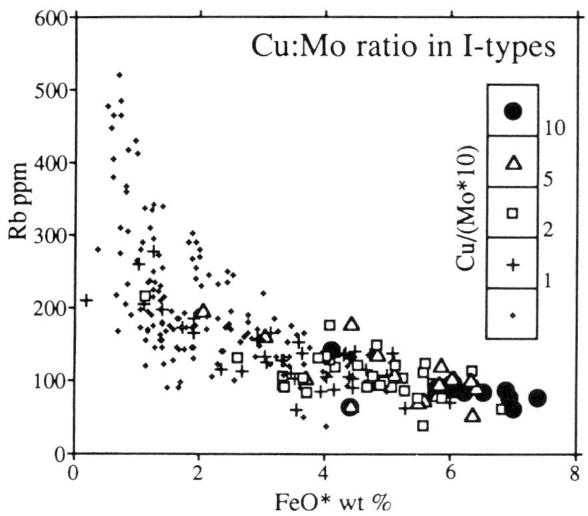

Figure 11 Cu/Mo ratio in I-types of the LFB; the pronounced decrease in the Cu/Mo ratio with decreasing FeO* is due to both a decrease in Cu and increase in Mo, particularly during fractional crystallisation; Cu mineralised granites in the LFB are higher in FeO* and lower in Rb than Mo mineralised granites (Fig. 5) and are also characterised by higher Cu/Mo ratios.

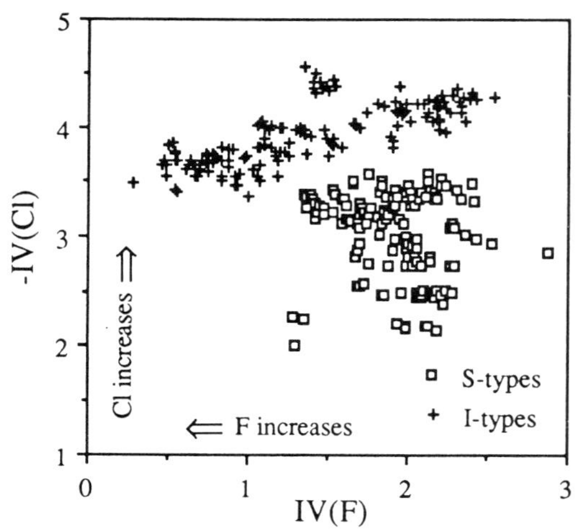

Figure 12 Concentration of F and Cl in biotite from I- and S-type granites from eastern Australia expressed as intercept values (IV) which correct F and Cl abundances in biotites for Fe–F and Mg–Cl avoidances (see Munoz 1984); for any value of IV(F), S-types have significantly lower Cl concentrations.

4.1.2. Reduced I- and S-type granites undergoing fractional crystallisation. Fractional crystallisation in a reduced magma appears to result in the early loss of Cu through the precipitation of sulfides. Mo may also be lost in that way, but also through its enhanced partitioning into ilmenite (or other Fe-Mg-Ti-bearing phases) under reducing conditions. Cl contents are less than that of equivalent I-type granites and that, allied with the inferred behaviour of S, will not encourage the efficient extraction of chalcophile elements from the melt into an exsolving aqueous phase. As fractional crystallisation continues, Sn and W would increase in amount in the melt phase, due to the absence of sphene and the low melt f_{O_2}, and partition into any exsolving aqueous phase. In such melts, the extraction of Sn and W would be enhanced by late exsolution of this aqueous fluid. Mineralisation would be typically poor in S (Burnham & Ohmoto 1980).

4.2. Unfractionated and restite-dominated I- and S-type granites

A result of this survey has been the observation of an extreme lack of significant magmatic-hydrothermal mineralisation associated with restite fractionated granite suites within the LFB. This result emphasises the importance of progressive crystallisation of melt-rich magmas in order to allow the concentration of elements into smaller melt volumes. The restite-fractionation mechanism has three important metallogenic implications. Firstly, trace element (and ore-element) concentrations within restite-fractionated suites will not vary significantly during differentiation. Assuming typical degrees of partial melting, the concentrations of these element within the melt fraction are not likely to be enriched more than a few times above their original concentration in the source protolith. Secondly, it is possible that ore elements present in the magma may be contained within restite minerals and not in the melt phase. Consequently those elements may never have been available for partitioning into an exsolving aqueous phase. Thirdly, mafic restite-rich magmas would be at a lower temperature and contain less available heat than the melt-rich equivalents. Thus there is less opportunity for fractional crystallisation to occur should the magma rid itself of its entrained restite, and less heat would be provided to the wall rocks or available for driving hydrothermal systems. Narrow contact metamorphic aureoles around relatively mafic restite-rich granodioritic plutons are a consequence of this property (Chappell *et al.* 1987).

4.3. W mineralised granites

The broad metallogenic association of W with Mo and Sn is a common feature (e.g. Einaudi *et al.* 1981; Kwak 1987). Many studies have attempted to ascertain the factors specific to the formation of W mineralisation (e.g. Newberry & Swanson 1986; Keith *et al.* 1989; Newberry *et al.* 1990; van Middelaar & Keith 1990). In the LFB no simple chemical discriminant appears to encompass W-dominant mineralised granites and the conclusion must be that W can behave incompatibly in a wide range of intermediate to felsic and fractionated magmas. At higher f_{O_2}, W mineralisation is accompanied by Mo, with increasing Mo/W ratios corresponding to higher f_{O_2} conditions. In porphyry

Mo-style deposits, and those within the LFB associated with very oxidised granites, Mo is overwhelmingly dominant over W. At lower f_{O_2} conditions, W is accompanied by Sn. W/Sn ratios in Sn–W ore deposits do not, however, always relate to simple chemical or oxidation parameters in the associated granites. However, differences in the hydrothermal geochemistry of W and Sn transport and deposition may be important in explaining some of these differences. Sn is transported dominantly by complex Sn–Cl species in hydrothermal fluids (Heinrich 1990) while W is transported dominantly as tungstates, and not as chlorides (Jaireth *et al.* 1990; Wood & Vlassopoulos 1989). W solubility in hydrothermal solutions is strongly temperature-dependent and falls dramatically over the range from 400°C to 300°C (Jaireth *et al.* 1990). Sn precipitation as cassiterite is more complex and proceeds via a number of physicochemical mechanisms (Heinrich 1990).

4.4. Why all prospective granites are not mineralised: other factors

Despite the conclusion here that granites have predictable ore-element associations, it is equally clear that not all prospective granites are mineralised. Besides more mundane reasons for this, such as the non-exposure or erosional removal of pre-existing mineralisation, other factors are also important if not critical in this regard. These relate to the timing of release of the magmatic aqueous phase in relation to the degree of crystallisation of the pluton (e.g. Candela & Holland 1986; Candela 1989). This timing may be dependent on a range of factors including initial H_2O content of the magma and the level of final emplacement. The level of emplacement, and thus confining pressure, will also influence the degree of fracturing that occurs in response to fluid exsolution from the magma (Burnham & Ohmoto 1980).

The chemical and structural nature of the wallrocks is also important. Skarn or carbonate replacement deposits will not form in the absence of wallrock carbonates. The largest Sn and W deposits within the LFB are carbonate-replacement bodies and skarns located in western Tasmania and King Island. Here, ore-prospective granites have been emplaced into late Proterozoic to Cambrian quartzose, pelitic and carbonate lithologies of variable metamorphic grade (Turner & Taheri 1990). At both Renison and King Island, faults acted to channel fluids from the respective granites to reactive carbonate horizons (Williams *et al.* 1989; Wesolowski *et al.* 1988). Although the size and grade characteristics of those ore deposits are consequences of their favourable lithostratigraphic and structural settings, it is the granite that predetermined the ore element assemblage observed in those deposits.

5. Conclusions: mineralisation and ore-element assemblages as extensions of magmatic processes

In this study, the basic association of Mo, Cu and Au with oxidised I-type granites is confirmed and Sn is shown to be associated with both reduced S- and I-types. While these associations are already established (e.g. Burnham & Ohmoto 1980; Ishihara 1981) the critical role of fractional crystallisation has not generally been emphasised. The importance of fractional crystallisation of melt-rich magmas is clearly demonstrated here by the direct comparison of granite suites showing contrasting fractionation mechanisms. Other more general factors such as differing Cl contents and the differences in the behaviour of S between I- and S-type granites may also be important, particularly in the relative behaviour of the chalcophile elements.

World-class porphyry-style Cu–Au and Mo mineralisation is absent from the Siluro-Devonian and Carboniferous granites of the LFB. The association of porphyry Cu–Au deposits with less fractionated oxidised magma, and that of porphyry Mo deposits with more evolved felsic and fractionated magmas, are directly analogous to the associations present in the LFB. However, the tectono-magmatic settings cited for the generation of these larger deposits (subduction and tensional rifting, respectively) are vastly different to those proposed for the LFB. However, it is clear that the basic determining processes that produce the observed ore-element assemblages operate at all scales to produce the deposits ranging from the largest to the smallest. These assemblages are the result of general magmatic processes whose control parameters are largely set by the physical conditions and the lithotypes at the source of the magma. Ore-element assemblages and associations should therefore be interpreted as a logical extension and product of magmatic sources and processes (Stanton 1972). Whether or not this potential is realised in the formation of an ore deposit, however, relies on additional factors not all of which are directly magmatically related.

6. Acknowledgements

PLB acknowledges the receipt of an Australian Research Council Research Fellowship. BWC wishes to acknowledge financial support for the study of granites in the Lachlan Fold Belt from the Australian Research Council, the ANU Faculties Research Fund, and a group of mineral exploration companies through the Australian Mineral Industries Research Association Ltd. Thoughtful reviews of this paper by Phil Candela and Allan White are gratefully acknowledged.

7. References

Arnold, G. O. & Sillitoe, R. H. 1989. Mount Morgan gold–copper deposit, Queensland, Australia: evidence for an intrusion-related replacement origin. ECON GEOL **84,** 1805–16.

Baker, E. M. & Andrew, A. S. 1991. Geologic, fluid inclusion, and stable isotope studies of the gold bearing breccia pipe at Kidston, Queensland, Australia. ECON GEOL **86,** 810–30.

Burnham, C. W. 1979. Magmas and hydrothermal fluids. *In* Barnes, H. L. (ed.) *Geochemistry of hydrothermal ore deposits* (2nd edn), 71–136. New York: Wiley.

Burnham, C. W. & Ohmoto, H. 1980. Late-stage processes of felsic magmatism. *In* Ishihara, S. & Takenouchi, S. (eds) *Granite magmatism and related mineralisation.* MIN GEOL SPEC ISSUE **8,** 1–11.

Candela, P. A. 1989. Felsic magmas, volatile, and metallogenesis. SOC ECON GEOL REV ECON GEOL **4,** 223–33.

Candela, P. A. & Bouton, S. L. 1990. The influence of oxygen fugacity on tungsten and molybdenum partitioning between silicate melts and ilmenite. ECON GEOL **85,** 633–40.

Candela, P. A. & Holland, H. D. 1986. A mass transfer model for copper and molybdenum in magmatic hydrothermal systems: The origin of porphyry-type ore deposits. ECON GEOL **81,** 1–19.

Carmichael, I. S. E. 1991. The redox state of basic and silicic magmas: a reflection of their source regions? CONTRIB MINERAL PETROL **106,** 129–41.

Carroll, M. R. & Rutherford, M. J. 1985. Sulphide and sulphate saturation in hydrous silicate melts. J GEOPHYS RES **90,** C601–12.

Chappell, B. W. 1984. Source rocks of I- and S-type granites in the Lachlan Fold Belt, southeastern Australia. PHILOS TRANS R SOC LONDON **A310,** 693–707.

Chappell, B. W. & Stephens, W. E. 1988. Origin of infracrustal (I-type) granite magmas. TRANS R SOC EDINBURGH EARTH SCI **79,** 71–86.

Chappell, B. W. & White, A. J. R. 1984. I- and S-type granites in the Lachlan Fold Belt, southeastern Australia. *In* Xu Keqin & Tu Guanchi (eds) *Geology of granites and their metallogenic relations*, 87–101. Beijing: Science Press.

Chappell, B. W. & White, A. J. R. 1992. I-type and S-type granites in the Lachlan Fold Belt. TRANS R SOC EDINBURGH EARTH SCI **83**, 1–26.

Chappell, B. W., White, A. J. R. & Wyborn, D. 1987. The importance of residual source material (restite) in granite petrogenesis. J PETROL **28**, 1111–38.

Chappell, B. W., White, A. J. R. & Hine, R. 1988. Granite provinces and basement terranes in the Lachlan Fold Belt, southeastern Australia. AUST J EARTH SCI **35**, 505–21.

Chappell, B. W., Wyborn, L. A. I., White, A. J. R., Burnham, C. W. & Wyborn, D. (in prep.). S-type granites of the Wagga Basement Terrane: an example of compositional variation resulting from sequential restite fractionation, fractional crystallisation, and hydrothermal alteration.

Cochrane, G. W. 1971. Tin deposits of Victoria. GEOL SURV VICTORIA BULL **60**.

Collins, W. J., Beams, S. D., White, A. J. R. & Chappell, B. W. 1982. Nature and origin of A-type granites with particular reference to southeastern Australia. CONTRIB MINERAL PETROL **80**, 189–200.

Einaudi, M. T., Meinert, L. D. & Newberry, R. J. 1981. Skarn deposits. ECON GEOL 75TH ANNIVERSARY VOL 317–91.

Flood, R. H. & Shaw, S. E. 1975. A cordierite-bearing granite suite from the New England Batholith, NSW. CONTRIB MINERAL PETROL **52**, 157–64.

Haughton, D. R., Roeder, P. L. & Skinner, B. J. 1974. Solubility of sulphur in mafic magmas. ECON GEOL **69**, 451–67.

Heinrich, C. A. 1990. The chemistry of hydrothermal tin (−tungsten) ore deposition. ECON GEOL **85**, 457–81.

Hine, R., Williams, I. S. & Chappell, B. W. 1978. Contrasts between I- and S-type granitoids of the Kosciusko Batholith. J GEOL SOC AUST **25**, 219–34.

Horton, D. J. 1978. Porphyry-type copper-molybdenum mineralisation belts in eastern Queensland. ECON GEOL **73**, 904–21.

Ishihara, S. 1981. The granitoid series and mineralisation. ECON GEOL 75TH ANNIVERSARY VOL, 458–84.

Ivanova, G. F. & Butuzova, Ye. G. 1968. Distribution of tungsten, tin and molybdenum in granites of eastern Transbaikalia. GEOKHIMIYA **6**, 689–700.

Jaireth, S., Heinrich, C. A. & Solomon, M. 1990. Chemical controls on hydrothermal tungsten transport and the precipitation of ferberite and scheelite. GEOL SOC AUST ABSTR **25**, 269–70.

Katsura, T. & Nagashima, S. 1974. Solubility of sulphur in some magmas at 1 atmosphere. GEOCHIM COSMOCHIM ACTA **38**, 517–31.

Keith, J. D., van Middelaar, W., Clark, A. H. & Hodgson, C. J. 1989. Granitoid textures, compositions, and volatile fugacities associated with the formation of tungsten-dominated skarn deposits. SOC ECON GEOL REV ECON GEOL **4**, 235–50.

Kilinc, I. A. & Burnham, C. W. 1972. Partitioning of chloride between a silicate melt and coexisting aqueous phase from 2 to 8 kilobars. ECON GEOL **67**, 231–5.

Kwak, T. A. P. 1987. *W–Sn skarn deposits and related metamorphic skarns and granitoids*. Amsterdam: Elsevier.

Lehmann, B. 1990. *Metallogeny of Tin*. Berlin: Springer.

Loiselle, M. C. & Wones, D. R. 1979. Characteristics and origin of anorogenic granites. GEOL SOC AM ABSTR PROG **11**, 468.

Lynton, S. J., Candela, P. A. & Piccoli, P. M. 1990. Experimental determination of copper partitioning between pyrrhotite and high silica rhyolite. GEOL SOC AM ABSTR PROG **22**, 181.

Markham, N. L. & Basden, H. 1974. *The mineral deposits of New South Wales*. Sydney: Geological Survey of New South Wales.

Munoz, J. L. 1984. F–OH and Cl–OH exchange in micas with applications to hydrothermal ore deposits. MIN SOC AM REV MINERAL **13**, 469–93.

Murray, C. G. 1990. Tasman Fold Belt in Queensland. *In* Hughes, F. E. (ed.) *Geology of the mineral deposits of Australia and Papua New Guinea*, 1431–50. Melbourne: Australasian Institute of Mining and Metallurgy.

Newberry, R. J. & Swanson, S. E. 1986. Scheelite skarn granitoids: an evaluation of the roles of magmatic source and process. ORE GEOL REV **1**, 57–81.

Newberry, R. J., Burns, L. E., Swanson, S. E. & Smith, T. E. 1990. Comparative petrologic evolution of the Sn and W granites of the Fairbanks-Circle area, interior Alaska. *In* Stein, H. J. & Hannah, J. L. (eds) *Ore-bearing granite systems: petrogenesis and mineralising processes*. PROF PAP US GEOL SURV **246**, 121–42.

Plimer, I. R. 1980. Exhalative tin and tungsten deposits associated with mafic volcanism as precursors to tin and tungsten deposits associated with granites. MINER DEPOSITA **15**, 275–89.

Stanton, R. L. 1972. *Ore petrology*. New York: McGraw-Hill.

Stanton, R. L. 1990. Magmatic evolution and the ore type-lava type affiliations of volcanic exhalative ores. *In* Hughes, F. E. (ed.) *Geology of the mineral deposits of Australia and Papua New Guinea*, 101–7. Melbourne: Australian Institute of Mining and Metallurgy.

Tacker, R. C. & Candela, P. A. 1987. Partitioning of molybdenum between magnetite and melt: a preliminary study of the partitioning between silicic magmas and crystalline phases. ECON GEOL **82**, 1827–38.

Taylor, J. R. & Wall, V. J. 1990. The behaviour of tin in magmatic-hydrothermal systems. GEOL SOC AUST ABSTR **25**, 270–1.

Turner, N. J. & Taheri, J. 1990. *Tin and tungsten deposits and related Devonian granitoids*. Excursion Guide E2, 10th Australian Geological Convention, Geological Society of Australia.

van Middelaar, W. T. & Keith, J. D. 1990. Mica cheimstry as an indicator of oxygen and halogen fugacities in the CanTung and other W-related granitoids in the North American Cordillera. *In* Stein, H. J. & Hannah, J. L. (eds) *Ore-bearing granite systems; petrogenesis and mineralising processes*. GEOL SOC AM PROF PAP **246**, 205–20.

Watchhorn, R. B. & Wilson, C. J. L. 1989. Structural setting of the gold mineralization at Stawell, Victoria, Australia. *In* Keays, R. R., Ramsay, W. R. H. & Groves, D. I. (eds) *The geology of gold deposits: The perspective in 1988.* ECON GEOL MONOGR **6**, 292–309.

Weber, C. R., Paterson, I. B. L. & Townsend, D. J. 1978. Molybdenum in New South Wales. GEOL SURV NEW SOUTH WALES MIN RES **43**.

Wesolowski, D., Cramer, J. & Ohmoto, H. 1988. Scheelite mineralsiation in skarns adjacent to Devonian granitoids at King Island, Tasmania. *In* Taylor, R. P. & Strong, D. F. (eds) *Recent Advances in the Geology of Granite-Related Mineral Deposits*. CAN INST MIN METALL SPEC VOL **39**, 234–51.

Whalen, J. B. & Chappell, B. W. 1988. Opaque mineralogy and mafic mineral chemistry of I- and S-type granites of the Lachlan Fold Belt, southeast Australia. AM MINERAL **73**, 281–96.

White, A. J. R. & Chappell, B. W. 1977. Ultrametamorphism and granitoid genesis. TECTONOPHYSICS **43**, 7–22.

White, A. J. R. & Chappell, B. W. 1988. Some supracrustal (S-type) granites of the Lachlan Fold Belt. TRANS R SOC EDINBURGH EARTH SCI **79**, 169–81.

Williams, E., McClenaghan, M. P. & Collins, P. L. F. 1989. Mid-Palaeozoic deformation, granitoids and ore deposits. *In* Burrett, C. F. & Martin, E. L. (eds) *Geology and Mineral Resources of Tasmania*. GEOL SOC AUST SPEC PUBL **15**, 238–92.

Wood, S. A. & Vlassopoulos, D. 1989. Experimental determination of the hydrothermal solubility and speciation of tungsten at 500°C and 1 kbar. GEOCHIM COSMOCHIM ACTA **53**, 303–12.

Wyborn, D., Turner, B. S. & Chappell, B. W. 1987. The Boggy Plain Supersuite: a distinctive belt of I-type igneous rocks of potential economic significance in the Lachlan Fold Belt. AUST J EARTH SCI **34**, 21–43.

PHILLIP L. BLEVIN, Department of Geology, The Australian National University, GPO Box 4, Canberra ACT 2601, Australia.

BRUCE W. CHAPPELL, Department of Geology, The Australian National University, GPO Box 4, Canberra ACT 2601, Australia.

Transactions of the Royal Society of Edinburgh: Earth Sciences, **83**, 317–326, 1992

Controls on ore metal ratios in granite-related ore systems: an experimental and computational approach

Philip A. Candela

ABSTRACT: Size and composition (bulk metal ratios) of magmatic hydrothermal mineral deposits are affected by a number of chemical and physical processes including the nature of the source region and mode of emplacement. At shallow levels, rising plumes of vapour bubbles + melt, and the advection of water through interconnected vapour bubbles, allows access of the magmatic aqueous phase to the upper reaches of a magma chamber. These processes are operative at shallow levels where low water solubility and high molar volume for water make these processes more efficient.

Partitioning experiments suggest that oxygen fugacity-dependent crystal/melt partitioning of ore metals leads to different efficiencies of removal of Cu, W, and Mo from silicate melts into ore-forming aqueous fluids. For example, the Mo/W ratio in magmatic hydrothermal deposits should increase as the oxygen fugacity of the magma increases. Further, Cu should behave as a crystal-compatible element in H_2O-undersaturated, sulfide-saturated felsic magmas with $f_{O_2} <$ NNO + 1 due to the strong partitioning of Cu from the melt into pyrrhotite.

Cycling of oxidised, hydrated, sulfidised and Cl-enriched oceanic crust into mantle can give rise to magmas that contain S but are oxidised (≥NNO). The combination of high oxidation state, relatively hydrous *but* shallow conditions and a high Cl/H_2O ratio leads to saturation with respect to H_2O early during crystallisation, and loss of a large proportion of magmatic Cu to the aqueous phase. Ores formed from these oxidised magmas also possess high Mo/W ratios due to the effect of oxygen fugacity on the sequestering of Mo vs W.

In less oxidised magmas, Cu and Mo are partitioned into sulfides and Ti-bearing phases, respectively, resulting in lower efficiencies of removal of Cu and Mo from melts into aqueous fluids. Further, the partitioning of W into crystallising phases is *reduced*, producing a more efficient removal of W into ore-forming fluids. This ultimately leads to mineral deposits with higher W/(Mo + Cu) ratios relative to deposits associated with more oxidised systems. Silicic, high-F magmas with $f_{O_2} \approx$ NNO can be found in tensional environments (e.g. rocks associated with the Climax-type deposits of the Colorado Mineral Belt). High HF/H_2O activity ratios in the source regions yield melts that evolve an aqueous phase late during crystallisation, leading to relatively low ratios of compatible/incompatible elements in the melt at H_2O saturation.

KEY WORDS: magmatic-hydrothermal processes, ore metals, granites.

Bulk metal ratios in magmatic hydrothermal mineral deposits are affected by a number of factors including the nature of the magmatic source region and the mode of pluton emplacement. The Magmatic-Hydrothermal Theory for the origin of porphyry, skarn and other types of deposits predicts that magma chemistry exerts strong controls on the composition of mineralised rocks. In accord with the magmatic-hydrothermal model, these deposits are the products of processes extending from magma generation through magmatic evolution and magma transport, to crystallisation, aqueous phase exsolution, aqueous transport and deposition, and plutonic consolidation. This paper is a synthesis of some recent experimental, field and theoretical research on the magmatic controls on mineralisation produced by magmatic-hydrothermal processes. In the first section of the paper, the physical aspects of plutonic consolidation and concomitant evolution of aqueous fluids according to my recently published model (Candela 1991) are reviewed and refined. In the second section, experimental constraints on ore-fluid chemistry are discussed in the light of mass transfer modelling of the chemistry of aqueous phase exsolution. In the final section the models of the first two sections are discussed in the light of the Henderson porphyry Mo deposits, and other porphyry-type deposits.

1. Crystallisation and aqueous phase evolution in epizonal plutons

Following the emplacement of a hydrous, felsic magma in the epizonal environment, crystallisation occurs due to the irreversible loss of heat to the surroundings. Heat passes out of the magma chamber at the margins of the body, and the solidus moves towards the interior of the chamber, defining an inwardly-progressing boundary. As crystallisation proceeds, H_2O increases in concentration in the residual melt. During the cooling of subliquidus, multicomponent liquids, in homogeneous equilibrium, aqueous phase saturation can occur due to crystallisation and/or decreasing pressure. Either process causes the H_2O concentration in the melt to approach the saturation H_2O concentration, $C_w^{l,s}$. As a melt rises toward the surface, the saturation H_2O concentration decreases with decreasing pressure. When the saturation H_2O concentration is lowered to the actual H_2O

concentration in the melt, first boiling occurs. As the melt continues to rise and H_2O is evolved, the solidus temperature increases and the melt is driven toward higher crystallinities. This line of reasoning has led to the truism that the H_2O content of a magma limits the ability of a granitic melt to rise to high levels. However, the argument is based on equilibrium considerations, and felsic melts that exhibit evidence of high initial H_2O contents (i.e. phenocrysts of biotite (Webster *et al.* 1987) or amphibole (Westrich *et al.* 1988)) erupt on the surface. Clearly, equilibrium phase proportions do not limit the dynamic rise of magmas. Complex interplay among the rate of crystallisation, the rate of heat flow out of the magma and the rate of rise of the body, can allow magmas to attain higher levels of intrusion. Evidence from rhyolitic volcanism demands this conclusion. Thus, for kinetic reasons, melts can intrude to levels higher than those dictated by the limitations of equilibrium solidus temperatures.

After a magma reaches a terminal level of intrusion, crystallisation can occur quasi-isobarically. Under these conditions, the ratio of initial to saturation H_2O concentrations, $C_w^{l,o}/C_w^{l,s}$, a master chemical variable for H_2O saturation, can be defined. This ratio indicates the fraction of melt remaining at the commencement of H_2O saturation for the limiting case of a magma that crystallises only anhydrous phases. At some points in space and time within the magma, bubbles of the magmatic aqueous phase(s) will nucleate and grow (second boiling). As crystallisation of the H_2O-saturated melt proceeds, H_2O and other substances that partition into the volatile phase(s) are progressively released from the magma under the constraints that, at equilibrium, the chemical potentials of the components in the volatile phases are equal to the chemical potentials of the components in the melt phase.

The physical style of volatile saturation in magma chambers is not well known. H_2O saturation may occur within a marginal crystallisation interval or within the adjacent chemical boundary layer as these zones progressively sweep through the magma chamber over time. Alternatively, H_2O saturation may be confined to inner regions of the magma chamber, with vapour saturation occurring late in the history of the magmatic system. According to Candela (1991), the master physical variable that controls the style of aqueous phase evolution is the Lewis number, Le, defined as the ratio of the diffusivity of heat, k, to the diffusivity of water, D_w. Either H_2O saturation is locally achieved (high Le) at the site of crystallisation (boundary zone H_2O saturation), or H_2O "back diffuses" into the residual melt (low Le). Given the range of H_2O diffusivities in rhyolitic melts (10^{-7} to $10^{-9}\,cm^2\,s^{-1}$) (Delaney & Karsten 1981) and reasonable thermal diffusivities for rock systems ($7\times10^{-3}\,cm^2\,s^{-1}$, Irvine 1970), Lewis numbers are orders of magnitude greater than unity. For $Le \gg 1$, the rate of diffusion of H_2O back into the magmatic residua is insignificant relative to the rate of crystallisation, and H_2O will either accumulate in the crystallisation interval or at the crystal/melt interface. In systems that are high in B, F, H_2O, or Li, and/or are highly peraluminous or peralkaline, the diffusivity of H_2O is significantly higher (Dingwell 1988), and Le would be closer to unity. In this case, H_2O saturation could be localised, and restricted to interior portions of the pluton, or if dyking occurs, the volatile and rare-element-charged melt may be intruded as a pegmatite (H_2O-saturated or not) (London *et al.* 1989). Under the high Lewis number conditions that prevail during the crystallisation of normal metaluminous granitic melts, vapour saturation should occur in all volume elements within the magma chamber, and the vapourous surface would sweep through the magma behind the liquidus and ahead of the solidus.

Candela (1991) has proposed that localisation of the ore fluid into the apical regions of a magma chamber could occur by buoyant rise of bubble-laden magmatic plumes, or by advection of aqueous fluids through spanning clusters of vapour bubbles. At the Le numbers that prevail in metaluminous felsic systems ($Le \gg 1$), vapour evolution will be confined to the crystallisation interval during quasi-isobaric crystallisation. According to Marsh (1988b), felsic magmas will be resistant to flow at crystallinities of a few tens of per cent. If vapour saturation and bubble growth occur before $\approx$20% of the magma crystallises, then the boundary layer will become mobile due to the buoyancy of the aqueous phase. If H_2O saturation occurs after approximately 20% crystallisation, then the crystallisation interval will be immobile, and the volume proportion of aqueous phase will increase, until the bubbles form what is termed, in percolation theory, a spanning cluster. This is achieved on a three-dimensional lattice when the volume proportion of the potentially percolating phase is 0·31. When this is achieved, the aqueous phase can advect through the spanning cluster of bubbles (Candela 1991).

Chemically, exsolution of the aqueous phase would follow a Rayleigh fractionation law as the fluid flows through the network of interconnected bubbles. Before the onset of critical percolation (whether in a crystallisation interval or in a rising, bubble-laden plume), exsolution of H_2O from the melt will not be fractional. In all cases, the "effective volume" of the melt will be limited by the slow diffusivity of H_2O that characterises felsic magmatic systems, and the difficulties inherent in the physical mixing of highly viscous fluids. Observational tests of whether H_2O exsolution is fractional can be obtained from studies of the deuterium/hydrogen (D/H) vs H_2O content systematics of felsic extrusive and intrusive rocks. Taylor (1988) published a review of H-isotopic studies of igneous rocks, and summarised work performed by a number of workers. Taylor drew the tentative conclusion that the relationship between D/H ratios and residual H_2O concentrations of the whole rock can best be explained by fractional vapourisation with D fractionating into the aqueous phase. Therefore, the chemical models presented in the next section of this paper are predicated on a fractional vapourisation model, consistent with the advective/spanning cluster model, and the D/H data from igneous rocks.

The achievement of critical percolation, which heralds the establishment of the spanning cluster of H_2O-bearing bubbles within a crystallisation interval, is possible if the initial H_2O concentration in the melt is above a certain value for a given pressure ($\approx 0{\cdot}35 \cdot C_w^{l,s}$ at 200 bars and $\approx C_w^{l,s}$ at 1750 bars) (Candela 1991). Buoyant plumes of bubbles + liquid ± crystals may form and transport vapour-saturated liquid to the apical portions of a magma at any pressure if the initial H_2O concentration is greater than $0{\cdot}8 \cdot C_w^{l,s}$. The buoyantly rising plumes may erupt, yielding permeable froths (Eichelberger *et al.* 1986), may drive dyke injection, or may lift cupolas which solidify to yield miarolitic aplites or granophyres. These processes may explain how large amounts of magmatic aqueous fluid, charged with mineralising agents, are focused into small regions. In systems that are too dry or deep to exhibit either plume rise or the establishment of spanning clusters of H_2O-rich bubbles, the magmatic aqueous phase will remain in a dispersed state, and can gain egress from the crystallised pluton only after the pluton has cracked during its subsolidus cooling. These unfocused, dispersed fluids may produce small mineral shows, but probably cannot be active in high temperature hydrothermal ore formation.

2. Partitioning of metals between melt and crystallising phases

Mineral/melt partitioning is important in controlling ore metal concentrations in the melt both before and after the saturation of the melt + crystal mixture with an aqueous phase during second boiling. Clearly, the partitioning of elements into an ore fluid will also be affected by aqueous phase/melt partition coefficients, and much effort has been expended in our laboratory and others to determine these important equilibrium constants. However, within reasonable limits of the aqueous phase/melt partition coefficients, if an element is perfectly incompatible, then 100% of the element in question will ultimately be removed into the magmatic aqueous phase. Conversely, the proportion of any given ore metal, originally present in solution in the melt phase, that is sequestered into crystallising phases is unavailable for partitioning into the magmatic aqueous phase(s) according to the orthomagmatic-hydrothermal theory for the origin of some mineral deposits. A quantitative model developed by the author (cf. Candela 1989a) can be used to calculate the instantaneous concentration of metals in ore fluids, or the integrated amount of ore metal that can be removed into a magmatic aqueous phase relative to the crystalline phases. A function (the efficiency function) is defined by the expressions (given, in the case of Mo)

$$\mathrm{E(Mo)} = \mathrm{M}^{a,\mathrm{T}}_{\mathrm{Mo}}/\mathrm{M}^{m,\mathrm{o}}_{\mathrm{Mo}}. \quad (1)$$

This equation expresses the efficiency of removal as the ratio of the total quantity of Mo removed into the aqueous phase, $\mathrm{M}^{a,\mathrm{T}}_{\mathrm{Mo}}$, divided by the initial quantity of Mo in the melt, $M^{m,\mathrm{o}}_{\mathrm{Mo}}$. The numerator of equation (1) is proportional to the integral of the instantaneous concentration of the element in question over reaction progress. The instantaneous concentration of an ore metal in the magmatic-aqueous phase ($\mathrm{C}^{\mathrm{a}}_{\mathrm{i}}$) is modelled as following a fractional boiling law (Candela 1989a). When this integral is combined with the definition in equation (1) (see Candela 1989a for the derivation of this expression) the final relation is

$$\mathrm{E(Mo)} = (\mathrm{C}^{\mathrm{l,o}}_{\mathrm{w}}/\mathrm{C}^{\mathrm{l,o}}_{\mathrm{Mo}}) \,.\, \int_0^1 \mathrm{C}^{\mathrm{a}}_{\mathrm{Mo}} \,.\, (\mathrm{Z}) \,.\, \mathrm{dZ}. \quad (2)$$

In the case of elements with constant Nernst partition coefficients (the conditions required include, depending on the element, constant temperature, pressure, oxygen fugacity, or activities of other constituents; see Candela 1990) the efficiency function (Candela 1986, 1989a) can be expressed as

$$\mathrm{E(Mo)} = (\mathrm{C}^{\mathrm{l,o}}_{\mathrm{w}}/\mathrm{C}^{\mathrm{l,s}}_{\mathrm{w}})^{\bar{\mathrm{D}}_{\mathrm{Mo}}} \,.\, (1 + (\bar{\mathrm{D}}_{\mathrm{Mo}}/\mathrm{C}^{\mathrm{l,s}}_{\mathrm{w}} \,.\, \mathrm{D}^{a/m}_{\mathrm{Mo}}))^{-1}. \quad (3)$$

In equation (3), E(Mo) is the efficiency of removal of Mo into the exsolving aqueous fluid from a silicate melt, $\mathrm{C}^{\mathrm{l,o}}_{\mathrm{w}}$ and $\mathrm{C}^{\mathrm{l,s}}_{\mathrm{w}}$ are the initial and saturation weight fraction of H_2O in the melt respectively, $\bar{\mathrm{D}}_{\mathrm{Mo}}$ is the bulk mineral/melt partition coefficient, and $\mathrm{D}^{a/m}_{\mathrm{Mo}}$ is a vapour/melt partition coefficient.

3. Ore metal partitioning the condensed systems: crystal/melt equilibria and ore formation

Experiments performed in our laboratory (Candela & Bouton 1990; Lynton *et al.* 1990) suggest that oxygen fugacity-dependent crystal/melt partitioning of ore metals leads to different efficiencies of Cu, W, and Mo removal from silicate melts into ore-forming aqueous fluids.

The partitioning of Mo and W between ilmenite and silicic melts was studied by Candela and Bouton (1990). The experiments were conducted at 800°C, 1 kbar (1 kbar = 100 MPa) and at two oxygen fugacities of the system. At the oxygen fugacity of the graphite–methane ($-H_2O$) buffer, the partition coefficient for Mo between ilmenite and melt (expressed as ilm/melt, in ppm) is $\mathrm{D}^{\mathrm{ilm/melt}}_{\mathrm{Mo}} = 0{\cdot}42$, and the partition coefficient for W between ilmenite and melt (expressed as ilm/melt, in ppm) is $\mathrm{D}^{\mathrm{ilm/melt}}_{\mathrm{W}} = 0{\cdot}39$. Near the Ni–NiO buffer, $\mathrm{D}^{\mathrm{ilm/melt}}_{\mathrm{Mo}} = 0{\cdot}22$ and $\mathrm{D}^{\mathrm{ilm/melt}}_{\mathrm{W}} = 0{\cdot}5$. These variations with respect to oxygen fugacity are critically significant at the 0·06 and 0·33 levels for Mo and W, respectively. The results for Mo are in agreement with the earlier work of Tacker and Candela (1987). Generalising these results for the partitioning of W and Mo between melt and crystalline oxides to other mineral/melt systems suggests that the oxygen fugacity prevailing during magmatic crystallisation plays a role in determining the Mo/W ratio in porphyry, skarn, or other hydrothermal mineral deposits. These variations should be seen in the behaviour of Mo and W during fractionation. Figure 1 shows behaviour of Mo and W in the Tanjungpandan Batholith, Indonesia (Lehmann & Harmanto 1990) and the behaviour of Mo in rhyolitic systems from the Great Basin (Haffty & Noble 1972). The Tanjungpandan Batholith is more reduced than the volcanic suite studied by Haffty and Noble (Lehmann 1990). This figure clearly shows that Mo behaves much more compatibly under reducing conditions than under oxidising conditions.

The effect of variable oxygen fugacity on the relative efficiencies of removal of W and Mo from silicate melts into ore-forming fluids can be expressed quantitatively by using the equation for the efficiency of removal of Nernstian elements (equation (3)) for each element under both oxidising (f_{O_2} = NNO) and reducing (f_{O_2} = GM) conditions. In the following calculations, the initial and saturation H_2O concentrations are set equal to 0·01 (1 wt%) and 0·04 (4 wt%, H_2O exsolution occurs at pressure ≈1 kbar), respectively. A model $\bar{\mathrm{D}}$ is calculated by assuming that the ferromagnesian minerals possess individual mineral/melt partition coefficients which are equal to the partition coefficients determined by Candela and Bouton (1990) for ilmenite. Furthermore, for the sake of illustration, it will be assumed that the ferromagnesian constituents make up 10% of the solidus assemblage. We therefore use $\mathrm{D_W} = 0{\cdot}05$ and $\mathrm{D_{Mo}} = 0{\cdot}02$ to model the efficiency of removal of metals under oxidising conditions, and $\mathrm{D_W} = 0{\cdot}039$ and $\mathrm{D_{Mo}} = 0{\cdot}05$ to model this process under reducing conditions. Although this calculation is rather crude, it can be used to illustrate the effect of oxygen fugacity on the efficiency of removal of W relative to Mo as a function of differential sequestration during plutonic consolidation. The Mo aqueous phase/melt partition coefficient used in this calculation is from Candela & Holland (1984) ($\mathrm{D}^{a/m}_{\mathrm{Mo}} = 2{\cdot}5$). The data on the aqueous paper by Manning and Henderson (1984) suggests that $\mathrm{D}^{a/m}_{\mathrm{W}}$ is of the order of unity. The ratio of the efficiency of removal of the order of unity. The ratio of the efficiency of removal of W to the efficiency of removal of Mo increases by 50% from 0·51 when f_{O_2} = NNO (representative of the oxidation state of arc magmas) to 0·77 when f_{O_2} = GM (representative of the oxidation state of magmas contaminated by or produced by the melting of C-bearing metasedimentary rocks). Clearly, oxygen fugacity can significantly affect the W/Mo ratio in a magmatic hydrothermal mineral deposit.

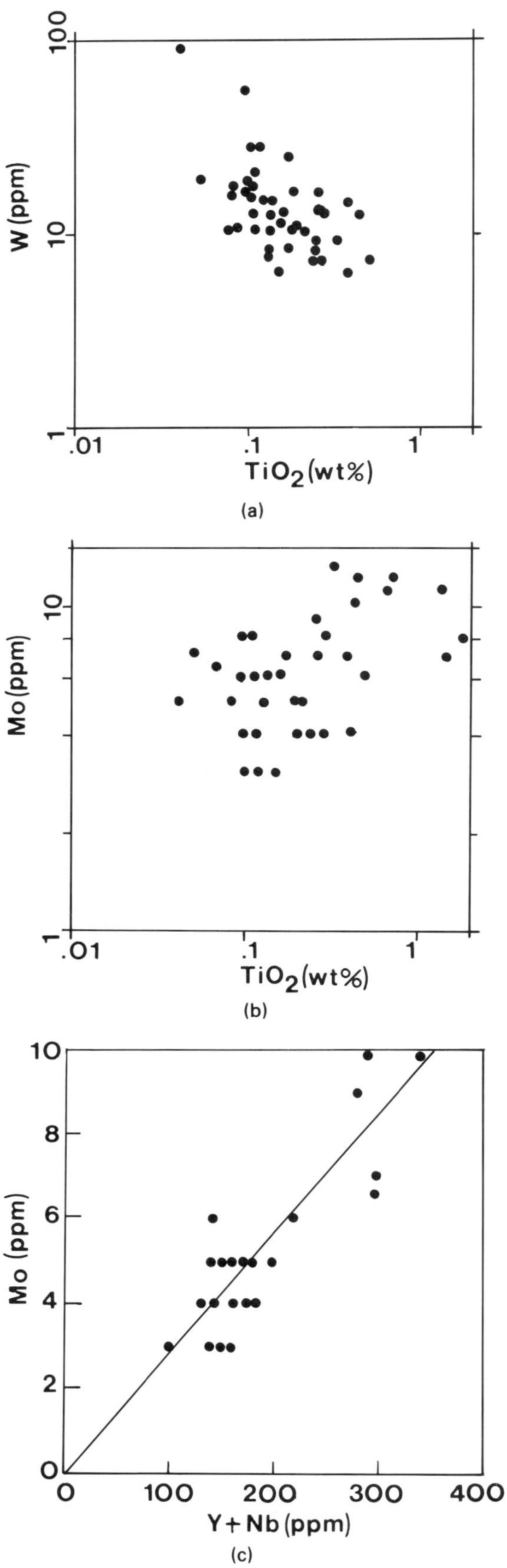

Figure 1 (a) Log W vs log TiO_2 for rocks of the ilmenite-series Tanjungpandan Batholith; W increases with fractionation.

(b) Log Mo vs log TiO_2 for the same rocks as (a); Mo decreases with fractionation.

((a) and (b) are redrawn after Lehmann & Harmanto 1990).

(c) Mo vs Y + Nb from magnetite-series lavas from Nevada; Mo increases with fractionation (Data of Haffty & Noble 1972, redrawn from Lehmann 1990.)

Table 1 Summary of experimental data on crystal/melt equilibria

System	Metal	f_{O_2}	$D^{(xl/melt)}$	Reference	Prob. value
Mt/Melt	Mo	NNO	0·21	1	<0·01
Mt/Melt	Mo	GM	0·52	1	
Ilm/Melt	Mo	NNO	0·22	2	0·06
Ilm/Melt	Mo	GM	0·42	2	
Ilm/Melt	W	NNO	0·50	2	0·33
Ilm/Melt	W	GM	0·39	2	
Po/Melt	Cu	NNO	566	3	0·15
Po/Melt	Cu	GM	908	3	

Sources: 1. Tacker & Candela 1987; 2. Candela & Bouton 1990; 3. Lynton, Candela & Piccoli 1990.

According to the study by Lynton *et al.* (1990) Cu partitions strongly into pyrrhotite from felsic melts of Bishop Tuff composition at 800°C, 1 kbar, S fugacities in the vicinity of 10^{-1} bar and oxygen fugacities of NNO and GM with $D_{Cu}^{a/m} \approx 600$ at NNO and ≈900 at GM (see Table 1 for a summary of Cu, Mo and W partition coefficients). Cu should behave as a crystal-compatible element in H_2O-undersaturated, sulfide-saturated felsic magmas with $f_{O_2} \leq$ NNO + 1 due to the strong partitioning of Cu into pyrrhotite from the melt. If these preliminary data are extrapolated to intermediate natural melts with of the order of 10^3 ppm S, rough estimates of bulk partition coefficients can be made. For a dacite crystallising at 950–900°C and 2 kbar, a pyrrhotite-saturated melt at $f_{O_2} \approx$ NNO would contain of the order of 1000 ppm S (Carroll & Rutherford 1987; Luhr 1990). If all the S precipitates as pyrrhotite, the final crystallisation products would contain approximately 0·275% pyrrhotite. If $D_{Cu}^{a/m} = 600$, then the bulk partition coefficient = 1·65, indicating that Cu would behave as an incompatible element under these conditions (Lynton *et al.*, in prep.).

4. Crystallisation of ore-related plutons at shallow levels

In the model presented in the first section of this paper, advection of an H_2O-bearing phase occurs by porous media flow through spanning clusters of bubbles. This model will be used here to show how the chemical systematics suggested in this paper can be applicable to systems such as the world-class Mo deposits of Climax-Henderson, Colorado, U.S.A. By using Darcy's law in the form

$$Q = (k/\nu) \, . \, (\Delta p/h) \qquad (4)$$

the flux (in $gm \, . \, cm^{-2} \, . \, s^{-1}$) of fluid flowing through the apical cap of a magma chamber can be calculated. Kinematic viscosities for the aqueous phase of the order of $10^{-3} \, cm^2 \, . \, s^{-1}$ at 800°C and 1 kbar are given by Norton and Knight (1977). Permeabilities of the bubble clusters equal to 10^{-8} to $10^{-12} \, cm^2$ can be estimated from measured permeability of pumice (Eichelberger *et al* 1986; Candela 1991). Further, after the spanning cluster is achieved, the pressure at any point in a column of vapour will be less than that in the adjacent crystal mush by a factor of $\Delta p = g \, . \, h \, . \, \Delta\rho$, where g is the acceleration due to gravity, h is characteristic height of the vapour column, and $\Delta\rho \approx 2 gm \, . \, cm^{-2}$. In this case,

$$Q = (k/\nu) \, . \, g \, . \, \Delta\rho \qquad (4a)$$

yields values for Q between 10^{-6} and $10^{-2} \, gm \, . \, cm^{-2} \, . \, s^{-1}$. However, the percolation network is limited to the supersolidus crystallisation interval. Once the interval has swept through the apical region, and these rocks cool to subsolidus temperatures, the mineralisation event may cease, or the mineralised zone will descend into the apical region. If the mineralisation is limited to a shell which has as its locus the intrusion (as is the case at Henderson), then the time available for operation of magmatic-hydrothermal processes is limited by the time-scale for crystallisation of the apical region. The primary process in the crystallisation of a magma body is the irreversible loss of heat to the surroundings by conduction. Experimental simulation (Brandeis & Marsh 1989) and theoretical calculations (Brandeis & Jaupart 1987; Norton & Taylor 1979) of the cooling and crystallisation of magmas suggest that these systems cool mainly by conduction, with convection relegated to a relatively minor role. This is borne out by observations of the Hawaiian Lava lakes (Peck *et al.* 1977; Marsh 1988a, b). Therefore, for all but small times (where transient effects related to nucleation delay dominate) the rate of advancement of the solidus can be given by an expression (after Jaeger 1968) of the form

$$x = 0{\cdot}28 \,.\, t^{1/2}, \tag{5}$$

where x is the position of the solidus (crystallisation proceeds in the positive x direction) in centimetres and t is in seconds. Equation (5) assumes a constant $\Delta T = 800°C$ (ΔT is the temperature of crystallisation minus the temperature of the country rock far from the intrusion) and assumes a fixed melting point. However, equation (5) yields a first-order estimate of the rate of advance of the crystallisation front. If x is taken as the radius of the apical region of the magma, then the cross-sectional area crystallised will be given by $A = \pi \cdot (0{\cdot}28)^2 \cdot t$. The total fluid flow before the completion of crystallisation is given by $Q \cdot A \cdot t$, where t is the time necessary to crystallise the cupola.

At the Henderson Porphyry Molybdenum Deposit, Carten *et al.* (1988a, b) demonstrate a close relationship between mineralisation-bearing veins and the associated intrusive rocks. Differences in the intensity of mineralisation associated with the stocks at Henderson correlate directly with differences in igneous rock textures. The lowest traceable levels of vein-filled fractures are limited to the upper 200–300 m of the apex of the intrusion, yet the size of the cupola, relative to the mass of ore, is impressively small. When the volume of the Seriate stock (volume $= 1{\cdot}3 \times 10^{-2}\,km^3$) is compared with the mass of Mo in the associated ore shell ($4{\cdot}3 \times 10^{11}$ g), concentrations of the order of 1 wt% Mo are calculated for the cupola magmas if this Mo is restored to the intrusion that has been shown to be genetically related to the mineralisation. According to the present model, the cupola associated with mineralisation was not the sole source of the Mo. Rather, fluids containing Mo passed from the underlying lower levels of the intrusion by advection through spanning clusters of volatile-rich bubbles that had achieved connectivity between the bulk of the intrusion and the overlying cupola (the Seriate stock). The cupola was the locus of mineralising fluids as long as some portion of it remained supersolidus. The cooling time for a stock the size of the Seriate stock, using equation (5), is of the order of 3×10^9 s (in 100 years). Assuming a model circular section for this irregularly shaped stock yields an area $= 4{\cdot}1 \times 10^8\,cm^2$. The total mass of H_2O which can pass through the interconnected bubbles in the crystallisation interval of the partially crystallised Seriate stock is then given by: $M_{H_2O} = Q \cdot A \cdot t = 10^{-2} \cdot 4{\cdot}1 \times 10^8 \cdot 3 \times 10^9 = 10^{16}$ g, or 10^{13} kg of H_2O, given a permeability of $10^{-8}\,cm^2$. This is the amount of H_2O that would be evolved from the crystallisation of 10^{15} kg of melt with 1% initial H_2O. A lower permeability of the spanning cluster, equal to $10^{-12}\,cm^2$, yields a mass of H_2O equal to 10^{11} kg. These lower and upper limits correspond (at densities of $3\,Mg/m^3$) to volumes of 0·03–333 km^3. Therefore, depending on the permeability of the spanning cluster, H_2O evolved from a volume of melt equal to the volume of the Seriate stock itself, at a minimum, to a volume four orders of magnitude larger can be tapped *through* the Seriate Stock. Figure 2 is a plot showing the required concentrations of Mo as a function of tappable volume to yield an ore shell the size of

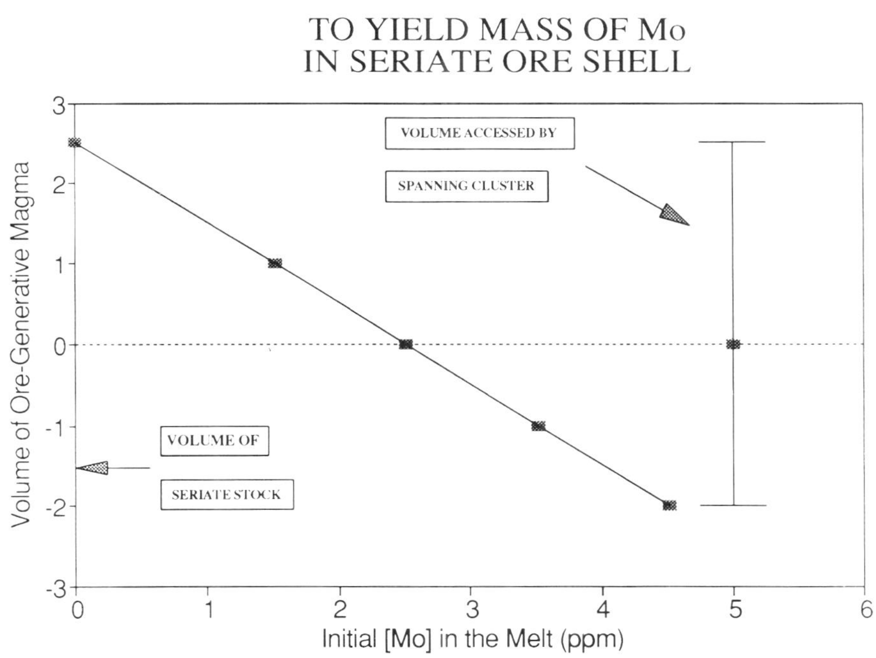

Figure 2 Inverse relationship between volume of magma and the initial Mo concentration (both shown on a log scale) in that magma that is required for the generation of the Seriate ore shell of the Henderson Porphyry Mo deposit, Colorado; possible range of magam volumes tappable by spanning clusters of vapour bubbles is shown by best estimate and associated error bars on the right-hand side of the figure.

the Seriate ore shell, assuming an efficiency of removal for Mo equal to 50% (Candela & Holland 1986; Candela & Bouton 1990). This model obviates the need for unrealistic (and indeed, nonuniformitarian) Mo concentrations in a granitic melt. Carten *et al.* (1988a, b) do argue for an unusual magma composition. However, the relatively simple, albeit highly fractionated, granite at Henderson speaks strongly against any arguments for unusual compositions. This discourse shows the feasibility of extracting metals from a larger body than is present in the cupola of a pluton. Even the largest of the magma bodies implied as a possible Mo source in Figure 2 is smaller than large ash eruptions such as the Bishop Event in eastern California, U.S.A. Further, the large amount of F fixed in some ore deposits as fluorite and other F-bearing minerals, as at the Climax-Henderson Mo deposits (Carten *et al.* 1988a, b), does not imply aqueous compatibility of F. Relatively high activities of F relative to H_2O can be achieved in aqueous fluids evolved from melts with F concentrations between a few hundred and approximately 10^4 ppm. Fluids derived from these magmas can be responsible for high-fluoridation state assemblages (Candela 1989a, b; Ague & Brimhall 1988). Further, the flow through spanning cluster model developed by Candela (1991) and expanded upon in this paper can solve problems of both Mo and F supply without appealing to unrealistically high metal or F concentrations or unusual and undemonstrated partition coefficients for ore metals and F.

5. Discussion

Answering questions such as why a porphyry Cu–Mo deposit forms in a given environment rather than a porphyry Mo–W deposit, requires a synthesis combining experimental and theoretical research with regional, petrological and ore-related field studies. The efficiency function (equation (2)) provides the link among these approaches. Ratios of calculated efficiencies of removal of metals from silicate melts into aqueous fluids, such as E(Cu)/E(Mo), can be calculated as a function of: (1) oxygen fugacity; (2) proportion of minerals in the crystallising assemblage (both of which vary the mineral/melt partition coefficients); (3) variable aqueous phase/melt partition coefficients; (4) Cl/H_2O ratios in the melt, and (5) initial and saturation H_2O concentrations in the melt phase. These variables can be related to a number of tectonic and petrological variables and the resulting efficiency ratios can be compared to the ratios of commodities found in mineral deposits.

Examination of the literature on Mo and W mineral deposits indicates that many deposits with high W/Mo ratios are rather reduced and exhibit high CH_4/CO_2 ratios in fluid inclusions, and low oxygen fugacities of the associated igneous rocks (see Table 2). On the other hand, many deposits with high Mo/W ratios are associated with rather oxidised magmatic systems. In general, it appears that oxygen fugacity-dependent sequestration of Mo and W in ferromagnesian phases during plutonic consolidation exerts a strong control on the relative efficiencies of removal of these metals into ore-forming fluids.

The ratio $C_w^{l,o}/C_w^{l,s}$ is also important in determining the efficiency of removal of metals from magmas into ore-forming fluids. A low ratio of $C_w^{l,o}/C_w^{l,s}$ indicates a large amount of crystallisation before H_2O saturation, because a significant amount of crystallisation is required to increase H_2O concentrations to the saturation level. This results in the sequestering of a large proportion of crystal-compatible elements in the minerals of the crystallising assemblage before the evolution of an H_2O-rich volatile phase. On the other hand, a high $C_w^{l,o}/C_w^{l,s}$ ratio allows compatible elements to gain access to the magmatic aqueous phase by inducing the early saturation of the melt with respect to H_2O. In this model, crystallising phases compete with the magmatic aqueous phase for ore metals. Whereas crystal-incompatibility is clearly important in determining whether an element will be available for ore formation by orthomagmatic-hydrothermal processes, it is not the case that, if an element is found in an orthomagmatic-hydrothermal mineral deposit, it must behave as an incompatible element before the melt reaches H_2O saturation. Cu is the type example, At oxygen fugacities higher than approximately NNO + 1, pyrrhotite is destabilised, and Cu would probably behave as an incompatible element during crystallisation. However, early pyrrhotite is not uncommon in arc magmas, and Cu is generally observed to behave as a compatible element (Andiambololona & Dupuy 1978). After the melt saturates with respect to an aqueous phase, it may become undersaturated with respect to pyrrhotite due to a reaction relationship among the pyrrhotite, melt and aqueous phases (Candela & Holland 1986). This hypothesis is consistent with the observation that pyrrhotite is rarely found in contact with the glass phase in

Table 2 Summaries of field relationships between Mo/W ratios and magmatix oxygen fugacity

Reference	Region	Evidence
Sato 1982	Japan	High Mo scheelite occurs in pyrite, magnetite and andraditic garnet-bearing skarn, characterised by $\delta^{34}S > 0$ in sulfides; associated granites are magnetite-series. Mo-free scheelite occurs in pyrrhotite-bearing skarns characterised by $\delta^{34}S < 0$ and low-andradite garnet; associated granites are ilmenite series.
Barton 1990	Great Basin western U.S.A.	Deep lithophile element mineralisation is associated with more reduced granites and alteration, and contains more W than shallower, more oxidised magmatic-hydrothermal deposite classes.
van Middelaar & Keith 1990	N American Cordillera	High-grade W deposits have CH_4-rich fluid inclusions, and are associated with granites characterised by low-ferric Fe-biotite.

lavas, indicating that melts become undersaturated with respect to pyrrhotite at some stage (Whitney & Stormer 1983).

Our laboratory data discussed above, concerning the partitioning of Cu, Mo and W between silicate melts and crystalline phases, suggests that under oxidising conditions ($f_{O_2} \approx$ NNO) E(Cu) and E(Mo) will be elevated, whereas E(W) will be modest. The converse is true under reduced conditions. Detailed experimental data are lacking, but simplistic arguments based on oxidation states (Candela 1989b), and geological inferences (Lehmann 1990) suggest that Sn behaves in a manner similar to W.

Cu, a chloride-aqueous phase compatible element, partitions in favour of the aqueous phase relative to the melt phase. Its Nernst partition coefficient is a function of the aqueous chloride concentration raised to the first power (i.e. it has a ligation number of one). On the other hand, Mo and probably W have partition coefficients that are independent of chloride (Candela & Holland 1984; Wood & Vlassopoulos 1989; Candela & Bouton 1990). These experimental results, together with my previous modelling efforts (Candela 1989a, b) suggest that E(Cu)/E(Mo or W) will increase as the Cl/H_2O ratio of the magma increases.

At f_{O_2} = NNO, Cu is crystal compatible, and Mo is crystal incompatible. Therefore, for a given Cl/H_2O ratio, E(Mo)/E(Cu) will vary inversely with $C_w^{l,o}/C_w^{l,s}$. That is, drier melts fractionate to a greater extent before H_2O saturation, Cu becomes sequestered, and Mo-rich, Cu-depleted ore fluids will be generated. Figure 3(a), (b) shows the efficiencies of removal for a number of elements. In both (a) and (b) the oxygen fugacity is set at the NNO buffer, and the initial Cl/H_2O weight ratio is set at 0·05, reasonable conditions for arc magmas. In Figure 3(a), the initial and saturation H_2O concentrations are low and equal, modelling H_2O saturation in a shallow environment. Under these conditions, the ratio of the efficiency of Cu removal to Mo removal is maximised (E(Cu)/E(Mo) = 3·3). In Figure 3(b), the initial H_2O concentration is half of the saturation H_2O concentration, and the melt evolves an aqueous fluid (and crystallises) at a deeper level. In this case, Cu partitioning is suppressed, and the ratio of E(Cu)/E(Mo) is only 0·33. In these calculations, Cu was modelled as a compatible

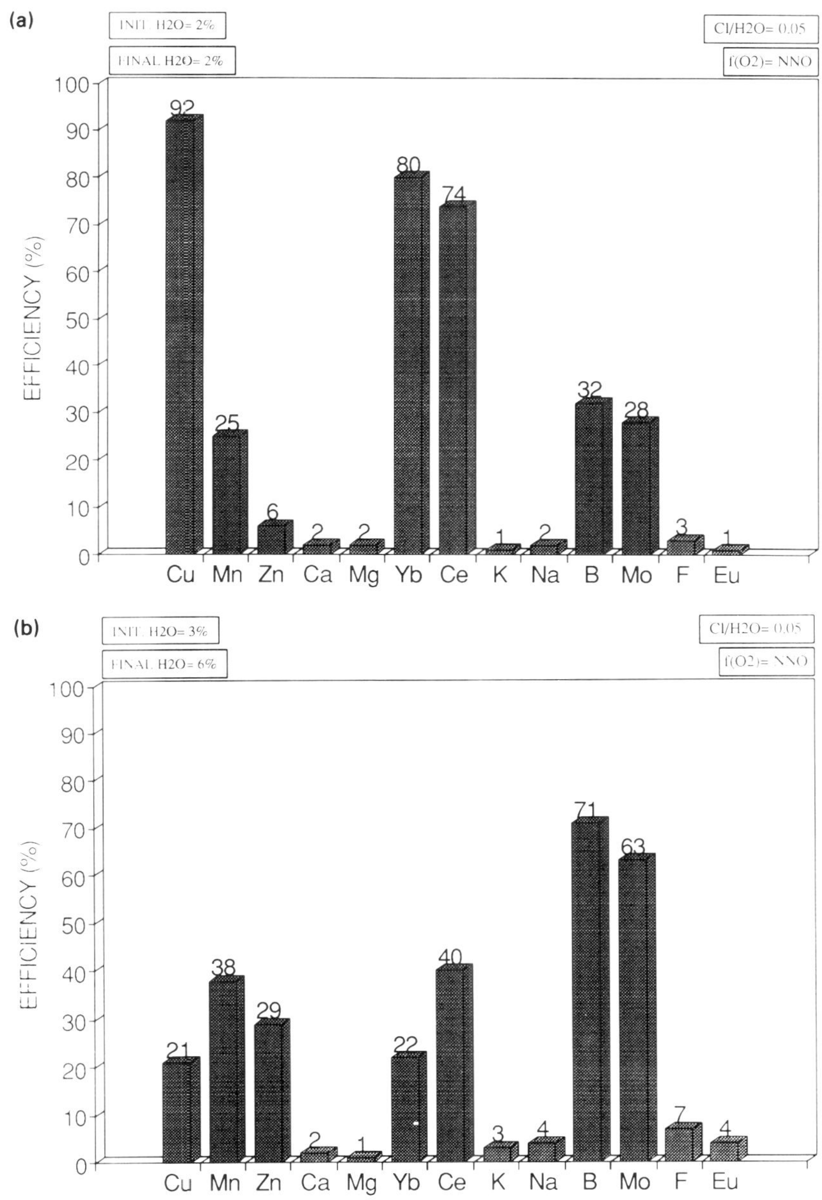

Figure 3(a), (b) Efficiencies of removal of elements, under the indicated conditions, from a silicate melt into an aqueous fluid during second boiling; see text for details.

element before H_2O saturation ($\bar{D} = 2{\cdot}0$), and as an incompatible element after H_2O saturation ($\bar{D} = 0{\cdot}1$). Mo was modelled as an incompatible element with a $\bar{D} = 0{\cdot}1$. Aqueous phase/melt partition coefficients, bulk partition coefficients for other elements, and methods of calculation are given in Candela (1989a). Values of $C_w^{l,o}/C_w^{l,s}$ close to unity (as in Fig. 3a) can be produced by H_2O saturation from a magma at a shallow level (<3–4 km). This is consistent with the general observation that porphyry Cu deposits are relatively shallow deposits. Some Mo and W deposits are deeper than porphyry Cu deposits (Candela 1989b). However, relatively shallow, silicic, high-F magmas with $f_{O_2} \approx$ NNO, hosting high (Mo ± W)/Cu deposits can be found in tensional environments (e.g. rocks associated with the Climax-Henderson ores). Whereas H_2O and F in hydrous minerals can flux melting in the source regions of these magmas, F does not promote aqueous phase saturation. High HF/H_2O activity ratios in the source regions yield melts with low $C_w^{l,o}$ that evolve an aqueous phase late during crystallisation, leading to relatively low ratios of compatible/incompatible elements in the melt at H_2O saturation, even at relatively shallow levels of intrusion. This effect may explain, in part, why F-rich systems yield ore fluids with high ratios of incompatible elements to compatible elements.

Porphyry Cu deposits are large tonnage (averaging $>10^6$ tons of ore), low grade (0·6% Cu) intrusion-related hydrothermal deposits that form in arc systems. According to both fluid inclusion data and stratigraphic reconstructions, the genetically related intrusions are emplaced to very shallow depths (<4 km, commonly 1–2 km). Cu is commonly concentrated by a single small pluton of an intrusive complex; this pluton is usually the youngest and most felsic of the suite (Titley & Beane 1981). Cycling of oxidised, hydrated, sulfidised, and Cl-enriched (Piccoli & Candela 1991) oceanic crust into the mantle can give rise to H_2O-rich slab contributions that contain Cl and S but are oxidised ($f_{O_2} \geq$ NNO) and which initiate melting of the overlying mantle wedge. Melts produced therein rise into the lower crust, causing repeated melting and the production of intermediate to silicic melts (Hildreth & Moorbath 1988). As Marsh has indicated (1982), any single batch of magma in an arc system would solidify long before reaching the surface if it moved through ductile rock with a temperature described by a normal geotherm. Marsh concludes that the first few bodies of magma do not reach the surface, but leave behind a "hot passageway" that insulates successive magmatic pulses. That is, for a given magmatic centre, the latest pulse of melt will experience the least crystallisation, and, because of the lowered average viscosity of the conduit due to preheating, will probably experience the most rapid ascent. The rapid ascent, from the crust/mantle boundary, of crystal-poor magmas with considerable H_2O contents allows the melts to reach shallow levels in arc environments. The combination of high oxidation state, relatively hydrous *but* shallow conditions and a high Cl/H_2O ratio leads to saturation with respect to H_2O early during crystallisation and loss of a large proportion of magmatic Cu to the aqueous phase. Ores formed from these oxidised magmas also possess high Mo/W ratios due to the effect of oxygen fugacity on the sequestering of Mo vs W.

In less oxidised magmas, Cu and Mo are partitioned into sulfide and Ti-bearing phases, respectively, resulting in lower efficiencies of removal of Cu and Mo from melts into aqueous fluids. Further, the partitioning of W into crystallising phases is *reduced*, producing a more efficient removal of W into ore-forming aqueous fluids. This ultimately leads to mineral deposits with higher W/(Mo + Cu) ratios relative to deposits associated with more oxidised systems. Sn probably behaves in a manner similar to W (Lehmann 1990).

In conclusion, it is quite tempting to ascribe the oxygen fugacity-related dualism of Cu–Mo deposits (high f_{O_2}) on the one hand, and Sn–W deposits (low f_{O_2}) on the other hand, to the effects of oxygen fugacity on crystal/melt partitioning. This forward modelling effort, based on mass transfer systematics and experimental data on metal partitioning in crystal-melt-aqueous phase systems, stands up well when compared with natural data. In a recent series of papers, Barton (1990) and Barton *et al.* (1988) summarise important data on the distribution of magmatic-hydrothermal mineral deposits in the western U.S.A. According to their studies "... copper deposits occur at the shallowest levels ... whereas tungsten skarns occur where exposures are deeper ..." (Barton *et al.* 1988, from the abstract). Further, Barton (1990) shows that, in the Great Basin, Cu ± Mo porphyry and skarn deposits are associated with oxidised igneous rocks and alteration, and occur at depths of 1–5 km. Quartz monzonite-type porphyry Mo deposits, also associated with oxidised rocks, occur at depths of 3–9 km. Associated skarns contain Mo and W. The deepest mineralisation (5–12 km) is classified by Barton as lithophile element mineralisation, and is associated with reduced igneous rocks and alteration, and contains more W than the shallower, and more oxidised classes of deposits in the Great Basin of the western U.S.A. As Barton (1990) points out, the melts associated with these individual classes of mineralisation are different, and other factors may also come into play in producing variations in commodity ratios. It should be noted that the factors cited here, namely oxidation state of the magmatic system during fractional crystallisation (or during AFC processes) and systematics related to the timing of crystallisation as determined by the master chemical variable, $C_w^{l,o}/C_w^{l,s}$, are but a few of the factors affecting the ultimate composition of the mineralised zone.

6. Acknowledgements

I would like to thank Phil Piccoli and Tom Williams for reading early versions of this manuscript, for drafting the figures, and for general assistance in assembling the paper. The comments of an anonymous reviewer are gratefully acknowledged. Special thanks to Dr Andrew R. Campbell of the New Mexico Institute of Mining and Technology, who reviewed a later version of this paper at short notice. I would like to thank Dr Bruce Chappell of the Australian National University for inviting me to submit a paper to this volume. My most sincere appreciation goes to my family, Mary, Emily, Alison and Nicholas, for their unwavering tolerance and understanding. This research was supported by the National Science Foundation (EAR-8706198 and EAR-9004154) and the Magma Energy Project of the Department of Energy.

7. References

Auge, J. J. & Brimhall, G. H. 1988. Magmatic arc asymmetry and distribution of anomalous plutonic belts in the batholiths of California: Effects of assimilation, crustal thickness, and depth of crystallization. BULL GEOL SOC AM **100,** 912–27.

Andiambololona, R. & Dupuy, C. 1978. Répartition et

comportment des éléments de transition dans les roches volcaniques. I. cuivre et zinc. BULL B.R.G.M. **2** (II), 121–38.

Barton, M. D. 1990. Cretaceous magmatism, metamorphism, and metallogeny in the east-central Great Basin. *In* Anderson, J.L. (ed.) *The nature and origin of Cordilleran magmatism: Boulder Colorado.* GEOL SOC AM MEM **174,** 283–302.

Barton, M. D. & 8 others 1988. Mesozoic contact metamorphism in the western United States. *In* W. G. Ernst (ed.) *Metamorphism and crustal evolution, western conterminous United States,* Rubey volume 7 110–178. Englewood Cliffs, New Jersey: Prentice-Hall.

Brandeis, G., Jaupart, C. & Allegre, C. J. 1984. Nucleation, crystal growth and the thermal regime of cooling magmas. J GEOPHYS RES **89,** 10161–77.

Brandeis, G. & Jaupart, C. 1987. Crystal sizes in intrusions of different dimensions: constraints on the cooling regime and the crystallization kinetics. *In* Mysen, B.O. (ed.) *Magmatic processes: physiochemical principles,* 307–318. New York.

Brandeis, G. & Marsh, B. D. 1989. The convective liquidus in a solidifying magma chamber: a fluid dynamic investigation. NATURE **339,** 613–16.

Candela, P. A. 1986. Generalized mathematical models for the fractional evolution of vapor from magmas in terrestrial planetary crusts. *In Advances in Physical Geochemistry* 6: *Chemistry and physics of the terrestrial planets,* 362–96. New York: Springer.

Candela, P. A. 1989a. Magmatic ore-forming fluids: thermodynamic and mass transfer calculations of metal concentrations. *In* Whitney, J. A. & Naldrett, A. J. (eds) *Ore Deposits Associated with Magmas, Reviews in Economic Geology* 4, 203–21. Society of Economic Geologists.

Candela, P. A. 1989b. Felsic magmas, volatiles, and metallogenesis. *In* Whitney, J. A. & Naldrett, A. J. (eds) *Ore Deposits Associated with Magmas, Reviews in Economic Geology* 4, 223–33. Society of Economic Geologists.

Candela, P. A. 1991. Physics of Aqueous Phase evolution in Plutonic environments. AM MINERAL **76,** 1081–91.

Candela, P. A. & Bouton, S. L. 1990. The influence of oxygen fugacity on tungsten and molybdenum partitioning between silicate melts and ilmenite. ECON GEOL **85,** 633–40.

Candela, P. A. & Holland, H. D. 1984. The partitioning of Cu and Mo between silicate melts and aqueous fluids. GEOCHIM COSMOCHIM ACTA **48,** 373–80.

Candela, P. A. & Holland, H. D. 1986. A mass transfer model for copper and molybdenum in magmatic hydrothermal systems: the origin of porphyry-type ore deposits. ECON GEOL **81,** 1–19.

Carroll, M. R. & Rutherford, M. J. 1987. The stability of igneous anhydrite: experimental results and implications for sulfur behavior in the 1982 El Chichón trachyandesite and other evolved magmas. J PETROL **28,** 781–801.

Carten, R. B., Walker, B. M., Geraghty, E. P. & Gunow, A. J. 1988a. Comparison of field-based studies of the Henderson porphyry molybdenum deposit, Colorado, with experimental and theoretical models of porphyry systems. *In* Taylor, R. P. & Strong, D. F. (eds) *Recent advances in the geology of granite-related mineral deposits,* 351–66. SPEC VOL **39** CIMM.

Carten, R. D., Geraghty, E. P., Walker, B. M. & Shannon, J. R. 1988b. Cyclic development of igneous features and their relationship to high-temperature hydrothermal features in the Henderson porphyry molybdenum deposit, Colorado. ECON GEOL **83,** 266–96.

Delaney, J. R. & Karsten, J. L. 1981. Ion microprobe studies of water in silicate melts: concentration-dependent diffusion in obsidian. EARTH PLANET SCI LETT **52,** 191–202.

Dingwell, D. B. 1988. The structures and properties of fluorine-rich magmas: a review of experimental studies. *in* Taylor, R. P. & Strong, D. F. (eds) *Recent advances in the geology of granite-related mineral deposits,* 1–12. SPEC VOL **39** CIMM.

Eichelberger, J. C., Carrigan, C. R., Westrich, H. R. & Price, R. H. 1986. Non-explosive silicic volcanism. NATURE **323,** 598–602.

Haffty, J. & Noble, D. C. 1972. Release and migration of molybdenum during the primary crystallization of peralkaline silicic volcanic rocks. ECON GEOL **70,** 857–912.

Hildreth, W. & Moorbath, S. 1988. Crustal contribution to arc magmatism in the Andes of Central Chile. CONTRIB MINERAL PETROL **98,** 455–89.

Irvine, T. N. 1970. Heat transfer during solidification of layered intrusions. I. Sheets and sills. CAN J EARTH PLANET SCI **7,** 1031–60.

Jaeger, J. C. 1968. Cooling and solidification of igneous rocks. *In* Hess, H. H. & Poldervaart, A. (eds) *Basalts,* 503–36. New York: Interscience Publishers.

Langmuir, C. H. 1989. Geochemical consequences of *In Situ* crystallization. NATURE **340,** 199–205.

Lehmann, B. 1990. *Metallogeny of Tin.* Heidelberg: Springer-Verlag.

Lehmann, B. & Harmanto, 1990. Large-scale tin depletion in the Tanjungpandan tin granite, Belitung Island, Indonesia. ECON GEOL **85,** 99–111.

London, D., Morgan, G. B., VI, & Hervig, R. L. 1989. Vapor-undersaturated experiments with Macusani glass + H_2O at 200 MPa and the internal differentiation of granitic pegmatites. CONTRIB MINERAL PETROL **102,** 1–17.

Luhr, J. F. 1990. Experimental phase relations of water- and sulfur-saturated arc magmas and the 1982 eruptions of El Chichón volcano. J PETROL **31,** 1071–114.

Luhr, J. F., Carmichael, I. S. E. & Varekamp, J. C. 1984. The 1982 eruptions of El Chichón volcano, Chiapas, Mexico: Mineralogy and petrology of the anhydrite-bearing pumices. J VOLCANOL GEOTHERM RES **23,** 69–108.

Lynton, S. J., Candela, P. A. & Piccoli, P. M. 1990. Experimental determination of copper partitioning between pyrrhotite and high silica rhyolite. GEOL SOC AM ABST PROG **22,** 181.

Manning, D. A. C. & Henderson, P. 1984. The behavior of tungsten in granitic melt-vapor systems. CONTRIB MINERAL PETROL **86,** 286–93.

Marsh, B. D. 1982. On the mechanics of igneous diapirism, stoping, and zone melting. AM J SCI **282,** 808–55.

Marsh, B. D. 1988a. Causes of magmatic diversity. NATURE **333,** 97.

Marsh, B. D. 1988b. Crystal capture, sorting, and retention in convecting magma. GEOL SOC AM BULL **100,** 1720–37.

Norton, D.L. & Knight, J. 1977. Transport Phenomena in Hydrothermal Systems: Cooling Plutons. AM J SCI **277,** 937–81.

Norton, D. & Taylor, H. P. Jr. 1979. Quantitative simulation of the hydrothermal systems of crystallizing magmas on the basis of transport theory and oxygen isotope data: an analysis of the Skaergaard intrusion. J PETROL **20,** 421–86.

Peck, D. L., Hamilton, M. S. & Shaw, H. R. 1977. Numerical analysis of lava lake cooling models: Part II, Application to Alae Lava Lake, Hawaii. AM J SCI **277,** 415–37.

Piccoli, P. M. & Candela, P. M. 1991. Spatial variations in model halogen concentrations in magmas based on apatite chemistry: examples from natural systems. GEOL SOC AM ABSTR PROG **23,** 5.

Sato, K. 1982. Chaacteristics of tungsten skarns in Japan, two constrasting types. *In* Hepworth, J. V. & Yu, H. Z. (eds) *Proc. Symp. W Geol. Jiangxi, China, 1981,* 203–209.

Tacker, R. C. & Candela, P. A. 1987. Partitioning of molybdenum between magnetite and melt: a preliminary experimental study of the partitioning of ore metals between silicic magmas and crystalline phases. ECON GEOL **82,** 1827–38.

Taylor, B. E. 1988. Degassing of rhyolitic magmas: Hydrogen isotope evidence and implications for magmatic-hydrothermal ore deposits. *In* Taylor, R. P. & Strong, D. F. (eds) *Recent Advances in the Geology of Granite-Related Mineral Deposits.* 33–49. SPEC VOL **39,** CIMM.

Titley, S. R. & Beane, R. E. 1981. Porphyry copper deposits. *In* Skinner, B. F. (ed.) *Economic Geology 75th Anniversary Volume.* Connecticut: Economic Geology.

Urabe, T. 1985. Aluminous granite as a source of hydrothermal ore deposits: an experimental study. ECON GEOL **80,** 148–57.

van Middelaar, W. T. & Keith, J. D. 1990. Mica chemistry as an indicator of oxygen and halogen fugacities in the CanTung and other W-related granitoids in the North American Cordillera. *In* Stein, H. J. & Hannah, J. L. (eds) *Ore-bearing granitic systems; Petrogenesis and mineralizing processes.* GEOL SOC AM SPEC PAPER **246,** 21–34.

Webster, J. D., Holloway, J. R. & Hervig, R. L. 1987. Phase equilibria of a Be, U, and F-enriched vitrophyre from Spor Mountain. Utah. GEOCHIM COSMOCHIM ACTA **51,** 389–402.

Webster, J. D. & Holloway, J. R., 1990. Partitioning of F and Cl between magmatic hydrothermal fluids and highly evolved granitic magmas. *In* Stein, H. J. & Hannah, J. L. (eds) *Ore-bearing granitic systems; Petrogenesis and mineralizing processes.* GEOL SOC AM SPEC PAP **246,** 21–34.

Westrich, H. R., Stockman, H. W. & Eichelberger, J. C. 1988. Degassing of rhyolitic magma during ascent and emplacement. J GEOPHYS RES **93,** 6503–11.
Whitney, J. A. & Stormer, J. C. Jr. 1983. Igneous sulfides in the Fish Canyon Tuff and the role of sulfur in calcalkaline magmas. GEOLOGY **11,** 99–102.
Wood, S. A. & Vlassopoulos, D. 1989. Experimental determination of the hydrothermal solubility and speciation of tungsten at 500 C and 1 kbar. GEOCHIM COSMOCHIM ACTA **53,** 303–12.

PHILIP A. CANDELA, Laboratory for Mineral Deposit Research, Department of Geology, University of Maryland, College Park, Maryland, 20742-4211, U.S.A

Transactions of the Royal Society of Edinburgh: Earth Sciences, **83,** 327–345, 1992

The Cretaceous volcanic-plutonic province of the central Queensland (Australia) coast—a rift related 'calc-alkaline' province

A. Ewart, R. W. Schon and B. W. Chappell

ABSTRACT: Silicic and minor intermediate and mafic pyroclastics, lavas, and dykes occupy a NW-trending zone through the Whitsunday, Cumberland and Northumberland Island groups, and locally areas on the adjacent mainland, over a distance of more than 300 km along the central Queensland coast. K–Ar and Rb–Sr data indicate an age range of 95–132 Ma, with the main activity approximately between 105–120 Ma; there is, however, evidence for easterly increasing ages. Comagmatic granites, some clearly intrusive into the volcanics, occur together with two localised areas of Triassic potassic granites (229 Ma), that form the immediate basement.

The volcanics are dominantly rhyolitic to dacitic lithic ignimbrites, with intercalated surge and bedded tuffs, accretionary lapilli tuffs, and lag deposits. Associated rock types include isolated rhyolitic and dacitic domes, and volumetrically minor andesite and rare basalt flows. The sequence is cut by abundant dykes, especially in the northern region and adjacent mainland, ranging from dolerite through andesite, dacite and rhyolite. Dyke orientations show maxima between NW–NNE. Isotope data, similarities in petrography and mineralogy, and alteration patterns all suggest dyke intrusion to be broadly contemporaneous with volcanism. The thickness of the volcanics is unconstrained, although in the Whitsunday area, minimum thicknesses of >1 km are inferred. Eruptive centres are believed to occur throughout the region, and include at least two areas of caldera-style collapse. The sequences are thus considered as predominantly intracaldera.

The phenocryst mineralogy is similar to modern "orogenic" volcanics. Phases include plagioclase, augite, hypersthene (uralitised), magnetite, ilmenite, with less common hornblende, and even rarer quartz, sanidine, and biotite. Fe-enriched compositions only develop in some high-silica rhyolites. The granites range from quartz diorite to granite s.s., and some contain spectacular concentrations of partially disaggregated dioritic inclusions.

Chemically, the suite ranges continuously from basalt to high-silica rhyolite, with calc-alkali to high-K affinities, and geochemical signatures similar to modern subduction-related magmas. Only the high-silica rhyolites and granites exhibit evidence of extensive fractional crystallisation (e.g. pronounced Eu anomalies). Variation within the suite can only satisfactorily be modelled in terms of two component mixing, with superimposed crystal fractionation. Nd and Sr isotope compositions are relatively coherent, with ε_{Nd} +2·2 to +7·3, and I_{Sr} (calculated at 110 and 115 Ma) 0·7031–0·7044. These are relatively primitive, and imply mantle and/or newly accreted crustal magma sources.

The two end-members proposed are within-plate tholeiitic melt, and ?low-silica rhyolitic melts generated by partial fusion of Permian (to ?Carboniferous) arc and arc basement. The arc-like geochemistry is thus considered to be source inherited. The tectonic setting for Cretaceous volcanism is correlated with updoming and basin rifting during the early stages of continental breakup, culminating in the opening of the Tasman Basin. Cretaceous volcanism is also recognised in the Maryborough Basin (S Queensland), the Lord Howe Rise, and New Caledonia, indicating the regional extent of volcanism associated with the complex breakup of the eastern Australasian continent margin.

KEY WORDS: rifting, silicic volcanism, pyroclastics, isotopes, trace elements, Cretaceous, crustal evolution, SW Pacific.

The Lower Cretaceous igneous provinces of the central Queensland coast were first mapped by Clarke *et al.* (1971). The southeasterly extension of the volcanics and intrusives had been previously recorded by Jensen *et al.* (1966), but mapped as Tertiary in age. Clarke *et al.* (1971) divided the volcanics into two units, the Whitsunday and Prosperpine volcanics; the former were defined as the thick (>1000 m) sequence of felsic to intermediate pyroclastics and minor flows which form most of the off-shore islands in the Proserpine map sheet area (Fig. 1). These authors considered their depositional environment to be subaqueous, in a land-locked intramontane basin. The Prosperine Volcanics were defined as the thick (>1000 m) sequence of rhyolite flows and minor pyroclastics and andesite overlying Lower Permian volcanics (Airlie Volcanics) on the mainland coast, and regarded as the terrestrial equivalents of the off-shore sequences, deposited on a relatively stable block of pre-existing volcanic basement.

This study has confirmed the southeasterly extension of the Whitsunday volcanics to the Percy Islands (Fig. 1), a strike length of approximately 300 km. Moreover, Early Cretaceous intermediate-silicic volcanism is known further south, most notably in the Maryborough region (Ellis & Whitaker 1976) comprising the Grahams Creek Formation. Further off-shore extensions of ?Late Cretaceous volcanism are recorded in the wells of Wreck Is. and Capricorn 1A (see Fig. 17; Davies & Marshall in press), and imply an extensive off-shore zone of volcanism of at least 900 km length. Within the study areas, Clarke *et al.* (1971) recognised the existence of associated granites, presumed to be coeval with the silicic volcanics, together with abundant dykes ranging from mafic to silicic in compositions, that cut most of the granites and volcanics alike (Figs 1, 2). This study has confirmed the Cretaceous age of many of the granites, but has shown some to be Triassic, most notably those of the Shaw Is. and Knight Is. groups (Fig. 1). A further complication in the region is the extent of faulting (Fig. 2), including the possible existence of 90 km of lateral movement along the Cumberland fault.

The main purpose of this paper is to present the first petrological and geochemical data on the Whitsunday Volcanics, a suite of dolerites, andesites, dacites and rhyolites, to interpret these data in terms of the tectonomagmatic affinities of the magmas, and to consider their significance in terms of the possible origin(s) of the magmas.

1. Tectonic environment

The Cretaceous was a period of intense plate divergence of East Gondwana which resulted in the isolation of the Australian continent, and the development of its modern outline (Veevers *et al.* 1991). Rifting along the modern eastern Australian coast is interpreted as commencing in the late Cretaceous (96 Ma), with separation underway by 84 Ma; termination of spreading within the Tasman Sea occurred by approximately 55 Ma (Veevers *et al.* 1991). The zone of divergence, however, did not pass close to the present eastern Queensland coast, having diverted eastwards along a transform fault, leaving the off-shore Queensland and Marian plateaux. The Whitsunday volcanics are situated along the western margin of the Marian Plateau (Falvey & Mutter 1981). These authors, however, infer the existence of a series of sediment-filled rift-valley sequences around and through the plateaux from the Early Cretaceous. Henderson (1980) and Smart and Senior (1980) also drew attention to the necessity of having a felsic-intermediate volcanic zone of considerable magnitude as provenance to Cretaceous sediments in the Great

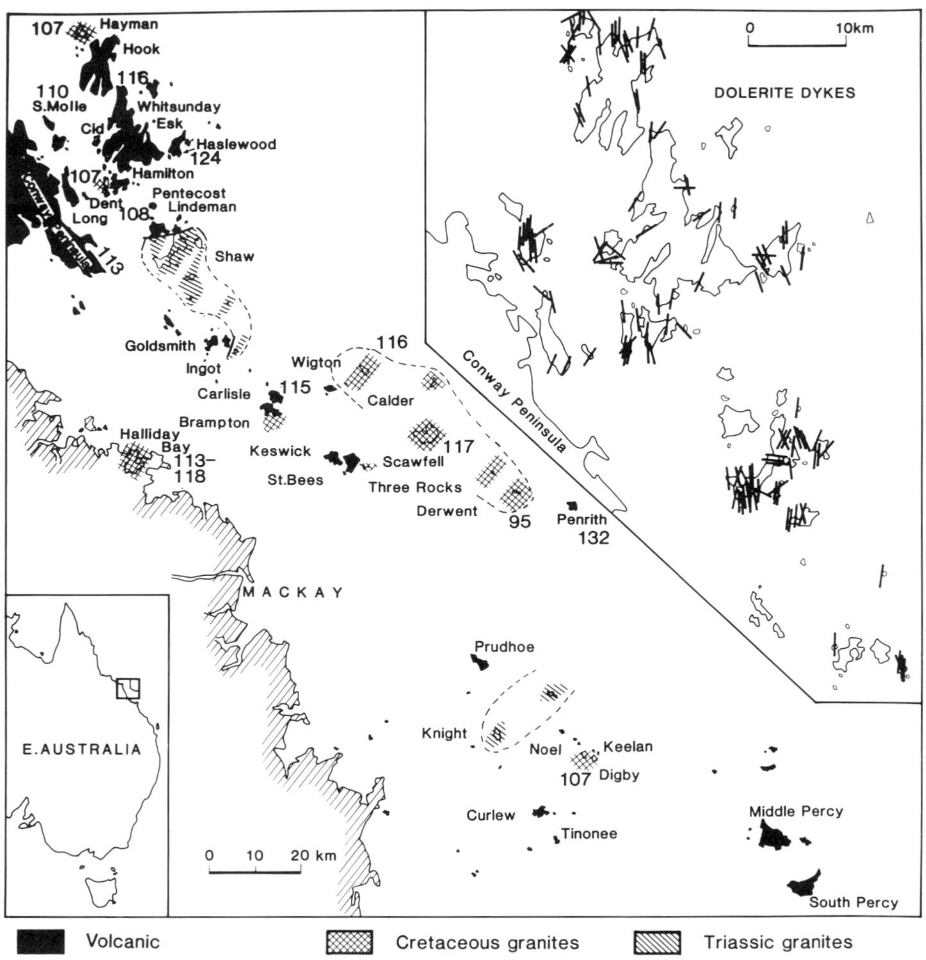

Figure 1 Generalised location map of the central Queensland coast showing the distribution of the Cretaceous volcanics (black) and the Triassic and Cretaceous granites; inset shows the generalised distribution of dolerite dykes in the northern islands; figures are K–Ar and Rb–Sr dates (Table 1) for lavas, ignimbrites, and granites; location of islands referred to in text is shown.

Artesian Basin. They infer the site of volcanism to be largely eroded and downfaulted on the present continental shelf, only a small fraction of which is preserved in the Whitsunday and Proserpine volcanics.

A divergence of interpretation, however, concerns the tectonic affiliation of the Cretaceous volcanics. Veevers (1984, e.g. pp. 136, 137, 205; see also Henderson 1980) consider the volcanics to be part of a magmatic arc associated with a convergent plate boundary which existed several hundred kilometres E of the present eastern Australian coast (see also Veevers *et al.* 1991). The reasoning seems to be based on the petrographically-determined calc-alkaline affinities of the volcanism, and thus the inferred correlation with modern convergent plate margin volcanism.

Other authors, however, clearly identify the Cretaceous eastern Australian margin as a passive divergent margin (e.g. Falvey & Middleton 1981; Falvey & Mutter 1981; Symonds *et al.* 1984, 1988; Symonds & Willcox 1989), with a detachment model being applied by Etheridge and Lister (1987) and Lister *et al.* (1988) to the margins of the Tasman Sea. These models imply no prior convergent phase in the late Jurassic–Early Cretaceous, and the Whitsunday Volcanics are interpreted as erupting within Lower Cretaceous rift basins or troughs developed prior to the actual rifting stage; in fact, it is possible that they occupy early failed rift systems. Falvey and Middleton (1981) suggest that basin subsidence may occur as rift grabens and half grabens flanking incipient continent–ocean boundary as much as 50 Ma before breakup; volcanism is noted to occur only close to the incipient boundary.

In conclusion, therefore, we see no compelling tectonic evidence for the existence of a convergent plate boundary association with the Whitsunday Volcanics. This is considered a significant aspect in the interpretation of the petrogenesis of the magmas.

2. Geochronology

Webb and McDougall (1968) established the Lower Cretaceous ages of the Whitsunday and Proserpine volcanics by means of an eight-point Rb–Sr isochron of 113 ± 5 Ma (corrected from 111 Ma using updated constants) for the Proserpine volcanics, plus four K–Ar dates. This is now supplemented by an additional nine K–Ar dates and 49 Rb–Sr data sets (Table 2) specifically on the Whitsunday Volcanics, and an additional 15 Rb–Sr data sets for the Proserpine Volcanics. The latter provide an identical isochron of 113 Ma to that noted above. A summary of dates is presented in Table 1.

Combined K–Ar and Rb–Sr data on the granites show the existence of both Triassic and Cretaceous granites. The Rb–Sr isochron of the former gives an age of $229 \pm 1{\cdot}1$ Ma (MSWD = 2·8), with initial $^{87}Sr/^{86}Sr$ of $0{\cdot}70353 \pm 0{\cdot}00007$. These granites comprise the Shaw Is. and Knight Is. groups, and the ages are confirmed by a K–Ar date from Knight Is.

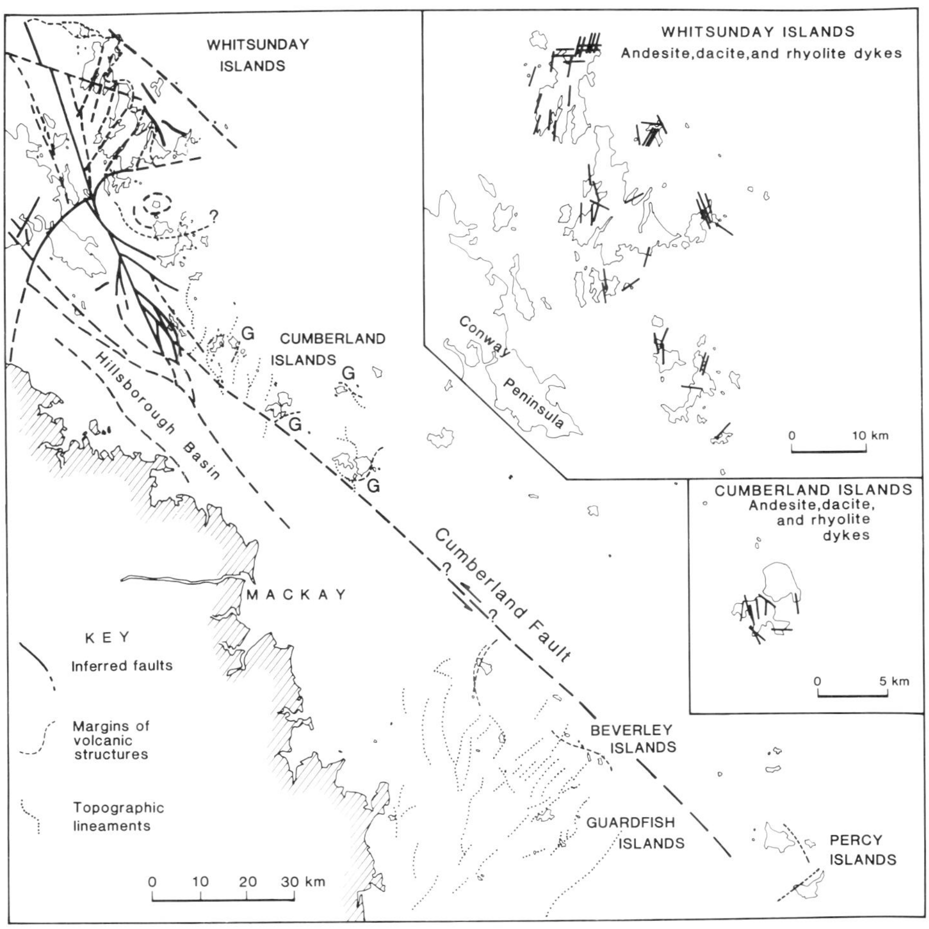

Figure 2 Map of the Whitsunday Volcanics and associated granites, showing inferred fault structures, submarine topographic lineament patterns and volcanic boundaries (the symbol G being used to indicate Cretaceous granite contacts); inset shows the generalised distribution of intermediate to silicic dykes in the northern and central island groups.

of 232 ± 3 Ma. The combined data for the Cretaceous granites indicate three discrete age groupings, namely: (i) Scawfell–Three Rocks (K–Ar 116·6 Ma), and the separate Wigton–Calder Is. (K–Ar 116·1, assumed to also apply to Calder Is.) groups; (ii) Derwent Is. (K–Ar 94·8); (iii) the Hayman–Dent–Noel Is. zone which define a single isochron of 107 ± 2·5 Ma. Field relationships show the Dent, Hayman and Noel-Digby granites to be intrusive into volcanics, whereas dolerite dykes are recorded as cutting all but the Derwent granite. The remaining granites have no volcanic contacts exposed.

The age range of the Cretaceous granites suggests that volcanics should also exhibit comparable age ranges. This is certainly confirmed (Table 1), although complicated by apparently varying age patterns between the ignimbrites, lavas and dykes. The ignimbrites define two, and possibly four age groupings, namely 132 Ma (the outlaying Penrith Is.), and 115–117 Ma (the Hook–Whitsunday–Carlisle–Pentecost–Goldsmith Is. group). A possible age grouping of approximately 125 Ma needs further confirmation for the ignimbrites on islands occurring E of Whitsunday Is. (and also E of a major tectonic, possibly collapse structure, Fig. 2), while a younger K–Ar age of 98 and 108 Ma is reported for alunite altered sample from Pentecost (Webb & McDougall 1968), the exact significance of which is uncertain; it is noted, however, that this ignimbrite possesses a distinctive mineralogy, and possibly represents a younger eruptive phase.

The rhyolitic to andesitic lavas post-date the ignimbrites, consistent with field relations, with Rb–Sr ages of 105, 110 and 113 Ma, the youngest group interestingly belonging to the southernmost group of islands. Only three K–Ar dyke ages are available, ranging from 128 Ma on Penrith Is. (containing the oldest ignimbrites), to 107 and 100 Ma. Dyke emplacement thus apparently occurred over an extended time interval, largely coeval with volcanism and granite emplacement. This is consistent with Rb–Sr dyke data which fail to give a geologically meaningful isochron (nominally 126 Ma).

Although still incomplete, the data do suggest a systematic age pattern (Fig. 1), with ages broadly increasing towards the E for the granites and ignimbrites. This pattern tends to reverse on the adjacent mainland for granite emplacement (coeval volcanics being absent). Moreover, within any one zone in the volcanic belt, granite emplacement appears to postdate associated volcanism by up to 8 Ma. In terms of the

Table 1 Summary of radiometric dating on Whitsunday and related islands

Rb–Sr	K–Ar
1. *Granites*	
Basement (Shaw Is. group, Knight Is., Conway Peninsula) 229 ± 1·1 (MSWD 2·8)	Knight Is. 232 ± 3 (M)
Hayman–Dent–Noel 107 ± 2·5 (MSWD 0·1)	Scawfell 116·6 ± 1·3 (H)
	Wigton 116·1 ± 1·3(H)
	Derwent 94·8 ± 1·1(H)
2. *Ignimbrites*	
Hook–Carlisle–Goldsmith–Pentecost 116·5 ± 2·2 (MSWD 1·4)	Penrith 131·8 ± 1·5 (H)
Esk–Haslewood ~ 124 (2 points)	Esk 125 ± 3 (H)
	Carlisle 115 ± 3 (H)
	Whitsunday 115 ± 2·5 (H)
	Pentecost 98 and 108 ± 3 (Alunite)
3. *Lavas*	
Shaw–Lindeman–Cid–Hamilton–Ingot 113·3 ± 1·2 (MSWD 1·8)	
S Molle–Long 110 ± 0·9 (MSWD 1·6)	
Tinonee–Curlew–Keelan–Keswick 105·4 ± 1·7 (MSWD 1·7)	
4. *Dykes*	
	Penrith (Hornblende andesite) 128·4 ± 1·4 (H)
	Lindeman (Dacite) 100 ± 3 (H)
	S Molle (gabbro/dolerite) 107 ± 2 (P)
5. *Mainland*	
Conway Peninsula 113 ± 3 (MSWD 12·5, 23 points)	Halliday Bay 113 and 118 ± 3 (H) (Separate granite phases)

Date sources: Clarke *et al.* 1971; Webb & McDougall 1968; Ewart & Schon, unpublished data. H = hornblende; P = plagioclase; M = muscovite.

Table 2 Frequency % occurrence of phenocryst phases in Whitsunday volcanics, compared to modern southwestern Pacific rhyolite and dacite occurrences

	Rhyolites and dacites Whitsunday Is.	Southwestern Pacific dacites*	Rhyolites of Taupo Volcanic Zone (New Zealand)*
Plagioclase	94	100	97
Quartz	15	14	62
Augite	82	89	9
± Augite + Hypersthene	(82)⁺⁺	82	94
Hornblende	15⁺	32	58
Biotite	8⁺	14	30
Samples	279	141	77

Notes: * After Ewart 1979; ⁺ Includes identifiable pseudomorphs after these phases; ⁺⁺ Hypersthene is altered throughout these volcanics, but is inferred to have been an originally common phase, based on pseudomorphs.

rifting model, the age patterns can be interpreted in terms of progressive rifting and volcanism above a linear heat source approximately coincident with the present central Queensland coastline, with net eastward movement of the rifted crustal blocks. The oldest date of 132 Ma (Penrith Is.) is consistent with the lower Cretaceous age of the more southerly Grahams Creek Formation (136 Ma), and ages of 136–145 Ma for associated intrusives (Green & Webb 1974; see also Fig. 18). The total age range for intrusive and extrusive activity is 37 Ma (95–132 Ma) for the Whitsunday region.

3. Outline of volcanic geology

The volcanics are dominated by rhyolitic to dacitic ignimbritic units (ranging from 10–100 m in thickness) throughout the volcanic zone, accompanied by surge and bedded tuff deposits, lenticulitic and pumiceous zones and, on most islands, containing lithic concentration zones and more locally, coarse lag deposits. In fact, a characteristic feature of the ignimbrites is their lithic-rich nature, the fragments comprising volcanic types ranging from andesite and basalt through to fluidal rhyolites, inferred to include Cretaceous and Triassic basement volcanics. In addition to the primary pyroclastics, local thick sequences of phreatomagmatic deposits, and high energy coarse volcanic sandstones are encountered. Lavas are mainly rhyolitic with minor andesites and dacites, and rarely basalts, occurring as isolated flows, lava domes, and basaltic scoria cones.

Dykes are common and characteristic throughout the region, ranging from dolerites to rhyolites, with widths from 0·5–100 m. The dolerites are most abundant in the northern part of the region (Fig. 1), being particularly conspicuous as swarms cutting the granites, and the volanics at the northern end of Hook Is.; there are also some particularly thick (up to 100 m) gabbroic dyke/sill complexes on S Molle Is. This latter locality is especially significant as the geochemically most "primitive" dykes are here. The intermediate to silicic dykes are likewise widespread, locally up to 40 m thick (these thicker units inferred to be volcanic feeders), the most intense concentration occurring again at the northern end of Hook Is. (Fig. 2). Dyke trends are dominantly northerly for all types, with only local northwesterly and southwesterly trends developed. Dolerites are noted to cut rhyolitic dykes in the rare dyke intersections observed.

3.1. Volcanic environments

The volcanic sequences are interpreted as subaerially deposited, with only localised shallow water deposits (cf. Clarke *et al.* 1971). The deposits within the majority of islands are interpreted as "intracaldera" facies. Supporting evidence includes: (i) Estimated thicknesses of >1 km for the sequences on at least Hook, Whitsunday, and Lindeman Is.; the occurrence of massive ignimbritic units on other islands likewise indicates total thicknesses greater than typical of outflow facies; (ii) pervasive low grade hydrothermal alteration; (iii) the abundance of intermediate to silicic dykes; dolerites seem most abundant either in the granites, or close to inferred caldera collapse margins; (iv) occurrences of very coarse lag deposits, some with clasts 3–6 m in diameter.

Figure 2 shows the inferred distribution of the tectonic boundaries and structures, based on limited available aeromagnetic data, with other structural lineaments from bathymetry, exposed basement-volcanic contact zones and the distribution of inferred near-vent pyroclastic facies. One major boundary lies along the eastern margin of Hook and Whitsunday Is., swings westward to the northern end of Hayman Is., with the southern margin passing through the Shaw–Lindeman Is. groups (Figs 1, 2); this boundary is believed to represent a volanic-collapse feature. The islands E of Whitsunday tend to preserve less-altered pyroclastics and are inferred to represent (older) outflow facies. This structure is the only "caldera-like" collapse structure so far recognised in the region, the southern volcanic centres being far more fragmented, and perhaps graben-controlled.

The overall regional pattern of volcanicity is thus interpreted as involving multiple vent eruptions, localised within at least four centres of collapse, and aligned along a complex volcano-tectonic rift depression, possibly of the style seen in the Taupo Volcanic Zone of New Zealand, although with much greater length.

4. Petrography

4.1. Volcanics

4.1.1. Rhyolites and dacites. Based on phenocryst assemblages, the following mineralogical groups are recognised:

(i) Plagioclase, augite, Ti-magnetite, ± uralitised orthopyroxene. This is the predominent mineralogical assemblage throughout the region. Phenocryst abundances range from <5% (commonly <2%) in lavas, to typically between 5–15% (with local crystal-rich lithologies) occurring within ignimbrites (eruptive crystal concentrations).

(ii) Plagioclase, quartz, augite, and Fe–Ti oxides, with 5–20% total phenocrysts. This assemblage is found within some ignimbrites of the central and southern islands, but is generally uncommon.

(iii) Plagioclase, augite, hornblende, and Fe–Ti oxides, with phenocryst abundances ranging from 5–25%, and rarely containing biotite. This assemblage occurs in dykes and commonly in ignimbrites from Hook, Esk, and Haslewood Is. within the northern area, and Penrith and Middle Percy in the southern area.

(iv) Plagioclase, quartz, hornblende, augite, and Fe–Ti oxides, with phenocryst contents 15–30%. This assemblage occurs rarely in dykes, being most characteristic of the Carlisle and some Long Is. ignimbrites.

(v) Plagioclase, quartz, biotite, augite, and Fe–Ti oxides, with phenocryst contents of 10–30%. This is found within dykes and the Pentecost Is. ignimbrites.

(vi) Plagioclase, sanidine, quartz, biotite, ± hornblende, Fe–Ti oxides; phenocryst contents 20–40%. This occurs in some dykes, and very rarely in high-silica rhyolitic ignimbrites.

4.1.2. Andesites, basalts and dolerites. Dolerite dykes and the rare basalts are usually nearly aphyric, with sparse augite and/or plagioclase phenocrysts. Andesite dykes and lavas characteristically contain phenocrystal plagioclase (15–30%), augite (3–10%), uralitised orthopyroxene (3–5%), and Ti-magnetite.

4.1.3. Discussion. Only relatively rare extrusive equivalents of the dolerite dykes are found, most notably as local basalt flows on S Molle and Lindeman Is. Dykes, plugs, and flows of andesite, however, occur sporadically throughout the region. The rhyolitic lavas with very low phenocryst abundances have no pyroclastic equivalents, although the pyroxene and hornblende-bearing rhyolitic dykes have equivalents among the ignimbrites. The sanidine-bearing rhyolite dykes apparently have only extremely rare ignimbritic equivalents. Relative dyke ages are poorly constrained, but dykes having assemblages (v) and (vi), above, cut rhyolite domes and an andesite plug, both of the latter inferred as among the youngest extrusive phases. The youngest dyke dated (100 Ma; Table 1), from Lindeman Is., has a type (iv) assemblage.

Comparison of the overall pattern of phenocryst abundances with modern southwestern Pacific "orogenic" silicic volanics (Table 2) shows the closest similarity with the orogenic dacites, but even here, the Whitsunday volanics apparently contain less common hornblende and biotite; they are clearly predominantly pyroxene rhyolites and dacites.

Secondary alteration. Low grade pervasive hydrothermal alteration is a regional characteristic of the volcanics, usually more extensively developed in the pyroclastics, but also occurring as microveining in lavas. Secondary phases include albite, chlorite, calcite, epidote, zoisite, sericite, actinolite, quartz, and sphene, with localised developments of prehnite, pyrite, laumontite, pyrophyllite and alunite.

4.2. Granites

These range from granite (s.s.) to quartz diorite, and petrographically fall into natural groupings:

Derwent. Strongly porphyritic, with phenocrystal K-feldspar + quartz + minor plagioclase + accessary Fe-hornblende, biotite, allanite, magnetite and rutile.

Scawfell and Three Rocks. Perthite + quarz + plagioclase (rimmed by K-feldspar) + accessory hornblende, biotite, allanite, ilmenite + magnetite.

Digby and Noel. Porphyritic, with plagioclase + quartz + perthite (interstitial to, and rims plagioclase) + accessary hornblende (rare augite cores), biotite, sphene, + ilmenite.

Dent. Porphyritic with plagioclase (K-feldspar rims) + quartz + K-feldspar + accessary pyroxene, magnetite, + rutile.

Hayman, Wigton and Calder. These contain abundant but irregularly distributed inclusions, and range from adamellite to granodiorite. They contain plagioclase (K-feldspar rims) + quartz + K-feldspar + accessary hornblende, biotite, sphene, allanite, magnetite, ilmenite, ± rutile. Hayman granites also contain some remarkably euhedral epidotes, suggesting primary crystallisation (although not readily reconciled with their shallow emplacement). The most significant aspect of these granites is the presence of abundant quartz diorite inclusions, locally forming spectacular zones of mingling. The inclusions are veined and injected by granodiorite-adamellite melt, and all stages of disaggregation are present from 20 m-long tabular "rafts" of quartz diorite to small diffuse "clots". The inclusions are fine to medium grained, consisting of plagioclase with interstitial quartz, K-feldspar plus accessory hornblende, ± augite, ± biotite, magnetite and ilmenite.

5. Mineralogy

Plagioclase. Compositions are commonly modified by albitisation, but remnant primary compositions are sufficiently often preserved to determine original compositional ranges. Within the dolerites (including basaltic andesite dykes), plagioclase (mainly coarse groundmass, less often phenocrystal) ranges from labradorite to bytownite, although within the coarse doleritic dyke/sill complexes on S Molle, plagioclases are zoned continuously through to oligoclase, due to *in-situ* fractionation. Phenocrystal andesite plagioclases are labradorite to median bytownite, whereas, in the dacitic ignimbrites, dominant compositions are andesine, although it is common to find individual crystals of labradorite–bytownite compositions. Low-silica rhyolitic dykes and ignimbrites also exhibit divergent plagioclase compositions (labradorite–andesine), suggesting that these are not simply the result of eruptive entrainment of andesitic lithic clasts into ignimbrites, and tend to point to a pre-eruptive mixing process. High-silica rhyolites contain oligoclase, in some examples coexisting with a sodic sanidine phase, but again sporadic isolated more calcic plagioclase occurs.

Pyroxene and amphiboles (Figs 3, 4). Only calcic pyroxenes are preserved, although orthopyroxene is inferred to have originally been a widespread phase in the andesitic and rholitic volcanics. Reference to the pyroxene compositional ranges shows that with the exception of high silica rhyolites, there is little variation in Fe–Mg–Ca ratios throughout the dolerite to low-silica rhyolite magma compositions. In fact, the general compositions are relatively magnesian, this being a common "orogenic" characteristic, and is emphasised by the overlap of the compositional fields with the respective modern Southwestern Pacific "orogenic" pyroxenes (shown by the circled fields in Fig. 3). Within the doleritic pyroxenes, the more calcic pyroxene (and higher Al and Ti) compositions are found within the *ne*- and *ol*-normative dolerites, indicating that the silica-undersaturated chemistry of these dolerites is a primary magmatic feature. Fe-enriched compositions only occur in the rhyolites, most notably the high-silica types, but significantly, bimodal pyroxene compositions are found in some low and high-silica rhyolites (Fig. 3, h-i). Evidence for mixed pyroxene compositions, most notably in some andesites, a trachyandesite, and dacites, also comes from either bimodal Al solid solution abundances, or wider ranges of Al than normally found in pyroxenes from such rocks. The high-silica rhyolite pyroxenes exhibit very limited Al solid solution ranges.

With the exception of rare Fe-rich hornblendes in high-silica rhyolites, the phenocrystal volcanic hornblendes are consistently relatively magnesian, and are again similar to compositions within "orogenic" silicic volcanics (e.g. Ewart 1982). In terms of major cation substitutions (Fig. 4), the amphiboles range from actinolitic hornblende in the rhyolites through hornblende in the rhyolites and dacites, towards hastingsite in the andesites and some dacites. Relict kaersutite is recorded in a single "mixed" andesite. The total data exhibit a very smooth trend through the compositional range, but as with the pyroxenes, bimodal compositional ranges are observed in dacites and low-silica rhyolitic ignimbrites and dykes. Application of the amphibole–plagioclase geothermometer (Blundy & Holland 1990) has not provided definitive temperature estimates, as taking extremes of compositional ranges found within relevant dacites and low-silica rhyolites gives an exaggerated temperature range between 620–1170°C. This is suggested to reflect disequilibrium among these compositions.

Amphibole compositions in the granites exhibit differences from those in the volcanics (Figs 3, 4). Firstly, there is a tendency for increasing Fe-enrichment, most pronounced in the late-stage Derwent porphyritic granite. Secondly, the compositions are offset towards more edenitic compositions, especially pronounced in the Derwent hornblendes. Other hornblendes, however, tend toward more actinolitic compositions, and this is attributed to solidus–subsolidus re-equilibration processes.

Fe–Ti oxides. Titaniferous magnetite is the dominant oxide phenocryst phase in the andesites and dacites, in some samples, coexisting with minor ilmenite. Coexisting oxide phases occur in the rhyolites, often exhibiting marked Mn enrichment. T°C and f_{O_2} estimates from a few available relatively homogeneous compositions suggest rhyolitic equilibration temperatures of 800–900°C, the hornblende-bearing types lying above NNO buffer, and the pyroxene rhyolites between NNO and FMQ.

6. Chemistry

6.1. Major elements

The overall range, and continuity of compositions of the volcanic and intrusive phases is readily apparent from the total alkalis–silica plot (Le Maitre 1989; Fig. 5). The bulk of

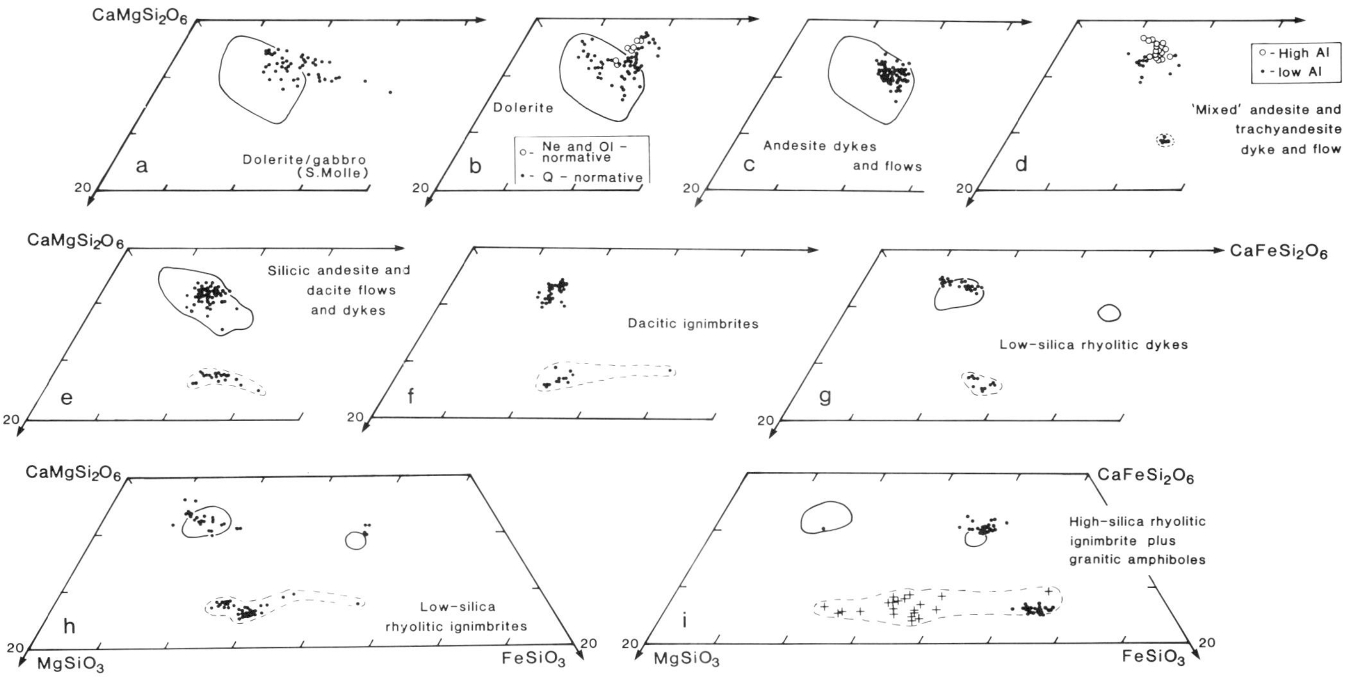

Figure 3 Pyroxene and amphibole phenocryst compositions in the Whitsunday Volcanics and Cretaceous granites ((i) only), expressed in terms of Ca, Mg, and Fe + Mn (atomic %); continuous circled fields are the pyroxene fields from southwestern Pacific "orogenic" lavas of the equivalent bulk rock compositions (after Ewart 1979, 1982); data lying in the dotted fields are amphibole compositions; in (i), the crosses represent granite amphiboles excluding Derwent Is., while the filled black dots represent Derwent Is. amphiboles.

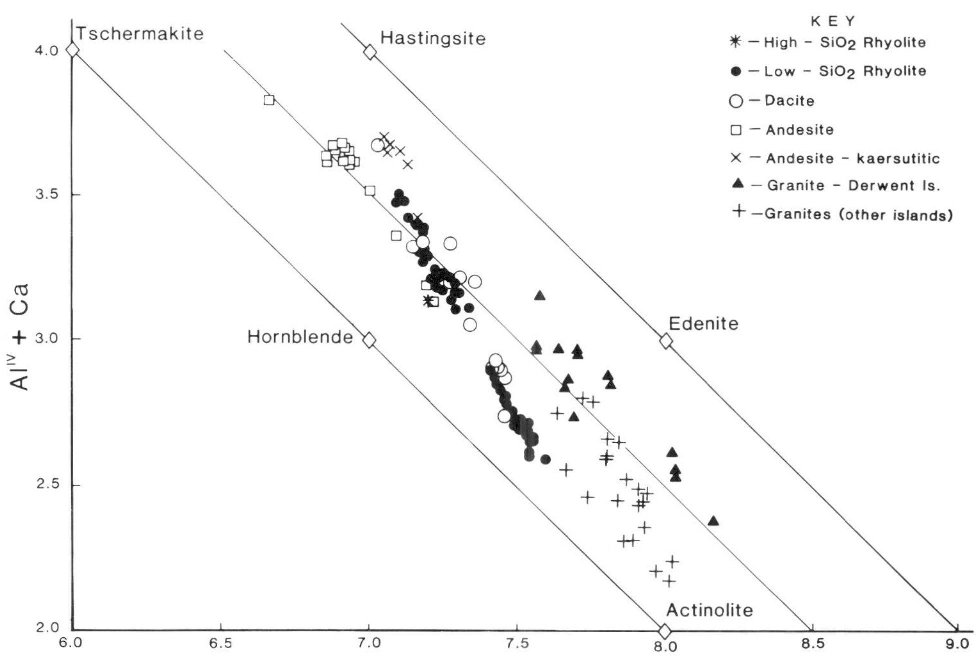

Figure 4 Amphibole phenocryst compositions in the Whitsunday volcanics and coeval granites.

the data plot within the basalt–basaltic andesite–andesite–dacite–rhyolite fields, with a relatively small percentage classifying as "alkalic", especially in the trachyandesite field. In this respect, the Cretaceous suite contrasts strongly with the alkaline eastern Australian Cainozoic intraplate provinces (Ewart *et al.* 1988), but is comparable with the overall calc-alkaline chemistry of many modern western Pacific island arc regions (e.g. Ewart 1979, 1982). The rhyolites are almost all peraluminous, averaging 1·2% normative corundum; this compares with 0·8% for the granites.

Figure 6a–d illustrates the silica distribution in relation to modes of occurrence. The ignimbrites and dykes exhibit continuous ranges from dolerites to rhyolites (dykes) and andesite to rhyolite (ignimbrites). The dykes exhibit a relatively constant frequency across the whole silica range

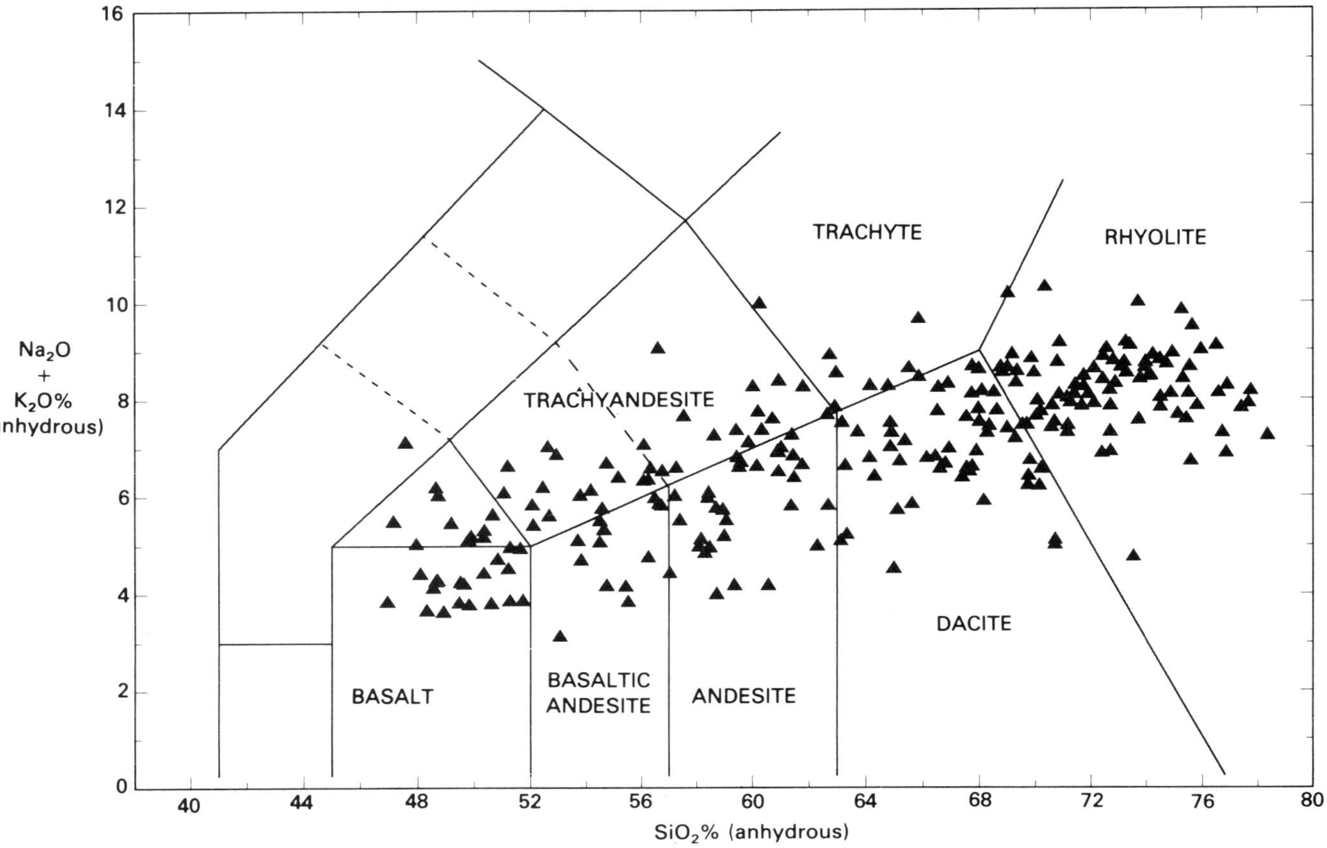

Figure 5 Total alkalis–silica plot of the Whitsunday volcanics and Cretaceous granites; field boundaries are after Le Maitre (1989).

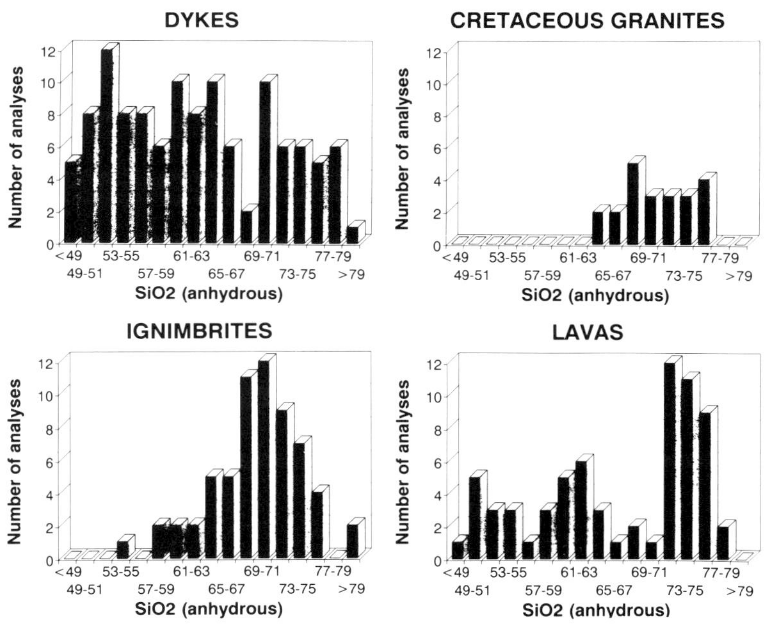

Figure 6 Silica histograms showing the distribution of compositions within the dykes, lavas, and ignimbrite, and coeval granites.

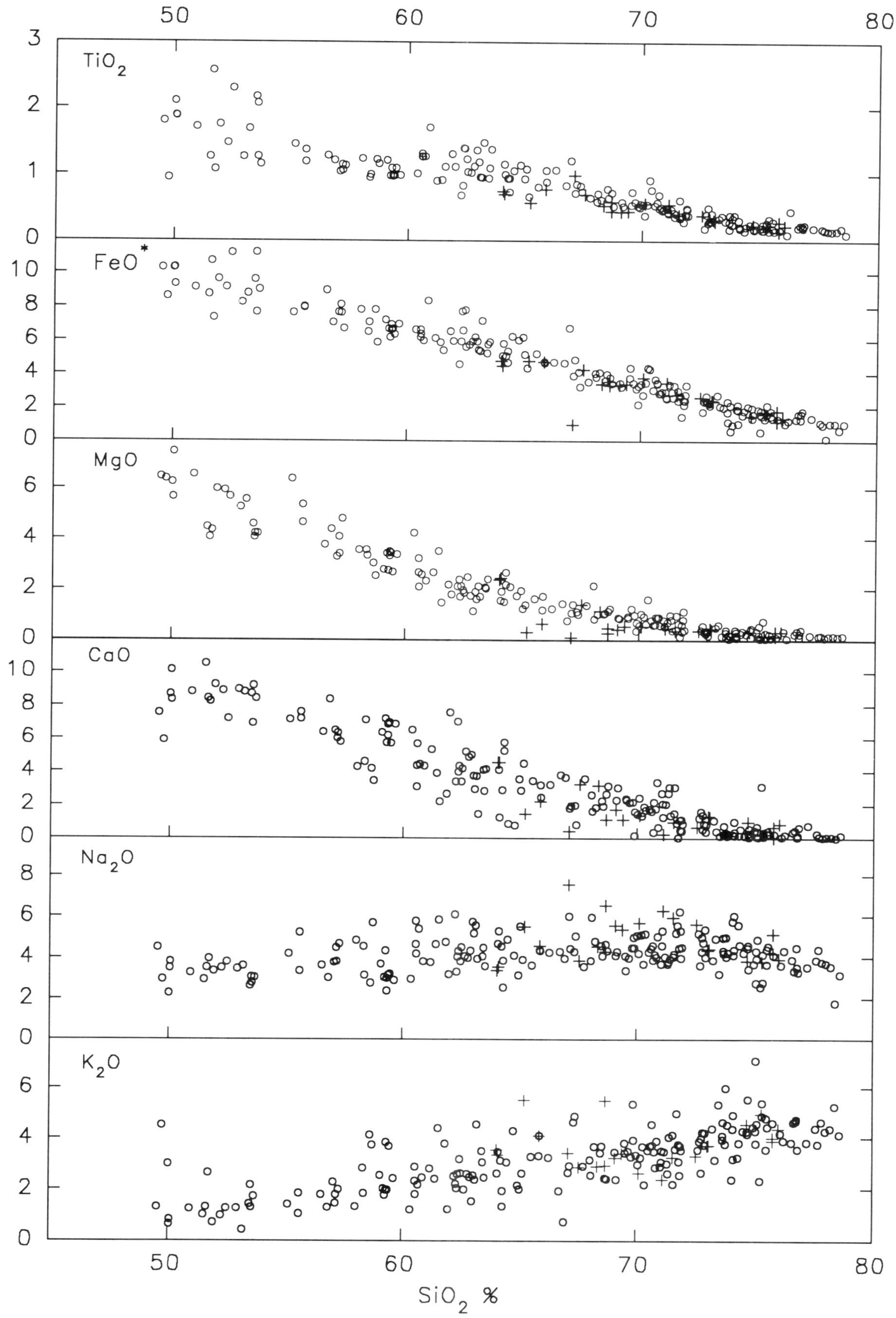

Figure 7 Composite Harker Diagrams for the Whitsunday Volcanics and (crosses) Cretaceous granites.

while the ignimbrites are dominantly dacitic to low-silica rhyolites. In contrast, the lavas tend to be trimodal, with distinct basalt, andesite and high silica rhyolite types.

Harker diagrams (Fig. 7a–i) show the expected trends with respect to silica, and are generally comparable to those of the New England granites (Shaw & Flood 1981), although differences exist with respect to MgO, CaO and FeO*. The Whitsunday data exhibit a wider scatter for the alkalis and CaO, possibly reflecting alteration effects. The TiO_2 and P_2O_5 data also exhibit some divergence at the mafic end of the range, almost certainly reflecting a range of doleritic compositions from *ne*- to *Q*-normative. As the majority of mafic dykes are labradorite-normative, they classify as alkali olivine basalts to tholeiites, although the majority are olivine to quartz tholeiites.

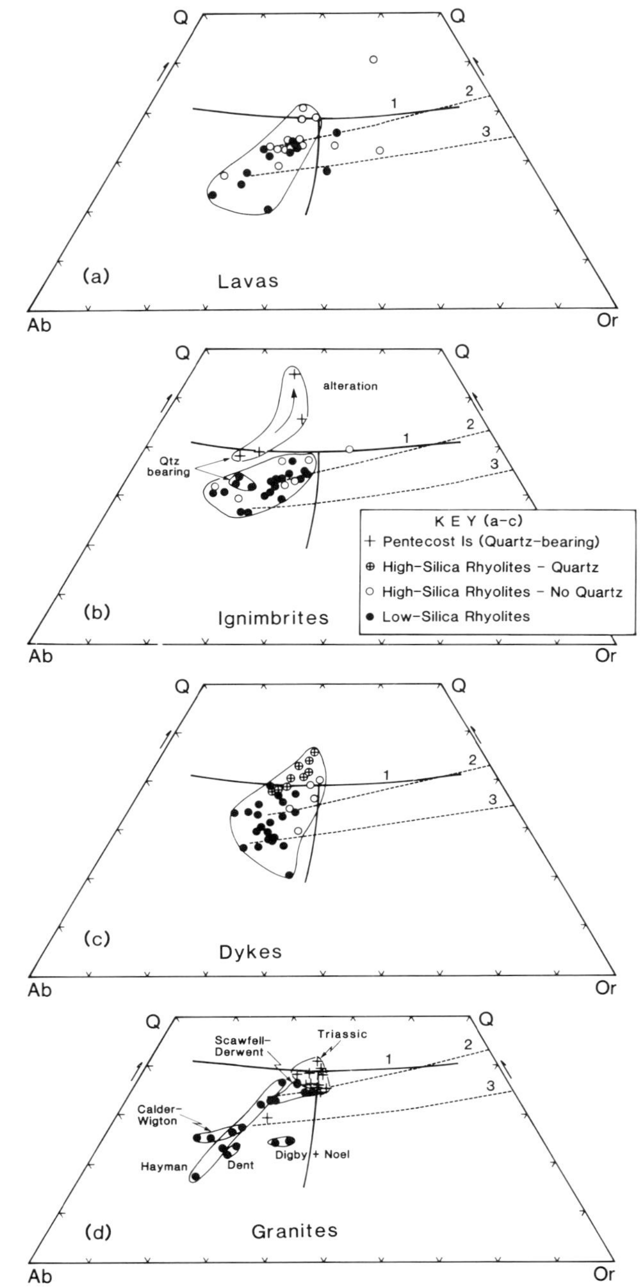

Figure 8 Normative Q–Ab–Or components of the rhyolite lavas, ignimbrites, and dykes, and the granites; boundary curves are: 1, water saturated curves, 100 MPa for composition An_3 after James and Hamilton (1969); curves 2 and 3 are the calculated 4-phase surfaces in the system Ab–An–Or–Q–H_2O for $a_w = 1{\cdot}0$ and 0·2 respectively, at 200 MPa (after Nakvasil 1988); in (a) to (c), the main groups of data are circled, excluding outlying points believed to be modified by alteration.

In terms of normative Ab-Or-Q components (Fig. 8), the silicic volcanics and granites are of sodic, rather than potassic affinities. The Triassic and Cretaceous granites are compositionally distinct, and with one exception, the former represent hypersolvus minimum-temperature melt granites whose compositions project near the 100 MPa water-saturated piercing point in the $[Ab\text{-}Or\text{-}Q]_{97}An_3$ system (James & Hamilton 1969). The Cretaceous granites, in contrast, extend from near minimum-temperature melt compositions (e.g. Scawfell–Derwent Is. group) to granodiorites, and cluster into distinct groupings usually reflecting similar chemistry between geographically closely-spaced islands. The trends of the data fields vary, with some apparently controlled by a four-phase surface as defined in the Ab-An-Or-Q-H_2O system (Nekvasil 1988), while other granite data fields are cross-cutting, possibly representing dioritic mixing trends (i.e. the inclusion-rich granites, previously noted).

The rhyolitic volcanics (Fig. 8a–c) define similar compositional fields to the coeval granites, although with variable trends. The high-silica rhyolites project closest to the minimum-temperature compositions, although phenocrystal quartz and sanidine are uncommon. In the plots, however, the quartz-bearing dykes and ignimbrites are distinguished, and are seen to define distinct fields. Considered overall, the rhyolitic compositions are consistent with magmas evolving under relatively low (<200 MPa) pressures. One possible exception is the group of phenocrystal quartz-bearing, low-silica rhyolitic to dacitic ignimbrites (Carlisle and Long Is.) which appear to have equilibrated at higher pressures. The mineralogically distinctive Pentecost Is. ignimbrites (plag–qtz–biot–augite) also define a unique compositional field, although the higher normative Q portion is alteration controlled. Apart from these exceptions, the bulk of the rhyolitic data for the dykes, lavas, and ignimbrites exhibit completely overlapping compositional groupings, and unlike the granites, cannot be divided into distinctive island groupings (i.e. volcanic centres).

6.2. Trace elements

6.2.1. Introduction. Data are based on XRF and neutron activation analyses, and representative data are presented in Table 3. In terms of general element behaviour, Rb and Th are incompatible throughout the compositional range, whereas Ba and Zr exhibit similar behaviour only up to the high-silica rhyolite range, indicating late stage control by feldspar and zircon fractionation, respectively. W, Ta, and LREE are broadly incompatible, whereas Nb and Y exhibit little systematic variation throughout the whole compositional range. Elements such as Sc, V, Ni, and Cr correlate with FeO* and MgO, behaving compatibly. The important overall point shown by the trace element data is again the continuity of element abundance patterns.

Mineralogical evidence suggests the possibility of mixing as an important process in these magmas. Routine plotting of many element ratio combinations has shown a number that apparently follow mixing trends, although not all can be easily distinguished from fractional crystallisation trends. One example is illustrated (Fig. 9), for Yb/Th–Ce/Th. Both sets of data (<69% and >69% SiO_2) lie along a linear mixing curve, and the insets show attempts at modelling the variation in terms of fractional crystallisation. These calculated curves, utilising basaltic andesite, granodiorite, and low-silica rhyolite starting compositions, suggest that neither the trends, nor more importantly, the extents of fractionation, match the observed data trends. It is noted here that attempts to model the overall trace element variations of the whole volcanic suite, using fractional crystallisation and AFC models, have generally been unsuccessful, even though satisfactory major element fits can be established by least squares method for fractional crystallisation only models. As will be further shown, mixing

Table 3 Selected chemical analyses of Whitsunday volcanics and granites

	SM11	Li06	Pr01	Es01	Hs17	No02	Sc04	Samples
SiO_2	50·07	59·38	60·6	69·12	74·10	64·08	74·68	SM11—Dolerite dyke/sill (100 m thick), S Molle Is.
TiO_2	1·87	0·97	1·25	0·59	0·21	0·75	0·24	Li06—andesite lava, Lindeman Is.
Al_2O_3	17·51	16·81	17·20	15·76	13·55	16·22	13·71	Pr01—Andesite dyke, Prudhoe Is.
Fe_2O_3	1·58	1·13	1·08	0·58	0·37	0·81	0·25	Es01—Ignimbrite, Esk Is.
FeO	7·89	5·66	5·41	2·91	1·83	4·04	1·25	Hs17—Ignimbrite, Haslewood Is.
MnO	0·18	0·14	0·14	0·11	0·08	0·08	0·05	No02—Granodiorite, Noel Is.
MgO	7·46	3·14	2·09	0·87	0·30	2·37	0·33	Sc04—Granite, Scawfell Is.
CaO	8·36	6·98	5·68	2·32	1·00	4·54	0·98	
Na_2O	3·81	3·15	4·19	4·11	4·12	3·37	3·85	
K_2O	0·81	1·98	1·80	3·46	4·38	3·53	4·54	
P_2O_5	0·46	0·37	0·56	0·16	0·07	0·21	0·11	
LOI	(2·89)	(2·66)	(2·77)	(3·15)	(2·42)	(2·37)	(2·37)	
Rb	20	60	47	97	134	115	143	
Cs	1·4	2·0	1·4	3·5	2·6	3·1	4·6	
Ba	197	416	393	566	710	365	449	
Sr	551	539	610	336	121	435	125	
Zr	199	167	231	278	319	218	130	
Hf	3·7	3·9	5·2	6·5	7·8	6·6	3·8	
Y	31	25	34	36	44	23	24	
Ta	0·73	1·4	1·2	1·5	2·0	2·3	1·7	
Nb	11	8	10	10	11	6	9	
Zn	109	92	99	93	77	78	42	
Cu	66	30	23	17	8	41	5	
Ni	109	14	15	22	1	68	17	
Cr	289	50	8	13	7	56	11	
V	183	172	132	38	9	131	13	
Sc	25·6	22·0	18	12	9·1	14	3·1	
Th	0·5	6·8	5·6	9·1	13·0	19·0	17·5	
La	14·2	20·0	27·0	27·0	33·0	25·5	25·5	
Ce	36·7	48·0	66	65	79	62	53	
Nd	23·6	26·7	39·0	35·5	42·0	32·5	21·5	
Sm	5·6	5·5	8·3	7·2	8·5	6·1	3·5	
Eu	1·89	1·48	2·20	1·78	1·44	1·23	0·58	
Gd	6·3	5·0	7·5	6·3	7·6	4·9	3·0	
Tb	0·92	0·88	1·15	1·10	1·30	1·00	0·65	
Ho	1·14	0·92	1·45	1·45	1·70	1·05	0·75	
Yb	2·95	2·58	3·45	3·65	4·45	2·45	1·90	
Lu	0·45	0·39	0·52	0·57	0·68	0·37	0·29	

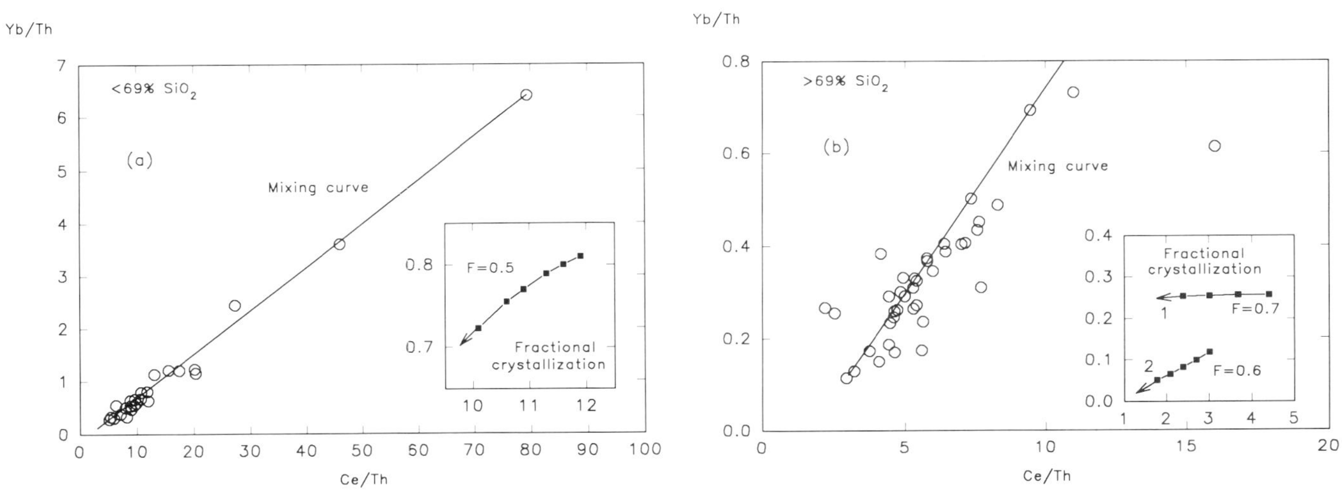

Figure 9 Yb/Th–Ce/Th plot for the Whitsunday volcanics and granites, divided into <69 and >69% SiO_2 groupings; inset in (a) represents the calculated fractional crystallisation curve for a basaltic andesite to granodiorite (dacite), extending to $F = 0{\cdot}5$ (Plag 0·59, Cpx 0·12, Opx 0·20, Mag 0·06, Ilm 0·02, Ap 0·01); inset in (b) represents calculated fractional crystallisation curves from granodiorite to low-silica rhyolite (Curve 2; Plag 0·63, Cpx 0·16, Opx 0·14, Mag 0·02, Ilm 0·03, Ap 0·02) and low-silica to high-silica rhyolite (Curve 1; Plag 0·83, Cpx 0·07, Mag 0·10).

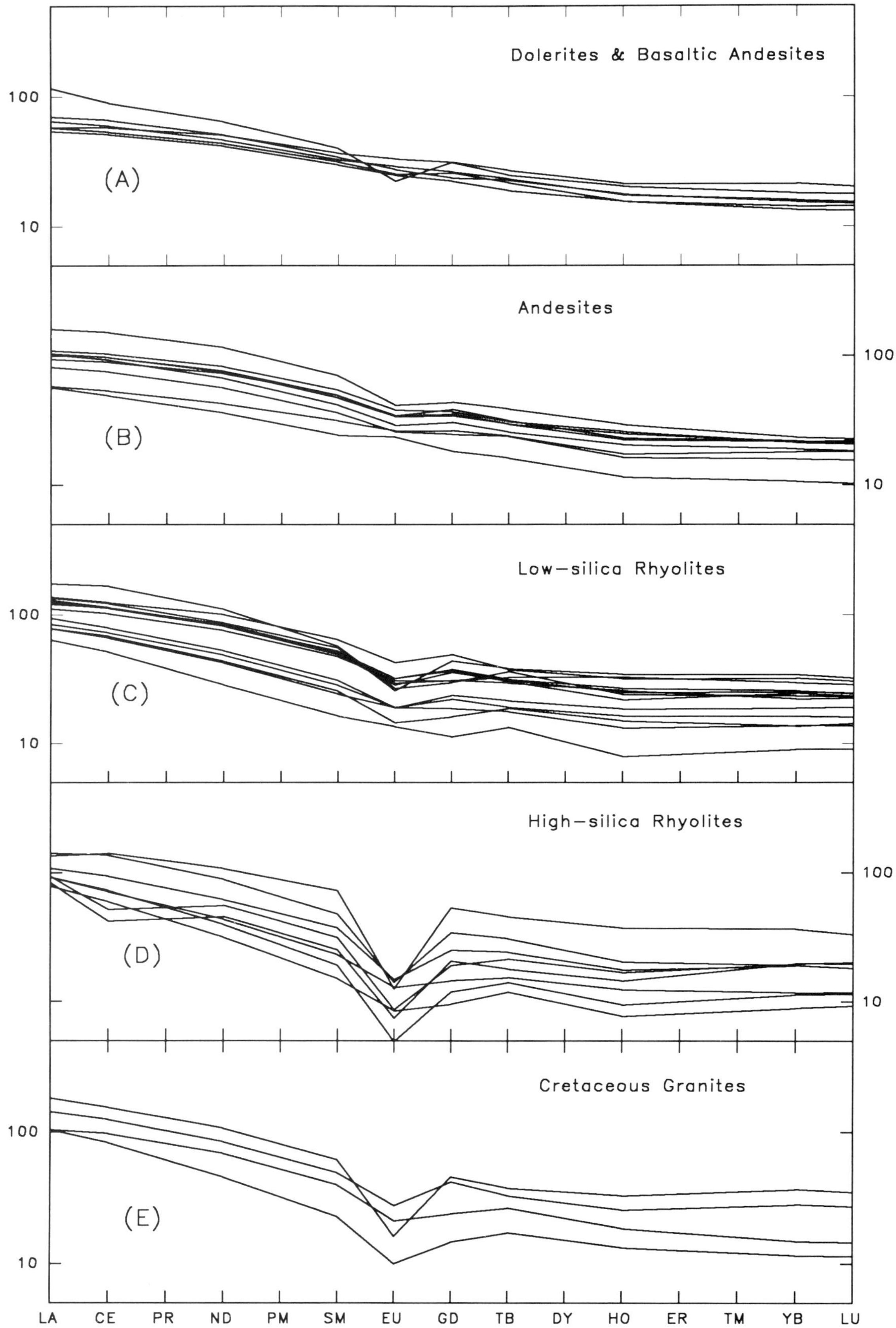

Figure 10 Chondrite-normalised REE abundances comparing the major bulk rock volcanic groupings, and the Cretaceous granites.

relationships have proved the most successful, although superimposed fractional crystallisation has certainly occurred.

6.2.2. Rare earth elements. Selected patterns are illustrated in Figure 10, for the major bulk rock volcanic groupings and granites. The overall data indicate that: (i) the dolerites, basaltic andesites, and andesites exhibit similar patterns, with La_N/Yb_N ratios between 3·2–5·4 (one sample 6·7), the majority between 4·0–5·4. A positive correlation is apparent between La_N and La_N/Yb_N, while very small negative Eu anomalies are present in a few patterns; (ii) the rhyolite patterns entirely overlap those of the andesites, with generally slightly higher LREE compared to dolerites and basaltic andesites. La_N/Yb_N ratios lie between 3·6–6·0, with the highest ratio extending to 8·9. The rhyolites are thus not strongly fractionated in terms of their REE abundances or LREE fractionation. Negative Eu anomalies are most strongly developed in the higher-silica rhyolites; (iii) the Cretaceous granites are intermediate between the low- and high-silica rhyolites.

As the Eu anomalies are indicative of feldspar fractionation, it is relevant to relate Eu/Eu* with Ba and Sr abundances (Fig. 11a, b). The Sr abundances show a systematic decrease with increasing Eu depletion, explicable in terms of plagioclase-dominated fractionation. Ba-Eu/Ea* relations are more complex, showing initially increasing Ba with decreasing Eu/Eu*, followed by limited Ba depletion with further Eu depletion (i.e. at Eu/Eu* $< 0{\cdot}6$). This again is consistent with feldspar fractionation, allowing for the changing value of K_D within increasingly siliceous magmas, but the more extreme late state Ba depletion seems to require the presence of sanidine in the fractionating assemblage in the high-silica rhyolites, a phase not commonly observed. Calculated mixing curves (not shown) do not duplicate these observed data trends.

7. Tectonomagmatic affinities

The association of dolerites, through andesites and rhyolites, and coeval granites, is a common association in modern subduction-related volcanic provinces (e.g. Ewart 1982). As previously discussed, however, the Whitsunday volcanics appear more feasibly related to rifting and crustal thinning, rather than a convergent tectonic environment.

The existence of a possible subduction-signature is explored in Figures 12–14. The Ba/Nb–La/Nb data (Fig. 12a, b) show a complete gradation from "intraplate" (as exemplified by the eastern Australian Cainozoic provinces), to arc-like (as exemplified by the Southwestern Pacific

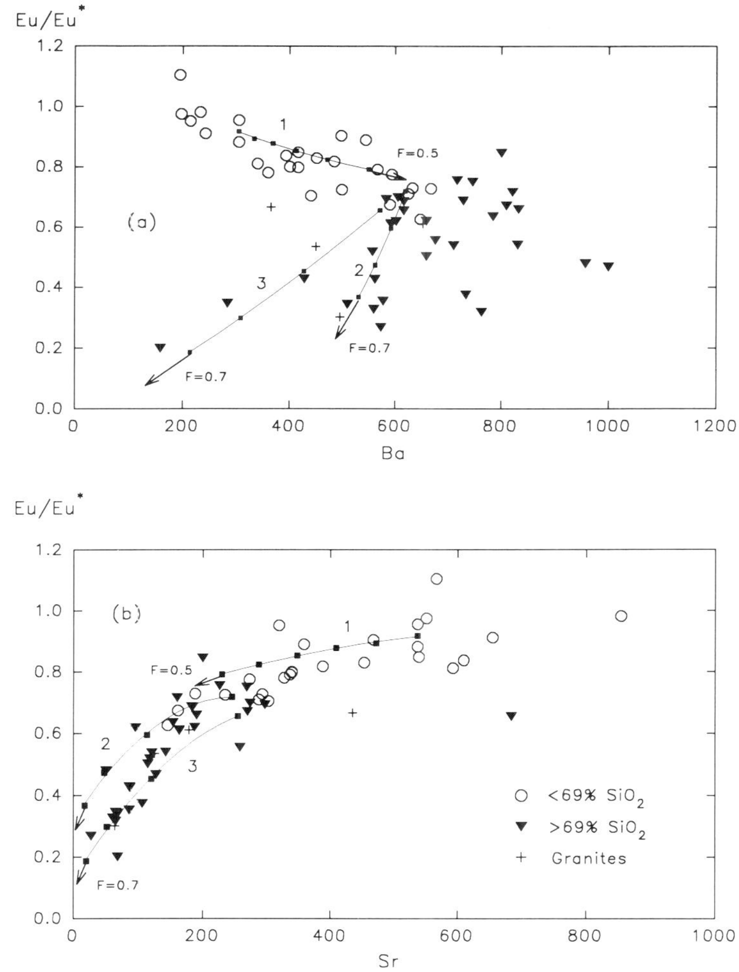

Figure 11 Eu anomalies versus Ba and Sr (ppm), the data divided into <69 and >69% SiO_2, and granite groupings; calculated fractional crystallisation paths are shown for: 1, basaltic andesite to granodiorite (dacite; Plag 0·59, Cpx 0·12, Opx 0·20, Mag 0·06, Ilm 0·02, Ap 0·01); 2 and 3, low-silica rhyolite (two different starting compositions) to high-silica rhyolite; curve 3 incorporates sanidine into the fractionating assemblage (Plag 0·83, Cpx 0·07, Mag 0·10 and Plag 0·60, San 0·32, Cpx 0·01, Mag 0·07, respectively); high-silica rhyolite partition coefficients used in curves 2 and 3.

andesites, and the Taupo Volcanic Zone rhyolites). In fact, the Whitsunday data appear to be in part transitional. The data trends are not readily explained in terms of fractional crystallisation, especially as the fields for the lower silica (<69% SiO_2) and higher silica volcanics and granites almost completely coincide. This suggests that these somewhat variable and transitional geochemical characteristics are source related. Significantly, many of the dolerites plot within the eastern Australian Cainozoic intra-plate fields.

Further comparisons between possible arc-like and within-plate signatures for the Whitsunday volcanics and granites are presented in terms of the ternary Hf–Th–Ta abundances (after Wood 1980; Fig. 13) and Th/Yb–Ta/Yb ratios (after Pearce 1983; Fig. 14). In Figure 13, the data plot as a remarkably linear array which is most readily explicable as a mixing array, as shown by the calculated mixing lines. Fractional crystallisation certainly cannot give rise to such an array. The mafic end of the spectrum is also informative, projecting in the E-type MORB and within-plate tholeiitic fields, whereas the silicic volcanics and granites plot predominantly within the calc-alkaline "destructive margin" volcanic field.

Further confirmation of the previous data trends is obtained in Figure 14, the data again defining an array consistent with a two-component mixing process. One end of the array, defined by dolerite and basaltic andesite dykes, lies in the mantle array (near average E-type MORB), whereas the main body of the data lie in the region defined by Pearce (1983) as characterising calc-alkaline to shoshonitic active continental margins, and alkaline oceanic

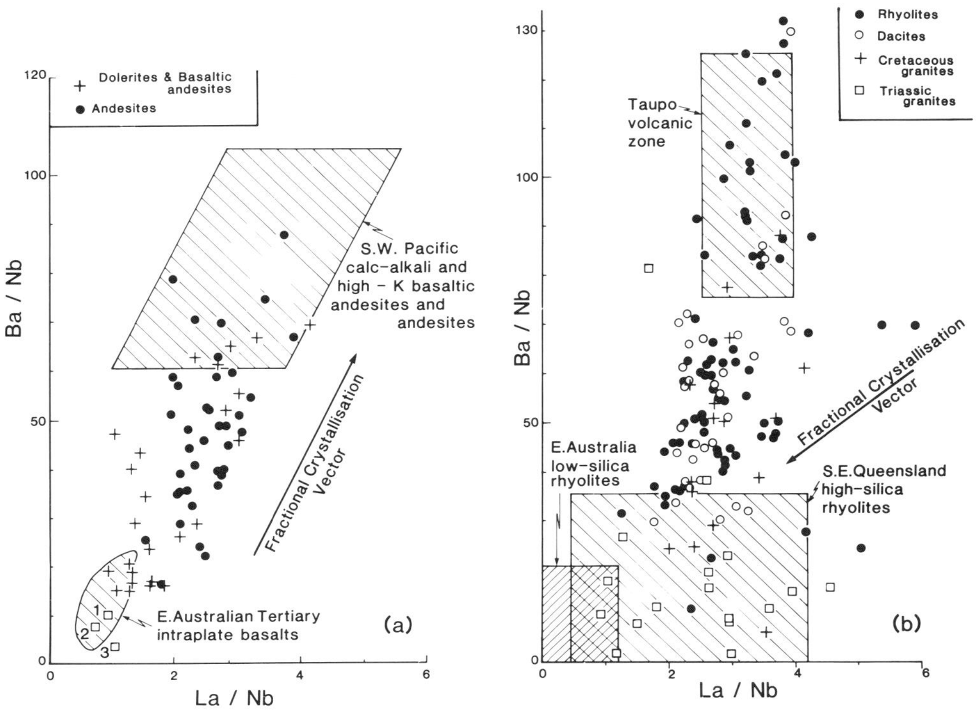

Figure 12 Ba/Nb–La/Nb plots for the Whitsunday Volcanics and Triassic and Cretaceous granites; intraplate fields are after Ewart and Chappell (1989), the averaged southwestern Pacific andesites after Ewart (1982), and the Taupo Volcanic Zone rhyolites after G. Corlett (pers. comm.); hollow squares labelled 1, 2, and 3 in (a) are the average primitive mantle, E-type MORB, and N-type MORB, respectively (after Sun and McDonough 1989).

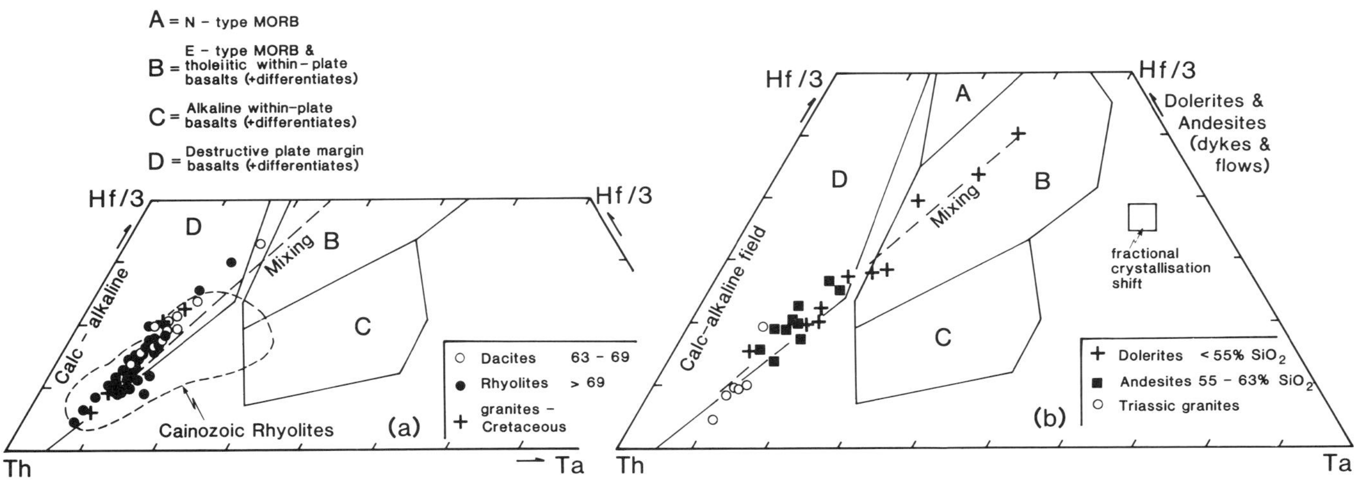

Figure 13 Hf–Ta–Th relationships of the rhyolites, dacites and Cretaceous granites (a), and andesites, dolerites and Triassic granites (b); field boundaries after Wood (1980).

arcs. The Whitsunday array again contrasts markedly with both the eastern Australian basalt array and high-silica rhyolite field.

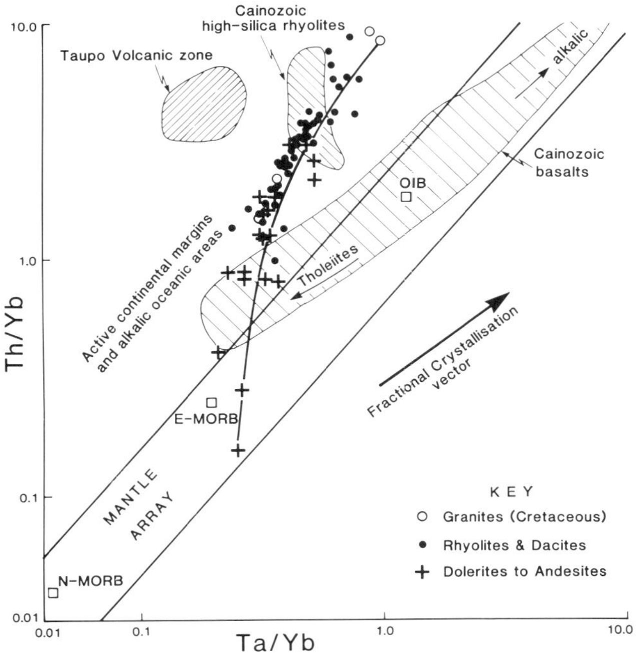

Figure 14 Th/Yb–Ta/Yb for the Whitsunday volcanics and coeval granites, based on plot by Pearce (1983); sources of comparative data as in Figure 12.

8. Isotope geochemistry

Figure 15 plots the Nd and Sr isotopic data for the volcanics and granites. The data are corrected to 115 Ma, except where specific K–Ar or Rb–Sr dates are available (Table 4). Also plotted are comparative fields for the southern New England (Australia) granites, the Cainozoic southern-central Queensland lava field basalts, and the granulite inclusions occurring in these lavas, together with limited data for Carboniferous and Triassic andesites from southern Queensland. Comparison of these data indicates the following points:

(i) The Whitsunday volcanics comprise a coherent data set with ε_{Nd} from +7·3 to +2·2 and initial Sr isotopic ratios from 0·70312–0·70436. The S Molle Is. dolerite, identified as the most primitive sample from trace element data, is among the most isotopically primitive of the data set. The rhyolites lie between ε_{Nd} from +2·2 to +5·7 and I_{Sr} 0·70332–0·70406. The remaining data, however, exhibit overlap between isotopic compositions of the andesites, dacites and rhyolites.

(ii) The Cretaceous granites have ε_{Nd} +6·5 to +3·4, and initial $^{87}Sr/^{86}Sr$ from 0·70312–0·70436, completely overlapping the volcanics, and confirming their comagmatic relationships. The most primitive sample is characterised by extremely abundant dioritic inclusions.

(iii) The Triassic granites are compositionally distinct, with ε_{Nd} + 1·5 to +4·2 and initial $^{87}Sr/^{86}Sr$ 0·7116–0·894 (one value at 0·70461), calculated to 115 Ma. These isotopic data thus preclude the Triassic granites as a possible source for the Cretaceous silicic magmas.

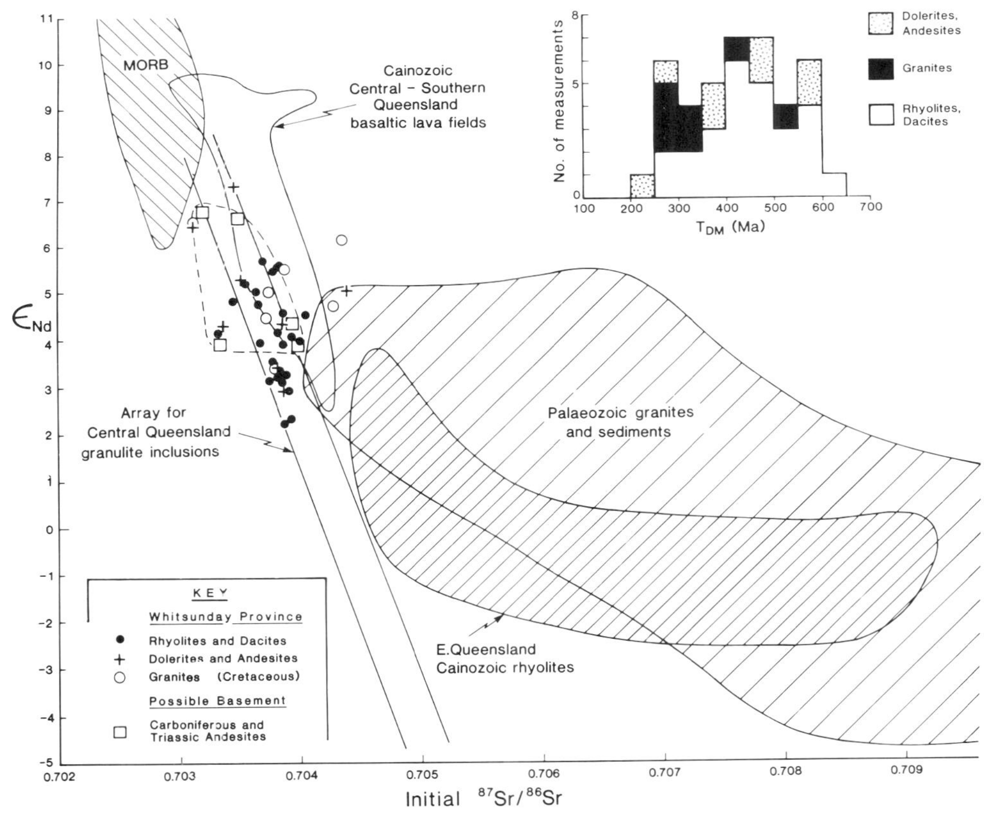

Figure 15 ε_{Nd}-initial $^{87}Sr/^{86}Sr$ ratios for the Whitsunday volcanics; lavas are corrected to 115 Ma, except where specific dates available; inset shows histogram of Nd T_{DM} model ages; comparative data fields presented for modern MORB glasses: Carboniferous and Triassic andesites (Ewart *et al.* 1988, and Table 3, corrected to 115 Ma); granulite inclusion array, (Ewart *et al.* 1988; O'Reilly *et al.* 1988; corrected to 115 Ma); intraplate silicic volcanics and basaltic lava fields (Ewart *et al.* 1988), New England granites (Hensel *et al.* 1985; corrected to 115 Ma).

(iv) The Cretaceous isotopic compositions overlap the fields of the Carboniferous–Triassic andesites, the Cainozoic basalts, and the granulite xenolith array. O'Reilly *et al.* (1988) interpret these inclusions in general as frozen melts, or re-equilibrated cumulates, at the crust–mantle boundary, although inferring that the eastern-central Queensland inclusions reflect interaction between basalt and an older mafic crust with time-integrated Rb and LREE depletion. The critical point with regard to the Cretaceous magmas is that their source(s) are isotopically very similar to the lower crust–lithosphere sources of the Cainozoic basalts and the Carboniferous–Triassic andesites.

The southern New England Batholith granites are generally more radiogenic in Sr than the Cretaceous magmas, implying a different and more evolved source. Palaeozoic–Mesozoic upper crustal sediments are also isotopically precluded as contributing to the Cretaceous magma evolution, noting the Cainozoic silicic magma data field in which such sediments are inferred to have been involved through AFC process (Ewart *et al.* 1988).

Figure 16 further compares the initial Sr isotopic compositions of the southern New England Batholith, the Triassic silicic volcanic provinces of southern Queensland (Stephens 1991), and the Cretaceous suite. Even the least radiogenic of the New England granites (also the least potassic, Shaw & Flood 1981) are mostly more radiogenic than the other two suites. The major significant point, however, is the close correspondence between the Triassic and Cretaceous initial Sr ratios, with the peak of the latter being slightly more radiogenic, possibly reflecting a slight isotopic composition change during the Triassic to Cretaceous interval in a common, or related magma source. If so, then a source Rb/Sr of 0·04–0·08 is implied (consistent with an andesitic source; see below). ε_{Nd} of the Triassic suite ranges from 5·5–2·7, a very similar range to the Cretaceous suite.

The isotopic data thus point to relatively primitive and/or newly accreted, unfractionated source for both the Triassic and Cretaceous magma suites. Stephens (1991) has proposed that the Triassic silicic volcanics originate from

Table 4 Sr and Nd isotopic analyses of the Whitsunday volcanics and granitoids

Sample**	Island location	$^{143}Nd/^{144}Nd^{++}$	$\varepsilon_{Nd(115)}$	$^{147}Sm/^{144}Nd$	Measured $^{87}Sr/^{86}Sr^{++}$	Initial $^{87}Sr/^{86}Sr^{*}$	Rb (ppm)	Sr (ppm)	Rock type	(Anhydrous) SiO_2%
Bo03	Border	0·512768 ± 14	3·52	0·130	0·705989 ± 7	0·70378	88	188	D	70·03
Ca01	Carlisle	0·512776 ± 14	3·89	0·115⁺	0·704769 ± 10	0·70386	69	358	I	68·78
Cd03	Calder	0·512924 ± 15	6·52	0·132	0·705090 ± 6	0·70312	58	140	G	70·13
Ci01	Cid	0·512766 ± 14	3·41	0·135	0·704167 ± 10	0·70382 (113)	47	623	L	63·02
Cu05	Curlew	0·512839 ± 12	5·00	0·123	0·703759 ± 8	0·70364 (105)	24	842	L	64·34
De12	Dent	0·512809 ± 10	4·45	0·121	0·706177 ± 14	0·70371 (107)	100	178	G	69·41
Dw01	Derwent	0·512866 ± 16	5·51	0·109	0·709947 ± 10	0·70388 (95)	104	67	G	75·29
Es01	Esk	0·512732 ± 12	2·89	0·125⁺	0·705407 ± 8	0·70392 (125)	97	336	I	69·12
Go05	Goldsmith	0·512791 ± 10	4·06	0·123⁺	0·709820 ± 2	0·70394	158	127	I	75·33
Ha04	Hayman	—	—	—	0·705912 ± 24	0·70371 (107)	75	150	G	68·70
Ha05	Hayman	—	—	—	0·709788 ± 38	0·70374 (107)	88	64	G	75·75
Hm02	Hamilton	0·512970 ± 10	7·30	0·141	0·703692 ± 12	0·70346	45	899	D	55·14
Hm07	Hamilton	0·512785 ± 10	3·95	0·123	0·716415 ± 34	0·70401 (113)	136	51	L	78·38
Ho03	Hook	0·512757 ± 20	3·35	0·127	0·705461 ± 14	0·70383	99	287	I	70·78
Ho15	Hook	0·512757 ± 16	3·26	0·132	0·706863 ± 13	0·70390	101	161	I	68·26
Ho19/1	Hook	0·512756 ± 12	3·24	0·132⁺	0·708009 ± 12	0·70385	80	91	I	74·69
Ho29	Hook	0·512811 ± 8	4·32	0·133	0·704710 ± 12	0·70386	70	387	I	62·53
Ho34	Hook	0·512832 ± 14	4·72	0·133⁺	0·704594 ± 7	0·70366	67	339	D	63·07
Ho39	Hook	0·512790 ± 18	3·93	0·131	0·705696 ± 20	0·70367	117	273	I	64·79
Hs17	Haslewood	0·512696 ± 10	2·21	0·123⁺	0·709572 ± 8	0·70387 (125)	134	121	I	74·10
In02/2	Ingot	0·512818 ± 10	4·48	0·131	0·704987 ± 8	0·70403 (113)	141	684	L	71·42
Keel	Keelan	0·512863 ± 12	5·69	0·108	0·711605 ± 9	0·70370 (105)	119	65	L	76·74
Kes8	Keswick	0·512865 ± 10	5·45	0·127⁺	0·705101 ± 8	0·70378 (105)	82	268	L	71·83
Li06	Lindeman	0·512739 ± 16	2·88	0·135	0·704391 ± 29	0·70387 (113)	60	539	L	59·38
Lo11	Long	0·512805 ± 14	4·26	0·128	0·704475 ± 10	0·70337 (110)	83	339	L	62·92
Mp16	Middle Percy	0·512840 ± 10	5·00	0·124	0·705537 ± 7	0·70440	57	235	D	60·57
No02	Noel	0·512831 ± 14	4·99	0·113⁺	0·704913 ± 10	0·70375 (107)	115	435	G	64·08
Pe01	Pentecost	0·512695 ± 18	2·31	0·115⁺	0·704589 ± 11	0·70393	53	356	I	71·60
Pr01	Prudhoe	0·512859 ± 10	5·28	0·131	0·703888 ± 16	0·70352	47	610	D	60·60
Sm08	South Molle	0·512791 ± 12	4·13	0·119	0·730512 ± 21	0·70332 (100)	162	27	L	75·33
Sm10	South Molle	0·512858 ± 16	5·19	0·135	0·706731 ± 18	0·70355 (110)	83	118	L	71·92
Sm11	South Molle	0·512929 ± 14	6·44	0·145	0·703287 ± 36	0·70312 (110)	20	551	D	50·07
Sc04	Scawfell	0·512808 ± 14	4·70	0·103⁺	0·709775 ± 11	0·70429 (117)	143	125	G	74·68
Sh08	Shaw	0·512798 ± 12	4·15	0·126	0·706791 ± 48	0·70382 (113)	122	191	L	72·93
Sh09	Shaw	0·512744 ± 14	3·09	0·127	0·705289 ± 10	0·70386	83	274	D	71·32
Sh10	Shaw	0·512731 ± 16	3·10	0·109	0·705907 ± 59	0·70375	83	182	D	76·55
Sh25	Shaw	0·512741 ± 18	3·19	0·116⁺	0·709328 ± 14	0·70381 (113)	183	154	L	73·50
TP01	Tinonee Peak	0·512857 ± 20	5·57	0·108⁺	0·705319 ± 11	0·70384 (105)	55	160	L	74·11
TP07	Tinonee Peak	0·512855 ± 18	5·53	(0·108)	0·706453 ± 10	0·70382 (105)	86	141	L	77·58
Tr01	Three Rocks	0·512882 ± 18	6·14	(0·103)	0·709973 ± 11	0·70436 (117)	137	117	G	75·99
Wg01	Wigton	0·512760 ± 14	3·38	0·129⁺	0·705202 ± 8	0·70380 (116)	83	282	G	68·37
Triassic Andesites (Neara Volcanics)										
37958		0·512775 ± 11	3·88	0·115	0·703705 ± 11	0·70334	79	1042		
37967		0·512939 ± 14	6·61	0·145	0·703621 ± 23	0·70349	24	754		
38645		0·512803 ± 16	4·31	0·112	0·704771 ± 12	0·70394	89	552		

Notes: ++ quoted uncertainties are ±2se; analyses performed at Sirotope, C.S.I.R.O., North Ryde. $^{147}Sm/^{144}Nd$ calculated from isotope dilution data, except samples with ⁺ (neutron activation), and figures in brackets (inferred), * calculated for 115 Ma except where indicated otherwise (see Table 1). D = dykes; G = granites; I = ignimbrites; L = lava flows. ** See Section 11 for University of Queensland rock collection numbers.

partial melting of an andesitic crust of early Triassic (or older) age. It is proposed here that a plausible source may be partial melting of mixed igneous and sediment components of the Permo–Triassic arc crust. The geographic plausibility of this is illustrated in Figure 17, in which the zones of both Permian and Triassic volcanism and plutonism are seen to run obliquely to the present coast, and are thus intersected by the zone of Cretaceous rifting. It is therefore expected that remelting of the older Mesozoic crust should contribute to Cretaceous magma genesis. The relatively primitive isotopic compositions of the Triassic and Cretaceous volcanics in the central Queensland region also implies an absence of an older, more geochemically evolved crust being present beneath the Permo–Triassic arcs, and strengthens the view for the progressive lateral accretion of the eastern Queensland crust by arc migration, at least to the early Mesozoic (e.g. Veevers *et al.* 1991). The involvement of Mesozoic to Palaeozoic crust in the Cretaceous magma genesis also receives support from the Nd model T_{DM} ages (inset, Fig. 15), most of which lie between 250–600 Ma.

9. Discussion

The trace element data point toward two main processes operating in the evolution of the Whitsunday magmas: (i) Two component mixing, one end-member inferred to be a within-plate tholeiitic magma (near E-type MORB), the second a silicic magma inferred to be rhyolitic, presumably low-silica rhyolite. (ii) Fractional crystallisation superimposed on the mixing trends, being especially pronounced in the high-silica rhyolites. These processes are interpreted to have produced the overall continuity of magma compositions from mafic to felsic. The trace element data also reveal the Cretaceous granites to be geochemically indistinguishable from the silicic volcanics.

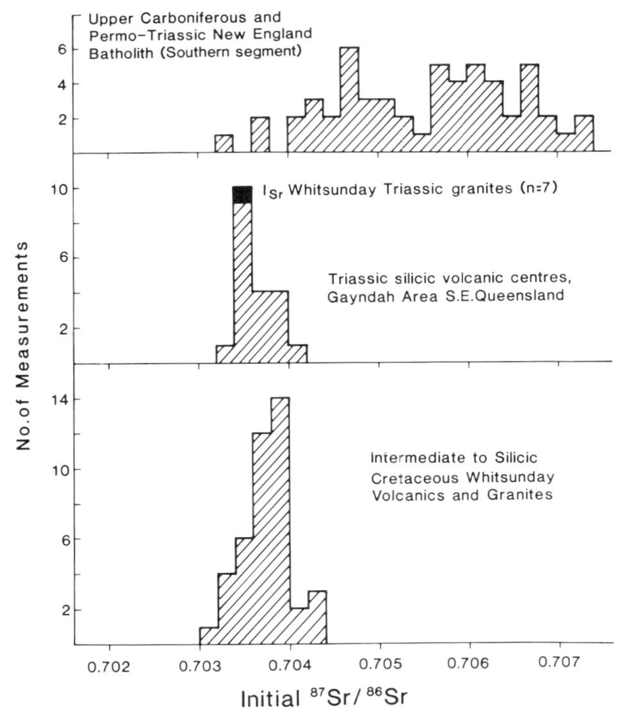

Figure 16 Histograms comparing the initial Sr isotope ratios of the southern New England Batholith (Shaw & Flood 1981), the central Queensland Triassic silicic volcanic centres (Gayndah region; Stephens 1991), and the Whitsunday Volcanics and granites (this paper).

The isotopic data clearly imply the involvement of a juvenile crust and lithosphere, the former being interpreted as most plausibly Permo–Triassic arc crust. Partial refusion of this crust is here inferred to have produced low-silica rhyolites, the remelting process developing in response to basaltic magma injection from upwelling lithosphere and asthenosphere. In fact, such remelting may also involve in part remelting of restite developed during the Triassic phase of volcanism, as Triassic granites are locally present underlying the Whitsunday volcanic region, as described. Triassic rhyolites also occur on the adjacent mainland (Conway Peninsula, Fig. 1). A plausible tectonic model is passive continental margin extension by detachment faulting (e.g. Lister & Etheridge 1989).

It is therefore proposed that the Whitsunday magmatism represents a calc-alkaline to high-K magma suite produced as a consequence of a divergent, rather than convergent plate margin. The development of intermediate magma compositions is interpreted to result largely from mixing processes, with superimposed fractional crystallisation. The "arc-like" signatures identified within the suite, especially in the intermediate to silicic compositions, are interpreted to be inherited from the Mesozoic arc source, through the production of rhyolitic magmas. Mixing of these magmas

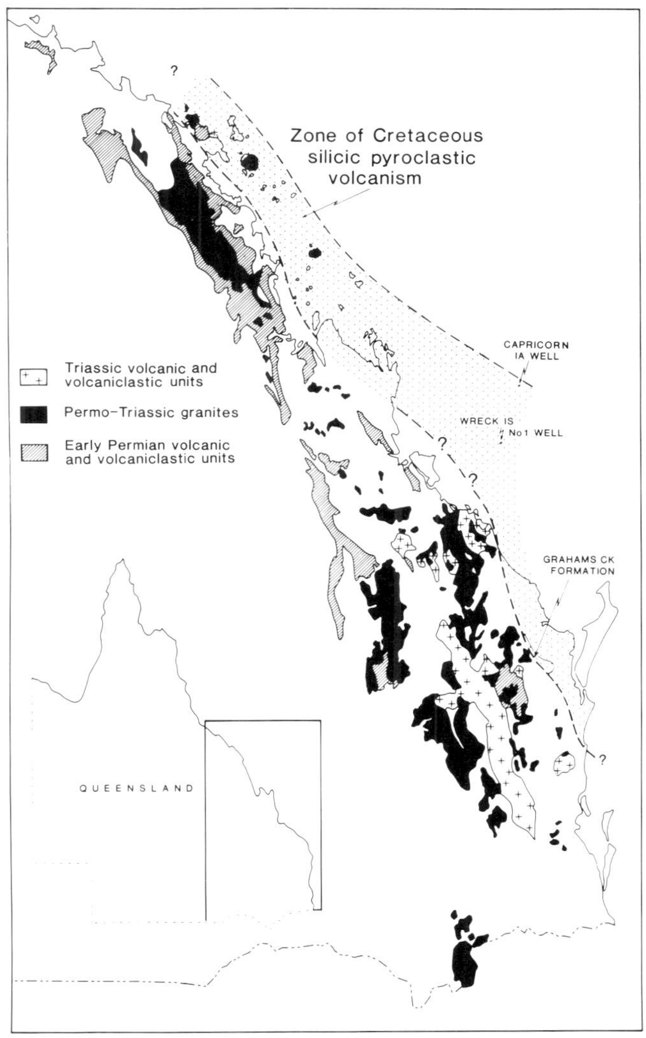

Figure 17 Map illustrating the generalised distribution in central and southern Queensland of Permian and Triassic volcanic and volcaniclastic units, and Permo-Triassic granites, in relation to the inferred distribution of the Cretaceous volcanic rift zone (see also Clarke *et al.* 1971); geological boundaries after Day *et al.* (1983).

with lithosphere-derived tholeiites has allowed these "arc" signatures to be transferred through much of the Whitsunday magma suite. Mixing is interpreted to dominantly involve magma mixing. The data of Frost and Mahood (1987) indicate that efficient mixing is only likely if either compositional differences between host and injected mafic magma are less than 10% SiO_2, or if the mass fraction of mafic magma exceeds approximately 0·5. The resulting mixed magmas correspond to tonalites to mafic granodiorites, i.e. corresponding to the compositions of the inclusions within some of the Cretaceous granites. The more silicic compositions are thus likely to represent localised magma mingling and assimilation with superimposed fractional crystallisation. The latter has demonstrably occurred, especially in the high silica rhyolites. It is also significant to recall the more complete continuity of rock compositions within the dykes (Fig. 6), perhaps reflecting more efficient mechanical stirring during dyke emplacement.

On a regional scale, as previously noted, Cretaceous volcanism was almost certainly originally more widespread along the eastern Australian coast, possibly now downfaulted along the continental shelf. In the context of the southwestern Pacific, Cretaceous volcanism occurs in New Zealand, the Lord Howe Rise, and New Caledonia (Fig. 18). In New Zealand, volcanic products of similar age (89 Ma) to the Whitsundays are found only in the Mt Somers region (Barley 1987). They range from basaltic andesite to high-silica rhyolite, with medium to high-K affinities. The rhyolites are peraluminous, and are interpreted by Barley (1987) to be partial melts of the quartzo-feldspathic Torlesse sediments (i.e. a more "mature" and more radiogenic source than that of the Whitsunday magmas). The intermediate composition is explained by AFC processes involving mantle derived tholeiites. Volcanicity is correlated with a change from compressional to extensional tectonic regimes after the Rangitata orogeny.

Cretaceous rhyolitic pyroclastics and lavas are present on the southern part of the Lord Howe Rise, recovered at the base of the DSDP Hole 207 (Burns *et al.* 1973; McDougall & van der Lingen 1974). Bimodal Cretaceous volcanism occurs extensively along the length of New Caledonia, part of the Norfolk Ridge (Paris 1981), and includes abundant rhyolitic and rhyodacitic tuffs interbedded in thick sediment sequences in the Noumea and Diahot basins (65–75 Ma). These widespread, although apparently localised, occurrences of Cretaceous silicic volcanism are thus of considerable significance in terms of the Cretaceous fragmentation of the eastern Australasian continent; the submarine Lord Howe and Norfolk ridges are regarded as rifted fragments of the former continent (e.g. Kroenke 1984). Of particular further interest is the recent recovery of pre-Upper Eocene rhyolites (exact age not yet known) in the lower 210 m of ODP Hole 841 on the Tonga forearc (ODP Leg 135 Scientific Party 1991), raising speculation that the Tonga and Lau ridges may also represent the easternmost rifted fragments of the former continental margin.

It is thus possible that rhyolitic volcanism accompanying the Cretaceous regional fragmentation was originally widespread in the southwestern Pacific, but is now largely buried in the off-shore ridges, and on the continental shelves. If so, the rhyolites will have sampled, by partial melting, segments of the former eastern Australasian continental margin of various ages and in various stages of evolution.

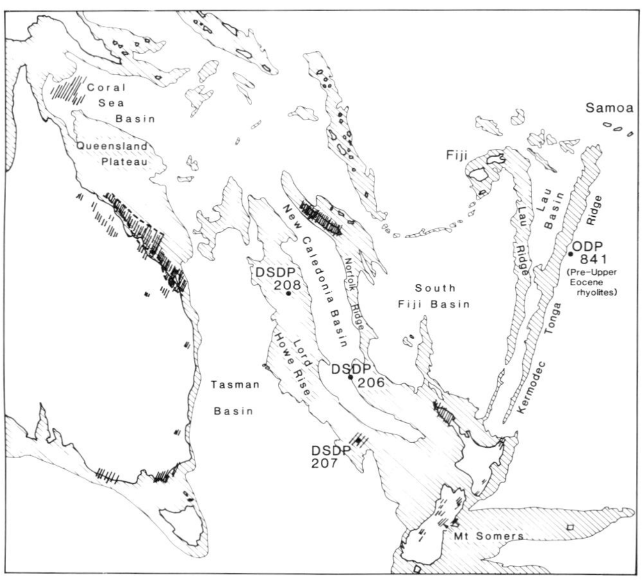

Figure 18 Map of southwestern Pacific illustrating the generalised distribution of continents and continental shelves, islands, and submarine ridges (based on 2000 m bathymetric contour), and showing the general distribution of inferred, drilled, and exposed Cretaceous volcanic and intrusive provinces.

10. Acknowledgements

This project was funded by an Australian Research Council grant, including post-doctoral position to RWS. All neutron activation analyses were carried out in the Department of Geology, Australian National University. Valuable field assistance was given by C. J. Stephens, A. Faulkner, and P. Colls, while much laboratory assistance was provided by F. Audsley and R. Hall (University of Queensland) and Dr D. J. Whitford and S. Craven (Sirotope, CSIRO). Stimulating discussion with J. Parianos, S. Bryan, and P. Downes, each completing theses on the region, are acknowledged. Dr S. D. Weaver and an anonymous reviewer provided helpful comments on the manuscript.

11. Note added in proof

The following are the University of Queenslnd official departmental museum numbers for the samples included in Table 4. UQ number is in parenthesis after sample.

Bo03 (49247), Ca01 (49269), Cd03 (49274), Ci01 (49275), Cu05 (49300), De12 (49312), Dw01 (49331), Es01 (49332), Go05 (49342), Ha04 (49346), Ha05 (49347), Hm02 (49356), Hm07 (49361), Ho03 (49378), Ho15 (49391), Ho19/1 (49396), Ho29 (49407), Ho34 (49412), Ho39 (49419), Hs17 (49477), In02/2 (49418), Keel (49484), Kes8 (49496), Li06 (49525), Lo11 (49555), Mp16 (49571), No02 (49575), Pe01 (49576), Pr01 (49597), Sm08 (49619), Sm10 (49621), Sm11 (49622), Sc04 (49640), Sh08 (49650), Sh09 (49651), Sh10 (49652), Sh25 (49667), TP01 (49702), TP07 (49708), Tr01 (49709), Wg01 (49726).

12. References

Barley, M. E. 1987. Origin and evolution of Mid-Cretaceous, garnet-bearing, intermediate and silicic volanics from Canterbury, New Zealand. J VOLCANOL GEOTHERM RES **32**, 247–67.

Blundy, J. D. & Holland, T. J. B. 1990. Calcic amphibole equilibria and a new amphibole–plagioclase geothermometer. CONTRIB MINERAL PETROL **104,** 208–24.

Burns, R. E. & Andrews *et al.* 1973. *Initial Reports of the Deep Sea Drilling Project, Washington* (*U.S. Government Printing office*) **21,** 197–269.

Clarke, D. E., Paine, A. G. L. & Jensen, A. R. 1971. Geology of the Proserpine 1:250,000 Sheet area, Queensland. BUR MINER RES GEOL GEOPHYS REP **144**.

Davies, P. J. & Marshall, J. F. (in press). The Great Barrier Reef—evolution and growth. SPEC PUB INT ASSOC SEDIMENTOL.

Day, R. W., Whitaker, W. G., Murray, G. C., Wilson, I. H. & Grimes, K. G. 1983. Queensland geology. A companion guide to the 1:250,000 scale geological map (1975). GEOL SURV QUEENSLAND PUB **383**.

Ellis, P. L. & Whitaker, W. G. 1976. Geology of the Bundaberg 1:250,000 sheet area. GEOL SURV QUEENSLAND REP **90**.

Etheridge, M. A. & Lister, G. S. 1987. Extensional models for passive margin evolution. *In Extended Abstracts—Applied Extension Tectonics, Bureau Mineral Resources Record 1987/51,* 87–94.

Ewart, A. 1979. A review of the mineralogy and chemistry of Tertiary–Recent dacitic, rhyolitic, and related salic volcanic rocks. *In* Barker, F. (ed.) *Trondhjemites, dacites and related rocks,* 13–121. Amsterdam: Elsevier.

Ewart, A. 1982. The mineralogy and petrology of Tertiary–Recent orogenic volcanic rocks: with special reference to andesitic–basaltic compositional range. *In* Thorpe, R. S. (ed.) *Andesites,* 25–95. New York: Wiley.

Ewart, A. & Chappell, B. W. 1989. Trace element geochemistry. *In* Johnson, R. W. (ed.) *Intraplate volcanism in Eastern Australia and New Zealand,* 219–35. Cambridge: Cambridge University Press.

Ewart, A., Chappell, B. W. & Menzies, M. A. 1988. An overview of the geochemical and isotopic characteristics of the Eastern Australian Cainozoic volcanic provinces. J PETROL SPEC LITHOSPHERE ISSUE, 225–73.

Falvey, D. A. & Middleton, M. F. 1981. Passive continental margins; evidence for a pre-breakup deep crustal metamorphic subsidence mechanism. *Oceanol. Acta, Proc. 26th Intern. Geol. Congress, Geology of continental margins symposium, Paris,* 7–17 July, 1980, 103–14.

Falvey, D. A. & Mutter, J. C. 1981. Regional plate tectonics and the evolution of Australia's passive continental margins. BMR J GEOL GEOPHYS **6,** 1–29.

Frost, T. P. & Mahood, G. A. 1987. Field, chemical, and physical constraints on mafic–felsic magma interaction in the Lamarck Granodiorite, Sierra Nevada, California. GEOL SOC AM BULL **99,** 272–91.

Green, D. C. & Webb, A. W. 1974. Geochronology of the northern part of the Tasman geosyncline. *In* Denmead, A. K., Tweedale, G. W. & Wilson, A. F. (eds), *The Tasman Geosyncline—A symposium,* 275–93. Brisbane: Geological Society of Australia, Queensland Division.

Henderson, R. A. 1980. Structural outline and summary geological history for northeastern Australia. *In* Henderson, R. A. & Stephenson, P. J. (eds) *The geology and geophysics of northeastern Australia,* 1–26. Brisbane: Geological Society of Australia, Queensland Division.

Hensel, H. D., McCulloch, M. T. & Chappell, B. W. 1985. The New England Batholith; constraints on its derivation from Nd and Sr isotopic studies of granitoids and country rocks. GEOCHIM COSMOCHIM ACTA **49,** 369–84.

James, R. S. & Hamilton, D. L. 1969. Phase relation in the system NaAlSi308–CaAl2Si208 at 1 Kilobar water vapour pressure. CONTRIB MINERAL PETROL **21,** 111–41.

Jensen, A. R., Gregory, C. M. & Forbes, V. R. 1966. Geology of the Mackay 1:250,000 Sheet area, Queensland. BUR MINER RES GEOL GEOPHYS REP 104.

Kroenke, L. W. 1984. Cenozoic tectonic development of the Southwest Pacific. UN ESCAP, CCOP/SOPAC TECH BULL **6**.

Le Maitre, R. W. (ed.) 1989. *A classification of igneous rocks and glossary of terms. Recommendations of the International Union of Geological Sciences Subcommission on the systematics of igneous rocks.* Oxford: Blackwell.

Lister, G. S., Etheridge, M. A. & Symonds, P. A. 1988. Extension history of the margins of the Tasman Sea. *Abstracts, No. 21, Ninth Aust. Geol. Convention, February 1–5, 1988, Brisbane,* 252–3.

Lister, G. S. & Etheridge, M. A. 1989. Detachment models for uplift and volcanism in the Eastern Highlands, and their application to the origin of passive margin mountains. *In* Johnson, R. W. (ed.) *Intraplate Volcanism in Eastern Australia and New Zealand,* 297–313. Cambridge: Cambridge University Press.

McDougall, I. & van der Lingen, G. J. 1974. Age of the rhyolites of the Lord Howe Rise and the evolution of the southwest Pacific Ocean. EARTH PLANET SCI LETT **21,** 117–26.

Nekvasil, H. 1988. Calculated effect of anorthite component on the crystallization paths of H_2O-undersaturated haplogranitic melts. AM MINERAL **73,** 966–82.

ODP Leg 135 Scientific Party 1991. A new view of arc/backarc systems. GEOTIMES **36(8),** 19–20.

O'Reilly, S. Y., Griffin, W. L. & Stabel, A. 1988. Evolution of Phanerozoic Eastern Australian lithosphere: Isotopic evidence for magmatic and tectonic underplating. J PETROL SPEC LITHOSPHERE ISSUE, 89–108.

Paris, J.-P. 1981. Geologie de la Nouvelle-Calédonie. BUR RECHERCHES GEOL MIN **113**.

Pearce, J. A. 1983. Role of the sub-continental lithosphere in magma genesis at active continental margins. *In* Hawkesworth, C. J. & Norry, M. J. (eds) *Continental basalts and mantle xenoliths* 230–49. Orpington: Shiva.

Shaw, S. E. & Flood, R. H. 1981. The New England Batholith, Eastern Australia: geochemical variations in space and time. J GEOPHYS RES **86,** 10,530–44.

Smart, J. & Senior, B. R. 1980. Jurassic–Cretaceous basins of northeastern Australia. *In* Henderson, R. A. & Stephenson, P. J. (eds) *The geology and geophysics of northeastern Australia,* 315–28. Brisbane: Geological Society of Australian, Queensland Division.

Stephens, C. J. 1991. *The Mungore Cauldron and Gayndah Centre. Late Triassic large-scale silicic volcanism in the New England Fold Belt near Gayndah, southeast Queensland.* Unpublished Ph.D. Thesis, University of Queensland, Brisbane.

Sun, S.-S. & McDonough, W. F. 1989. Chemical and isotopic systematics of oceanic basalts. *In* Saunders, A. D. & Norry, M. J. (eds) *Magmatism in the ocean basins,* 313–45. GEOL SOC SPEC PUBL **42**.

Symonds, P. A., Fritsch, J. & Schluter, H.-U. 1984. Continental margin around the western Coral Sea Basin: Structural elements, seismic sequences, and petroleum geological aspects. *In* Watson, S. T. (ed.) *Trans. Third Circum-Pacific Energy & Mineral Resources Conf., Hawaii,* 243–52. Tulsa: American Association of Petroleum Geologists.

Symonds, P., Davies, P. J., Feary, D. A. & Pigram, C. J. 1988. Geology of the northeastern Australian margin basins. *In* PESA (Qld)—ODCAA SPE Petroleum Symp., Queensland 1988; Exploration Development: 61–77.

Symonds, P. A. & Willcox, J. B. 1989. Australia's petroleum potential in areas beyond an Exclusive Economic zone. BMR J AUST GEOL GEOPHYS **11,** 11–36.

Veevers, J. J. (ed.) 1984. *Phanerozoic earth history of Australia.* Oxford: Clarendon Press.

Veevers, J. J., Powell, C. McA. & Roots, S. R. 1991. Review of seafloor spreading around Australia. I. Synthesis of the patterns of spreading. AUST J EARTH SCI **38,** 373–89.

Webb, A. W. & McDougall, I. 1968. The geochronology of the igneous rocks of Eastern Queensland. J GEOL SOC AUST **15,** 313–46.

Wood, D. A. 1980. The application of a Th–Hf–Ta diagram to problems of tectonomagmatic classification and to establishing the nature of crustal contamination of basaltic lavas of the British Tertiary volcanic province. EARTH PLANET SCI LETT **50,** 11–30.

A. EWART and R. W. SCHON, Department of Earth Sciences, University of Queensland, St Lucia, Queensland 4072, Australia.
B. W. CHAPPELL, Department of Geology, Australian National University, Canberra, A.C.T. 2601, Australia.

Transactions of the Royal Society of Edinburgh: Earth Sciences, **83**, 347–359, 1992

Rhyolites as probes of the crust: examples from Tertiary volcanism in Arizona, southwestern U.S.A.

Sonia Esperança and Thomas C. Moyer

ABSTRACT: Geochemical and isotopic data obtained from three bimodal basalt-rhyolite fields in west-central Arizona point to significant variation in the sources of the siliceous products despite petrological similarities between them. The studied fields, which straddle the boundary between the Basin and Range and Colorado Plateau physiographic provinces, include Castaneda Hills (CH) in the Basin and Range, Kaiser Spring (KS) in the Transition Zone, and Mount Floyd (MF) in the Colorado Plateau. Two types of rhyolite (high-silica [HSR] and low-silica [LSR]) occur as lavas in the KS and MF fields, whereas all analysed CH samples are HSR. These lavas generally post-date low-angle extension of the Basin and Range crust and become generally younger towards the Colorado Plateau. Our isotopic data illustrate that the CH rhyolites require a source with comparatively radiogenic Sr, that most rhyolites require a source with comparatively unradiogenic $^{206}Pb/^{204}Pb$, and that the MF–HSR require a source with comparatively radiogenic Nd and $^{206}Pb/^{204}Pb$. The isotopic data clearly indicate that the western Arizona rhyolites contain a large crustal component. Importantly, the Pb isotopic compositions of the rhyolites illustrate the transition between two crustal provinces identified by other workers on the basis of Nd and Pb isotopic studies of Proterozoic granites. The Sr, Nd, and Pb isotopic compositions of our rhyolites can be modelled by mixing a basaltic component with three hypothetical crustal end-members which themselves may be formed by mixing two crustal components. One crustal component, which is characterised by unradiogenic Pb and Sr, may be similar to the mafic gneiss xenoliths of Tule Tank on the Colorado Plateau. The second crustal component, which has more radiogenic Sr and Pb, may be similar to the Proterozoic Fenner Gneiss of southeastern California. This conceptual model indicates an apparent increase in the contribution of the depleted Tule Tank source from the Basin and Range onto the Colorado Plateau and accommodates the notion that there is an underlying consistency to the petrogeneses of the western Arizona rhyolites as suggested by their chemistry and mineralogy. Although the CH and KS lavas require only one crustal end-member, the MF–HSR and LSR require two isotopically distinct sources. This difference may be attributed to the homogenisation of diverse crustal lithologies by ductile stretching during crustal extension.

KEY WORDS: rhyolites; bimodal volcanic fields; Sr, Nd, Pb isotopic compositions; Arizona crust

Late Cenozoic volcanism in Arizona (southwestern U.S.A.) manifests the heating episode that preceded and accompanied Tertiary crustal thinning and extension of the Basin and Range province. The style and scale of volcanism in Arizona has varied dramatically since 30 Ma; earlier products (18–30 Ma) are dominantly large-volume, intermediate composition lavas and pyroclastics, whereas subsequent deposits are either bimodal (basalt-rhyolite) or primarily basaltic lavas (Nealey & Sheridan 1989).

Geophysical studies show that as a consequence of crustal extension, the present crustal thickness varies from about 25 km under the Basin and Range province to about 45 km beneath the adjoining Colorado Plateau (e.g. Hauser *et al.* 1987; McCarthy & Thompson 1988). The extensional process was complex: while the upper crust was uplifted and brittlely deformed along low-angle normal faults between about 14 and 20 Ma (Suneson & Lucchita 1983), the middle to lower crust was intruded by mafic magma and stretched laterally in ductile fashion (e.g. Gans 1987; however, cf. Wernicke 1981, 1985). Recent reflection seismology traverses in western Arizona reveal a regionally complex crustal structure that supports the notion of lateral flow at depth (McCarthy *et al.* 1991). Subsequent deformation along high-angle normal faults presumably has had less effect on the deep crust, but has created the distinctive physiography of the Basin and Range and delineated the margins of the Colorado Plateau. Obviously, the mid-Tertiary volcanic and extensional episode profoundly affected the crust of Arizona, but the processes by which changes were effected, and the extent to which the crust was modified by stretching and intrusion, remain generally unknown.

We investigated the products of three bimodal basalt-rhyolite fields to address the question of lateral variations in crustal magma sources from the Basin and Range to the Colorado Plateau. These centres, which generally postdate low-angle crustal extension, define a 200 km-long transect that crosses the two tectonic provinces at a sharp angle to their boundary (Fig. 1). The transect includes, from SW to NE respectively, the volcanic fields of: (1) Castaneda Hills (CH) in the Basin and Range; (2) Kaiser Spring (KS) in the structural transition zone between the two provinces; and (3) Mount Floyd (MF) on the southwestern margin of the Colorado Plateau. The volcanic centres are aligned with the southerly projection of the Bright Angel fault, a major Precambrian fracture system that was reactivated during

Tertiary extension (Huntoon & Sears 1975; Shoemaker *et al.* 1978). The transect also lies close to an area identified, on the basis of isotopic data, as a boundary between crust formed in the Proterozoic (Nd model ages of about 1·7 Ga; Bennett & DePaolo 1987) to the E and older crust with an Archaean component (Nd model ages of 2·0–2·3 Ga) to the W (cf. Wooden *et al.* 1988; Fig. 2).

Previous studies show that two types of rhyolite occur in the bimodal fields: high-silica rhyolite (HSR), which has in excess of 75% SiO_2 (normalised to 100% volatile-free) and low-silica rhyolite (LSR), which has from 68 to 74% SiO_2. Moyer and Nealey (1989) summarised the regional compositional variations of rhyolites in the CH, KS, and MF fields and speculated on the probable effects of crustal thickness and source heterogeneity on the geochemistry of the silicic rocks. Radiogenic isotope data (Sr, Nd, and Pb) were reported for basalts and rhyolites from KS by Moyer and Esperança (1989a). They concluded that the rhyolites contain a large portion of crustal material and that some basalts could have been contaminated at lower crustal levels (cf. Moyer & Esperança 1989b).

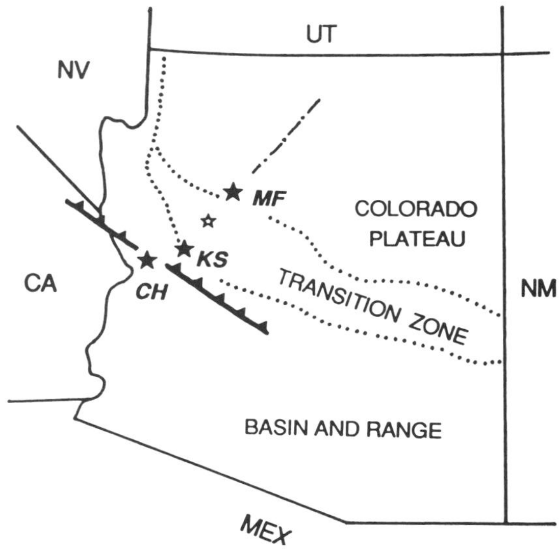

Figure 1 Stars locate the bimodal volcanic fields in western Arizona: MF = Mount Floyd; KS = Kaiser Spring; CH = Castaneda Hills; the Mount Hope field (open star) is not included in this study; saw-toothed line denotes the northern limit of detachment faulting and metamorphic core complex uplift in the Basin and Range; dash-dot line indicates the Bright Angel fault; physiographic province boundaries are shown by dotted lines; neighbouring states and countries are indicated by abbreviations.

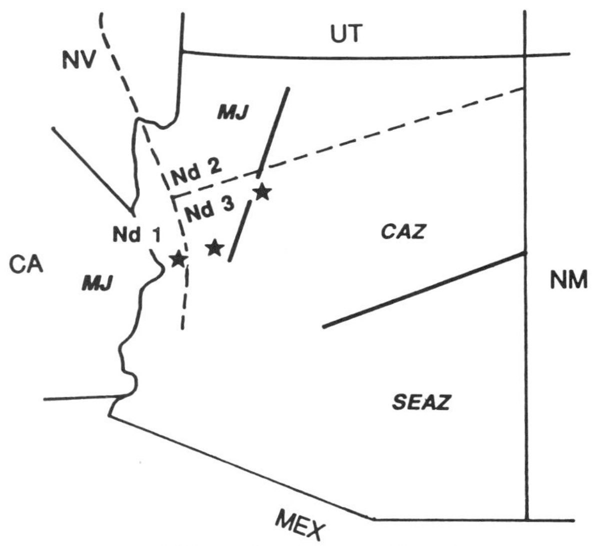

Figure 2 Bimodal fields of the study are shown in relation to isotopic crustal province boundaries; dashed lines designate Nd isotopic provinces from Bennett and DePaolo (1987): Nd 1 = T_{DM} 2.0–2.3 Ga, Nd 2 = T_{DM} 1.8–2.0 Ga, Nd 3 = T_{DM} 1·7–1·8 Ga; solid lines denote Pb isotopic crustal provinces from Wooden and Miller (1990): MJ = Mojavia terrane, CAZ = Central Arizona terrane, SEAZ = Southeast Arizona terrane.

The present study was designed to measure the Sr, Nd, and Pb isotopic compositions of the felsic end-members in the MF and CH fields to help contrain possible petrogenetic models for the evolution of these silicic magmas and to probe for differences in the crustal sources of these lavas across the tectonic boundary. The available geophysical data regarding the probable crustal structure beneath the area, coupled with isotopic data for crustal and mantle xenoliths, basalts, and crustal rocks in the region, enable us to derive a more accurate petrogenetic view of rhyolite genesis.

1. Field relations

We summarise below salient field descriptions of the three volcanic fields of interest. Additional descriptions can be found in Nealey (1980) and Goff *et al.* (1983) for the MF lavas, in Moyer (1986, 1990) and Moyer & Esperança (1989a) for the KS lavas, and in Suneson (1980), Suneson & Lucchitta (1979, 1983), Lucchitta & Suneson (1985, 1988, 1989), and Sherrod (1988) for the CH field. Important field observations include the similar character of the KS and CH high-silica rhyolite lavas which differs from that of the MF–HSR, the similarity of the KS and MF low-silica rhyolites, and the regional variations in the style of faulting and the degree to which the lavas in each field are structurally disrupted.

1.1. Mount Floyd

The silicic centres at MF occur as isolated hills surrounded by coeval flows of basaltic lava. Low-silica rhyolites (LSR) erupted from four vents and high-silica rhyolites (HSR) were extruded from at least five additional events (Nealey 1980; Moyer & Nealey 1989; Nealey, unpublished mapping). LSR occur as isolated domes or complexes of overlapping domes that were generally preceded by pyroclastic emission. One of the LSR lavas (sample MF-1) was extruded adjacent to an earlier basaltic lava flow and may have used the same conduit (Nealey 1980). Aphyric HSR occur as low, obsidian-capped mounds that are difficult to discern in the field, whereas phenocryst-bearing HSR lavas form distinct buttes that cap pyroclastic deposits. HSR sample MF-6 was collected from a dome adjacent to a basalt cinder cone but the temporal relationships between the two lavas are equivocal (Nealey 1980). Although variably devitrified and altered, none of the MF silicic lavas contain lithophysae. The available radiometric ages suggest that silicic volcanism migrated from S to N. LSR domes have been dated from 9·8 ± 0·7 Ma (Goff *et al.* 1983) to 2·7 ± 0·7 Ma (Nealey, unpublished data). Most of the mafic end-members in this field are alkali-olivine basalts or hawaiites, some of which contain cumulate xenoliths and abundant megacrysts. Nealey (1980) additionally reported spinel and amphibole in some xenolith samples. The mafic lavas erupted from cinder and spatter cones scattered throughout the field (Nealey 1980). Geological mapping by Nealey (1980 and unpublished data) indicates that the MF lavas are generally unfaulted; however, the field is bounded by NW-striking high-angle normal faults, one of which is still active. Mafic vents form both northwesterly and northeasterly alignments, suggesting control by basement fractures (Nealey 1980).

1.2. Kaiser Spring

The KS-silicic lavas erupted from at least 25 different vents that form two subparallel N–S-trending belts (Moyer 1986, 1990). Lavas of the older eastern belt are monogenetic HSR domes and coulees, whereas the younger western belt extruded flows of LSR from possibly 12 vents and HSR from one vent. HSR lavas occur as isolated or overlapping domes and flows preceded by pyroclastic activity. Their devitrified and altered upper portions contain abundant lithophysae lined with vapour-phase silica, garnet, and rare topaz (Burt *et al.* 1981). LSR lavas were extruded as domes and (comparatively) widespread lava flows that followed initial pyroclastic activity. Although variably devitrified and crystallised, the LSR lavas are not lithophysal. Both HSR and LSR pyroclastic deposits are interbedded with mafic lavas in the southern portion of the field; in one instance, LSR lava includes quenched xenoliths of mafic magma. K–Ar ages reported for HSR lavas of the eastern belt are 12·0 and 11·3 Ma, whereas K–Ar ages for LSR lavas of the western belt vary from 9·6 to 8·8 Ma (Moyer & Esperança 1989a). The mafic volcanism at KS has a compositional range that includes tholeiites, quartz-bearing basaltic andesites and alkali-olivine basalts. Most of these lavas were erupted from vents peripheral to the silicic volcanism. The KS volcanics were deposited above Precambrian crystalline basement that was uplifted and eroded along NW-striking normal faults (Moyer 1990). The volcanic pile is cut by numerous small displacement normal faults that strike NW and NNE, but evidence for significant syn-volcanic faulting is lacking.

1.3. Castaneda Hills

Silicic lavas were erupted from three centres in the CH field: Aubrey Peaks, Castaneda Hills, and Bill Williams Mountains (Suneson 1980; Suneson & Lucchita 1983; Sherrod 1988). Multiple domes, flows and dykes occur in these centres with at least 25 vents recognised in the field. Rhyolitic lavas and their preceding pyroclastic deposits are interbedded with basaltic lavas and sedimentary units (Suneson & Lucchitta 1983), although evidence for widespread basalt-rhyolite interaction is absent. The lavas have basal vitrophyres that are variably hydrated and spherulitically devitrified. The exposed upper surfaces of many of these rhyolites, especially at Aubrey Peaks, contain abundant lithophysae lined with silica phases and, at least in one case, garnet. Our study of glassy samples does not corroborate the report of a single LSR lava in the CH region (Suneson 1980; Moyer & Nealey 1989), so we conclude that either such lavas are absent or that they comprise only a minor rhyolite component in this field. HSR lavas were erupted over an age range of 10·3–15·1 Ma, with most activity occurring between 12·4 and 13·7 Ma (average of 8 K–Ar ages yields 13·0 ± 1·0 Ma; Suneson & Lucchitta 1979, 1983). Extrusions of tholeiitic basalts and quartz-bearing basaltic andesites were contemporaneous with silicic volcanism at CH (Suneson & Lucchitta 1983). These lavas were succeeded by flows of megacryst-bearing alkali-olivine basalt from dykes and spatter cones (Suneson & Lucchitta 1983). The rhyolites and older basalt lavas are cut by numerous normal faults, some of which are rotational faults associated with the waning stages of detachment faulting and uplift of the adjacent Rawhide metamorphic core complex (Shackelford 1980; Suneson & Lucchitta 1983). Listric normal faulting was succeeded by an episode of NW-striking high-angle normal faulting that occurred from about 14 to 8 Ma (Suneson & Lucchitta 1983). Megacryst-bearing basalt lavas, which yield radiometric ages between 5·4 and 8·6 Ma, are generally unfaulted (Suneson & Lucchitta 1983).

2. Petrography

Petrographic descriptions and locations of the rhyolite samples discussed herein are summarised in Appendix 1. Additional data are given in Suneson (1980), Moyer (1986 and in prep.), Moyer & Esperança (1989a) and Nealey (1980). The petrographic similarity of the lavas from these fields, especially the LSR, is striking.

High-silica rhyolite lavas have variable phenocryst contents that range from 0 to about 20 vol%. Most samples contain alkali feldspar, oligoclase to albite feldspar, quartz, Fe-rich biotite, and Fe–Ti oxide. Some samples additionally contain hornblende. Accessory minerals include zircon, apatite, monazite, and allanite; the latter two phases are quite common in the CH lavas. Xenocrysts are minor to absent in the HSR lavas.

Low-silica rhyolite lavas generally contain from 10 to 30 vol% phenocrysts. Andesine to labradorite feldspar dominates a variable mafic-silicate phase assemblage and Fe–Ti oxide. LSR lavas are subdivided into three general groups based on the mafic-silicate phases present: biotite plus hornblende, hornblende plus orthopyroxene, and hornblende plus ortho- and clinopyroxene. The latter group is restricted to the KS field. Zircon and apatite are accessory phases. Xenocrysts, which are common in these lavas, are predominantly plagioclase with spongy to sieved textures and anhedral, deeply embayed quartz.

3. Geochemistry

3.1. Analytical techniques

Kilogram samples, trimmed of weathering products, were crushed to centimetre size in a hardened steel roller mill. Chips were washed in distilled water and approximately 50 g alteration-free material was hand-selected and powdered in an alumina ceramic shatterbox. All samples analysed in this study have fresh, glassy matrices.

Major and minor elements for MF, CH, and some KS samples were analysed by commercial X-ray fluorescence (XRF); additional analyses for KS were obtained at Arizona State University. Analyses were made on fused glass discs by wavelength XRF spectrometry. Analytical precision determined by duplicate analyses of US Geological Survey standards G-2 and RGM-1 is better than 10% relative for all elements except Nb (30%).

Isotopic analyses were determined at the Department of Terrestrial Magnetism, Carnegie Institution of Washington. Sr, Nd, and Pb were separated for isotopic analysis using ion exchange techniques. Samples were dissolved using a mixture of HF and HNO_3 in tightly-capped PFA vessels. Sr and Nd were separated on HCl and alpha-hydroxy isobutyric acid cation exchange columns, respectively. Pb was separated using a modification of the procedure described by Manhes *et al.* (1978). The Sr and Pb isotopic measurements were performed on a multi-collector VG-354 mass spectrometer. All results were fractionation corrected to $^{87}Sr/^{86}Sr = 0{\cdot}1194$ and are reported relative to $^{87}Sr/^{86}Sr = 0{\cdot}7080$ for the Eimer and Amend standard. Pb results were corrected for fractionation using the difference between measured values for the NBS 981 Pb standard and those reported by Catanzaro *et al.* (1968). Nd isotopic compositions were measured on a 15 in mass spectrometer and fractionation corrected to $^{146}NdO/^{144}NdO = 0{\cdot}722250$ ($^{146}Nd/^{144}Nd = 0{\cdot}7219$). Values reported are relative to $^{143}Nd/^{144}Nd = 0{\cdot}511860$ for the La Jolla standard. ε_{Nd} values were calculated using the average chondritic value of $^{143}Nd/^{144}Nd = 0{\cdot}51264$.

3.2. Elemental geochemistry

In Table 1 we report new whole-rock major and minor element data for 22 rhyolite samples from the KS, CH and MF fields. Additional analyses of rhyolites from these fields are reported in Moyer & Esperança (1989a), Moyer & Nealey (1989), Moyer (1986), Suneson & Lucchitta (1983) and Nealey (1980).

Table 1 X-Ray fluorescence analyses and CIPW norms of rhyolites

	Castaneda Hills							Kaiser Spring						
	CH89-4	CH90-4	CH90-5	CH90-13	CH90-16	CH90-21	CH90-33	EM-6	EM-21	EM-33	BC-4	KS-44	EM-83	EM-84
SiO_2	77·3	77·3	77·5	77·4	77·2	76·7	77·3	75·6	76·4	77·3	76·8	76·8	71·7	69·7
TiO_2	0·12	0·12	0·09	0·09	0·10	0·21	0·10	0·18	0·16	0·08	0·05	0·05	0·41	0·37
Al_2O_3	12·6	12·6	12·6	12·7	12·8	12·6	12·7	13·2	12·9	12·9	13·4	13·2	14·4	15·5
Fe_2O_3	1·06	1·07	1·08	0·94	0·94	1·45	1·00	1·21	1·15	0·78	0·71	0·79	2·91	3·08
MnO	0·03	0·03	0·04	0·05	0·05	0·04	0·05	0·08	0·08	0·06	0·13	0·12	0·07	0·06
MgO	0·04	0·07	0·03	0·08	0·05	0·24	0·09	0·33	0·26	0·06	0·01	0·02	0·97	1·01
CaO	0·62	0·46	0·47	0·42	0·51	0·77	0·49	0·87	0·78	0·72	0·52	0·30	2·21	3·02
Na_2O	3·40	3·70	4·02	3·81	3·25	3·23	3·66	3·77	3·92	3·72	4·93	4·48	3·48	3·75
K_2O	4·87	4·61	4·11	4·55	5·08	4·75	4·58	4·66	4·31	4·29	3·49	4·28	3·70	3·34
P_2O_5	0·02	0·02	0·02	0·02	0·02	0·04	0·02	0·04	0·03	0·02	0·02	0·02	0·15	0·12
LOI	4·77	3·93	4·31	3·31	3·23	3·32	2·62	2·62	3·16	5·31	4·23	1·00	3·77	2·85
Anal Σ	100·27	99·02	100·39	100·27	100·23	99·34	100·40	99·40	99·78	100·37	99·97	100·40	99·59	99·63
Rb	221	201	231	202	206	169	225	284	338	289	485	384	118	73
Sr	19	28	10	<10	43	133	<10	99	91	34	13	4	270	388
Y	35	47	43	56	38	36	41	67	74	53	109	88	30	25
Zr	109	119	112	96	92	123	92	106	107	86	84	78	160	147
Nb	23	18	32	25	16	21	34	27	34	34	57	56	31	13
Ba	<10	<10	<10	<10	<10	88	<10	185	150	14	3	10	944	1175
Q	37·4	36·9	37·2	36·7	37·5	37·6	37·1	33·4	34·9	37·5	33·5	33·5	31·0	26·8
Or	28·8	27·2	24·3	26·9	30·0	28·1	27·1	27·5	25·5	25·4	20·6	25·3	21·9	19·8
Ab	28·8	31·3	34·0	32·3	27·5	27·4	31·0	31·9	33·2	31·5	41·7	37·9	29·5	31·8
An	3·0	2·2	2·2	2·0	2·4	3·6	2·3	4·1	3·7	3·4	2·5	1·4	10·0	14·2
Hy	0·66	0·74	0·72	0·77	0·67	1·29	0·81	1·48	1·29	0·65	0·65	0·71	3·80	4·06
Di														
Ap	0·04	0·04	0·04	0·04	0·04	0·09	0·04	0·09	0·07	0·04	0·04	0·04	0·33	0·26
Il	0·23	0·23	0·17	0·17	0·19	0·40	0·19	0·34	0·30	0·15	0·10	0·10	0·78	0·71
Mt	0·62	0·62	0·63	0·55	0·55	0·84	0·58	0·70	0·67	0·45	0·41	0·46	1·69	1·79
C	0·66	0·74	0·73	0·79	1·08	0·84	0·88	0·47	0·44	0·88	0·61	0·70	1·01	0·51

	Mount Floyd							
	MF-6	MF-8	MF-10	MF-15	MF-1	MF-4	MF-14	MF-21
SiO_2	76·9	76·2	77·2	73·5	73·1	69·3	69·6	72·0
TiO_2	0·06	0·06	0·07	0·30	0·24	0·37	0·37	0·32
Al_2O_3	13·1	13·3	12·8	13·7	15·1	15·5	15·6	14·5
Fe_2O_3	0·92	1·06	0·88	2·04	1·74	3·00	2·96	2·76
MnO	0·08	0·10	0·07	0·05	0·06	0·06	0·06	0·07
MgO	0·01	0·04	0·02	0·59	0·52	1·39	1·23	0·95
CaO	0·21	0·19	0·22	1·45	2·02	3·50	3·32	2·45
Na_2O	4·30	4·52	4·51	4·02	4·27	4·30	4·43	4·00
K_2O	4·32	4·47	4·21	4·26	2·85	2·46	2·27	2·85
P_2O_5	0·02	0·02	0·02	0·06	0·09	0·13	0·12	0·12
LOI	3·39	2·85	1·08	0·77	2·54	2·31	1·70	1·93
Anal Σ	99·98	99·90	100·05	99·97	99·88	100·02	99·73	99·66
Rb	256	338	211	96	38	38	41	44
Sr	<10	<10	<10	223	818	992	982	685
Y	55	75	12	<10	<10	<10	<10	11
Zr	85	78	81	135	127	113	107	145
Nb	57	74	64	20	<10	19	16	<10
Ba	<10	<10	<10	1287	1738	1442	1497	1584
Q	34·7	32·1	34·2	29·6	32·2	25·2	25·9	30·9
Or	25·5	26·4	24·9	25·2	16·9	14·6	13·4	16·9
Ab	36·4	38·3	38·2	34·1	36·2	36·4	37·5	33·9
An	0·9	0·8	1·0	6·8	9·4	15·8	15·7	11·4
Hy	0·68	0·89	0·65	2·40	2·16	4·65	4·53	3·80
Di				0·04		0·63		
Ap	0·04	0·04	0·04	0·13	0·20	0·29	0·26	0·26
Il	0·11	0·11	0·13	0·57	0·46	0·71	0·71	0·61
Mt	0·53	0·62	0·51	1·19	1·01	1·75	1·72	1·61
C	1·02	0·73	0·47		1·54		0·11	0·67

Notes: Analyses recalculated to 100 wt% volatile-free; original analytical summation including loss on ignition is shown as Anal Σ. Total Fe given as Fe_2O_3. Rb, Sr, Y, Zr, and Ba for Kaiser Spring samples analysed at Arizona State University. Minor elements adjusted by factor used to normalise major element oxides to 100%.

HSR lavas are chemically evolved (molar Mg values 2–36), weakly peraluminous (0·1–1·1 wt% normative corundum), and have normative compositions that plot near the haplogranite eutectic (Tuttle & Bowen 1958; Luth *et al.* 1964; Ebadi & Johannes 1991). They are characterised by high concentrations of incompatible elements (Rb, U, Th, Nb) and low concentrations of feldspar- and mafic-silicate-compatible elements (Sr, Ba, Eu, Ti, Mg) (ref. Moyer & Esperança 1989a; Moyer & Nealey 1989; Fig. 3). In contrast, the LSR lavas are more primitive (molar Mg values 31–48), metaluminous to weakly peraluminous and contain significant normative An (8–16 wt%). These lavas have low concentrations of incompatible elements and high concentrations of compatible elements. Microprobe analyses show that the matrix glasses of many LSR lavas from KS have HSR compositions (Moyer, unpublished data). The limited data of Moyer & Nealey (1989) led them to suggest that the compositional gap separating the HSR and LSR lavas of a single field widens from the Basin and Range onto the Colorado Plateau. Our data corroborate that observation, with the exception that we have not found fresh samples of LSR in the CH field.

The CH high-silica rhyolites have a comparatively restricted compositional range. The most mafic of our samples (CH 90-21) has a Ba and Sr content similar to "primitive" HSR samples from KS. CH compositions bridge the small gap between the HSR and LSR samples of KS in terms of their Rb, Y, Zr, and Nb contents (Fig. 3b), but when compared to HSR lavas from KS and MF, the CH lavas clearly have lower Rb, Nb, and Th contents and higher LREE contents (Fig. 4a). The latter observation accords with the ubiquitous allanite and less common monazite in the CH rocks.

Moyer and Esperança (1989a) showed that the KS HSR and LSR lavas have strikingly different trace element compositions (Fig. 4b). HSR lavas have high concentrations of incompatible elements and relatively flat REE patterns ($La_N/Lu_N < 2$) with large, negative Eu anomalies. Their high, but variable, Rb and HREE contents distinguish them in the transect (Moyer & Nealey 1989). In contrast, LSR lavas have LREE-enriched patterns ($La_N/Lu_N = 8{\cdot}7–12{\cdot}0$), higher concentrations of compatible elements, and negative Nb anomalies (Fig. 4b). Minor element variation diagrams for the KS lavas indicate the presence of two trends of HSR, both evolving from two samples of hornblende-two pyroxene rhyodacite with slightly different bulk chemical composition (EM 83 and EM 39). The first trend includes a biotite HSR (KSP 70) and aphyric lavas KSP 56 and KS 44, whereas the second trend includes the biotite-hornblende HSR samples of Elephant Mountains (EM 6, EM 21, EM 17) and aphyric lava BC 4 (Fig. 5).

There is a larger compositional gap between most members of the HSR and LSR groups at MF than is observed at KS (Moyer & Nealey 1989; Fig. 4c). One of our samples (MF-15) is an aphyric lava that is compositionally intermediate to the two groups. As discussed below, this sample has the isotopic characteristics of other HSR lavas and is considered a "primitive" member of that group. Chemically evolved HSR lavas from MF have trace element compositions generally similar to evolved HSR lavas from KS, with the exception of having lower abundances of Rb and HREE (Fig. 4a) and higher incompatible element abundances (Nb, Cs, U) at a given Rb content (Fig. 3b). MF LSR lavas are richer in Ba and Sr and have lower contents of Rb, Th, and HREE than LSR lavas from KS (Fig. 4d).

3.3. Isotope geochemistry

New isotopic analyses for 13 rhyolite samples are presented in Table 2 and shown in Figures 6 and 7. Additional isotopic

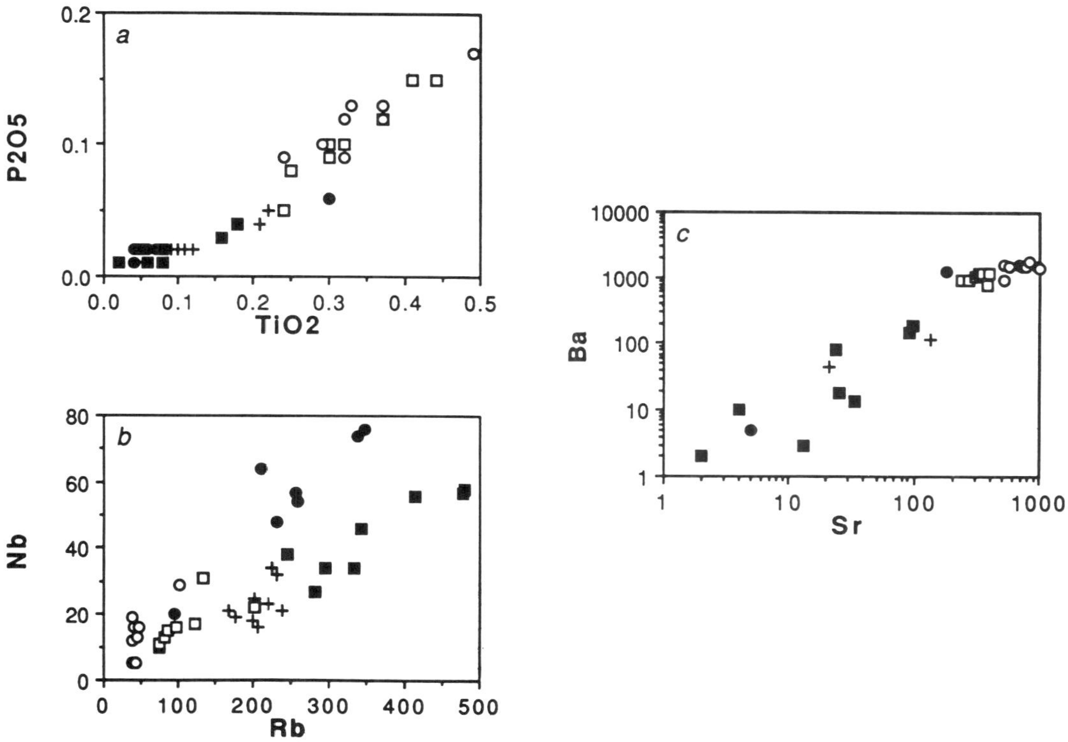

Figure 3 Elemental variations for HSR and LSR samples.
Symbols: crosses = Castaneda Hills; solid squares = Kaiser Spring HSR; open squares = Kaiser Spring LSR; solid circles = Mount Floyd HSR; open circles = Mount Floyd LSR. Data are from Table 1, Moyer & Esperança (1989a), and Moyer & Nealey (1989).

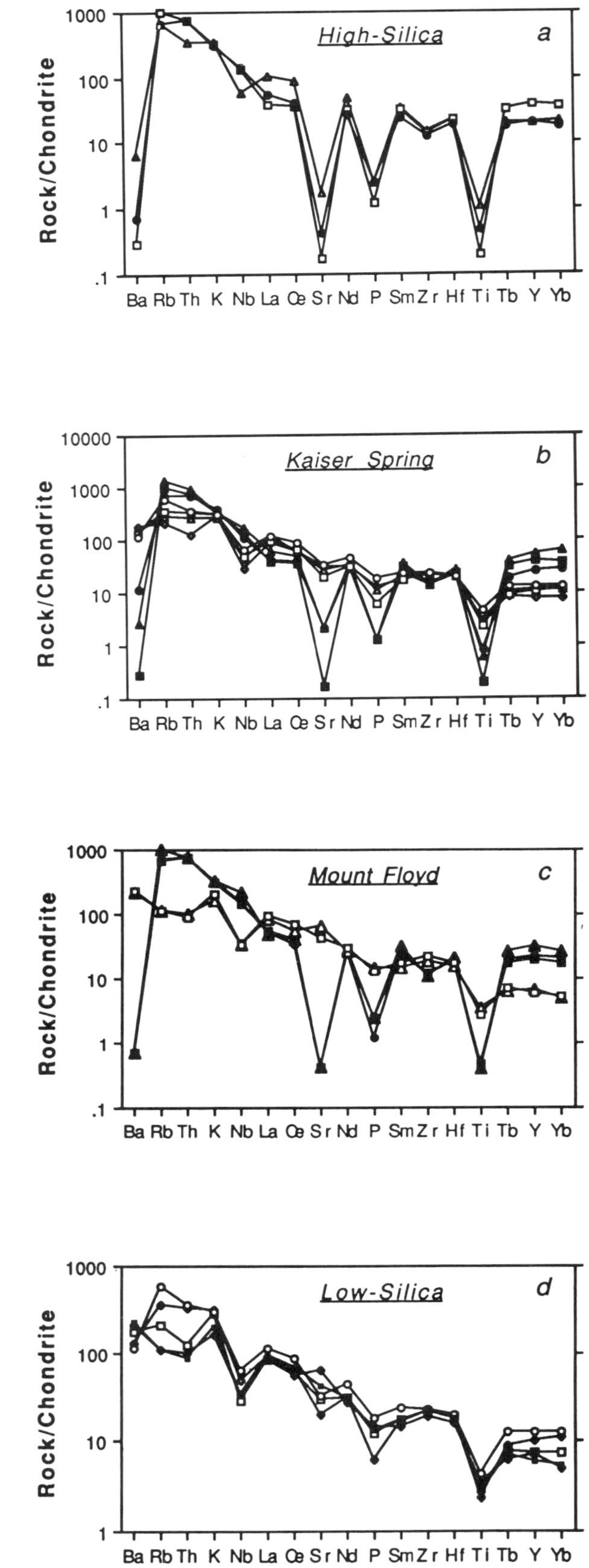

Figure 4 Chondrite-normalised spidergrams (chondrite values from Thompson *et al.* 1984). (a) High-silica rhyolites. Triangle—Mount Floyd (NMF-86-14); square—Kaiser Spring (KSP-56), solid circle—Castaneda Hills (CH-42). (b) Kaiser Spring rhyolites. Open symbols—HSR; solid symbols—LSR. (c) Mount Floyd rhylites. open symbols—HSR; solid symbols—LSR. (d) Low-silica rhyolites. Open symbols—Kaiser Spring; solid symbols—Mount Floyd. Data are from Moyer & Esperança (1989a) and Moyer & Nealey (1989).

data are given in Moyer & Esperança (1989a), Moyer & Nealey (1989), and Suneson & Lucchitta (1983).

The rhyolites from CH have the most radiogenic Sr isotopic compositions of all the samples analysed and yield a poorly constrained Rb–Sr isochron with an initial Sr isotopic composition of 0·71080 and a slope that corresponds to an age of 13·0 Ma ($r^2 = 0{\cdot}979$). The isochron age agrees with K–Ar ages of these lavas reported by Suneson and Lucchitta (1983). Whole-rock initial Sr ratios calculated at 13 Ma for our samples (Table 2) are in general agreement with two determinations on sanidine separates of 0·7093 and 0·7141 reported by Suneson and Lucchitta (1983). Interestingly, the CH rhyolite sample that has the highest ε_{Nd} and $^{206}Pb/^{204}Pb$ ratio (CH 90-5) also has the least "primitive" chemical characteristics such as the lowest Mg# and Sr content, and the highest Rb concentration.

The LSR from KS have heterogeneous isotopic compositions. Two-pyroxene LSR are more mafic (higher Mg#) and plot closer to basalts in the Sr-Nd isotopic plane (Moyer & Esperança 1989a, and this work). In contrast, hornblende-biotite LSR lavas have more negative ε_{Nd} and slightly more radiogenic Sr isotopic compositions. Age-corrected Sr isotopic compositions of the HSR lavas are less radiogenic than values for several tholeiitic basalts in this locality and approximate those of the LSR lavas. The LSR and HSR lavas from this field have similar ε_{Nd} values and $^{206}Pb/^{204}Pb$ ratios; however, the HSR have distinctly more radiogenic $^{208}Pb/^{204}Pb$ compositions (Fig. 7).

The MF high-silica rhyolites have a small isotopic compositional range that is closer to that of coeval basalts than to any of the rhyolite samples analysed from either the KS or CH fields. A linear regression of Rb–Sr data for all MF HSR samples yields an age of $6{\cdot}2 \pm 0{\cdot}1$ Ma with an

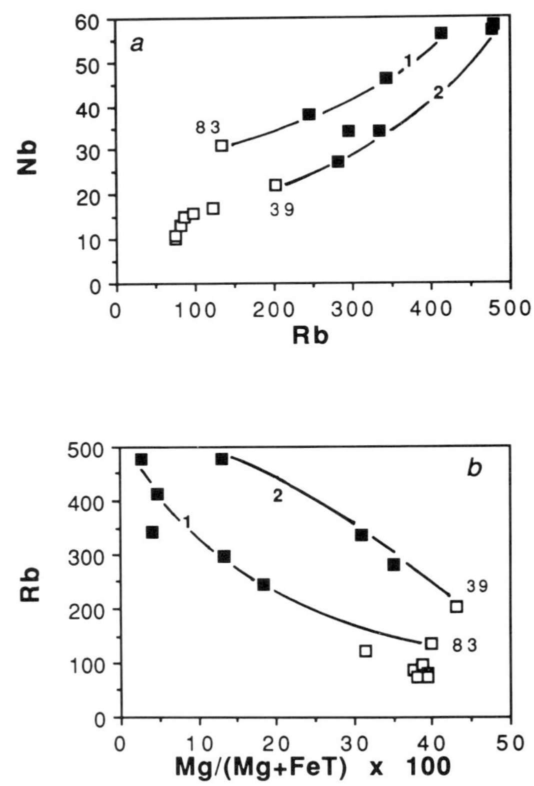

Figure 5 Elemental variations for the Kaiser Spring rhyolites; symbols as in Figure 3; samples EM-39 and EM-83 are hbld-opx-cpx LSR (Moyer & Esperança 1989a); Mg/(Mg + Fe_T) computed on an atomic basis; solid lines labelled 1 and 2 are trends discussed in text.

initial Sr isotopic composition of 0·7066. Because radiometric ages from this field provide good evidence for LSR volcanism over an extended period of time (Nealey 1980), the isochron age could be inaccurate. Considering the errors, these ages agree with an $^{40}Ar/^{39}Ar$ age of approximately 6 Ma for a HSR sample (L.D. Nealey, pers. comm.). The age-corrected values for the LSR samples at MF show unradiogenic Sr isotope compositions, lower than many basalts as well as the HSR in the field. The LSR samples also have much lower ε_{Nd} values and $^{206}Pb/^{204}Pb$ ratios than the HSR and basalts.

Except for the HSR lavas from MF, all of our rhyolite

Table 2 Isotopic composition of rhyolites

Sample	$^{87}Rb/^{86}Sr$	$^{87}Sr/^{86}Sr_m$	$^{87}Sr/^{86}Sr_i$	$^{147}Sm/^{144}Nd$	$^{143}Nd/^{144}Nd$	ε_{Nd}	$^{206}Pb/^{204}Pb$	$^{207}Pb/^{204}Pb$	$^{208}Pb/^{204}Pb$	Rb	Sr	Sm	Nd	Age (Ma)
					Castaneda Hills									
CH 89-4	27·28	0·716540	0·711504		0·512227	−8·0	17·432	15·495	37·924	203·5	21·6		27·94	13·0
CH 90-5	48·35	0·721353	0·712427	0·1506	0·512282	−6·9	17·478	15·518	38·023		13·3	5·45	21·88	13·0
CH 90-13	113·7	0·731030	0·710047				17·343	15·498	37·799		5·0			13·0
CH 90-16	17·08	0·713778	0·710625								33·9			13·0
CH 90-21	4·77	0·710401	0·709520		0·512118	−10·1	17·293	15·490	37·743		98·2		23·69	13·0
					Kaiser Spring									
EM-6	8·66	0·706610	0·705220		0·512175	−9·0	17·310	15·490	37·869	297	99·3		26·33	11·3
EM-80	1·51	0·706647	0·706443	0·1042	0·512278	−7·0	17·365	15·476	37·514	123	235	3·36	19·5	9·5
					Mount Floyd									
MF-1	0·210	0·704462	0·704446				17·561	15·494	37·538	42·3	581	2·6	18·1	5·4
MF-6	458·6	0·744495	0·705421		0·512488	−2·9	17·996	15·539	38·159	259	1·64		16·51	6·0
MF-8	208·6	0·723296	0·705523	0·1940	0·512516	−2·4	18·098	15·539	38·093	342	4·75	6·03	18·8	6·0
MF-10	455·5	0·751321	0·712511		0·512486	−3·0	18·034	15·536	38·136	232	1·48		17·4	6·0
MF-14	0·138	0·703889	0·703884	0·1029	0·512180	−8·9	17·518	15·482	37·399	35	735	2·79	17·3	2·7
MF-15	1·38	0·706738	0·706621		0·512483	−3·0	17·864	15·531	38·160	83·9	176·1		21·64	6·0
MF-21	0·186	0·704473	0·704447		0·512088	−10·7	17·277	15·461	37·143				20·18	9·8

Notes: Rb, Sr, Sm, and Nd by isotope dilution, ε_{Nd} computed using the average chondritic value for $^{143}Nd/^{144}Nd$ of 0·51264. Initial Sr computed using the ages shown. $^{87}Rb/^{86}Sr$ for MF-21 computed using Rb and Sr by XRF (Table 1).

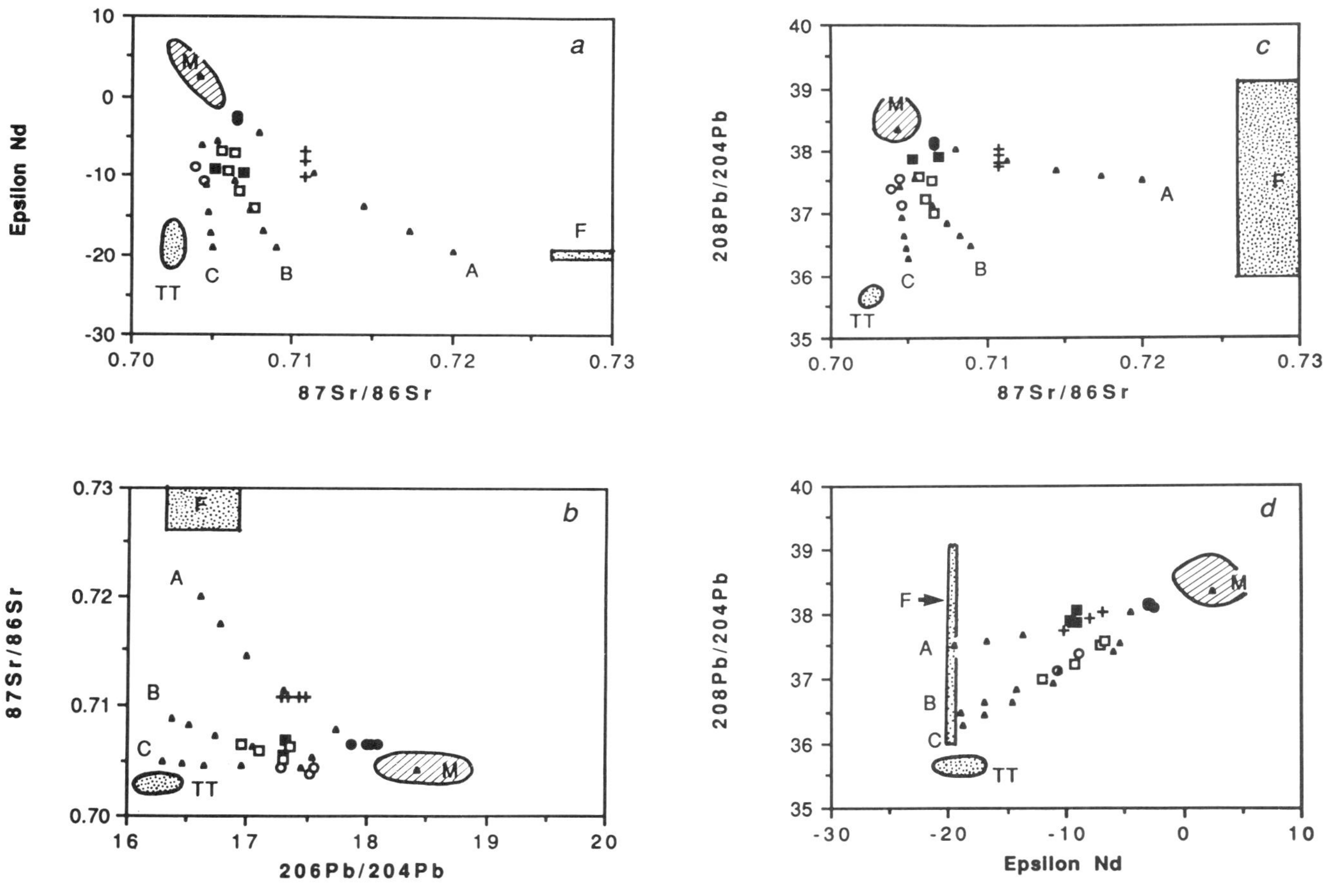

Figure 6 Isotopic variations: (a) ε_{Nd} vs $^{87}Sr/^{86}Sr_i$; (b) $^{87}Sr/^{86}Sr_i$ vs $^{206}Pb/^{204}Pb$; (c) $^{208}Pb/^{204}Pb$ vs $^{87}Sr/^{86}Sr_i$; (d) $^{208}Pb/^{204}Pb$ vs ε_{Nd}. Symbols as in Figure 3; data are from Table 2 and Moyer & Esperança (1989a); small triangles are mixing curves at intervals of 20% computed between a basalt component (M) and three crustal components (A, B, C); endmember compositions are given in Table 3; diagonally ruled field encloses alkali-olivine basalt samples from the bimodal fields (Moyer & Esperança 1989a and unpublished data); stippled fields labelled TT and F enclose analyses reported for the Tule Tank mafic gneiss xenoliths (Nealey & Unruh, in press) and Fenner Gneiss (Bennett & DePaolo 1987; Bender 1990).

samples are characterised by unradiogenic Pb isotopic compositions ($^{206}Pb/^{204}Pb < 17{\cdot}52$) that are distinct from the majority of the alkali basalts in the three fields (Fig. 7). In general, the LSR lavas have less radiogenic $^{208}Pb/^{204}Pb$ compositions than the HSR lavas.

4. Discussion

The variation in petrological and chemical characteristics of the rhyolites from the three fields is quite large even for specific types within each individual locality, attesting to the complexity of the petrogenetic process and to the myriad possible sources involved in the formation of these siliceous magmas. Several important features of the isotopic data are illustrated in Figures 6 and 7: (1) the CH rhyolites require a source with comparatively radiogenic Sr, (2) most rhyolites require a source with comparatively unradiogenic $^{206}Pb/^{204}Pb$, and (3) the MF–HSR require a source with comparatively radiogenic Nd and $^{206}Pb/^{204}Pb$. Given what is known about the composition and structure of the basement beneath these areas and of the geochemistry of the likely crustal contributors sampled as xenoliths in several volcanic localities in the region from as early as 25 Ma, it is possible to establish a working model with some petrogenetic constraints for these rhyolites.

Our data clearly indicate that the western Arizona rhyolites contain a large crustal component. The isotopic

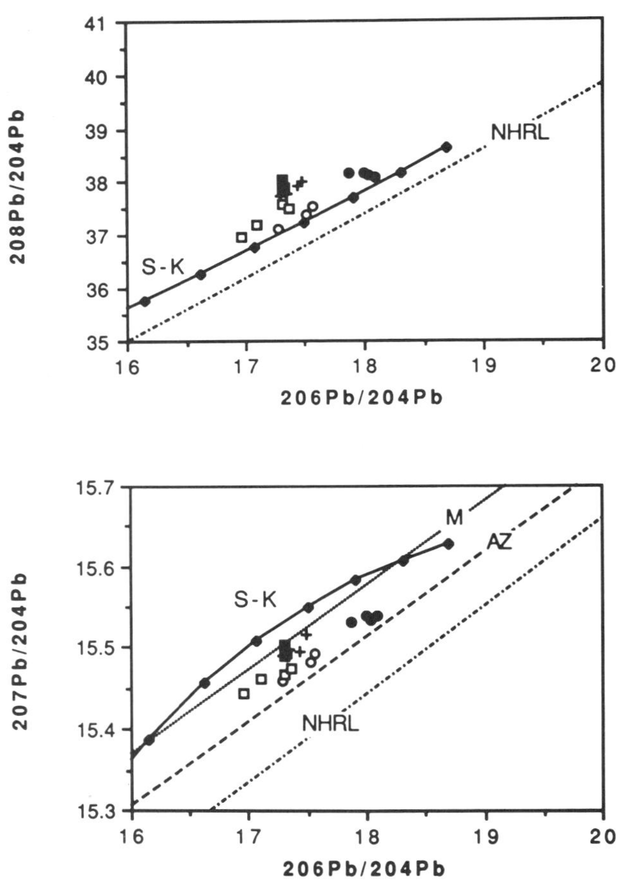

Figure 7 Pb isotopic compositions: symbols as in Figure 3; data are from Table 2 and Moyer & Esperança (1989a); solid line is Stacey and Kramers' (1975) model curve for average crust (dots are at intervals of 0·25 Ga); dash-dot line labelled NHRL is the northern hemisphere regression line for oceanic basalts given by Hart (1984); dashed line labelled AZ and dotted line labelled M are trends for Proterozoic crust in the Arizona and Mojave crustal provinces, respectively (Wooden & Miller 1990).

compositions of the mafic and felsic end-members of these bimodal suites do not overlap, which precludes a simple crystal fractionation model for the rhyolites studied and indicates that variable amounts of crustal material were involved in their generation. The depleted mantle Nd model ages (T_{DM}) for LSR samples from KS vary from 1·2–1·6 Ga and those from MF vary from 1·3–1·4 Ga. We point out that the T_{DM} model ages for the LSR are maximum values. Although we do not know to what extent the Sm/Nd ratios of the LSR magmas were modified by crystallisation during ascent and storage before eruption, crystal fractionation of the observed phase assemblage would produce greater LREE-enrichment and consequently younger model ages. Considering the approximate age of cratonisation of the region (1·6–1·7 Ga; Condie 1981), this suggests that a large percentage of old crustal material was involved in rhyolite genesis. To further quantify the percentage of reworked crustal material is difficult because the isotopic compositions of deep crustal xenoliths from the region are highly variable and because we lack adequate knowledge of the isotopic compositional range of regional Proterozoic upper crustal lithologies. The problem is exacerbated by debate about the "real" isotopic composition of the lower crust in a volcanically active area (see Esperança *et al.* 1990 and Johnson 1990).

The Pb isotopic data (Fig. 7) corroborate the interpretation that the rhyolites are derived largely from Proterozoic crustal materials. Wooden *et al.* (1988) and Wooden and Miller (1990) showed that the Proterozoic crust beneath central and southern Arizona has Pb isotopic characteristics that are distinct from those of adjacent southeastern California. Our Pb isotopic data overlap the range for Proterozoic upper crust, suggesting that most, if not all, of the Pb in our samples has a crustal origin. In addition, the Pb–Pb data lie along a poorly constrained isochron that yields an age of 1·7–1·9 Ga.

The rhyolite Pb data display subtle variations that can be interpreted in terms of the Pb isotopic crustal provinces defined by Wooden *et al.* (1988) and Wooden and Miller (1990) for Proterozoic samples. Our volcanic fields occur near the boundary between the Mojave and Central Arizona provinces and their Pb isotopic compositions contain characteristics of both (Fig. 2). All of our samples plot below Stacey and Kramers' (1975) two-stage Pb evolution line on $^{207}Pb/^{204}Pb$ vs $^{206}Pb/^{204}Pb$, indicating that they are derived from sources that evolved with U/Pb ratios less than Stacey and Kramers' crustal average of 9·7 (Fig. 7). However, KS–HSR and CH samples plot at higher $^{207}Pb/^{204}Pb$ values, near the trend for Mojave province rocks. The MF samples plot along Stacey and Kramers' evolution curve for $^{208}Pb/^{204}Pb$ vs $^{206}Pb/^{204}Pb$, whereas the KS and CH samples plot above this model curve. The higher than average Th/U ratios modelled for the latter samples is characteristic of the Mojave province, whereas the MF samples are more similar to Proterozoic crust from southeastern Arizona (Wooden & Miller 1990).

The combined Sr, Nd, and Pb isotopic compositions of the western Arizona rhyolites do not point to a single crustal endmember; instead these data require that multiple crustal components and a basaltic component be involved in magma genesis. Figures 6 and 7 illustrate that mixes of basalt with three hypothetic crustal endmembers (A, B, and C; Table 3) can explain the isotopic variability that we observe. In this manner, the CH and MF–HSR represent lavas derived from sources that are mixtures of basalt with crustal endmember A, the KS–HSR and LSR lavas are formed from a source region that is a hybridisation of basalt and crustal endmember B, and the MF–LSR are lavas derived from a source that is a mix of basalt and crustal endmember C. Despite the complex processes of magma genesis and source region compositional variations that are indicated by the isotopic data, the mineralogical similarities of the lavas and their progressive chemical variations from the Basin and Range onto the Colorado Plateau (Moyer & Nealey 1989) suggests some underlying consistency to the origin of these rhyolites.

Importantly, rocks or xenoliths with the isotopic compositions of our three hypothetical endmembers have not been described in the literature. Suites of lower crustal xenoliths from the region for which data are available invariably have Pb isotopic compositions that are too radiogenic ($^{206}Pb/^{204}Pb > 18{\cdot}0$) to serve as suitable crustal endmembers (e.g. Camp Creek, AZ granulites and amphibolites (Esperança *et al.* 1988); Geronimo, AZ granulites and cumulates (Kempton *et al.* 1990); Piute Mountains, CA quartzofeldspathites (Hanchar, 1991)). Likewise, most Proterozoic and younger granitoids from Arizona and many from the eastern Mojave Desert have unsuitably radiogenic Pb (Wooden *et al.* 1988; Wooden & Miller 1991).

Although our simple models do not indicate involvement of known lower crustal sources, their contribution is not precluded in more complex models that utilise several endmembers. Considering the Proterozoic tectonic evolution of the region, which includes the opening and closing of a small ocean basin succeeded by large volume melting and plutonism (Anderson 1989), it is possible that appropriate crustal sources exist at depth that have not been sampled by subsequent volcanism. The model outlined below is, therefore, a conceptual framework within which to conduct future studies.

We are aware of only two crustal lithologies that can serve as suitable endmembers for our rhyolites. These are the Proterozoic granites with low $^{206}Pb/^{204}Pb$ ratios ($<17{\cdot}0$) that occur in the eastern Mojave Desert (e.g. granodioritic Fenner Gneiss, Wooden *et al.* 1988; Bender 1990) and the mafic gneisses found as xenoliths in Colorado Plateau lavas at Tule Tank, AZ ($^{206}Pb/^{204}Pb < 16{\cdot}5$; Nealey & Unruh in press; L.D. Nealey, pers. comm. 1991). Whereas the Fenner Gneiss has the Pb isotopic characteristics of the Mojave crustal province, the Tule Tank xenoliths plot along the trend for Proterozoic granites from the Arizona crustal province. Moreover, the Fenner Gneiss provides the radiogenic Sr component required for modelling the CH rhyolites, whereas the Tule Tank xenoliths have the unradiogenic Sr isotopic composition that is necessary for modelling the MF–LSR.

We suggest that our hypothetical endmembers are themselves "mixtures" of two crustal components that have the isotopic characteristics similar to those of the Fenner Gneiss and Tule Tank xenoliths. Because both have variable elemental compositions, mixing of the Fenner Gneiss and Tule Tank components forms a range of isotopic compositions that include those indicated by our rhyolite data. As illustrated, crustal endmember A contains a large proportion of Fenner Gneiss (90–95%), crustal endmember B is composed of approximately 60–65% Fenner Gneiss, whereas crustal endmember C has the smallest proportion of Fenner Gneiss (35–40%). Thus, the hypothesised crustal endmembers increase in Tule Tank component from the Basin and Range onto the Colorado Plateau.

Table 3 Mixing endmember compositions

	M	A	B	C
$^{87}Sr/^{86}Sr$	0·7042	0·7200	0·7090	0·7050
ε_{Nd}	+2·5	−19·5	−19·0	−18·8
$^{206}Pb/^{204}Pb$	18·43	16·60	16·36	16·30
$^{207}Pb/^{204}Pb$	15·55	15·44	15·36	15·33
$^{208}Pb/^{204}Pb$	38·35	37·50	36·50	36·30
Sr (ppm)	500	620	650	675
Nd (ppm)	30	55	70	80
Pb (ppm)	5	12	15	17

We caution that the simple models utilised herein generalise the petrogenetic conditions and probably do not portray accurately the details of magma genesis. Importantly, we point out that the proportion of basalt mixed with crustal material in our models is very high (50–85%) and is at odds with the T_{DM} and Pb isotopic data discussed previously. If the basalt proportions are correct, then the primary magmas for rhyolites from all fields are mafic andesites (51–55 wt% SiO_2). The absence of lavas with silica contents between 54 and 67 wt% SiO_2 argues against this condition and suggests more complex models such as AFC or the involvement of additional, possibly lower crustal, sources. Detailed modelling of this type is beyond the scope of this paper, however.

Fractional crystallisation certainly plays an important role in the genesis of the HSR lavas from all fields as evidenced by their extremely low concentrations of compatible elements and high abundances of incompatible elements (Moyer & Esperança 1989a; Moyer & Nealey 1989). Whereas the HSR from KS probably formed from LSR parent magmas, the isotopic data indicate that the same is not true at MF unless large quantities of crust were also assimilated during the transformation. Since, due to their low Sr contents, the $^{87}Sr/^{86}Sr$ ratios of the HSR lavas can be easily altered by contamination, it is conceivable that the difference between the KS and CH HSR is merely a result of greater assimilation of radiogenic upper crust by the CH magmas.

Our present working model agrees with the petrological characteristics of the western Arizona rhyolites from which we interpret their crustal source components as meta-igneous. In the absence of extreme differentiation, the preponderance of hornblende and pyroxene in the LSR and some "primitive" HSR samples suggests that the lavas were derived from a source that contains hornblende (Fenner Gneiss) and possibly pyroxene (Tule Tank). In this aspect, amphibolitic rocks, such as the xenoliths of Camp Creek (Esperança *et al.* 1988) and the metavolcanics near Bagdad (Anderson *et al.* 1955) are also attractive candidates. Moreover, the metaluminous to weakly peraluminous character of the rhyolites taken in concert with the absence of aluminous phases such as cordierite argues against a major contribution from a metasedimentary source component.

The idea that two crustal sources contribute in varying proportions to the western Arizona rhyolites is appealing for many reasons. Most importantly, it accommodates the notion that there is an underlying consistency to the petrogenesis of the western Arizona rhyolites. Moreover, it explains the transition from Mojave province Pb isotopic ratios in the CH field to more Arizona-like values in the MF field and supports the transitional nature of the crust in western Arizona. This model also predicts higher Ba, Sr, and Mg values and lower Rb, U, and incompatible element abundances for rocks developed from sources that contain the greatest proportion of depleted, mafic Tule Tank component. Moyer and Nealey (1989) described exactly these relations for LSR lavas from these fields.

Finally, our working model indicates that the CH and KS

rhyolites require only a single crustal source, whereas two isotopically distinct crustal sources are required to explain the MF lavas. This may suggest that melting of the comparatively thick crust of the Colorado Plateau occurred at different levels, perhaps as basalt ponded at various locations dictated by density changes within the crust (e.g. Glazner & Ussler 1989). Alternatively, it may indicate that the crust beneath the Basin and Range has become more homogenised as a result of the ductile stretching that accompanied crustal extension (e.g. Gans 1987; McCarthy *et al.* 1991). In either case, basalt input into the crust to form the rhyolites seems an inescapable conclusion, although the amount of basalt required remains an open question.

5. Conclusions

Our study of the petrological and geochemical characteristics of rhyolites from three bimodal fields in Arizona located in the Basin and Range, Transition Zone, and the Colorado Plateau provinces leads to the following conclusions: (1) Both HSR and LSR are products of partial melting of crustal materials with an addition of mafic melts. (2) Two crustal sources can be identified from their distinct imprints on the rhyolites: (a) crust characterised by unradiogenic Sr and Pb isotopic compositions that is exemplified by the Tule Tank mafic gneiss xenoliths, and (b) crust with more radiogenic Pb and Sr isotopic compositions that is exemplified by the Fenner Gneiss. (3) There is an apparent increase in the contribution of the depleted Tule Tank crustal source from the Basin and Range onto the Colorado Plateau. (4) The Castaneda Hills and Kaiser Spring rhyolites require only one crustal source, whereas the Mount Floyd rhyolites require two isotopically distinct sources.

6. Acknowledgements

This work was supported by the collaborative NSF grants EAR-8904688 to SE and EAR-8904445 to TCM. We are indebted to the staff at DTM (Carnegie Institution) for allowing SE's use of their research facilities. Dave Nealey and Dan Unruh (US Geological Survey) graciously shared their data and preprints of their work on the Tule Tank xenoliths. We also thank Joe Wooden and Calvin Miller for helpful discussions of the Proterozoic and younger granites of the region. The paper was improved as a result of two anonymous reviews.

7. Appendix 1—Sample descriptions

CH-89-4 Aphyric rhyolite (34°23′15″N 113°47′05″W). Upper flow (of three) on Aubrey Peak Ridge. Cut by listric normal faults. *Phenocrysts*: ($\ll$1 vol%) qtz $\gg$ opaq > plag > K-spar. Subhedral to anhedral (embayed) quartz, sized to 1 mm; subhedral, unzoned plagioclase, <0·2 mm; subhedral, unzoned alkali feldspar, <0·2 mm; subhedral Fe-Ti oxide, common with zircon, <0·3 mm. *Accessory*: biotite, zircon. No xenocrysts. *Groundmass*: clear, slightly pumiceous glass with microphenocrysts of plag, bio, oxide. Abundant microlites, rare spherulites (<0·1 mm) and veins of devitrified material that are visible in hand specimen and were avoided in powders for geochemistry.

CH-90-4 Aphyric rhyolite (34°23′22″N 113°48′53″W). Aubrey Peaks region. *Phenocrysts* (<1vol%): qtz > opaq. Anhedral quartz, <0·5 mm; subhedral Fe-Ti oxide, <0·3 mm. *Accessory*: zircon. *Xenoliths*: altered, oxidised metagranitic assemblage up to 1 mm diameter. *Groundmass*: light brown glass with scattered microlites along vesicles. Few microphenocrysts of plagioclase.

CH-90-5 Biotite rhyolite (34°23′03″N 113°48′39″W). Prominent dome west of Aubrey Peak ridge. *Phenocrysts*: (2–5 vol%) qtz > K-spar > plag $\gg$ bio = opaq. Euhedral to anhedral, embayed quartz, sized to 1·25 mm; euhedral/subhedral, unzoned plagioclase, sized from 0·4–1·0 mm; euhedral to subhedral alkali feldspar, occurs with Fe-Ti oxide, sized to 1 mm; euhedral, yellowish brown to dark brown biotite, rare with oxides, <0·5 mm; subhedral Fe-Ti oxide, rarely found with apatite, <0·3 mm. *Accessory*: apatite. No xenocrysts. *Groundmass*: light brown glass with rare microphenocrysts of plag and few microlites. Some perlitic racks, veins, and tiny spherulites.

CH-90-13 Biotite rhyolite (34°20′07″N 113°50′58″W). Dome north of Aubrey Peaks. *Phenocrysts*: (1–3 vol%) qtz > K-spar ≥ plag. Euhedral to subhedral quartz, <0·75 mm; euhedral to subhedral, unzoned alkali feldspar, rarely with oxide, 0·2–0·5 mm; sharply euhedral to subhedral, unzoned to slightly zoned plagioclase, rarely with oxide, 0·2–0·75 mm; euhedral to subhedral biotite, dark olive brown to extremely dark brown, rarely with oxides, <0·5 mm; subhedral Fe-Ti oxide rarely with zircon, K-spar, plag, <0·3 mm. *Accessory*: allanite, zircon. No xenocrysts. *Groundmass*: clear, pumiceous/vesicular glass with microphenocrysts of plag, bio, opaq, allanite. Microlites are somewhat aligned by flow.

CH-90-16 Rhyolite (34°28′56″N 113°49′29″W). Dome south of Aubrey Peaks. *Phenocrysts*: (2–5 vol%) qtz = plag = K-spar $\gg$ opaq. Euhedral to anhedral quartz, 0·2–0·7 mm; euhedral to subhedral, unzoned plagioclase, rarely with oxide, 0·2–0·6 mm; euhedral to subhedral, unzoned alkali feldspar, 0·2–0·6 mm; subhedral Fe-Ti oxide, rarely with zircon, <0·25 mm. *Accessory*: zircon. No xenocrysts. *Groundmass*: clear, with microphenocrysts of plag, bio, opaq. Rare microlites, some perlitic fractures.

CH-90-21 Biotite-Hornblende Rhyolite (34°27′09″N 113°56′35″W). Faulted lava flow in the Castaneda Hills. Same unit as CHNW-155 collected by Suneson (1980). *Phenocrysts*: (15–20 vol%) plag = Kspar = qtz $\gg$ hbl. Euhedral to subhedral, slightly to moderately zoned plagioclase, occurs with bio, k-spar, oxide, zirc, hbld, 0·3–1·5 mm; euhedral to subhedral, unzoned alkali feldspar, occurs with plag, bio, oxide, 0·2–0·8 mm; anhedral to embayed, unzoned alkali feldspar, with bio, oxide, rare allanite, 0·7–1·5 mm; euhedral to subhedral, olive-brown to reddish brown biotite, occurs with k-spar, plag, oxide, zirc, apat, allanite, 0·2–1·25 mm, most < 0·6 mm; euhedral olive-brown to honey brown hornblende, occurs with plag, oxide, rare zirc, 0·3–1·0 mm; subhedral to anhedral, embayed quartz, rarely occurs with bio, oxide, sized to 2·0 mm; subhedral Fe-Ti oxide, occurs with any phase, <0·3 mm. *Accessory*: zircon (up to 0·15 mm), apatite, monazite, allanite. Note: this is the most accessory-rich sample found in this study. *Xenocrysts*: K-spar (?), qtz (?). *Groundmass*: clear, pumiceous glass, aligned microlites, equant opaques.

CH-90-33 Biotite rhyolite (34°25′59″N 113°54′06″W). Dome adjacent to Castaneda Well. *Phenocrysts*: (15–20 vol%) K-spar = qtz > plag $\gg$ bio. Euhedral to anhedral, embayed quartz, 0·2–1·5 mm; euhedral to subhedral, weakly zoned alkali feldspar, occurs with plag, bio, sized to 1·5 mm; euhedral to subhedral, weakly to moderately zoned plagioclase, occurs with k-spar, bio, rare oxide, sized to

1·5 mm; euhedral to subhedral, reddish-brown to dark brown biotite, occurs with oxide, k-spar, plag, 0·2–0·5 mm; euhedral to subhedral Fe-Ti oxide, occurs with zirc, apat, <0·3 mm. *Accessory*: allanite (common), rare zircon, apatite. *Groundmass*: clear glass, with few microlites. Vesicular with cristobalite common in cavities. Spherulites and devitrification around vesicle walls and some crystal boundaries.

BC-4 Aphyric HSR (34°38′24″N 113°21′26″W) Negro Ed dome. This dome contains vapour phase garnet and topaz. Refer to Moyer (1986) for sample description.

KS-44 Aphyric HSR (34°36′07″N 113°25′38″W). This dome and flow contain vapour-phase garnet. Refer to Moyer (1986) for sample description.

EM-6 Biotite-Hornblende HSR (34°40′03″N 113°24′47″W). Dome on eastern side of Elephant Mountains. Refer to Moyer (1986) for sample description.

EM-21 Biotite-Hornblende HSR (34°39′12″N 113°27′17″W). Flow that extends SW from the Elephant Mountains. Refer to Moyer (1986) for sample description.

EM-33 Aphyric HSR (34°37′48″N 113°27′06″W). Lava flow dismembered by extensive normal faulting. Refer to Moyer (1986) for sample description.

EM-80 Hornblende-Orthopyroxene-Biotite LSR (34°40′07″N 113°28′24″W). Refer to Moyer & Esperança (1989) for sample description and elemental analysis.

EM-83 Hornblende-Orthopyroxene-Clinopyroxene LSR (34°40′19″N 113°28′00″W). Lava flow that underlies Neale Mesa LSR. Refer to Moyer (1986) for sample description.

EM-84 Hornblende-Orthopyroxene LSR (34°40′31″N 113°28′02″W). Neale Mesa dome and lava flow. One of the youngest units in the northwestern part of the volcanic field. Refer to Moyer (1986) for sample description.

MF-1 Biotite-Hornblende LSR (35°20′28″N 112°43′30″W). Mt Rushmore unit of Nealey (1980). *Phenocrysts*: plag > hbl = bio > opaq. Euhedral plagioclase, moderately zoned, sized to 1 mm; euhedral hornblende, broken, olive green, present with oxides and plag, sized to 0·75 mm; euhedral biotite, brown to reddish brown, present with plag, less common Fe-Ti oxide, and rare apatite, sized to 1 mm. *Accessory*; apatite, zircon. *Xenocrysts*: qtz, plag (spongy). *Groundmass*: abundant microphenocrysts in clear, microlite-poor glass.

MF-4 Hornblende-Orthopyroxene LSR (35°22′25″N 112°43′28″W). Base of Radio Tower hill. *Phenocrysts*: (20–25 vol%) plag ≥ hbl > opx > opaq. Euhedral plagioclase, moderately zoned to 1·5 mm; euhedral to subhedral hornblende, broken and occasionally hollow, olive green to reddish brown, present with plag, Fe-Ti oxide, rarely apatite, some grains enclose opx, sized to 1·5 mm; euhedral/subhedral orthopyroxene, pale green to pink, with oxide, rare apatite, some grains reacting to hornblende, sized to 0·5 mm. *Accessory*: apatite, zircon. *Xenocrysts*: plag (spongy). *Groundmass*: clear, microlite-rich glass, local incipient devitrification.

MF-6 Biotite HSR (35°22′59′N 112°41′31″W). York dome unit of Nealey (1980); same unit as sample NMF 86-11 of Moyer & Nealey (1989). *Phenocrysts*: (5–10 vol%) plag = K-spar > qtz ≫ bio ≥ opaq. Euhedral/subhedral plagioclase, slightly zoned, present with K-spar, sized to 1·5 mm; euhedral to subhedral alkali-feldspar, unzoned, present with plag, sized to 1.5 mm; subhedral to anhedral/embayed quartz, as isolated crystals, to 1 mm; euhedral biotite, light brown to honey brown, to 0·25 mm. *Accessory*: apatite, rare zircon. *Xenocrysts*: quartz?. *Groundmass*: clear, pumiceous glass with scarce microlites.

MF-8 Biotite HSR (35°27′06″N 112°44′00″W). Round Mountain dome; same unit as sample NMF86-13 of Moyer & Nealey (1989). *Phenocrysts*: (about 15 vol%) K-spar ≥ plag > qtz ≫ bio ≥ opaq. Euhedral to subhedral alkali-feldspar, crystal edge fractured to a depth of 0·05 mm, present with plag and rarely quartz and biotite, sized to 1 mm; euhedral/subhedral plagioclase, slightly zoned, present with K-spar and rarely bio and qtz, sized to 1 mm; subhedral to anhedral/embayed quartz, commonly as single crystals, sized to 1 mm; euhedral biotite, very dark brown to reddish brown, present with K-spar, plag, qtz, sized to 0·5 mm. *Accessory*: rare apatite, allanite?. *Xenocrysts*: none. *Groundmass*: clear, vesicular glass with scattered mirolites.

MF-10 Aphyric HSR (35°5′24″N 112°43′36′W). Smith Tank/Partridge Creek unit of Nealey (1980); same unit as sample NMF86-14 of Moyer & Nealey (1989). Only two crystals observed: brown euhedral hornblende, 0·25 mm; anhedral granular quartz, 1·5 mm. No accessories or xenocrysts seen. *Groundmass*: clear glass with abundant microphenocrysts of plag (<0·05 mm) and oxides, many microlites.

MF-14 Hornblende-Orthopyroxene LSR (35°23′32″N 112°42′14″W). Mt Floyd dome; same unit as sample NMF86-5 of Moyer & Nealey (1989). *Phenocrysts*: (20–25 vol%) plag > hbl > opx > opaq. Euhedral plagioclase, moderately zoned, present with hbl, opx, or oxide, generally <1 mm; euhedral to subhedral hornblende, broken and occasionally hollow, olive green to olive brown, present with plag, Fe-Ti oxide, rarely >1 mm; euhedral/subhedral orthopyroxene, clear to pale green, present with oxide, rare apatite, rarely surrounded by hornblende, sized from 0·2–0·4 mm. *Accessory*; apatite, zircon. *Xenocrysts*: plag (spongy). *Groundmass*: clear, microlite-rich glass, local incipient devitrification.

MF-15 & MF-17 Aphyric HSR/LSR (35°27′09″N 112°38′00″W). Presley Tank unit of Nealey (unpublished mapping). No phenocrysts. *Accessories*: very rare zircon, very tiny cpx?, very rare apatite. *Xenocrysts*: qtz (rounded), plag (rounded/embayed), opx (anhedral), oliv (rounded with partial opx overgrowth). *Groundmass*: clear to brown glass with skeletal plag, K-spar, and abundant microlites.

MF-21 Hornblende-Biotite LSR (35°13′55″N 112°44′12″W). Picacho Butte; same unit as sample NPB of Moyer & Nealey (1989). *Phenocrysts*: (10–20 vol%) hbl > plag ≫ bio ≥ opaq. Euhedral to subhedral (hollow core) hornblende, yellowish to olive green to reddish brown, present with plag, sized to 1·75 mm; euhedral/subhedral plagioclase, moderately zoned, present with hbl, oxide, sized to 1 mm; euhedral biotite, olive brown, present with oxide, sized <0·6 mm; euhedral equant Fe-Ti oxide, present with zircon, apatite, sized <0·3 mm. *Accessory*: apatite, zircon. *Xenocrysts*: qtz, plag (spongy). *Groundmass*: clear glass with microphenocrysts of plag (swallowtail), bio, oxide, hbl (?) and aligned microlites. Locally devitrified with clay (?) alteration.

8. References

Anderson, C. A., Scholz, E. A. & Strobell, J. D. Jr. 1955. Geology and ore deposits of the Bagdad area, Yavapai County, Arizona. US GEOL SURV PROF PAP **278**.

Anderson, P. 1979. Proterozoic plate tectonic evolution of Arizona. ARIZONA GEOL SOC DIG **17,** 17–55.

Bender, E. E. 1990. *Geology of the Fenner Gneiss, Piute and Old Woman Mountains, San Bernadino County, California.* M.Sc. thesis, Vanderbilt University, Tennessee.

Bennett, V. C. & DePaolo, D. J. 1987. Proterozoic crustal history of the western United States as determined by neodymium isotopic mapping. GEOL SOC AM BULL **99,** 674–85.

Burt, D. M., Moyer, T. C. & Christiansen, E. H. 1981. Garnet- and topaz-bearing rhyolites from near Burro Creek, Mohave County, Arizona—Possible exploration significance. ARIZONA GEOL SOC DIG **13,** 1–4.

Catanzaro, E. J., Murphy, T. J., Shields, W. R. & Garner, E. L. 1968. Absolute isotopic ratios of common equal-atom and radiogenic lead isotopic standards. J RES NAT BUR STAND SECT A **72,** 261–7

Condie, K. C. 1981. Precambrian rocks of the southwestern United States and adjacent areas of Mexico. NEW MEXICO BUR MINES MINER RESOUR RESOUR MAP 13, Scale 1:1,000,000.

Ebadi, J. & Johannes, W. 1991. Beginning of melting and first composition of melts in the system Qz-Ab-Or-H_2O-CO_2. CONTRIB MINERAL PETROL **106,** 286–95.

Esperança, S., Carlson, R. W. & Shirey, S. B. 1988. Lower crustal evolution under central Arizona: Sr, Nd and Pb isotopic and geochemical evidence from the mafic xenoliths of Camp Creek. EARTH PLANET SCI LETT **90,** 26–40.

Esperança, S., Carlson, R. W. & Shirey, S. B. 1990. Reply to "Comment on: Lower crustal evolution under central Arizona: Sr, Nd and Pb isotopic and geochemical evidence from the mafic xenoliths of Camp Creek" by C.M. Johnson. EARTH PLANET SCI LETT **99,** 406–9.

Gans, P. B. 1987. An open-system, two-layer stretching model for the eastern Great Basin. TECTONICS **6,** 1–12.

Glazner, A. F. & Ussler, W. III, 1989. Crustal extension, crustal density, and the evolution of Cenozoic magmatism in the Basin and Range of the western United States, J GEOPHYS RES **94,** 7952–60.

Goff, F. E., Eddy, A. C. & Arney, B. H. 1983. Reconnaissance geologic strip map from Kingman to south of Bill Williams Mountain, Arizona. LOS ALAMOS NAT LAB PUB LA-9202-MAP.

Hanchar, J. M. 1991. *Lower crustal xenoliths in Tertiary dikes, Old Woman Mountains area, southeastern California.* M.Sc. Thesis, Vanderbilt University, Tennessee.

Hart, S. R. 1984. A large-scale isotope anomaly in the southern hemisphere mantle. NATURE **309,** 753–7.

Hauser, E. C., Gephart, J., Latham, T., Oliver, J., Kaufman, S., Brown, L. & Lucchitta, I. 1987. COCORP Arizona transect: Strong crustal reflections and offset Moho beneath the transition zone. GEOLOGY **15,** 1103–6.

Huntoon, P. W. & Sears, J. W. 1975. Bright Angel and Eminence faults, eastern Grand Canyon, Arizona. GEOL SOC AM BULL **86,** 465–72.

Johnson, C. M. 1990. Comment on "Lower crustal evolution under central Arizona: Sr, Nd and Pb isotopic and geochemical evidence from the mafic xenoliths of Camp Creek" by Esperança, S., Carlson, R. W. and Shirey, S. B. EARTH PLANET SCI LETT **99,** 400–5.

Kempton, P. D., Harmon, R. S., Hawkesworth, C. J. & Moorbath, S. 1990. Petrology and geochemistry of lower crustal granulites from the Geronimo Volcanic Field, southeastern Arizona. GEOCHIM COSMOCHIM ACTA **54,** 3401–26.

Lucchitta, I. & Suneson, N. H. 1985. Geologic map of the Artillery Peak Northwest quadrangle, Mohave County, Arizona. US GEOL SURV OPEN-FILE REP OF-85-277.

Lucchitta, I. & Suneson, N. H. 1988. Geologic map of the Planet 2 SW quadrangle, Mohave County, Arizona. US GEOL SURV OPEN-FILE REP OF-88-547.

Lucchitta, I. & Suneson, N. H. 1989. Geologic map of the Planet 2 NW quadrangle, Mohave County, Arizona. US GEOL SURV OPEN-FILE REP OF-89-673

Luth, W. C., Jahns, R. H. & Tuttle, O. F. 1964. The granite system at pressures of 4 to 10 kilobars. J GEOPHYS RES **69,** 759–73.

Manhes, G., Minster, J. F. & Allegre, C. J. 1978. Comparative uranium-thorium-lead and rubidium-strontium study of the Saint Severin amphoterite: Consequences for early solar system chronology. EARTH PLANET SCI LETT **39,** 14–24.

McCarthy, J. & Thompson, G. A., 1988. Seismic imaging of extended crust with emphasis on the western United States. GEOL SOC AM BULL **100,** 1361–74.

McCarthy, J., Larkin, S. P., Fuis, G. S., Simpson, R. W. & Howard, K. A. 1991. Anatomy of a metamorphic core complex: Seismic refraction/wide-angle reflection profiling in southeastern California and western Arizona. J. GEOPHYS RES **96,** 12,259–92.

Moyer, T. C. 1986. *The Pliocene Kaiser Spring (AZ) bimodal volcanic field: Geology, geochemistry, and petrogenesis.* Ph.D. Dissertation, Arizona State University Tempe.

Moyer, T. C. 1990. Generalized geologic map of the Kaiser Spring volcanic field, Mohave County, Arizona. ARIZONA GEOL SURV CONTRIB MAP CM-90-C, SCALE 1:50,000.

Moyer, T. C. & Esperança, S. 1989a. Geochemical and isotopic variations in a bimodal magma system; The Kaiser Spring volcanic field, Arizona. J GEOPHYS RES **94,** 7841–59.

Moyer, T. C. & Esperança, S. 1989b. Petrologic and geochemical variations in basalts associated with bimodal volcanism in the Arizona transition zone: Preliminary results from Kaiser Spring. NEW MEXICO BUR MINES MINERAL RESOUR BULL **131,** 198.

Moyer, T. C., & Nealey, L. D. 1989. Regional compositional variations of late Tertiary bimodal rhyolite lavas across the Basin and Range/Colorado Plateau boundary in western Arizona. J GEOPHYS RES **94,** 7799–816.

Nealey, L. D. 1980. *Geology of Mount Floyd and vicinity, Coconino County, Arizona.* M.S. Thesis, North. Arizona University, Flagstaff.

Nealey, L. D. & Sheridan, M. F. 1991. Post-Laramide volcanic rocks of Arizona and northern Sonora, Mexico, and their inclusions. ARIZONA GEOL SOC DIG **17,** 609–47.

Nealey, L. D. & Unruh, D. M. in press. Geochemistry and isotopic characteristics of deep crustal xenoliths from Tule Tank, San Francisco Volcanic Field, Northern Arizona. ARIZONA GEOL SOC GUIDE: *The Proterozoic of Arizona.*

Shackelford, T. J. 1980. Tertiary tectonic denudation of a Mesozoic-early Tertiary(?) gneiss complex, Rawhide Mountains, Arizona. GEOLOGY **8,** 190–4.

Sherrod, D. R. 1988. Preliminary geologic map of the Monkeys Head quadrangle, Mohave and La Paz Counties, Arizona. US GEOL SURV OPEN-FILE REP OF-88-597.

Shoemaker, E. M., Squires, R. L. & Abrams, M. J. 1978. Bright Angel and Mesa Butte fault systems of northern Arizona. In: Smith, R. B. & Eaton, G. P., Cenozoic *Tectonics and Regional Geophysics of the Western Cordillera,* 341–68. MEM GEOL SOC AM **152**.

Stacey, J. S. & Kramers, J. D. 1975. Approximation of terrestrial lead isotope evolution by a two-stage model. EARTH PLANET SCI LETT **26,** 207–21.

Suneson, N. H. 1980. *The origin of bimodal volcanism, west-central Arizona.* Ph.D. Dissertation, University of California, Santa Barbara.

Suneson, N. H. & Lucchitta, I. 1979. K/Ar ages of Cenozoic volcanic rocks, west-central Arizona. ISOCHRON WEST **24,** 25–9.

Suneson, N. H. & Lucchitta, I. 1983. Origin of bimodal volcanism, southern Basin and Range province, west-central Arizona. GEOL SOC AM BULL **94,** 1005–19.

Thompson, R. N., Morrison, M. A., Hendry, G. L. & Parry, S. J. 1984. An assessment of the relative roles of crust and mantle in magma genesis: An elemental approach. PHILOS TRANS R SOC LONDON **A310,** 549–90.

Tuttle, O. F. & Bowen, N. L. 1958. Origin of granite in the light of experimental studies in the system $NaAlSi_3O_8$-$KAlSi_3O_8$-SiO_2-H_2O. GEOL SOC AM MEM **74**.

Wernicke, B., 1981. Low-angle normal faults in the Basin and Range province–Nappe tectonics in an extending orogen. NATURE **291**, 645–8.

Wernicke, B. 1985. Uniform-sense simple shear of the continental lithosphere. CAN J EARTH SCI **22,** 108–25.

Wooden, J. L. & Miller, D. M. 1990. Chronologic and isotopic framework for early Proterozoic crustal evolution in the eastern Mojave Desert region, SE California. J GEOPHYS RES **95,** 20, 133–46.

Wooden, J. L., Stacey, J. S., Doe, B., Howard, K. A. & Miller, D. M. 1988. Pb isotopic evidence for the formation of Proterozoic crust in the southwestern United States. In Ernst, W. (ed.) *Metamorphism and crustal evolution of the western United States, Rubey Volume III,* 68–86. Englewood Cliffs, NJ: Prentice-Hall.

SONIA ESPERANÇA, Department of Heritage and Resource Management, Victoria College-Rusden, Clayton, Victoria 3168, Australia.
THOMAS C. MOYER, Department of Geology, Vanderbilt University, Nashville, TN 37235, U.S.A.

Transactions of the Royal Society of Edinburgh: Earth Sciences, **83,** 361–375, 1992

The Papoose Flat Pluton of eastern California: a reassessment of its emplacement history in the light of new microstructural and crystallographic fabric observations

R. D. Law, S. S. Morgan, M. Casey, A. G. Sylvester and M. Nyman

ABSTRACT: One of the most outstanding apparent examples in N America of a forcibly emplaced pluton is the Papoose Flat Pluton of eastern California. Sideways expansion of this granitic pluton, during emplacement into a series of Cambrian shelf strata, has been regarded by early workers as resulting in the observed intense crystal plastic deformation of the pluton's mylonitic border facies and surrounding country rocks. This deformation is evidenced by up to 90% thinning of individual stratigraphic layers within the pluton's metamorphic aureole, although such intense penetrative deformation of the country rocks is not observed outside the aureole.

Previously published quartz **c**-axis fabrics associated with this deformation (and presented on projection planes oriented perpendicular to lineation) were interpreted as being symmetrical with respect to foliation and lineation, implying almost coaxial deformation histories. Such fabrics could be interpreted as indicating that the pluton evolved by "ballooning" as a result of new magma being intruded into its core during emplacement. However, a major problem with applying the strict ballooning model to the Papoose Flat Pluton is that while oblate strains would be expected to develop in association with a ballooning mechanism, the mylonitic rocks of this elongate WNW–ESE-trending pluton and its aureole are characterised by both a strongly developed foliation, which is concordant with the pluton's margin, and an intense, NW–SE trending, shallow plunging stretching lineation.

Previously published fabrics from the Papoose Flat Pluton and its metamorphic aureole have been rotated on to a projection plane oriented parallel to lineation and perpendicular to foliation. Examination of the fabrics in this projection plane has revealed that they are in fact dominantly asymmetric, and that a constant sense of asymmetry is detected across the pluton, suggesting a consistent (top-to-the-SE) shear-sense. This new interpretation is strongly supported by microstructural and petrofabric analysis of additional L-S tectonites collected, during recent fieldwork, from both the aureole and quartz veins within the pluton's gneissic border facies. Thus mylonite formation around the Papoose Flat Pluton could have involved large-scale consistently oriented translation and associated shearing, rather than passive "blister-like" coaxial deformation associated with pluton ballooning. It should be noted that mylonitic deformation is restricted to the western half of the pluton, features indicative of a more "permitted" emplacement mechanism being found in the eastern portion of the pluton.

The detected top-to-the-SE shear-sense could be interpreted as indicating that the granitic material forming the western part of the pluton was forcibly intruded in a northwestward direction from the pluton source as a nearly solidified wedge beneath a static cover of sedimentary rocks. Alternatively, the detected shear sense could also be interpreted as indicating SE-directed thrusting of the cover rocks over the underlying pluton, the western margin of the pluton suffering intense mylonitic deformation, while the eastern margin was located in a "stress-shadow" region. If this alternative interpretation is correct, then the deformation temperatures indicated by the pattern of quartz **c**-axis fabrics dictate that thrusting must either be synchronous with pluton emplacement, or at least have commenced during the early stages of pluton cooling.

KEY WORDS: calcite, contact metamorphism, emplacement mechanisms, granite, mylonite, quartz, shear sense, texture goniometry.

In recent years there has been a resurgence of interest in the mechanisms and timing of pluton emplacement relative to regional deformation (e.g. Brun & Pons 1981; Hutton 1982; Bateman 1984; Castro 1986; Hollister & Crawford 1986; Courrioux 1987; Guineberteau *et al.* 1987; Hibbard 1987; Hutton 1988a,b; Paterson & Tobisch 1988; Paterson *et al.* 1989; Ramsay 1989; Brun *et al.* 1990; Paterson *et al.* 1991a,b). Pluton emplacement mechanisms have traditionally been divided into two basic types: forcible emplacement, including doming, diapirism and ballooning; and "permitted" emplacement, involving stoping, cauldron subsidence and sheeting. Hutton (1988a) argued that the critical factor controlling the emplacement mechanism of an individual pluton is the relationship between the rate of buoyant uprise of magma and the rate of tectonic cavity opening. For forcible emplacement, the rate of magma uprise is greater than the rate of tectonic cavity opening, whereas for permitted intrusions, the rate of cavity opening is greater than the rate of magma uprise.

The spectacularly exposed Papoose Flat Pluton of eastern California (Fig. 1) has been regarded as a classical example of a forcibly emplaced pluton (Sylvester *et al.* 1978). On palinspastic reconstruction (Nelson *et al.* 1977) the pluton is roughly elliptical-rectangular in plan view. Sideways expansion of the granitic pluton, during emplacement into a series of Cambrian shelf strata, was regarded by Sylvester *et al.* (1978) as resulting in the intense crystal plastic deformation of the western half of the pluton's margin and surrounding country rocks. This localised (or blister-like) deformation of the country rock (Sylvester *et al.* 1978) around the concordantly intruded western part of the pluton caused up to 90% thinning of individual stratigraphic layers. Mylonitic foliation within the deformation zone defines a concentric pattern about the pluton and is parallel to both bedding and the pluton walls. The grain shape stretching lineation lying within the foliation (Fig. 2) is horizontal-gently plunging and, unlike the concentric foliation pattern, is oriented sub-parallel to the long axis of the pluton on palinspastic reconstruction (Nelson *et al.* 1977). Penetrative deformation associated with crystal plasticity is not observed outside the pluton's metamorphic aurole.

As pointed out by Ramsay (1989, p. 203), a pluton expanding ("ballooning") as a result of new magma being intruded into the pluton core should be characterised by oblate shape fabrics in the outer (previously solidified) parts of the pluton and adjacent country rocks. The presence of a strong linear element in the grain shape fabric associated with the mylonite in the aureole of the Papoose Flat Pluton is, therefore, difficult to reconcile with a ballooning mechanism of emplacement for the pluton.

It could, of course, be argued that the grain shape lineation in these tectonites is not a true stretching lineation, but has been produced by the superposition of several deformation phases. This alternative explanation may be readily tested through examination of the **c**-axis pattern of preferred orientation in the quartz-rich tectonites. In crystal-plastic deformation of quartz aggregates, a clear relationship has been documented between strain symmetry and **c**-axis fabric patterns (see reviews by Schmid & Casey 1986; Law 1990). While plane strain tectonites are characterised by crossed-girdle fabrics, true oblate strains are associated with small-circle girdles of **c**-axes centred about the pole of mylonitic foliation (Fig. 3). In this paper, we demonstrate that in all quartz-rich tectonites examined from the Papoose Flat Pluton and its aureole, the grain shape lineation is a true stretching lineation.

Possible supporting evidence for a ballooning emplacement mechanism for the Papoose Flat Pluton was originally thought to be provided by the crystallographic fabrics present within the marble and quartzite of the pluton's metamorphic aureole. The **c**-axis fabrics originally measured by Sylvester and Christie (1968), and Sylvester (1969) within these plastically deformed and dynamically recrystallised rocks were regarded as being almost symmetrically oriented with respect to foliation and lineation. By analogy with recent experimental and numerical fabric studies (e.g. Tullis *et al.* 1973; Lister & Hobbs 1980; Wenk *et al.* 1987; Jessell 1988) the apparently symmetric cross-girdle quartz and calcite **c**-axis fabrics described by Sylvester and Christie (1968) and Sylvester (1969) could be interpreted as indicating almost coaxial (i.e. pure shear) deformation histories (Fig. 3). This interpretation would tend to support, at least in general terms, the pluton emplacement model proposed by Sylvester *et al.* (1978) in which the country rocks deformed coaxially as a passive "blister" on the

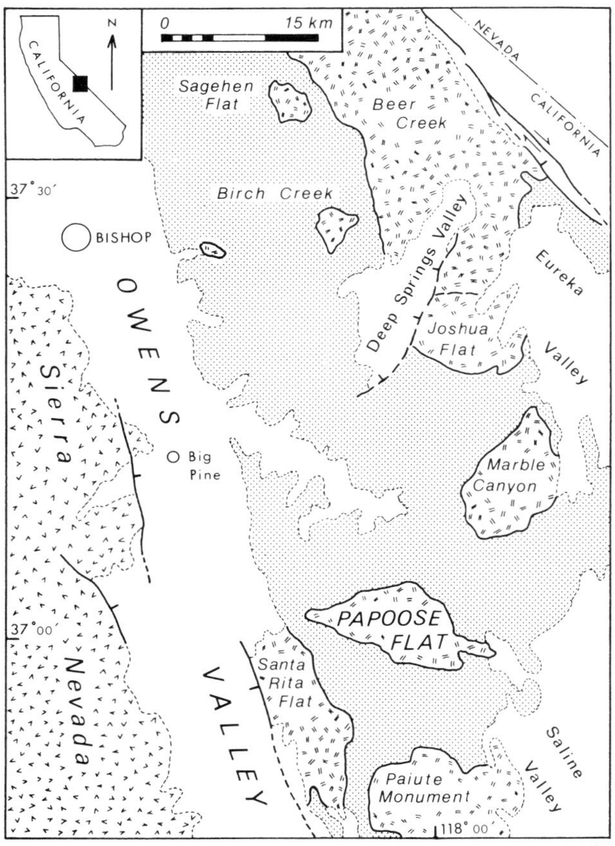

Figure 1 Regional geological setting of the Papoose Flat Pluton and other plutons of the White-Inyo Range (adapted from Sylvester *et al.* 1978); stippled pattern = pre-plutonic rocks; v-shaped pattern = granitic rocks of the Sierra Nevada, no pattern = Cainozoic volcanic rocks and alluvium; pluton ages: Papoose Flat = 75–81 Ma, Birch Creek = 77–80 Ma, Sagehen Flat = 130–138 Ma, Beer Creek = 150–171 Ma, Josuha Flat = 153–184 Ma, Santa Rita Flat = 156–160 Ma, Paiute Monument 160–190 Ma, Marble Canyon—undated (see Sylvester *et al.* 1978 for details).

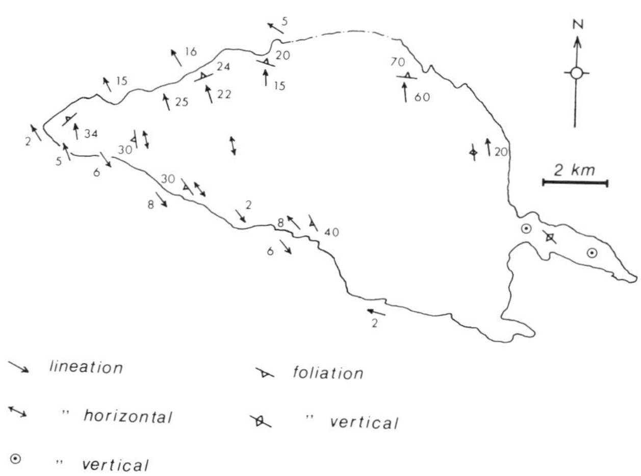

Figure 2 Schematic map of mylonitic foliation and lineation in the Papoose Flat Pluton and its metamorphic aureole; simplified from Neson *et al.* (1977).

surface of a sidways-expanding pluton, although deformation is clearly associated with plane strain rather than oblate strains.

The quartz **c**-axis fabrics described by Sylvester (1966), Sylvester and Christie (1968) and Sylvester (1969) were displayed on projection planes oriented perpendicular to both foliation and lineation. On such projection planes, typical **c**-axis fabrics (see review by Price 1985) plot around the periphery of the pole figure and are easily misinterpreted. Replotting of the original quartz **c**-axis fabric data described by Sylvester on projection planes oriented perpendicular to foliation and parallel to lineation has revealed (Law *et al.* 1990) that the majority of fabrics are in fact assymmetric with respect to foliation and lineation (Fig. 4). By analogy with numerical fabric studies (e.g. Lister & Hobbs 1980; Etchecopar & Vasseur 1987) this asymmetrical relationship is considered to be indicative of non-coaxial (approximately simple shear) strain histories (Fig. 3). Further, a consistent sense of fabric asymmetry is detected across the pluton (Fig. 4), suggesting a constant shear sense. Viewed toward the N and E, a dominantly dextral shear sense is indicated by these fabrics (cf. Figs 3, 4). This top-to-the-SE shear sense has also recently been detected over the pluton by S. R. Paterson and his co-workers (Godin & Paterson 1991; Paterson *et al.* 1991a) using microstructures observed within the pluton and its aureole rocks.

As independently outlined by both Law *et al.* (1990) and Paterson *et al.* (1991a), therefore, it is possible that pluton evolution and deformation could have involved large-scale, consistently oriented, translation associated with top-to-the-SE shearing rather than passive "blister-like" coaxial deformation associated with an expanding pluton.

In the present paper, a preliminary report is given on the microstructures and crystallographic fabrics contained within a new suite of samples collected from the Papoose Flat Pluton and its aureole. These deformation features are interpreted in the light of recent advances in our understanding of microstructural and crystal fabric evolution (see reviews by Bouchez *et al.* 1983; Simpson & Schmid 1983; Lister & Snoke 1984; Simpson 1986; Law 1990; Hanmer & Passchier 1991), and their implications for the tectonic history of the area are discussed. Details of the grain-scale microstructural, metamorphic, and crystal fabric deformation features in these tectonites are the subject of on-going research by the authors, and will form the subject of a separate paper.

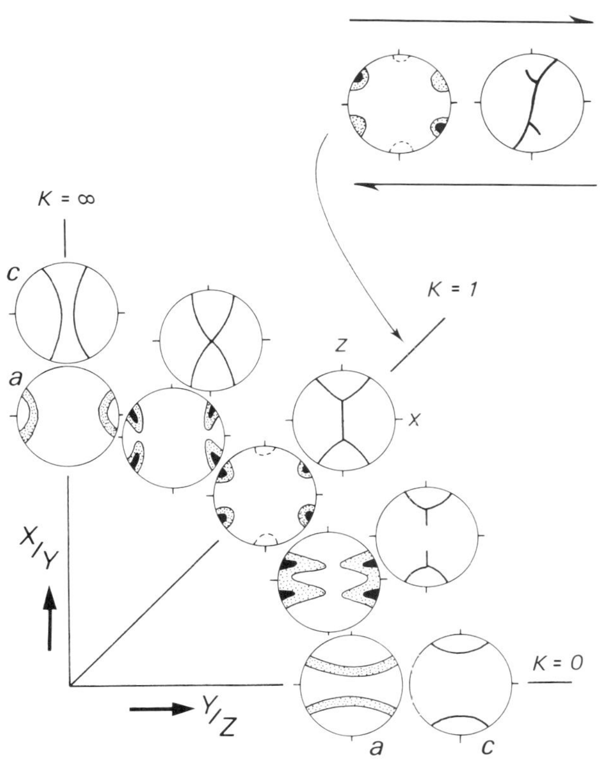

Figure 3 Theoretical relationships (for coaxial deformation) between strain symmetry (expressed by Flinn Plot) and quartz **c**- and **a**-axis fabrics; oblate strain ellipsoids indicated by K = 0, plane strain ellipsoids by K = 1 and prolate strains by K = ∞; quartz **c**-axis fabrics represented by fabric skeletons; **a**-axis fabrics represented by contours; lineation = X, pole to foliation = Z; also shown are **c**- and **a**-axis fabrics associated with non-coaxial plane strain (dextral simple shear) deformation; adapted from Schmid & Casey (1986).

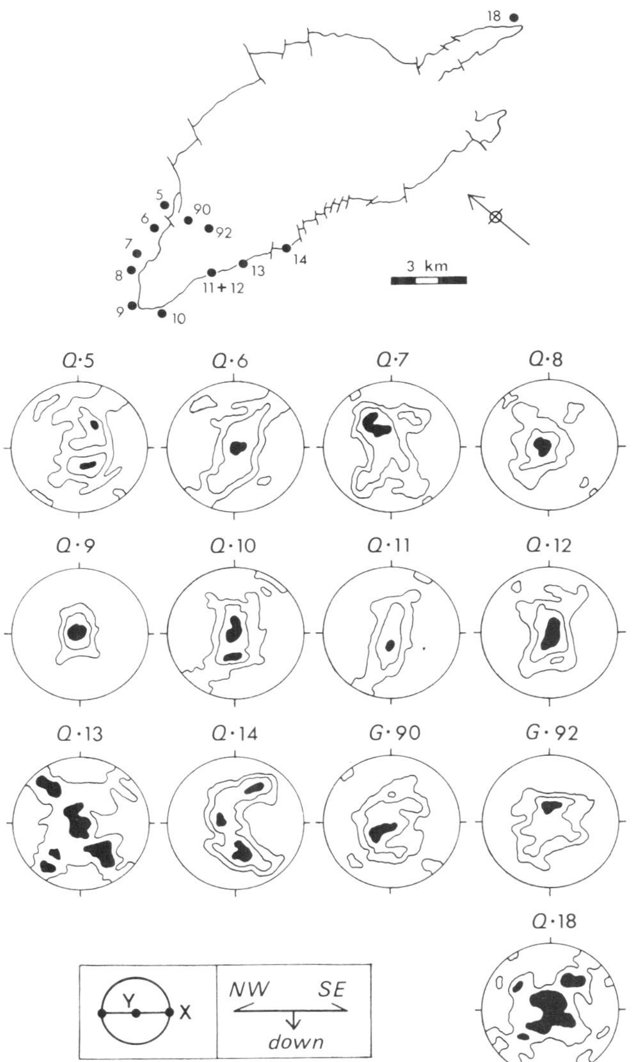

Figure 4 Optically measured quartz **c**-axis fabrics from the Harkless quartzite (Q) and granitic rocks of the Papoose Flat pluton (G) presented on projection planes parallel to lineation (X) and perpendicular to foliation (XY); these fabrics were originally presented by Sylvester & Christie (1968, fig. 1) on YZ projection planes and have now been rotated onto the XZ plane; contour intervals: 1, 2 & 5 times uniform distribution; 300 grains measured in each sample; areas of maximum intensity indicated by solid ornament; location of individual sampling sites indicated; note the dominant sense of fabric asymmetry indicating a top-to-the-SE shear-sense.

1. Geological setting

The Papoose Flat Pluton is one of several granitic bodies situated within the White-Inyo Range of eastern California (Fig. 1) that are regarded as satellites of the Sierra Nevada Batholith (e.g. Bateman *et al.* 1963; Ross 1965, 1969). The pluton, which is composed mainly of biotite quartz monzonite with K-feldspar megacrysts, crops out as an elongated W–E-trending dome, which is 16 km long and 8 km wide (Ross 1967; Nelson *et al.* 1977; Sylvester *et al.* 1978). The Papoose Flat Pluton is one of the youngest plutons in the White-Inyo Range, yielding K–Ar dates on biotite from two samples of 75 and 81 Ma (Kistler *et al.*

The White-Inyo Range is an elongate, NNW-trending horst block which is tilted to the E and bounded by high-angle Cainozoic normal and dextral oblique slip faults in Owens Valley to the W, and in the Fish Lake, Eureka and Saline valleys to the E (Fig. 1). Most of the range consists of virtually unmetamorphosed late Precambrian and Palaeozoic strata (Nelson 1962; Ross 1965) that are warped into broad folds. Evidence for the Mesozoic structural evolution of the White and Inyo mountains has been summarised by Dunne *et al.* (1978), who recognised periods of late Permian, middle to late Jurassic, and Cretaceous deformation. In the vicinity of the Papoose Flat Pluton, the dominant structural feature of the Inyo Mountains is the Inyo anticline, a major upright SE-plunging fold (? belonging to the F3 fold phase of Sylvester & Babcock 1975) which is cut by NNW- and NE-striking faults (Nelson 1966).

1.1. Structure of the pluton

Detailed accounts of the petrology and structure of the Papoose Flat Pluton have been given by Ross (1965) and Sylvester *et al.* (1978). Detailed maps, cross-sections and palinspastic reconstructions of the pluton have been presented by Nelson *et al.* (1977).

The Papoose Flat Pluton has a relatively simple outcrop pattern (Fig. 1) with a long narrow "tail" at its eastern end that Sylvester *et al.* (1978) refer to as the apophysis. Both the pluton and its aureole are well exposed. Contacts with the country rocks are typically sharp, and dip radially outward at about 30–60°. Around the apophysis, however, the contact is irregular and locally gradational with small screens and inclusions of metamorphosed country rocks (Sylvester *et al.* 1978).

Sylvester *et al.* (1978) demonstrated that the pluton is situated structurally within the SW-dipping limb of the Inyo anticline. The pluton is regarded as having cut upward and discordantly across several formations near the core of the fold until it encountered the Poleta Formation, against which the contact is concordant for a distance of 10 km around the western half of the pluton (Sylvester *et al.* 1978, fig. 9). Here the Poleta and younger formations were shown by Sylvester *et al.* (1978) to be deflected to the W and thinned to as little as 10% of their regional stratigraphic thickness, although in the eastern part of the pluton thinning is not observed where the contact is discordant. Sylvester *et al.* (1978) subdivided the granitic rocks into three textural types distinguished by mesoscopic fabric elements: (1) the gneissic border facies, (2) the homogeneous core facies, and (3) the apophysis facies.

The gneissic border facies is most strongly developed within 10–20 m of the contact around the western half of the pluton, and is characterised by an intensely-developed foliation and lineation. Isolated patches of the border facies are also found where topographically high outcrops are near the inferred position of the pluton's roof. The border facies around the rest of the pluton is also foliated and lineated, although not nearly as strongly as at the western end.

The core facies is a relatively homogeneous quartz monzonite with hypidiomorphic-granular texture, locally containing K-feldspar megacrysts that are invariably euhedral (Sylvester *et al.* 1978). The apophysis facies is distinguished from the core facies by the elongate habit of about 10% of the K-feldspar megacrysts. Sub-parallel alignment of these megacrysts imparts a vertical lineation to the granite in the apophysis, possibly indicating magmatic flow; foliation is commonly absent (Sylvester *et al.* 1978).

1.2. Structure of the aureole

Attenuation of the deflected Cambrian strata around the western margin of the Papoose Flat Pluton has been achieved by intense crystal plastic deformation and dynamic recrystallisation of quartz and calcite (Sylvester 1966, 1969; Sylvester & Christie 1968). This intense plastic deformation has locally resulted in formation of strongly-developed grain shape fabrics (foliation and lineation) and associated crystallographic preferred orientation.

Detailed mapping (Nelson 1966, 1971; Ross 1965, 1967a; Nelson *et al.* 1977) has revealed that the deflected strata are progressively thinned along the western and southwestern margins of the pluton to less than 10% of their original stratigraphic thickness (Sylvester *et al.* 1978). In spite of such extreme attenuation, all major stratal subunits of each formation, which can be recognised in the undeformed, unmetamorphosed sections elsewhere in the White Inyo Range, can also be recognised with confidence in the attenuated section. Such stratigraphic correlations have clearly demonstrated that while stratigraphic continuity is maintained in the attenuated section, strata are not uniformly thinned along strike, but pinch and swell within the aureole (Sylvester 1966; Sylvester *et al.* 1978).

In contrast to the western part of the Papoose Flat Pluton, intense crystal plastic deformation is rarely observed around the eastern margin of the pluton (i.e. the apophysis). There, discordant contacts at the pluton margin, lack of penetrative deformation features, and hornfels textures in the surrounding country rocks suggest a passive (permitted) rather than forcible emplacement mechanism.

1.3. Metamorphic petrology of the aureole rocks

Sylvester (1966) and Sylvester *et al.* (1978) demonstrated that contact metamorphism of the country rocks surrounding the Papoose Flat Pluton produced a metamorphic aureole of metapelitic and calcareous hornfels and schistose mylonite having mineral assemblages typical of the albite-epidote and hornblende hornfels facies of metamorphism. Two metamorphic zones, which are separated by a gradational contact, were distinguished by Sylvester (1966) and Sylvester *et al.* (1978): (1) an outer zone, characterised by muscovite- and chlorite-bearing spotted slate and phyllite, reflecting albite-hornfels facies grade of contact metamorphism, and (2) an inner zone, characterised by the presence of biotite and andalusite in pelitic schist and semi-hornfels, reflecting hornblende hornfels grade of contact metamorphism.

The aureole has been mapped by Sylvester (1966, plate 4) as ranging (in plan view) from less than 100 m to 600 m in width N of the pluton, and up to 1,000 m in width at the eastern end. Inasmuch as the pluton contact dips from 30° to 45° away from the pluton, the actual thickness of the aureole is of the order of 100–200 m measured normal to the pluton contact (Sylvester *et al.* 1978).

2. Techniques of quartz and calcite petrofabric analysis

In this paper, all microstructural and crystal fabric data are presented on section planes oriented perpendicular to the mylonitic foliation (XY) and parallel to lineation (X). In all figures, these XZ section planes are, in a geographical sense, viewed toward the N and E. Petrofabric analyses have been made of: (1) quartz veins within the gneissic border facies, (2) quartzite and marble within the Poleta Formation located immediately above the gneissic border facies, and (3) quartzite within the overlying Harkless Formation.

Petrofabric analyses of quartz **c**-axis preferred orientation were carried out on one thin section from each sample collected using an optical microscope and universal stage, a general minimum of 600 **c**-axis being measured in each thin section. The **c**-axis data are displayed on equal area, upper hemisphere stereographic projections, whose plane of projection contains the sample lineation (X) and pole (Z) to foliation; in all these stereographic projections the foliation is vertical and lineation within the foliation horizontal. The **c**-axis plots were contoured using the computer program developed by Starkey (1989).

Petrofabric analyses of both calcite and quartz crystallographic preferred orientation in the samples collected were also carried out at ETH, Zurich using an automated Scintag XDS 2000 X-ray texture goniometer operating in combined reflection and transmission modes. Quartz **c**-axes cannot be measured directly with the X-ray texture goniometer, but must be calculated through the Orientation Distribution Function (ODF), from other pole figures obtained by X-ray texture goniometry (Casey 1981). The calculated quartz and calcite **c**-axis fabrics of representative samples are included in this paper. In representative samples, quartz **c**-axis fabrics are also compared with their corresponding **a**-axis fabrics derived from X-ray texture goniometry. Pole figures obtained using X-ray texture goniometry are displayed using the same orientation convention as that previously described for the optically-derived quartz **c**-axis fabrics.

3. Deformation features within the gneissic border facies

3.1. Granitic gneiss

The granitic rocks of the gneissic border facies are characterised by an intensely-developed foliation. This foliation, which is defined by parallelism of mica flakes and plastically deformed quartz aggregates, anastomoses around the large (0·5–2·0 cm) euhedral-rounded K-feldspar megacrysts present within the gneiss. The foliation is parallel to both the pluton-country rock contact and to lithologic layering and tectonic foliation in the locally surrounding country rock strata. The main lineation in the granitic rocks is defined by a shape preferred orientation of elongate mica blebs and quartz aggregates. In general, the lineation trends NW–SE and plunges gently. All features so far observed within the gneissic border facies are indicative of solid state deformation. No unequivocal evidence for a magmatic foliation (see review by Paterson *et al.* 1989) has been observed within these rocks either during the present study or by previous workers (Sylvester *et al.* 1978; Paterson *et al.* 1991a).

Few unequivocal macroscopic shear sense indicators have been observed to date within the gneissic border facies in our study (cf. Paterson *et al.* 1991a, p. 325). In joint faces oriented perpendicular to foliation (XY) and parallel to lineation (X), the gneiss appears to be characterised by a symmetric relationship between feldspar megacrysts, anastomosing foliation, and pressure-shadow development (Fig. 5(a)). Such an apparent lack of unequivocal shear sense indicators within these L-S tectonites suggests an approximately coaxial bulk deformation history (Morgan *et al.* 1991).

3.2. Quartz veins in the gneissic border facies

The granitic rocks of the gneissic border facies are cut by a series of thin (10–50 cm) aplitic sills and dykes which, at least locally, are intensely folded. An axial planar foliation defined by the preferred alignment of feldspar and quartz grains is present in the folded aplites. This axial planar foliation is oriented parallel to foliation in the surrounding gneiss. Both the aplites and the granitic rocks of the gneissic

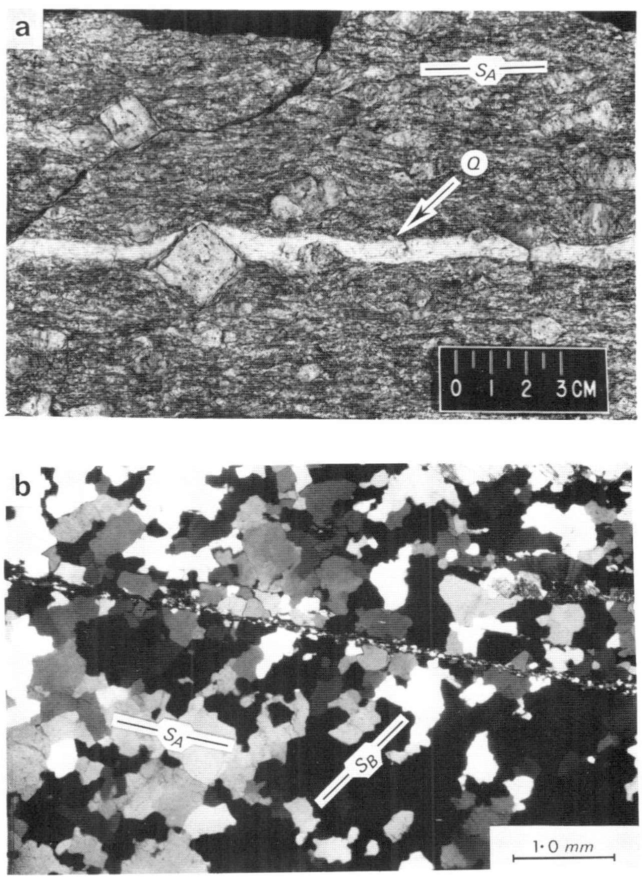

Figure 5 Mylonitic rocks from the gneissic border facies.
(a) Joint face displaying typical macroscopic fabric elements in a hand sample (not *in situ*) from the gneissic border facies (sample PF.38); joint face is oriented parallel to stretching lineation and perpendicular to mylonitic foliation; note lack of any obvious shear-sense indicators at the hand sample scale; Q = Mylonitic quartz vein.
(b) Photomicrograph (viewed toward the NE) of a mylonitic quartz vein (sample PF.1) within the pluton's gneissic border facies; elongate dynamically recrystallised quartz grains display a preferred alignment (S_B) which is oblique to mylonitic foliation (S_A); sense of obliquity suggests a dextral (top-to-the-SE) shear sense; angle of obliquity is greater than 45°; scale bar = 1 mm (1000 microns).

border facies are cut by a series of thin (10–30 cm) quartz veins, which are commonly oriented parallel to the foliation in the surrounding gneiss. The quartz veins are also intensely foliated and lineated, the foliation and lineation being generally oriented parallel to foliation and lineation in the immediately surrounding gneiss.

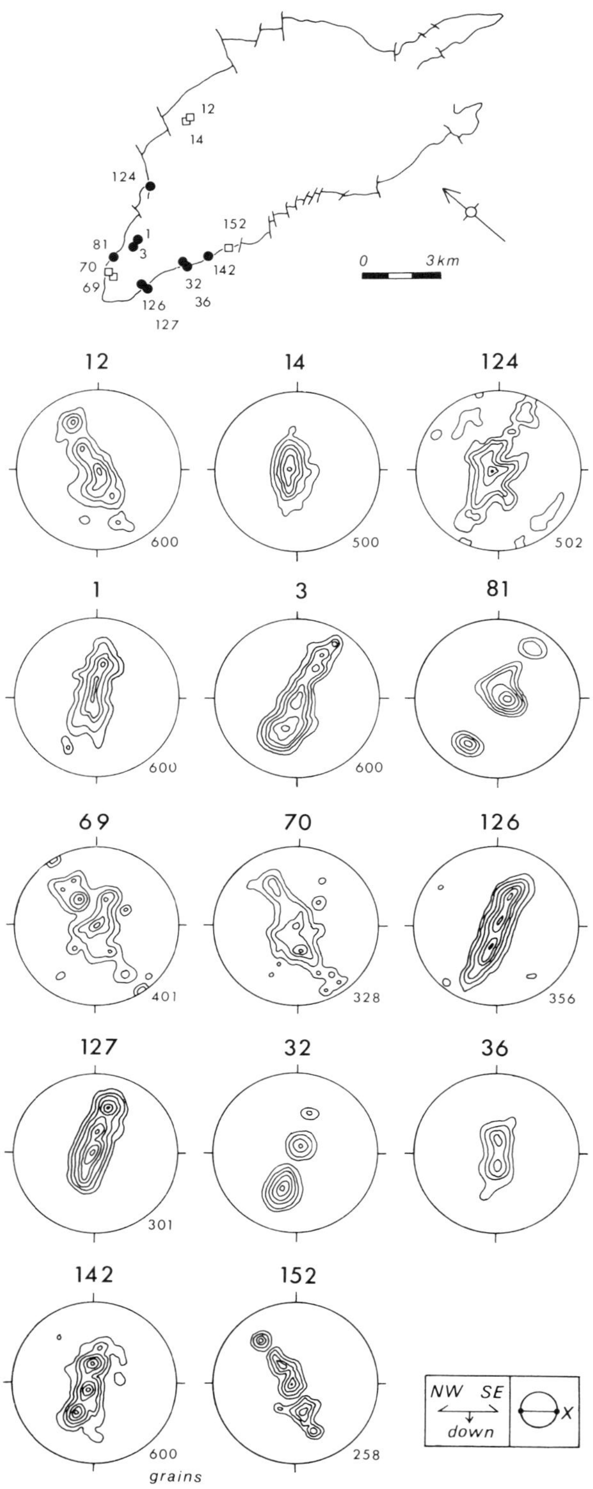

Figure 6 C-axis fabrics (viewed towards the NE) from selected mylonitic quartz veins within the gneissic border facies; fabrics in samples PF.32, 36 and 81 measured by X-ray texture goniometry and ODF analysis, all other fabric measured optically; locations of samples indicating dextral (top-to-the-SE) shear-sense indicated by circles; location of samples indicating sinistral shear sense indicated by squares; contour intervals: 1, 2, 3, 5, 10, 15 and 20 times uniform distribution.

In XZ thin sections, the quartz veins are composed of equant to elongate recrystallised grains ranging in size from 400–1,000 μm. The elongate quartz grains are commonly aligned with their long axes oblique to the macroscopic foliation (Fig. 5b). Adopting the nomenclature introduced by Law *et al.* (1984), this preferred alignment of elongate recrystallised grains is referred to as S_B. The sense of obliquity between S_B and mylonitic foliation (S_A) may be used as a shear sense indicator (see reviews by Bouchez *et al.* 1983; Simpson & Schmid 1983). In the majority of veins examined, the sense of obliquity between S_A and S_B indicates a top-to-the-SE shear sense (e.g. Fig. 5b). Local exceptions to this general observation are described in Section 4 of this paper.

Internally, the grains display either straight or slightly undulatory extinction. Subgrains are rarely observed in the optical microscope. Grain boundaries are commonly lobate, suggesting the importance of grain boundary migration. Taken together, these microstructural features clearly indicate the importance of dynamic recrystallisation at relatively high homologous temperatures within the quartz veins (see reviews by Urai *et al.* 1986; Knipe & Law 1987).

All quartz veins within the gneissic border facies are characterised by intensely-developed asymmetric single girdle **c**-axis fabrics (Fig. 6). In ten of the fourteen quartz vein samples so far analysed by optical and X-ray texture goniometry methods, a top-to-the-SE shear sense is indicated by obliquity between the **c**-axis single girdle and foliation and lineation (cf. Figs 3, 4, 6). For all individual quartz veins examined, the same shear sense is indicated by both crystal fabric asymmetry and obliquity between S_A and S_B.

In Figure 7, the optically-measured **c**-axis fabric from sample PF.1, collected from a quartz vein located on the NW margin of the pluton (Fig. 6), is compared with the sample's corresponding **a**-axis pole figure measured by X-ray texture goniometry. This **a**-axis fabric is representative of all quartz veins analysed in the present study. The **a**-axis fabric is composed of three point maxima aligned within the XZ plane and, by analogy with quartz fabric modelling studies (Fig. 3), clearly indicates that the quartz veins have undergone plane strain deformation, and that the grain shape lineation is a true stretching lineation.

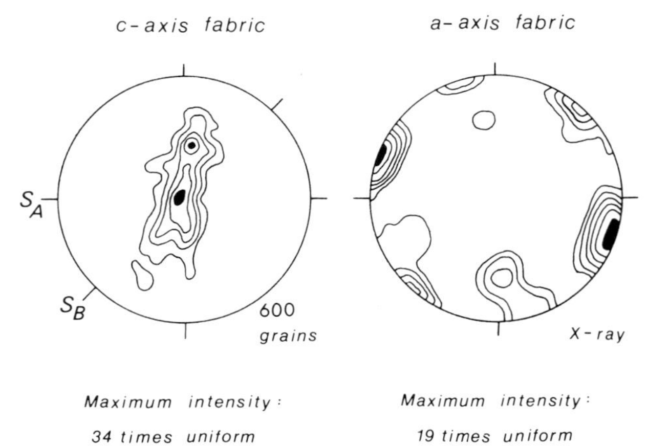

Figure 7 Comparison of optically measured **c**-axis fabric from a mylonitic quartz vein (sample PF.1) and the corresponding **a**-axis fabric measured by X-ray texture goniometry; the **a**-axis fabric consists of three strong point maxima positioned close to the XZ plane (cf. Fig. 3); contour intervals: 1, 2, 5, 10 and 20 times uniform distribution (**c**-axis fabric); 1, 2, 3, 5, 7, 9 and 11 times uniform distribution (**a**-axis fabric).

4. Plastically deformed rocks of the metamorphic aureole

Five detailed sampling traverse have been made around the western margin of the Papoose Flat Pluton outward from the gneissic border facies into the penetratively deformed metasedimentary rocks of the aureole (Fig. 8). In all of these measured sections, microstructural analysis indicates that both quartzite and marble in the Poleta, Harkless and Saline Valley formations have suffered intense crystal plastic deformation (Sylvester & Christie 1968; Sylvester 1969). This intense deformation caused formation of a strong mylonitic foliation oriented parallel to both lithologic layering and the pluton margin, and a mineral stretching lineation that is particularly well developed in the quartzite.

In the following sections, each sampling traverse is described in turn with particular emphasis on the observed microstructural and crystal fabric shear sense indicators. Cross-sections illustrating these traverses are all drawn parallel to the local dip direction of mylonitic foliation and lithologic layering.

4.1. Traverse A

The complete stratigraphic sequence of the Poleta Formation is exposed in this measured section (Fig. 9). Here the Poleta Formation has been tectonically thinned to 10% of its regional stratigraphic thickness (Sylvester *et al.* 1978). This extreme tectonic thinning is expressed in outcrop by an intensely-developed mylonitic foliation in the Poleta and Harkless formations, as well as the underlying gneissic border facies. Within the mylonitic foliation, a mineral elongation lineation is strongly developed in both quartz-rich units of the Poleta and Harkless formations, and in mylonitic quartz veins in the underlying gneissic border facies. This mineral lineation plunges gently to the SE approximately perpendicular to the cross-section in Figure 9.

Both the mylonitic quartzite within the Harkless and Poleta formations, and the mylonitic quartz veins (PF.32 & 36) in the gneissic border facies, display intensely developed **c**-axis fabrics (Fig. 9). While the mylonitic aureole rocks are characterised by cross-girdle **c**-axis fabrics, the mylonitic quartz veins are characterised by single girdle fabrics. All quartz **c**-axis fabrics are asymmetric with respect to foliation and lineation in terms of either skeletal outline or density distribution. The sense of asymmetry indicates a dextral (top-to-the-SE) shear-sense. This inferred shear-sense is confirmed by the microstructures in the Poleta and Harkless quartzite and schist (Fig. 10(a)–(d)). Shear bands (extensional crenulation cleavage) are well developed in the pelitic layers of both the Harkless and Poleta formations. These shear bands intersect the penetrative foliation in a line oriented perpendicular to the stretching lineation and consistently indicate a top-to-the SE shear-sense.

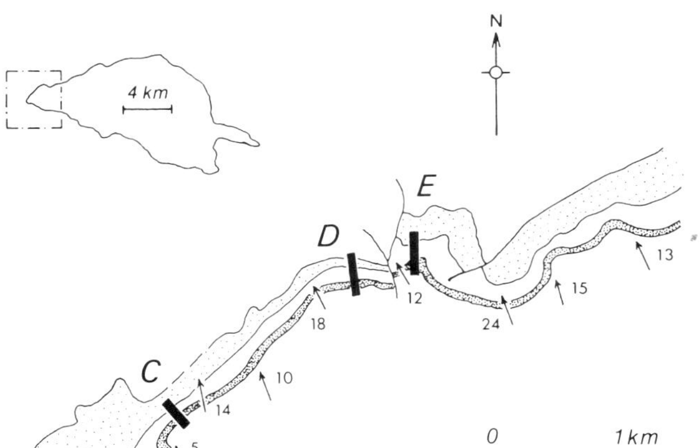

Figure 8 Locations of the five sampling traverses (A–E): all traverses are oriented parallel to the local dip direction of mylonitic foliation; local trend and plunge of lineation indicated by arrows; Poleta Formation = dense stipple pattern, Harkless Formation = unornamented, Saline Valley formation = light stipple.

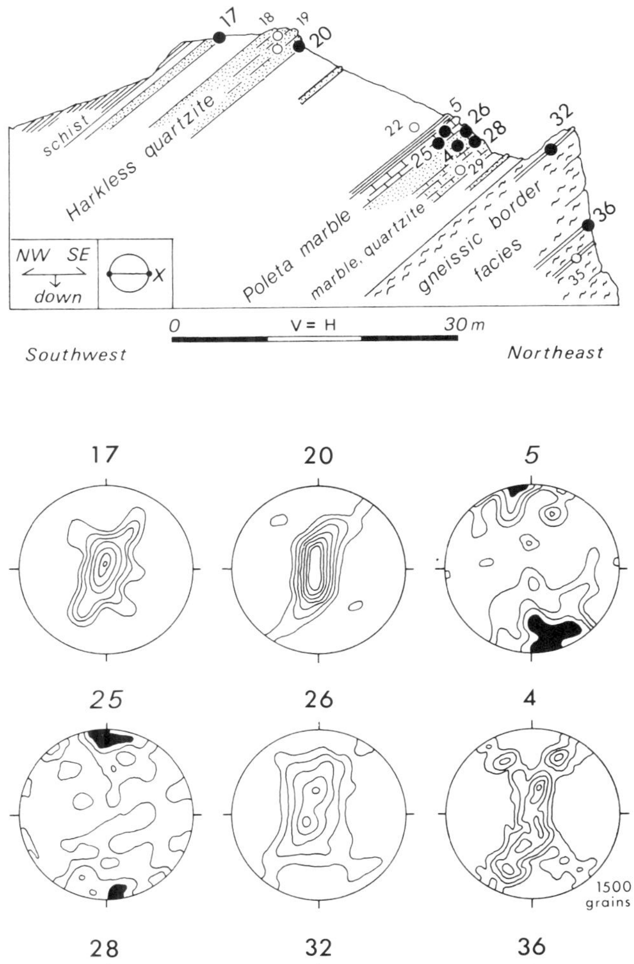

Figure 9 Cross-section along Traverse A: location of sampling sites indicated; PF.5 and 25: calcite **c**-axis fabrics; all other samples: quartz **c**-axis fabrics; with the exception of sample PF.4, all **c**-axis fabrics measured by X-ray texture goniometry and ODF analysis; 1500 quartz grains measured optically in PF.4; note: mineral stretching lineation along this traverse is sub-perpendicular to the measured cross-section; contour intervals: 1, 3, 5, 7, 9, 11, 13 and 15 times uniform distribution (quartz fabrics), 1, 2, 3 and 4 times uniform (calcite fabrics), areas of maximum intensity indicated by solid ornament.

All the **c**-axis fabrics measured by X-ray texture goniometry are characterised by an intense point maximum (up to ten times uniform distribution in the quartzite, up to 25 times uniform in the quartz veins) oriented within the foliation perpendicular to lineation. In geological terranes where crystal fabric evolution is associated with regional metamorphism, such Y axis point maxima fabrics, which are inferred to indicate the dominance of prism slip, are almost always associated with either amphibolite facies or higher conditions of metamorphism (see reviews by Schmid & Casey 1986; Law 1990).

Strong calcite crystallographic fabrics are also present in this section (Fig. 9). The Poleta marble (samples PF.5 and 25) is characterised by a calcite **c**-axis point maximum oriented within the XZ specimen plane at a low angle to the foliation pole (Z). Calcite-dolomite geothermometry of sample PF.5 indicates *minimum* temperatures of metamorphism of 530–540°C and 480–490°C using the models of Rice (1977) and Annovitz and Essene (1987), respectively. Shear-sense interpretation of calcite fabric asymmetry is, due to the possibility of twinning, more complicated than in the case of quartz (see reviews by Schmid *et al.* 1987; Law 1990). The sense of asymmetry in the calcite **c**-axis fabrics from samples PF.5 and 25 is indicative of sinistral (top-to-the-NW) shear, assuming the fabrics are produced by dislocation creep rather than later twinning deformation.

Quartz **a**-axis fabrics for Traverse A are shown in Figure 11. With the exception of sample PF.29, all quartzites and quartz veins measured in this section are characterised by three **a**-axis point maxima oriented within the XZ plane. These fabrics indicate plane strain deformation conditions (cf. Figs 3, 11). In contrast, the **a**-axis fabric for sample PF.29 (situated 3·6 m above the pluton margin) consists of a broad small-circle girdle of **a**-axes centred about the pole (Z) to mylonitic foliation. Such small-circle girdle **a**-axis fabrics indicate deformation in the flattening field (Fig. 3). This strain symmetry interpretation agrees with the observation that, particularly around the NW margin of the pluton, the lowermost pelitic units of the Poleta Formation, although lineated, also contain foliation-parallel garnet-rich skarn deposits with chocolate-tablet boudinage structures (Sylvester & Christie 1968, plate 1).

4.2. Traverse B

In this measured section (Fig. 12) the Poleta Formation is even more intensely thinned than in Traverse A. The **c**-axis

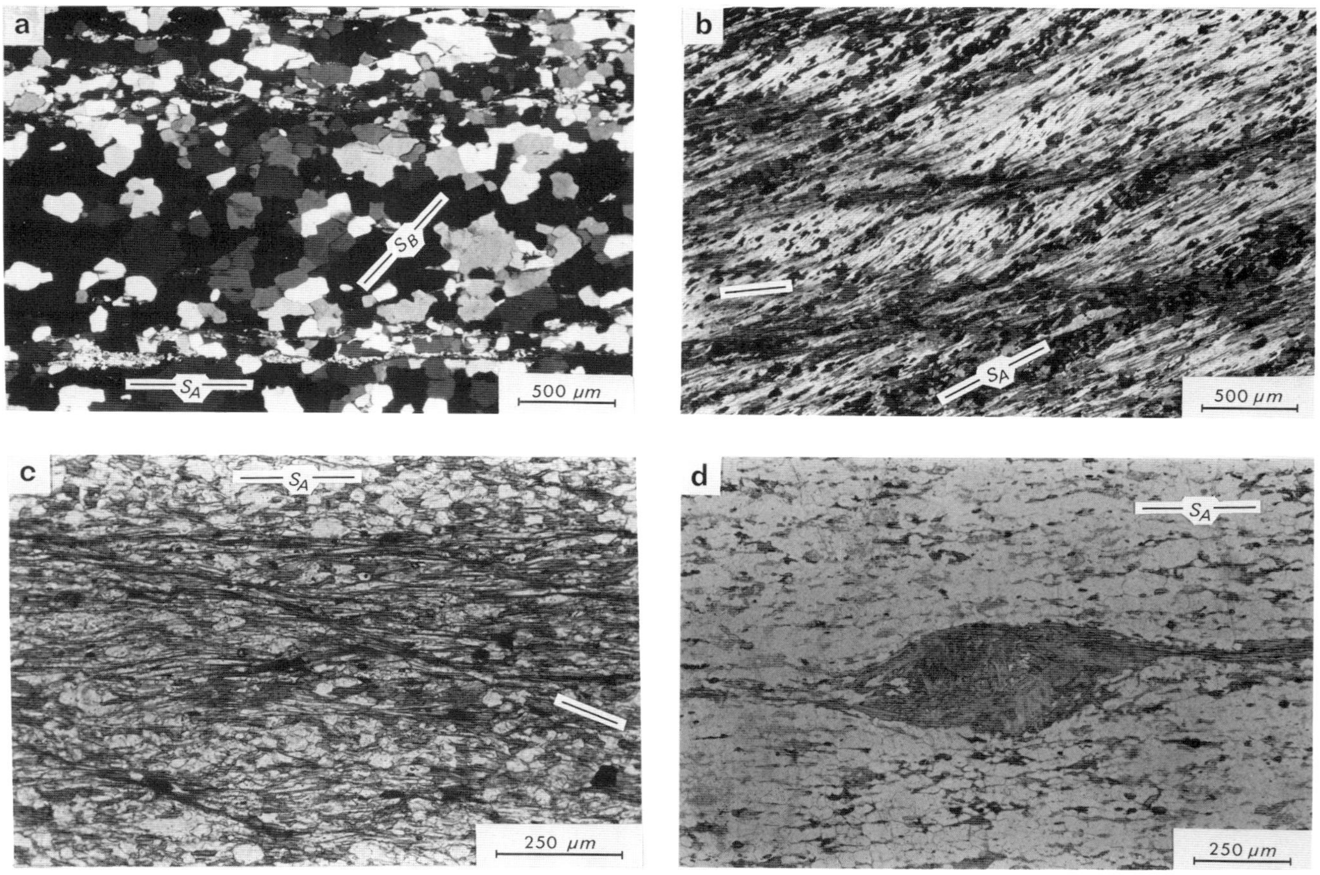

Figure 10 Photomicrographs (viewed toward the NE) of schist and mylonitic quartzite from the Harkless Formation exposed in Traverse A; in all cases a dextral (top-to-the-SE) shear sense is indicated by the observed microstructures.

(a) Preferred alignment (S_B) of elongate, dynamically recrystallised quartz grains which dip more steeply to the NW than mylonitic foliation (S_A) defined by phyllosilicates (sample PF.18); cross-polarised light.

(b) Shear bands (extensional crenulation cleavage) within Harkless black schist (sample PF.22); foliation (S_A) defined by phyllosilicates; orientation of shear bands indicated by bar; cross-polarised light.

(c) Shear band (extensional crenulation cleavage) within thin micaceous quartzite layer from the Poleta Formation (sample PF.26); foliation (S_A) defined by phyllosilicates and elongate quartz grains; orientation of shear bands indicated by bar; plane-polarised light.

(d) Asymmetric overgrowths on phyllosilicate clast within thin micaceous quartzite layer from the Poleta Formation (sample PF.29); overgrowths are parallel to mylonitic foliation (S_A); plane-polarised light.

preferred orientation of three mylonitic quartzite samples from the Harkless Formation, and two mylonitic quartz veins from the underlying gneissic border facies, have been optically measured in this section. All are asymmetric fabrics. A top-to-the-SE shear sense is indicated within both the mylonitic quartz veins (PF126 and 127) and the quartzite samples. This shear sense is also indicated by shear bands (extensional crenulation cleavage) in the overlying Harkless schist.

4.3. Traverse C

A domainal distribution of quartz **c**-axis fabric asymmetry has been detected in the Harkless quartzite of this section (Fig. 13). At a structural distance of 23·2 m above the pluton contact (sample PF.90), fabric asymmetry indicates a sinistral (top-to-the-NW) shear sense. In contrast, at 25·3 m above the contact, the measured **c**-axis fabric from sample PF.89 is almost symmetric with respect to foliation and lineation, indicating approximately coaxial (pure shear) deformation. At 39·6 m above the contact, the **c**-axis fabric (PF.88), although symmetric in terms of skeletal outline, is asymmetric in terms of density distribution. This density asymmetry may indicate a weak component of dextral (top-to-the-SE) shear. No clearly unambiguous microstructural shear sense indicators have been observed within either the Harkless quartzite or schist exposed along this section.

All three **c**-axis fabrics are characterised by an intense Y point maximum with **c**-axis intensity values ranging between ten and twelve times uniform distribution. As discussed in Section 4.1, these point maxima indicate relatively high temperatures of deformation.

The Poleta marble (sample PF.93) in this section also displays a strong crystallographic preferred orientation (Fig. 13). The calcite **c**-axis fabric is asymmetric with respect to foliation and lineation and consists of two intense point maxima, the dominant point maximum (ten times uniform distribution) being oriented within the XZ plane at a low angle to the pole to foliation (Z). Calcite-dolomite geothermometry of sample PF.93 indicates *minimum* temperatures of metamorphism of 420–500°C and 380–460°C using the models of Rice (1977) and Annovitz and Essene (1987), respectively. The sense of asymmetry in the calcite **c**-axis fabric from sample PF.93 indicates dextral (top-to-the-SE) shear assuming the fabric is produced by dislocation creep rather than later twinning deformation.

4.4. Traverse D

A domainal distribution of shear sense regimes has also been revealed by microstructural and crystal fabric analysis

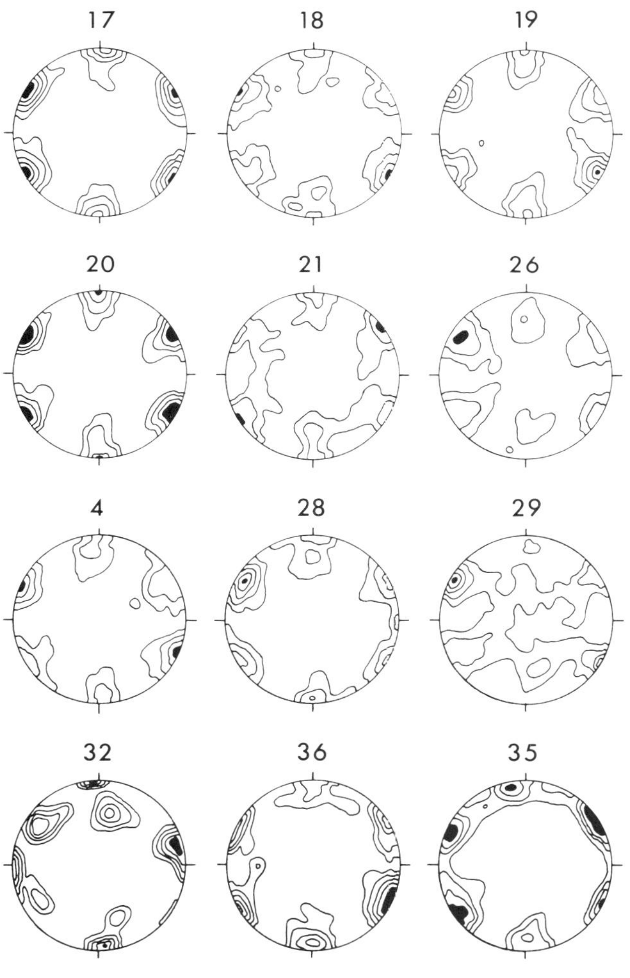

Figure 11 Quartz **a**-axis fabrics from Traverse A viewed toward the NE; see Figure 9 for sample localities and orientation convention; all fabrics measured by X-ray texture goniometry; only sample PF.29 displays a small circle girdle distribution of **a**-axes centred about the pole (Z) to mylonitic foliation; contour intervals: 1, 2, 4, 6 and 7 times uniform; areas of maximum intensity indicated by solid ornament.

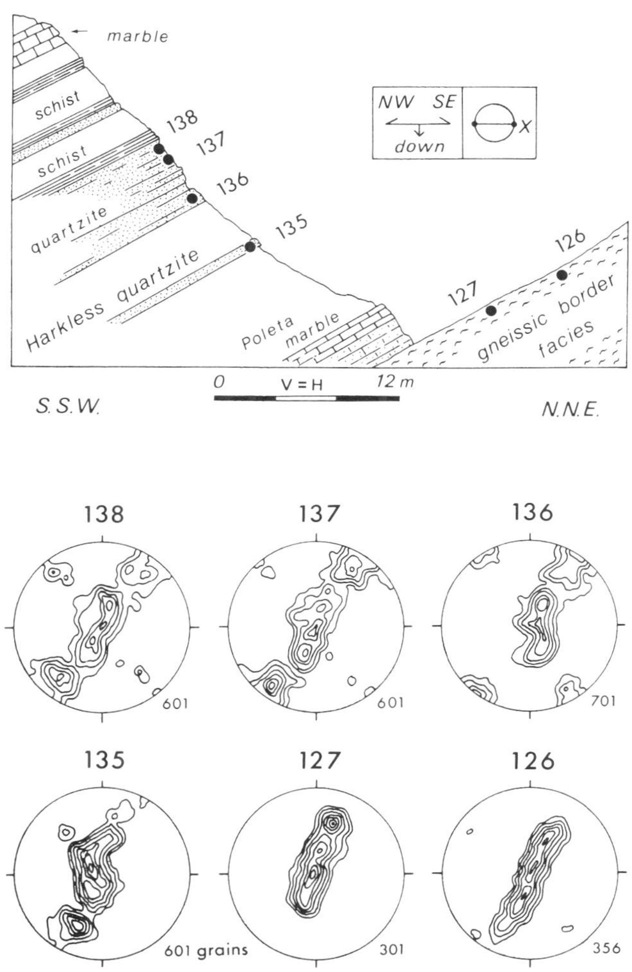

Figure 12 Cross-section along Traverse B; location of sampling sites indicated; all fabrics are optically measured quartz **c**-axis fabrics; contour intervals: 1, 2, 4, 6, 10, 15 and 20 times uniform; note: mineral stretching lineation along this traverse is sub-perpendicular to the measured cross-section.

of samples collected along Traverse D (Fig. 14). A top-to-the-NW shear-sense is indicated by asymmetric **c**-axis fabrics in two quartz veins (samples PF.69 and PF.70) in the gneissic border facies at distances of 0·5 and 3·5 m beneath the pluton's margin. This shear sense is confirmed by the preferred alignment (S_B) of elongate recrystallised quartz grains oriented oblique to mylonitic foliation (S_A). Similarly, in the surrounding gneiss a weak component of top-to-the-NW shear is indicated by rare asymmetric fibrous overgrowths on feldspar porphyroclasts. In contrast, within the Harkless quartzite above the pluton margin, a top-to-the-SE shear-sense is indicated by asymmetric quartz **c**-axis fabrics. Macroscopic shear bands (extensional crenulation cleavage) within the overlying Harkless schist also clearly indicate a top-to-the-SE shear-sense. All **c**-axis quartz fabrics are characterised by Y-axis point maxima indicating relatively high temperatures of deformation.

In this section, calcite-dolomite geothermometry has been carried out on two samples of Poleta marble. For sample PF.71 (collected 0·5 m above the pluton margin), *minimum* temperatures of metamorphism of 510–530°C and 470–490°C are indicated using the models of Rice (1977) and Annovitz and Essene (1987), respectively. Similarly, at a structurally higher level (11·6 m above the pluton margin), minimum temperatures of metamorphism of 490–530°C and 450–490°C are estimated for sample PF.72.

Sample PF.71 is characterised by a calcite **c**-axis fabric which is markedly asymmetric with respect to foliation and lineation (Fig. 14). The sense of asymmetry in this fabric is indicative of dextral (top-to-the-SE) shear if the fabric was produced by dislocation creep rather than later twinning deformation.

4.5. Traverse E

Traverse E, on the northern margin of the pluton, extends in a S–N direction from the gneissic border facies, upward through the Poleta and Harkless formations into the marble of the Saline Valley Formation (Fig. 15). The stretching lineation in this series of outcrops plunges NW, sub-parallel to the plane of cross-section.

Calc-silicate skarn deposits are well developed in the schist of the Poleta formation located immediately above the pluton. These foliation-parallel, layered skarn deposits clearly display three-dimensional pinch and swell structures (chocolate-tablet boudinage), the more rigid garnet and epidote-rich skarn layers being segmented into a series of oblate spheroids around which the schistose layering in the pelitic schist anastomoses (Sylvester & Christie 1968, plate 1). The grain shape lineation lying within the foliation in

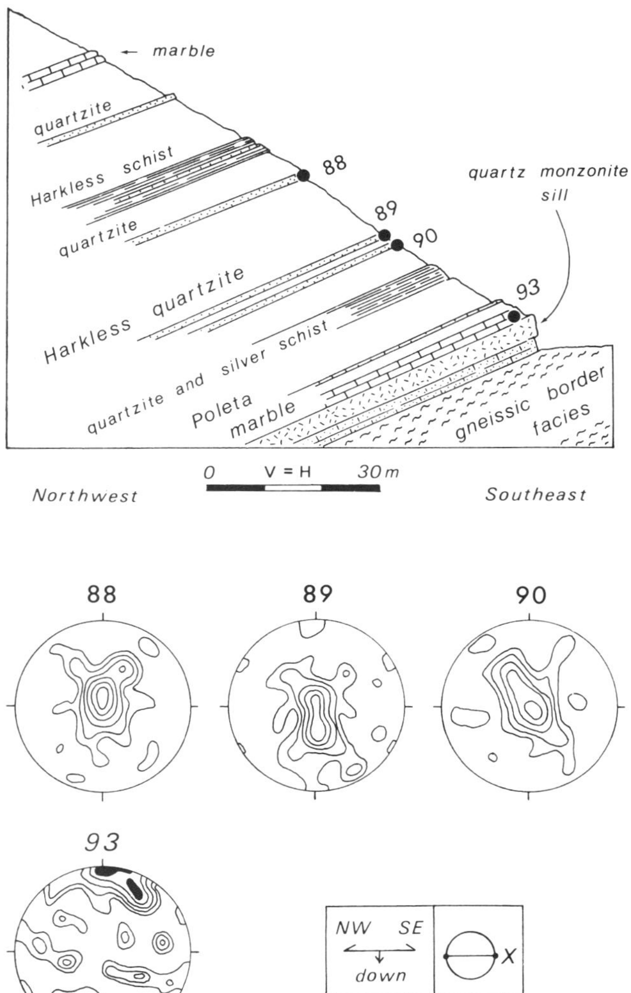

Figure 13 Cross-section along Traverse C; location of sampling sites indicated; sample PF.93: calcite **c**-axis fabric; all other samples: quartz **c**-axis fabrics; all fabrics measured by X-ray texture goniometry and ODF analysis; contour intervals: 1, 2, 3, 4, 5, 6 and 7 times uniform distribution.

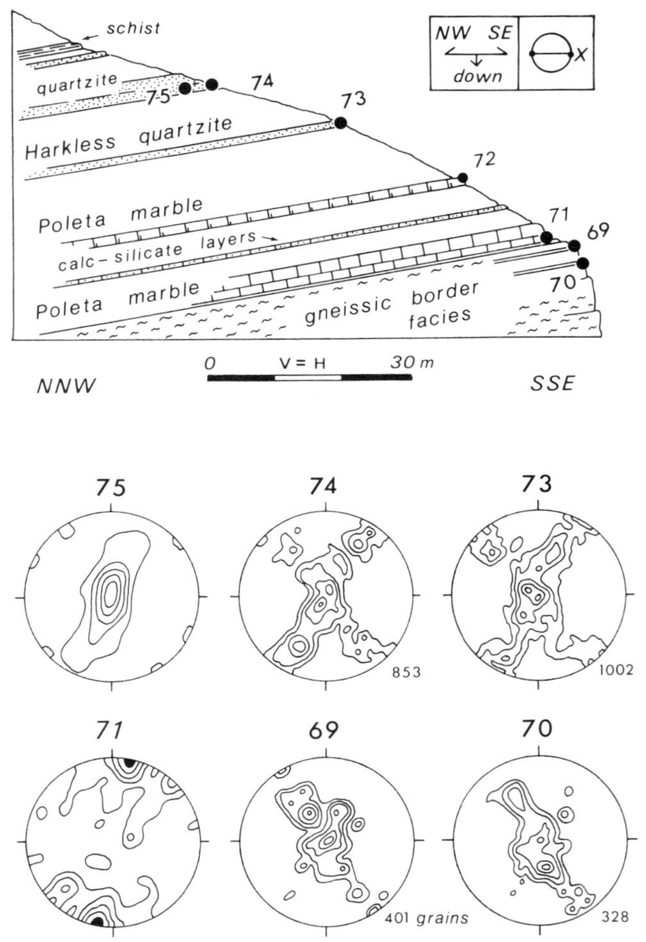

Figure 14 Cross-section along Traverse D; location of sampling sites indicated; sample PF.71: calcite **c**-axis fabric; all other samples: quartz **c**-axis fabrics; samples PF.75 and 71 measured by X-ray texture goniometry and ODF analysis; all other fabrics measured optically (number of grains measured in each sample is indicated); contour intervals: 1·0, 1·5, 2·0, 2·5 and 3·0 times uniform distribution (calcite fabric—PF.71); 1, 2, 3, 5, 7 and 9 times uniform (all quartz fabrics).

these schists trends parallel to the stretching lineation in both the underlying gneissic border facies and the overlying mylonitic marble and quartzite.

An intense **c**-axis preferred orientation is detected in sample PF.81 collected from a mylonitic quartz vein 3·0 m beneath the pluton-cover rock contact within the gneissic border facies. The **c**-axis fabric consists of three point maxima aligned on a single girdle whose asymmetry with respect to foliation and lineation indicates a top-to-the-SE shear-sense (Fig. 15). The same shear sense is indicated by the asymmetric cross-girdle **c**-axis fabric measured within sample PF.77 collected from the Harkless quartzite at a measured distance of 26·0 m above the pluton's margin.

Asymmetric calcite crystal fabrics have also been detected in the marble layers exposed along this traverse. Sample PF.80, collected within the Poleta Formation at a distance of 11·0 m above the pluton's margin, exhibits a calcite **c**-axis fabric (Fig. 15) whose sense of asymmetry is consistent with a top-to-the-SE shear-sense, assuming that intracrystalline deformation is associated with dislocation processes rather than twinning (see review by Law 1990).

Within marble of the structurally higher Saline Valley Formation, a top-to-the-SE shear-sense is indicated by elongate, dynamically recrystallised calcite grains that dip more steeply to the NNW than the main mylonitic foliation. In thin section, these oblique grain shapes defining S_B are extremely elongate (aspect ratios of between 3:1 and 5:1) and are interpreted as being produced by dynamic recrystallisation and associated dislocation processes rather than by twinning. In some layers, this oblique grain shape alignment is visible at the hand sample scale (Fig. 16). The main mylonitic foliation (S_A) and the oblique grain shape alignment (S_B) intersect in a line that is oriented perpendicular to the mineral stretching lineation.

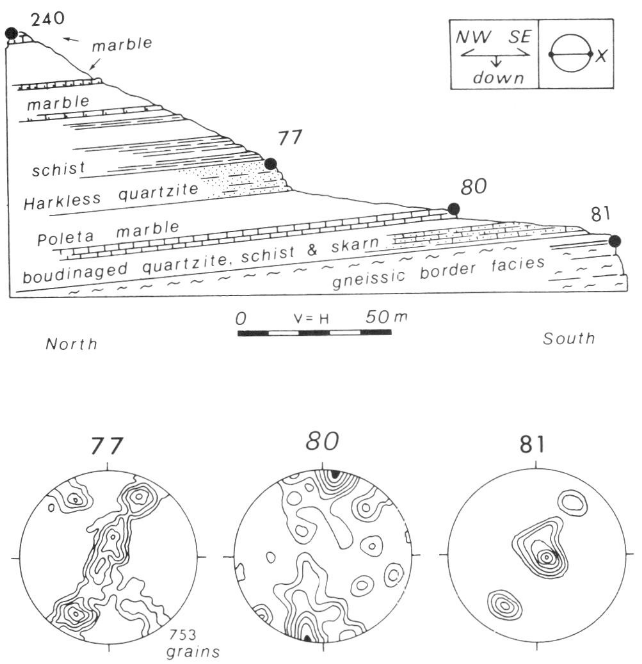

Figure 15 Cross-section along Traverse E; location of sampling sites indicated; sample PF.80: calcite **c**-axis fabric; PF.77 and 81; quartz **c**-axis fabrics; PF.77 measured optically, samples PF.80 and 81 measured by X-ray texture goniometry and ODF analysis; contour intervals: 1, 2, 3, 4, 5, 6, and 7 times uniform distribution (PF.77 and 80); 4, 5, 6, 7, 10 and 13 times uniform (PF.81).

5. Discussion

5.1. Strain path and temperature of deformation

Both microstructural and crystal fabric analysis of the mylonitic cover rocks situated in the contact aureole surrounding the western part of the Papoose Flat Pluton indicate that deformation was dominated by strongly non-coaxial, plane strain conditions. Within the underlying pluton's gneissic border facies, similar strain path conditions are indicated by microstructures and crystal fabrics in the mylonitic quartz veins. In contrast, the apparent rarity of unequivocal shear sense indicators in the gneiss would appear to indicate a more coaxial strain path. Only locally within calc-silicate rocks located immediately above the pluton margin have we observed both shape fabrics and crystallographic fabrics indicative of deformation within the flattening field.

The microstructural and crystal fabric kinematic indicators reported in this study clearly indicate a dominant top-to-the-SE shear sense associated with crystal plastic deformation. Exceptions to this general observation are present, but appear to be restricted to fairly localised structural domains.

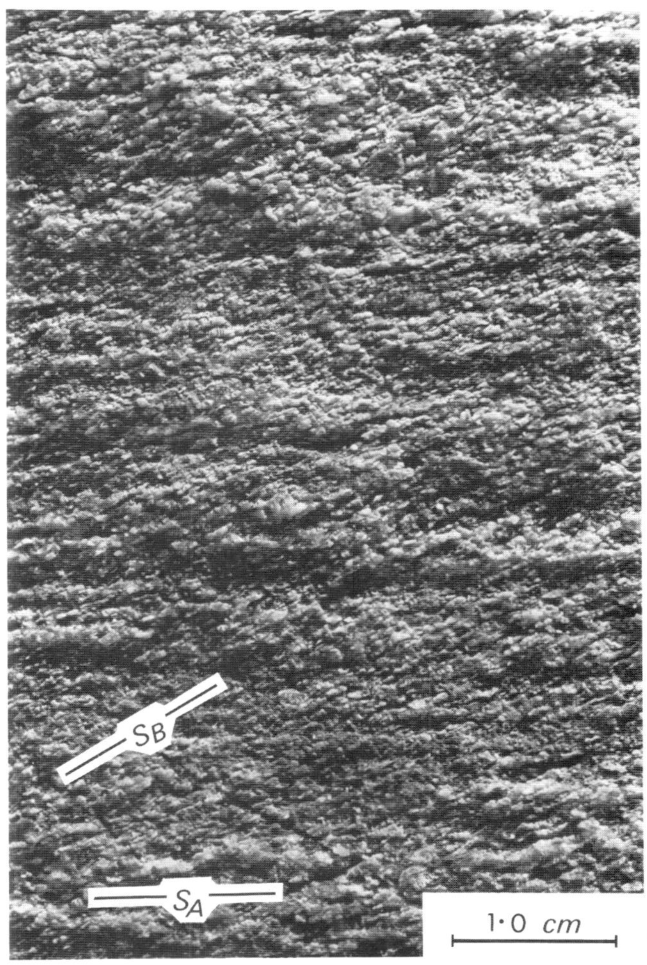

Figure 16 Elongate dynamically recrystallised calcite grains in Saline Valley marble defining a grain shape fabric (S_B) which is oriented oblique to mylonitic foliation (S_A)—sample PF.240; exposure surface (viewed toward the NE) is oriented sub-perpendicular to S_A and parallel to stretching lineation; area shown measures 6 cm × 4 cm; sample collected at a structural height of approximately 70·0 m above the pluton margin; see Figure 15 for sample location.

We consider the presence of a strongly-developed point maximum in the quartz **c**-axis fabrics in these tectonites (oriented within the mylonitic foliation plane perpendicular to lineation) to indicate relatively high temperatures of crystal plastic deformation. In deformation associated with regional metamorphism, such Y-axis parallel point maxima are generally associated with at least amphibolite facies conditions (see review by Schmid & Casey 1986). Relatively high deformation temperatures are also indicated by lobate quartz grain boundaries (particularly in the quartz veins), suggesting a high degree of grain boundary mobility during dynamic recrystallisation (Urai *et al.* 1986). Further supporting evidence for this inferred relatively high temperature of deformation is provided by preliminary calcite-dolomite geothermometry studies of marble from the Poleta Formation situated at distances of less than 12 m above the pluton margin. For the localities reported in this paper, minimum estimates for temperatures of metamorphism range between 380 and 540°C (Nyman & Law, unpublished data, 1991).

We must also ask whether the deformation features observed within the gneissic border facies may have developed under magmatic conditions. Criteria for the identification of magmatic and tectonic foliations have recently been reviewed by Paterson *et al.* (1989). One crystal fabric criterion that may be applied is the presence or absence of quartz prism ⟨**c**⟩ slip in association with igneous intrusions such as granites. Prism ⟨**c**⟩ slip, evidenced by a **c**-axis point maximum oriented sub-parallel to the stretching lineation, may indicate plastic deformation at temperatures near the granite solidus (Blumenfeld *et al.* 1986; Gapais & Barbarin 1986; Blumenfeld & Bouchez 1988). Deformation at least within the quartz veins from the gneissic border facies, however, is clearly dominated by lower temperature ⟨**a**⟩ slip (see reviews of slip system criteria by Schmid & Casey 1986; Law 1990). No evidence for quartz ⟨**c**⟩ slip has been found within the rocks of the gneissic border facies. Similarly, no microstructural or chemical evidence for magmatic foliations have been observed within the pluton (Sylvester *et al.* 1978; Paterson *et al.* 1991a). Deformation within the gneissic border facies would clearly appear to have taken place in the solid state.

5.2. Deformation models

Any model for the deformation features associated with the Papoose Flat Pluton's gneissic border facies and surrounding aureole rocks must take into account the following observations:

(1) Intense crystal plastic deformation is restricted to the western part of the pluton and its surrounding aureole. In contrast, the eastern part of the pluton is characterised by discordant contacts at the pluton margin, lack of penetrative deformation features and hornfels textures in the surrounding country rocks (Sylvester *et al.* 1978).

(2) Mylonite formation in: (a) the quartzites and at least some marble layers within the aureole, and (b) quartz veins within the gneissic border facies, is clearly a plane strain deformation associated with a dominant top-to-the-SE shear sense. In contrast, only rarely are unequivocal examples of shear sense indicators observed within granitic rocks of the gneissic border facies (cf. Paterson *et al.* 1991a, p. 325).

(3) The pattern of quartz **c**-axis preferred orientation within both the mylonitic aureole rocks and quartz veins within the gneissic border facies indicates relatively high (>450°C) temperatures of deformation.

Three deformation models, which are potentially consistent with the above observations, are illustrated in Figure 17.

In the first model (Fig. 17(a)) the granite forming the western part of the pluton was forcibly intruded in a northwestward direction from the pluton source as a nearly solidified wedge beneath a static cover of sedimentary rocks. In this model, sideways magmatic wedging toward the NW, rather that *in situ* ballooning associated with flattening

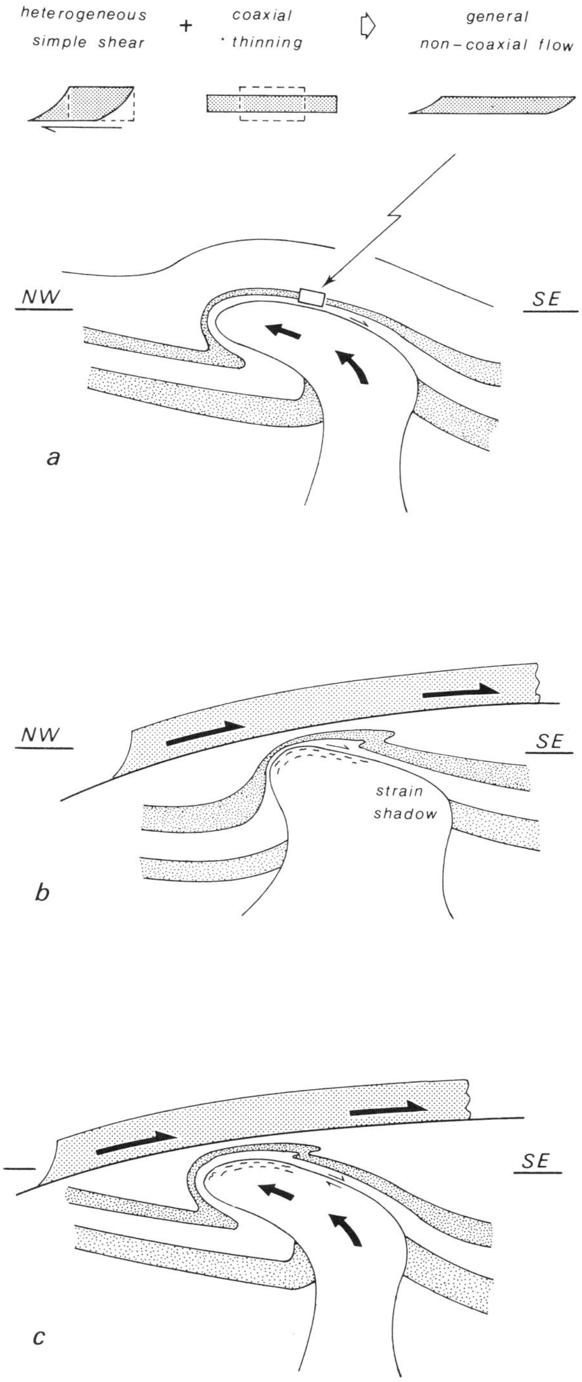

Figure 17 Diagrammatic explanation of tectonic models for the observed localisation of intense non-coaxial strain associated with crystal plastic deformation on the western part of the Papoose Flat Pluton and its metamorphic aureole.

(a) Strain due to "magmatic wedging": observed 90% thinning of individual stratigraphic units indicates that simple shear deformation must have been accompanied by layer parallel extension (i.e. coaxial thinning) unless the shear zone boundaries were oriented oblique to bedding.

(b) Regional SE-directed thrusting of cover rocks over the pluton.

(c) Combined "magmatic wedging" and regional overthrusting.

strains in the cover rocks, is proposed. However, as recently pointed out by Paterson *et al.* (1991a), within the White and Inyo mountains a number of other plutons of similar age to the Papoose Flat Pluton also show more intense solid-state deformation on their western or northwestern margins than on their eastern margins. We agree with Paterson *et al.* (1991a) that it seems at least unlikely that so many of these Cretaceous-age plutons should all have been forcibly intruded towards the W or NW.

In the second model (Fig. 17(b)) the mylonitic deformation observed around the western margin of the Papoose Flat Pluton is associated with regional SE-directed thrusting of the cover rocks over the underlying pluton (Paterson *et al.* 1991a). A similar model for the late Cretaceous age Pellisier Flat Pluton, located in the northern part of the White-Inyo Range, was also proposed by Dunne *et al.* (1978). From study of foliation development around cordierite porphyroblasts in the Papoose Flat Pluton's aureole, Paterson *et al.* (1991a) suggested that much of the deformation may have occurred after pluton emplacement and associated contact metamorphism. Although the timing of deformation and pluton emplacement remains enigmatic, we maintain that the relatively high deformation temperatures (>450°C) indicated by the quartz mylonite **c**-axis fabrics dictate that the pluton must have been acting as a heat source at the time of the proposed thrusting-related mylonite formation. Nowhere outside the pluton's aureole has evidence been found for either mylonite formation or amphibolite facies regional metamorphism. If the regional thrusting model (Fig. 17(b)) is applicable, then the over-riding thrust sheet must have been moving during the pluton's early stages of cooling.

Taken in conjunction with the quartz mylonite results reported in the present paper, we interpret the observation by Paterson *et al.* (1991a) that foliation development in the pelitic units of the pluton's aureole is (at least locally) syn-post cordierite porphyroblast growth, as an indication that thrusting may have occurred over an appreciable interval of the pluton's cooling history. In this model the high temperature crystal fabrics in the quartz mylonite are interpreted as recording a relatively early stage in the pluton's thrusting-related cooling history, whereas the post-cordierite deformation in the schistose units may be recording continued thrusting-related deformation at progressively lower temperatures during pluton cooling. Because the Papoose Flat Pluton has yielded K–Ar ages on biotite of 75–81 Ma (Kistler *et al.* 1965), these lines of argument clearly limit the postulated thrusting to be late Cretaceous in age. It should be borne in mind, however, that regional mapping (Nelson 1966, 1971; Ross 1965, 1967a,b) revealed only local evidence for thrusting within the strata of the Papoose Flat area. In contrast, considerable evidence for large-scale thrusting has been recorded in the northern part of the White-Inyo Range (e.g. Dunne *et al.* 1978; Hanson *et al.* 1987).

The third model (Fig. 17(c)) takes structural elements from both of the previously described models. In this compromise model, we envisage magmatic wedging as operating at the same time as regional overthrusting. With the evidence available, it is difficult to assess the merits of this compromise model in relation to the second model. One potentially fruitful line of future research may be to map out in detail the change in foliation orientation traced outward from the pluton. For example, as illustrated by Brun and Pons (1981) and Brun (1983, fig. 5), the pattern of foliation development at the pluton scale may be diagnostic of synchronous forcible pluton emplacement and regional overthrusting.

The apparently dominant coaxial nature of deformation within the granitic rocks of the pluton's gneissic border facies remains to be explained. One possibility is that the scarcity of unequivocal shear-sense indicators in the granitic gneiss reflects a rheological contrast between the feldspar-rich granite and the rheologically weaker quartzite, marble and schist. As pointed out by Lister and Williams (1983, p. 13) a domainal distribution of materials with differing rheological properties within an evolving shear zone may result in strain path partitioning, strain accumulating coaxially in the rheologically stronger material while, at the same time, the weaker material undergoes non-coaxial flow. Alternatively, the inferred coaxial deformation in the gneissic border facies could reflect the influence of forcible pluton emplacement operating either during or before regional thrusting.

6. Conclusions

Microstructural and crystallographic fabric analysis indicates that mylonite formation around the western margin of the Papoose Flat Pluton is dominantly associated with plane strain deformation involving, in at least the pelitic and quartz-rich units, top-to-the-SE shearing. Crystal plastic deformation is restricted to the pluton's gneissic border facies and the overlying aureole rocks surrounding the western margin of the pluton. With the information presently available, these deformation features may be best explained by a tectonic model involving regional SE-directed thrusting of the cover rocks over the underlying pluton. This overthrusting may, or may not, have been accompanied by synchronous forcible emplacement of the underlying Papoose Flat Pluton. However, although the timing of deformation and pluton emplacement remains enigmatic, we maintain that the relatively high deformation temperatures (>450°C) indicated by the **c**-axis fabrics within the mylonites dictate that the pluton must have been acting as a heat source at the time of the proposed thrusting-related mylonite formation. Nowhere outside the pluton and its aureole has evidence been found for either mylonite formation or equivalent temperature regional metamorphism. Therefore, if the regional thrusting model is applicable, then the over-riding thrust sheet must have been moving during the pluton's early stages of cooling.

7. Acknowledgments

This study was supported by: (1) National Science Foundation Grant EAR-9018929 to Law and Sylvester, (2) a grant from the Creative Match Grants Program of Virginia Tech to Law, and (3) a Sigma Xi grant to Morgan. The authors wish to express their gratitude to Clem Nelson for his continued support during the course of this work, and to Robert Miller for his constructive review of an earlier version of the manuscript. Law particularly wishes to thank Rick Sibson for introducing him to the geology of the Papoose Flat Pluton and for providing logistic support during the early stages of the project.

8. References

Anovitz, L. M. & Essene, E. J. 1987. Phase equilibria in the system $CaCo_3$–$MgCo_3$–$FeCO_3$. J PETROL **28,** 389–414.

Bateman, P. C., Clark, L. C., Huber, N. K., Moore, J. G. & Rinehart, C. D. 1963. The Sierra Nevada batholith: a synthesis of recent work across the central part. US GEOL SURV PROF PAP **414-D**.

Bateman, R. 1984. On the role of diapirism in the segregation, ascent and final emplacement of granitoids. TECTONOPHYSICS **110**, 211–31.

Blumenfeld, P. & Bouchez, J.-L. 1988. Shear criteria in granite and migmatite deformed in the magmatic and solid states. J STRUCT GEOL **10**, 361–72.

Blumenfeld, P., Mainprice, D. & Bouchez, J. L. 1986. C-slip in quartz from subsolidus deformed granite. TECTONOPHYSICS **127**, 97–115.

Bouchez, J.-L., Lister, G. S. & Nicolas, A. 1983. Fabric asymmetry and shear sense in movement zones. GEOL RDSCH **72**, 401–19.

Brun, J. P. 1983. L'origine des domes gneissiques: modeles et tests. BULL GEOL SOC FRANCE **25**, 219–26.

Brun, J. P. & Pons, J. 1981. Strain patterns of pluton emplacement in a crust undergoing non-coaxial deformation. J STRUCT GEOL **3**, 219–29.

Brun, J. P., Gapais, D., Cogne, J. P., Ledru, P. & Vigneresse, J. L. 1990. The Flamanville Granite (NW France): an unequivocal example of a syntectonically expanding pluton. GEOL J **25**, 271–86.

Casey, M. 1981. Numerical analysis of X-ray texture data: an implementation in FORTRAN allowing triclinic or axial specimen symmetry and most crystal symmetries. TECTONOPHYSICS **78**, 51–64.

Castro, A. 1986. Structural pattern and ascent model in the Central Extramadura batholith, Hercynian belt, Spain. J STRUCT GEOL **8**, 645–63.

Courrioux, G. 1987. Oblique diapirism: the Criffel granodiorite-granite zoned pluton (SW Scotland). J STRUCT GEOL **9**, 313–30.

Dunne, G. C., Gulliver, R. M. & Sylvester, A. G. 1978. Mesozoic evolution of rocks of the White, Inyo, Argus, and Slate ranges, eastern California. *In* Howell, D. G. & McDougall, K. A. (eds) *Mesozoic paleogeography of the western United States*. Pacific Section, Society of Economic Paleontologists and Mineralogists, Pacific Coast Paleogeography Symposium 2, 189–208, Los Angeles, California.

Etchecopar, A. & Vasseur, G. 1987. A 3-D kinematic model of fabric development in polycrystalline aggregates: comparisons with experimental and natural examples. J STRUCT GEOL **9**, 705–17.

Gapais, D. & Barbarin, B. 1986. Quartz fabric transition in a cooling syntectonic granite (Hermitage Massif, France). TECTONOPHYSICS **125**, 357–70.

Godin, P. D. & Paterson, S. R. 1991. Re-evaluation of a blistering pluton: post-emplacement regional deformation at the Cretaceous Papoose Flat Pluton, White-Inyo Mountains, California. GEOL SOC AM ABSTR **23**(2), A29.

Guineberteau, B., Bouchez, J.-L. & Vignneresse, J.-L. 1987. The Mortagne granite pluton (France) emplaced by pull-apart along a shear zone: structural and gravimetric arguments, regional implication. BULL GEOL SOC AM **99**, 763–70.

Hanmer, S. & Passchier, C. 1991. *Shear-sense indicators: a review*. Geological Survey of Canada, Paper 90-17. Ottawa: Ministry of Supply and Services.

Hanson, R. B., Saleeby, J. B. & Fates, D. G. 1987. Age and tectonic setting of Mesozoic metavolcanic and metasedimentary rocks, northern White Mountains, California. GEOLOGY **15**, 1074–78.

Hibbard, M. J. 1987. Deformation of incompletely recrystallised magma systems: granite gneisses and their tectonic implications. J GEOL **95**, 543–61.

Hollister, L. & Crawford, M. L. 1986. Melt enhanced deformation: a major tectonic process. GEOLOGY **14**, 558–61.

Hutton, D. W. 1982. A tectonic model for the emplacement of the Main Donegal granite, NW Ireland. J GEOL SOC LONDON **139**, 615–31.

Hutton, D. W. 1988a. Granite emplacement mechanisms and tectonic controls: inferences from deformation studies. TRANS R SOC EDINBURGH EARTH SCI **79**, 245–55.

Hutton, D. W. 1988b. Igneous emplacement in a shear zone termination: the biotite granite at Strontian, Scotland. BULL GEOL SOC AM **100**, 779–94.

Jessell, M. W. 1988. Simulation of fabric development in recrystallizing aggregates—II: example model runs. J STRUCT GEOL **10**, 779–94.

Kistler, R. W., Bateman, P. C. & Branock, W. W. 1965. Isotopic ages of minerals from granitic rocks of the central Sierra Nevada and Inyo Mountains. BULL GEOL SOC AM **76**, 155–64.

Knipe, R. J. & Law, R. D. 1987. The influence of crystallographic orientation and grain boundary migration on microstructural and textural evolution in a S-C mylonite. TECTONOPHYSICS **135**, 155–69.

Law, R. D. 1990. Crystalographic fabrics: a selective review of their applications to research in structural geology. *In* Knipe, R. J. & Rutter, E. H. (eds) *Deformation mechanisms, rheology and tectonics*. SPEC PUBL GEOL SOC LONDON **54**, 335–52.

Law, R. D., Knipe, R. J. & Dayan, H. 1984. Strain path partitioning within thrust sheets: microstructural and petrofabric evidence from the Moine thrust zone at Loch Eriboll, northwest Scotland. J STRUCT GEOL **6**, 477–97.

Law, R. D., Casey, M. & Knipe, R. J. 1986. Kinematic and tectonic significance of microstructures and crystallographic fabrics within quartz mylonites from the Assynt and Eriboll regions of the Moine thrust zone, NW Scotland. TRANS R SOC EDINBURGH EARTH SCI **77**, 99–123.

Law, R. D., Morgan, S. S. & Sylvester, A. G. 1990. Evidence for large scale shearing during emplacement of the Papoose Flat Pluton, California: a re-examination of the microstructures and crystallographic fabrics. GEOL SOC AM ABST **22**(7), A183.

Lister, G. S. & Hobbs, B. E. 1980. The simulation of fabric development during plastic deformation: the effect of deformation history. J STRUCT GEOL **2**, 355–70.

Lister, G. S. & Snoke, A. 1984. S-C mylonites. J STRUCT GEOL **6**, 617–38.

Lister, G. S. & Williams, P. F. 1983. The partitioning of deformation in flowing rock masses. TECTONOPHYSICS **92**, 1–33.

Morgan, S. S., Law, R. D. & Sylvester, A. G. 1991. Strain-path partitioning during contact metamorphism and emplacement of the Papoose Flat pluton, eastern California. GEOL SOC AM ABSTR PROG **23**(5), A175.

Nelson, C. A. 1962. Lower Cambrian–Precambrian succession, White Inyo Mountains, California. BULL GEOL SOC AM **73**, 139–44.

Nelson, C. A. 1966. *Geologic map of the Waucoba Mountain quadrangle, Inyo County, California*. U.S. Geological Survey Quadrangle Map GQ-528, scale 1:62,500. Reston, Virginia: U.S. Geological Survey.

Nelson, C. A. 1971. *Geologic map of the Waucoba Spring quadrangle, Inyo County, California*. U.S. Geological Survey Quadrangle Map GQ-921, scale 1:62,500. Reston, Virginia: U.S. Geological Survey.

Nelson, C. A., Oertel, G., Christie, J. M. & Sylvester, A. G. 1977. *Geologic map, structure sections and palinspastic map of the Papoose Flat pluton, Inyo Mountains, California*. GEOL SOC AM MAP & CHART SERIES MC-20.

Paterson, S. R., Vernon, R. H. & Tobisch, O. T. 1989. A review of criteria for the identification of magmatic and tectonic foliations in granitoids. J STRUCT GEOL **11**, 349–63.

Paterson, S. R., Brudos, T., Fowler, K., Carlson, C., Bishop, K. & Vernon, R. H. 1991a. Papoose Flat Pluton: forceful expansion or postemplacement deformation. GEOLOGY **19**, 324–7.

Paterson, S. R., Vernon, R. H. & Fowler, T. K. 1991b. Aureole tectonics. REV MINERAL **26**, 673–722.

Paterson, S. R. & Tobisch, O. T. 1988. Using pluton ages to date regional deformations: problems with recently used criteria. GEOLOGY **16**, 1108–11.

Price, G. P. 1985. Preferred orientations in quartzites. *In* Wenk, H.-R. (ed.) *Preferred orientations in deformed deformed metals and rocks: an introduction to modern texture analysis*, 385–406. Orlando: Academic Press.

Ramsay, J. G. 1989. Emplacement kinematics of a granite diapir: the Chindamora batholith, Zimbabwe. J STRUCT GEOL **11**, 191–209.

Rice, J. M. 1977. Contact metamorphism of impure dolomite limestone in the Boulder aureole, Montana. CONTRIB MINERAL PETROL **59**, 237–59.

Ross, D. C. 1965. Geology of the Independence quadrangle, Inyo County, California. BULL US GEOL SURV **1181-0**, 64.

Ross, D. C. 1967a. *Geologic map of the Waucoba Wash quadrangle, Inyo County, California*. U.S. Geological Survey, Geological Quadrangle Map GQ-612, scale 1:62,500.

Ross, D. C. 1967b. *Generalised map of the Inyo Mountains region*. U.S. Geological Investigations Map I-506, scale 1:125,000.

Ross D. C. 1969. Description petrography of three large granitic bodies in the Inyo Mountains, California. U.S. Geological Survey, Professional Paper 601, 47p.

Schmid, S. M., Panozzo, R. & Bauer, S. 1987. Simple shear experiments on calcite rocks: rheology and microfabric. J

STRUCT GEOL **9,** 747–78.

Schmid, S. M. & Casey, M. 1986. Complete fabric analysis of some commonly observed quartz **c**-axis patterns. *In* Heard, H. C. & Hobbs, B. E. (eds) *Mineral and Rock Deformation: Laboratory Studies, the Paterson Volume.* AM GEOPHYS UNION, GEOPHYS MONOGR **36,** 263–86.

Simpson, C. 1986. Determination of movement sense in mylonites. J GEOL EDUC **34,** 246–61.

Simpson, C. & Schmid, S. M. 1983. An evaluation of criteria to deduce the sense of movement in sheared rocks. BULL GEOL SOC AM **94,** 1281–8.

Starkey, J. 1989. *Microcomputer Workshop: Quantitative Petrofabric Analysis,* 9th Annual Meeting, Structural Geology & Tectonics Division, Geological Association of Canada, London, Ontario, Canada.

Sylvester, A. G. 1966. *Structural and metamorphic petrology of the contact aureole of the Papoose Flat Pluton, Inyo Mountains, California.* Unpublished Ph.D. thesis, University of California at Los Angeles.

Sylvester, A. G. 1969. A Microfabric study of calcite, dolomite and quartz around the Papoose Flat Pluton, California. BULL GEOL SOC AM **80,** 1311–28.

Sylvester, A. G., Oertel, G., Nelson, C. A. & Christie, J. M. 1978. Papoose Flat pluton: a granitic blister in the Inyo Mountains, California. BULL GEOL SOC AM **89,** 1205–19.

Sylvester, A. G. & Babcock, J. W. 1975. Significance of multiphase folding in the White-Inyo range, eastern california. GEOL SOC AM ABSTR PROG **7,** 1289.

Sylvester, A. G. & Christie, J. M. 1968. The origin of crossed-girdle orientations of optic axes in deformed quartzites. J GEOL **76,** 571–80.

Tullis, J., Christie, J. M. & Griggs, D. T. 1973. Microstructures and preferred orientations of experimentally deformed quartzites. BULL GEOL SOC AM **84,** 297–314.

Urai, J. L., Means, W. D. & Lister, G. S. 1986. Dynamic recrystallization of minerals. *In* Heard, H. C. & Hobbs, B. E. (eds) *Mineral and rock deformation: laboratory studies, the Paterson Volume.* AM GEOPHYS UNION, GEOPHYS MONOGR **36,** 161–99.

Wenk, H.-R., Takeshita, T., Bechler, E., Erskine, B. G. & Matthies, S. 1987. Pure shear and simple shear calcite textures. Comparison of experimental, theoretical and natural data. J. STRUCT GEOL **9,** 731–46.

R. D. LAW, S. S. MORGAN & M. NYMAN, Department of Geological Sciences, Virginia Polytechnic Institute & State University, Blacksburg, Virginia 24061, U.S.A.

M. CASEY, Geologisches Institut, ETH-Zentrum, Zurich CH 8092, Switzerland.

A. G. SYLVESTER, Department of Geological Sciences, University of California at Santa Barbara, Santa Barbara, California 93106, U.S.A.

Granite sheeted complexes: evidence for the dyking ascent mechanism

Donald H. W. Hutton

ABSTRACT: A review of granite emplacement mechanisms in transcurrent, extensional and contractional (thrust sense) shear zones reveals that in all three tectonic settings the plutons have been constructed by multiple granite sheeting parallel to the shear zone walls and deformation fabrics. The sheets and plutons are non-Andersonian in type and were emplaced obliquely to the principal stress directions. Their shape and orientation is more likely to reflect the exploitation of faults and shear zones which were active during emplacement. Sheeting (dyking) is therefore also likely to be the mechanism of ascent along fault zones in the crust.

KEY WORDS: dykes, emplacement, ascent, granite viscosity.

The mechanism by which granitic magma ascends through the continental crust is a fundamental problem because there is a lack of clear observational data on the subject. The process has not been observed and it is unclear what features, if any, which are preserved in plutons relate to their transport or ascent mechanism. Of the few physically possible ascent mechanisms (stoping, melting, fracture transport, diapirism; see Miller *et al.* (1988) for a review) by the late 1960s diapirism had emerged as the most attractive candidate. It appeared to be physically feasible and could be mathematically modelled; analogues could be seen in other low-density geological materials (serpentinite and salt); it was physically reproducible in analogue centrifuge experiments; and, perhaps most importantly, tear-drop shaped plutons, interpreted as diapirs frozen in the act, could be seen in the geological record. By the early 1980s however, two things had happened to question diapirism as the general granite ascent mechanism. Firstly, it was appreciated by a number of authors that diapirism was just as much a thermal process as it was a buoyancy process, since granitic diapirs needed to heat and soften the country rocks around them for ascent to continue. Thus the track length or ascent distance of a thermally isolated, convecting diapir is now considered by many to be largely confined to the hot and ductile lower crust and to be of limited applicability to the cooler and more brittle upper crust (Marsh 1982; Daly & Raefsky 1985; Brun *et al.* 1990). Secondly, a number of key studies on the deformation associated with probable granitic diapirs (Ardara (Holder 1979), Chindamora (Ramsay 1989)) showed that these plutons had gained their shapes by *in situ* expansion at the emplacement level (ballooning). Thus, although evidence of diapirism (independent upward ascent or piercement of a spherical or subspherical granitic body driven by buoyancy) has subsequently been identfied in a number of plutons (e.g. Criffel (Courrioux 1987), Arran (England 1990)) it is now appreciated, more than ever before, that in plutons features related to the ascent phase may be easily confused with or even entirely overprinted by features related to the emplacement mechanism. Thus what we see at the emplacement level may only relate to the emplacement process and may have little or nothing to do with the ascent mechanism. This somewhat nihilistic approach to the field data, combined with the objections raised by the thermal modelling, has tended to dampen enthusiasm for this ascent mechanism (Bateman 1984; Brun *et al.* 1990). There are probably a number of plutons in the world which retain evidence of the vertical ascent and thus are technically diapirs, but a great many others are neither diapirs nor balloons, and cannot possibly qualify for inclusion in this category. Of the remaining processes, ascent by fracture transport (Turcotte & Emerman 1985) is physically possible, but has had few adherents probably because of a lack of observational data. In what follows, the author presents data from the emplacement level, and therefore possibly one stage removed, for ascent by a sheeting or dyking process. This has arisen from work on emplacement in tectonically controlled settings, which will be reviewed before considering the implications for ascent.

1. Emplacement in tectonic settings

During the last ten years or so, much of the research in granite emplacement mechanisms has concentrated on intrusion in, or associated with, major tectonic structures. This has particularly been the case with emplacement in transcurrent fault systems; a number of models have been developed to account for the fabric patterns, strain variations, shear sense and pluton geometries found in the various situations. Many of the models derived from these data bear an obvious resemblance to structural geometries in other geological situations (basins, volcanic centres, etc.) and especially to smaller-scale vein-type analogues. Thus granite emplacement is seen in rhombohedral pull-aparts (Mortagne (Guineberteau *et al.* 1987); Boulder and Tobacco Root (Schmidt *et al.* 1990); Rehamma (Lagarde *et al.* 1990); in fault releasing bends (Wyangala (Morand 1988); in tension gashes (Extramadura (Castro 1986); Varmala (Nironen 1989)); in shear zone terminations (Strontian (Hutton 1988a); Easter Jebilet (Lagarde *et al.* 1990); Fanad (White & Hutton 1985)); and in dilational volumes produced by other longitudinal shear strain gradients (Main Donegal Granite (Hutton 1982)). In all of these models, a strong appeal is made for space creation within active, developing fault systems. A similar theme has arisen from a fewer number of studies of syntectonic emplacement in extensional regimes (Hutton *et al.* 1990; Koukoukevlas & Pe-Piper 1991) where space is created by the extensional displacement across ramp-flat geometries.

Very little structurally based work has been carried out on emplacement in thrust regimes. De Wit *et al.* (1987) record the emplacement of tabular sill-like intrusions along thrusts during the late stages of movement in the Barberton Belt of southern Africa. Similarly John (1988) describes sill-like intrusions occupying undulating major flat-lying shear zones in the Chemenheuvi Mountains, California. Schmidt *et al.* (1990), based on seismic data, propose that the main mass of the Boulder Batholith, Montana, is bounded at depth by a major flat-lying thrust of the Rocky Mountain System. Probably the most studied pluton of this type is the Great Tonalite Sill of Southeastern Alaska and British Columbia (Hutton & Ingram 1992; Brew & Ford 1981; Crawford & Hollister 1982). This pluton, which is a series of narrow lensoid bodies, which are steeply inclined and concordant with their envelope rocks, runs for over 800 km along strike and yet is usually only less than 20 km wide. Although it is strongly deformed, the preservation of magmatic features within it demonstrates that the long narrow geometry is largely original and not solely the result of imposed strain. Detailed structural studies (summarised in Hutton & Ingram 1992) show that the pluton was first deformed before it had completely crystallised and that emplacement occurred totally within a strike-parallel, steeply continentward-inclined, high-angle, reverse shear zone, in which there was a considerable amount of additional across-strike shortening.

Hutton and Ingram have not detected any evidence of early PFC (pre-full crystallisation, Hutton 1988b) extensional deformation in the pluton which could be related to space-creating mechanisms. On the contrary, the earliest PFC "lock up" shears are contractional and can be related to across-strike shortening along the entire zone. The principal conclusion from the Great Tonalite Sill studies is that emplacement occurred along the median plane of an active contractional (high-angle reverse thrust) shear zone. Clearly the Great Tonalite Sill was able to overcome contractional orogenic stresses during its emplacement and space must have been created by some type of forceful mechanism. Yet the pluton is clearly not a series of diapirs or balloons. The Great Tonalite Sill, as an example of emplacement in a contractional orogenic setting, therefore stands in marked constrast to emplacement models in strike-slip and extensional regimes, where the inferred primary space-creating mechanism is dilation along or associated with the major structures. Given this contrast, is there anything which is common to all three tectonic settings in terms of emplacement process? For the plutons with which the author is most familiar, or those whose internal geometries are relatively well described, it is believed that the unifying factor is that they are internally sheeted bodies. Before discussing the significance of this, in the following section a brief description of a number of sheeted plutons will be given.

2. Sheeted granite plutons

Perhaps the best and clearest example of a sheeted granite pluton is the Ox Mountains Granite, which has been described by McCaffrey (1989, 1992). This 27 km-long and 6 km-wide Caledonian pluton lies astride the postulated trace of the Highland Boundary Fault in western Ireland (Hutton & Dewey 1986). Although very strongly deformed by sinistral movements associated with this major transcurrent shear zone, the internal geometries of its petrographic components are well preserved. These are units of diorite, granodiorite and tonalite which occur in strike-parallel, steeply-inclined sheets intruded into each other. Individual sheets reach up to ~2 km long and 300 m wide, although many are much smaller than this. In places, overlapping sheet tips and bridges are preserved, suggesting original *en echelon* emplacement geometries. Chilled margins are absent, indicating that the host granites were still hot during the emplacement of later phases.

In terms of the extensional tectonic setting, the basic sheet-like form of these plutons is readily apparent (Hutton *et al.* 1990; Koukoukevlas & Pe-Piper 1991). In Greenland, the gently and steeply-inclined sheets are up to 500 m thick and their flat segments may have aerial extents of tens or even hundreds of square kilometres (Hutton *et al.* 1990; Bridgewater *et al.* 1974). With the Great Tonalite Sill, the long thin sheet-like geometry of the individual intrusions and the entire body is particularly obvious. In addition, the external country rock contacts are typically sheeted by tonalite bodies up to 50 m thick. Ingram (per. comm.) has also discovered an internally sheeted complex in one part of the pluton, the sheet boundaries being differentiated by chilling. In less well-constrained situations, de Wit *et al.* (1987) depict tabular sheet-like intrusions along gently inclined thrusts in the Barberton Belt and Schmidt *et al.* (1990) display a similar geometry for the central part of the Boulder Batholith, Montana.

While in a number of such situations internal petrographical variations allow the composite sheeted nature of the intrusions to be identified, this may not be very common. Plutons which are constructed by sheets exhibiting little petrographical variation may render the identification of these sheets very difficult. Such cryptic sheeted bodies may only be recognised by the existence of "ghost stratigraphy". Probably the best described example of this in the world is in the Main Donegal Granite.

As first described by Pitcher and Read (1959) and later discussed by Pitcher (1970) and Pitcher and Berger (1972), within the Main Donegal Granite there are a number of strike-parallel zones of country rock xenoliths (Figs 1, 2). Within each of these zones, there is an integrity of country rock lithology, and even sequence, which in many cases can be traced for tens of kilometres along strike. Remarkably, most of these "raft trains" can be traced back into the country rocks at the pluton margins and, in one spectacular case, into a major downward-protruding prong or roof pendant of country rock. Thus the arrangement of the variable and well-differentiated stratigraphy of the late Precambrian country rock metasediments (pelites, quartzites, limestones) and an older Donegal granite is preserved as xenoliths within the pluton itself. Since petrogenesis by *in situ* melting can be ruled out for this pluton (Pitcher & Berger 1972), and the xenoliths are thus unlikely to be relict "resistors", the only sensible interpretation must be that the xenoliths are very largely *in situ*. That being the case, and given that many of the xenoliths can be shown to be surrounded on the top, bottom and sides by granite, then the granite must have emplaced itself around them as a series of interdigitating and coalescing sheets. The existence in parts of the pluton of *en echelon*, right-stepping raft arrangements (Pitcher & Read 1959, Fig. 10) may be a reflection of *en echelon* sheeting, supporting a sinistral component of shear associated with emplacement (Hutton 1982). The width of the thickest, first-order sheets in the pluton is presumably reflected by the separations of the major raft trains and these vary between 0·5 and 2 km.

The Main Donegal Granite may be typical of a large number of plutons which have been emplaced by a multiple sheeting process, but which exhibit little petrographical variation. It is also possible (Pitcher & Berger 1972) that petrographically similar sheets with high remaining melt contents, when emplaced into each other, may undergo a degree of physical and chemical mixing across their boundaries. Thus to some extent the multiple sheets of such plutons may coalesce and the plutons may appear to be relatively homogeneous. The existence of ghost stratigraphy and separated raft trains may therefore be one of the few clues available for this type of emplacement process.

3. The significance of sheeted complexes: non-Andersonian bodies

Anderson (1951) developed a model of dyke emplacement by wedging (Fig. 3) in which the slight excess pressure of the magma over the compressive (tectonic) wall-rock stress causes a tension at the tip of the wedge. As pointed out by Pollard and others (Pollard & Johnston 1973; Roberts 1970) such a force acting across a narrow tip creates very high localised tensile and shear stresses which easily overcome the wall-rock strength and allow the rapid propagation of the dyke tip. Because of the need for the slight excess

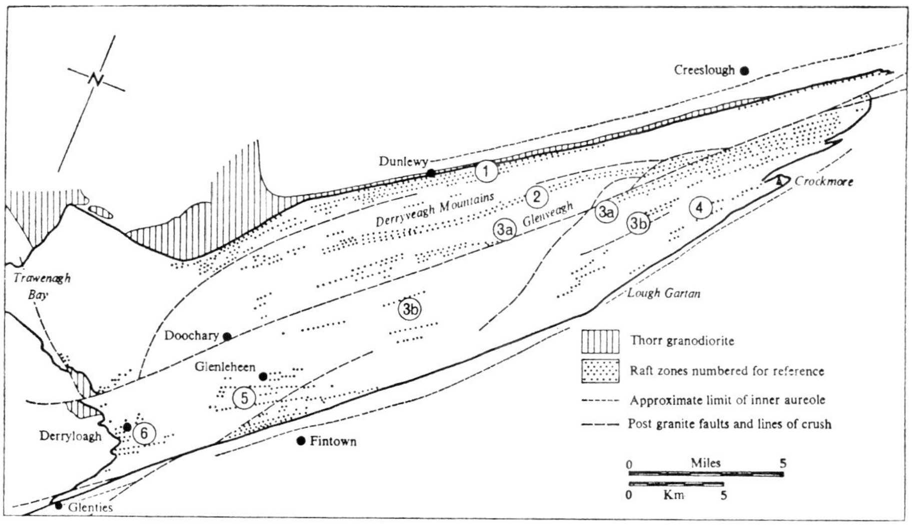

Figure 1 Map of the Main Donegal Granite taken from Pitcher (1970) and reproduced with permission: this shows a simplified representation of the distribution of the xenolith trains in the pluton; in each of the trains (numbered 1–6) the xenoliths have a range of lithology specific to the train; the trains therefore preserve a ghost stratigraphy.

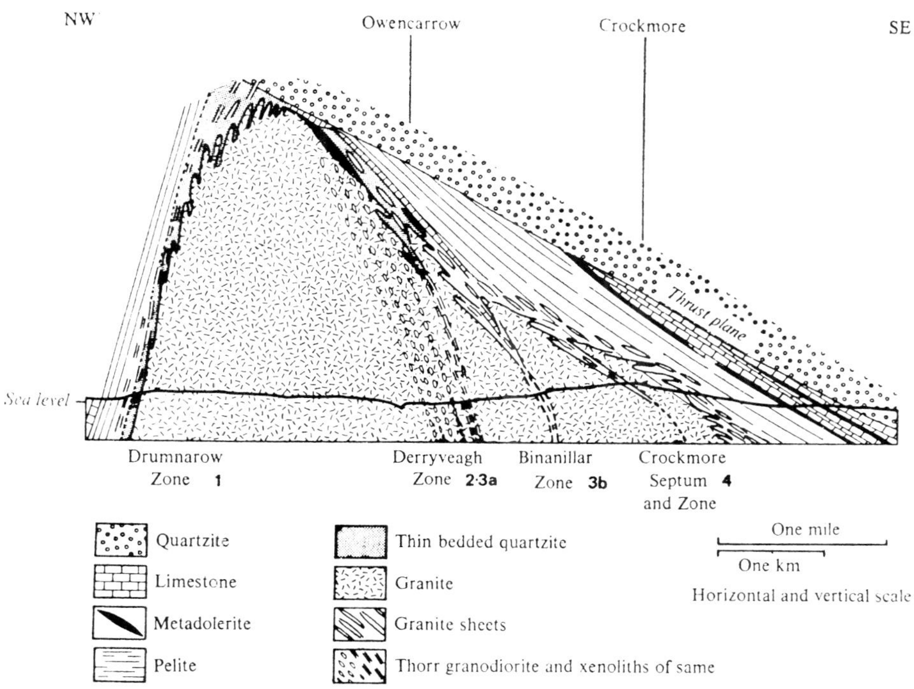

Figure 2 Schematical and generalised cross-section of the Main Donegal Granite taken from PItcher (1970) and reproduced by permission.

pressure to allow propagation to occur, Anderson argued that dykes would tend to align themselves perpendicular to the least compressive stress and parallel to the maximum compressive stress. Thus dykes should, in the presence of a regional tectonic stress, form swarms parallel to σ_1. At high levels in the crust, the same control was held to apply and close to the surface the weight of overlying rock diminishes, the least compressive stress becomes vertical, and the igneous sheets become flat and sill-like.

In the cases of the Great Tonalite Sill, the Main Donegal Granite and the S Greenland Rapakivi granites (Fig. 4) these are all steep or (in the case of S Greenland) on average moderately inclined. In no case do they lie at right angles to the minimum compressive stress or parallel to the maximum compressive stress (Fig. 4). In the case of the Great Tonalite Sill, the body lies nearly at right angles σ_1, and for both the Main Donegal Granite and the Greenland Rapakivi bodies the sheets are oblique to the inferred directions of both the minimum and maximum compressive stresses. The first conclusion therefore of this section is that the igneous sheets involved in these plutons do not behave according to the well-tested and accepted models of E. M. Anderson. This author proposes that their shapes and geometries are generally unrelated to the orientations and magnitudes of crustal stress systems but they are most closely related to the shapes and geometries of the active shear zones which they intrude.

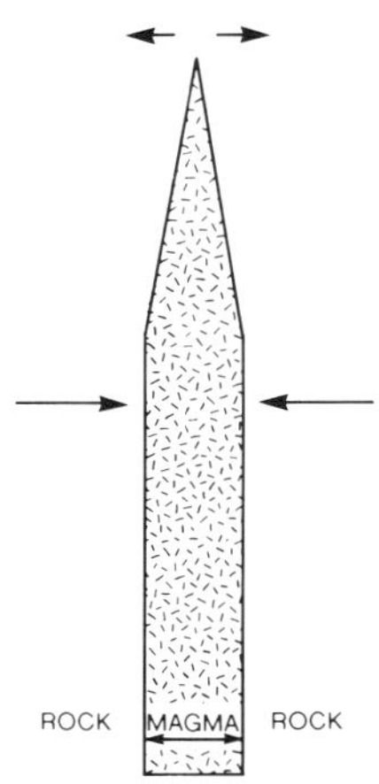

Figure 3 Magmatic wedging; after Anderson (1951); see text for details.

The proposal is therefore that the sheets exhibit exploitation and accommodation geometries which are the energetically most favourable way for a body of magma to occupy long, narrow zones of weakness or potential dilation space. They are likely to emplace along the median planes of these zones because this is where high continuing strains occur, where the rocks are rendered weakest for the longest periods of time and where sheet-tip propagation stresses can be most effective. In essence, this model deals with the natural consequences of intruding magma into crust which contains active faults and shear zones, rather than into crust which is isotropic and flawless as is perhaps more appropriate to the Anderson model. Multiple granite sheeting as a method of constructing plutons in tectonically controlled settings is a new mechanism of emplacement which must be at least as important as other classical intrusion mechanisms (diapirism, ballooning, cauldrons, stoping, etc.). When it operates with magmas which change little during the time taken for pluton construction, it will produce rather homogeneous-looking plutons, whose only real evidence of their emplacement mechanism is the presence of ghost stratigraphy: such cryptically sheeted plutons may be much more common than has been appreciated.

4. Implications for magmatic ascent

If sheet propagation stresses can exploit planes of weakness at the emplacement level, so as to build up granitic plutons, and since the shear and fault zones in all cases cut deep into the crust either vertically or obliquely, then propagation stresses can exploit these anisotropies below the level of emplacement and so they are also a mechanism for magmatic ascent. Thus although there is a philosophical point to be made about confusing the identification of

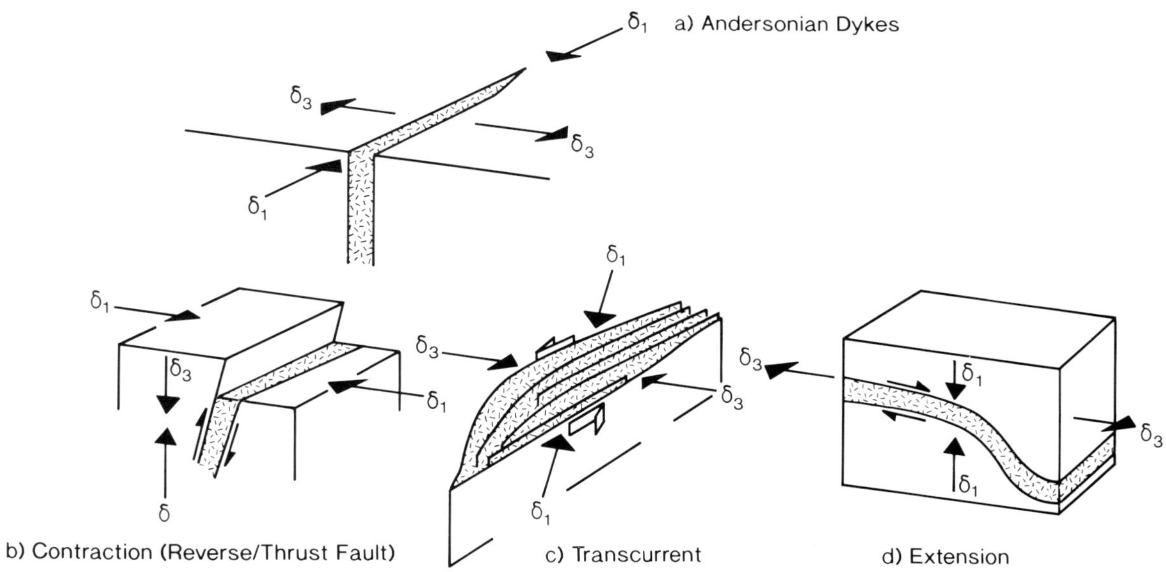

Figure 4 (a) Andersonian dykes emplaced in a regional stress field at right angles to the minimum principal stress and parallel to the maximum principal stress. (b)–(d) Non-Andersonian sheets typical of emplacement mechanisms in tectonically controlled situations: (b) Contraction (reverse/thrust situations) (Great Tonalite Sill, S.E. Alaska); (c) Transcurrent (Main Donegal Granite); (d) Extension (Rapakivi Granite, S Greenland). All show emplacement oblique to and often at high angles to the principal stresses.

emplacement and ascent mechanisms (based mainly on the problem of the confusion of balloons and diapirs) the author believes that, in this example, a strong case can be made for the preservation of the essential ascent mechanism at the emplacement level. It seems reasonable to view the whole process as one in which magma ascends as sheets along deep-reaching faults and shear zones which intercept the melt regions. At certain levels in the crust and positions within these shear zones, dilation zones occur and at these points the sheets begin to exploit the situation by repeated injection until whole plutons are constructed. The dilation, however, need not be a necessary component and, as the Great Tonalite Sill so vividly illustrates, magma, if produced in sufficient quantity, will simply find space for itself by wedging the walls of the shear zone apart, even in a strictly contractional regime.

One of the restraints on transporting magma up narrow dyke-like conduits is the viscous drag encountered against the dyke walls. Opposing this and driving the magma upwards (Lister & Kerr 1991) is the buoyancy head of the magma and the pressure difference between source and emplacement level. In general, the narrower the dyke, the lower must be the viscosity in order for ascent, at a rate fast enough to prevent premature crystallisation in the conduit, to occur. Current textbooks (e.g. Hall 1987) quote large ranges in granite viscosities, with the highest viscosities many orders of magnitude larger than that of basaltic magmas. Very high viscosities for granites would simply prevent their ascent along the sorts of sheet widths which have been described in this paper. Therefore, the final conclusion is that the identification of granitic ascent-related sheets whose widths are up to only three to four orders of magnitude larger than the widths of typical basaltic sheets places constraints on the viscosities of granitic magmas, and implies that these may be much lower than is generally held to be the case.

5. Acknowledgements

These data and ideas have arisen through nearly ten years of research on granite emplacement mechanisms in tectonic settings that has been done in conjunction with past and present research students, research assistants and close colleagues. I thank Dunstan Cooper, Gabriel Courrioux, Richard England, Gary Ingram, John Jacques, Chris Jones, Ken McCaffrey, Michele McErlean, Bob Miller, and John Reavy for their contributions to this programme and for their discussion. Wally Pitcher is thanked for his continuing encouragement of this work, and finally I would like to acknowledge the hard work, kindness and hospitality shown by Bruce Chappell, Alan White, Ian Williams and Charlotte Allen at the Canberra meeting in 1991.

6. References

Anderson, E. M. 1951. *The dynamic of faulting*. Edinburgh: Oliver & Boyd.

Bateman, R. 1984. On the role of diapirism in the segregation, ascent and final emplacement of granitoids. TECTONOPHYSICS **110,** 210–31.

Brew, D. A. & Ford, A. B. 1981. The Coast Plutonic Complex Sill, Southeastern Alaska. *In* Albert, N. R. D. & Hudson, T. (eds) *The United States Geological Survey in Alaska—accomplishments during* 1979, USGS CIRC 823-B, 96–9.

Bridgewater, D., Sutton, J. & Watterson, J. 1974. Crustal downfolding associated with igneous activity. TECTONOPHYSICS **21,** 57–77.

Brun, J. P., Gapais, D., Cogne, J. P., Ledru, P. & Vigneresse, J. L. 1990. The Flamanville Granite (Northwest France): an unequivocal example of a syntectonically expanding pluton. GEOL J **25,** 271–86.

Castro, A. 1986. Structural pattern and ascent model in the Central Extramadura batholith, Hercynian belt, Spain. J STRUCT GEOL **8,** 633–45.

Courrioux, G. 1987. Oblique diapirism: the Criffel granodiorite/granite zoned pluton (southwest Scotland). J STRUCT GEOL **9,** 89–124.

Crawford, M. L. & Hollister, L. S. 1982. Contrasts of metamorphic and structural histories across the Work Channel Lineament, Coast Plutonic Complex, British Columbia. J GEOPHYS RES **87,** 3847–60.

Daly, S. F. & Raefsky, A. 1985. On the penetration of a hot diapir through a strongly temperature-dependent viscosity medium. GEOPHYS J R ASTRON SOC **83,** 657–82.

de Wit, M. J., Armstrong, R., Hart, R. J. & Wilson, A. H. 1987. Felsic igneous rocks within 3·3 to 3·5 Ga Barberton Greenstone Belt: high crustal level equivalents of the surrounding tonalite-trondjemite terrain, emplaced during thrusting. TECTONICS **6,** 529–49.

England, R. W. 1990. The identification of granitic diapirs. J GEOL SOC LONDON **147,** 931–3.

Guineberteau, B., Bouchez, J. L. & Vigneresse, J. L. 1987. The Mortagne Granite pluton (France) emplaced by pull-apart along a shear zone: structural and gravimetric arguments, regional implications. BULL GEOL SOC AM **99,** 763–70.

Hall, A. 1987. *Igneous petrology*. Harlow: Longman.

Holder, M. T. 1979. An emplacement mechanism for post tectonic granites and its implications for their geochemical features. *In* Atherton, M. J. (eds) *Origin of granite batholiths: geochemical evidence*, 116–128. Nantwich: Shiva.

Hutton, D. H. W. 1982. A tectonic model for the emplacement of the Main Donegal Granite, NW Ireland. J GEOL SOC LONDON **139,** 615–31.

Hutton, D. H. W. 1988a. Igneous emplacement in a shear-zone termination: the biotite granite at Strontian, Scotland. GEOL SOC AM BULL **100,** 1392–9.

Hutton, D. H. W. 1988b. Granite emplacement mechanisms and tectonic controls: inferences from deformation studies. TRANS R SOC EDINBURGH EARTH SCI **79,** 245–55.

Hutton, D. H. W. & Dewey, J. F. 1986. Palaeozoic terrane accretion in the Western Irish Caledonides. TECTONICS **5,** 1115–24.

Hutton, D. H. W. & Ingram, G. M. 1992. The Great Tonalite Sill of South East Alaska and British Columbia: emplacement into an active contractional high angle reverse shear zone. TRANS R SOC EDINBURGH EARTH SCI **83,** 383–6.

Hutton, D. H. W., Dempster, T. J., Brown, P. E. & Becker, S. M. 1990. A new mechanism of granite emplacement: intrusion in active extensional shear zones. NATURE **343,** 452–5.

John, B. E. 1988. Structural reconstruction and zonation of a tilted mid-crustal magma chamber: the felsic Chemehuevi Mountains plutonic suite. GEOLOGY **16,** 613–7.

Koukoukevlas, I. & Pe-Piper, G. 1991. The Oligocene Xanthi pluton, northern Greece: a granodiorite emplaced during regional extension. J GEOL SOC LONDON **148,** 749–58.

Lagarde, J. L., Omar, S. A. & Roddaz, B. 1990. Structural characteristics of granitic plutons emplaced during weak regional deformation: examples from late Carboniferous plutons, Morocco. J STRUCT GEOL **12,** 805–21.

Lister, J. R. & Kerr, R. C. 1991. Fluid-mechanical models of crack propagation and their application to magma transport in dykes. J GEOPHYS RES **96** (B6), 10,049-77.

Marsh, B. D. 1982. On the mechanics of igneous diapirism, stoping and zone melting. AM J SCI **282,** 808–55.

McCaffrey, K. J. W. 1989. *The emplacement and deformation of granitic rocks in a transpressional shear zone: the Ox Mountains Igneous Complex*. Unpublished Ph.D. Thesis, University of Durham.

McCaffrey, K. J. W. 1992. Igneous emplacement in a transpressive shear zone: the Ox Mountains igneous complex. J GEOL SOC LONDON (in press).

Miller, C. F., Watson, E. B. & Harrison, T. M. 1988. Perspectives on the source, segregation and transport of granitoid magmas. TRANS R SOC EDINBURGH EARTH SCI **79,** 135–56.

Morand, V. J. 1988. Emplacement and deformation of the Wyangala Batholith, New South Wales. AUS J EARTH SCI **35,** 339–53.

Nironen, M. 1989. Emplacement and structural setting of granitoids in the early Proterozoic Tampere and Savo schist belts, Finland—implications for contrasting crustal evolution. GEOL SURV FINLAND BULL **346**.

Pitcher, W. S. 1970. Ghost stratigraphy in granites: a review. *In* Newall, G & Rast, N. (eds) *Mechansims of igneous intrusion.* GEOL J SPEC PUBL **2,** 123–140.

Pitcher, W. S. & Berger, A. R. 1972. *The Geology of Donegal: a study of granite emplacement and unroofing.* London: Wiley.

Pitcher, W. S. & Read, H. H. 1959. The Main Donegal Granite. Q J GEOL SOC LONDON **114,** 259–305.

Pollard, D. D. & Johnston, A. M. 1973. Mechanics of growth of some laccolithic intrusions in the Henry Mountains, Utah. II Bending and failure of overburden layers and sill formation. TECTONOPHYSICS **18,** 311–45.

Ramsay, J. G. 1989. Emplacement Kinematics of a granite diapir: the Chindamora batholith, Zimbabwe. J. STRUCT GEOL **11,** 191–209.

Roberts, J. L. 1970. The intrusion of magma into brittle rocks. *In* Newall, G & Rast, N. (eds) *Mechanisms of igneous intrusion.* GEOL J SPEC PUBL **2,** 287–338.

Schmidt, C. J., Smedes, H. W. & O'Neill, J. M. 1990. Syncompressional emplacement of the Boulder and Tobacco Root Batholiths (Montana, U.S.A.) by pull apart along old fault zones. GEOL J **25,** 305–18.

Turcotte, D. L. & Emerman, S. H., 1985. Magma fracture as a mechanism for magma migration. EOS **66,** 361.

White, N. J. & Hutton, D. H. W. 1985. The structure of the Dalradian rocks in West Fanad, County Donegal. IRISH J EARTH SCI **7,** 79–92.

DONALD H. W. HUTTON, Department of Geological Sciences, University of Durham, Durham DH1 3LE, England, U.K.

Transactions of the Royal Society of Edinburgh: Earth Sciences, **83**, 383–386, 1992

The Great Tonalite Sill of southeastern Alaska and British Columbia: emplacement into an active contractional high angle reverse shear zone (extended abstract)

Donald H. W. Hutton and Gary M. Ingram

The Great Tonalite Sill (GTS) of southeastern Alaska and British Columbia (Brew & Ford 1981; Himmelberg *et al.* 1991) is one of the most remarkable intrusive bodies in the world: it extends for more than 800 km along strike and yet is only some 25 km or less in width. It consists of a belt of broadly tonalitic sheet-like plutons striking NW–SE and dipping steeply NE, and has been dated between 55 Ma and 81 Ma (J. L. Wooden, written communication to D. A. Brew, April 1990) (late Cretaceous to early Tertiary). The sill (it is steeply inclined and rather more like a "dyke") is emplaced along the extreme western margin of the Coast Plutonic and Metamorphic Complex (CPMC), the high grade core of the Western Cordillera. The CPMC forms the western part of a group of tectonostratigraphic terranes including Stikine and Cache Creek, collectively known as the Intermontane Superterrane (Rubin *et al.* 1990). To the W of the GTS, rocks of the Insular Superterrane, including the Alexander and Wrangellia terranes and the Gravina belt, form generally lower metamorphic grade assemblages. The boundary between these two superterranes is obscure but it may lie close to, or be coincident with, the trace of the GTS.

The rocks of the GTS are heterogeneous but a medium- to coarse-grained hornblende-biotite tonalite is the dominant rock type. Garnet is a commonly occurring accessory phase. Medium-grained diorites are also found and microdiorite enclaves and synplutonic dykes are common. Granodiorite and late stage granite, trondhjemite, and alaskitic pegmatite sheets are less abundant. Coarse, euhedral hornblende-bearing rocks, of dominantly ultramafic composition (appinite in the British usage) occur as small bodies often along contacts between the intrusion and the country rock and these are sometimes magmatically disrupted and mingled with other phases.

The significance and origin of this remarkably-shaped body are intriguing questions. This paper summarises a new set of results concerning the structure and deformational history of the pluton and its western wall rocks. Our data come from some 18 detailed traverses which we have carried out across the pluton between Haines in Alaska and the Skeena River in British Columbia (Fig. 1) together with the detailed mapping of Ingram in the northern Juneau area. Although a number of studies on the structural setting and deformation of the GTS have been published, we believe that we are the first to have systematically examined these features along its entire length. The data that we have collected are voluminous and in many areas there are significant variations in structure and structural history for either local or more regional tectonic reasons. In what we present below, these are overlooked and we attempt to describe a generalised picture of these features. We present a typical traverse through the GTS which we feel is applicable to it along its entire length.

1. Structure and deformation of the country rocks

In general terms the bedding and early rock fabrics of the Insular Superterrane W of the GTS have variable but low dips and are generally of low metamorphic grade. Approaching the GTS, young, steeply inclined (usually NE-dipping), intense crenulation cleavages and schistosities develop. These fabrics contain a down dip (nearly 90° pitch) stretching lineation. Strains always increase eastwards towards the GTS. Strain measurements in the Taku Inlet area confirm this qualitative impression and show x/z increasing from 4·5, at 4 km W of the GTS, to x/z values in excess of 15 at the western contact. Particularly well documented in the northern (Petersburg to Haines) area (Himmelberg *et al.* 1991) is an increase in metamorphic grade towards the western margin of the GTS. Over distances of about 8–10 km a complete Barrovian system of inverted isograds parallel to the tectonic strike and GTS contact is crossed, with sillimanite-bearing anatectites occurring closest to the intrusive body. High grade minerals associated with this sequence (especially kyanite and sillimanite) are aligned parallel to the down dip lineation, indicating, together with other microscopic observations, that this metamorphism and deformation are synchronous.

Associated with the intense fabric are folds (of earlier fabrics and bedding) which have horizontal to steeply plunging hinges and SW to sinistral vergence. The steeply-plunging folds are believed to have rotated into this orientation parallel to the stretching lineation in areas of highest strain.

Sense of shear indicators include pressure shadows around individual grains (often garnet), sheared crystals (especially plagioclase laths in deformed granite veins), extensional shears and larger discrete shears often >10 m long. In general, all of these features show ambiguity in the dominant direction of shear, with the exception of the latest-stage large, flat-lying shears. The ambiguity implies that considerable amounts of pure shear deformation have affected these rocks. However, when a dominant sense of shear is indicated by these features (particularly at the most penetrative, grain scale) it is always top towards the SW. We therefore conclude that the zone of high deformation immediately W of the GTS is a high-angle reverse shear zone in which large amounts of additional pure shear shortening have been accommodated.

2. Structure of the Great Tonalite Sill

The contact between the GTS and the country rocks is sheeted, both to the W and to the E, with individual sheet widths varying between a few tenths of a metre to 20–30 m. Although much of the GTS is highly deformed, there are some areas where strains are low enough to reveal original, intrusive features. Thus, for example, in Ford's Terror Inlet and in Taku Inlet, we have observed primary magmatic scouring structures. In the Mount Juneau pluton, primary magmatic sheets some 3 m thick and at least 30 m long, with one-way chilled contacts, have been discovered. This could be termed a granitic sheeted dyke complex and its existence, together with the nature of the pluton country rock contacts, implies that the GTS intruded as a series of injecting and coalescing sheets or dykes of tonalite. The perservation of these original, pre-deformational features indicates that the bulk strain affecting the GTS is not excessive and therefore that the intrusion was originally emplaced as a long and relatively thin series of bodies.

The deformation of the GTS is generally pervasive and often intense. The fabrics are all steeply NE-inclined and parallel to the fabrics in the western country rocks (Fig. 2). The ealiest deformation is a magmatic state or PFC (Hutton 1988) fabric of euhedral hornblende and plagioclase laths with undeformed interstitial phases. This has down-dip stretching lineations and is affected by small discrete extensional shears which we consider to have formed during the latest stage of magmatic state deformation ("lock-up" shears). This PFC deformation is in general restricted to the central parts of the pluton.

Towards the margins of the intrusion, solid state or CPS (Hutton 1988) fabrics are developed involving the plastic deformation of all the magmatic phases. Despite this, it is

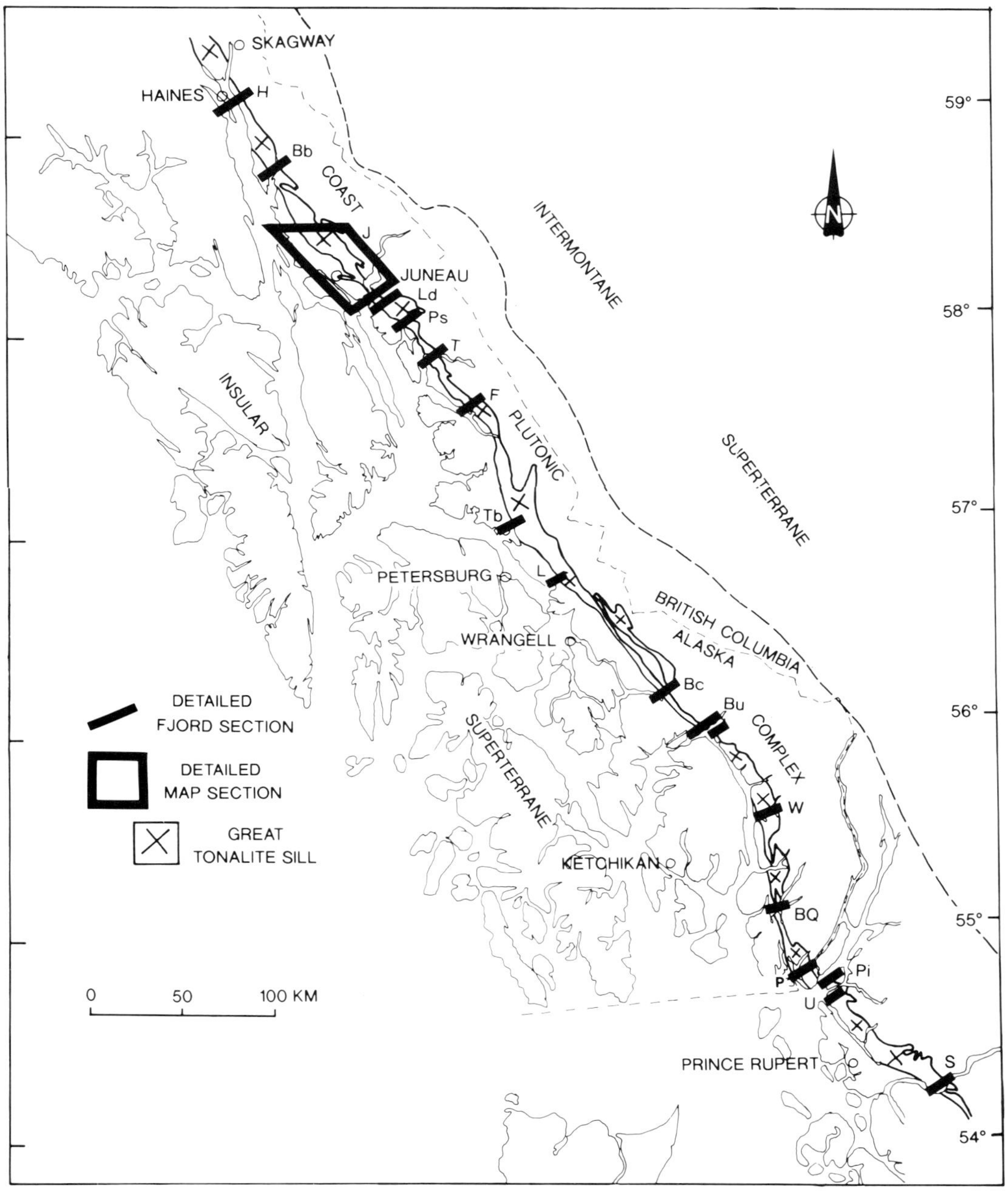

Figure 1 The Great Tonalite and its setting in South East Alaska and British Columbia. Traverse locations and study areas are indicated: H = Haines, B = Berners Bay, J = Juneau (including Taku Inlet), Ld = Lake Dorothy, PS = Port Snettisham, T = Tracy Arm, F = Ford's Terror, Tb = Thomas Bay, L = Le Conte Bay, W = Walker Cove, BG = Boca de Quadro, P = Pearse Canal, Pi = Portland Inlet, U = Union Inlet, S = Skeena River.

still possible to see preserved aligned euhedral phenocrysts, indicating that PFC fabrics are pervasive throughout the plutons but have been modified, often strongly, in generally marginal regions by deformation continuing down temperature in the solid state. Lineations associated with CPS fabrics are down dip with a high pitch angle, identical in geometry to that seen in the PFC fabrics. As with the western country rocks, strong and often mylonitic CPS fabrics have associated shear sense indicators. Plastic deformation, pressure shadows, and shearing of plagioclase phenocrysts and hornblende laths, are particularly common. Garnet, which may be magmatic and primary, is also useful as a shear sense indicator. On a slightly larger scale there are extensional crenulations cleavages and, where mylonitic zones are found at the pluton contacts, SC fabrics are present. As in the country rocks, these structures are all cut by discrete ~1 m-long shears and larger (~10 m-long) flat-lying or gently NE dipping top to the SW shears, often with granitic and trondhjemitic melts or pegmatites present. In general, with all of the small-scale structures, there is amgibuity in shear sense and implied pure shear with NE to SW horizontal shortening and vertical extension. However, as with the country rocks, whenever there is a dominant shear sense, especially at the smallest grain scale, this is top towards the SW. At much lower temperatures deformation continued sporadically in generally narrow zones within the GTS. Mylonites showing considerable grain reduction but not cataclasis, are common and dominant in some sections. The large low-angle shears cut these fabrics. The bulk strain within the GTS is large and in some sections we have measured values from deformed xenoliths which are in excess (x/z) of 80:1.

Our data show that the GTS began to be deformed before it had completely crystallised and that the deformation continued down temperature into the solid state and even to mylonites. The observed deformation was dominantly of pure shear type (with K values between 0 and 1) with a NE–SW shortening and a vertical extension. Combined with this shortening event is a (generally) poorly registered simple shear which has a top to the SW. Only the very latest structures (the large gently dipping shears) show a 100% consistent shear sense of top to the SW. Precisely the same deformation is seen in the western country rocks and thus a deformation zone or shear zone exists here in the country rocks and in the site of the GTS. This, as with the GTS itself, is a remarkably long (>800 km) and narrow (<30 km) zone of continental and plate tectonic dimension. Our observation of generally top to the SW shear for this shear zone is backed by data from metamorphic rocks (Crawford & Hollister 1982) which shows uplift of rocks to the E of the GTS (relative to the W) of ~15 km between 60 and 50 Ma.

3. Summary

1. The steeply-inclined GTS is a syntectonic intrusion (Brew & Ford 1981).
2. The GTS was emplaced along the axis of a long (>800 km) and relatively narrow (<30 km) high-angle reverse shear zone which has a top to the SW simple shear component and steeply plunging lineations.
3. The shear zone was active when the GTS was emplaced into it.
4. The GTS was emplaced as a series of steeply inclined sheets by dyking and wedging mechanisms.
5. There is little or no sign of transcurrent shear within this zone.
6. The GTS was emplaced into the middle of an active contractional orogenic belt and therefore its wedging action and magmatic body force was able to overcome orogenic compressional stresses.
7. The GTS does not obey the Andersonian model of dykes

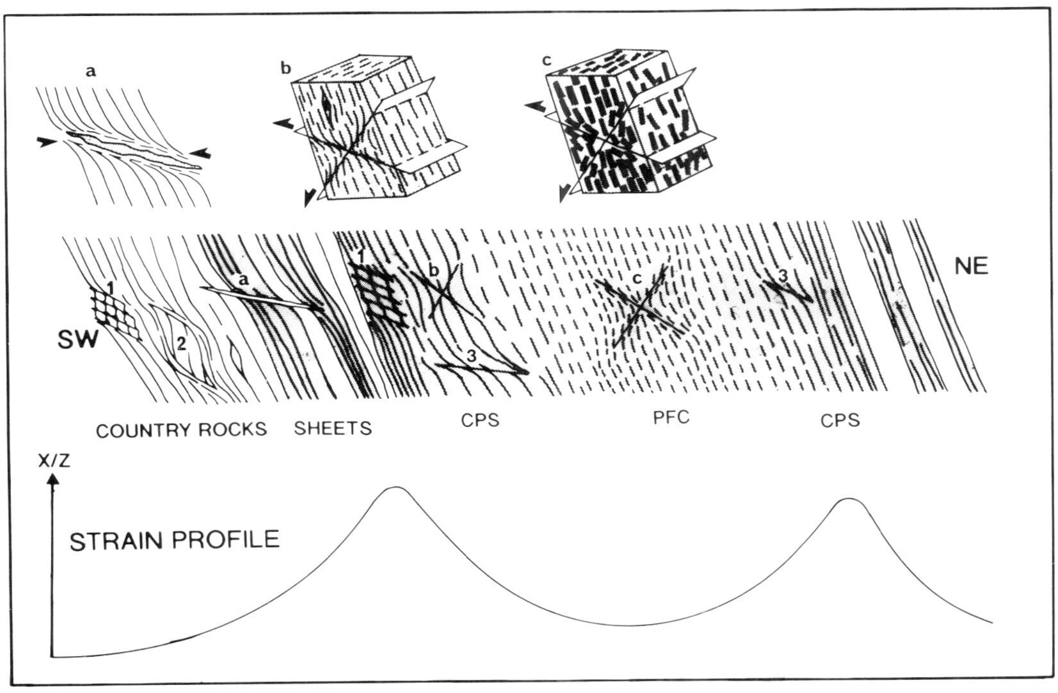

Figure 2 Generalised cross-section of the Great Tonalite Sill: (a) late-stage, top-to-SW, ductile shears, intruded by granitic melt; (b) secondary CPS foliations, with down-dip lineations, cut by conjugate ductile shears. Asymmetric tails on phenocrysts; (c) primary PFC fabrics, with down-dip lineations, cut by PFC lock-up shears.
Top-to-SW shear sense criteria: (1) extensional crenulation cleavages; (2) asymmetric boudins; (3) low-angle, late, ductile shears.

emplaced parallel to the maximum principal stress and at right angles to the minimum principle stress. Rather, it was emplaced at right angles to σ_1.

8. This is an example of syntectonic granite emplacement in a contractional high-angle reverse/thrust regime and as such it complements earlier work on syntectonic emplacement in transcurrent and extensional tectonic regimes (Hutton 1982; Hutton *et al.* 1990). In all three settings, the detailed mechanism of emplacement is one of granitic sheeting. This must now be seen as a distinctive category of pluton construction along with diapirism and cauldrons.

9. If the GTS does represent the boundary between the Intermontane and Insular superterranes, then accretion of these terranes must have been orthogonal, not strike slip, in the late Cretaceous to early Palaeocene.

4. Acknowledgements

We are particularly grateful to David Brew of the U.S.G.S. who originally introduced us to the GTS and who has continued to encourage and support us logistically. DHWH is grateful to Weecha Crawford, Linc Hollister, Bill Crawford, and Rick Matthews who provided a wonderful trip to the GTS and other rocks of NW British Columbia. Funding for field work has been provided by the Shell International Petroleum Company, the Royal Society of London and the United States Geological Survey.

5. References

Brew, D. A. & Ford, A. B. 1981. The Coast Plutonic Complex Sill, Southeastern Alaska. *In* Albert, N. R. D. & Hudson, T. (eds) *The United States Geological Survey in Alaska—Accomplishments during* 1979. U.S.G.S. CIRCULAR **823-B,** 96–9.

Crawford, M. L. & Hollister, L. S. 1982. Contrasts of metamorphic and structural histories across the Work Channel Lineament, Coast Plutonic Complex, British Columbia. J GEOPHYS RES **87,** 3849–60.

Himmelberg, G. R., Brew, D. A. & Ford, A. B. 1991. Development of inverted metamorphic isograds in the western metamorphic belt, Juneau, Alaska. J METAMORPHIC GEOL **9,** 165–80.

Hutton, D. H. W. 1982. A tectonic model for the emplacement of the Main Donegal Granite, NW Ireland. J GEOL SOC LONDON **139,** 615–31.

Hutton, D. H. W. 1988. Granite emplacement mechanisms and tectonic controls: inferences from deformation studies. TRANS R SOC EDINBURGH EARTH SCI **79,** 245–55.

Hutton, D. H. W., Dempster, T. J., Brown, P. E. & Becker, S. M. 1990. A new mechanism of granite emplacement: intrusion in active extensional shear zones. NATURE **343,** 452–5.

Rubin, C. M., Saleeby, J. B., Cowan, D. S., Brandon, M. T. & McGroder, M. F. 1990. Regionally extensive mid-Cretaceous west-vergent thrust system in the northwestern Cordillera: implications for continent-margin tectonism. GEOLOGY **18,** 276–80.

D. H. W. HUTTON, Department of Geological Sciences, Durham University, Durham DH1 3LE, U.K.
G. M. INGRAM, Department of Geological Sciences, Durham University, Durham DH1 3LE, U.K.

Transactions of the Royal Society of Edinburgh: Earth Sciences, **83**, 387–397, 1992

Calculated melt and restite compositions of some Australian granites

C. Wayne Burnham

ABSTRACT: The thermodynamic relations embodied in the Quasicrystalline Model of Burnham and Nekvasil (1986), as recently extended by the author, have been used to quantitatively assess the feldspar-quartz liquidus relations in two I-type (Jindabyne and Moruya) and two S-type (Bullenbalong and Dalgety) suites of Australian granites, using analytical data provided by B. W. Chappell and co-workers. Among the more notable results obtained from these calculations at a constant pressure of 5·0 kbar and $X_w^m = 0{\cdot}30$ ($\approx 2{\cdot}8$ wt% H_2O), for purposes of comparison, are that: (1) felsic melts of remarkably uniform, but distinctive composition can be extracted from each suite, leaving solid residues in amounts up to 65 mol%; (2) all melts from both S-type suites have two feldspars plus quartz on their liquidii, whereas both I-type suites have only plagioclase plus quartz on their liquidii; (3) the total solid residue ranges from 27–63% in the Jindabyne suite, from 15–62% in the Moruya suite, from 30–65% in the Bullenbalong suite, and from 27–65% in the Dalgety suite; (4) liquidus temperatures of the S-type Bullenbalong and Dalgety melts are similar (856° and 860°C), reflecting similar feldspar compositions of An_{53}, Or_{75} and An_{60}, Or_{77}, respectively; (5) liquidus temperatures of the I-type Jindabyne and Moruya melts, however, are distinctly different (950° and 894°C), reflecting correspondingly different plagioclase compositions of An_{80} and An_{52}; (6) the calculated liquidus plagioclase composition throughout a given suite is very uniform ($\pm 1\%$) and amounts to as much as 46% of the total rock; and (7) these calculated liquidus and residual plagioclase compositions are also the same, within the uncertainty of measurement, as those of the plagioclase crystal-cores determined optically by A. J. R. White. The only plausible explanation for this remarkable consanguinity in plagioclase liquidus, residue, and crystal-core compositions, hence liquidus temperatures, is that the bulk of the residue is restite, in accordance with the model of White and Chappell (1977). This explanation is corroborated by the very systematic variations in the amounts of individual restite minerals with respect to total restite contents. Accordingly, those members of each suite that contain more than 60% total restite probably closely represent the bulk composition of the source rock, which is dioritic or andesitic for the Jindabyne suite, tonalitic or dacitic for the Moruya suite, pelitic metagreywacke for the Bullenbalong suite, and feldspathic metagreywacke for the Dalgety suite. As a corollary, those members with less than 60% restite must have undergone melt–restite segregation (unmixing), probably during ascent and emplacement.

KEY WORDS: cryoscopic, equilibrium, liquidus, mixing, phases, plagioclase, pressure, Quasi-crystalline, speciation, suites, temperature, thermodynamic

Considerable controversy has arisen recently (cf. Clemens 1989) over the restite model of White and Chappell (1977), as recently expounded by Chappell *et al.* (1987). The basic argument appears to be whether the more mafic members of certain Australian granite suites reflect a greater accumulation of precipitates from an initially near-liquidus melt or a greater retention of restite phases.

In an effort to shed light on this critical petrological problem and, perhaps, contribute to the resolution of the controversy, the thermodynamically-based Quasi-crystalline Model (Burnham & Nekvasil 1986), as modified by Nekvasil (1990) and extended here by the author, has been applied to the calculation of the liquidus phase and melt compositions and temperatures in several suites of Australian granites. The suites selected initially as representative of I-type granites were the Jindabyne (Hine *et al.* 1978) and Moruya (Griffin *et al.* 1978) Suites and those representative of S-type granites were the Bullenbalong and Dalgety Suites (White *et al.* 1977; Chappell 1984; White & Chappell 1989). Subsequently, analytical data from several other suites have been investigated to determine the extent to which they differ from the four initial suites. The data base used in all calculations on Australian granites is that established by B. W. Chappell and co-workers at the Australian National University and the sample numbers referred to in this communication are those in that data base.

1. Basic features of the thermodynamic formulations

1.1. Limitations on treatment of thermodynamic relations

A comprehensive treatment of the thermodynamic relations used in the calculations presented here is beyond the scope of this communication, but a brief discussion of the more salient features may serve to place the results in clearer perspective. A more comprehensive treatment of both the thermodynamic relations and the melt speciation model may be found in Burnham and Nekvasil (1986) and in Burnham (1981).

1.2. The cryoscopic equation of phase quilibrium

The fundamental cryoscopic equation for crystal–melt equilibrium, as it is used in the calculations presented here, may be expressed as

$$a_i^m/a_i^s = \exp\left[-\left(\Delta G^\circ_{mi} + \int_1^P \Delta V_{mi} dP\right)\Big/RT\right]$$

where superscripts m and s refer to melt and solid phases, respectively, and subscript i refers to component i. The subscript m refers to the melting process, R is the gas constant, T is temperature in Kelvins, and P is pressure in bars.

The first term on the right-hand side of this equation, ΔG°_{mi}, represents the standard state Gibbs free energy of fusion (melting) of crystalline substance i at T (K) and one bar and is graphed for the feldspars, quartz, and diopside (for comparison) in Figure 1(a). The $\int_1^P \Delta V_{mi} dP$ term represents the pressure integral of the volume change upon melting of crystalline i and takes the standard state cryoscopic relations in Figure 1(a) up to the pressure of interest. Its value, as shown in Figure 1(b), is dependent upon both temperature and pressure. Both of these terms on the right-hand side of the equation are fundamental properties of pure substance i and have been evaluated from essentially all of the relevant published experimental and calorimetric data (cf. Burnham & Nekvasil 1986). Except for the choice of a formula unit of each non-volatile component that contains eight oxygen atoms, these terms are not model dependent.

In contrast, the activity terms on the left-hand side of the above equation represent properties of substance i in solution, either melt (a_i^m) or solid (a_i^s), and as such their values depend not only upon their respective mol fractions, but upon their mixing behaviour with other substances in the solutions. The activity relations of the *an, ab,* and *or* components in the ternary feldspar solid solutions, for example, are those derived by Nekvasil (1990). Those for the various quadrilateral and non-quadrilateral components in the clinopyroxenes (di) and orthopyroxenes (hy) have been closely approximated from published experimental data and the assumption of ideal mixing in both pyroxenes at liquidus temperatures. The activity relations of the

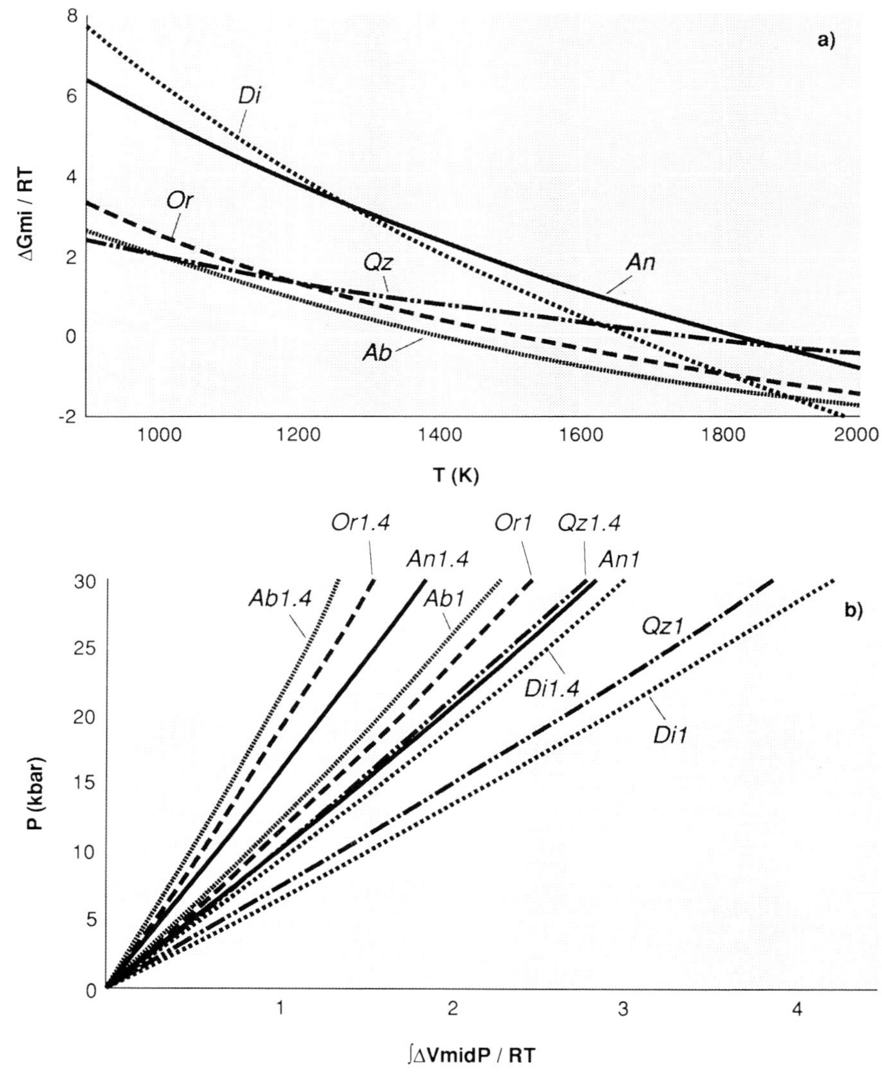

Figure 1 Cryoscopic relations for albite (Ab), sanidine (Or), quartz (Qz), anorthite (An), and diopside (Di): (a) the standard state (1·0 bar) cryoscopic curves as a function of temperature (T K); (b) the integral of $\Delta V_{mi}/RT$ with respect to P; the numbers 1 and 1·4 following the mineral symbols refer to the 1000 K and 1400 K isotherms, respectively, as examples. In these figures all mineral components are defined on the basis of eight gram atoms of oxygen per formula unit.

various components in aluminosilicate melts, on the other hand, are much more complicated and, to a certain extent, are model dependent.

1.3. The Quasi-crystalline Model

The Quasi-crystalline Model is perhaps best summarised by the following quotation from Burnham and Nekvasil (1986, p. 240):

> The model treats igneous rock melts as multicomponent solutions of chemically discrete and thermodynamically distinct neutral complexes or species that are correlated with thermodynamic components. Many of these species mimic the solid phases that crystallise from the melts in stoichiometry and presumably in the structure of their polyhedral elements. Those species that do not mimic the crystalline phases are linked to those that do through homogeneous speciation equilibria of two types: (1) simple dissociation, as in the incongruent melting of the feldspars, and (2) interaction between two or more reactive species (components) to produce new species with different stoichiometries, hence different thermodynamic properties . . . .
>
> The basis for this speciation model was the discovery that the activities (a_a^{am}) of the aluminosilicate components (feldspar-like species) in most anhydrous silicate melts calculated from the basic cryoscopic equations . . . are either equal to or *less than* their corresponding mole fractions (X_a^{am}), when all non-volatile components are chosen on the basis of equal numbers of oxygen atoms (eight) per formula unit and, therefore, of nearly equal molal volumes.

On this basis of equal numbers of oxygen atoms per formula unit, the feldspar (*an, ab, or*) and silica (*qz*) components are found to mix ideally (Burnham & Nekvasil 1986), as are the products of speciation. To illustrate the nature and extent of speciation, as well as mixing behaviour of speciation products, the melting relations of anorthite and albite, as a function of pressure, are now briefly analysed.

1.4. Speciation in anorthite melts

The experimentally determined melting curve (x) of pure anorthite (Wood 1978; Goldsmith 1980; Boettcher *et al.* 1984) is compared with that obtained from the cryoscopic equation for anorthite (c), assuming no speciation by dissociation, in Figure 2. The experimentally-determined curve at pressures beyond the 9·4 kbar (1 kbar = 100 MPa) singular point (San1), where corundum first appears, takes on a negative dT/dP (reciprocal Clapeyron) slope, which steepens with increasing pressure to the invariant point (Ian) at 32·3 kbar, 1682 K (1409°C), involving anorthite (An, grossularite (Gr), kyanite (Ky), quartz (Qz), and melt (M), but not corundum (Wood 1978). The activity of the *an* component (a_{an}^{am}) at the invariant point, calculated from the cryoscopic equation (cf. Fig. 1), is 0·108. In accordance with the Quasi-crystalline Model, in which the true value of $X_i^{am}(tr) = a_i^{am}$, the fact that the liquidus temperature of anorthite at 32·3 kbar has fallen approximately 375°C from the $a_{an}^{am} = 1{\cdot}0$ curve (c) is thus attributed to 89% of the *an* melt component having dissociated in accordance with the speciation reaction

$$\underset{\text{Anorthite}}{CaAl_2Si_2O_8} \Leftrightarrow 0{\cdot}108\,\underset{an}{CaAl_2Si_2O_8} + 0{\cdot}446\,\underset{gr}{Ca_2Al_{1{\cdot}333}Si_2O_8} + 0{\cdot}372\,\underset{as}{Al_{3{\cdot}2}Si_{1{\cdot}6}O_8} + 0{\cdot}074\,\underset{qz}{Si_4O_8}.$$

Thus, according to the stoichiometry of this reaction, the speciation model predicts that $X_{qz}^{am} = a_{qz}^{am} = 0{\cdot}074$. The cryoscopic equation for quartz (cf. Fig. 1), which was derived entirely independently from that for anorthite, yields $a_{qz}^{zm} = 0{\cdot}072$ at this invariant point! This remarkably precise agreement between the stoichiometrically-derived X_{qz}^{am} and the cryoscopically-derived a_{qz}^{am} at this high pressure is surprising, but not unique, as will now be demonstrated for the melting of albite.

1.5. Speciation in albite melts

The experimentally-determined melting curve (x) of pure albite (Boettcher & Wyllie 1968; Bell & Roseboom 1969; Boettcher *et al.* 1982) is also compared with the calculated melting curve (c)—assuming no speciation—in Figure 2. After passing through a singular point involving the appearance of jadeite at approximately 31·5 kbar, the melting curve of albite then terminates in an invariant point (Iab) at 33·3 kbar and 1640 K (1367°C), where albite,

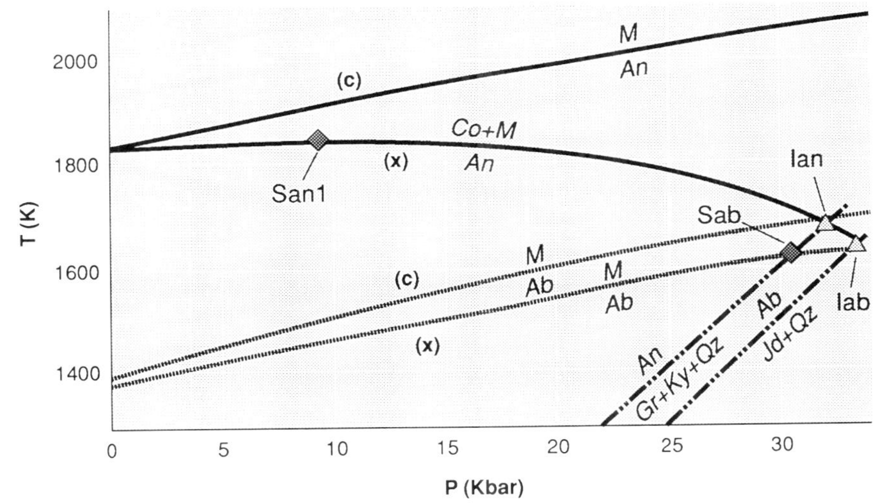

Figure 2 Melting temperatures (T K) of anorthite (An) and albite (Ab) as functions of pressure. Curves labelled (x) are experimentally determined (see text for references), whereas those labelled (c) are calculated on the assumption of no melt speciation ($a_i^m = X_i^m = 1{\cdot}0$). Ian and Iab are invariant points and San1 and Sab are singular points on the melting curves of anorthite and albite, respectively. A second singular point on the melting curve of anorthite at approximately 32·1 kbar, where corundum (Co) disappears, is not shown for clarity.

jadeite, quartz, and melt are in equilibrium. At this point, which is only approximately 61°C below the calculated $a_{ab}^{am} = 1{\cdot}0$ curve, $a_{ab}^{am} = X_{ab}^{am}(tr) = 0{\cdot}771$. The stoichiometry of the melting reaction

$$\underset{\text{Albite}}{NaAlSi_3O_8} \Leftrightarrow \underset{ab}{0{\cdot}771NaAlSi_3O_8} + \underset{jd}{0{\cdot}172Na_{1{\cdot}333}Al_{1{\cdot}333}Si_{2{\cdot}667}O_8} + \underset{qz}{0{\cdot}057Si_4O_8}$$

yields $X_{qz}^{am} = 0{\cdot}057$; the independently-dervied cryoscopic equation for quartz (cf. Fig. 1) yields precisely the same value for a_{qz}^{am}! This remarkable agreement between a_{qz}^{zm} and X_{qz}^{zm} in both albite and anorthite melts attests not only to the internal consistency and accuracy of the cryoscopic equations, but also to the ideal mixing behaviour of the *an*, *ab*, and *qz* components, as well as the nature and extents of the speciation reactions.

1.6. $X_i^{am} - a_i^{am} = f(P, X_j^{am}, X_w^m)$ as a measure of speciation

It will be observed in the melt speciation reactions above that $X_{ab}^{am} - a_{ab}^{am} = 1{\cdot}0 - 0{\cdot}771 = 0{\cdot}229$, whereas $X_{an}^{am} - a_{an}^{am} = 1{\cdot}0 - 0{\cdot}108 = 0{\cdot}892$ at nearly the same pressure. This almost fourfold difference in the extent of speciation is the fundamental cause for the experimental observation of Green (1969) and others more recently that the liquidus plagioclase in mafic melts of fixed bulk composition becomes more albitic with increasing pressure. This effect is enhanced, moreover, by the stronger tendency for the *an* component to further speciate by interaction with the olivine- and pyroxene-like components of mafic melts, as mentioned briefly by Burnham (1981, p. 201). The enhancement of the albite content of liquidus plagioclase as a result of the differential lowering of a_{an}^{am} with respect to X_{an}^{am} by these two types of speciation reaction has important implications for the origin of massif anorthosites, as well as tholeitic and alkali basalts. More important in the present context, it makes it highly unlikely that granite magmas with substantial quantities of An_{80} or higher liquids plagioclase compositions can be produced by admixing olivine basalt magma, even at mid-crustal pressures.

Before proceeding to a discussion of the feldspar-quartz-melt equilibrium relations in some Australian granites, the question of conformity of the melt speciation (Quasi-crystalline) model to the Gibbs Phase Rule should be mentioned. Boettcher and Burnham (1983) pointed out that identifying melt species with thermodynamic components is fully compatible with the Gibbs Phase Rules as propounded by J. Willard Gibbs (cf. Gibbs 1961).

2. Feldspar-quartz-melt equilibrium

2.1. Input data

The calculations embodied in the results presented here were carried out on a personal computer with the aid of the subroutine "EQUIL", which was originally written by Nekvasil (1986) and subsequently updated by Nekvasil (1990) and, more recently by the author. For a given bulk composition, the program calculates liquidus temperatures and compositions of the various minerals, including the feldspars, quartz, clinopyroxene, and orthopyroxene. Very few data are available for magnetite, ilmenite, and apatite, hence their solubilities in granitic melts as functions of temperature, pressure, and H_2O content are estimated.

The actual input data consist of: (1) the mol fraction (eight oxygen atoms per formula unit) of each non-volatile melt component derived mostly from the CIPW Norm of the analysed rock, (2) pressure in kilobars, and (3) either the activity of H_2O (a_w^m) or its mol fraction (X_w^m) in the melt. For purposes of comparison within each suite and especially between suites of granites, a pressure of 5·0 kbar and an H_2O mol fraction of 0·30, which is equivalent to approximately 2·8wt%, have been chosen as appropriate for the conditions prevailing during fluid-absent partial melting of the granite source rocks (cf. Burnham 1979).

Modest changes in the values of these independent variables have correspondingly modest effects on the plagioclase liquidus relations, but large effects on the quartz liquidus relations. Decreasing pressure to 2·0 kbars at $X_w^m = 0{\cdot}30$, as during magma emplacement, for example, decreases the plagioclase liquidus temperature by only 12°C, as can be deduced by linear extrapolation of the plagioclase liquidii in Figures 3 and 6. This same decrease in pressure, on the other hand, decreases the quartz liquidus temperature by 90°C, as also can be deduced from Figures 3 and 6. In order to restore both plagioclase and quartz on the liquidus at this lower pressure, therefore, it is necessary to resorb quartz restite, if available, and to lower the temperature and precipitate plagioclase. This very large effect of pressure on quartz liquidus temperatures, as shown in Figures 3 and 6, is a direct result mainly of the very small negative slope of the cryoscopic curve for quartz in Figure 1(a). The impact of these grossly different Clapeyron slopes of the quartz and the feldspar liquidii, which are a direct measure of the enthalpies of fusion, on melt–restite equilibria will be discussed at greater length in Section 3.

If now pressure is fixed as the independent variable and X_w^m is increased to 0·40 (≈4·3wt% H_2O), for example, the plagioclase liquidus temperature is depressed by roughly 54°C and the quartz liquidus temperature is depressed by 80°C, hence an additional 3% quartz must be added—or an

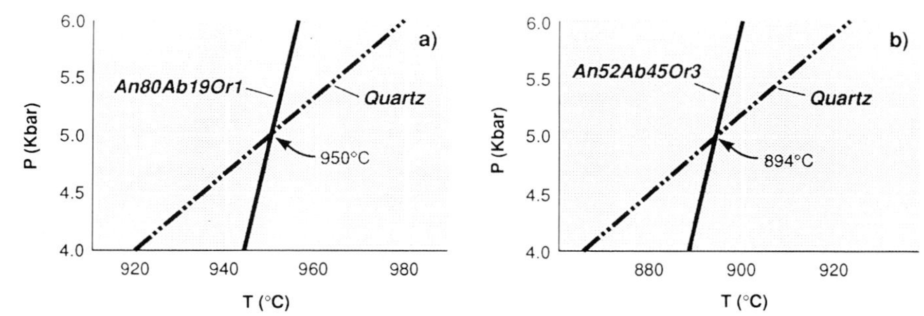

Figure 3 Calculated liquidus relations of the Jindabyne (a) and Moruya (b) Suite melts at 5·0 ± 1·0 kbar and $X_w^m = 0{\cdot}30$ (≈2·8 wt% H_2O); differences in slope of the plagioclase and quartz liquidus curves play an important role in phase relations during magma decompression, as discussed in the text.

approximately equivalent amount of plagioclase removed—to re-establish plagioclase-quartz-melt equilibrium. The substantial lowering even of the plagioclase liquidus upon addition of only 1·5wt% H_2O is due, in part, to the large differences between the gram formula weights of H_2O and the non-volatile melt components (eight oxygen atoms per formula weight), which results in a large value of X_w^m for a relatively small weight fraction of H_2O. It is also due, in part, to the manner in which H_2O dissolves in the melts, which gives rise to the squared term in the following expression for a_i^{hm} in melts where $X_w^m \leq 0{\cdot}50$ (Burnham & Davis 1974):

$$a_i^{hm} = a_i^{am}(1 - X_w^m)^2.$$

Here the superscripts hm and am refer to hydrous and anhydrous melts, respectively. It will be observed that it is this expression, which contains no term for the activity of H_2O (a_w^m), that is substituted for a_i^m in the cryoscopic equation of 1·2 to calculate equilibria in hydrous melts. Thus, inasmuch as the activities of all anhydrous melt components are lowered to the same proportionate extent by a given mol fraction of H_2O, the much larger effect on quartz equilibrium temperatures noted above is again the direct result of the much lower negative slope of the cryoscopic curve for quartz in Figure 1(a).

To place these differential effects of pressure and H_2O content on quartz-feldspar-melt equilibrium in perhaps a more familiar petrological perspective, increasing pressure preferentially increases the quartz liquidus temperature, thereby shifting the quartz-feldspar cotectic in the haplogranite system (cf. Tuttle & Bowen 1958) *away* from the quartz corner. Increasing X_w^m, on the other hand, preferentially decreases the quartz liquidus temperature, thereby shifting the cotectic *toward* the quartz corner.

2.2. Normative minerals and melt components

Among the CIPW normative minerals in granitic melts, only corundum (*co*) and wollastonite (*wo*) do not appear as melt components. In hypersthene-free quartz-normative melts, it has been shown experimentally (Voigt 1983; Joyce 1985; Burnham & Nekvasil 1986) that normative corundum, as $Al_{5\cdot333}O_8$, combines with *qz* (Si_4O_8) to form *dpy* ($Al_{1\cdot455}Si_{2\cdot91}O_8$), a dehydroxylated pyrophyllite-like species (component), in amounts up to 7·7 mol%. In hypersthene normative melts, however, normative corundum first combines with *hy* and *qz* in equal proportions to produce *cd* ($(Mg, Fe)_{0\cdot89}Al_{1\cdot78}Si_{2\cdot22}O_8$], a cordierite-like component, in amounts up to 8·0 mol% (T. Chacko, unpublished experimental results, 1985). In these same experiments, it was found that, in the presence of excess sillimanite and almandine garnet, cordierite-like, but very little almandine-like species appeared to exist in the melt at pressures up to 6·0 kbar. The amount of almandine-like species presumably increases at higher pressures, however, in response to the increasing tendency for Al to shift from fourfold to sixfold coordination in aluminosilicate melts. The principal significance of the existence of such large amounts of Fe-rich *cd* component in melts produced from ferruginous peraluminous sediments is that the total normative hypersthene content, as well as the siderophile element contents of the melts, are greatly enhanced, as shown by White and Chappell (1988). With respect to normative *wo*, of course, it combines with normaitve *en* and *fs* to make *di* and *hd* melt components.

Finally, it should be mentioned that the micas and amphiboles do not exist as melt species (components). Instead, they appear to precipitate by the inverse of their incongruent melting reactions when X_w^m reaches a certain level and their critical anhydrous components either approach or reach saturation. For common hornblende, $X_w^m \geq 0{\cdot}30$ (Burnham 1979) and the critical components appear to be $hy + di$ ($\pm hd$) and $an + ab$; for biotite, $X_w^m \geq 0{\cdot}35$ and the critical components are *or* and *hy* or *cd*; and for muscovite, $X_w^m \geq 0{\cdot}59$ and the critical components are *or* and *dpy*.

2.3. Calculation of liquidus relations

The fundamental assumption that underlies the present scheme for calculating the melting relations in all granitic magmas is that *the melts produced at the site of magma generation were in equilibrium with plagiolcase feldspar and quartz*. As a consequence of this basic assumption, the first step in determining liquidus relations is to iteratively calculate how much plagioclase of the most anorthitic composition possible must be removed as solid residue from the bulk composition to put plagioclase of the same anorthite content on the liquidus with quartz. Concomitantly, other phases are removed as necessary to avoid supersaturation at the calculated plagioclase-quartz liquidus temperature. Convergence on the maximum anorthite content possible of a plagioclase in equilibrium with quartz on the liquidus for a given bulk composition is crucial because: (1) it is that plagioclase of fixed An content which is in equilibrium with melt at the site of partial melting under a particular set of P-T-X_w^m conditions, (2) it is that plagioclase of fixed An content that must be removed as restite in generally varying proportions from every unfractionated member of a given suite, and (3) it is also that plagioclase of fixed An content which serves as a reference for the detection of post-generation plagioclase-melt fractionation.

Plagioclase fractionation is clearly indicated when convergence can be obtained only on an anorthite content that is significantly lower (≥3·0% An) than in the more mafic members of a given suite. Among the 25 (originally 26) compositions investigated from the Jindabyne, Moruya, Bullenbalong, and Dalgety suites, which ranged from the most felsic to the most mafic within each suite, none was found to be fractionated, although the plagioclase residue was calculated to be as little as 5·6 mol (≅volume)% in one sample from the Moruya Suite (MG20) and as much as 45·6% in one sample from the Jindabyne Suite (KB139), both of them I-type (cf. Fig. 7 below). In comparison, the smallest amount of plagioclase residue removed from S-type bulk compositions was 7·4 mol% from a sample in the Bullenbalong Suite (KB45) and the most removed was 26·9% from a sample in the Dalgety Suite (BG37). The exact percentage of plagioclase residue, hereinafter identified with restite, is perhaps not that significant; the important point is that the amounts are substantial in all samples investigated and their compositions are the same throughout each suite.

2.4. The Jindabyne Suite (I-type)

The calculated liquidus relations for all eight investigated members of the Jindabyne Suite at 5·0 kbar and $X_w^m = 0{\cdot}30$ are shown in the P-T projection of Figure 3(a). The calculated liquidus temperature at which both plagioclase ($An_{80}Ab_{19}Or_1$) and quartz are in equilibrium with melt of trondhjemitic granite composition, as shown in Figure 4, is 950 ± 4°C. The calculated An content of the restite plagioclase, it should be noted, is precisely that of the plagioclase cores in KB22 (Jindabyne Tonalite, 30% restite plagioclase) reported by Chappell *et al.* (1990).

Although quartz is calculated to be on the liquidus for all compositions, the amounts of restite quartz ranges from essentially zero (just saturated with quartz) in the most mafic composition (KB139) to 7·0 mol% in the more felsic KB35 sample. Owing to the high temperature required for the anorthitic plagioclase equilibrium (An_{80}), the amount of *qz* component in the melt required for coexistence of quartz with plagioclase and melt is correspondingly high, as shown in Figure 4. It is noteworthy that if KB139 is assumed to have been completely molten at its source, its calculated liquidus temperature at 5·0 kbar and $X_w^m = 0{\cdot}30$ would have been 1250°C! Moreover, the calculated liquidus phases would have been An_{95} plagioclase and an enstatitic pyroxene.

The observations that: (1) the calculated liquidus temperatures are virtually the same throughout the suite, (2) the calculated liquidus plagioclase compositions are the same (An_{80}) for all members of the suite, and (3) plagioclase of An_{80} composition is found as crystal cores throughout the suite (Chappell *et al.* 1990; A. J. R. White pers. comm. 1991), together mitigate against assimilation, liquidus phase accumulation, or magma mixing as the cause of compositional variations within the Jindabyne Suite. If wallrock assimilation was involved in producing any of the analysed members of the Jindabyne Suite, then the assimilated rock would have to have been derived from a virtually identical source rock; otherwise, the melt composition would be substantially different from the norm and the composition of the calculated liquidus plagioclase would have been substantially different from that of the plagioclase cores. This last constraint is also applicable if liquidus phase accumulation is hypothesised. Finally, the magma mixing hypothesis is totally inconsistent with all three of the above enumerated observations. These same three observations, it should be noted, are equally applicable to the other three suites investigated and, when combined with additional evidence presented in Section 3 below, provide strong support for the restite model of White and Chappell (1977).

A feature of the melt compositions of the Jindabyne Suite in Figure 4, which is shared to a lesser extent by the Moruya melts, is the substantial variability in the *or, ab, hy + di,* and *cd* contents. These variations appear to reflect fractionation of hornblende from the more felsic magmas of these suites during ascent and emplacement, as they contain none of the *di* component that is found in the more mafic members, they are corundum normative, and they contain increasing amounts of restite quartz at 5·0 kbar. The S-type melts, on the other hand, exhibit less variation apparently because hornblende is not a stable phase in these systems. Thus, unlike the minor elements such as Sr, which tend to exhibit greater uniformity relative to SiO_2 in I-type granites as opposed to S-type granites (Chappell *et al.* 1990), the melt components in the I-type compositions of Figure 4 clearly show a wider variation than in the S-type compositions. The greater variability in S-type granites observed by Chappell *et al.* (1990), on the other hand, is reflected in the much greater variability in the proportions of restite minerals, as shown by comparison of Figures 7 and 8 in Section 3 below.

2.5. The Moruya Suite (I-type)

The calculated liquidus relations of the four members of the Moruya Suite, ranging from the most mafic (MG42) to the most felsic (MG20; Chappell & Stevens 1988, table 2), are shown in the P-T projection of Figure 3(b). Owing to the much higher *ab* and lower *an* contents of Moruya melts, as shown in Figures 4 and 5, the calculated An content of the liquidus plagioclase (An_{52}) and the liquidus temperature (894 ± 2°C) of this trondhjemitic granite melt are correspondingly much lower than for the Jindabyne melts at 5·0 kbar and $X_w^m = 0{\cdot}30$. Despite this overall more felsic nature of the Moruya Suite, however, its most mafic member (MG42, Tuross Head Tonalite), the composition of which is presented in Figure 5, still is calculated to contain 62% restite, of which 43% is plagioclase of the above composition.

If, as with the most mafic member of the Jindabyne Suite, it is assumed that the Tuross Head Tonalite was completely molten at its source, then the calculated liquidus temperature would have been 1170°C and the liquidus plagioclase would have been $An_{88}Ab_{11}Or_{11}$ in composition. Obviously, this temperature is still much too high for a crustal environment and clearly indicates that roughly 43% plagioclase in the Tuross Head Tonalite is restite in origin. Indeed, Chappell and Stephens (1988), on independent grounds, argued that "... the composition of MG42 is close to that of the source rocks of the suite". With 62% total restite, it is unlikely that any appreciable amount of melt

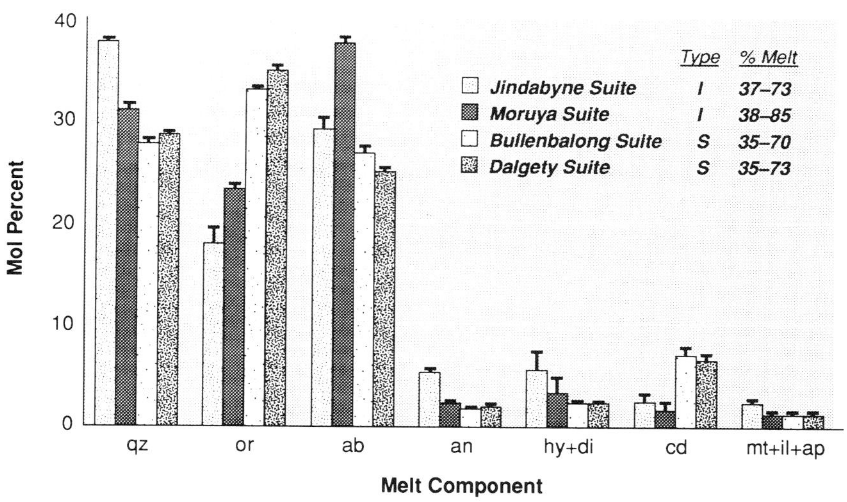

Figure 4 Calculated liquidus melt compositions (mol percent) at 5·0 kbar, $X_w^m = 0{\cdot}30$ of the I-type Jindabyne and Moruya Suites, and the S-type Bullenbalong and Dalgety Suites; error bars represent mean deviations from the average compositions; melt components and their total percentages are as described in the text.

segregation has occurred (cf van der Molen & Paterson 1979), hence the present findings are in complete accord and the composition of MG42 is therefore shown in Figure 5 as representative of the Moruya Suite source rocks.

2.6. The Bullenbalong Suite (S-type)

The calculated liquidus relations in the five (excluding KB118, as discussed below) members of the Bullenbalong Suite melts investigated are shown in the P-T projection of Figure 6(a) and the melt composition, which is that of a potassic granite, is shown in Figure 4. It will be noticed in Figure 6(a) that both plagioclase ($An_{53}Ab_{40}Or_7$) and K-feldspar ($Or_{75}Ab_{21}An_4$) are calculated to be in equilibrium with quartz and melt at 856 ± 1°C, 5·0 kbar, and $X_w^m = 0{\cdot}30$. The noticeable differences in the Clapeyron slopes of the plagioclase and K-feldspar liquidii, incidentally, are due, in part, to a combination of a greater extent of speciation of the *an* melt component and a lesser extent of speciation of the *or* component (less incongruency of melting) with increasing pressure.

In the most mafic member of this suite (KB32, Jillamatong Granodiorite), the normative composition of which is presented in Figure 5, Chappell *et al.* (1990) report "... prominent cores near An_{55} ..." in the plagioclase crystals; the calculated liquidus and restite plagioclase is An_{53}, as indicated above, in very close agreement. The Jillamatong Granodiorite is calculated to contain 65% restite, whereas the most felsic member of the suite (KB45; White & Chappell 1988, Table 1) contains 30% restite, including 19% quartz.

One of the samples originally chosen as representative of the Bullenbalong Suite was KB118 from the Amboyne unit. It turned out to have distinctly different liquidus phase relations: (1) the calculated liquidus temperature is 919°C, (2) the plagioclase composition is $An_{81}Ab_{17}Or_2$, and (3) K-feldspar is not on the liquidus. These properties, which are grossly different from the rest of the Bullenbalong Suite, are precisely those of the Cootralantra Granodiorite (BB83, Chappell *et al.* 1990) and other related granites (BB38 and BB36). The facts that the calculated liquidus temperatures and An contents of the plagioclase are so much higher and K-feldspar is not on the liquidus, clearly indicate that the source rocks of the Cootralantra Suite were substantially more calcareous than those of the Bullenbalong Suite and therefore should not be included in that suite. Indeed, insofar as liquidus phase relations are concerned, there are more cogent reasons for separating the Cootralantra group from the Bullenbalong suite than there are for separating the latter suite from the Strathbogie Suite, as now discussed.

The Strathbogie Suite (White & Chappell 1988) is another group of S-type granites in central Victoria that resemble, in certain respects, the Bullenbalong Suite. In fact, they have nearly the same liquidus relations at 5·0 kbar and $X_w^m = 0{\cdot}30$ (T = 850°C, An_{53} plagioclase, and Or_{75} K-feldspar), but they are more reduced and have higher Ba contents. In the most felsic member of this suite (CV114, White & Chappell 1988, table 1), the most anorthitic plagioclase obtainable on the liquidus is $An_{46}Ab_{47}Or_7$, which indicates that the more felsic members of this suite, at least, tend to be fractionated, unlike those of the Bullenbalong Suite. It also should be

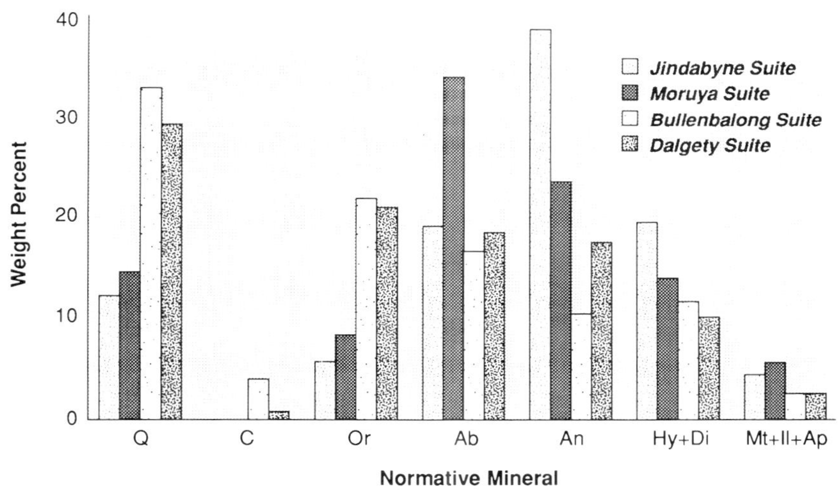

Figure 5 CIPW normative compositions of the most mafic members of the Jindabyne, Moruya, Bullenbalong, and Dalgety suites: inasmuch as the melt contents total only 35–38% (cf. Fig. 4), it is assumed that the bulk compositions of these members are closely representative of the source rocks of their respective suites.

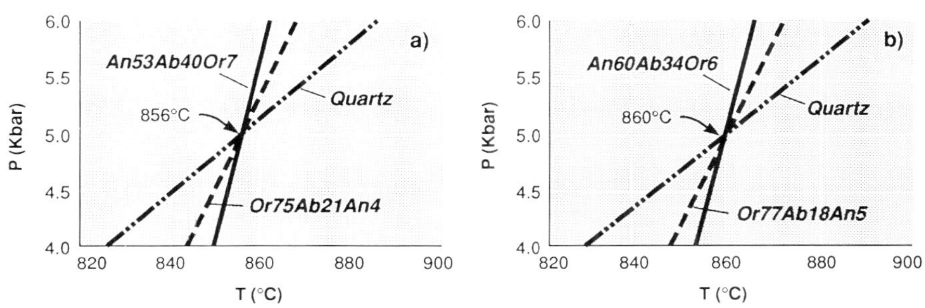

Figure 6 Calculated liquidus relations of the Bullenbalong (a) and Dalgety (b) Suite melts at 5·0 ± 1·0 kbar and $X_w^m = 0{\cdot}30$ (≈2·8 wt% H_2O); substantial differences in the Clapeyron slopes of the plagioclase and alkali feldspar liquidii are explained in the text.

mentioned that 17 of the 18% crystal residue in CV114 is quartz, which is consistent with the restite quartz content of the more felsic members of the Bullenbalong Suite.

2.7. The Dalgety Suite (S-type)

The P-T diagram in Figure 6(b) shows the calculated liquidus relations in the Dalgety Suite. For all eight samples of this suite investigated, whose potassic granite melt compositions are shown in Figure 4, the calculated liquidus temperature is 860 ± 1°C and the liquidus phases are plagioclase ($An_{60}Ab_{34}Or_6$), K-feldspar ($Or_{77}Ab_{18}An_5$), and quartz.

In comparison with the Bullenbalong Suite, the slightly higher liquidus temperature, as well as the slightly more anorthitic plagioclase, are indicative of a somewhat more feldspathic source rock, as shown in Figure 5. The most mafic member of the suite (BB12), the normative composition of which is shown in Figure 5, contains 65% restite, whereas the most felsic member (BB2) contains 27%, including 16% quartz. Ninety per cent of the remaining restite in BB2 is plagioclase of An_{60} composition, precisely the same as the cores of the crystals reported by Chappell *et al.* (1990).

3. Restite in perspective (R.I.P.)

3.1. Source rocks of the Jindabyne, Moruya, Bullenbalong and Dalgety Suites

On the premise that a granitic magma body at crustal levels which contains less than approximately 35 vol% melt is incapable of segregating melt from solids (van der Molen & Paterson 1979) during diapiric ascent, the most mafic members of each of the four suites, calculated to contain from 35–38% melt, are presumed to closely represent the bulk compositions of their respective source rocks. The CIPW norms of these four bulk rock compositions are therefore presented as representatives of their source rocks in Figure 5.

Accordingly, the source rock of the Jindabyne Suite is here regarded as dioritic or andesitic, whereas the source rock of the Moruya Suite is regarded as tonalitic or dacitic. A comparison of the two S-type source rocks in Figure 5, on the other hand, indicates that the Bullenbalong Suite was derived from a more pelitic metagreywacke, whereas the Dalgety Suite was derived from a more feldspathic metagreywacke.

3.2. Restite systematics

The uniformity of calculated melt and plagioclase compositions throughout all members of a given suite, except for the effects already mentioned and attributed to minor hornblende fractionation during emplacement of the Jindabyne and Moruya suites, clearly rules out magma mixing to explain the compositional variations within a given suite. It also rules out contamination or cumulates from extraneous sources, unless the extraneous contaminants or cumulates are in equilibrium with the melts. To test this possibility that part of what is regarded as restite in a given member of a suite was actually derived from a source extraneous to that particular body during magma ascent and emplacement, the calculated mol per cents of individual restite minerals are plotted against the mol per cent total restite in the I-type and S-type suites in Figures 7 and 8, respectively.

In these figures, it is assumed for the purposes of discussion and further comparison that all magmas of all suites ascended essentially along the isenthalpic adiabat from their source regions at a depth of approximately 18 km (5·0 kbar) to a depth of 7·5 km (2·0 kbar). During this adiabatic ascent, it is further assumed that quartz-plagioclase-melt equilibrium was maintained; hence, owing to the drastically different Clapeyron slopes of quartz and the feldspars (Figs 3, 6), as much as 10 mol% quartz per mol of melt must have been resorbed in the process. The heat required for this resorption, which is essentially the heat of fusion of quartz, was obtained by cooling and precipitation of plagioclase, the molal heat of crystallisation of which is approximately twice that required to resorb an equimolal amount of quartz. The net result of closed system decompression from 5·0 to 2·0 kbar, therefore, was to resorb up to 10% quartz, precipitate up to 5% plagioclase, increase the total melt content by up to 5%, and lower the liquidus temperature along the isenthalpic adiabat, which is likely to be somewhat closer to the K-feldspar liquidus than the plagioclase liquidus in Figure 6. Herein, incidentally, lies the explanation of why, in shallow seated porphyries and pyroclastics, quartz phenocrysts very commonly show evidence of resorption, whereas plagioclase, the liquidus slope of which is close to the isentropic (not isenthalpic) adiabat, very rarely, if ever, does.

2.3. The Jindabyne Suite

The remarkably linear array, with very little scatter, of decreasing restite mineral percentages with decreasing total restite shown in Figure 7(a) clearly demonstrates that virtually nothing of extraneous origin has been added to the magmas of this suite, unless the contaminant had the same bulk composition as the Jindabyne source rocks (Fig. 5). Moreover, if magma mixing has occurred in any member of this suite, both magmas must necessarily have been derived from the same source rock, i.e. the Jindabyne source rock.

3.4. The Moruya Suite

Although only four samples from the Moruya Suite were investigated, as shown in Figure 7(b), the conclusions drawn for the Jindabyne Suite are equally applicable to this suite. The very minor scatter in the relative proportions of restite minerals in both I-type suites is not only indicative of a purely restite origin for these minerals, but of the very uniform and orderly process of restite-melt segregation (unmixing), in which the relative proportions of restite minerals remain remarkably constant during unmixing. This latter feature, which is not as apparent in the Bullenbalong Suite discussed below, may be due to the higher temperatures of the I-type magmas (cf Fig. 3 with Fig. 6), but it is more likely a reflection of the greater uniformity in the bulk compositions of the I-type source rocks as compared with the S-type source rocks (Chappell *et al.* 1990).

3.5. The Bullenbalong Suite

As indicated above and shown in Figure 8(a), the S-type Bullenbalong Suite shows greater scatter than either of the I-type suites (cf Fig. 7), despite the expanded scale of the ordinate in Figure 8(a). This greater scatter, produced mainly by samples KB90 and KB121, conceivably could be due to greater retention at lower magma temperatures of restite K-feldspar in KB90 and of restite plagioclase in KB121 during ascent and emplacement at the presumed depth of 7·5 km. It is also possible, on the other hand, that the metasedimentary source rocks of KB90 locally were slightly enriched in K-feldspar (5·0%), whereas those of

KB121 were half as much enriched in plagioclase (2·5%). For reasons discussed more fully below, this latter interpretation is preferred.

3.6. The Dalgety Suite

The variation in restite mineral proportions with total restite in the Dalgety Suite is significantly more uniform than in the Bullenbalong Suite, despite a temperature difference of only 4°C. Except for one sample (BG7), as shown in Figure 8(b), the uniformity is comparable to that in the I-type suites.

The fact that the restite quartz content of BG7 is several per cent higher than that of the presumed source rock in Figure 5 clearly precludes preferential retention of quartz as the mechanism for producing this anomaly. The most viable alternative, as with the feldspars in the two samples from the Bullenbalong Suite, is that the excess quartz (7·0%) is a reflection of a local stratigraphic heterogeneity in the Dalgety source rocks. Support for this hypothesis comes from the fact that removal of this 7·0% excess quartz places the percentages of all of the restite minerals in BG7 close to the linear trends for the rest of the suite, as indicated in Figure 8(b).

3.7. Restite unmixing systematics

A comparison of the regression lines for the restite minerals in Figures 7(a) through to 8(b) reveals that plagioclase (An + Ab) is a restite mineral in all members of all four suites and is the dominant restite mineral in both I-type suites (Fig. 7). Hypersthene is the second most abundant restite mineral in the I-type suites, but disappears from the assemblage at total restite contents less than approximately 18%, presumably as a result of hornblende formation and unmixing. Quartz is the most abundant restite in the Bullenbalong Suite, whereas it is second to plagioclase in the Dalgety Suite.

4. Concluding summary statement

The basic premise of this investigation has been that all members of each granite suite examined contained plagioclase feldspar and quartz on their liquidii. The first task, therefore, was to determine, through application of the thermodynamically-based Quasi-crystalline Model (Burnham & Nekvasil 1986), the composition of the plagioclase and the temperature at which this plagioclase was in equilibrium with quartz on the liquidus at P = 5·0 kbar and $X_w^m = 0{\cdot}30$. In accomplishing this task, it was found that: (1) all members of each suite investigated were calculated to contain plagioclase of the same composistion on the liquidus; (2) this plagiclase composition was the same as that of unaltered crystal cores determined optically; (3) the calculated liquidus temperatures of all members of each

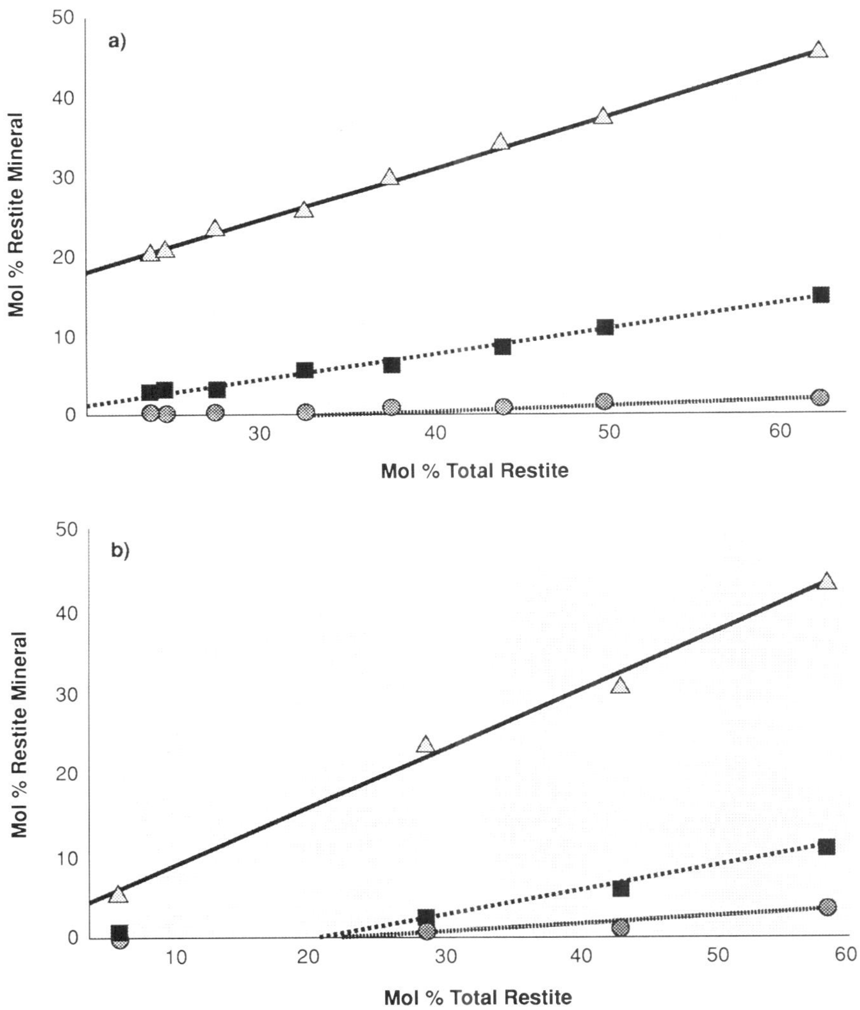

Figure 7 Calculated percentages of restite minerals versus total restite in the Jindabyne (a) and Moruya (b) Suite rocks at 2·0 kbar, $X_w^m = 0{\cdot}30$; triangles represent plagioclase (An + Ab), squares represent hypersthene (Hy), and circles represent magnetite plus ilmenite plus apatite (Mt + Il + Ap).

suite were the same within, at most, ±4°C; (4) the liquidus melts in all members of each suite were of the same composition, except for the effects of minor hornblende fractionation in some members of the I-type suite; (5) the most mafic member of each suite contained only 35–38 mol (≈volume)% melt, which has been interpreted as an indication of close identity with the magma source rock; and (6) with only three minor exceptions in the two S-type suites, the trends in the individual residual (restite) mineral contents in proceeding from the most mafic to the most felsic member of each suite are highly linear.

Together, these six findings virtually preclude assimilation of extraneous wallrock, accumulation of minerals precipitated earlier at higher temperatures than the calculated liquidus, or mixing of magmas from different sources as processes by which the bulk compositional variations within each of the suites investigated could have been produced. The only viable process, which is fully consistent with all of these findings, is that the observed compositional variations are the result of very systematic variations in the extent of restite unmixing from magmas derived by partial melting of unambiguously identifiable source rocks. Thus, the present results, contrary to the prior prejudices of the author, fully corroborate the basic tenets of the Restite Model of White and Chappell (1977).

As a closing caveat, the present findings that only melt-restite equilibria were involved in the origin of the Australian granites investigated does not necessarily imply that all granites worldwide originated in precisely this way. However, each suite of granites, wherever it occurs, must be examined primarily in terms of its phase equilibrium relationships if processes other than melt-restite segregation are to be revealed.

5. Acknowledgements

Special thanks are due to Hanna Nekvasil, whose "EQUIL" subroutine made possible the calculated phase equilibrium relations presented here. Special thanks also are due to A. J. R. White and B. W. Chappell for providing the data base, the selection of the most appropriate suites to be investigated, and for providing logistical support in the initial stages of this project (including computer access) during a visit to Australia early in 1991. The review comments of H. Ohmoto and A. J. R. White, which have led to substantial improvements in this presentation, are greatly appreciated.

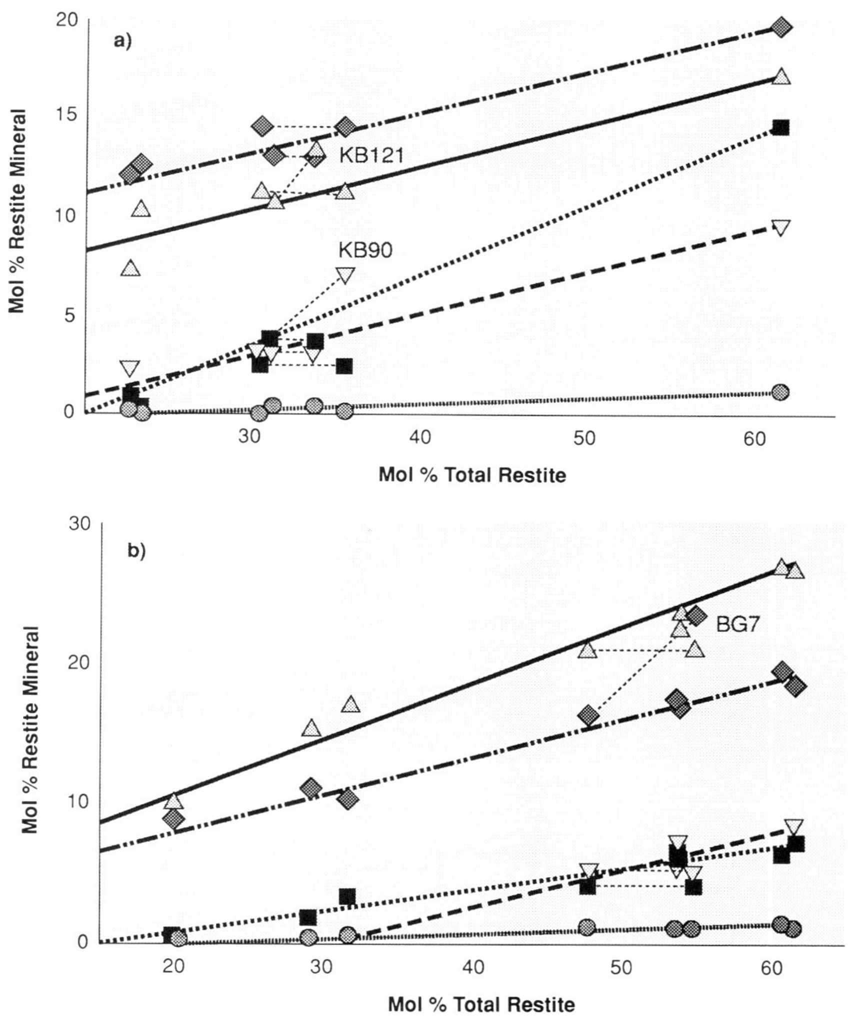

Figure 8 Calculated percentages of restite minerals versus total restites in the Bullenbalong (a) and Dalgety (b) Suite rocks at 2·0 kbar, $X_w^m = 0{\cdot}30$; upright triangles represent plagioclase (An + Ab), squares represent hypersthene plus cordierite (Hy + Cd), diamonds represent quartz (Qz), inverted triangles represent K-feldspar (Or), and circles represent magnetite plus ilmenite plus apatite (Mt + Il + Ap); lines connecting points indicate the effects of subtracting 2·5% An + Ab from KB121, 5·0% Or from KB90, and 7·0% Qz from BG7.

6. References

Bell, P. M. & Roseboom, E. H. Jr. 1969. Melting relationships of jadeite and albite to 45 kilobars with comments on melting diagrams of binary systems at high pressures. *In* Papike, J. J. (ed) *Pryoxenes and amphiboles: crystal chemistry and phase petrology,* 151–61. MINER SOC AM SPEC PAP **2**.

Boettcher, A. L. & Burnham, C. W. 1983. The structures of silicate liquids: A note. J GEOL **91,** 607–8.

Boettcher, A. L. & Wyllie, P. J. 1969. Phase relationships in the system $NaAlSiO_4$–SiO_2–H_2O to 35 kilobars pressure. AM J SCI **267,** 875–909.

Boettcher, A. L., Burnham, C. W., Windom, K. E. & Bohlen, S. R. 1982. Liquids, glasses, and the melting of silicates to high pressures. J GEOL **90,** 127–38.

Boettcher, A. L., Guo, Q., Bohlen, S. R. & Hansen, B. 1984. Melting in feldspar-bearing systems to high pressures and the structures of aluminosilicate liquids. GEOLOGY **12,** 202–4.

Burnham, C. W. 1979. Magmas and hydrothermal fluids. *In* Barnes, H. L. (ed.) *The geochemistry of hydrothermal ore deposits,* 71–136. New York: John Wiley.

Burnham, C. W. 1981. The nature of multicomponent aluminosilicate melts. *In* Rickard, D. T. & Wickman, F. E. (eds) *Geochemistry of solutions at high temperatures and pressures.* PHYS CHEM EARTH **13** &**14,** 191–227.

Burnham, C. W. & Davis, N. F. 1974. The role of H_2O in silicate melts: Thermodynamic and phase relations in the system $NaAlSi_3O_8$—H_2O to 10 kilobars and 1100°C. AM J SCI **274,** 902–40.

Burnham, C. W. & Nekvasil, H. 1986. Equilibrium properties of granite pegmatite magmas. AM MINERAL **71,** 239–63.

Chappell, B. W. 1984. Source rocks of I- and S-type granites in the Lachlan Fold Belt, southeastern Australia. PHILOS TRANS R SOC LONDON **A310,** 693–707.

Chappell, B. W. & Stevens, W. E. 1988. Origin of infracrustal (I-type) granite magmas. TRANS R SOC EDINBURGH: EARTH SCI **79,** 71–86.

Chappell, B. W., White, A. J. R. & Wyborn, D. 1987. The importance of residual source material (restite) in granite petrogenesis. J PETROL **28,** 1111–38.

Chappell, B. W., Williams, I. S., White, A. J. R. & McCulloch, M. T. 1990. Granites of the Lachlan Fold Belt. EXCURSION GUIDE A-2, ICOG-7. Canberra: Bureau Mineral Resources.

Clemens, J. D. 1989. The importance of residual source material (restite) in granite petrogenesis. J PETROL **30,** 1313–16.

Gibbs, J. W. 1961. *The Scientific Papers of J. Willard Gibbs,* Vol. 1, *Thermodynamics.* New York: Dover.

Goldsmith, J. R. 1980. The melting and breakdown reactions of anorthite at high pressures and temperatures. AM MINERAL **65,** 272–84.

Green, T. H. 1969. High-pressure experimental studies on the origin of anorthosite. CAN J EARTH SCI **6,** 427–40.

Griffin, T. J., White, A. J. R. & Chappell, B. W. 1978. The Moruya Batholith and geochemical contrasts between the Moruya and Jindabyne Suites. J GEOL SOC AUST **25,** 235–47.

Hine, R., Williams, I. S., Chappell, B. W. & White, A. J. R. 1978. Contrasts between I- and S-type granitoids of the Kosciusko Batholith. J GEOL SOC AUST **25,** 219–34.

Joyce, D. 1985. *A phase equilibrium study in the system $NaAlSi_3O_8$–SiO_2–Al_2SiO_5–H_2O at 2 kilobars, and petrogenetic implications.* M.S. Thesis, Pennsylvania State University, University Park, PA.

Nekvasil, H. 1986. *A theoretical thermodynamic investigation of the system Ab–Or–An–Qz(–H_2O) and implications for melt speciation.* Ph.D. Thesis, Pennsylvania State University, University Park, PA.

Nekvasil, H. 1990. Reaction relations in the granite system: Implications for trachytic and syenitic magmas. AM MINERAL **75,** 560–1.

Tuttle, O. F. & Bowen, N. L. 1958. Origin of granite in light of experimental studies. MEM GEOL SOC AM **74**.

van der Molen, I. & Paterson, M. S. 1979. Experimental deformation of partially-melted granite. CONTRIB MINERAL PETROL **70,** 299–318.

Voigt, D. E. 1983. *The solubility of Al_2SiO_5 in the system $KAlSi_3O_8$–SiO_2–H_2O at 2 kbar, and its implications for melt speciation.* M.S. Thesis, Pennsylvania State University, University Park, PA.

White, A. J. R. & Chappell, B. W. 1977. Ultrametamorphism and granitoid genesis. TECTONOPHYSICS **43,** 7–22.

White, A. J. R. & Chappell, B. W. 1988. Some supracrustal (S-type) granites of the Lachlan Fold Belt. TRANS R SOC EDINBURGH: EARTH SCI **79,** 169–81.

White, A. J. R. & Chappell, B. W. 1989. *Geology of the Numbia 1:100,000 Sheet (8624).* Sydney: Geological Survey of N.S.W.

White, A. J. R., Williams, I. S. & Chappell, B. W. 1977. *Geology of the Berridale 1:100,000 Sheet (8625).* Sydney: Geological Survey of N.S.W.

Wood, B. J. 1978. Reaction involving anorthite and $CaAl_2SiO_6$ pyroxenes at high pressures and temperatures. AM J SCI **278,** 930–942.

C. WAYNE BURNHAM, Department of Geosciences, The Pennsylvania State University, University Park, PA 16802, U.S.A.

Transactions of the Royal Society of Edinburgh: Earth Sciences, **83**, 399–407, 1992

Feldspar crystallisation in felsic magmas: a review

Hanna Nekvasil

ABSTRACT: Understanding the controls on the evolution of natural feldspars is greatly assisted by coupling experimental determinations of feldspar/melt equilibria with thermodynamic modelling and the calculation of crystallisation paths. Such a combined approach permits the evaluation of the influence of intensive and extensive variables on feldspar compositions. Feldspar compositional paths are influenced only to a minor degree by pressure. The presence of H_2O or other melt component incompatible in feldspar has a more major effect, not only in increasing the temperature interval over which feldspar crystallises, but also in decreasing the amount of Ab enrichment of the feldspar which occurs during crystallisation. The amount and behaviour of H_2O in the magma has a pronounced influence on feldspar compositions when two feldspars are stable. Under conditions of high bulk H_2O content or constant activity of H_2O, plagioclase and alkali feldspar compositions evolve by increasing Ab content similar to the behaviour manifested in the simple binary systems. If the bulk H_2O content is low, however, and the H_2O content increases during crystallisation, plagioclase evolves by increasing Ab content until alkali feldspar is stabilised. From that point on, plagioclase compositions change mainly by decreasing Or content, while alkali feldspar evolves mainly by increasing Or content.

KEY WORDS: ternary feldspar, crystallisation paths, trachyte, granite, thermodynamic modelling, phase relations

Feldspars are ubiquitous in felsic rocks, ranging from low temperature two-mica granites to high temperature quartz syenites. They form both binary and ternary solid solutions and, because of the sluggish kinetics of their homogenisation, commonly show compositional zoning. This zonation reflects the history of the magma and may provide important clues as to the source rock and magmatic processes leading to the development of felsic magmas. In order to be able to use feldspar zoning to understand the magmatic history of any particular body, however, we must first understand the controls on feldspar composition imposed by magmatic intensive and extensive variables. The article summarises our understanding of the behaviour of feldspars under equilibrium conditions in the simplest to the more complex compositional systems as a stepping stone towards the understanding of more complex kinetic phenomena. Because summaries of experimental work are available elsewhere (e.g. Johannes 1978), this work emphasises the information obtained through modelling efforts.

1. The binary feldspar systems: the effects of pressure and H_2O

The characteristics of the binary plagioclase system are now reasonably well understood. The solidus and liquidus in this system form the well-known continuous melting loop at 1 atm (Bowen 1913). This melting loop has provided us with the basic tenet of feldspar crystallisation which has been applied widely to natural rocks: isobarically decreasing temperature results in an increase in the Ab content of plagioclase. This tenet must be modified for alkali feldspar. The early work of Schairer (1950) and Bowen and Tuttle (1950) demonstrated the effect of the strong non-ideality of Or–Ab solid solutions in producing the 1 atm azeotrope in this system. This has led to the conclusion that bulk compositions on the Or side of the thermal minimum (azeotrope at high temperatures, eutectic at low temperatures) crystallise by decreasing Or content with isobarically decreasing temperature, while bulk compositions on the Ab side of the thermal minimum evolve by increasing Or content during an isobaric temperature drop. However, most natural alkali feldspars in Ca-bearing bulk compositions lie on the Or side of the azeotrope and thus show the same evolutionary trends of increasing Ab content with decreasing temperature as plagioclase.

The effects of many intensive and extensive variables on feldspar compositions have not been determined experimentally. However, Nekvasil (1986) and Nekvasil and Burnham (1987) conducted a computational investigation of the effects of such intensive variables. Their phase relations at constant H_2O contents for H_2O-undersaturated conditions were predicted. Without available experimental data for comparison, these results cannot be checked for accuracy but only evaluated for internal consistency. However, comparisons of predicted H_2O-saturated phase relations against H_2O-saturated experimental data for a variety of compositions (Nekvasil & Burnham 1987) and comparisons against recent anhydrous data (Nekvasil 1991) indicate that the calculation methodology has good predictive capabilities in these systems. Until more internally consistent experimental data are available, however, a detailed error analysis cannot be conducted on such calculations. Hence, the validity of the temperatures and compositions calculated at any set of specific conditions remains unknown. Importantly, however, because investigations of the phase relations with different solid and liquid thermodynamic models gives rise to similar trends, these *trends* should be robust and useful for understanding the controls on feldspar compositional evolution.

As magmas form at pressures above ambient, the effect of pressure on the feldspar melting relations must be considered. Its effect on the binary melting loops at constant

H_2O content has not been determined experimentally; however, Nekvasil and Burnham (1987) investigated these effects computationally. The calculated effects of pressure on the plagioclase melting loop are summarised in Figure 1. The effect of increasing pressure is minor (as shown by arrows **a** in Fig. 1), inducing only a slight increase in Ab content of plagioclase at a constant melt composition and a small amount of thermal stabilisation with increasing pressure.

H_2O has long been considered a very important volatile in the formation and evolution of felsic magmas. Its strong effect on reducing solidus and liquidus temperatures has permitted the interpretation that most granitic magmas arise from crustal anatexis. However, based on phase equilibria studies of natural granites, dehydration/melting reactions of hydrous minerals and textural and field observations, it is now generally accepted that most granitic magmas do not reach H_2O-saturation until late in their evolutionary history, if at all. For magmas which do not have a hydrous mineral on the liquidus, the H_2O content in the melt probably increases continually during crystallisation of anhydrous phases until a hydrous mineral is stabilised. As long as all of the melt components needed to make this hydrous mineral are available, the H_2O activity will remain essentially constant and continued crystallisation of anhydrous minerals will simply result in further crystallisation of the hydrous mineral. (This does not, however, imply completely invariant H_2O content of the melt, because the temperature and the composition of the melt are continuously changing.) Once this is no longer the case, however, the H_2O content of the melt will begin to build up once more and may reach the saturation value. H_2O-saturation will take place only

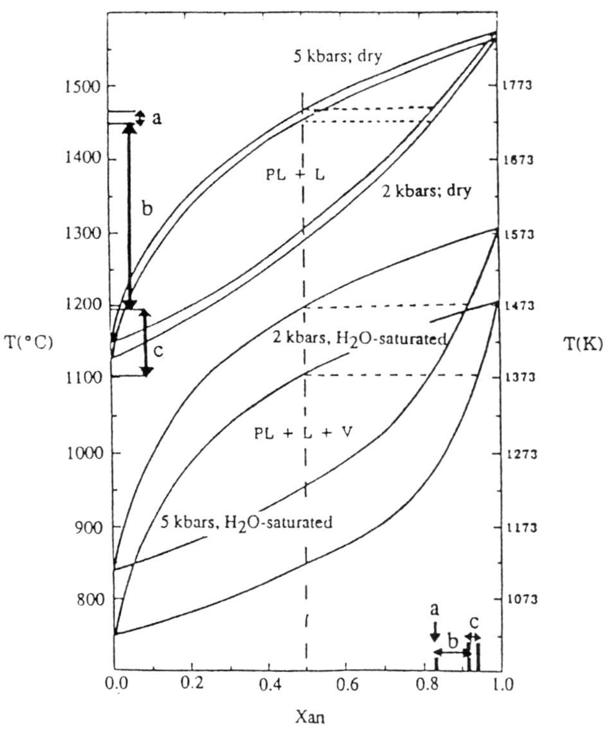

Figure 1 The effects of pressure and H_2O content on the plagioclase melting loop, modified from the calculations of Nekvasil (1986); arrows **a**, **b** and **c** indicate the change in temperature and plagioclase composition as (a) pressure is decreased from 5 kbar to 2 kbar under anhydrous conditions, (b) H_2O content is increased from 0 to 6 wt% (H_2O-saturated) isobarically at 2 kbar total pressure and (c) the pressure is increased from 2 to 5 kbar under H_2O-saturated conditions (which results in an increase in melt H_2O content of ~3 wt%).

very close to the solidus for "dry" magmas, but at an earlier stage for more H_2O-rich bulk compositions.

The effect of isobarically changing H_2O content on the binary feldspar systems is perhaps better constrained experimentally than the effect of pressure. The strong destabilisation of plagioclase and alkali feldspar with increasing H_2O content can be seen by comparison of the 1 atm melting loops (Bowen 1913; Schairer 1950) and the 2 kbar (~6 wt% H_2O) and 5 kbar (~10 wt% H_2O) H_2O-saturated melting loops (1 kbar = 100 MPa; Yoder *et al.* 1957; Johannes 1978; Erikson 1979). (This comparison is valid only because of the small effect of pressure alone.) Figure 1 shows the calculated H_2O-saturated plagioclase melting loops at 2 and 5 kbar. These calculated melting loops agree with available experimental data, as discussed by Nekvasil and Burnham (1987), and can be used as a basis for determining the effect of H_2O. Comparison of these H_2O-saturated melting loops with the predicted anhydrous melting loops at the same pressure indicates a widening of the melting loops with increased H_2O content and hence lower temperature. The widening of the H_2O-bearing melting loops is a reflection of the slight non-ideality of the plagioclase solid solution with increases in magnitude with decreasing temperature (Saxena & Ribbe 1972; Seil & Blencoe 1979; Henry *et al.* 1982). The widening of the melting loops results in a higher An content of plagioclase and Or content of alkali feldspar (on the Or side of the thermal minimum) for a given melt composition than is found at lower H_2O contents (arrows **b** in Fig. 1).

Much experimental effort has been expended in determining the effect of changing P_{H_2O} on the feldspar melting loops (e.g. Yoder *et al.* 1957; Johannes 1978; Erikson 1979). Increasing P_{H_2O} results in combined increase of both total pressure and H_2O content (because of the increased H_2O solubility with increased pressure). The results of combination of the effects of these two variables can be seen by comparing the H_2O-saturated melting loops at 2 and 5 kbar total pressure. The net result of increasing P_{H_2O} is a slight expansion of the melting loop and hence a more An-rich plagioclase and Or-rich alkali feldspar for the same anhydrous melt composition as at lower P_{H_2O}.

From the above analysis, it is evident that if crystallisation takes place under conditions of continually increasing H_2O content due to the precipitation of anhydrous minerals, then the relative strengths of two competing factors will dictate the evolution of the feldspar compositions. First, decreasing temperature drives the melt composition towards Ab which will move the feldspar composition also towards Ab. Countering this tendency is the increase in H_2O content which should induce enrichment of the feldspars in An or Or. The effect of decreasing temperature on the melt composition is much stronger than the expansion of the melting loop and there is a net increase in the amount of Ab in the plagioclase with dropping temperature. However, a new melting loop will be followed during crystallisation, one which is much steeper in T–X space than that at constant H_2O content.

2. The effect of dilution by additional melt components

When an additional melt component which does not partition into feldspar is added to the feldspar systems, Qz for example, the feldspar melt components are diluted, thereby depressing the liquidus temperatures. Using the plagioclase-quartz system as an example, Figure 2a shows the calculated liquidus surface at 2 kbar, under H_2O-

saturated conditions from Nekvasil and Burnham (1987), indicating the temperature depression of plagioclase induced by the addition of Qz component in the melt. The effect of the Qz component on crystal/melt partitioning of Ab and An can perhaps be best seen by a series of sections through the system at constant Qz content (Fig. 2b) projected back into the plagioclase system. The projected isoplethal plagioclase melting loops show very little change other than temperature depression and there is little flattening of the loops induced simply by dilution. The flatter loop projected from the cotectic is simply due to the variation in Qz content along this curve. Crystallisation paths of Qz-normative (but Qz-undersaturated) magmas cut across numerous Qz-isopleths. Therefore, the influence of Qz component on the feldspar loop must be considered along a crystallisation path. Figure 2a shows an example of such a path which follows a curved trend from the bulk composition towards the cotectic and saturation with quartz. The melt compositions along this path have been projected from quartz (and H_2O) into the plagioclase system in Figure 2b in order to permit direct visualisation of the effect of continual buildup of quartz on the plagioclase melting loop. As can be seen in this figure, the increase in Qz content of the melt induces continual temperature depression and inhibits the plagioclase from attaining the albitic compositions of the pure melting loop. This is the same effect that would by induced by continual build-up of H_2O as both of these components act more or less as dilutants of the plagioclase components.

Phase relations in the alkali feldspar–quartz system (i.e. the haplogranite system) were studied by Tuttle and Bowen (1958) and Luth *et al.* (1964), who concluded that the alkali feldspar in this system showed the same behaviour as in the simple feldspar system. However, with the temperature

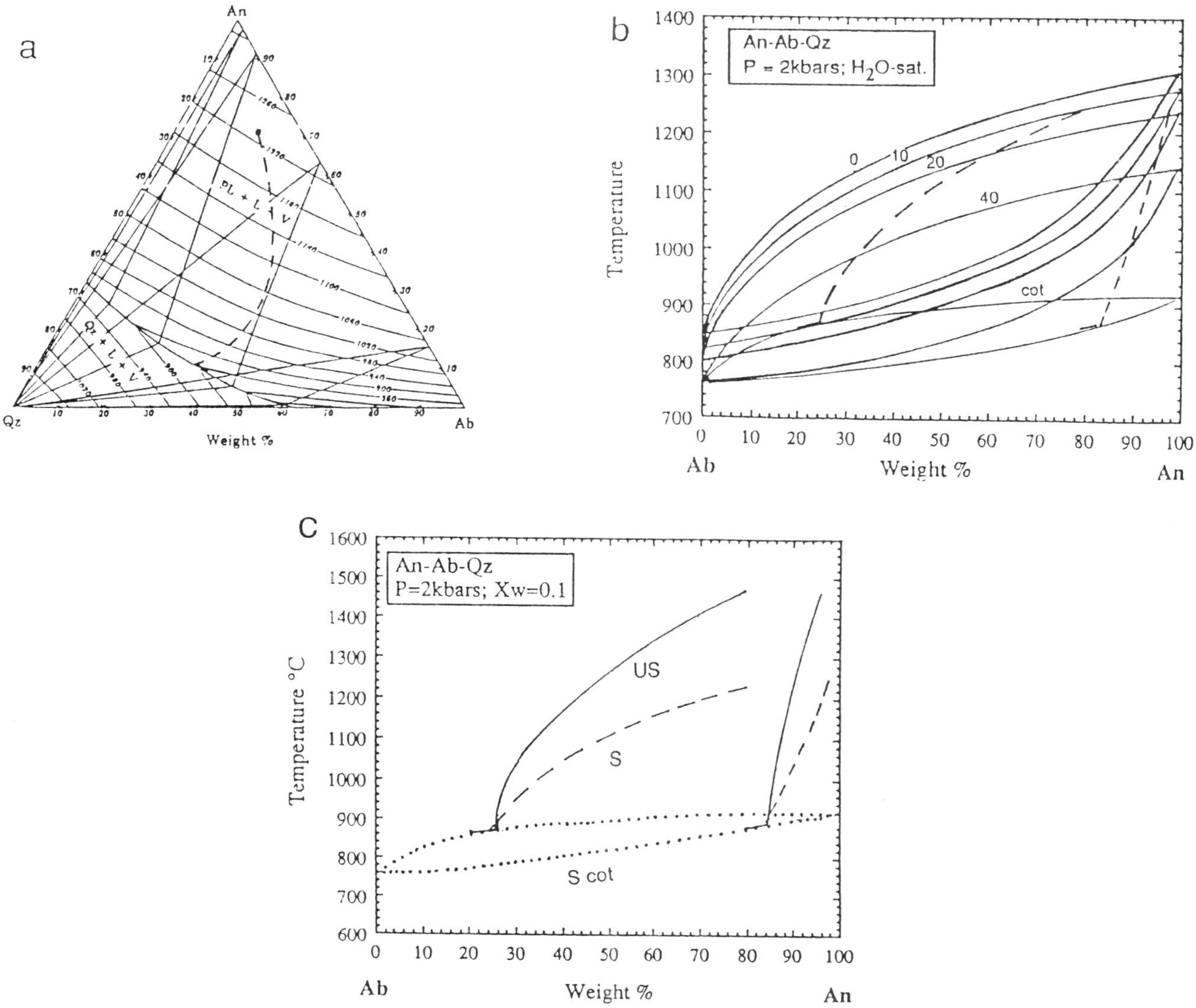

Figure 2 The effects of increasing silica content and H_2O content on plagioclase compositional evolution: (a) The equilibrium crystallisation path for the composition $An_{71\cdot8}Ab_{19\cdot3}Qz_{8\cdot9}$ (wt%) (dashed curve) plotted onto the calculated liquidus surface in the plagioclase-quartz system at 2 kbar under H_2O-saturated conditions from Nekvasil (1986). (b) Projections of the liquidus relations along Qz-isopleths (numbered in wt% Qz) and along the cotectic in the plagioclase-quartz system into the plagioclase system, showing the effect of excess silica on the H_2O-saturated plagioclase melting loop; dashed curves represent the melt compositions of the path in (a) (curve closest to Ab), and the coexisting plagioclase compositions (curve closest to An) during equilibrium crystallisation and concomitant increase in the silica content of the melt. (c) Plagioclase compositions and projected melt compositions for the H_2O-unbuffered and H_2O-undersaturated (US) crystallisation path (solid curves) of the same anhydrous bulk composition as in (a) but with 0·75 wt% H_2O added to the bulk composition; shown for comparison is the H_2O-saturated (S) path (dashed curves) of (b) and the projected plagioclase loop at the H_2O-saturated cotectic (S cot).

depression induced by the addition of quartz, the intersection of the alkali feldspar solvus by the solidus occurs at a lower P_{H_2O} than in the binary, resulting in a ternary eutectic. As discussed by Boettcher (1980), until the temperatures in the binary are low enough to intersect the solvus, the two-feldspar + L cotectic, which intersects the ternary eutectic, does not extend to the binary. This complicates the alkali feldspar projections in the haplogranite system, but the basic trends seen in Figure 2 also hold for this system.

In order to be able to understand feldspar evolution in natural felsic magmas, the combined effects of increasing silica as well as increasing H_2O content must be understood for compositions in which one feldspar or two feldspars crystallise. For these natural magmas, therefore, H_2O must be considered an independent compositional variable.

3. H_2O as an independent compositional variable

Experimental investigation of phase equilibria under H_2O-undersaturated conditions is not a simple matter. Using H_2O in the absence of a fluid results in numerous experimental problems such as sluggish kinetics of equilibration (Johannes 1978), and inhomogeneous distribution of H_2O in the melt. Mixed volatile experiments using a second volatile which partitions only minimally into felsic melts are now quite popular (e.g. Webster *et al.* 1987; Holtz *et al.* 1988). Although it is possible to obtain some information about the effect of H_2O on magmatic evolution from such experiments, it is difficult to obtain detailed information on the evolution of melts in which the H_2O content is allowed to increase freely. In order to elucidate the trends arising from such H_2O-unbuffered crystallisation, Nekvasil (1988b) conducted a computational investigation using the methodology of Nekvasil (1988a).

Adding H_2O as an independent compositional variable to the plagioclase–quartz system changes the plagioclase/melt compositional relations yet again. Figure 2c shows the evolution of the projected melt and feldspar compositions during equilibrium crystallisation for the same bulk composition as in Figure 2b, but with 10 mol% (0·75 wt%) bulk H_2O content, indicating the combined effect of increasing H_2O and excess silica in the melt. Because at its final state the melt is H_2O-saturated, yet the early stages of crystallisation take place under H_2O-undersaturated conditions, the crystallisation temperature interval is significantly greater for H_2O-unbuffered conditions than H_2O-buffered conditions. As shown in Figure 1, the low initial H_2O content of the melt produces a more albitic liquidus plagioclase (for the same initial melt composition) than is initially produced under H_2O-saturated conditions. This plagioclase should then evolve towards Ab with decreasing temperature in response to the changing melt composition. Competing against this, however, is the increase in H_2O content which drives the plagioclase composition towards increasing An content. Because of the two competing effects, the resulting path of evolving plagioclase compositions is now concave towards An, moving closer to the H_2O-saturated path with decreasing temperature.

Fractional crystallisation paths enhance these trends. Because the build-up of H_2O in the melt is more efficient during fractional crystallisation than during equilibrium crystallisation, the H_2O-saturated cotectic is reached at an earlier stage of crystallisation in the former case and the concavity of the path of evolving feldspars shown in Figure 2c is more pronounced. Additionally, once the H_2O-saturated cotectic is reached, the melt will follow this cotectic to its termination at the eutectic in the system Ab–Qz–H_2O.

From the analysis of systems such as those discussed above, it is readily apparent that increasing abundances of components incompatible in feldspars result in feldspar compositional paths which deviate significantly from the simple binary systems. Feldspar precipitation in natural felsic magmas, however, is not only affected by such incompatible components, but also by the precipitation of a second feldspar.

4. Simultaneous evolution of two feldspars

4.1. Controls on the compositions of coexisting feldspars

The simplest system which simulates the phase relations of two simultaneously crystallising feldspars is the system Ab–Or–An. Phase relations in this system under H_2O-saturated conditions were determined by Yoder *et al.* (1957). Within this system is a two-feldspar cotectic which separates the very small Af + L + V field from the dominant Pl + L + V field. The shape of this cotectic has not been well constrained by experiments and the compositions of the two feldspars coprecipitating are only poorly known. In order to understand the nature of this system, several workers (e.g. Stewart & Roseboom 1962; Carmichael 1963, 1965) used geometric approaches and inferences from naturally-occurring feldspars to attempt to constrain the topology of the two-feldspar + L curve at high and low temperatures and pressures. Such approaches can produce, however, only general constraints as to the nature of these systems. More recently, Nekvasil (1990), and Nekvasil and Lindsley (1990) undertook a computational investigation of this system using a modification of the method of Nekvasil (1988a), which was based on the thermodynamic mixing properties of the melt components in the hydrous granite system (Nekvasil 1986; Burnham & Nekvasil 1986), the thermodynamic properties of the crystalline feldspar components of ternary and binary plagioclase and alkali feldspar (Lindsley & Nekvasil 1989), as well as mass balance constraints. The modified method permits calculation of the detailed relations between the ternary feldspar solvus and the two-feldspar + L surface in the system Ab–Or–An–H_2O, as shown by Nekvasil (1990) and Nekvasil and Lindsley (1990).

Figure 3 shows the curve bounding the plagioclase + L(+V) and the alkali feldspar + L(+V) surfaces in this system at high and low H_2O contents. Melts in equilibrium with two feldspars will evolve along a curve such as these curves if the activity of H_2O (and the H_2O content) remains essentially constant. At low H_2O contents, such curves do not extend to the Ab–Or boundary. At some point (the "neutral point" of Abbott 1978), therefore, they begin to exhibit peritectic-like characteristics. In the peritectic region of each curve, one feldspar (most commonly plagioclase—see Nekvasil (1990) for further discussion) will react with the melt to produce a second feldspar and therefore will decrease in abundance as the melt evolves. Depending upon the bulk composition, it is possible to lose one feldspar completely before the solidius temperature is reached. Innumerable two-feldspar + L curves at constant H_2O content together describe a two-feldspar + L surface across which a melt must evolve if the H_2O content increases during crystallisation (of anhydrous phases). This surface has a cotectic-like (even) region as well as a smaller peritectic-like (odd) region as shown in Figure 3. Equilibrium paths of increasing H_2O content may include early crystallisation in the even region, then evolution into the odd region and finally re-entrance into the even region before the solidus temperature is reached.

When the three-feldspar melt components are considered, the feldspars' composition must also be considered relative to the three components instead of only the binary components. In this case, the ternary feldspar solvus becomes a very important control in the evolution of the feldspars. Figure 4 shows the ternary feldspar solvus as a series of isothermal solvus sections computed using the Lindsley and Nekvasil (1989) model for ternary feldspar solid solutions. Importantly, as for the binary systems, the solvus expands with decreasing temperature, although not symmetrically. Compositions around the nose of the solvus change more drastically with temperature when compared with those near the limbs of the solvus. The tielines in Figure 4 show the compositions of coexisting feldspars at three temperatures for a given bulk feldspar composition. With decreasing temperature, the plagioclase composition in equilibrium with alkali feldspar undergoes a significant decrease in Or content with little change in the An content. The alkali feldspar compositions undergo a major *increase* in Or content with a slight decrease in An content.

The co-evolution of magmatic feldspars is dictated by the combined effect of temperature on expanding the solvus (for any given bulk feldspar composition), as shown in Figure 4, as well as by the change in melt composition towards Ab (which will change the bulk feldspar compositions towards Ab). If the dominant effect on the feldspar compositions is the increase in Ab content of the melt and change in the bulk feldspar composition towards albite, then plagioclase will follow a path with the characteristics of the arrow **b** in Figure 4, and alkali feldspar the characteristics of path **c**. This was shown to the case for H_2O-unbuffered crystallisation of plagioclase in the plagioclase-quartz system (Fig. 2c). Because of the shallower T–X slope around the nose of the solvus, however, it may be possible that for high temperature magmas, the expansion of the solvus with decreasing temperature is dominant over the changing melt composition and paths such as **a** for plagioclase and **d** for the alkali feldspar might be followed. In this case, should a step-wise fractional crystallisation process be followed by the crystallising feldspars, the plagioclase may become normally zoned with respect to Or content instead of An content, while the alkali feldspar develops reverse zoning in Or content. Evaluation of this possibility requires an understanding of the details of crystallisation paths in the two-feldspar system.

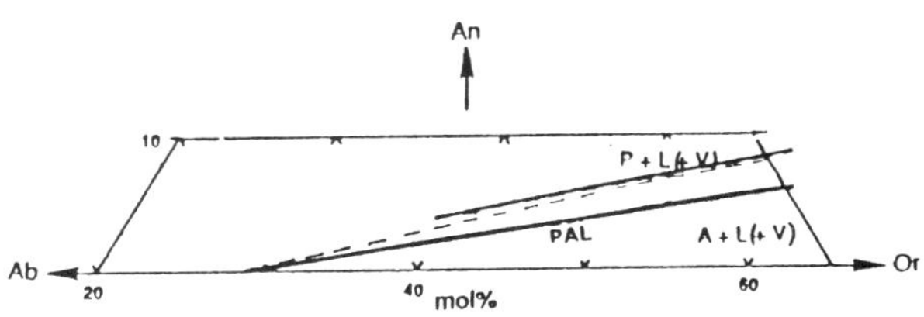

Figure 3 A portion of the calculated 5 kbar plagioclase + alkali feldspar + L (PAL) surface in the system An–Ab–Or–H_2O shown by the H_2O-saturated cotectic (lowermost solid curve) and the 0·1 H_2O isoactivity two-feldspar + L curve (uppermost solid curve); the dashed curve outlines the boundary between the peritectic-like (odd) region of this surface (above the dashed curve) and the cotectic-like (even) region (below the dashed curve); during H_2O-buffered equilibrium crystallisation, melts entering the odd region along an H_2O-isoactivity curve will never leave the odd region and therefore will undergo some reaction with plagioclase which will lead to partial or complete resorption of plagioclase; during H_2O-unbuffered equilibrium crystallisation, melts may enter the odd region, but will later leave it as the H_2O content of the melt builds up. During H_2O-buffered crystallisation, if a melt reaches an H_2O-isoactivity two-feldspar + L curve in the odd region during fractional crystallisation, it will immediately leave this surface and only a single alkali feldspar will precipitate. In contrast, intersection of the curve in the even region will result in melt evolution along the curve as two feldspars precipitate until the "neutral" point is reached. At this point the melt will leave the curve and only a single feldspar will precipitate. During H_2O-unbuffered fractional crystallisation, however, after the melt leaves the surface it will eventually reach the surface once again at a point with higher H_2O content.

4.2. Crystallisation paths in the quartz-free two-feldspar system

4.2.1. H_2O-buffered paths. Feldspar evolutionary paths in the feldspar system differ during equilibrium crystallisation depending upon the behaviour of H_2O in the system. Figure 5a shows the calculated compositions of feldspars crystallising at equilibrium at 5 kbar under conditions of constant H_2O activity (nearly constant H_2O content) equal to 0·1 for a trachyte composition from Moll (1981) projected through quartz (39·3 wt% Or, 52·0 wt% Ab, 8·67 wt% An). In a natural setting, buffering of H_2O at such low activities would probably require unrealistically large volumes of a CO_2-rich fluid to be present; however, the characteristics of H_2O-buffered paths at low H_2O activities give an important insight into the complex topology of the two-feldspar system at low H_2O contents. For the bulk composition chosen, plagioclase is the liquidus phase and its composition decreases in An content with dropping temperature as the composition of the melt decreases in An content. At the inflection point, the solvus is reached and two feldspars, of compositions shown by the

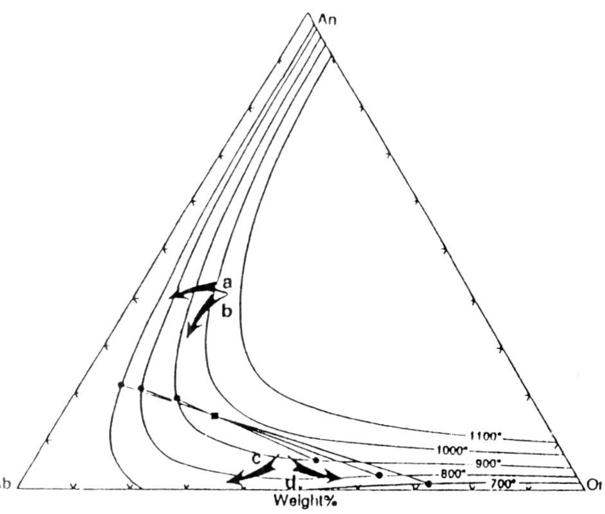

Figure 4 Ternary feldspar solvus at 5 kbar calculated using the model of Lindsley and Nekvasil (1990) as a series of isothermal solvus sections (temperature in °C): three tielines (thin lines) are shown for the same bulk feldspar composition (solid square) at 900, 800 and 700°C, indicating that the equilibrium plagioclase compositions become more Or-poor and the co-existing alkali feldspar compositions more Or-rich with lower temperatures; arrows indicate possible compositional paths of feldspars during crystallisation with decreasing temperature; if the effect of decreasing temperature is dominant over the effect of driving the bulk feldspar composition towards Ab because of the increase in Ab content of the melt during crystallisation, plagioclase compositions may evolve along a path such as **a**, while alkali feldspar evolves along a path such as **d**; if instead, the changing melt composition is dominant, path **b** may be followed by plagioclase and path **c** by alkali feldspar; from Nekvasil (in press).

highest temperature tieline, become stable. As shown in Figure 5b, the solidus temperature is reached in less than 20° from this point and therefore only a small segment of the solvus is traversed by the feldspars. Importantly, the path produced because of the small temperature interval to the solidus gives rise to feldspar compositions which follow a path very similar to an isothermal solvus section and the main effect on the composition of the feldspars is the increasing Ab content of the melt. Thus, the feldspars, although becoming more ternary in nature with dropping temperature, show the basic compositional trends of the binary system of decreasing An content of plagioclase and decreasing Or in the alkali feldspar.

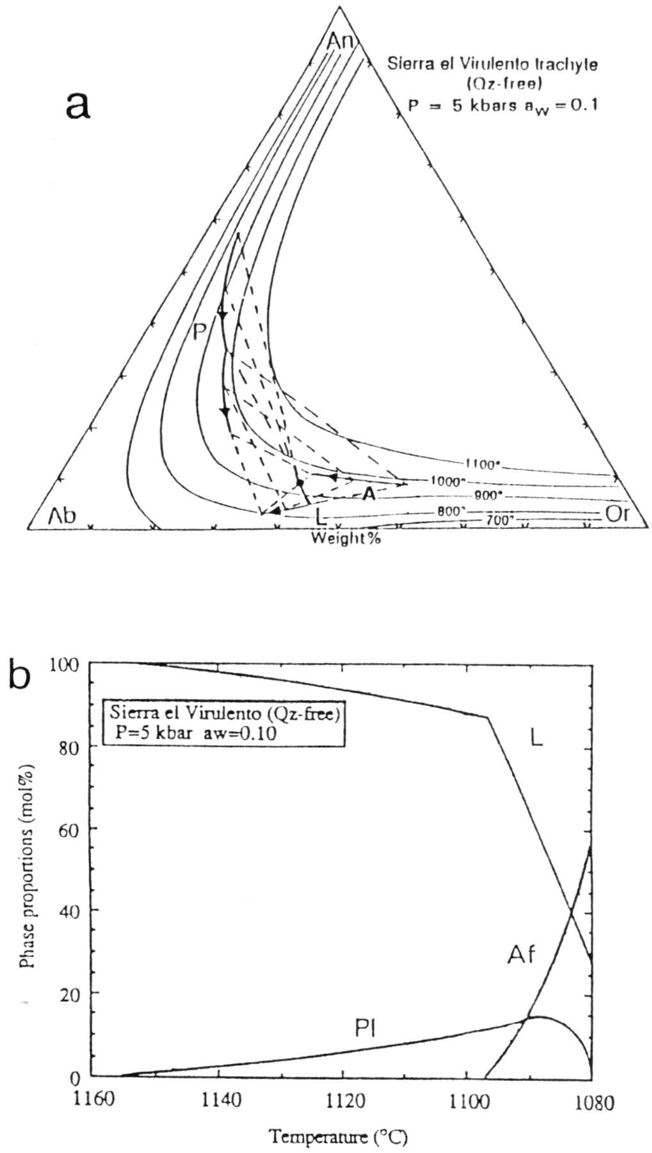

Figure 5 (a) Equilibrium crystallisation path for the Qz-free renormalised Sierra el Virulento composition (39·3 wt% Or, 52·0 wt% Ab, 8·87 wt% An) at 5 kbar and $a_w = 0{\cdot}1$ (~1·5 wt% H_2O), showing the compositional evolution of plagioclase (P), alkali feldspar (A) and melt (L); superimposed upon the path are the isotherms of the solvus from Figure 4; the solvus is reached at the inflection in the plagioclase path, at which point alkali feldspar is stabilised; tie lines and three phase triangles (dashed lines) indicate the compositions of coexiting phases; the bulk composition lies outside the polythermal solvus section for this H_2O content, hence, although plagioclase is the liquidus phase, all of the plagioclase will react with the melt to produce more alkali feldspar and hence the subsolidus assemblage will be a single alkali feldspar of the bulk composition.
(b) Variation in phase proportions (mol%) during equilibrium crystallisation of the bulk composition of (a) above; the melt intersects the two-feldspar + L curve in its odd (peritectic-like) region and, soon afterwards, plagioclase undergoes reaction with liquid to produce alkali feldspar; this reaction goes to completion before the solidus temperature is reached; after this point, further cooling yields only a single alkali feldspar; modified from Nekvasil (1990).

Figure 5b shows the variations in phase proportions along the crystallisation path in Figure 5a. Because of the odd region of the two-feldspar + L H_2O-isoactivity curve, crystallisation of alkali feldspar is accompanied by resorption of plagioclase in a reaction relation. Under these conditions of H_2O content, this reaction eventually proceeds to completion, but there is a temperature interval over which both feldspars coexist at equilibrium. The complete loss of plagioclase is not a requirement for all bulk compositions in this system at low H_2O activities. Instead, this behaviour is very dependent upon bulk composition (see Nekvasil (1990) for more details).

Fractional crystallisation paths can differ considerably from equilibrium paths such as that in Figure 5a. The most pronounced difference occurs once the two-feldspar + L H_2O-isoactivity curve is intersected by the melt. If this intersection occurs in the odd region of the curve during perfect fractional crystallisation, the melt immediately leaves the curve and moves into the alkali feldspar + L field, as is normal peritectic behaviour. If this intersection occurs in the even region, then two feldspars will coprecipitate until the neutral point on the curve is reached. The melt will then leave the curve at this point and only a single feldspar will precipitate. As for equilibrium crystallisation, once this stage is reached, any plagioclase crystals which remained entrained in the magma during fractionation would undergo resorption. The commonly described corroded plagioclase grains in sanidine of syenites and trachytes may well represent accidental mantling of such residual grains by alkali feldspar.

The compositions of feldspars forming by equilibrium crystallisation under H_2O-buffered conditions at higher H_2O activities show similar features to Figure 5a, although with less ternary solution. With increasingly higher H_2O activities (and hence H_2O contents), the reaction relation (peritectic-like behaviour), which results in complete resorption of plagioclase during equilibrium crystallisation for $a_w = 0{\cdot}1$, does not go to completion and only partial resorption of plagioclase takes place. Eventually, at even higher H_2O activities, no partial resorption may occur at any stage because of the expansion of the cotectic-like region of the two-feldspar + L curve at higher H_2O activity (as seen in Fig. 3) and, once the solvus is intersected, the feldspars will coprecipitate to the solidus temperature.

4.2.2. H_2O-unbuffered paths. If the H_2O content of the melt is free to increase during crystallisation of the feldspars, the path is unbuffered with respect to H_2O content. In this case, the crystallisation temperature interval will be greatly expanded due to the continuous depression of the crystalline phase stabilities by the increasing H_2O content. The relative strengths of the two competing factors, shown by both sets of arrows in Figure 4, will dictate the actual paths taken by the feldspars.

Figure 6a shows the evolution of the two feldspars +L for 1·2 and 2·2 wt% bulk H_2O contents (Xw (molefraction of H_2O) = 0·15 and 0·25, respectively) during equilibrium crystallisation. The continual build-up of H_2O in the melt to saturation induces the large crystallisation temperature interval shown in Figure 6b when compared to crystallisation under conditions of constant H_2O activity. The high temperatures of the dry melts result in intersection of the solvus in the region around the nose of the solvus where, as shown in Figure 4, the solvus expands to the greatest extent per interval of temperature drop. For the dry case, the increase in Ab content of the melt is a secondary effect, being overshadowed by the expansion of the solvus. The trends shown by arrows **a** and **d** in Figure 4, therefore, describe the evolving feldspar compositions. Importantly, under such "H_2O-unbuffered" conditions, the lower the bulk H_2O content (i.e the drier the magma), the more strongly plagioclase will evolve by decreasing Or content rather than decreasing An content, and the alkali feldspar will evolve by *increasing* rather than decreasing Or content with dropping temperature. Unlike the case of crystallisation under H_2O-buffered conditions (Fig. 5b), the melt

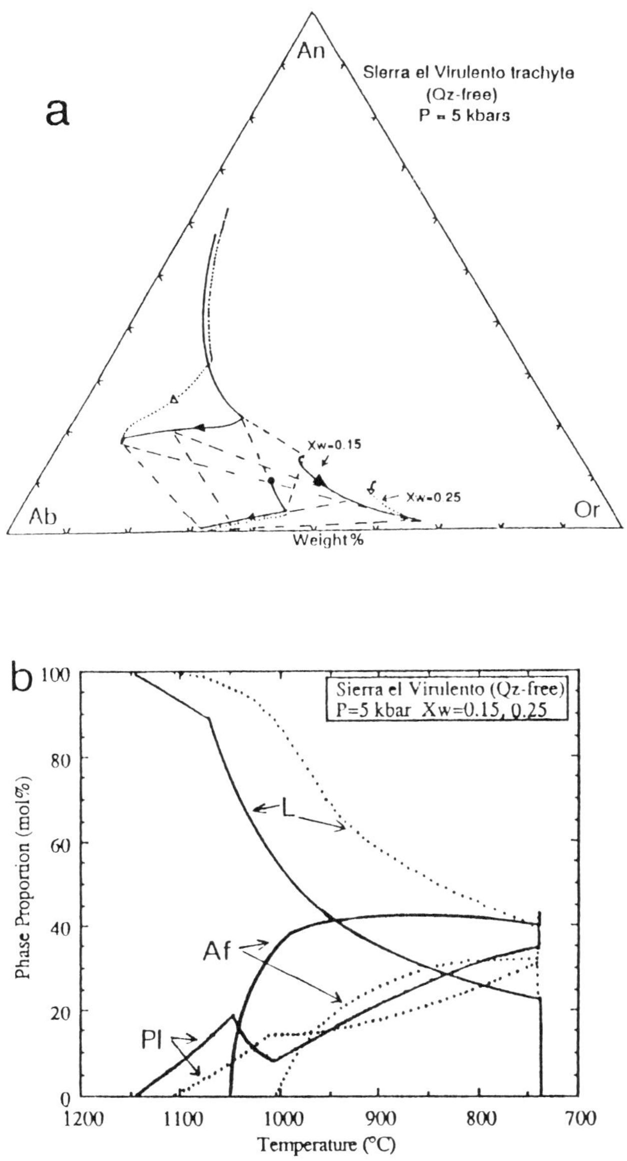

Figure 6 Calculated equilibrium crystallisation paths for the projected Qz-free trachyte composition (of Fig. 5a) at 5 kbar and for bulk H_2O contents of 1·2 (Xw = 0·15; solid curves) and 2·2 (Xw = 0·25; dotted curves) wt%. For these H_2O-unbuffered paths, the melt H_2O content is allowed to build up to saturation because of the precipitation of anhydrous phases:
(a) Compositional evolution of the feldspars and liquid; arrows indicate the direction of decreasing temperature; bulk composition is shown by the black dot.
(b) Variation in phase proportions along the paths in (a). Modified from Nekvasil (in press).

will eventually reach the even region of the two-feldspar + L surface and, instead of losing all of the plagioclase by peritectic-like reaction before the solidus temperature is reached, plagioclase will eventually begin coprecipitating with alkali feldspar once more (as shown in Fig. 6b). The amount of plagioclase which will undergo resorption during equilibrium crystallisation under H_2O-unbuffered conditions depends on the bulk H_2O content. The higher the bulk H_2O content, the smaller the amount of resorption of plagioclase which would occur before coprecipitation commenced again (shown by the more minor dip in plagioclase abundance for 2·2 wt% bulk H_2O content when compared with 1·2 wt% H_2O in Fig. 6b).

As for the H_2O-buffered paths, H_2O-unbuffered fractional crystallisation paths can differ significantly from equilibrium paths. The most significant difference is once again the movement of the liquid off the two-feldspar + L *surface* and into the alkali feldspar + L field as soon as the melt reaches the odd portion of the surface. This would eventually be followed, however, by precipitation of two feldspars once the H_2O content built up sufficiently that the even portion of the surface caught up with the changing melt composition.

4.3. Quartz-normative compositions

Although the feldspar system appears at first pass to be a good analogue for syenites and trachytes, the ubiquitous presence of quartz in granites and quartz syenites and the presence of normative Qz in rhyolites and Qz-trachytes indicates that the effect of Qz in the melt should be considered when evaluating differentiation trends of the natural compositions. Curves **a** in Figure 7a show the path of feldspar evolution for the Sierra el Virulento Qz-normative trachyte composition for 0·75 wt% bulk H_2O content undergoing equilibrium crystallisation. For this composition, not only does the H_2O content of the melt continue to build up to saturation with vapour (fluid), but also the Qz content of the melt builds up to saturation with quartz. As can be seen by these curves, the path taken is similar to the Qz-free composition (Fig. 6a); however, the temperatures attained are lower by dilution and the feldspars are not as ternary in composition. In many respects, the addition of quartz as a diluting component acts similarly to the addition of H_2O to make a higher bulk H_2O content. The curve dividing the odd and even regions of the two-feldspar + quartz + L surface lies at higher Ab contents than that dividing these regions on the two-feldspar + L surface (shown in Fig. 3). Because of this, the partial resorption of plagioclase seen for the Qz-free composition (Fig. 6b) is not manifested (Fig. 7b), although there is a considerable decrease in the amount of plagioclase precipitated per unit temperature interval (as seen just below the temperature of appearance of alkali feldspar). These characteristics are enhanced with further increases in Qz content. Curves **b** in Figure 7a show the feldspar paths for a rhyolite composition (obtained by addition of silica to the Sierra el Virulento trachyte composition until the bulk composition lay within the granite field of Tuttle & Bowen 1958) with the same bulk H_2O as for curves **a**. Importantly, as for the trachyte, for a dry granite or rhyolite, plagioclase evolves first by decreasing An content and then predominantly by decreasing Or content, while alkali feldspar evolves by increasing Or content.

Upon further increases in bulk H_2O content of the rhyolite (3·5 wt%, shown by arrows **c** in Fig. 7a), a more normal evolutionary path is obtained. Plagioclase evolves by decreasing An content and the alkali feldspar composition remains essentially constant. This type of behaviour readily explains the normal zoning found in the plagioclase of many granitic rocks as well as the common occurrence of homogeneous alkali feldspar of Or_{76-80} composition in many granites. The dotted arrow in Figure 7b shows the variations in abundances of the feldspars of paths **c** in Figure 7a. The strong temperature depression induced by the additional H_2O and the coprecipitational behaviour of the two feldspars are both readily apparent.

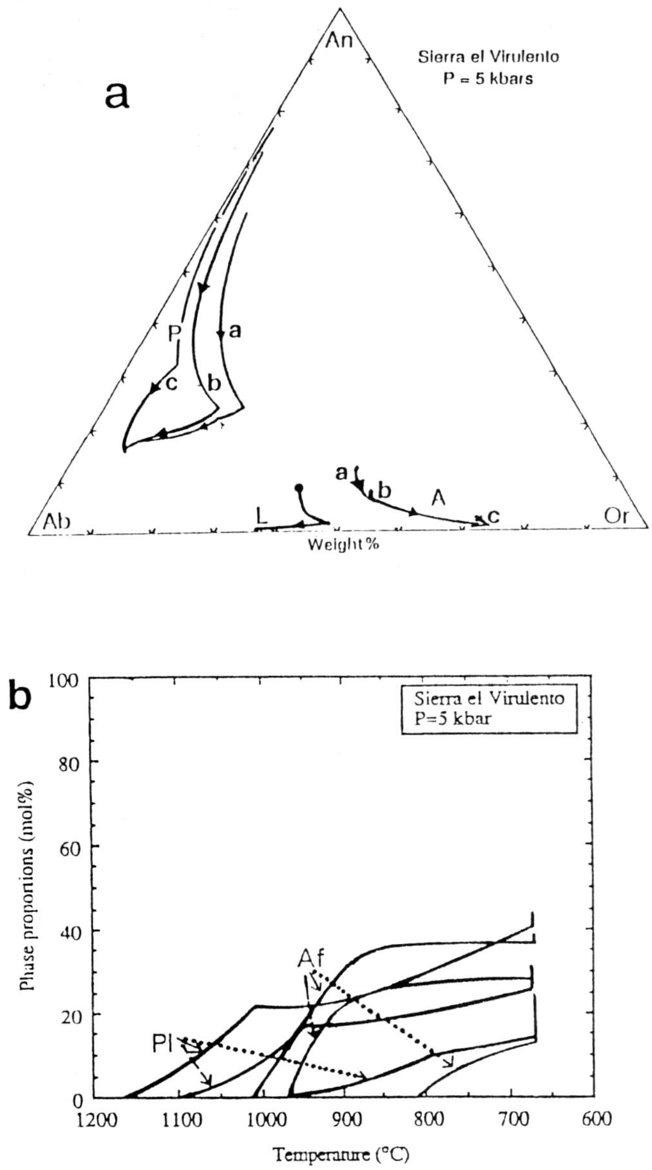

Figure 7 Equilibrium crystallisation paths for the Qz-normative Sierra el Virulento trachyte composition at 5 kbar:

(a) Plagioclase (P), alkali feldspar (A), and liquid (L) compositional evolution during crystallisation; curves **a** indicate the feldspar evolutionary path for the Qz-normative Sierra el Virulento trachyte ($An_{7\cdot1}Ab_{45\cdot2}Or_{32\cdot3}Qz_{15\cdot4}$) with 0·75 wt% bulk H_2O content, which can be compared to the paths in Figure 6a for the renormalised Qz-free composition; the inflection in the plagioclase path indicates the onset of appearance of quartz and its effect in further concentrating the feldspathic components in the melt; curves **b** show the feldspar evolutionary trends for a fictive rhyolite with 0·75 wt% bulk H_2O content obtained by increasing the Qz content of the Sierra el Virulento trachyte composition until the bulk composition ($Or_{28\cdot5}Ab_{37\cdot5}An_{5\cdot88}Qz_{30}$) lies in the field of common granites as shown by Tuttle and Bowen (1958); curves **c** show the feldspar evolutionary paths for the same rhyolite as for curves **b** but with 3·5 wt% bulk H_2O content.

(b) Variations in proportions of the two feldspars during crystallisation (the curves for quartz and liquid have been omitted for clarity); solid arrows are for curves **a** in (a), the dashed arrows are for curves **b** in (a), and the dotted arrows are for curves **c** in (a); modified from Nekvasil (in press).

5. Conclusions

The compositional evolution of feldspars in multicomponent magmas is considerably more complex than in the binary systems. The combined effects of pressure, H_2O content, the behaviour of H_2O during crystallisation, the abundance of non-feldspathic melt components, and the presence of a second feldspar all play a role in dictating the evolutionary path of a crystallising feldspar. When each of these variables is considered and their combined effects evaluated, their relative importance can be evaluated. The effect of pressure alone appears to be the least significant, in that the solidus compositions and liquidus temperatures are affected only minimally by pressure. The effect of higher H_2O content, in contrast, is very strong, resulting in major temperature depression and an increase in the An or Or component in the solid for a given melt composition. Evolution of feldspars under H_2O-undersaturated conditions, therefore, is a combined effect of the increase in An or Or content induced by increasing H_2O content, as well as the increases in Ab content induced by the evolving melt composition. The diluting effect seen by the addition of H_2O can also be induced by the addition of an additional melt component relatively incompatible in feldspars, such as Qz (its diluting effect is also enhanced by the negative deviations from ideality).

When two feldspars are stable, the competing effects of solidus temperature depression (induced by the buildup of H_2O during H_2O-unbuffered crystallisation) on widening the solvus section and the effects of the changing melt compositions on the bulk feldspar compositions, result in very different paths depending on the H_2O content (and the presence of additional components in the melt). Feldspars in "wet" felsic magmas evolve similarly to the binary systems, that is, both plagioclase and alkali feldspar evolve by increasing Ab content with dropping temperature. In "dry", high temperature magmas, in contrast, plagioclase evolves to ternary compositions by decreasing its An but increasing its Or content until alkali feldspar appears. Further evolution of plagioclase is predominantly by decreasing Or and not decreasing An content. In fact, with sufficiently low H_2O contents, it is possible for plagioclase to continue its evolution essentially along an An isopleth. Alkali feldspar also behaves quite differently in "dry" magmas, evolving by increasing Or content with dropping temperature. This type of behaviour is not restricted to trachytes and syenites in that it can occur even in a silica-rich rhyolite, provided that the diluting effect of the excess silica is compensated for by even lower H_2O contents.

If, as is reasonably concluded, natural magmas crystallise most commonly under H_2O-unbuffered conditions, then the zoning trends seen in Figure 7a (for curves **a** and **b**) should be found in phonolites, trachytes (syenites) and high temperature rhyolites (granites). However, high temperature ternary feldspars commonly exsolve during cooling (e.g. Fuhrman *et al.* 1988), thereby obliterating any existing zoning patterns. Careful reconstruction of zoning profiles as conducted by Parsons and Brown (1988) for the Klokken syenite may reveal these patterns. For this syenite, three main compositional clusters were identified: a standard plagioclase, a ternary oligoclase, and a sodic sanidine, with compositions which are very consistent with a path such as shown in Figure 7a (curves **b** and **c**). The zoning pattern in the sanidine is toward Or with dropping temperature; in the oligoclase the data are permissive, but not definitive, of zoning by decreasing Or content.

Reverse zoning of alkali feldspar in Qz-normative trachytes, pantellerites, comendites and phonolites have been reported widely (e.g Carmichael 1965; Nash *et al.* 1969; Nicholls & Carmichael 1969) and the origin of such zoning has remained a source of debate. Because many of these rock types show peralkaline characteristics, it was natural for early workers to link this zoning to peralkalinity, which resulted in the postulation of complex fractionation curves for the feldspars in such magmas (e.g. Nash *et al.* 1969). The phase relations in the two-feldspar + Qz system indicate that reversely zoned alkali feldspar may result from as simple a process as stepwise fractional crystallisation under H_2O-unbuffered conditions, where equilibrium was maintained between crystals and melt between the steps and does not require peralkalinity. Such H_2O-unbuffered behaviour of dry magmas may also explain the zoning pattern in rocks such as pantellerites, in which only a single feldspar crystallises (alkali feldspar) because the Or enrichment is linked to the expansion of the solvus and the An-free solvus also expands with decreasing temperature.

6. Acknowledgements

This work was supported in part by NSF grant EAR 8916050 which is gratefully acknowledged. The reviews of J. G. Blencoe and an anonymous reviewer greatly improved the manuscript and were appreciated.

7. References

Abbott, R. N. Jr. 1978. Peritectic relations in the system An–Ab–Or–Qz–H_2O. CAN MINERAL **16,** 245–56.

Boettcher, A. L. 1980. The systems albite–orthoclase–water and albite–orthoclase–quartz–water. J GEOPHYS RES **85**(B12), 6955–62.

Bowen, N. L. 1913. The melting phenomena of the plagioclase feldspars. AM J SCI **35,** 577–99.

Bowen, N. L. & Tuttle, O. F. 1950. The system $NaAlSi_3O_8$–$KAlSi_3O_8$–H_2O. J GEOL **58,** 489–511.

Burnham, C. W. & Nekvasil, H. 1986. Equilibrium properties of granite pegmatite magmas. AM MINERAL **71,** 239–63.

Carmichael, I. S. E. 1963. The crystallization of feldspar in volcanic acid liquids. Q J R SOC LONDON **119,** 95–131.

Carmichael, I. S. E. 1965. Trachytes and their phenocrysts. MINERAL MAG **34,** 107–25.

Erikson, R. L. 1979. *An experimental and theoretical investigation of plagioclase melting relations.* M.S. Thesis, The Pennsylvania State University.

Fuhrman, M., Frost, R. & Lindsley, D. H. 1988. Crystallization conditions of the Sybille Monzosyenite, Laramie Anorthosite Complex, Wyoming. J PETROL **29,** 699–729.

Henry, D. J., Navrotsky, A. & Zimmermann, H. D. 1982. Thermodynamics of plagioclase-melt equilibria in the system albite-anorthite-diopside. GEOCHIM COSMOCHIM ACTA **46,** 381–91.

Johannes, W. 1978. The melting of plagioclase in the system Ab–An–H_2O and Qz–Ab–An–H_2O at P_{H_2O} = 5 kbar, an equilibrium problem. CONTRIB MINERAL PETROL **66,** 295–303.

Lindsley, D. H. & Nekvasil, H. 1989. A ternary feldspar model for all reasons. EOS **70**(15), 506.

Luth, W. C., Jahns, R. & Tuttle, F. 1964. The granite system at pressures of 4 to 10 kbar. J GEOPHYS RES **69,** 759-73.

Moll, E. J. 1981. Geochemistry and petrology of mid-Tertiary ash flow tuffs from the Sierra la Virulent area, Eastern Chihuahua, Mexico. J GEOPHYS RES **86**(B11), 10321–34.

Nash, W. P., Carmicahel, I. S. E. & Johnson, R. W. 1969. The mineralogy and petrology of Mount Suswa, Kenya. J PETROL **10,** 409–39.

Nekvasil, H. 1986. *A theoretical thermodynamic investigation of the system Ab–Or–An–Qz(–H_2O) and implications for melt speciation.* Ph.D. Thesis, The Pennsylvania State University,

Nekvasil, H. 1988a. Calculation of equilibrium crystallization paths of compositionally simple hydrous felsic melts. AM MINERAL **73**(9), 956–65.

Nekvasil, H. 1988b. Calculated effect of anorthite component on the crystallization paths of H_2O-undersaturated haplogranitic melts. AM MINERAL **73**(9), 966–82.

Nekvasil, H. 1990. Reaction relations in the granite system: Implications for trachytic and syenitic magmas. AM MINERAL **75,** 560–71.

Nekvasil, H. 1991. New experimental constraints on anhydrous phase equilibria in the granite system. EOS **72,** 303.

Nekvasil, H. & Burnham, C. W. 1987. The calculated individual effects of pressure and water content on phase equilibria in the granite system. *In* Mysen, B. O. (ed.) *Magmatic processes: Physicochemical principles,* GEOCHEM SOC SPEC PUB **1**. University Park, PA: The Geochemical Society.

Nekvasil, H. & Lindsley, D. H. 1990. Termination of the 2 feldspar + liquid curve in the system Ab–Or–An–H_2O at low H_2O contents. AM MINERAL **75,** 1071–9.

Nicolls, J. & Carmichael, I. S. E. 1969. Peralkaline liquids: A petrological study. CONTRIB MINERAL PETROL **20,** 268–94.

Parsons, I. & Brown, W. L. 1988. Sidewall crystallization in the Klokken intrusion: zoned ternary feldspars and coexisting minerals. CONTRIB MINERAL PETROL **98,** 431–43.

Saxena, S. K. & Ribbe, P. H. 1972. Activity-composition relations in in feldspars. CONTRIB MINERAL PETROL **37,** 131–8.

Schairer, J. F. 1950. The alkali feldspar join in the system $NaAlSiO_4$–$KAlSiO_4$–SiO_2. J GEOL **58,** 512–7.

Seil, M. K. & Blencoe, J. G. 1979. Activity-composition relations of $NaAlSi_3O_8$–$CaAl_2Si_2O_8$ feldspars at 2 kbars, 600–800°C. GEOL SOC AM ABSTR PROG **11,** 513.

Stewart, D. B. & Roseboom, E. H., Jr. 1962. Lower temperature terminations of the three-phase region plagioclase-alkali felspar-liquid. J PETROL **3,** 280–315.

Tuttle, O. F. & Bowen, N. L. 1958. Origin of granite in light of experimental studies in the system $NaAlSi_3O_8$–$KAlSi_3O_8$–SiO_2–H_2O. GEOL SOC AM **74**.

Webster, J. D., Holloway, J. H. & Hervig, R. L. 1987. Phase equilibria of a Be-, U- and F-enriched vitrophyre from Spor Mountain, Utah. GEOCHIM COSMOCHIM ACTA **51,** 389–402.

Yoder, H. S., Stewart, D. B. & Smith, J. R. 1957. Ternary feldspars. CARNEGIE INST WASHINGTON YEARB **55,** 206–14.

HANNA NEKVASIL, Department of Earth and Space Sciences, State University of New York, Stony Brook, NY 11794–2100, U.S.A.

Transactions of the Royal Society of Edinburgh: Earth Sciences, **83**, 409–416, 1992

Peraluminous granites: the effect of alumina on melt composition and coexisting minerals

François Holtz, Wilhelm Johannes and Michel Pichavant

ABSTRACT: Liquidus phase relationships at H_2O-saturated and -undersaturated conditions and 2 kbar in the systems Qz–Or–Ab (SiO_2–$KAlSi_3O_8$–$NaAlSi_3O_8$), Qz–Or–Ab–Al_2O_3, and subsystems are compared and discussed. In the peraluminous systems (i.e. when melts are saturated with respect to mullite) the liquidus temperatures are lowered by 40–55°C for compositions in the quartz primary field and by 15–25°C for cotectic compositions. The composition of the Qz–Ab eutectic and of the minimum are slightly shifted towards more Qz-rich compositions (minimum composition at $P(H_2O) = 2$ kbar in the system Qz–Or–Ab–Al_2O_3 saturated with respect to mullite: $Qz_{40}Or_{23}Ab_{37}$). In melts saturated with sillimanite or mullite, the effect of high Al content may be lower for the Qz–Or than for the Qz–Ab eutectic.

The depression of the liquidus temperatures may be partly related to the higher H_2O solubility in melts saturated with respect to mullite. The solubility of H_2O in a melt with a composition of $Qz_{28}Or_{34}Ab_{38}$ at 2 kbar and 800°C is $5{\cdot}77 \pm 0{\cdot}15$ wt% H_2O and $6{\cdot}36 \pm 0{\cdot}30$ wt% H_2O in a melt with the same Qz/Or/Ab proportions but saturated with respect to mullite.

The effect of high Al contents on the Mg and Fe contents of Ca-free granite melts was investigated at 775°C–3 kbar (melts coexisting with phlogopite), and at 820°C–2 kbar (melts coexisting with biotite and spinel), under NNO buffer conditions. Less than 0·15 wt% MgO is incorporated in subaluminous melts coexisting with phlogopite, whereas peraluminous melts (2·9 wt% normative corundum) contain 0·6–0·7 wt% MgO. A similar behaviour of the MgO content is observed for melts coexisting with biotite. In contrast, no significant effect of high Al contents on the FeO content of melt coexisting with biotite was observed. This suggests that the Fe/Mg ratio may be significantly lower in peraluminous than in subaluminous granitic melts.

KEY WORDS: granite, H_2O, melt, peraluminous, phase relations.

Crustal granites are mostly generated by partial melting of a large variety of rocks occurring in the middle to lower continental crust. The highly variable composition of crustal granites is partly explained by large differences in source rock materials. For example, elements occurring in small amounts (e.g. F, B, Li) may have a significant influence on the composition of the generated melts and on the mineralogy of the granites crystallising from these melts. The effects of some of these minor elements were investigated in the last ten years (e.g. Chorlton & Martin 1978; Manning 1981; Pichavant 1981, 1987; Murrell 1983), but the influence of some major features such as reduced H_2O activity or high Al content, common to most of the granites generated in regions of highly evolved continental crust, was not considered in detail. The effect of reduced H_2O activities on solidus and liquidus phase relations in the subaluminous haplogranite system were investigated very recently (Keppler 1989; Ebadi & Johannes 1991; Holtz *et al.* 1992b); however, the influence of the high normative corundum content of melts was not determined systematically for granite compositions.

In this study, data obtained from new experimental results showing the effects of high Al contents on phase relations in the system Qz–Or–Ab–Al_2O_3 and its subsystems, on the Mg and Fe contents of melts, and on the H_2O solubility in granite melts are presented and discussed. The results are obtained by different experimental techniques and analytical methods but a detailed description of experimental or analytical problems is not given here.

1. Effect of alumina on phase relations in the system Qz–Or–Ab–Al_2O_3(–H_2O–CO_2)

1.1. Effect at $P(H_2O) = P(fluid) = 2$ kbar

Liquidus phase relationships were investigated at 2 kbar (1 kbar = 100 MPa) in the system Qz–Or–Ab–Al_2O_3–H_2O, using dry peraluminous glasses (5 wt% normative corundum) +10 wt% H_2O as starting materials. The Qz/Or/Ab proportions of the investigated compositions are represented by the dots in Figure 1. The results are given in Table 1 and shown on Figure 1, which is a projection of the system Qz–Or–Ab–Al_2O_3–H_2O onto the anhydrous Qz–Or–Ab plane. In all experiments, the melt was coexisting with mullite, so that the temperatures of beginning of crystallisation of quartz of alkali feldspar, illustrated by the isotherms on Figure 1, are not really liquidus temperatures. The compositions of coexisting alkali feldspar and melt (defined by microprobe analysis, see analytical method in Holtz *et al.* 1992b) or of coexisting alkali feldspar, quartz and melt are indicated by the conjugation lines or three phase triangles in Figure 1, respectively. When compared to the liquidus phase relationships defined by Tuttle and Bowen (1958) and Holtz *et al.* (1992b) at $P(H_2O) = 2$ kbar (Fig. 2a), it can be noted that the quartz-feldspar cotectic curve for peraluminous compositions is shifted towards more Qz-rich compositions. Our preliminary results, combined with those of Voigt and Burnham (1983) and Voigt and Joyce (1991), suggest that the displacement of the cotectic curve is more pronounced for Ab-rich than for

Or-rich compositions. However, more binary compositions should be investigated in order to confirm this observation, the importance of the shift of the eutectic points (Qz–Ab and Qz–Or) varying as a function of the subaluminous data used for comparison (Tuttle & Bowen 1958; Shaw 1963; Voigt & Burnham 1983). It is worth noting that Voigt and Burnham (1983) and Voigt and Joyce (1991) used crystals of sillimanite (always present in the run products) in order to obtain peraluminous synthetic melts. This should not, however, affect significantly the liquidus temperature when compared to our experimental method leading to the crystallisation of mullite as the aluminous phase (Holtz *et al.* 1992c).

Along the cotective curve, the minimum temperature is 25 ± 10°C lower than the minimum temperature in the subaluminous system (660°C in this study; 685°C in Tuttle & Bowen 1958; see also Figure 3a); the minimum melt composition is $Qz_{40}Or_{23}Ab_{37}$ (Table 2). Considering the available experimental and analytical data, there is a relatively high uncertainty on the Or/Ab ratio (±2% normative) of the minimum composition defined from Figure 1. However, the Qz content is well constrained between Qz_{39} and Qz_{41}.

The results obtained in the system $Qz–Or–Ab–Al_2O_3–H_2O$ are consistent with previous data of Voigt and Burnham (1983) and Voigt and Joyce (1991), investigating liquidus phase relationships at $P(H_2O) = 2$ kbar. In the presence of sillimanite, the eutectic temperature in the system $Qz–Or–Al_2SiO_5–H_2O$ is 30 ± 5°C to 13 ± 5°C lower (depending on the data used for comparison, see Table 2) than the eutectic temperature in the system $Qz–Or–H_2O$. The quartz content of the eutectic composition is slightly higher (+1·5 ± 1 wt% SiO_2) when compared with data of Shaw (1963) or similar when compared with the data of Tuttle and Bowen (1958) or Holtz *et al.* (1992b). Very recent results obtained in the system $Qz–Or–Ab–Al_2SiO_5–H_2O$ indicate a decrease of the minimum liquidus temperature of 15–20°C when the melts are saturated with

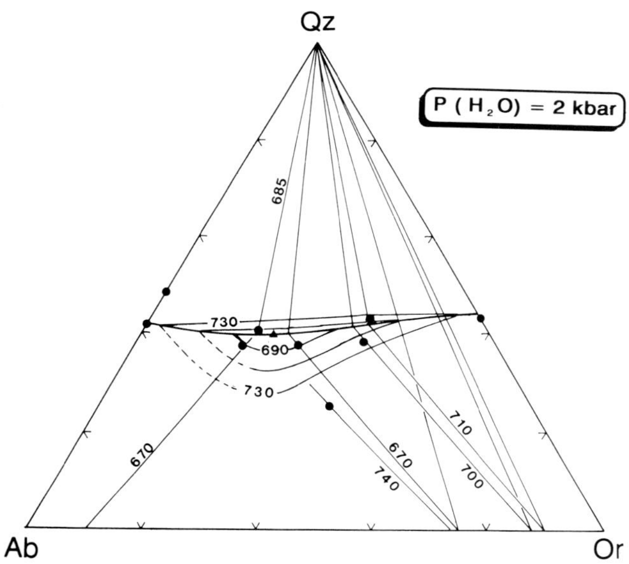

Figure 1 Liquidus phase relations in the system $Qz–Or–Ab–Al_2O_3–H_2O$ at 2 kbar, projected onto the anhydrous Qz–Or–Ab plane; circles represent the investigated starting compositions; in all run products, melts (±crystals) coexist with mullite; isotherms correspond to the temperature of beginning of crystallisation of quartz or alkali feldspar; the position and temperature of the binary eutectic points Qz–Ab and Qz–Or are from Voigt & Burnham (1983) and Voigt & Joyce (1991); in this case, melts coexist with sillimanite.

Table 1 Experimental results at $P(H_2O) = 2$ kbar. The label of the synthetic glasses (no.) indicates the expected normative Qz/Or/Ab proportions (i.e. the amounts of Si, Al, Na and K given in the glass $3_{75}42_{25}$ were calculated to correspond to normative proportions of 37·5% Qz, 40% Or and 22·5% Ab). The two temperatures applied for reversal experiments are given in parentheses. Run durations were always more than 10 days for direct experiments and more than 20 days for two steps reversal experiments. Abbreviations: L = liquid; Mul = mullite; Qtz = quartz; Sa = sanidine; phases in parentheses denote very low amounts; (*) = composition of coexisting melt and mineral phase(s) reported in Figure 1

Glass no.	T (°C)	run products	
442	725	L + Mul	
	710 (690–710)	L + Mul + (Qtz) + (Sa)	(*)
	690	L + Mul + Qtz + Sa	
424	710	L + Mul	
	710 (685–710)	L + Mul	
	700 (685–700)	L + Mul + (Qtz)	
	685	L + Mul + Qtz	(*)
$3_{75}42_{25}$	720 (700–720)	L + Mul + ((Sa))	
	710	L + Mul + (Sa)	
	710 (690–710)	L + Mul + (Sa)	
	700	L + Mul + Sa + (Qtz)	(*)
	685	L + Mul + Sa + Qtz	
$3_5 33_5$	700	L + Mul	
	700 (670–700)	L + Mul	
	690 (670–690)	L + Mul + (Sa)	
	685	L + Mul	
	680 (660–680)	L + Mul + (Sa) + ((Qtz))	
	670	L + Mul + (Sa) + ((Qtz))	(*)
$3_5 24_5$	700	L + Mul	
	700 (670–700)	L + Mul	
	690 (670–690)	L + Mul + Sa	
	670	L + Mul + Sa	(*)
$2_5 43_5$	750	L + Mul + Sa	
	740 (725–740)	L + Mul + Sa	(*)
$4_2 5_8 0$	770 (750–770)	L + Mul + (Sa)	
	760	L + Mul + Sa	
$3_7 06_3$	745	L + Mul	
	730	L + Mul + Sa	
$4_4 05_6$	730	L + Mul + Qtz	
$3_7 06_3$	735	L + Mul + Qtz	

respect to silimanite (Voigt & Joyce 1991; see also Table 2). A lower liquidus temperature for the minimum melt composition and a displacement of the cotectic curve towards the Qz apex were also observed by Dimitriadis (1978), using quartz, alkali feldspar and muscovite as solid starting materials. The lowering of the liquidus temperature of the minimum melt composition is finally in good agreement with theoretical chemographic relationships (Thompson & Algor 1977; Abbott & Clarke 1979; Abbott 1985).

1.2. Effect at $P(H_2O) < P(fluid) = 2$ kbar

Liquidus phase relationships were investigated in the systems $Qz–Or–Ab–H_2O–CO_2$ and $Qz–Or–Ab–Al_2O_3–H_2O–CO_2$ (Holtz *et al.* 1992b, c). CO_2 was added to the systems in order to reduce the H_2O activity (aH_2O). For all compositions, experiments were performed using 90 wt% anhydrous glass and 10 wt% fluid phase (composed of 70 mol% H_2O and 30 mol% CO_2) as starting material (for more details about the experimental method, see Holtz *et al.* 1992b, c). With such proportions, the H_2O content of the subaluminous melts during the experiment (H_2O content of the quenched glass) is 3·4 ± 0·2 wt% H_2O for compositions close to the minimum melt composition, corresponding to a H_2O activity of approximately 0·5 (using Burnham's model; Burnham 1979).

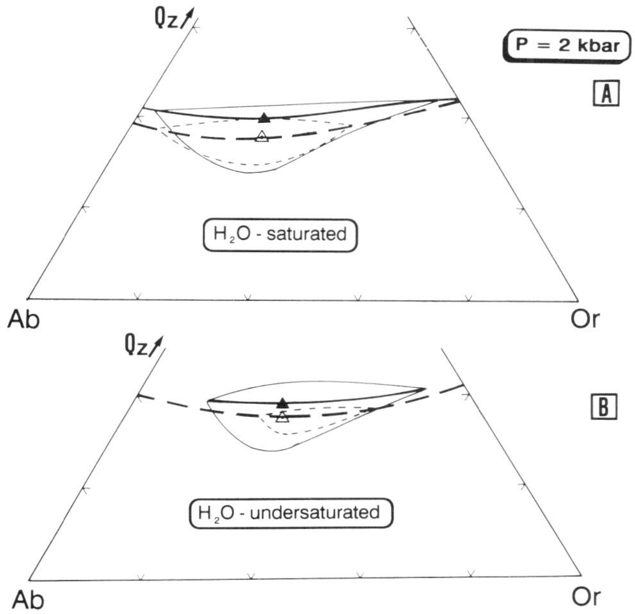

Figure 2 Projection of the minima (triangles), quartz–alkali feldspar cotectic lines (bold curves) and liquidus isotherms (light curves) of the subaluminous Qz–Or–Ab system (dashed lines and open triangle) and peraluminous Qz–Or–Ab–Al_2O_3 system (full lines and black triangle) at $P = 2$ kbar:
(a) 730°C liquidus isotherm is drawn, H_2O-saturated conditions (source of data: Tuttle & Bowen 1958; Holtz *et al.* 1992b, for the subaluminous system; Voigt & Burnham 1983; Voigt & Joyce 1991; this study, for the peraluminous system).
(b) 830°C liquidus isotherm is drawn, H_2O activity of approximately 0·5 (source of data: Holtz *et al.* 1992b for the subaluminous system; Holtz *et al.* 1992c for the peraluminous system).

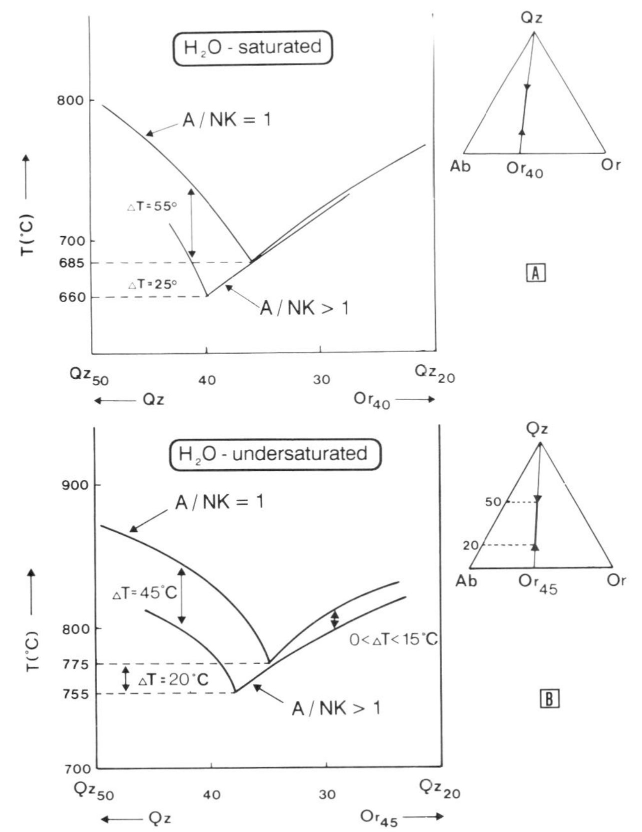

Figure 3 Section through the liquidus surfaces in the subaluminous (A/NK = 1) Qz–Or–Ab and peraluminous (A/NK > 1) Qz–Or–Ab–Al_2O_3 systems at P = 2 kbar:
(a) section along the join Qz–Or_{40}, H_2O-saturated conditions;
(b) section along the join Qz–Or_{45}, H_2O activity approximately 0·5 (from Holtz *et al.* 1992c).

Table 2 Composition of eutectic and minimum points in the subaluminous and peraluminous Qz–Ab–Or system at 2 kbar and $P(H_2O) = 2$ kbar

System	T (°C)	Qz/Or/Ab normative proportions	Source of data
Qz–Ab	750	39/0/61	1
	745	40/0/60	6
Qz–Ab–Al_2O_3	736	42/0/58	4, 7
	>730		8
Qz–Or	755	43/57/0	1
	767	42/58/0	2
	750	43·5/56·5/0	6
Qz–Or–Al_2O_3	737	43·5/56·5/0	3
Qz–Ab–Or	685	35/26/39	1
	685	36/25/39	5
Qz–Ab–Or–Al_2O_3	663		4
	660	40/23/37	8

Notes: (1) Tuttle and Bowen (1958), data given in Luth (1976)
(2) Shaw (1963)
(3) Voigt and Burnham (1983)
(4) Voigt and Joyce (1991)
(5) Holtz *et al.* (1992b).
(6) Pichavant *et al.* (1992)
(7) Joyce (pers. comm., 1991)
(8) This study

Compared to H_2O-saturated conditions, the liquidus temperatures are higher at reduced H_2O activity. The main results are reported in Figures 2b and 3b. When compared to the H_2O-undersaturated phase relations in the subaluminous system (Figs 2b and 3b), the quartz-feldspar cotectic curve of the peraluminous system is displaced towards the Qz apex and the cotectic temperatures are lowered, in the same way as at H_2O-saturated conditions. However, it can be noted that the effects of alumina on the phase relationships are slightly lower at H_2O-undersaturated conditions. At H_2O-undersaturated conditions, the decrease of the liquidus temperature of the minimum melt composition is 20°C, the minimum melt composition contains 3 wt% more normative Qz (instead of 25°C and 4 wt% normative quartz, respectively, at H_2O-saturated conditions). Although the small differences observed here lie within the uncertainty range of the determined values, other features confirm the lower effect of alumina at H_2O-undersaturated conditions. For example, the decrease of the liquidus temperatures for compositions situated in the quartz field of the peraluminous system is often more important at H_2O-saturated conditions (see Fig. 3).

1.3. Petrogenetic implications

The higher Qz content of the minimum melt composition in melts saturated with respect to sillimanite or mullite may have petrogenetic consequences and may partially explain the relatively high Qz content of peraluminous leucogranites, when compared to the Qz content of the subaluminous minimum melt composition (see for example the high Himalaya leucogranites; Le Fort 1981; Castelli & Lombardo 1988; France-Lanord & Le Fort 1988; Scaillet *et al.* 1990). However, the displacement of the minimum remains relatively low (for a given aH_2O). Other factors such as the presence of Ca, F, Li, probably produce a more significant shift of the minimum composition when projected onto the Qz–Or–Ab plane (James & Hamilton 1969; Manning 1981; Murrell 1983). More important is the effect of excess alumina on liquidus temperatures and on the evolution of magmas during their crystallisation history. Granite melts are H_2O-undersaturated at the beginning of their crystallisation history and the results show that, for H_2O-undersaturated aluminosilicate melts with composition in the quartz primary field, the temperature of beginning of crystallisation may be reduced considerably (45 ± 10°C at $aH_2O \approx 0{\cdot}5$). Residual, almost H_2O-saturated, peraluminous quartzo-feldspathic melts will reach the solidus at lower temperatures than predicted from the phase relations in the subaluminous Qz–Or–Ab system. The solidus may be lowered more significantly if the effect of alumina is combined with that of elements usually concentrated in residual melts (F, B, Li), explaining the very low solidus of some natural rocks such as Macusani glass or Harding pegmatite, containing 5% and 2·4% normative corundum, respectively (Pichavant *et al.* 1987; London *et al.* 1989; Burnham & Jahns 1962).

2. Effect of alumina on solubility of H_2O in the system Qz–Or–Ab–Al_2O_3

The shift of the cotectic line towards more Qz-rich compositions with increasing alumina content of the melt observed in the Qz–Ab–Or system has been explained by the formation of a stable complex in the melt with a structure close to dehydroxylated pyrophyllite (Burnham & Nekvasil 1986). However, the lower liquidus temperatures observed for the minimum melt composition and the Qz–Ab and Qz–Or eutectics in peraluminous systems, when compared to the subaluminous ones (Voigt & Joyce 1991; this study), may also suggest that the H_2O solubility is higher in peraluminous melts.

Experiments were conducted in order to define the H_2O solubility in peraluminous melts. Since the H_2O solubility of haplogranite melts is significantly dependent on their anhydrous composition (e.g. Tuttle & Bowen 1958; Luth 1976; Voigt *et al.* 1981; Holtz *et al.* 1992a), H_2O solubility data of subaluminous and peraluminous melts can only be compared if their normative Qz/Or/Ab proportions are similar. In addition, the starting material (gel, powdered glass, block of glass), the experimental method and the method used for the determination of the water content may have a significant influence on the H_2O solubility data (see e.g. McMillan & Holloway 1987; Day & Fenn 1982; Dingwell *et al.* 1984). Thus, comparison of data obtained by different methods may be inappropriate to define small H_2O solubility variations. In this study, the H_2O solubility in a subaluminous and a peraluminous melt with a composition $Qz_{28}Or_{34}Ab_{38}$ (i.e. approximately the minimum melt composition at 5 kbar, for a H_2O activity of 0·5) was investigated between 0·5–3 kbar and 800–900°C. The starting material was a single piece of bubble-free anhydrous glass. After the experiments, the hydrous subaluminous glasses were free of crystals, whereas the peraluminous glasses always contained mullite. The crystallisation of mullite is the result of the high Al_2O_3 content of the starting peraluminous glass (6·2% normative corundum). The determination of the H_2O contents was made by Karl Fischer titration (for more details, see Holtz *et al.* 1992a) and the results are given in Table 3.

The solubility of H_2O in the peraluminous glasses had to be calculated, since they contained significant amounts of mullite. Previous data showed that Al_2O_3 solubility in Qz–Or–Ab melts (composition close to the minimum point) coexisting with sillimanite or mullite at 2 kbar and 700–800°C corresponds approximately to 2·2% normative corundum at H_2O-saturated conditions (Voigt & Joyce 1991) and to 1·8–2·2% normative corundum at H_2O-undersaturated conditions (Holtz *et al.* 1992c). In the calculation of the H_2O solubility of the peraluminous melts, it was considered that 2·2 wt% Al_2O_3 was used to form mullite (with a composition $2SiO_2/3Al_2O_3$), implying that 5·6 wt% anhydrous mullite is present in the run products. The calculated H_2O solubility data are given in Table 3. In addition, a minimal and maximal H_2O solubility value are given. These two extreme values were calculated considering a Al_2O_3 solubility corresponding to 4·2 wt% normative corundum (i.e. 2·6 wt% mullite with a composition $SiO_2/2Al_2O_3$ is present in the run product) and 1·2 wt% normative corundum (7 wt% mullite with a composition $2SiO_2/3Al_2O_3$) in the melt, respectively. Due to the crystallisation of mullite, the normative Qz content of the

Table 3 Results of H_2O measurements (Karl Fischer titration) and H_2O solubilities in subaluminous and peraluminous glasses with similar starting Qz/Ab/Or proportions ($Qz_{28}Ab_{38}Or_{34}$). An approximative H_2O solubility value of the peraluminous glasses is calculated considering that 5·6 wt% mullite is present in the run product. Maximal and minimal H_2O solubility values are given, considering that respectively 7 and 2·6 wt% mullite occur in the run products

	P (kbar)	T (°C)	Duration (days)	measured H_2O contents (wt%)	H_2O solubility in wt% Min.		Max.
				Subaluminous composition			
*	0·5	800	22	2·68 2·70		2·69	
	1	800	12	4·07		4·07	
*	1	800	21	4·05 4·05		4·10	
				4·11 4·20			
*	2	800	14	5·80 5·75		5·77	
	3	800	11	7·01		7·01	
	3	800	15	7·28		7·28	
	3	800	15	7·35		7·35	
				Peraluminous composition			
	0·5	900	10	2·57	2·64	2·69	2·76
	0·5	800	11	2·72	2·79	2·88	2·92
	1	850	10	3·92	4·02	4·15	4·22
	1	800	10	4·04	4·15	4·27	4·34
	2	800	12	5·90	6·05	6·25	6·34
	2	800	14	6·07	6·23	6·43	6·53
	3	800	12	7·25	7·44	7·68	7·80
	3	800	12	7·20	7·39	7·62	7·74

Note: * = data from Behrens (in prep).

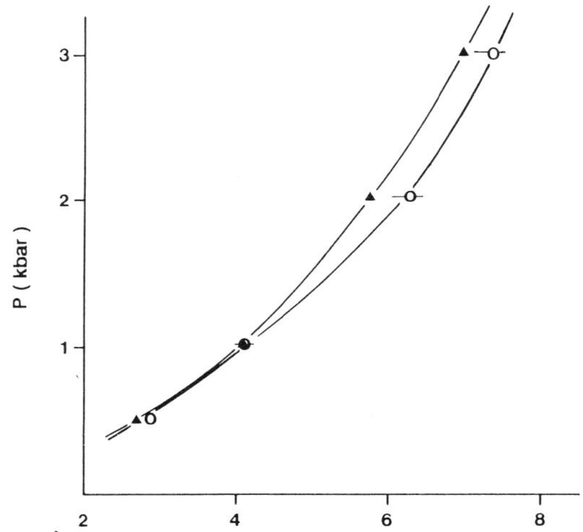

Figure 4 Evolution of H_2O solubility at 800°C as a function of pressure for a subaluminous melt (triangles; starting glass composition $Qz_{28}Or_{34}Ab_{38}$; 0% normative corundum) and a peraluminous melt saturated with respect to mullite (dots; similar starting Qz/Or/Ab proportions; 6·2% normative corundum). The H_2O solubility of the peraluminous melt is calculated considering that 5·6 wt% mullite is coexisting with the melt; horizontal lines correspond to the calculated maximum and minimum possible variation of H_2O solubility for peraluminous melts (see text and Table 3 for more details).

peraluminous melt is somewhat lowered when compared to that of the starting composition (1·2–1·6 wt% SiO_2 is used to form mullite if the melt contains 2·2% normative corundum). However, a variation of 1–2% normative Qz will not affect significantly the H_2O solubility values for haplogranite compositions close to the minimum point (Holtz *et al.* 1992a).

The variation of H_2O solubility of the subaluminous and peraluminous granite compositions as a function of P is illustrated in Figure 4. The pressure dependence of H_2O solubility is given for 800°C. At 0·5 and 1 kbar, H_2O solubility values are very slightly higher in peraluminous melts than in subaluminous melts. However, the variations are within the uncertainty range and they cannot be considered as significant. In contrast, at higher pressures (i.e. at higher H_2O contents of the melt) peraluminous melts contain more H_2O when compared to subaluminous melts (+0·3–0·5 wt% at 2 and 3 kbar).

2.1. Petrogenetic implications

The higher H_2O solubility in peraluminous melts at pressures higher than 2 kbar may explain in part the lower liquidus temperatures observed in peraluminous granite compositions (see before) and may affect chemical (diffusivity, solubility of elements) and physical properties (viscosity, density) of melts. As an example, the viscosity of residual H_2O-saturated peraluminous melts may be lowered by half an order of magnitude at 2 kbar. If the effect of alumina on phase relationships is more important for Ab-rich compositions, as suggested from preliminary experimental data, the H_2O solubility difference between Ab-rich subaluminous and peraluminous compositions could be even higher.

3. Effect of alumina on the composition of granitic melts

The chemical properties of granite melts may be dependent on their alumina content. As an example, Montel *et al.* (1988) showed that the solubility of apatite in granitic melts (with 75 wt% SiO_2) may be approximately ten times higher in peraluminous than in subaluminous melts. At 2 kbar, the P_2O_5 content of peraluminous melts coexisting with apatite was determined to be 0·41 wt% (Montel *et al.* 1988), whereas experimentally-determined values for met-aluminous melts are 0·06 wt% (Harrison & Watson 1983). The P_2O_5 values obtained by Montel *et al.* (1988) are of the same order of magnitude as (or even higher than) the P_2O_5 contents of most of the peraluminous granites (see review in Bea *et al.* 1992). This, in turn, has important petrogenetic implications and the widespread conception of the restitic nature of apatite in peraluminous granites has to be reconsidered.

In this study preliminary investigations were performed in order to define the effect of high alumina content of granitic melts on the solubility of phlogopite and biotite, and hence, on the Mg and Fe content of the melts. The investigations focused on Mg and Fe since these two elements mostly control the amount of ferromagnesian minerals in granites.

3.1. Effect of alumina on the Mg content of granite melts

The experiments were performed using a mixture of 70·4 wt% anhydrous granite glass, 17·6 wt% synthetic phlogopite and 12 wt% H_2O as starting material. The Qz/Or/Ab proportions in the starting peraluminous and subaluminous glasses were $Qz_{38}/Or_{25}/Ab_{37}$, a composition close to the 2 kbar minimum melt composition (at H_2O-saturated conditions). The experimental temperature and pressure (775°C, 3 kbar) were chosen in order to obtain melts in equilibrium with phlogopite (Bohlen *et al.* 1983; Peterson & Newton 1989) and the run durations were seven days. The experimental products were composed of glass and small crystals of phlogopite (identified by X-ray techniques). The solid starting material being composed of 20 wt% phlogopite and 80 wt% anhydrous glass, the fraction of melt is always higher than 80 wt% in the experiments, allowing a microprobe analysis of the glass with a defocused beam of more than 15 μm.

Table 4 Composition of quenched glasses from experiments conducted with glass + phlogopite + H_2O or glass + biotite + H_2O as starting materials. Microprobe analyses are recalculated on an anhydrous basis

	Starting material and experimental conditions					
crystal	Phl	Phl	Bt	Bt	Bt	Bt
norm. C in glass	0·2	6·2	0·2	1·7	3·2	6·2
P (kbar)	3	3	2	2	2	2
T (°C)	775	775	820	820	820	820
	Microprobe analysis (average of 3 to 8 analyses) of quenched glasses					
K_2O	4·13	4·01	6·28	5·75	5·51	5·21
SiO_2	79·82	77·14	75·06	74·66	73·98	74·81
Na_2O	4·16	4·12	3·95	4·25	3·72	3·39
Al_2O_3	11·76	14·06	13·28	13·84	14·38	14·81
FeO	—	—	1·28	1·25	1·25	1·17
MgO	0·13	0·67	0·16	0·18	0·32	0·59
Total	100·00	100·00	100·00	100·00	100·00	100·00
Tot. of microprobe analysis	94.39	90·42	92·99	94·19	93·43	91·57
	CIPW norms					
Qz	39·7	36·8	26·94	26·82	30·16	33·45
Or	24·4	23·7	37·12	34·03	32·86	30·81
Ab	35·2	34·9	33·23	35·94	31·70	28·63
C	0·4	2·9	—	0·61	2·30	3·58
Hy	0·3	1·7	2·43	2·40	2·77	3·32
Mt	—	—	0·14	0·20	0·20	0·19

The compositions of the glasses obtained after the experiments are given in Table 4. Since no quartz or alkali feldspar were present in the run products, the normative Qz/Or/Ab proportions of the glasses do not vary significantly when compared to the starting glasses ($Qz_{38}Or_{25}Ab_{37}$). The MgO content of the subaluminous glass is very low (less than 0·15 wt% MgO). In contrast, the MgO content of the peraluminous glass is 0·67 ± 0·05 wt% (squares in Fig. 5a). The normative corundum content of the peraluminous glass after the experiment is lower than that of the starting glass (2·9 wt%). Since only phlogopite (and glass) was determined in the run products, this mineral must correspond to an Al-rich phlogopite. In the case of peraluminous melts with a similar Qz/Or/Ab composition but without Mg, the normative corundum content at 2 kbar and 700–800°C never exceeds 2·2% (see above). Thus, high corundum content of the Mg-bearing peraluminous melt suggests that some Al forms complexes with Mg in the melt.

3.2. Effect of alumina on the Fe content of granite melts

The experimental method and the starting granite glasses were similar to those used in experiments conducted with phlogopite (see above). Two additional glass compositions were investigated; they were obtained by mixing the subaluminous and the peraluminous glasses in the following proportions: 75 wt% subaluminous glass +25 wt% peraluminous glass; 50 wt% subaluminous glass +50 wt% peraluminous glass (normative corundum of the starting glass mixture: 1·7 and 3·2, respectively, see Table 4). The composition of the starting biotite was $K_2Mg_2Fe_4(Si_6Al_2O_{20})OH_4$. The experimental temperature and pressure were 820°C and 2 kbar. Oxygen fugacity was controlled by N–NO buffer (technique described by Puziewicz & Johannes 1988). The run products are composed of biotite, magnetite (in metaluminous compositions) or Al-rich magnetite (in peraluminous compositions) and more than 80 wt% glass. Because of the small size of the minerals, microprobe analyses of biotite and magnetite are not precise (contamination by the surrounding glass) and are not discussed in detail here.

The composition of the quenched melts coexisting with biotite and magnetite is given in Table 4. The positive correlation between normative corundum and MgO content of the melt, observed in the phlogopite-bearing granite system, is also clearly observed in the biotite-bearing system (dots in Fig. 5a). In contrast, the FeO contents are not significantly different in peraluminous and subaluminous melts (dots in Fig. 5b) and correspond to 1·2–1·3 wt% FeO (recalculated for an anhydrous composition). The K_2O content of the melt increases with decreasing Al_2O_3 content (or normative corundum content), suggesting that a more important amount of biotite is present in experiments with peraluminous melts. Although the Na_2O content may vary from one sample to the other, no sodic phase was found in the run products and this variation could be due to a loss of Na during the analysis of some samples (although measurements were performed with a defocused beam and

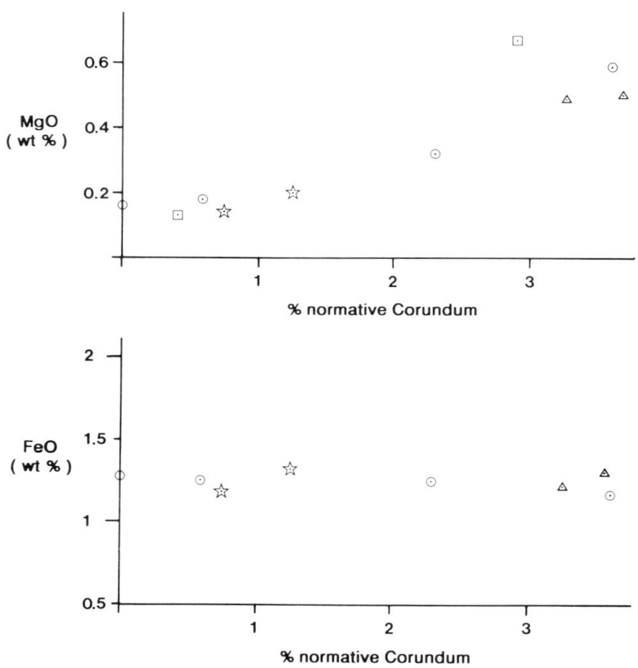

Figure 5 MgO (Fig. 5a) and FeO (Fig. 5b) contents of Ca-free granite melts coexisting with phlogopite (squares) and biotite + magnetite (circles) as a function of the normative corundum content of the melt. Additional data: stars = data of Puziewicz and Johannes (1988) at $a_{H_2O} = 1$, 820 and 850°C, 2 kbar, for melts coexisting with cordierite and magnetite; triangles = data of Puziewicz and Johannes (1990) at $a_{H_2O} < 1$, 830°C, 2 kbar, for melts coexisting with biotite, pyroxene and magnetite.

2 s counting time for Na and K, for more detail on microprobe analytical method see Puziewicz & Johannes 1988). This will, however, not affect significantly the measured Fe and Mg content of the glasses, but could artificially increase the normative corundum content of the melts. Thus, the normative corundum content values reported in Figure 5 must be considered as maximum values. Assuming 80 wt% melt in the run products and taking into account that the starting glasses contain 4·24–3·99 wt% Na_2O, a normative corundum content of 0, 0·6, 1·7 and 2·5% should be expected for the four starting compositions (instead of 0, 0·6, 2·3 and 3·6% obtained by microprobe analysis, see Table 4).

3.3. Comparison with other data and petrogenetic implications

Care must be taken when comparing our data with those of previous studies, the Fe/Mg ratio of melts being controlled by the nature and composition of the coexisting phases (e.g. Le Breton & Thompson 1988) and by the prevailing $f(O_2)$. In addition, the FeO, and especially the MgO, content of silicate melts coexisting with ferromagnesian phases increase with increasing T (e.g. Clemens & Wall 1981; Conrad *et al.* 1988), so that Fe- and Mg-bearing melts analysed in this study can only be compared to melts obtained between 800–850°C.

The data obtained by Puziewicz and Johannes (1988, 1990) for peraluminous melts and subaluminous melts show a similar behaviour for Mg (see Fig. 5a). Although the peraluminous melts coexist with cordierite and spinel (instead of biotite + spinel in our case), the MgO content is higher than 0·49 wt% for similar *P–T* conditions (2 kbar, 820 and 850°C, composition I, Puziewicz & Johannes 1988). Melts with lower normative corundum contents and coexisting with biotite, orthopyroxene and magnetite only contain 0·14–0·20 wt% MgO (at 830°C and 2 kbar, Puziewicz & Johannes 1990), a value close to those determined in the subaluminous melts in this study. Experiments performed at similar temperatures but at higher pressures (8 to 10 kbar) in synthetic or natural systems confirm the low MgO content of melts in subaluminous compositions (Naney 1983; Conrad *et al.* 1988). Although higher than those recorded at 2 kbar, the MgO contents (0·2–0·3 wt% MgO) defined by Naney (1983) and Conrad *et al.* (1988) remain lower than those observed for peraluminous compositions at 2 kbar.

The comparison of the FeO contents of our glasses with that given in other studies is difficult, since it is dependent on the oxygen fugacity which may vary with the experimental apparatus and method. We only outline here that the data obtained by Puziewicz and Johannes (1988, 1990), using the same pressure vessels, are close to those obtained in this study (Fig. 5b) and that the FeO content of experimental melts obtained at the same temperature in granite or granodiorite systems is mostly between 1 and 2 wt% FeO (e.g. Naney 1983; Miller *et al.* 1985; Clemens *et al.* 1986; Conrad *et al.* 1988; Puziewicz & Johannes 1988, 1990; except Clemens & Wall 1981, with 2·85 wt% FeO).

Petrogenetic implications of the low Fe and Mg content of granitic melts generated at temperature lower than 850°C have been discussed in detail (e.g. Clemens & Wall 1981; Clemens *et al.* 1986; Wall *et al.* 1987; Chappell *et al.* 1987). The most important implication concerns the low amount of ferromagnesian minerals which can be dissolved in granite melts at this temperature. Melts extracted from the source and almost free of residual minerals will have a leucogranite composition. More interesting is the effect of high alumina contents on the MgO content in melts. Although the MgO contents of the glasses only reflect a minimum solubility value for magnesium in Qz–Or–Ab granitic melts coexisting with phlogopite or biotite (in order to confirm this value, experiments in which phlogopite or biotite crystallises should be performed), the analytical results show that high Al contents enhance the incorporation of Mg in granitic melts. Peraluminous melts may be more Mg-rich than subaluminous ones. The Fe/Mg ratio of peraluminous melts can be significantly lower, thus favouring the crystallisation of an Al- and Mg-rich phase such as cordierite. However, although cordierite-biotite-bearing leucogranites exist, this is not the case in all peraluminous leucogranites (Himalayan leucogranites have usually high Fe/Mg; e.g. Le Fort 1981; Castelli & Lombardo 1988; Scaillet *et al.* 1990) and additional conditions must be required, such as the participation of an Mg–rich phase in the fusion in the source area.

4. Summary and conclusions

Phase relationships in the Qz–Or–Ab–Al_2O_3 system and the solubility of elements like Mg, P are dependent on the Al content of the melt. At 2 kbar, liquidus surfaces in the Qz–Or–Ab–Al_2O_3 system saturated with respect to sillimanite or mullite are shifted towards the Qz apex, thus lowering the liquidus temperatures in the quartz primary field by 40–55°C and producing a slight shift of the minimum composition when compared to the subaluminous Qz–Or–Ab system. The lowering of the liquidus temperature for all cotectic compositions may be partly explained by the slightly higher H_2O solubility observed in peraluminous melts at 2 kbar. The higher H_2O solubility in peraluminous melts may also lower the viscosity of such melts and enhance the emplacement of peraluminous granites as small bodies. The incorporation of Mg and P is enhanced by high Al contents of the melt, and this may influence the composition (especially the Fe/Mg ratio), the amount and the nature of mineral phases (biotite, cordierite, apatite) crystallising from granite melts.

This work is a very general review of the possible effects of high alumina contents on the composition of granitic melts. The effect of light elements (F, B, Li), combined with that of Al, was not studied experimentally and is a further important field to be investigated. Though more detailed investigations have to be performed, especially in Fe-Mg-bearing and Ca-bearing granite systems, the data compiled in this study show that chemical and physical properties of granite melts may change significantly when the Al concentration exceeds that required to form feldspar-like units in the melt (when the melt contains normative corundum).

5. Acknowledgements

This work was partly supported by Procope (German-French Cooperation Program N° 91162) and SFB 173 (German Science Foundation). Technical assistance was provided by O. Diedrich, J. Koepcke and F. Schulze. Thanks are due to C. W. Burnham for a critical review of the manuscript.

6. References

Abbott, R. N. Jr. 1985. Muscovite-bearing granites in the AFM liquidus projection. CAN MINERAL **23,** 553–61.

Abbott, R. N. Jr & Clarke, D. B. 1979. Hypothetical liquidus relationships in the subsystem Al_2O_3–FeO–MgO projected from quartz, alkali feldspar and plagioclase for $a(H_2O) \leq 1$. CAN MINERAL **17,** 549–60.

Bea, F., Fershtater, G. & Corretgé, L. G. 1992. The geochemistry of phosphorus in granite rocks. LITHOS (in press).

Bohlen, S. R., Boettcher, A. L., Wall, V. J. & Clemens, J. D. 1983. Stability of phlogopite-quartz and sanidine-quartz: a model for melting in the lower crust. CONTRIB MINERAL PETROL **83,** 270–7.

Burnham, C. W. 1979. The importance of volatile constituents. *In* Yoder, H. S. (ed) *The evolution of the igneous rocks: fiftieth anniversary perspectives,* 439–82. Princeton: Princeton University Press.

Burnham, C. W. & Jahns, R. H. 1962. A method for determining the solubility of water in silicate melts. AM J SCI **260,** 721–45.

Burnham, C. W. & Nekvasil, H. 1986. Equilibrium properties of granite pegmatite magmas. AM MINERAL **71,** 239–63.

Castelli, D. & Lombardo, B. 1988. The Gophu La and Western Lunana granites: Miocene muscovite leucogranites of the Bhutan Himalaya. LITHOS **21,** 211–25.

Chappell, B. W., White, A. J. R. & Wyborn, D. 1987. The importance of residual source material (restite) in granite petrogenesis. J PETROL **28,** 1,111–38.

Chorlton, L. B. & Martin, R. F. 1978. The effect of boron on the granite solidus. CAN MINERAL **16,** 239–44.

Clemens, J. D. & Wall, V. J. 1981. Origin and crystallization of some peraluminous (S-type) granitic magmas. CAN MINERAL **19,** 111–31.

Clemens, J. D., Holloway, J. R. & White, A. J. R. 1986. Origin of an A-type granite: experimental constraints. AM MINERAL **71,** 317–24.

Conrad, W. K., Nicholls, I. A. & Wall, V. J. 1988. Water-saturated and -undersaturated melting of metaluminous and peraluminous crustal compositions at 10 kb: evidence for the origin of silicic magmas in the Taupo volcanic zone, New Zealand, and other occurrences. J PETROL **29,** 765–803.

Day, H. W. & Fenn, P. M. 1982. Estimating the P–T–X(H_2O) conditions during crystallization of low calcium granites. J GEOLOGY **90,** 485–507.

Dimitriadis, S. 1978. Some liquid compositions in the peraluminous haplo-granite system. N JAHRB MINER MONATSH **1978,** 377–83.

Dingwell, D. B., Harris, D. M. & Scarfe, C. M. 1984. The solubility of H_2O in melts in the system SiO_2–Al_2O_3–Na_2O–K_2O at 1 to 2 kbars. J GEOLOGY **92,** 387–95.

Ebadi, A. & Johannes, W. 1991. Beginning of melting and composition of first melts in the system Qz–Ab–Or–H_2O–CO_2. CONTRIB MINERAL PETROL **106,** 286–95.

France-Lanord, C. & Le Fort, P. 1988. Crustal melting and granite genesis during the Himalayan collision orogenesis. TRANS R SOC EDINBURGH EARTH SCI **79,** 197–207.

Hamilton, D. L. & Oxtoby, S. 1986. Solubility of water in albite melt determined by the weight-loss method. J GEOLOGY **94,** 626–30.

Harrison, T. M. & Watson, E. B. 1983. The behavior of apatite during crustal anatexis: equilibrium and kinetic considerations. GEOCHEM COSMOCHIM ACTA **48,** 1467–77.

Holtz, F., Behrens, H., Dingwell, D. B. & Taylor, R. 1992a. Water solubility in aluminosilicate melts of haplogranitic compositions at 2 kbar. CHEM GEOL (in press).

Holtz, F., Pichavant, M., Barbey, P. & Johannes, W. 1992b. Effects of H_2O on liquidus phase relations in the haplogranite system at 2 and 5 kbar. AM MINERAL (submitted).

Holtz, F., Johannes, W. & Pichavant, M. 1992c. Effect of excess aluminium on phase relations in the system Qz–Ab–Or. Experimental investigation at 2 kbar and reduced H_2O-activity. EUR J MINERAL **4,** 137–52.

James, R. S. & Hamilton, D. L. 1969. Phase relations in the system $NaAlSi_3O_8$–$KAlSi_3O_8$–$CaAlSi_3O_8$–SiO_2 at 1 kilobar water vapour pressure. CONTRIB MINERAL PETROL **84,** 355–64.

Keppler, H. 1989. The influence of the fluid phase composition on solidus temperatures in the haplogranite system $NaAlSi_3O_8$–$KAlSi_3O_8$–SiO_2–H_2O–CO_2. CONTRIB MINERAL PETROL **102,** 321–7.

Le Breton, N. & Thompson, A. B. 1988. Fluid-absent (dehydration) melting of biotite metapelites in the early stages of crustal anatexis. CONTRIB MINERAL PETROL **99,** 226–37.

Le Fort, P. 1981. Manaslu leucogranite: a collision signature of the Himalaya, a model for its genesis and emplacement. J GEOPHYS RES **86,** 10,545–68.

London, D., Hervig, R. L. & Morgan, VI, G. B. 1989. Melt-vapor solubilities and elemental partitioning in peraluminous granite–pegmatite systems: experimental results with Macusani glass at 200 Mpa. CONTRIB MINERAL PETROL **99,** 360–73.

Luth, W. C. 1976. Granitic rocks. *In* Bailey, D. K. & MacDonalds, R. (eds) *The evolution of the crystalline rocks,* 335–417. London: Academic Press.

Manning, D. A. C. 1981. The effect of fluorine on liquidus phase relationships in the system Qz–Ab–Or with excess water at 1 kbar. CONTRIB MINERAL PETROL **76,** 206–15.

McMillan, P. F. & Holloway, J. R. 1987. Water solubility in aluminosilicate melts. CONTRIB MINERAL PETROL **97,** 320–32.

Miller, C. F., Watson, A. B. & Rapp, R. P. 1985. Experimental investigation of mafic mineral–felsic liquid equilibria: preliminary results and petrogenetic implications. EOS **66,** 1130.

Montel, J. M., Mouchel, R. & Pichavant, M. 1988. High apatite solubility in peraluminous melts. TERRA ABSTR **8,** 71.

Murrell, J. S. 1983. An experimental study of the effects of lithium on the granite system. PROC USSHER SOC **5,** 417–20.

Naney, M. T. 1983. Phase equilibria of rock-forming ferromagnesian silicates in granitic systems. AM J SCI **283,** 993–1033.

Peterson, J. W. & Newton, R. C. 1989. Reversed experiments on biotite–quartz–feldspar melting in the system KMASH: implications for crustal anatexis. J GEOLOGY **97,** 465–85.

Pichavant, M. 1981. An experimental study of the effect of boron on a water saturated haplogranite at 1 kbar vapour pressure. CONTRIB MINERAL PETROL **76,** 430–9.

Pichavant, M. 1987. Effects of B and H_2O on liquidus phase relations in the haplogranite system at 1 kbar. AM MINERAL **72,** 1056–70.

Pichavant, M., Valencia, Herrera, J., Boulmier, S., Briqueu, L., Joron, J. L., Juteau, M., Marin, L., Michard, A., Sheppard, S. M. F., Treuil, M. & Vernet, M. 1987. The Macusani glasses, SE Peru: evidence of chemical fractionation in peraluminous magmas. *In* Mysen, B. O. (ed.) *Magmatic processes: physicochemical principles,* 359–73. GEOCHEM SOC SPEC PUBL **1**.

Pichavant, M., Holtz, F. & McMillan, P. 1992. Phase relations and compositional dependence of H_2O solubility in quartz-feldspar melts. CHEM GEOL (in press).

Puziewicz, J. & Johannes, W. 1988. Phase equilibria and compositions of Fe–Mg–Al minerals and melts in water-saturated peraluminous granitic systems. CONTRIB MINERAL PETROL **100,** 156–68.

Puziewicz, J. & Johannes, W. 1990. Experimental study of a biotite-bearing system under water-saturated and water-undersaturated conditions. CONTRIB MINERAL PETROL **104,** 397–406.

Scaillet, B., France-Lanord, C. & Le Fort, P. 1990. Badrinath-Gangotri plutons. Petrological and geochemical evidence for fractionation processes in a high Himalayan leucogranite. J VOLC GEOTHERM RES **44,** 163–88.

Shaw, H. R. 1963. The four-phase curve sanidine–quartz–liquid–gas between 500 and 4000 bars. AM MINERAL **48,** 883–96.

Thompson, A. B. & Algor, J. R. 1977. Model systems for anatexis of pelitic rocks. I. Theory of melting reactions in the system $KAlO_2$–$NaAlO_2$–Al_2O_3–SiO_2–H_2O. CONTRIB MINERAL PETROL **63,** 247–69.

Tuttle, O. F. & Bowen, N. L. 1958. Origin of granite in the light of experimental studies in the system $NaAlSi_3O_8$–$KAlSi_3O_8$–SiO_2–H_2O. GEOL SOC AM MEM **74**.

Voigt, D. E., Bodnar, R. J. & Blencoe, J. G. 1981. Water solubility in melts of alkali feldspar composition at 5 kbar, 950°C. EOS **62,** 428.

Voigt, D. E. & Burnham, C. W. 1983. The solubility of Al_2SiO_5 in the system $KAlSi_3O_8$–SiO_2–H_2O at 2 kbar, and its implication for melt speciation. EOS **64,** 342.

Voigt, D. E. & Joyce, D. B. 1991. Depression of the granite minimum by the addition of sillimanite. EOS **72,** 304.

Wall, V. J., Clemens, J. D. & Clarke, D. B. 1987. Models for granitoid evolution and source compositions. J GEOL **95,** 731–49.

FRANCOIS HOLTZ, Institut für Mineralogie, Universität Hannover, Welfengarten 1, D-3000 Hannover 1, Germany; present address: CRSCM-CNRS, 1A rue de la Férollerie, 45071 Orléans Cedex 02, France.

WILHELM JOHANNES, Institut für Mineralogie, Universität Hannover, Welfengarten 1, D-3000 Hannover 1, Germany.

MICHEL PICHAVANT, CRSCM-CNRS, 1A rue de la Férollerie, 45071 Orléans Cedex 02, France.

Transactions of the Royal Society of Edinburgh: Earth Sciences, **83**, 417–422, 1992

Melting of plagioclase in granite and related systems: composition of coexisting phases and kinetic observations

Wilhelm Johannes and François Holtz

ABSTRACT: Melting experiments with plagioclases were performed in the systems Ab–An, Ab–An–H_2O, Qz–Ab–An–H_2O, and Qz–Or–Ab–An–H_2O–CO_2. The experimental products were analysed by electron microprobe, and the kinetics of the reactions were studied qualitatively.

Melting of the plagioclase in the system Ab–An ($P = 1$ atm, $T = 1420$°C) is very fast in the first minutes but becomes slower with increasing run duration and is incomplete even after 1000 hours. The Ab/An fractionation between melt and residual plagioclase is similar to that described by Bowen (1913).

Melting kinetics of plagioclase in the system Ab–An–H_2O ($P_{H_2O} = 5$ kbar, $T = 1000$°C) is controlled by the diffusion of water into the plagioclase structure. Melting is especially fast parallel to the *a*-axis. The experimental products show separation of melts and crystals.

In the tonalite system Qz–Ab–An–H_2O, equilibrium melting could be observed down to 830°C ($P_{H_2O} = 2$ kbar) but not a lower temperatures. The kinetics of the reaction is enhanced by deficiency or excess of alumina in the aluminosilicate melt surrounding the plagioclase crystals. The fractionation of Ab and An between melt and plagioclase crystals is more pronounced in the presence of quartz than in the Ab–An–H_2O system. The ratio An/An + Ab is approximately 0·35 in the melt and 0·85 in the coexisting plagioclase ($T = 880$°C).

In the haplogranodiorite system Qz–Or–Ab–An–H_2O–CO_2, melting reactions were performed at $P = 0{\cdot}5$ kbar, $T = 880$°C, and a_{H_2O} of approximately 0·5. It is assumed that near equilibrium compositions of melt and coexisting residual plagioclase could be obtained in long duration runs (run time = 60 days). The distribution of Ab and An between melt and minerals is similar to that observed in the tonalite system. The partial melt coexisting with an An-rich plagioclase and Or-rich K-feldspar is relatively poor in An.

KEY WORDS: granite, melting, plagioclase, reaction kinetics, tonalite

Feldspars are the most abundant minerals in crustal rocks; however, our knowledge of the numerous reactions in which feldspars are involved is poor. This is especially true for plagioclase-melt interactions. There is a need for additional information concerning (1) the distribution of the plagioclase components Ab ($NaAlSi_3O_8$) and An ($CaAl_2Si_2O_8$) between melts and crystals, and (2) the compositions of partial melts in the tonalite and granite systems.

Dry melting of plagioclase at $P = 1$ atm was described by Bowen (1913) and by Tsuchiyama and Takahashi (1983). The solidus and liquidus temperatures determined in 1913 and 1983 are in very good agreement. According to Tsuchiyama and Takahashi (1983), the melting of plagioclase is initiated by heterogeneous nucleation of melt and the An content of unmelted portions increases with time until the equilibrium composition is reached. The authors concluded that the melting process is controlled by chemical diffusion in the crystal.

Yoder *et al.* (1957) investigated solidus and liquidus temperatures in the system Ab–An–H_2O at 5 kbar water pressure (1 kbar = 100 MPa). The liquidus curve was reproduced by Johannes (1978); the solidus was found to be at lower temperatures (up to 60°C when compared to Yoder's data). Yoder (1968) determined liquidus temperatures and established the cotectic line in the tonalite system Qz–Ab–An–H_2O at $P_{H_2O} = 5$ kbar. The solidus surface of this system was determined by Johannes (1978) ($P_{H_2O} =$ 5 kbar). Johannes (1978) also studied the kinetics of plagioclase–quartz–melt interactions at 730°C and $P_{H_2O} =$ 5 kbar and found that, at this temperature, about 100,000 years are needed to reach equilibrium compositions. Equilibrium compositions could be obtained in the tonalite systems in experiments performed at $P_{H_2O} = 2$ kbar and high temperatures (830°C and more) by Johannes (1989). The results show a difference between solidus and liquidus temperatures of more than 100°C for intermediate Ab/An ratios of the bulk composition.

There have been several attempts to obtain equilibrium between melt, quartz, and coexisting feldspars in experiments with natural and synthetic granitic starting materials (for summary see Johannes 1980, 1985). Equilibrium compositions could not be proved in these attempts. In time-dependent studies performed in the system Qz–Or–Ab–An–H_2O at $P_{H_2O} = 5$ kbar and 660°C it could be shown that plagioclases melt almost stoichiometrically and do not reach equilibrium compositions at the applied conditions (Johannes 1980). Metastable melting was not only considered as an experimental problem but also as a rock-forming process.

The tonalite system Qz–Ab–An–H_2O was reinvestigated at $P_{H_2O} = 2$ kbar by Johannes (1989). In that study, single crystals of plagioclase surrounded by powdered quartz and water were used as the starting materials. The crystals of the experimental products contained volumes of melt and new plagioclase big enough to be analysed with an electron microprobe. Experimental temperatures were high, because equilibrium compositions could be observed at high temperatures only (≥830°C). The results obtained in the tonalite system Qz–Ab–An–H_2O encouraged us to repeat experiments in the binary and ternary systems Ab–An and Ab–An–H_2O and to try to obtain equilibrium data in the granite system Qz–Or–Ab–An (–H_2O). The first results are summarised in this paper.

1. Experimental results

1.1. Melting of plagioclase An_{66} at 1420°C

Tsuchiyama and Takahashi (1983) performed partial melting experiments on plagioclase feldspars at 1 atm to investigate textures and kinetics of the melting. The authors observed that melting was initiated by heterogeneous nucleation of melt at the surface and inside the crystal. The melt composition was homogeneous and had approximately the liquidus composition (at given T), irrespective of run time. The An content of the crystals increased with time until the equilibrium composition was reached. It was concluded that the melting process is controlled by diffusion processes in the crystal. In contrast to this conclusion, Johannes (1978) assumed that, in his experiments, the reaction rate is controlled by surface (interface) processes.

New kinetic studies were performed at 1420°C. Starting materials were single crystals of plagioclase from Lake County, Oregon. These crystals, which were also used in the more complex systems described below, have small amounts of K_2O and FeO (See Table 1).

Run times were 0·01, 0·1, 1, 10, 100 and 1000 hours. First melt could be observed after a run duration of approximately one minute, appearing as isolated or chain-forming drops. The reaction rates were high within the first hour, but decreased rapidly with increasing run duration. Unmolten volumes could be observed even after 1000 hours. K diffuses out of the unmolten plagioclase portions into the melt and Fe also leaves the crystal.

The melt areas are surrounded by new An-rich plagioclase. This new plagioclase is An-rich from the very beginning. Back-scattered electron images of experimental products show sharp boundaries between new and old plagioclase volumes (Fig. 1). Microprobe profiles obtained with help of a well-focused beam (diameter approximately 1 μm) also yielded abrupt step-like changes of element concentration and no diffusion profiles. We conclude that melting kinetics is controlled by rearrangement of atoms at the crystal-melt interface (interface kinetics) and not by chemical diffusion. There is no explanation for the existence of old unmolten plagioclase portions after experimental durations of 1000 hours.

The composition of the melt fits in the liquidus determined by Bowen (1913) but the composition of the new plagioclase indicates a solidus curve approximately 30°C below that of Bowen (Fig. 9). This difference is probably due to the small amounts of K (and Fe?) present in the starting plagioclase.

Table 1 Compositions of starting plagioclases

Label	An_{45} (1)	An_{60} (2)	An_{66} (2)
K_2O	1·19	0·30	0·12
Na_2O	5·41	4·16	3·78
Al_2O_3	26·62	29·22	30·24
FeO	0·03	0·48	0·46
CaO	9·27	12·04	13·50
SiO_2	56·67	53·65	51·22
Total	99·45	99·84	99·32
Ab	47·8	37·8	33·4
Or	6·9	1·8	0·7
An	45·3	60·4	65·9
$\Delta 2\theta$ (131–1$\bar{3}$1)	1·82	2·03	2·08

Notes: (1) from Tellnes, Norway
(2) from Lake County, Oregon, U.S.A.
Compositions given in wt% oxides; single analyses obtained by electron microprobe. Ab, Or, An proportions given in mol%. The Tellnes plagioclase is almost ordered, the Lake County plagioclases are almost disordered (see $\Delta 2\theta$ (131–1$\bar{3}$1) values above).

1.2. Melting of plagioclase at 1000°C and 5 kbar water pressure

Starting materials were plagioclase plates of 1 × 2 × 3 mm having 60 mol% An component and almost the same K and Fe contents as the material used for dry melting (see Table 1). After an experimental duration of ten hours, the plagioclase had completely reacted. After six hours the composition of the plagioclase core remained unchanged (An_{60}). Around this core there was a deep reaction zone and further outside a zone of clear glass. The reaction zone was 0·6–0·7 mm deep parallel to the a-axis and almost 0·4 mm vertical to that direction.

The interface reaction front between the unchanged plagioclase and the reaction zone consisted of up to 0·1 mm long "needles" (Fig. 2). These needles were Ab-rich H_2O-saturated glass. The reaction rate of wet plagioclase melting seems to be related to the diffusion of water in the plagioclase structure. At 900°C the diffusion coefficient for water in plagioclase is almost 10^{-9} (cm^2/s), determined from dehydration experiments (Behrens, pers. comm.). The preferred propagation of the melt parallel to the a-axis and, thus, parallel to the fracture planes (010) and (001), is obviously controlled by the structure of plagioclases (low concentration of chemical bonds across (010) and (001)).

The reaction zone (zone between reaction front and clear glass surrounding the reaction zone) is composed of new An-rich plagioclase crystals of nearly square shape forming a

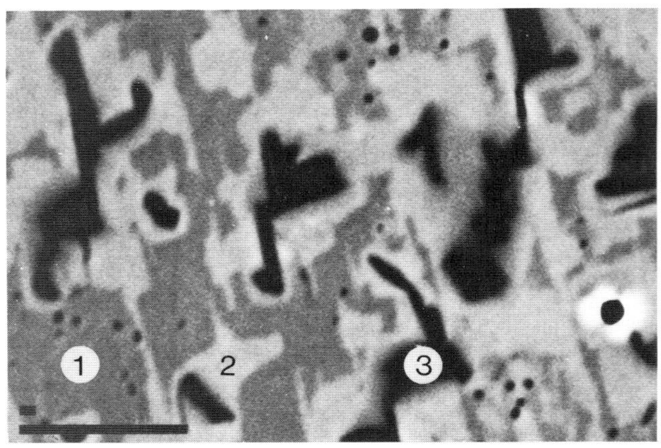

Figure 1 Back-scattered electron image of an experimental product. Starting material: plagioclase An_{60}, $T = 1420$°C, $P =$ 1 atm, run duration (d) = 100 hours. 1 (grey areas): unchanged plagioclase of composition An_{60}; 2 (light grey areas): new plagioclase (An_{80}); 3 (dark areas): Ab-rich melt (bar = 10 μm).

mosaic structure (Fig. 3). The crystals are approximately 50 μm long and have almost the same optical orientation. The new plagioclases are An-rich (almost An_{80}; detailed information about the compositions of coexisting phases will be given in Koepke *et al.* in prep.).

The space between the plagioclase squares is empty. During the experiment it was probably filled by supercritical water. This water replaced the water-saturated plagioclase melt which was formed in the reaction front adjacent to the An-rich crystals. The experimental product may be described as the result of experimental melt segregation. In the three-phase assemblage supercritical H_2O-rich gas phase $+H_2O$ saturated Ab-rich melt + An-rich plagioclase crystal, the contact between gas phase and plagioclase crystals is obviously more stable than melt-plagioclase contacts.

1.3. Melting of plagioclase in the tonalite system Qz–Ab–An at $P_{H_2O} = 2$ kbar

Results of plagioclase–quartz–melt interactions at $P_{H_2O} =$ 2 kbar and $T = 820$–880°C are described in detail by Johannes (1989). In this paper, additional results obtained at 880°C (see Fig. 9), and findings achieved in kinetic studies performed at this temperature are discussed.

In the kinetic studies, components not incorporated in quartz and plagioclase (FeO, MgO, HF, HCl) were added to the starting assemblage quartz + plagioclase An_{66}. In further experiments Al_2O_3 (as gel) or synthetic phlogopite (K Mg_3 $(OH)_2$ Si_3AlO_{10}) was part of the starting material. Run durations were 1–30 days.

There was no noticeable change of the reaction rate when additional oxides (MgO, FeO) or acids (HF, HCl) were present in the samples. But reaction rates increased when additional Al_2O_3 or phlogopite was present. The reaction was approximately ten times faster in the presence of phlogopite and five times faster in the presence of excess (10 wt%) Al_2O_3. In this paper only the results with excess Al_2O_3 will be presented.

Excess Al_2O_3 crystallises as Al_2O_3-rich mullite at the beginning of an experiment. This mullite forms long and thin needles (Figs 4, 5). Some needles are more than 50 μm long but they are never thicker than 1 μm. Only semi-quantitative compositions could be obtained using the electron microprobe. The Al_2O_3 content of the first needles and of those deep in plagioclase is approximately 95 wt%. The Al_2O_3 content of the needles in the outer portions of the former plagioclase is lower (approximately 80 wt%).

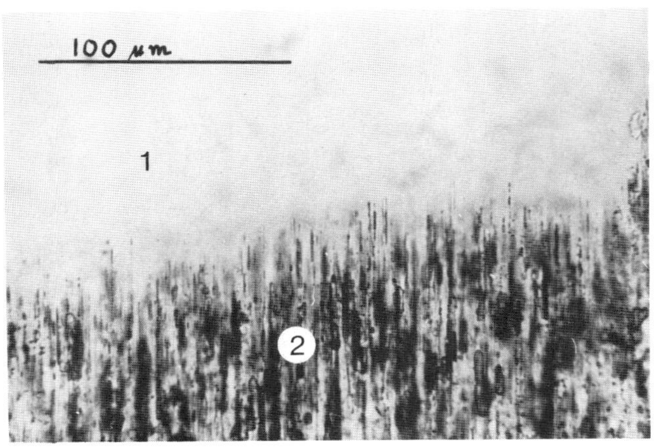

Figure 2 Thin section photograph of an experimental product obtained in the system Ab–An–H_2O at $P(H_2O) = 5$ kbar, $T =$ 1000°C, $d = 6$ hours. 1 (grey area): unchanged plagioclase of composition An_{60}; 2: "needles" consisting of melt oriented along the a-axis.

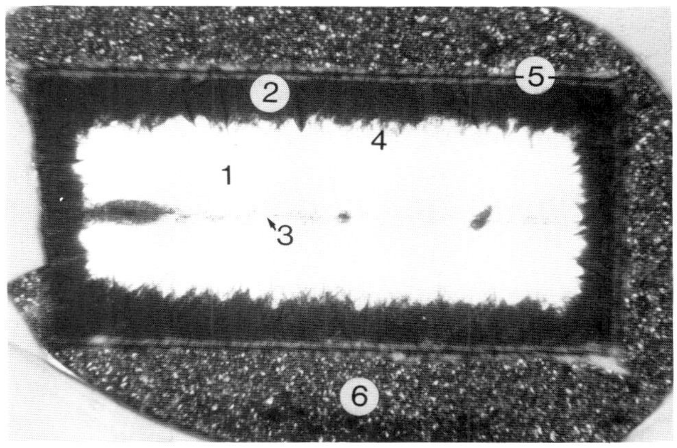

Figure 4 Thin section photograph of an experimental product obtained in the system Qz–Ab–An–Al_2O_3–H_2O at $P(H_2O) =$ 5 kbar, $T = 880$°C, $d = 6$ days, section is cut perpendicular to (010). 1: unchanged plagioclase An_{66}; 2: reaction zone; 3: fracture parallel to (010); 4: reaction front; 5: original border of plagioclase sample; 6: quartz cemented by glass.

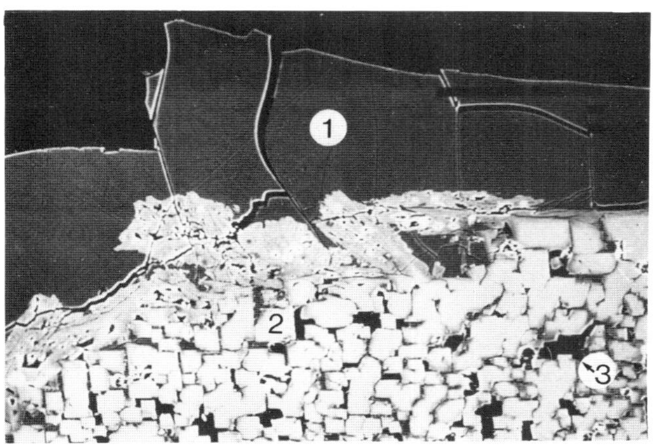

Figure 3 Back-scattered electron image of an experimental product (same sample as described in Fig. 2). 1: Ab-rich H_2O-saturated glass; 2 (white square): An-rich new plagioclase; 3 (dark space between the white squares): empty, probably filled by H_2O during the run (bar = 100 μm).

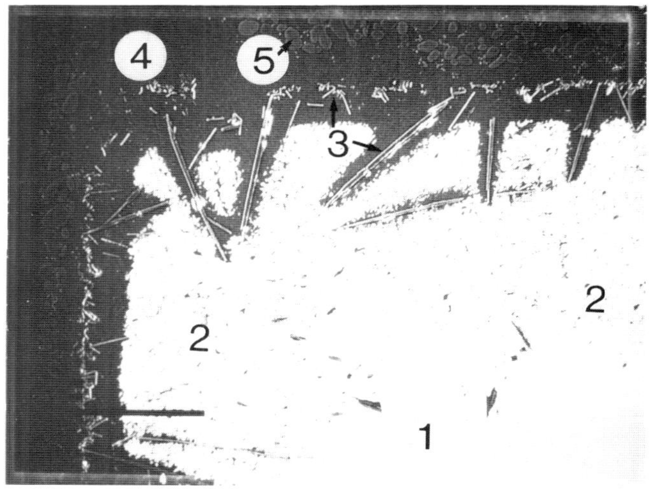

Figure 5 Back-scattered electron image of an experimental product (same sample as in Fig. 4). 1: unchanged plagioclase An_{66}; 2: reaction zone; 3: mullite; 4: Ab-rich H_2O-saturated tonalitic glass; 5 (small grey grains): unreacted quartz (bar = 100 μm).

The mullite needles seem to penetrate into the plagioclase crystal. They can always be observed at the forefront of the reaction zone. There is obviously an interaction between the end of the needle and the plagioclase (Fig. 6). The An content of the new plagioclase formed together with the quartz–feldspar melt is high. The new plagioclase seems to have a near-equilibrium composition from the very beginning. There are, not yet understood, interactions in the mullite-plagioclase interface controlling the kinetics of the quartz-plagioclase melting reaction.

When phlogopite is present in the starting material, the phlogopite is a sink for Al_2O_3. The original Al_2O_3 content of the synthetic phlogopite of 12 wt% increases during the run up to 18 wt%. The surroundings of the reacting phlogopite become undersaturated with respect to Al_2O_3. This undersaturation appears to be accelerating the quartz–plagioclase melting reaction. So far we do not have convincing explanations for the observation that both over- and undersaturation with Al_2O_3 is stimulating quartz-plagiocalse melting.

1.4. Melting in the granite system Qz–Or–Ab–An(–H_2O)

There have been numerous attempts to obtain equilibrium compositions in the cotectic granitic assemblage quartz–plagioclase–alkali feldspar–melt. They were unsuccessful because either run temperature or water activity was too low. According to solidus temperatures published by Ebadi and Johannes (1991), there is obviously only a chance at low pressure ($P \leq 1$ kbar) of getting a cotectic quartz–alkali feldspar–plagioclase–melt assemblage at relatively high water activities ($a_{H_2O} \geq 0{\cdot}5$) and high temperature. First experiments yielding probably near equilibrium compositions were performed at 880°C, $P = 0{\cdot}5$ kbar, and a_{H_2O} of approximately 0·5, with a run duration of 60 days. The solid starting material was a mixture of plagioclase crystals (An_{45}, Table 1) from Tellnes, Norway, (size of crystals between 0·2 × 0·2 × 0·2 mm and 0·2 × 0·2 × 0·5 mm) and a glass of composition 40 wt% Qz and 60 wt% Or. The gas phase with a_{H_2O} of approximately 0·5 was composed of 45 mol% H_2O and 55 mol% CO_2. The modified Redlich–Kwong equation (Kerrick & Jacobs 1981) was used to calculate the a_{H_2O}.

The experimental result was as follows: unchanged plagioclase with irregular borders is surrounded by newly crystallised alkali feldspar (normative composition: Or_{78}, Ab_{19}, An_3). New An-rich plagioclase (composition An_{65}, Ab_{30}, Or_5) was found in the form of "schlieren" in both old plagioclase and new alkali feldspar (Fig. 7). Clear glass surrounds the alkali feldspars. A few needles of wollastonite and rare tiny rods of clinopyroxene could be detected in the glass.

The melt surrounding the crystals was not homogeneous. SiO_2 decreased and Al_2O_3, Na_2O, and CaO increased in profiles from outside towards the feldspars. Quartz was not present. So the run product was not a complete cotectic quartz–plagioclase–alkali feldspar–melt assemblage. But the melts (now glasses), analysed near the alkali feldspar overgrowing the plagioclase may have near equilibrium composition and were probably only too low in SiO_2. The composition of the glass near the feldspars was (in terms of wt% normative components): 16·3% Qz, 47·0% Or, 30·2% Ab, 6·5% An. The results (normalised feldspar components) are shown and compared with the starting material in Figure 8.

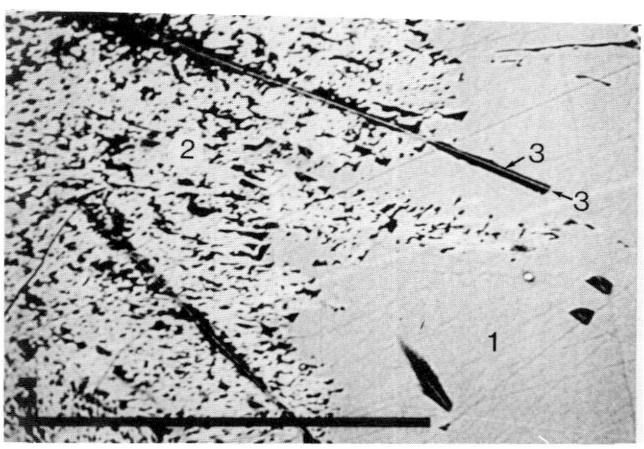

Figure 6 Back-scattered image of an experimental product (same sample as in Figs 4, 5). 1: unchanged plagioclase; 2: reaction zone; 3: mullite needle in contact with plagioclase at its very end (bar = 100 μm).

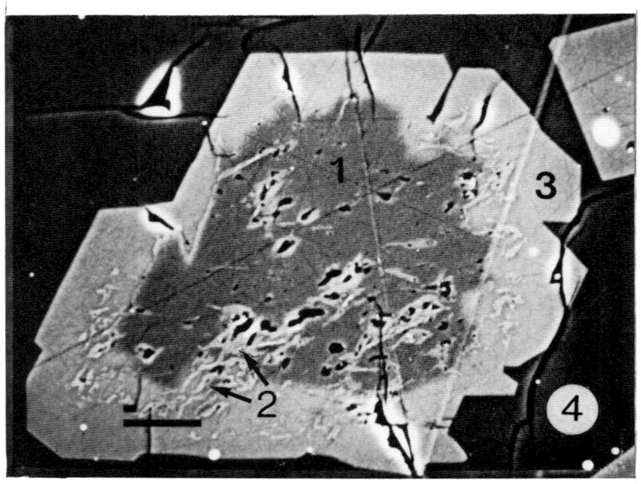

Figure 7 Back-scattered electron image of an experimental product obtained in the system Qz–Or–Ab–An–H_2O–CO_2. Starting material: plagioclase An_{45} surrounded by a glass of composition 40 wt% Qz and 60 wt% Or. $T = 880$°C; $p = 0{\cdot}5$ kbar; $X_{H_2O} = 0{\cdot}45$; $d = 60$ days. 1: unchanged plagioclase An_{45}; 2: new plagioclase (An_{65}, Ab_{30}, Or_5); 3: new alkali feldspar (Or_{78}, Ab_{19}, An_3); 4: glass ($Qz_{16{\cdot}3}$, $Or_{47{\cdot}0}$, $Ab_{30{\cdot}2}$, $An_{6{\cdot}5}$). All data in wt%. Bar = 10 μm.

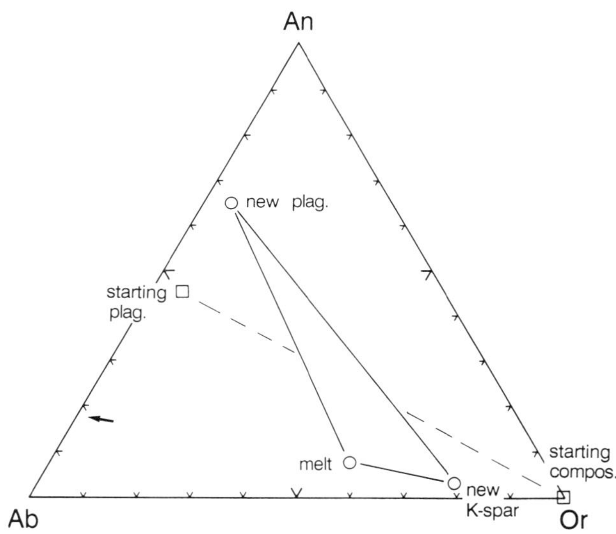

Figure 8 Composition of starting materials and experimental products projected onto the An–Ab–Or plane (system and run conditions as described in text and Fig. 7). Squares: composition of starting materials; circles: composition of run products; arrow: indication of the An/Ab ratio of the melt after the experiment.

2. Interpretation of the results

The experimental results confirm that the melting kinetics of plagioclase are strongly temperature-dependent. At high temperatures, excess or deficiency of alumina in aluminosilicate melts surrounding plagioclase crystals enhanced the reaction rate, suggesting that alumina activity is a kinetic controlling parameter. Thus, complete reaction of plagioclase crystals with a surrounding melt should be attained more rapidly in peralkaline or strongly peraluminous melts than in subaluminous ones.

The An/An + Ab ratios of melts and coexisting plagioclases observed in the systems Ab–An, Qz–Ab–An–H_2O, and Qz–Or–Ab–An–H_2O–CO_2 are compared in Figure 9. The difference between the An/An + Ab ratios of melt and coexisting plagioclase increases with the increasing number of components taking part in the melting reaction. This experimental finding means that there is always a considerable difference in the composition of partial melt and coexisting minerals in granitic systems as far as the plagioclase components are concerned.

Because of the flattening of the plagioclase loop in granitic systems (Fig. 9), very calcic plagioclase is expected as a liquidus phase. An-rich plagioclase may be formed by partial melting of more sodic bulk composition. This calcic restitic plagioclase can be preserved during crystallisation of the magma, implying a relatively fast cooling history prohibiting equilibration between this plagioclase and the crystallising melt. The restite An-rich plagioclase may become surrounded by more Ab-rich zones crystallising from the melt. Zoned plagioclases observed in granitic rocks of the Lachlan Fold Belt (Hine *et al.* 1978; Chappell *et al.* 1991) are assumed to be the result of such a process. Plagioclases in some granites of the Lachlan Fold Belt have very calcic cores (An_{80}) and sodic rims (An_{25-35}). Such a pronounced difference between cores and rims agrees with the solidus-liquidus relationships (melting loop) shown in Figure 9 for granitic compositions. The An-rich cores may be the restitic parts of the plagioclases occurring in the magma at maximum melting, thus supporting the restite model of Chappell *et al.* (1987).

An-rich plagioclases may often occur as restite minerals in partly molten rocks, e.g. in migmatities. The leucosomes of migmatites are assumed to be crystallised partial melts (Gupta & Johannes 1982; Johannes & Gupta 1982) which often contained restitic plagioclases. But An-rich cores indicating restite material are only rarely to be seen in leucosomes (Mehnert & Büsch 1982). The reasons for the absence of zoned plagioclases in migmatite leucosomes are probably low cooling rates and high water activities during the last stage of migmatite formation. Under such conditions, the former composition of protolith plagioclase could obviously be re-established. On the other hand, fast cooling and low water contents of a magma seem to enhance the formation and persistence of zoned plagioclases. High (low) water activities seem to be, besides temperature and cooling history, important factors controlling formation and persistence of homogeneous (zoned) plagioclases.

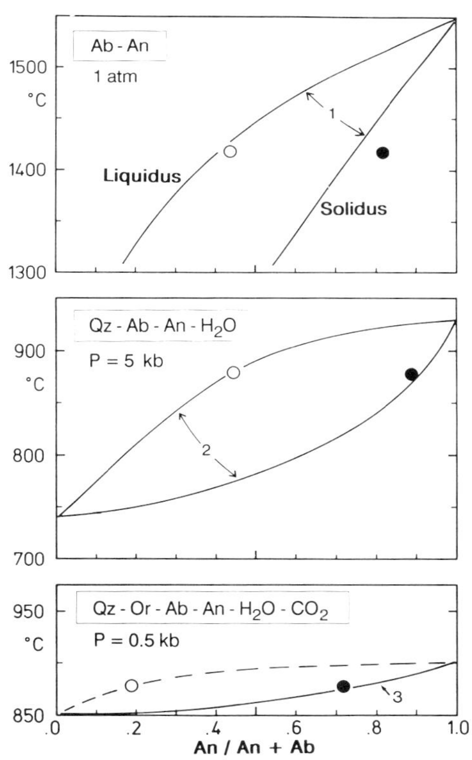

Figure 9 Comparison of An/An + Ab ratios determined in the melts (circles) and new plagioclases of experimental products (dots) in the systems Ab–An, Qz–Ab–An–H_2O, and Qz–Or–Ab–An–H_2O–CO_2. The results obtained in the granite system Qz–Or–Ab–An–H_2O–CO_2 are first preliminary (near equilibrium?) data for $a_{H_2O} \approx 0{\cdot}5$. Data in the Ab–An system are given in weight fractions. All other data are in mole fractions. 1: Bowen (1913); 2: Johannes (1989); 3: solidus extrapolated from data given by Johannes (1984) and Ebadi & Johannes (1991).

3. Acknowledgements

This work was supported by the German Science Foundation (DFG, Projekt O/26-1) and SFB 173. Comments by J. Blencoe and an unknown reviewer were very helpful and are appreciated.

4. References

Bowen, N. L. 1913. The melting phenomena of the plagioclase feldspars. AM J SCI **35,** 577–99.

Chappell, B. W., White, A. J. R. & Wyborn, D. 1987. The importance of residual source material (restite) in granite petrogenesis. J PETROL **28,** 1,111–38.

Chappell, B. W., White, A. J. R. & Williams, I. S. 1991. A transverse section through granites of the Lachlan Fold Belt. Second Hutton Symposium on Granites and Related Rocks, Excursion guide. REC BUR MINER RESOUR GEOL & GEOPHYS AUST, 1991/22.

Ebadi, A. & Johannes, W. 1991. Beginning of melting and composition of first melts in the system Qz–Ab–Or–H_2O–CO_2. CONTRIB MINERAL PETROL **106,** 286–95.

Gupta, L. N. & Johannes, W. 1982. Petrogenesis of a stromatic migmatite (Nelaug, Southern Norway). J PETROL **23,** 548–67.

Hine, R., Williams, I. R., Chappell, B. W. & White, A. J. R. 1978. Contrasts between I- and S-type granitoids of the Kosciusko Batholith. J GEOL SOC AUST **25,** 219–34.

Johannes, W. 1978. Melting of plagioclase in the system Ab–An–H_2O and Qz–Ab–An–H_2O at P_{H_2O} = 5 kbars, an equilibrium problem. CONTRIB MINERAL PETROL **66,** 295–303.

Johannes, W. 1980. Metastable melting in the granite system Qz–Or–Ab–An–H_2O. CONTRIB MINERAL PETROL **72,** 73–80.

Johannes, W. 1984. Beginning of melting in the granite system Qz–Or–Ab–An–H_2O. CONTRIB MINERAL PETROL **86,** 264–73.

Johannes, W. 1985. The significance of experimental studies in the formation of migmatites. *In* Ashworth, J. R. (ed.) *Migmatites,* 36–85. Glasgow: Blackie and Son.

Johannes, W. 1989. Melting of plagioclase-quartz assemblages at 2 kbar water pressure. CONTRIB MINERAL PETROL **103,** 270–6.

Johannes, W. & Gupta, N. L. 1982. Origin and evolution of a migmatite. CONTRIB MINERAL PETROL **79,** 14–23.

Kerrick, D. M. & Jacobs, G. K. 1981. A Modified Redlich-Kwong equation for H_2O, CO_2 and H_2O–CO_2 Mixtures at elevated pressures and temperatures. AM J SCI **281,** 735–67.

Koepke, J., Behrens, H. & Johannes, W. in prep. Melting kinetics of plagioclase dry and at 5 kbar water pressure. CONTRIB MINERAL PETROL (in prep.)

Mehnert, K. R. & Büsch, W. 1982. The initial stage of migmatite formation. N JB MINER ABH **145,** 211–38.

Tsuchiyama, A. & Takahashi, E. 1983. Melting kinetics of plagioclase feldspar. CONTRIB MINERAL PETROL **84,** 345–54.

Yoder, H. S. 1968. Albite–anorthite–quartz–water at 5 kb. CARNEGIE INST WASHINGTON YEARB **66,** 477–8.

Yoder, H. S., Stewart, D. B. & Smith, J. R. 1957. Feldspars. CARNEGIE INST WASHINGTON YEARB **56,** 206–14.

WILHELM JOHANNES, Institut für Mineralogie, Universität Hannover, Welfengarten 1, D-3000 Hannover 1, Germany

FRANCOIS HOLTZ, Institut für Mineralogie, Universität Hannover, Welfengarten 1, D-3000 Hannover 1, Germany; present address: CRSCM-CNRS, 1A rue de la Férollerie, 45071 Orléans Cedex 02, France.

Transactions of the Royal Society of Edinburgh: Earth Sciences, **83**, 423–428, 1992

A two-parameter Margules method for modelling the thermodynamic mixing properties of albite-water melts

James G. Blencoe

ABSTRACT: A two-parameter Margules method has been developed for modelling the thermodynamic mixing properties and hypersolidus phase relations of albite-water (*ab-w*) melts. The method is based largely on phase-equilibrium and calorimetric data that are either currently available or readily obtainable; *P-V-T* data are not required.

The new modelling method has been applied to calculate: (1) thermodynamic mixing properties for *ab-w* melts at 2·5 kbar, 815°C; and (2) hypersolidus phase relations for the *ab-w* system at 2·5 kbar. Results are uniformly positive—activity-composition relations are similar to those predicted by equations based on *P-V-T* data, excess enthalpies are in good agreement with data obtained from calorimetry, and calculated phase relations are fully compatible with phase-equilibrium data for *ab-w* melts acquired at 2·5 kbar.

It is concluded that the Margules method for thermodynamic modelling of *ab-w* melts is a practical alternative to the *P-V-T* modelling method.

KEY WORDS: igneous petrology, granite, magmas, silicate melts, thermodynamics, albite, water

Igneous petrologists generally accept the view that albite-water (*ab-w*) melts are physicochemical analogues for granitic magmas. (Thermochemical terms used in this paper are defined in Table 1.) This idea, borne long ago in the pioneering studies of Goranson (1938) and Tuttle and Bowen (1958), has prompted numerous investigations of the phase relations and thermodynamics of the $NaAlSi_3O_8$—H_2O system. In recent years, increased emphasis has been placed on formulating comprehensive thermodynamic models for *ab-w* melts (Burnham & Davis 1974; Stolper 1982; Silver & Stolper 1985, 1989). These efforts have been successful to a large degree, but all models developed to date are marred by discrepancies with one or more key sets of phase-equilibrium or spectroscopic data for high-temperature phases of the $NaAlSi_3O_8$—H_2O system.

It is universally acknowledged that the *P-V-T* method (Burnham & Davis 1971, 1974) is a fundamentally sound approach to thermodynamic modelling of *ab-w* melts. However, *P-V-T* experiments on hydrous silicate melts are extremely difficult to perform, and there are numerous potential sources of error in the resulting volumetric data. Thus, it is appropriate to investigate the possibility that an alternative modelling approach might yield results of comparable or superior accuracy.

It has not been recognised previously that the two-parameter Margules formulation can be applied in modelling the thermodynamic mixing properties and hypersolidus phase relations of *ab-w* melts. The method is based on the equilibria $\mu_{ab}^{L(Ab)} = \mu_{ab}^{Ab(L)}$ and $\mu_w^{L(V)} = \mu_w^{V(L)}$, which apply at all *P-T* conditions between the hydrous solidus and the anhydrous liquidus of the *ab-w* system. This paper reports the results of calculations conducted to assess the performance of the Margules method in modelling the thermodynamic mixing properties and phase relations of *ab-w* melts at 2·5 kbar.

1. Thermodynamic theory

Thermodynamic mixing properties for *ab-w* melts can be calculated from two-parameter Margules equations based on the equilibria

$$\mu_{ab}^{L(Ab)} = \mu_{ab}^{Ab(L)} \tag{1}$$

and

$$\mu_w^{L(V)} = \mu_w^{V(L)}, \tag{2}$$

which apply at all *P-T* conditions between the reactions $Ab + V \leftrightarrow L$ and $Ab \leftrightarrow L^*$. Equations (1)–(2) and the definitions

$$\mu_{ab}^{L} = \mu_{ab}^{\circ,L} + RT \ln a_{ab}^{L} \tag{3}$$

and

$$\mu_{w}^{L} = \mu_{w}^{\circ,L} + RT \ln a_{w}^{L} \tag{4}$$

yield the identities

$$a_{ab}^{L(Ab)} = \exp(-\Delta G_{fus,ab}^{\circ,Ab}/RT) \tag{5}$$

and

$$a_w^{L(V)} = a_w^{V(L)} \cong 1{\cdot}0. \tag{6}$$

Substituting the generalised two-parameter Margules equations

$$a_{ab}^{L} = X_{ab}^{L} \exp[(X_w^L)^2(W_{G,ab} + 2(W_{G,w} - W_{G,ab})X_{ab}^L)] \tag{7}$$

and

$$a_{w}^{L} = X_{w}^{L} \exp[(X_{ab}^L)^2(W_{G,w} + 2(W_{G,ab} - W_{G,w})X_{w}^L)] \tag{8}$$

into equations (5) and (6), we obtain two equations that can be solved simultaneously for the mixing parameters $W_{G,ab}$ and $W_{G,w}$. Algebraic manipulation of these equations yields

$$W_{G,ab} = \frac{[1 - 4X_w^{L(V)} + 5(X_w^{L(V)})^2 - 2(X_w^{L(V)})^3]A + 2RT[(X_w^{L(V)})^2 - (X_w^{L(V)})^3]B}{RT[(X_w^{L(V)})^2 - 2(X_w^{L(V)})^3 + (X_w^{L(V)})^4]} \tag{9}$$

and

$$W_{G,w} = \frac{2[-X_w^{L(V)} + 2(X_w^{L(V)})^2 - (X_w^{L(V)})^3]A + RT[(X_w^{L(V)})^2 - 2(X_w^{L(V)})^3]B}{RT[(X_w^{L(V)})^2 - 2(X_w^{L(V)})^3 + (X_w^{L(V)})^4]}, \tag{10}$$

Table 1 Notation

General thermochemical symbols
a = activity
G = molal Gibbs free energy
H = molal enthalpy
P = pressure (1 kbar = 100 MPa)
R = universal gas constant, 1·98726 cal/gfw-deg
T = temperature
V = volume
X = mole fraction
γ = activity coefficient
μ = chemical potential

Mixing model symbols
$W_{G,ab}$ = Margules mixing parameter for the $NaAlSi_3O_8$ component
$W_{G,w}$ = Margules mixing parameter for the H_2O component

Superscripts
Ab = analbite
ex = excess
L = liquid (silicate melt)
V = H_2O-rich vapour
° = in the Raoult's law standard state (Blencoe *et al.* 1982)
* = anhydrous

Subscripts
ab = $NaAlSi_3O_8$ component
fus = fusion
w = H_2O component

Miscellaneous
$L(Ab), L(V)$, etc. = a phase coexisting with the phase(s) in parentheses
Δ = change

where

$$A = \Delta G^{\circ,Ab}_{fus,ab} + RT \ln \left(\frac{1}{X^{L(Ab)}_w}\right)$$

and

$$B = \ln X^{L(V)}_w - \ln a^{L(V)}_w.$$

If $X^{L(Ab)}_w$, $X^{L(V)}_w$, $\Delta G^{\circ,Ab}_{fus,ab}$ and $a^{L(V)}_w$ are known quantities for a given set of P-T conditions, equations (9) and (10) yield values for $W_{G,ab}$ and $W_{G,w}$ that can be substituted into equations (7) and (8) to evaluate a^L_{ab} and a^L_w. Values for γ^L_{ab}, γ^L_w, and G^{ex} are obtained subsequently from the relations

$$\gamma^L_{ab} = a^L_{ab}/X^L_{ab}, \quad (11)$$

$$\gamma^L_w = a^L_w/X^L_w, \quad (12)$$

and

$$G^{ex} = RT(X^L_{ab} \ln \gamma^L_{ab} + X^L_w \ln \gamma^L_w). \quad (13)$$

Finally, if $X^{L(Ab)}_w$, $X^{L(V)}_w$, $\Delta G^{\circ,Ab}_{fus,ab}$ and $a^{L(V)}_w$ are known quantities for a series of temperatures at a given pressure, H^{ex} for ab-w melts can be calculated using the expression

$$H^{ex} = -RT^2\left(X^L_{ab}\frac{\partial \ln \gamma^L_{ab}}{\partial T} + X^L_w \frac{\partial \ln \gamma^L_w}{\partial T}\right) \quad (14)$$

(see Section 4).

2. Input data

It has been noted that Margules parameters for ab-w melts ($W_{G,ab}$ and $W_{G,w}$) can be evaluated if $X^{L(Ab)}_w$, $X^{L(V)}_w$, $\Delta G^{\circ,Ab}_{fus,ab}$ and $a^{L(V)}_w$ are known quantities for a fixed set of P-T conditions. In the sections below, it is shown that: (1) values for $X^{L(Ab)}_w$ and $X^{L(V)}_w$ can be obtained from empirical equations for the $L(Ab)$ and $L(V)$ boundary curves (Fig. 1), respectively; and (2) $\Delta G^{\circ,Ab}_{fus,ab}$ and $a^{L(V)}_w$ can be computed directly from explicit equations for these parameters.

2.1. $X^{L(Ab)}_w$

An empirical equation for the hydrous liquidus (the $L(Ab)$ boundary curve) yields values for $X^{L(Ab)}_w$. On an isobaric T-X_{H_2O} diagram (e.g. Fig. 1), this phase boundary extends from the hydrous solidus to the anhydrous liquidus, which are defined, respectively, by the reactions $Ab + V \leftrightarrow L$ and $Ab \leftrightarrow L^*$. Between the temperatures of these two reactions, the $L(Ab)$ curve separates the $Ab + L$ and L phase fields. It is thus evident that $Ab + L$ and L phase-equilibrium data do not define the $L(Ab)$ boundary curve, they delimit it. Consequently, an equation for the $L(Ab)$ equilibrium must be developed empirically, and the graphical expression of this function must pass *between* the phase-equilibrium data for the $Ab + L$ and L phase fields.

The hypersolidus phase-equilibrium data obtained by Fenn (1973) indicate that, for 2·5 kbar, the $L(Ab)$ equilibrium can be represented graphically by a gently-sloping, slightly convex-upward curve (Fig. 1). A curve with that shape is well represented by an equation of the form

$$X^{L(Ab)}_w = a + bT + cT^2. \quad (15)$$

The coefficients in equation (15) can be evaluated in a straightforward manner by assigning T-X_{H_2O} coordinates to three "control points" along the $L(Ab)$ equilibrium. Two of these points are the low- and high-temperature end-points of the hydrous liquidus, which have the T-X_{H_2O} coordinates $\{T^{Ab(L,V)}_{fus}, X^{L(Ab,V)}_w\}$ and $\{T^{Ab(L^*)}_{fus}, 0\}$, respectively. These points are selected to achieve complete consistency with values for $T^{Ab(L,V)}_{fus}$, $X^{L(Ab,V)}_w$ and $T^{Ab(L^*)}_{fus}$ that are derived by independent means (e.g. a value for $T^{Ab(L^*)}_{fus}$ is obtained directly from a $\Delta G^{\circ,Ab}_{fus,ab}$ equation—see discussion below). Appropriate values for the coordinates of the third point are obtained by trial-and-error estimation. This approach ultimately yields an equation that is a good "fit" to the $Ab + L$ and L phase-equilibrium data.

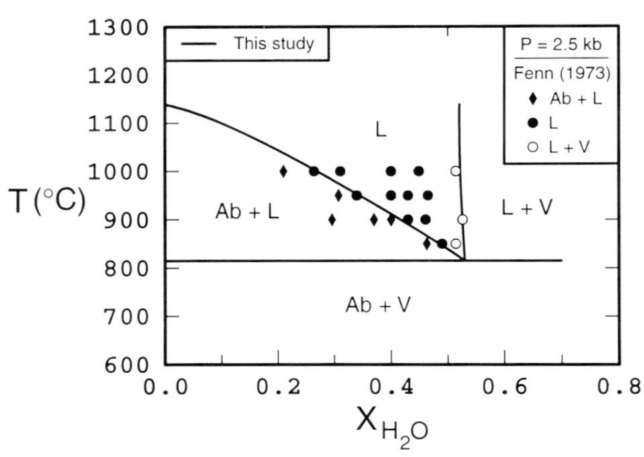

Figure 1 Hypersolidus phase-equilibrium data (Fenn 1973) and calculated phase relations (this study) for the ab-w system at 2·5 kbar; Ab = pure crystalline analbite, L = silicate melt, V = H_2O-rich vapour.

2.2. $X_w^{L(V)}$

Values for $X_w^{L(V)}$ are obtained using a procedure similar to the one described for evaluating $X_w^{L(Ab)}$. The task at hand is to devise a temperature-dependent $X_w^{L(V)}$ equation that accurately represents the $L(V)$ boundary curve at the pressure under consideration. In view of the observation that this phase boundary is essentially a straight line to pressures as high as 8 kbar (Hamilton & Oxtoby 1986), it is appropriate to obtain calculated values for $X_w^{L(V)}$ using an equation of the form

$$X_w^{L(V)} = a + bT. \tag{16}$$

The b coefficient in equation (16) will always be a small value because, in addition to being nearly straight, the $L(V)$ curve is subvertical to pressures as high as 8 kbar (Hamilton & Oxtoby 1986).

2.3. $\Delta G^{\circ,Ab}_{fus,ab}$

Values for $\Delta G^{\circ,Ab}_{fus,ab}$ can be obtained from any one of several competing published equations (Burnham 1979; Burnham & Nekvasil 1986; Kress *et al.* 1988). In the present study, $\Delta G^{\circ,Ab}_{fus,ab}$ was estimated using the relation

$$\Delta G^{\circ,Ab}_{fus,ab} = -6760{\cdot}1 + 152{\cdot}682T + 0{\cdot}003912T^2 - 21{\cdot}195T\ln T + 0{\cdot}17(P - 1{\cdot}0) \tag{17}$$

(T in Kelvins, P in bars), which is based on: (1) the 1 atm heat capacity data of Stebbins *et al.* (1983), and (2) the assumption that $\Delta V^{\circ,Ab}_{fus,ab}$ is essentially constant to pressures as high as 10 kbar.

2.4. $a_w^{L(V)}$

A value for $a_w^{L(V)}$ is obtained automatically by assuming that $V(L)$ is pure H_2O, which gives $a_w^{L(V)} = 1{\cdot}0$. However, in the present study, the small amount of silicate material dissolved in $V(L)$ was accounted for thermodynamically by stipulating that: (1) it is albitic in composition, and (2) it exhibits Henrian thermodynamic behaviour in the vapour. With these assumptions, $a_w^{L(V)}$ was evaluated from the identity

$$a_w^{L(V)} = 1{\cdot}0 - X_{ab}^{V(L)} \tag{18}$$

using the relation

$$X_{ab}^{V(L)} = 6{\cdot}8206 \times 10^{-5}P + 7{\cdot}53365 \times 10^{-5}P^2 \tag{19}$$

(P in kilobars). Equation (19) was developed by the writer using data presented by Clark (1966).

3. Calculations for $T_{fus}^{Ab(L,V)}$: a special case

A general case and a special case are encountered in evaluations of $W_{G,ab}$ and $W_{G,w}$ for a given pressure. The general case holds when $T > T_{fus}^{Ab(L,V)}$; in this circumstance, the procedures described in Sections 2.1 and 2.2 are used to evaluate $X_w^{L(Ab)}$ and $X_w^{L(V)}$, respectively, and calculations performed for temperatures at regular intervals between $T_{fus}^{Ab(L,V)}$ and $T_{fus}^{Ab(L^*)}$ yield values for $W_{G,ab}$ and $W_{G,w}$ which are used to develop isobaric, polythermal equations for these parameters. The special case holds when $T = T_{fus}^{Ab(L,V)}$; in this circumstance $X_w^{L(Ab)} = X_w^{L(V)} = X_w^{L(Ab,V)}$, a value for which can be obtained directly from experimental data on the solubility of water in *ab-w* melts. The latter observation is important, because it means that experimental data on $X_w^{L(Ab,V)}$ and calculated values for $\Delta G^{\circ,Ab}_{fus,ab}$ and $a_w^{L(V)}$ are sufficient to evaluate a-X_w^L, γ-X_w^L and G^{ex}-X_w^L relations for *ab-w* melts at P-T conditions along the hydrous solidus.

4. Calculated thermodynamic mixing properties for albite-water melts

The thermodynamic procedures described in Sections 2 and 3 have been used to quantify thermodynamic mixing properties for *ab-w* melts at 2·5 kbar. The equations required to perform these calculations were derived in four steps. First, the procedures described in Section 2 and the experimental data obtained by Fenn (1973) were used to evaluate $X_w^{L(Ab)}$, $X_w^{L(V)}$, $\Delta G^{\circ,Ab}_{fus,ab}$, $a_w^{L(V)}$, $W_{G,ab}$ and $W_{G,w}$ for temperatures at regular intervals between $T_{fus}^{Ab(L,V)}$ and $T_{fus}^{Ab(L^*)}$. Subsequently, the values for $W_{G,ab}$ and $W_{G,w}$ obtained in step 1 were used to develop least-squares-fit polythermal equations for these mixing parameters, with the following results:

$$W_{G,ab} = 36{\cdot}1744 - 5576{\cdot}07/T - 4{\cdot}07643\ln T \tag{20}$$

and

$$W_{G,w} = 411{\cdot}011 - 67835{\cdot}3/T - 50{\cdot}2746\ln T \tag{21}$$

(T in Kelvins). Next, equations (20) and (21) were substituted into equations (7) and (8) to obtain polythermal equations for a_{ab}^L and a_w^L at 2·5 kbar. Finally, equations (7) and (8) were substituted into equations (11) and (12), respectively, to obtain isobaric, polythermal equations for γ_{ab}^L and γ_w^L. The latter equations yield values for $\partial \ln \gamma_{ab}^L/\partial T$ and $\partial \ln \gamma_w^L/\partial T$, which are required to calculate H^{ex} using equation (14).

Discussion below focuses on results obtained for 2·5 kbar, 815°C—nominally a set of P-T conditions along the hydrous solidus of the *ab-w* system (cf. Goldsmith & Jenkins 1985). The estimate $X_w^{L(Ab,V)} = 0{\cdot}53$ for 2·5 kbar, 815°C, was obtained from the water solubility data presented by Burnham and Jahns (1962) and Hamilton and Oxtoby (1986). Calculated values for $\Delta G^{\circ,Ab}_{fus,ab}$ and $a_w^{L(V)}$ were obtained from equations (17) and (18), respectively. Thermodynamic mixing properties were computed from equations (7), (8), (11)–(14), (20) and (21). Results are plotted in Figures 2–4. Also shown for comparison are results obtained from the thermodynamic model developed by Burnham and Davis (1974).

Examining Figures 2a and 2b, it is evident that calculated a-X_w^L relations obtained from the Margules method are in close agreement with corresponding a-X_w^L relations yielded by the Burnham and Davis model. The small discrepancies between the results are largely attributable to the different mathematical forms of the Margules equations for a_{ab}^L and a_w^L and corresponding equations in the Burnham and Davis model. Fundamentally, the Margules equations are third-order functions (cubic in X_w^L), whereas the Burnham and Davis expressions are second-order equations (quadratic in X_w^L). Thus, it is not surprising to observe in Figures 2a and 2b that the curves derived from the Margules method have sigmoidal forms, whereas corresponding results obtained from the Burnham and Davis model (i.e. two sets of results, one for $X_w^L \leq 0{\cdot}5$ and one for $X_w^L > 0{\cdot}5$) are represented by monotonic curves.

Figures 3a, b and 4a illustrate calculated values for γ_{ab}^L, γ_w^L and G^{ex}, respectively, obtained from the Margules modelling performed in the present study and from the Burnham and Davis model. As in the case of calculated a-X_w^L relations, there is good agreement between the results obtained from the two models.

Calculated and calorimetrically measured H^{ex}-X_w^L relations for *ab-w* melts are illustrated in Figure 4b, whereupon it is evident that there is substantial agreement between the H^{ex}-X_w^L relations derived from the Margules method (this study) and excess enthalpies reported by

Clemens and Navrotsky (1987). Both sets of results indicate a negative (exothermic) H^{ex} for *ab-w* melts. Each of the three curves representing Margules H^{ex}-X^L_w relations are convex downward, with minima at $X^L_w = 0{\cdot}35$; the calorimetry-based curve for metastable *ab-w* melts at 1 bar, 25°C, is also convex downward, with a minimum at $X^L_w = 0{\cdot}27$. Finally, both the Margules H^{ex}-X^L_w relations and the results obtained from high-*P* drop calorimetry indicate that H^{ex} becomes less exothermic with increasing temperature.

In contradistinction to the results obtained from the Margules method and from calorimetric measurements, values for H^{ex} derived from the Burnham and Davis model are only slightly exothermic, and there is no indication of a minimum in H^{ex} vs. X^L_w. Furthermore, while H^{ex}-X^L_w relations derived from the Burnham and Davis model become less exothermic with increasing temperature, changes from 815°C to 1033°C are negligibly small. These observations suggest that the Burnham and Davis model does not yield accurate calculated H^{ex}-X^L_w relations for *ab-w* melts at 2·5 kbar.

5. Calculated phase relations for the albite-water system at 2·5 kbar

5.1. Procedures for calculating hypersolidus phase relations

Isobaric, hypersolidus T-X_{H_2O} phase relations for the *ab-w* system are readily calculated in stepwise fashion as follows: first, quantify $T^{Ab(L^*)}_{fus}$; next, evaluate $T^{Ab(L,V)}_{fus}$; finally, calculate values for $X^{L(Ab)}_w$ and $X^{L(V)}_w$ in the range $T^{Ab(L,V)}_{fus} < T < T^{Ab(L^*)}_{fus}$. The writer uses simple iterative numerical algorithms to complete these calculations. $T^{Ab(L^*)}_{fus}$ is quantified from equation (17) by iterative calculation of the temperature at which $\Delta G^{\circ,Ab}_{fus,ab} = 0$ for the pressure under consideration. $T^{Ab(L,V)}_{fus}$ is evaluated by selecting trial values for this temperature until the condition $X^{L(Ab)}_w = X^{L(V)}_w$ is satisfied. At a given temperature, values for $X^{L(Ab)}_w$ and $X^{L(V)}_w$ are obtained by solving the identities

$$X^{L(Ab)}_w = 1{\cdot}0 - \exp\left(-\Delta G^{\circ,Ab}_{fus,ab}/RT\right)/\gamma^{L(Ab)}_{ab} \quad (22)$$

and

$$X^{L(V)}_w = (1{\cdot}0 - X^{V(L)}_{ab})/\gamma^{L(V)}_w, \quad (23)$$

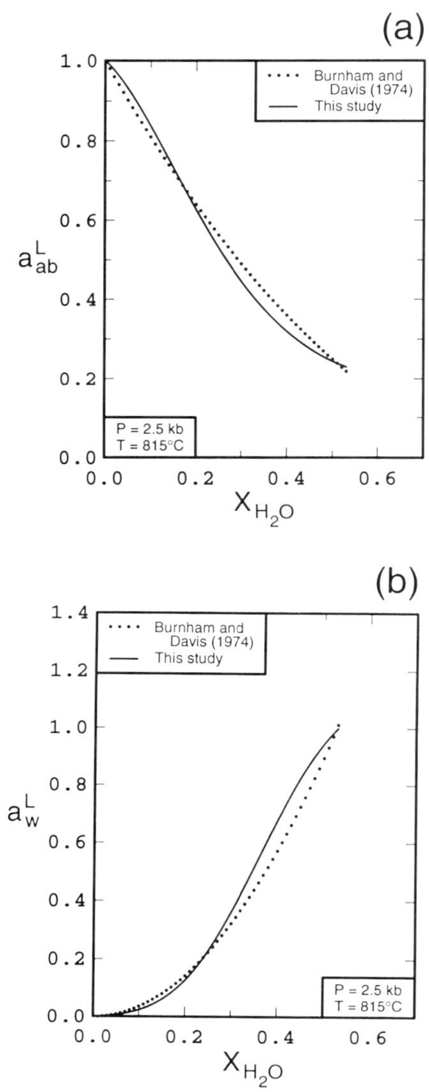

Figure 2 Calculated a^L_{ab}-X^L_w and a^L_w-X^L_w relations for *ab-w* melts at 2·5 kbar, 815°C.

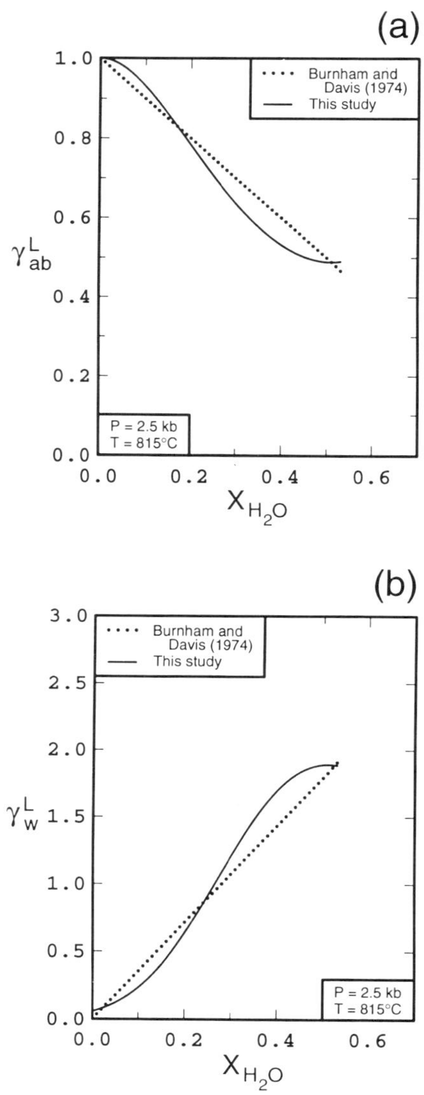

Figure 3 Calculated γ^L_{ab}-X^L_w and γ^L_w-X^L_w relations for *ab-w* melts at 2·5 kbar, 815°C.

the values for $\gamma_{ab}^{L(Ab)}$ and $\gamma_w^{L(V)}$ being supplied by equations (7), (8), (11), (12), (20) and (21). In the process of calculating $T_{fus}^{Ab(L,V)}$, the interim result $X_w^{L(Ab)} > X_w^{L(V)}$ indicates that the trial value of $T_{fus}^{Ab(L,V)}$ is too low; the opposite circumstance—$X_w^{L(Ab)} < X_w^{L(V)}$—indicates that the trial value of $T_{fus}^{Ab(L,V)}$ is too high. Finally, it is evident that equations (22) and (23) also supply calculated values for $X_w^{L(Ab)}$ and $X_w^{L(V)}$ at temperatures between $T_{fus}^{Ab(L,V)}$ and $T_{fus}^{Ab(L^*)}$.

5.2. Results

The thermodynamic procedures described in Section 5.1 were used to calculate hypersolidus phase relations for the *ab-w* system at 2·5 kbar. Results are illustrated in Figure 1, whereupon it can be seen that the calculated phase relations provided by the Margules method are fully accordant with the experimental data obtained by Fenn (1973). This favourable result exemplifies one of the key advantages of the Margules method *vis-à-vis* alternative methods for thermodynamic modelling of *ab-w* melts: namely that, because the Margules method is based solidly on hypersolidus phase-equilibrium data, good agreement between these data and calculated phase relations is virtually assured.

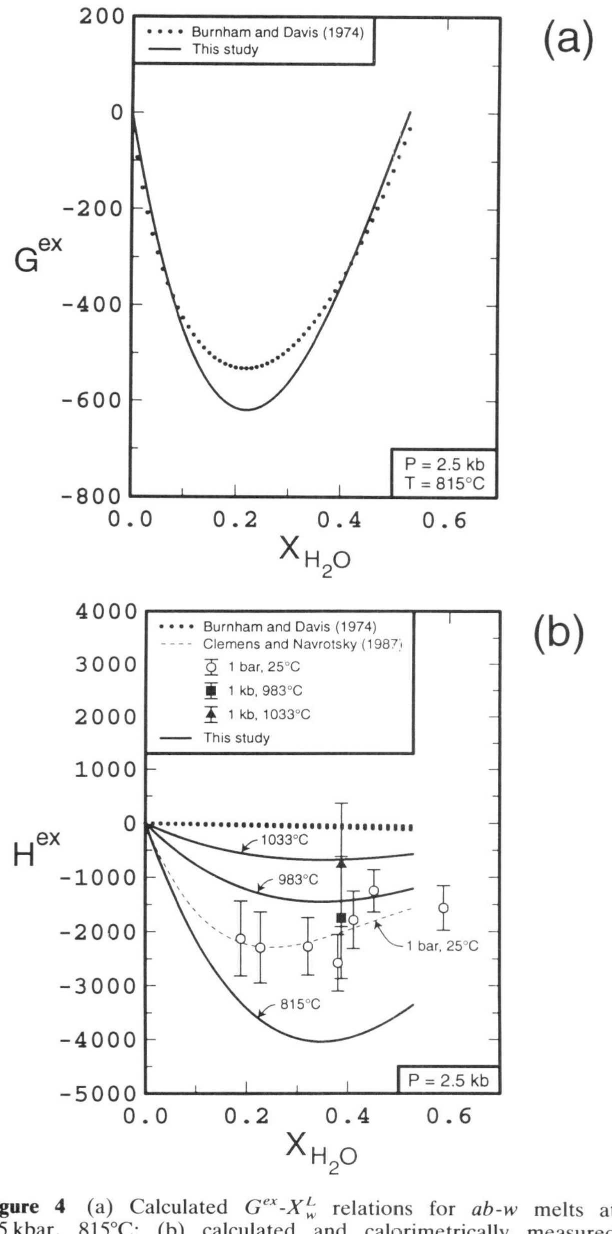

Figure 4 (a) Calculated G^{ex}-X_w^L relations for *ab-w* melts at 2·5 kbar, 815°C; (b) calculated and calorimetrically measured H^{ex}-X_w^L relations for *ab-w* melts at various pressures and temperatures. Dotted and solid lines represent calculated H^{ex}-X_w^L relations for 2·5 kbar, 815, 983 and 1033°C (Burnham & Davis 1974; this study); H^{ex} data were derived from solution calorimetry (1 bar, 25°C) and high-*P* drop calorimetry (1·0 kbar, 983 and 1033°C) (Clemens & Navrotsky 1987, Tables 5 and 7); dashed curve illustrates H^{ex}-X_w^L relations calculated from a fourth-order polynomial equation that was fit to the 1 bar, 25°C data using the method of least squares.

6. Discussion

An important question pertaining to the immediate application of the Margules method in thermodynamic modelling of *ab-w* melts concerns the availability and accuracy of phase-equilibrium and calorimetric data that are required to apply the method to pressures below and above 2·5 kbar. It has been shown that the method relies on four input thermochemical quantities: $X_w^{L(Ab)}$, $X_w^{L(V)}$, $\Delta G_{fus,ab}^{\circ,Ab}$ and $a_w^{L(V)}$. Arguments advanced in previous sections of this paper support the suggestion that it is presently possible to calculate accurate values for $\Delta G_{fus,ab}^{\circ,Ab}$ and $a_w^{L(V)}$ to pressures as high as 10 kbar. However, obtaining accurate values for $X_w^{L(Ab)}$ and $X_w^{L(V)}$ is more problematical. This subject merits further discussion.

Key observations concerning $X_w^{L(Ab)}$ are illustrated in Figure 1. The phase-equilibrium data plotted in this figure (Fenn 1973) indicate that the hydrous liquidus is nearly a linear function of X_{H_2O} at 2·5 kbar. The morphology obtained from thermodynamic modelling (this study) is a slightly convex upward curve. To the extent that this quasi-linear behaviour persists to pressures below and above 2·5 kbar, values for $X_w^{L(Ab)}$ at these pressures could be estimated with fair accuracy using experimental data on the hydrous solidus (Goldsmith & Jenkins 1985) and on the solubility of water in *ab-w* melts (Burnham & Jahns 1962; Hamilton & Oxtoby 1986). Such estimates would be hazardous for pressures above 4 kbar, however, because: (1) there are significant discrepancies between the published data for $X_w^{L(Ab,V)}$ (Burnham & Jahns 1962; Hamilton & Oxtoby 1986), and (2) $T_{fus}^{Ab(L^*)}$ is known with less certainty. An additional potential complicating factor is the possibility that the T-X_{H_2O} trend of the hydrous liquidus becomes more convex upward with increasing pressure.

There are two main considerations in estimating values for $X_w^{L(V)}$: the accuracy of experimental data for this parameter, and the dX_{H_2O}/dT slope of the $L(V)$ equilibrium. The good accord between the $X_w^{L(Ab,V)}$ data obtained by Burnham and Jahns (1962) and Hamilton and Oxtoby (1986) for pressures below 4 kbar indicates that values for $X_w^{L(Ab,V)}$ can be estimated with good accuracy for pressures up to 4 kbar. However, the water solubility data obtained by these investigators diverge markedly for pressures above 4 kbar. The Burnham and Jahns data indicate that $X_w^{L(Ab,V)}$ increases from ~0·59 at 4 kbar to ~0·69 at 8 kbar; by contrast, the data obtained by Hamilton and Oxtoby indicate that $X_w^{L(Ab,V)}$ increases from ~0·58 at 4 kbar to only ~0·63 at 8 kbar.

A nearly linear and approximately vertical trend of $L(V)$ at pressures from 0·56 kbar to 4 kbar is strongly supported by water solubility data for *ab-w* melts obtained by Hamilton and Oxtoby (1986). Water solubility data acquired by these investigators at pressures above 4 kbar are also consistent with a linear, subvertical morphology of $L(V)$ in isobaric T-X_{H_2O} section, but there is sufficient scatter in the data to permit a slightly arcuate, subvertical trend. In this regard, it is noteworthy that the thermodynamic model for *ab-w* melts developed by Silver and Stolper (1985) predicts $L(V)$ to be essentially linear with a strongly positive dX_{H_2O}/dT slope at pressures above 8 kbar.

It is evident from the foregoing discussion that full implementation of the Margules method is contingent upon obtaining additional experimental data on : (1) the hydrous liquidus of the *ab-w* system, and (2) the solubility of water in *ab-w* melts at pressures above 4 kbar.

7. Conclusions

Results of calculations performed for this paper indicate that the Margules method yields accurate calculated thermodynamic mixing properties and hypersolidus phase relations for *ab-w* melts at 2·5 kbar. It is also evident that: (1) the Margules method relies on thermochemical data for high temperature phases of the $NaAlSi_3O_8$—H_2O system that are either currently available or readily obtainable, and (2) the thermodynamic procedures applied to perform the modelling employ ordinary methods of numerical evaluation. Finally, it should be recognised that the Margules method has the practical advantage that *P-V-T* data are not required for modelling calculations. Difficulties associated with performing *P-V-T* experiments on hydrous silicate melts and gleaning accurate volumetric data therefrom make it highly unlikely that additional *P-V-T* data for *ab-w* melts will be forthcoming in the near future. It is concluded from these observations that the Margules method is an attractive alternative to the *P-V-T* method for thermodynamic modelling of *ab-w* melts.

8. Acknowledgements

The following individuals have assisted the writer in his efforts to develop a new thermodynamic model for albite-water melts: Drs C. Wayne Burnham, I. S. E. Carmichael, D. L. Hamilton, D. B. Joyce, V. C. Kress, G. A. Merkel, H. Nekvasil, E. M. Stolper and D. J. Wesolowski. Reviews of the draft manuscript by Drs W. Johannes and P. I. Nabelek are gratefully acknowledged. The research performed for this paper was supported financially by the Division of Engineering and Geosciences, Office of Basic Energy Sciences, U.S. Department of Energy under contract DE-AC05-84OR21400 with Martin Marietta Energy Systems, Inc.

9. References

Blencoe, J. G., Merkel, G. A. & Seil, M. K. 1982. Thermodynamics of crystal-fluid equilibria, with applications to the system $NaAlSi_3O_8$–$CaAl_2Si_2O_8$–SiO_2–NaCl–$CaCl_2$–H_2O. *In* Saxena, S. K. (ed.) *Advances in Physical Geochemistry* **2,** 282–311. New York: Springer.

Burnham, C. W. 1979. The importance of volatile constituents. *In* Yoder, H. S. Jr. (ed.) *The Evolution of the Igneous Rocks,* 439–82. Princeton: Princeton University Press.

Burnham, C. W. & Davis, N. F. 1971. The role of H_2O in silicate melts: I. *P-V-T* relations in the system $NaAlSi_3O_8$—H_2O to 10 kilobars and 1000°C. AM J SCI **270,** 54–79.

Burnham, C. W. & Davis, N. F. 1974. The role of H_2O in silicate melts: II. Thermodynamic and phase relations in the system $NaAlSi_3O_8$—H_2O to 10 kilobars, 700° to 1100°C. AM J SCI **274,** 902–40.

Burnham, C. W. & Jahns, R. H. 1962. A method for determining the solubility of water in silicate melts. AM J SCI **260,** 721–45.

Burnham, C. W. & Nekvasil, H. 1986. Equilibrium properties of granite pegmatite magmas. AM MINERAL **71,** 239–63.

Clark, S. P. Jr. 1966. Solubility. *In* Clark, S. P. Jr. (ed.) *Handbook of Physical Constants,* 415–36. GEOL SOC AM MEM **97**.

Clemens, J. D. & Navrotsky, A. 1987. Mixing properties of $NaAlSi_3O_8$ Melt–H_2O: New calorimetric data and some geological implications. J GEOL **95,** 173–86.

Fenn, P. M. 1973. *Nucleation and growth of alkali feldspars from melts in the system $NaAlSi_3O_8$—$KAlSi_3O_8$—H_2O.* Ph.D thesis, Stanford University.

Goldsmith, J. R. & Jenkins, D. M. 1985. The hydrothermal melting of low and high albite. AM MINERAL **70,** 924–33.

Goranson, R. W. 1938. Silicate-water systems: phase equilibria in the $NaAlSi_3O_8$—H_2O and $KAlSi_3O_8$—H_2O systems at high temperatures and pressures. AM J SCI **35A,** 71–91.

Hamilton, D. L. & Oxtoby, S. 1986. Solubility of water in albite-melt determined by the weight-loss method. J GEOL **94,** 626–30.

Kress, V. C., Williams, Q. & Carmichael, I. S. E. 1988. Ultrasonic investigation of melts in the system Na_2O—Al_2O_3—SiO_2. GEOCHIM COSMOCHIM ACTA **52,** 283–93.

Silver, L. A. & Stolper, E. M. 1985. A thermodynamic model for hydrous silicate melts. J GEOL **93,** 161–78.

Silver, L. A. & Stolper, E. M. 1989. Water in albitic glasses. J PETROL **30,** 667–709.

Stebbins, J. F., Carmichael, I. S. E. & Weill, D. E. 1983. The high temperature liquid and glass heat contents and the heats of fusion of diopside, albite, sanidine, and nepheline. AM MINERAL **68,** 717–30.

Stolper, E. M. 1982. The speciation of water in silicate melts. GEOCHIM COSMOCHIM ACTA **46,** 2609–20.

Tuttle, O. F. & Bowen, N. L. 1958. Origin of granite in light of experimental studies in the system $KAlSi_3O_8$—$NaAlSi_3O_8$—SiO_2—H_2O. GEOL SOC AM MEM **74**.

JAMES G. BLENCOE, Geosciences Group, Chemistry Division, Oak Ridge National Laboratory, Oak Ridge, Tennessee 37831-6110, U.S.A.

Transactions of the Royal Society of Edinburgh: Earth Sciences, **83**, 429–438, 1992

Experimental phase equilibrium studies of garnet-bearing I-type volcanics and high-level intrusives from Northland, New Zealand

T. H. Green

ABSTRACT: Rare garnet phenocrysts and garnet-bearing xenoliths occur in high-silica, metaluminous to peraluminous andesites and dacites (and their high-level intrusive quartz diorite equivalents) from a Miocene calc-alkaline province in Northland, New Zealand. These garnets are among the most Ca-rich (17–28 mol% grossular) garnets of igneous origin so far recorded in calc-alkaline suite rocks. Associated minerals are dominant hornblende and plagioclase and minor augite, occurring as phenocrysts in xenoliths and as inclusions in the garnet. This mineralogy points to the I-type character of the garnet-bearing host magma compositions, and contrasts this garnet occurrence with the more frequently recorded grossular-poor (3–10 mol%) garnets with hypersthene, plagioclase, biotite and cordierite, found in S-type volcanic and intrusive host rocks.

Detailed experimental work on a glass prepared from one of the garnet-bearing dacites closely constrains the conditions under which the natural phenocryst and xenolith mineral assemblages formed. This work was conducted over a pressure-temperature range of 8–20 kbar, 800–1050°C with 3–10 wt% of added H_2O, defining overall phase relationships for these conditions. Importantly, amphibole only appears at temperatures of 900°C or less and clinopyroxene at >900°C (with 3wt% H_2O). Orthopyroxene occurs with garnet at lower pressure (~15 kbar with 3 wt% H_2O; ~<10 kbar with 5 wt% H_2O). Absence of orthopyroxene from the natural garnet-bearing assemblages indicates pressures above these limits during crystallisation. Plagioclase is markedly suppressed (with respect to temperature) with increasing H_2O content, and for pressures of 10–15 kbar, the maximum H_2O content possible in the magma with retention of clinopyroxene and plagioclase together (as evident in xenoliths) is 5–6 wt%. Finally, the lack of quartz in any of the xenoliths suggests magma H_2O content higher than 3% (where quartz appears with amphibole at 900°C), since the quartz liquidus temperature decreases with increasing H_2O content, and with decreasing pressure. In experiments with 5 wt% H_2O, a quartz-free field of crystallisation of garnet–clinopyroxene–amphibole–plagioclase occurs between 10 and 15 kbar and temperatures between 850 and 900°C. In addition, detailed experimentally-determined garnet compositional trends, together with ferromagnesian mineral compositional data for specific experiments with 5 wt% H_2O added and run at 10–13 kbar and ~900°C, suggest that the natural assemblages formed at these conditions. This implies that the parental dacitic magma must have been derived at mantle depths (the Northland crust is ~25 km thick), and any basaltic or basaltic andesite precursor must have contained ~2–3 wt% H_2O.

The unique nature of the Northland volcanics and high-level intrusives, preserving evidence of relatively grossular-rich garnet fractionation in the high-pressure crystallisation history of an originally mantle-derived magma, is attributed to a combination of unusually hydrous conditions in the source region, complex tectonic history involving obduction and subduction, possible incorporation of crustal slivers in a mantle-crust interaction zone, and relatively thin (~25 km) crust.

KEY WORDS: calc-alkaline, high-pressure, lower-crust, mantle, petrogenesis, subduction, xenoliths

Rare almandine-rich garnets have been known in silicic volcanic and plutonic rocks for a long time (e.g. Green 1915), but they only received sporadic detailed attention (Miyashiro 1955; Oliver 1956) before the publication of experimental results which showed that almandine-rich garnet is a liquidus mineral in silicic compositions at high pressure (Green & Ringwood 1968a,b). These and later experimental studies (Green 1972; Green & Ringwood 1972) incorporated the suggestion that garnet in silicic igneous rocks provides evidence of crystallisation of magma at high pressure, possibly linked with eclogite melting in a subduction zone context. However, a difficulty with this interpretation is that the experimentally-obtained garnets are almandine-rich with subordinate pyrope *and* grossular (>15 mol%), while the natural garnets are almandine-rich with subordinate pyrope and only ~5 mol% grossular (neglecting spessartine-rich garnets present in very late-stage melts) (Green & Ringwood 1968b, 1972; Fitton 1972; Birch & Gleadow 1974; Wood 1974).

As part of Chappell and White's (1974) subdivision of silicic igneous rocks into I and S types, it was recognised that almandine-rich grossular-poor garnet is characteristic of rocks with S-type affinities, i.e. strongly corundum-normative character. This garnet was then attributed to being either a residual or phenocrystic mineral in a magma derived from partial melting of a pelitic source at moderate crustal depths (Green 1976; Clemens & Wall 1981; Vielzeuf & Holloway 1988). Garnet associated with I-type silicic igneous rocks has been recorded relatively rarely (Ujike & Onuki 1976; Zen & Hammarstrom 1984; Embey-Isztin *et al.* 1985; Barth & Ehlig 1988), and has been attributed to crystallisation at moderate pressure (6–10 kbar) (1 kbar = 100 MPa). Significantly, these garnets are richer in grossular (10–46 mol%) than the garnet characterising S-type igneous rocks, and they occur with hornblende ± epidote. Similar grossular content (14–19 mol%) has been recorded in garnets from strongly peraluminous rhyolitic lavas and ignimbrites from Spain (Gilbert & Rogers 1989). In that case, the high grossular content was attributed to the depth of crystallisation, following the suggestion of Green (1977) that the grossular content of near-liquidus garnet increased with increasing pressure. However, the strongly peraluminous character of these rocks may be due, at least in part, to alkali-loss from glass during post-crystallisation alteration (Gilbert & Rogers 1989). Thus the apparent association of phenocrystic almandine garnets with significant grossular content with an S-type magma, in that case, is equivocal.

More recently, Day *et al.* (1991) reported an unusual garnet-bearing phenocryst and xenolith assemblage from I-type silica-rich andesites and dacites from Northland, New Zealand. These garnets contain 17–28 mol% grossular and are associated with *both* augite and hornblende in host rocks which are metaluminous to slightly peraluminous (1·6% normative diopside to 1·1% normative corundum). These contrast with the I-type volcanics from Hungary and Japan referred to earlier, which contained garnet associated with hornblende and orthopyroxene.

Applying available experimentally determined high-pressure phase relationships for andesite, Day *et al.* (1991) concluded that the garnet bearing hornblende ± augite assemblages formed at 10–20 kbar and ~900–1000°C. Application of geothermobarometry to a single orthopyroxene-bearing garnet–hornblende xenolith indicated conditions of crystallisation for the garnet and orthopyroxene at 10–12 kbar and 950–1000°C.

The phenocryst and xenolith mineral assemblage in the Northland andesites–dacites lends itself to experimental duplication. This enables us to determine, more precisely than is possible from generalised andesite phase diagrams, the physical conditions (pressure, temperature and water content) under which the host or related magma might have crystallised the observed mineral assemblage. Thus the purpose of this paper is to present experimental data, for varying pressure, temperature and water contents, on the crystallisation of a Northland dacite, and to make detailed comparison of compositional trends observed in experimental run products with trends recorded in the natural minerals (Day *et al.* 1991). Duplication or near-duplication of minerals and their compositions, and the resultant delineation of appropriate physical conditions, also enable some depth and temperature limits to be placed on the source region for the host magma. Furthermore, the results have application to the possible crystallisation and fractionation of I-type magmas (volcanic and plutonic) at depth, with particular reference to bridging the thermal divide between metaluminous and peraluminous compositions (Ellis & Thompson 1986; Conrad *et al.* 1988).

1. Petrological and mineralogical background

The Miocene calc-alkaline suite from Northland includes strongly porphyritic basaltic andesite, andesite and dacite and weakly porphyritic rhyolite. Garnet is recorded only in high-silica andesite and dacite (Day *et al.* 1991). Discrete garnet crystals are argued to be cognate (and not xenocrystic) because of their euhedral form, oscillatory zoning and correlation between garnet and host rock composition (Day *et al.* 1991). Garnet in the xenoliths exhibits similar compositional features to the garnet phenocrysts, together with igneous microstructures (Day *et al.* 1991). Thus these xenoliths (type 1a of Day *et al.* 1991) are also interpreted as cognate. Consistent compositional correlation between garnet-bearing xenoliths and phenocryst minerals and their host-rock composition suggests that the latter reflects the overall composition of the bulk magma from which the observed phenocrysts and igneous-origin xenoliths crystallised. Thus the choice of a bulk composition for high-pressure experimental work is justified, but cannot be proved to be the *exact* composition from which all the phenocrysts and xenoliths formed, with progressive crystallisation.

A garnet phenocryst- and xenolith-bearing dacitic composition from a small sub-volcanic stock at Taipa [NZMS 1:63000 N7/973832], Northland, New Zealand, was selected for the experimental study. Specific characteristic features of the phenocryst-xenolith assemblages in the Taipa dacite which should be sought after in the experimental attempts at duplication are summarised as follows (from Day *et al.* 1991):

(a) Garnet phenocrysts ($mg = 32–46$) ($mg = 100\,Mg/(Mg + Fe)$ atomic proportions) may have oscillatory, normal or reverse zoning (in mg) and some are resorbed, with reaction coronas of hornblende and/or plagioclase.

(b) Other phenocrysts, associated with garnet, are hornblende ($mg = 50–70$) and plagioclase (An = 40–56), with minor ilmenite. Orthopyroxene occurs as microphenocrysts, never in contact with garnet.

(c) Included minerals in garnet are augite ($mg = 67–81$), hornblende ($mg = 52–63$), plagioclase (An = 69–76), ilmenite and apatite, but never quartz or orthopyroxene.

(d) Amphibole is zoned from pargasitic cores to hornblende rims.

(e) Garnet ($mg = 37–54$)-bearing xenoliths are typically hornblende-rich ($mg = 56–64$) with plagioclase (An = 46–82). Similar xenoliths from other Northland localities contain rare augite ($mg = 69–75$). Quartz is not present, and bronzite ($mg = 71–80$) occurs in only one garnet ($mg = 35–47$)-bearing xenolith; the bronzite is enclosed in poikilitic hornblende ($mg = 72–75$). Significantly, this xenolith is more magnesian than is characteristic for the garnet + hornblende + plagioclase ± augite xenoliths.

(f) The Taipa dacite contains a complex and varied xenolith population, in addition to the garnet-bearing type (1a of Day *et al.* 1991). These other xenoliths include more magnesian hornblende + bronzite ± augite (+garnet, in one example referred to in the previous point) (type 1b), meta-igneous xenoliths (type 1c), metamorphic xenoliths (type 2) interpreted as fragments of regionally metamorphosed basalts and volcaniclastic sediments, and finally composite xenoliths consisting of a meta-igneous portion (type 1c) in sharp contact with a metamorphic portion (type 2).

The features summarised in points (a) to (e) reflect

Table 1 Taipa dacite starting composition (XRF analysis of bulk rock composition compared with electron microprobe analysis of glass chips); recalculated to 100% anhydrous.

	XRF	Electron probe (n = 16)		CIPW Norms XRF	CIPW Norms Electron probe
SiO_2	64·76	64·50	Q	18·7	18·8
TiO_2	0·53	0·56	Or	13·3	12·6
Al_2O_3	16·93	17·06	Ab	29·7	29·2
FeO*	4·64	4·65	An	23·8	24·8
MnO	0·09	0·10	Di	0·7	0·4
MgO	2·01	2·09	Hy	12·5	12·8
CaO	5·14	5·30	Il	1·0	1·1
Na_2O	3·51	3·45	Ap	0·3	0·4
K_2O	2·25	2·13			
P_2O_5	0·14	0·16			

Note: Total Fe expressed as FeO.

cooling and progressive crystallisation processes producing non-equilibrium features, which are inevitably more complex than those to be expected in equilibrium experiments. The latter can only set out to reproduce specific, hopefully recognisable, stages in the overall crystallisation history.

2. Experimental procedure

The bulk composition of the Taipa dacite is given in Table 1. The rock powder was melted at ~1400°C for 30 minutes in an inert atmosphere, and quenched to a glass (twice, with fine grinding in between). This glass powder (see Table 1 for electron microprobe analysis of glass fragments) was used as the starting composition for a series of high-pressure experiments conducted in a 12·7 mm piston-cylinder apparatus, using talc/pyrex pressure cells with air-fired boron nitride internal components. A −10% pressure correction was applied to nominal piston-in pressures (Green *et al.* 1966), and temperatures were measured with a Pt, Pt 10% Rh thermocouple. No correction was applied for any possible pressure effect on the electromotive force of the thermocouple. Temperatures were controlled to within ±5°C of the set point, and allowing for temperature gradients within the pressure cell, are accurate to ±15°C, while pressures in the 8–20 kbar range are believed accurate to within ±1 kbar. About 10–15 mg of sample was encapsulated within Ag_{70}–Pd_{30} tubing and 3 or 5 wt% water was added, using a microsyringe, to most runs, although some reconnaissance experiments were conducted with higher water contents (6–10 wt%). Oxygen fugacity for hydrous runs in these pressure cell assemblies is suggested to be close to the magnetite–wustite buffer (Green & Pearson 1985). Run times generally varied from ~24 hours at 1050°C to ~72 hours at 800°C. At the conclusion of a run, the sample was crushed and examined optically, before mounting in epoxy and polishing for electron microprobe analysis.

3. Analytical procedure

Electron microprobe analyses of phases in the experimental runs were obtained using an ETEC wavelength dispersive autoprobe, with an accelerating voltage of 15 kV and current of 50 nA. Correction procedures followed those of Bence and Albee (1968). Analysis of crystals from 5–50 μm in size involved a finely-focused beam (~1 μm). Where possible, several grains of each mineral were analysed and, if large enough, checked for homogeneity via analyses of rims and cores. Representative analyses are presented in Table 2.

4. Approach to equilibrium

The results obtained are for synthesis experiments only, but are argued to closely approach equilibrium for the run times adopted, because of the relatively hydrous character of the melts (3 wt% or greater added H_2O) promoting diffusion, and because of the consistent pattern of crystallisation. This pattern is consistent between the present work using a glass starting composition, and other experimental studies on broadly similar bulk compositions, using natural rock compositions, not glasses (e.g. Wyllie & Rutter 1988, and references therein). The only obvious equilibrium problem in the present work relates to the nucleation of garnet. This is clearly illustrated in experiments with 3 wt% H_2O, where a glass starting composition with 1% garnet (separated from the original rock) seeds added, showed partial (rim) re-equilibration of the seed garnet, and garnet growth at up to about 5 kbar lower pressure than in runs without garnet seeds. The 1% garnet seeds were successfully dissolved in experiments interpreted to be on the low pressure side of the garnet stability field (see Fig. 1a,b).

Finally, Fe-loss to noble metal capsules, a general problem in hydrous melting experiments with relatively high degrees of melting, restricted a complete approach to equilibrium, because of changing composition. This problem was minimised by using talc/pyrex pressure cell assemblages, Ag_{70}–Pd_{30} capsules (Ellis 1986) and the shortest run time possible (e.g. to allow resorption of garnet seeds). Iron loss from the charge was commonly measured (via broad beam, average of three mobile analyses, covering wide, representative areas) to be about 10% of the amount present. This means that precise duplication of natural assemblages in terms of *mg* is not possible. Nevertheless, the iron-loss problem of the order indicated should not greatly affect reasonable interpretation of pressure (*P*) temperature (*T*) conditions based on experiments, for crystallisation of the observed phenocryst/xenolith assemblage.

5. Results

5.1. Phase equilibrium

The phase equilibrium results are represented in Figure 1 for 3 wt% and 5 wt% of added H_2O. This represents the bulk H_2O content in the charge and is not the H_2O content of the melt, which will increase with increasing crystallisation. The most significant points are summarised as follows:

(a) Garnet is stable at 8 kbar, 850°C, 3 wt% H_2O in garnet-seeded runs, and at slightly higher pressure in runs with 5 wt% H_2O. In unseeded runs, garnet does not appear until pressures of 13 kbar or greater are reached for temperatures of 850–1050°C.

(b) Orthopyroxene coexists with garnet for a relatively small pressure range above the incoming of garnet (~0·5–1·5 kbar) in garnet-seeded runs. Orthopyroxene is absent at higher pressure.

(c) With falling temperature, above 10 kbar, liquidus clinopyroxene is joined by garnet, then plagioclase, and is finally replaced by amphibole and quartz.

Table 2 Representative analyses of mineral products from experimental runs, together with analyses of selected natural phenocryst/xenolith minerals from sample 38629 for comparison (see Day *et al.* 1991 for details of 38629); abbreviations: n = number of analyses averaged; seeds refers to 1% garnet seed material added (y = yes, n = no); c = core, r = rim; incl = including.

(a)

	garnet								amphibole					
Run No.	38629 I	1354	1362	1327	1318c	1318r	1251c	1251r	38629 I	1354	1327	1298	1162	1139
P (kbar)		10	10	12·5	15	15	17·5	17·5		10	12·5	17·5	17·5	17·5
T (°C)		900	950	850	950	950	850	850		900	850	850	850	850
Time (hrs)		29	48	48	30	30	30	30		29	48	30	30	30
%H_2O		5	3	5	5	5	5	5		5	5	3	5	8
Seeds		Y	Y	N	N	N	N	N		Y	N	N	N	N
n	43	1	2	2	5	6	1	1	11	2	3	10	5	5
SiO_2	38·57	39·45	40·36	38·92	38·23	38·47	39·19	39·03	41·19	44·09	43·57	42·54	43·63	42·84
TiO_2	0·36	0·83	1·10	1·29	1·78	1·41	1·86	1·64	2·69	2·66	2·00	2·03	1·95	1·81
Al_2O_3	21·31	21·23	21·31	21·33	21·16	21·27	20·77	21·45	14·12	13·95	15·17	15·37	14·93	15·05
FeO_T	23·12	22·05	23·05	22·16	22·51	21·59	23·57	22·53	14·17	15·59	15·64	18·29	16·10	14·83
MnO	0·95	0·89	0·90	0·80	0·75	0·61	1·50	0·95	0·16	0·18	0·27	0·26	0·27	0·21
MgO	8·05	7·41	7·94	6·14	7·86	8·23	4·09	5·22	11·41	10·86	9·21	8·91	8·98	9·85
CaO	7·95*	8·20	8·37	10·03	8·85	8·76	11·09	11·21	11·26	9·46	9·79	8·85	9·42	9·24
Na_2O	—	—	0·07	0·05	0·04	0·06	0·08	0·09	2·20*	2·07	2·16	1·86	1·99	2·27
K_2O	—	—	—	—	—	—	—	—	0·77**	0·83	0·85	0·92	0·89	0·74
	100·31	100·06	103·1	100·72	101·18	100·41	102·15	102·12	97·97	99·69	98·66	99·03	98·16	96·84
mg	38·3 +	37·5	38·0	33·0	38·3	40·4	23·6	29·2	58·9 +	55·4	51·2	46·5	49·9	54·2
Ca:Mg	21·4:30·1	22·1:31·6	22·3:29·5	28·0:23·8	23·7:29·3	23·6:30·9	31·5:16·2	31·1:20·1	29·5:41·6	25·8:41·1	28·1:36·8	24·9:34·9	27·3:36·2	26·8:39·7
:Fe	:48·5	:46·3	:48·1	:48·2	:47·0	:45·5	:52·3	:48·8	:29·0	:33·1	:35·1	:40·2	:36·4	:33·5
Al^{IV}									1·89	1·63	1·61	1·73	1·58	1·64
Al^{VI}									0·58	0·75	1·00	0·94	1·01	1·00
Na + K									0·78	0·74	0·77	0·70	0·73	0·79
Na/K									4·34	3·81	3·88	3·07	3·40	4·66

Garnet 38629 I: *6·9–8·6 (c r); +31–41 (c r)

Amphibole 38629 I: *1·83–2·48 range; **0·71–0·84 range; +56–62 range

(b)

	clinopyroxene				orthopyroxene		
Run No.	38629 K	1342	1318	1333	38629X	1342	1351
P (kbar)		10	15	11·5		10	10
T (°C)		950	950	850		950	1050
Time (hrs)		48	31	50		48	29
%H_2O		5	5	5		5	5
Seeds		Y	N	Y		Y	Y
n	22ϕ	5ϕ	3ϕ	1	7ϕ	1	4ϕ
SiO_2	49·30	49·46	49·34	49·80	52·56	50·94	52·12
TiO_2	0·54	0·68	0·76	0·64	0·10	0·32	0·21
Al_2O_3	4·73*	5·60	8·91	6·12	1·70*	4·98	3·42
FeO_T	9·83	11·31	8·50	9·81	17·16	21·04	16·81
MnO	0·24	0·31	0·08	0·12	0·34	0·39	0·42
MgO	12·74	13·33	10·91	11·63	25·98	20·66	24·90
CaO	20·90	17·03	19·93	21·65	1·05	1·66	1·87
Na_2O	0·43**	0·59	1·20	0·54	0·05	—	0·09
K_2O	—	0·05	0·09	0·08	—	—	—
	98·88	98·50	99·79	100·38	99·30	99·99	100·26
mg	69·8 +	67·7	69·6	67·9	72·9 +	63·6	72·5
Ca:Mg	45·1:38·3	38·5:41·7	47·7:36·4	47·6:35·6	2·1:71·5	3·6:61·4	3·7:69·8
:Fe	:16·6	:19·8	:15·9	:16·8	:26·5	:35·1	:26·4
Al^{IV}	0·13	0·14	0·17	0·14	0·07	0·11	0·10
Al^{VI}	0·08	0·13	0·12	0·13	—	0·11	0·05
Na + K							
Na/K							
	ϕ incl. 0·17% Cr_2O_3	ϕ incl. 0·14% Cr_2O_3	ϕ incl. 0·07% Cr_2O_3		ϕ incl. 0·36% Cr_2O_3		ϕ incl. 0·42% Cr_2O_3
	*2·67–5·63 range				*1·31–2·26 range		
	**0·36–0·51 range				+65·6–78·4 range		
	+68·8–71·8 range						

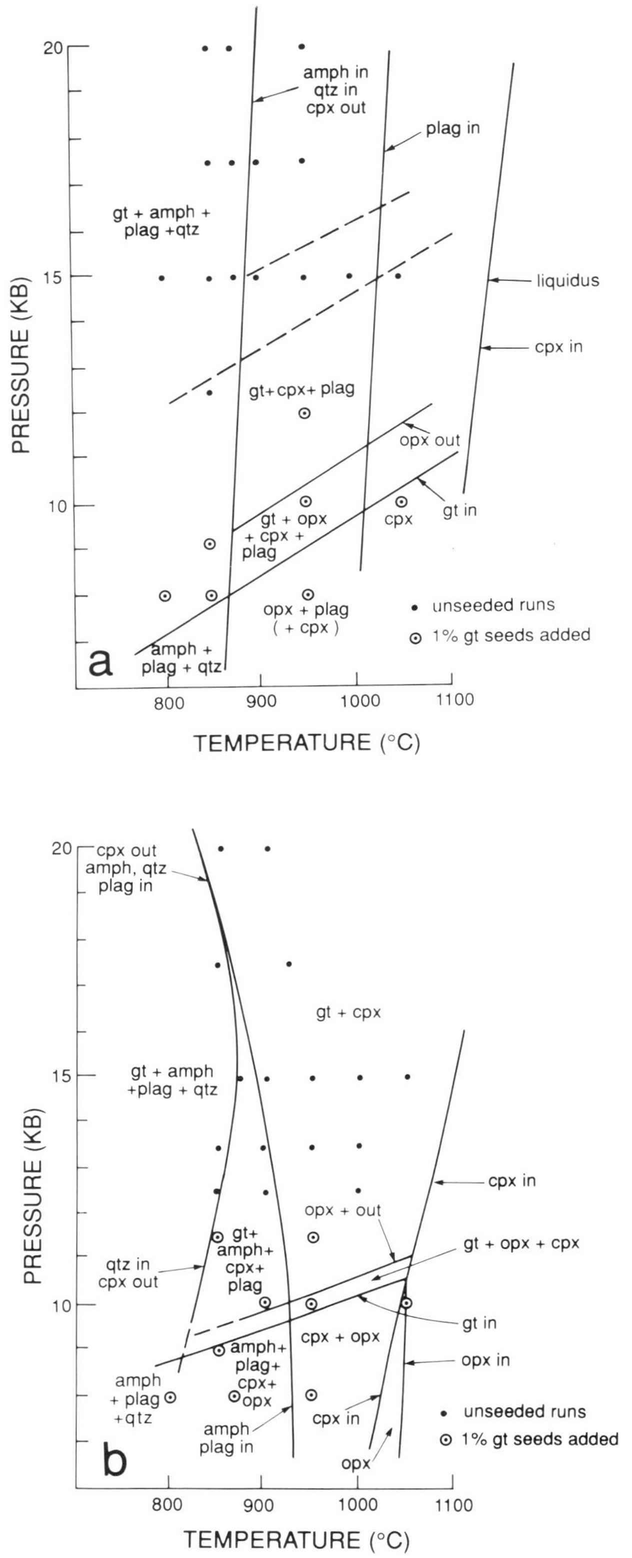

Figure 1 Pressure-temperature phase diagram for garnet-bearing dacite (AU38629—housed in Geology Department, Auckland University, New Zealand) with (a) 3% and (b) 5 wt% of H_2O added; abbreviations: gt = garnet, cpx = Ca-clinopyroxene, opx = orthopyroxene, plag = plagioclase, amph = Ca-amphibole, qtz = quartz; phase boundaries are approximate, representing the best fit for the spacing of the experimental runs; position of the liquidus is estimated only (it is not critical to the interpretation of the results); glass was present in all runs; dashed boundaries in (a) refer to gt in, opx out (respectively, with increasing *P*) boundaries obtained when no garnet seeds were added; detailed experimental conditions and results are available on request.

(d) In 3 wt% H_2O runs, quartz and amphibole appear together, as clinopyroxene disappears at ~875–900°C. Plagioclase appears below 1000°C and there is a relatively wide temperature interval (~120°C) where garnet, plagioclase and clinopyroxene coexist. Accessory ilmenite is present at 12·5 kbar, 850°C and 15 kbar, 900°C, while rutile occurs at 20 kbar, 875°C.

(e) In 5 wt% H_2O runs, the appearance of quartz and plagioclase is suppressed to lower temperatures (compared with the 3 wt% H_2O runs). A small temperature interval is recognisable, in the lower pressure range (e.g. 10–13 kbar), where clinopyroxene and amphibole coexist with plagioclase and garnet. Plagioclase and amphibole appear together at ~900°C in this pressure range.

(f) In additional experiments (not plotted in Fig. 1) with >5 wt% H_2O (8 and 10 wt%), the plagioclase liquidus is suppressed (e.g. below 800°C at 17·5 kbar), and the garnet-amphibole field expanded.

5.2. Compositional features

Representative analyses are given in Table 2. It was difficult to obtain high-quality analyses of plagioclase because of its thin lath-like habit and the unavoidable Na-loss resulting from use of a finely focused electron beam on these crystals. Hence only ferromagnesian mineral analyses are presented and discussed.

5.2.1. Garnet. Most garnet cores are rich in inclusions of other phases in the experimental runs, while the rims are inclusion-free, and wherever possible rim and core compositions of garnets have been obtained. All experimentally crystallised garnet compositions plot within the designated field in Figure 2. Care was taken to avoid analysis of any relict seed garnet (if evident in garnet cores) where garnet-seeded runs were conducted. In Figure 3 representative compositional trends, from core to rim, and for varying pressure (*P*) at two fixed temperatures (*T*) are presented. The core to rim variation in the garnet seed crystals is also shown, for comparison. Noteworthy points on garnet compositional features are:

(i) *mg* increases from core to rim in runs at 850–950°C, shows little variation at 1000°C and decreases slightly at 1050°C. This is broadly similar to *mg* zoning variations observed in natural garnets (Day *et al.* 1991) and, although manifestly a non-equilibrium effect, it appears to correlate with garnet-silicate liquid Mg/Fe partition relations noted by Green (1977) and Ellis (1986).

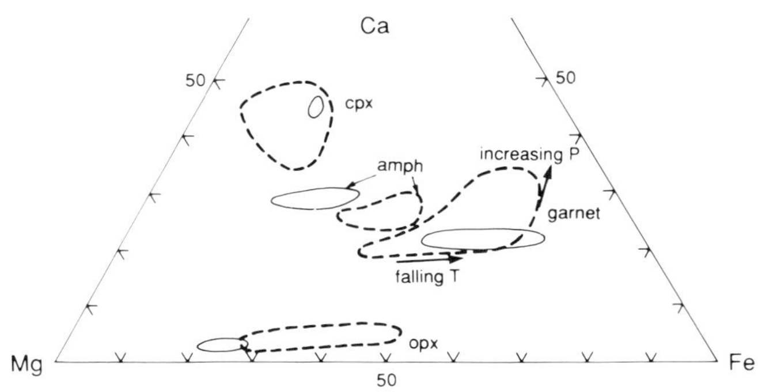

Figure 2 Ca–Mg–Fe composition plot showing compositional fields for pyroxenes, amphibole and garnet, comparing 38629 phehocryst/xenolith compositions with those obtained experimentally; dashed lines denote experimentally produced compositions, and solid lines define variation in the natural compositions; abbreviations as for Figure 1.

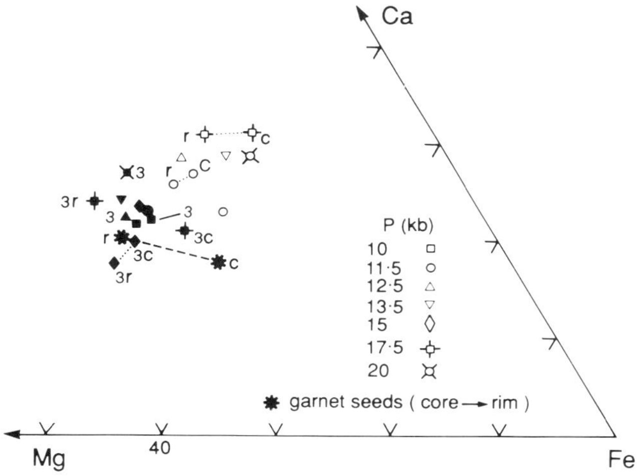

Figure 3 Enlargement of a portion of the Ca–Mg–Fe plot showing garnet compositions plotted for various pressures; points marked with 3 denote experiments with 3 wt% H_2O added run at 950°C; un-numbered points are all experiments with 5 wt% H_2O added; solid symbols 950°C, open symbols 850°C, c = core, r = rim; garnet seeds denote the core/rim compositions of garnet separated from a xenolith from 38629, and added to the glass starting composition.

(ii) Grossular content increases with increasing *P* (cf. Green 1977), although the pattern is complicated by common, but variable, zoning in grossular. In most cases, grossular decreases from core to rim (variation of 0–8 mol%). However, in some runs, most notably at 17·5 kbar, grossular increases from core to rim (0–5 mol%).

(iii) Grossular content increases with increasing H_2O content from 3 to 5 wt% in runs at 15 kbar (e.g. from 18·4 to 23·6 mol% grossular, garnet rims, 950°C) and 17·5 kbar

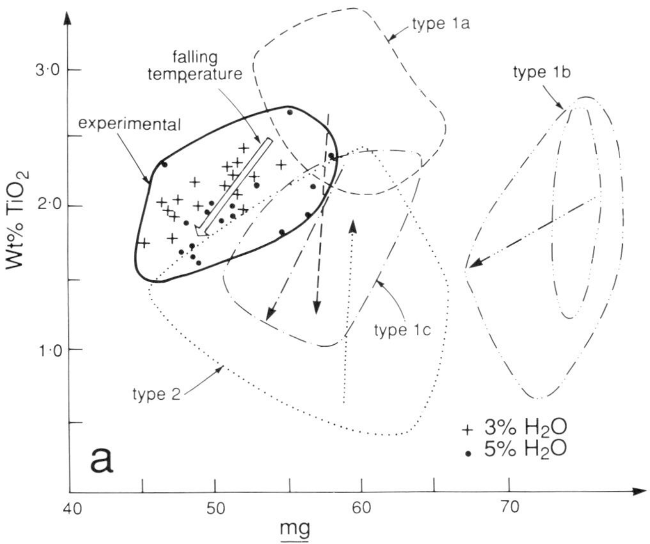

Figure 4 Compositional fields for experimentally obtained Ca-amphiboles compared with amphiboles occurring in the various xenolith types (types 1a–1c cognate, igneous; type 2 accidental, metamorphic); arrows associated with the various xenolith types denote compositional variation (core to rim) found in amphiboles on the margins of the xenoliths; small field inside the type 1b field denotes the compositional field for poikilitic amphibole in the single garnet-bearing type 1b xenolith.

(a) TiO_2 vs *mg* where *mg* = 100 Mg/(Mg + Fe) atomic proportions;

(b) Al^{VI} vs *mg*;

(c) Al^{IV} vs Na + K where Al, Na, K in atomic proportions.

For clarity, the field for type 2 amphiboles is not plotted in (b) and (c).

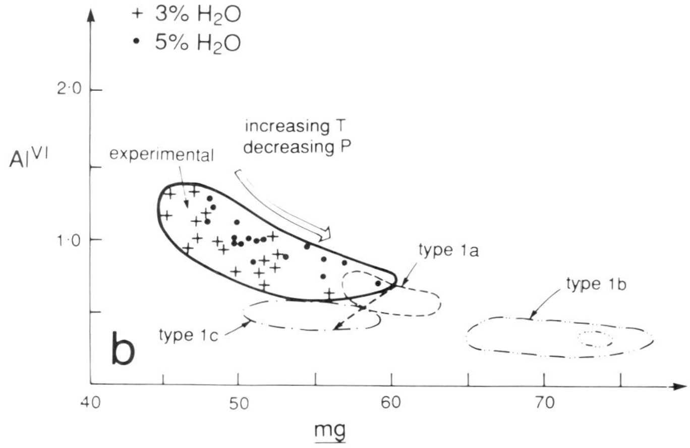

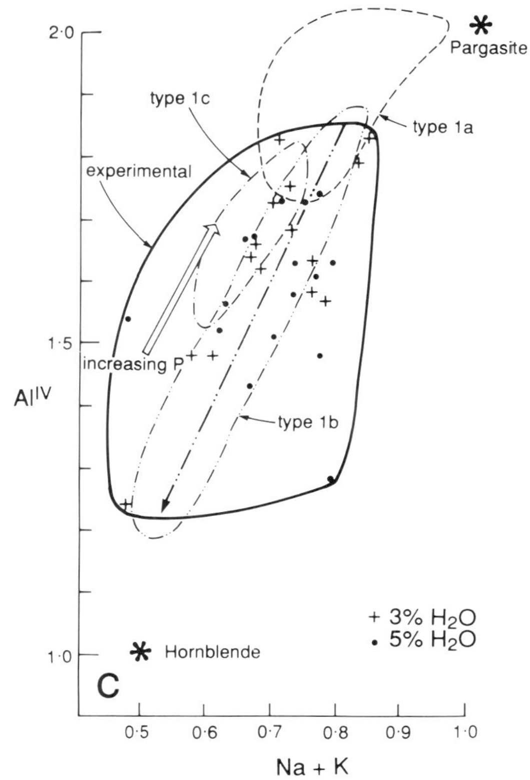

Figure 4—(*Continued*)

(e.g. from 26·7 to 31·1 mol% grossular, garnet rims, 850°C), but shows little change at 10 kbar.

(iv) Closest compositional matching between experimental and natural garnets (from 38629) was obtained for garnet rims from a run (1354) at 10 kbar, 900°C and 5 wt% added H_2O, and a run (1362) at 10 kbar, 950°C and 3 wt% added H_2O.

5.2.2. Amphibole. Experimentally obtained amphibole compositions are plotted in Figure 4. They are unzoned and vary from ferroan pargasite through ferroan pargasitic hornblende to edenitic hornblende (Hawthorne 1981) with 1·5–2·6 wt% TiO_2 (0·16–0·29 cations). Na/K varies from 1·9–4·8, increasing with increasing H_2O content and *T* and decreasing *P*. Al^{VI} increases with decreasing *T* and increasing *P*, while Al^{IV} and (Na + K) increase with increasing *P* and *T*. Figure 4(a) and (b) show only slight overlap between experimental and natural amphibole compositions and point to crystallisation of the natural amphiboles at slightly higher temperatures or from a parent composition with a slightly higher *mg* than the experimental starting composition. Figure 4(c) shows a greater overlap between experimental amphiboles and amphiboles from the higher *mg* type 1b xenoliths, rather than the garnet-bearing type 1a xenoliths. The Ca–Mg–Fe plot (Fig. 2) does not show overlap for the natural and experimental compositions. The Mg/Fe discrepancy can be explained by differing temperatures of crystallisation or *mg* of the bulk composition, but no consistent trend of varying Ca with *T*, *P* or H_2O content was observed in the experiments.

Overall, amphibole crystallised at 900°C, 10 kbar with 5 wt% H_2O (run 1354) approximates the natural amphibole compositions apart from lower *mg*, and Ca, as already noted (see also Table 2).

5.2.3. Clinopyroxene. Compositions from experimental runs are plotted in Figure 2, and compared with the natural clinopyroxene found as inclusions in garnet from 38629. The clinopyroxenes crystallised over a *P*, *T* range of 850–1050°C, 10–17·5 kbar with 3–10 wt% H_2O. Figure 2 shows the wide compositional variation of the experimentally produced clinopyroxene, which encompasses the field of the natural clinopyroxenes. The closest compositions to the natural pyroxene formed between 850°C, 11·5 kbar and 950°C, 10 kbar with 5 wt% H_2O, although the *mg* is low. Higher pressure clinopyroxenes from 3–5 wt% H_2O runs have distinctly higher Na and Al than the natural examples.

5.2.4. Orthopyroxene. Although orthopyroxene does not occur as phenocrysts or in type 1a garnet-bearing xenoliths, it is common in the higher *mg* type 1b xenoliths, one of which contains garnet. Analyses plotted in Figure 2 and given in Table 2 show that in the experiments a liquidus *T* of 1050°C at 10 kbar, 5 wt% H_2O is required for experimental orthopyroxene *mg* to be close to the natural orthopyroxene. At temperatures where orthopyroxene occurs with garnet and clinopyroxene (e.g. run 1342), the *mg* (63·6) is much lower than the natural orthopyroxene. Thus the latter probably crystallised from a more *mg*-rich composition (cf. Day *et al.* 1991). Al_2O_3 content of 1·3–2·3 wt% suggests *P* of crystallisation between 8 and 10 kbar at a probable *T* of ~950°C with ~5 wt% H_2O (experimental orthopyroxene at 8 kbar, 950°C, 3 wt% H_2O contained 1·2% Al_2O_3, while at 10 kbar, 1050°C, 5 wt% H_2O it contained 3·4% Al_2O_3).

6. Discussion of results

6.1. Conditions of crystallisation

The matching of the phase assemblage garnet + clinopyroxene then garnet (±clinopyroxene) + hornblende + plagioclase, with the notable absence of quartz and orthopyroxene phenocrysts, indicates that a bulk composition such as the dacite (or high-silica andesite) 38629 must have crystallised at >10 kbar at 950°C, cooling through 900°C to ~850°C (see Fig. 1). This is consistent with the conclusions of Day *et al.* (1991). The H_2O content of the magma must have been close to 5 wt%, as higher amounts suppress plagioclase crystallisation too much and lower amounts both reduce the amphibole + clinopyroxene field and expand the quartz field. For the degree of crystallisation involved at 850–900°C, the water content in the actual melt approximates 6–7 wt%.

Compositional matching and compositional trends with changing *P*, *T* and H_2O content in experimentally produced minerals are internally consistent in suggesting beginning of crystallisation at *T* slightly higher than 900°C at 10 kbar and/or from a bulk composition with a slightly higher *mg* than 38629. Any iron loss from the dacite to the capsule walls (see earlier) would cause higher *mg* in the crystallising minerals and so cannot be responsible for the *mg* difference between the naturally occurring and experimentally obtained hornblende and clinopyroxene. Thus the suggestion of a bulk composition with a higher *mg* than 38629 is reinforced by any iron-loss effect.

6.2. Significance of depth of origin

The experimental data supporting the crystallisation of the Northland, New Zealand dacite at ~900°C and ~10 kbar (equivalent to >30 km depth) places the process at sub-crustal depths in Northland, since seismic data suggest the crust is only about 25 km thick in this region (Stern *et al.* 1987). This therefore provides evidence for the genesis of relatively silicic magma compositions in the mantle. The tectonic setting for the New Zealand garnet-bearing dacite is complex, with subduction apparently occurring soon after an obduction event (Day *et al.* 1991). This point, allied with initial $^{87}Sr/^{86}Sr$ ratios of 0·7065–0·7070 (Day *et al.* 1991), suggests that sialic crustal slivers may have been caught up in the mantle melting process at depths equal to or greater than those indicated by the crystallisation of the garnet-bearing assemblage.

The H_2O content of ~5 wt% interpreted for the dacite magma implies, through simple mass-balance assuming closed system fractionation, that any mantle derived basaltic to basaltic andesite precursor must have contained ~2–3 wt% H_2O. In turn, the mantle source region for these magmas would be required to have ~0·3–0·5 wt% H_2O, unless a significant contribution was obtained from the crustal contamination suggested by the isotopic data.

6.3. Role and significance of garnet

In a recent review, Kay and Kay (1990) emphasised the widespread occurrence of xenoliths and frequently associated high-pressure phenocrysts in convergent plate boundary igneous rocks. They pointed to the unique evidence these provided on the nature of petrogenetic processes in the lower crust and upper mantle in these regions, and noted that garnet-bearing xenoliths did not occur in oceanic island arcs, but that deformed garnet-bearing xenoliths, of proposed lower crustal origin, were present in continental marginal regions. They concluded that in the latter case the garnet formed as a result of post-magmatic recrystallisation. Thus it follows from the evidence considered by them, that garnet would not have a role in crystal–liquid evolution of the convergent plate boundary magmas. An alternative view put forward by Drummond and Defant (1990) argued that a particular convergent-plate boundary magmatic sub-group (trondhjemite–tonalite–dacite) was derived by melting of eclogitic subducted oceanic crust at 23–26 kbar. Thus, following this view, garnet would have a major role in the genesis of these particular rocks.

The Northland, New Zealand, garnet-bearing andesite–dacite occurrence does not have the chemical characteristics of the trondhjemite–tonalite–dacite suite of Drummond and Defant (1990), and it appears to fall between the oceanic island arc and continental margin convergent plate boundary situations discussed by Kay and Kay (1990). The Northland province is not a true oceanic island arc, and has developed in a "protocontinental" situation with a crustal thickness equal to that found in oceanic arcs without active back-arc spreading (Leeman 1983). Thus these new data on the Northland rocks do provide clearcut evidence of a role for garnet in the fractional crystallisation of magmas in convergent plate boundary regions. However, this role appears to be restricted to evolved compositions, and any mantle-derived precursor melt must have undergone considerable ferromagnesian mineral (and subordinate plagioclase) fractional crystallisation in the mantle before the appearance of garnet.

It is possible that a necessary additional feature, promoting the role of garnet crystallisation in the sub-crustal or crust/mantle interaction zone, is the modification of the mantle-derived magma by assimilation of relatively Al-rich crustal slivers. This has the effect of decreasing the normative di/hy ratio and shifting the bulk composition to lower normative diopside (i.e. less metaluminous) and higher normative quartz. This allows crystallisation of garnet as a near-liquidus mineral at lower pressure than in a strongly metaluminous, more mafic composition (Green 1982; Day *et al.* 1991). This is also borne out by the experiments of Conrad *et al.* (1988), who obtained garnet above the solidus in a "greywacke" composition (2·8% normative corundum) at 10 kbar, but did not observe garnet for similar conditions in a dacite (3·3% normative diopside).

Finally, the same argument applies to explaining the rarity of evidence for a role for garnet in the crystallisation of I-type plutonic and volcanic rocks, compared with S-type equivalents. For the latter strongly corundum-normative S-type compositions, garnet may crystallise at pressures as low as 4–5 kbar (Clemens & Wall 1981, 1984), giving a considerable depth range in the crust where *S*-type magmas may be generated leaving a garnet-bearing residue. In contrast, metaluminous to mildly corundum-normative I-type magmas require pressures of ~10 kbar for garnet to be stable, restricting origin of these rocks by direct crustal melting to regions where the crust is ~35 km thick. In this case, there is a greater likelihood that any mineralogical evidence for early garnet involvement will be destroyed as the magmas rise to higher levels in the crust, compared with S-type magmas where garnet remains stable to much shallower levels, and in part at least has a refractory source origin (Munksgaard 1985). Spessartine-rich garnets are excluded from the foregoing discussion as they are stable in both S- and I-type compositions to very low pressures (Green 1977; Miller & Stoddard 1981).

6.4. Application to granite genesis

The interpretation of a sub-crustal origin for high silica andesitic to dacitic extrusive and high-level intrusive igneous rocks provides direct evidence of possible controls on crystal fractionation involved in the derivation of M-type granites (Whalen 1985; Chappell & Stephens 1988). Coincidentally, the composition of the dacite studied here is remarkably similar to the average composition of the I-type tonalitic Peninsular Ranges Batholith (Silver & Chappell 1988). Thus, if deep-level (~30–40 km) crystal fractionation is involved in explaining the spectrum of compositions in this or other tonalitic suites, the mineral phase equilibria and compositional data presented here will have direct application. Such fractionation, involving Ca-clinopyroxene, Ca-amphibole and possibly ultimately grossular-bearing garnet, bridges the low-pressure thermal divide between metaluminous and peraluminous compositions (Ellis & Thompson 1986; Conrad *et al.* 1988), and allows a melt composition range of diopside-normative to corundum-normative derivatives in these suites. The same arguments could apply to I-type granodioritic suites where these are derived at sufficient depth in the crust. In the case of the Peninsular Ranges Batholith, geochemical evidence in the form of steep chondrite-normalised rare-earth element patterns (depleted in the heavy rare-earth elements) is consistent with a role for garnet in the derivation of the Central Tonalites, although no direct mineralogical evidence for garnet involvement is preserved (Gromet & Silver 1987).

7. Conclusions

High pressure experimental duplication of the garnet (17–28% grossular, 16–35% pyrope), Ca-clinopyroxene, Ca-amphibole, plagioclase phenocryst/xenolith mineral assemblage found in a Northland, New Zealand, dacite points to crystallisation from a relatively hydrous high-silica andesite to dacite magma at ~35 km depth and temperature of ~900°C, placing the process at mantle depths for Northland. This provides evidence for the possible crystal fractionation controls involved in the evolution of M-type and some I-type tonalitic magmas, which is rarely preserved in the mineralogy of the end-products. It also verifies the link, via fractional crystallisation of Ca-clinopyroxene and Ca-amphibole, between metaluminous and peraluminous compositions. This demonstrates appropriate conditions in the deep crust or upper mantle where M- and I-type granitic rocks may crystallise these typically metaluminous minerals together with garnet, a typical peraluminous mineral. In this situation, the latter mineral is noteworthy for its grossular content which is significantly higher than found in the garnet characterising S-type granitic rocks.

8. Acknowledgements

This research has been supported by grants from the Australian Research Council and Macquarie University. Skilled assistance from Dr R. A. Day in the early stages and from Dr J. Adam in the later stages is gratefully acknowledged. J. Cleasby, A. Wilde and A. Williams assisted with manuscript preparation, while Drs R. H. Flood, F. Holtz, C. E. Miller and E. P. Vicenzi provided constructive comments on the content.

9. References

Barth, A. P. & Ehlig, P. L. 1988. Geochemistry and petrogenesis of the marginal zone of the Mount Lowe Intrusion, Central San Gabriel Mountains, California. CONTRIB MINERAL PETROL **100,** 192–204.

Bence, A. E. & Albee, A. L. 1986. Empirical correction factors for the electron microanalysis of silicates and oxides. J GEOL **76,** 382–403.

Birch, W. D. & Gleadow, A. J. W. 1974. The genesis of garnet and cordierite in acid volcanics: evidence from the Cerberean Cauldron, Central Victoria, Australia. CONTRIB MINERAL PETROL **45,** 1–13.

Chappell, B. W. & Stephens, W. E. 1988. Origin of infracrustal (I-type) granite magmas. TRANS R SOC EDINBURGH EARTH SCI **79,** 71–86.

Chappell, B. W. & White, A. J. R. 1974. Two contrasting granite types. PAC GEOL **8,** 173–4.

Clemens, J. D. & Wall, V. J. 1981. Crystallization and origin of some peraluminous (S-type) granitic magmas. CAN MINERAL **19,** 111–32.

Clemens, J. D. & Wall, V. J. 1984. Origin and evolution of a peraluminous silicic ignimbritic suite: the Violet Town Volcanics. CONTRIB MINERAL PETROL **88,** 354–71.

Conrad, W. K., Nicholls, I. A. & Wall, V. J. 1988. Water-saturated and undersaturated melting of metaluminous and peraluminous crustal compositions at 10 kb: evidence for the origin of silicic magmas in the Taupo Volcanic Zone, New Zealand, and other occurrences. J PETROL **29,** 765–803.

Day, R. A., Green, T. H. & Smith, I. E. M. 1992. The origin and significance of garnet phenocrysts and garnet-bearing xenoliths in Miocene calc-alkaline volcanics from Northland, New Zealand, J PETROL **33** (in press).

Drummond, M. S. & Defant, M. J. 1990. A model for trondhjemite–tonalite–dacite genesis and crustal growth via slab melting: Archaean to modern comparisons. J GEOPHYS RES **95,** 21, 503–21.

Ellis, D. J. 1986. Garnet-liquid Fe^{2+}–Mg equilibria and implications for the beginning of melting in the crust and subduction zones. AM J SCI **286,** 765–91.

Ellis, D. J. & Thompson, A. B. 1986. Subsolidus and partial melting reactions in the quartz-excess $CaO + MgO + Al_2O_3 + SiO_2 + H_2O$ system under water-excess and water-deficient conditions to 10 kb: some implications for the origin of peraluminous melts from mafic rocks. J PETROL **27,** 91–121.

Embey-Isztin, A., Nozke-Fazekas, G., Kurat, G. & Brandstatter, F. 1985. Genesis of garnets in some magmatic rocks from Hungary. TSCHERMAKS MINERAL PETROG MITT **34,** 49–66.

Fitton, J. B. 1972. The genetic significance of almandine-pyrope phenocrysts in the calc-alkaline Borrowdale Volcanic Group, Northern England. CONTRIB MINERAL PETROL **36,** 231–48.

Gilbert, J. S. & Rogers, N. W. 1989. The significance of garnet in the Permo-Carboniferous volcanic rocks of the Pyrenees. J GEOL SOC LONDON **146,** 477–90.

Green, J. F. N. 1915. The garnets and streaky rocks of the English Lake District. MINERAL MAG **17,** 207–17.

Green, T. H. 1972. Crystallization of calc-alkaline andesite under controlled high-pressure hydrous conditions. CONTRIB MINERAL PETROL **34,** 150–66.

Green, T. H. 1976. Experimental generation of cordierite- or garnet-bearing granitic liquids from a pelitic composition. GEOLOGY **4,** 85–8.

Green, T. H. 1977. Garnet in silicic liquids and its possible use as a P-T indicator. CONTRIB MINERAL PETROL **65,** 59–67.

Green, T. H. 1982. Anatexis of mafic crust and high pressure crystallization of andesite. *In* Thorpe, R. S. (ed.) *Andesites: Orogenic Andesites and Related Rocks,* 465–487. Chichester: Wiley.

Green, T. H. & Pearson, N. J. 1985. REE partitioning between clinopyroxene and silicate liquid at moderate to high pressure. CONTRIB MINERAL PETROL **91,** 24–36.

Green, T. H. & Ringwood, A. E. 1968a. Genesis of the calc-alkaline igneous rock suite. CONTRIB MINERAL PETROL **18,** 105–62.

Green, T. H. & Ringwood, A. E. 1968b. The origin of garnet phenocrysts in calc-alkaline rocks. CONTRIB MINERAL PETROL **18,** 163–74.

Green, T. H. & Ringwood, A. E. 1972. Crystallization of garnet-bearing rhyodacite under high-pressure hydrous conditions. J GEOL SOC AUST **19,** 203–12.

Green, T. H., Ringwood, A. E. & Major, A. 1966. Friction effects and pressure calibration in a piston-cylinder apparatus at high pressure and temperature. J GEOPHYS RES **71,** 3589–94.

Gromet, L. P. & Silver, L. T. 1987. REE variations across the Peninsular Ranges batholith; implications for batholithic petrogenesis and crustal growth in magmatic arcs. J PETROL **28,** 75–125.

Hawthorne, F. C. 1981. Crystal chemistry of the amphiboles. *In* Veblen, D. R. (ed.) *Amphiboles and other hydrous pyriboles—mineralogy.* REV MINERAL, MINERAL SOC AM **9A,** 1–102.

Kay, R. W. & Kay, S. M. 1990. Basaltic composition xenoliths and the formation, modification, and preservation of lower crust. *In* Salisbury, M. H. & Fountain, D. M. (eds) *Exposed cross-sections of the continental crust,* 401–420, Dordrecht: Kluwer.

Leeman, W. P. 1983. The influence of crustal structure on compositions of subduction-related magmas. *In* Aramaki, S. & Kushiro, I. (eds) *Arc volcanism.* J VOLCANOL GEOTHERM RES **18,** 561–88.

Miller, C. F. & Stoddard, E. F. 1981. The role of manganese in the paragenesis of magmatic garnet: an example from the Old Woman-Piute Range, California. J GEOL **89,** 233–46.

Miyashiro, A. 1955. Pyralspite garnets in volcanic rocks. J GEOL SOC JAPAN **61,** 463–70.

Munksgaard, N. C. 1985. A non-magmatic origin for compositionally zoned euthedral garnet in silicic Neogene volcanics from SE Spain. N JB MINERAL MH **85,** 73–82.

Oliver, R. L. 1956. The origin of garnets in the Borrowdale Volcanic Series and associated rocks, English Lake District. GEOL MAG **93,** 121–39.

Silver, L. T. & Chappell, B. W. 1988. The Peninsular Ranges Batholith: an insight into the evolution of the Cordilleran batholiths of southwestern North America. TRANS R SOC EDINBURGH EARTH SCI **79,** 105–21.

Stern, T., Smith, E. G. C. & Davey, F. J. 1987. Crustal and upper mantle structure of the northwestern North Island, New Zealand, from seismic reflection data. GEOPHYS J R ASTRON SOC **91,** 913–36.

Ujike, O. & Onuki, H. 1976. Phenocrystic hornblendes from Tertiary andesites and dacites, Kagawa Prefecture, Japan. J JAPAN ASSOC MINERAL PETROL ECON GEOL **71,** 389–99.

Vielzeuf, D. & Holloway, J. R. 1988. Experimental determination of the fluid-absent melting relations in a pelitic system. Consequences for crustal differentiation. CONTRIB MINERAL PETROL **98,** 257–76.

Whalen, J. B. 1985. Geochemistry of an island-arc plutonic suite: the Masilau-Yau Yau Intrusive complex, New Britain, P.N.G. J PETROL **26,** 603–32.

Wood, C. P. 1974. Petrogenesis of garnet-bearing rhyolites from Canterbury, New Zealand. NZ J GEOL GEOPHYS **17,** 759–87.

Wyllie, P. J. & Rutter, M. J. 1988. Melting sites at convergent plate boundaries with experimental data on vapor-absent tonalite gneiss. REND SOC ITAL MINERAL PETROL **43,** 1391–306.

Zen, E. & Hammarstrom, J. M. 1984. Mineralogy and a petrogenetic model for the tonalite pluton at Bushy Point, Revillagigedo Island, Ketchikan 1° × 2° Quadrangle, Southeastern Alaska. US GEOL SURV CIRCULAR **939,** 118–23.

T. H. GREEN, School of Earth Sciences, Macquarie University, N.S.W. 2109, Australia.

Transactions of the Royal Society of Edinburgh: Earth Sciences, **83**, 439–446, 1992

Crustal versus mantle sources of granitic magmas: a two-parameter model based on Nd isotopic studies

Donald J. DePaolo, Frank V. Perry and W. Scott Baldridge

ABSTRACT: Temporal and spatial variations in the Nd isotopic compositions of Tertiary caldera-forming rhyolite tuffs, and Cretaceous and Tertiary granites of the western U.S.A. are used as a basis for a model that accounts for the observed proportions of crustal versus mantle contributions to silicic magmas in terms of two parameters: the ambient crustal temperature and the rate of supply of basaltic magma from the mantle. The crustal contribution to silicic igneous rocks is measured in terms of the Neodymium Crustal Index (NCI). The relationships between crustal temperature, basalt supply and NCI are quantified using a model of a magma chamber undergoing continuous recharge, wall-rock assimilation and fractional crystallisation. From the model, a critical value of the ratio of basalt recharge-to-assimilation, $(r/a)_c$, is deduced, which increases with decreasing crustal temperature. The r/a value must exceed $(r/a)_c$ to allow the volume of differentiated magma to increase, a prerequisite for developing large volumes of silicic magma. Strongly peraluminous (or S-type) magmas (NCI = 0·8–1), form under conditions of high crustal temperature and low basalt supply. Metaluminous or I-type granites form over a wide range of conditions (NCI = 0·1–1), generally where basalt supply is substantial. In individual long-lived volcanic centres, the large-volume high-silica ignimbrites are associated with the highest r/a and lowest NCI values, indicating that these magmas are typically differentiates of mantle-derived basaltic parents.

KEY WORDS: Assimilation, fractional crystallisation, ignimbrite, magma chamber, rhyolite.

Although granitic rocks are among the most common rock types in the continental crust, there continues to be vigorous debate about the sources of granitic magma. Many consider the continental crust to be the primary or sole source. Those holding this view grant that heat for crustal fusion could be provided by basaltic magma entering or underplating the lower crust. A radically different view is that many granitic magmas, as well as many high silica rhyolite magmas, are formed by fractional crystallisation from basaltic precursor magmas which are in turn formed by melting in the mantle. This fractional crystallisation process is generally accompanied by assimilation of crustal rock material, so the resulting granitic magma is a mixture of so-called "mantle-derived" and "crustally-derived" components.

Isotopic ratios of the elements Nd, Sr, Pb, O and Ca are particularly suited to distinguishing mantle from crustal components in magmas; they have consequently been the basis for much of the recent discussion of the problem. Neodymium isotopes in particular are useful because in areas of Precambrian continental crust there is typically a substantial difference in Nd isotopic composition between the continental crustal rocks and the subjacent mantle magma sources.

Given that the source of granitic magma is in general some mixture of crustal and mantle components, a pure crustal origin representing an end-member case, then we seek to understand what parameters control the proportion of crustally-derived versus mantle-derived components in granitic magmas. An attempt at this was made by Farmer and DePaolo (1983, 1984) using data on Nd isotopic compositions of Mesozoic and Tertiary granites from the western U.S.A. Their model was that two parameters determined the amount of crustal material present in the end-product granitic liquids: the thickness of the continental crust at the start of granitic magmatism, and the flux of basalt entering the crust. Thin crust and high basalt fluxes are associated with granites that have the isotopic characteristics of mantle-derived magmas; thick crust and low basalt fluxes are associated mainly with granites that have the isotopic characteristics of the local lower or middle crust.

Recent work by several investigators on the Cainozoic large-volume rhyolite occurrences of the western U.S.A. partially confirm the Farmer and DePaolo model, but also suggest that an important modification is necessary. The Nd isotopic compositions of Cainozoic rhyolites indicate that the crustal contribution to the magmas varies considerably from very large (>80%) to small (<10%) in centres that are built in regions of similar-thickness continental crust (Perry *et al.* 1991, 1992). The new data indicate that the ambient crustal temperature, or geothermal gradient, at the onset of magmatism is more important than the crustal thickness in determining the amount of crustal material that is incorporated into magmas differentiating in the crust. The effect of basaltic magma flux, on the other hand, is confirmed by data on three separate volcanic systems.

The new observations confirm that granitic magmas span almost the complete range from purely crustally-derived to mostly mantle-derived, and suggest a simple parameterisation of the granite source problem in terms of crustal temperature and basalt magma flux. This paper reviews the evidence for the importance of these two controlling parameters, and describes a model that is an attempt at quantifying the effects of temperature and basalt flux.

1. The Neodymium Crustal Index (NCI)

The thesis of this paper is that Nd isotopic compositions of continental igneous rocks can be used to determine the fraction of a magma that is derived from the crust as opposed to that derived from the underlying mantle. This information is expressed as the Neodymium Crustal Index (NCI), the general definition of which is given in terms of an ε_{Nd} value of the crustal component (CC) and an ε_{Nd} value of the mantle component (MC):

$$NCI = \frac{\varepsilon_{Nd\,(Rock)} - \varepsilon_{Nd\,(MC)}}{\varepsilon_{Nd\,(CC)} - \varepsilon_{Nd\,(MC)}}.$$

The NCI is zero when the rock has the same ε_{Nd} value as the mantle component, and is unity when the rock has the same ε_{Nd} value as the crustal component. The NCI value does not directly give the volume or mass fraction of material derived from the crust, rather it is the fraction of the Nd derived from the crust. The volume and mass fractions can be estimated from the NCI value if the Nd concentrations in the two components are known or can be estimated.

The difficulty in the application of the NCI concept is in determining the appropriate values of the mantle and crustal ε_{Nd} values. The method proposed here is empirical. We use the example of the southwestern U.S.A., where there are abundant data on both crustal rocks and basalts from which to construct a model for the isotopic structure of the lithosphere. The regional crustal isotopic structure is described by Bennett and DePaolo (1988) and by DePaolo *et al.* (1990). There are three major provinces (Fig. 1), each of which is characterised by a particular average value of the Sm–Nd model age, which in turn corresponds to a particular average value of the ε_{Nd} value of the "crust". The ε_{Nd} values of crustal rocks of a given model age are variable; they can be described using the average model age in the province (T_{DM}) and the Sm/Nd ratio of the rocks:

$$\varepsilon_{Nd}(t) = \varepsilon_{Nd}(T_{DM}) + f_{Sm/Nd}Q_{Nd}(t - T_{DM}), \quad (1)$$

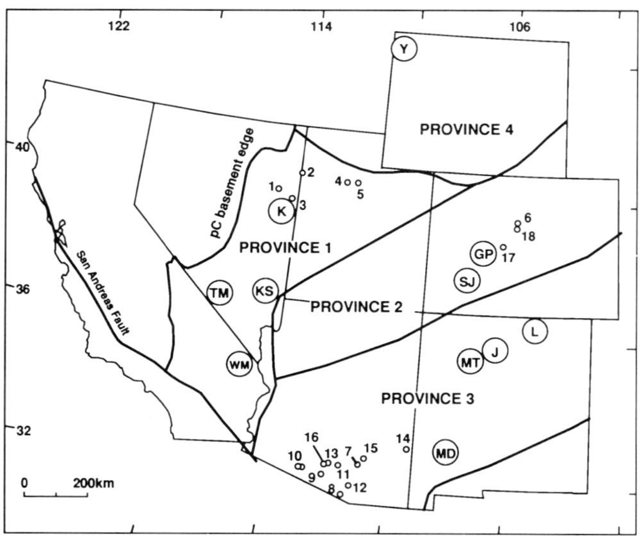

Figure 1 Locations of Tertiary rhyolite caldera complexes (large circles labelled with letters) and late Cretaceous and Tertiary granite samples (numbered small circles) listed in Table 1; also shown are approximate Precambrian basement age-province boundaries based on Sm-Nd model age (T_{DM}) values; model ages are 2·2 Ga, 2·0 Ga, 1·8 Ga and 2·7 Ga, for Provinces 1, 2, 3, and 4, respectively (Bennett & DePaolo 1988).

where $Q_{Nd} = 25{\cdot}13\ Gy^{-1}$ (DePaolo 1988) and:

$$f_{Sm/Nd} = \frac{Sm/Nd_{rock}}{Sm/Nd_{CHUR}} - 1. \quad (2)$$

Because typical intermediate-to-high-silica crustal rocks have a restricted range of $f_{Sm/Nd}$ values, for a given value of T_{DM}, an average value of $\varepsilon_{Nd}(t)$ can be specified within about ±2 units. The crustal $f_{Sm/Nd}$ values average about −0.4, with a range between −0·3 and −0·5 depending on the model age of the crust (DePaolo 1988). We have recently proposed the following, which is an explicit formulation for the ε_{Nd} value at time t (denoted $\varepsilon_{Nd}(t)$) in typical continental crust having a model age of T_{DM} (DePaolo *et al.* 1990):

$$\varepsilon_{Nd}(t) = \varepsilon_{Nd(CC)}(t) = \varepsilon_{Nd}(T_{DM}) + f^{*}_{Sm/Nd}Q_{Nd}(t - T_{DM}), \quad (3)$$

$$f^{*}_{Sm/Nd}(T_{DM}) = -0{\cdot}25 - 0{\cdot}08T_{DM}. \quad (4)$$

This formulation is used for predicting the $\varepsilon_{Nd}(t)$ value of the crustal component of granitic and rhyolitic rocks (Table 1), based on the average T_{DM} age of the local Precambrian basement (Fig. 1).

An obvious criticism of this approach would be that equations (3) and (4) are not likely to be meaningful because the Sm/Nd ratios in crustal rocks are too variable. There is ample evidence from xenoliths that the $f_{Sm/Nd}$ value of rocks in the crust is quite variable (e.g. Rudnick 1990; Wendlandt & DePaolo 1992). In particular, mafic lower crustal rocks typically have high Sm/Nd ratios ($f_{Sm/Nd}$ values near zero), and correspondingly high ε_{Nd} values, higher than the values calculated from equations (3) and (4). The existing data, however, indicate that the crustal material most likely to participate in granitic magma genesis is not mafic, but siliceous ($SiO_2 > 60\%$), and that the crustal component (CC) can be characterised by the low Sm/Nd ratios given by equation (3) and (4), which pertain to intermediate and silicic rocks in the crust. If the pertinent crustal rocks really were as variable in isotopic composition as the xenolith data suggests, then crustally-derived granitic magmas would also show a large variation in isotopic composition. In this regard, Nd and Ca isotopic data on strongly peraluminous granites, which are probably the most clearly crustal-anatectic in origin, are particularly significant. Strongly peraluminous granites in the southwestern U.S.A. always have ε_{Nd} values that are very close to the average value of their intermediate-to-silicic country rocks (Farmer & DePaolo 1983, 1984; Bennett & DePaolo 1988; DePaolo 1988). The conclusion from this observation is that granitic magmas, when melted from continental crust, are typically melted from intermediate-to-silicic source rocks and not from mafic source rocks. This conclusion is supported by K-Ca isotopic data (Marshall & DePaolo 1989). Granitic rocks from the western U.S.A. that are inferred to be melted from Precambrian basement rocks (based on Nd isotopes), regardless of whether they are peraluminous or metaluminous, also have elevated initial $^{40}Ca/^{42}Ca$ ratios that indicate that their sources had K/Ca weight ratios of about unity. In general, only silicic rock types, with SiO_2 contents of *c.* 65% or more, have such high K/Ca ratios. Mafic lower crustal rocks typically have K/Ca ratios of <0·1. Thus, for reasons not yet fully understood, mafic rocks in the crust do not readily participate in granitic magma genesis.

The NCI concept also relies on determination of the ε_{Nd} value of the mantle component. This must also be done empirically because the basalt-producing mantle reservoirs have a wide range of ε_{Nd} values. Our approach is to use

Table 1 Age, isotopic data and NCl values for Cretaceous and Tertiary granites and rhyolite

Locality	Age (Ma)	Crust age (Ga)	ε_{Nd} values (CC)[1]	(Rock)[2]	(MC)[3]	NCI[4]	$(r/a)_{ss}$[5]	REF[6]	MAP
Jemez (Upper Unit)	1·1	1·8	−12·6	−3·0	1·0	0·29	2·41	1	J
Jemez (Lower Unit)	1·5	1·8	−12·6	0·0	1·0	0·07	12·65	1	J
Yellowstone	2	2·7	−28·2	−9·0	−5·0	0·17	4·79	2	Y
Mt Taylor	3	1·8	−12·6	1·5	5·0	0·20	4·04	3	MT
Timber Mtn	13	2·2	−19·0	−11·0	−9·0	0·20	4·01	4	TM
Kane Springs	14	2·2	−19·0	−7·0	−6·0	0·08	12·00	5	KS
Woods Mtns	16	2·2	−19·0	−7·0	−6·0	0·08	11·98	6	WM
Latir	26	1·8	−12·4	−7·0	1·0	0·60	0·68	5	L
San Juan	27	2·0	−15·5	−7·0	1·0	0·48	1·07	5	SJ
Mogollon-Datil	34	1·8	−12·3	−9·0	1·0	0·75	0·33	5	MD
Grizzly Peak	34	2·0	−15·5	−12·0	1·0	0·79	0·27	5	GP
Kalamazoo	35	2·2	−18·8	−17·0	−6·0	0·86	0·16	5	K
Egan Range	32	2·2	−18·8	−16·9	−6·0	0·85	0.18	7	1
Gold Hill	38	2·2	−18·7	−18·7	−6·0	1·00	0·01	7	2
Kern Mtns	58	2·2	−18·5	−19·1	−6·0	1·05	0·01	7	3
Oquirrh Mtns	40	2·2	−18·7	−18·3	−6·0	0·97	0·03	7	4
Wasatch Mtns	28	2·2	−18·9	−18·3	−6·0	0·96	0·05	7	5
Wasatch Mtns	32	2·2	−18·8	−14·2	−6·0	0·64	0·56	7	5
Henderson	24	2·0	−15·6	−9·9	1·0	0·66	0·52	8	6
Henderson	24	2·0	−15·6	−9·6	1·0	0·64	0·56	8	6
Henderson	24	2·0	−15·6	−9·5	1·0	0·63	0·58	8	6
Santa Catalina Mtns	47	1·8	−12·2	−10·0	1·0	0·83	0·20	8	7
Baboquivari Mtns	58	1·8	−12·1	−12·1	1·0	1·00	0·01	8	8
Baboquivari Mtns	58	1·8	−12·1	−9·8	1·0	0·83	0·21	8	8
Sierra Blanca	58	1·8	−12·1	−12·4	1·0	1·02	0·01	8	9
Santa Catalina Mtns	26	1·8	−12·4	−6·7	1·0	0·57	0·74	8	7
Bingham	38	2·2	−18·7	−18·4	−6·0	0·97	0·03	8	5
Ajo	63	1·8	−12·0	−6·9	1·0	0·61	0·65	8	10
Ajo	63	1·8	−12·0	−8·3	1·0	0·71	0·40	8	10
Silverbell	67	1·8	−12·0	−8·1	1·0	0·70	0·43	8	11
Sierrita	61	1·8	−12·1	−7·7	1·0	0·67	0·50	8	12
Lakeshore	67	1·8	−12·0	−8·6	1·0	0·74	0·35	8	13
Morenci	55	1·8	−12·1	−6·0	1·0	0·53	0·87	8	14
San Manuel	70	1·8	−12·0	−5·6	1·0	0·51	0·96	8	15
Santa Catalina Mtns	70	1·8	−12·0	−7·6	1·0	0·66	0·51	8	7
Cimar Mtns	68	1·8	−12·0	−6·5	1·0	0·58	0·73	8	16
Sawatch Range	50	2·0	−15·3	−5·4	1·0	0·39	1·55	8	17
Front Range	39	2·0	−15·4	−8·4	1·0	0·57	0·75	8	18

Notes: 1. CC: crustal component; 2. Rock: measured value in the rock; 3. MC: mantle component; 4. Neodymium Crustal Index, defined in text; 5. Steady state recharge-to-assimilation ratio; 6. References: (1) DePaolo *et al* 1991 (2) Hildreth *et al.* 1991 (3) Perry *et al.* 1990 (4) Farmer *et al.* 1991 (5) Johnson 1991 (6) Musselwhite *et al.* 1989 (7) Farmer & DePaolo 1983 (8) Farmer & DePaolo 1984.

basalts erupted in spatial and temporal proximity to the granitic and rhyolitic magmas to determine the value of $\varepsilon_{Nd(MC)}(t)$. With the western U.S.A. province considered here, the $\varepsilon_{Nd(MC)}(t)$ value varies between −10 and +6 (Perry *et al.* 1987; Farmer *et al.* 1989). For pre-Oligocene granites we do not have sufficient data on contemporaneous basaltic rocks from which to estimate the ε_{Nd} value of the mantle component; for these rocks we use the values for the younger Tertiary basalts from the same area.

2. Sources of Tertiary granites and rhyolites of the western U.S.A.: crustal temperature and basalt supply as controlling parameters

NCI values have been calculated for Late Cretaceous-to-Pleistocene granitic rocks and large-volume rhyolites of the western U.S.A. (Table 1, Fig. 2). The NCI values vary between wide limits, *c.* 0·1–1·0, the conclusion being that granitic and rhyolitic magmas form both as crustal anatectic products (NCI = 0·8–1·0) and as differentiates of mantle-derived (presumably basaltic) magma (NCI = 0·1–0·2), as well as with subequal amounts of crustally-derived and mantle-derived components. In addition, the data show a

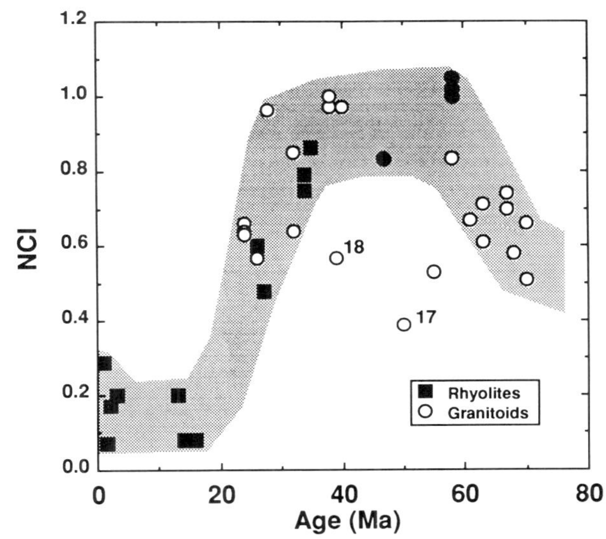

Figure 2 Variation of the NCI value with eruption age or emplacement age for the caldera-forming-rhyolites and granites of the western U.S.A., showing the inferred trend with shading; solid circles are peraluminous granites, which are restricted in time to the early part of the period of highest regional NCI values (about 40–60 Ga); two of the samples that do not fit the trend are from central Colorado (samples 17 and 18).

distinct trend when considered as a function of age. For the middle-to-late Tertiary rhyolites, there is a marked, and apparently gradual, decrease in NCI values between *c.* 35 and 17 Ma. In rocks younger than 17 Ma, all of the NCI values are low. When the data set is extended to include the Late Cretaceous-to-middle Tertiary granitic rocks, there is a suggestion of a peak in NCI values at 40–60 Ma, with somewhat lower values before, and a precipitous drop at *c.* 30–20 Ma.

The NCI data from the western U.S.A. indicate that the crustal contribution shows systematic trends with time over large areas; this observation suggests that the control on the crustal contribution must be a parameter that is relatively uniform regionally. We suggest that the primary controlling parameter is crustal temperature, or geothermal gradient. We interpret the data as indicating that crustal temperatures peaked at about 60–50 Ma, and then decreased rather markedly between about 35 and 15 Ma. Below we attempt to quantify the temperature changes. However, with regard to magma genesis, other data suggest that there is another important parameter controlling the NCI, and that is the flux of mantle-derived magma entering the crust from the mantle.

The data in Table 1 for the Jemez system in northern New Mexico show that a single system can exhibit a range of NCI values. A more complete representation of the data for the Jemez system is given in Figure 3. Shown as panels (a), (b), and (d) are ε_{Nd} value, NCI and approximate volume rate of eruption. When the system produced the largest volume of rhyolitic magma (*c.* 1,000 km^3 in less than 1 million years, with major rhyolitic ignimbrite eruptions at 1·4 and 1·1 Ma), the erupted rhyolitic tuffs had the lowest values of NCI of any of the lavas (including more mafic adesites and dacites) erupted over the entire 10+ million year history of the centre. Furthermore, lavas erupted both before and after the low-NCI caldera-forming eruptions, at 1·1 and 1·4 Ma, have higher NCI values. The Jemez data indicate that (1) the large eruption rate is caused by a large rate of infusion of mantle-derived (low-NCI) basalt to the crust, and (2) the same low $-\varepsilon_{Nd}$ crustal component was available to the magma system over a period of *c.* 10 million years.

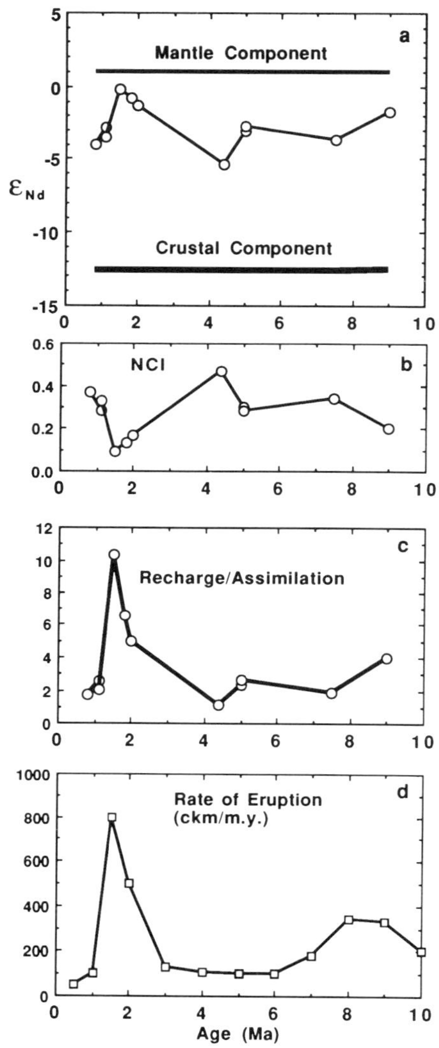

Figure 3 Variation of isotopic parameters and eruption rate with age for lavas and tuffs of the Jemez volcanic field, northern New Mexico (labelled "J" on Fig. 1):

(a) Inferred ε_{Nd} values for the mantle and crustal components of the magmas (Table 1), and the values in the eruptive products.

(b) Calculated NCI values.

(c) Steady-state recharge-to-assimilation ratio that corresponds to the ε_{Nd} values of (a). (See eqn. 9 in text.)

(d) Estimated volume rate of eruption, in km^3/m.y., for the Jemez system (adapted from Gardner & Goff 1984); peak eruption rate corresponds to the two large rhyolitic ashflow units at 1·4 and 1·1 Ma; the eruption rate was smoothed with the constraint that the area under the curve is approximately the total volume of the two ashflow eruptions.

3. The flux-temperature model

Our interpretation of the data summarised above is that the crustal contribution to silicic magmas is controlled by two parameters: the ambient crustal temperature and the rate of supply of basalt to the system. When the crustal temperature is high, it favours melting of the crust and the formation of granitic magma with a large crustal component. At any given crustal temperature, an increase in the rate of supply of basalt magma dilutes the crustal contribution and leads to granitic magmas with lower NCI values. Purely crustally-derived magmas must be formed in situations where the crust is very hot and the supply of mantle magma is low. Granitic magmas that are pure differentiates of basaltic magma are rare; we have no examples among the three dozen or so systems studied in the western U.S.A. However, many systems are strongly dominated by fractional crystallisation of basalt, albeit almost always with an admixture of crustal material at the level of at least 10%.

We have modelled the process of granitic magma formation in terms of a simple magma chamber undergoing recharge, assimilation and fractional crystallisation (Fig. 4). The complete derivation is not reproduced here (Perry *et al.* 1992), but the following expressions result, which give the rate of change of magma temperature with time:

$$M\frac{dT}{dt} = \dot{M}_c\left\{r(T_r - T) - a\left[\frac{L_a}{c} + (T - T_a)\right] + \frac{L_c}{c} - \mathbf{J}(T - T_c)\right\}, \tag{5}$$

and the rate of change of magma temperature with amount crystallised:

$$M\frac{dT}{dM_c} = \left\{r(T_r - T) - a\left[\frac{L_a}{c} + (T - T_a)\right] + \frac{L_c}{c} - \mathbf{J}(T - T_c)\right\}, \tag{6}$$

where: $a = \dot{M}_a/\dot{M}_c$ = assimilation rate/crystallisation rate,
$r = \dot{M}_r/\dot{M}_c$ = basalt recharge rate/crystallisation rate,
T = temperature of the magma in the chamber,
T_r = temperature of the recharging basalt magma,

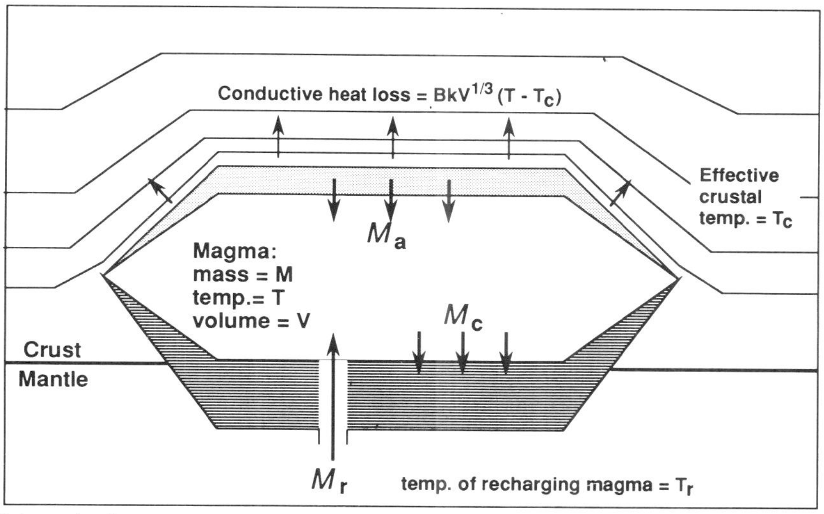

Figure 4 Schematic diagram of the parameters used in the recharge-assimilation-fractional crystallisation magma chamber model; light lines surrounding the chamber are schematic isotherms; rate of wall-rock assimilation is M_a, the rate of crystallisation is M_c and the rate of recharge with mafic magma is M_r.

T_c = effective ambient crustal temperature,
T_a = initial temperature of assimilated rock material ($\geq T_c$),
c = specific heat of solid and liquid crustal material (assumed equal to the specific heat of magma in the chamber),
L_a = specific heat of fusion of the crustal rock,
L_c = specific heat of crystallisation of the magma in the chamber

and $\mathbf{J} = (BkV^{1/3})/(cM_c)$, where:
V = volume of magma in the chamber ($=M/\rho$),
k = thermal conductivity of wall rock,
B = scaling parameter.

The dimensionless parameter **J** relates the rate of heat loss through the walls of the system to the rate of crystallisation. When **J** is large, there is a relatively large amount of heat loss per unit crystallisation. This corresponds in a closed system to a liquidus surface with a large slope with respect to temperature. When **J** is small, there is a relatively large amount of crystallisation per unit heat loss through the walls. For an open system of the type considered here, it is difficult to determine the value of **J**. Assimilation and recharge change the composition of the magma, so there is no simple relation between temperature and amount crystallised. We use values of **J** of 0·4 and 1·0, which span most of the range of values for closed systems.

To maintain or grow the chamber, we require that:

$$\frac{dT}{d(M_c/M)} > \zeta, \text{ and } (a + r - 1) > 0.$$

The strict steady state (constant magma mass and temperature) would correspond to $\zeta = 0$ and $a + r = 1$. We use a more general and less stringent condition, where the mass of the magma is constant but the temperature is allowed to decrease ($\zeta < 0$) at a small but non-zero rate. These requirements lead to:

$$(r/a)_c = \frac{(T - T_a) - \frac{L_c - L_a}{c} + \mathbf{J}(T - T_c) + \zeta}{(T_r - T) - \mathbf{J}(T - T_c) + \frac{L_c}{c} - \zeta}. \quad (7)$$

The parameter $(r/a)_c$, is the critical or minimum ratio of the recharge (with basalt) to the assimilation rate, needed to maintain or increase magma mass. This ratio is dependent mainly on the temperature difference between the magma and the surrounding wall rock ($T - T_c$), since the other parameters take on similar values from system to system. If the r/a value falls below the critical value, the system decreases in volume and eventually completely solidifies; if the r/a value is larger than the critical value, the system increases in volume. In order to develop a large magma body, r/a must exceed the critical value for an extended time period. Thus, for a given value of T_c, there is a minimum basalt influx required to grow a large magma body, and this minimum rate of supply increases as T_c decreases.

Figure 5 shows curves of $(r/a)_c$ for the parameters given in

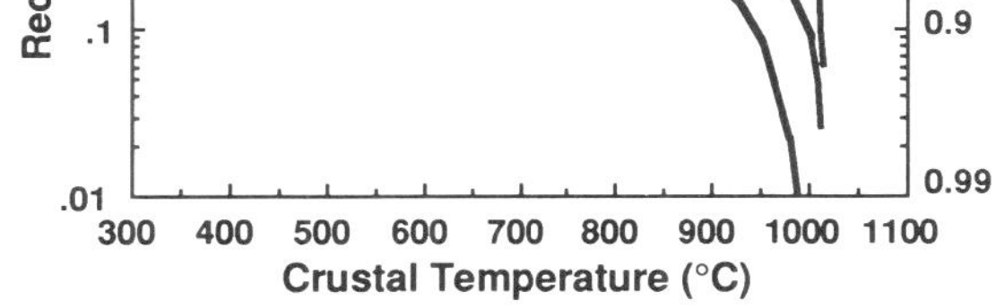

Figure 5 Calculated values of the critical recharge-to-assimilation ratio, $(r/a)_c$, versus effective crustal temperature, T_c, for the following parameters: $T_r = 1200°C$, $T_m = 1100°C$, $T_a = T_c$ or $T_a = 0{\cdot}5\ (T_m - T_c)$, $L_c = 100$ cal/g, $L_a = 70$ cal/g, $c = 0{\cdot}32$ cal/g°C, $\zeta = -10°C$; values chosen for L_c, L_a, c, and T_r, correspond to those used by Huppert and Sparks (1988); values of r/a and crustal temperature that fall above the lines correspond to conditions under which the magma volume increases with time; values that fall below the lines correspond to conditions under which the magma volume decreases with time; location of the lines is sensitive to the choice of the parameter **J**, defined in the text, which relates the rate of heat loss from the system to the rate of crystallisation; data from the layered intrusions Skaergard (Stewart & DePaolo 1990), Kiglapait (DePaolo 1985), and Muskox (Stewart & DePaolo 1992) suggest that values of **J** near 0·4–0·5 are more likely to be applicable than values approaching unity.

the caption. We consider one case where $T_a = T_c$ and another where $T_a > T_c$. The latter case is presumably applicable because material incorporated into the magma must come from near the wall of the magma body and therefore has been preheated to a temperature above the initial wall-rock temperature.

For comparison with the model results, a value for r/a for the rhyolites and granites can be calculated from the isotopic data using the equation for an RAFC magma chamber (DePaolo 1985):

$$M\frac{d\varepsilon_m}{dt} = M_a\frac{c_a}{c_m}(\varepsilon_a - \varepsilon_m) + M_r\frac{c_r}{c_m}(\varepsilon_r - \varepsilon_m), \quad (8)$$

where the subscripts m, a and r represent the magma (rock), assimilated material (crustal component) and recharging magma (mantle component), C is the concentration of Nd and ε is the ε_{Nd} value. For a system at steady state, $d\varepsilon_m/dt = 0$, then:

$$(r/a)_{ss} = \frac{C_a(\varepsilon_a - \varepsilon_m)}{C_r(\varepsilon_m - \varepsilon_t)} = \frac{C_a}{C_r}\left(\frac{1 - NCI}{NCI}\right). \quad (9)$$

In general, because r/a changes with time, most systems are not precisely in steady state. However, when the magma volume is increasing, it can be shown that the ε_{Nd} value of the magma rapidly becomes close to the steady state value, and therefore that $(r/a)_{ss}$ gives useful information on the state of the magma system (Fig. 3(c)). For simplicity, the ratio C_a/C_r is assumed to be unity for the $(r/a)_{ss}$ values displayed in Table 1 and Figures 3 and 6; it could normally take on values from slightly less than one, to perhaps as high as two.

The early Oligocene centres are characterised by r/a values less than 0·3. Our model predicts that crustal temperatures in the neighbourhood of 900–1,000°C are necessary to have large magma systems supported by such small recharge rates (Fig. 6). The late Oligocene centres have r/a values of between 0·5 and 1; these values would require crustal temperatures of 700–900°C. The Miocene and younger centres, were r/a values are typically >2·5, require crustal temperatures of about 500–700°C.

The Nd isotopic data of Figure 2 therefore can be interpreted in terms of a decreasing average wall-rock temperature for the magma systems during the period from 35–15 Ma. Because crustal temperature is approximately proportional to depth below the surface, decreasing temperature could be interpreted to be caused either by decreasing magma chamber depth or decreasing geothermal gradient, or by a combination of the two. If we assume that the depth of the magma bodies is constant and centred at roughly 30 km depth, the wall-rock temperatures indicated above correspond to a change in regional geothermal gradient from >30°C/km in the early Oligocene (35 Ma) to about 20°C/km by 17 Ma. This in turn corresponds to an increase in lithospheric thickness of approximately 50% over this 20 million year time interval. However, part of the difference could be explained by changes in the depth to the magma chambers. Magma chambers would be expected to become established at a depth where the intruding magma is neutrally buoyant. Because alkaline magmas tend to be inherently less dense than tholeiitic magmas (Kushiro 1982) they may pond at shallower crustal depths in contact with cooler wall rocks. Thus, to the extent that there might be systematic compositional changes from (more dense) tholeiitic to (less dense) alkaline in the basaltic magmas during the middle and late Tertiary, this effect could contribute to the trend of Figure 2, and decrease the amount of inferred lithospheric cooling.

The second part of the model is the dependence on basalt recharge rate. The graph of Figure 5 shows only the minimum value of r/a needed to maintain or grow the system. However, r/a can exceed this value, which will lead to a growing magma mass. It is this effect that we believe accounts for the data of the Jemez caldera-forming rhyolites. A similar trend is also observed at Yellowstone (Hildreth *et al.* 1991), and for the Long Valley system in eastern California (Halliday *et al.* 1989). The overall model, and the inferred conditions of formation for generic magma types, are summarised in Figure 6.

4. The crustal component: nature and entrainment mechanism

The data summarised here indicate that a substantial amount of crustal material can be incorporated into magma whose ultimate source is the mantle. However, the mechanisms by which this incorporation takes place are not well known. There are two constraints, however, that can be derived from the isotopic data. One is that, for most systems studied so far, the NCI value is almost constant over a silica range from *c.* 55–75% SiO_2 (Musselwhite *et al.* 1989; Johnson 1991). Therefore, if the magma bodies are layered, with the most mafic magma at the bottom, the incorporation of crustal wall-rock material must be directly into the base of the system. The generation of higher silica magma apparently occurs by fractional crystallisation from more

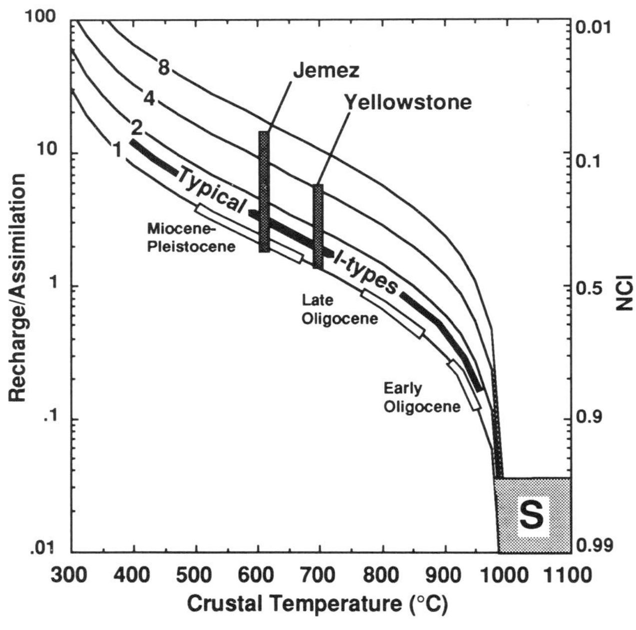

Figure 6 Summary diagram of the predictions of the RAFC model for the sources of typical metaluminous (or I-type) and strongly peraluminous (or S-type) granites and rhyolites: line labelled "1" corresponds to $(r/a)_c$ using the parameters of Figure 5, **J** = 0·5 and $T_a = T_c$; other lines are for r/a values equal to 2, 4 and 8 times $(r/a)_c$; field labelled "S" in the lower right corner of the diagram corresponds to the conditions inferred for the formation of S-type or strongly peraluminous granites; minimum temperature shown for S-type formation is controlled by the parameters chosen; a somewhat lower temperature of formation would result from a change in parameters, but the overall geometry of the diagram would remain the same; typical I-type or metaluminous granites and rhyolite are inferred to be generated under conditions near or above the $(r/a)_c$ line; data from the Jemez and Yellowstone systems indicate that r/a values can exceed the critical value by several times; scale of the right gives the NCI values that correspond to $C_a/C_r = 1$ (see equation (9)).

mafic magma, with very little further assimilation occurring along the way. The other observation that needs to be satisfied is that the Nd isotopic composition of the crustal component corresponds to the intermediate-to-silicic part of the crust. This is clearly indicated by the fact that ε_{Nd} values extend down to the low values characteristic of the more siliceous rocks of the crust (Table 1). The important aspect of this observation is that in many parts of the continental crust, the lower part of the crust is impoverished in high-silica rocks. Consequently, the data suggest that siliceous rocks from the upper or middle part of the crust are typically incorporated into the base of the magma system, which is in the lower crust.

Figure 7 shows a possible mechanism for incorporation of silicic crustal material into the base of the system. The magma system is envisioned as broadly diapiric in shape, with compositional layering from mafic at the base to highly silicic at the top. The diapir is maintained by the input of hot basaltic magma to the base, and if the r/a value is large enough, the magma body increases in volume with time. Because the upper part of the magma body has a low density, the diapir moves slowly upward in the crust, displacing material from above it. This material moves around the periphery of the body and is heated as it moves downward. When this heated wall-rock material moves under the magma system, the temperature is probably high enough that it is partially molten. Because small melt fractions will not escape easily if they are silica rich (McKenzie 1985), small-percentage melts from mafic rocks will be retained in the wall rock and not contribute to contaminating the main magma body. More siliceous domains of the wall rock, however, achieve high melt fractions that move upward into the magma body. In this manner, the contaminants to the magma body are restricted to be derived from relatively large silicic domains in the crust. Also, the material incorporated into the lower part of the magma system can be derived from crustal levels that were originally much closer to the surface.

Another conceivable mechanism for incorporation of relatively shallow crustal material into the deeper parts of the magma system is by stoping of the roof material and the sinking of the stoped blocks to the bottom of the chamber before there is substnatial reaction with the magma. The stoping mechanism is not preferred, however, because more contamination would be expected in the upper parts of the chamber than is observed. Also, large-scale stoping would probably tend to disrupt the compositional layering in the chamber.

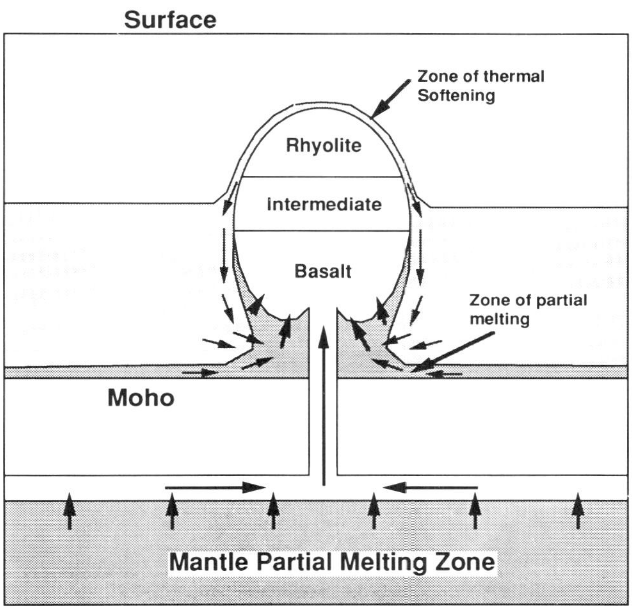

Figure 7 Schematic representation of the inferred geometry of a layered silicic magma system of the type that produces large-volume caldera-forming rhyolitic tuff eruptions. Within the chamber, stepwise fractional crystallisation causes the generation of layers of magma with upward-increasing silica content and decreasing density; these layers grow in volume with time if the r/a value is sufficiently high. The observation that relatively low-silica magmas (*c.* 55% SiO_2) have approximately the same NCI value as associated high-silica magmas (Musselwhite *et al.* 1989; Johnson 1991) leads to the inference that entrainment of crustal material occurs near the base of the system, possibly by upward movement of liquids derived from wall-rock material that is flowing downward around the slowly rising and inflating magma body (stoped blocks from the roof may also fall to the base of the system before being assimilated). Because only the more silicic parts of the wall rock become sufficiently molten to escape into the magma chamber, this model could account for the observation that the crustal component of continental rhyolites and granites generally has the Nd isotopic composition of the intermediate-to-silicic components of the Precambrian basement wall rock.

5. Summary

It is proposed that the amount of crustally-derived versus mantle-derived materials in granitic magmas is controlled mainly by two parameters: the ambient temperature of the crustal wall rocks at the level of the lower parts of the developing magma chamber, and the rate of influx of mantle-derived basaltic magma into the system. This "flux-temperature model" is deduced from spatial and temporal patterns of Nd isotopic values in Tertiary and late Cretaceous granites and rhyolitic ignimbrites of the southwestern U.S.A. The amount of crustally-derived material present in a magma system is parametrised as the Neodymium Crustal Index (NCI) which is calculated using the empirically-determined Nd isotopic compositions of local mantle and crustal magma sources.

The relationships between crustal temperature, basalt flux and NCI values are quantified using a relatively simple model of a magma chamber undergoing recharge, assimilation and fractional crystallisation. From this model, a critical value of the ratio of recharge-to-assimilation, $(r/a)_c$, can be derived, which is dependent mainly on the ambient crustal temperature outside the magma system. In long-lived systems, the r/a value deduced from Nd isotopic data varies between $(r/a)_c$ and much higher values. In a given system, large-volume high silica rhyolite ignimbrites are typically associated with the highest r/a values, indicating that these magmas are the result of increased basaltic influx and form primarily by fractional crystallisation from basaltic parents.

In terms of this model, S-type granites (or strongly peraluminous granites) form in a regime where crustal temperatures are high but mantle magma fluxes are low. This presumably corresponds to a regime where crustal heating occurs without crustal thinning, perhaps in a backarc setting where lithospheric thinning occurs as a result of corner flow above the downgoing slab. Metaluminous granites and rhyolites occur over a wide range of conditions, but generally are associated with larger basalt fluxes and lower crustal temperatures.

6. Acknowledgements

This work was supported by the Director, Office of Energy Research, Office of Basic Energy Sciences, Engineering and

Geosciences Division of the U.S. Department of Energy under contract No. DE-AC03-76SF00098. This synthesis draws on research supported also by the U.S. National Science Foundation, and the Institute of Geophysics and Planetary Physics of the Los Alamos National Laboratory.

7. References

Bennett, V. C. & DePaolo, D. J. 1988. Proterozoic crustal history of the western United States as determined by Neodymium isotopic mapping. GEOL SOC AM BULL **99,** 674–85.

DePaolo, D. J. 1985. Isotopic studies of processes in mafic magma chambers. J PETROL **26,** 925–51.

DePaolo, D. J. 1988. *Neodymium isotope geochemistry*. Heidelberg: Springer.

DePaolo, D. J., Linn, A. M. & Schubert, G. 1990. The continental crustal age distribution: Methods of determining mantle separation ages from Sm-Nd isotopic data and application to the western United States. J GEOPHYS RES **96,** 2,071–88.

DePaolo, D. J., Perry, F. V. & Baldridge, W. S. 1991. Neodymium isotopic monitor of basalt influx rates and eruption potential in continental magmatic systems. GEOL SOC AM ABSTR **23,** A396.

Farmer, G. L. & DePaolo, D. J. 1983. Origin of Mesozoic and Tertiary granite in the western U.S. and implications for pre-Mesozoic crustal structure. I. Nd and Sr isotopic studies in the geocline of the northern Great Basin. J GEOPHYS RES **88,** 3,379–401.

Farmer, G. L. & DePaolo, D. J. 1984. Origin of Mesozoic and Tertiary granite in the western U.S. and implications for pre-Mesozoic crustal structure. II. Nd and Sr isotopic studies of Cu- and Mo-mineralized granite in the Precambrian craton. J GEOPHYS RES **89,** 10,141–60.

Farmer, G. L., Perry, F. V., Semken, S., Crowe, B., Curtis, D. B. & DePaolo, D. J. 1989. Isotopic evidence on the structure and origin of subcontinental lithospheric mantle in southern Nevada. J GEOPHYS RES **94,** 7,885–98.

Farmer, G. L., Broxton, D. E., Warren, R. G. & Pickthorn, W. 1991. Nd, Sr, and O isotopic variations in metaluminous ash-flow tuffs and related volcanic rocks at the Timber Mountain/Oasis Valley caldera complex, SW Nevada: Implications for the origin and evolution of large volume silicic magma bodies. CONTRIB MINERAL PETROL **109,** 53–68.

Gardner, J. N. & Goff, F. 1984. Potassium-Argon dates from the Jemez volcanic field: Implications for volcanic activity in the north-central Rio Grande Rift. *In* Baldridge, W. S., Dickerson, P. W., Riecker, R. E. & Zidek, J. (eds) *RioGrande Rift: Northern New Mexico, New Mexico Geological Society, Socorro, 35th Annual Field Conference, Oct.* 11–13, 1984, Socorro: New Mexico Geological Society.

Hildreth, W., Halliday, A. N. & Christiansen, R. L. 1991. Isotopic and chemical evidence concerning the genesis and contamination of basaltic and rhyolitic magma beneath the Yellowstone Plateau volcanic field. J PETROL **32,** 63–138.

Halliday, A. N., Mahood, G. A., Holden, P., Metz, J. M., Dempster, T. J. & Davidson, J. P. 1989. Evidence for long residence times of rhyolitic magma in the Long Valley magmatic system: the isotopic record in precaldera lavas of Glass Mountain EARTH PLANET SCI LETT **94,** 274–90.

Huppert, H. E. & Sparks, R. S. J. 1988. The generation of granitic magmas by intrusion of basalt into continental crust. J PETROL **29,** 599–624.

Johnson, C. M. 1991. Large-scale crust formation and lithosphere modification beneath Middle to Late Cenozoic calderas and volcanic fields, western North America. J GEOPHYS RES **96,** 13,485–507.

Kushiro, I. 1982. Density of tholeiite and alkali basalt magmas at high pressures. YEARB CARNEGIE INST WASHINGTON **81,** 305–9.

Marshall, B. D. & DePaolo, D. J. 1989. Calcium isotopes in igneous rocks and the origin of granite. GEOCHIM COSMOCHIM ACTA **53,** 917–22.

McKenzie, D. 1985. The extraction of magma from the crust and mantle. EARTH PLANET SCI LETT **74,** 81–91.

Musselwhite, D. S., DePaolo, D. J. & McCurry, M. 1989. The evolution of a silicic magma system: Isotopic and chemical evidence from the Woods Mountains volcanic center, eastern California. CONTRIB MINERAL PETROL **101,** 19–29.

Perry, F. V., Baldridge, W. S. & DePaolo, D. J. 1987. Role of asthenosphere and lithosphere in the generation of late Cenozoic basaltic rocks from the Rio Grande rift and adjacent regions of the southwestern United States. J GEOPHYS RES **92,** 9,193–213.

Perry, F. V., Baldridge, W. S., DePaolo, D. J. & Shafiqullah, M. 1990. Evolution of a magmatic system during continental extension: The Mount Taylor volcanic field, New Mexico. J GEOPHYS RES **95,** 19,327–48.

Perry, F. V., DePaolo, D. J. & Baldridge, W. S. 1991. Isotopic evidence for a decline in crustal contributions to caldera-forming rhyolites of the western United States during the middle to late Cenozoic. GEOL SOC AM ABSTR **23,** A441.

Perry, F. V., DePaolo, D. J. & Baldridge, W. S. 1992. Isotopic evidence for a decline in crustal contributions to caldera-forming rhyolites of the western United States during the middle to late Cenozoic. GEOL SOC AM BULL (submitted).

Rudnick, R. L. 1990. Nd and Sr isotopic compositions of lower-crustal xenoliths from North Queensland, Australia: Implications for Nd model ages and crustal growth processes. CHEM GEOL **83,** 195–208.

Stewart, B. W. & DePaolo, D. J. 1990. Isotopic studies of processes in mafic magma chambers: II. The Skaergard intrusion, East Greenland. CONTRIB MINERAL PETROL **104,** 125–41.

Stewart, B. W. & DePaolo, D J. 1992. Diffusive isotopic contamination of mafic magma by coexisting silicic liquid in the Muskox intrusion. SCIENCE **255,** 708–11.

Wendlandt, E. & DePaolo, D. J. 1992. Nd and Sr isotope chronostratigraphy of Colorado Plateau lithosphere: Implications for magmatic and tectonic underplating of the continental crust. EARTH PLANET SCI LETT (submitted).

DONALD J. DePAOLO, Center for Isotope Geochemistry, Department of Geology and Geophysics, University of California, Berkeley, California 94720, U.S.A.; and Earth Sciences Division, Lawrence Berkeley Laboratory, Berkeley, California 94720, U.S.A.

FRANK V. PERRY, Department of Geology, University of New Mexico, Albuquerque, NM 87131, U.S.A.

W. SCOTT BALDRIDGE, Earth and Environmental Sciences Division, Los Alamos National Laboratory, Los Alamos, NM 87545, U.S.A.

Transactions of the Royal Society of Edinburgh: Earth Sciences, **83**, 447–458, 1992

Some observations on the use of zircon U–Pb geochronology in the study of granitic rocks

Ian S. Williams

ABSTRACT: *In situ*, microscale, U–Pb isotopic analyses of zircon using the SHRIMP ion microprobe demonstrate both the potential and the limitations of zircon U–Pb geochronology. Most zircons, whether from igneous or metamorphic rocks, need to be considered as mixed isotopic systems. In simple, young igneous rocks the mixing is principally between isotopically disturbed and undisturbed zircon. In polymetamorphic rocks, several generations of zircon growth can coexist, each with a different pattern of discordance. A similar situation exists for igneous rocks rich in inherited zircon, as these contain both melt-precipitated zircon and inherited components of several different ages. Microscale analysis by ion probe makes it possible to sample the record of provenance, age and metamorphic history commonly preserved within a single zircon population. It also indicates how the interpretation of conventionally-measured bulk zircon isotopic compositions might be improved.

KEY WORDS: Dalgety, discordance, inheritance, ion probe, Kameruka, monazite, Mount Sones, radiogenic Pb, Watergums, Watersmeet.

Of the several isotopic systems that have been explored as a means of determining the absolute ages of granitic rocks, the U–Pb system in zircon currently enjoys a reputation as one of the most broadly applicable and reliable. Zircon is ubiquitous in granites, and is more resistant to isotopic disturbance by secondary geological processes such as metamorphism, alteration and weathering than most other minerals. Also, the U–Pb system itself has the unique feature that, because it consists of two U isotopes that decay at different rates to two Pb isotopes, any isotopic disturbance that does occur can be detected by a simple test for internal consistency between the radiogenic Pb isotopic composition and the Pb/U ratios.

As important as these factors are, it is unlikely that zircon dating would have achieved its present popularity had it not been for the major advances that have been made in U–Pb microchemistry and mass spectrometry, particularly over the last 20 years (e.g. Krogh 1973; Lancelot *et al.* 1975; Manhes *et al.* 1978). These have resulted in its being possible to make increasingly precise analyses of ever smaller zircon samples.

Much slower to develop, however, has been an understanding of the properties of the zircon U–Pb system as such. Despite the enormous amount of zircon U–Pb data now available, some very basic questions about the nature and causes of isotopic disturbance (discordance) and the preservation of older isotopic systems in younger zircon populations (inheritance) have yet to be answered.

1. Microscale zircon analysis

The difficulty in understanding zircon U-Pb systems stems from the fact that even in rapidly-cooled igneous rocks the crystals of zircon commonly vary greatly in their contents of trace elements, including the radioactive elements U and Th. Complicating matters further, individual crystals generally are strongly zoned, with internal differences in trace element contents of two to three orders of magnitude over distances of a few micrometres not being unusual. With the passing of time, the decay of the radioactive elements in such crystals not only generates a correlated zonation in the concentration of radiogenic Pb but also, if the temperature remains below that at which zircon anneals, creates a zonation in the extent to which the crystal structure becomes damaged by radiation. The effect of that damage is to cause a differential expansion of the crystal lattice (Holland & Gottfried 1955), resulting in the fracturing of the lower Th–U, less damaged zones and a physical weakening of the whole crystal. If the radiation damage is sufficiently severe, all trace of the zircon crystal lattice can be destroyed; the zircon is metamict.

This varied structural damage within the zircon crystals from any one rock means that the response of the zircon U–Pb system to events causing isotopic disturbance differs from grain to grain. This feature was recognised and exploited very early in the history of zircon U–Pb geochronology. Within the population of zircon crystals from a single rock there tends to be a correlation between mean radioactivity and U–Pb discordance (Silver 1963). There also tends to be a correlation between radioactivity and other physical properties of individual grains, for example grain size and magnetic susceptibility. Those properties have been utilised to subdivide zircon populations according to their radioactivity levels and hence relative discordance, thereby defining the pattern of discordance and allowing the concordant (undisturbed) isotopic composition to be inferred (Silver & Deutsch 1961, 1963; Silver 1963).

This approach to dating zircon has changed little in nearly 30 years. Certainly the physical techniques for separating low radioactivity, undamaged crystals from the remainder of the zircon population have been improved, particularly with the development of the air abrasion and high gradient magnetic separation techniques by Krogh (1982a, 1982b). However, little real progress has been made in understanding how the zircon U–Pb system behaves under different geological conditions, and in exploiting that behaviour not only to improve the accuracy (as distinct from the precision) of zircon age determinations but also to extract more chronological information from "disturbed" isotopic systems.

One of the reasons for the slow progress is the very popularity of the zircon U–Pb method as a geochronometer. In many laboratories the push to isolate the most concordant zircon in order to measure geological ages with maximum precision takes precedence over studies of discordance, which all too often give only ambiguous results. However, a far more fundamental reason is that, as advanced as the techniques of microchemistry are, the smallest samples analysed are generally a single zircon crystal, or at best a fragment of a crystal, weighing in the order of a microgram. Given the micrometre scale of zircon zoning, most such analyses are as much an average of a variety of compositions as are analyses of whole zircon populations. With the scale of the analysed sample being several orders of magnitude larger than the scale of the isotopic heterogeneity within individual crystals, the detailed study of that heterogeneity is made almost impossible.

Over the last decade, an alternative method of zircon analysis has been developed with the express intention of reducing the sample size to such an extent that it becomes possible to analyse the interior of single crystals at a scale approaching that on which the internal isotopic heterogeneity occurs. The technique uses a sensitive, high mass-resolution ion microprobe (SHRIMP) built for the purpose (Compston *et al.* 1982) and is based on an analytical approach developed by researchers such as Andersen and Hinthorne (1972) and Hinthorne *et al.* (1979).

The SHRIMP is a large mass spectrometer with which small areas (about 30 μm in diameter) on the surface of a target crystal can be analysed. The crystal is bombarded by a finely focused ion beam and the ionised portion of the material eroded is transferred directly to the mass spectrometer for determination of its chemical and isotopic composition. The interiors of crystals are accessed by sectioning them by polishing before analysis.

Ion microprobe analysis has been accepted only slowly by geoscientists at large. There are a number of reasons for this, some valid concerns, some misconceptions and some unfounded prejudices.

One valid concern is that of the accuracy of isotope ratio measurements. The mass spectrum produced by an ion probe is a complex mixture of atomic and molecular species. Some of the more ambitious of the early ion probe studies (e.g. Lovering *et al.* 1981) suffered from the low mass resolution of the probes then available, which necessitated large, difficult corrections for the interferences between species that could not be separated by mass. SHRIMP largely bypasses this problem by using high mass resolution to reduce interferences to levels where the corrections required are minimal. As for other mass spectrometers, the corrections for machine-induced mass dependent mass fractionation are determined by analysing standards.

A second concern is the problem of measuring interelement ratios. Although ion probes, like other mass spectrometers, measure isotopic ratios directly, the ratios between elements can only be determined relative to those in standards. Calibration procedures have been refined to the extent that some element ratios in some minerals can now be measured to a precision of about 2% per determination. However, those procedures are sensitive to the exact mineralogy of the target and are therefore not necessarily equally accurate over the whole compositional range even of one mineral. This sensitivity, and the limited homogeneity of available standards, mean that it will be some time yet before ion probe analysis is as precise or accurate as isotope dilution, but the present accuracy is nevertheless more than adequate to solve some very basic scientific problems.

It is through the common tendency to confuse precision and accuracy that the principal misconception regarding ion probe analysis has arisen. Because an ion probe analysis consumes about one thousand times less zircon than the smallest conventional analyses, individual ion probe analyses have larger analytical uncertainties, i.e. are less precise, than conventional zircon analyses, generally by about a factor of ten. It does not follow, however, that the ion probe analysis is any the less correct, i.e. accurate. In fact, a suite of relatively imprecise ion probe analyses on a single zircon crystal, distinguishing differences between core and rim ages and regions of isotopic discordance, is likely to be far more informative than a single high precision conventional analysis of the average composition of the whole grain.

The small-scale isotopic analyses made possible by the SHRIMP have led to a much clearer understanding of the way zircon and zircon U–Pb systems can behave under different geological conditions. On the one hand, this understanding has resulted in the development of much more effective strategies for the microscale sampling of the record of provenance, age and metamorphic history that commonly is preserved within a zircon population. On the other, it has indicated a number of ways in which the interpretation of conventionally-determined zircon discordance patterns might be improved. The present paper draws together a few examples from both previously published and new ion probe zircon studies, the latter to be published in full elsewhere, that illustrate some of the present potential and limitations of zircon U–Pb geochronology in the study of granites.

2. The Watergums Granite—simple discordance

The Watergums Granite is one of several Devonian A-type granites that occur along the eastern margin of the large Silurian-Devonian Bega Batholith in southeastern Australia. The geology and petrology of the granite have been described in detail by Collins *et al.* (1982) and its petrogenesis has been studied experimentally by Clemens *et al.* (1986).

In terms of zircon geochronology, the Watergums Granite is a relatively simple case for a number of reasons. It forms a small pluton (28 km^2) intruded into low-grade Ordovician turbidites and surrounded by a narrow contact aureole; there is no evidence for significant post-emplacement metamorphism. The granite is less felsic than many A-types (73·6% SiO_2), but nevertheless shows the high Zr, Nb, Y, REE, Sc, Zn and Ga characteristic of such rocks (Collins *et al.* 1982). The sample discussed here, GI1, has a moderately low initial $^{87}Sr/^{86}Sr$ (0·70516) and a slightly positive ε_{Nd} (+0·3), giving it a depleted mantle Nd model age of 1,050 Ma (Chappell *et al.* 1990). It is probable, therefore, that the magma is derived from a source rock containing components that are Proterozoic in age or older.

Field evidence suggests that the Watergums Granite was emplaced to a very high crustal level, probably in an almost entirely molten state; the rock is medium-grained and equigranular, virtually free of phenocrysts, xenocrysts or mafic enclaves and contains microgranophyric patches and miarolitic cavities. On this assumption, Clemens *et al.* (1986)

calculated a minimum temperature for the magma of 830°C. If all the apatite in the rock precipitated from the melt, the measurements of P solubility in granitic melts by Harrison and Watson (1984) indicate that the temperature exceeded 910°C.

Clemens *et al.* (1986) argued on textural grounds that zircon was one of the first minerals to crystallise in the magma and was entirely melt precipitated. The fact that zircon was present in all their experimental runs and they were not able to locate the zircon saturation boundary was cited as supporting evidence. Using the Zr solubility data of Watson and Harrison (1983) they calculated that the zircon saturation temperature for the magma would be about 940°C.

In this context, the zircon from the Watergums Granite was analysed with the expectation that it would be of uniform age and undisturbed isotopically by post-emplacement processes. This expectation was based also on the morphology of the zircons. Most of the crystals consist of {100} prisms terminated by simple {101} pyramids, similar to type D of Pupin (1980), a type of zircon that rarely contains inheritance. A concordia plot of the SHRIMP U–Pb analyses of 32 areas on 29 grains (Fig. 1) shows that these expectations were largely fulfilled; all but four of the analyses are clustered close to the concordia at an age of about 400 Ma.

It is the few analyses that are not in the cluster that initially are of most interest from the point of view of discordance. Two of those analyses have the same $^{207}Pb/^{206}Pb$ age as the main population, but are significantly lower in $^{206}Pb/^{238}U$. The other two have very high $^{207}Pb/^{206}Pb$ ages, $1{,}811 \pm 60$ and $2{,}481 \pm 40$ Ma respectively. These older ages were obtained on the centres of grains in which there was no obvious core, just a faint irregular shadow in transmitted light to suggest a break in the grains' growth histories. Analyses of the zircon near the margin of both grains plot within the main population close to 400 Ma.

This is the manner in which inherited zircon commonly occurs in a granite, not as discrete older grains but as nuclei on which zircon has later precipitated from the melt. The cores and overgrowths are rarely distinguishable optically, particularly in thin section, making visual inspection of zircon crystals a totally inadequate means of assessing whether or not inheritance is present, let alone its abundance. On the other hand, the structural breaks within zircons that commonly accompany inheritance are splendidly revealed in backscattered electron or cathodoluminescence images of sectioned grains (Paterson *et al.* 1992; Vavra 1990).

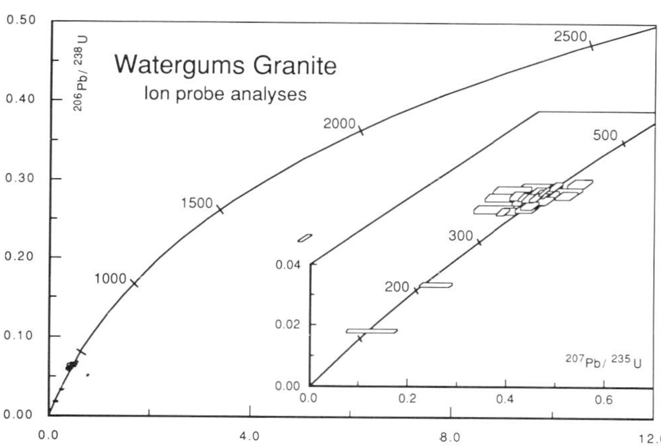

Figure 1 Concordia plot of ion probe analyses of zircons from the Watergums Granite; inset shows an enlargement of the lower end of the concordia; size of each point illustrates the respective 1σ analytical uncertainties.

There are exceptions to the occurrence of inherited zircon as cores. The Parks Pond Pluton, Maine, U.S.A., contains a zircon population which is bimodal. Most of the zircon grains are colourless and structureless and were precipitated from the melt at 410 Ma. Rare grains are red-brown, strongly zoned and 2·1 and 3·1 Ga old (Eriksson & Williams 1992). No overgrowths of younger zircon on the older grains have been found, despite a determined search. How the older zircons have been transported in the magma without acting as nuclei for further zircon precipitation has yet to be explained. None have been identified in thin section so it has not been possible to test one hypothesis, that they were armoured in refractory minerals or enclaves. Another possibility is that they were picked up by the magma after most zircon precipitation from the melt had finished. There are late Precambrian gabbros (now retrograded to eclogite) in the Central Alps, Switzerland, in which inherited Archaean zircon dominates the zircon population (Gebauer *et al.* 1988). There, too, there are discrete melt-precipitated zircons but no discernible melt-precipitated rims on the inherited grains.

The two Watergums Granite zircon cores with high $^{207}Pb/^{206}Pb$ ages are strongly discordant, indicating a major disturbance to their Pb/U. From the fact that the $^{206}Pb/^{238}U$ age of the core with lower $^{207}Pb/^{206}Pb$ is less than 400 Ma, at least part of the Pb/U disturbance occurred after the pluton was emplaced. This being so, it is invalid to estimate the primary age of the other core by assuming single stage Pb loss at 400 Ma. All that can be inferred with reasonable confidence is that the cores are at least as old as their $^{207}Pb/^{206}Pb$ ages, and even that conclusion is not necessarily correct, as will be seen in the example of the Mount Sones granulite discussed below. Had the cores been larger, the question of their ages might have been resolved by obtaining additional analyses.

The two strongly discordant analyses from ~400 Ma-old grains are unequivocal evidence for very young disturbance of the zircons' Pb/U; the disturbance must have been at least as late as the youngest $^{206}Pb/^{238}U$ age, 115 Ma. The question is, why have those two analysed areas been affected so much more than the remainder?

Following the work of Silver (1963) it has become generally accepted that discordance in zircon is principally a function of its radioactivity. As more single crystal and ion probe work is done, however, this is found to be true only in the most general sense, so that on a grain by grain, zone by zone scale, discordance obviously is controlled by other factors as well.

The relationship between radioactivity (approximated by U content) and discordance (measured by $^{206}Pb/^{238}U$) for the Watergums Granite zircons, excluding the two inherited cores, is plotted in Figure 2. Radioactivity does not control discordance in this case. The control instead appears to be Th/U (Fig. 3); even though both the Th and U contents of the discordant areas are close to the averages for the population, both have unusually high Th/U ratios. They also have very low Pb–Th ages. This illustrates a phenomenon that has been found, through ion probe analysis, to be quite common in zircons; high Th, and high Th/U in particular, significantly increases a zircon's susceptibility to isotopic disturbance. Whether this is a characteristic of the zircon as such or related to the presence of small amounts of strongly discordant thorite has yet to be investigated.

The next question is whether any of the analyses in the

cluster at ~400 Ma are discordant. A limitation of ion probe data is that individual measurements of radiogenic $^{207}Pb/^{206}Pb$, particularly in young zircons, are too imprecise to allow a rigorous assessment of discordance on the basis of the difference between the measured $^{206}Pb/^{238}U$ and $^{207}Pb/^{206}Pb$ ages. The internal variation in $^{206}Pb/^{238}U$ can be checked, however, against the precision expected from ion counting statistics and the concurrent replicate measurements of the zircon standard. Such a procedure has been found to be a very sensitive detector of discordance, especially when the sample and standard are analysed 20 to 50 times, and has now been adopted in studies of the Phanerozoic time-scale, for which the best possible precision and accuracy is required (Compston *et al.* 1992; Claoué-Long *et al.* 1992).

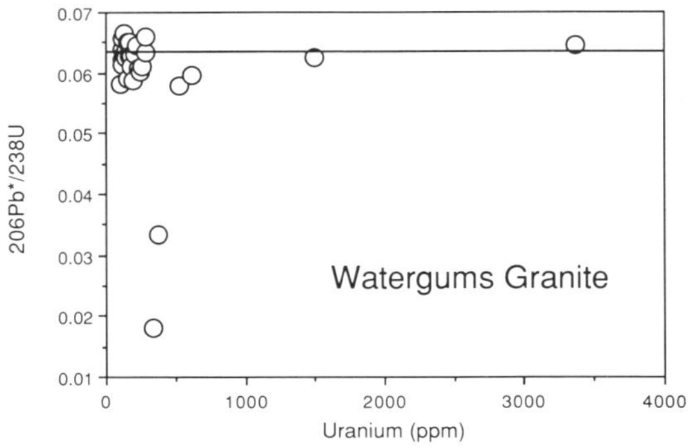

Figure 2 Plot showing that in the zircons from the Watergums Granite, discordance (measured by radiogenic $^{206}/^{238}U$) is not a function of U concentration; analytical uncertainties are less than the size of the points.

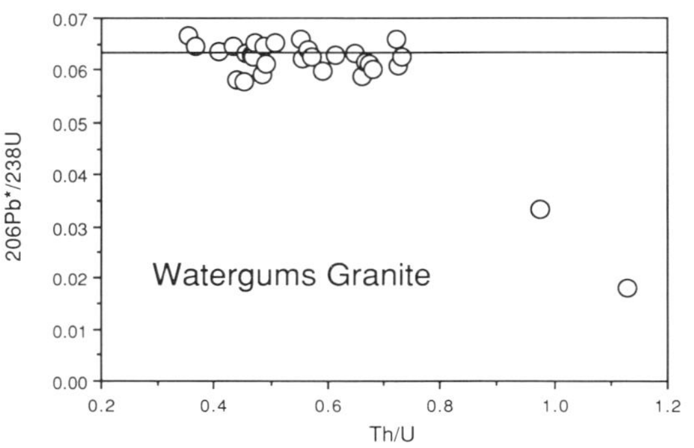

Figure 3 Plot of discordance (measured by radiogenic $^{206}Pb/^{238}U$) in the zircons from the Watergums Granite as a function of Th/U; analytical uncertainties are less than the size of the points.

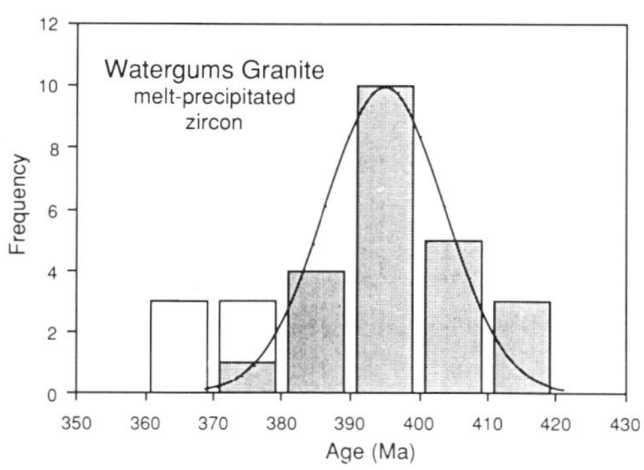

Figure 4 A histogram of the $^{206}Pb/^{238}U$ ages measured on individual melt-precipitated zircons from the Watergums Granite, with a Gaussian distribution for the uncertainty in Pb/U measurements obtained concurrently on a standard zircon superimposed; shaded analyses are consistent with that distribution, unshaded analyses are skewed towards low $^{206}Pb/^{238}U$.

The standards analysed concurrently with the Watergums Granite zircons showed a coefficient of variation in $^{206}Pb/^{238}U$ of 2·2%, the cluster of granite zircon analyses a coefficient of variation of 3·9%, the latter much greater than expected. When the analyses of the granite zircons are plotted as a histogram (Fig. 4) they are clearly skewed towards low $^{206}Pb/^{238}U$, as would be expected if there was a small amount of discordance in some areas due to radiogenic Pb loss. Removing the five lowest values leaves 23 analyses equal in $^{206}Pb/^{238}U$ within error, the mean age being 395 Ma. The standard deviation of the population is 8·6 Ma, giving a standard error of the mean of 1·8 Ma and a 95% confidence limit for the mean of ±3·7 Ma ($t = 2{\cdot}07$). This is the pattern of discordance found in the zircons from most unaltered, unmetamorphosed igneous rocks, namely the great bulk of the zircon is isotopically a closed system, some parts of some grains show a small deficiency in radiogenic Pb and in some parts the Pb deficiency is extreme. The controlling factor, particularly in older and trace-element-rich grains, is probably localised breakdown of the zircon crystal structure. Krogh and Davis (1974, 1975) found that Precambrian zircon crystals commonly contain patches or zones of alteration that are hydrated, enriched in Fe, Al, and Ca, and depleted in radiogenic Pb. Such alteration, they concluded, was "a contributing factor if not the major cause of discordance". Sommerauer (1979) has shown that the second "phase" in zircons is an unstable mixture of amorphous Zr and Si-rich subsystems in which Pb is mobile even at low temperatures.

Despite their relative simplicity, the data from the Watergums Granite illustrate the need for caution in the interpretation of conventional zircon analyses from such rocks. More often than not, inheritance cannot be detected optically, so the analysis of any one grain must be treated as though it were a mixture of the contributions from an old core and young overgrowth, both of which almost certainly have a different history of isotopic disturbance. Even if the analyses cluster close to the concordia, the uncertainty remains as to whether they are correct, slightly old due to unrecognised inheritance, or slightly young due to minor Pb loss. This applies even to the very best of modern analytical work (e.g. Tucker *et al.* 1990).

3. The Watersmeet gneiss—discordance associated with amphibolite-grade metamorphism

Metamorphism can affect a zircon population in a number of ways. Factors such as metamorphic grade, fluid flux and composition, the duration of metamorphism and previous thermal history determine whether zircon is stable, recrystallises or dissolves, whether new zircon is precipitated on old or forms new crystals and whether and how the isotopic systems in the different generations of zircon respond.

At Watersmeet, in northern Michigan, there is an early Archaean tonalitic gneiss that has had a complex metamorphic history, yet illustrates quite well some aspects of zircon's response to metamorphism. The gneiss is part of the oldest sialic crust in the Superior province. It outcrops as a small dome surrounded and overlain by early Proterozoic metasedimentary and metavolcanic rocks.

The Watersmeet gneiss was first dated by Peterman *et al.* (1980) as part of their Rb–Sr and U–Pb study of the

Marenisco–Watersmeet area. Even though the Rb–Sr whole-rock and mineral ages of the rock were about 1·75 Ga, conventional zircon analyses indicated an age of at least 3·41 Ga. This identified the gneiss as one of the oldest rocks in N America, so it has since been the subject of a number of studies including Sm–Nd analyses (McCulloch & Wasserburg 1980; Futa 1981), further conventional zircon analyses (Peterman *et al.* 1986), Hf isotopic studies both by conventional techniques (Patchett 1983) and ion microprobe (Kinny *et al.* 1991) and zircon analyses by ion microprobe (Kinny *et al.* 1991). As a result, the protolith of the gneiss is known to be an igneous rock of tonalitic composition 3,636 ± 6 Ma old with an earlier crustal history possibly extending to 3·8 Ga. It was deformed and metamorphosed to amphibolite grade in the late Archaean (2·75–2·6 Ga) and again, also to amphibolite grade, during the Penokean orogeny, a major regional tectonothermal event at 1·8–1·7 Ga (Sims *et al.* 1984).

Figure 5 shows the zircon U–Pb analyses for sample M83 (Peterman *et al.* 1980), tonalitic augen gneiss from the northern part of the Watersmeet dome. The conventional analyses are those of Peterman *et al.* (1980, 1986) and the ion probe data are from Kinny *et al.* (1991). There are several features of the data that are of interest in understanding discordance. Firstly, the range in Pb/U and Pb isotopic composition measured by the ion probe is enormously greater than that measured conventionally, a direct consequence of the much smaller volume of zircon sampled for each ion probe analysis. Secondly, the upper and lower intercepts of the discordance trend defined by the conventional analyses do not coincide with the principal clusters of ion probe data. The conventional analyses are, however, wholly explicable as different mixtures of those compositions that the ion probe has distinguished. Thirdly, the ion probe data do not define a single discordance line, but instead there is a large cluster of analyses near the concordia in the region of 3·6–3·5 Ga, a band of analyses trending from ~3·6 Ga towards ~2·5 Ga and a second band of analyses trending from ~2·5 Ga towards the origin.

The explanation for this complex discordance pattern is to be found in the zircons themselves. The zircons were derived dominantly from the gneiss's igneous precursor; they have well-developed igneous growth zoning and a wide range of U and Th contents. A very few crystals are overgrown by a thin mantle of metamorphic zircon, generally less than 20 μm thick (see Fig. 4, Kinny *et al.* 1991). The discordance pattern is therefore composite; it shows the compositions of igneous zircon variously affected by metamorphism, of metamorphic zircon, and of some mixtures of the two. Superimposed on all are the sometimes severe effects of post-metamorphic isotopic disturbance.

The feature of greatest interest in the present context is the distribution of the analyses of igneous zircon that cluster near the concordia at 3·6–3·5 Ga (Fig. 5). The analyses are not collinear but, despite a significant range in Pb/U, they do have a very sharply defined upper limit in radiogenic $^{207}Pb/^{206}Pb$. It is from this limiting value of $^{207}Pb/^{206}Pb$ that Kinny *et al.* (1991) inferred the zircon's crystallisation age to be 3,636 ± 6 Ma. They also noted that the zircons showed a general decrease in Pb/U with increasing U content. If $^{207}Pb/^{206}Pb$ rather than Pb/U is plotted against U content (Fig. 6), some very interesting features of the zircon's response to metamorphism emerge. Firstly, the zircons with lowest U contents are those with highest $^{207}Pb/^{206}Pb$. The converse is not true, however; the zircons with highest $^{207}Pb/^{206}Pb$ have a significant range of U contents. Secondly, as U content increases, the range of measured $^{207}Pb/^{206}Pb$ increases. Thirdly, above a threshold of about 700 ppm U all of the analyses show reduced $^{207}Pb/^{206}Pb$.

The trajectory of the discordance trend for the igneous zircon data, and the analyses of the few metamorphic rims wide enough to be probed, strongly suggest that the dispersion in $^{207}Pb/^{206}Pb$ reflects the zircon's response to the first amphibolite-grade metamorphism at ~2·6 Ga. In fact the $^{207}Pb/^{206}Pb$ ages measured on two rims and two highly discordant igneous zircon centres are equal within error at 2,642 ± 15 Ma (Fig. 6), exactly the age proposed independently for that metamorphism (Peterman *et al.* 1986). There is no evidence in any of the zircon analyses for any response to the second amphibolite grade metamorphism at ~1·75 Ga, despite its having totally reset the gneiss's whole-rock and mineral Rb–Sr and strongly disturbed the whole-rock and mineral Sm–Nd.

Using $^{207}Pb/^{206}Pb$ therefore as a measure of the effects of metamorphism, some interesting conclusions follow. In this case, metamorphism has had the least effect on the lowest U parts of the zircon grains. As U content increases, the likelihood of metamorphic effects and their possible magnitude also increase, but some domains can remain isotopically closed. Over a threshold U content, in this case 700 ppm, isotopic disturbance is inevitable and generally profound, resulting, in most such zircon, in complete loss of previously accumulated radiogenic Pb. Given the strength of the zircon's zoning, even the weak effects seen below 700 ppm U could be due to those analyses being of mixtures including disturbed, high-U zones.

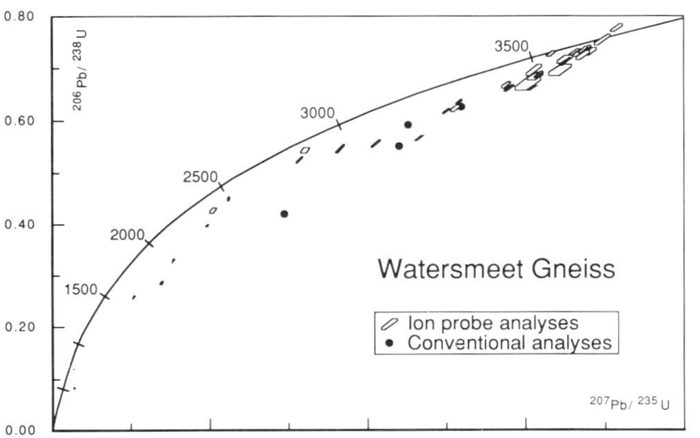

Figure 5 Concordia plot of ion probe and conventional analyses of zircons from the Watersmeet gneiss; size of each ion probe data point illustrates the respective 1σ analytical uncertainties.

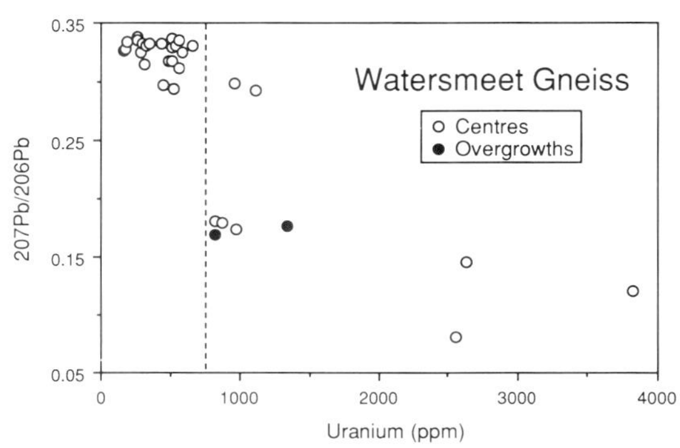

Figure 6 Plot of discordance (measured by radiogenic $^{207}Pb/^{206}Pb$) in the zircons from the Watersmeet gneiss as a function of U concentration; analyses of two younger metamorphic zircon overgrowths are shown separately; analytical uncertainties are less than the size of the points.

Such behaviour provides an explanation for the relationship between radioactivity and discordance so commonly seen in conventional zircon analyses (Silver 1963); if many zircon populations affected by Pb loss (whether due to metamorphism, alteration, weathering or dilation) are mixtures of lower U zircon that is virtually undisturbed isotopically and higher U zircon in which disturbance is extreme, such a relationship must follow. While this suggestion is not new (e.g. Krogh & Davis 1975) the example of the Watersmeet gneiss shows that this really is the pattern of Pb loss at the smallest scale that currently can be analysed. The concept of a threshold U content below which Pb loss can be minimal also is not new (e.g. Silver *pers. comm.* 1978; Ludwig & Cooper 1984; Fanning *et al.* 1988), but once again, the present example shows the effect particularly clearly.

One feature of conventional zircon data for which the Watersmeet analyses do offer a new explanation is the apparent curvature of discordance trends towards upper concordia intersections, a feature that has been observed more commonly as sample sizes are reduced and air abrasion of zircon is used to minimise discordance (e.g. Krogh 1982b; Fanning *et al.* 1988). This curvature usually simply consists of the most concordant analyses falling off the trend line defined by the more discordant grains or fractions. The Watersmeet data suggest that the explanation for this effect is that within the least discordant domains in zircon there is a small range in Pb/U but a constant radiogenic Pb isotopic composition. With the analysis of very small, extreme zircon fractions, this range in Pb/U becomes evident because discordance is no longer dominated by admixture with highly discordant domains. An important corollary to this observation is that in cases where concordia curvature is observed, the age of the zircon should be determined not by extrapolation of the main discordance trend to intersect the concordia, but rather from the highest $^{207}Pb/^{206}Pb$ age measured, particularly if the same high $^{207}Pb/^{206}Pb$ is measured on more than one zircon fraction. Ages inferred from the extrapolation of the near-linear portion of discordance trends in such cases risk being overestimates.

4. The Mount Sones gneiss—discordance associated with granulite-grade metamorphism

Zircons respond to the extreme conditions of granulite-grade metamorphism in a number of different ways, leading at times to very complex discordance patterns. A particularly good example of this is the discordance in the Mount Sones orthogneiss from Enderby Land, Antarctica (Williams *et al.* 1984; Black *et al.* 1986).

The Mount Sones gneiss is part of the Napier complex, a major Archaean granulite terrane that has had a protracted history of high pressure, very high temperature metamorphism. The precursor to the gneiss was probably a granodiorite or tonalite. Three principal episodes of deformation and metamorphism have been recognised in the complex, the first isoclinal folding at very high granulite grade (950°C, ~1 GPa), the second further isoclinal folding still at high grade and the third, following cooling to ~700°C, open folding accompanied by a further thermal pulse (Sheraton *et al.* 1980). From Rb–Sr, U–Pb and Sm–Nd studies of the region the first metamorphism is thought to have occurred at about 3·0 Ga (Black & James 1979; Sheraton & Black 1983; McCulloch & Black 1984). Whether the second high grade event was at 2·9 Ga (Black & James 1983) or 2·5 Ga (Grew & Manton 1979; De Paolo *et al.* 1982) is a matter under debate. Black *et al.* (1983) and McCulloch and Black (1984) argue that the event at 2·5 Ga was geologically relatively minor, despite its widespread effect on isotopic systems. The southern margin of the Napier complex, 100 km from Mount Sones, suffered a further high grade metamorphism along with the adjacent Rayner complex at about 1·0 Ga (Grew 1978; Black *et al.* 1983).

This complex metamorphic history is reflected in the compound discordance pattern for the zircons from the Mount Sones granulite, described in detail by Williams *et al.* (1984) and Black *et al.* (1986). The zircons show a variety of structures (Black *et al.* 1986, fig. 2). Most of the grains have pronounced euhedral growth zoning, with large compositional differences between the zones. In some crystals this zoned zircon surrounds a core that is relatively structureless. Metamorphism has had at least three different effects on the zircons: many grains have been partly recrystallised, with the compositional differences between the zones sufficiently homogenised that only "ghost" zoning remains; on many grains there are structureless overgrowths that both crosscut and penetrate the euhedral zoning; and there is a minor population of equant, structureless zircons that appear to be newly grown.

Conventional and ion probe analyses of these zircons (Williams *et al.* 1984; Black *et al.* 1986) are plotted in Figure 7. Even more so than for the gneiss at Watersmeet, the conventional analyses from the Mount Sones granulite fail to show the full range of zircon compositions present and to identify the principal ages recorded. The ion probe analyses fall into four separate fields: cores, zoned zircon, overgrowths and new grains, each zircon type defining a different discordance trend.

The value of this example in understanding zircon discordance in granites is that it illustrates very clearly how complex zircon behaviour can be and yet how robust the zircon U–Pb system is under extreme geological conditions, preserving a record of a multistage history that can be accessed provided that analyses are made on a small-enough scale.

Consider first the cores. Most of the core analyses plot above the concordia. Such reverse discordance is rarely seen

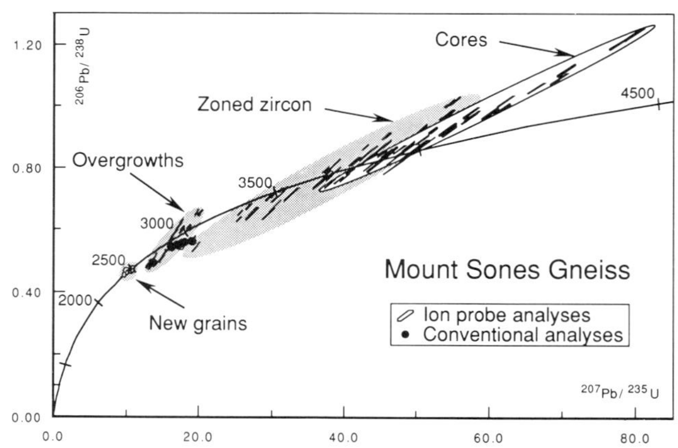

Figure 7 Concordia plot of ion probe and conventional analyses of zircons from the Mount Sones gneiss; four generations of zircon growth are distinguishable from the crystal morphology: cores, zoned zircon, overgrowths and recrystallised zircon, and new metamorphic grains; differences in isotopic composition between these generations, detected only by the ion probe, are highlighted; size of each ion probe data point illustrates the respective 1σ analytical uncertainties.

in conventional analyses because in most zircons Pb loss dominates. When reverse discordance is seen, it can be an indication of analytical problems, such as failure to achieve complete equilibration between dissolved zircon and U spike during sample preparation. In ion probe analyses also it can signal failure in calibration procedures due, for example, to matrix effects related to the partial breakdown of the zircons' crystal structure (McLaren *et al.* 1990).

Nearly all the reversely discordant core analyses from the Mount Sones granulite, however, are from a single low-U core (Williams *et al.* 1984) in which the Pb isotopic composition varies as a function of Pb concentration on a sub-micrometre scale, but the U and Th concentrations remain constant. This pattern suggests that Pb has migrated into the core from surrounding high-U zircon and is now very irregularly distributed. The trend of the analyses implies that this occurred during the metamorphism at 2·5 Ga. The age of the cores is about 3·9 Ga.

The significance of this example in the interpretation of other zircon data is that it shows that it is possible for zircons to contain radiogenic Pb that is, or has at some time been, unsupported by U. Such Pb has a higher $^{207}Pb/^{206}Pb$ than would be the case had the production of radiogenic Pb (dominantly ^{206}Pb) continued uninterrupted to the present day. The $^{207}Pb/^{206}Pb$ age calculated from such Pb gives an overestimate of the age of the zircon, the $^{207}Pb/^{206}Pb$ ages of the core in question are up to 200 Ma greater than the age that the zoned zircons indicate for the Mount Sones gneiss. Where there is evidence for unsupported radiogenic Pb, the commonly accepted interpretation of $^{207}Pb/^{206}Pb$ ages as minimum age estimates is invalid.

The analyses of the dominant zircon type in the gneiss, the zoned zircon, plot in a band that broadly extends the trend defined by the cores (Fig. 7). Most of the analyses are normally discordant, although there are some exceptions. The trend of the band shows that, as in the cores, the dispersion in composition is due largely to the 2·5 Ga event but, given that the points are not collinear, some of the dispersion must have been generated earlier. There is no indication that the zoned zircon is significantly younger than the cores.

The analyses of the overgrowths plot in a separate cluster close to 3·0 Ga, indicating that they represent either new zircon growth or the recrystallisation of pre-existing zircon with complete loss of radiogenic Pb at that time. Black *et al.* (1986) favoured the latter interpretation. An age of 3·0 Ga was proposed for the first granulite-grade metamorphism by Black and James (1979). There is significant dispersion in the analyses of the overgrowths, with a mixture of normal and reverse discordance, the trend of which suggests isotopic disturbance at about 1·0 Ga, the age of the granulite-grade event which affected the Napier complex 100 km S of Mount Sones and the adjacent Rayner complex.

The equant, structureless grains plot concordantly at 2·5 Ga, the age of pervasive isotopic resetting in the region, the grade of which is disputed. The composition of the structureless zircons themselves suggests that the grade might have been relatively high; they have a much higher Th/U than the other zircon in the gneiss and the growth of new zircon with extreme Th/U, either high or low, is commonly associated with upper amphibolite and granulite grade metamorphism (e.g. Williams & Claesson 1987).

This example illustrates a number of salient points. Several generations of zircon can co-exist in one rock, each with its own pattern of discordance. Contrary to the general experience from conventional analyses, zircon can contain some unsupported radiogenic Pb, meaning that $^{207}Pb/^{206}Pb$ ages cannot automatically be interpreted as minimum ages. It is possible for zircon to recrystallise during high-grade metamorphism, resulting in the loss of all previously accumulated radiogenic Pb. This recrystallisation can be localised, however, and radiogenic Pb be retained elsewhere in the same grain. During metamorphism whole new zircons can grow.

5. The Kameruka Granodiorite—discordance in an I-type granite containing significant inheritance

The production of a granitic magma can be considered as the final stage of the ultrametamorphism of its source rocks (White & Chappell 1977) and the discordance in granitic zircons needs to be evaluated with similar precepts in mind to those used in interpreting the zircons from the metamorphic rocks discussed above. The difference between the two is principally one of degree; in most metamorphic rocks the zircon is dominated by that from the protolith, while in most granitic rocks the zircon is dominantly melt-precipitated. There are nevertheless some exceptions and the relative proportions of the two components vary widely.

The amount of protolith zircon that survives in a granitic magma depends largely upon whether or not the magma is saturated in Zr, a function of melt composition and magma temperature (Watson & Harrison 1983). Reaction kinetics also play a role, however, there being cases where zircon is known to have survived in, for example, basalt magmas in which it almost certainly was chemically unstable (Compston *et al.* 1985/86).

Even more so than in metamorphic rocks, protolith zircons in granitic rocks are likely to be of a variety of ages and to have had different thermal histories. The problem is to establish what those ages, and the various responses to those histories, might have been.

The Kameruka Granodiorite is one of the larger granitic bodies (570 km^2) towards the eastern side of the Silurian-Devonian Bega Batholith. Its petrography and chemical features have been described by Chappell *et al.* (1990). Like the majority of the granites of the Bega Batholith, the Kameruka Granodiorite is I-type. The sample discussed here, AB40, has a moderately low initial $^{87}Sr/^{86}Sr$ (0·70560) and slightly negative ε_{Nd} (−3·7), giving it a depleted-mantle Nd model age of 1,190 Ma (Chappell *et al.* 1990). The granite's source must therefore be at least in part of Precambrian age.

Ion probe analyses of zircons from the Kameruka Granodiorite are plotted in Figure 8. No conventional analyses have been made. The most striking feature of the data is the very wide range in composition and the strong tendency of the analyses to plot close to the concordia. This pattern reflects the fact that the granodiorite contains a relatively large amount of inherited zircon. As in the Watergums Granite, the inheritance consists entirely of older nuclei within zircon precipitated from the melt; no inherited grains were identified that did not have a melt-precipitated overgrowth. Grains appearing to be wholly melt-precipitated are common.

A feature of the Kameruka Granodiorite is that the inherited cores within many of the zircon crystals are clearly visible in an optical microscope. They are even more obvious in backscattered electron images (Paterson *et al.* 1992). This makes targeting the inherited components easier (in 16 of the 42 grains probed the core was found to be inherited) but it also means that inheritance is grossly over-represented in the data set. With targeting entirely at the discretion of the analyst and the number of analyses relatively small, ion probe data sets always tend to be strongly biased towards selected compositions, particularly if there has not been a conscious effort to ensure that the analyses are representative of the population. This is not a weakness of the method, as there are great advantages in being able to select those components in a zircon population that are most informative, but it does invalidate the use of most ion probe data in assessing total-population characteristics, for example the zircon's gross chemical composition or the relative abundance of inheritance. That can only be done through experiments designed for the purpose, using large-sample multigrain conventional analyses.

The majority of the zircon data from the Kameruka Granodiorite plot in a cluster close to the concordia at about 400 Ma (Fig. 8). This group includes the analyses of a number of melt-precipitated overgrowths and clearly is dating the final crystallisation of zircon from the magma. The distribution of the analyses is very similar to those of the melt-precipitated zircon in the Watergums Granite, namely there is a range in Pb/U that is slighly greater than that expected from the concurrent analyses of the zircon standard, with the scatter due to a few analyses that are lower than average. If the three analyses with lowest Pb/U are omitted from consideration, the remaining 26 agree within analytical uncertainty at an age of 421 ± 3 ($t\sigma$-95% confidence) Ma. This is the best estimate of the emplacement age of the granite.

Unlike the Watergums Granite, there are sufficient analyses of inherited cores (18) to show that there are clusters in the data close to 600 Ma and again close to 1·1 Ga. This, plus the three analyses giving $^{207}Pb/^{206}Pb$ ages in the range 2·7–2·0 Ga, is a clear indication that there are inherited components of more than one age or that, if there is only a single inherited component, it has had a multistage history. The former is much the more likely. With the complexity of zircon behaviour in high-grade metamorphic rocks in mind, the principal question is: how best to use the analyses to determine the ages of those inherited components?

There is no unique solution to the problem. For the youngest of the inherited cores, those with $^{206}Pb/^{238}U$ ages of about 600 Ma, the precision of the measurements of radiogenic $^{207}Pb/^{206}Pb$ is insufficient to determine whether or not the data are concordant. The clustering in $^{206}Pb/^{238}U$ is very tight, however, and there is a high probability that, like the melt-precipitated zircon, many of these cores have remained closed isotopic systems. If that is so, then the crystallisation age of the majority of the youngest cores is probably in the range 630–560 Ma.

The composition of the radiogenic Pb in the older cores at about 1·1 Ga can be measured with greater precision and the near concordance of the analyses is more readily demonstrable. There is undoubtedly a correlation between $^{206}Pb/^{238}U$ and $^{207}Pb/^{206}Pb$, indicating that the dispersion in Pb/U is not the result of a recent isotopic disturbance. The analytical precision is not sufficient, however, to determine whether the five analyses represent a true age range of the order of 1,260–930 Ma, or Pb loss during the 420 Ma magmatic event from cores all about 1,260 Ma old. An argument in favour of the former is that the two highest analyses, with $^{206}Pb/^{238}U$ ages of 1,263 and 1,212 Ma, respectively, are from adjacent areas on the same core, suggesting that Pb/U in the cores might be relatively constant. In lieu of a better interpretation, the $^{206}Pb/^{238}U$ ages of these cores are therefore considered to be close approximations to their crystallisation ages.

Such an argument cannot be made for the cores with $^{207}Pb/^{206}Pb$ ages greater than 2·0 Ga, one of the analyses being grossly discordant. Also, with only a single analysis for each core, it cannot be determined whether the discordance was caused recently, at 420 Ma, earlier than that time, or by a combination of events. All that can be inferred is that each core is probably at least as old as its $^{207}Pb/^{206}Pb$ age, being aware nevertheless of the possibility that some of the radiogenic Pb might at some time have been unsupported by U.

A histogram of the individual ages inferred from all the analyses is plotted in Figure 9. This emphasises the polymodal distribution of the ages. Melt-precipitated zircon dominates, even though the analytical strategy was to target inheritance. There is a pronounced cluster of ages at 630–560 Ma, a second cluster at 1,260–930 Ma, and three ages in the late Archaean to early Proterozoic.

The youngest inherited zircons are exactly the age of the dominant zircon inheritance in a variety of microdioritic enclaves from several other granites of the Bega Batholith (Chen & Williams 1990). This, and similar ages of inheritance in many other Bega Batholith granites, suggests strongly that the immediate protolith to the Bega Batholith is a 650–500 Ma-old lower crustal igneous underplate. From the other inherited zircons, it can be inferred that the underplate possibly was derived from a 1,250–900 Ma-old precursor and that a small amount of sediment also was part

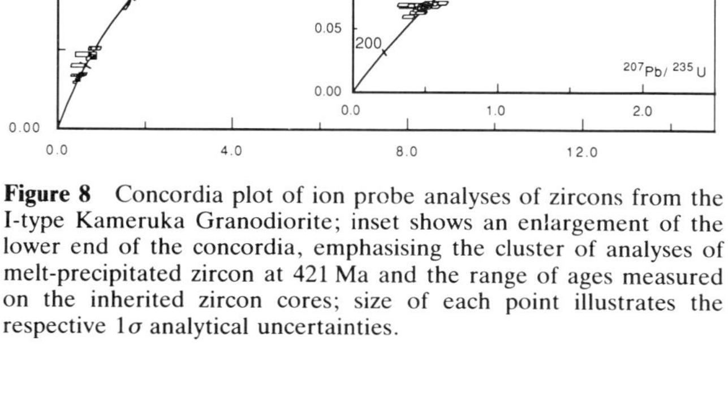

Figure 8 Concordia plot of ion probe analyses of zircons from the I-type Kameruka Granodiorite; inset shows an enlargement of the lower end of the concordia, emphasising the cluster of analyses of melt-precipitated zircon at 421 Ma and the range of ages measured on the inherited zircon cores; size of each point illustrates the respective 1σ analytical uncertainties.

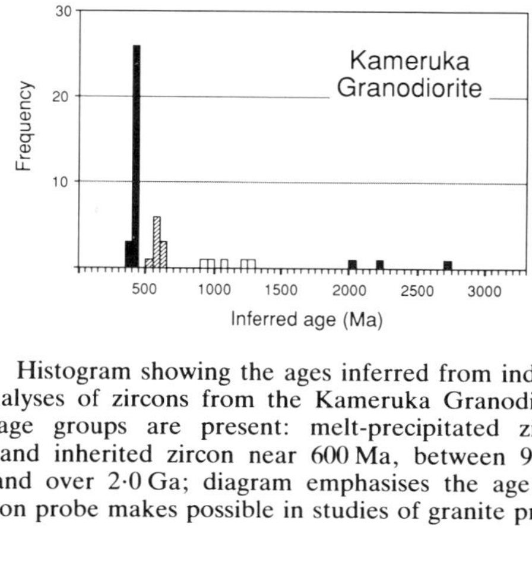

Figure 9 Histogram showing the ages inferred from individual ion probe analyses of zircons from the Kameruka Granodiorite; four distinct age groups are present: melt-precipitated zircon near 420 Ma, and inherited zircon near 600 Ma, between 900 Ma and 1·3 Ga, and over 2·0 Ga; diagram emphasises the age resolution that the ion probe makes possible in studies of granite protoliths.

of the granite's source (Williams *et al.* 1988). In the one granite zircon population, therefore, there is a record of at least three generations of igneous activity, dating the granite, its precursor, and possibly its precursor in turn.

6. The Dalgety Granodiorite—discordance in an S-type granite

A feature of all but the easternmost batholiths in southeastern Australia is the great abundance of S-type granites. Such granites, characterised by relatively low Na, Ca, Sr and oxidation state, relatively high Cr, Ni and $\delta^{18}O$, radiogenic initial Sr and unradiogenic initial Nd, are thought to have been derived from source rocks dominated by material that has been weathered at the Earth's surface, namely sediment (Chappell & White 1974, 1984, 1992). A corollary is that the S-type granites are to be expected to be rich in zircon inherited from that sediment, and that is indeed the case.

In the Berridale Batholith of the Lachlan Fold Belt, S- and I-type granites are present over about equal areas, with the intrusion relationships showing that in most cases the S-type granites are the older. A number of attempts have been made to measure the emplacement ages accurately, but with only limited success. Studies utilising Rb–Sr, K–Ar and $^{40}Ar–^{39}Ar$ have given very consistent results on the later plutons, showing them to be mostly 415–410 Ma old, but there has been a longstanding concern that the ages measured on the earlier plutons, particularly the S-type granites, are underestimates due to partial isotopic resetting (Williams *et al.* 1975, 1982; Tetley 1979). Crosschecks using the more robust Rb–Sr whole-rock method have not been possible because S-type granites commonly preserve inherited Rb–Sr isochrons (Roddick & Compston 1977). The inheritance in the zircon populations is so pronounced that emplacement ages cannot be determined accurately using conventionally-measured zircon U–Pb (Williams 1978) and doubt has even been cast on the validity of emplacement ages inferred from monazite U–Pb (Sawka *et al.* 1986).

Ion microprobe analyses have been made of zircon from a number of S-type granites from the Berridale, and adjacent Kosciusko, batholiths. In all those granites the abundance of inherited zircon is far greater than in any of the associated I-type granites, to the extent that in some S-type granites zircon precipitated from the melt fraction of the magma is so rare as to be difficult to date (Chappell *et al.* 1990). The Dalgety Granodiorite is a rock in which the relative proportion of melt-precipitated zircon is one of the highest.

The Dalgety Granodiorite forms the largest S-type pluton ($310\ km^2$) in the Berridale Batholith, although the several plutons comprising the more mafic Cootralantra Granodiorite are exposed over a slightly greater total area ($>360\ km^2$). The chemical and mineralogical features of the granodiorite have been summarised by Chappell *et al.* (1990). The Dalgety Granodiorite is not typical of the S-type granites of the Kosciusko-Berridale region; it contains neither the cordierite nor aluminosilicate minerals common in the other S-type granites, it has very few enclaves and it is relatively high in Ca and Na, and therefore lower in normative corundum. These features suggest that the sedimentary source for the granodiorite was slightly more feldspar-rich, or slightly less mature, than the sources for the other associated S-type granites. The distinctiveness of the granodiorite extends to its isotopic composition, in that the sample under consideration here (BB9) has the lowest initial $^{87}Sr/^{86}Sr$ (0·7099) and least negative ε_{Nd} (−6·1) of any analysed S-type granite in the region (McCulloch & Chappell 1982).

Ion probe analyses of zircon from Dalgety Granodiorite sample BB9 are plotted in Figure 10. Also shown are the conventional analyses of a number of zircon fractions and two fractions of monazite from the same rock (Williams 1978). Inherited zircon is very abundant as cores overgrown by thin mantles of zircon precipitated from the melt. In fact, of the 56 crystals sampled, only nine of those in which cores were sought proved not to contain one. The likely presence of inheritance was signalled by the morphology of the zircon crystals. The majority have relatively low aspect ratios and complex pyramidal terminations, with well-developed {211} and {110} faces (similar to morphology type S_{11} of Pupin 1980). This morphology very commonly is developed in zircon populations rich in inheritance.

As in the other examples discussed in this paper, the dispersion in the ion probe analyses is far greater than in the conventional, the ion probe having identified a number of distinct co-existing zircon components that are simply mixed together in different proportions in each of the conventionally analysed fractions. There is a great similarity between the components recognisable in this rock and those in the I-type Kameruka Granodiorite from 95 km away. Analyses of a number of melt-precipitated overgrowths and some structureless grains cluster close to the concordia at about 400 Ma, there is a group of analyses, also close to the concordia, that have $^{206}Pb/^{238}U$ ages between 550 and 450 Ma, there are near-concordant analyses dispersed about 1·0 Ga and there are a number of discordant analyses with $^{207}Pb/^{206}Pb$ ages between 2·0 and 3·3 Ga.

The interpretation of the analyses can be treated exactly as it was for the Kameruka Granodiorite zircons. The 22 analyses of melt-precipitated zircon show a slightly greater range in $^{206}P/^{238}U$ than is expected from the individual analytical uncertainties, all due to one low value. If that analysis is excluded, the remainder are equal within error at a mean age of 426 ± 6 ($t\sigma$) Ma. This becomes the best estimate available for the emplacement age of the granodiorite. The ages of the youngest of the inherited zircon are also best inferred from $^{206}Pb/^{238}U$, indicating a range of values between 580 and 420 Ma, with the principal

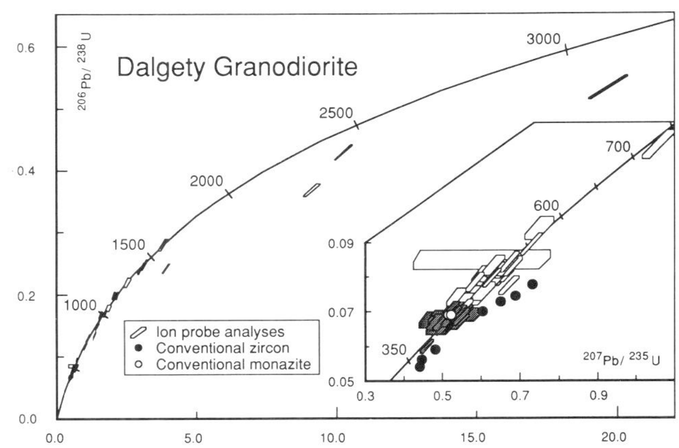

Figure 10 Concordia plot of ion probe and conventional analyses of zircons, and conventional analyses of monazites, from the S-type Dalgety Granodiorite; inset shows an enlargement of the lower end of the concordia; ion probe analyses of melt-precipitated zircon are distinguished by shading to emphasise their similarity to the conventionally-measured monazite isotopic composition; size of each ion probe data point illustrates the respective 1σ analytical uncertainties.

cluster close to 500 Ma. Some of these ages are indistinguishable from the granodiorite's emplacement age. All are distinguished by having been measured on distinct cores but, as in the Kameruka Granodiorite, it is not possible to be sure that all "cores" are inherited.

The range in the near-concordant analyses older than 600 Ma is much greater than in the Kameruka Granodiorite. In one sense this simplifies their interpretation because the dispersion is such that the analyses obviously are not collinear and therefore cannot represent the 426 Ma disturbance of a single inherited component about 1·6 Ga old. On the other hand, there is little sign of any tendency to cluster in $^{206}Pb/^{238}U$, leaving open the possibility that the dispersion is due to slight discordance in zircons of several similar ages.

When the inferred ages of all the zircons are plotted on a histogram (Fig. 11) the similarity between the ages of the inherited components in the Kameruka and Dalgety Granodiorites is emphasised (see also Fig. 9). This not only suggests that in these two very different granites the inheritance is ultimately derived from the same thermal events, but also lends support to the arguments above by which the ages of the inheritance have been deduced.

As well as being of great interest for their own sake, the data from the Dalgety Granodiorite have an important bearing on the interpretation of conventional U–Pb analyses of both zircon and monazite. The four conventional analyses of zircon fractions that are nearly collinear (Fig. 10) closely resemble the patterns of analyses very commonly obtained for rocks rich in inherited zircon (e.g. Grauert & Arnold 1968; Gulson & Krogh 1973; Pidgeon & Aftalion 1978). A popular interpretation of such a discordance pattern is as a mixing line between igneous or metamorphic zircon and an inherited component, the age of the former given by the lower concordia intersection and the average age of the latter by the upper intersection (e.g. Pidgeon & Aftalion 1978). This concept continues to be used, despite it having been known for many years that the lower intersection age is likely to be a serious underestimate because of subsequent Pb loss (Pasteels & Michot 1968). Such a line for the Dalgety Granodiorite would have concordia intersection ages of 380 ± 40 Ma and 1,750 ± 300 Ma, respectively.

The ion probe analyses, and the three lowest conventional analyses, show that such interpretations are invalid. Firstly, it is simplistic, and in the present case clearly incorrect, to assume that all the inheritance will necessarily be of one age or even a limited range of ages. Given this and the variable discordance in the inherited components, the age obtained by upward extrapolation of such discordance trends is of little geological value. It is relatively safe to infer, however,

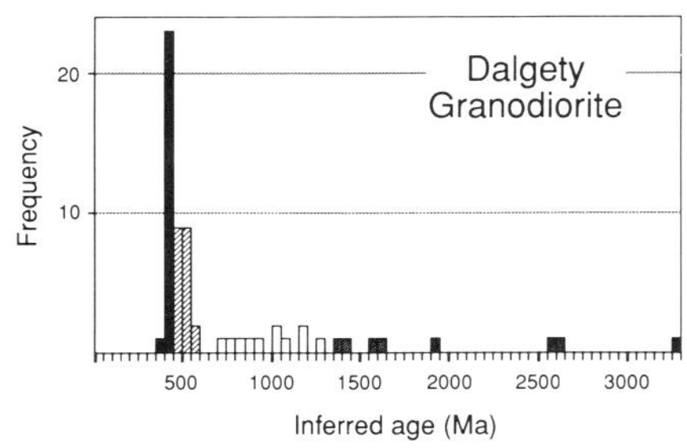

Figure 11 Histogram showing the ages inferred from individual ion probe analyses of zircons from the Dalgety Granodiorite: the same four age groups are distinguished as in the histogram for the Kameruka Granodiorite (Fig. 9), illustrating the similarity of the age of inheritance in these two very different granites.

that at least one of the inherited components is at least as old as the highest $^{207}Pb/^{206}Pb$ age measured.

The downward extrapolation of the discordance trend also has little age significance because of the downward "droop" caused by recent Pb loss. In the present case, the intersection age of 380 Ma is more than 30 Ma younger than the K–Ar and Rb–Sr biotite ages measured on the same rock (Williams *et al.* 1982) and nearly 50 Ma younger than the age of melt-precipitated zircon measured by the ion probe. The effects of Pb loss are very evident in the three lowest conventional analyses but not so much in the ion probe analyses. This is because the parts of zircon grains to be analysed by ion probe are selected to be free of obvious crystal defects or alteration.

The ion probe zircon data also cast the conventional monazite analyses in a whole new light. Even though it has always been considered possible that the monazite ages might correctly record the emplacement ages of the S-type granites (Williams 1978), the consistent discrepancy of several million years between the monazite ages and those measured by other methods has been used to argue that the monazite ages reflect a pre-emplacement stage of magma genesis or inheritance (Sawka *et al.* 1986; Sawka & Harrison 1986). In the case of the granites of southeastern Australia at least, that argument can no longer be sustained. The ion probe zircon age of 426 ± 6 Ma and conventionally measured monazite age of 430 ± 3 Ma for the Dalgety Granodiorite are indistinguishable. The existence of inherited monazite is a well-documented phenomenon (Copeland *et al.* 1988) but in the Lachlan Fold Belt granites the present evidence is that both monazite and melt-precipitated zircon accurately record crystallisation on emplacement.

7. Overview

The microscale isotopic analyses made possible by the SHRIMP ion microprobe have had, and will continue to have, a major impact on zircon geochronology, both in realising its potential and in recognising its limitations. They have proved some long-held precepts, revealed the flaws in some dogma, explained some apparent inconsistencies, and vastly expanded the scope of the technique as a whole. The foregoing examples were selected to demonstrate particular aspects of zircon's behaviour, but they also raise a number of points that are relevant to all zircon work, no matter which analytical technique is employed.

Zircons generally are composites on a grain-to-grain and intragrain scale. This cannot be ignored. Several generations of zircon can co-exist within one population or crystal, each affected differently from an isotopic viewpoint by its post-crystallisation history.

The melt-precipitated zircon in many igneous rocks is dominantly a closed U–Pb isotopic system. The small fraction that is isotopically disturbed usually has suffered severe Pb loss. All else being equal, zircon's susceptibility to disturbance increases with increasing damage to its crystal structure. This varies on a sub-micrometre scale, increasing, for example, with increasing U content (but becoming significant only above a threshold concentration of a few hundred parts per million that is different from rock to rock), increasing Th/U, increasing age and increasing time since the zircon was last annealed.

For relatively young zircons (younger than mid Proterozoic) the most accurate measure of age is generally obtained from radiogenic $^{206}Pb/^{238}U$. This is least susceptible to errors in the determination of radiogenic $^{207}Pb/^{206}Pb$, such as arise from errors in estimating the amount or composition of the initial Pb or unsuspected inheritance. In older zircons, domains with undisturbed Pb/U are scarce, but more common are weakly disturbed domains in which an accurate record of age is still preserved by the radiogenic Pb isotopic composition. In rare cases even the upper limit provided by this age is demonstrably unreliable due to the presence of radiogenic Pb unsupported by U.

It is simplistic to treat conventionally-defined discordance trends as dominantly two-component systems. Even in simple metamorphosed rocks, the combined effects of new zircon growth and isotopic disturbance invalidate the use of such discordance lines to infer either the age of the protolith or of the metamorphism in any but the most general terms. In multiply-metamorphosed rocks, conventional analyses yield little information at all.

The same applies to zircon populations rich in inheritance. Individual inherited zircon cores commonly preserve an accurate isotopic record of their age, but the range in analyses of populations obtained by mixing these cores, plus the melt-precipitated zircon, plus the variable discordance in both, generate mixing arrays that define neither the age of magmatism nor the age of any one inherited component.

The future in U–Pb geochronology, of zircon and many other minerals, lies in microscale analyses. Only in this way will mixed isotopic systems be adequately deconvoluted, the reasons for and implications of discordance be understood, and the full potential of the U–Pb technique be realised.

8. Acknowledgements

This paper has benefited greatly from reviews by Drs D. Gebauer, B. A. Paterson and L. T. Silver, and detailed criticism by Drs T. R. Ireland and J. E. H. Williams. The ideas discussed are distilled from innumerable exchanges with a great many colleagues, foremost among them Bill Compston, Bruce Chappell, Lee Silver, and Allan White.

9. References

Andersen, C. A. & Hinthorne, J. R. 1972. U, Th, Pb and REE abundances and $^{207}Pb/^{206}Pb$ ages of individual minerals in returned lunar material by ion microprobe mass analysis. EARTH PLANET SCI LETT **14,** 195–200.

Black, L. P. & James, P. R. 1979. Preliminary isotopic ages from Enderby Land, Antarctica. J GEOL SOC AUST **26,** 266–7.

Black, L. P. & James, P. R. 1983. Geological history of the Archaean Napier complex of Enderby Land. *In* Oliver, R. L., James, P. R. & Jago, J. B. (eds) *Antarctic Geoscience,* 11–5. Canberra: Australian Academy of Science.

Black, L. P., James, P. R. & Harley, S. L. 1983. The geochronology, structure and metamorphism of early Archaean rocks at Fyfe Hills, Enderby Land, Antarctica. PRECAMB RES **21,** 197–222.

Black, L. P., Williams, I. S. & Compston, W. 1986. Four zircon ages from one rock: the history of a 3930 Ma-old granulite from Mount Sones, Enderby Land, Antarctica. CONTRIB MINERAL PETROL **94,** 427–37.

Chappell, B. W. & White, A. J. R. 1974. Two contrasting granite types. PAC GEOL **8,** 173–4.

Chappell, B. W. & White, A. J. R. 1984. I- and S-type granites in the Lachlan Fold Belt, southeastern Australia. *In* Xu Keqin & Tu Guanchi (eds) *Geology of granites and their metallogenic relations,* 87–101, Beijing: Science Press.

Chappell, B. W. & White, A. J. R. 1992. I- and S-type granites in the Lachlan Fold Belt. TRANS R SOC EDINBURGH EARTH SCI **83,** 1–26.

Chappell, B. W., Williams, I. S., White, A. J. R. & McCulloch, M. T. 1990. *Granites of the Lachlan Fold Belt,* ICOG-7 Excursion Guide A2, BMR RECORD 1990/48.

Chen, Y. D. & Williams, I. S. 1990. Zircon inheritance in mafic inclusions from Bega Batholith granites, southeastern Australia: an ion microprobe study. J GEOPHYS RES **95,** 17,787–96.

Claoué-Long, J. C., Jones, P. J. & Roberts, J. 1992. The age of the Devonian-Carboniferous boundary. *In* Paproth, E. (ed.) *Final Report of the IUGS Working Group on the Devonian-Carboniferous boundary* (in press).

Clemens, HJ.D., Holloway, J. R. & White, A. J. R. 1986. Origin of an A-type granite: experimental constraints. AM MINERAL **71,** 317–24.

Collins, W. J., Beams, S. D., White, A. J. R. & Chappell, B. W. 1982. Nature and origin of A-type granites with particular reference to southeastern Australia. CONTRIB MINERAL PETROL **80,** 189–200.

Compston, W., Williams, I. S. & Clement, S. W. 1982. U–Pb ages within single zircons using a sensitive high mass-resolution ion microprobe. ABSTR 30TH AM SOC MASS SPECTROM CONF, 593–5.

Compston, W., Williams, I. S., Campbell, I. H. & Gresham, J. J. 1985/86. Zircon xenocrysts from the Kambalda volcanics: age constraints and direct evidence for older continental crust below the Kambalda-Norseman greenstones. EARTH PLANET SCI LETT **76,** 299–311.

Compston, W., Williams, I. S., Kirschvink, J. L., Zhang Zichao & Ma Guogan 1992. Zircon U–Pb ages for the Early Cambrian time scale. J GEOL SOC LONDON (in press).

Copeland, P., Parrish, R. R. & Harrison, T. M. 1988. Identification of inherited radiogenic Pb in monazite and its implications for U–Pb systematics. NATURE **333,** 760–3.

De Paolo, D. J., Manton, W. I., Grew, E. S. & Halpern, M. 1982. Sm–Nd, Rb–Sr and U–Th–Pb systematics of granulite facies rocks from Fyfe Hills, Enderby Land, Antarctica. NATURE **298,** 614–8.

Eriksson, S. C. & Williams, I. S. 1992. Inherited zircons from the Paleozoic Parks Pond pluton, Maine: direct evidence for Archean components in the Avalon basement. BULL GEOL SOC AM (in press).

Fanning, C. M., Flint, R. B., Parker, A. J., Ludwig, K. R. & Blissett, A. H. 1988. Refined Proterozoic evolution of the Gawler craton, South Australia, through U–Pb zircon geochronology. PRECAMBRIAN RES **40/41,** 363–86.

Futa, K. 1981. Sm–Nd systematics of a tonalite augen gneiss and its constituent minerals from northern Michigan. GEOCHIM COSMOCHIM ACTA **45,** 1,245–9.

Gebauer, D., Quandt, A., Compston, W., Williams, I. S. & Grünenfelder, M. 1988. Archaean zircons in a retrograded, Caledonian eclogite of the Gotthard Massif (Central Alps, Switzerland). SCHWEIZ MINERAL PETROGR MITT **68,** 485–90.

Grauert, B. & Arnold, A. 1968. Deutung diskordanter Zirconalter der Silvrettadecke und des Gotthardmassivs (Schweizer Alpen). CONTRIB MINERAL PETROL **20,** 34–56.

Grew, E. S. 1978. Precambrian basement at Molodezhnaya station, East Antarctica. BULL GEOL SOC AM **89,** 801–13.

Grew, E. S. & Manton, W. I. 1979. Archean rocks in Antarctica: 2·5 billion year uranium lead ages of pegmatites in Enderby Land. SCIENCE **206,** 443–5.

Gulson, B. L. & Krogh, T. E. 1973. Old lead components in the young Bergell Massif, south-east Swiss Alps. CONTRIB MINERAL PETROL **40,** 239–52.

Harrison, T. M. & Watson, E. B., 1984. The behaviour of apatite during crustal anatexis: equilibrium and kinetic considerations. GEOCHIM COSMOCHIM ACTA **48,** 1,464–77.

Hinthorne, J. R., Andersen, C. A., Conrad, R. L. & Lovering, J. F. 1979. Single-grain $^{207}Pb/^{206}Pb$ and U/Pb age determinations with a 10-μm spatial resolution using the ion microprobe mass analyser (IMMA). CHEM GEOL **25,** 271–303.

Holland, H. D. & Gottfried, D. 1955. The effect of nuclear radiation on the structure of zircon. ACTA CRYSTALLOGR **8,** 291–300.

Kinny, P. D., Compston, W. & Williams, I. S. 1991. A reconnaissance ion-probe study of hafnium isotopes in zircons. GEOCHIM COSMOCHIM ACTA **55,** 849–59.

Krogh, T. E. 1973. A low-contamination method for hydrothermal decomposition of zircon and extraction of U and Pb for isotopic age determinations. GEOCHIM COSMOCHIM ACTA **37,** 485–94.

Krogh, T. E. 1982a. Improved accuracy of U–Pb zircon dating by selection of more concordant fractions using a high gradient magnetic separation technique. GEOCHIM COSMOCHIM ACTA **46,** 631–5.

Krogh, T. E. 1982b. Improved accuracy of U–Pb zircon ages by the creation of more concordant systems using an air abrasion technique. GEOCHIM COSMOCHIM ACTA **46,** 637–49.

Krogh, T. E. & Davis, G. L. 1974. Alteration in zircons with discordant U–Pb ages. CARNEGIE INST WASHINGTON YEARB **73,** 560–7.

Krogh, T. E. & Davis, G. L. 1975. Alteration in zircons and differential dissolution of altered and metamict zircon. CARNEGIE INST WASHINGTON YEARB **74,** 619–23.

Lancelot, J., Vitrac, A. & Allègre, C. J. 1975. Uranium and lead isotopic dating with grain-by-grain zircon analysis: a study of complex geological history with a single rock. EARTH PLANET SCI LETT **29,** 357–66.

Lovering, J. F., Travis, G. A., Comaford, D. J. & Kelly, P. R. 1981. Evolution of the Gondwana Archaean shield: zircon dating by ion microprobe, and relationships between Australia and Wilkes Land. *In* Glover, J. E. & Groves, D. I. (eds) *Archean Geology.* GEOL SOC AUST SPEC PUBL **7,** 193–203.

Ludwig, K. R. & Cooper, J. A. 1984. Geochronology of Precambrian granites and associated U–Th–Ti mineralisation, northern Olary Province, South Australia. CONTRIB MINERAL PETROL **86,** 298–308.

Manhes, G., Minster, J. F. & Allègre, C. J. 1978. Comparative Uranium-thorium-lead and rubidium-strontium study of the Saint Sèverin amphoterite: consequences for early solar system chronology. EARTH PLANET SCI LETT **39,** 14–24.

McCulloch, M. T. & Black, L. P. 1984. Sm–Nd isotopic systematics of Enderby Land granulites and evidence for redistribution of Sm and Nd during metamorphism. EARTH PLANET SCI LETT **71,** 46–58.

McCulloch, M. T. & Chappell, B. W. 1982. Nd isotopic characteristics of S- and I-type granites. EARTH PLANET SCI LETT **58,** 51–64.

McCulloch, M. T. & Wasserburg, G. J. 1980. Early Archean Sm–Nd model ages from a tonalitic gneiss, northern Michigan. SPEC PAP GEOL SOC AM **182,** 135–8.

McLaren, A. C., Fitz Gerald, J. D. & Williams, I. S. 1990. The microstructure of zircon and its influence on the Pb/U age measured by the ion microprobe. GEOL SOC AUST ABSTR **27,** 65.

Pasteels, P. & Michot, J. 1968. Nouveaux résultats géochronologiques obtenus par la méthode U-Pb sur les zircons des Monts Sör-Rondane (Antarctique). ANN SOC GÉOL DE BELGIQUE **91,** 283–303.

Patchett, P. J. 1983. Importance of the Lu–Hf isotopic system in studies of planetary chronology and chemical evolution. GEOCHIM COSMOCHIM ACTA **47,** 81–91.

Paterson, B. A., Rogers, G., Stephens, W. E., Williams, I. S., Hinton, R. W. & Herd, D. A. 1992. The nature of zircon inheritance in two granite plutons. TRANS R SOC EDINBURGH EARTH SCI **83,** 459–71.

Peterman, Z. E., Zartman, R. E. & Sims, P. K. 1980. Tonalitic gneiss of early Archean age from northern Michigan. SPEC PAP GEOL SOC AM **182,** 125–34.

Peterman, Z. E., Zartman, R. E. & Sims, P. K. 1986. A protracted Archean history in the Watersmeet gneiss dome, northern Michigan. BULL US GEOL SURV **1622,** 51–64.

Pidgeon, R. T. & Aftalion, M. 1978. Cogenetic and inherited zircon U–Pb systems in granites: Palaeozoic granites of Scotland and England. *In* Bowes, D. R. & Leake, B. E. (eds) *Crustal evolution in northwestern Britain and adjacent regions.* GEOL J SPEC ISSUE **10,** 183–220.

Pupin, J. P. 1980. Zircon and granite petrology. CONTRIB MINERAL PETROL **73,** 207–20.

Roddick, J. C. & Compston, W. 1977. Strontium isotopic equilibration: a solution to a paradox. EARTH PLANET SCI LETT **34,** 238–46.

Sawka, W. N., Banfield, J. F. & Chappell, B. W. 1986. A weathering-related origin of widespread monazite in S-type granites. GEOCHIM COSMOCHIM ACTA **50,** 171–5.

Sawka, W. N. & Harrison, T. M. 1986. The interpretation of monazite ages from I- and S-type granitoids. GEOL SOC AM PROG ABSTR **18,** 64.

Sheraton, J. W., Offe, L. A., Tingey, R. J. & Ellis, D. J. 1980. Enderby Land, Antarctica–an unusual Precambrian high-grade metamorphic terrain. J GEOL SOC AUST **27,** 1–18.

Sheraton, J. W. & Black, L. P. 1983. Geochemistry of Precambrian gneisses: relevance for the evolution of the East Antarctic shield. LITHOS **16,** 273–96.

Silver, L. T. 1963. The relation between radioactivity and discordance in zircons. NAT ACAD SCI PUBL **1075,** 34–9.

Silver, L. T. & Deutsch, S. 1961. Uranium lead method on zircons. ANN NEW YORK ACAD SCI **91,** 279–83.

Silver, L. T. & Deutsch, S. 1963. Uranium-lead isotopic variations in zircons: a case study. J GEOL **71,** 721–58.

Sims, P. K., Peterman, Z. E., Prinz, W. C. & Benedict, F. C. 1984. Geology, geochemistry, and age of Archean and early Proterozoic rocks in the Marenisco-Watersmeet area, northern Michigan. PROF PAP US GEOL SURV **1292–A,** A1-A41.

Sommerauer, J. 1979. Physico-chemical stability of zircon and its U–Pb system. GEOL SOC AM PROG ABSTR **11,** 521.

Tetley, N. W. 1979. *Geochronology by the* $^{40}Ar/^{39}Ar$ *technique using the HIFAR reactor.* Unpublished Ph.D. Thesis, The Australian National University, Canberra.

Tucker, R. D., Krogh, T. E., Ross, R. J. Jr. & Williams, S. H. 1990. Time-scale calibration by high-precision U–Pb zircon dating of interstratified volcanic ashes in the Ordovician and Lower Silurian stratotypes of Britain. EARTH PLANET SCI LETT **100,** 51–8.

Vavra, G. 1990. On the kinematics of zircon growth and its petrogenetic significance: a cathodoluminescence study. CONTRIB MINERAL PETROL **106,** 90–9.

Watson, E. B. & Harrison, T. M. 1983. Zircon saturation revisited: temperature and composition effects in a variety of crustal magma types. EARTH PLANET SCI LETT **64,** 295–304.

White, A. J. R. & Chappell, B. W. 1977. Ultrametamorphism and granitoid genesis. TECTONOPHYSICS **43,** 7–22.

Williams, I. S. 1978. *The Berridale Batholith: a lead and strontium isotopic study of its age and origin.* Unpublished Ph.D. Thesis, The Australian National University, Canberra.

Williams, I. S. & Claesson, S. 1987. Isotopic evidence for the Precambrian provenance and Caledonian metamorphism of high grade paragneisses from the Seve Nappes, Scandinavian Caledonides, II, Ion microprobe zircon U–Th–Pb. CONTRIB MINERAL PETROL **97,** 205–17.

Williams, I. S., Compston, W., Chappell, B. W. & Shirahase, T. 1975. Rubidium-strontium age determinations on micas from a geologically controlled, composite batholith. J GEOL SOC AUST **22,** 497–505.

Williams, I. S., Tetley, N. W., Compston, W. & McDougall, I. 1982. A comparison of K–Ar and Rb–Sr ages of rapidly cooled igneous rocks: two points in the Paleozoic time scale re-evaluated. J GEOL SOC LONDON **139,** 557–68.

Williams, I. S., Compston, W., Black, L. P., Ireland, T. R. & Foster, J.J. 1984. Unsupported radiogenic Pb in zircon: a cause of anomalously high Pb–Pb, U–Pb and Th–Pb ages. CONTRIB MINERAL PETROL **88,** 322–7.

Williams, I. S., Chen, Y. D., Chappell, B. W. & Compston, W. 1988. Dating the sources of Bega Batholith granites by ion microprobe. GEOL SOC AUST ABSTR **21,** 424.

IAN S. WILLIAMS, Research School of Earth Sciences, The Australian National University, GPO Box 4, Canberra City, ACT 2601, Australia.

Transactions of the Royal Society of Edinburgh: Earth Sciences, **83**, 459–471, 1992

The nature of zircon inheritance in two granite plutons

B. A. Paterson, W. E. Stephens, G. Rogers, I. S. Williams, R. W. Hinton and D. A. Herd

ABSTRACT: Using zircons taken from two granite plutons, Strontian (Caledonian, northwestern Scotland) and Kameruka (Bega Batholith, southeastern Australia), this study presents observations that have a bearing on refractory zircons as provenance indicators. Two broad textural types of refractory zircon were identified: (1) those which show simple two-stage growth histories; and (2) those which have apparently undergone repeated periods of growth, resorption, mechanical abrasion, fracturing and fracture-healing. SHRIMP U–Pb ages obtained from the Kameruka zircons indicate that the cores are the textural manifestation of inheritance. The shapes of refractory cores are not unambiguously indicative of their ultimate origin, since they may also be modified by processes that occur before and after incorporation into the magma. The cores within the two populations show a great diversity in types and styles of zoning, and in composition, implying that they have not chemically equilibrated internally, or externally with their host melt.

KEY WORDS: Bega Batholith, Caledonian, core, electron microprobe, growth histories, Hf, ion microprobe, mineral zoning, Nd isotopes, REE, rim, SHRIMP, Th, U.

The occurrence of old, so-called inherited radiogenic Pb (Pb*) within zircons from otherwise younger granites is a commonly reported phenomenon (Grauert & Arnold 1968; Pasteels 1970; Gulson & Krogh 1973; Pidgeon & Aftalion 1978; and many others). This indicates: (1) that it is possible for zircons to remain undissolved within crustal melts (Watson & Harrison 1983); and (2) that at the temperatures attained by crustal melts (700–900°C) the diffusion of Pb in zircon is very slow.

Despite the extensive reporting within the geological literature of occurrences of inherited Pb* within zircon, and the acknowledged isotopic complexity of some zircon populations (Harrison *et al.* 1987; Williams *et al.* 1988; Chen & Williams 1990), surprisingly little is known about the textural characteristics of refractory zircon. Moreover, if refractory zircons also remain closed with respect to the exchange of other elements, then they could be a bountiful source of age, isotopic and chemical information, characteristic of their previous environment(s) of growth. In combination with experimental data on intra-zircon element diffusion rates, they could also yield information on the time-temperature history of crustal melting episodes (Watson & Harrison 1984).

This paper presents textural, compositional and isotopic observations on zircons from two granite plutons that are known to contain a component of inherited Pb*. The principal aims are to demonstrate: (1) that granite zircons often have complex, protracted growth histories; (2) that refractory zircons in crustal melts apparently undergo little exchange with respect to Hf, Y, the REE, Th and U; and (3) that an understanding of zircon growth histories is important to studies which use high spatial resolution analysis techniques.

1. Zircon inheritance in granites

Zircons that contain a component of inherited Pb* do not plot on the U–Pb concordia line and are said to be discordant. Early workers (e.g. Grauert & Arnold 1968; Pidgeon & Aftalion 1978) observed that different zircon fractions from the same rock often defined a linear pattern of discordance which had upper and lower intercepts with concordia. The lower intercept was usually interpreted as representing the crystallisation age of the granite, whereas the upper intercept was interpreted as representing the age of the inherited component. A necessary, although often unstated, assumption in the extrapolation of reverse discordia is that there is only one age of inherited zircon. However, there is mounting evidence (e.g. Harrison *et al.* 1987; Williams *et al.* 1988; Foster *et al.* 1989; Chen & Williams 1990; Rogers & Dunning 1991) that suggests that this assumption is wrong in many cases, and that multiple age (or mixed) inheritance (Miller *et al.* 1988) in granites is common. Mixed inheritance in granites has only recently been unambiguously identified, largely through the *in situ* analysis of U–Pb isotopes using SHRIMP (Compston *et al.* 1984). The existence of mixed inheritance is not altogether surprising for the following reasons:

(1) Melting of a heterogeneous crust containing rocks of a number of different ages would result in a granite with mixed zircon inheritance. Williams *et al.* (1988) and Chen and Williams (1990) interpreted the mixed inheritance they identified within the I-type granites of the Bega Batholith (southeastern Australia) as resulting from the melting of igneous rocks with a variety of ages plus a small component of sediment in the source.

(2) Many individual granite source rocks (igneous, sedimentary and metamorphic) may themselves contain zircon populations with multiple ages. Sedimentary provenance studies using single-grain dating of zircons have shown that detrital zircons commonly have disparate ages and therefore disparate original sources (Schärer & Allègre 1982; Krogh 1986; Rogers *et al.* 1990). Harrison *et al.* (1987) explained the presence of mixed inheritance in two peraluminous granites from New Hampshire as being caused by the original magmas being derived from sedimentary source rocks.

Williams and Claesson (1987) showed that detrital zircon populations with multiple ages are highly resistant to metamorphism, even to granulite facies conditions; detrital zircons from paragneisses of the Seve Nappes (Scandinavian Caledonides) retained a variety of ages indicative of their original sources. Other studies of high-grade gneisses with both sedimentary and igneous protoliths have shown that repeated high-grade metamorphism can create zircon populations with multiple ages (Black *et al.* 1986; Kinny *et al.* 1988).

In summary, therefore, we would expect granites commonly to contain mixed zircon inheritance. A corollary of this is that zircon inheritance should be complex in terms of its textural, compositional and isotopic characteristics.

2. Case studies and methods

2.1. Samples

The zircon populations from two samples were studied: one from the Strontian pluton (Caledonian, northwestern Scotland), the other from the Kameruka pluton (Lachlan Fold Belt, southeastern Australia).

Strontian is a composite pluton consisting of two main intrusions (Sabine 1963), namely the outer Loch Sunart Granodiorite and the inner, cross-cutting Glen Sanda Granodiorite (Paterson *et al.* 1992). Pidgeon and Aftalion (1978) and Halliday *et al.* (1979) reported that the Loch Sunart Granodiorite did not contain inherited zircon. However, using the same sample as that analysed by Halliday *et al.* (1979), and by carefully abrading selected euhedral prisms and prism tips, Rogers and Dunning (1991) demonstrated that the Loch Sunart Granodiorite does contain zircon inheritance. A concordant U–Pb zircon age of 425 ± 3 Ma (2σ) was obtained by Rogers and Dunning (1991) for this intrusion. The later Glen Sanda Granodiorite is known to contain substantial zircon inheritance (Halliday *et al.* 1979) and zircons from this intrusion (sample BPSRG1) have been used for this study.

Kameruka is the larger of two plutons which comprise the Kameruka Supersuite of the Siluro–Devonian Bega Batholith. Like the majority of the plutons of the Bega Batholith, Kameruka is an I-type granite in the sense of Chappell and White (1974). Its petrography and chemical features have been described by Chappell *et al.* (1991). The age of the Kameruka sample (AB40), based on an average of 26 SHRIMP U–Pb analyses obtained from melt-precipitated zircon rims, is 421 ± 3 Ma (2σ) (Williams this volume).

2.2. Methods

The internal zoning textures of zircons were observed using backscattered electron (BSE) imaging on a JEOL JCXA-733 electron microprobe at St Andrews University; the images obtained primarily reflect the compositional contribution to electron backscattering and are termed atomic number (Z) contrast images (ZCI) (see Paterson *et al.* (1989) and Paterson & Stephens (1992) for details of the method used). Studies were carried out on separated, hand-picked grains mounted in epoxy.

Quantitative analyses of zircons were obtained using a wavelength dispersive electron microprobe (JEOL JCXA-733 at St Andrews University) and an ion microprobe (Cameca ims-4f at Edinburgh University). A combined approach using both techniques was employed in order to maximise the spatial and compositional information. The electron microprobe, while having good spatial resolution (2–3 μm), generally has poor detection limits (for the REE, Th and U the detection limits are in the order of 200–400 ppm). The ion microprobe has considerably better detection limits (generally <1 ppm) but poorer spatial resolution (~25 μm).

Electron microprobe analyses were obtained using a focused beam of 20 kV and 50 nA. Interferences in the REE L-series X-ray spectra were avoided by the use of Lα lines for some of the REE elements and Lβ lines for others (Exley 1980). Peaks were considered significant if the count-rate of the net peak was more than three times the standard deviation of the error of that net peak. Peak and background positions are given in Table 1. Count times of 100 s were used for both peak and background positions for all elements on standards and unknowns.

Table 1 Analysis conditions used in the electron microprobe analysis of zircon

Element	Analysing crystal	X-ray line	Peak position (nλ)	Background positions (nλ)	(nλ)	Standard	C_{MDL} (wt%)	Standard errors in % at two concentrations (0·1 wt%)	(0·5 wt%)
Mg	TAP	Kα_1	9·89000	9·61394	10·16560	Periclase	0·004	—	—
Al	TAP	Kα_1	8·33934	7·78800	8·58540	Corundum	0·002	—	5·1
Si	TAP	Kα_1	7·12542	6·52772	7·40110	Zircon	—	—	—
P	PET	Kα_1	6·15700	6·12630	6·21925	Apatite	0·014	—	—
Ca	PET	Kα_1	3·35839	3·29565	3·42124	Wollastonite	0·005	6·2	—
Mn	LiF	Kα_1	2·10182	2·07204	2·15354	Mn metal	0·009	—	—
Fe	LiF	Kα_1	1·93604	1·86410	2·01500	Fe metal	0·008	7·3	—
Y	TAP	Lα_1	6·44880	6·35590	6·54039	REE-3	0·018	29	9·8
Zr	PET	Lα_1	6·07050	6·00720	6·13280	Zircon	—	—	—
La	LiF	Lα_1	2·66570	2·63700	2·69404	REE-3	0·033	—	—
Ce	LiF	Lα_1	2·56150	2·53276	2·59050	REE-3	0·026	27	—
Pr	LiF	Lβ_1	2·25880	2·23710	2·28042	REE-3	0·039	—	—
Nd	LiF	Lβ_1	2·16690	2·15397	2·18160	REE-2	0·037	—	—
Sm	LiF	Lβ_1	1·99806	1·86854	2·01687	REE-2	0·036	—	—
Dy	LiF	Lβ_1	1·71062	1·69868	1·83141	REE-4	0·039	—	—
Er	LiF	Lα_1	1·78425	1·69868	1·83141	REE-4	0·021	33	—
Yb	LiF	Lα_1	1·67189	1·63560	1·69868	REE-2	0·021	21	—
Hf	TAP	Mα_1	7·53900	7·40110	7·67697	Zircon	—	—	—
Th	PET	Mα_1	4·13810	4·07592	4·20049	ThO_2	0·038	20	5·6
U	PET	Mα_1	3·91000	3·84870	4·00248	UO_2	0·035	21	6·0

Notes: REE-1, 2, 3 and 4 are REE-doped glass standards of Drake and Weill (1972). Minimum detectable limits (C_{MDL}) were calculated from peak and background counts obtained on the relevant standard using the method of Scott and Love (1983). Values of analytical precision calculated from the standard errors obtained from both standards and unknowns.

Ion microprobe analyses were obtained using a 14.5 kV primary O^- beam, focused to spot of approximately 25 μm in diameter; for complete details of the ion microprobe analysis procedure, see Hinton and Upton (1991).

U–Pb isotopic analyses were performed on the SHRIMP using a 10 kV primary O^- beam focused to a spot of 25 μm in diameter; details of the analysis procedure have been given by Compston *et al.* (1984), Williams and Claesson (1987) and Chen and Williams (1990).

3. Textural observations

All the zircon grains described here contain core structures which have an epitaxial relationship with a euhedral rim. The terms "core" and "rim" are used to denote parts of a crystal which are distinct in terms of their composition or texture, and which have a generally concentric relationship. This definition allows for the occurrence of several core/rim structures within a single crystal. The nature of the interface between core and rim or between two cores can be described, firstly, by its shape, and secondly, by whether the change in composition across the interface is continuous (gradual) or discontinuous (sharp). The ability to judge whether a compositional boundary is continuous or discontinuous is determined by the resolution of the imaging technique. In the case of BSE imaging, this is determined by the electron beam operating conditions, the width of the electron beam, the mean atomic number ($\bar{Z}$) of the specimen and detector characteristics. The resolution of the Z-contrast images shown here is estimated to be better than 1 μm.

3.1. Examples of core structures

The examples described have been specifically chosen to illustrate the range of zoning textures seen in the two zircon populations. It is our experience that the zircon zoning textures from the Strontian and Kameruka plutons are by no means unusual.

3.1.1. Example 1 (Fig. 1(a)). This zircon contains many of the basic features by which core structures are defined. It contains a relatively small, rounded core within an otherwise euhedral crystal. The core/rim interface is convex outwards and is defined by a distinct change in grey-level at the rounded core/rim interface. The truncation of linear compositional discontinuities within the core (they would be planar in three dimensions) also defines the core/rim interface. The arrangement of face-parallel zoning within the rim immediately adjacent to the core/rim interface is clearly controlled by the external shape of the core.

3.1.2. Example 2 (Fig. 1(b)). The core/rim interface, which is defined principally by changes in grey-level, has a very irregular shape with both convex and concave segments. As with example 1, the face-parallel zoning within the rim is controlled by the shape of the core. The zoning within the core is complex; there are both planar and non-planar compositional discontinuities, and there are numerous small apatite inclusions.

3.1.3. Example 3 (Fig. 1(c)). The core in this example is different from those of examples 1 and 2. The zoning within the centre of the grain is truncated by a non-planar boundary towards one end of the crystal; note that part of this boundary is concave outwards. The position of the other core boundaries is less obvious; they are euhedral and do not truncate the zoning within the core. The rim material of this grain is concentrated at the terminations of the crystal.

3.1.4. Example 4 (Fig. 1(d)). This example has two approximately concentric cores. The interfaces between the inner and outer cores and the outer core and the rim are generally rounded and defined principally on changes in grey-level. Note that the outer core/rim interface truncates both the inner core/outer core interface and the zoning within the inner core. The two interfaces within this grain are compositionally discontinuous, which is particularly noticeable at the interface between the inner and outer cores. The fact that the rim to this grain is not entirely euhedral is probably due to fracturing during mineral separation.

3.1.5. Example 5 (Fig. 1(e)). The core/rim boundary is defined by the truncation of the marked face-parallel zoning within the core. In addition to this, the zoning within the core is cross-cut by several vein-like areas with generally higher $\bar{Z}$ (lighter grey). However, these areas are truncated by the zoning within the rim of the crystal. The margins to the vein-like areas are compositionally continuous, and the boundaries are not as sharp as other compositional boundaries within this grain; compare the boundaries of the vein-like areas with the boundaries within the face-parallel zoning of the crystal rim.

3.1.6. Example 6 (Fig. 1(f)). The zoning within this euhedral zircon is complex, and in places the textural relationships are ambiguous and thus difficult to describe objectively. The zoning consists of a number of approximately concentric core structures separated by non-planar boundaries. Some of these non-planar compositional discontinuities truncate planar zoning structures, whereas others separate areas of distinct grey-level. Although some of the cracks seen in zircon grains are artefacts of sample preparation, most of those in this example are integral features of the grain. The pattern of cracks appears to be controlled by the zoning; many are seen to terminate at non-planar boundaries. There are only a few cracks in the outermost rim and in the innermost core.

3.2. Zoning within zircon cores

Here we are concerned with the zoning inside the outermost core/rim boundary of otherwise euhedral zircons. The cores within both zircon populations that were studied display a wide variety of types and styles of zoning. Figures 1(a), 1(b), 1(e), 1(f), 2 and 3 show 11 zircons from the Glen Sanda Granodiorite (Strontian) sample, each with a distinctive core structure. Some cores are comparatively devoid of any internal structure (Fig. 3(a)), whereas others contain strong face-parallel zoning markedly different in spacing and style from the zoning within the rim (Fig. 2(b)).

3.3. Summary

Although zircon cores display a wide variety of types and styles of zoning, for the purposes of the interpretation of individual growth histories, two broad textural types can be identified: (1) those which have simple two-stage growth histories; and (2) those which have multi-stage growth histories.

4. Compositional observations

The great diversity in types and styles of zoning displayed by the zircon cores from the Glen Sanda Granodiorite indicates diversity of composition, both within individual cores and between cores. However, since BSE grey-levels are not unique to a particular composition (many different compositions may give rise to the same BSE intensity), it is necessary to use microanalysis techniques to document the

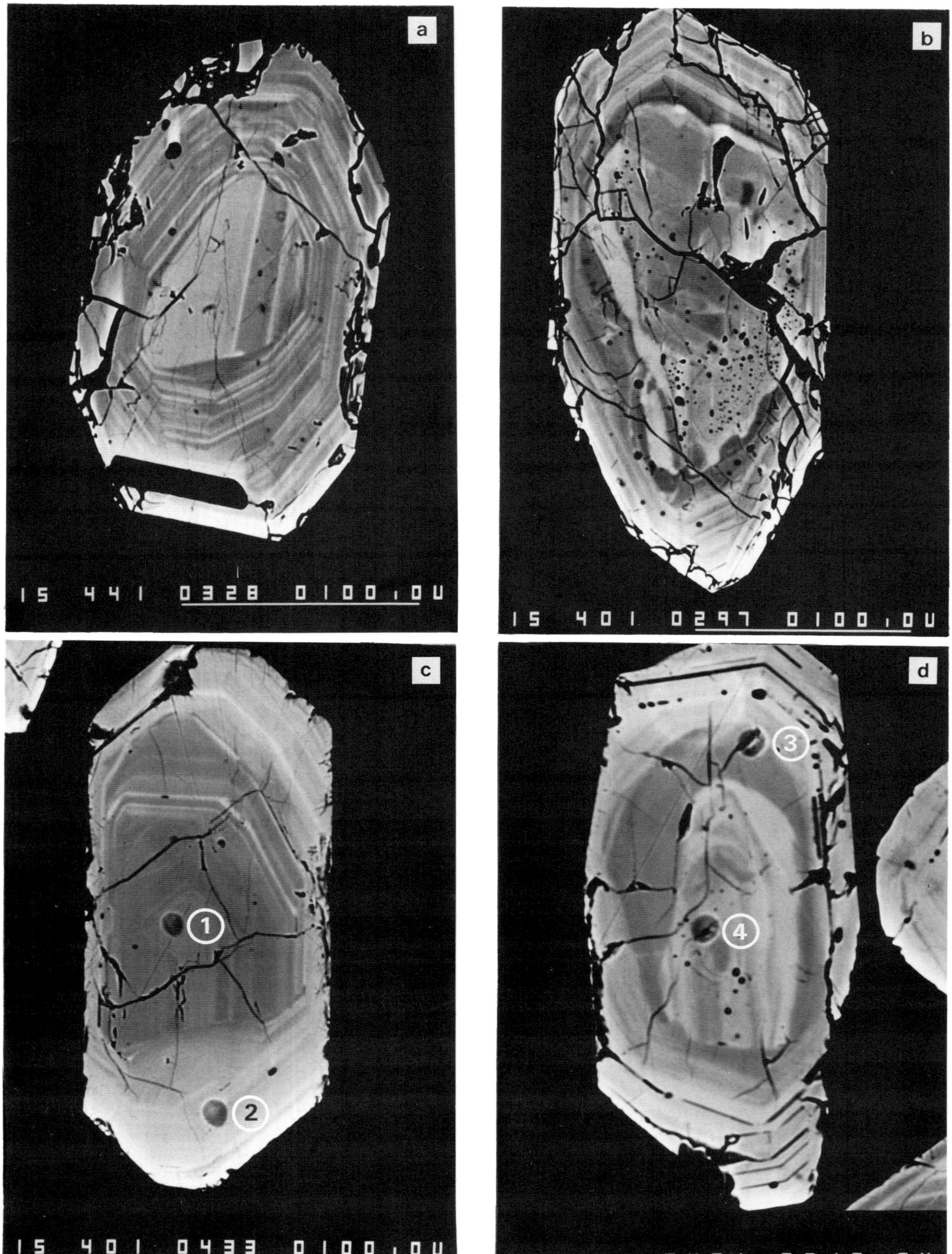

Figure 1 Z-constrast images of zircons from the Strontian and Kameruka plutons. See text for description: (a) example 1, Strontian, scale 100 μm; (b) example 2, Strontian, scale 100 μm; (c) example 3, Kameruka, scale 100 μm. Numbered pits are SHRIMP analysis areas; (d) example 4, Kameruka, scale 10 μm. Numbered pits are SHRIMP analysis areas; (e) example 5, Strontian, scale 10 μm; (f) example 6, Strontian, scale 100 μm.

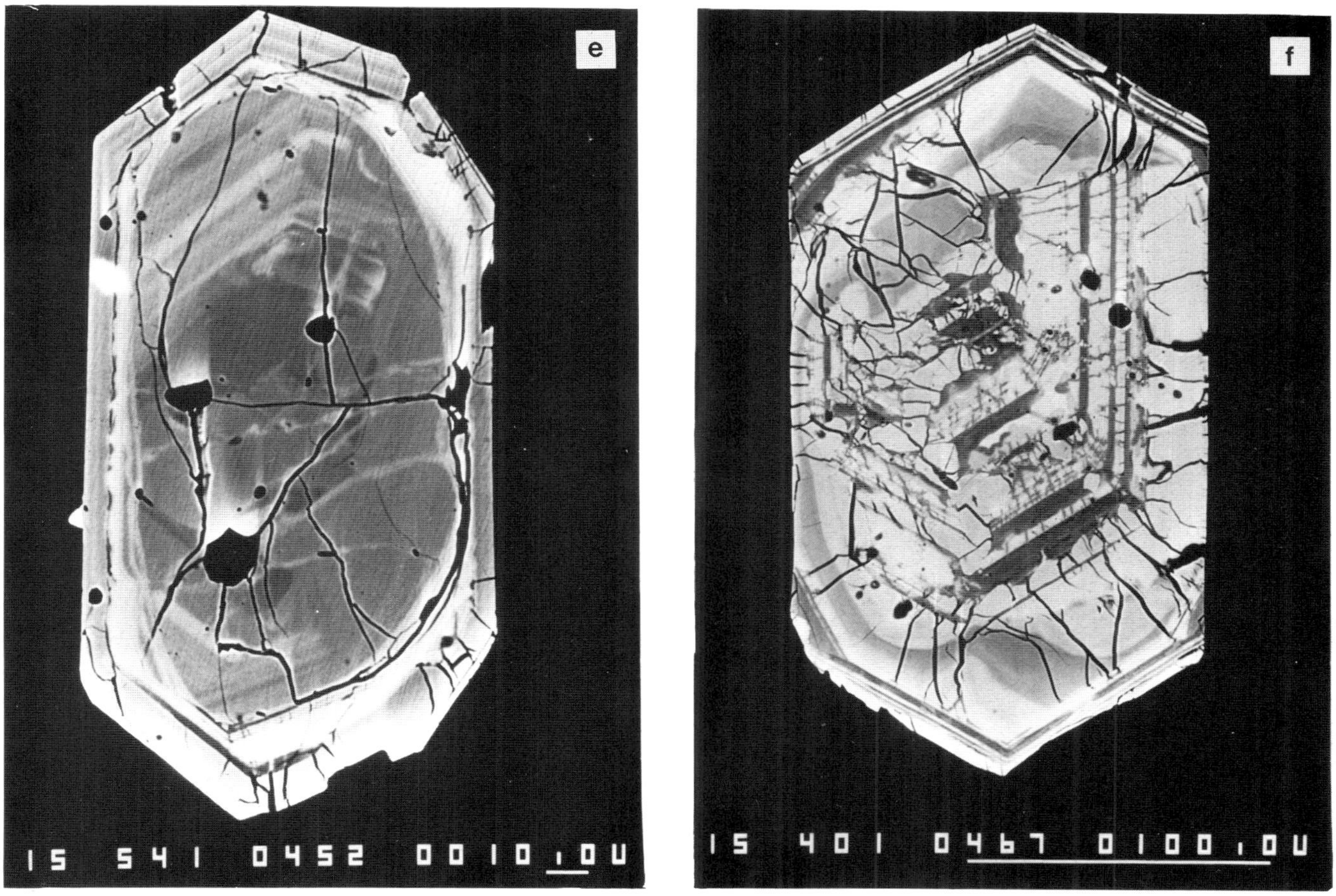

Figure 1 (Continued)

within-grain compositional variations indicated by the Z-contrast images.

Many of the apparent variations in the trace element chemistry of zircon that have been documented can be explained by the presence of mineral inclusions, the effects of alteration and mineral zoning. For example, Hinton and Upton (1991) recently demonstrated that the relatively flat chondrite-normalised LREE patterns often observed in bulk analyses of zircons (e.g. Nagasawa 1970; Mahood & Hildreth 1983; Heaman *et al.* 1990) are due to the presence of inclusions, and do not accurately reflect the partitioning of REE into the zircon lattice. The use of *in situ* microanalysis techniques means that it is possible to avoid these problems. With the exception of positive Ce anomalies (due to minor Ce^{4+}) and negative Eu anomalies (due to minor Eu^{2+}), zircons are commonly HREE-enriched with patterns that are uniformly steep from Lu down to La (Hinton & Upton 1991). The REE patterns obtained from refractory zircons may give information about parental melt chemistries (Hinton & Upton 1991) and, in the case of Ce and Eu anomalies, oxygen fugacities.

Owing to the relatively large diameter (~25 μm) primary beam of an ion microprobe, the analyses presented represent averages of relatively large volumes. However, using electron microprobe analysis, which has a spatial resolution of 2–3 μm, and step-scans on the Cameca ion microprobe (~2 μm), it can be shown that Y, the REE, U and Th concentrations in zircons tend to behave coherently: their spatial distribution is responsible for the fine-scale zoning seen on Z-contrast images (Paterson 1990; Hinton & Upton 1991). Furthermore, Hinton and Upton (1991) also showed that the chondrite-normalised REE *patterns* of adjacent zones within a single crystal are similar. This means that while absolute element abundances within zircons may show variations due to the presence of fine-scale zoning, element ratios and REE patterns may potentially contain geologically useful information.

4.1. Electron microprobe analyses

Four cores (Fig. 2) from the Glen Sanda Granodiorite were analysed by electron microprobe, the purpose being to illustrate the diversity of compositional variation typically present within zircon cores. The analysis numbers on the photographs are cross-referenced with Table 2. The analysis areas, two or three per core, were chosen to reflect both the style of zoning within the core and the observed range of $\bar{Z}$.

4.1.1. Example 7 (Fig. 2(a)). The core of this example generally has a much lower $\bar{Z}$ than the thin euhedral rim. Apart from Si and Zr, only Hf was detectable in significant quantities; other trace elements such as Y, the REE, Th and U were either not detected or are present in quantities that are very close to the minimum detectable limit. The analyses indicate that this core has a relatively restricted range of HfO_2 concentration from 1·73–1·93 wt%.

4.1.2. Example 8 (Fig. 2(b)). The core in this zircon also shows a restricted range of HfO_2 concentration, significantly lower than those of example 6. Analysis 5, which was obtained from an area within the core with the highest $\bar{Z}$, also contains significant quantities of Y and the HREE (Er and Yb). The high $\bar{Z}$ area (analysis 6) that truncates the zoning within the core of this grain has much higher HfO_2 (2·15 wt%) and UO_2 concentrations (0·16 wt%) than the inner core, although no Y or HREE were detected.

4.1.3. Example 9 (Fig. 2(c)). Texturally, this grain

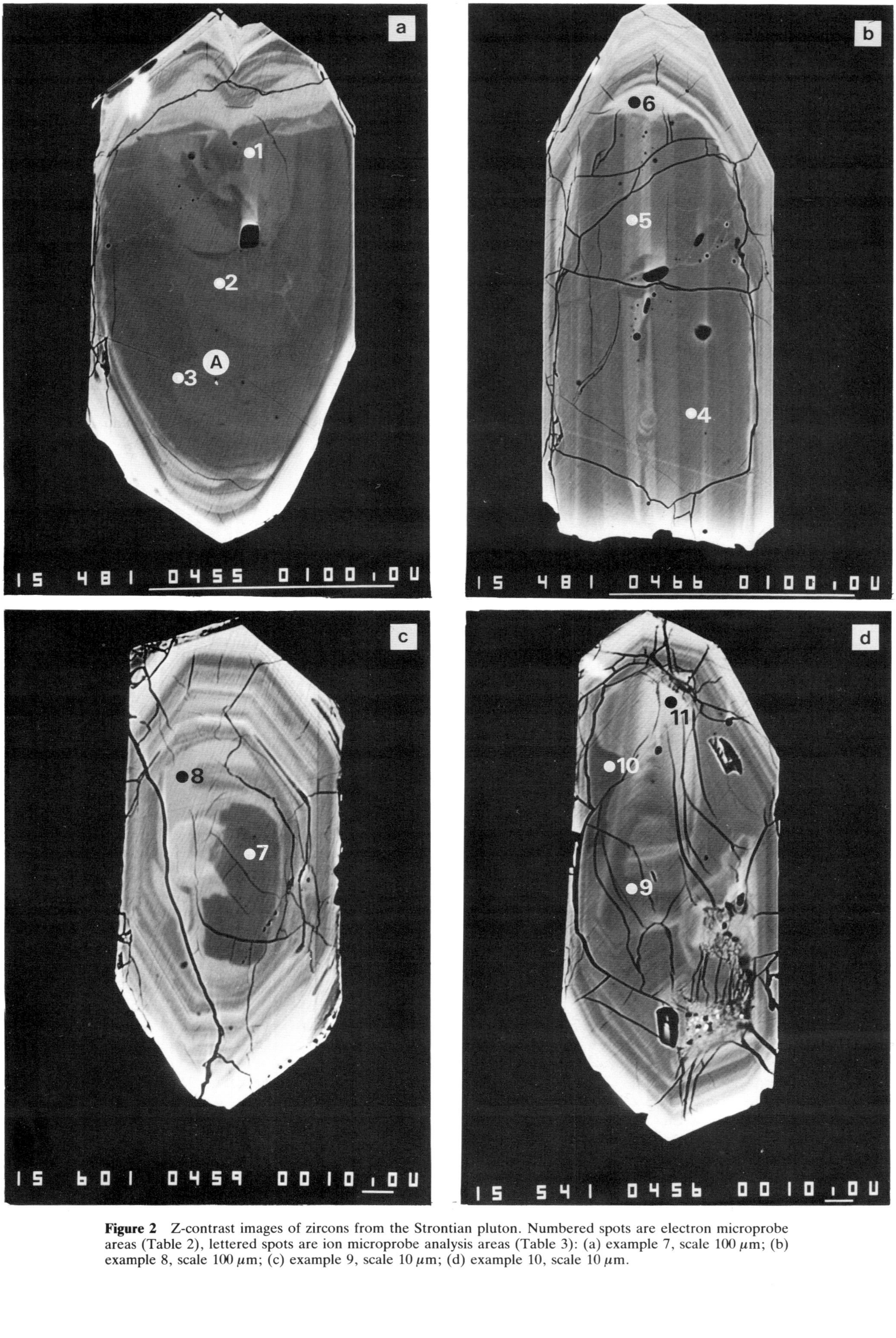

Figure 2 Z-contrast images of zircons from the Strontian pluton. Numbered spots are electron microprobe areas (Table 2), lettered spots are ion microprobe analysis areas (Table 3): (a) example 7, scale 100 μm; (b) example 8, scale 100 μm; (c) example 9, scale 10 μm; (d) example 10, scale 10 μm.

Table 2 Electron microprobe analyses of zircon cores from Strontian. Only element detected in significant quantities are shown

	1	2	3	4	5	6	7	8	9	10	11
SiO_2	32·10	32·49	32·43	31·65	31·88	32·00	32·48	32·50	32·11	32·16	31·80
ZrO_2	65·18	66·09	65·21	66·03	65·75	66·35	65·19	65·19	65·97	65·86	64·58
HfO_2	1·83	1·93	1·73	1·14	1·12	2·15	1·54	1·92	1·91	1·43	1·20
Y_2O_3	n.d.	0·04	n.d.	0·19	0·88	n.d.	0·13	0·03	n.d.	n.d.	0·51
UO_2	0·04	0·04	n.d.	0·05	0·05	0·16	n.d.	0·18	n.d.	n.d.	0·43
TOTAL	99·15	100·59	99·37	99·13	100·04	100·71	99·39	99·86	99·99	99·45	99·24

Notes: All concentrations in wt%. Analysis numbers cross-referenced with Figure 2. n.d. not detected.
Analysis 4 includes: 0·07 wt% Yb_2O_3.
Analysis 5 includes: 0·12 wt% Er_2O_3; 0·19 wt% Yb_2O_3; 0·05 wt% ThO_2.
Analysis 6 includes: 0·05 wt% Yb_2O_3.
Analysis 7 includes: 0·05 wt% Yb_2O_3.
Analysis 8 includes: 0·04 wt% Yb_2O_3.
Analysis 11 includes: 0·04 wt% Al_2O_3; 0·08 FeO; 0·07 CaO; 0·05 Ce_2O_3; 0·07 Er_2O_3; 0·12 wt% Yb_2O_3; 0·29 wt% ThO_2.

appears to consist of two approximately concentric cores. The area with low $\bar{Z}$ (analysis 7) contains 1·54 wt% HfO_2 and 1·3 wt% Y_2O_3 and a small amount of Yb_2O_3 close to the minimum detectable limit. This contrasts with the outer high $\bar{Z}$ area (analysis 8) which has 1·92 wt% HfO_2, and significant quantities of UO_2 (0·18 wt%).

4.1.4. Example 10 (Fig. 2(d)). The core of this zircon is texturally complex. The pattern of cracks is closely related to the pattern of $\bar{Z}$ variations. Many of the cracks appear to originate in three distinct areas of the grain, and at least one of these areas (bottom right) contains distinct high $\bar{Z}$ phases, including thorite and uraninite. Only Si, Zr and Hf were present in detectable quantities in analyses 9 and 10, which, however, gave significantly different concentrations of HfO_2. Analysis 11 detected the presence of relatively large quantities of Y_2O_3, Yb_2O_3, Er_2O_3, ThO_2 and UO_2. Unusually, analysis 11 also detected the presence of significant concentrations of elements such as Al, Fe, Ca and Ce. The HfO_2 concentration of analysis 11 (1·20 wt%) is significantly lower than the concentrations of analyses 9 and 10.

4.2. Ion microprobe analyses

Four cores were also analysed by ion microprobe (Figs 2(a), 3, Table 3), the main purpose being to obtain complete REE patterns for four distinctly different zircon cores (Fig. 4). The areas analysed were selected without obvious fine-scale structures in view of the limited spatial resolution of the primary ion beam. With the exception of the Eu and Ce anomalies and variations in absolute REE abundances, the cores have very similar chondrite-normalised REE patterns. In addition to variations in the REE, the four cores also show marked differences in the absolute concentrations of Hf, Th, and U.

5. Isotopic observations

5.1. U–Pb ages obtained by SHRIMP

Using SHRIMP, the different growth stages of the zircons from the Kameruka Granodiorite were analysed for their U–Pb isotopic composition (see Figs 1(c), 1(d), 5 and Table 4). These analyses were obtained as part of a much larger study of zircons from the granites of the Bega Batholith (Williams *et al.* 1988; Chen & Williams 1990). The results in Table 4 do not represent the complete data set available for Kameruka; only the analyses that are cross-referenced with the photographs in this paper are given.

U–Pb ages obtained from zircon rims (Figs 1(c)-analysis 2; 5(a)-analysis 7) give values which are interpreted as representing the emplacement age of Kameruka (421 ± 3 Ma). A variety of ages was obtained from the cores, ranging from 509 Ma to 1263 Ma. The zircon in Figure 1(d) contains two approximately concentric cores; the inner core gave an age of 929 Ma (analysis 4), whereas the outer core gave an age of 626 Ma (analysis 3). These are thought to be real; the 626 Ma age does not represent loss of Pb* nor a mixed analysis between the 929 Ma core and the 421 Ma melt-precipitated overgrowth.

5.2. Sm–Nd isotopic composition of granite zircons

As part of a study of the influence that accessory minerals have on the behaviour of REE and Sm–Nd isotopes in magmatic processes (Paterson *et al.* 1992), REE-bearing phases from the Glen Sanda Granodiorite (Strontian) were analysed for their Sm–Nd isotopic composition by conventional techniques. The ε_{Nd425} values obtained from apatite, −4·0 and −4·9, and monazite, −5·2, are in equilibrium with each other and are similar to their host rock, −4·4. However, the ε_{Nd425} values of −7·8 and −7·9 for zircon are not in equilibrium with the apatite, monazite or whole-rock values, a difference beyond that expected from analytical uncertainty. The zircon contains only a small proportion of the whole-rock Nd (<0·3%), and thus it cannot cause a measurable shift in the whole-rock Nd isotopic composition.

6. Interpretation

The subhedral and anhedral zircon core structures from Strontian and Kameruka are interpreted as representing refractory material that was incorporated into the magma either during partial melting, or between that time and final crystallisation, and were subsequently the sites of new, melt-precipitated zircon growth. Although this has been the assumption of many U–Pb zircon studies, it has only been through the selection of grains on the basis of morphology (e.g. Schärer & Allègre 1983) and analysis by SHRIMP that the basic truth of this assumption has been proved. There are several lines of evidence which support the interpretation.

(1) The subhedral and anhedral core structures in the Kameruka zircons all give pre-emplacement U–Pb ages when analysed by SHRIMP, whereas the rims give ages consistent with emplacement. However, it should be noted that one of the Kameruka zircons with inheritance, example 3 (Fig. 1(c)), had a euhedral core with a euhedral overgrowth, indicating that zircon inheritance need not always occur as anhedral cores (see also Pidgeon & Aftalion 1978). Although there are no SHRIMP U–Pb ages from the Glen Sanda Granodiorite zircons, this part of the Strontian pluton is known from a previous U–Pb study (Halliday *et al.* 1979) to contain a substantial component of inherited Pb*.

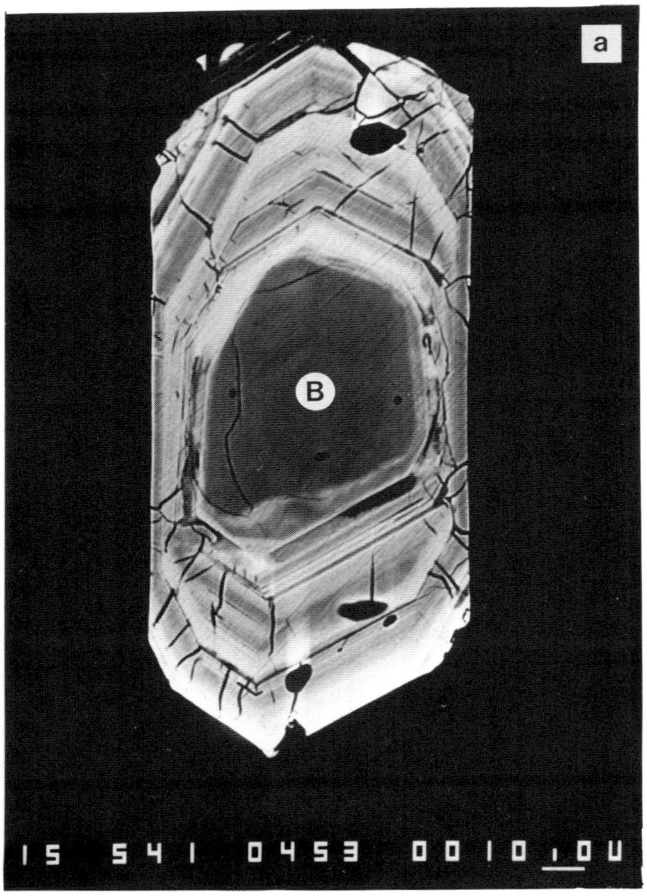

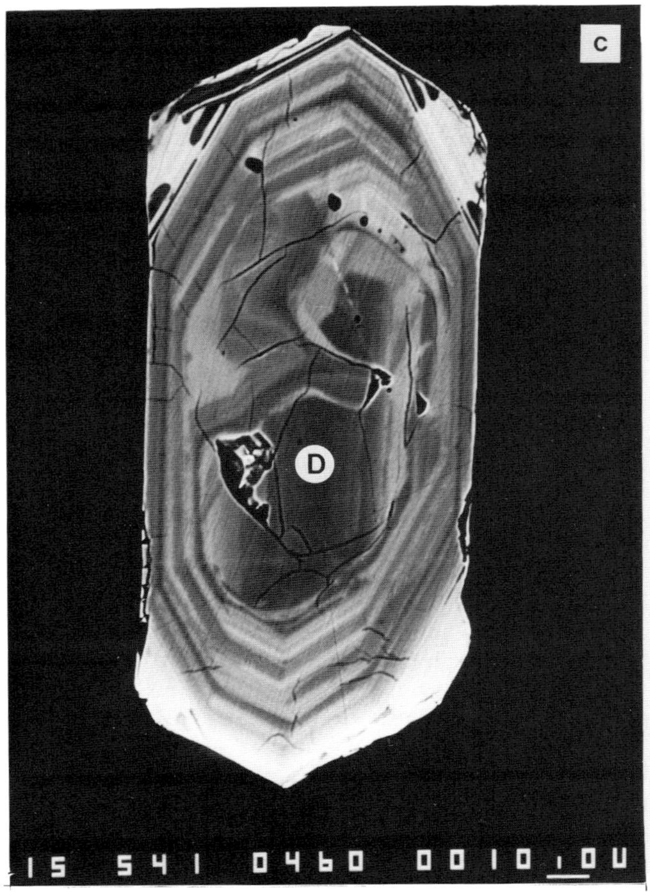

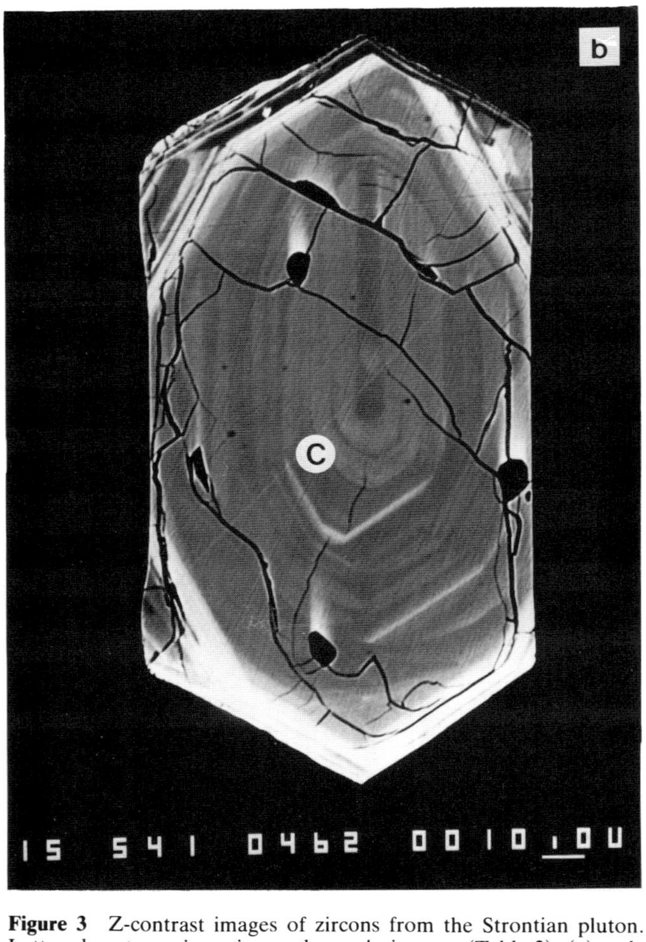

Figure 3 Z-contrast images of zircons from the Strontian pluton. Lettered spots are ion microprobe analysis areas (Table 3): (a) scale 10 μm; (b) scale 10 μm; (c) scale 10 μm.

(2) Internal zoning structures are commonly truncated by core/rim boundaries, indicating that the core has, at some time, undergone a reduction in size, due to dissolution, mechanical abrasion or fracturing. It is important to note that it is theoretically possible for non-planar boundaries to develop under conditions of crystal *growth* rather than dissolution, and that the existence of non-planar crystal interfaces should never be considered *prima facie* evidence for crystal dissolution (Donaldson 1985a) and/or mechanical abrasion; other evidence is required.

(3) The fact that the arrangement of the face-parallel zoning within the zircon rims often reflects the shape of the

Table 3 Ion microprobe analyses of zircons from Strontian. Concentrations in ppm unless stated. Analysis letters cross-referenced with Figures 2(a) and 3

	A	B	C	D
Y	518	1020	1020	1580
La	0·13	0·14	0·37	0·35
Ce	21·8	30·3	42·6	40·4
Pr	0·13	0·23	0·53	0·33
Nd	0·81	1·50	4·72	4·14
Sm	1·59	3·03	4·99	5·85
Eu	0·52	0·73	1·85	0·97
Gd	9·36	17·2	26·5	30·3
Tb	3·36	6·78	8·24	10·89
Dy	46·8	90	100	151
Ho	17·8	35·1	36·5	58
Er	90	174	175	295
Tm	20·4	38·0	38·8	70
Yb	179	325	334	594
Lu	46·3	77	79	115
Th	109	73	146	103
U	398	138	226	385
Hf	1·37%	1·20%	1·10%	1·42%

core, implies that the core/rim boundary represents a previous exterior surface on which the rim material was grown. The core/rim boundaries are in effect small-scale unconformities.

(4) There is usually a marked contrast between the core and rim in terms of the type and style of zoning present. This would be expected if the core and rim grew in different environments.

(5) The textures obviously give no direct indication of the actual time-gap between different growth stages. However, many of the cracks within cores terminate at the core/rim boundary (see Fig. 1(e) and (f)), indicating that the cracking occurred prior to growth of the rim. Such cracking is caused by the differential expansion of the zircon structure due to slowly accumulated α-particle damage.

For the zircons shown here, only the high spatial resolution U–Pb analyses obtained using SHRIMP can be considered to be *proof* that the cores represent material that was refractory during magmatism. The other lines of evidence do not have a unique interpretation, but are characteristics that might be expected of an inherited zircon population. It should be stressed that conventional U–Pb isotopic analyses of zircons selected on the basis of morphology also indicate that zircon cores commonly represent the textural manifestation of inherited zircon. The presence of visible cores, however, is not proof of the presence of inheritance.

7. Discussion

7.1. Zircon dissolution in silicate melts: theoretical considerations

The initial shape of an older zircon incorporated into a melt will depend on the crystallisation history of the parent rock from which it was derived. In general zircons from sub-alkaline rocks develop crystal faces (see Pupin 1980), whereas those from alkaline rocks are generally more irregular. The shapes of zircons found in sedimentary rocks are determined by their original source and the effects of sedimentary transport (Robson 1987).

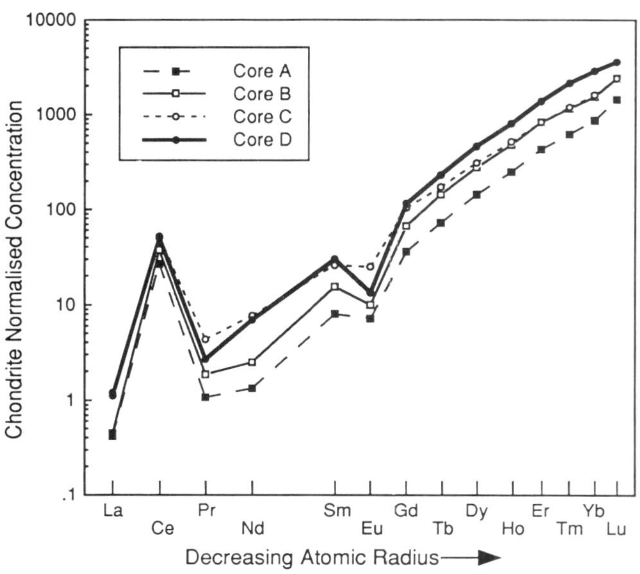

Figure 4 Chondrite-normalised REE patterns obtained from four zircon cores (analyses A, B, C and D–Figs 2(a), 3 and Table 3).

When zircons are incorporated into a melt, dissolution will take place only if the melt is undersaturated with respect to Zr (Watson & Harrison 1983). The rate of crystal dissolution in a melt may be controlled either by the rate of interface reaction or by the rate at which elements are transported, either by diffusion or bulk flow, away from the dissolving interface and replaced by components in the bulk melt. If element transport in the melt is more rapid than interface reaction, then interface reaction will be the rate-controlling step, and *vice versa.* Where transport is the rate-controlling step, dissolution will be more rapid at crystal corners and edges, due to better bulk melt access, and rounded morphologies will result (see Donaldson (1985a) for discussion about dissolution morphologies). Where interface reaction is the rate-controlling step and the interface kinetics are anisotropic in different crystallographic directions, then, theoretically, faceted morphologies could develop during dissolution (Donaldson 1985a), although there are few documented cases involving dissolution of silicate minerals in silicate melts. In the case of zircon dissolution, other, non-crystallographic, anisotropies (e.g. compositional zoning, cracks, fission tracks and metamict areas) may be important where interface reaction is the rate-controlling step (Harrison & Watson 1983), and complex dissolution shapes could result.

Harrison and Watson (1983) concluded from their dissolution experiments, that in the absence of bulk flow, zircon dissolution rates are controlled by Zr diffusion rather than interface kinetics. If this is commonly the case, then crystal anisotropies would be unimportant and rounded morphologies favoured. A number of studies (Harrison & Watson 1983; Donaldson 1985b; Zhang *et al.* 1989), have observed that, in general, interface reaction is unimportant in determining the dissolution rates of silicate minerals in silicate melts. Interface reaction may only become important either during bulk magma flow—in which a crystal is in continual contact with the bulk melt—or where the melt is only slightly undersaturated with respect to the components of the dissolving crystal (Donaldson 1985a; Zhang *et al.* 1989).

7.2. Zircon dissolution in silicate melts: textural observations

The majority of the zircons in the Strontian and Kameruka samples have cores with smooth, rounded outlines which truncate zoning within the cores. Such core/rim boundaries must have been produced by dissolution and/or mechanical abrasion. If dissolution were responsible for the truncation of core zoning structures and the rounded outlines, then transport must have been the rate-determining step in the dissolution reaction. If mechanical abrasion were responsible for the rounded shapes, then the zircons did not dissolve when incorporated into the melt. Extensive dissolution of a zircon which had been rounded during sedimentary transport would modify its shape, with the resultant shape being dependent on dissolution kinetics.

A small number of zircon cores from the Strontian and Kameruka plutons do not have smooth, rounded outlines. The cores with highly irregular outlines *could have* developed during a regime of dissolution in which interface reaction was the rate-determining step, controlled by heterogeneities such as zoning and state of metamictisation. If this is the case, then it would indicate dissolution under conditions where interface reaction is the rate-controlling step, possibly due to low degrees of Zr undersaturation, after the bulk of the zircons had partially dissolved. An

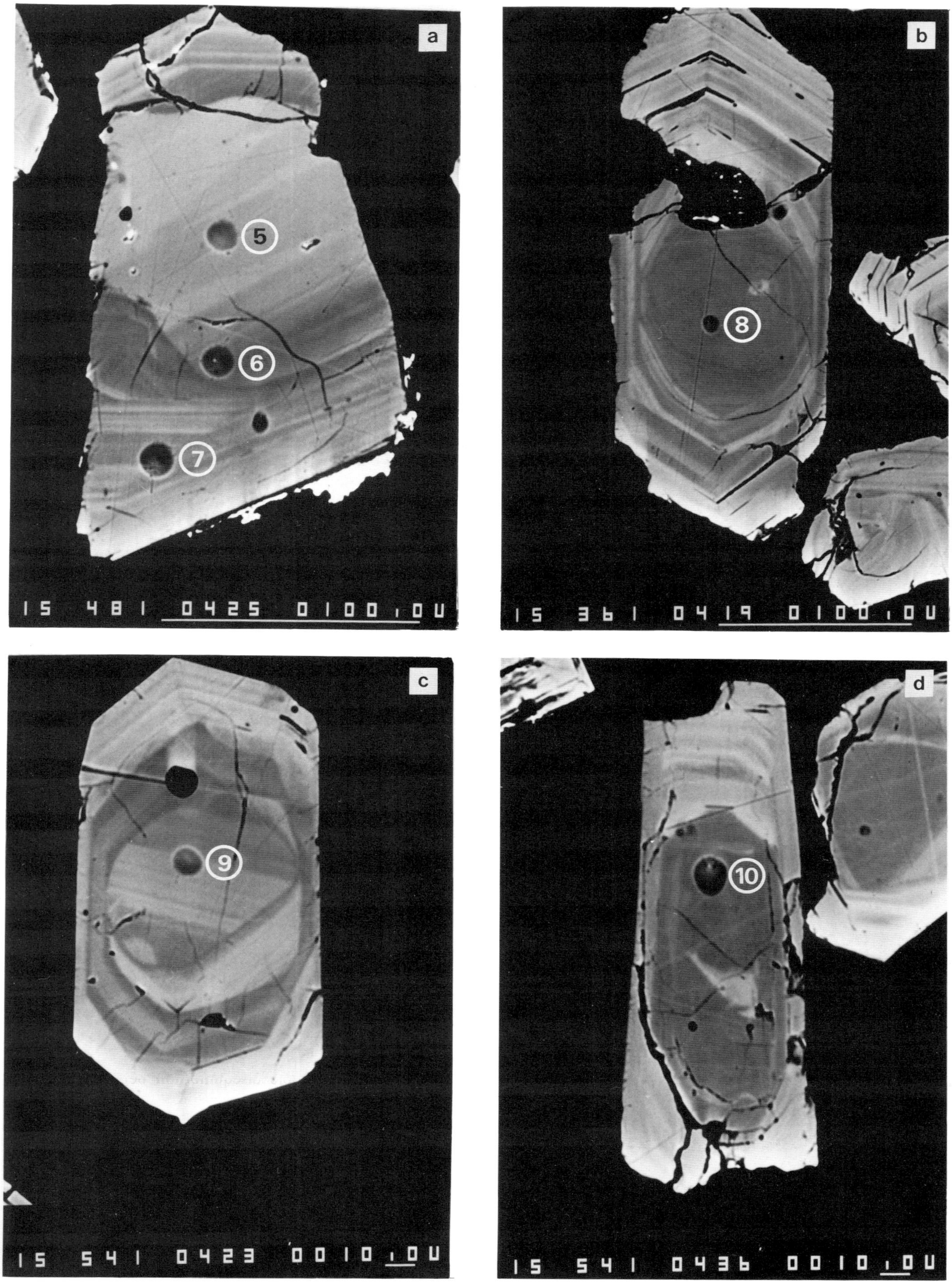

Figure 5 Z-contrast images of zircons from the Kameruka pluton. Numbered pits are SHRIMP analysis areas, see Table 4: (a) scale 100 μm; (b) scale 100 μm; (c) scale 10 μm; (d) scale 10 μm.

Table 4 SHRIMP U–Pb ages from the Kameruka grains. Analyses cross-referenced with Figures 1(c), (d) and 6. Ages calculated from the measured $^{206}Pb/^{238}U$ ratios. Analyses 2 and 7, from zircon rims, give ages which are interpreted as the emplacement age of Kameruka (421 ± 3 Ma). Errors are 2σ

Analysis number	Age (Ma)
1	509 ± 18
2	emplacement
3	626 ± 22
4	929 ± 32
5	1263 ± 42
6	1212 ± 40
7	emplacement
8	633 ± 22
9	592 ± 20
10	590 ± 22

alternative possibility is that irregular shapes are a primary pre-incorporation feature and no dissolution occurred because the melt was already Zr-saturated.

7.3. Textural, compositional and isotopic disequilibrium

The zircon cores from both the Strontian and Kameruka plutons appear to retain the primary zoning structures and chemistries developed in their original growth environment(s). The preservation of compositional heterogeneities implies that there was no significant internal diffusion of the elements responsible for the zoning. The strong similarities in the chondrite-normalised REE *patterns* obtained from four texturally diverse cores implies that they originally grew from melts with similar REE patterns, although the variety of zoning styles suggests that the environments of growth were different in other respects. The only marked differences in the REE patterns of the cores are in the magnitudes of the Eu and Ce anomalies. These probably reflect variations in the prevailing redox conditions in the environments in which the inherited cores crystallised.

Complete chemical equilibration of the refractory zircon component with the bulk magma would have resulted in homogeneous cores, the composition of which would have been determined by prevailing zircon/melt partition coefficients. The marked difference in the compositions of individual cores in the Strontian sample also argues that extensive chemical equilibration did not occur.

The Sm–Nd isotopic disequilibrium between zircon and coexisting apatite and monazite from the Glen Sanda Granodiorite (Strontian) is thought to be due to the presence of an inherited Nd isotopic component in the zircon, rather than being due to a post-crystallisation isotopic disturbance (Paterson *et al.* 1992). The disequilibrium is consistent with the preservation of zoning structures within the cores, and the presence of an inherited U–Pb zircon component in the Glen Sanda Granodiorite (Halliday *et al.* 1979). If the Nd isotopic disequilibrium in the zircons is due to an inherited Nd isotopic component, it would imply that zircon did not completely equilibrate and that diffusion of Sm and Nd (and presumably the other REE) at temperatures in excess of 700°C is very slow (Paterson *et al.* 1992).

There is evidence that at high metamorphic grades, zircons remain closed to exchange of Lu–Hf isotopes (Pettingill & Patchett 1981; Stille & Steiger 1991; Kinny *et al.* 1991). The work of Kinny *et al.* (1991), which used SHRIMP, demonstrated that the core and rim of individual zircons can have different Hf isotopic compositions. Although no similar studies have been carried out on zircons with known U–Pb inheritance, it seems likely that refractory zircons also inherit their original Hf isotopic composition.

8. Conclusions

(1) In terms of growth histories, two broad textural types of zircon can be identified in the Strontian and Kameruka plutons: (a) those that contain a single core and a euhedral overgrowth (i.e. grains which have simple two-stage growth histories); and (b) those which have undergone repeated periods of resorption/mechanical abrasion and growth (i.e. grains which have multi-stage growth histories). Individual cores can be defined by their shape, which can range from rounded to highly irregular, and by the zoning on either side of the core/rim interface.

(2) U–Pb isotopic analyses by SHRIMP demonstrate that in general anhedral zircon cores from the Kameruka intrusion represent the textural manifestation of inheritance, although one of the inherited grains contained a euhedral core structure. The fact that the SHRIMP U–Pb data obtained from the zircon cores have a range of ages, also shows that the Kameruka zircons have mixed inheritance.

(3) On textural and compositional criteria, it is argued that the U–Pb inheritance that is known to exist in the Glen Sanda Granodiorite (Strontian pluton), is present in the form of subhedral and anhedral core structures. The compositional and textural diversity of individual zircon cores and the presence of grains with multi-stage growth histories is consistent with the zircon population having mixed inheritance. Although this interpretation can only be proved by selective U–Pb analysis, it is worth noting that Rogers and Dunning (1991) interpreted the pattern of inheritance in the outer Loch Sunart Granodiorite of the Strontian pluton as being due to mixed inheritance.

(4) The shapes of refractory cores are not unambiguously indicative of their ultimate origin, since shape is also a function of dissolution kinetics. For example, the rounded shapes of cores could result from partial dissolution where transport in the melt is the rate-determining step, or they could be due to mechanical abrasion during sedimentary transport sometime before their incorporation in the melt.

(5) The preservation of zoning within refractory zircon cores, the compositional diversity that is indicated by the variety of zoning styles, and the quantitative electron and ion microprobe analyses, imply that the cores have neither chemically equilibrated internally, nor externally with their host melt. If the refractory cores had completely equilibrated, they would be unzoned and compositionally identical.

(6) The Sm–Nd isotopic disequilibrium between zircon and coexisting apatite and monazite from Strontian is interpreted as being due an inherited Nd isotopic component in the zircon, rather than post-crystallisation isotopic disturbance (Paterson *et al.* 1992). The results imply that zircons remain closed systems with respect to Sm, Nd and REE diffusion at the elevated temperatures of crustal melts.

(7) The knowledge derived from studies of refractory zircon populations which combine textural, compositional, geochronological and isotopic information, should prove a powerful way of determining granite provenance. However, for this ideal to be a realistic objective there needs to be a

better understanding of element partitioning and diffusivity in zircon at high temperatures. Part of this understanding will come from laboratory experiments, and part from isotopic and chemical studies of zircons from well-characterised magmatic systems. In parallel to this, there need to be technical advances in the ability to measure fine-scale isotopic and chemical variations within single zircons.

9. Acknowledgements

Part of this work was carried out whilst BP held a NERC studentship (GT4/86/GS/93) at the University of St Andrews. NERC also supported this work through the grants to GR (GR3/7222), WES (GR3/4633) and the Edinburgh University ion microprobe laboratory. SURRC is also supported by the Scottish Universities. Two constructive reviews greatly improved the final manuscript.

10. References

Black, L. P., Williams, I. S. & Compston, W. 1986. Four zircon ages from one rock: the history of a 3930 Ma-old granulite from Mount Sones, Enderby Land, Antarctica. CONTRIB MINERAL PETROL **94,** 427–37.

Chappell, B. W. & White, A. J. R. 1974. Two contrasting granite types. PAC GEOL **8,** 173–4.

Chappell, B. W., White, A. J. R. & Williams, I. S. 1991. A tranverse section through granites of the Lachlan Fold Belt. Excursion Guide: Second Hutton Symposium on Granites and Related Rocks. Canberra 1991. BUR MINER RES, GEOL GEOPHYS, REC **1991/92.**

Chen, Y. D. & Williams, I. S. 1990. Zircon inheritance in mafic inclusions from the Bega Batholith granites, southeastern Australia: an ion microprobe study. J GEOPHYS RES **95,** (suppl B) 17,787–96.

Compston, W., Williams, I. S. & Meyer, C. 1984. Geochronology of zircons from the lunar breccia 73217 using a sensitive high mass-resolution ion microprobe. J GEOPHYS RES **89,** (suppl B) 525–34.

Donaldson, C. H. 1985a. A comment on crystal shapes resulting from dissolution in magmas. MIN MAG **49,** 129–32.

Donaldson, C. H. 1985b. The dissolution rate of olivine, plagioclase, and quartz in a basaltic melt. MIN MAG **49,** 683–93.

Drake, M. J. & Weill, D. F. 1972. New rare earth element standards for electron microprobe analysis. CHEM GEOL **10,** 179–81.

Exley, R. A. 1980. Microprobe studies of REE-rich accessory minerals: implications for Skye granite petrogenesis and REE mobility in hydrothermal systems. EARTH PLANET SCI LETT **48,** 97–110.

Foster, D. A., Harrison, T. M. & Miller, C. F. 1989. Age, inheritance, and uplift history of the Old-Woman-Piute batholith, California and implications for K-feldspar age spectra. J GEOL **97,** 232–43.

Grauert, B. & Arnold, A. 1968. Duetung diskordanter Zirkonalter der Silvrettadecke und des Gotthardmassivs (Schweizer Alpen). CONTRIB MINERAL PETROL **20,** 34–56.

Gulson, B. L. & Krogh, T. E. 1973. Old lead component in the young Bergell Massif, South-east Swiss Alps. CONTRIB MINERAL PETROL **40,** 239–52.

Halliday, A. N., Aftalion, M., van Breemen, O. & Jocelyn, J. 1979. Petrogenetic significance of Rb–Sr and U–Pb isotopic systems in the 400 Ma-old British Isles granitoids and their hosts. *In* Harris, A. L., Holland, C. H. & Leake, B. E. (eds) *The Caledonides of the British Isles-Reviewed.* GEOL SOC LONDON SPEC PUBL **8,** 653–61.

Harrison, T. M. & Watson, E. B. 1983. Kinetics of zircon dissolution and zirconium diffusion in granitic melts of variable water content. CONTRIB MINERAL PETROL **84,** 66–72.

Harrison, T. M., Aleinikoff, J. N. & Compston, W. 1987. Observations and controls on the occurrence of inherited zircon in Concord-type granitoids, New Hampshire. GEOCHIM COSMOCHIM ACTA **51,** 2549–58.

Heaman, L. M., Bowins, R. & Crockett, J. 1990. The chemical composition of igneous zircon suites: implications for geochemical tracer studies. GEOCHIM COSMOCHIM ACTA **54,** 1597–607.

Hinton, R. W. & Upton, B. G. J. 1991. The chemistry of zircon: variations within and between large crystals from a syenite and alkali basalt xenoliths. GEOCHIM COSMOCHIM ACTA **55,** 3287–302.

Kinny, P. D., Williams, I. S., Froude, D. O., Ireland, T. R. & Compston, W. 1988. Early Archaean zircon ages from orthogneisses and anorthosites at Mount Narryer, Western Australia. PRECAMBRIAN RES **38,** 325–41.

Kinny, P. D., Compston, W. & Williams, I. S. 1991. A reconnaissance ion-probe study of hafnium isotopes in zircons. GEOCHIM COSMOCHIM ACTA **55,** 849–59.

Krogh, T. E. 1986. Single zircon analysis and the provenance of Meguma and Spruce Brook sediments in eastern Canada. TERRA COGNITA **6,** 152.

Mahood, G. & Hildreth, W. 1983. Large partition coefficients for trace elements in high-silica rhyolites. GEOCHIM COSMOCHIM ACTA **47,** 11–30.

Miller, C. F., Watson, E. B. & Harrison, T. M. 1988. Perspectives on the source, segregation and transport of granitoid magmas. TRANS R SOC EDINBURGH EARTH SCI **79,** 135–56.

Nagasawa, H. 1970. Rare earth concentrations in zircons and apatites and their host dacites and granites. EARTH PLANET SCI LETT **9,** 359–64.

Pasteels, P. 1970. Uranium-lead radioactive ages of monazite and zircon from the Vire-Carolles granite (Normandy). A case study of zircon-monazite discrepancy. ECLOGAE GEOL HELV **63,** 231–37.

Paterson, B. A. 1990. *Accessory mineral growth histories: implications for granitoid petrogenesis* (Unpublished Ph.D. Thesis, University of St Andrews, St Andrews, Scotland, U.K.).

Paterson, B. A. & Stephens, W. E. 1992. Kinetically induced compositional zoning in titanite: implications for accessory-phase/melt partitioning of trace elements. CONTRIB MINERAL PETROL **109,** 373–85.

Paterson, B. A., Stephens, W. E. & Herd, D. A. 1989. Zoning in granitoid accessory minerals as revealed by backscattered electron imagery. MIN MAG **53,** 55–61.

Paterson, B. A., Rogers, G. & Stephens, W. E. 1992. Evidence for inherited Sm–Nd isotopes in granitoid zircons. CONTRIB MINERAL PETROL in press.

Pettingill, H. S. & Patchett, P. J. 1981. Lu–Hf total-rock age for the Amîtsoq gneisses, West Greenland. EARTH PLANET SCI LETT **55,** 150–6.

Pidgeon, R. T. & Aftalion, M. 1978. Cogenetic and inherited zircon U–Pb systems in granites: Palaeozoic granites of Scotland and England. *In* Bowes, D. R. & Leake, B. E. (eds) *Crustal evolution in Northwestern Britain and adjacent regions.* GEOL J SPEC ISSUE **10,** 183–220.

Pupin, J. P. 1980. Zircon and granite petrology. CONTRIB MINERAL PETROL **73,** 207–20.

Robson, D. A. 1987. The persistence of zircons in the clastics of the sedimentary succession of northern Britain: an SEM study. *In* Marshall, J. R. (ed.) *Clastic particles: scanning electron microscopy and shape analysis of sedimentary and volcanic clasts,* 51–64. New York: Van Nostrand Reinhold.

Rogers, G., Krogh, T. E., Bluck, B. J. & Kwok, Y. Y. 1990. Provenance ages of the Torridonian sandstone of NW Scotland using single grain U–Pb zircon analysis. GEOL SOC AUST ABSTR (ICOG 7) **27,** 84.

Rogers, G. & Dunning, G. R. 1991. Geochronology of appinitic and related granitic magmatism in the W Highlands of Scotland: constraints on the timing of transcurrent fault movement. J GEOL SOC LONDON **148,** 17–27.

Sabine, P. A. 1963. The Strontian granite complex, Argyllshire. BULL GEOL SURV GB **20,** 6–42.

Schärer, U. & Allègre, C. J. 1982. Investigation of the Archaean crust by single-grain dating of detrital zircon in a graywacke of the Slave Province, Canada. CAN J EARTH SCI **94,** 438–51.

Schärer, U. & Allègre, C. J. 1983. The Palung granite (Himalaya); high-resolution U–Pb systematics in zircon and monazite. EARTH PLANET SCI LETT **63,** 423–32.

Scott, V. D. & Love, G. 1983. *Quantitative electron-probe microanalysis.* Chichester: Ellis Horwood Limited.

Stille, P. & Steiger, R. H. 1991. Hf isotope systematics in granitoids from the central and southern Alps. CONTRIB MINERAL PETROL **107.** 273–8.

Watson, E. B. & Harrison, T. M. 1983. Zircon saturation revisited:

temperature and composition effects in a variety of crustal magma types. EARTH PLANET SCI LETT **64,** 295–304.
Watson, E. B. & Harrison, T. M. 1984. Accessory minerals and the geochemical evolution of crustal magmatic sytems: a summary and prospectus of experimental approaches. PHYS EARTH PLANET INT **35,** 19–30.
Williams, I. S. & Claesson, S. 1987. Isotopic evidence for the Precambrian provenance and Caledonian metamorphism of the high grade paragneisses from the Seve Nappes, Scandinavian Caledonides. II. Ion microprobe zircon U–Th–Pb. CONTRIB MINERAL PETROL **97,** 205–17.
Williams, I. S., Chen, Y., Chappell, B. W. & Compston, W. 1988. Dating sources of Bega Batholith granites by ion microprobe. GEOL SOC AUST ABSTR **21,** 424.
Zhang, Y., Walker, D. & Lesher, C. E. 1989. Diffusive crystal dissolution. CONTRIB MINERAL PETROL **102,** 492–513.

BRUCE A. PATERSON, Isotope Geology Unit, Scottish Universities Research and Reactor Centre, East Kilbride, Glasgow G75 0QU, Scotland and Department of Geography and Geology, Division of Geology, University of St Andrews, St Andrews, Fife KY16 9ST, Scotland; present address: Isotope Geology Unit, SURRC.
W. EDRYD STEPHENS, Department of Geography and Geology, Division of Geology, University of St Andrews, St Andrews, Fife KY16 9ST, Scotland.
GRAEME ROGERS, Isotope Geology Unit, Scottish Universities Research and Reactor Centre, East Kilbride, Glasgow G75 0QU, Scotland.
IAN S. WILLIAMS, Research School of Earth Sciences, Australian National University, GPO Box 4, Canberra ACT 2601, Australia.
RICHARD W. HINTON, Department of Geology and Geophysics, University of Edinburgh, Edinburgh EH9 3JW, Scotland.
DONALD A. HERD, Department of Geography and Geology, Division of Geology, University of St Andrews, St Andrews, Fife KY16 9ST, Scotland.

Transactions of the Royal Society of Edinburgh: Earth Sciences, **83**, 473–483, 1992

A SHRIMP ion microprobe study of inherited and magmatic zircons from four Scottish Caledonian granites

R. T. Pidgeon and W. Compston

ABSTRACT: Using the ion microprobe SHRIMP we have analysed zircons from the Ben Vuirich, Glen Kyllachy, Inchbae and Vagastie Bridge granites from the Scottish Caledonides, in an attempt to resolve the ages of inherited zircons shown to be present in these granites by previous conventional multigrain analyses. Middle Proterozoic age components were found in inherited zircons from all four granites. Late Proterozoic (900–1,100 Ma) components have been identified in zircons from the Glen Kyllachy and Ben Vuirich granites in the Grampian Highlands. A Late Archaean age has only been detected in one zircon from the Glen Kyllachy granite. The distribution of inherited components in the granite zircon populations could reflect fundamental divisions in the age composition of granite source rocks; however, detailed assessment of this possibility must await further ion microprobe analyses on zircons from many more granites.

SHRIMP isotopic and U, Th and Pb analyses were made on successive shells of zoned zircon surrounding inherited cores from the Glen Kyllachy granite to monitor chemical changes during magmatic zircon growth. Results show that zircon shells have characteristic but significantly different Th, U and Pb concentrations. Magmatic zircon from the Vagastie Bridge granite also forms as clearly defined oscillatory zoned shells around unzoned nuclei of inherited zircon. However, the distinction between magmatic and inherited zircon in zircons from the Inchbae granite is less clear. Zircons from the Ben Vuirich granite occur as euhedral, magmatic zircons, or as rounded, subhedral, inherited zircon grains. A SHRIMP age of $597 \pm 11\,(2\sigma)$ Ma for euhedral magmatic zircon from this granite is identical, within the uncertainty, to the conventional multigrain zircon age of $590 \pm 2\,(2\sigma)$ Ma reported by Rogers *et al.* (1989) and confirms the conclusions of those authors that sedimentation of the Dalradian sequence took place in the Precambrian.

KEY WORDS: geochronology, U–Pb ages, zircon ages, zircon Th–U–Pb systems in granites.

In this paper we report the results of an ion microprobe study of individual zircons from four Caledonian granites from the Scottish Highlands. It incorporates a reconnaissance study, made in 1983 and 1984, of inherited and magmatic zircon from the three granites and a detailed investigation, made in 1991, of the age of magmatic zircons from the Ben Vuirich granite. Three of the granites, the Inchbae, Vagastie Bridge and Ben Vuirich granites, are pre-tectonic "early" granites of Read (1961). The fourth, Glen Kyllachy granite, is a late-tectonic granite (van Breemen & Piasecki 1983). Locations and descriptions of the investigated granitic rocks and their zircons are given by Pidgeon and Johnson (1974), Pankhurst and Pidgeon (1976), and Pidgeon and Aftalion (1978), and are shown in Figure 1. Conventional zircon U–Pb studies by these authors and others have demonstrated the presence of inherited zircon in the four granites. The inherited zircon often forms nuclei for growth of magmatic zircon, and direct investigation of the isotopic systems of such complex zircon grains is not possible using conventional zircon analyses. This limitation has been overcome by the development of the ANU ion microprobe SHRIMP, which provides a means of measuring the U–Pb isotopic characteristics of 20–40 μm diameter areas on the polished internal surfaces of individual zircon grains (Compston *et al.* 1984). Our initial objective was to use this spatial resolution of SHRIMP to determine the ages of inherited zircon cores. These cores are generally encased in magmatic zircon rims. Chemical changes across the rims reflect evolving magma conditions, and our second objective was to examine the potential for using the spatial resolution of SHRIMP to identify the Th–U–Pb isotopic and chemical patterns across well-developed zircon rims as potential monitors of magmatic processes. This initial study was made on well-developed, concentric, zircon shells in zircons from the Glen Kyllachy granite.

Finally, we have applied SHRIMP techniques as described by Compston *et al.* (1992) to date zircons from the Ben Vuirich granite, to provide an independent age determination for this granite. The Ben Vuirich granite intrudes the Dalradian rocks of the Scottish Highlands and has until recently been considered to be Ordovician in age (Pankhurst & Pidgeon 1976). However, the recently reported conventional multigrain zircon age determination of 590 ± 2 Ma (Rogers *et al.* 1989) has resulted in a reassignment of the age of this granite to the Late Proterozoic and a reinterpretation of the age of Dalradian sedimentation, which must also be Precambrian.

1. Analytical techniques

Zircon grains were mounted in "Epirez Epimount" epoxy resin and polished to approximately half their thickness.

U–Th–Pb isotopic analyses of the exposed zircon were performed on the ANU ion microprobe SHRIMP. A detailed description of the operation of this instrument is contained in Compston *et al.* (1984) and Williams *et al.* (1984). Analyses reported in Table 1 were made during 1983 and 1984. The analytical procedure at that time has been described by Williams *et al.* (1984). The Pb isotope ratios, and the ratios UO^+/U^+, ThO^+/UO^+ and UO^+/Zr_2O^+, were measured by magnetic field stepping using an ion counting collector. Data were combined and reduced in sets of seven cycles and two such sets were measured in succession per spot. Standard samples were analysed at the beginning and the end of each day's run and generally in the middle of the day, to monitor variations in sensitivity and discrimination between different ion species. Discrimination in the raw, measured $^{206}Pb^+/^{238}U^+$ was corrected using UO^+/U^+ as described previously (Compston *et al.* 1984) and the reduced $^{206}Pb/^{238}U$ ratios were normalised to the value of 0·0928 for the SL3 standard (equivalent to 572 Ma) determined to within 2% from replicate isotope dilution analyses (Williams *et al.* unpublished; Black *et al.* 1986). Discrimination between Pb isotopes was less than 0·1% per mass unit and no correction has been applied to Pb isotope ratios. Data recorded on Table 1 are the mean values of each set of two consecutive analyses for each spot, corrected for common Pb using the "^{208}Pb method" of Compston *et al.* (1984) and a common Pb composition determined from the model of Cumming and Richards (1975) appropriate to the age of the analysed zircon area. The precision was estimated from the combined, corrected, duplicate analyses. This takes into account ion counting statistics and additional uncertainty relating to unspecified instrumental effects. However, this is also vulnerable to inhomogeneity in the zircon on the scale of the depth of the analysis pit. Significant inhomogeneity was identified by conspicuously large variation between repeat measurements, and these have not been used in the error estimation. The precision of the calibration curve of the standard also contributes to the uncertainty in $^{206}Pb^*/^{238}U$ ratios and we have combined this uncertainty with that determined from the duplicate analyses to give a total coefficient of variation of 2% for $^{206}Pb^*/^{238}U$ and 2·3% for $^{207}Pb^*/^{235}U$. Determinations of U, Th and Pb concentrations are estimated to be within 20%, which is the uncertainty estimate reflecting the inhomogeneity of the zircon standard.

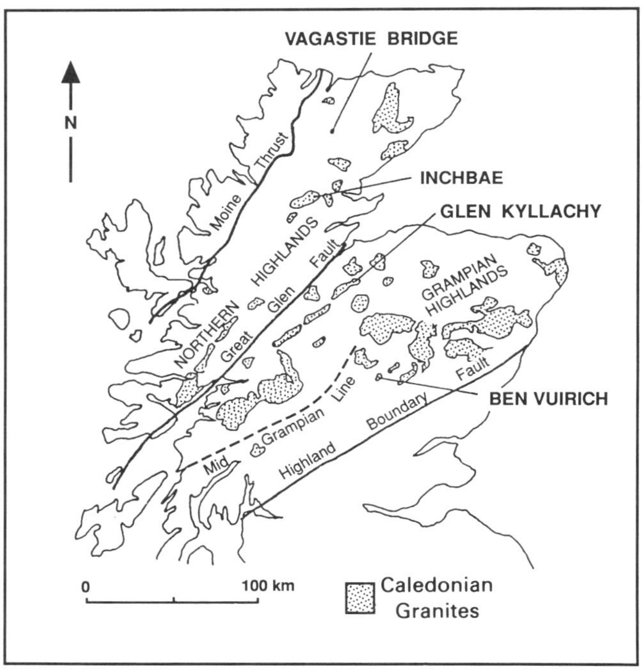

Figure 1 Map of northern Scotland showing location of the four granites investigated.

For analyses of euhedral, magmatic zircons from the Ben Vuirich granite (Table 1) the procedure followed, and the method of data reduction, is that described by Compston *et al.* (1992). Data for each spot were collected in sets of either four or five scans through the mass range of $Zr_2{}^{16}O$, Pb isotopes, ^{238}U, $^{238}U^{16}O$, $^{232}Th^{16}O$, at a mass resolution of 5500. The analysed spot diameter was 25 μm and sensitivity ranged from 7–30 cps per ppm Pb. The raw, measured $^{206}Pb^+/^{238}U^+$ was corrected and normalised to the SL3 standard as described previously. The data set was completed in one day and each sample analysis was alternated with an analysis of the standard (SL3) to closely control the calibration of $^{206}Pb/^{238}U$ with UO^+/U^+. After forming the calibration curve for the session, each standard analysis was processed as an "unknown" to obtain an empirical measure of the overall reproducibility. This represents the limiting precision in the session for the unknown zircons and has been added to the errors for each analysis arising from its counting statistics. Results on Table 2 were corrected for common Pb using the ^{208}Pb method.

Uncertainties in ages reported in this paper are at the 2σ level. Decay constants used are those recommended by the Subcommission on Geochronology (Steiger & Jäger 1977).

2. Analytical results

2.1. The Vagastie Bridge granite

SHRIMP analyses are reported for zircons from a sample of foliated, hornblende-bearing, augen granite from Vagastie Bridge, Sutherland (RC662) (Fig. 1). This rock and its zircons are described in Pidgeon and Aftalion (1978). The zircons are composite grains consisting of an irregularly shaped, unzoned zircon core surrounded by an oscillatory zoned rim. A polished section of a typical grain (grain A), which has been etched with HF to reveal internal structure, is shown on Figure 2 (A) and (B). The irregularly shaped central core is clearly identified in both the transmitted and reflected light photomicrographs as having no obvious internal structure, whereas the surrounding rim is composed of numerous fine, euhedral, concentric zones. This morphology is characteristic of all zircons in the population and is interpreted as a central inherited zircon nucleus, rounded by mechanical abrasion or chemical corrosion, surrounded by a shell of zoned zircon formed during crystallisation of the granite magma. Although our objectives were to resolve the age structure of the inherited cores and to determine the age of the magmatic zircon rims, this is not feasible from the eight analyses reported in Table 1 and the present results are considered to represent a preliminary survey of the within-grain U–Pb systems of these zircons.

Results are reported for core-rim pairs on three grains (3, A and B) and the separate core and rim of other grains (Table 1). Data points are shown on a concordia plot in Figure 3.

The four analyses on zircon cores confirm these as inherited zircon (Fig. 3(A)). One data point (3-1) is concordant within the uncertainty at *c.* 1,603 Ma ($^{207}Pb/^{206}Pb$ age). The other three points are discordant and

Table 1 Ion microprobe analyses made in 1983–84

Grain-area	U ppm	Th ppm	^{204}Pb ppb	f%	$^{232}Th/^{238}U$	$^{208}Pb^*/^{206}Pb^*$	$^{208}Pb^*/^{232}Pb$	$^{206}Pb^*/^{238}U$	$^{207}Pb^*/^{235}U$	$^{207}Pb^*/^{206}Pb^*$
Inchbae granite (RC317)										
1-1	470	190	6	1·0	0·402	0·140	0·0300 ± 7	0·0890 ± 17	0·731 ± 17	0·0597 ± 7
2-1	540	195	5	0·9	0·361	0·125	0·0319 ± 7	0·0915 ± 17	0·775 ± 18	0·0615 ± 7
3-1 centre	66	66	8	2·9	0·994	0·200	0·0185 ± 5	0·0916 ± 176	1·145 ± 220	0·091 ± 18
4-1 centre	1,035	195	19	0·3	0·189	0·056	0·0289 ± 7	0·0971 ± 18	0·818 ± 19	0·0611 ± 7
5-1	98	31	7	5·1	0·315	0·172	0·0523 ± 12	0·0923 ± 18	0·866 ± 20	0·0681 ± 8
6-1 centre	1,330	510	10	0·26	0·383	0·114	0·0918 ± 21	0·3034 ± 57	4·49 ± 10	0·1074 ± 12
6-2 edge	124	50	4	3·3	0·411	0·180	0·0425 ± 10	0·0945 ± 18	0·813 ± 19	0·0622 ± 7
6-3 centre	1,000	380	3	0·38	0·380	0·119	0·0710 ± 16	0·2260 ± 42	3·215 ± 74	0·1031 ± 12
Vagastie Bridge granite (RC662)										
2-1 rim	540	125	83	1·9	0·230	0·018	0·005 ± 1	0·0676 ± 12	0·502 ± 11	0·0535 ± 6
3-1 core	195	68	176	3·8	0·350			0·2940 ± 56	4·008 ± 92	0·0989 ± 11
3-2 rim	820	208	215	6·2	0·255			0·0710 ± 13	0·514 ± 12	0·0525 ± 6
A-1 core	150	45	53	4·6	0·290	0·092	0·042 ± 1	0·1376 ± 26	1·711 ± 39	0·0902 ± 10
A-2 rim	1,050	275	80	2·2	0·263	0·075	0·018 ± 1	0·0617 ± 11	0·470 ± 11	0·0552 ± 6
B-1 core	615	205	90	1·2	0·333	0·098	0·0735 ± 17	0·2483 ± 47	3·660 ± 83	0·1070 ± 12
B-2 rim	960	245	95	2·0	0·257	0·062	0·0145 ± 5	0·0603 ± 11	0·461 ± 11	0·0554 ± 6
E-1 core	180	50	65	6·6	0·283	0·097	0·0398 ± 9	0·1009 ± 19	0·984 ± 23	0·0708 ± 8
Ben Vuirich granite (RC992)										
1-1	315	107	97	1·9	0·341	0·87	0·0713 ± 16	0·2809 ± 53	4·050 ± 77	0·1046 ± 14
1-2	335	102	60	0·35	0·307	0·090	0·0528 ± 12	0·2640 ± 50	3·587 ± 68	0·0988 ± 11
6-1	1,115	310	30	0·33	0·279	0·083	0·0495 ± 11	0·1720 ± 32	1·715 ± 32	0·0723 ± 8
6-2	603	150	27	0·55	0·248	0·073	0·0444 ± 10	0·1582 ± 30	1·540 ± 29	0·0706 ± 8
15-1	370	13	17	0·38	0·038	0·011	0·0449 ± 10	0·1596 ± 30	1·707 ± 32	0·0776 ± 9
16-1	320	260	17	1·4	0·830	0·243	0·0265 ± 6	0·0932 ± 18	0·643 ± 12	0·0487 ± 6
19-1	1,365	295	61	0·4	0·216	0·063	0·0520 ± 12	0·1817 ± 34	1·808 ± 34	0·0722 ± 8
Glen Kyllachy granite (RC722)										
2-1 core (A)	170	40	55	1·7	0·240	0·0613	0·0704 ± 16	0·2786 ± 52	4·203 ± 97	0·1100 ± 12
2-2 rim (D)	5,110	205	3,900	23·0	0·0407			0·0517 ± 10	0·425 ± 10	0·0592 ± 7
3-1 core (A)	360	95	45	1·5	0·268	0·0838	0·0592 ± 14	0·1893 ± 36	2·408 ± 55	0·0923 ± 11
7-1 core (A)	95	54	7	1·7	0·574	0·1929	0·1103 ± 25	0·3240 ± 61	5·03 ± 11	0·1126 ± 13
7-2 core (A)	145	55	20	1·6	0·388	0·1291	0·0981 ± 22	0·2910 ± 55	4·268 ± 98	0·1064 ± 12
7-3 inner rim (B)	2,860	200	220	2·3	0·0695	0·0225	0·0226 ± 5	0·0697 ± 13	0·563 ± 13	0·0588 ± 6
7-4 outer rim (C)	14,190	965	217	0·55	0·0675	0·0242	0·0276 ± 6	0·0777 ± 15	0·593 ± 14	0·0553 ± 6
7-5 outer rim (C)	11,910	805	345	0·76	0·0673	0·0220	0·0275 ± 6	0·0840 ± 16	0·650 ± 15	0·0562 ± 6
7-6 inner rim (B)	2,520	195	26	0·20	0·0768	0·0216	0·0195 ± 4	0·0695 ± 13	0·546 ± 13	0·0570 ± 6
7-7 inner rim (B)	2,180	160	20	0·14	0·0733	0·0204	0·0190 ± 4	0·0698 ± 13	0·546 ± 13	0·0567 ± 6
7-8 outer rim (C)	13,330	1,140	130	0·26	0·0810	0·0271	0·0265 ± 6	0·0838 ± 16	0·650 ± 15	0·0561 ± 6
10-1 inner rim (B)	2,030	145	21	0·20	0·0712	0·0194	0·0192 ± 4	0·0706 ± 13	0·552 ± 13	0·0567 ± 6
10-2 core (A)	360	85	28	0·23	0·238	0·0642	0·0800 ± 18	0·2979 ± 56	4·40 ± 10	0·1072 ± 12
10-3 outer rim (D)	2,680	210	2,650	36·0	0·0813	0·075	0·030 ± 1	0·634 ± 12	0·137 ± 10	0·0287 ± 38
11-1 core edge (B)	1,310	170	32	0·27	0·128	0·0376	0·0514 ± 12	0·1755 ± 33	1·920 ± 44	0·0776 ± 9
11-2 core centre (B)	745	275	29	0·28	0·365	0·104	0·0525 ± 12	0·1850 ± 35	1·990 ± 46	0·0779 ± 9
12-1 core (A)	645	345	85	1·8	0·528	0·121	0·067 ± 2	0·2912 ± 55	7·42 ± 17	0·1885 ± 22
12-2 inner rim (B)	1,110	90	66	0·9	0·084	0·002	0·0571 ± 13	0·4324 ± 82	10·68 ± 24	0·1790 ± 21
12-3 outer rim (D)	1,890	180	1,550	27·1	0·095	0·0618	0·0275 ± 6	0·0418 ± 8	0·421 ± 10	0·0730 ± 8

Notes: * radiogenic isotope; f % is the % common ^{206}Pb in the total ^{206}Pb; uncertainties are 1σ.

Table 2 Ion microprobe analyses on euhedral zircons from the Ben Vuirich Granite (RC992)

Grain-area	U ppm	Th ppm	^{204}Pb ppb	f%	measured ratios $^{206}Pb/^{238}U$	measured ratios $^{207}Pb/^{206}Pb$	measured ratios $^{208}Pb/^{206}Pb$	corrected for common Pb $^{206}Pb^*/^{238}U$	corrected for common Pb $^{207}Pb^*/^{235}U$	corrected for common Pb $^{207}Pb^*/^{206}Pb^*$
21-1	123	32	4	0·6	0·0984 ± 23	0·0664 ± 15	0·0247 ± 33	0·0978 ± 23	0·827 ± 36	0·0613 ± 21
22-1	179	68	11	1·2	0·1011 ± 24	0·0626 ± 12	0·0334 ± 22	0·0998 ± 24	0·720 ± 35	0·0523 ± 21
23-1	196	106	5	0·6	0·0956 ± 22	0·0649 ± 12	0·0308 ± 12	0·0950 ± 22	0·788 ± 38	0·0601 ± 23
24-1	88	36	6	1·3	0·1037 ± 25	0·0668 ± 19	0·0378 ± 18	0·1023 ± 25	0·790 ± 50	0·0560 ± 32
24-2	114	55	3	0·6	0·0993 ± 24	0·0632 ± 17	0·0315 ± 16	0·0987 ± 24	0·797 ± 48	0·0585 ± 30
25-1	89	27	12	3·0	0·0920 ± 22	0·0796 ± 25	0·0261 ± 68	0·0893 ± 22	0·679 ± 52	0·0552 ± 32
25-2	172	42	49	5·8	0·1025 ± 24	0·1025 ± 20	0·0445 ± 78	0·0986 ± 23	0·734 ± 45	0·0551 ± 30
26-1	109	33	32	6·1	0·0985 ± 24	0·1036 ± 21	0·0502 ± 63	0·0924 ± 22	0·677 ± 48	0·0531 ± 34
27-1	376	337	22	1·2	0·0972 ± 22	0·0613 ± 9	0·0317 ± 10	0·0960 ± 22	0·674 ± 39	0·0510 ± 26
28-1	182	79	10	0·8	0·1448 ± 34	0·0727 ± 11	0·0441 ± 22	0·1437 ± 34	1·318 ± 51	0·0665 ± 19
29-1	174	58	27	3·3	0·0980 ± 23	0·0861 ± 15	0·0336 ± 40	0·0947 ± 22	0·768 ± 37	0·0588 ± 23
30-1	210	25	2	1·2	0·1031 ± 24	0·0629 ± 12	0·0306 ± 35	0·1029 ± 24	0·873 ± 31	0·0615 ± 14
31-1	136	30	1	0·1	0·0978 ± 23	0·0657 ± 15	0·0300 ± 26	0·0977 ± 23	0·870 ± 35	0·0646 ± 19
32-1	459	375	43	2·1	0·0946 ± 22	0·0713 ± 9	0·0322 ± 11	0·0927 ± 21	0·695 ± 36	0·0544 ± 24
33-1	92	55	2	5·4	0·0966 ± 23	0·0673 ± 20	0·0265 ± 26	0·0961 ± 23	0·833 ± 57	0·0629 ± 38
34-1	168	114	39	4·7	0·1009 ± 24	0·0960 ± 18	0·0376 ± 23	0·0962 ± 23	0·761 ± 52	0·0574 ± 36

Notes: f% is percent non radiogenic ^{206}Pb in total ^{206}Pb; uncertainties are 1σ.

fall along an approximate trend from about 400 Ma to 1,600–1,700 Ma. The discordant points A-1, B-1 and E-1 are interpreted as having experienced one or more isotopic disturbances since their initial crystallisation. The essentially concordant 1,603 Ma age of grain 3 may indicate the age of a component within the granite source. The lowest point (E-1) falls essentially on the reverse discordia defined by the conventional multigrain analyses (Pidgeon & Aftalion 1978) (see points on Fig. 3(A)). It is conceivable that the inherited zircons were originally older than middle Proterozoic, and have been strongly isotopically disturbed about 1,600 Ma and then during granite formation about 407 Ma, resulting

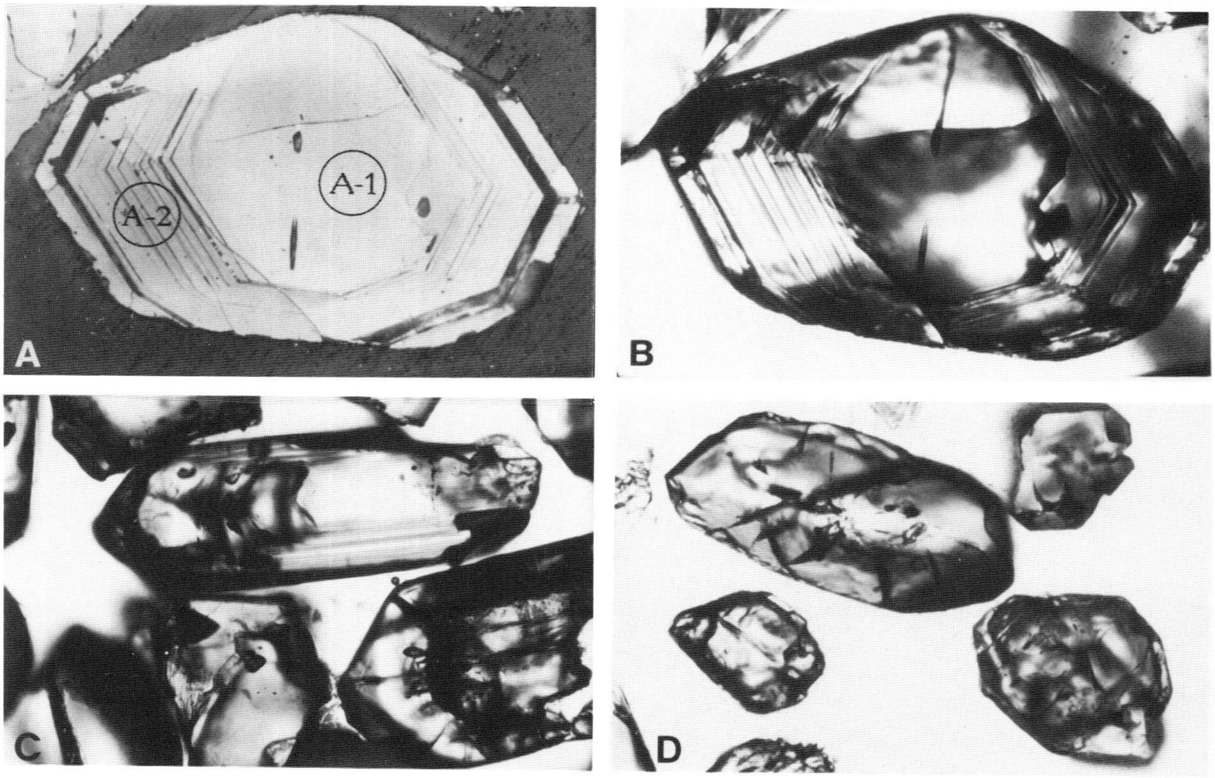

Figure 2 (A) and (B) Reflected and transmitted light photomicrographs of a typical zircon (grain A) from the Vagastie Bridge granite (RC622); grain is approximately 110 μm in length and consists of an unzoned, rounded xenocryst core encased in a shell of fine, concentric, euhedral zones distinguished by semi-transparent layers which are prone to etching with HF; areas where the oscillatory zoning has been replaced by unzoned zircon, such as at the top left part of this crystal, can be explained by adsorption and reprecipitation during magmatism or by inhomogeneous recrystallisation of zoned zircon after crystallisation; locations of SHRIMP analysed areas (see Table 1) are shown.

(C) Transmitted light photomicrograph of a group of euhedral, generally faintly zoned zircons from the Inchbae granite (RC317); grains are approximately 120 μm in length; central unzoned areas of these grains might be interpreted as xenocryst cores, as in the case of zircons from the Vagastie Bridge granite (A and B), but this is not supported by microprobe analyses which show that most unzoned central areas do not have significant older radiogenic Pb.

(D) Typical rounded, irregularly shaped zircons from the Ben Vuirich granite (RC992); these grains have no later overgrowths of magmatic zircon; euhedral, magmatic zircons from the Ben Vuirich granite (not shown), described by Rogers *et al.* (1989), appear to have crystallised separately from the zircon xenocrysts.

in the present discordant trend. However, a simpler and more feasible explanation is that the inherited zircon population is dominated by middle Proterozoic zircons which have been isotopically disturbed to varying degrees by events associated with formation of the granite magma.

The four analyses of zircon rims fall close to concordia at between 390 Ma and 420 Ma (Fig. 3(B)). This falls within the age limits proposed for this granite by Pidgeon and Aftalion (1978) but the present data are not sufficient to provide a precise age estimate. These data show no definite evidence of inherited radiogenic Pb, although this might be proposed for the two slightly reversely discordant points 2-1 and 3-2. The spread in $^{206}Pb/^{238}U$ ratios exceeds the uncertainty at the 2σ level, suggesting that analysed areas in grains B-2 and A-2 may have undergone minor, recent, isotopic disturbance.

2.2. The Inchbae granite

The Inchbae granite forms the main rock-type of the complex Carn Chuinneag granite (Fig. 1). It is a strongly foliated, coarse-grained, biotite granite and has been described by Johnson and Shepherd (1970) and Shepherd (1973). A conventional multigrain zircon U–Pb age of 560 ± 10 Ma has been reported for the Carn Chuinneag granite by Pidgeon and Johnson (1974). These authors also showed that the zircon population from the Inchbae granite contained a significant component of inherited radiogenic Pb, and zircons analysed in the present work were from the same population (sample RC317) described by Pidgeon and Johnson (1974). The zircons are generally euhedral, but HF etching of polished surfaces does not show well-developed oscillatory zones. A typical grain, shown in Figure 2(C), has a weak oscillatory zoned margin surrounding an unzoned centre. Such grains can be interpreted as consisting of an inherited core surrounded by a magmatic rim, but this is not always confirmed by the SHRIMP analyses which have shown suspected cores to have ages close to the age of the surrounding rim. A few irregularly shaped, highly conchoidally fractured, almost opaque grains (e.g. grain 6) stand out as markedly different in origin and are suspected as being inherited. These grains have narrow rims of clear zircon which can be interpreted as magmatic growth.

Ion microprobe analyses on six grains are reported in Table 1. These include two clear, featureless grains (grains 1 and 5), representing the majority of crystals, possible cores from faintly oscillatory zoned grains (2, 3 and 4) and a possible core and rim from translucent grain 6. The analysis of grain 3 has an excessive error and has been discarded. Analyses on grain 1 and faintly zoned grains 2 and 4 fall, with overlapping errors, closed to concordia between about 540 Ma and 620 Ma. The analysis of the clear zircon rim on grain 6 also falls within this group. The displacement to the right of this group of the data point from area 5-1 (Fig. 3(D)) is attributed to an underestimate of the uncertainty in the determined radiogenic $^{207}Pb/^{206}Pb$ ratio, which might be due to undetected hydride interference at mass 207. The centre of gravity of this group coincides with a cluster of conventional data points (from Pidgeon & Johnson 1974) which are essentially concordant *c.* 580 Ma. These results raise the question of whether the previous age of 568 ± 3 Ma reported for this granite (Pidgeon & Johnson 1974) is too young. A second explanation, which is not excluded by the precision of the data points, is that the 568 Ma-age is correct and that areas 6-2 and 4-1 have a small component of

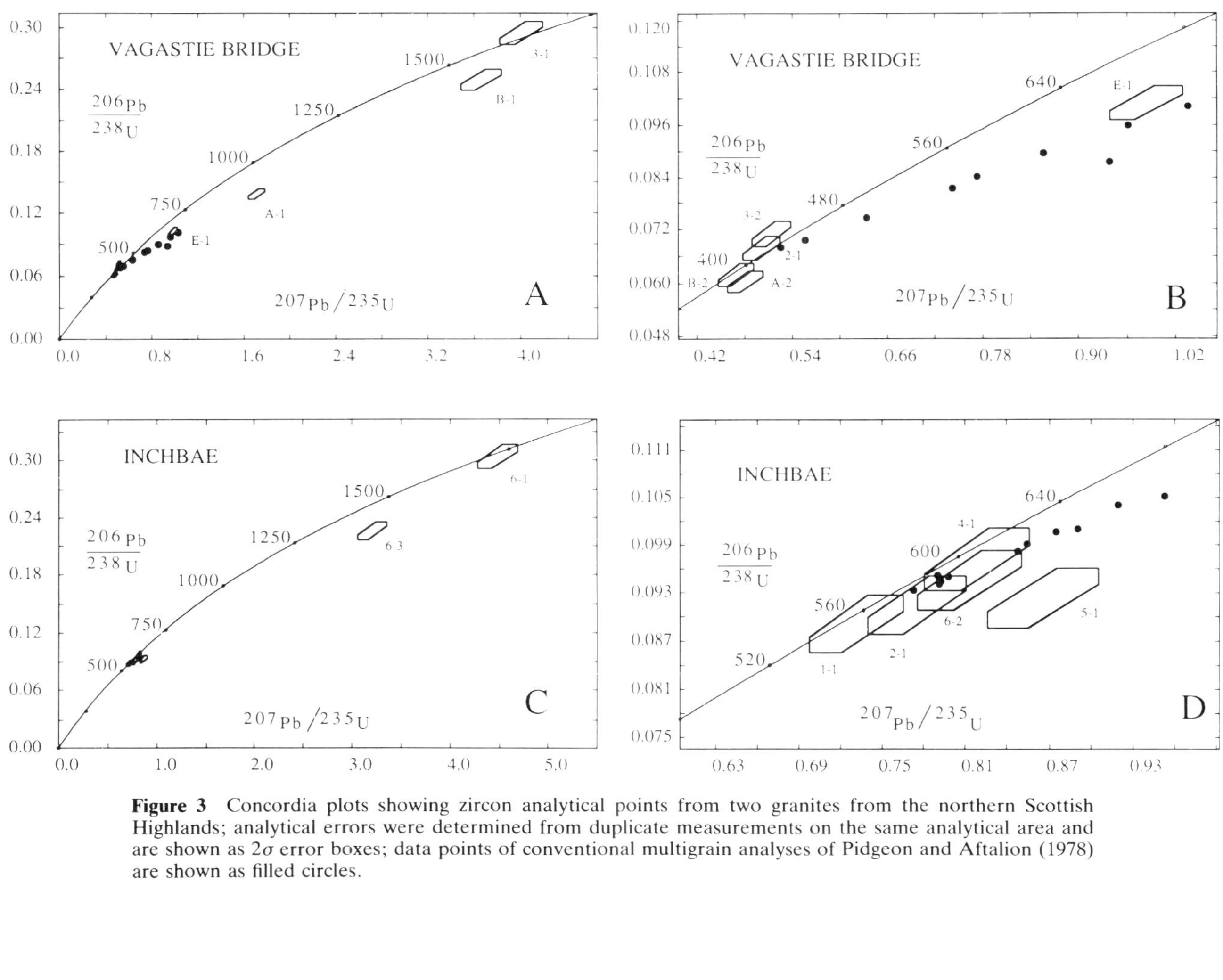

Figure 3 Concordia plots showing zircon analytical points from two granites from the northern Scottish Highlands; analytical errors were determined from duplicate measurements on the same analytical area and are shown as 2σ error boxes; data points of conventional multigrain analyses of Pidgeon and Aftalion (1978) are shown as filled circles.

inherited radiogenic Pb. High-precision conventional analyses on clear zircon crystals, or a more detailed ion microprobe study, are needed to resolve these possibilities.

Two analyses from the centre of grain 6 (6-1, 6-3) confirm this to be an inherited core (Fig. 3(C)). One point is essentially concordant at about 1,750 Ma. The second point (6-3) is about 35% discordant (Fig. 3), and falls close to a reference chord joining 1,750 Ma to 560 Ma. This variable discordance within the same grain is interpreted as a consequence of inhomogeneous response of domains within the zircon to later isotopic disturbances. The conventional data points (Pidgeon & Johnson 1974) represent bulk averages of magnetically selected size fractions of the original population. Analysed aliquots could include grains of a number of ages. The present limited number of ion microprobe analyses is insufficient to test the interpretation of Pidgeon and Aftalion (1978), based on the trend of the reverse discordia through the multigrain analytical points, that the inherited zircon population is dominated by zircons of middle Proterozoic age.

2.3. The Glen Kyllachy granite

The Glen Kyllachy granite is a weakly foliated granodiorite approximately 8 km^2 in area (Fig. 1) which intrudes the Moine metasediments in the Central Highlands late in the tectonic history of this part of the Caledonian orogeny (van Breemen & Piasecki 1983). Those authors reported a Rb–Sr whole rock isochron age of 443 ± 4 Ma (initial $^{87}Sr/^{86}Sr$ of 0·7176 ± 0·0004) for the granite and also a Rb–Sr age of 431 ± 5 Ma for primary muscovite. This age is in agreement with the 433 Ma age reported by Halliday *et al.* (1979) for muscovite from granite sample RC722. Zircons from this sample were analysed by Pidgeon and Aftalion (1978) using conventional U–Pb multigrain techniques and are the subject of the present ion microprobe study. In Pidgeon and Aftalion (1978) this sample was incorrectly assigned to the Moy granite.

These grains have distinct, rounded, unzoned, conchoidally fractured zircon cores surrounded by a mantle of zoned zircon. Conventional U–Pb isotopic analyses (Pidgeon & Aftalion 1978) show these zircons to have a significant inherited component (Fig. 4(A) and (B)). Photomicrographs of two typical grains (7 and 10) are shown in Figure 5. On a concordia plot, data-points scatter about a "reference" reverse discordia which intersects concordia at 300–350 Ma and 1,500–1,600 Ma. The scatter of points about this discordia suggests that the inherited zircon population is composed of zircons of significantly different ages. The main objective of the ion microprobe study was to resolve this age structure. The surrounding mantle of magmatic zircons can be divided optically into a number of distinct shells (Fig. 5) which represent stages in the growth of magmatic zircon, and our second objective was to examine the U–Pb chemistry of successive shells as potential monitors of magmatic conditions during progressive crystallisation.

2.3.1. The zircon cores. Analyses of cores within the six grains show three ages of inherited zircon (Fig. 4(A), (B)).

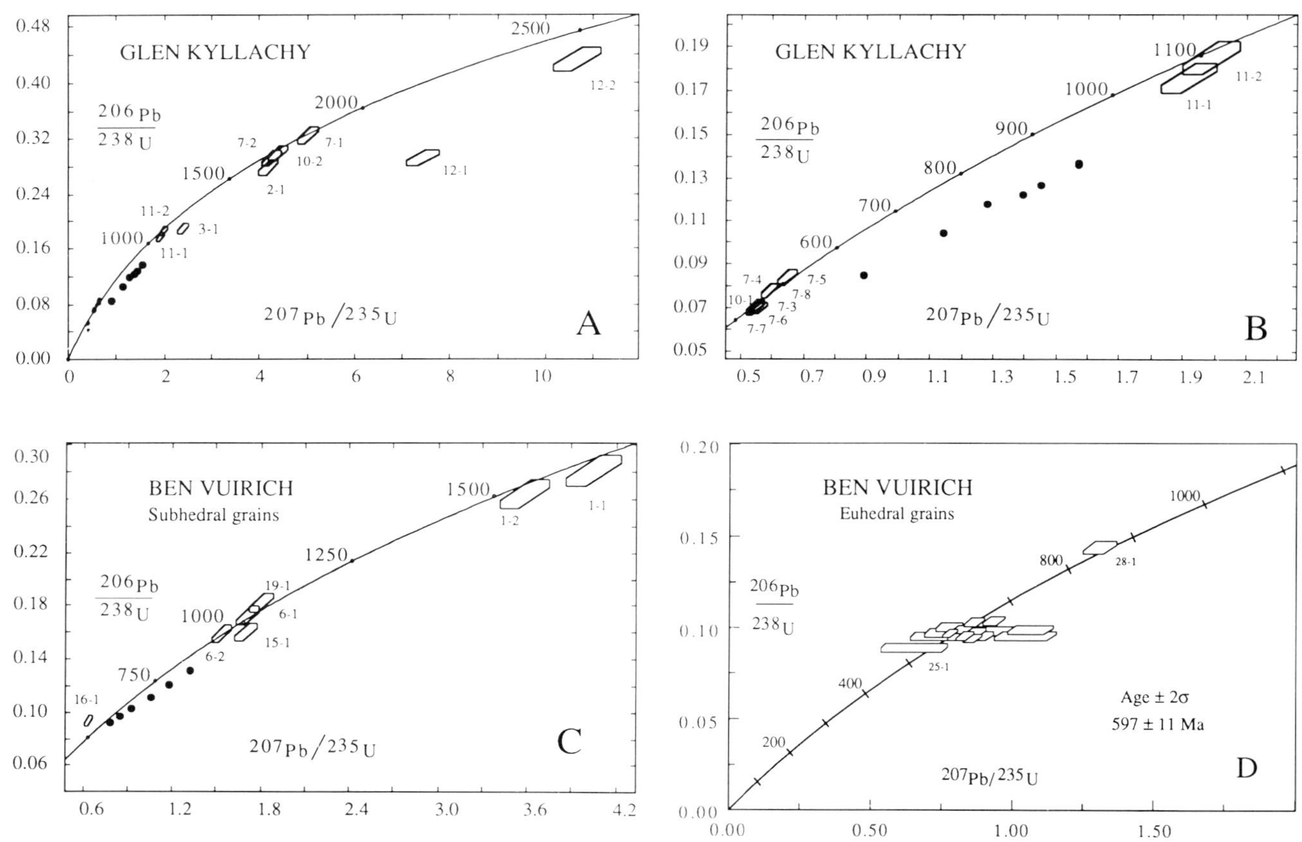

Figure 4 Concordia plots showing zircon analytical points from two granites from the Grampian Highlands; error boxes for analyses made in 1983–84 on core and rims of zircons from the Kyllachy granite and subhedral zircons from the Ben Vuirich granite were determined from duplicate analyses on the same analytical area and are drawn at the 2σ level (A, B and C); error boxes for 1991 analyses on euhedral zircon from the Ben Vuirich analyses determined from counting statistics are drawn at the 1σ level (D); data points of conventional multigrain analyses of Pidgeon and Aftalion (1978) are shown as filled circles on A, B and C.

The oldest core was identified in grain 12 (Fig. 4(A)). The two analysed areas in this grain have $^{207}Pb/^{206}Pb$ ages of 2,730 Ma and 2,640 Ma, respectively. The two data points are discordant but fall approximately on a chord joining 2,750 Ma to the cluster of analytical points between 400 Ma and 500 Ma. A second group of data points (2-1, 7-1, 7-2, 10-2) fall close to concordia between about 1,700 Ma and 1,850 Ma. The two measurements of areas on the inherited core of grain 7 cover the total spread in age, suggesting that this inherited core has undergone a heterogeneous response to later isotopic disturbance. Analytical points from grain 11 (11-1, 11-2) cluster near concordia about 1,100 Ma (Fig 4(A), (B)). Point 3-1 is located a little above and to the right of the other two points (Fig. 4(A)). This can be interpreted as a second *c.* 1,100 Ma zircon which has a minor component of older radiogenic Pb, or, alternatively, as an older zircon with an original middle Proterozoic age (*c.* 1,700 Ma), which has been strongly isotopically disturbed about 590 Ma (the age of the granite), resulting in the present discordant point which is only apparently related to the *c.* 1,100 Ma age of grain 11.

2.3.2. The zircon rims. The rounded zircon cores are surrounded by complex mantles of oscillatory zoned and mottled zircon (see typical examples in Fig. 5). The mantles

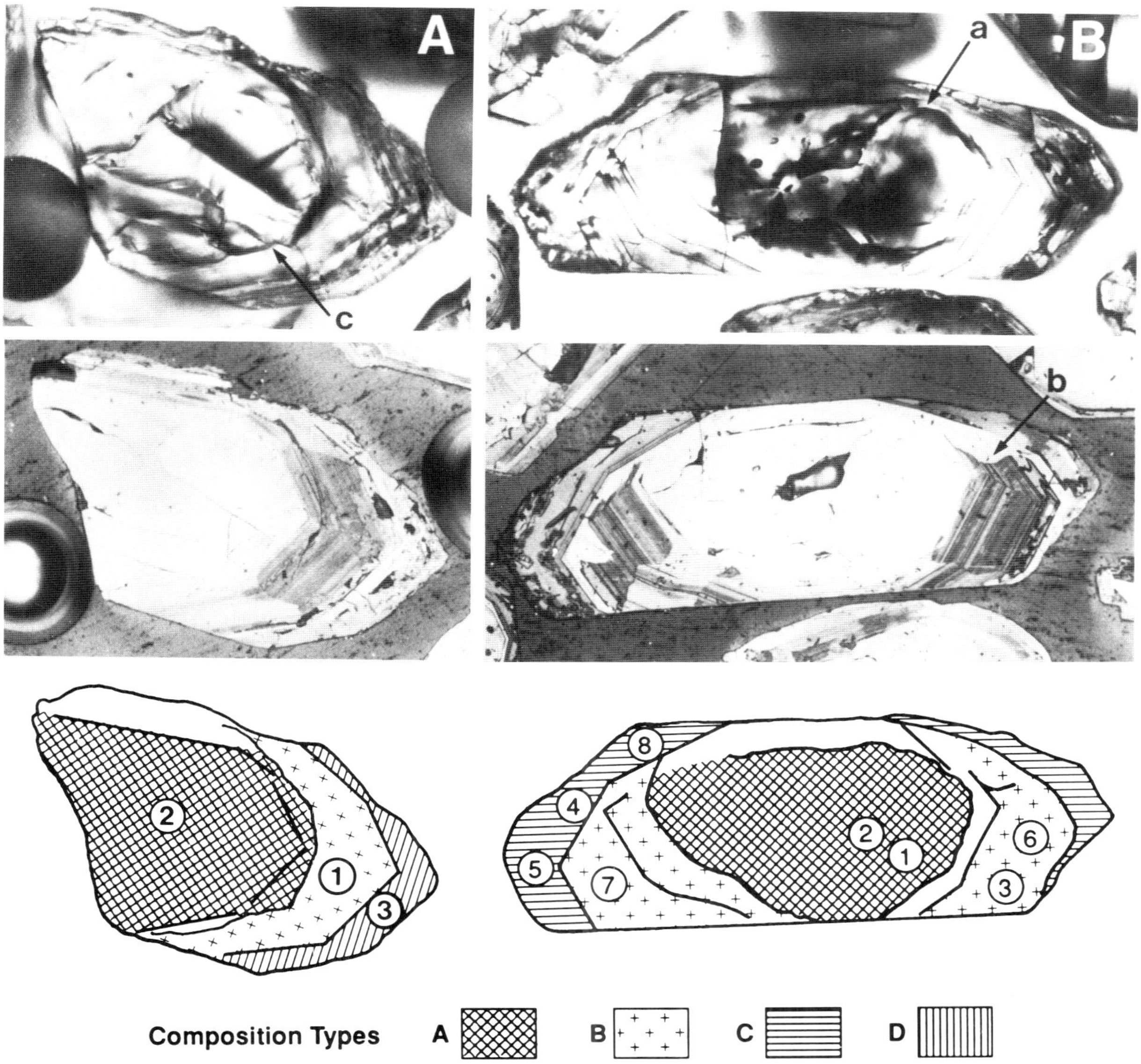

Figure 5 Reflected and transmitted photomicrographs of zircon grains 10(A) and 7(B) from the Glen Kyllachy granite; these show that grains consist of rounded, unzoned zircon cores surrounded by separate shells of oscillatory and irregularly zoned zircon; shells can be distinguished by characteristic HF etching patterns in the reflected light photomicrographs and by transparency contrast in the transmitted light photomicrographs; the rounded surfaces of the inherited zircon nuclei are indicated at (c) for grain 10 and (a) for grain 7 on the transmitted light photomicrographs; corrosive replacement, or recrystallisation, of magmatic zircon is shown at (b) for grain 7; successive shells of magmatic zircon have characteristic U–Th–Pb abundances which are indicated by sketches of the two grains; also shown on these sketches are the positions of analytical areas listed in Table 1; type A zircon, which refers to the unzoned, zircon xenocryst cores, has an inherited isotopic age and low Th (34–55 ppm), U (95–145 ppm) and common Pb (7–20 ppm ^{204}Pb); zircon type B has a magmatic age and Th of 160–200 ppm, U of 2,180–2,860 ppm and common Pb of 20–200 ppb ^{204}Pb; zircon type C has a similar U–Pb age to type B but much higher Th (805–1,140 ppm), U (11,910–14,190 ppm) and ^{204}Pb (130–345 ppb); zircon type D has a similar content of Th and U to type B but has a significantly higher content of ^{204}Pb (1,550–3,900 ppb) and an apparently discordant U–Pb age.

of these zircons consist of a number of shells, identifiable by HF etching, which are broad enough (greater than the approximately 30 μm diameter of analysed areas) to be investigated by ion microprobe. In this part of the study we are concerned with the U, Th and Pb chemistry of the magmatic zircon rims as well as the isotopic ratios, and we have used the analytical ability of the ion microprobe to monitor changes in the chemistry of these elements within and between morphologically identifiable zircon shells.

The most detailed study has been made on grain 7 (Table 1) which consists of an unzoned, inclusion-free, fractured zircon core separated by an irregular rounded margin from an inner shell of oscillatory zoned zircon. This shell has been resorbed and partially annealed and is unconformably surrounded by an outer rim of mottled zircon (Fig. 5(B)). Analyses of these three main morphological subdivisions (Table 1 and Fig. 5(B)) show that each subdivision has a characteristic isotopic discordance pattern and Th–U-common Pb chemistry. Analyses made for each subdivision over a number of days gave consistent results, which reinforces our confidence in the ability of the ion probe to determine Th–U–Pb concentrations of zircon unknowns to within the uncertainty estimated at ±20%. The inherited unzoned core is characterised by low Th (34–55 ppm), U (95–145 ppm) and common Pb (7–20 ppb ^{204}Pb). This composition, characterised by low U, Th and common Pb, is referred to as composition type A. Analyses of the inner shell of oscillatory zoned zircon, immediately surrounding the zircon core (7-3, 7-6, 7-7) show this to have almost concordant U–Pb ages at about 500 Ma and contents of Th of 160–200 ppm, U of 2,180–2,860 ppm, and common Pb of 20–200 ppb ^{204}Pb. This is referred to as composition type B. The outer rim on this grain has a concordant U–Pb age similar to type B but much higher Th (805–1,140 ppm), U (11,910–14,190 ppm) and ^{204}Pb (130–345 ppb) contents and is referred to as composition type C (Fig. 5(B)).

Unzoned cores within the other six grains have inherited U–Pb systems with similar Th–U-common Pb contents to the type A composition defined from the core in grain 7. In grains 10 (Figs. 5(A)) and 11, the surrounding oscillatory zoned zircon has Th–U–Pb contents similar to the type B zircon. The zoned inner rim of grain 12 also has a type B composition, but the inherited ^{207}Pb/^{206}Pb ratio, equivalent to an age of about 2·6 Ga, indicates that this is part of an inherited grain and cannot be equated with the magmatic zircon rims. The outer rim of grain 10 (Fig. 5(A)) has a similar Th and U content to type B, but has a significantly higher content of common Pb (2,650 ppb ^{204}Pb) and a much more discordant U–Pb system. This apparent discordance is believed to be an artefact, related to enhanced uncertainty in determining Pb/U as a result of the large common Pb content. This distinctive zircon composition is referred to as type D. As well as in grain 10, Type D zircon is found on the outer rims of grains 2 and 12, suggesting that these three outer rims crystallised under similar magmatic conditions.

The ability of the ion probe to resolve the distinct chemical compositions of the successive zircon growth stages gives information on crystallisation history of the zircons. The progression of magmatic zircon growth is shown on Figure 5(A) and (B). On grain 7 the inherited core of composition type A is overlain by type B and then by C, whereas in grain 10, type A is overlain by B which is overlain by D. On grain 2 (not shown) the original nucleus of type A zircon is overlain directly by type D. The growth of overlying zircon shells appears to take place as a pulse with a characteristic composition. This is followed by a hiatus in zircon growth, sometimes accompanied by a stage of resorption, seen as a corrosion surface on oscillatory zoned zircon (Fig. 5(B)) followed by further deposition of zircon with a different chemical composition. An important observation is that zircons do not show a consistent stratigraphy of chemical layering but, in some grains, certain chemical types appear to be missing, as if these grains experienced a hiatus in deposition during a specific growth event compared to other grains, which continued to grow at this time. An example is illustrated in Figure 5, where grain 10 shows an outward transition from composition types B to D, whereas grain 7 shows a transition from composition types B to C.

2.4. The Ben Vuirich granite

The Ben Vuirich granite is a small stock which intruded Dalradian metasediments of the Grampian Highlands (Fig. 1) after major regional (D1/D2) deformations, but before or during the main D3 deformation and peak of regional metamorphism is the Grampian terrane (Rogers *et al.* 1989).

Zircons analysed in this study (Table 1) were from the same population investigated by Pankhurst and Pidgeon (1976). The sample RC992 (National Grid NH988679) is located at position BV3 in figure 1 of that paper. The zircon population is composed mainly of irregularly shaped, rounded grains together with a minor amount of euhedral crystals (Pankhurst & Pidgeon 1976). The euhedral crystals characteristically have fine oscillatory zoning and contain a few fine rod- and needle-shaped inclusions. The rounded, irregularly shaped grains are heavily conchoidally fractured, have no obvious zones and contain small irregularly shaped opaque inclusions. Typical grains of this type are shown in Figure 2(D). These grains are interpreted as inherited xenocrysts, whereas the euhedral grains are considered to represent new magmatic zircon that crystallised as discrete crystals, without the aid of old zircon nuclei. Optical examination of grains from this population failed to reveal evidence of the presence of complex grains, with inherited cores surrounded by magmatic zircon rims.

Zircon data points from conventional multigrain analyses (Pankhurst & Pidgeon 1976) fall on a reverse discordia with intersections with concordia at 514 ± 7 Ma and 1,316 ± 26 Ma. Those authors interpreted the reverse discordia as evidence for the presence of older inherited zircon. Although the complexity of the age structure of the inherited zircons cannot be determined from the conventional analyses, Pankhurst and Pidgeon (1976) were strongly influenced by the linearity of the reverse discordia and tended to the view that it indicated a source of restricted age range. Our 1983–84 ion microprobe study was made to resolve the question of the age structure of the inherited zircon population.

Pankhurst and Pidgeon (1976) interpreted the lower intersection of the reverse discordia with concordia at 514 ± 7 Ma as indicating the emplacement age. However, the recent conventional zircon age of 590 ± 2 Ma for magmatic zircon from the Ben Vuirich granite (Rogers *et al.* 1989) has demonstrated that the previous multigrain age estimate was too low, probably as a result of significant discordance of the U–Pb systems of a number of grains included in the analysed samples. This result has such geological importance that a SHRIMP study of the magmatic zircon was made in October 1991 to provide an independent estimate of the age.

In the following, ion microprobe results are described in two parts. The first part describes analyses made in 1983–84 on subhedral zircons with rounded forms which are interpreted as inherited zircon. The second part describes analyses made in 1991 on euhedral zircons interpreted as having formed during crystallisation of the granite magma.

2.4.1. Subhedral zircon. The oldest ages determined were from analyses on two areas within irregularly shaped grain 1 (Fig. 4(C)). The dispersion of the two data points is interpreted to be due partly to uncertainties in the measurements and partly to a real difference in the degree of discordance of areas within the crystal. Analysis spot 1-1 is interpreted to be less disturbed than spot 1-2 and the position of point 1-1 near concordia suggests that the inherited zircon age is close to 1,700 Ma. In contrast to this, two analyses on Grain 6 and analyses on grains 15 and 19 fall in a group near concordia between about 940 Ma and 1,100 Ma (Fig. 4(C)). The spread of these points can also be explained in terms of the combined effects of experimental error, possible small differences in the original ages of the three cores, and a component of Pb loss due to a later event, most likely melting and granite formation. This may also explain the small discrepancy in the two analyses on the core of grain 6. The analytical point on irregularly shaped grain 16 (Table 1, Fig. 4(C)) is different from the others. The $^{206}Pb/^{238}U$ age of this grain is similar to ages reported below for euhedral zircons, indicating that it does not have an inherited component, and it is interpreted as a misidentified magmatic zircon. However, the data point is reversely discordant (Fig. 4(C)), suggesting that ^{207}Pb has been underestimated, although the reason for this is not known. The advantages of the ion microprobe in resolving the age structure of the inherited zircon population are clearly illustrated in this figure. Although the conventional zircon multigrain points of Pankhurst and Pidgeon (1976), shown as filled circles on Figure 4(C), show an almost perfect linear trend, there is no doubt that this trend is composed of zircons of at least two populations with ages of approximately 1,700 Ma and 950–1,100 Ma. On this evidence, the upper intersection age of $1{,}316 \pm 26$ Ma (Pankhurst & Pidgeon 1976) is an apparent age reflecting the mixture of at least these two ages of inherited zircon together with new magmatic zircon.

2.4.2. Euhedral zircon. SHRIMP analyses were made on fourteen grains selected, on the basis of their euhedral, elongate forms, as magmatic zircon. These grains were distinguished from the rounded, anhedral grains that were interpreted as inherited. The analytical procedure used has been described in the analytical methods section and by Compston *et al.* (1992). Basic to the procedure is the close control of the calibration of Pb/U by alternating measurements of the sample and the standard, and a total of 13 standard measurements were made alternating with samples over one continuous analysis session. The coefficient of variation of 12 of the 13 standard determinations of $^{206}Pb/^{238}U$ (one measurement was discarded as an outlier) was 2·3%. Results of analyses of samples, corrected for common Pb by the ^{208}Pb method referred to earlier, are shown on Table 2 and on a concordia plot in Figure 4(D). Uncertainties of ratios on Table 2 are 1σ errors based on counting statistics. Of these points, 18-1 (Table 2) is conspicuously anomalous and is interpreted as an inherited grain (Fig. 4). The other points fall within a group, except for analysis 15-1 which falls significantly below the main group of analysis points (Fig. 4(D)). This zircon area is interpreted to have undergone isotopic disturbance since crystallisation, whereas all other analysed areas have $^{206}Pb/^{238}U$ ratios within 3σ of a common value and are interpreted to have behaved as closed U–Pb systems. The weighted mean $^{206}Pb/^{238}U$ age of the remaining 14 analyses is 597 Ma. Combining the error in the age with that of the standard gives a final age estimate of 597 ± 11 Ma (2σ). This age is not significantly different from the conventional, multigrain determination of 590 ± 2 Ma reported by Rogers *et al.* (1989).

One analysis (28-1) is clearly an outlier (Table 2). This has a concordant age of about 860 Ma and may represent an earlier zircon crystallisation event which was not detected in the 1984–85 study of anhedral inherited grains. This is too young to be related to the "Grenville ages" determined for some inherited anhedral zircons from this population (Fig. 4). Sufficient work was not done to determine whether this grain was entirely this age or whether the area analysed was part of a poorly defined older zircon core. The fourth multigrain sample analysed by Rogers *et al.* (1989) also contained an inherited component. This may indicate the presence of complex zircons, with nuclei of inherited zicron, surrounded by a mantle of magmatic zircon growth. Equally, the analysed multigrain sample may have contained one or more anhedral, inherited zircon xenocrysts that were not detected during sample selection.

3. Discussion and conclusions

3.1. Significance of the inherited zircon components

The presence of inherited zircon in the four granites had previously been demonstrated by conventional multigrain U–Pb isotopic analyses. However, it is generally not possible to determine specific ages of inherited zircon using this technique, as the analysed multigrain samples are complex mixtures of inherited and magmatic zircon. Whereas this appeared to be an intractable problem 15 years ago, the advent of the SHRIMP ion microprobe has radically changed this perspective. SHRIMP analysis spots can be targeted on to the inherited cores themselves and ages measured directly, without contamination by magmatic zircon or other generations of inherited zircon. This has opened up new possibilities for examining the distribution of ages of inherited zircon in granites, with implications for the nature of the source rocks, and provides a means of investigating the internal stability of the U–Th–Pb system within individual inherited zircons. Following their initial crystallisation, inherited zircons may have survived a range of geological conditions such as weathering, transport, deposition, metamorphism and melting of their host rock, exposure to a granite melt, recrystallisation and possibly later metamorphism. As a consequence, it is not unexpected that analysed areas of some inherited zircon have been isotopically disturbed during one or more of these events (e.g. A-1 and B-1 from Vagastie Bridge and 12-1 and 12-2 from Glen Kyllachy). However, what is surprising is that our results show that it is not uncommon to find areas of inherited zircon that are concordant at what may be their original age or possibly an age of essentially complete resetting. For example, zircon area 3-1 from the Vagastie Bridge granite, zircon area 6-1 from the Inchbae granite, zircon areas 7-1 and 11-2 from the Glen Kyllachy granite and areas 6-1 and 19-1 from the Ben Vuirich granite are concordant within the uncertainties (Figs 3(A); 4(A), (C)). Our results also indicate that some inherited zircons show an inhomogeneous response to later isotopic disturbance. For example, two analyses within the inherited core of grain 12 from the Glen Kyllachy granite show different degrees of discordance (Fig. 4(A)). This is also shown for the inherited core in grain 6 from the Inchbae granite and to a lesser extent for grain 7 from Glen Kyllachy (Fig. 4(A)) and the inherited cores in grains 1 and 6 from Ben Vuirich (Fig. 4(C)). The observation that some zircon cores are essentially concordant, or only slightly discordant, has made it possible to identify three broad ages of inherited zircon within the four granites. The oldest age is indicated from the two discordant points on grain 12 from the Glen Kyllachy granite. An extrapolation of a line through these two data points gives a late Archaean intersection with concordia. This intersection may reflect multiple disturbances of this zircon and represent a minimum age. On the other hand, the intersection age may be close to the true age and correlate with Lewisian (possibly Scourian) events identified in the Precambrian basement of western Scotland. A second grouping of middle Proterozoic inherited zircons has been identified from concordant and discordant inherited zircons from all four granites. The range of $^{207}Pb/^{206}Pb$ ages of concordant points is from 1,604 Ma (3-1 Vagastie Bridge) to 1,756 Ma (1-1 Ben Vuirich). The third group is of Late Proterozoic age. The spread in ages of *c*. 900–1,140 Ma may represent the real distribution of zircon ages in this group, or it may be an overestimate of the original age spread, due to minor discordance of some analysed areas (e.g. area 6-2 from Ben Vuirich and the apparent minor reverse discordance of area 19-1 from the same population). However, this group of points is particularly significant as it demonstrates the presence of Grenville-aged zircons in the granite source rocks and confirms previous interpretative evidence for the presence of Grenville material in the sub-Grampian lower crust (e.g. van Breemen & Hawkesworth 1980; Piasecki 1980; Clayburn *et al.* 1983).

Although specific ages of inherited zircon have been identified, the limited amount of available data prevents determination of the statistical distribution of age components within the inherited zircon populations. Nevertheless, a suggestion of a regional pattern of inherited zircon ages has emerged. This needs to be considered within the geological framework of the four investigated granites, which are distributed along a N–S transect of the Highlands (Fig. 1). The Vagastie Bridge and Inchbae granites occur in the North Highlands Terrane of Hutton (1987), whereas the Glen Kyllachy and Ben Vuirich granites occur in Hutton's Grampian Terrane. The Ben Vuirich granite lies S of the Mid-Grampian Line of Halliday *et al.* (1985) (Fig. 1), whereas the Glen Kyllachy granite lies to the N of the line. All four granites have a component of middle Proterozoic inherited zircons, whereas a Late Archaean-aged component has been identified only in one zircon from the Glen Kyllachy granite. Inherited zircons of Grenville age are presently found only in the Glen Kyllachy and Ben Vuirich granites from the Grampian Terrane. Speculation on the significance of this preliminary pattern is premature and must await further confirmation, or otherwise, of this distribution of inherited zircon ages by a comprehensive ion microprobe investigation. Once firmly established, regional trends in the ages of inherited components in the granite zircons will have profound implications for questions on the nature of the granite source rocks and the deep crust underlying the Caledonian Fold Belt.

3.2. The age of the Ben Vuirich granite

The SHRIMP age of 597 ± 11 Ma (2σ) for the Ben Vuirich granite confirms the age of 590 ± 2 Ma reported by Rogers *et al.* (1989). The age of 597 ± 11 Ma is interpreted as the age of crystallisation during emplacement of the granite. This revises the previous multigrain zircon age of $514 \pm 6/7$ Ma by Pankhurst and Pidgeon (1976), based on the lower intersection with concordia of a discordia through analyses of mixtures of inherited and magmatic zircon. The new age requires that the deposition of the Dalradian sequence be assigned to the Precambrian. This also requires that the D1/D2 nappe structures, which preceded the emplacement of the Ben Vuirich granite, also be Precambrian in age.

3.3. Crystallisation of zircon in the granite magmas

Crystallisation of zircon from the Glen Kyllachy granite magma involved a number of growth stages, revealed by HF etching patterns as a series of symmetrical shells surrounding a central inherited zircon nucleus. The separate growth stages have characteristic chemical signatures, identified by SHRIMP analyses, from one zircon to another. The consistency of the chemistry of consecutive shells suggests pulses of crystallisation under both stable conditions and magma composition, followed by a rapid change in composition, followed by a new period of stable crystal growth. The inconsistency in the stratigraphy of growth zones in some grains (Fig. 5) is evidenced by the absence of characteristic growth stages. This cannot be explained by the progressive incorporation of zircons in mineral hosts as the magma crystallised, as this would result in an interrupted stratigraphy rather than missing layers. Another possible explanation is that zircons experienced episodic stages of growth in a uniformly evolving magma, but corrosion during adsorption stages preferentially removed shells from some grains before crystallisation resumed. The difficulty with this explanation is that the main boundaries between shells appear euhedral (Fig. 5). It would be expected that strong corrosion would produce rounded contacts between consecutive zircon shells. A third and favoured explanation is that crystallisation conditions, and possibly the composition of the magma itself, were inhomogeneous over an unspecified scale and that zircons experienced different crystallisation histories in different parts of the magma before being finally brought together in the volume of granite represented by the present sample.

Euhedral zircons from the Ben Vuirich granite appear to have crystallised separately from the inherited grains, allowing these to be used directly to date the age of crystallisation using conventional techniques. However, direct dating of magmatic and inherited zircon from the Glen Kyllachy granite will be extremely difficult, and may be impossible, using conventional techniques, even where single crystals can be analysed, as almost all zircons are complex grains consisting of inherited cores and magmatic rims. This is also suspected to be the case for the Vagastie Bridge granite. Zircons from this granite appear to be dominantly complex grains, consisting of rounded cores of inherited zircon encased in magmatic zircon (Fig. 2). On the other hand, zircons from the Inchbae granite are euhedral and weakly zoned. Rounded, unzoned areas located at the centre of grains and surrounded by a weakly zoned margin (shown on Fig. 2(C)), can be postulated to be inherited cores but analytical results on these areas (1-1, 2-1, 3-1 and 4-1 in Table 1) show no convincing evidence of inherited radiogenic Pb. Within the uncertainty, these data points can be interpreted as concordant (Fig. 3(D)), suggesting a zircon age of about 580 Ma, which is older than the published zircon age of 568 ± 3 Ma (Pidgeon & Johnson 1974). Alternatively, points can also be interpreted as falling along the reverse discordia defined by the conventional analyses points, suggesting that the analysed areas contain a minor component of inherited radiogenic Pb.

4. Acknowledgements

We are pleased to acknowledge Drs Ian Williams and John Foster for assistance with the early ion microprobe analyses. We are also very grateful to Robert Reinfrank for analysing the euhedral zircons from the Ben Vuirich granite and reducing the data. We thank Dr David Nelson and an anonymous referee for valuable comments on the paper. This research has been supported by the Research School of Earth Sciences, ANU, and by a grant from the Australian Research Council.

5. References

Black, L. P., Williams, I. S. & Compston, W. 1986. Four zircon ages from one rock: the history of a 3930 Ma-old granulite from Mount Sones, Enderby Land, Antarctica. CONTRIB MINERAL PETROL, **94,** 427–37.

Clayburn, J. A. B., Harmon, R. S., Pankhurst, R. J. & Brown, J. F. 1983. Sr, O, and Pb isotope evidence for origin and evolution of Etive Igneous Complex, Scotland. NATURE **303,** 492–7.

Compston, W., Williams, I. S. & Meyer, C. 1984. U–Pb geochronology of zircons from Lunar breccia 73217 using a sensitive high mass-resolution ion microprobe. Proc. XIV Lunar Planetary Science Conference. J GEOPHYS RES **89,** supplement B525.

Compston, W., Williams, I. S., Kirschvink, J. L., Zhang Zichao & Ma Guogan 1992. Zircon U–Pb ages for the Early Cambrian time-scale. J GEOL SOC LONDON (in press).

Cumming, G. L. & Richards, J. R. 1975. Ore lead isotope ratios in a continuously changing earth. EARTH PLANET SCI LETT **28,** 155–71.

Halliday, A. N., Aftalion, M., van Breemen, O. & Jocelyn, J. 1979. Petrogenetic significance of Rb–Sr and U–Pb isotopic systems in the 400 Ma old British Isles granitoids and their hosts. GEOL SOC SPEC PUBL **8,** 653–61.

Halliday, A. N., Stephens, W. E., Hunter, R. H., Menzies, M. A., Dickin, A. P. & Hamilton, P. J. 1985. Isotopic and chemical constraints on the building of the deep Scottish lithosphere. SCOTT J GEOL **21,** 465–91.

Hutton, D. H. W. 1987. Strike-slip terranes and a model for the evolution of the British and Irish Caledonides. GEOL MAG **124,** 405–25.

Johnson, M. R. W. & Shepherd, J. 1970. Note on the age of metamorphism of the Moinian. SCOTT J GEOL **6,** 228–30.

Pankhurst, R. J. & Pidgeon, R. T. 1976. Inherited isotope systems and the source region pre-history of early Caledonian Granites in the Dalradian Series of Scotland. EARTH PLANET SCI LETT **31,** 35–68.

Piasecki, M. A. J. 1980. New light on the Moine rocks of the Central Highlands of Scotland. J GEOL SOC LONDON **137,** 41–59.

Pidgeon, R. T. & Aftalion, M. 1978. Cogenetic and inherited zircon U–Pb systems in Palaeozoic granites from Scotland and England. *In* Bowes, D. R. & Leake, B. E. (eds) *Crustal processes and evolution in N.W. Britain and adjacent regions*. GEOL J SPEC ISSUE **10,** 183–220.

Pidgeon, R. T. & Johnson, M. R. W. 1974. A comparison of zircon U–Pb and whole-rock Rb–Sr systems in three phases of the Carn Chuinneag granite, northern Scotland. EARTH PLANET SCI LETT **24,** 105–12.

Read, H. H. 1961. Aspects of Caledonian magmatism in Britain. LIVERPOOL AND MANCHESTER GEOL J **2,** 653–83.

Rogers, G., Dempster, T. J., Bluck, B. J. & Tanner, P. W. G. 1989. A high precision U–Pb age for the Ben Vuirich granite: implications for the evolution of the Scottish Dalradian supergroup. J GEOL SOC LONDON **146,** 789–98.

Shepherd, J. 1973. The structure and structural dating of the Carn Chuinneag Intrusion, Ross-shire. SCOTT J GEOL **9,** 63–88.

Steiger, R. H. & Jager, E. 1977. Subcommission on geochronology: convention on the use of decay constants in geo- and cosmochronology. EARTH PLANET SCI LETT **36,** 359–62.

van Breemen, O. & Hawkesworth, C. J. 1980. Sm–Nd isotopic study of garnets and their metamorphic host rocks. TRANS R SOC EDINBURGH EARTH SCI **75,** 65–70.

van Breemen, O. & Piasecki, M. A. J. 1983. The Glen Kyllachy Granite and its bearing on the nature of the Caledonian Orogeny in Scotland. J GEOL SOC LONDON **140,** 47–62.

Williams, I. S., Compston, W., Black, L. P., Ireland, T. R. & Foster, J. J. 1984. Unsupported radiogenic Pb in zircon: a cause of anomalously high Pb–Pb, U–Pb, and Th–Pb ages. CONTRIB MINERAL PETROL **88,** 322–7.

R. T. PIDGEON, School of Applied Geology, Curtin University of Technology, GPO Box U 1987, Perth WA 6001, Australia

W. COMPSTON, Research School of Earth Sciences, The Australian National University, GPO Box 4, Canberra City, ACT 2601, Australia

Transactions of the Royal Society of Edinburgh: Earth Sciences, **83**, 485–506, 1992

Abstracts

Age of cassiterite mineralisation at the Zaaiplaats Tin Mine, South Africa: implications for the age of the Bushveld Complex

Anita S. Andrew, Brian L. Gulson, Mel T. Jones and Peter J. Pollard

Tin-tungsten-rare-earth-element mineralisation at Zaaiplaats is hosted by the Bobbejaankop and Lease granites: miarolitic, brick-red alkali feldspar granites which form at high levels within the Lebowa Granite Suite of the Bushveld Complex. Hydrothermal minerals occur as pervasive alteration and within miarolitic cavities. Cassiterite occurs as a cavity filling, as a replacement mineral in tabular, subhorizontal zones and in shallowly-plunging and branching pipes, and as disseminated mineralisation unrelated to fracture systems.

The age of mineralisation at Zaaiplaats has been determined by U–Pb dating of cassiterites. Bulk fractions of cassiterite have $^{206}Pb/^{204}Pb$ ratios from 30–3,500. Uranium–Pb data for five cassiterites with the highest $^{206}Pb/^{204}Pb$ ratios lie on a chord with intersections at $2{,}099 \pm 3$ Ma and $\approx 300 \pm 20$ Ma (MSWD = 0·44). Four samples are from pipe systems and one from cavity fill, suggesting a common age within 6 Ma. The $2{,}099 \pm 3$ Ma age is at the upper limit of published $^{87}Rb/^{86}Sr$ estimates for emplacement of both the Lebowa Granite Suite ($2{,}046 \pm 55$ Ma) and the Rustenberg Layered Suite ($2{,}061 \pm 27$ Ma) and identical within error of an $^{40}Ar/^{39}Ar$ age of $2{,}096 \pm 12$ Ma for the Upper Zone of the Rustenburg Layered Suite.

The precise U/Pb dating of cassiterites at Zaaiplaats suggests that mineralisation occurred early in the cooling history of the Bobbejaankop and Lease granites. This is consistent with fluid inclusion and stable isotope data which show that cassiterite and the related brick-red alkali feldspar alteration result from high temperature (>250°C) interaction of magmatic fluids with the cooling granite.

The origin and geotectonic setting of Andean granites

Michael P. Atherton and Nicholas Petford

The Andean magmatic cycle began in Peru with the formation of a major depositional structure, the West Peruvian Trough, the westerly component of which has a volcanic facies indicating it formed very rapidly in a relatively deep-sea spreading/ridge system during the Albian. Beneath this major deposystem is an arch-like structure (density of $3{\cdot}00\,g\,cm^{-3}$) made up of mafic, mantle-derived material. Fracturing and splitting of the crust above this structure was associated with intensive dyking and intrusion/extrusion of basaltic material during the Albian.

The Coastal Batholith was intruded (101–37 Ma) into this linear edifice, and is confined mainly within it. Intrusion was permissive at 3–4 km from the surface, and was accompanied by large synplutonic basaltic dyke swarms, emphasising further the role of extension during batholith emplacement. The tonalitic–granitic rocks of the batholith formed on melting of this "new" basaltic crust at shallow depth on adiabatic decompression during the extension following basin inversion. About 20 Ma, granitic magmatism jumped to the E, producing small stocks within coeval "plateau" basalts (52·5–12 Ma). Intrusion occurred during a major faulting phase and took the form of stocks, domes, breccia pipes and dykes.

The last magmatic events in northern Peru include the Cordillera Blanca Batholith (12–13 Ma) which was intruded into a major strike-slip transtensional zone. Chemical, mineralogical and isotopic data indicate a much deeper source than that for the Coastal Batholith, which is considered to be newly accreted lower crust beneath the batholith.

Thus, the tectonic environment during the Andean in Peru was dominantly extensional, the magmatism younging to the E. All extensional systems are orthogonal to the Andean trend and relate to rifting associated with spreading systems opening up in the same direction. The relation to subduction is unclear.

The Grayback Pluton: magmatism in a Jurassic back-arc environment, Klamath Mountains, Oregon

Calvin G. Barnes, Kenneth Johnson, Melanie A. Barnes, Tore Prestvik and R. W. Kistler

The Late Jurassic Grayback Pluton was emplaced in a back-arc environment (Mesozoic remnant arc) behind a contemporaneous Late Jurassic arc. The main stage of the pluton consists of an early unit of reversely zoned tonalite to gabbro that was subsequently intruded by synplutonic noritic and gabbroic magmas. Late-stage tonalitic and granitic dykes typically contain mafic enclaves and zones of hybrid tonalite and quartz diorite. The mafic rocks in the pluton are predominantly calc-alkaline, with magnesian clinopyroxene, calcic cores in plagioclase, and elemental abundances similar to H_2O-rich arc basalts. Some mafic samples contain relatively Fe-rich clinopyroxene, lack calcic plagioclase and are compositionally similar to evolved high-alumina tholeiite. Both groups of mafic rocks lack depletions of HFSE that are typical of arc magmas. Intermediate rocks span a wide range of elemental compositions that can be explained by mixing of calc-alkaline or tholeiitic mafic magmas with tonalitic or granitic magmas, combined with variable degrees of crystal fractionation. Initial $^{87}Sr/^{86}Sr$, ε_{Nd}, and $\delta^{18}O$ range from 0·7028–0·7045, 4–9, and 7·8–14‰, respectively. Late-stage felsic rocks have the highest $^{87}Sr/^{86}Sr$ and $\delta^{18}O$ and display arc-like depletions of HFSE. These data suggest depleted mantle sources for the mafic magmas but a heterogeneous (young remnant arc) crustal source for both the tonalitic and granitic magmas. Mixing of mantle-derived basaltic magmas with isotopically variable crustal melts gave rise to the observed range of compositions. The heterogeneity and compositional range of the Grayback Pluton suggest a complex crustal plumbing system that allowed for injection of batches of mafic magma (derived from at least two types of mantle source rocks), mixing with pods of crustal melts (some tonalitic and some granitic) and partial homogenisation of the mixtures to form the main stage of the pluton.

Forceful emplacement in the southern Adamello Massif, Italian Alps: intrusive history of the Lago della Vacca Suite

Jonathan D. Blundy and Barbara E. John

The Lago della Vacca Suite (LVS) is a small plutonic body, exposed over $20\,km^2$, in the southern Adamello Massif, Italy. The LVS was emplaced 40 Ma at ~200 MPa, into warm but consolidated precursor granitic and ultramafic rocks, and Triassic dolomites. The suite exhibits crude concentric zonation from dioritic margins through tonalite, to a granodiorite core, each rock type representing a discrete magmatic pulse, emplaced before the previous pulse was fully solidified. Synplutonic dykes and mafic inclusions are widespread.

High-temperature foliations are well-developed in all units, and define steeply inward-dipping concentric trajectories defined by the planar alignment of plagioclase, amphibole and biotite, and flattened (oblate) mafic inclusions. Lineations are rare and unsystematic. Foliation is deflected around a large rigid body of mafic/ultramafic country rock, but passes continuously into warm precursor granites, consistent with radial expansion against wall rocks of variable rheology. Thin aplite dykes and small-scale shear zones also have orientations consistent with radial expansion.

Mafic inclusions used as strain markers indicate a general trend of decreasing strain toward the centre of the suite. In detail, the strain distribution is very heterogeneous, with decametre wide zones of high strain ($9 \leq X/Z \leq 12$), within broader zones of lower strain ($X/Z \leq 3$). The high-strain zones are concentric, laterally discontinuous and characterised by fine-grained dioritic rocks. Compositionally, these rocks resemble undeformed synplutonic mafic dykes.

Amphibole-plagioclase thermometry constrains the temperature of deformation to 850–750°C. There is no evidence of a low temperature component, despite evidence of grain-size reduction and recrystallisation in the most deformed diorites. Amphibole in these rocks is characterised by high K contents and low *mg#* relative to amphibole in undeformed rocks of comparable composition, suggesting interaction of the deforming rocks with evolved interstitial melts. We conclude that the syn-intrusive deformation of the LVS occurred at magmatic temperatures, and propose that strain partitioning reflects local variations in temperature.

A-type granite magmatism in the western Mediterranean

Bernard Bonin

The Permo-Triassic period was characterised in the western Mediterranean area by abundant magmatic activity just after the end of the Variscan orogenesis. Volcanic–plutonic complexes were emplaced mainly in the internal parts of the orogen and constitute a 4,000 km-long province stretching from Morocco through Corsica and the Alps to Austria. The magmatic activity has taken place during two discrete episodes:

(i) The mid-Permian (*c.* 270 ± 5 Ma) was marked by the emplacement of the first A-type formations, with ignimbritic traps, and ring complexes made up by a gabbro–monzonite pink to red subsolvus biotite granite association.

(ii) At the Permian-Triassic boundary (245 ± 10 Ma), numerous A-type caldera-related ring complexes were emplaced along the future Alpine belt under an incipient rift regime. The suite is characterised by a basalt–trachyte–rhyolite (comendite) association, syenites, and a large variety of hypersolvus to subsolvus, peraluminous to peralkaline, grey-green to pink A-type granites.

Sources and conditions of differentiation of A-type granite magmatism changed with time. At the end of the Variscan orogenesis, a new mantle source replaced the old complex system of mixed oceanic-continental crust-mantle sources. Primary mantle-derived melts were probably trapped at the crust-mantle boundary where they evolved through high-pressure fractionation to intermediate compositions. Intermediate melts were collected in shallow magma chambers, placed at the ductile-brittle transition in the crust, where they evolved through low-pressure fractionation to granitic residual melts. Felsic melts were subsequently less and less contaminated by crustal host rocks. Strong correlations between hydrothermal events, mineralisations and crustal isotope signatures suggest that crustal contribution relates essentially to percolating fluids.

Magmatism and hydrothermal metasomatism in Mesozoic Damaraland granites, northern Namibia

Peter Bowden, Judith Kinnaird, Michael Diehl and Franco Pirajno

The fracturing of Gondwanaland as part of the larger Pangaean fragmentation has been chronicled by numerous geological features, including examples of magmatism oriented along major zones of weakness in the continental lithospheric segments of S America and Africa. Following the widespread outpouring of the Karoo basalts, Jurassic-Cretaceous magmatism was finally concentrated along several major lineaments linked to the development of a series of transformed fracture systems in the S Atlantic Ocean. This activity was marked on the southwestern African continent by the emplacement of a number of felsic-mafic complexes, granite centres, carbonatites and kimberlites.

The Damaraland zone of anorogenic complexes in northern Namibia extends more than 400 km as a well-defined series of NE-oriented lineaments between the Ugab and Khan rivers. The lineament zone coincides with the axis of the Damaran (Pan-African) orogen. Twenty Mesozoic granitic centres have been identified along this zone. Based on evidence from these centres, the marginal zones of granite intrusions are the best places where hydrothermal overprinting of an original magmatic assemblage can be observed. The fluids of magmatic origin causing endogenetic hydrothermal metasomatism by intergranular exchange are derived from the consanguineous crystallisation residua of the host pluton. There is also the possibility, in multicentred anorogenic granites, that residual fluids are expelled from later adjacent intrusions whose fluid chemistry may be out of equilibrium with the bulk chemistry of the earlier crystallised pluton. This involves mass transfer of components to form new subsolidus assemblages.

Origin and distribution of granitic and related rocks in the Coast Plutonic-Metamorphic Complex, N American Cordillera, southeastern Alaska, U.S.A.

David A. Brew

Discrete chrono-lithological granitic rock units occur in both continuous and discontinuous belts that run the 1000 km-plus length of the late Mesozoic and Cainozoic Coast Plutonic-Metamorphic Complex (CPMC) in southeastern Alaska and the adjacent parts of British Columbia. The CPMC has four main zones that are characterised by specific granitic units and to some extent by specific metamorphic protoliths: the western metamorphic zone or belt, the central metamorphic and granitic zone, the central granitic zone, and the eastern metamorphic zone.

There are four major granitic belts in the CPMC. The tectonic setting of the oldest, pre-Late Triassic, belt is uncertain; the two youngest major belts, the 85–55 Ma Great Tonalite Sill belt and the 55–45 Ma Coast Mountains belt, are generally related to the alternately transpressive and transtensive accretion of the Wrangellia-Alexander and Chugach tectonostratigraphic terranes (Insular Superterrane) to the previously emplaced Yukon prong and Stikine terranes (Intermontane Superterrane). The tectonic settings of the next-to-oldest major belt, the Admiralty-Revillagigedo belt of magmatic-epidote-bearing tonalite, is probably similar. The tectonic setting of the minor 110–100 Ma Klukwan Duke Alaskan-type ultramafic belt is enigmatic. Calcalkalic, island-arc affinity rocks are predominant in all belts. However, each has a somewhat different origin that is reflected in its structural, modal compositional and major, minor, trace, and rare-earth-element compositional characteristics. Three minor granitic belts, of Late Triassic and Early Jurassic, early Early Cretaceous, and of Late Oligocene age, are also present in the CPMC.

Transpression and granite genesis

Michael Brown, Richard S. D'Lemos and Robin A. Strachan

Transpression involves horizontal shortening across and vertical lengthening along a zone of transcurrent shear. This thickens crust and inverts sedimentary basins, which leads to high-temperature metamorphism and anatexis. Shear zones provide convenient paths of ingress for volatiles as well as ascent paths for magma, and deformation enhances melt segregation during anatexis. Mantle-deep terrane boundary faults provide access for mantle-derived magma to intrude and interact with the crust. In the Cadomian Belt of northwestern France, arc-related outboard terranes have been amalgamated with inboard terranes, interpreted to represent elements of a behind-arc basin sedimentary sequence, during sinistral transpression. In the outboard terranes, post-kinematic plutonic complexes comprise diorite-to-granite with evidence for local co-mingling of magmas. More extensive mixing between depleted and enriched components is evidenced by the geochemistry and in the isotopic signature of the granitic rocks. By contrast, in the inboard terranes, thickening of the Brioverian succession during transpressional terrane accretion has resulted in fluid-present anatexis in the St Malo migmatite belt. The St Malo migmatite belt is imbricated by steep, sinistral strike-slip shear zones which contain syn-kinematic granites, suggesting that shear zone deformation and anatexis were broadly coeval. Further inboard, the high-level Mancellian granites are intruded into Brioverian rocks. By analogy with the shear zone system to the W, these granites may have been emplaced into pull-apart structures along a jog in the E–W-trending zone of sinistral shear at the S margin of the belt of transpression. The St Malo migmatites/anatectic granites and the Mancellian granites have like petrographical features, statistically indistinguishable chemical compositions, and similar Nd isotopic characteristics. These features suggest a common origin for the St Malo migmatites/anatectic granites and the Mancellian granites, which is interpreted to be by anatexis of the immature arc-derived turbidites of the Brioverian succession.

Three-dimensional analysis of the crystallisation sequence and textural development in two Peruvian granitoid rocks

David Bryon

Microscopic geometrical and textural characteristics of crystallisation sequences, and inferences regarding the rheological properties of crystallising granites, cannot be properly quantified without three-dimensional textural analysis. Three-dimensional analysis utilising serial sectioning and image digitisation has been used to interpret textural development in two granodiorites from the Linga superunit of the Coastal Batholith, Peru. The high level of emplacement of these granodiorites (Atherton 1984) and limited deformation allows excellent preservation of the primary textures. Despite their very similar modal and normative compositions, the two rocks' different textures reflect the variation in *nucleation density, growth rate* and *timing* of crystallisation of individual phases. Serial geometries produced for all the major phases identify the individual role that each phase plays during the three main stages of crystallisation, namely: (i) early crystallisation from a melt containing a high proportion of melt, where crystal growth proceeded relatively uninhibited; (ii) development of a touching three-dimensional crystal framework (R.C.M.P. of Arzi 1978); (iii) subsequent interstitial crystallisation where the growth of individual crystals was controlled by the already crystalline surrounding material. All the geometries support the modelled crystallisation path deduced from the An : Ab : Qz : Or system, and suggest that all the major phases had started to crystallise by stage (ii), i.e. the development of a crystal framework.

Arzi, A. A. 1978. TECTONOPHYSICS **44,** 173–84.
Atherton, M. P. 1984. *In* (Harmon & Barrerio) (eds). *Andean magmatism,* 168–179. Nantwich: Shiva.

Geochemical contrasts between intersecting Devonian and Carboniferous magmatic belts in the eastern Lachlan Fold Belt

Paul F. Carr, Bruce W. Chappell and **Brian G. Jones**

The subvolcanic Marulan Batholith is the same general age as most batholiths in the Lachlan Fold Belt and constitutes part of a world-wide magmatic event about 400 Ma. The Marulan Batholith is closely associated in the field with several Carboniferous contact-aureole plutons and, therefore, marks the intersection of a southern Devonian magmatic belt with a northern Carboniferous magmatic belt.

Differences in the sources for the Devonian and Carboniferous granites are reflected in pronounced geochemical differences between these granites. The Devonian plutons are characterised by a continuous and fairly uniformly distributed range of compositions from 56–73% SiO_2. In contrast, the Carboniferous plutons have a weak bimodal distribution of compositions and have significantly higher average Na_2O and Sr contents. Rare-earth-element patterns for the Carboniferous granites show considerably more fractionation and enrichment in LREE than the geographically associated Devonian plutons.

Isotopic data for Devonian plutons form three groups, two of which have parallel isochrons with early Devonian ages and initial ratios of 0·7062 and 0·7071, respectively. Variations within these two groups may be explained by feldspar fractionation. Plutons in the third group do not fit an isochron, have the highest initial $^{87}Sr/^{86}Sr$ ratios (0·7077–0·7100 at 395 Ma), and have a linear array on a plot of initial $^{87}Sr/^{86}Sr$ versus 1/Sr. Variations within this third group possibly reflect derivation from a heterogeneous source.

Initial $^{87}Sr/^{86}Sr$ ratios for the Carboniferous plutons are appreciably lower and range from 0·7047–0·7052. The isotopically more primitive source for the Carboniferous magmas may have underplated the crust during the early Devonian and initiated partial melting to produce the Marulan Batholith.

Evolution of the Indian Peak Volcanic field, western U.S.A.: implications for the origin of A- and I-type granites in subduction settings

Eric H. Christiansen, Myron G. Best and **Joaquin Ruiz**

The Indian Peak caldera complex erupted 10,000 km^3 of magma 32–27 Ma. Nd-Sr isotopic data and temporal-spatial patterns show that these high-K calc-alkaline rocks formed in an open magma system, driven by mafic magma derived from a southward-migrating thermal anomaly. The migration may be associated with progressive foundering of subducting lithosphere away from the overriding continental plate. As the thermal anomaly stagnated in the southern Great Basin, the high power input optimised magma generation in the crust. Small volumes of andesitic lavas are derivatives of this mantle power supply. They are compositionally diverse and not co-magmatic, but all have low TiO_2, negative Nb-anomalies and REE patterns which lack Eu anomalies. Three large-volume crystal-rich dacite tuffs have all of the characteristics of I-type granites, but isotopic and elemental mixing trends suggest that they formed by mixing of andesite and rhyolite magmas rather than by melting a distinctive source in the crust. The death of the magma system was marked by the eruption of a trachytic tuff with plagioclase, clino- and orthopyroxene, and oxides. Nd isotope ratios are higher and Sr isotope ratios lower than in the dacites but similar to andesites. The trachyte has many of the characteristics of A-type granites, including a high eruption temperature, high alkalies, Ba, Zr and Ga, but it can be modelled as a product of clinopyroxene-dominated fractionation from andesitic magma at relatively high temperatures and does not require a distinctive crustal source.

Contact metamorphosed granites: their metamorphic zonation and implications upon emplacement mechanisms

A. P. Clare

Petrological and structural studies of the contact metamorphic effects of one granite upon another and the surrounding hornfels are presented for the Bendemeer Adamellite (41 km NE of Tamworth, N.S.W.) and the Gwydir River Adamellite (30 km W of Armidale, N.S.W.) in the southern part of the New England Batholith. The Permian Bendemeer Adamellite intrudes the Carboniferous S-type microcline and cordierite-bearing Banalasta Adamellite, the late Permian Gwydir River Adamellite intrudes the middle Permian I-type orthoclase-bearing Yarrowyck Granodiorite. Both aureoles are shown to have essentially identical metamorphic zonations.

A mineralogical and microstructural examination of the

metagranites away from the later intrusives reveals three distinct zones: (1) a regional unaffected zone (to within 4 km of the Bendemeer Adamellite and to within 1·5 km of the Gwydir River Adamellite); (2) a low-grade strained zone (between 4 and 1·5 km and between 1·2 and 0·35 km, respectively, for the Bendemeer and Gwydir River plutons) characterised by high K-feldspar triclinicities and deformational microstructures; and (3) a high-grade annealed zone (within 1·5 and 0·35 km, respectively, of the Bendemeer and Gwydir River plutons), characterised by low to zero K-feldspar triclinicities and recovery and recrystallisation microstructures. The hornfelsic rocks in both areas record metamorphic changes up to the hornblende-hornfels (brown, granoblastic tschermakitic hornblende) in the metabasalts and cordierite-K-feldspar facies in the pelitic rocks.

This study has revealed two distinct, emplacement-induced domains within the aureoles: an outer strained envelope recording minor ductile deformation and an inner thermally annealed envelope. The discordant nature of the intruding pluton's contacts, low contact temperatures (650°C) and negligible observed shortening in the aureole preclude emplacement mechanisms by doming, melt zoning or ballooning diapirism. The meridonal trend of the plutonic suites in the New England Batholith (sub-parallelling major regional faults and a once-convergent plate margin), the metamorphic zonation outlined and its implications upon emplacement-induced deformation and thermal annealing and negligible shortening, favour an emplacement mechanism involving dyke propagation at depth and, at high levels, stoping with associated minor late-stage diapiric accommodation.

Mid-crustal contamination of granitic magmas

W. J. Collins and **R. H. Vernon**

In the Anmatjira Range, Arunta Inlier, central Australia, Proterozoic (1,820–1,760 Ma) megacrystic granitic magmas have intruded as sheet-like bodies at mid-crustal levels (10–20 km depth). The sheets show a remarkable mineralogical and geochemical coherency with their enclosing country rocks; cordierite-rich, garnet-rich and orthopyroxene-rich (charnockitic) granitic sheets, all containing megacrystic K-feldspar, occur within cordierite-rich, garnet-rich and orthopyroxene-rich gneisses/granofelses, respectively. The distinctive mineralogy of the granite sheets in the Anmatjira Range is reflected geochemically. Cordierite granites have high MgO, Ni, Cr and Zn contents, but low Na/K ratios, typical of S-type granites. Garnet granites have high Fe/Mg ratios and are typically silicic (>73% SiO_2). Orthopyroxene granites, which usually are hornblende-bearing, have higher CaO contents and Na/K ratios, and an extended silica range (65–75% SiO_2). The latter are typical of I-type granites and they form a *continuous* geochemical lineage from 50–75% SiO_2 with the enclosing mafic granofelses.

The 10–20 km-depth of emplacement for the granite sheets is that estimated for magma chambers in magmatic arcs. Thus, it is suggested that the Anmatjira Range is an analogue for the internal zones of a continental magmatic arc, and that considerable contamination of granites can occur in this environment. With repeated emplacement of granites at this level, and passage of granitic magmas to higher levels, the resultant low-pressure–high-temperature metamorphism in the mid-crustal "aureole" reaches upper amphibolite facies conditions and migmatites develop, which sufficiently reduce the viscosity of the host rocks to allow physical disaggregation and incorporation into the granite magmas. Either S- or I-type granites can be produced, depending on the chemistry of the enclosing country rock. The data suggest that many granites acquire their geochemical characteristics at mid-crustal levels.

The multiple origin of peraluminous granites: consequences for their metallogenic potential

Michel Cuney and **Jean-Marc Stussi**

Peraluminous granites represent about 70% of the plutonic activity of the European Variscan. They are characterised by a high peraluminous index, but show distinctive mineralogical, geochemical and genetic characteristics when compared with the typical Australian S-type granites. They have been subdivided into three groups, each being marked by either increasing or decreasing peraluminous index (I.A. = Al − K − Na − 2Ca) with differentiation (I.D. = Fe + Mg + Ti). Modelling by the least squares method of these chemical variations gives a quantitative evaluation of the amount of fractionating minerals or of magma mixing:

(i) S-type granites typically show a strong decrease of their peraluminous index IA with differentiation (ID). They are best modelled by the fractionation of biotite, cordierite and/or garnet, quartz and plagioclase which may represent restitic minerals (resitite unmixing).

(ii) AKG granites (Aluminopotassic granites similar to the Gueret Batholith (5,000 km^2) in the northern French Massif Central (FMC), taken as a reference) are characterised by a slight increase of the peraluminous index despite large variations of ID. The peraluminous minerals cordierite and muscovite do not present, or present very limited, fractionation. This type of chemical variation is best modelled by mixing the most peraluminous facies with an intermediate metaluminous hornblende-bearing magma. No significant metal deposit is related to them.

(iii) AKL granites (Aluminopotassic granites, leucocratic and from Limousin, western FMC, where they are well exposed) are muscovite ± sillimanite ± andalusite ± garnet-bearing. Their peraluminous index shows a slight to strong increase for a limited variation of ID. The best model implies the fractionation of biotite, plagioclase, orthoclase and quartz. Most Variscan economic U, Sn, W, Li, Ta, Be deposits are related to this type of granite.

Redeposited and primary volcanic facies in the middle to late Silurian Tumut Trough: deposition on an unstable continental edge

Kelsie Dadd

Late Silurian rocks of the Tumut Trough in the southeastern Lachlan Fold Belt of N.S.W., Australia, comprise a conformable sequence of dacitic volcaniclastic rocks, sedimentary rocks and pillow basalt intruded by mafic and felsic igneous bodies. Massive dacitic volcaniclastic rocks (MDV) in the sequence have been interpreted as subaerial ash flow and ash fall deposits. However, facies analysis of the MDV and surrounding rock units indicates that this interpretation is unlikely. The MDV are considered to be very proximal, submarine, mass-flow deposits resulting from the redeposition of large volumes of unconsolidated pyroclastic material soon after eruption.

In the W, the MDV is interstratified with marine sedimentary rocks in both massive and graded beds, and units of chaotic limestone boulder conglomerate. Pillow basalt occurs at the top of the sequence. To the E, the MDV occurs with bedded volcaniclastic rocks that grade upward from dacitic pebble conglomerate to laminated siltstone, deposited from turbidity currents and large allochthonous blocks of limestone.

The MDV lacks stratification, implying very rapid accumulation from dense flows. Phenocryst content in the dacite is typically 30–50%. This high crystal content probably added to the buoyancy of the flows, increasing their competence and mobility and also inhibiting the development of grading. There is no evidence for hot emplacement.

Previous interpretations for the Tumut Trough include a continental rift setting (Lightner 1977; Wyborn *et al.* 1981), a pull-apart basin in a strike-slip fault system (Stuart-Smith 1990), a fore-arc basin (Crook 1980), a back-arc rift (Crook & Powell 1976; Ashley *et al.* 1979) and the upwelling of hot mantle and crustal stretching along a transform fault (Packham 1987). The interstratification of dacitic volcaniclastic rocks, turbidites, chaotic limestone boulder conglomerate and marine basalt favour deposition at the foot of an unstable slope in an arc environment. The S-type geochemistry indicates this had a substantial sialic basement.

Ashley, P. M., Brown, P. F., Franklin, B. J., Ray, A. S. & Scheibner, E. 1979. J GEOL SOC AUST **26,** 45–60.

Crook, K. A. W. 1980. J GEOL SOC AUST **27,** 215–32.

Crook, K. A. W. & Powell, C. McA. 1976. *Field guide for Excursion 17A, 25th International Geological Congress, Australia, 1976.*

Lightner, J. D. 1977. Unpublished MSc. Thesis, Australian National University.

Packham, G. H. 1987. AM GEOPHYS UN GEODYN SER **19,** 67–82.
Stuart-Smith, P. G. 1990. J STRUCT GEOL **12,** 621–38.
Wyborn, D., Chappell, B. W. & Johnston, R. M. 1981. J GEOPHYS RES **86,** 10,335–48.

The Middle Proterozoic rhyolite-hosted Pea Ridge iron and rare-earth-element deposit: a magmatic source for Olympic Dam-type deposits in the midcontinent region of the U.S.A.

W. C. Day, G. B. Sidder, R. O. Rye, L. M. Nuelle and **E. B. Kisvarsanyi**

An origin for the ore fluids that deposited iron, copper, rare-earth elements (REE), and gold in Olympic Dam-type deposits remains enigmatic. However, recent research on one such deposit in Missouri, U.S.A., indicates that the ore fluids were magmatic and may have formed from immiscible liquids derived from magmas cogenetic with the host terrane.

The Pea Ridge deposit is one of several apatite-bearing magnetite bodies that intrude Middle Proterozoic anorogenic high-silica rhyolite in southeastern Missouri, U.S.A. At Pea Ridge, REE-rich breccia pipes cut the ore body and exhibit fluidisation textures, with clasts of iron ore and host rock supported in a matrix of monazite, barite, xenotime, apatite and rock flour. Evidence for boiling is seen in the primary fluid inclusions in quartz from the breccia pipes.

$\delta^{18}O$ quartz-magnetite and $\delta^{34}S$ barite-pyrite equilibrium temperatures show that the early ore solutions were hot (680°C), clearly in the temperature range of magmatic systems, but cooled through time (300°C). The primary ore fluids had a heavy ^{18}O signature (+15‰), controlled by equilibrium with a heavy ^{18}O magma of crustal origin.

Trace element modelling indicates that the ore deposit formed from an immiscible melt generated from the iron-rich trachyte suite. The magnetite ore and REE-bearing breccia pipes formed as a coherent system that underwent secondary boiling and caused catastrophic brecciation during emplacement of the REE-bearing breccia pipes.

Characterisation and petrogenetic subdivision of A-type granites

G. Nelson Eby

A data bank of published and unpublished analyses has been established for the A-type granites (and associated felsic volcanics) from a variety of geological settings. All of the suites plot within the WPG field of Pearce *et al.* (1984) and the A-type fields of Whalen *et al.* (1987). Compared to the I- and S-type granites, the A-types have greater FeO/MgO ratios and higher absolute abundances of a number of incompatible trace elements. Utilising various indicators of magmatic evolution, of which the Eu/Eu* ratio (a measure of feldspar fractionation) is one of the most useful, A-type granites generally have higher abundances of Ga, Nb, Y, REE (except Eu), Ta, Zr, Hf and Th than I- and S-type granites at a similar stage of magmatic evolution. These observations suggest that most, if not all, A-type granites are not the product of extreme fractionation of I-type magmas.

On the basis of the chemical data, the A-type granites can be divided into two groups (A_1 and A_2). The A_1 group consists of suites from oceanic islands, intracontinental rift zones and regions of crustal doming. These suites often have associated mafic rocks, thus showing a bimodal character. They are characterised by a number of trace element ratios (such as Y/Nb and Yb/Ta) similar to those of oceanic island basalts. The A_2 group consists of suites which have higher Y/Nb and Yb/Ta ratios (similar to those of continental margin basalts), are generally less enriched in incompatible elements and were usually emplaced in post collision-post orogenic environments. These suites appear to have originated either in the subcontinental mantle or within the continental crust. On the basis of the chemical characteristics, it is suggested that the A_1 group represents granites derived by fractionation of magmas similar to those which give rise to oceanic island basalts and thus have a mantle source. The A_2 group, which includes the Gabo and Mumbulla suites of the Lachlan Fold Belt, may represent granites derived from the remelting of lithospheric crust (or mantle?) as proposed by Collins *et al.* (1982).

A number of elemental screens have been developed to facilitate the identification of the A_1 and A_2 groups. The most useful screens are those utilising Y/Nb, Yb/Ta, Ga/Al and Rb/Sr ratios and three component diagrams utilising Y, Nb, Zr, Hf, Th, Ce, Yb, Ta and Ga.

Collins, W. J., Beams, S. D., White, A. J. R. & Chappell, B. W. 1982. CONTRIB MINERAL PETROL **80,** 189–200.
Pearce, J. A., Harris, N. B. W. & Tindle, A. G. 1984. J PETROL **25,** 956–83.
Whalen, J. B., Currie, K. L. & Chappell, B. W. 1987. CONTRIB MINERAL PETROL **95,** 407–19.

Magmatic evolution of Late Proterozoic ultrapotassic syenites in northeastern Brazil and metasomatic mantle source

Valderez P. Ferreira, Alcides N. Sial and **James A. Whitney**

Ultrapotassic peralkalic silica-saturated plutons (580 Ma) are widespread in the Cachoeirinha-Salgueiro foldbelt, northeastern Brazil. They consist of alkali-feldspar syenites carrying pyroxenite as co-magmatic inclusions, syn-plutonic or late-stage dykes, which have the same mineralogical phases as the host syenites, (aegirine-augite, microcline, sphene, apatite, ±blue amphibole, ± magnetite), only proportions differ. Rare inclusions of a "mixed" rock (about 50% syenite +50% pyroxenite in an emulsion-like texture) are also observed. Pyroxenes in the three rocks are all only slightly zoned, silica-saturated and extremely low in Al (0·2–1·4%), suggesting crystallisation at low pressures. Amphiboles are mostly K-rich richterite, characterised by high SiO_2, low Al_2O_3 and TiO_2 contents and low Mg#.

The three rock types are characterised by an overall enrichment in incompatible elements, with negative spikes at Nb, Sr, P and Ti in spidergram-type diagrams. They have similar REE chondrite-normalised patterns, with negative slopes and lack of Eu anomaly, with the total REE in the pyroxenites greater than the total REE in the syenites. The pattern for the mixed rock is between the ones for the pyroxenites and syenites in both REE and spidergram diagrams.

The rocks have similar, high $\delta^{18}O$ values (average w.r. +8‰ SMOW, corrected from pyroxene), similar, high $\delta^{34}S$ (average +11‰ CDT), similar, high $^{87}Sr/^{86}Sr$ initial ratios (about 0·710) and similar, low $^{143}Nd/^{144}Nd$ ratios (average 0·51104).

Altogether, field and geochemical characteristics indicate chemical equilibrium among the three rock types and suggest liquid immiscibility between syenite and pyroxenite, the mixed rock representing magma composition before splitting.

The unusual isotopic signatures as well as the enrichment in LIL elements point to an anomalous enriched mantle source. Phlogopite pyroxenite xenoliths, assumed to be source fragments, display carbonate minerals and replacement textures, suggesting CO_2/H_2O metasomatism. Sm/Nd model age indicates that the enrichment took place at about 2.1 Ga ago.

Structures in metasedimentary inclusions as "windows" into granite source regions

Peter D. Fleming

Deformation structures preserved in metasedimentary restite inclusions can provide valuable information on the deformation history in the granite source area up to the time of melt generation, particularly if the inclusions record more deformations than undergone by the country rock by emplacement time. The data can add details to the view through the "window" to the source that is given by current petrological and geochemical arguments. The improved view can give extra information about the nature of the source rocks and their structural/tectonic history.

Pelitic restite inclusions from the Yabba Adamellite in the northeastern part of the Palaeozoic Lachlan Fold Belt (LFB) of southeastern Australia illustrate the strategy. The 415 Ma-adamellite intruded before and during the first discernible deformation in the Early Ordovician low-grade country rocks, yet the inclusions record at least four deformations in their microstructure. The inclusions and the adamellite have therefore been derived from a source with a tectonic history significantly longer than is recorded in the exposed Early Ordovician rocks. The source is therefore most likely to be a complexly deformed, metasedimentary terrane of pre-Ordovician age. The House Creek Granodiorite and the Cobaw Granite Complex occur respectively 40 km and 250 km to the W across the LFB from the Yabba pluton. Though details vary, similar conclusions can be drawn from studies of the structures recorded in their restite inclusions.

The examples discussed provide support to independent assertions, based on petrological and geochemical grounds, that at least in places the Palaeozoic rocks of the LFB rest structurally over pre-Ordovician continental crust rather than lie directly on oceanic crust. A less favoured alternative is that the complexly deformed sources are of approximately similar age to the exposed country rocks, but have early structural histories developed more or less independently in a tectonically separate terrane now overthrust by the country rock sequence.

Textural and chemical zoning in the François Pluton, Newfoundland

R. H. Flood and **S. E. Shaw**

The François zoned Pluton of southern Newfoundland exhibits only minor silica variation (72–76%) but is strongly zoned microstructurally. The microstructural zones have distinctive U, Th and K values as shown by published airborne radiometric maps.

The François Pluton outcrops as two lobes, with the southwestern lobe truncating the northeastern lobe. Rb/Sr bulk rock data from both lobes indicate an age of $383{\cdot}0 \pm 8{\cdot}7$ Ma, while the average of three biotite/bulk rock pairs give a younger age of 371 ± 4 Ma with an initial $^{87}Sr/^{86}Sr$ ratio of $0{\cdot}7048 \pm 0{\cdot}0006$. The pluton is massive, has sharp cross-cutting contacts and has few enclaves. Microcline is the K-feldspar in all microstructural zones regardless of grain-size and is the result of minor deformation.

Both lobes of the François Pluton are concentrically zoned, with an outer medium-grained marginal granite in each lobe grading inward to a coarse-grained leucogranite. The leucogranite in turn has a sharp contact with an inner microgranite that gradationally coarsens to the centre of the pluton. Although the bulk-rock chemistry changes only slightly, the coarser-grained variants have a higher K-feldspar to plagioclase ratio than the finer-grained parts of the pluton.

Changes in microstructure, mineralogy and chemistry result from changing H_2O pressures during crystallisation. Initial H_2O pressure increase during the formation of the outer medium-to-coarse-grained leucogranite produced a grain-size increase and a shift in the feldspar cotectic, causing a progressive increase in K-feldspar. The second stage of crystallisation, that is the microgranite in the sharp contact with the coarse-grained leucogranite against microgranite, is produced by the sudden loss of H_2O pressure. This pressure-quench event produces the narrow zone of microgranite with microstructures similar to those of microgranitic enclaves. The coarser-grained granite in the centre of the pluton indicates a second increase in H_2O pressure.

Magma mingling and the origin of the Mannum A-type granite, S Australia

J. D. Foden and **S. P. Turner**

The 485 Ma Mannum pluton is a high-level, potassic, rapakivi (A-type) granite. It is part of a late regional bimodal magmatic association in the Delamerian Orogeny. Swarms of contained dioritic enclaves result from mingling of a mafic magma when both phases were partially liquid. Mingling of melts, rather than inclusion of solid xenoliths, is demonstrated by the amoeboid physical shape of the enclaves with plastically deformed outlines. These include and react with alkali feldspar phenocrysts from the host granite and have hybrid margins. Geochemical transects from enclave core to host granite show sigmoidal shapes due to diffusional exchange. Although the profiles for many elements (e.g. Rb) are modelled as a result of diffusional process, those for some elements (REE, Y, Zr) are more complex due to reaction "front" migration, controlled by the preferential growth of sphene, apatite and zircon. The abundance of these minerals is promoted by diffusion-controlled activity gradients between mafic and felsic magma. The mafic enclaves and granite show initial Sr-isotopic equilibrium, indicating efficient diffusional exchange between mafic and felsic liquids. The "screen" of mafic droplets is either the draw-up of a lower layer of a layered magma chamber, or the intersection of the cooling granite body by a dyke. Our evidence is that these compositionally distinct mafic and felsic components did not mix freely, and interaction was restricted to diffusional exchange after mingling. After thermal equilibration, the viscosity contrast between the two magmas was too great for unrestricted mixing. During mingling, fluid motion of the granite was restricted to the regime of laminar flow.

Our evidence for temporal and spatial association of mafic and felsic magmas provides a clear solution to the thermal budgetary problems that may seem to be associated with post-orogenic granites. In fact, our results including Sr and Nd-isotopic data, are quite consistent with a common origin of both granite and enclaves as the products of fractionation of a parental continental tholeiitic basalt magma.

Early Proterozoic calcalkaline magmatism in the Hooper Complex, West Kimberley, Western Australia

T. J. Griffin and **I. M. Tyler**

The Hooper Complex in the W Kimberley region of Western Australia contains the third largest area (5,500 km^2) of Early Proterozoic felsic igneous rocks in Northern Australia, after Halls Creek (9,600 km^2) and Pine Creek (8,700 km^2). This I-type suite, which consists of comagmatic volcanic, hypabyssal and granite rocks dominated by monzogranite, is calcalkaline on an AFM plot.

The felsic igneous activity at ?1,840–1,860 Ma is part of the main cratonising event in the W Kimberley region, the Early Proterozoic Hooper Orogeny. The igneous phase followed moderate to high-grade metamorphism which affected deformed, interlayered, turbiditic sedimentary rocks and mafic sills, and was synchronous with the second deformation (D_2). Small areas of anatectic granite formed during the high-grade metamorphism of the sedimentary rocks have S-type characteristics, containing less than half the levels of CaO and Na_2O of the later, dominant I-type granites. A pronounced tectonic foliation is widespread throughout the felsic igneous suite, and is associated with metamorphism under greenschist and amphibolite facies conditions during the compressional Middle Proterozoic Yampi Orogeny.

The Hooper Complex felsic igneous suite is similar to other Early Proterozoic suites in Northern Australia, with high levels of K_2O and incompatible elements such as La, Ce and Rb, and low levels of MgO, CaO and Ni. The majority of rocks are enriched in alkalis; however, the associated mafic and intermediate rocks define the calcalkaline trend. The large volume of felsic igneous rocks of the Hooper Complex appear to be significantly more diverse than other Early Proterozoic orogenic belts of Northern Australia for which there are abundant data, and they exhibit many similarities with calcalkaline rocks in modern convergent continental margins. The Hooper Complex magmatism has sutured the Kimberley Craton, to the N, with the craton to the S which now underlies the Canning Basin. An origin by subduction-related partial melting of underplated material, previously fractionated from the mantle, is proposed for the Hooper Complex felsic igneous suite.

Petrological characteristics of magmatic epidote-bearing granites of the western cordillera of America

Jane M. Hammarstrom and **E-an Zen**

Magmatic epidote-bearing plutons occur in a discontinuous belt within the accreted terranes of the western N American cordillera. The Admiralty-Revillagigedo plutonic belt, southeastern Alaska comprises a regionally extensive suite of diorites, quartz diorites, tonalites and granodiorites. Similar magmatic epidote-bearing rock types are present in the Ecstall Pluton, British Columbia and in the Tenpeak and Oval Peak plutons, northern Cascades, Washington. A more evolved suite (tonalite to trondhjemite) is present along the western margins of the Idaho Batholith. The characteristic mineral assemblage is hornblende, biotite, epidote (±allanite), plagioclase (An_{15}–An_{50}), quartz, orthoclase, ilmenite ± magnetite ± sphene, ±garnet, and apatite. These rocks are predominantly metaluminous (molar A/CNK = 0·6–1·3). Limited trace element data show that these rocks are high in Sr (300–1,404 ppm) and low in Rb (3–·79 ppm). Oxidation ratios are low (whole-rock FeO/Fe_2O_3 ranges from ~1–16), and the relatively reduced nature of the rocks is reflected by the facts that: (1) ilmenite > magnetite; (2) sphene rims some ilmenite, but is not ubiquitous as a primary constituent; and (3) Fe^{3+}/Fe_{Total} for biotite (~0·06–0·07) and hornblende (~0·19–0·31) and pistacite contents of epidote (10–30 mol%) indicate redox states near or slightly more oxidising than those defined by the nickel-bunsenite buffer. Hornblendes are aluminous ferroan pargasites. For both biotite and hornblende, mg# (Mg/(Mg + Fe)) ranges from about 0·4–0·6, but hornblendes have slightly lower mg# than coexisting biotite. Garnet (grossular-almandine; 3–20 mol% spessartine) is present in parts of most of the complexes. Magmatic epidote, hornblende geobarometry, garnet compositions, field relations and pressure estimates from associated metamorphic assemblages all suggest that these plutons crystallised at mid- to lower crustal depths. The presence of magmatic epidote alone is not an unambiguous indicator of pressure. The collective chemical and mineralogical characteristics of this subset of calcalkaline I-type granites may be useful in recognising such plutons elsewhere.

The effect of boron and fluorine, in mineral additives, on the production of granitic partial melts from greywacke sources

M. J. Harvey, C. M. B. Henderson and **D. A. C. Manning**

The granitic melt compositions produced by experimentally induced partial melting of greywackes, with volatile bearing additives, are investigated within a framework of variable pressure, temperature and volatile composition, in an attempt to shed further light on the processes of granite formation. Partial melts have been produced from "dry" and "wet" greywacke systems, at pressures of 0·1 and 0·4 GPa and temperatures between 750°C and 900°C. These systems were then further perturbed by the addition of B and F, added as natural minerals: tourmaline, cryolite, or ulexite. Further runs were executed using simple laboratory reagents to act as control experiments.

The hydrous, alkali-rich, partial melts produced in these experiments are chemically very complex and require particularly careful analysis. The loss of alkalis during electron microprobe analysis has been avoided by the use of a freezing stage attached to a JEOL JXA 6400 analytical S.E.M., allowing the specimen to be cooled to −193°C and permitting analysis to be carried out using a reduced beam current (1.5 nA). Analysis of specimens at a range of temperatures has enabled optimum analytical conditions to be determined; these represent a considerable improvement on standard electron microprobe analysis at room temperature.

Melts produced in this study plot close to the appropriate minima in Qz–Ab–Or, but within the quartz liquidus field. This may reflect the low potassium content of the greywacke starting materials; residual crystalline phases include: cordierite, ilmenite, apatite, zircon, plagioclase feldspar and pyroxene, all K-free phases. The degree of melting increases from "dry", through hydrous to B and F-bearing melts. Tourmaline (added as 10 wt%, approximately equal to 1 wt% B_2O_3) is completely consumed, as are equivalent quantities (with respect to the atomic proportions of F in the charge) of cryolite. These experiments demonstrate the probable importance of natural volatile-rich minerals, particularly tourmaline, for the processes of anatectic melting.

The shoshonite-associated, zone gold skarn system at Junction Reefs, New South Wales, Australia

R. Hine, A. Mandyczewsky and **N. Gray**

Within the Junction Reefs and Burnt Yards goldfields there are several different styles of gold mineralisation that we suggest are genetically associated with a suite of Ordovician shoshonitic intrusives ranging from high-K diorite (48% SiO_2 1·4% K_2O) to monzonite (58% SiO_2, 3·6% K_2O). Ore styles include retrogressed skarns, intrusive hosted sheeted veins, fissure veins and hydrothermal breccia pipes. The ores exhibit similar geochemical character, in that they have a strong arsenic-gold association coupled with a low base metal and silver content. Age determinations indicate that the gold skarns and the sheeted veins formed at essentially the same time (i.e. around 400 Ma).

The gold skarns, which are economically the most significant ore type in the district, are hosted by a 39 m-thick, gently tilted sequence of Early Ordovician limestone, siltstone and chert. The skarn is centred on a quartz monzodiorite stock from which the silicate mineralogy varies outwards from a garnet-quartz core zone assemblage, through a pyroxene zone, to a pyroxene-ferrohastingsite zone and an outer retrogressed calcite-chlorite-quartz zone with semi-massive pyrrhotite, arsenopyrite, pyrite and minor chalcopyrite. All of the gold ore bodies are found within the retrogressed zone. The quartz monzodiorite is fractionated and forms part of an intrusive complex centred on a large high-K diorite pluton. The diorite is pre-ore as it hosts both sheeted sulphide-quartz vein and breccia pipe ores.

Emplacement of the diorite, which is chemically the most primitive member of the suite, was not accompanied by any major metasomatic activity. Rather, there is good evidence that the gold mineralising events were initiated, in each instance, by the evolution of magmatic hydrothermal solutions generated by the crystallisation of felsic fractionates of the shoshonitic magma suite in the closing stages of its 30–40 Ma history.

High level A-type sheet granites of the Wichita Mountains Igneous Province, Oklahoma, U.S.A.

J. P. Hogan, M. C. Gilbert, B. L. Weaver and **J. D. Myers**

The Wichita Mountains Igneous Province consists of Cambrian-age mafic and felsic igneous rocks intruded during rifting of the southern margin of the N American craton. Layered gabbros form a ~3–4 km-thick substrate for a ~1·5 km-thick rhyolite pile. Intrusive into the rhyolite are a series of high SiO_2 (71·0–77·6 wt%) alkali-feldspar granites of the Wichita Granite Group (WGG). The granites have A-type characteristics (e.g. metaluminous-subalkaline, magmatic fluorite, Zr > 500 ppm) and a "within plate" trace element signature. They occur as thin (~0·5 km) but laterally extensive (20–55 km) sheets. The WGG has been subdivided into three chemical suites. The Mount Scott Suite is impressive for its remarkable chemical homogeneity (i.e. SiO_2 71·0–73·6%; CaO 1·0–1·5%, Rb/Sr ~1·3%) over a distance of at least 55 km. The other suites show more petrographical and chemical variability. Coarse-grained (~1 cm) seriate and fine-grained granophyric microgranite are characteristic. In general, coarser intrudes finer-grained granite, which commonly occurs as stoped blocks. Miaroles at the base of stoped blocks provide evidence for upward volatile migration. Segregation of hornblende ± magnetite in layers exhibiting "crossbedding" is commonly observed near granite contacts. Abundant granophyre, miaroles and intrusion into an approximately coeval volcanics suggest emplacement pressures of 10^2–10^3 bars. Comparison of crystallisation histories of WGG with 1 kbar (1 kbar = 100 MPa) T − xH_2O phase relationships for the A-Type Watergums Granite (Clemens *et al.* 1986) imply liquidus temperatures of 900–975°C and initial xH_2O of <2·8 wt%. Temperatures of ~900°C are indicated by Zr geothermometry (Watson & Harrison 1981). Early crystallisation of hornblende-sphene-magnetite and quartz indicate intrinsic f_{O_2} values greater than FMQ (Wones 1989). Ubiquitous salmon-red alkali feldspars record a pervasive late-stage oxidation. Within granite sheets, hornblende and biotite exhibit antipathetic modal variation. Both phases can be replaced by the assemblage magnetite-fluorite, a trend consistent with increasing f_{F_2} and f_{O_2} during crystallisation.

Gravitational settling, flow sorting and volatile migration are processes suggested to have produced limited compositional and mineralogical variability. In contrast to the high viscosities predicted for "dry" granite melts, these processes and the laterally extensive nature of individual sheets suggest crystallisation from relatively fluid melts. We suggest that high fluorine activities implied for these melts significantly lowered the viscosity of these magmas, enhancing the recorded emplacement and crystallisation processes. However, the chemical distinctiveness of the granite suites is not adequately explained by high-level differentiation processes alone and thus must reflect chemical heterogeneities intrinsic to the source regions of these granites.

Clemens, J. D., Hollaway, J. R. & White, A. J. R. 1986. AM MINERAL **71**, 317–24.
Watson, E. B. & Harrison, T. M. 1983. EARTH PLANET SCI LETT **64**, 295–304.
Wones, D. R. 1989. AM MINERAL **74**, 744–49.

Granites of the Ravenswood Batholith, northern Australia

L. J. Hutton, I. P. Rienks, and **D. Wyborn**

The Ravenswood Batholith divides into four major suites using evidence of timing of intrusion relative to deformation, radiometric data and geochemistry. The first major suite is pretectonic and was probably intruded between about 480 Ma and 470 Ma. The suite includes calc-alkaline granite and granodiorite, and tholeiitic gabbro. Some granites may have been formed by mixing and mingling of magma types. Small peraluminous garnet- and muscovite-bearing biotite granites are the only granites in the batholith where feldspar fractionation has played a major role in their genesis. They are included in the pretectonic suite. The second suite is syndeformational. Deformed and undeformed phases occur along a major shear zone and are chemically indistinguishable. Radiometric data indicate an Early Silurian age. The third major suite postdates deformation and was intruded between 425 Ma and 400 Ma. It includes voluminous calc-alkaline granodiorite and tonalite, and minor granite and diorite. Fractionation and crustal contamination do not appear to have played a major role in the genesis of this suite. Compositional variation is probably caused by source rock variation and restite unmixing. Rocks of the fourth suite are from the extensive and chemically distinct Carboniferous to Permian N Queensland Volcanic and Plutonic Province. Migmatitic and peraluminous two-mica granites occur within the Early Palaeozoic basement rocks. The age of migmatisation is not known. Gravity values over the batholith are higher than expected over deep-seated granites, suggesting that most of the granites yield to denser gabbro between 5 km and 10 km deep. Two plutons have low gravity values, however, and are probably more deeply seated. We speculate that the bimodal pre-tectonic suite resulted from the intrusion of a tholeiitic magma into continental crust. The post-tectonic suite is probably derived from partial melting of a mafic source rock, possibly the older mafic underplate.

Tertiary extension related granite magmatism in the Rhodope Massif, northern Greece

Catrin E. Jones and **John Tarney**

Voluminous Tertiary granites intrude the Rhodope Massif in northern Greece, exhibiting an age sequence broadly younging from N to S. The largest, the Skaloti Granite complex, displays rock types varying from strongly foliated hornblende-biotite-tonalites exposed at its southern and western margins to unfoliated garnetiferous corundum-normative leucogranites in upper/central portions. This and other Tertiary granites (Lefkogia, Vrondou) were intruded as large-scale, sub-horizontal N–NE-dipping sheets broadly concordant along major Alpine structural horizons. Smaller, later, unfoliated Tertiary intrusions (Xanthi, Granitis) have more boss-like forms. Fabrics are not related to the waning stages of Alpine deformation, but were developed during or shortly after emplacement of the magmas into active extensional shear zones. Granite magma generation may have been a consequence of the return to a normal thermal regime following subduction and orogeny, but substantially enhanced by the onset in Tertiary times in northern Greece of the extensional regime now affecting the whole of the Aegean region.

The granites have Na/K ratios >1, high K/Rb ratios, high absolute Ba, Sr, P, moderate Ti and low Y and HREE contents. Their mantle-normalised spider-diagrams have slightly positive or no Eu anomalies and large negative Nb anomalies. Many of these characteristics cannot be accounted for by magmatic fractionation, AFC or crustal assimilation processes, but must reflect the geochemical composition of a major mantle source component. Moreover, these chemical signatures are inherent in many sodic granites worldwide. These features may be associated with incongruent melting of amphibole in lithospheric mantle. The generation of potassic granites may instead require their parental magma genesis from a phlogopite-bearing source.

We suggest that the granites of Rhodope were generated under conditions of pure shear lithospheric stretching aided by the breakdown of hydrous mineral phases (mainly amphibole) in the lithosphere that owed their origin to the long pre-history of subduction and fluid activity in the region.

Deformation structures and intrusion tectonics of Late Cretaceous Obara Granite distributed in the Ryoke Belt of central Japan

Yuji Kanaori, Shin-ichi Kawakami, Sayuri Onishi, Tomohiro Shinkai, Kenji Yairi and **Yutaka Nakai**

The Late Cretaceous Obara granite exhibits an open V-shaped geometry, 3–5 km wide and 50 km in length. The body consists of coarse-grained, porphyritic hornblende-biotite granodiorite and medium-grained hornblende-biotite granodiorite, and contains distinctive, elongate microgranitic encalves oriented parallel to a magmatic foliation. The enclave elongation is also parallel to the boundary between the Obara body and other surrounding granites.

Cataclasite zones approximately 50 m wide occur parallel to the magmatic foliation at the northern margin of the pluton. The relationship between the magmatic foliation and enclave elongation suggests that the elongation of the enclaves probably resulted from ductile deformation of melt and crystalline mixtures at high temperatures. Microscopic observations reveal that polygonisation commonly occurred in quartz, and biotite and quartz fill intergranular fractures in plagioclase. This evidence suggests that the quartz and biotite were deformed ductilely while the plagioclase deformed brittlely. The deformation occurred at temperatures between 400°C and 650°C, consistent with the existence of deformation lamellae in quartz and the bending of biotite. The microstructures of cataclasites were also examined, with the quartz and feldspars found to be shattered into fragments. However, the biotite was not shattered and exhibited a kinked or elongated texture. The existence of brittlely-deformed quartz and feldspars, and plastically-deformed biotite implies that the deformation occurred over a temperature range of 250–400°C. The observance of such variously deformed structures suggests that the Obara body has been extensively deformed under a regional stress field over a wide range of temperatures: high temperatures at an intrusive stage to a low-temperature solid stage after cooling.

Two contrasting styles of plutonism of the Ryukyu Arc, Japan

Yoshinobu Kawano and **Hiroo Kagami**

Cretaceous (70 Ma) to Tertiary (19 Ma) plutonic rocks occur on several islands and range from gabbro to granite. From age data, plutonic rocks in the Ryukyu Arc are divided into older and younger groups. Intrusion of the older group took place at 70–54 Ma and that of the younger group at 40–19 Ma. Plutonic rocks from the northern part of the Ryukyu Arc belong to the older group and the others to the younger one. The older group is characterised by higher FeO* and low Al_2O_3 contents compared with the younger group and the range of Y content of the former is wider than the latter. In Rb-(Y + Nb) and Nb-Y diagrams, the older group mainly belongs to WPG (within plate granite) and the younger group to VAG (volcanic arc granite).

In Cretaceous times, the Ryukyu Arc was located to the NW of the present position. At that time, the crust underneath the Ryukyu Arc was thick and felsic igneous activity took place in the northern part of the arc and E China and S Korea (Fukken-Ryeongnam

Belt). It is considered that magmas caused by partial melting of the lower crust were affected by upper crustal material during their ascent through the thick crust. In Tertiary times, the crust underneath the Ryukyu Arc thinned by extension. As igneous activity of the second group took place under these conditions, it was not affected by upper crustal material.

Magmatism, uranium mineralisation and hydrothermal activity: granitic examples from the Damaran Orogen of Namibia

Judith A. Kinnaird, P. Bowden and **G. J. H. Oliver**

The Pan-African Damaran Orogen forms one of the extensive late Proterozoic thermal, tectono-thermal and orogenic belts in the southwestern part of the African continent. The belt comprises: a N–S-oriented zone along the Namibian-Angolan coastline and a 400 km-wide NE-oriented zone between the Congo and Kalahari cratons. This latter belt has been divided into four main zones, of which the central zone is characterised by medium to high grades of metamorphism and extensive granite plutonism. Over 200 plutons occur, covering about 74,000 km^2 in area, ranging from over 5,000 km^2 in size to small veins and sheets. More than 95% of the plutons are granitic in composition, dominated by monzogranites.

Within the central zone, late-stage leucocratic granitic sheets carry important enrichments of uranium. On a regional scale, these uraniferous leucogranitic sheets and small plutons are spatially associated with domal structures found only along the axial zone of the Damaran belt. They are restricted to a structural zone approximately 50 km wide and extending NE for more than 100 km. Early, post-F1, pre-F2 tectonised (boudinaged) granitic sheets are non-uraniferous. These are cross-cut by post-F3 non-tectonised coalescing uranium-bearing granitic sheets.

Textures in the late leucogranite veins, which originally consisted of granophyric intergrowths of quartz and feldspar, have been modified into a range of textural facies which vary from coarse pegmatitic to micrographic. Petrographical evidence suggests that following primary magmatic crystallisation of uranium-rich minerals, there were several hydrothermal episodes. The earliest can be related to residual high-temperature magmatic fluids concentrated during the crystallisation of post-F3 granitic intrusions. A lower temperature hydrothermal episode (possibly related to post-Damaran tectonics) was responsible for fluid movement along faulted contact zones between granitic sheets and country rock. A range of secondary minerals dominated by beta-uranophane is related to the latest meteoric fluid circulation along joints.

Origin and significance of mafic enclaves in peraluminous A-type granites: examples from northeastern New South Wales, Australia

B. Landenberger

The Chaelundi Complex of the New England Orogen of northeastern New South Wales comprises an older I-type and a younger A-type suite of granites. The A-type suite is dominated by a leucoadamellite similar to other leucoadamellites in the region. However, the suite is compositionally extended, ranging from quartz monzonite (66% SiO_2) to leucoadamellite (76% SiO_2). The A-type suite is distinguished from the I-type by higher alkali and high-field-strength element contents, and lower CaO, MgO and Fe^{3+}/Fe^{3+}. Systematic major, trace and REE variation within the suite indicates that the leucoadamellite is derived from the quartz monzonite by 60% fractionation, involving removal of plagioclase, K-feldspar and orthopyroxene, along with minor hornblende and ilmenite.

Volumetrically minor mafic mineral clusters, comprising hypersthene, augite and calcic plagioclase, also occur in the quartz monzonite. Such clusters are better preserved as enclaves in the Woodlands Quartz Monzonite. Mineral chemistry and petrographical relations indicate that these enclaves are quenched basaltic liquids. Whole-rock geochemistry reveals that the enclaves have a high-K, high-alumina basaltic composition and have undergone minor mixing with the host granite. Underplating of similar mantle-derived magmas is likely to have initiated melting in the lower crust, producing both the I- and A-type granites present in the region. Subtle differences between these granites suggest a similar, if not identical, source rock. Previously underplated arc basalts would have a composition similar to, but more primitive than, the basalt represented by the enclaves, and are a likely source. Volatile concentrations during partial melting, particularly that of F, probably determines the initial composition, and hence the affinity of these granitic melts.

Textural evolution of the two-phase granite pluton at Tai Lam, western New Territories, Hong Kong

R. L. Langford

The Tai Lam granite pluton is one of many intensely complex, high-level, two-phase plutons that characterise large areas of Hong Kong and the S China coastal granite belt. It is a structurally distinct, leucocratic biotite monzogranite of late Yenshanian (Mesozoic) age, characterised by small relict bodies of coarse- and medium-grained primary-texture granite surrounded by finer-grained, megacrystic granites displaying a wide and complex range of textures produced by remobilisation and recrystallisation. These textures form a continuum from the primary-texture granite through to granites with an entirely derived texture and geochemistry.

Major and trace element abundances show similar trends to those associated with tin mineralisation in granites in Peninsular Malaysia. Textural modification of the coarser primary-texture granite results in an increase in SiO_2 and Na_2O, while TiO_2, Al_2O_3, total Fe_2O_3, MgO, CaO and K_2O are progressively depleted. Trace-element abundances show a progressive enrichment in Rb, Pb, Cr, Ga, Nb, Sn and W, and there is depletion in Sr, Ce, Ba, La and Zr.

The extensive areas of megacrystic fine-grained and fine- to medium-grained granite represent the syn-intrusive remobilisation and resorption of a partially cooled, deep-seated phase of coarser-grained, weakly evolved granite in response to rapid pressure release during emplacement at a higher level in the crust. The progressive remobilisation of the primary-texture granite by the addition of late-stage, high-level, fractionated residual magmatic fluids appears to be dominated by fluidisation, brecciation and alkali metasomatism. Quartz veins with associated Sn-W mineralisation suggest that hydrothermal activity was important in the later stages of pluton development.

Timing of emplacement and crystallisation of the Himalayan leucogranites

Patrick Le Fort, Arnaud Pêcher and **Stéphane Guillot**

The idea that the High Himalayan Granites are linked to the tectonic evolution of the Himalaya was suggested long ago. Thus, dating of the granite is a choice way of dating the mountain-building. However, detailed investigations of the deformation characteristics and petrology of the granite and surrounding rocks suggest that the relationship is complicated.

In the country rocks of the granites, the early fabric of the Main Central Thrusting (MCT) is overprinted by the structures related to the N Himalayan Shear Zone (NHSZ). Large-scale structures, such as the N-vergent Annapurna fold of central Nepal, belong to this second deformation that may be in part coeval with the continuation of the movement along the MCT at lower levels. In the granitic rocks, the magmatic fabric is often overprinted by a sub-solidus, high-temperature foliation and/or lineation.

The migmatitic source area of the granite has undergone a decompressive phase with development of sillimanite often associated with shear planes preferentially collecting the anatectic material. The plutonic bodies cut the regional structures of collapse type such as the N-vergent Annapurna fold or the N Himalayan Shear Zone. In the thin upper contact aureoles, the metamorphic minerals are associated with a cleavage due to the dynamic emplacement of the granite.

Absolute dating of the granite has proved to be a quite disappointing and debatable matter. U-Pb and whole-rock Rb-Sr ages extend from around 25 Ma or less, to 18 Ma; the time-scale for the granite emplacement and crystallisation is of the order of a few Ma. Absolute dating of the periods of deformation is essentially lacking, but they should have lasted around 20 Ma for the first (MCT), and from around one to several Ma for the second (NHSZ). This may explain the complex relationships between the granite and the deformation patterns. Altogether, the emplacement and crystallisation of the High Himalayan Granites is a rather long process that occurs somewhat late in the tectonic evolution of the mountain range.

Granite emplacement and deformation in the Old Woman Mountains, southeastern California

K. J. W. McCaffrey, C. F. Miller, K. A. Howard, K. E. Karlstrom and **C. Simpson**

Thd Old Woman-Piute Batholith was emplaced into orogenic continental crust and provides an insight into the interplay between plutonism and deformation in the Mojave Desert. Individual plutons are generally sheet-like and concordant to the major regional structure, the Scanlon Nappe, which placed inverted Proterozoic basement and lower Palaeozoic strata on upright Palaeozoic to lower Mesozoic strata. The 73 Ma composite batholith includes both metaluminous granodiorite, which underlies the nappe, and slightly younger peraluminous granite plutons. These intrusions clearly postdate nappe development. Pre- and post-pluton emplacement fabrics were distinguished using strain analysis techniques on Proterozoic gneiss and Mesozoic dykes.

Syn- to post-intrusion deformation is demonstrated by: granodiorite fabrics that are parallel to, but weaker than, those in adjacent country rocks; melt-filled shear zones in granodiorite; and folded gneissic xenoliths and schlieren associated with a strong magmatic fabric in the granodiorite. This deformation continued far below granitic solidus temperatures in shear zones on the pluton margins.

Field relations, strain and kinematic data permit reconstruction of the pluton emplacement style and the role played by the magmatic events in the post-thrusting evolution of the Old Woman Mountains. The cooling history indicated by $^{40}Ar/^{39}Ar$ data suggests rapid denudation of the plutons ending by 65 Ma. Syn-magmatic deformation and later deformation may have contributed to unroofing during this interval. It is plausible that magmatism occurred in response to crustal thickening and profoundly influenced the subsequent extensional collapse of the thrust pile.

Pb isotopic evidence for deep crustal-scale fluid transport during granite petrogenesis

M. T. McCulloch, J. D. Woodhead and **B. W. Chappell**

The initial Pb isotopic composition has been determined in K-feldspars from Palaeozoic granites of the Lachlan Fold Belt (LFB) of southeastern Australia. In contrast to the wide range in initial Nd and Sr isotopic compositions ($\varepsilon_{Nd(I)} = +3$ to $-8{\cdot}9$ and $^{87}Sr/^{86}Sr_{(I)} = 0{\cdot}70408–0{\cdot}71206$), the Pb isotopic compositions for K-feldspars from the Bega and Berridale batholiths in the LFB of southeastern Australia exhibit an extremely limited variation in Pb isotopic composition with $^{206}Pb/^{204}Pb$ ratios of 18·142–18·185, $^{207}Pb/^{204}Pb$ ratios of 15·581–15·630 and $^{208}Pb/^{204}Pb$ ratios of 38·042–38·207. Despite this, there is a relatively good correlation with ε_{Nd} values, for example, samples having low, and hence more primitive, $^{207}Pb/^{204}Pb$ ratios also having more positive ε_{Nd} values. There is also an excellent correlation between single-stage Pb-Pb and T^{Nd} model ages. The Pb-Pb model ages, however, have a significantly reduced range of from ~320–440 Ma compared to the T^{Nd} model ages which range from ~800–1,770 Ma.

Previous Pb isotopic studies, particularly of ore bodies, have shown that Pb isotopic compositions are relatively uniform and tend to reflect average values of terranes or upper-crustal segments. Thus in high-level supergene deposits, Pb appears to be particularly mobile in fluid systems. In the LFB, detailed isotopic and geochemical studies have shown that the Palaeozoic (400–420 Ma) granites are derived from complex, dominantly intracrustal, source rocks with a wide range of Sm-Nd depleted mantle model ages (600–1,800 Ma). Furthermore, two chemically distinct types of source rocks have been recognised, infracrustal and supracrustal, producing, respectively, I-type and S-type granites. The Pb isotopic composition of the LFB granites would be expected to reflect this complex source history. The extremely limited range in Pb isotopic compositions and consequent limited range in Pb-Pb model ages, despite the large range in T^{Nd} crust formation ages, is therefore attributed to the mobile behaviour of Pb in deep, crustal-scale fluid circulation systems that were established immediately before or during granite plutonism. The fluid circulation system probably affected whole crustal segments and operated over length scales of $\sim 10^3$ km, and may play an important role in the redistribution of soluable elements as well as the convective transport of heat within the crust.

A new shear criterion for rocks deformed in the magmatic state: examples from the Thorr Granite, Co. Donegal

Michele A. McErlean

A fundamental concept when studying deformation fabrics in granitic rocks is that of relating the time of deformation to the crystallisation state of the magma, hence the subdivision of granite fabrics into magmatic-state and solid-state types. The development of both types of fabric can be related to the operation of non-coaxial strains, including simple shear. Observations about the sense of shear are obviously critical to any structural study. While there are abundant useful shear sense indicators in rocks deformed in the solid state, those deformed in the magmatic state often lack documented sense of shear indicators, such as tiling fabrics. By applying a concept proposed by Fernandez and Laporte, it is now possible to use the subfabrics often observed in granites deformed in the magmatic state to determine the sense of shear during deformation.

The concept relies on the fact that different shapes of grains rotate at different rates when subjected to non-coaxial shear. Crystals with low grain-shape ratios rotate faster than those with higher grain-shape ratios because they have a smaller period of rotation about a fixed axis, thereby resulting in the development of polymodal and asymmetric fabrics.

This has been tested on the Thorr Granite, the oldest member of the Donegal Batholith (418 Ma). The pluton carries a fairly well-developed and pervasive magmatic-state deformation fabric which is defined mainly by feldspar phenocrysts. In areas throughout the pluton, but especially in the central Gola facies, this is markedly bimodal in nature, the subfabrics being defined by two families of feldspar phenocrysts. By visually rotating the subfabric defined by those feldspars with a higher grain-shape ratio through the acute angle towards the subfabric with the lower grain-shape ratios, the sense of shear can be determined, since this is how the crystals would have continued to rotate had the magma stayed hot.

Results from Thorr are predominantly anticlockwise rotations, consistent with sinistral shear along a northerly trend. This supports other data from the margins of the pluton and together they indicate that the Thorr granite lay in the site of a N-trending sinistral shear zone which entirely pre-dated the NE-trending Main Donegal Granite shear zone.

Emplacement of mid-crustal plutons in the Ross Lake Fault Zone, North Cascades, northwestern U.S.A.

Robert B. Miller

The 10 km-wide Ross Lake Fault Zone (RLFZ) exemplifies the relationships between pluton emplacement and strike-slip faults. This terrane boundary was intruded by numerous plutons that record spatial and temporal variations in emplacement mechanisms. Emplacement was particularly influenced by stepover zones between dextral strike-slip fault segments and by a regional change from transpression to transtension.

Plutons were intruded at 87–91 Ma, 60–68 Ma, and 48–50 Ma.

The oldest body, the Black Peak Batholith, shows evidence for both stoping and forcible emplacement. It intruded across a left-stepping jog or bend in the RLFZ at its southern end and a right-stepping one at its northern end. Slip across the southern, contractional stepover was transferred through the pluton by reverse-slip ductile shear zones that strike at a high angle to the strike-slip segments.

The 60–68 Ma plutons were emplaced during regional transpression in which dextral strike-slip and thrusting occurred on different segments of the RLFZ. Tectonic loading occurred before and/or during intrusion. Hornblende barometry for the epidote-bearing Oval Peak Batholith (65 Ma) records crystallisation at ~6 kbar (1 kbar = 100 MPa), considerably greater than the <3 kbar for the nearby Black Peak Batholith. The Oval Peak was intruded syntectonically as an asymmetric expanding diapir along a thrust segment. In contrast to this diapiric pluton, variably deformed 60–68 Ma rocks in a 2 km-wide belt in the thrust footwall intruded as moderately dipping sheets that are typically 1–10 m wide.

The 48–50 Ma plutons were intruded during regional transtension and oblique (dextral-normal) slip in the RLFZ. They lack solid-state deformation, but show strong structural control where numerous sheets and irregular masses form, in map view, a 10 km-long wedge within an extensional stepover.

Collectively, these patterns demonstrate how terrane boundaries may serve as a locus of protracted plutonism during varying tectonic regimes and indicate the importance of stepovers in localising emplacement.

The contribution of early Proterozoic basement to Caledonian magma genesis in northwestern Britain and Ireland

R. J. Muir, W. R. Fitches and **A. J. Maltman**

The source of the inherited Proterozoic lead component in the Caledonian granites has been the subject of much speculation. The apparent absence of a significant volume of early Proterozoic crust in Scotland and Ireland has led to three alternative suggestions for the origin of this isotopic signature: (1) Lewisian crust (2·9–2·7 Ga) thoroughly reworked during the Laxfordian tectonothermal event (1·9–1·7 Ga); (2) Laxfordian granitic material (1·7 Ga); (3) a mixture of Lewisian (2·9–2·7 Ga) and Grenville crust (1·0 Ga). A more likely source for the Proterozoic component is the newly recognised Rhinns Complex in the southern Inner Hebrides.

Precambrian basement on the southern side of the Great Glen Fault zone is exposed on the southern half of the Rhinns of Islay, in northeastern Colonsay and on Inishtrahull. Previous workers have generally correlated this basement with the late Archaean Lewisian complex. However, detailed field, petrographical and geochemical studies have revealed that the basement rocks comprise a deformed igneous association of syenite and gabbro, with minor mafic and felsic intrusions. This association, collectively termed "The Rhinns Complex" has an alkalic composition. Isotopic data indicate that the complex represents new addition of material to the crust at *c.* 1·8 Ga. The igneous protolith was juvenile mantle-derived material, not reworked Archaean crust.

The inherited zircons in the Caledonian granites may reflect the presence of early Proterozoic basement (Rhinns Complex) beneath Scotland and northwestern Ireland. It is important to note, however, that the granites have passed through Moine and Dalradian metasedimentary rocks which also have early Proterozoic Nd model ages and contain Proterozoic zircon. Moreover, at least part of the Dalradian sediment was derived from the Rhinns Complex. Consequently, an alternative explanation of the granite data is contamination by the country rocks. In support of this, many of the granites have isotopic compositions indicative of mixing of early Proterozoic, Dalradian and mantle-derived components.

The petrogenesis of the Proterozoic Harney Peak leucogranite, Black Hills, South Dakota, U.S.A.

Peter I. Nabelek and **Carol Russ-Nabelek**

The peraluminous Harney Peak Granite is associated with one of the largest rare-element pegmatite fields in N America. The granite was emplaced at 3–4 kbar (1 kbar = 100 MPa) as multiple sills and dykes at the culmination of a regional high-temperature, low-pressure metamorphic event. Principally along the periphery of the main pluton and in the satellite intrusions, the sills segregated into granite-pegmatite couplets. The minerals include quartz, K-feldspar, sodic plagioclase, muscovite ± garnet. The predominant ferromagnesian mineral in the granite's core is biotite, whereas at the periphery of the main pluton and in the satellite intrusions it is tourmaline. Oxygen isotopes in most samples equilibrated at magmatic temperatures between >800°C and 650°C. The biotite-containing granites have $\delta^{18}O$ of 11·5 ± 0·6‰ and tourmaline-rich granites 13·2 ± 0·8‰. The difference is best explained by generation from two different sources, consistent with radiogenic isotope data (Walker *et al.* 1986; Krogstad *et al.* 1992). The schists have an identical average $\delta^{18}O$ value to that of the high-$\delta^{18}O$ granites.

SiO_2, CaO, MgO, FeO, Sr, Zr, W of the low- and high-$\delta^{18}O$ suites overlap, whereas the concentration of Ba and TiO_2 is higher and of MnO and B lower in the low-$\delta^{18}O$ granites. Most samples are more potassic than the H_2O-saturated minima, consistent with melting at $a_{H_2O} < 1$ (Holtz & Johannes 1991; Pichavant 1987). These features suggest that the low-$\delta^{18}O$ granites were generated by high-extent, biotite-dehydration melting of more immature metasedimentary source deeper in the crust, whereas low-extent, muscovite-dehydration melting of schists higher in the crust resulted in the peripheral tourmaline granites. The results of this study indicate that chemical and mineralogical variations in pegmatitic leucogranites may be largely due to conditions prevailing in the source regions during melting.

Holtz, F. & Johannes, W. 1991. J PETROL **32,** 935–58.
Krogstad, E. J., Walker, R. J., Nabelek, P. I. & Russ-Nabelek, C. 1992. GEOLOGY (in review).
Pichavant, M. 1987. AM MINERAL **72,** 1,056–70.
Walker, R. J., Hanson, G. N. & Papike, J. J. 1986. GEOCHIM COSMOCHIM ACTA **50,** 2833–46.

Along-arc migration of Cretaceous granitic magmatism and low-P metamorphism in the Southwest Japan Arc: what caused the Cordilleran-type orogeny along the Eurasian Continental Margin?

Takashi Nakajima

During the late Mesozoic, before the opening of the Japan Sea, the Pacific plate had subducted beneath the Eurasian continent and caused a Cordilleran-type orogeny including the formation of accretionary prism, low-pressure and high-pressure regional metamorphism and granitic magmatism along the continental margin. These orogenic products are widely exposed in the Southwestern Japan Arc, especially in the northern part of southwestern Japan, where Cretaceous granites and the related volcanics occupy more than half of the ground surface.

The granitic rocks are mainly of I-type and ilmenite series. They are divided into two groups in terms of field occurrence and lithology. "Ryoke-type" granites are associated with low-pressure-type regional metamorphic rocks and are sometimes gneissose, whereas "Sanyo-type" granites intrude cogenetic volcanic rocks and are all massive. They are regarded as the deep and shallow parts of the granitic province, respectively.

Extensive geochronological work has revealed that the isotopic ages of these granites have a systematic lateral variation. Rb-Sr whole-rock isochron ages of the Sanyo-type granites become younger eastwards from 100–70 Ma over 700 km. Rb-Sr/K-Ar biotite ages basically follow the eastward-younging trend with a 5–10 Ma time-lag from the Rb-Sr whole-rock ages for the entire area. The ages of the Ryoke low-pressure metamorphic rocks are fairly consistent with that lateral age variation trend.

From the systematic along-arc age variation of these granitic and metamorphic rocks, we infer the migration of a tectonic setting in which the granitic magmatism and low-P metamorphism took place. The tectonic models for granitic magmatism based on the situation of steady-state subduction cannot explain these geochronological features.

RTT (ridge-trench-trench) migration involving the subduction of the Kula-Pacific Ridge beneath the Eurasia plate might have driven the Cretaceous Cordilleran-type orogeny along the eastern margin of the Eurasian continent. Eastward-younging MORB-type basaltic volcanism within the Cretaceous accretionary prism, as well as widespread occurrence of high-magnesian andesites at the early stage of the granitic magmatism, could be the results of ridge subduction.

Role of mafic magmas in granite genesis: evidence from enclaves in Silurian Devonian plutons of southeastern Australia

I. A. Nicholls, C. Carson, C. Legg, R. Maas and **G. Eberz**

Petrographic and major/trace element studies of abundant microgranular enclaves in the early Devonian I-type Swifts Creek Pluton and middle Devonian S-type Wilson's Promontory Batholith of eastern Victoria support their origin as mafic magma globules which, after injection into pre-existing felsic magma chambers, mingled mechanically with crystal-rich, viscous host magmas and underwent variable mixing. Within some enclaves, "trails" of quartz and feldspar megacrysts derived from the host demonstrate extensive mixing. Other non-megacrystic enclaves show little evidence for mixing, and zoned examples with fine-grained margins indicate rapid cooling against host magma and internal differentiation. Swifts Creek Pluton Sr and Nd isotopic ratios fall into almost distinct "enclave" and "host" fields, which together define a two-component "mixing" trend compatible with involvement of primitive (meta-igneous) and evolved (metasedimentary) crustal magma sources. Trace element systematics are also in broad agreement with a two-component mixing model (Eberz *et al.* 1990).

The late Silurian S-type Deddick Granodiorite in northeastern Victoria, a member of the Bullenbalong Suite of the Kosciusko Batholith, contains abundant gneissic metasedimentary enclaves, ranging from fertile examples with quartzofeldspathic layers to highly residual types dominated by cordierite + biotite + sillimanite + spinel ± andalusite ± corundum assemblages. Microgranular enclaves with igneous microstructures are rare (<10% of enclaves) and comprise biotite-rich microtonalitic types with plagioclase to An_{85} and orthopyroxene to Mg_{87}, and microgranodiorites. The host granodiorite remote from enclaves is isotopically uniform (initial $\varepsilon_{Nd} \sim -10$, $^{87}Sr/^{86}Sr \sim 0{\cdot}715$) and contrasts strongly with the enclaves themselves, which show broad variation in isotopic ratios within and between groups. Gneissic enclaves are isotopically more evolved than the host (−10–12, 0·717–0·722), while microgranodiorite enclaves (−4–8, 0·709–0·713) are more primitive. Microtonalite enclaves (−6–12, 0·713–0·717) overlap with the host, but show much greater petrographic and isotopic diversity. Host and enclaves together define a curved $\varepsilon_{Nd}-{}^{87}Sr/^{86}Sr$ array compatible with two-component mixing, but trace-element systematics are not readily reconciled with this. The simplest interpretations appear to be that gneissic enclaves (at least some originating at shallow depths compatible with andalusite stability) represent fragments of a previously partially melted metasedimentary succession encountered during uprise of host magma, while microtonalite enclaves may represent partially hybridised products of mingling between small volumes of mafic magma and host granodiorite magma already carrying material derived from metasedimentary enclaves. Microgranodiorite enclaves probably represent distinct magmas which intruded the host pluton as small dykes late in its emplacement history and also became involved in limited magma mingling.

Eberz *et al.* 1990. CHEM GEOL **85**, 119.

Interpretation of the chemistry of the mineral apatite from granitic rocks: a preliminary model calculation for the estimation of initial halogen contents in the Tuolumne Intrusive Suite, Sierra Nevada Batholith, California

P. M. Piccoli and **P. A. Candela**

The concentration of the halogens in the mineral apatite can be used to deduce changes in fugacity ratios during the crystallisation of granites. In addition, the initial concentration of Cl and F in the melt and the associated magmatic aqueous phase can be calculated if the temperature of apatite crystallisation can be estimated. Model apatite saturation temperatures (AST) can be calculated given the apatite solubility function determined by Harrison and Watson (1984). Our calculations suggest the following AST for the units of the Tuolumne Intrusive Suite (in °C); Kuna Crest (KC) 928, Half Dome Equigranular (HDE) 901, Half Dome Porphyry (HDP) 915, Cathedral Peak (CP) 925 and Johnson Porphyry (JP) 795. In addition, over half of the apatite crystallises within 60°C of the AST for the KC, HDE, HDP and CP, and within 30°C for the JP. The proportion of crystallisation before apatite saturation has also been estimated: KC (35%), HDE (9%), HDP (9%), CP (8%), and JJP (16%). The ratio of the mole fraction of chlorapatite to hydroxapatite and fluorapatite to hydroxyapatite within the host rocks decreases from the outer to the inner units of the Tuolumne Intrusive Suite: KC (0·080, 1·70), HDE (0·064, 4·57), HDP (0·019, 2·71), CP (0·030, 5·00) and JP (0·016, 8·00).

The composition of apatite can be used in conjunction with the AST to calculate initial F and Cl in the melt and magmatic aqueous phase. Given this information, the following concentrations for Cl (wt%) and F(ppm), respectively, in the aqueous phase have been estimated: KC (44; 45), HDE (32; 100), HDP (10; 65), CP (16; 130), and JP (5; 60). The coexisting magma is estimated to contain the following concentration of Cl (ppm) and F (ppm), respectively: KC (8,200; 190), HDE (5,800; 410), HDP (1,800; 270), CP (3,000; 540) and JP (800; 300). This drop in initial Cl in the melt from the outer to the inner units of the Tuolumne Intrusive Suite corresponds well with the increase in $(Sr^{87}/Sr^{86})^{\circ}$ determined by Kistler *et al.* (1986). The relationship between decreasing Cl and increasing $(Sr^{87}/Sr^{86})^{\circ}$ suggests that variations in initial Cl in the melt are controlled by variable amounts of a subduction-related component.

Harrison, E. B. & Watson, T. M. 1984. GEOCHIM COSMOCHIM ACTA **48,** 1,467–77.

Kistler, R. W., Chappell, B. W., Peck, D. L. & Bateman, P. C. 1986. CONTRIB MINERAL PETROL **94,** 205–20.

Mantle-crust interaction in the European Hercynides: isotopic and geochemical evidence from Spanish Central System granites

L. Pinarelli, R. Petrini and **A. Rottura**

The Spanish Central System (SCS) is a major granite complex in the Hercynian Iberian Massif. It consists largely of granitic occurrences associated with minor mafic to intermediate intrusives, emplaced at shallow level in late Hercynian times.

The granites from Bejar in the western SCS consist of porphyritic biotite-monzongranite-granodiorite (BG), often cordierite-bearing (CMG), containing quartz dioritic to tonalitic mafic enclaves. Both BG and CMG are moderately peraluminous (A/CNK = 1·10–1·33) and LREE-enriched; they show pronounced negative Eu anomalies (0·57–0·23) and variable HREE fractionation ($(Gd/Yb)_N = 1{\cdot}5$–2·8), with geochemical characteristics similar to those of volcanic arc granites.

The initial (at 300 Ma) Sr and Nd isotopic compositions are variable in both granite types ($(^{87}Sr/^{86}Sr)_i = 0{\cdot}70802$–0·70924 in BG and 0·70799–0·70897 in CMG; $\varepsilon_{Nd} = -3{\cdot}9$–−8·0 in BG and −4·7–−7·8 in CMG). Mafic enclaves have slightly less evolved isotopic signatures ($(^{87}Sr/^{86}Sr)_i = 0{\cdot}70718$–0·70797; $\varepsilon_{Nd} = -4{\cdot}3$–−5·5), similar to those of gabbroics occurring in the eastern SCS.

Trace-element patterns and isotopic features of the SCS granites, along with the close association in space and time with mafic rocks, suggest an involvement of mantle derivatives, through a complex mechanism of mantle-crust interaction. Simple two-component mixing calculations (depleted mantle model) would indicate a range of 40–60% mantle component in the granites. The narrow range of Sr and Nd isotopic composition in gabbroic to granitic rocks suggests two alternative conclusions: either the granites are simply fractionated from the hybrid gabbroic magma, or, more probably, all the gabbro-granite suite is produced by a complex steady-state balance between crystallisation, crustal assimilation and new magma input, causing the magma to evolve chemically but not isotopically. In the last case, the buffering effect on the isotopes would impede a realistic estimate of the increasing crustal contribution in the progressively more evolved lithologies.

The Rosses multi-pulse Pluton: fractures and fractals

W. S. Pitcher

The Rosses Pluton of Donegal, Ireland, is a typical example of a nested grouping of near-circular intrusions, here representing four consecutive magma pulses, G_1, G_2, G_3 and G_4. Advantage has been taken of the excellent exposure on glaciated pavements to re-examine the contacts, metre by metre. All have been found to be sharp, with fining directions and hanging-wall pegmatite providing ample evidence of younging.

The linear, angular outline of the contacts is clear at all scales down to a few centimetres, and each intrusion has an overall polygonal outline. The contact planes of each of the intrusions conforms to a simple, structural pattern, seemingly controlled by a coeval fracture system akin to that of jointing, a view confirmed by the spalling off of rectangular joint blocks of the country-rock granodiorite at outer contacts, and of earlier pulse members at internal contacts. Though there are some common features, a contact-fracture pattern is specific to each intrusion event.

Overall, the emplacement is envisaged as being due to the collapse, during domal uplift, of arches of just consolidated granite which diminished in diameter with time.

All this is merely a refinement of the original work of Pitcher (1953, 1972), but the re-examination was motivated by the possibility of applying fractal analysis to this almost ideal example of igneous fracturing. On the gross scale, i.e. with a "stride" greater than about 250 m, the fractal dimensions of the contacts turn out to be self-similar and self-affine, but the D-values (degree of roughness) are generally low, expressing an approach of the contact planes to a Euclidean line. Of course, mathematical precision is difficult to obtain in the field from field measurements, but the preliminary results are: Thorr Pluton (the country rock) against G_1, $D = 1{\cdot}059$; G_1 against G_2, $D = 1{\cdot}016$, a very low value possibly expressing a low rigidity of G_1 at the time of the intrusion of G_1; a view consonant with other field observations; G_2 against G_3, $D = 1{\cdot}064$, the same as the outer contact yet surprisingly low considering the obvious angularity. On a stride scale smaller than 250 m the fractal dimensions turn out to be even lower in value, yet still self-similar. The reason for this is obvious in the field, where the variation in degree of angularity or roughness is seen to be very variable in the sense that long linears are broken by short, discontinuous lengths of small-scale, angular re-entrants. It is difficult to understand why this should be so in reasonably homogeneous materials, but perhaps disorder is more natural than order!

As earlier suspected by Pitcher, certain principal linears parallel the early regional joint system and it is envisaged that the overall shapes of the contacts were controlled by the interplay of a regional and a local, intrusion-centred, stress system. At the joins of the tangential linears, the change in direction required to follow an overall arcuate form results only after a number of "trail" fractures filled by granitic apophyses.

It is hoped that this very elementary attempt to use fractals will encourage their more general use and engender discussion.

Pitcher, W. S. 1953. PROC GEOL ASSOC **64,** 153–82.
Pitcher, W. S. & Berger, A. R. 1972. *The geology of Donegal: A study of granite emplacement and unroofing.* New York: Wiley-Interscience.

A peraluminous passing to metaluminous intrusive suite by means of a complex mixing process: geochemical evidence from the Miocene-Pliocene intrusive rocks from southern Tuscany and the Tyrrenian Sea, Italy

G. Poli, P. Manetti and S. Tommasini

The Miocene-Pliocene intrusive rocks of the Tuscan Magmatic Province (TMPI) have been found on the mainland, on islands of the Tuscan archipelago, as well as in boreholes. TMPI range in composition from granodiorite to alkaligranite, with a strong predominance of monzogranites. The rocks can be divided into five facies on the basis of field and petrological constraints, whereas from a geochemical viewpoint three trends exist in any Harker's diagram using major and trace elements. Aluminium Saturation Index (ASI) increases with differentiation and TMPI straddle the I-type and S-type fields. Genetic models that relate these chemical variations, as a whole and within individual plutons, to the segregation of major and accessory mineral phases are inconsistent with major, trace and rare-earth element abundance data. A simple two-end-member mixing process fails also to explain the geochemical characteristics of the plutons, either as a group or singly. The petrogenetic model that is able to explain all the petrographical and geochemical features shown by TMPI involves a three-stage process as suggested by Poli and Tommasini (1991). The model consists of three main stages concerning injection, evolution by means of a contamination and fractional crystallisation (CFC) process, and mixing processes which the mafic magmas underwent when injected into anatectic crustal environments.

The CFC process involved crustal peraluminous magma and mafic magma of probable subcrustal origin which formed microgranular enclaves, and a mixing process between the same crustal peraluminous magma and different evolved magmas derived from the mafic magma. The felsic peraluminous end member could be the leucocratic facies of the Giglio Pluton, whereas the mafic end member could be similar to the Capraia volcanic rocks. Thus, the petrogenetic model proposed for TMPI is able to explain the dichotomy between S- and I-type granite characteristics.

Structural controls on granite genesis in transpressional shear zones

R. John Reavy

Existing attempts to classify and account for granite magmatism from a genetic or environmental standpoint still do not explain fully the diversities encountered in various settings. This is because such schemes only loosely address the geological effects of tectonism and do not consider in detail how tectonic processes could actually control melt generation, magma ascent and pluton emplacement. In this contribution, the Hercynian belt of Iberia is compared to the Caledonian of the British Isles. Although there are marked contrasts of granite type between these two orogens (the causes of these differences are addressed here), they are both belts in which transpressional tectonics prevailed at the time of magmatism.

As transpression thickens the crust, there is the potential to generate granitic magma in such zones. In the Hercynian transpressional shear zones of Iberia, thickening together with hydrous fluxing created intracrustal wet melting of fertile Gondwanan sediments to produce syntectonic peraluminous two-mica granites. In the northern part of the British and Irish Caledonides, the association of compositionally-expanded granites with a major mantle input and transpressional shear zones may be explained by melting of continental crust at the lower limits of transpressional faults detaching into the Moho.

Individual transcurrent shear zones may, therefore, not only simply control the ascent paths, siting and emplacement mechanisms of plutons, but more fundamentally may also, due to the thickening inherent in the geometry of transpression, be ultimately responsible for granite genesis in such zones.

Alteration effects of trioctahedral micas as a contribution to petrogenesis of granitic rocks from the Elbe lineament (Saxonia)

A. Renno and W. Schmidt

Within the comprehensive examination of the Variscan and Prevariscan granites of the Elbe lineament in southeastern Saxonia (Schmidt 1992) a detailed investigation of the trioctahedral micas has been carried out. The mineral separates were analysed using classical wet chemical and trace-element methods and were characterised by X-ray diffraction and Mössbauer spectroscopy. Such investigations give valuable hints about petrogenesis. But one has to keep in mind the fundamental methodical difficulties, particularly the impossibility of separation of single mineral generations and the significant influence, on chemical analysis, of inclusions of accessory minerals and mineral alteration. The micas from different rock types were classified as Mg^- and Fe^{2+} biotites following Foster (1960).

With the help of statistical methods, we examined the effects of alteration to the major and trace-element concentrations in micas. Factor analysis shows that the formation of hydrobiotite has the most significant influence (30·6 variance%). The factor scores show a congruent behaviour of Rb, MnO, Li, Fe_2O_3, Nb, H_2O, B and L.O.I. Discriminant analysis was used to examine which element contents support the empirical division into normal and "changed" (hydro) biotites. We controlled the influence of the biotite classification in the first discriminant analysis, and proved that H_2O, Ba, Pb, L.O.I., Cr and Sn support our division. We can point out that even changes of micas, which are scarcely visible and difficult to identify by chemical methods, have a decisive influence on the starting values of petrogenetic interpretation (particularly Fe^{2+} and Fe^{3+}).

According to our results of major element analysis and X-ray diffraction, we can distinguish two different alteration processes (chloritisation, and the formation of hydrobiotite-vermiculitisation). The chloritisation is found in variable degrees in all examined rock types, but the formation of hydrobiotite is limited to a spatial and temporal section, which are additionally characterised by the martitisation of magnetite. The process seems independent of the rock type.

Schmidt, W. 1992. LITHOS (submitted).
Foster, M. D. 1960. US GEOL SURV PROF PAP **354B**.

Lachlan orogen pluton-emplacement structures

M. J. Rickard

There are three types of granite body in the Lachlan Fold Belt. Large homogeneous batholiths, elliptical plutons in strings or isolated, and narrow lenticular bodies. There is little deformation effect in the country rock adjacent to plutons or batholiths, and most have a sharp steep contact with a narrow contact hornfels aureole. Only a few small plutons show evidence for forceful intrusion. The S-type granites tend to be foliated (biotite and xenoliths), and this probably results from "ballooning pressure" against the country rock during intrusion a short distance from source. Some I-type plutons show marginal "flow foliation" that traces parallel to the contact around elliptical bodies. A tectonic mylonitic fabric is in places superimposed on both the above foliations. It is recognised by flaser quartz strings and new cross-cutting growth of biotite. A particularly broad and intense zone parallels the eastern edge of the Murrumbidgee Batholith along the trace of the S-I line. Evidence that aplite dykes intrude parallel to later diagonal fault directions suggests that emplacement was in a wrench regime. Some of the large thin lens-like bodies may have intruded up major extensional Reidel shears.

The similarity and regularity of pluton and fault spacing suggest that magma was ponded at the base of the brittle upper part of the crust about 8–10 km thick and that this carapace warped and faulted, permitting plutons to emplace passively. (Rickard & Ward 1981; Rickard (in press).) Deformation was initially and subsequently by compression and strike-slip faulting, but the huge extent of granitic material indicates considerable extension during the time of emplacement.

Rickard, M. J. & Ward, P. 1981. J GEOL SOC AUST **28,** 19–32.
Rickard, M. J. (in press). PROC 7TH INT BASEMENT TECTONICS CONF. Amsterdam: Kluwer.

S-type granites of the Lachlan Fold Belt: are Proterozoic source rocks really necessary?

Allan G. Rossiter

A new model for the generation of S-type granite suites within the LFB eliminates the need for Proterozoic source rocks and the geological difficulties which accompany it. It is envisaged that melting of early Cambian tholeiitic basalts (and dolerites) and overlying middle-late Cambrian sediments took place when mantle-sourced shoshonitic magma ponded at the interface between them. The anatectic process was controlled by the breakdown of hornblende in the (meta)tholeiites and biotite in the (meta)sediments. Experimental data and the observed restite phases are consistent with the melt zone lying between depths of about 10 km and 18 km. Mixing of mantle material and crustal melts generated near the interface (which was at a depth of some 15 km) produced the low-Si endmember of each suite. As more and more shoshonitic magma entered the system, more and more tholeiite and sediment melted so that the low-Si endmember was limited to $SiO_2 > 63\%$ and initial $^{87/86}Sr \geq 0{\cdot}708$. Variation in the mantle component was responsible for most of the chemical differences between suites. Mixing of the low-Si endmember and melt derived purely from sedimentary rocks (the high-Si endmember) occurred as the entire contents of the melt zone moved towards the upper crust. The chemical (including isotopic) variation in S-type granites containing up to 72–73% SiO_2 can be explained in this way. Crystal fractionation in high-level magma chambers appears to have played a part in the formation of more siliceous varieties. Some of the I-type granitic rocks of the Lachlan Fold Belt were probably derived by partial melting of early Cambrian(?) calc-alkaline, continental-margin rocks. The I-type granites contain more Al and less Fe than LFB andesites and mantle melts may have been involved in their genesis as well.

A re-evaluation of granite types of the Calabrian Arc, Italy in the context of the southwestern Hercynian belt

Alessandro Rottura and **Raffaellamaria Campana**

The late-Hercynian granites of the southern Calabrian Arc have been referred in the past to two different suites, one predominantly calc-alkaline, weakly peraluminous (biotite ± hornblende quartz gabbro/diorite to granodiorite) and the other strongly peraluminous (two-mica ± Al-silicate-bearing trondhjemite to granite). In particular the Villa S. Giovanni (VSG) and Capo Rasocolmo (CR) strongly peraluminous granites, coeval but not spatially related to the calc-alkaline granites, have been considered S-type.

A geomathematical approach to their geochemical composition indicates that the VSG and CR granites are distinguishable from the calc-alkaline granites, as well as from the classic S-type granites of southeastern Australia. The calc-alkaline granites show basically I-type compositional characteristics, and have trace-element patterns suggestive of subduction zone processes during development of their source, and of a plagioclase-dominated mineralogical control. They have recently been interpreted as having been generated by interaction of subcontinental mantle-derived magmas with crustal components, in a context dominated by crustal recycling. The strongly peraluminous granites have an S-type signature in terms of enclaves and mineralogy, while their geochemical characteristics are similar to those of volcanic arc granites. They have high Sr and Ba, low Rb and highly-fractionated REE patterns ($La_n/Yb_n = 78–27$; $Yb_n < 4$), suggestive of hornblende (±garnet)-dominated fractionation processes and similar to those of the Archaean tonalite-trondhjemite-granodiorite suites. Therefore, they cannot be directly related to the associated calc-alkaline granites by high-level fractionation, and must represent separate melts.

Geochemical and petrographical evidence indicates a trondhjemite-dominated composition for the VSG and CR granites, with crystal-liquid mixing coupled with metasediment assimilation as the leading process for their genesis. Whether the trondhjemitic magma is due to mantle input at the time of granite production or represents a crustal melt derived from a meta-igneous source (gt-amphibolite), is still an open question. Nd_{DM} model ages (1·2–1·8 Ga) are not conclusive.

Volcanic plutonic connections, Uralla Igneous Centre, N.S.W.

S. E. Shaw and **R. H. Flood**

The late Permian Uralla Igneous Centre consists of K-rich andesitic lavas and pyroclastic flows (Harnham Hill volcanic rocks), the

Terrible Vale Pluton which is microstructurally zoned from dacite-tuffasite to quartz-monzodiorite, the Uralla granodiorite-diorite Pluton, and the Kentucky diorite Pluton. The Harnham Hill volcanic rocks and underlying sedimentary rocks are intruded and contact metamorphosed by the Terrible Vale Pluton, which is crescent-shaped, the dyke-like extensions of which follow possible ring fractures. The preservation of the outlier of volcanic and sedimentary rocks suggests cauldron subsidence.

The Terrible Vale Pluton, although uniform in composition, is microstructurally zoned from a narrow marginal two-pyroxene dacite-tuffasite zone, to an inner coarser-grained porphyritic quartz-monzodiorite facies. The dacite-tuffasite which occurs along intrusive margins has elongate fragmented phenocrysts and shard-like devitrified glass fragments aligned in a sub-vertical foliation. Superficially, the tuffasite has many features in common with ignimbrite flows. In the quartz-monzodiorite facies both the groundmass and phenocryst plagioclase grains are zoned and rectangular, but quartz occurs as polygonal aggregates rather than as interstitial grains, and biotite forms poikilitic grains that enclose or partly enclose quartz.

The Terrible Vale Pluton appears to represent the preserved parts of a high-level magma chamber that vented to the surface during a major roof-fracture event but then crystallised more slowly to produce the observed microstructural zonation. The volcanic, subvolcanic and plutonic rocks are inferred to be parts of a geochemically coherent igneous centre, one of many such short-lived centres within the New England Batholith that formed during the late Permian and early Triassic.

The Walcha Road adamellite: an example of a large restite-poor zoned pluton

S. E. Shaw and **R. H. Flood**

The Walcha Road Pluton of southern New England is 30 km long, 20 km wide and is elliptical in outline except where minor faults off-set the margin. Biotite/bulk-rock Rb/Sr ages for the pluton and two quartz micromonzonite stocks along the southern and western margins give late Permian ages of approximately 247 Ma and an initial $^{87}Sr/^{86}Sr$ ratio of 0·7046.

A foliation defined by the preferred orientation of the long axes of both feldspar crystals and microgranitic enclaves is generally margin-parallel and is sub-vertical along the southern margin. The pluton is concentrically and smoothly zoned from a discontinuous outer margin of quartz monzonite (colour index = 35) with hornblende in excess of biotite, to two central areas of hornblende-free biotite-adamellite (colour index = 6). The marginal quartz-monzonite is even-grained and grades over a few tens of metres into a porphyritic mafic adamellite with more than 8% pink K-feldspar megacrysts up to 20 mm in length. This zone grades inwards over several hundred metres into a more leucocratic adamellite with larger (50 mm) but less abundant pink K-feldspar megacrysts. Closer to the pluton centre the abundance of K-feldspar megacrysts reduces to near zero, although the total modal K-feldspar changes little from margin to core.

The zoning results from the crystallisation of the more mafic rocks along the margin as cumulates, and the felsic residual melt collecting in the centre and near the top of the magma chamber. Calculations based on pluton shape as deduced from outcrop mapping and feldspar foliation measurements suggest that the felsic central parts of the pluton are volumetrically small compared with the more mafic marginal rocks. The quartz-micromonzonite stocks along the southern and northwestern margins of the pluton are sufficiently fine-grained to show that intermediate-composition magmas with little restite were present at the time.

Late Precambrian epidote-bearing calc-alkalic intrusive suite in northeastern Brazil: oxygen isotopes and depth of crystallisation

A. N. Sial, V. P. Ferreira and **R. V. Fodor**

Over 50 late Precambrian (~600 Ma) Conceiçao-type meta- to peraluminous tonalites and granodiorites intruded phyllites of the Cachoeirinha-Salgueiro Fold Belt (CSF) in northeastern Brazil, and contain primary epidote (based on criteria of Zen & Hammarstrom 1984). Petrographically equivalent epidote-bearing granites (EBG) intruded schists of the Serido Fold Belt (SFB), a northward extension of the CSF, and amphibolite-grade metasediments of the Riacho do Pontal Fold Belt (RPF), a southern extension of the CSF. Magmatic epidote is also found in trondhjemitic plutons that intruded Salgueiro schists as well as in plutons in three other Fold Belts. This phase is observed in four textural relationships, two of them indisputably magmatic and two from sub-solidus reactions.

In the CSF, contemporaneous EBG solidified around 6–7 kbar (1 kbar = 100 MPa) if all Al variation in hornblende is ascribed to pressure. Lower pressure for quartz diorite enclaves results from Al loss by subsolidus reaction with plagioclase, generating granular epidote. Temperatures of equilibration for hornblende-plagioclase assemblages are in the range 700–800°C. They differ from Mesozoic EBG in N America in that they intruded phyllites, likewise Palaeozoic plutons in Argentina, New England and New Zealand.

In the SFB, Al in hornblende indicates emplacement around 2·5–4·0 kbar, and 5·5 kbar when they only intruded basement rocks. Temperatures for equilibration of hornblende-plagioclase assemblages in this case are in the range 680–760°C.

In the CSF, EBG exhibit moderate Sr, Ba and Zr, low Nb (<20 ppm), are LREE-enriched and HREE-depleted, with discrete negative Eu anomaly. In the SFB, they are higher in Sr, with Ba contents equivalent to CSF granites, Zr slightly higher and low Nb. They exhibit less steep REE-patterns with negative slope, lacking or exhibiting discrete Eu anomaly. EBG in the CSF display high $\delta^{18}O$ (+11–+13‰) and amphibolite xenoliths, probably from the source, have $\delta^{18}O$ values around +10.5‰. Equivelent plutons in the SFB display $\delta^{18}O$ values slightly lower (+6·2–+8·2‰) which are lower than equivalent epidote-bearing tonalites and trondhjemites in northwestern N America (+7·5–+9·0‰, e.g. Hazard Creek Complex, Idaho). This demonstrates that EBG form from more than one kind of source material, always with the presence of a component in the magma derived from sedimentary or altered volcanic rocks. Processes leading to their formation are repeated through geological time and magmas intrude different crustal levels.

Zen, E-an & Hammarstrom, J. M. 1984. GEOLOGY **12**, 515–8.

Tracking magmatic arc evolution across contrasting lithosphere settings in southwestern N America

Leon T. Silver

Two great batholiths of the southern Cordillera, the Peninsular Ranges and the Sonoran, are tectonically dislocated elements of a formerly continuous magmatic arc produced in a single cycle. The cycle was initiated in earliest Cretaceous time and continued into the Palaeocene. (The commonly employed distinction between coastal batholiths and "Laramide" plutons of the interior must be reassessed.) The magmatic arc transgresses time and space northeastward, continuing across contrasting lithospheric settings ranging from Mesozoic oceanic lithosphere in the W to early Proterozoic crust-lithosphere in the E. Petrological, chemical and isotopic properties of the plutonic rocks show dramatic changes in many forms. This particular geological setting provides an opportunity to investigate the relation of various initial isotopic signatures ($^{87}Sr/^{86}Sr$, $^{143}Nd/^{144}Nd$, Pb isotopes, $\delta^{18}O$) and trace-element indices to tectonic setting. The volume of plutonic rocks exposed in the Sonoran Batholith, which extends from southern Arizona to southern Sinaloa and from the Gulf of California to Chihuahua, is fully comparable to the Peninsular Ranges Batholith, although plutons of the former are more dispersed to the E. These volumes provide a major challenge in modelling various source regions and petrogenetic mechanisms. Further, in the Sonoran Batholith region, earlier Jurassic and Triassic magmatic arcs had extensively taxed the putative source regions for significant quantities of volcanic and plutonic differentiates. From where has all of this material been derived? How much is from the mantle? How much is recycled? Where did the energy for mobilisation come from and how was it applied? Finally, from where did the mid-Tertiary magmatic cycle represented by the great Sierra Madre Occidental volcano-plutonic arc, which is superimposed on the Sonoran Batholith, derive its materials? Observations, discussion and some possible insights are offered; all of the answers are not yet in hand.

Granites in the late Cainozoic arc-type magmatic associations of the southwestern Pacific

Ian E. M. Smith and **Donald S. Clarke**

Granites, as individual plutons and as small-scale clusters of plutons, are a subordinate component of the suites of dominantly andesitic volcanic rocks which comprise the late Cainozoic arc-type magmatic associations of the southwestern Pacific. This association of magmatic rocks dominates the late Cainozoic geological record of the southwestern Pacific margin from Papua New Guinea to northern New Zealand.

High-level granite intrusives (porphyries) are commonly associated with the volcanics. In only a few areas, notably in Papua New Guinea and New Zealand, granitic plutons represent deeper levels of the magmatic systems; these are temporally but not always spatially associated with the arc-type volcanism. In New Caledonia, granites of comparable type and age occur without coeval volcanics.

The structural setting of these granites shows considerable diversity. In Papua New Guinea, granites occur as discrete pluton complexes with or without associated volcanics and as pluton within metamorphic core complexes. In New Caledonia, they occur as isolated plutons intruding an ophiolite. In New Zealand, granitic plutons constitute a part of arc-type volcanic associations, although they are usually spatially separate from coeval volcanic rocks.

The granites of these southwestern Pacific arc systems are mainly diorite and granodiorite, but include adamellite and minor mafic diorite and monzonite. In southeastern Papua New Guinea there are also shoshonitic pluton complexes dominated by monzonite and syenite, and including gabbro and cumulate pyroxenite; these are associated with volcanic rocks of similar geochemical character. Variation in chemical compositions correlates with tectonic setting and also with the complexity of earlier geological history. In general terms, the rocks are mainly M-type granites. Petrogenetic hypotheses involve a subduction-related component as well as a chemical component linked to deep crustal levels.

Mantle and crust involvement in the genesis of subduction-related granites: evidence from the calc-alkaline magmatism of central Peru during the Andean orogenesis

Pierre Soler

During the Andean orogenesis (Albian to Present), the space and time distributions of the calc-alkaline magmatic rocks of central Peru have been tightly controlled by the features of the subduction of the Nazca (previously Farallon) plate. Over this time, the calc-alkaline volcanic rocks and I-type granites display complex variations in their geochemical features.

For a given period, overall similarities in their space distribution and compositions imply that both types of rock were derived from the same source, an enriched mantle modified by fluids extracted from the subducted slab. Subordinate chemical differences indicate, however, that the crustal histories of the magmas leading to plutonic or volcanic rocks diverge at an early stage, and that the plutons are not simply the magma chambers of the volcanoes.

The variations of their compositions in time correspond mostly to crustal processes in deep magma chambers which include: (1) diffusive exchange between magmas and crust, a selective and self-limiting process which accounts in particular for the variation from medium- to high-K contents; (2) assimilation of the country rock at an early stage of differentiation; (3) modifications of the mineralogy of the crystallising phases, with the appearance of garnet at the liquidus and the later crystallisation of plagioclase which lead to lower HREE, Sc, Mn, and higher Sr and LREE in the intermediate and felsic rocks.

The observed evolution is a succession of compositional jumps which are contemporaneous with the tectonic events and associated crustal thickening and are interpreted as the result of an increasing degree of crustal assimilation and an evolution under higher pressure at the early stage of differentiation.

The evolution of the chemical and isotopic features of the calc-alkaline magmatic rocks of central Peru may be regarded as the geochemical fingerprint of the orogeny upon mantle-derived magmas which do not appear to have changed in time.

The origin of Cooma Supersuite granites: protoliths and early magmatic processes

D. A. Steele, R. C. Price, P. D. Fleming and **C. M. Gray**

The Siluro-Devonian Yabba adamellite is a residue/inclusion-rich, syn-tectonic pluton located between the late Cambrian(?) Gundowring migmatite terrane and lower-grade early Ordovician Lockhart terrane in the high-*T*, low-*P* Wagga Metamorphic Belt, southwestern Australia. This body is compositionally and isotopically similar to the Cooma Supersuite (CS) granites of the Lachlan Fold Belt (LFB). Both are characterised by regional-style contact aureoles and by low $CaO–Na_2O$, highly peraluminous compositions; the latter has been considered indicative of derivation from the extensive feldspar-poor Ordovician turbidites exposed in southeastern Australia (e.g. Chappell 1984). These characteristics distinguish the Yabba and CS granites from the voluminous contact aureole S-type granites of the LFB.

Petrographical, structural and geochemical comparison of metasedimentary enclaves in the Yabba body with the two "host" tectonometamorphic terranes indicates: (1) the Yabba adamellite and, by implication, the CS granites were derived from the late Cambrian(?) Gundowring terrane, *not* from the Lockhart terrane or the other exposed Ordovician flysch; (2) one of the quartzofeldspathic lithologies found in the Gundowring terrane is absent from the inclusion population and represents the "fertile" source component.

The partial melting and virtually complete extraction of the fertile quartzofeldspathic gneisses gave rise to residue-rich magmas that almost perfectly image their sources. The chemical variation observed within the Yabba adamellite reflects the incorporation of various residual and non-melted source components during magma genesis. The metasedimentary enclaves represent non- or partially melted (infertile) protolith.

The Gundowring terrane and/or its tectonic/compositional equivalents probably underlie the monotonous Ordovician turbidite sequences and thus form part of the basement to the LFB. The CS granites were probably derived from this basement. Geochemical evidence precludes the derivation of the voluminous contact aureole S-type granites (e.g. the Bullenbalong Suite) from the Gundowring terrane by simple partial melting. A more immature (feldspathic) metasedimentary protolith or a multi-component process (e.g. Gray 1984) are required to produce the chemical characteristics of these S-type granites.

Chappell, B. W. 1984. PHILOS TRANS R SOC LOND **A-310,** 693–707.

Gray, C. M. 1984. EARTH PLANET SCI LETT **70,** 47–60.

Geochemical constraints on fractionation and source models for an A-type granite: the Mt Walsh granite in the New England Fold Belt, eastern Australia

C. J. Stephens

Recent reviews of A-type granites (Eby 1990) defined two sub-groups based on geochemical criteria: a group that have distinct OIB-like trace-element characteristics, and a second group that display trace-element characteristics closer to calc-alkaline volcanics, and which are often associated with I-type granite plutonism. This paper describes the geology and geochemistry of a late Triassic A-type granite in eastern Australia that is of this latter type.

The Mt Walsh granite is a discrete, circular intrusion that straddles the margin of a coeval silicic I-type cauldron. The intrusion is 60 km^2 in area, flat topped, and exposes 570 m of relief. Dissection of the intrusion along fractures provides an opportunity to sample vertical and horizontal variations within the stock. The intrusion is predominantly metaluminous to weakly peralkaline, high-silica alkali granite displaying granophyric textures. Low silica granite occurs high within the core of the stock, and alaskite forms a thin (<30 m) cap across the top of the stock. A high level of intrusion is indicated by the abundance of miarolitic cavities, and by intrusion of the stock 400 m into intracaldera ignimbrite. The mineralogy comprises embayed quartz and perthitic potassium feldspar, with minor sodic plagioclase (especially in low silica granite), interstitial annite-rich biotite, phenocrystic edenitic amphibole, and accessory magnetite, zircon, apatite and allanite. The biotites contain up to 2·2% F, and show trends of decreasing Al^{IV} and increasing Fe/Fe + Mg with increasing host rock SiO_2 and alkalinity. This trend contrasts with the increasing Al^{IV} with Fe/Fe + Mg observed in peraluminous granites. Biotite and amphibole Fe/Fe + Mg ratios are typical of ilmenite series granites, consistent with an inferred f_{O_2} close to NNO. P_{H_2O} is inferred to be less than 1 kbar (1 kbar = 100 MPa).

Trace-element abundances define a zone depleted in Ba, Sr and Zr, and enriched in Rb and La, across the top and extending down the margins of the stock. High values for Rb/Ba define a thin zone across the top of the stock corresponding to alaskite. These variations are consistent with crystal fractionation occurring via sidewall crystallisation, and can be modelled in terms of Rayleigh fractionation of an assemblage comprising potassium feldspar-plagioclase-allanite-zircon ± quartz. The isotopic composition is indistinguishable from the associated I-type granites ($\varepsilon_{Sr} = -10$, $\varepsilon_{Nd} = +4{\cdot}3$, $\delta^{18}O = -8‰$) and defines an identical age (221·8 Ma) and I_{Sr} (0·70354 ± 0·00001). Modelling of REE data shows that partial melting of restite or tonalite sources is incapable of generating the REE compositions of low-silica A-type granite. Partial melting of an anhydrous (?granulite) source is unlikely because modelling of the I-type granite sources excludes the presence of all but recently accreted crust within the source area. Low-silica A-type granite cannot be derived by (A)FC from any of the I-type granites, but can be modelled closely by Rayleigh fractionation from associated mafic volcanics that have primitive mantle-like geochemistry (Ba/La < 10). The mafic volcanics, nevertheless, show evidence for contamination by a calc-alkaline component, a feature reflected in the calc-alkaline spidergrams shown by all igneous rocks of the complex.

It is interpreted that the A-type Mt Walsh granite was produced by crystal fractionation from a basaltic parent derived from a primitive mantle source. The basaltic magma was contaminated with a calc-alkaline source, probably within the crust, before the evolution of the Mt Walsh silicic magmas. The silicic magmas underwent crystal fractionation after emplacement into a high-level magma chamber, with REE-enriched magma produced by sidewall crystallisation ponding at the top of the stock, and crystallising as alaskite.

Eby, G. N. 1990. LITHOS **26,** 115–34.

The geology and petrology of arfvedsonite granites, Hinchinbrook Island, N Queensland

P. J. Stephenson, B. W. Chappell, M. McCulloch, M. T. Frost and E. Reid

Permian arfvedsonite granites occur within a 16 km-long batholith over a vertical range of 1,100 m, in rugged wilderness. Some earlier and later felsic dyke units are related chemically. The granites are dominantly hypersolvus with relatively low modal biotite and/or arfvedsonite. Near-horizontal layering occurs in numerous places and some indicates downward solidification (Stephenson 1990). Geochemically, the rocks are A-type (sense of Whalen *et al.* 1987), and strongly fractionated. They contrast with country rock I-type granites and volcanics. Four traverses from summits to sea were used to test vertical fractionation, but no consistent variations were found. Considerable trace element fluctuations occur. Few rocks are peralkaline in the sense of agpaitic index, but rather peraluminous, consistent with the low arfvedsonite modes, less than 2%. Classic textures suggest that most of the rock features are primary, with amphibole-biotite growth (commonly in cavities) from late residual fluids.

An Rb-Sr isochron gives 275 ± 5 Ma (Sr* 0·701138 ± 0·008, MSWD 13·5), and this isochron appears to be coincident for the I-type granites in the nearby Palm Islands. Sm-Nd results are more dispersed and give a considerable range of model ages (ε_{Nd} − 1·3–8·0). Evidence suggests that the arfvedsonite granites evolved almost simultaneously with the I-type Palms granites, with close consanguity.

The batholith is interpreted to contain a thick "roof" facies, passing down into slightly more calcic and aluminous rocks beneath. Of possible alternatives, partial melting of crustal tonalite to granodiorite (Creaser *et al.* 1991) is preferred to a residual source model (Collins *et al.* 1982). Considerable fractionation occurred, but details are not established.

Collins, W. J., Beams, S. D., White, A. J. R. & Chappell, B. W. 1982. CONTRIB MINERAL PETROL **80,** 189–200.
Creaser, R. A. *et al.* 1991. GEOLOGY **19,** 163–66.
Stephenson, P. J. 1990 GEOL J **125,** 407–19.
Whalen, J. B., Currie, K. L. & Chappell, B. W. 1987. CONTRIB MINERAL PETROL **95,** 407–19.

Classification and nomenclature for Hong Kong granites

P. J. Strange, R. L. Langford and R. J. Sewell

The first systematic geological survey of Hong Kong in the late 1960s recognised four main phases of Upper Jurassic syntectonic plutonic intrusion. Widely mapped lithological units were assigned to a phase according to their appearance and mutual cross-cutting relationships, but no attempt was made to delineate individual plutons. Considerable textural variation within some bodies resulted in an unworkable system of naming and classification.

Recent field mapping by the Hong Kong Geological Survey at scales of 1:20,000 and 1:5,000, together with geochemistry and whole-rock Rb-Sr geochronology, has resulted in the establishment of a pluton-based nomenclature for the granite, granodiorite and syenite of Hong Kong. Fifteen intrusive units have been identified ranging in age from late Jurassic to early Cretaceous, and these have been subdivided into two suites: the Lamma Suite and the Lion Rock Suite.

The Lamma Suite comprises two small relatively unfractionated batholithic bodies, Tai Po Granodiorite and Sung Kong Granite, which represent the oldest intrusive units in the Territory. The Lion Rock Suite encompasses several cross-cutting, well-defined, normal to highly fractionated granite plutons which are assigned to three subgroups. Subgroup I includes two large elliptical plutons of Tsing Shan and Tai Lam, both in excess of 20 km. Subgroup II embraces the large (>20 km) elliptical plutons of Lantau and Sha Tin, plus two bodies of quartz syenite: Cape D'Aguilar and Lantau syenites. The almost circular plutons of Kowloon, Stanley, Chi Ma Wan, South Lamma, Mt Butler, Kwun Tong and the small (2 km-diameter) King's Park Pluton make up subgroup III.

Petrological diversity in I-type granites due to variation in H_2O fugacity during melting of amphibolite

J. H. Tepper

Petrological diversity (quartz diorite-granite) is characteristic of granites in I-type batholiths, and is generally ascribed to crystal fractionation, restite separation, and/or magma mixing. However, data from the Chilliwack Batholith (North Cascades, Washington) support an alternative hypothesis, in which the diversity reflects differences in H_2O fugacity during melting of amphibolitic lower crust.

REE patterns of Chilliwack granites vary systematically with SiO_2 content (Tepper *et al.*, 1989). Intermediate plutons have Eu/Eu* < 0·85, whereas siliceous plutons have little or no Eu anomaly. All samples from the same pluton have similar patterns. The lack of a Eu anomaly in the siliceous plutons precludes their having undergone significant crystal/restite fractionation. Geochemical data for these plutons are modelled by 20–30% melting at elevated f_{H_2O} of amphibolite having trace element and Sr/Nd isotopic characteristics of Chilliwack gabbros. Data for intermediate plutons are modelled by 30–50% melting of a similar source, but at lower f_{H_2O}. The main effect of f_{H_2O} variation is to change the proportions of plagioclase versus amphibole in the residuum (cf. Beard & Lofgren 1991). This strongly influences melt composition, with higher f_{H_2O} (residuum = amph + cpx ± plag) resulting in magmas with high SiO_2 and low MgO and FeO (and no Eu anomaly) compared to magmas generated at low f_{H_2O} (residuum = plag + cpx + opx). The f_{H_2O} differences probably reflect additions of H_2O to the lower crust from crystallising basaltic magmas having a range of H_2O contents. Individual granitic plutons are interpreted as discrete magma batches that underwent limited modification by fractionation, etc. during ascent. Based on geochemical and lithological similarities, this model may be applicable to other I-type batholiths.

Beard, & Lofgren, 1991. J PETROL.
Tepper, J. H. *et al.* 1989. EOS **70,** 43.

Genesis of A-type granites: trace element and isotope constraints from Oligo-Miocene Yemen alkaline granites

S. Tommasini, P. Manetti, G. Poli and **S. Conticelli**

Yemen was the site, in Oligo-Miocene times, of extensive magmatic activity which brought about the formation of a basalt plateau and the emplacement of several granitic plutons. This activity developed in two phases which climaxed at 30–26 Ma and 22–20 Ma, respectively. During the second phase, alkaline granites and granophyres were emplaced. They consist of light grey to pinkish, medium to coarse-grained rocks, having characteristics from subsolvus to hypersolvus with emplacement depths in the range of 0·05–0·2 GPa. Three main groups have been distinguished on the basis of the presence of different ferromagnesian minerals: alkali amphibole granite, alkali amphibole plus biotite granite, and calcic amphibole plus biotite granite.

Yemen granites are emplaced in anorogenic settings, and have typical physicochemical characteristics such as high temperature, high halogen and low H_2O contents, and high HFSE abundances of A-type granites. They fall well outside VAG and CG fields, indicating that crustal partial melting did not play a leading role in their genesis. Moreover, the ratios of HFSE of basaltic rocks belonging to the Yemen plateau relate to those of the granites. This fact suggests a genetic link between basalts and granites. However, the variation in radiogenic isotope ratios recorded from the Yemen granites precludes an origin solely through closed-system processes. The absence of crustal xenoliths, magmatic enclaves and contact aureoles supports an assimilation plus fractional crystallisation (AFC) process, where the heat required for assimilation was provided by the latent heat of crystallisation. The genesis of the most and least alkaline granites can be accounted for by FC (ol + pl + cpx+/−mt) process from a parental alkali and transitional basalt, respectively, accompanied by different degrees of assimilation at different levels in the crust.

Derivation of A-type magma by fractionation of basaltic magma with an example from the Padthaway Ridge, S Australia

Simon P. Turner, John D. Foden and **Robert S. Morrison**

Various petrogenetic schemes have been proposed for A-type granite and volcanic rocks, a popular one being re-melting of a melt-depleted source. However, simple modelling shows that the high incompatible/compatible element ratios characteristic of A-types are not readily obtainable from such a source, because the first melting event has already lowered this ratio. Instead, extended fractionation of a mantle-derived basaltic magma will produce high incompatible to compatible element ratios and also explains the high-temperature nature of A-types, their typically low initial $^{87}Sr/^{86}Sr$ ratios and their bimodal association with mafic rocks. Granophyres from layered mafic intrusions provide evidence for this origin of A-type magma, as do new data presented here for a suite of A-type granites and volcanics from the Padthaway Ridge in S Australia. Mineralogical, chemical and isotopic arguments show that the Padthaway Suite evolved from the same basaltic magma that formed contemporaneous gabbroic plutons which themselves contain A-type granophyres. Olivine, pyroxene and hypersolvus feldspar assemblages in the granites and volcanics document temperatures of 900–1,000°C and H_2O-undersaturated conditions with $H_2O < 3\%$. Curvilinear geochemical trends and large negative Eu anomalies indicate a history of protracted fractionation involving pyroxene and feldspar. High $\varepsilon_{Nd}(+2-3)$ and low initial $^{87}Sr/^{86}Sr$ (0·703–0·706) for the whole suite of granites, volcanics, gabbros and granophyres indicate a mantle-derived parental magma. Major- and trace-element modelling confirms that these A-type magmas can be produced by fractionation (*c.* 90%) of such a magma, though it must be somewhat enriched, indicating either a lithospheric mantle source (K-rich layer) or limited crustal contamination. While there may be various kinds of A-types, their high temperatures dictate that mantle magmas be invoked on thermal grounds in any petrogenetic model. In the fractionation model outlined here, the ratio of mafic to silicic magma is likely to be ~9:1. This suggests that A-type suites, which typically intrude in extensional or post-orogenic settings, mark episodes of crustal growth in which considerable mantle material is added to the crust.

Some thoughts on cordierite-bearing granites: are host-rocks the clue?

J. M. Ugidos and **C. Recio**

Cordierite-bearing granites (CBG) form batholiths in some collisional domains. Determining the original I/S character of such granites may be controversial. The possibility of assimilation should be considered based on the following facts: (1) the high content of prismatic cordierite in high-grade hornfelses, migmatitic hornfelses and contact migmatites in thermal aureoles (TA); (2) syn-orogenic metamorphism is of the intermediate P type; (3) frequently CBG are late orogenic. Point (1) indicates the high capacity of some host-rocks to produce cordierite and that high temperatures have been reached at low pressure. This resulted from the addition of heat input to the pre-heated host-rocks. Then, if pressure drops, fluid absent reactions are expected in the host. If the metasediments involved have pelitic composition, the reactions will result in cordierite/garnet + Kfd + H_2O or melt. In the first case, subsequent increments of T may result in the equilibrium q + ab + kfd + cord + H_2O = melt; in the second case, melt is produced directly because of the fluid-absent reaction. The amount of melt is small, but the important point is the generation of intercrystalline melts in the TA of an intruding granite. Point (2) implies possible restites being left behind after regional anatexis. If restites are affected also by a lowering of pressure, reactions transforming the restitic parageneses into lower-P parageneses are to be expected. The interaction of a granite magma with any of these possible rock-types developed in its TA can result in granites with apparent S character, since even small assimilation percentages may drastically modify the original granite mineralogy and normative corundum content. Andalusite-bearing, sillimanite-bearing or garnet-bearing cordierites, or other kinds of cordierite-rich rocks found in TA, are important points that must be taken into account when considering the significance of S-type mineralogy in some granites.

Subdivision of I-type granitic rocks in the Archaean Murchison Province, Western Australia

Liang G. Wang, Neal J. McNaughton and **David I. Groves**

Three major suites of granitic rocks can be identified in the Archaean Murchison Province of Western Australia on the basis of this work (Wang in prep) and that of Watkins and Hickman (1990): Suite I comprises monzogranites and granodiorites which occur as homogeneous regional batholiths; Suite II consists of trondhjemite and tonalite plutons; and Suite III includes A-type (Loiselle & Wones 1979) monzogranite and syenogranite plutons. Each of the three suites have coherent but distinct geochemical and isotopic (Pb, Sr and Nd) characteristics indicative of their derivation from distinct source regions. Suite I can be interpreted as being derived from partial melting of underlying tonalitic continental crust. Suite II from partial melting of subducted/underplated oceanic crust and Suite III from partial melting of dehydrated siliceous granulite.

According to Chappell and White (1974), both Suite I and Suite II granites, which have contrasting compositions and origins, have to be classified as I-type granites. It is proposed here that granites derived from earlier granites or gneisses comprising continental crust should be termed G-type granites (e.g. Suite I), and that granites derived from a basaltic source should be termed I-type granites (e.g. Suite II).

The above subdivision of the I-type granites of Chappell and White (1974) may have important tectonic implications. The G-type granites may be generated during anatexis of continental crust within various tectonic environments, whereas the newly-defined I-type granites, with more restricted compositions and specific source, are likely to be indicators of subduction/underplating tectonic regimes.

Chappell B. W. & White, A. J. R. 1974. PAC GEOL **8,** 173–4.
Loiselle, M. C. & Wones, D. R., 1979. GEOL SOC AM ABSTR PROG **11,** 468.
Wang, L. G. Ph.D. thesis, in preparation.
Watkins & Hickman 1990. GEOL SURV WESTERN AUST BULL **137**.

Proterozoic granites, with some characteristics of A-type magmas, in the Arunta Block, central Australia

R. G. Warren and **S.-S. Sun**

Assessment of 800 analyses in the BMR ROCKCHEM database of Proterozoic rocks from the Arunta Block shows that most felsic igneous rocks fall into one of three classes: two showing typical low-Sr, high-Y Proterozoic signatures in multi-element plots, and the third with *flat profiles* (high-Sr, low-Y). The low-Sr, high-Y classes subdivide into a *normal* Proterozoic type (cf. Kalkadoon-Leichhardt-Ewen association), and an *enriched* type. Both have the characteristics of Sr-depleted felsic magmas, especially high incompatible elements. Enriched types have higher U, Th, K, Rb, REE and Y relative to normal types but, commonly, lower K/Rb and higher Th/U. High LREE and Th are contained in allanite or monazite and, in some, high F in fluorite. Some chemical characteristics of the enriched suites, particularly the high incompatible elements, hint at affinities with true A-type magmas (rare in the Arunta Block); but some enriched types have high Ba, not all have high F, and Al/Ga is in the same range as in normal types. The normal and enriched types occur throughout the block, the flat-profile type mainly in the SE. Distinct suites, delineated within both major types from subtle chemical variations, perhaps reflecting source, have an areal distribution with probable time-tectonic significance. Regional variations among enriched suites mimic those of the normal types in the same areas, suggesting a evolutionary link, supported by sparse Nd isotope model age data. Field evidence shows that the enriched suites were emplaced later, either syn- or post-tectonic. Their low Cs and high Rb/Cs dictate a refractory source, while low-Sr, high-Y dictate a shallow source. Overall, the evidence suggests enriched magmas formed through regional metamorphism of high-*T*, low-*P* style, possibly with variable addition of second-stage melts from source regions, either during renewed extension or through A-subduction.

A granite geochemical and isotopic transect of the southern Canadian Appalachian orogen

Joseph B. Whalen, Ernst Hegner, George A. Jenner and **Frederick J. Longstaffe**

From W to E, the four major Appalachian tectonostratographic zones are Humber (HZ), Dunnage (DZ), Gander (GZ) and Avalon (AZ). Recent deep seismic reflection studies have defined three main lower crustal blocks (LCB) beneath this orogen: Grenville (under HZ and western DZ), Central (under eastern DZ and GZ) and Avalon (under AZ). All zones are intruded by late to post-orogenic Siluro-Devonian granites. In addition, GZ and AZ contain pre- to syn-orogenic Ordovician and pre-Iapetus Precambrian granites, respectively. Based on Na_2O and AI index values (generally $>3{\cdot}2$ wt% and $<1{\cdot}1$) and high-field strength element contents, most granites are I-types; some have "A-type affinities".

ε_{Nd} data map three contrasting groups of granites: HZ + DZ, GZ and AZ. Positive ε_{Nd} values for HZ, DZ and AZ granites indicate melting of predominately juvenile sources. Higher $^{207}Pb/^{204}Pb$ and $\delta^{18}O$ values of AZ granites could reflect incorporation of a minor radiogenic supercrustal component which has produced elevated Pb and O but has had little effect on their juvenile Nd signature. Negative ε_{Nd} and elevated $^{207}Pb/^{204}Pb$ and $\delta^{18}O$ values for GZ granites are consistent with reworking of older (1·1–1·8 Ga) crust containing a significant supercrustal component.

Granite data suggest that different protoliths are juxtaposed at the DZ–GZ and GZ–AZ boundaries. As Nd data effectively rule out Grenville basement as a source for HZ + DZ granites, the seismically defined Grenville LCB is either not "Grenville-like" compositionally and/or it is composite. The geochemical and isotopic similarity between Ordovician and Siluro–Devonian GZ granites indicates that no major post-middle Ordovician basement-cover detachments have occurred in GZ. GZ, therefore, probably is the surface expression of the underlying Central LCB. Contrasting ε_{Nd} signatures from AZ and GZ granites help substantiate the seismic interpretation that GZ and AZ are underlain by different crustal blocks; i.e. Precambrian Avalon basement is not a suitable protolith for GZ granites.

Inherited and detrital zircons—vital clues to the granite protoliths and early igneous history of southeastern Australia

Ian S. Williams, Bruce W. Chappell, Yadong D. Chen and **Keith A. W. Crook**

The SHRIMP ion microprobe has been used to measure the U–Th–P isotopic composition of over 1,000 zircons from more than 20 plutons and enclaves from the Palaeozoic batholiths of southeastern Australia. The ages of the principal components in the granites' sources have been established and a start made on determining the areal variation in the relative abundances of those components.

The granites range in age from 440–380 Ma. Zircons with inherited cores are rare in the I-type granites, but virtually every zircon in the S-types contains an older core. In most granites, most of the cores are 650–450 Ma old, there is a lesser group 1,075–800 Ma old, and a lesser group again with ages up to 3,350 Ma. Not all age groups are represented in every granite, and there is a marked lack of the two older groups in the I-type granite enclaves. There is little difference between the I- and S-type granites in the relative abundances of cores of different ages. The inherited zircon becomes older and more abundant westwards across the I-type Bega Batholith.

Zircons from the Bega Batholith's Ordovician sedimentary country rock show the same age groups as the inherited zircons in the granites, but the 650–450 Ma age group is significantly less abundant in the sediments than in the granites.

It appears that 1,075–800 Ma ago, and again 650–450 Ma ago, there was considerable igneous activity in southeastern Australia and nearby regions that produced both the immediate sources for the I-type granites and their precursors (probably by underplating of the crust) and also generated, through volcanism, much of the detritus which became the source for the S-type granites. Recycling of those sediments produced much of the early Palaeozoic flysch. The presence of 1,100–3,350 Ma inherited zircon in many of the I-type granites is evidence for a small component of sediment in those magmas. The chemical and isotopic trends across the Bega Batholith suggest that the amount of that sediment increases westwards.

Geochemical and isotopic study of granites in the Arunta Inlier, central Australia: implications for Proterozoic crustal evolution

J.-X. Zhao, M. T. McCulloch and **R. G. Warren**

The Arunta Inlier, the largest in central Australia (~200,000 km^2), is situated on the southern margin of the early Proterozoic Northern Australian Orogenic Province. Systematic geochemical variations are observed in granites from the Arunta Inlier. Overall, 20 granite suites identified are classified into three broad geochemical groups, a high-Na, Sr calcalkaline-trondhjemitic group (CAT), a "normal" granite group, and a high-heat-production group (HHP). The CAT group occurs only in the southern and southeastern margin, and consists of two subgroups, i.e. gabbro-diorite-tonalite-trondhjemite suites analogous to modern continental margin calcalkaline intrusions, and tonalite-trondhjemite-granodiorite suites similar to Archaean TTG suites. The HHP group is confined mainly within the northwestern inland section, where it intrudes coexisting normal granites. It is exceptionally enriched in Th, U, K, Rb, Pb and LREE and depleted in MgO, CaO, Ba and Sr. The above geochemical diversity suggests that the Barramundi Igneous Association is insignificant in the Arunta Inlier.

Although geochemically diverse, the Arunta granites are relatively uniform in terms of Nd isotopic signatures. Most of the granites analysed yield T_{DM}^{Nd} model ages ranging from 1·95–2·35 Ga, which are significantly older than those obtained from the Musgrave Inlier to the S, suggesting two distinct provinces forming during separate crustal formation events.

It is considered that the inland section of the Arunta Inlier was dominated by plume-driven ensialic rifting processes, while the southeastern margin was controlled by subduction-driven plate tectonics. The origin of the CAT group is related to the subduction processes. The normal granite group was derived by partial melting from a plume-derived mafic underplate. The younger HHP group was generated by partial melting of the normal group during anorogenic events.

List of Abstract authors' names and addresses

Andrew, Anita S., CSIRO Division of Exploration Geoscience, PO Box 136, North Ryde, NSW 2113, Australia

Atherton, Michael P., Department of Earth Sciences, University of Liverpool, Herdman Building, Brownlow Street, Liverpool, L69 3BX, U.K.

Barnes, Calvin G., Department of Geosciences, Texas Tech University, Lubbock, TX 79409, U.S.A.

Barnes, Melanie A., Department of Geosciences, Texas Tech University, Lubbock, TX 79409, U.S.A.

Best, Myron G., Department of Geology, Brigham Young University, Provo, Utah 84602, U.S.A.

Blundy, Jonathan D., Department of Geology, Bristol University, Bristol, U.K.

Bonin, Bernard, Laboratoire de Pétrographie-Volcanologie, Bâtiment 504, Université de Pari-Sud, F 91405 Orsay Cedex, France

Bowden, Peter, Department of Geography and Geology, University of St Andrews, Fife, KY16 9ST, Scotland, U.K.

Brew, David A., U.S. Geological Survey, Menlo Park, CA 94025-3591, U.S.A.

Brown, Michael, Department of Geology, University of Maryland, College Park, MD 20742, U.S.A.

Bryon, David, 114 Leopold Road, Kensington, Liverpool 7, Merseyside L7 8SS, U.K.

Campana, Raffaellamaria, Dipartimento di Scienze Mineralogiche, Università di Bologna, Piazza Porta S. Donata 1, I-40126 Bologna, Italy

Candela, P. A., Department of Geology, University of Maryland, College Park, MD 20742, U.S.A.

Carr, Paul F., Department of Geology, University of Wollongong, PO Box 1144, Wollongong, NSW 2500, Australia

Carson, C., Department of Earth Sciences, Monash University, Clayton, VIC 3168, Australia

Chappell, Bruce W., Department of Geology, Australian National University, GPO Box 4, Canberra, ACT 2601, Australia

Chen, Yadong D., School of Earth Sciences, Macquarie University, Sydney, NSW 2113, Australia

Christiansen, Eric H., Department of Geology, Brigham Young University, Provo, Utah 84602, U.S.A.

Clare, A. P., Esso Australia Ltd, 127 Kent Street, Sydney, NSW 2000, Australia

Clarke, Donald S., Department of Geology, University of Auckland, Private Bag, Auckland, New Zealand

Collins, W. J., Department of Geology, University of Newcastle, Newcastle, NSW 2308, Australia

Conticelli, S., Dipartimento di Scienze della Terra, Firenze, Italy

Crook, Keith A. W., Department of Geology, Australian National University, GPO Box 4, Canberra, ACT 2601, Australia

Cuney, Michel, Centre de Recherches sur la Geologie des Matières Premières Minerales et Energetiques, CREGU et GDR CNRS-CREGU, BP 23, 54 501 Vandoeuvre-les-Nancy Cedex, France

D'Lemos, Richard S., Department of Geology, Oxford Polytechnic, OX3 0BP, U.K.

Dadd, Kelsie, Department of Applied Geology, University of Technology Sydney, P.O. Box 123, Broadway, N.S.W. 2007, Australia

Day, W. C., U.S. Geological Survey, Mail Stop 905, Denver, CO 80225, U.S.A.

Diehl, Michael, Geological Survey, PO Box 2168, Windhoek, Namibia

Eberz, G., Redlands Technologies, Horsham, West Sussex, U.K.

Eby, G. Nelson, Department of Earth Sciences, University of Massachusetts at Lowell, Lowell, MA 01854, U.S.A.

Ferreira, Valderez P., Federal University of Pernambuco, Department of Geology, Recife, PE, 50000, Brazil

Fitches, W. R., Institute of Earth Studies, University of Wales, Aberystwyth, U.K.

Fleming, P. D., Department of Geology, La Trobe University, Bundoora, VIC 3083, Australia

Flood, R. H., School of Earth Sciences, Macquarie University, Sydney, NSW 2109, Australia

Foden, J. D., Department of Geology and Geophysics, University of Adelaide, GPO Box 498, Adelaide, SA 5001, Australia

Fodor, R. V., Department of Marine, Earth and Atmospheric Sciences, North Carolina State University, Raleigh, NC 27695-8208, U.S.A.

Frost, M. T., CSIRO, Pt. Melbourne, Australia

Gilbert, M. C., School of Geology and Geophysics, University of Oklahoma, Norman, Oklahoma 73019, U.S.A.

Gray, N., Climax Mining Ltd., PO Box 3595, Weston Creek, ACT 2611, Australia

Gray, C. M., Department of Geology, La Trobe University, Bundoora, VIC 3083, Australia

Griffin, T. J., Geological Survey of Western Australia, 100 Plain Street, East Perth, WA 6004, Australia

Groves, David I., Key Centre for Strategic Mineral Deposits, Department of Geology, The University of Western Australia, Nedlands, WA 6009, Australia

Guillot, Stéphane, Geodynamique des chaines alpines, C.N.R.S., Institut Dolomieu, 15 rue Maurice Gignoux, F-38000 Grenoble, France

Gulson, Brian L., CSIRO Division of Exploration Geoscience, PO Box 136, North Ryde, NSW 2113, Australia

Hammarstrom, Jane M., U.S. Geological Survey, 959 National Center, Reston, VA 22092, U.S.A.

Harvey, M. J., Department of Geology, University of Manchester, Manchester, U.K.

Hegner, Ernst., Geological Survey of Canada, 601 Booth Street, Ottawa, K1A OE8, Canada

Henderson, C. M. B., Department of Geology, University of Manchester, Manchester, U.K.

Hine, R., Climax Mining Ltd., PO Box 3595, Weston Creek, ACT 2611, Australia

Hogan, J. P., School of Geology and Geophysics, University of Oklahoma, Norman, Oklahoma 73019, U.S.A.

Howard, K. A., US Geological Survey, Menlo Park, CA 94025, U.S.A.

Hutton, L. J., Senior Geologist, Queensland Department of Resource Industries, GPO Box 194, Brisbane, QLD 4001, Australia

Jenner, George A., Department of Earth Sciences, Memorial University, St. John's, Newfoundland A1C 3X5, Canada

John, Barbara E., Department of Earth Sciences, University of Cambridge, Downing Street, Cambridge, U.K.

Johnson, Kenneth, Department of Geosciences, Texas Tech University, Lubbock, TX 79409, U.S.A.

Jones, Brian G., Department of Geology, University of Wollongong, PO Box 1144, Wollongong, NSW 2500, Australia

Jones, Mel T., Pancontinental Mining Limited, Gateway, 1 Macquarie Place, Sydney, NSW 2000, Australia

Jones, Catrin E., Department of Geology, University of Leicester, University Road, Leicester, U.K.

Kagami, Hiroo, Institute foı Study of the Earth's Interior, Okayama University

Kanaori, Yuji, Department of Geological Sciences, Faculty of General Education, Gifu University, 1-1 Yanagido, Gifu City 501-11, Japan

Karlstrom, K. E., Department of Geology, Northern Arizona University, Flagstaff, AZ 86011, U.S.A.

Kawakami, Shin-ichi, Department of Earth and Planetary Sciences, Faculty of Education, Gifu University, 1-1 Yanagido, Gifu City 501-11, Japan

Kawano, Yoshinobu, Graduate School of Science and Technology, Niigata University, Ikarashi 8050, Niigata 950-21, Japan (present address: Geological Survey of Japan, Hokkaido Branch)

Kinnaird, Judith, Department of Geology, University College, Cork, Eire

Kistler, R. W., U.S. Geological Survey, 345 Middlefield Road, Menlo Park, CA 94025, U.S.A.

Kisvarsanyi, E. B., Missouri Department of Natural Resources, P.O. Box 250, Rolla, MO 65401, U.S.A.

Landenberger, B., Department of Geology, University of Newcastle, Callaghan, NSW 2308, Australia

Langford, R. L., Hong Kong Geological Survey, Geotechnical Engineering Office, 11/F Civil Engineering Building, 101 Princess Margaret Road, Homantin, Kowloon, Hong Kong

Le Fort, Patrick, Géodynamique des chaines alpines, C.N.R.S., Institut Dolomieu, 15 rue Maurice Gignoux, F-38000 Grenoble, France

Legg, C., Department of Earth Sciences, Monash University, Clayton, VIC 3168, Australia

Longstaffe, Frederick J., Department of Geology, University of Western Ontario, London, Ontario, N6A 5B7, Canada

Maas, R., Department of Applied Physics, Curtin University of Technology, GPO Box U1987, Perth, WA 6001, Australia

Maltman, A. J., Institute of Earth Studies, University of Wales, Aberystwyth, U.K.

Mandyczewsky, A., Climax Mining Ltd., PO Box 3595, Weston Creek, ACT 2611, Australia

Manetti, P., Dipartimento di Scienze della Terra, Firenze, Italy

Manning, D. A. C., Department of Geology, University of Manchester, Manchester, U.K.

McCaffrey, K. J. W., Department of Earth and Planetary Sciences, Johns Hopkins University, Baltimore, MD 21218, U.S.A. and Department of Geology, Vanderbilt University, Nashville, TN 37235, U.S.A.

McCulloch, M. T., Research School of Earth Sciences, Australian National University, GPO Box 4, Canberra, ACT 2601, Australia

McCulloch, M., CSIRO, Pt. Melbourne, Australia

McErlean, Michele A., Department of Geological Sciences, University of Durham, South Road, Durham, DH1 3LE, U.K.

McNaughton, Neal J., Key Centre for Strategic Mineral Deposits, Department of Geology, the University of Western Australia, Nedlands, WA 6009, Australia

Miller, Robert B., Department of Geology, San Jose State University, San Jose, CA 95192-0102, U.S.A.

Miller, C. F., Department of Geology, Vanderbilt University, Nashville, TN 37236, U.S.A.

Morrison, Robert S., Geological Survey, Department of Mines and Energy, GPO Box 2655, Alice Springs, NT 5750 Australia

Muir, R. J., Department of Geology, University of Canterbury, Christchurch 1, New Zealand

Myers, J. D., Department of Geology and Geophysics, University of Wyoming, Laramie, Wyoming 82071, U.S.A.

Nabelek, Peter I., Department of Geological Sciences, University of Missouri, Columbia, MO 65211, U.S.A.

Nakai, Yutaka, Department of Earth Sciences, Aichi University of Education, 1 Hirosawa, Igayo-Cho, Kariya City 448, Japan

Nakajima, Takashi, Geochemistry Department, Geological Survey of Japan, 1-1-3 Higashi, Tsukuba 305, Japan

Nicholls, I. A., Department of Earth Sciences, Monash University, Clayton, VIC 3168, Australia

Nuelle, L. M., Missouri Department of Natural Resources, P.O. Box 250, Rolla, MO 64501, U.S.A.

Oliver, G. J. H., Department of Geography and Geology, University of St Andrews, St Andrews KY16 9ST, Scotland, U.K.

Onishi, Sayuri, Department of Earth and Planetary Sciences, Faculty of Education, Gifu University, 1-1 Yanagido, Gifu City 501-11, Japan

Pêcher, Arnaud, Geodynamique des chaines alpines, C.N.R.S., Institut Dolomieu, 15 rue Maurice Gignoux, F-38000 Grenoble, France

Petford, Nicholas, Department of Earth Sciences, University of Liverpool, Herdman Building, Brownlow Street, Liverpool, L69 3BX, U.K.

Petrini, R., Istituto di Geocronologia e Geochimica Isotopica, CNR, via C. Maffi 36, I-56100 Pisa, Italy

Piccoli, P. M., Department of Geology, University of Maryland, College Park, MD 20742, U.S.A.

Pinarelli, L., Istituto di Geocronologia e Geochimica Isotopica, CNR, via C. Maffi 36, I-56100 Pisa, Italy

Pirajno, Franco, Western Mining, Perth, Australia

Pitcher, W. S., 14 Church Road, Upton, Wirral, Merseyside L49 6JZ, U.K.

Poli, G., Dipartimento di Scienze della Terra, Piazza University, I-06100 Perugia, Italy

Pollard, Peter J., Department of Geology, James Cook University of North Queensland, Townsville, Australia

Prestvik, Tore, Department of Geology, University of Trondheim, Norwegian Institute of Technology, N 7034 Trondheim, Norway

Price, R. C., Department of Geology, La Trobe University, Bundoora, VIC 3083, Australia

Reavy, R. John, Department of Earth Sciences, University of Oxford, Parks Road, Oxford OX1 3PR, UK

Recio, C., Servicio General de Analisis de Isótopos Estables, Fac. de Ciencias, Universidad de Salamanca, 37008-Salamanca, Spain

Reid, E., University of Tasmania

Renno, A., Inst. für Mineralogie, Geochemie und Lagerstattenlehre, Bergakademie Frieberg, Brennhausgasse 14, 9200 Frieberg, Germany

Rickard, M. J., Geology Department, Australian National University, GPO Box 4, Canberra, ACT 2601, Australia

Rienks, I. P., Geologist, Queensland Department of Resource Industries, GPO Box 194, Brisbane, QLD 4001, Australia

Rossiter, Allan G., Department of Geology, La Trobe University, Bundoora, Victoria 3083, Australia

Rottura, Alessandro, Dipartimento di Scienze Mineralogiche, Università di Bologna, Piazza Porta S. Donata 1, I-40126 Bologna, Italy

Ruiz, Joaquin, Department of Geosciences, University of Arizona, Tucson, Arizona 85721, U.S.A.

Russ-Nabelek, Carol, Department of Geological Sciences, University of Missouri, Columbia, MO 65211, U.S.A.

Rye, R. O., U.S. Geological Survey, Mail Stop 905, Denver, CO 80225, U.S.A.

Schmidt, W., Inst. für Mineralogie, Geochemie und Lagerstattenlehre, Bergakademie Frieberg, Brennhausgasse 14, 9200 Frieberg, Germany

Sewell, R. J., Hong Kong Geological Survey, Geotechnical Engineering Office, 11/F Civil Engineering Building, 101 Princess Margaret Road, Homantin, Kowloon, Hong Kong

Shaw, S. E., School of Earth Sciences, Macquarie University, Sydney, NSW 2109, Australia

Shinkai, Tomohiro, Department of Earth and Planetary Sciences, Faculty of Education, Gifu University, 1-1 Yanagido, Gifu City 501-11, Japan

Sial, Alcides N., Federal University of Pernambuco, Department of Geology, Recife, PE, 50000, Brazil

Sidder, G. B., U.S. Geological Survey, Mail Stop 905, Denver, CO 80225, U.S.A.

Silver, Leon T., Division of Geological and Planetary Sciences, California Institute of Technology, Pasadena, CA 91125, U.S.A.

Simpson, C., Department of Earth and Planetary Sciences, Johns Hopkins University, Baltimore, MD 21218, U.S.A.

Smith, Ian E. M., Department of Geology, University of Auckland, Private Bag, Auckland, New Zealand

Soler, Pierre, ORSTOM, 213 rue Lafayette, 75010 Paris, France, and Department of Geological Sciences and INSTOC, Snee Hall, Cornell University, Ithaca, NY 14853-1504, U.S.A.

Steele, D. A., Department of Geology, Australian National University, GPO Box 4, Canberra, ACT 2601, Australia

Stephens, C. J., Department of Geology and Mineralogy, University of Queensland, St Lucia, QLD 4072, Australia

Stephenson, P. J., James Cook University, Townsville, QLD 4811, Australia

Strachan, Robin A., Department of Geology, Oxford Polytechnic, OX3 0BP, U.K.

Strange, P. J., Hong Kong Geological Survey, Geotechnical Engineering Office, 11/F Civil Engineering Building, 101 Princess Margaret Road, Homantin, Kowloon, Hong Kong

Stussi, Jean-Marc, CRPG, BP 20, 45 501 Vandoeuvre-les-Nancy Cedex, France

Sun, S. S., Bureau of Mineral Resources, GPO Box 378, Canberra, ACT 2601, Australia

Tarney, John, Department of Geology, University of Leicester, University Road, Leicester, U.K.

Tepper, J. H., Department of Geology, Lafayette College, Easton, PA 18042, U.S.A.

Tommasini, S., Dipartimento di Scienze della Terra, Piazza University, I-06100 Perugia, Italy

Turner, S. P., Department of Geology and Geophysics, University of Adelaide, GPO Box 498, Adelaide, SA 5001, Australia

Tyler, I. M., Geological Survey of Western Australia, 100 Plain Street, East Perth, WA 6004, Australia

Ugidos, J. M., Departamento de Geología, Fac. de Ciencias, Universidad de Salamanca, 37008—Salamanca, Spain

Vernon, R. H., School of Earth Sciences, Macquarie University, Sydney, NSW 2109, Australia

Wang, Liang G., Key Centre for Strategic Mineral Deposits, Department of Geology, The University of Western Australia, Nedlands, WA 6009, Australia

Warren, R. G., Bureau of Mineral Resources, GPO Box 378, Canberra, ACT 2601, Australia

Weaver, B. L., School of Geology and Geophysics, University of Oklahoma, Norman, Oklahoma 73019, U.S.A.

Whalen, Joseph B., Geological Survey of Canada, 601 Booth Street, Ottawa, K1A 0E8, Canada

Whitney, James A., Department of Geology, University of Georgia, Athens, GA 30602, U.S.A.

Williams, Ian S., Research School of Earth Sciences, Australian University, GPO Box 4, Canberra, ACT 2601, Australia

Woodhead, J. D., Research School of Earth Sciences, Australian National University, GPO Box 4, Canberra, ACT 2601, Australia

Wyborn, D., Senior Research Scientist, Bureau of Mineral Resources, GPO Box 378, Canberra, ACT 2601, Australia

Yairi, Kenji, Department of Geological Sciences, Faculty of General Education, Gifu University, 1-1 Yanagido, Gifu City 501-11, Japan

Zen, E-an, Department of Geology, University of Maryland, College Park, MD 20742, U.S.A.

Zhao, J. X., Research School of Earth Sciences, Australian National University, GPO Box 4, Canberra, ACT 2601, Australia

INDEX – Authors of papers

Adams, C. J., 281
Allen, C. A., 179

Bacon, C. R., 27
Baldridge, W. S., 439
Barbarin, B., 145
Barbero, L., 127
Bennett,V. C., 49
Blencoe, J. G., 423
Blevin, P. L., 305
Brown, P. E., 173
Burnham, C. W., 387

Campbell, I. H., 211
Candela, P. A., 317
Casey, M., 361
Champion, D. C., 115
Chappell, B. W., 1, 115, 211, 305, 327
Compston, W., 473
Cornejo, P. C., 63

Darbyshire, D. P. F., 269
Dempster, T. J., 173
DePaolo, D. J., 439
Didier, J., 145
Downes, H., 235
Drummond, B. J., 201

Ellis, D. J., 95, 155
Esperança, S., 347
Ewart, A., 327

Fallick, A. E., 227, 247
Foden, J., 83
Foster, D. A., 49

Gibson, I. L., 281
Green, T. H., 429

Haapala, I., 167
Hanchar, J. M., 49
Harrison, S. M., 291
Harrison, T. M., 49, 173
Herd, D. A., 459
Hill, R. I., 211
Hinton, R. W., 459
Holtz, F., 409, 417
Hutton, D. H. W., 173, 377, 383

Iiyama, J. T., 259
Ingram, G. M., 383

Johannes, W., 409, 417

Kilpatrick, J. A., 155
Komatsu, M., 259

Langford, R. L., 269
Law, R. D., 361
Leeman, W. P., 71

McCormick, A. G., 227
Mahood, G. A., 63
Meighan, I. G., 227
Mertzman, S. A., 71
Miller, C. F., 49
Morgan, S. S., 361
Moyer, T. C., 347

Nekvasil, H., 399
Norman, M. D., 71
Nyman, M., 361

Obata, M., 95

Pankhurst, R. J., 281, 291
Paterson, B. A., 459
Perry, F. V., 439
Pichavant, M., 409
Pidgeon, R. T., 473

Rämö, O. T., 167
Rapela, C. W., 291
Recio, C., 247
Rogers, G., 459

Sandiford, M., 83
Schon, R. W., 327
Sewell, R. J., 269
Shimura, T., 259
Stephens, W. E., 191, 459
Strange, P. J., 269
Sylvester, A. G., 361

Thirlwall, M. F., 235
Turner, S., 83

Ugidos, J. M., 247

Villaseca, C., 127

Wark, D. A., 49
Warren, R. G., 201
Weaver, S. D., 281
White, A. J. R., 1
Williams, I. S., 447, 459
Williamson, B. J., 235
Wooden, J. L., 49
Wyborn, D., 201
Wyborn, L. A. I., 201

Zeck, H. P., 139
Zen, E-an, 107
Zhou, S., 83

Printed in Northern Ireland